THE BOOK OF COMMON LOGARITHMS

J.S. GILES

ARTHUR H. STOCKWELL LTD.
Elms Court Ilfracombe
Devon

First published in Great Britain, 1973

SBN 7223 0215-0
PRINTED IN GREAT BRITAIN BY
ARTHUR H. STOCKWELL LTD.
Elms Court - Ilfracombe
Devon

THE BOOK OF COMMON LOGARITHMS

A table responsive to limited or extended digit application . . . simple in operation.

ACKNOWLEDGEMENTS

The Author wishes to thank Friends who showed an interest, provided paper and books, and helped with the proofs; the Publisher, who almost drove one frantic but for his part bore with the writer; and the Printer, for a presentable book. The research of that Mind or those Minds of the seventeenth century or earlier, which first apprehended the 'principle' of Logarithms, is acknowledged and appreciated.

INTRODUCTORY NOTES

The procedure in working logarithms has been explained in many text books previously, and basically it is the same here. Apart therefore from passing indications included in the examples, a repetition would appear unnecessary. So now for the present table.

In essence, two columns only, consisting of seven digit marginal numbers, in four digit 'leader' form, stretch throughout the book, columns 3 and 4; 5 and 6; etc., being the continuation of one and two. The difference between an applied number and the nearest marginal less than itself constitutes an 'excess' on that marginal.

No excess has more than two digits, and most, or many of those practiced in figure-work will be able to work them mentally. The requisite excess table is the first found after the figure worked. To keep the place if the table is overleaf, or splits on the page, it is perhaps advisable to jot down the number in the opposing column, or at least its three last digits.

The excess table stops (except the last one in the book) at the number immediately preceding, so any excess on that marginal should be applied to the next later table.

The first column, without heading, is initially to be regarded as the 'Application' column. Sections of this column are indicated by the large numbers at the bottom of every sixteenth page which give the range. The second column is in the first place to be treated as the 'Logarithm' column.

The sums are worked in the usual manner, and now the second column is to be regarded as the 'Antilogarithm' column. (Strictly speaking numbers with an established decimal point, relative to the first column are the Antilogarithms, but for the present purpose *only* I am using the term for figures in the second column which together with the characteristic, via the first column, give the antilogarithm). Sections of this second column used as such are indicated by the second section of the larger numbers in the index.

We now apply the result of our working to the second column, apply any excess to the 'Anti-Log' Excess table, and use its equivalent to adjust the first column, retranslate the Characteristic and receive the answer.

An example of multiplication – 6000·801 X 109·01

The characteristic of the first number, one less than the numbers of digits before the decimal is 3. By aid of the index 1st larger number we find 6000750 on page 363. The difference is 51. This applied to following excess table gives equivalent 37 which is added to the marginal in the second column making 3·7782092. The second number 109·01 characteristic 2. By way of index 1st larger number again, we find 1090100 on page 29. There is no excess so the log is 2·0374659. The two logarithms added together become 5·8156751 now by way of index 2nd number we find ·8156711 on page 389. The difference of 40 applied to Anti-Log Excess table leaves 60 to be added to the first column making the answer, with decimal one place more than the characteristic 654146·7 as against the true 654147·31701.

ROUGHER METHODS AND IMPERFECTIONS

(a) Find the nearest marginal and if the applied figure is not more than five above or below take the opposed marginal. Otherwise, consult the numbers below the excess tables. These numbers represent rough half-way measures of the margins they cover. So mentally add the first number to the first column marginal and if the applied falls near add the half measure equivalent to the logarithm marginal. If however, you consider the applied number falls nearer the quarter or three-quarter difference then halve the equivalent, and in the first case add to the lower logarithm margin, or in the second, deduct from the higher.

Repeat at each stage, though of course in finding the antilogarithm we mentally add the half measure (first on left 'Anti Log' Excesses) to the second column and make adjustment accordingly. All this is done quite roughly, much less in fractions. Imperfection – 10 times the rate of 'C' or within a unit to 100,000.

(b) Taking the nearest marginal at a guess, but adding the half measure equivalent when obviously indicated. Repeat at each stage. Imperfection – within say 2 in 100,000 or ·2 to 10,000.

Do not use these last two methods for multiplication or division of 'powers' or the imperfection in the majority of cases will be unduly aggravated (Squares and Square roots may be worked) but work under 'C'.

(c) The main system – worked in the manner of the earlier example.

			Unit(s)
Imperfections	100 to 1,000	within	$\frac{1}{1,000}$
	1,000 to 10,000	within	$\frac{1}{100}$
	10,000 to 100,000	within	$\frac{1}{10}$
	100,000 to 1,000,000	within	1
	1,000,000 to 10,000,000	within	10

and in such terms, with slight qualification for extensions (see later) for any amount upwards or downwards.

Or to memorise, without studying a chart, the imperfection is within one millionth of the sum represented by the next higher characteristic or we may put the indicator over the sixth place of the first positive.

Therefore: $1{,}000{,}\overset{1}{0}00{,}000$ plus is within 10,000 so also are all numbers up to $9{,}999{,}999{,}999 \cdot \overset{1}{\dot{9}}$

The "ceiling" round number (here 10^{10}) is included in all cases but in these isolated instances the indicator is over the seventh place.

Coming down: $100{,}00\overset{1}{0}$ plus to $999{,}999 \cdot \overset{1}{\dot{9}}$ is within a unit, while a fraction conclusion – $.00100000\overset{1}{0}$ plus, to $.00999999\overset{1}{\dot{9}}$ is within one hundred millionth of the unit. Out of well over a hundred practical tests spread throughout the series, ten only had imperfections greater than .3 of the imperfection shown in the chart, the furthest out being the example worked ·61701 of the unit short. The imperfection may be either way, short, or over.

Respecting numbers having more than seven digits. After taking the characteristic, cut back and adjust to the nearest seventh place; where no integer, seventh from the first positive. The difference arising hereby cannot, even if both numbers extend, reach ·55 of the imperfection shown in the chart, so that if the example worked is anywhere near the extreme, the two discrepancies taken together will not break bounds overmuch, nor, considering that compensations are active, will the "curtailment" necessarily increase the imperfection. It may reduce it, and probably in practice it would be rare even for an extension to take the imperfection beyond the limit given.

Further possibilities – very large products (decimals apart) defy logarithms by reason of the imperfection, nevertheless they may be tamed somewhat by a partial multiplication sufficient to provide all the digits after the first five:

e.g. 672431 x 49432

Working the logarithm first will also indicate the number of digits:

672,431 = LOG 5·8276477
49,432 = LOG 4·6940082
10·5216559 apply ·5216559

which gives antilogarithm with decimal point one place further than positive characteristic –

33,239,620,000.

Eleven digits, so we require the last six.

Therefore:

672431
49432

344862
17293
9724
879
24

609192

Which perfects at 33,239,609,192

A product complete in seven digits –

4123 x 1815 merely needs

23
15

15
3

45 as adjustment

Eight digits – 6324 x 3165

324
165
———
620
44
4
———
460 as adjustment, and so on.

In making the adjustment, the sixth place of the antilogarithm may not change at all, but if altered, can only alter by one unit up or down. Judgment will be exercised when the sixth place of the Antilog is a 9, or a 0 as these can effect the previous digit(s).

Illustrative of this last point: $3{,}000^2$ by the book is

1
8,999,998 but really it is 9,000,000. On the other hand

1
24,365,020 may require 24364983 as correct. Of course numbers with a decimal are included but decimals expand the digits rather than the imperfection and should the digits be extensive by reason of decimals I can only suggest they be cut back to the nearest at a place considered sufficient to produce a relative approximation. However, it may be claimed that by the use of the table and only a negligible 'tail end' multiplication to give up to a four figure adjustment, perfection may be obtained for products up to nine digits, or for whole numbers to an inclusive 1000 million, (with a very remote chance? of a two unit change in the antilog sixth following an extension 'cut back', but this could only be certain 8 digits x 2 digits and we would not use logarithms here).

The number of digits in a product derived from decimal numbers cannot be determined from the position of the decimal point in the antilogarithm, and here we must assess the initial terms. Quite simple, for this purpose 4·123 x 181·5 is treated as 4000 x 2000 = 7 digits, 63·24 x 3·165 as 6000 x 3000 = 8 digits, ·006324 x 3·165 = 9 digits.

In doubtful cases, e.g. 1.245 x 8003 or 31.68 x 317.3 whether there is a carry forward making another digit or not, a glance at the amount of the first positive of the antilog will decide, (If low, carry forward, if high, no carry forward).

Adjustments are only to be made after counting five places starting with the *First Positive.* Where there is no integer (antilog) noughts immediately following a decimal point are not included in the count, ·006324 x 3·165 (above) adjustment is 460 the last three digits of the *nine.*

To resume. With regard to accuracy. Other than those 'unassailable' surds, etc., and their manipulation, I have checked all the marginals by running down the columns, restricting the progressive variation to a unit on the seventh. Where there is a jump in one column (allowed by the two digit excess) there is a corresponding jump in the other. These leaps occasion the isolated halfway measures found earlier in the book; the detailed excesses not being affected follow later.

Although the figures are simple to work, a couple of warnings may save a lot of errors –

1. Concentrate, for which purpose you are using a particular column, and be careful to use the 'Excesses' table for the first column, and the 'Antilog Excesses' for the second.

2. Seeing it is often easier when working under 'C' to take the higher marginal and *deduct* the equivalent difference from the opposed marginal remember when addition is the rule, or subtraction is required.

December, 1973. J.S.G.

1000000	·0000000	1001476	·0006407	1002951	·0012794	1004422	·0019162	1005893	·0025513
023	099	499	505	973	892	445	260	915	610
046	197	522	603	996	990	467	358	938	708
068	296	544	702	1003019	·0013088	490	456	960	805
091	395	567	800	041	186	513	554	983	903
114	493	590	898	064	284	535	652	1006006	·0026000
136	592	613	997	087	382	558	749	028	098
159	690	635	·0007095	109	480	581	847	051	195
182	789	658	193	132	578	603	945	073	293
205	888	681	292	154	676	626	·0020043	096	390
227	986	704	390	177	774	648	140	118	488
250	·0001085	726	488	200	872	671	238	141	585
273	184	749	587	222	970	694	336	163	683
295	282	772	685	245	·0014068	716	434	186	780
318	381	794	783	268	167	739	532	209	878
341	479	817	881	290	265	762	629	231	975
364	578	840	980	313	363	784	727	254	·0027072
386	677	863	·0008078	336	461	807	825	276	170
409	775	885	176	358	559	830	923	299	267
432	874	908	275	381	657	852	·0021020	321	365
455	973	931	373	404	755	875	118	344	462
477	·0002071	954	471	426	853	898	216	366	560
500	170	976	570	449	951	920	314	389	657
523	269	999	668	471	·0015049	943	411	411	755
545	367	1002022	766	494	147	965	509	434	852
568	466	044	865	517	245	988	607	457	949
591	564	067	963	539	343	1005011	705	479	·0028047
614	663	090	·0009061	562	441	033	802	502	144
636	761	112	159	585	539	056	900	524	242
659	860	135	258	607	637	078	998	547	339
682	958	157	356	630	735	101	·0022095	569	437
705	·0003057	180	454	653	833	124	193	592	534
727	156	203	552	675	931	146	291	615	631
750	254	225	651	698	·0016029	169	388	637	729
773	353	248	749	721	127	191	486	660	826
795	451	271	847	743	225	214	584	682	924
818	550	293	946	766	323	236	681	705	·0029021
841	648	316	·0010044	789	421	259	779	727	119
864	747	339	142	812	519	282	877	750	216
886	846	361	240	834	617	304	975	773	313
909	944	384	339	857	715	327	·0023072	795	411
932	·0004043	407	437	880	813	349	170	818	508
955	141	429	535	902	911	372	268	840	606
977	240	452	633	925	·0017009	395	365	863	703
1001000	338	474	732	948	107	417	463	886	800
023	437	497	830	970	205	440	561	908	898
045	535	520	928	993	303	462	658	931	995
068	634	542	·0011026	1004016	401	485	756	953	·0030093
091	732	565	125	038	499	508	854	976	190
113	831	588	223	061	597	530	951	999	287
136	929	610	321	083	694	553	·0024049	1007021	385
159	·0005028	633	419	106	792	576	146	044	482
181	126	656	517	129	890	598	244	066	579
204	225	678	616	151	988	621	342	089	676
227	323	701	714	174	·0018086	643	439	112	774
249	422	724	812	196	184	666	537	134	871
272	520	746	910	219	282	689	634	157	968
295	619	769	·0012008	241	380	711	732	179	·0031065
318	717	792	107	264	477	734	830	202	163
340	816	815	205	287	575	757	927	224	260
363	914	837	303	309	673	779	·0025025	247	357
386	·0006013	860	401	332	771	802	122	270	455
408	111	883	499	354	869	825	220	292	552
431	210	905	598	377	967	847	317	315	649
454	308	928	696	400	·0019065	870	415	337	746

1007360	·0031844
383	941
405	·0032038
428	135
450	233
473	330
496	427
518	524
541	622
563	719
586	816
608	913
631	·0033011
653	108
676	205
699	302
721	399
744	497
766	594
789	691
811	788
834	886
856	983
879	·0034080
901	177
924	274
947	372
969	469
992	566
1008014	663
037	760
059	857
082	954
104	·0035052
127	149
149	246
172	343
194	440
217	537
240	634
262	731
285	828
307	925
330	·0036022
352	120
375	217
397	314
420	411
443	508
465	605
487	702
510	799
532	896
555	993
577	·0037090
600	187
622	284
645	381
667	478
690	575
712	672
735	769
757	866
780	963
802	·0038060

1008825	·0038157
847	254
870	351
892	448
915	545
937	642
960	739
983	836
1009005	933
028	·0039030
050	127
073	224
095	320
118	417
140	514
163	611
186	708
208	805
231	902
253	999
276	·0040096
298	193
321	290
343	386
366	483
388	580
411	677
434	774
456	871
478	968
501	·0041065
523	161
546	258
568	355
591	452
613	549
636	646
658	742
681	839
703	936
726	·0042033
748	130
771	226
793	323
816	420
838	517
861	614
883	711
906	807
928	904
951	·0043001
973	098
996	194

Excesses

10 = 43	1 = 4
20 = 86	2 = 9
	3 = 13
	4 = 17
	5 = 22
	6 = 26
	7 = 30
	8 = 35
	9 = 39

11 = 48

Anti Log Excesses

10 = 2	1 = 0
20 = 5	2 = 0
30 = 7	3 = 1
40 = 9	4 = 1
50 = 12	5 = 1
60 = 14	6 = 1
70 = 16	7 = 2
80 = 18	8 = 2
90 = 21	9 = 2

48 = 11

1010018	·0043291
041	388
063	485
086	581
108	678
131	775
153	872
176	968
198	·0044065
221	162
243	258
266	355
288	452
311	549
333	645
356	742
378	839
401	936
423	·0045032
446	129
468	226
491	322
513	419
536	516
558	612
581	709
603	805
626	902
648	999
671	·0046095
693	192
716	289
738	385
761	482
783	579
806	675
828	772

1010851	·0046869
873	965
896	·0047062
918	158
941	255
963	351
986	448
1011008	544
031	641
053	737
076	834
098	930
121	·0048027
143	123
165	220
188	316
210	413
233	509
255	606
278	702
300	799
323	895
345	992
368	·0049088
390	185
412	281
435	378
457	474
480	571
502	667
525	764
547	860
570	957
592	·0050053
615	150
637	246
659	342
682	439
704	535
727	632
749	728
772	825
794	921
817	·0051018
839	114
862	211
884	307
906	403
929	500
951	596
974	693
996	789
1012019	885
041	982
064	·0052078
086	175
109	271
131	367
154	464
176	560
198	656
221	753
243	849
266	945
288	·0053042

1012311	·0053138
333	234
356	331
378	427
400	524
423	620
445	716
468	812
490	909
513	·0054005
535	101
558	197
580	294
603	390
625	486
648	582
670	679
692	775
715	871
737	967
760	·0055063
782	160
805	256
827	352
850	448
872	545
894	641
917	737
939	833
962	929
984	·0056025
1013007	122
029	218
052	314
074	410
097	506
119	602
142	698
164	795
186	891
209	987
231	·0057083
254	179
276	275
299	371
321	468
344	564
366	660
388	756
411	852
433	948
456	·0058044
478	140
501	236
523	332
545	429
568	525
590	621
613	717
635	813
658	909
680	·0059005
702	101
725	197
747	293

1013770	·0059389
792	486
814	582
837	678
859	774
882	870
904	966
926	·0060062
949	158
971	254
994	350
1014016	446
038	542
061	638
083	734
106	830
128	926
151	·0061021
173	117
195	213
218	309
240	405
263	501
285	597
307	693
330	789
352	885
375	981
397	·0062077
419	173
442	269
464	365
487	460
509	556
531	652
554	748
576	844
599	940
621	·0063036
644	131
666	227
688	323
711	419
733	515
756	611
778	707
800	802
823	898
845	994
868	·0064090
890	186
912	282
935	378
957	473
979	569
1015002	665
024	761
047	857
069	953
091	·0065048
114	144
136	240
158	336
181	432
203	527

1015225	·0065623
248	719
270	815
292	911
315	·0066007
337	102
360	198
382	294
404	390
427	485
449	581
472	677
494	772
516	868
539	964
561	·0067059
584	155
606	251
629	346
651	442
673	538
696	634
718	729
741	825
763	921
785	·0068016
808	112
830	208
853	303
875	399
897	495
920	590
942	686
964	781
987	877
1016009	973
032	·0069068
054	164
076	259
099	355
121	450
143	546
166	642
188	737
210	833
233	928
255	·0070024
277	120
300	215
322	311
345	406
367	502
389	598
412	693
434	789
456	884
479	980
501	·0071075
523	171
546	266
568	362
590	457
613	553
635	648
657	744

1016679	·0071839
702	935
724	·0072030
746	126
769	221
791	317
813	412
836	508
858	603

Excesses

11 = 47

Anti Log Excesses

47 = 11

1016881	·0072703
905	803
928	903
952	·0073003
975	103
999	203
1017022	303
045	403
069	503
092	603
116	702
139	802
163	902
186	·0074002
209	102
233	202
256	302
280	402
303	502
327	602
350	702
373	802
397	902
420	·0075002
444	101
467	201
490	301
514	401
537	501
560	601
584	701
607	800
631	900
654	·0076000
677	100
701	200
724	300
747	400
771	499
794	599
818	699
841	799
864	899
888	999
911	·0077098
935	198
958	298

1017981	·0077398
1018005	498
028	597
051	697
075	797
098	897
122	996
145	·0078096
168	196
192	296
215	396
238	495
262	595
285	695
309	795
332	894
355	994
379	·0079093
402	193
426	293
449	392
472	492
496	592
519	691
542	791
566	891
589	990
613	·0080090
636	190
659	289
683	389
706	489
729	588
753	688
776	788
800	887
823	987
846	·0081087
870	186
893	286
917	386
940	485
963	584
987	684
1019010	784
033	883
057	983
080	·0082082
104	182
127	281
150	381
174	481
197	580
220	680
244	780
267	879
291	978
314	·0083078
337	177
361	277
384	376
407	476
431	575
454	675
477	774

1019501	·0083873
524	973
547	·0084072
571	172
594	271
617	371
641	470
664	569
687	669
711	768
734	868
757	967
781	·0085067
804	166
827	265
851	365
874	464
898	564
921	663
944	762
968	862
991	961

Excesses

10 = 43	1 = 4
20 = 86	2 = 9
	3 = 13
	4 = 17
	5 = 21
	6 = 26
	7 = 30
	8 = 34
	9 = 39

12 = 50

Anti Log Excesses

10 = 2	1 = 0
20 = 5	2 = 0
30 = 7	3 = 1
40 = 9	4 = 1
50 = 12	5 = 1
60 = 14	6 = 1
70 = 16	7 = 2
80 = 19	8 = 2
90 = 21	9 = 2

50 = 12

1020014	·0086060
038	160
061	259
085	359
108	458
131	557
155	657
178	756
201	855
225	955
248	·0087054
272	154
295	253
318	352

1020342	·0087452	1021857	·0093898	1023370	·0100326	1024882	·0106734	1026390	·0113123
365	551	880	997	394	424	905	833	414	222
388	650	904	·0094096	417	523	928	931	437	320
412	750	927	195	440	622	951	·0107030	460	418
435	849	950	294	463	721	974	128	483	516
458	948	974	393	487	819	997	227	507	614
482	·0088048	997	492	510	918	1025021	325	530	712
505	147	1022020	591	533	·0101017	044	423	553	810
528	246	043	690	556	116	067	522	576	909
552	346	067	789	580	214	090	620	600	·0114007
575	445	090	888	603	313	113	719	623	105
598	544	113	987	626	412	137	817	646	203
622	644	137	·0095086	649	511	160	916	669	301
645	743	160	185	673	609	183	·0108014	692	399
668	842	183	284	696	708	206	112	715	497
692	942	207	383	719	807	229	211	739	595
715	·0089041	230	482	742	906	253	309	762	693
738	140	253	581	766	·0102004	276	407	785	791
762	240	276	680	789	103	299	506	808	889
785	339	300	779	812	201	322	604	831	987
808	438	323	878	835	300	346	702	854	·0115085
832	537	346	977	859	399	369	801	877	184
855	637	369	·0096076	882	497	392	899	901	282
878	736	393	174	905	596	415	997	924	380
901	835	416	273	929	695	439	·0109096	947	478
925	934	439	372	952	793	462	194	970	576
948	·0090033	463	471	975	892	485	292	993	674
971	133	486	570	998	990	508	391	1027016	772
995	232	509	669	1024022	·0103089	532	489	039	870
1021018	331	532	768	045	188	555	587	063	968
041	430	556	867	068	286	578	686	086	·0116066
064	529	579	966	092	385	601	784	109	164
088	628	602	·0097065	115	484	625	882	132	262
111	728	626	163	138	582	648	981	155	360
134	827	649	262	162	681	671	·0110079	178	458
158	926	672	361	185	779	694	177	202	556
181	·0091025	696	460	208	878	717	275	225	654
204	124	719	559	231	977	741	374	248	752
228	224	742	658	254	·0104075	764	472	271	850
251	323	765	757	278	174	787	570	294	948
274	422	789	856	301	272	810	668	318	·0117046
297	521	812	954	324	371	833	767	341	144
321	620	835	·0098053	347	469	857	865	364	242
344	719	858	152	371	568	880	963	387	340
367	818	882	251	394	666	903	·0111061	410	438
391	917	905	350	417	765	926	160	433	536
414	·0092016	928	448	440	863	949	258	457	634
437	115	952	547	464	962	972	356	480	732
461	214	975	646	487	·0105060	996	454	503	830
484	313	998	745	510	159	1026019	553	526	928
507	413	1023021	844	533	257	042	651	549	·0118026
531	512	045	943	557	356	065	749	573	124
554	611	068	·0099041	580	454	088	847	596	222
577	710	091	140	603	553	112	946	619	320
601	809	115	239	626	651	135	·0112044	642	418
624	908	138	338	650	750	158	142	665	516
647	·0093007	161	437	673	848	181	240	689	614
671	106	185	535	696	947	204	338	712	711
694	205	208	634	719	·0106045	228	436	735	809
717	304	231	733	742	144	251	535	758	907
741	403	254	832	766	242	274	633	781	·0119005
764	502	277	931	789	341	297	731	805	103
787	601	301	·0100029	812	439	321	829	828	201
811	700	324	128	835	538	344	927	851	299
834	799	347	227	858	636	367	·0113025	874	396

1027897	·0119494
920	592
944	690
967	788
990	886
1028013	984
036	·0120081
060	179
083	277
106	375
129	473
152	571
175	668
199	766
222	864
245	962
268	·0121059
291	157
314	255
337	353
361	450
384	548
407	646
430	744
453	841
476	939
499	·0122037
523	135
546	232
569	330
592	428
615	526
638	623
661	721
685	819
708	917
731	·0123014
754	112
777	210
800	307
823	405
847	503
870	601
893	698
916	796
939	894
962	991
985	·0124089
1029009	187
032	284
055	382
078	480
101	578
124	675
147	773
171	870
194	968
217	·0125065
240	163
263	261
286	358
309	456
333	553
356	651
379	748

1029402	·0125846
425	944
448	·0126041
471	139
495	236
518	334
541	431
564	529
587	627
610	724
633	822
656	919
680	·0127017
703	114
726	212
749	309
772	407
795	504
818	602
841	699
864	797
887	894
911	992
934	·0128089
957	187
980	284

Excesses

10 = 42	1 = 4
20 = 85	2 = 8
	3 = 13
	4 = 17
	5 = 21
	6 = 25
	7 = 30
	8 = 34
	9 = 38
12 = 49	

Anti Log Excesses

10 = 2	1 = 0
20 = 5	2 = 0
30 = 7	3 = 1
40 = 9	4 = 1
50 = 12	5 = 1
60 = 14	6 = 1
70 = 17	7 = 2
80 = 19	8 = 2
90 = 21	9 = 2
49 = 12	

1030003	·0128382
026	479
049	577
072	674
095	772
118	869
141	967
164	·0129064
188	162
211	259

1030234	·0129356
257	454
280	551
303	649
326	746
349	844
372	941
396	·0130038
419	136
442	233
465	331
488	428
511	526
534	623
557	720
580	818
603	915
626	·0131012
649	110
673	207
696	304
719	402
742	499
765	596
788	694
811	791
834	888
857	986
880	·0132083
904	180
927	278
950	375
973	472
996	570
1031019	667
042	764
065	861
088	959
111	·0133056
134	153
158	250
181	347
204	445
227	542
250	639
273	736
296	833
319	930
342	·0134028
365	125
389	222
412	319
435	416
458	514
481	611
504	708
527	805
550	902
573	999
596	·0135097
620	194
643	291
666	388
689	485
712	582

1031735	·0135679
758	777
781	874
804	971
827	·0136068
850	165
874	262
897	359
920	457
943	554
966	651
989	748
1032012	845
035	942
058	·0137039
081	136
104	233
127	330
150	427
173	524
196	621
220	719
243	816
266	913
289	·0138010
312	107
335	204
358	301
381	398
404	495
427	592
450	689
473	786
496	883
519	980
542	·0139077
565	174
588	271
611	368
634	465
657	562
680	659
703	756
727	853
750	950
773	·0140047
796	144
819	241
842	338
865	435
888	532
911	629
934	726
957	823
980	920
1033003	·0141017
026	114
049	210
072	307
095	404
118	501
141	598
164	695
187	792
211	888

1033234	·0141985
257	·0142082
280	179
303	276
326	373
349	470
372	566
395	663
418	760
441	857
464	954
487	·0143051
510	147
533	244
556	341
579	438
602	534
625	631
648	728
672	825
695	921
718	·0144018
741	115
764	212
787	308
810	405
833	502
856	599
879	695
902	792
925	889
948	986
971	·0145082
994	179
1034017	276
040	372
063	469
086	566
109	662
132	759
155	856
179	952
202	·0146049
225	146
248	242
271	339
294	436
317	532
340	629
363	726
386	822
409	919
432	·0147016
455	112
478	209
501	305
524	402
547	498
570	595
593	692
616	788
639	885
662	981
685	·0148078
708	174

1034731	·0148271
754	368
777	464
800	561
823	657
846	754
869	850
892	947
915	·0149044
938	140
961	237
984	333
1035007	430
030	526
053	623
076	719
099	816
122	912
145	·0150009
168	105
191	202
214	298
237	395
260	491
283	588
306	684
329	781
352	877
375	974
398	·0151070
421	167
444	263
467	360
490	456
513	552
536	649
559	745
582	841
605	938
628	·0152034
651	131
674	227
697	323
720	420
743	516
766	612
789	709
812	805
835	902
858	998
881	·0153094
904	191
927	287
950	383
973	480
996	576
1036019	672
042	769
065	865
087	961
110	·0154058
133	154
156	250
179	347
202	443

1036225	·0154539
248	636
271	732
294	828
317	925
340	·0155021
363	117
386	213
409	310
432	406
455	502
478	598
501	695
524	791
547	887
570	983
593	·0156080
616	176
639	272
662	368
685	464
708	561
731	657
754	753
777	849
800	946
823	·0157042
846	138
869	234
892	330
915	427
938	523
961	619
984	715
1037007	811
030	907
052	·0158003
075	100
098	196
121	292
144	388
167	484
190	580
213	676
236	773
259	869
282	965
305	·0159061
328	157
351	253
374	349
397	445
420	541
443	637
466	733
489	829
512	925
534	·0160022
557	118
580	214
603	310
626	406
649	502
672	598
695	694

1037718	·0160790
741	886
764	982
787	·0161078
810	174
833	270
856	366
879	462
901	558
924	654
947	750
970	846
993	942
1038016	·0162038
039	134
062	230
085	326
108	422
130	518
153	614
176	710
199	806
222	902
245	998
268	·0163094
291	190
314	286
337	382
360	477
383	573
406	669
429	765
452	861
475	957
497	·0164053
520	148
543	244
566	340
589	436
612	532
635	628
658	724
681	819
704	915
727	·0165011
750	107
773	203
796	299
819	394
842	490
864	586
887	682
910	778
933	873
956	969
979	·0166065
1039002	161
025	257
048	353
071	448
093	544
116	640
139	736
162	832
185	927

1039208	·0167023
231	119
254	215
277	310
300	406
323	502
345	597
368	693
391	789
414	884
437	980
460	·0168076
483	171
506	267
529	363
552	458
574	554
597	650
620	745
643	841
666	937
689	·0169032
712	128
735	224
758	319
781	415
804	510
826	606
849	702
872	797
895	893
918	989
941	·0170084
964	180
987	275

Excesses

10 = 42	1 = 4
20 = 84	2 = 8
	3 = 13
	4 = 17
	5 = 21
	6 = 25
	7 = 29
	8 = 34
	9 = 38

12 = 49

Anti Log Excesses

10 = 2	1 = 0
20 = 5	2 0
30 = 7	3 = 1
40 = 9	4 = 1
50 = 12	5 = 1
60 = 14	6 = 1
70 = 17	7 = 2
80 = 19	8 = 2
90 = 21	9 = 2

49 = 12

1040010	·0170371
033	467
056	562
078	658
101	754
124	849
147	945
170	·0171040
193	136
216	232
239	327
262	423
284	518
307	614
330	709
353	805
376	900
399	996
422	·0172091
444	187
467	282
490	378
513	473
536	569
559	664
582	760
604	855
627	951
650	·0173046
673	142
696	237
719	333
742	428
765	524
787	619
810	715
833	810
856	905
879	·0174001
902	096
925	192
948	287
971	383
994	478
1041016	573
039	669
062	764
085	860
108	955
131	·0175050
154	146
177	241
200	337
223	432
245	528
268	623
291	718
314	814
337	909
360	·0176004
383	100
405	195
428	291
451	386
474	481

1041497	·0176577
520	672
543	767
565	863
588	958
611	·0177054
634	149

Excesses

12 = 49

Anti Log Excesses

49 = 12

1041658	·0177249
682	349
706	449
730	549
754	649
778	749
802	849
826	949
850	·0178049
874	149
898	249
922	349
946	449
970	549
994	649
1042018	749
042	849
066	949
090	·0179049
114	149
138	249
162	349
186	449
210	549
234	649
258	749
282	849
306	949
330	·0180049
354	149
377	248
401	348
425	448
449	548
473	648
497	748
521	848
545	948
569	·0181048
593	148
617	248
641	348
665	448
689	547
713	647
737	747
761	847
785	947
809	·0182047

1042833	·0182147
857	246
881	346
905	446
929	546
953	646
977	746
1043001	845
025	945
049	·0183045
073	145
097	245
121	344
145	444
169	544
193	644
217	743
241	843
265	943
289	·0184042
312	142
336	242
360	341
384	441
408	541
432	641
456	740
480	840
504	940
528	·0185039
552	139
576	239
600	338
624	438
648	538
672	637
696	737
720	837
744	936
768	·0186036
791	136
815	235
839	335
863	434
887	534
911	634
935	733
959	833
983	933
1044007	·0187032
031	132
055	232
079	331
103	431
127	530
151	630
175	730
199	829
223	929
247	·0188028
270	128
294	228
318	327
342	427
366	526

1044390	·0188626
414	726
438	825
462	925
486	·0189024
510	124
534	223
558	323
582	422
606	522
630	621
654	721
678	820
702	920
726	·0190019
750	119
773	218
797	317
821	417
845	516
869	616
893	715
917	815
941	914
965	·0191014
989	113
1045013	212
037	312
061	411
085	511
108	610
132	709
156	809
180	908
204	·0192008
228	107
252	206
276	306
300	405
324	505
347	604
371	703
395	803
419	902
443	·0193002
467	101
491	200
515	300
539	399
563	498
586	597
610	697
634	796
658	895
682	994
706	·0194094
730	193
754	292
778	391
802	491
825	590
849	689
873	788
897	888
921	987

1045945	·0195086
969	185
993	284
1046017	384
041	483
064	582
088	681
112	780
136	880
160	979
184	·0196078
208	177
232	276
256	376
280	475
303	574
327	673
351	772
375	872
399	971
423	·0197070
447	169
471	268
495	367
519	467
542	566
566	665
590	764
614	863
638	962
662	·0198062
686	161
710	260
734	359
758	458
781	557
805	656
829	756
853	855
877	954
901	·0199053
925	152
949	251
973	350
996	449
1047020	548
044	647
068	746
092	845
116	944
139	·0200043
163	142
187	241
211	340
235	439
259	538
283	637
306	736
330	835
354	934
378	·0201033
402	132
426	231
450	330
473	429

1047497	·0201528
521	627
545	726
569	825
593	924
616	·0202023
640	121
664	220
688	319
712	418
736	517
760	616
783	715
807	814
831	913
855	·0203012
879	111
903	210
927	308
951	407
974	506
998	605
1048022	704
046	802
070	901
094	·0204000
118	099
142	198
166	296
190	395
213	494
237	593
261	692
285	790
309	889
333	988
357	·0205087
381	186
404	284
428	383
452	482
476	581
500	679
523	778
547	877
571	976
595	·0206074
619	173
642	272
666	371
690	469
714	568
738	667
761	766
785	864
809	963
833	·0207062
857	160
881	259
904	358
928	456
952	555
976	654
1049000	752
024	851

Number	Log
1049047	·0207950
071	·0208048
095	147
119	245
143	344
167	443
191	541
214	640
238	739
262	837
286	936
310	·0209035
334	133
357	232
381	330
405	429
429	528
453	626
476	725
500	823
524	922
548	·0210021
572	119
595	218
619	316
643	415
667	514
691	612
714	711
738	809
762	908
786	·0211007
810	105
834	204
857	302
881	401
905	499
929	598
953	696
977	795

Excesses

10 = 42	1 = 4
20 = 83	2 = 8
	3 = 12
	4 = 17
	5 = 21
	6 = 25
	7 = 29
	8 = 33
	9 = 37

12 = 49

Anti Log Excesses

10 = 2	1 = 0
20 = 5	2 = 0
30 = 7	3 = 1
40 = 10	4 = 1
50 = 12	5 = 1
60 = 14	6 = 1
70 = 17	7 = 2
80 = 19	8 = 2
90 = 22	9 = 2

49 = 12

Number	Log
1050001	·0211893
024	992
048	·0212090
072	189
096	287
120	386
144	484
167	583
191	681
215	780
239	878
263	976
287	·0213075
310	173
334	272
358	370
382	468
406	567
429	665
453	764
477	862
501	960
525	·0214059
548	157
572	256
596	354
620	452
644	551
667	649
691	748
715	846
739	944
762	·0215043
786	141
810	239
834	337
857	436
881	534
905	632
929	730
952	829
976	927
1051000	·0216025
024	123
047	222
071	320
095	418
119	516
142	615
166	713
190	811

Number	Log
1051214	·0216909
238	·0217008
261	106
285	204
309	302
333	401
357	499
380	597
404	695
428	794
452	892
476	990
499	·0218088
523	187
547	285
571	383
595	481
618	580
642	678
666	776
690	874
713	972
737	·0219070
761	168
785	267
808	365
832	463
856	561
880	659
903	757
927	855
951	953
975	·0220051
998	149
1052022	248
046	346
070	444
093	542
117	640
141	738
165	836
188	934
212	·0221032
236	130
260	228
283	326
307	424
331	522
355	620
378	719
402	817
426	915
450	·0222013
473	111
497	209
521	307
545	405
568	503
592	601
616	699
640	797
663	895
687	993
711	·0223091
735	189

Number	Log
1052758	·0223287
782	385
806	483
830	581
853	679
877	776
901	874
925	972
948	·0224070
972	168
996	266
1053020	364
043	462
067	560
091	658
115	756
138	854
162	952
186	·0225049
210	147
233	245
257	343
281	441
305	539
328	637
352	734
376	832
400	930
423	·0226028
447	126
471	224
495	321
518	419
542	517
566	615
590	713
613	811
637	908
661	·0227006
684	104
708	202
732	299
756	397
779	495
803	593
827	690
850	788
874	886
898	984
921	·0228081
945	179
969	277
993	375
1054016	472
040	570
064	668
087	765
111	863
135	961
159	·0229059
182	156
206	254
230	352
254	449
277	547

Number	Log
1054301	·0229645
325	742
349	840
372	938
396	·0230036
420	133
444	231
467	329
491	426
515	524
539	622
562	719
586	817
610	914
633	·0231012
657	110
681	207
705	305
728	402
752	500
776	598
799	695
823	793
847	890
870	988
894	·0232086
918	183
942	281
965	378
989	476
1055013	574
036	671
060	769
084	866
107	964
131	·0233061
155	159
178	256
202	354
226	451
249	549
273	646
296	744
320	841
344	939
367	·0234036
391	134
415	231
438	329
462	426
486	523
509	621
533	718
557	816
581	913
604	·0235010
628	108
652	205
675	303
699	400
723	497
746	595
770	692
794	790
817	887

1055841	·0235984
865	·0236082
889	180
912	277
936	374
960	471
983	569
1056007	666
031	763
054	861
078	958
102	·0237055
126	153
149	250
173	348
197	445
220	542
244	640
268	737
291	834
315	932
339	·0238029
363	126
386	224
410	321
434	418
457	516
481	613
505	710
528	807
552	905
576	·0239002
599	099
623	196
646	294
670	391
694	488
717	585
741	683
765	780
788	877
812	974
836	·0240072
859	169
883	266
907	363
930	460
954	557
978	655
1057001	752
025	849
049	946
072	·0241043
096	140
119	238
143	335
167	432
190	529
214	626
238	723
261	820
285	918
309	·0242015
332	112
356	209

1057380	·0242306
403	403
427	500
450	597
474	695
498	792
521	889
545	986
568	·0243083
592	180
616	277
639	374
663	471
686	568
710	666
734	763
757	860
781	957
804	·0244054
828	151
852	248
875	345
899	442
923	539
946	636
970	733
994	830
1058017	927
041	·0245024
064	121
088	217
112	314
135	411
159	508
183	605
206	702
230	799
254	896
277	993
301	·0246090
325	187
348	284
372	381
395	478
419	575
443	671
466	768
490	865
513	962
537	·0247059
561	156
584	253
608	350
631	447
655	544
679	640
702	737
726	834
749	931
773	·0248028
797	125
820	222
844	318
868	415
891	512

1058915	·0248609
939	706
962	802
986	899
1059009	996
033	·0249093
057	190
080	286
104	383
128	480
151	577
175	674
199	770
222	867
246	964
270	·0250061
293	158
317	254
340	351
364	448
387	545
411	641
434	738
458	835
481	932
505	·0251028
529	125
552	222
576	319
599	415
623	512
646	609
670	706
693	802
717	899
741	996
764	·0252092
788	189
811	286
835	382
859	479
882	576
906	672
929	769
953	866
977	962

Excesses

10 = 41	1 = 4
20 = 82	2 = 8
	3 = 12
	4 = 16
	5 = 21
	6 = 25
	7 = 29
	8 = 33
	9 = 37

12 = 49

Anti Log Excesses

10 = 2	1 = 0
20 = 5	2 = 0
30 = 7	3 = 1
40 = 10	4 = 1
50 = 12	5 = 1
60 = 14	6 = 1
70 = 17	7 = 2
80 = 19	8 = 2
90 = 22	9 = 2

49 = 12

1060000	·0253059
024	155
047	252
071	349
095	445
118	542
142	639
165	735
189	832
213	929
236	·0254025
260	122
283	218
307	315
331	411
354	508
378	604
401	701
425	798
449	894
472	991
496	·0255087
519	184
543	280
567	377
590	473
614	570
637	666
661	763
685	859
708	956
732	·0256052
755	149
779	245
802	342
826	438
849	535
873	631
896	728
920	824
944	920
967	·0257017
991	113
1061014	210
038	306
061	403
085	499
108	596
132	692
156	788
179	885

1061203	·0257981
226	·0258078
250	174
273	270
297	367
320	463
344	560
367	656
391	752
415	849
438	945
462	·0259042
485	138
509	234
532	331
556	427
579	524
603	620
627	716
650	813
674	909
697	·0260005
721	102
744	198
768	294
791	390
815	487
838	583
862	679
886	776
909	872
933	968
956	·0261065
980	161
1062003	257
027	353
050	450
074	546
098	642
121	738
145	835
168	931
192	·0262027
215	123
239	219
262	316
286	412
309	508
333	604
357	700
380	797
404	893
427	989
451	·0263085
474	181
498	278
521	374
545	470
568	566
592	662
615	758
639	854
662	951
686	·0264047
709	143

								Anti Log Excesses 48 = 12	
1062733	·0264239	1064260	·0270479	1065787	·0276701	1067310	·0282905	1068670	·0288431
756	335	284	575	810	796	334	·0283000	694	531
780	431	307	671	834	892	357	095	719	631
803	527	331	766	857	988	380	191	743	731
827	623	354	862	881	·0277083	404	286	768	831
850	719	378	958	904	179	427	381	792	931
874	815	401	·0271054	928	274	451	476	817	·0289031
897	912	425	150	951	370	474	572	842	131
921	·0265008	448	246	975	465	497	667	866	231
944	104	472	342	998	561	521	762	891	331
968	200	495	437	1066022	656	544	857	915	431
991	296	519	533	045	752	568	953	940	531
1063015	392	542	629	068	847	591	·0284048	965	631
038	488	566	725	092	943	614	143	989	731
062	584	589	821	115	·0278038	638	238	1069014	831
085	680	613	916	139	134	661	334	038	931
109	776	636	·0272012	162	229	685	429	063	·0290031
132	872	660	108	186	325	708	524	087	131
156	968	683	204	209	420	731	619	112	231
179	·0266064	707	300	233	516	755	714	137	331
203	160	730	395	256	611	778	810	161	431
226	256	754	491	280	707	802	905	186	531
250	353	777	587	303	802	825	·0285000	211	631
273	449	801	683	326	897	849	095	235	731
297	545	824	779	350	993	872	190	260	831
320	641	848	874	373	·0279088	896	286	284	931
344	737	871	970	397	184	919	381	309	·0291031
367	833	895	·0273066	420	279	942	476	334	131
391	929	918	162	443	375	966	571	358	231
414	·0267025	942	257	467	470	989	666	383	330
438	121	965	353	490	566	1068013	762	408	430
461	217	989	449	514	661	036	857	432	530
485	313	1065012	545	537	757	060	952	457	630
508	409	036	640	560	852	083	·0286047	481	730
532	505	059	736	584	947	107	142	506	830
555	601	083	832	607	·0280043	130	238	531	930
579	697	106	927	631	138	154	333	555	·0292030
602	793	129	·0274023	654	234	177	428	580	130
626	889	153	119	677	329	200	523	605	230
649	985	176	214	701	425	224	618	629	330
673	·0268081	200	310	724	520	247	713	654	430
696	177	223	406	748	616	271	809	678	530
720	273	247	502	771	711	294	904	703	629
743	368	270	597	794	806	317	999	727	729
767	464	294	693	818	902	341	·0287094	752	829
790	560	317	789	841	997	364	189	777	929
814	656	341	884	865	·0281093	388	284	801	·0293029
837	752	364	980	888	188	411	380	826	129
861	848	387	·0275076	912	283	434	475	850	229
884	944	411	171	935	379	458	570	875	329
908	·0269040	434	267	959	474	481	665	900	429
931	136	458	362	982	570	505	760	924	528
955	232	481	458	1067006	665	528	855	949	628
978	328	505	554	029	760	551	950	973	728
1064002	424	528	649	052	856	575	·0288046	998	828
025	520	552	745	076	951	598	141		
049	616	575	840	099	·0282047	622	236		
072	712	599	936	123	142	645	331		
096	807	622	·0276032	146	237				
119	903	646	127	170	333	Excesses			
143	999	669	223	193	428				
166	·0270095	693	318	217	524	12 = 48			
190	191	716	414	240	619				
213	287	740	510	263	714				
237	383	763	605	287	810				

Excesses

10 = 41	1 = 4
20 = 82	2 = 8
	3 = 12
	4 = 16
	5 = 20
	6 = 24
	7 = 29
	8 = 32
	9 = 37

12 = 48

Anti Log Excesses

10 = 2	1 = 0
20 = 5	2 = 0
30 = 7	3 = 1
40 = 10	4 = 1
50 = 12	5 = 1
60 = 15	6 = 1
70 = 17	7 = 2
80 = 20	8 = 2
90 = 22	9 = 2

48 = 12

1070022	·0293928
047	·0294028
072	128
096	228
121	327
145	427
170	527
194	627
219	727
244	826
268	926
293	·0295026
317	126
342	225
367	325
391	425
416	525
440	625
465	724
489	824
514	924
539	·0296024
563	123
588	223
612	323
637	422
661	522
686	621
711	721
735	821
760	920
784	·0297020
809	120
834	219
858	319
883	418
907	518

1070932	·0297618
956	717
981	817
1071006	917
030	·0298016
055	116
079	216
104	315
128	415
153	514
178	614
202	714
227	813
251	913
276	·0299013
301	112
325	212
350	311
374	411
399	511
423	610
448	710
473	809
497	909
522	·0300008
546	108
571	207
595	307
620	406
645	506
669	605
694	705
718	804
743	904
768	·0301003
792	103
817	202
841	302
866	401
890	500
915	600
940	699
964	799
989	898
1072013	998
038	·0302097
062	197
087	296
111	395
136	495
160	594
185	694
209	793
234	892
258	992
283	·0303091
307	191
332	290
357	390
381	489
406	588
430	688
455	787
479	886
504	986

1072528	·0304085
553	185
577	284
602	383
626	483
651	582
675	681
700	781
724	880
749	980
773	·0305079
798	178
823	278
847	377
872	476
896	575
921	675
945	774
970	873
994	972
1073019	·0306071
043	171
068	270
092	369
117	468
141	568
166	667
190	766
215	865
239	964
264	·0307064
289	163
313	262
338	361
362	460
387	560
411	659
436	758
460	857
485	956
509	·0308056
534	155
558	254
583	353
607	453
632	552
656	651
681	750
705	849
730	949
755	·0309048
779	147
803	246
828	345
852	444
877	543
901	642
926	741
950	840
975	939
999	·0310038
1074024	138
048	237
073	336
097	435

1074122	·0310534
146	633
171	732
195	831
219	930
244	·0311029
269	128
293	227
318	326
342	425
367	524
391	623
416	722
440	821
465	920
489	·0312019
514	118
538	217
563	316
587	415
612	514
636	613
661	712
686	811
710	910
734	·0313009
759	108
783	207
808	306
832	404
857	503
881	602
906	701
930	800
955	899
979	998
1075004	·0314097
028	196
053	294
077	393
102	492
126	591
150	690
175	789
199	888
224	987
248	·0315086
273	184
297	283
322	382
346	481
371	580
395	679
420	777
444	876
469	975
493	·0316074
518	173
542	272
567	370
591	469
615	568
640	667
664	766
689	864

1075713	·0316963
738	·0317062
762	161
787	259
811	358
836	457
860	556
885	654
909	753
934	852
958	951
983	·0318049
1076007	148
032	247
056	346
080	444
105	543
129	642
154	740
178	839
203	938
227	·0319036
252	135
276	233
300	332
325	431
349	529
374	628
398	727
422	825
447	924
471	·0320022
496	121
520	220
545	318
569	417
593	516
618	614
642	713
667	811
691	910
716	·0321008
740	107
764	206
789	304
813	403
838	501
862	600
886	699
911	797
935	896
960	994
984	·0322093
1077009	191
033	290
057	388
082	487
106	585
131	684
155	782
180	881
204	979
229	·0323078
253	176
278	275

Number	Log
1077302	·0323373
327	472
351	570
376	669
400	767
425	866
449	964
473	·0324062
498	161
522	259
547	358
571	456
596	555
620	653
645	752
669	850
693	948
718	·0325047
742	145
767	244
791	342
815	440
840	539
864	637
889	736
913	834
938	933
962	·0326031
986	129
1078011	228
035	326
059	424
084	522
108	621
133	719
157	817
181	915
206	·0327014
230	112
254	210
279	308
303	407
328	505
352	603
376	701
401	800
425	898
449	996
474	·0328095
498	193
523	291
547	389
572	488
596	586
620	684
645	782
669	881
694	979
718	·0329077
742	175
767	274
791	372
816	470
840	568
865	667

Number	Log
1078889	·0329765
913	863
938	961
962	·0330059
986	157
1079011	256
035	354
060	452
084	550
108	648
133	746
157	844
181	942
206	·0331040
230	139
255	237
279	335
303	433
328	531
352	629
376	727
401	825
425	923
449	·0332021
474	120
498	218
523	316
547	414
571	512
596	610
620	708
644	806
669	904
693	·0333003
718	101
742	199
766	297
791	395
815	493
839	591
864	689
888	787
912	885
937	983
961	·0334081
986	179

Excesses

10 = 40	1 = 4
20 = 81	2 = 8
	3 = 12
	4 = 16
	5 = 20
	6 = 24
	7 = 28
	8 = 32
	9 = 36

12 = 49

Anti Log Excesses

10 = 2	1 = 0
20 = 5	2 = 0
30 = 7	3 = 1
40 = 10	4 = 1
50 = 12	5 = 1
60 = 15	6 = 1
70 = 17	7 = 2
80 = 20	8 = 2
90 = 22	9 = 2

49 = 12

Number	Log
1080010	·0334277
034	375
059	473
083	570
107	668
132	766
156	864
181	962
205	·0335060
229	158
254	256
278	354
302	452
327	550
351	648
375	746
400	843
424	941
449	·0336039
473	137
497	235
522	333
546	431
570	529
595	627
619	724
644	822
668	920
692	·0337018
717	116
741	214
765	312
790	410
814	508
838	605
863	703
887	801
912	899
936	997
960	·0338095
985	192
1081009	290
033	388
058	486
082	584
107	682
131	779
155	877
180	975
204	·0339073
228	171

Number	Log
1081253	·0339268
277	366
301	464
326	561
350	659
374	757
399	854
423	952
447	·0340050
472	147
496	245
520	343
544	441
569	538
593	636
617	734
642	831
666	929
690	·0341027
715	124
739	222
763	320
788	417
812	515
836	612
861	710
885	808
909	905
934	·0342003
958	101
982	198
1082006	296
031	393
055	491
079	589
104	686
128	784
152	882
177	979
201	·0343077
225	174
250	272
274	369
298	467
323	565
347	662
371	760
396	857
420	955
444	·0344053
468	150
493	248
517	345
541	443
566	540
590	638
614	735
639	833
663	930
687	·0345028
712	125
736	223
760	320
785	418
809	515

Number	Log
1082833	·0345613
858	710
882	808
906	905
930	·0346003
955	100
979	198
1083003	295
028	392
052	490
076	587
101	685
125	782
149	879
173	977
198	·0347074
222	172
246	269
270	366
295	464
319	561
343	658
367	756
392	853
416	951
440	·0348048
464	145
489	243
513	340
537	437
562	535
586	632
610	729
635	827
659	924
683	·0349021
708	119
732	216
756	313
781	411
805	508
829	605
853	702
878	800
902	897
926	994
951	·0350092
975	189
999	286
1084024	384
048	481
072	578
096	675
121	773
145	870
169	967
193	·0351064
218	162
242	259
266	356
290	453
315	551
339	648
363	745
387	842

1084412	·0351940
436	·0352037
460	134
485	231
509	328
533	425
557	523
582	620
606	717
630	814
654	911
679	·0353008
703	105
727	202
751	299
775	397
800	494
824	591
848	688
873	785
897	882
921	979
946	·0354076
970	173
994	270
1085018	368
043	465
067	562
091	659
115	756
140	853
164	950
188	·0355047
212	144
237	242
261	339
285	436
309	533
334	630
358	727
382	824
406	921
431	·0356018
455	115
479	212
503	309
527	406
552	503
576	600
600	697
624	793
649	890
673	987
697	·0357084
721	181
745	278
770	375
794	472
818	569
842	666
867	763
891	860
915	957
939	·0358053
964	150

1085988	·0358247
1086012	344
036	441
061	538
085	635
109	732
133	829
158	925
182	·0359022
206	119
230	216
255	313
279	410
303	507
327	604
352	700
376	797
400	894
424	991
448	·0360088
473	184
497	281
521	378
545	475
570	571
594	668
618	765
642	862
666	959
691	·0361055
715	152
739	249
763	346
787	442
812	539
836	636
860	733
884	829
908	926
933	·0362023
957	120
981	216
1087005	313
030	410
054	507
078	603
102	700
126	797
151	894
175	990
199	·0363087
223	184
247	280
272	377
296	474
320	570
344	667
368	764
393	860
417	957
441	·0364054
465	150
490	247
514	344
538	441

1087562	·0364537
586	634
611	731
635	827
659	924
683	·0365021
707	117
731	214
756	310
780	407
804	503
828	600
852	697
876	793
901	890
925	986
949	·0366083
973	180
997	276
1088021	373
046	469
070	566
094	662
118	759
142	855
166	952
191	·0367048
215	145
239	241
263	338
287	434
312	531
336	627
360	724
384	820
409	917
433	·0368013
457	110
481	206
505	303
530	399
554	495
578	592
602	688
626	785
650	881
675	977
699	·0369074
723	170
747	267
771	363
795	459
820	556
844	652
868	748
892	845
916	941
940	·0370038
965	134
989	230
1089013	327
037	423
061	519
085	616
109	712

1089134	·0370808
158	905
182	·0371001
206	098
230	194
254	290
279	387
303	483
327	579
351	676
375	772
399	869
424	965
448	·0372061
472	158
496	255
520	351
544	447
568	543
593	639
617	735
641	831
665	927
689	·0373024
713	120
738	216
762	312
786	409
810	505
834	601
858	697
883	793
907	890
931	986
955	·0374082
979	178

Excesses

10 = 40	1 = 4
20 = 80	2 = 8
	3 = 12
	4 = 16
	5 = 20
	6 = 24
	7 = 28
	8 = 32
	9 = 36

12 = 48

Anti Log Excesses

10 = 2	1 = 0
20 = 5	2 = 0
30 = 7	3 = 1
40 = 10	4 = 1
50 = 12	5 = 1
60 = 15	6 = 1
70 = 17	7 = 2
80 = 20	8 = 2
90 = 22	9 = 2

48 = 12

1090003	·0374274
027	371
051	467
076	563
100	659
124	755
148	852
172	948
196	·0375044
220	140
244	237
268	333
293	429
317	525
341	621
365	718
389	814
413	910
437	·0376006
461	102
485	198
510	294
534	390
558	486
582	582
606	678
630	774
655	871
679	967
703	·0377063
727	159
751	255
775	351
800	447
824	543
848	639
872	735
896	831
920	927
944	·0378023
968	119
993	215
1091017	311
041	407
065	503
089	599
113	695
137	791
161	887
185	983
210	·0379079
234	175
258	271
282	367
306	463
330	559
354	655
378	751
402	847
426	943
451	·0380039
475	135
499	231
523	327
547	423

1091571	·0380519
595	614
619	710
643	806
668	902
692	998
716	·0381094
740	190
764	286
788	382
812	478
836	574
860	670
884	765
909	861
933	957
957	·0382053
981	149
1092005	245
029	340
053	436
077	532
101	628
126	724
150	820
174	915
198	·0383011
222	107
246	203
270	299
294	394
318	490
342	586
367	682
391	777
415	873
439	969
463	·0384065
487	160
511	256
535	352
559	448
584	543
608	639
632	735
656	831
680	926
704	·0385022
728	118
752	213
776	309
800	405
824	500
848	596
872	692
896	787
920	883
944	979
969	·0386074
993	170
1093017	266
041	362
065	457
089	553
113	649

1093137	·0386744
161	840
185	936
209	·0387031
233	127
257	222
281	318
305	413
329	509
353	605
377	700
401	796
426	891
450	987
474	·0388083
498	178
522	274
546	369
570	465
594	560
618	656
642	752
666	847
690	943
714	·0389038
738	134
762	229
786	325
810	420
834	516
858	611
883	707
907	802
931	898
955	993
979	·0390089
1094003	184
027	280
051	376
075	471
099	566
123	662
147	757
171	853
195	948
219	·0391044
243	139
267	234
291	330
315	425
340	521
364	616
388	711
412	807
436	902
460	998
484	·0392093
508	189
532	284
556	379
580	475
604	570
628	665
652	761
676	856

1094700	·0392952
724	·0393047
748	142
772	238
797	333
821	428
845	524
869	619
893	715
917	810
941	905
965	·0394001
989	096
1095013	191
037	287
061	382
085	477
109	572
133	668
157	763
181	858
205	953
230	·0395049
254	144
278	239
302	334
326	430
350	525
374	620
398	715
422	811
446	906
470	·0396001
494	096
518	192
542	287
566	382
590	477
614	572
638	668
662	763
686	858
710	953
734	·0397049
758	144
782	239
806	334
830	429
854	525
878	620
902	715
926	810
950	905
974	·0398000
998	095
1096022	191
046	286
070	381
094	476
118	571
142	666
166	761
190	856
214	951
238	·0399047

1096262	·0399142
286	237
310	332
334	427
358	522
382	617
406	712
430	807
454	902
478	997
502	·0400092
526	187
550	282
574	377
598	473
622	568
646	663
670	758
694	853
718	948
742	·0401043
766	138
790	233
814	328
838	423
862	518
886	613
910	708
934	803
958	898
982	993
1097006	·0402088
030	183
054	278
078	373
102	468
126	563
150	658
174	753
198	848
222	943
246	·0403038
270	133
294	228
318	323
342	418
366	512
390	607
414	702
438	797
462	892
486	987
510	·0404081
533	176
557	271
581	366
605	461
629	556
653	650
677	745
701	840
725	935
749	·0405030
773	125
797	219

1097821	·0405314
845	409
869	504
893	599
917	693
941	788
965	883
989	978
1098013	·0406072
037	167
061	262
085	357
109	452
133	546
157	641
181	736

Excesses

12 = 48

Anti Log Excesses

48 = 12

1098206	·0406836
232	936
257	·0407036
282	136
307	236
333	336
358	436
383	536
409	636
434	736
459	836
484	936
510	·0408036
535	136
560	236
585	336
611	436
636	536
661	636
687	736
712	836
737	936
762	·0409036
788	136
813	236
838	336
864	435
889	535
914	635
939	735
965	835
990	935
1099015	·0410035
040	135
066	235
091	335
116	435
142	535
167	634
192	734

1099217	·0410834
243	934
268	·0411034
293	134
319	233
344	333
369	433
394	533
420	633
445	733
470	832
495	932
521	·0412032
546	132
571	232
596	331
622	431
647	531
672	631
697	730
723	830
748	930
773	·0413029
798	129
823	229
849	329
874	428
899	528
924	628
950	728
975	827

Excesses

10 = 40	1 = 4
20 = 79	2 = 8
	3 = 12
	4 = 16
	5 = 20
	6 = 24
	7 = 28
	8 = 32
	9 = 36

12 = 48

Anti Log Excesses

10 = 3	1 = 0
20 = 5	2 = 1
30 = 8	3 = 1
40 = 10	4 = 1
50 = 13	5 = 1
60 = 15	6 = 2
70 = 18	7 = 2
80 = 20	8 = 2
90 = 23	9 = 2

48 = 12

1100000	·0413927
025	·0414027
051	126
076	226
101	326

1100126	·0414425
152	525
177	625
202	724
228	824
253	924
278	·0415023
303	123
329	223
354	322
379	422
404	522
430	621
455	721
480	821
505	920
531	·0416020
556	119
581	219
606	318
632	418
657	517
682	617
707	717
732	816
758	916
783	·0417015
808	115
833	214
859	314
884	413
909	513
934	613
959	712
985	812
1101010	911
035	·0418011
060	110
086	210
111	309
136	409
161	508
186	608
212	707
237	807
262	906
287	·0419006
313	105
338	205
363	304
388	403
413	503
439	602
464	702
489	801
514	900
540	·0420000
565	099
590	199
615	298
640	397
666	497
691	596
716	695
741	795

1101767	·0420894
792	994
817	·0421093
842	192
867	292
893	391
918	490
943	590
968	689
994	788
1102019	888
044	987
069	·0422086
094	186
120	285
145	384
170	484
195	583
221	682
246	782
271	881
296	980
321	·0423080
347	179
372	278
397	377
422	477
447	576
472	675
498	775
523	874
548	973
573	·0424072
598	172
623	271
649	370
674	469
699	569
724	668
749	767
774	866
800	966
825	·0425065
850	164
875	263
901	362
926	461
951	561
976	660
1103001	759
027	858
052	957
077	·0426056
102	156
128	255
153	354
178	453
203	552
228	651
254	750
279	849
304	948
329	·0427047
354	146
379	245

1103405	·0427345
430	444
455	543
480	642
505	741
530	840
556	939
581	·0428038
606	137
631	236
656	335
681	434
707	533
732	632
757	731
782	830
807	929
832	·0429028
858	127
883	226
908	325
933	424
958	523
983	622
1104009	721
034	820
059	919
084	·0430018
109	117
134	216
160	315
185	414
210	513
235	612
260	711
285	810
311	908
336	·0431007
361	106
386	205
411	304
436	403
462	502
487	601
512	700
537	799
562	898
587	997
612	·0432095
637	194
663	293
688	392
713	491
738	590
763	688
788	787
813	886
838	985
863	·0433084
889	183
914	281
939	380
964	479
989	578
1105014	677

1105039	·0433776
065	874
090	973
115	·0434072
140	171
165	269
190	368
216	467
241	566
266	665
291	763
316	862
341	961
367	·0435060
392	158
417	257
442	356
467	455
492	553
517	652
542	751
568	849
593	948
618	·0436047
643	145
668	244
693	343
718	441
743	540
768	639
794	737
819	836
844	935
869	·0437033
894	132
919	231
944	329
969	428
994	526
1106020	625
045	724
070	822
095	921
120	·0438019
145	118
170	217
195	315
220	414
246	513
271	611
296	710
321	808
346	907
371	·0439006
396	104
421	203
446	301
472	400
497	498
522	597
547	695
572	794
597	892
622	991
647	·0440089
672	188

1106698	·0440286
723	385
748	483
773	582
798	680
823	778
848	877
873	975
898	·0441074
923	172
948	271
973	369
998	468
1107024	566
049	664
074	763
099	861
124	960
149	·0442058
174	157
199	255
224	353
249	452
274	550
299	649
324	747
350	846
375	944
400	·0443042
425	141
450	239
475	338
500	436
525	534
550	633
576	731
601	829
626	928
651	·0444026
676	125
701	223
726	321
751	420
776	518
801	616
826	714
851	813
876	911
901	·0445009
927	108
952	206
977	304
1108002	402
027	501
052	599
077	697
102	795
127	894
152	992
177	·0446090
202	188
227	287
252	385
277	483
302	581

1108327	·0446679
352	777
378	876
403	974
428	·0447072
453	170
478	268
503	366
528	465
553	563
578	661
603	759
628	857
653	955
678	·0448054
703	152
728	250
753	348
778	446
803	544
829	643
854	741
879	839
904	937
929	·0449035
954	133
979	232
1109004	330
029	428
054	526
079	624
104	722
129	820
154	918
179	·0450016
204	114
229	212
254	310
280	408
305	506
330	604
355	702
380	800
405	898
430	996
455	·0451094
480	192
505	290
530	388
555	486
580	584
605	682
630	780
655	878
680	976
705	·0452074
730	171
755	269
780	367
805	465
830	563
855	661
880	759
905	857
930	955

1109955	·0453053
980	151

Excesses

10 = 39	1 = 4
20 = 79	2 = 8
	3 = 12
	4 = 16
	5 = 20
	6 = 24
	7 = 28
	8 = 31
	9 = 35

12 = 48

Anti Log Excesses

10 = 3	1 = 0
20 = 5	2 = 1
30 = 8	3 = 1
40 = 10	4 = 1
50 = 13	5 = 1
60 = 15	6 = 2
70 = 18	7 = 2
80 = 20	8 = 2
90 = 23	9 = 2

48 = 12

1110005	·0453249
030	347
055	445
080	542
105	640
130	738
155	836
181	934
206	·0454032
231	130
256	228
281	326
306	423
331	521
356	619
381	717
406	815
431	913
456	·0455011
481	108
506	206
531	304
556	402
581	500
606	598
631	695
656	793
681	891
706	989
731	·0456087
756	185
781	282
806	380
831	478

1110856	·0456576
881	674
906	771
931	869
956	967
981	·0457065
1111006	162
031	260
056	358
081	455
106	553
131	651
156	749
181	846
206	944
231	·0458042
256	140
281	237
306	335
331	433
356	530
381	628
406	725
431	823
456	921
481	·0459018
506	116
531	213
556	311
581	409
606	506
631	604
656	702
681	799
706	897
731	994
756	·0460092
781	190
806	287
831	385
856	482
881	580
906	678
931	775
956	873
981	970
1112006	·0461068
031	166
056	263
081	361
106	459
131	556
156	654
181	751
206	849
231	946
256	·0462044
281	141
306	239
331	336
356	434
381	531
406	629
430	726
455	823

1112480	·0462921
505	·0463018
530	116
555	213
580	311
605	408
630	505
655	603
680	700
705	798
730	895
755	993
780	·0464090
805	187
830	285
855	382
879	480
904	577
929	674
954	772
979	869
1113004	966
029	·0465064
054	161
079	259
104	356
129	453
154	551
179	648
204	745
229	843
254	940
279	·0466037
304	135
328	232
353	329
378	427
403	524
428	621
453	719
478	816
503	913
528	·0467011
553	108
578	205
603	303
628	400
653	497
678	594
703	692
728	789
753	886
777	984
802	·0468081
827	178
852	275
877	373
902	470
927	567
952	664
977	762
1114002	859
027	956
052	·0469053
077	151

1114102	·0469248
127	345
152	442
177	539
202	636
226	734
251	831
276	928
301	·0470025
326	122
351	219
376	317
401	414
426	511
451	608
476	705
501	802
526	899
551	996
576	·0471093
601	190
626	287
651	384
675	482
700	579
725	676
750	773
775	870
800	967
825	·0472064
850	161
875	258
900	355
925	452
950	549
975	646
1115000	743
024	840
049	937
074	·0473034
099	131
124	229
149	326
174	423
199	520
224	617
248	714
273	811
298	908
323	·0474005
348	102
373	199
398	296
423	393
448	490
472	587
497	684
522	781
547	878
572	974
597	·0475071
622	168
647	265
672	362
696	459

1115721	·0475556
746	653
771	750
796	847
821	944
846	·0476041
871	137
896	234
920	331
945	428
970	525
995	622
1116020	718
045	815
070	912
095	·0477009
120	106
144	203
169	299
194	396
219	493
244	590
269	687
294	784
319	880
344	977
368	·0478074
393	171
418	267
443	364
468	461
493	558
518	655
543	751
568	848
592	945
617	·0479042
642	138
667	235
692	332
717	429
742	525
767	622
792	719
816	816
841	912
866	·0480009
891	106
916	202
941	299
966	396
991	493
1117016	589
040	686
065	783
090	880
115	976
140	·0481073
165	170
190	266
214	363
239	459
264	556
289	653
314	749

1117339	·0481846
363	942
388	·0482039
413	136
438	232
463	329
488	426
512	522
537	619
562	715
587	812
612	909
637	·0483005
661	102
686	198
711	295
736	391
761	488
786	584
810	681
835	778
860	874
885	971
910	·0484067
935	164
959	260
984	357
1118009	454
034	550
059	646
084	743
109	839
134	936
158	·0485032
183	129
208	225
233	322
258	418
283	514
308	611
333	707
358	804
382	900
407	997
432	·0486093
457	189
482	286
507	382
532	479
556	575
581	672
606	768
631	864
656	961
681	·0487057
705	154
730	250
755	346
780	443
805	539
830	635
854	732
879	828
904	925
929	·0488021

1118954	·0488117
979	214
1119003	310
028	406
053	503
078	599
102	695
127	792
152	888
177	984
202	·0489081
226	177
251	273
276	370
301	466
325	562
350	659
375	755
400	851
425	948
449	·0490044
474	140
499	236
524	333
549	429
574	525
598	622
623	718
648	814
673	910
698	·0491007
723	103
747	199
772	295
797	392
822	488
847	584
872	680
896	777
921	873
946	969
971	·0492065
995	161

Excesses

10 = 39	1 = 4
20 = 78	2 = 8
	3 = 12
	4 = 16
	5 = 19
	6 = 23
	7 = 27
	8 = 31
	9 = 35

12 = 47

Anti Log Excesses

10 = 3	1 = 0
20 = 5	2 = 1
30 = 8	3 = 1
40 = 10	4 = 1
50 = 13	5 = 1
60 = 15	6 = 2
70 = 18	7 = 2
80 = 21	8 = 2
90 = 23	9 = 2

47 = 12

1120020	·0492257
045	354
070	450
095	546
119	642
144	738
169	834
194	931
218	·0493027
243	123
268	219
293	315
318	411
342	507
367	603
392	699
417	795
442	891
467	987
491	·0494084
516	180
541	276
566	372
591	468
616	564
640	660
665	756
690	852
715	948
740	·0495044
765	140
789	236
814	332
839	428
864	524
888	620
913	716
938	812
963	908
988	·0496004
1121012	100
037	196
062	292
087	388
111	484
136	580
161	676
186	772
210	868
235	964

1121260	·0497060	1122869	·0503288	1124476	·0509499	1126081	·0515691	1127682	·0521867
285	156	894	384	500	594	105	786	707	962
309	252	918	479	525	689	130	881	732	·0522056
334	348	943	575	550	785	155	976	756	151
359	444	968	671	575	880	179	·0516072	781	246
383	539	992	766	599	976	204	167	805	341
408	635	1123017	862	624	·0510071	229	262	830	436
433	731	042	958	649	167	253	357	855	531
458	827	067	·0504053	674	262	278	452	879	626
482	923	091	149	698	357	303	547	904	720
507	·0498019	116	245	723	453	327	642	929	815
532	115	141	340	748	548	352	737	953	910
557	211	165	436	772	644	377	832	978	·0523005
581	307	190	532	797	739	401	928	1128003	099
606	403	215	627	822	834	426	·0517023	027	194
631	499	240	723	846	930	451	118	052	289
656	595	264	819	871	·0511025	475	213	077	384
680	690	289	914	896	120	500	308	101	479
705	786	314	·0505010	920	216	525	403	126	573
730	882	339	106	945	311	549	498	151	668
755	978	363	201	970	406	574	594	175	763
779	·0499074	388	297	994	502	598	688	200	858
804	170	413	393	1125019	597	623	783	225	952
829	265	437	488	044	692	648	878	249	·0524047
854	361	462	584	068	788	672	973	274	142
879	457	487	679	093	883	697	·0518068	299	237
903	553	512	775	118	978	721	163	323	331
928	649	536	870	142	·0512074	746	259	348	426
953	745	561	966	167	169	771	354	372	521
978	840	586	·0506061	192	264	795	449	397	616
1122002	936	610	157	216	359	820	544	422	710
027	·0500032	635	253	241	455	845	639	446	805
052	128	660	348	266	550	869	734	471	900
077	224	685	444	290	645	894	829	495	994
102	320	709	539	315	740	918	924	520	·0525089
126	415	734	635	340	836	943	·0519019	545	184
151	511	759	730	364	931	968	114	569	279
176	607	784	826	389	·0513026	992	209	594	373
201	703	808	921	414	121	1127017	304	619	468
225	798	833	·0507017	438	216	042	399	643	563
250	894	858	112	463	312	066	494	668	658
275	990	882	208	488	407	091	589	692	752
300	·0501086	907	303	512	502	116	684	717	847
325	182	932	399	537	597	140	779	742	942
349	277	957	494	562	693	165	873	766	·0526036
374	373	981	590	586	788	190	968	791	131
399	469	1124006	685	611	883	214	·0520063	815	226
424	565	031	781	636	978	239	158	840	320
448	660	055	876	660	·0514073	264	253	864	415
473	756	080	971	685	169	288	348	889	510
498	852	105	·0508067	710	264	313	443	913	604
523	948	130	162	735	359	338	538	938	699
547	·0502043	154	258	759	454	362	633	963	794
572	139	179	353	784	549	387	728	987	888
597	235	204	449	809	644	412	823	1129012	983
622	331	229	544	833	740	436	918	036	·0527078
646	426	253	639	858	835	461	·0521013	061	172
671	522	278	735	883	930	485	108	085	267
696	618	303	830	908	·0515025	510	202	110	362
720	713	327	926	932	120	535	297	135	456
745	809	352	·0509021	957	215	559	392	159	551
770	905	377	117	982	311	584	487	184	646
795	·0503001	402	212	1126007	406	608	582	208	740
819	096	426	308	031	501	633	677	233	835
844	192	451	403	056	596	658	772	257	929

1129282	·0528024
307	118
331	213
356	307
380	402
405	497
430	591
454	686
479	780
504	875
528	969
553	·0529064
577	159
602	253
627	348
651	442
676	537
700	631
725	726
750	820
774	915
799	·0530009
823	104
848	199
873	293
897	388
922	482
947	577
971	671
996	766
1130020	861
045	955

Excesses

10 = 39	1 = 4
20 = 77	2 = 8
	3 = 12
	4 = 15
	5 = 19
	6 = 23
	7 = 27
	8 = 31
	9 = 35

12 = 47

Anti Log Excesses

10 = 3	1 = 0
20 = 5	2 = 1
30 = 8	3 = 1
40 = 10	4 = 1
50 = 13	5 = 1
60 = 16	6 = 2
70 = 18	7 = 2
80 = 21	8 = 2
90 = 23	9 = 2

47 = 12

1130071	·0531055
097	155
123	255
149	355

1130175	·0531455
201	555
227	655
253	755
279	854
305	954
331	·0532054
357	154
383	254
409	354
435	454
461	554
487	654
513	754
539	854
565	954
591	·0533054
617	154
643	254
669	354
695	454
721	554
747	653
773	753
799	853
825	953
851	·0534053
877	153
903	253
929	353
955	453
981	553
1131007	652
033	752
059	852
085	952
111	·0535052
137	152
163	251
189	351
215	451
241	551
267	651
293	750
319	850
345	950
371	·0536050
397	150
423	250
449	349
475	449
501	549
527	649
553	748
579	848
605	948
631	·0537048
657	147
683	247
709	347
735	447
761	546
787	646
813	746
839	846

1131865	·0537945
891	·0538045
917	145
943	244
969	344
995	444
1132021	543
047	643
073	743
099	842
125	942
151	·0539041
177	141
203	241
229	340
255	440
281	540
307	639
333	739
359	838
385	938
411	·0540038
437	137
463	237
488	336
514	436
540	535
566	635
592	735
618	834
644	934
670	·0541033
696	133
722	233
748	332
774	432
800	531
826	631
852	730
878	830
904	929
930	·0542029
956	128
982	228
1133008	327
034	427
060	526
086	626
112	725
138	825
164	924
190	·0543024
216	123
242	223
268	322
294	422
320	521
346	621
371	720
397	820
423	919
449	·0544019
475	118
501	218
527	317

1133553	·0544417
579	516
605	615
631	715
657	814
683	913
709	·0545013
735	112
761	211
787	311
812	410
838	510
864	609
890	708
916	808
942	907
968	·0546006
994	106
1134020	205
046	304
072	404
098	503
124	602
150	701
176	801
202	900
228	999
253	·0547099
279	198
305	297
331	396
357	496
383	595
409	694
435	794
461	893
487	992
513	·0548091
539	191
565	290
590	389
616	488
642	587
668	686
694	786
720	885
746	984
772	·0549083
797	182
823	282
849	381
875	480
901	579
927	678
953	777
979	877
1135005	976
031	·0550075
057	174
083	273
109	372
134	472
160	571
186	670
212	769

1135238	0550868
264	968
290	·0551067
316	166
342	265
368	364
394	463
420	562
446	661
471	760
497	859
523	958
549	·0552057
575	157
601	256
627	355
653	454
678	553
704	652
730	751
756	850
782	949
808	·0553048
834	147
860	246
886	345
911	444
937	543
963	642
989	741
1136015	840
041	938
067	·0554037
093	136
118	235
144	334
170	433
196	532
222	631
248	730
274	829
300	928
326	·0555027
351	126
377	225
403	324
429	423
455	521
481	620
507	719
533	818
558	917
584	·0556016
610	115
636	214
662	313
688	412
714	511
740	609
766	708
791	807
817	906
843	·0557005
869	104
895	202

Number	Log
1136921	·0557301
947	400
973	499
998	598
1137024	696
050	795
076	894
102	993
128	·0558092
154	190
180	289
206	388
231	487
257	585
283	684
309	783
335	881
361	980
387	·0559079
413	178
438	276
464	375
490	474
516	572
542	671
568	770
594	868
619	967
645	·0560066
671	165
697	263
723	362
749	461
774	559
800	658
826	757
852	856
878	954
903	·0561053
929	152
955	250
981	349
1138007	448
033	546
058	645
084	743
110	842
136	941
162	·0562039
188	138
213	236
239	335
265	433
291	532
317	631
342	729
368	828
394	926
420	·0563025
446	124
472	222
498	321
524	419
549	518
575	616

Number	Log
1138601	·0563715
627	813
653	912
679	·0564010
705	109
731	207
756	306
782	404
808	503
834	601
860	700
886	798
912	897
937	995
963	·0565094
989	192
1139015	291
041	389
067	488
092	586
118	685
144	783
170	882
196	980
221	·0566079
247	177
273	276
299	374
325	472
351	571
376	669
402	767
428	866
454	964
479	·0567062
505	161
531	259
557	358
582	456
608	554
634	653
660	751
685	849
711	948
737	·0568046
763	144
789	243
814	341
840	439
866	537
892	636
918	734
944	832
969	931
995	·0569029

Excesses

10 = 38	1 = 4
20 = 77	2 = 8
	3 = 11
	4 = 15
	5 = 19
	6 = 23
	7 = 27
	8 = 31
	9 = 34

13 = 50

Anti Log Excesses

10 = 3	1 = 0
20 = 5	2 = 1
30 = 8	3 = 1
40 = 10	4 = 1
50 = 13	5 = 1
60 = 16	6 = 2
70 = 18	7 = 2
80 = 21	8 = 2
90 = 23	9 = 2

50 = 13

Number	Log
1140021	·0569127
047	225
073	324
098	422
124	520
150	619
176	717
202	815
228	913
253	·0570012
279	110
305	208
331	306
356	405
382	503
408	601
434	699
459	798
485	896
511	994
537	·0571092
562	191
588	289
614	387
640	485
666	583
691	681
717	779
743	878
769	976
795	·0572074
821	172
846	270
872	368
898	466
924	564
950	663

Number	Log
1140975	·0572761
1141001	859
027	957
053	·0573055
079	153
105	251
130	349
156	447
182	545
208	643
233	741
259	839
285	938
311	·0574036
336	134
362	232
388	330
414	428
439	526
465	624
491	722
517	820
543	918
568	·0575016
594	114
620	212
646	310
671	408
697	506
723	604
749	702
774	800
800	898
826	996
852	·0576094
877	192
903	290
929	388
955	486
980	584
1142006	682
032	780
058	878
083	976
109	·0577074
135	172
160	269
186	367
212	465
238	563
263	661
289	759
315	857
340	955
366	·0578053
392	151
418	249
443	346
469	444
495	542
521	640
546	738
572	836
598	933
624	·0579031

Number	Log
1142649	·0579129
675	227
701	325
727	422
752	520
778	618
804	716
830	814
856	912
881	·0580009
907	107
933	205
959	303
984	400
1143010	498
036	596
062	694
087	791
113	889
139	987
165	·0581085
190	182
216	280
242	378
268	476
293	573
319	671
345	769
371	867
396	964
422	·0582062
448	160
473	257
499	355
525	453
551	551
576	648
602	746
628	844
653	941
679	·0583039
705	137
730	234
756	332
782	429
808	527
833	625
859	722
885	820
910	917
936	·0584015
962	112
988	210
1144013	308
039	405
065	503
090	600
116	698
142	796
167	893
193	991
219	·0585088
245	186
270	284
296	381

1144322	·0585479
347	576
373	674
399	771
425	869
450	967
476	·0586064
502	162
527	259
553	357
579	454
604	552
630	649
656	747
682	844
707	942
733	·0587039
759	137
784	234
810	332
836	429
862	527
887	624
913	722
939	819
964	917
990	·0588014
1145016	111
041	209
067	306
093	403
118	501
144	598
170	695
195	793
221	890
247	988
272	·0589085
298	182
323	280
349	377
375	474
400	572
426	669
452	766
477	864
503	961
529	·0590058
555	156
580	253
606	350
632	448
657	545
683	643
709	740
735	837
760	935
786	·0591032
812	129
837	227
863	324
889	421
914	518
940	616
966	713

1145991	·0591810
1146017	907
043	·0592005
068	102
094	199
120	296
145	394
171	491
196	588
222	685
248	783
273	880
299	977
325	·0593074
350	171
376	269
402	366
427	463
453	560
479	657
504	754
530	852
556	949
581	·0594046
607	143
632	240
658	338
684	435
709	532
735	629
761	726
786	823
812	920
838	·0595017
863	114
889	211
915	308
940	405
966	503
992	600
1147017	697
043	794
068	891
094	988
120	·0596085
145	182
171	279
197	376
222	473
248	570
274	667
299	764
325	861
351	958
376	·0597055
402	153
427	250
453	347
479	444
504	541
530	638
556	735
581	832
607	929
633	·0598026

1147658	·0598123
684	220
709	317
735	414
761	511
786	608
812	705
837	801
863	898
888	995
914	·0599092
940	189
965	286
991	383
1148016	480
042	577
068	674
093	771
119	868
145	965
170	·0600061
196	158
222	255
247	352
273	449
298	546
324	643
350	740
375	836
401	933
427	·0601030
452	127
478	224
504	321
529	418
555	514
580	611
606	708
632	805
657	901
683	998
708	·0602095
734	192
759	288
785	385
811	482
836	579
862	675
887	772
913	869
939	966
964	·0603063
990	159
1149015	256
041	353
067	450
092	546
118	643
143	740
169	837
194	933
220	·0604030
246	127
271	224
297	320

1149322	·0604417
348	514
374	611
399	707
425	804
450	901
476	997
502	·0605094
527	191
553	287
578	384
604	481
629	577
655	674
681	770
706	867
732	964
757	·0606060
783	157
809	254
834	350
860	447
885	543
911	640
937	737
962	833
988	930

Excesses

10 = 38	1 = 4
20 = 76	2 = 8
	3 = 11
	4 = 15
	5 = 19
	6 = 23
	7 = 27
	8 = 30
	9 = 34

13 = 49

Anti Log Excesses

10 = 3	1 = 0
20 = 5	2 = 1
30 = 8	3 = 1
40 = 11	4 = 1
50 = 13	5 = 1
60 = 16	6 = 2
70 = 18	7 = 2
80 = 21	8 = 2
90 = 24	9 = 2

49 = 13

1150013	·0607026
039	123
064	219
090	316
116	413
141	509
167	606
192	702
218	799

1150244	·0607896
269	992
295	·0608089
320	185
346	282
371	378
397	475
422	571
448	668
473	764
499	861
524	957
550	·0609054
575	150
601	247
626	343
652	440
678	536
703	633
729	729
754	826
780	922
806	·0610018
831	115
857	211
882	308
908	404
933	501
959	597
985	693
1151010	790
036	886
061	983
087	·0611079
113	175
138	272
164	368
189	465
215	561
240	657
266	754
291	850
317	947
342	·0612043
368	140
393	236
419	332
444	429
470	525
495	622
521	718
547	814
572	911
598	·0613007
623	103
649	199
674	296
700	392
725	488
751	585
776	681
802	777
827	873
853	970
878	·0614066

1151904	·0614162	1153563	·0620410	1155219	·0626640	1156872	·0632853	1158524	·0639048
929	259	588	506	245	736	898	949	549	143
955	355	614	602	270	832	923	·0633044	575	239
981	451	639	698	296	928	949	140	600	334
1152006	547	665	794	321	·0627023	974	235	625	429
032	644	690	890	347	119	1157000	331	651	524
057	740	716	986	372	215	025	426	676	619
083	836	741	·0621082	398	310	050	521	701	714
108	932	767	178	423	406	076	617	727	809
134	·0615028	792	274	448	502	101	712	752	904
159	124	818	370	474	597	127	808	778	·0640000
185	221	843	466	499	693	152	903	803	095
210	317	869	562	525	789	177	998	828	190
236	413	894	658	550	884	203	·0634094	854	285
261	509	920	753	575	980	228	189	879	380
287	605	945	849	601	·0628076	254	284	904	475
312	702	971	945	626	171	279	380	930	570
338	798	996	·0622041	652	267	304	475	955	665
363	894	1154022	137	677	363	330	570	981	760
389	990	047	233	703	458	355	666	1159006	855
415	·0616086	073	329	728	554	381	761	031	950
440	182	098	425	753	649	406	857	057	·0641045
466	279	124	521	779	745	432	952	082	140
491	375	149	617	804	841	457	·0635047	108	236
517	471	175	713	830	936	482	143	133	331
542	567	200	809	855	·0629032	508	238	159	426
568	663	226	905	880	128	533	333	184	521
593	759	251	·0623000	906	223	559	429	209	616
619	856	277	096	931	319	584	524	235	711
644	952	302	192	957	414	609	619	260	806
670	·0617048	328	288	982	510	635	715	286	901
695	144	353	384	1156008	606	660	810	311	996
721	240	379	480	033	701	686	905	336	·0642091
746	337	404	576	059	797	711	·0636000	362	186
772	433	430	672	084	892	736	096	387	281
797	529	455	767	110	988	762	191	412	376
823	625	481	863	135	·0630083	787	286	438	471
849	721	506	959	161	179	813	382	463	566
874	817	532	·0624055	186	275	838	477	488	661
900	913	557	151	212	370	864	572	514	756
925	·0618009	582	247	237	466	889	667	539	851
951	105	608	343	263	561	914	763	564	946
976	201	633	438	288	657	940	858	590	·0643041
1153002	297	659	534	313	753	965	953	615	136
027	393	684	630	339	848	991	·0637049	641	231
053	490	710	726	364	944	1158016	144	666	326
078	586	735	821	390	·0631039	041	239	691	421
104	682	761	917	415	135	067	334	717	516
129	778	786	·0625013	440	230	092	430	742	611
155	874	812	109	466	326	118	525	767	706
180	970	837	204	491	421	143	620	793	801
206	·0619066	863	300	517	517	168	715	818	896
231	162	888	396	542	612	194	811	843	991
257	258	914	492	568	708	219	906	869	·0644086
282	354	939	587	593	803	245	·0638001	894	181
308	450	965	683	618	899	270	096	919	276
333	546	990	779	644	994	296	192	945	371
359	642	1155015	875	669	·0632090	321	287	970	465
384	738	041	970	695	185	346	382	995	560
410	834	066	·0626066	720	281	372	477		
435	930	092	162	745	376	397	573		
461	·0620026	117	258	771	472	423	668		
486	122	143	353	796	567	448	763		
512	218	168	449	822	663	473	858		
537	314	194	545	847	758	499	953		

Excesses

10 = 38	1 = 4
20 = 75	2 = 8
	3 = 11
	4 = 15
	5 = 19
	6 = 23
	7 = 26
	8 = 30
	9 = 34

13 = 49

Anti Log Excesses

10 = 3	1 = 0
20 = 5	2 = 1
30 = 8	3 = 1
40 = 11	4 = 1
50 = 13	5 = 1
60 = 16	6 = 2
70 = 19	7 = 2
80 = 21	8 = 2
90 = 24	9 = 2

49 = 13

1160021	·0644655
046	750
072	845
097	940
122	·0645035
148	130
173	225
198	320
224	415
249	510
274	605
300	699
325	794
350	889
376	984
401	·0646079
426	174
452	269
477	364
503	458
528	553
553	648
579	743
604	838
629	933
655	·0647028
680	122
705	217
731	312
756	407
781	502
807	597
832	691
857	786
883	881
908	976
934	·0648071

1160959	·0648165
984	260
1161010	355
035	450
060	545
086	640
111	734
136	829
161	924
187	·0649019
212	113
237	208
263	303
288	398
313	492
338	587
364	682
389	777
414	871
440	966
465	·0650061
490	156
516	250
541	345
566	440
592	534
617	629
642	724
668	818
693	913
718	·0651008
744	102
769	197
795	291
820	386
845	481
871	575
896	670
921	765
947	859
972	954
997	·0652048
1162022	143
048	238
073	332
098	427
124	521
149	616
174	710
199	805
225	900
250	994
275	·0653089
301	183
326	278
351	372
377	467
402	561
427	656
452	750
478	845
503	939
528	·0654034
554	128
579	223

1162604	·0654317
629	412
655	506
680	601
705	695
731	790
756	884
781	978
807	·0655073
832	167
857	262
882	356
908	451
933	545
958	640
984	734
1163009	829
034	923
059	·0656018
085	112
110	207
135	301
161	396
186	490
211	584
237	679
262	773
287	867
312	962
338	·0657056
363	150
388	245
414	339
439	434
464	528
489	622
515	717
540	811
565	905
591	·0658000
616	094
641	188
666	283
692	377
717	471
742	565
767	660
793	754
818	848
843	943
868	·0659037
894	131
919	225
944	320
969	414
995	508
1164020	603
045	697
070	791
096	885
121	980
146	·0660074
171	168
197	262
222	357

1164247	·0660451
273	545
298	639
323	734
348	828
374	922
399	·0661016
424	111
450	205
475	299
500	393
525	487
551	582
576	676
601	770
626	864
652	958
677	·0662052
702	147
727	241
753	335
778	429
803	523
828	618
854	712
879	806
904	900

Excesses

13 = 49

Anti Log Excesses

49 = 13

1164931	·0663000
958	100
984	200
1165011	300
038	400
065	500
092	600
118	700
145	799
172	899
199	999
226	·0664099
253	199
279	299
306	399
333	499
360	599
387	699
413	799
440	899
467	998
494	·0665098
521	198
547	298
574	398
601	498
628	598
655	698
682	797

1165708	·0665897
735	997
762	·0666097
789	197
816	297
842	397
869	497
896	596
923	696
950	796
976	896
1166003	996
030	·0667096
057	196
084	296
111	395
137	495
164	595
191	695
218	795
245	895
271	994
298	·0668094
325	194
352	294
379	393
405	493
432	593
459	693
486	792
513	892
540	992
566	·0669092
593	191
620	291
647	391
673	490
700	590
727	690
754	789
780	889
807	989
834	·0670089
861	188
887	288
914	388
941	487
968	587
994	687
1167021	786
048	886
075	986
101	·0671085
128	185
155	284
182	384
208	483
235	583
262	682
289	782
315	882
342	981
369	·0672081
396	180
422	280

1167449	·0672379
476	479
503	579
530	678
556	778
583	877
610	977
637	·0673076
664	176
690	275
717	375
744	475
771	574
798	674
825	773
851	873
878	972
905	·0674072
932	171
958	271
985	370
1168012	470
039	569
065	669
092	768
119	868
146	967
172	·0675066
199	166
226	265
253	365
279	464
306	564
333	663
360	762
386	862
413	961
440	·0676061
467	160
493	260
520	359
547	459
574	558
600	657
627	757
654	856
681	956
707	·0677055
734	155
761	254
788	353
814	453
841	552
868	651
894	750
921	850
948	949
975	·0678048
1169001	147
028	247
055	346
081	445
108	544
135	644
161	743

1169188	·0678842
215	941
241	·0679041
268	140
295	239
322	338
348	438
375	537
402	636
429	735
455	835
482	934
509	·0680033
536	132
562	232
589	331
616	430
643	529
669	628
696	728
723	827
749	926
776	·0681025
803	124
830	224
856	323
883	422
910	521
936	620
963	719
990	819

Excesses

10 = 37	1 = 4
20 = 75	2 = 7
	3 = 11
	4 = 15
	5 = 19
	6 = 22
	7 = 26
	8 = 30
	9 = 34

13 = 48

Anti Log Excesses

10 = 3	1 = 0
20 = 5	2 = 1
30 = 8	3 = 1
40 = 11	4 = 1
50 = 13	5 = 1
60 = 16	6 = 2
70 = 19	7 = 2
80 = 21	8 = 2
90 = 24	9 = 2

48 = 13

1170016	·0681918
043	·0682017
070	116
096	215
123	314

1170150	·0682414
176	513
203	612
230	711
257	810
283	909
310	·0683008
337	107
363	207
390	306
417	405
443	504
470	603
497	702
523	801
550	900
577	999
604	·0684098
630	197
657	296
684	395
711	494
737	593
764	692
791	791
818	890
844	989
871	·0685088
898	187
925	286
951	385
978	484
1171005	583
031	682
058	781
085	880
112	979
138	·0686078
165	177
192	276
218	375
245	474
272	573
298	672
325	771
352	870
378	969
405	·0687068
431	167
458	265
485	364
511	463
538	562
565	661
591	760
618	859
644	958
671	·0688056
698	155
724	254
751	353
778	452
804	551
831	649
858	748

1171884	·0688847
911	946
938	·0689045
965	143
991	242
1172018	341
045	440
071	539
098	638
125	736
151	835
178	934
205	·0690033
231	132
258	230
284	329
311	428
338	527
364	625
391	724
418	823
444	922
471	·0691020
497	119
524	218
551	317
577	415
604	514
631	613
657	711
684	810
711	909
737	·0692007
764	106
791	204
818	303
844	402
871	500
898	599
924	698
951	796
978	895
1173004	993
031	·0693092
058	191
084	289
111	388
137	487
164	585
191	684
217	782
244	881
271	980
297	·0694078
324	177
350	276
377	374
404	473
430	571
457	670
484	769
510	867
537	966
563	·0695064
590	163

1173617	·0695261
643	360
670	458
697	557
723	655
750	754
776	852
803	951
830	·0696049
856	148
883	246
910	344
936	443
963	541
989	640
1174016	738
043	837
069	935
096	·0697034
123	132
149	230
176	329
202	427
229	526
256	624
282	723
309	821
336	919
362	·0698018
389	116
415	215
442	313
469	411
495	510
522	608
549	706
575	805
602	903
628	·0699002
655	100
682	198
708	297
735	395
762	493
788	592
815	690
841	788
868	887
894	985
921	·0700083
947	181
974	280
1175001	378
027	476
054	575
080	673
107	771
133	870
160	968
187	·0701066
213	165
240	263
266	361
293	459
319	558

1175346	·0701656
372	754
399	852
426	951
452	·0702049
479	147
505	245
532	344
558	442
585	540
612	638
638	736
665	835
691	933
718	·0703031
745	129
771	227
798	326
825	424
851	522
878	620
904	718
931	816
958	915
984	·0704013
1176011	111
038	209
064	307
091	405
117	504
144	602
170	700
197	798
223	896
250	994
277	·0705092
303	190
330	289
356	387
383	485
409	583
436	681
463	779
489	877
516	975
542	·0706073
569	171
595	269
622	367
648	465
675	563
702	661
728	759
755	857
781	955
808	·0707053
834	151
861	249
888	347
914	445
941	543
967	641
994	739
1177020	837
047	935

1177073	·0708033
100	130
127	228
153	326
180	424
206	522
233	620
259	718
286	816
312	914
339	·0709012
365	110
392	208
418	305
445	403
471	501
498	599
524	697
551	795
577	893
604	991
630	·0710088
657	186
683	284
710	382
737	480
763	578
790	675
816	773
843	871
869	969
896	·0711067
922	164
949	262
976	360
1178002	458
029	556
055	654
082	751
108	849
135	947
161	·0712045
188	143
214	240
241	338
267	436
294	534
320	631
347	729
373	827
400	925
426	·0713022
453	120
479	218
506	316
532	413
559	511
585	609
612	706
638	804
665	902
691	999
718	·0714097
744	195
771	293

1178797	·0714390
824	488
850	586
877	683
903	781
930	879
956	976
983	·0715074
1179009	172
036	269
062	367
089	465
115	562
142	660
168	757
195	855
221	953
248	·0716050
274	148
301	246
327	343
354	441
380	538
407	636
433	734
460	831
486	929
513	·0717026
539	124
566	221
592	319
619	416
645	514
672	611
698	709
725	806
751	904
778	·0718001
804	099
831	196
857	294
884	391
910	489
937	586
963	684
990	781

Excesses

10 = 37	1 = 4
20 = 74	2 = 7
	3 = 11
	4 = 15
	5 = 18
	6 = 22
	7 = 26
	8 = 30
	9 = 33

13 = 48

Anti Log Excesses

10 = 3	1 = 0
20 = 5	2 = 1
30 = 8	3 = 1
40 = 11	4 = 1
50 = 13	5 = 1
60 = 16	6 = 2
70 = 19	7 = 2
80 = 22	8 = 2
90 = 24	9 = 2

48 = 13

1180016	·0718879
043	976
069	·0719074
096	171
122	269
149	366
175	464
202	561
228	659
255	756
281	853
308	951
334	·0720048
361	146
387	243
414	340
440	438
467	535
493	632
519	730
546	827
572	925
599	·0721022
625	119
652	217
678	314
704	411
731	509
757	606
784	703
810	801
837	898
863	995
890	·0722092
916	190
943	287
969	384
996	482
1181022	579
049	676
075	774
102	871
128	968
155	·0723066
181	163
208	260
234	357
261	455
287	552
314	649
340	746

1181366	·0723844
393	941
419	·0724038
446	135
472	233
499	330
525	427
551	524
578	621
604	719
631	816
657	913
684	·0725010
710	107
737	205
763	302
789	399
816	496
842	593
869	690
895	788
922	885
948	982
974	·0726079
1182001	176
027	273
054	371
080	468
107	565
133	662
160	759
186	856
212	953
239	·0727050
265	148
292	245
318	342
345	439
371	536
397	633
424	730
450	827
477	924
503	·0728021
530	118
556	215
583	312
609	410
635	507
662	604
688	701
715	798
741	895
768	992
794	·0729089
820	186
847	283
873	380
899	477
926	574
952	671
979	768
1183005	865
031	961
058	·0730058

Erratum: the top two numbers in col. 5 should read 1181366
393

1183084	·0730155	1184800	·0736449	1186513	·0742725	1188224	·0748982	1189932	·0755221
110	252	827	546	539	821	251	·0749078	958	317
137	349	853	643	566	918	277	174	984	413
163	446	880	739	592	·0743014	303	270		
190	543	906	836	618	110	329	366		
216	640	932	933	645	207	356	462		
242	737	959	·0737029	671	303	382	558		
269	834	985	126	697	399	408	654		
295	931	1185011	223	724	496	434	750		
322	·0731028	038	319	750	592	461	846		
348	125	064	416	777	688	487	942		
375	222	090	512	803	784	513	·0750039		
401	319	116	609	829	881	539	135		
428	416	143	706	856	977	566	231		
454	512	169	802	882	·0744073	592	327		
480	609	195	899	908	170	618	423		
507	706	222	996	935	266	644	519		
533	803	248	·0738092	961	362	671	615		
560	900	274	189	988	459	697	711		
586	997	301	285	1187014	555	723	807		
613	·0732094	327	382	040	651	750	903		
639	191	353	479	067	748	776	999		
665	288	380	575	093	844	802	·0751095		
692	385	406	672	119	940	829	191		
718	481	432	768	146	·0745037	855	287		
744	578	459	865	172	133	881	383		
771	675	485	961	198	229	907	479		
797	772	512	·0739058	224	325	934	575		
824	869	538	154	251	422	960	671		
850	965	564	251	277	518	986	767		
876	·0733062	591	348	303	614	1189013	863		
903	159	617	444	330	711	039	959		
929	256	643	541	356	807	065	·0752055	1190011	·0755509
955	353	670	637	382	903	092	151	037	604
982	450	696	734	409	·0746000	118	247	063	700
1184008	546	723	830	435	096	144	343	090	796
035	643	749	927	461	192	170	439	116	892
061	740	775	·0740023	488	288	197	535	142	987
087	837	802	120	514	385	223	631	168	·0756083
114	934	828	216	540	481	249	727	195	179
140	·0734030	854	313	567	577	275	823	221	275
167	127	881	409	593	673	302	919	247	370
193	224	907	506	619	769	328	·0753015	274	466
220	321	934	602	645	866	354	110	300	562
246	417	960	699	672	962	380	206	326	658
273	514	986	795	698	·0747058	407	302	353	753
299	611	1186013	891	724	154	433	398	379	849
325	708	039	988	751	250	459	494	405	945
352	804	065	·0741084	777	346	485	590	431	·0757041
378	901	092	181	803	443	512	686	458	136
405	998	118	277	830	539	538	782	484	232
431	·0735095	145	374	856	635	564	878	510	328
458	191	171	470	882	731	590	974	536	424
484	288	197	567	909	827	617	·0754070	563	519
510	385	224	663	935	924	643	166	589	615
537	482	250	760	961	·0748020	669	262	615	711
563	578	276	856	988	116	695	358	641	807
589	675	303	953	1188014	212	722	454	668	902
616	772	329	·0742049	040	308	748	550	694	998
642	869	355	146	066	405	774	645	720	·0758094
669	965	381	242	093	501	800	741	746	190
695	·0736062	408	339	119	597	827	837	773	285
721	159	434	435	145	693	853	933	799	381
748	256	460	532	172	789	879	·0755029	825	477
774	352	487	628	198	885	905	125	851	572

Excesses

10 = 37	1 = 4
20 = 73	2 = 7
	3 = 11
	4 = 15
	5 = 18
	6 = 22
	7 = 26
	8 = 29
	9 = 33

13 = 48

Anti Log Excesses

10 = 3	1 = 0
20 = 5	2 = 1
30 = 8	3 = 1
40 = 11	4 = 1
50 = 14	5 = 1
60 = 16	6 = 2
70 = 19	7 = 2
80 = 22	8 = 2
90 = 25	9 = 2

48 = 13

1190878	·0758668	1192582	·0764880	1194284	·0771074	1195984	·0777249	1197680	·0783408
904	764	608	975	310	169	1196010	344	706	503
930	859	634	·0765071	336	264	036	439	733	598
956	955	660	166	363	359	062	534	759	692
982	·0759051	687	262	389	454	088	629	785	787
1191009	147	713	357	415	549	114	724	811	882
035	242	739	452	441	644	140	818	837	976
061	338	765	548	467	740	167	913	863	·0784071
087	434	791	643	493	835	193	·0778008	889	165
113	529	817	739	520	930	219	103	915	260
139	625	844	834	546	·0772025	245	198	941	355
166	721	870	930	572	120	271	293	967	449
192	816	896	·0766025	598	215	297	388	993	544
218	912	922	120	624	310	323	482	1198019	638
244	·0760008	948	216	650	405	349	577	046	733
270	103	975	311	676	500	376	672	072	827
297	199	1193001	406	703	595	402	767	098	922
323	295	027	502	729	690	428	862	124	·0785016
349	390	053	597	755	785	454	956	150	111
375	486	079	692	781	880	480	·0779051	176	205
402	581	105	787	807	976	506	146	202	300
428	677	132	883	833	·0773071	532	241	228	394
454	773	158	978	859	166	558	336	255	489
480	868	184	·0767073	885	261	585	431	281	583
507	964	210	169	912	356	611	525	307	678
533	·0761060	236	264	938	451	637	620	333	772
559	155	263	359	964	546	663	715	359	867
585	251	289	455	990	641	689	810	385	961
612	346	315	550	1195016	736	715	905	411	·0786056
638	442	341	645	042	831	741	999	437	150
664	538	367	741	068	926	767	·0780094	463	245
690	633	394	836	094	·0774021	793	189	489	339
717	729	420	931	121	116	819	284	515	434
743	824	446	·0768027	147	211	845	378	541	528
769	920	472	122	173	306	871	473	568	623
795	·0762015	498	217	199	401	898	568	594	717
822	111	525	312	225	496	924	663	620	812
848	206	551	408	251	591	950	757	646	906
874	302	577	503	277	686	976	852	672	·0787001
900	398	603	598	303	781	1197002	947	698	095
927	493	629	694	330	876	028	·0781042	724	190
953	589	655	789	356	971	054	136	750	284
979	684	682	884	382	·0775066	080	231	776	378
1192005	780	708	980	408	161	106	326	802	473
032	875	734	·0769075	434	256	132	420	828	567
058	971	760	170	460	351	158	515	854	662
084	·0763066	786	265	487	446	184	610	880	756
110	162	813	361	513	541	211	704	906	850
137	257	839	456	539	636	237	799	932	945
163	353	865	551	565	731	263	894	958	·0788039
189	448	891	646	591	826	289	989	985	133
215	544	917	741	618	921	315	·0782083	1199011	228
241	639	944	837	644	·0776015	341	178	037	322
268	735	970	932	670	110	367	273	063	417
294	830	996	·0770027	696	205	393	367	089	511
320	925	1194022	122	722	300	420	462	115	605
346	·0764021	048	217	748	395	446	557	141	700
372	116	074	312	775	490	472	651	167	794
398	212	101	408	801	585	498	746	193	888
425	307	127	503	827	680	524	841	219	983
451	403	153	598	853	775	550	935	245	·0789077
477	498	179	693	879	870	576	·0783030	271	172
503	593	205	788	905	965	602	125	297	266
529	689	232	883	931	·0777059	628	219	323	360
556	784	258	979	958	154	654	314	349	455

1199375	·0789549
402	643
428	738
454	832
480	927
506	·0790021
532	115
558	210
584	304
610	398
636	493
662	587
688	681
714	776
740	870
766	964
792	·0791058
818	153
844	247
870	341
896	436
922	530
948	624
974	719

Excesses

10 = 36	1 = 4
20 = 73	2 = 7
	3 = 11
	4 = 15
	5 = 18
	6 = 22
	7 = 25
	8 = 29
	9 = 33

13 = 48

Anti Log Excesses

10 = 3	1 = 0
20 = 6	2 = 1
30 = 8	3 = 1
40 = 11	4 = 1
50 = 14	5 = 1
60 = 17	6 = 2
70 = 19	7 = 2
80 = 22	8 = 2
90 = 25	9 = 2

48 = 13

1200000	·0791813
026	907
052	·0792001
078	096
104	190
130	284
156	378
182	472
208	567
235	661
261	755
287	849

1200313	·0792943
339	·0793037
365	132
391	226
417	320
443	414
469	508
495	602
521	697
547	791
573	885
599	979
625	·0794073
652	167
678	261
704	355
730	450
756	544
782	638
808	732
834	826
860	920
886	·0795014
912	108
938	203
964	297
990	391
1201016	485
042	579
068	673
094	767
120	861
146	956
172	·0796050
198	144
224	238
250	332
276	426
302	520
328	614
354	708
380	802
406	896
432	990
458	·0797084
484	178
510	272
536	366
562	460
588	554
614	648
640	742
666	836
692	930
718	·0798024
744	118
770	212
796	306
822	400
848	494
874	588
900	681
926	775
952	869
978	963

1202004	·0799057
030	151
056	245
082	339
108	433
134	527
160	621
186	715
212	809
238	903
264	997
290	·0800091
316	184
342	278
368	372
394	466
420	560
446	654
472	748
498	842
524	936
550	·0801030
576	123
602	217
628	311
654	405
680	499
706	592
732	686
758	780
784	874
810	968
836	·0802062
862	155
888	249
914	343
940	437
966	531
992	624
1203018	718
044	812
070	906
096	999
122	·0803093
147	187
173	281
199	374
225	468
251	562
277	656
303	749
329	843

Excesses

13 = 47

Anti Log Excesses

47 = 13

1203357	·0803943
384	·0804043
412	143
440	243

1203468	·0804343
495	443
523	543
551	642
579	742
606	842
634	942
662	·0805042
689	142
717	242
745	342
773	442
800	542
828	642
856	741
883	841
911	941
939	·0806041
966	141
994	241
1204022	341
049	441
077	540
105	640
132	740
160	840
188	940
215	·0807040
243	139
271	239
298	339
326	439
354	539
381	638
409	738
437	838
464	938
492	·0808038
520	137
547	237
575	337
603	437
630	536
658	636
686	736
713	836
741	935
769	·0809035
796	135
824	235
852	334
879	434
907	534
935	633
962	733
990	833
1205018	933
045	·0810032
073	132
101	232
128	331
156	431
184	531
211	630
239	730

1205267	·0810830
294	929
322	·0811029
350	129
377	228
405	328
433	428
460	527
488	627
516	726
543	826
571	926
599	·0812025
626	125
654	224
682	324
709	424
737	523
765	623
792	722
820	822
848	922
875	·0813021
903	121
930	220
958	320
986	419
1206013	519
041	618
068	718
096	817
124	917
151	·0814016
179	116
206	215
234	315
262	414
289	514
317	613
345	713
372	812
400	912
428	·0815011
455	111
483	210
511	310
538	409
566	509
594	608
621	708
649	807
677	906
704	·0816006
732	105
759	205
787	304
815	403
842	503
870	602
897	702
925	801
953	900
980	·0817000
1207008	099
035	199

1207063	·0817298
091	397
118	497
146	596
173	695
201	795
229	894
256	993
284	·0818093
311	192
339	291
367	391
394	490
422	589
449	689
477	788
505	887
532	987
560	·0819086
587	185
615	284
643	384
670	483
698	582
725	681
753	781
781	880
808	979
836	·0820079
863	178
891	277
919	376
946	475
974	574
1208001	674
029	773
057	872
084	971
112	·0821070
139	169
167	268
195	367
222	467
250	566
277	665
305	764
333	863
360	962
388	·0822061
415	161
443	260
471	359
498	458
526	557
553	656
581	755
609	854
636	954
664	·0823053
691	152
719	251
747	350
774	449
802	548
829	647

1208857	·0823746
884	845
912	944
939	·0824044
967	143
994	242
1209022	341
049	440
077	539
104	638
132	737
160	836
187	935
215	·0825034
242	133
270	232
298	331
325	430
353	529
380	628
408	727
436	826
463	925
491	·0826024
518	123
546	222
574	321
601	420
629	519
656	618
684	717
711	816
739	915
766	·0827013
794	112
821	211
849	310
876	409
904	508
931	607
959	706
987	805

Excesses

10 = 36	1 = 4
20 = 72	2 = 7
	3 = 11
	4 = 14
	5 = 18
	6 = 21
	7 = 25
	8 = 29
	9 = 32

14 = 50

Anti Log Excesses

10 = 3	1 = 0
20 = 6	2 = 1
30 = 8	3 = 1
40 = 11	4 = 1
50 = 14	5 = 1
60 = 17	6 = 2
70 = 19	7 = 2
80 = 22	8 = 2
90 = 25	9 = 2

50 = 14

1210014	·0827904
042	·0828002
069	101
097	200
124	299
152	398
179	496
207	595
234	694
262	793
289	892
317	990
344	·0829089
372	188
400	287
427	386
455	484
482	583
510	682
537	781
565	880
592	978
620	·0830077
647	176
675	275
702	374
730	472
757	571
785	670
813	769
840	867
868	966
895	·0831065
923	164
950	262
978	361
1211005	460
033	559
060	657
088	756
115	855
143	953
170	·0832052
198	151
226	250
253	348
281	447
308	545
336	644
363	743
391	841

1211418	·0832940
446	·0833038
473	137
501	236
528	334
556	433
583	531
611	630
639	729
666	827
694	926
721	·0834024
749	123
776	222
804	320
831	419
859	517
886	616
914	715
941	813
969	912
996	·0835010
1212024	109
051	208
079	306
106	405
134	503
161	602
189	700
216	799
244	897
271	996
299	·0836094
326	193
354	291
381	390
409	488
436	587
463	685
491	784
518	882
546	981
573	·0837079
601	178
628	276
656	375
683	473
711	572
738	670
766	769
793	867
821	966
848	·0838064
876	162
903	261
931	359
958	457
986	556
1213013	654
041	752
068	851
096	949
123	·0839047
151	146
178	244

1213206	·0839342
233	441
261	539
288	637
316	736
343	834
371	932
398	·0840031
426	129
453	227
481	326
508	424
536	522
563	621
591	719
618	817
646	916
673	·0841014
700	112
728	211
755	309
783	407
810	505
838	604
865	702
893	800
920	898
948	997
975	·0842095
1214003	193
030	292
058	390
085	488
112	586
140	684
167	782
195	881
222	979
249	·0843077
277	175
304	273
332	371
359	469
386	567
414	666
441	764
469	862
496	960
523	·0844058
551	156
578	254
606	353
633	451
661	549
688	647
716	745
743	843
771	941
798	·0845039
826	138
853	236
881	334
908	432
935	530
963	628

1214990	·0845726
1215018	824
045	922
072	·0846020
100	118
127	217
155	315
182	413
209	511
237	609
264	707
292	805
319	903
346	·0847001
374	099
401	197
429	295
456	393
484	491
511	589
539	686
566	784
594	882
621	980
649	·0848078
676	176
704	274
731	372
758	470
786	568
813	666
841	764
868	862
895	960
923	·0849058
950	155
978	253
1216005	351
032	449
060	547
087	645
115	743
142	841
169	939
197	·0850037
224	135
252	232
279	330
306	428
334	526
361	624
389	722
416	820
443	918
471	·0851015
498	113
526	211
553	309
580	407
608	505
635	602
663	700
690	798
717	896
745	994

1216772	·0852091
800	189
827	287
854	385
882	483
909	580
937	678
964	776
991	874
1217019	971
046	·0853069
074	167
101	264
128	362
156	460
183	557
211	655
238	753
265	850
293	948
320	·0854046
348	143
375	241
402	339
430	436
457	534
484	632
512	729
539	827
566	925
594	·0855022
621	120
648	218
676	315
703	413
730	511
758	608
785	706
812	804
840	901
867	999
895	·0856096
922	194
949	292
977	389
1218004	487
032	584
059	682
086	780
114	877
141	975
169	·0857072
196	170
223	267
251	365
278	462
305	560
333	657
360	755
387	852
415	950
442	·0858047
469	145
497	242
524	340

1218551	·0858437
579	535
606	632
633	729
661	827
688	924
715	·0859022
743	119
770	217
797	314
825	412
852	509
879	607
907	704
934	802
961	899
989	997
1219016	·0860094
043	191
071	289
098	386
126	484
153	581
180	678
208	776
235	873
263	971
290	·0861068
317	165
345	263
372	360
400	458
427	555
454	652
482	750
509	847
536	944
564	·0862042
591	139
618	236
646	334
673	431
700	528
728	626
755	723
782	820
810	918
837	·0863015
864	112
892	210
919	307
946	404
973	501

Excesses

10 = 36	1 = 4
20 = 71	2 = 7
	3 = 11
	4 = 14
	5 = 18
	6 = 21
	7 = 25
	8 = 29
	9 = 32

14 = 50

Anti Log Excesses

10 = 3	1 = 0
20 = 6	2 = 1
30 = 8	3 = 1
40 = 11	4 = 1
50 = 14	5 = 1
60 = 17	6 = 2
70 = 20	7 = 2
80 = 22	8 = 2
90 = 25	9 = 2

50 = 14

1220001	·0863599
028	696
055	793
082	890
110	988
137	·0864085
164	182
192	280
219	377
246	474
273	571
301	668
328	765
355	863
383	960
410	·0865057
437	154
465	251
492	348
519	445
547	542
574	640
601	737
629	834
656	931
683	·0866028
711	125
738	222
765	320
793	417
820	514
847	611
875	708
902	805
929	902
957	999
984	·0867097

1221011	·0867194
039	291
066	388
093	485
121	582
148	679
175	776
202	873
230	970
257	·0868067
284	165
311	262
339	359
366	456
393	553
421	650
448	747
475	844
502	941
530	·0869038
557	135
584	232
611	329
639	426
666	523
693	620
720	717
748	814
775	911
802	·0870008
830	105
857	202
884	299
911	396
939	493
966	590
993	687
1222020	784
048	881
075	978
102	·0871074
129	171
157	268
184	365
211	462
239	559
266	656
293	753
320	850
348	947
375	·0872044
402	140
429	237
457	334
484	431
511	528
538	625
566	722
593	819
620	915
648	·0873012
675	109
702	206
729	303
757	400

1222784	·0873496
811	593
838	690
866	787
893	884
920	980
947	·0874077
975	174
1223002	271
029	368
057	464
084	561
111	658
138	755
166	851
193	948
220	·0875045
247	142
275	238
302	335
329	432
356	529
384	625
411	722
438	819
466	915
493	·0876012
520	109
547	206
575	302
602	399
629	496
656	592
684	689
711	786
738	882
765	979
793	·0877076
820	172
847	269
875	366
902	462
929	559
956	656
983	752
1224011	849
038	945
065	·0878042
092	139
119	235
147	332
174	428
201	525
228	622
255	718
283	815
310	911
337	·0879008
364	105
391	201
419	298
446	394
473	491
500	587
527	684

1224555	·0879780
582	877
609	973
636	·0880070
663	166
691	263
718	359
745	456
772	553
800	649
827	746
854	842
881	939
909	·0881035
936	132
963	228
990	325
1225018	421
045	518
072	614
100	711
127	807
154	904
181	·0882000
208	097
236	193
263	290
290	386
317	482
344	579
372	675
399	772
426	868
453	964
480	·0883061
508	157
535	254
562	350
589	446
616	543
644	639
671	735
698	832
725	928
752	·0884024
780	121
807	217
834	313
861	410
888	506
916	602
943	699
970	795
997	891
1226024	988
051	·0885084
079	180
106	276
133	373
160	469
187	565
214	661
241	758
268	854
296	950

1226323	·0886047
350	143
377	239
404	335
431	432
459	528
486	624
513	720
540	817
567	913
595	·0887009
622	105
649	202
676	298
703	394
731	491
758	587
785	683
812	779
839	875
866	971
894	·0888068
921	164
948	260
975	356
1227002	452
029	548
056	644
083	740
111	837
138	933
165	·0889029
192	125
219	221
246	317
274	413
301	509
328	605
355	701
382	797
410	894
437	990
464	·0890086
491	182
518	278
546	374
573	470
600	566
627	662
654	758
681	854
709	950
736	·0891046
763	142
790	238
817	335
844	431
871	527
898	623
926	719
953	815
980	911
1228007	·0892007
034	103
061	199

1228088	·0892295
116	391
143	487
170	583
197	679
224	774
251	870
278	966
305	·0893062
333	158
360	254
387	350
414	446
441	542
468	638
495	734
523	830
550	926
577	·0894022
604	118
631	213
658	309
685	405
712	501
740	597
767	693
794	789
821	885
848	981
875	·0895077
902	172
930	268
957	364
984	460
1229011	556
038	651
065	747
092	843
119	939
147	·0896035
174	130
201	226
228	322
255	418
282	514
309	609
337	705
364	801
391	897
418	993
445	·0897088
472	184
499	280
526	376
554	472
581	567
608	663
635	759
662	855
689	950
716	·0898046
743	142
770	238
797	333
824	429

1229852	·0898525
879	621
906	716
933	812
960	908
987	·0899003

Excesses

10 = 35	1 = 4
20 = 71	2 = 7
	3 = 11
	4 = 14
	5 = 18
	6 = 21
	7 = 25
	8 = 28
	9 = 32

14 = 50

Anti Log Excesses

10 = 3	1 = 0
20 = 6	2 = 1
30 = 8	3 = 1
40 = 11	4 = 1
50 = 14	5 = 1
60 = 17	6 = 2
70 = 20	7 = 2
80 = 23	8 = 2
90 = 25	9 = 3

50 = 14

1230014	·0899099
041	195
068	291
095	386
122	482
150	578
177	673
204	769
231	865
258	960
285	·0900056
312	152
339	247
367	343
394	439
421	534
448	630
475	726
502	821
529	917
556	·0901012
583	108
610	203
637	299
665	394
692	490
719	585
746	681
773	776
800	872

1230827	·0901967	1232586	·0908170	1234342	·0914355	1236096	·0920522	1237848	·0926672
854	·0902063	613	265	369	450	123	617	875	766
881	159	640	361	396	545	150	712	902	861
908	254	667	456	423	640	177	806	929	955
935	350	694	551	450	735	204	901	955	·0927050
962	445	721	646	477	830	231	996	982	144
989	541	748	742	504	925	258	·0921091	1238009	239
1231016	636	775	837	531	·0915020	285	185	036	333
043	732	802	932	558	115	312	280	063	428
071	827	829	·0909028	585	210	339	375	090	522
098	923	856	123	612	305	366	470	117	617
125	·0903018	883	218	639	400	393	564	144	711
152	114	910	313	666	495	420	659	171	805
179	209	937	408	693	590	447	754	198	900
206	305	964	504	720	684	474	848	225	994
233	400	991	599	747	779	501	943	252	·0928089
260	496	1233018	694	774	874	528	·0922038	279	183
287	591	045	789	801	969	555	133	306	277
314	687	072	884	828	·0916064	582	227	333	372
341	782	100	980	855	159	609	322	359	466
368	878	127	·0910075	882	254	636	416	386	561
395	973	154	170	909	349	663	511	413	655
422	·0904069	181	265	936	444	690	606	440	749
449	164	208	360	963	539	717	700	467	844
477	260	235	456	990	634	743	795	494	938
504	355	262	551	1235017	729	770	889	521	·0929033
531	451	289	646	044	824	797	984	548	127
558	546	316	741	071	919	824	·0923079	575	221
585	642	343	836	098	·0917014	851	173	602	316
612	737	370	931	125	108	878	268	629	410
639	833	397	·0911027	152	203	905	362	656	504
666	928	424	122	179	298	932	457	683	599
693	·0905023	451	217	206	393	959	552	710	693
720	119	478	312	233	488	986	646	737	787
747	214	505	407	260	583	1237013	741	763	882
774	310	532	502	287	678	040	835	790	976
801	405	559	597	314	773	067	930	817	·0930070
828	500	586	692	341	868	094	·0924025	844	165
855	596	613	788	368	963	121	119	871	259
883	691	640	883	395	·0918057	147	214	898	353
910	787	667	978	422	152	174	308	925	448
937	882	694	·0912073	449	247	201	403	952	542
964	977	721	168	476	342	228	498	979	636
991	·0906073	748	263	503	437	255	592	1239006	731
1232018	168	775	358	530	531	282	687	033	825
045	264	802	454	557	626	309	781	059	919
072	359	829	549	584	721	336	876	086	·0931013
099	454	856	644	611	816	363	971	113	108
126	550	883	739	638	911	390	·0925065	140	202
153	645	910	834	665	·0919005	417	160	167	296
180	740	937	929	692	100	444	254	194	390
207	836	964	·0913024	719	195	471	349	221	485
234	931	991	119	746	290	498	443	248	579
261	·0907026	1234018	215	773	385	525	538	274	673
289	122	045	310	800	479	551	632	301	768
316	217	072	405	827	574	578	727	328	862
343	312	099	500	854	669	605	821	355	956
370	408	126	595	881	764	632	916	382	·0932050
397	503	153	690	908	859	659	·0926010	409	145
424	598	180	785	934	953	686	105	436	239
451	694	207	880	961	·0920048	713	199	463	333
478	789	234	975	988	143	740	294	490	427
505	884	261	·0914070	1236015	238	767	388	517	522
532	980	288	165	042	333	794	483	544	616
559	·0908075	315	260	069	427	821	577	570	710

1239597	·0932804
624	899
651	993
678	·0933087
705	182
732	276
759	370
786	464
813	558
840	652
866	747
893	841
920	935
947	·0934029
974	123

Excesses

10 = 35	1 = 4
20 = 70	2 = 7
	3 = 11
	4 = 14
	5 = 18
	6 = 21
	7 = 25
	8 = 28
	9 = 32

14 = 50

Anti Log Excesses

10 = 3	1 = 0
20 = 6	2 = 1
30 = 9	3 = 1
40 = 11	4 = 1
50 = 14	5 = 1
60 = 17	6 = 2
70 = 20	7 = 2
80 = 23	8 = 2
90 = 26	9 = 3

50 = 14

1240001	·0934217
028	311
055	405
081	500
108	594
135	688
162	782
189	876
216	970
243	·0935064
269	158
296	252
323	346
350	440
377	535
404	629
431	723
458	817
484	911
511	·0936005
538	099

1240565	·0936193
592	287
619	381
646	475
672	569
699	663
726	757
753	851
780	946
807	·0937040
834	134
861	228
887	322
914	416
941	510
968	604
995	698
1241022	792
049	886
075	980
102	·0938074
129	168
156	262
183	355
210	449
237	543
264	637
290	731
317	825
344	919
371	·0939013
398	107
425	201
451	295
478	389
505	483
532	577
559	671
585	764
612	858
639	952
666	·0940046
693	140
719	234
746	328
773	422
800	516
827	610
854	704
880	797
907	891
934	985
961	·0941079
988	173
1242015	267
042	361
069	455
095	548
122	642
149	736
176	830
203	924
230	·0942018
256	111
283	205

1242310	·0942299
337	393
364	487
390	580
417	674
444	768
471	862
498	956
524	·0943049
551	143
578	237
605	331
632	424
658	518
685	612
712	706
739	799
766	893
792	987
819	·0944081
846	174
873	268
900	362
926	455
953	549
980	643
1243007	737
034	830
061	924
087	·0945018
114	111
141	205
168	299
195	392
222	486
249	580
276	673
302	767
329	861
356	954
383	·0946048
410	142
437	235
463	329
490	422
517	516
544	610
571	703
597	797
624	890
651	984
678	·0947078
705	171
731	265
758	358
785	452
812	546
839	639
865	733
892	826
919	920
946	·0948013
973	107
999	200
1244026	294

1244053	·0948387
080	481
107	574
133	668
160	761
187	855
214	948
241	·0949042
267	135
294	229
321	322
348	416
375	509
401	603
428	696
455	790
482	883
509	977
535	·0950070
562	164
589	257
616	350
642	444
669	537
696	631
723	724
749	817
776	911
803	·0951004
830	097
856	191
883	284
910	378
936	471
963	565
990	658
1245017	751
044	845
070	938
097	·0952032
124	125
151	218
178	312
204	405
231	499
258	592
285	685
312	779
338	872
365	966
392	·0953059

Excesses

14 = 49

Anti Log Excesses

49 = 14

1245421	·0953159
449	259
478	359
507	459
535	559

1245564	·0953659
592	759
621	858
650	958
678	·0954058
707	158
736	258
764	358
793	458
822	558
850	658
879	758
908	858
937	958
965	·0955058
994	158
1246023	257
051	357
080	457
109	557
138	657
166	757
195	857
224	957
252	·0956057
281	157
310	256
338	356
367	456
395	556
424	656
453	756
481	856
510	955
539	·0957055
567	155
596	255
625	355
653	455
682	554
711	654
739	754
768	854
796	954
825	·0958053
854	153
882	253
911	353
940	452
968	552
997	652
1247026	752
054	851
083	951
112	·0959051
140	151
169	250
197	350
226	450
255	549
283	649
312	749
341	849
369	948
398	·0960048

1247427	·0960148
455	247
484	347
513	447
541	546
570	646
598	746
627	845
656	945
684	·0961044
713	144
742	244
770	343
799	443
828	543
856	642
885	742
913	841
942	941
970	·0962040
999	140
1248028	240
056	339
085	439
113	538
142	638
170	737
199	837
228	937
256	·0963036
285	136
314	235
342	335
371	434
399	534
428	634
457	733
485	833
514	932
543	·0964032
571	131
600	231
629	330
657	430
686	529
714	629
743	728
771	828
800	927
829	·0965026
857	126
886	225
914	325
943	424
971	524
1249000	623
029	722
057	822
086	921
114	·0966021
143	120
171	220
200	319
229	418
257	518

1249286	·0966617
314	717
343	816
371	916
400	·0967015
429	114
457	214
486	313
514	412
543	511
571	611
600	710
629	809
657	909
686	·0968008
714	107
743	206
771	306
800	405
829	504
857	604
886	703
914	802
943	901
971	·0969001

Excesses

10 = 35	1 = 3
20 = 70	2 = 7
	3 = 10
	4 = 14
	5 = 17
	6 = 21
	7 = 24
	8 = 28
	9 = 31

14 = 49

Anti Log Excesses

10 = 3	1 = 0
20 = 6	2 = 1
30 = 9	3 = 1
40 = 11	4 = 1
50 = 14	5 = 1
60 = 17	6 = 2
70 = 20	7 = 2
80 = 23	8 = 2
90 = 26	9 = 3

49 = 14

1250000	·0969100
029	199
057	299
086	398
114	497
143	596
171	696
200	795
229	894
257	993
286	·0970093

1250314	·0970192
343	291
371	390
400	489
429	589
457	688
486	787
514	886
543	986
571	·0971085
600	184
629	283
657	382
686	481
714	581
743	680
771	779
800	878
829	977
857	·0972076
886	175
914	275
943	374
971	473
1251000	572
029	671
057	770
086	869
114	968
143	·0973067
171	166
200	265
229	365
257	464
286	563
314	662
343	761
371	860
400	959
428	·0974058
457	157
485	256
514	355
542	454
571	553
599	652
628	751
656	850
685	949
713	·0975048
742	147
770	246
799	345
828	444
856	543
885	642
913	741
942	840
970	939
999	·0976038
1252028	136
056	235
085	334
113	433
142	532

1252170	·0976631
199	730
227	829
256	928
284	·0977027
313	126
341	225
370	324
398	423
427	521
455	620
484	719
512	818
541	917
569	·0978016
598	115
626	214
655	313
683	411
712	510
740	609
769	708
797	807
826	905
854	·0979004
883	103
911	202
940	300
968	399
997	498
1253025	597
054	696
082	794
111	893
139	992
168	·0980091
196	190
225	288
253	387
282	486
310	585
339	683
367	782
396	881
424	980
453	·0981078
481	177
510	276
538	375
567	473
595	572
624	671
652	769
681	868
709	967
738	·0982066
766	164
795	263
823	362
852	460
880	559
909	658
937	756
966	855
994	953

1254022	·0983052
051	151
079	249
108	348
136	447
165	545
193	644
221	743
250	841
278	940
307	·0984038
335	137
364	235
392	334
421	433
449	531
478	630
506	728
535	827
563	925
592	·0985024
620	123
649	221
677	320
706	418
734	517
763	615
791	714
820	812
848	911
877	·0986009
905	108
934	206
962	305
991	403
1255019	501
048	600
076	698
105	797
133	895
162	994
190	·0987092
218	190
247	289
275	387
304	486
332	584
361	683
389	781
417	879
446	978
474	·0988076
503	174
531	273
560	371
588	470
616	568
645	666
673	765
702	863
730	961
759	·0989060
787	158
815	256
844	355

1255872	·0989453
901	551
929	650
958	748
986	847
1256014	945
043	·0990043
071	142
100	240
128	338
157	437
185	535
213	633
242	732
270	830
299	928
327	·0991026
356	125
384	223
412	321
441	420
469	518
498	616
526	714
555	813
583	911
611	·0992009
640	107
668	205
697	304
725	402
754	500
782	598
810	696
839	794
867	892
896	991
924	·0993089
953	187
981	285
1257009	383
038	481
066	579
095	678
123	776
152	874
180	972
208	·0994070
237	168
265	266
294	365
322	463
351	561
379	659
407	757
436	855
464	953
492	·0995051
521	149
549	247
578	345
606	444
634	542
663	640
691	738

1257719	·0995836
748	934
776	·0996032
804	130
833	228
861	326
890	424
918	522
947	620
975	718
1258003	817
032	915
060	·0997013
089	111
117	209
146	307
174	405
202	503
231	601
259	699
287	797
316	895
344	993
373	·0998091
401	188
429	286
458	384
486	482
514	580
543	678
571	776
599	874
628	972
656	·0999070
684	167
713	265
741	363
770	461
798	559
826	657
855	755
883	852
911	950
940	·1000048
968	146
996	244
1259025	342
053	440
081	537
110	635
138	733
167	831
195	929
223	·1001027
252	125
280	222
308	320
337	418
365	516
393	614
422	712
450	809
478	907
507	·1002005
535	103

1259564	·1002201
592	298
620	396
649	494
677	592
705	689
734	787
762	885
790	983
819	·1003080
847	178
875	276
904	374
932	471
961	569
989	667

Excesses

10 = 35	1 = 3
20 = 69	2 = 7
	3 = 10
	4 = 14
	5 = 17
	6 = 21
	7 = 24
	8 = 28
	9 = 31

14 = 49

Anti Log Excesses

10 = 3	1 = 0
20 = 6	2 = 1
30 = 9	3 = 1
40 = 12	4 = 1
50 = 14	5 = 1
60 = 17	6 = 2
70 = 20	7 = 2
80 = 23	8 = 2
90 = 26	9 = 3

49 = 14

1260017	·1003764
046	862
074	960
102	·1004058
131	155
159	253
187	351
216	448
244	546
272	644
301	741
329	839
358	936
386	·1005034
414	132
443	229
471	327
499	425
528	522
556	620

1260584	·1005718
613	815
641	913
669	·1006010
698	108
726	205
755	303
783	401
811	498
840	596
868	693
896	791
925	888
953	986
981	·1007084
1261010	181
038	279
066	376
094	474
123	571
151	669
179	766
208	864
236	961
264	·1008059
292	156
321	254
349	351
377	448
406	546
434	643
462	741
490	838
519	936
547	·1009033
575	130
604	228
632	325
660	423
688	520
717	618
745	715
773	812
802	910
830	·1010007
858	105
887	202
915	300
944	397
972	494
1262000	592
029	689
057	787
085	884
114	982
142	·1011079
170	176
199	274
227	371
255	468
283	566
312	663
340	761
368	858
397	955

1262425	·1012053
453	150
481	247
510	345
538	442
566	539
595	636
623	734
651	831
679	928
708	·1013025
736	122
764	220
793	317
821	414
849	511
877	609
906	706
934	803
962	900
990	997
1263019	·1014095
047	192
075	289
103	386
131	483
160	581
188	678
216	775
244	872
273	970
301	·1015067
329	164
357	261
386	358
414	456
442	553
470	650
499	747
527	844
555	942
584	·1016039
612	136
640	233
668	331
697	428
725	525
753	622
782	719
810	816
838	913
866	·1017010
895	107
923	204
951	302
980	399
1264008	496
036	593
064	690
093	787
121	884
149	981
177	·1018078
206	175
234	272

1264262	·1018369
290	466
318	563
347	660
375	757
403	854
431	951
460	·1019048
488	145
516	242
544	339
573	436
601	533
629	630
657	727
686	824
714	921
742	·1020018
771	115
799	212
827	309
855	406
884	503
912	600
940	697
968	794
997	891
1265025	988
053	·1021085
081	182
109	279
138	375
166	472
194	569
222	666
251	763
279	860
307	957
335	·1022054
363	151
392	248
420	344
448	441
476	538
504	635
533	732
561	829
589	926
617	·1023022
646	119
674	216
702	313
730	410
758	507
787	603
815	700
843	797
871	894
899	991
928	·1024087
956	184
984	281
1266012	378
041	474
069	571

1266097	·1024668
125	765
153	861
182	958
210	·1025055
238	152
266	248
294	345
323	442
351	538
379	635
407	732
436	829
464	925
492	·1026022
520	119
548	215
576	312
605	409
633	505
661	602
689	698
717	795
745	892
773	988
802	·1027085
830	182
858	278
886	375
914	472
942	568
971	665
999	762
1267027	858
055	955
083	·1028051
112	148
140	245
168	341
196	438
225	535
253	631
281	728
309	825
337	921
366	·1029018
394	114
422	211
450	307
478	404
507	501
535	597
563	694
591	790
620	887
648	983
676	·1030080
704	176
732	273
760	369
789	466
817	562
845	659
873	755
901	851

1267929	·1030948
957	·1031044
986	141
1268014	237
042	334
070	430
098	526
126	623
154	719
183	816
211	912
239	·1032009
267	105
295	201
323	298
351	394
380	491
408	587
436	684
464	780
492	876
520	973
548	·1033069
577	165
605	262
633	358
661	455
689	551
717	647
745	744
774	840
802	936
830	·1034033
858	129
886	225
914	322
942	418
971	514
999	611
1269027	707
055	804
083	900
111	996
139	·1035093
168	189
196	285
224	382
252	478
280	574
308	670
336	767
365	863
393	959
421	·1036055
449	151
477	248
505	344
533	440
562	536
590	633
618	729
646	825
674	921
702	·1037017
730	114

1269759	·1037210
787	306
815	402
843	498
871	595
899	691
927	787
956	883
984	980

Excesses

10 = 34	1 = 3
20 = 69	2 = 7
	3 = 10
	4 = 14
	5 = 17
	6 = 21
	7 = 24
	8 = 27
	9 = 31

14 = 49

Anti Log Excesses

10 = 3	1 = 0
20 = 6	2 = 1
30 = 9	3 = 1
40 = 12	4 = 1
50 = 15	5 = 1
60 = 17	6 = 2
70 = 20	7 = 2
80 = 23	8 = 2
90 = 26	9 = 3

49 = 14

1270012	·1038076
040	172
068	268
096	364
124	460
152	557
180	653
208	749
236	845
265	941
293	·1039037
321	133
349	230
377	326
405	422
433	518
461	614
489	710
517	806
546	902
574	998
602	·1040094
630	190
658	287
686	383
714	479
743	575

1270771	·1040671
799	767
827	863
855	959
883	·1041055
911	151
939	247
967	343
995	439
1271023	535
052	631
080	727
108	823
136	919
164	·1042015
192	111
220	207
248	303
276	399
304	495
333	591
361	687
389	783
417	879
445	974
473	·1043070
501	166
530	262
558	358
586	454
614	550
642	646
670	742
698	838
726	934
754	·1044030
782	126
810	222
839	317
867	413
895	509
923	605
951	701
979	797
1272007	893
035	989
063	·1045085
091	180
119	276
147	372
175	468
203	564
232	659
260	755
288	851
316	947
344	·1046042
372	138
400	234
428	330
456	426
484	521
512	617
540	713
568	809

1272596	·1046905	1274419	·1053121	1276239	·1059319	1278057	·1065501
625	·1047000	447	217	267	415	085	596
653	096	475	312	295	510	112	690
681	192	503	408	323	605	140	785
709	288	531	503	351	700	168	880
737	383	559	599	379	795	196	975
765	479	587	694	407	891	224	·1066070
793	575	615	790	435	986	252	165
821	671	643	885	463	·1060081	280	260
849	766	671	981	491	176	308	355
877	862	699	·1054076	519	271	336	450
905	958	727	172	547	367	364	545
933	·1048054	755	267	575	462	392	640
961	149	783	362	603	557	420	735
989	245	811	458	631	652	448	830
1273017	341	839	553	659	748	476	925
045	436	867	649	687	843	504	·1067019
073	532	895	744	715	938	532	114
101	628	923	840	743	·1061033	560	209
129	724	951	935	771	128	588	304
157	819	980	·1055030	799	223	616	399
185	915	1275008	126	827	319	644	494
213	·1049011	036	221	855	414	672	589
241	106	064	317	883	509	700	684
269	202	092	412	911	604	728	779
297	298	120	508	939	699	756	873
325	393	148	603	967	794	784	968
353	489	176	698	995	889	812	·1068063
381	584	204	794	1277023	985	840	158
410	680	232	889	051	·1062080	868	253
438	776	260	984	079	175	895	347
466	871	288	·1056080	107	270	923	442
494	967	316	175	135	365	951	537
522	·1050063	344	271	163	460	979	632
550	158	372	366	191	555	1279007	726
578	254	400	461	219	650	035	821
606	350	428	557	247	745	063	916
634	445	456	652	275	840	091	·1069011
662	541	484	747	303	935	119	106
690	636	512	843	330	·1063031	147	200
718	732	540	938	358	126	175	295
746	827	568	·1057033	386	221	203	390
774	923	596	129	414	316	231	485
803	·1051019	624	224	442	411	259	580
831	114	652	319	470	506	286	674
859	210	680	414	498	601	314	769
887	305	708	510	526	696	342	864
915	401	736	605	554	791	370	959
943	496	764	700	582	886	398	·1070053
971	592	792	796	610	981	426	148
999	688	820	891	638	·1064076	454	243
1274027	783	848	986	666	171	482	338
055	879	876	·1058081	694	266	510	432
083	974	904	177	721	361	538	527
111	·1052070	932	272	749	456	565	622
139	165	960	367	777	551	593	717
167	261	988	462	805	646	621	811
195	357	1276016	558	833	741	649	906
223	452	044	653	861	836	677	·1071001
251	548	072	748	889	931	705	095
279	643	100	843	917	·1065026	733	190
307	739	128	938	945	121	760	285
335	834	155	·1059034	973	216	788	380
363	930	183	129	1278001	311	816	474
391	·1053026	211	224	029	406	844	569

1279872	·1071664
900	758
928	853
956	948
984	·1072043

Excesses

10 = 34	1 = 3
20 = 68	2 = 7
	3 = 10
	4 = 14
	5 = 17
	6 = 20
	7 = 24
	8 = 27
	9 = 31

14 = 48

Anti Log Excesses

10 = 3	1 = 0
20 = 6	2 = 1
30 = 9	3 = 1
40 = 12	4 = 1
50 = 15	5 = 1
60 = 18	6 = 2
70 = 21	7 = 2
80 = 24	8 = 2
90 = 26	9 = 3

48 = 14

1280012	·1072137
040	232
067	327
095	421
123	516
151	611
179	706
207	800
235	895
263	990
291	·1073084
319	179
346	273
374	368
402	462
430	557
458	652
486	746
514	841
541	935
569	·1074030
597	124
625	219
653	314
681	408
709	503
737	597
765	692
793	786
821	881
848	976

1280876	·1075070	1282687	·1081206	1284496	·1087326	1286302	·1093427	1288105	·1099513
904	165	715	300	524	420	330	521	133	606
932	259	743	395	552	514	358	615	161	700
960	354	771	489	580	608	386	709	188	793
988	448	799	583	607	702	413	802	216	887
1281016	543	827	678	635	796	441	896	244	980
044	638	855	772	663	890	469	990	271	·1100074
072	732	882	866	691	984	497	·1094084	299	167
100	827	910	960	719	·1088078	524	177	327	261
127	921	938	·1082055	746	171	552	271	355	354
155	·1076016	966	149	774	265	580	365	382	448
183	110	994	243	802	359	608	459	410	541
211	205	1283022	337	830	453	635	552	438	634
239	299	049	432	857	547	663	646	465	728
267	394	077	526	885	641	691	740	493	821
295	488	105	620	913	735	718	833	521	914
322	583	133	714	941	829	746	927	549	·1101008
350	677	161	808	969	923	774	·1095021	576	101
378	772	188	903	996	·1089017	802	115	604	195
406	866	216	997	1285024	111	829	208	632	288
434	960	244	·1083091	052	205	857	302	659	381
462	·1077055	272	185	080	299	885	396	687	475
490	149	299	280	108	393	913	489	715	568
517	244	327	374	135	486	940	583	743	661
545	338	355	468	163	580	968	677	770	755
573	433	383	562	191	674	996	770	798	848
601	527	411	656	219	768	1287024	864	826	941
629	621	439	750	246	862	052	957	853	·1102035
657	716	466	845	274	956	079	·1096051	881	128
685	810	494	939	302	·1090050	107	145	909	221
712	905	522	·1084033	330	144	135	238	937	315
740	999	550	127	358	238	163	332	964	408
768	·1078094	578	221	385	332	190	426	992	502
796	188	606	315	413	425	218	519	1289020	595
824	282	634	409	441	519	246	613	047	688
852	377	661	504	469	613	274	707	075	782
880	471	689	598	497	707	301	800	103	875
907	565	717	692	524	801	329	894	131	968
935	660	745	786	552	895	357	987	158	·1103062
963	754	773	880	580	989	385	·1097081	186	155
991	849	801	974	608	·1091082	412	174	214	248
1282019	943	828	·1085068	635	176	440	268	241	342
047	·1079037	856	162	663	270	468	362	269	435
075	132	884	256	691	364	495	455	297	528
102	226	912	350	719	458	523	549	325	621
130	320	940	444	747	552	551	642	352	715
158	415	967	539	774	645	579	736	380	808
186	509	995	633	802	739	606	829	408	901
214	603	1284023	727	830	833	634	923	435	995
242	698	051	821	858	927	662	·1098017	463	·1104088
270	792	078	915	886	·1092021	689	110	491	181
297	886	106	·1086009	913	114	717	204	519	274
325	980	134	103	941	208	745	297	546	368
353	·1080075	162	197	969	302	773	391	574	461
381	169	190	291	997	396	800	484	602	554
409	263	218	385	1286024	489	828	578	629	648
437	358	245	479	052	583	856	671	657	741
465	452	273	573	080	677	883	765	685	834
492	546	301	667	108	771	911	858	713	927
520	640	329	761	136	865	939	952	740	·1105021
548	735	357	856	163	958	967	·1099045	768	114
576	829	385	950	191	·1093052	994	139	796	207
604	923	413	·1087044	219	146	1288022	232	823	301
632	·1081018	440	138	247	240	050	326	851	394
660	112	468	232	275	334	077	419	879	487

1289907	·1105580
934	674
962	767
990	860

Excesses

10 = 34	1 = 3
20 = 68	2 = 7
	3 = 10
	4 = 14
	5 = 17
	6 = 20
	7 = 24
	8 = 27
	9 = 30

14 = 47

Anti Log Excesses

10 = 3	1 = 0
20 = 6	2 = 1
30 = 9	3 = 1
40 = 12	4 = 1
50 = 15	5 = 1
60 = 18	6 = 2
70 = 21	7 = 2
80 = 24	8 = 2
90 = 27	9 = 3

47 = 14

1290017	·1105953
045	·1106046
073	140
100	233
128	326
155	419
183	512
211	605
238	698
266	792
294	885
321	978
349	·1107071
377	164
404	257
432	350
460	444
488	537
515	630
543	723
571	816
598	909
626	·1108002
654	096
682	189
709	282
737	375
764	468
792	561
820	654
848	747
875	840

1290903	·1108933
930	·1109026
958	120
986	213
1291013	306
041	399
069	492
096	585
124	678
152	771
179	864
207	957
235	·1110050
262	143
290	236
317	329
345	422
373	515
400	608
428	701
456	794
483	887
511	980
539	·1111073
566	166
594	259
622	352
649	445
677	538
704	631
732	724
760	817
787	910
815	·1112003
843	096
870	189
898	282

Excesses

14 = 47

Anti Log Excesses

47 = 14

1291928	·1112382
958	482
987	582
1292017	682
047	782
077	882
106	982
136	·1113082
166	182
196	282
225	382
255	482
285	582
315	682
345	782
374	882
404	982
434	·1114082
464	182

1292493	·1114282
523	382
553	482
583	582
612	682
642	782
672	882
702	982
732	·1115082
761	182
791	282
821	382
851	482
880	581
910	681
940	781
970	881
999	981
1293029	·1116081
059	181
089	281
119	381
148	481
178	581
208	681
238	781
267	880
297	980
327	·1117080
357	180
386	280
416	380
446	480
476	580
505	680
535	779
565	879
594	979
624	·1118079
654	178
684	278
713	378
743	478
773	577
802	677
832	777
862	877
891	977
921	·1119076
951	176
980	276
1294010	376
040	475
070	575
099	675
129	775
159	874
188	974
218	·1120074
248	174
278	273
307	373
337	473
367	572
397	672

1294426	·1120772
456	872
486	971
516	·1121071
545	171
575	270
605	370
635	470
664	569
694	669
724	768
753	868
783	968
813	·1122067
843	167
872	267
902	366
932	466
961	565
991	665
1295021	765
050	864
080	964
110	·1123063
139	163
169	263
199	362
229	462
258	562
288	661
318	761
347	860
377	960
407	·1124059
436	159
466	258
496	358
525	457
555	557
585	656
615	756
644	855
674	955
704	·1125054
733	154
763	253
793	352
822	452
852	551
882	651
911	750
941	850
971	949
1296001	·1126049
030	148
060	248
090	347
119	447
149	546
179	645
208	745
238	844
267	944
297	·1127043
327	142

1296356	·1127242
386	341
416	440
445	540
475	639
504	739
534	838
564	937
593	·1128037
623	136
653	236
682	335
712	434
742	534
772	633
801	732
831	832
861	931
890	·1129031
920	130
950	229
979	329
1297009	428
038	527
068	627
098	726
127	825
157	924
187	·1130024
216	123
246	222
275	322
305	421
335	520
364	619
394	718
424	818
453	917
483	·1131016
513	115
543	214
572	313
602	413
632	512
661	611
691	710
721	809
750	908
780	·1132008
809	107
839	206
869	305
898	405
928	504
958	603
987	702
1298017	802
046	901
076	·1133000
106	099
135	198
165	297
194	396
224	495
254	594

1298283	·1133694
313	793
343	892
372	991
402	·1134090
431	189
461	288
491	387
520	486
550	585
579	684
609	783
639	882
668	981
698	·1135080
728	179
757	278
787	377
816	476
846	575
876	674
905	773
935	872
964	971
994	·1136070
1299024	169
053	268
083	367
113	466
142	565
172	664
201	763
231	862
261	961
290	·1137060
320	159
349	258
379	357
409	456
438	554
468	653
498	752
527	851
557	950
586	·1138049
616	148
646	247
675	346
705	445
734	544
764	643
793	742
823	840
852	939
882	·1139038
911	137
941	236
970	335

Excesses

10 = 34	1 = 3
20 = 67	2 = 7
	3 = 10
	4 = 13
	5 = 17
	6 = 20
	7 = 23
	8 = 27
	9 = 30

15 = 50

Anti Log Excesses

10 = 3	1 = 0
20 = 6	2 = 1
30 = 9	3 = 1
40 = 12	4 = 1
50 = 15	5 = 1
60 = 18	6 = 2
70 = 21	7 = 2
80 = 24	8 = 2
90 = 27	9 = 3

50 = 15

1300000	·1139434
030	533
059	632
089	730
118	829
148	928
178	·1140027
207	125
237	224
267	323
296	422
326	520
355	619
385	718
415	817
444	916
474	·1141014
503	113
533	212
563	311
592	409
622	508
652	607
681	706
711	804
740	903
770	·1142002
800	101
829	199
859	298
888	397
918	495
947	594
977	693
1301006	792
036	890
065	989

1301095	·1143088
124	186
154	285
184	384
213	482
243	581
272	679
302	778
331	877
361	975
390	·1144074
420	173
449	271
479	370
508	468
538	567
568	666
597	764
627	863
656	961
686	·1145060
715	159
745	257
774	356
804	455
833	553
863	652
892	750
922	849
952	947
981	·1146046
1302011	144
040	243
070	341
099	440
129	538
158	637
188	735
217	834
247	932
276	·1147031
306	129
336	227
365	326
395	424
424	523
454	621
483	720
513	818
542	917
572	·1148015
601	114
631	212
660	311
690	409
720	507
749	606
779	704
808	803
838	901
867	999
897	·1149098
926	196
956	294
985	393

1303015	·1149491
044	590
074	688
103	786
133	885
162	983
192	·1150082
221	180
251	278
280	377
310	475
339	573
369	672
398	770
428	869
457	967
487	·1151065
516	164
546	262
575	360
605	459
634	557
664	655
693	753
723	852
752	950
782	·1152048
811	147
841	245
870	343
900	441
929	540
959	638
988	736
1304018	834
047	933
077	·1153031
106	129
136	227
165	326
195	424
224	522
254	620
283	718
313	816
342	915
372	·1154013
401	111
431	209
460	307
490	405
519	504
549	602
578	700
608	798
637	896
667	994
696	·1155092
726	190
755	288
785	386
814	485
844	583
873	681
903	779

1304932	·1155877
962	975
991	·1156073
1305020	171
050	269
079	367
109	465
138	563
168	661
197	760
227	858
256	956
286	·1157054
315	152
345	250
374	348
403	446
433	544
462	642
492	740
521	838
551	936
580	·1158034
610	132
639	230
669	328
698	426
728	524
757	622
786	720
816	818
845	916
875	·1159014
904	112
934	210
963	307
993	405
1306022	503
052	601
081	699
111	797
140	895
169	993
199	·1160091
228	189
258	287
287	385
317	483
346	580
376	678
405	776
435	874
464	972
494	·1161070
523	168
552	266
582	364
611	461
641	559
670	657
699	755
729	852
758	950
787	·1162048
817	146

1306846	·1162243
876	341
905	439
934	537
964	635
993	732
1307023	830
052	928
082	·1163026
111	123
141	221
170	319
200	417
229	514
259	612
288	710
317	808
347	906
376	·1164003
406	101
435	199
464	297
494	394
523	492
552	590
582	688
611	785
641	883
670	981
699	·1165079
729	176
758	274
788	371
817	469
846	567
876	664
905	762
934	860
964	957
993	·1166055
1308023	152
052	250
081	348
111	445
140	543
170	640
199	738
229	836
258	933
288	·1167031
317	129
347	226
376	324
406	421
435	519
464	617
494	714
523	812
553	909
582	·1168007
611	104
641	202
670	299
699	397
729	494

1308758	·1168592
788	689
817	787
846	884
876	982
905	·1169079
935	177
964	274
993	372
1309023	469
052	567
081	664
111	762
140	859
170	957
199	·1170054
228	151
258	249
287	346
317	444
346	541
375	639
405	736
434	834
463	931
493	·1171029
522	126
552	224
581	321
610	418
640	516
669	613
698	711
728	808
757	905
786	·1172003
815	100
845	197
874	295
903	392
933	490
962	587
991	684

Excesses

10 = 33	1 = 3
20 = 67	2 = 7
	3 = 10
	4 = 13
	5 = 17
	6 = 20
	7 = 23
	8 = 27
	9 = 30

15 = 49

Anti Log Excesses

10 = 3	1 = 0
20 = 6	2 = 1
30 = 9	3 = 1
40 = 12	4 = 1
50 = 15	5 = 2
60 = 18	6 = 2
70 = 21	7 = 2
80 = 24	8 = 2
90 = 27	9 = 3

49 = 15

1310021	·1172782
050	879
080	976
109	·1173074
138	171
168	268
197	365
226	463
256	560
285	657
315	755
344	852
373	949
403	·1174046
432	144
461	241
491	338
520	435
549	533
578	630
608	727
637	824
666	922
696	·1175019
725	116
754	213
784	310
813	407
843	505
872	602
901	699
931	796
960	893
989	990
1311019	·1176088
048	185
078	282
107	379
136	476
166	573
195	670
224	768
254	865
283	962
312	·1177059
341	156
371	253
400	351
429	448
459	545
488	642

1311517	·1177739
547	836
576	933
605	·1178030
635	127
664	224
693	322
722	419
752	516
781	613
810	710
840	807
869	904
898	·1179001
928	098
957	195
986	292
1312016	389
045	486
074	584
103	681
133	778
162	875
191	972
221	·1180069
250	166
279	263
309	360
338	457
367	554
397	651
426	748
455	844
484	941
514	·1181038
543	135
572	232
602	329
631	426
660	523
690	620
719	717
748	814
778	911
807	·1182008
836	104
865	201
895	298
924	395
953	492
983	589
1313012	686
041	783
071	880
100	977
129	·1183073
159	170
188	267
217	364
246	461
276	558
305	655
334	751
364	848
393	945

1313422	·1184042
452	139
481	236
510	332
540	429
569	526
598	623
627	720
657	817
686	914
715	·1185010
745	107
774	204
803	301
832	398
862	494
891	591
920	688
949	785
979	881
1314008	978
037	·1186075
066	172
096	268
125	365
154	462
183	559
213	655
242	752
271	849
301	945
330	·1187042
359	139
388	236
418	332
447	429
476	526
506	622
535	719
564	816
593	912
623	·1188009
652	105
681	202
710	299
740	395
769	492
798	589
827	685
857	782
886	878
915	975
944	·1189072
973	168
1315003	265
032	361
061	458
090	555
120	651
149	748
178	845
207	941
237	·1190038
266	134
295	231

1315324	·1190327
353	424
383	520
412	617
441	713
470	810
500	906
529	·1191003
558	099
587	196
617	292
646	389
675	485
704	582
733	678
763	775
792	871
821	968
850	·1192064
880	161
909	257
938	354
967	450
997	547
1316026	643
055	740
084	836
113	933
143	·1193029
172	126
201	222
230	318
260	415
289	511
318	607
347	704
377	800
406	897
435	993
464	·1194089
493	186
523	282
552	379
581	475
610	571
640	668
669	764
698	860
727	957
757	·1195053
786	150
815	246
844	342
873	438
902	535
932	631
961	727
990	823
1317019	920
048	·1196016
077	112
106	208
136	305
165	401
194	497

1317223	·1196593
252	690
282	786
311	882
340	979
369	·1197075
399	171
428	267
457	364
486	460
516	556
545	653
574	749
603	845
632	941
661	·1198037
691	134
720	230
749	326
778	422
807	518
836	614
865	711
895	807
924	903
953	999
982	·1199095
1318011	191
041	287
070	384
099	480
128	576
158	672
187	768
216	864
245	961
275	·1200057
304	153
333	249
362	345
391	441
420	537
450	633
479	729
508	825
537	922
566	·1201018
595	114
624	210
654	306
683	402
712	498
741	594
770	690
799	786
829	882
858	978
887	·1202074
916	170
945	266
974	362
1319003	458
033	554
062	650
091	746

1319120	·1202842
149	938
178	·1203034
208	130
237	226
266	322
295	418
324	514
353	610
382	706
412	802
441	898
470	994
499	·1204090
528	186
557	282
587	378
616	474
645	570
674	665
703	761
732	857
761	953
791	·1205049
820	145
849	241
878	337
907	433
936	529
966	624
995	720

Excesses

10 = 33	1 = 3
20 = 66	2 = 7
	3 = 10
	4 = 13
	5 = 17
	6 = 20
	7 = 23
	8 = 26
	9 = 30

15 = 49

Anti Log Excesses

10 = 3	1 = 0
20 = 6	2 = 1
30 = 9	3 = 1
40 = 12	4 = 1
50 = 15	5 = 2
60 = 18	6 = 2
70 = 21	7 = 2
80 = 24	8 = 2
90 = 27	9 = 3

49 = 15

1320024	·1205816
053	912
082	·1206008
111	104
140	200

1320170	·1206295
199	391
228	487
257	583
286	679
315	774
344	870
373	966
402	·1207062
432	157
461	253
490	349
519	445
548	540
577	636
606	732
635	828
664	924
693	·1208019
723	115
752	211
781	307
810	402
839	498
868	594
897	690
927	785
956	881
985	977
1321014	·1209073
043	168
072	264
101	360
130	455
159	551
189	647
218	743
247	838
276	934
305	·1210030
334	125
363	221
392	317
421	412
450	508
480	604
509	699
538	795
567	891
596	987
625	·1211082
654	178
684	274
713	369
742	465
771	561
800	656
829	752
858	847
887	943
916	·1212038
946	134
975	229
1322004	325
033	420

1322062	·1212516
091	611
120	707
149	803
178	898
207	994
236	·1213089
265	185
294	280
324	376
353	471
382	567
411	662
440	758
469	853
498	949
527	·1214044
556	140
585	235
614	331
643	426
672	522
702	617
731	713
760	808
789	904
818	999
847	·1215095
876	190
905	285
934	381
963	476
992	572
1323021	667
050	763
080	858
109	954
138	·1216049
167	145
196	240
225	336
254	431
283	526
312	622
341	717
370	813
399	908
428	·1217003
458	099
487	194
516	289
545	385
574	480
603	576
632	671
661	766
690	862
719	957
748	·1218052
777	148
806	243
836	338
865	433
894	529
923	624

1323952	·1218719
981	815
1324010	910
039	·1219005
068	100
097	196
126	291
155	386
184	481
213	577
242	672
271	767
300	862
329	958
358	·1220053
387	148
416	243
445	338
474	434
503	529
532	624
561	719
591	815
620	910
649	·1221005
678	100
707	196
736	291
765	386
794	481
823	576
852	671
881	767
910	862
939	957
968	·1222052
997	147
1325026	242
055	338
084	433
113	528
142	623
171	718
200	813
229	908
258	·1223003
287	098
316	193
345	289
374	384
403	479
432	574
461	669
490	764
519	859
548	954
577	·1224049
606	144
635	239
664	334
693	429
722	525
751	620
780	715
809	810

1325838	·1224905
867	·1225000
896	095
925	190
954	285
983	380
1326012	475
041	570
070	665
099	759
128	854
157	949
186	·1226044
215	139
244	234
273	329
302	424
331	519
360	614
389	709
418	804
447	899
476	993
505	·1227088
534	183
563	278
592	373
621	468
650	563
679	658
708	753
737	848
766	943
795	·1228038
824	133
853	227
882	322
911	417
940	512
969	607
998	702
1327027	797
056	892
085	987
114	·1229082
143	176
172	271
201	366
230	461
259	556
288	651
317	746
346	840
375	935
404	·1230030
433	125
462	220
491	314
520	409
549	504
578	599
607	693
636	788
665	883
694	978

1327723	·1231072
752	167
781	262
810	357
839	451
868	546
897	641
926	735
955	830
983	925
1328012	·1232020
041	114
070	209
099	304
128	398
157	493
186	588
215	682
244	777
273	872
302	966
331	·1233061
360	156
389	251
418	345
447	440
476	535
505	629
534	724
563	819
592	913
621	·1234008
650	102
679	197
708	292
736	386
765	481
794	576
823	670
852	765
881	859
910	954
939	·1235049
968	143
997	238
1329026	332
055	427
084	521
112	616
141	710
170	805
199	899
228	994
257	·1236088
286	183
315	278
344	372
373	467
402	561
431	656
460	750
488	845
517	939
546	·1237034
575	128

1329604	·1237223
633	317
662	412
691	506
720	601
749	695
778	790
807	884
836	978
864	·1238073
893	167
922	261
951	356
980	450

Excesses

10 = 33	1 = 3
20 = 66	2 = 7
	3 = 10
	4 = 13
	5 = 16
	6 = 20
	7 = 23
	8 = 26
	9 = 29

15 = 49

Anti Log Excesses

10 = 3	1 = 0
20 = 6	2 = 1
30 = 9	3 = 1
40 = 12	4 = 1
50 = 15	5 = 2
60 = 18	6 = 2
70 = 21	7 = 2
80 = 24	8 = 2
90 = 27	9 = 3

49 = 15

1330009	·1238545
038	639
067	733
096	828
125	922
154	·1239017
183	111
212	206
240	300
269	395
298	489
327	584
356	678
385	773
414	867
443	961
472	·1240056
501	150
530	244
559	339
588	433
616	527

1330645	·1240621
674	716
703	810
732	904
761	999
790	·1241093
819	187
848	282
877	376
905	470
934	565
963	659
992	753
1331021	847
050	942
079	·1242036
107	130
136	225
165	319
194	413
223	507
252	602
281	696
310	790
339	884
367	979
396	·1243073
425	167
454	261
483	356
512	450
541	544
570	638
599	732
628	826
656	921
685	·1244015
714	109
743	203
772	297
801	391
830	486
858	580
887	674
916	768
945	862
974	956
1332003	·1245050
032	145
061	239
090	333
118	427
147	521
176	615
205	710
234	804
263	898
292	992
321	·1246086
350	180
379	274
407	368
436	462
465	556
494	651

1332523	·1246745	1334398	·1252851	1336270	·1258939	1338139	·1265011
552	839	427	945	299	·1259033	168	105
580	933	455	·1253039	327	126	197	198
609	·1247027	484	132	356	220	225	291
638	121	513	226	385	313	254	385
667	215	542	320	414	407	283	478
696	309	571	414	443	501	312	571
725	403	600	507	472	594	341	664
754	497	628	601	500	688	369	758
782	591	657	695	529	781	398	851
811	685	686	788	558	875	427	944
840	779	715	882	587	968	456	·1266037
869	873	743	976	615	·1260062	484	131
898	967	772	·1254070	644	155	513	224
927	·1248061	801	163	673	249	542	317
955	155	830	257	702	342	571	410
984	249	858	351	730	436	599	504
1333013	343	887	444	759	529	628	597
042	437	916	538	788	623	657	690
071	531	945	632	817	716	686	783
100	625	974	726	846	810	714	876
129	719	1335003	819	874	903	743	969
157	813	031	913	903	997	772	·1267063
186	907	060	·1255007	932	·1261090	800	156
215	·1249001	089	101	961	184	829	249
244	095	118	194	989	277	858	342
273	189	147	288	1337018	371	887	435
302	283	176	382	047	464	915	528
330	377	204	476	076	558	944	622
359	471	233	569	104	651	973	715
388	565	262	663	133	745	1339001	808
417	659	291	757	162	838	030	901
446	753	320	851	191	931	059	994
475	847	349	944	220	·1262025	087	·1268087
504	941	377	·1256038	248	118	116	180
532	·1250035	406	131	277	212	145	274
561	129	435	225	306	305	173	367
590	223	464	319	335	398	202	460
619	316	492	412	363	492	231	553
648	410	521	506	392	585	260	646
677	504	550	600	421	678	288	739
705	598	579	693	450	772	317	833
734	692	607	787	478	865	346	926
763	786	636	880	507	959	374	·1269019
792	880	665	974	536	·1263052	403	112
821	974	694	·1257068	565	145	432	205
850	·1251068	723	161	593	239	461	298
879	162	751	255	622	332	489	391
907	255	780	348	651	425	518	484
936	349	809	442	679	519	547	577
965	443	838	536	708	612	576	670
994	537	866	629	737	705	604	764
1334023	631	895	723	766	798	633	857
052	725	924	817	794	892	662	950
081	819	953	910	823	985	691	·1270043
109	912	981	·1258004	852	·1264078	719	136
138	·1252006	1336010	097	880	172	748	229
167	100	039	191	909	265	777	322
196	194	068	285	938	358	806	415
225	288	097	378	967	452	834	508
254	382	126	472	995	545	863	601
282	475	154	565	1338024	638	892	694
311	569	183	659	053	732	920	787
340	663	212	752	082	825	949	880
369	757	241	846	110	918	978	974

Excesses

10 = 33	1 = 3
20 = 65	2 = 7
	3 = 10
	4 = 13
	5 = 16
	6 = 20
	7 = 23
	8 = 26
	9 = 29

15 = 49

Anti Log Excesses

10 = 3	1 = 0
20 = 6	2 = 1
30 = 9	3 = 1
40 = 12	4 = 1
50 = 15	5 = 2
60 = 18	6 = 2
70 = 22	7 = 2
80 = 25	8 = 2
90 = 28	9 = 3

49 = 15

1340007	·1271067
035	160
064	253
093	346
121	439
150	532
179	625
207	718
236	811
265	904
293	997
322	·1272090
351	182
380	275
408	368
437	461
466	554
494	647
523	740
552	833
580	926
609	·1273019
638	112
666	205
695	298
724	390
753	483
781	576
810	669
839	762
867	855
896	948
925	·1274041
953	134
982	227
1341011	320
039	413

1341068	·1274506
097	598
126	691
154	784
183	877
212	970
240	·1275063
269	156
298	249
326	342
355	434
383	527
412	620
441	713
469	805
498	898
527	991
555	·1276084
584	176
612	269
641	362
670	455
698	548
727	641
756	733
784	826
813	919
842	·1277012
871	105
899	198
928	291
957	383
985	476
1342014	569
043	662
071	754
100	847
128	940
157	·1278032
186	125
214	218
243	311
272	403
300	496
329	589
357	681
386	774
415	867
443	959
472	·1279052
501	145
529	237
558	330
587	423
616	516
644	608
673	701
702	794
730	886
759	979
788	·1280072
816	164
845	257
873	349
902	442

1342931	·1280535
959	627
988	720
1343017	813
045	905
074	998
102	·1281090
131	183
160	276
188	368
217	461
245	553
274	646
303	738
331	831
360	923
389	·1282016
417	108
446	201
474	293
503	386
532	479
560	571
589	664
617	756
646	849
675	941
703	·1283034
732	126
761	219
789	311
818	404
846	496
875	589
904	681
932	774
961	866
990	959
1344018	·1284051
047	144
076	236
105	329
133	421
162	514
191	606
219	699
248	791
277	883
305	976
334	·1285068
362	161
391	253
419	345
448	438
476	530
505	622
533	715
562	807
590	900
619	992
648	·1286084
676	177
705	269
733	362
762	454

1344791	·1286546
819	639
848	731
877	823
905	916
934	·1287008
962	101
991	193

Excesses

14 = 46

Anti Log Excesses

46 = 14

1345022	·1287293
053	393
084	493
115	593
146	693
177	793
208	893
239	993
270	·1288093
301	193
332	293
363	393
394	493
425	593
456	693
487	793
518	893
549	993
579	·1289092
610	192
641	292
672	392
703	492
734	592
765	692
796	792
827	892
858	992
889	·1290092
920	192
951	291
982	391
1346013	491
044	591
075	691
106	791
137	891
168	991
199	·1291090
230	190
261	290
292	390
322	490
353	590
384	689
415	789
446	889
477	989

1346508	·1292089
539	189
570	288
601	388
632	488
663	588
694	687
725	787
756	887
787	987
818	·1293086
849	186
880	286
911	386
942	485
973	585
1347004	685
035	785
065	884
096	984
127	·1294084
158	184
189	283
220	383
251	483
282	582
313	682
344	782
375	881
406	981
436	·1295081
467	180
498	280
529	380
560	479
591	579
622	679
653	778
684	878
715	978
746	·1296077
777	177
807	277
838	376
869	476
900	576
931	675
962	775
993	875
1348024	974
055	·1297074
086	173
117	273
148	372
178	472
209	571
240	671
271	770
302	870
333	969
364	·1298069
395	168
426	268
457	367
488	467

1348519	·1298566
549	666
580	765
611	865
642	964
673	·1299064
704	163
735	263
766	362
796	462
827	561
858	661
889	760
920	860
951	959
981	·1300059
1349012	158
043	258
074	357
105	456
136	556
167	655
198	755
229	854
260	954
290	·1301053
321	152
352	252
383	351
414	451
445	550
476	649
507	749
537	848
568	947
599	·1302047
630	146
661	245
692	345
722	444
753	543
784	643
815	742
846	841
877	941
908	·1303040
939	139
970	238

Excesses

10 = 32	1 = 3
20 = 65	2 = 6
30 = 97	3 = 10
	4 = 13
	5 = 16
	6 = 19
	7 = 23
	8 = 26
	9 = 29

15 = 50

Anti Log Excesses		1351573	·1308396	1353576	·1314825	1355575	·1321236	1357572	·1327627
		604	495	606	923	606	334	603	725
10 = 3	1 = 0	635	594	637	·1315022	637	433	633	823
20 = 6	2 = 1	666	693	668	121	667	531	664	922
30 = 9	3 = 1	697	792	699	220	698	630	695	·1328020
40 = 12	4 = 1	728	891	730	318	729	728	725	118
50 = 15	5 = 2	758	990	761	417	760	826	756	216
60 = 19	6 = 2	789	·1309089	791	516	790	925	787	314
70 = 22	7 = 2	820	188	822	615	821	·1322023	817	412
80 = 25	8 = 2	851	287	853	713	852	122	848	511
90 = 28	9 = 3	882	386	884	812	883	220	879	609
		913	485	915	911	913	319	909	707
50 = 15		943	584	945	·1316010	944	417	940	805
		974	683	976	108	975	515	971	903
1350001	·1303338	1352005	782	1354007	207	1356006	614	1358001	·1329001
031	437	036	881	038	306	036	712	032	099
062	536	067	980	068	405	067	811	063	197
093	635	098	·1310079	099	503	098	909	093	295
124	735	128	178	130	602	129	·1323007	124	393
155	834	159	277	161	701	159	106	155	492
186	933	190	376	191	800	190	204	185	590
217	·1304032	221	475	222	898	221	302	216	688
248	132	252	574	253	997	251	401	247	786
278	231	283	673	284	·1317096	282	499	277	884
309	330	313	772	314	194	313	597	308	982
340	429	344	871	345	293	343	696	339	·1330080
371	529	375	970	376	391	374	794	369	178
402	628	406	·1311069	407	490	405	892	400	276
433	727	437	168	437	588	435	991	431	374
463	826	467	267	468	687	466	·1324089	461	472
494	926	498	366	499	786	497	187	492	570
525	·1305025	529	465	530	884	527	286	523	669
556	124	560	564	560	983	558	384	553	767
587	223	590	663	591	·1318081	589	482	584	865
618	322	621	761	622	180	620	581	615	963
648	422	652	860	653	279	650	679	645	·1331061
679	521	683	959	683	377	681	777	676	159
710	620	713	·1312058	714	476	712	876	707	257
741	719	744	157	745	574	743	974	737	355
772	818	775	256	776	673	773	·1325072	768	453
803	917	806	355	806	771	804	171	799	551
833	·1306017	837	454	837	870	835	269	829	649
864	116	867	552	868	969	866	367	860	747
895	215	898	651	899	·1319067	896	466	891	844
926	314	929	750	929	166	927	564	921	942
957	413	960	849	960	264	958	662	952	·1332040
988	512	991	948	991	363	989	760	983	138
1351018	612	1353022	·1313047	1355022	462	1357019	859	1359013	236
049	711	052	145	052	560	050	957	044	334
080	810	083	244	083	659	081	·1326055	075	432
111	909	114	343	114	757	112	153	105	530
142	·1307008	145	442	145	856	142	252	136	628
173	107	176	541	175	954	173	350	167	726
203	207	206	640	206	·1320053	204	448	197	824
234	306	237	738	237	152	235	546	228	922
265	405	268	837	268	250	265	645	259	·1333020
296	504	299	936	298	349	296	743	289	118
327	603	329	·1314035	329	447	327	841	320	216
358	702	360	133	360	546	357	939	351	314
388	801	391	232	391	645	388	·1327038	381	412
419	900	422	331	421	743	419	136	412	510
450	999	452	430	452	842	449	234	443	608
481	·1308098	483	528	483	940	480	332	473	706
512	198	514	627	514	·1321039	511	431	504	804
543	297	545	726	544	137	541	529	535	902

1359565	·1334000
596	098
627	195
657	293
688	391
719	489
749	587
780	685
811	783
841	881
872	978
903	·1335076
933	174
964	272
995	370

Excesses

10 = 32	1 = 3
20 = 64	2 = 6
30 = 96	3 = 10
	4 = 13
	5 = 16
	6 = 19
	7 = 22
	8 = 26
	9 = 29
15 = 49	

Anti Log Excesses

10 = 3	1 = 0
20 = 6	2 = 1
30 = 9	3 = 1
40 = 12	4 = 1
50 = 16	5 = 2
60 = 19	6 = 2
70 = 22	7 = 2
80 = 25	8 = 2
90 = 28	9 = 3
49 = 15	

1360025	·1335468
056	565
087	663
117	761
148	859
179	957
209	·1336055
240	152
270	250
301	348
331	446
362	543
393	641
423	739
454	837
484	934
515	·1337032
546	130
576	228
607	325
638	423

1360668	·1337521
699	619
730	716
760	814
791	912
822	·1338010
852	107
883	205
914	303
944	400
975	498
1361005	596
036	693
066	791
097	889
128	986
158	·1339084
189	182
219	279
250	377
281	475
311	572
342	670
372	768
403	865
433	963
464	·1340061
495	158
525	256
556	354
586	451
617	549
648	647
678	744
709	842
739	939
770	·1341037
800	134
831	232
862	330
892	427
923	525
953	622
984	720
1362015	818
045	915
076	·1342013
106	110
137	208
167	305
198	403
229	500
259	598
290	695
320	793
351	890
382	987
412	·1343085
443	182
473	280
504	377
534	475
565	572
596	669
626	767

1362657	·1343864
687	962
718	·1344059
749	157
779	254
810	352
840	449
871	547
901	644
932	742
963	839
993	937
1363024	·1345034
054	132
085	229
116	326
146	424
177	521
207	618
238	716
268	813
299	910
330	·1346008
360	105
391	202
421	300
452	397
483	494
513	592
544	689
574	786
605	884
635	981
666	·1347078
697	176
727	273
758	370
788	468
819	565
849	662
880	760
910	857
941	954
971	·1348051
1364002	149
032	246
063	343
093	440
124	538
154	635
185	732
216	829
246	927
277	·1349024
307	121
338	218
368	316
399	413
430	510
460	607
491	705
521	802
552	899
582	996
613	·1350093

1364643	·1350191
674	288
704	385
735	482
765	579
796	676
826	774
857	871
887	968
918	·1351065
948	162
979	259
1365009	356
040	453
070	550
101	647
131	745
162	842
192	939
223	·1352036
253	133
284	230
315	327
345	424
376	521
406	618
437	715
467	812
498	909
529	·1353006
559	103
590	200
620	297
651	394
681	491
712	588
742	685
773	782
803	879
834	976
864	·1354073
895	170
925	267
956	364
986	461
1366017	558
047	655
078	752
108	849
139	946
169	·1355043
200	140
230	237
261	334
291	431
322	528
352	625
383	722
413	819
444	916
474	·1356012
505	109
535	206
566	303
596	400

1366626	·1356497
657	593
687	690
718	787
748	884
778	981
809	·1357078
839	175
870	272
900	369
931	466
961	562
992	659
1367022	756
053	853
083	950
114	·1358047
144	144
175	241
205	337
236	434
266	531
297	628
327	724
358	821
388	918
419	·1359015
449	111
480	208
510	305
541	402
571	498
602	595
632	692
663	789
693	885
724	982
754	·1360079
785	176
815	272
846	369
876	466
907	562
937	659
968	756
998	852
1368029	949
059	·1361046
089	142
120	239
150	336
181	432
211	529
241	626
272	722
302	819
333	916
363	·1362012
394	109
424	206
454	302
485	399
515	496
546	592
576	689

1368606	·1362786
637	882
667	979
698	·1363075
728	172
759	268
789	365
820	462
850	558
881	655
911	751
942	848
972	945
1369003	·1364041
033	138
064	234
094	331
125	427
155	524
185	621
216	717
246	814
277	910
307	·1365007
337	103
368	200
398	296
429	393
459	489
490	586
520	682
550	778
581	875
611	971
642	·1366068
672	164
702	261
733	357
763	454
794	550
824	647
855	743
885	840
915	936
946	·1367033
976	129

Excesses

10 = 32	1 = 3
20 = 64	2 = 6
30 = 95	3 = 10
	4 = 13
	5 = 16
	6 = 19
	7 = 22
	8 = 25
	9 = 29

15 = 48

Anti Log Excesses

10 = 3	1 = 0
20 = 6	2 = 1
30 = 9	3 = 1
40 = 13	4 = 1
50 = 16	5 = 2
60 = 19	6 = 2
70 = 22	7 = 2
80 = 25	8 = 3
90 = 28	9 = 3

48 = 15

1370007	·1367226
037	322
067	418
098	515
128	611
159	707
189	804
220	900
250	996
280	·1368093
311	189
341	285
372	382
402	478
432	574
463	671
493	767
524	863
554	960
585	·1369056
615	152
645	249
676	345
706	441
737	538
767	634
797	730
828	827
858	923
888	·1370019
919	115
949	212
979	308
1371010	404
040	500
070	597
101	693
131	789
161	885
192	982
222	·1371078
253	174
283	270
314	367
344	463
374	559
405	655
435	752
466	848
496	944
526	·1372040

1371557	·1372136
587	233
618	329
648	425
679	521
709	617
739	713
770	810
800	906
831	·1373002
861	098
891	194
922	290
952	387
982	483
1372013	579
043	675
073	771
104	867
134	964
164	·1374060
195	156
225	252
255	348
286	444
316	540
346	636
377	732
407	828
437	924
468	·1375020
498	116
528	212
559	308
589	404
619	500
650	596
680	692
711	788
741	884
772	980
802	·1376077
832	173
863	269
893	365
924	461
954	557
984	653
1373015	749
045	845
075	941
106	·1377037
136	133
166	228
197	324
227	420
257	516
288	612
318	708
348	804
379	900
409	996
439	·1378092
470	188
500	284

1373530	·1378379
561	475
591	571
621	667
652	763
682	859
712	955
743	·1379051
773	147
803	243
834	339
864	435
894	530
925	626
955	722
985	818
1374016	914
046	·1380010
076	106
106	202
137	297
167	393
197	489
227	585
258	681
288	777
318	872
348	968
379	·1381064
409	160
439	256
470	352
500	447
530	543
561	639
591	735
621	830
652	926
682	1382022
712	118
743	213
773	309
803	405
834	501
864	596
894	692
925	788
955	884
985	979
1375016	·1383075
046	171
076	267
107	362
137	458
167	554
197	649
228	745
258	841
288	936
318	·1384032
349	128
379	223
409	319
439	415
470	510

1375500	·1384606
530	702
561	797
591	893
621	988
652	·1385084
682	179
712	275
743	371
773	466
803	562
834	657
864	753
894	849
924	944
955	·1386040
985	135
1376015	231
045	326
076	422
106	518
136	613
166	709
197	804
227	900
257	996
287	·1387091
318	187
348	282
378	378
408	473
439	569
469	664
499	760
529	855
560	951
590	·1388046
620	142
651	237
681	333
711	428
742	524
772	619
802	715
833	810
863	906
893	·1389001
924	097
954	192
984	287
1377014	383
045	478
075	574
105	669
135	765
166	860
196	955
226	·1390051
256	146
287	242
317	337
347	432
377	528
408	623
438	718

1377468	·1390814
498	909
529	·1391004
559	100
589	195
619	290
650	386
680	481
710	576
740	672
771	767
801	862
831	958
861	·1392053
892	148
922	244
952	339
982	434
1378013	530
043	625
073	720
103	816
134	911
164	·1393006
194	101
224	197
254	292
284	387
315	482
345	578
375	673
405	768
435	863
465	959
496	·1394054
526	149
556	244
586	340
617	435
647	530
677	625
707	721
738	816
768	911
798	·1395006
828	101
859	197
889	292
919	387
949	482
980	577
1379010	672
040	768
070	863
101	958
131	·1396053
161	148
191	243
221	338
252	433
282	528
312	623
342	719
372	814
403	909

1379433	·1397004
463	099
493	194
523	289
553	384
584	479
614	574
644	669
674	764
705	860
735	955
765	·1398050
795	145
826	240
856	335
886	430
916	525
947	620
977	715

Excesses

10 = 32	1 = 3
20 = 63	2 = 6
30 = 95	3 = 9
	4 = 13
	5 = 16
	6 = 19
	7 = 22
	8 = 25
	9 = 28

15 = 48

Anti Log Excesses

10 = 3	1 = 0
20 = 6	2 = 1
30 = 9	3 = 1
40 = 13	4 = 1
50 = 16	5 = 2
60 = 19	6 = 2
70 = 22	7 = 2
80 = 25	8 = 3
90 = 28	9 = 3

48 = 15

1380007	·1398810
037	905
067	·1399000
097	095
128	190
158	285
188	380
218	475
248	570
278	665
309	760
339	855
369	950
399	·1400045
429	139
459	234
490	329

1380520	·1400424
550	519
580	614
610	709
640	804
671	899
701	994
731	·1401089
761	184
791	278
821	373
852	468
882	563
912	658
942	753
972	848
1381002	943
033	·1402038
063	133
093	228
123	323
153	417
183	512
214	607
244	702
274	797
304	892
334	987
364	·1403082
395	176
425	271
455	366
485	461
515	556
545	651
576	745
606	840
636	935
666	·1404030
696	125
726	220
757	314
787	409
817	504
847	599
877	693
907	788
938	883
968	978
998	·1405072
1382028	167
058	262
088	357
119	451
149	546
179	641
209	736
239	830
269	925
300	·1406020
330	115
360	209
390	304
420	399
450	493

1382480	·1406588
510	683
540	777
570	872
601	967
631	·1407061
661	156
691	251
721	345
751	440
781	535
811	629
842	724
872	818
902	913
932	·1408007
962	102
992	197
1383023	291
053	386
083	480
113	575
143	670
173	764
203	859
233	953
263	·1409048
293	142
324	237
354	332
384	426
414	521
444	615
474	710
504	805
534	899
565	994
595	·1410088
625	183
655	277
685	372
715	466
746	561
776	655
806	750
836	844
866	939
896	·1411033
926	128
956	222
986	317
1384016	411
047	506
077	600
107	695
137	789
167	884
197	978
227	·1412072
257	167
287	261
317	356
347	450
377	545
408	639

1384438	·1412733
468	828
498	922
528	·1413017
558	111
588	205
618	300
648	394
678	488
708	583
738	677
769	771
799	866
829	960
859	·1414054
889	149
919	243
949	337
979	432
1385009	526
039	620
069	715
099	809
130	903
160	998
190	·1415092
220	186
250	281
280	375
310	469
340	564
370	658
400	752
430	846
460	941
491	·1416035
521	129
551	223
581	318
611	412
641	506
671	600
701	695
731	789
761	883
791	977
821	·1417072
852	166
882	260
912	354
942	449
972	543
1386002	637
032	731
062	825
092	919
122	·1418014
152	108
182	202
213	296
243	390
273	484
303	579
333	673
363	767

1386393	·1418861
423	955
453	·1419049
483	144
513	238
543	332
573	426
603	520
633	614
663	709
693	803
723	897
753	991
783	·1420085
813	179
843	273
873	367
903	461
934	555
964	649
994	743
1387024	837
054	931
084	·1421025
114	119
144	213
174	307
204	401
234	495
264	589
294	684
324	778
354	872
384	966
414	·1422060
444	154
474	248
504	342
534	436
564	530
594	624
624	718
655	811
685	905
715	999
745	·1423093
775	187
805	281
835	375
865	469
895	563
925	657
955	751
985	845
1388015	938
045	·1424032
075	126
105	220
135	314
165	408
195	502
225	596
255	690
285	784
315	878

1388345	·1424972
375	·1425065
405	159
435	253
465	347
495	441
525	535
555	629
585	723
615	816
645	910
675	·1426004
705	098
735	192
765	286
795	379
825	473
855	567
885	661
915	755
945	849
975	942
1389005	·1427036
035	130
065	224
095	318
125	412
155	505
185	599
215	693
245	787
275	881
305	975
335	·1428068
365	162
395	256
425	350
455	443
485	537
515	631
545	725
575	818
605	912
635	·1429006
665	099
695	193
725	287
755	380
785	474
815	568
845	661
875	755
905	849
935	942
965	·1430036
995	130

Excesses

10 = 31	1 = 3
20 = 63	2 = 6
30 = 94	3 = 9
	4 = 13
	5 = 16
	6 = 19
	7 = 22
	8 = 25
	9 = 28

15 = 47

Anti Log Excesses

10 = 3	1 = 0
20 = 6	2 = 1
30 = 10	3 = 1
40 = 13	4 = 1
50 = 16	5 = 2
60 = 19	6 = 2
70 = 22	7 = 2
80 = 26	8 = 3
90 = 29	9 = 3

47 = 15

1390025	·1430223
055	317
085	410
115	504
145	597
174	691
204	785
234	878
264	972
294	·1431065
324	159
354	253
384	346
414	440
444	533
474	627
504	720
534	814
564	908
594	·1432001
624	095
654	188
684	282
714	376
744	469
774	563
804	656
834	750
864	843
893	937
923	·1433031
953	124
983	218
1391013	311
043	405
073	499
103	592

1391133	·1433686
163	779
193	873
223	966
253	·1434060
283	153
313	247
343	340
373	434
403	527
433	620
463	714
493	807
523	901
553	994
583	·1435088
612	181
642	274
672	368
702	461
732	555
762	648
792	741
822	835
852	928
882	·1436022
912	115
942	209
971	302
1392001	395
031	489
061	582
091	676
121	769
151	862
181	956
211	·1437049
241	142
271	236
301	329
330	422
360	516
390	609
420	702
450	796
480	889
510	982
540	·1438076
570	169
600	262
630	356
660	449
689	542
719	636
749	729
779	822
809	916
839	·1439009
869	102
899	196
929	289
959	382
989	475
1393019	569
048	662

1393078	·1439755
108	848
138	942
168	·1440035
198	128
228	221
258	314
288	408
318	501
348	594
378	687
407	780
437	873
467	967
497	·1441060
527	153
557	246
587	339
617	432
647	526
677	619
707	712
737	805
766	898
796	991
826	·1442085
856	178
886	271
916	364
946	457
976	550
1394006	643
036	736
066	829
096	922
125	·1443016
155	109
185	202
215	295
245	388
275	481
305	574
335	667
364	760
394	853
424	946
454	·1444039
484	133
514	226
543	319
573	412
603	505
633	598
663	691
693	784
723	877
753	970
783	·1445063
813	156
842	249
872	342
902	435
932	528
962	621
992	714

1395022	·1445807	1396961	·1451843	1398899	·1457862	1400000	·1461280	1401934	·1467273
052	900	991	935	929	954	030	372	963	365
081	993	1397021	·1452028	959	·1458046	060	465	993	458
111	·1446086	051	121	988	139	089	557	1402023	550
141	179	080	214	1399018	231	119	649	053	642
171	272	110	306	048	324	149	742	082	734
201	365	140	399	078	416	179	834	112	826
231	458	170	492	107	509	209	926	142	918
260	551	200	584	137	601	239	·1462019	172	·1468010
290	644	229	677	167	693	268	111	201	102
320	737	259	770	197	786	298	203	231	194
350	830	289	862	226	878	328	296	261	286
380	923	319	955	256	971	358	388	290	378
410	·1447016	349	·1453048	286	·1459063	388	480	320	470
439	109	379	140	316	155	417	573	350	562
469	202	408	233	345	248	447	665	379	654
499	295	438	326	375	340	477	757	409	746
529	388	468	418	405	433	507	849	439	838
559	480	498	511	435	525	536	942	468	930
589	573	528	604	464	618	566	·1463034	498	·1469022
618	666	558	696	494	710	596	126	528	114
648	759	587	789	524	802	626	218	557	206
678	852	617	882	554	895	655	311	587	298
708	945	647	974	583	987	685	403	617	390
738	·1448038	677	·1454067	613	·1460080	715	495	647	482
768	131	707	160	643	172	745	587	676	574
797	223	737	252	673	264	774	679	706	665
827	316	766	345	702	357	804	772	736	757
857	409	796	438	732	449	834	864	766	849
887	502	826	530	762	541	864	956	795	941
917	595	856	623	792	634	893	·1464048	825	·1470033
947	688	886	716	821	726	923	140	855	125
976	780	916	808	851	818	953	232	885	217
1396006	873	945	901	881	911	983	325	914	309
036	966	975	993	911	·1461003	1401012	417	944	401
066	·1449059	1398005	·1455086	940	095	042	509	974	493
096	152	035	178	970	188	072	601	1403003	585
126	245	065	271			102	693	033	677
156	337	095	364			131	786	063	769
186	430	124	456			161	878	092	861
216	523	154	549			191	970	122	953
246	616	184	641			220	·1465062	152	·1471045
275	709	214	734			250	155	181	137
305	802	244	827			280	247	211	229
335	894	273	919			309	339	241	321
365	987	303	·1456012			339	431	270	413
395	·1450080	333	104			369	524	300	504
425	173	363	197			398	616	330	596
455	266	392	289			428	708	359	688
485	359	422	382			458	800	389	780
514	451	452	474			487	892	419	872
544	544	482	567			517	984	448	964
574	637	511	659			547	·1466076	478	·1472055
604	730	541	752			577	168	508	147
634	822	571	844			606	260	537	239
664	915	601	937			636	353	567	331
693	·1451008	631	·1457029			666	445	597	423
723	101	660	122			696	537	626	515
753	193	690	214			725	629	656	606
783	286	720	307			755	721	686	698
813	379	750	399			785	813	716	790
842	472	780	492			815	905	745	882
872	564	810	584			844	997	775	974
902	657	839	677			874	·1467089	805	·1473066
932	750	869	769			904	181	835	157

Excesses

10 = 31	1 = 3
20 = 62	2 = 6
	3 = 9
	4 = 12
	5 = 16
	6 = 19
	7 = 22
	8 = 25
	9 = 28

15 = 47

Anti Log Excesses

10 = 3	1 = 0
20 = 6	2 = 1
30 = 10	3 = 1
40 = 13	4 = 1
50 = 16	5 = 2
60 = 19	6 = 2
70 = 22	7 = 2
80 = 26	8 = 3
90 = 29	9 = 3

47 = 15

1403864	·1473249
894	341
924	433
954	525
983	617
1404013	708
043	800
072	892
102	984
132	·1474076
161	168
191	259
221	351
250	443
280	535
310	627
339	718
369	810
399	902
428	993
458	·1475085
488	177
517	268
547	360
577	452
606	543
636	635
666	727
695	819
725	910
755	·1476002
784	094
814	186
844	277
873	369
903	461
933	553
962	644
992	736

Excesses

15 = 46

Anti Log Excesses

46 = 15

1405024	·1476836
057	936
089	·1477036
121	136
153	236
186	336
218	436
250	536
282	636
315	736
347	836
379	936
412	·1478036
444	136
476	236
509	336
541	435
1405574	·1478535
606	635
638	735
671	835
703	935
735	·1479035
768	135
800	234
832	334
865	434
897	534
930	634
962	734
994	833
1406027	933
059	·1480033
091	133
124	233
156	332
188	432
220	532
253	632
285	732
317	832
349	931
382	·1481031
414	131
446	231
479	331
511	430
543	530
576	630
608	730
641	830
673	930
705	·1482029
738	129
770	229
802	329
835	428
867	528
899	628
931	728
964	827
996	927
1407028	·1483027
060	127
093	226
125	326
157	426
190	525
222	625
254	725
287	825
319	924
352	·1484024
384	124
416	224
449	323
481	423
513	523
546	622
578	722
610	821
642	921
1407675	·1485020
707	120
739	219
771	319
804	419
836	518
868	618
901	717
933	817
965	917
997	·1486016
1408030	116
062	215
094	315
126	415
159	514
191	614
223	714
256	813
288	913
320	·1487012
352	112
385	211
417	311
449	410
481	510
514	610
546	709
578	808
611	908
643	·1488007
675	107
707	206
740	306
772	405
804	505
836	604
869	704
901	803
933	902
966	·1489002
998	101
1409030	200
062	300
095	399
127	499
159	598
191	697
224	797
256	896
288	995
320	·1490095
353	194
385	293
417	393
449	492
481	592
513	691
546	790
578	890
610	989
642	·1491088
675	188
707	287
739	386
1409771	·1491486
804	585
836	685
868	784
900	883
933	983
965	·1492082
997	181

Excesses

10 = 31	1 = 3
20 = 62	2 = 6
30 = 93	3 = 9
	4 = 12
	5 = 15
	6 = 19
	7 = 22
	8 = 25
	9 = 28

16 = 50

Anti Log Excesses

10 = 3	1 = 0
20 = 6	2 = 1
30 = 10	3 = 1
40 = 13	4 = 1
50 = 16	5 = 2
60 = 19	6 = 2
70 = 23	7 = 2
80 = 26	8 = 3
90 = 29	9 = 3

50 = 16

1410030	·1492281
062	380
094	479
126	578
159	678
191	777
223	876
255	975
288	·1493075
320	174
352	273
384	373
417	472
449	571
481	670
513	770
545	869
577	968
610	·1494067
642	167
674	266
706	365
738	464
771	564
803	663
835	762
867	861
899	960
1410931	·1495059
964	159
996	258
1411028	357
060	456
093	555
125	654
157	753
189	852
222	952
254	·1496051
286	150
318	249
351	348
383	447
415	546
447	645
480	744
512	843
544	942
576	·1497042
608	141
640	240
673	339
705	438
737	537
769	636
801	735
834	834
866	933
898	·1498032
930	131
962	230
994	329
1412027	428
059	527
091	626
123	725
155	824
188	923
220	·1499022
252	121
284	220
316	319
348	418
381	517
413	616
445	715
477	814
509	913
542	·1500012
574	111
606	210
638	308
670	407
702	506
735	605
767	704
799	803
831	902
863	·1501001
896	099
928	198
960	297
992	396

1413024	·1501495
056	594
089	692
121	791
153	890
185	989
217	·1502088
250	187
282	286
314	385
346	483
378	582
410	681
443	780
475	879
507	978
539	·1503077
571	175
603	274
635	373
667	472
700	570
732	669
764	768
796	867
828	965
860	·1504064
892	163
924	261
957	360
989	459
1414021	558
053	656
085	755
117	854
150	953
182	·1505051
214	150
246	249
278	347
311	446
343	545
375	644
407	742
439	841
471	940
504	·1506039
536	137
568	236
600	335
632	433
664	532
696	631
728	729
761	828
793	926
825	·1507025
857	124
889	222
921	321
953	420
985	518
1415017	617
049	715
081	814

1415114	·1507912
146	·1508011
178	109
210	208
242	307
274	405
306	504
338	602
371	701
403	799
435	898
467	996
499	·1509095
531	193
564	292
596	391
628	489
660	587
692	686
724	784
756	883
788	981
821	·1510080
853	178
885	277
917	375
949	474
981	572
1416013	670
045	769
077	867
109	966
141	·1511064
174	163
206	261
238	360
270	458
302	557
334	655
366	753
398	852
430	950
462	·1512048
494	147
527	245
559	344
591	442
623	540
655	639
687	737
719	835
751	934
783	·1513032
815	130
847	229
880	327
912	426
944	524
976	622
1417008	721
040	819
072	917
104	·1514016
136	114
168	212

1417200	·1514310
232	409
264	507
296	605
328	703
360	802
392	900
424	998
456	·1515097
488	195
520	293
552	391
585	490
617	588
649	686
681	784
713	883
745	981
777	·1516079
809	177
841	276
873	374
905	472
938	570
970	668
1418002	766
034	865
066	963
098	·1517061
130	159
162	257
194	355
226	453
258	551
290	650
322	748
354	846
386	944
418	·1518042
450	140
482	238
514	336
546	434
578	532
610	630
643	729
675	827
707	925
739	·1519023
771	121
803	219
835	317
867	415
899	513
931	611
963	709
995	807
1419027	905
059	·1520003
091	101
123	199
155	297
187	395
219	493
251	591

1419283	·1520689
315	787
347	885
379	983
411	·1521081
443	179
475	277
507	375
539	473
571	571
603	669
635	767
667	865
700	963
732	·1522061
764	159
796	257
828	355
860	453
892	551
924	649
956	747
988	845

Excesses

10 = 31	1 = 3
20 = 61	2 = 6
30 = 92	3 = 9
	4 = 12
	5 = 15
	6 = 18
	7 = 21
	8 = 25
	9 = 28

16 = 49

Anti Log Excesses

10 = 3	1 = 0
20 = 7	2 = 1
30 = 10	3 = 1
40 = 13	4 = 1
50 = 16	5 = 2
60 = 20	6 = 2
70 = 23	7 = 2
80 = 26	8 = 3
90 = 29	9 = 3

49 = 16

1420020	·1522943
052	·1523040
084	138
116	236
148	334
180	432
212	530
244	628
276	726
308	823
340	921
372	·1524019
404	117

1420436	·1524215
468	313
500	410
532	508
564	606
596	704
628	802
660	899
692	997
724	·1525095
756	193
788	291
820	389
852	486
884	584
916	682
948	780
980	877
1421012	975
044	·1526073
076	171
108	268
140	366
172	464
204	562
236	659
268	757
300	855
332	952
364	·1527050
396	148
428	246
459	343
491	441
523	539
555	637
587	734
619	832
651	930
683	·1528027
715	125
747	223
779	320
811	418
843	515
875	613
907	711
939	808
971	906
1422003	·1529004
035	101
067	199
099	297
131	394
163	492
195	589
227	687
259	785
291	882
323	980
355	·1530078
387	175
419	273
451	370
483	468

1422514	·1530565
546	663
578	760
610	858
642	956
674	·1531053
706	150
738	248
770	345
802	443
834	540
865	638
897	735
929	833
961	930
993	·1532028
1423025	125
057	222
089	320
121	417
153	515
185	612
217	710
249	807
281	905
313	·1533002
345	100
377	197
409	294
441	392
473	489
505	587
537	684
568	782
600	879
632	977
664	·1534074
696	172
728	269
760	366
792	464
824	561
856	658
888	756
919	853
951	951
983	·1535048
1424015	145
047	243
079	340
111	437
143	535
175	632
207	729
239	826
270	924
302	·1536021
334	118
366	215
398	313
430	410
462	507
494	605
526	702
558	799

1424590	·1536896
621	994
653	·1537091
685	188
717	285
749	383
781	480
813	577
845	675
877	772
909	869
941	966
972	·1538064
1425004	161
036	258
068	355
100	453
132	550
164	647
196	744
228	842
260	939
292	·1539036
323	133
355	230
387	327
419	425
451	522
483	619
515	716
547	813
579	910
611	·1540007
643	104
674	202
706	299
738	396
770	493
802	590
834	687
866	784
898	881
929	978
961	·1541075
993	172
1426025	270
057	367
089	464
120	561
152	658
184	755
216	852
248	949
280	·1542046
312	143
344	240
375	337
407	434
439	531
471	628
503	725
535	822
567	919
599	·1543016
630	113

1426662	·1543210
694	307
726	404
758	501
790	598
821	695
853	792
885	889
917	986
949	·1544083
981	180
1427013	277
045	374
076	470
108	567
140	664
172	761
204	858
236	955
268	·1545052
300	149
331	246
363	343
395	440
427	536
459	633
491	730
522	827
554	924
586	·1546021
618	118
650	215
681	311
713	408
745	505
777	602
809	699
841	796
872	892
904	989
936	·1547086
968	183
1428000	280
031	376
063	473
095	570
127	667
159	764
191	861
222	957
254	·1548054
286	151
318	248
350	344
381	441
413	538
445	635
477	731
509	828
541	925
572	·1549022
604	118
636	215
668	312
700	408

1428731	·1549505
763	602
795	699
827	795
859	892
891	989
922	·1550086
954	182
986	279
1429018	376
050	472
081	569
113	666
145	762
177	859
209	955
241	·1551052
272	149
304	245
336	342
368	439
400	535
431	632
463	729
495	825
527	922
559	·1552018
591	115
622	212
654	308
686	405
718	502
749	598
781	695
813	791
845	888
876	984
908	·1553081
940	177
972	274

Excesses

10 = 30	1 = 3
20 = 61	2 = 6
30 = 91	3 = 9
	4 = 12
	5 = 15
	6 = 18
	7 = 21
	8 = 24
	9 = 27

16 = 49

Anti Log Excesses

10 = 3	1 = 0
20 = 7	2 = 1
30 = 10	3 = 1
40 = 13	4 = 1
50 = 16	5 = 2
60 = 20	6 = 2
70 = 23	7 = 2
80 = 26	8 = 3
90 = 30	9 = 3

49 = 16

1430003	·1553371
035	467
067	563
099	660
130	756
162	853
194	949
226	·1554046
258	142
290	239
321	335
353	432
385	528
417	624
449	721
480	817
512	914
544	·1555010
576	107
608	203
640	300
671	396
703	493
735	589
767	685
798	782
830	878
862	975
894	·1556071
925	168
957	264
989	361
1431021	457
052	554
084	650
116	746
147	843
179	939
211	·1557035
243	132
274	228
306	325
338	421
370	517
401	614
433	710
465	806
497	903
528	999
560	·1558095
592	191

1431624	·1558288	1433686	·1564539	1435746	·1570773	1437802	·1576988
656	384	718	635	777	869	834	·1577084
688	480	749	731	809	964	865	179
719	576	781	827	840	·1571060	897	275
751	673	813	923	872	156	928	370
783	769	844	·1565019	904	252	960	466
815	865	876	115	935	347	992	561
846	962	908	211	967	443	1438023	657
878	·1559058	939	307	999	539	055	752
910	154	971	403	1436030	634	087	847
942	250	1434003	499	062	730	118	943
973	347	035	595	094	826	150	·1578038
1432005	443	066	691	125	922	181	134
037	539	098	787	157	·1572017	213	229
069	635	130	883	188	113	244	325
100	732	162	979	220	209	276	420
132	828	193	·1566075	252	305	307	516
164	924	225	171	283	400	339	611
195	·1560021	257	267	315	496	370	707
227	117	288	363	347	592	402	802
259	213	320	459	378	687	434	897
291	309	352	555	410	783	465	993
322	406	384	651	442	879	497	·1579088
354	502	415	747	473	974	529	184
386	598	447	843	505	·1573070	560	279
418	694	479	939	536	165	592	375
449	791	511	·1567035	568	261	623	470
481	887	542	131	600	357	655	566
513	983	574	227	631	452	687	661
544	·1561079	606	323	663	548	718	757
576	176	637	419	695	644	750	852
608	272	669	515	726	739	782	947
640	368	701	611	758	835	813	·1580043
671	464	732	707	790	931	845	138
703	560	764	802	821	·1574026	876	233
735	656	795	898	853	122	908	328
767	753	827	994	884	217	939	424
798	849	859	·1568090	916	313	971	519
830	945	890	186	948	409	1439002	614
862	·1562041	922	282	979	504	034	709
893	137	954	378	1437011	600	065	805
925	233	985	474	043	696	097	900
957	329	1435017	569	074	791	129	995
989	425	049	665	106	887	160	·1581091
1433020	522	081	761	138	982	192	186
052	618	112	857	169	·1575078	224	281
084	714	144	953	201	173	255	377
116	810	176	·1569049	232	269	287	472
147	906	208	144	264	364	318	568
179	·1563002	239	240	296	460	350	663
211	098	271	336	327	556	382	758
242	194	303	432	359	651	413	854
274	290	334	528	391	747	445	949
306	386	366	623	422	842	477	·1582044
338	482	398	719	454	938	508	140
369	579	429	815	486	·1576033	540	235
401	675	461	911	517	129	571	330
433	771	492	·1570007	549	224	603	425
465	867	524	103	580	320	634	521
496	963	556	198	612	415	666	616
528	·1564059	587	294	644	511	697	711
560	155	619	390	675	607	729	806
591	251	651	486	707	702	760	902
623	347	682	581	739	797	792	997
655	443	714	677	770	893	824	·1583092

1439855	·1583187
887	283
918	378
950	473
981	568

Excesses

10 = 30	1 = 3
20 = 61	2 = 6
30 = 91	3 = 9
	4 = 12
	5 = 15
	6 = 18
	7 = 21
	8 = 24
	9 = 27

16 = 48

Anti Log Excesses

10 = 3	1 = 0
20 = 7	2 = 1
30 = 10	3 = 1
40 = 13	4 = 1
50 = 17	5 = 2
60 = 20	6 = 2
70 = 23	7 = 2
80 = 26	8 = 3
90 = 30	9 = 3

48 = 16

1440013	·1583663
044	758
076	854
107	949
139	·1584044
171	139
202	234
234	329
266	424
297	519
329	615
360	710
392	805
424	900
455	995
487	·1585090
519	185
550	280
582	376
613	471
645	566
676	661
708	756
739	851
771	947
802	·1586042
834	137
866	232
897	327
929	422
960	517

1440992	·1586612	1443040	·1592783	1445086	·1598936	1447129	·1605072
1441023	707	072	878	118	·1599030	160	166
055	802	103	972	149	125	192	260
086	897	135	·1593067	181	219	223	354
118	992	166	162	212	314	255	449
149	·1587087	198	257	244	408	286	543
181	182	229	351	275	503	318	637
212	277	261	446	306	598	349	731
244	372	292	541	338	692	380	826
275	467	324	635	369	786	412	920
307	562	355	730	401	881	443	·1606014
338	657	387	825	432	975	475	108
370	753	418	920	464	·1600070	506	203
401	848	450	·1594014	495	164	537	297
433	943	481	109	527	259	569	391
464	·1588038	513	204	558	353	600	485
495	133	544	299	590	448	632	579
527	228	575	393	621	542	663	673
559	323	607	488	652	637	694	768
590	418	638	583	684	731	726	862
622	513	670	677	715	825	757	956
653	608	701	772	747	920	788	·1607050
685	703	733	867	778	·1601014	820	144
716	798	764	962	810	109	851	239
748	893	796	·1595056	841	203	883	333
779	988	827	151	873	298	914	427
811	·1589083	859	246	904	392	946	521
842	178	890	341	936	487	977	615
874	273	921	435	967	581	1448009	709
906	368	953	530	998	676	040	804
937	463	985	625	1446030	770	071	898
969	558	1444016	719	061	864	103	992
1442000	653	048	814	093	959	134	·1608086
032	748	079	909	124	·1602053	166	180
063	842	111	·1596003	155	147	197	274
095	937	142	098	187	242	228	368
126	·1590032	174	192	218	336	260	462
158	127	205	287	250	431	291	557
189	222	237	382	281	525	323	651
221	317	268	476	312	619	354	745
253	412	300	571	344	714	385	839
284	507	331	666	375	808	417	933
316	601	363	760	406	902	448	·1609027
347	696	394	855	438	997	479	121
379	791	426	950	469	·1603091	511	215
410	886	457	·1597044	501	185	542	309
442	981	489	139	532	280	573	403
473	·1591076	520	233	564	374	605	497
505	170	552	328	595	469	636	592
536	265	583	423	627	563	668	686
568	360	614	517	658	657	699	780
599	455	646	612	690	752	730	874
631	550	677	707	721	846	762	968
662	645	709	801	752	940	793	·1610062
694	740	740	896	784	·1604035	824	156
725	835	772	990	815	129	856	250
757	929	803	·1598085	846	223	887	344
788	·1592024	835	179	878	317	918	438
820	119	866	274	909	412	950	532
851	214	898	368	941	506	981	626
882	309	929	463	972	600	1449013	720
914	404	960	558	1447003	694	044	814
946	499	992	652	035	789	075	908
977	593	1445023	747	066	883	107	·1611002
1443009	688	055	841	097	977	138	096

1449169	·1611190
201	284
232	378
263	472
295	566
326	660
358	754
389	848
420	942
452	·1612036
483	130
514	224
546	318
577	412
608	506
640	600
671	693
703	787
734	881
765	975
797	·1613069
828	163
859	257
891	351
922	445
953	539
985	633

Excesses

10 = 30	1 = 3
20 = 60	2 = 6
30 = 90	3 = 9
	4 = 12
	5 = 15
	6 = 18
	7 = 21
	8 = 24
	9 = 27

16 = 47

Anti Log Excesses

10 = 3	1 = 0
20 = 7	2 = 1
30 = 10	3 = 1
40 = 13	4 = 1
50 = 17	5 = 2
60 = 20	6 = 2
70 = 23	7 = 2
80 = 27	8 = 3
90 = 30	9 = 3

47 = 16

1450016	·1613726
048	820
079	914
110	·1614008
142	102
173	196
204	290
236	384
267	477

1450298	·1614571	1452334	·1620664	1454367	·1626739	1456397	·1632796	1458424	·1638837
330	665	365	757	398	832	428	889	455	930
361	759	396	851	429	925	460	982	486	·1639023
393	853	428	944	460	·1627018	491	·1633075	518	115
424	947	459	·1621038	492	112	522	168	549	208
455	·1615040	490	131	523	205	553	262	580	301
487	134	521	225	554	298	584	355	611	394
518	228	553	319	585	391	615	448	642	487
549	322	584	412	617	485	647	541	673	580
581	416	615	506	648	578	678	634	705	672
612	509	647	599	679	671	709	727	736	765
643	603	678	693	711	765	740	820	767	858
674	697	709	786	742	858	771	913	798	951
706	791	741	880	773	951	803	·1634006	829	·1640043
737	885	772	973	804	·1628044	834	099	861	136
768	979	804	·1622067	836	138	865	192	892	229
799	·1616072	835	160	867	231	896	285	923	322
831	166	866	254	898	324	927	378	954	414
862	260	898	348	929	417	958	471	985	507
893	354	929	441	961	511	990	564	1459016	600
925	448	960	535	992	604	1457021	657	048	693
956	541	992	628	1455023	697	052	750	079	785
987	635	1453023	722	055	791	083	843	110	878
1451019	729	054	815	086	884	114	936	141	971
050	823	085	909	117	977	146	·1635029	172	·1641063
082	917	117	·1623002	148	·1629070	177	122	203	156
113	·1617011	148	096	180	164	208	215	234	249
144	104	179	189	211	257	239	307	265	342
176	198	210	283	242	350	270	400	297	434
207	292	242	377	273	443	301	493	328	527
238	386	273	470	305	537	333	586	359	620
270	479	304	563	336	630	364	679	390	713
301	573	336	657	367	723	395	772	421	805
332	667	367	750	398	816	426	865	452	898
363	760	398	843	430	910	457	958	483	991
395	854	429	937	461	·1630003	489	·1636051	514	·1642083
426	947	461	·1624030	492	096	520	144	546	176
457	·1618041	492	124	523	189	551	237	577	269
488	135	523	217	554	282	582	330	608	361
520	228	554	310	585	375	613	423	639	454
551	322	586	404	617	469	644	516	670	546
582	416	617	497	648	562	676	609	701	639
614	509	648	590	679	655	707	702	733	732
645	603	680	684	710	748	738	795	764	824
676	697	711	777	741	841	769	888	795	917
708	791	742	871	773	934	800	981	826	·1643010
739	884	773	964	804	·1631027	832	·1637073	857	102
771	978	805	·1625058	835	120	863	166	888	195
802	·1619072	836	151	866	214	894	259	919	288
833	166	867	245	897	307	925	352	950	380
865	259	898	338	928	400	956	445	982	473
896	353	930	432	960	493	987	538		
927	447	961	525	991	586	1458019	630		
959	540	992	618	1456022	679	050	723		
990	634	1454023	712	053	772	081	816		
1452021	728	055	805	085	865	112	909		
052	821	086	898	116	958	143	·1638002		
084	915	117	992	147	·1632051	175	094		
115	·1620008	148	·1626085	178	144	206	187		
146	102	179	179	210	238	237	280		
177	196	210	272	241	331	268	373		
209	289	242	365	272	424	299	466		
240	383	273	459	303	517	330	559		
271	477	304	552	335	610	362	651		
303	570	335	645	366	703	393	744		

Excesses

10 = 30	1 = 3
20 = 60	2 = 6
30 = 90	3 = 9
	4 = 12
	5 = 15
	6 = 18
	7 = 21
	8 = 24
	9 = 27

16 = 47

Anti Log Excesses									
		1461600	·1648283	1463620	·1654281	1465637	·1660263	1467652	·1666228
		631	376	651	374	668	355	683	320
10 = 3	1 = 0	662	468	682	466	699	447	714	411
20 = 7	2 = 1	694	561	713	558	730	539	745	503
30 = 10	3 = 1	725	653	744	650	761	631	776	595
40 = 13	4 = 1	756	746	775	742	792	722	807	686
50 = 17	5 = 2	787	838	806	834	823	814	837	778
60. = 20	6 = 2	818	931	837	927	854	906	868	869
70 = 23	7 = 2	849	·1649023	868	·1655019	885	998	899	961
80 = 27	8 = 3	880	115	899	111	916	·1661090	930	·1667053
90 = 30	9 = 3	911	208	930	203	947	182	961	144
		942	300	961	295	978	274	992	236
47 = 16		973	392	992	387	1466009	366	1468023	328
		1462004	484	1464023	479	040	457	054	419
1460013	·1643565	036	577	054	571	071	549	085	511
044	658	067	669	086	664	102	641	116	603
075	751	098	761	117	756	133	733	147	694
106	843	129	853	148	848	164	825	178	786
137	936	160	946	179	940	195	917	209	877
168	·1644029	191	·1650038	210	·1656032	226	·1662008	240	969
199	121	222	130	241	124	257	100	271	·1668061
231	214	253	223	272	216	288	192	302	152
262	306	284	315	303	308	319	284	333	244
293	399	315	407	334	400	350	376	364	336
324	491	346	500	365	492	381	467	395	427
355	584	378	592	396	584	412	559	426	519
386	676	409	685	427	676	443	651	457	610
418	769	440	777	458	768	474	743	488	702
449	862	471	869	489	860	505	835	518	793
480	954	502	962	520	952	536	927	549	885
511	·1645047	533	·1651054	551	·1657044	567	·1663018	580	976
542	139	564	146	582	136	598	110	611	·1669068
573	232	595	239	613	228	629	202	642	160
604	324	626	331	644	320	660	294	673	251
635	417	657	423	675	412	691	385	704	342
667	509	688	515	706	504	722	477	735	434
698	602	719	608	737	596	753	569	766	525
729	694	750	700	769	689	784	661	797	617
760	787	781	792	800	781	815	752	828	708
791	880	812	884	831	873	846	844	859	800
822	972	843	977	862	965	877	936	890	891
853	·1646064	874	·1652069	893	·1658057	908	1664028	921	983
884	157	905	161	924	149	939	119	952	·1670074
915	249	936	253	955	241	970	211	983	166
946	342	967	346	986	333	1467001	303	1469014	257
977	434	998	438	1465017	425	032	394	045	348
1461009	527	1463029	530	048	517	063	486	076	440
040	619	061	622	079	609	094	578	107	531
071	712	092	714	110	700	125	670	138	623
102	804	123	806	141	792	156	761	169	714
133	897	154	899	172	884	187	853	199	806
164	989	185	991	203	976	218	945	230	897
195	·1647081	216	·1653083	234	·1659068	249	·1665037	261	989
226	174	247	175	265	160	280	128	292	·1671080
258	266	278	267	296	252	311	220	323	172
289	359	309	360	327	344	342	312	354	263
320	451	340	452	358	436	373	403	385	354
351	544	371	544	389	528	404	495	416	446
382	636	403	636	420	620	435	587	447	537
413	729	434	728	451	711	466	678	478	629
445	821	465	820	482	803	497	770	509	720
476	914	496	913	513	895	528	861	539	812
507	·1648006	527	·1654005	544	987	559	953	570	903
538	098	558	097	575	·1660079	590	·1666045	601	995
569	191	589	189	606	171	621	136	632	·1672086

1469663	·1672178
694	269
725	360
756	452
787	543
818	634
849	726
880	817
911	909
942	·1673000
973	091

Excesses

10 = 30	1 = 3
20 = 59	2 = 6
30 = 89	3 = 9
	4 = 12
	5 = 15
	6 = 18
	7 = 21
	8 = 24
	9 = 27

15 = 45

Anti Log Excesses

10 = 3	1 = 0
20 = 7	2 = 1
30 = 10	3 = 1
40 = 13	4 = 1
50 = 17	5 = 2
60 = 20	6 = 2
70 = 24	7 = 2
80 = 27	8 = 3
90 = 30	9 = 3

45 = 15

1470004	·1673183
035	274
066	365
097	457
128	548
159	639
190	730
220	822
251	913
282	·1674004
313	095
344	187
375	278
406	369
437	461
468	552
499	643
530	735
560	826
591	918
622	·1675009
653	100
684	192
715	283
746	374

1470777	·1675465
808	557
839	648
870	739
900	830
931	921
962	·1676012
993	104
1471024	195
055	286
086	377
117	468
147	560
178	651
209	742
240	833
271	924
302	·1677015
332	107
363	198
394	289
425	380
456	471
487	563
518	654
549	745
579	836
610	927
641	·1678018
672	110
703	201
734	292
765	383
796	474
827	565
858	656
889	747
919	839
950	930
981	·1679021
1472012	112
043	203
074	294
105	385
136	476
166	568
197	659
228	750
259	841
290	932
321	·1680023
351	115
382	206
413	297
444	388
475	479
506	570
537	661
568	752
598	843
629	934
660	·1681025
691	116
722	207
753	298

1472784	·1681389
815	480
845	571
876	662
907	753
938	844
969	935
1473000	·1682026
030	117
061	208
092	299

Excesses

15 = 45

Anti Log Excesses

45 = 15

1473126	·1682399
160	499
194	599
228	699
262	799
296	899
330	999
364	·1683099
398	199
432	299
466	399
500	499
534	599
568	699
601	799
635	899
669	999
703	·1684099
737	199
771	299
805	399
839	499
873	599
907	699
940	799
974	899
1474008	999
042	·1685099
076	199
110	299
144	399
178	499
212	599
246	699
280	799
314	898
348	998
382	·1686098
416	198
450	298
484	398
518	498
552	597
586	697
619	797

1474653	·1686897
687	997
721	·1687096
755	196
789	296
823	396
857	496
891	595
925	695
958	795
992	895
1475026	995
060	·1688094
094	194
128	294
162	394
196	494
230	593
264	693
297	793
331	893
365	993
399	·1689092
433	192
467	292
501	392
535	491
568	591
602	691
636	791
670	890
704	990
737	·1690090
771	189
805	289
839	389
873	488
907	588
941	688
974	788
1476008	887
042	987
076	·1691087
110	186
144	286
178	386
212	485
246	585
280	684
313	784
347	884
381	983
415	·1692083
449	182
483	282
517	382
551	481
584	581
618	680
652	780
686	880
720	979
753	·1693079
787	178
821	278

1476855	·1693378
889	477
923	577
957	676
990	776
1477024	876
058	975
092	·1694075
126	174
160	274
194	374
228	473
261	573
295	672
329	772
363	871
397	971
430	·1695070
464	170
498	269
532	368
566	468
600	567
634	667
667	766
701	865
735	965
769	·1696064
803	164
837	263
871	362
905	462
938	561
972	661
1478006	760
040	859
074	959
107	·1697058
141	158
175	257
209	356
243	456
276	555
310	655
344	754
378	853
412	953
445	·1698052
479	152
513	251
547	350
581	450
614	549
648	648
682	748
716	847
750	946
783	·1699045
817	145
851	244
885	343
919	443
952	542
986	641
1479020	741

1479054	·1699840
088	939
121	·1700038
155	138
189	237
223	336
257	435
290	535
324	634
358	733
392	832
426	931
459	·1701031
493	130
527	229
561	328
595	427
628	527
662	626
696	725
730	824
764	923
797	·1702023
831	122
865	221
899	320
933	419
966	518

Excesses

10 = 29	1 = 3
20 = 59	2 = 6
30 = 88	3 = 9
	4 = 12
	5 = 15
	6 = 18
	7 = 21
	8 = 24
	9 = 27

17 = 50

Anti Log Excesses

10 = 3	1 = 0
20 = 7	2 = 1
30 = 10	3 = 1
40 = 14	4 = 1
50 = 17	5 = 2
60 = 20	6 = 2
70 = 24	7 = 2
80 = 27	8 = 3
90 = 31	9 = 3

50 = 17

1480000	·1702617
034	717
068	816
102	915
135	·1703014
169	113
203	212
237	311
271	410

1480304	·1703509
338	608
372	708
406	807
440	906
473	·1704005
507	104
541	203
575	302
609	401
642	500
676	599
710	698
744	797
778	896
811	995
845	·1705094
879	193
913	292
946	391
980	490
1481014	589
047	688
081	787
115	886
149	985
182	·1706084
216	183
250	282
284	381
317	480
351	579
385	678
419	777
453	876
486	975
520	·1707074
554	173
588	272
621	371
655	470
689	569
722	668
756	766
790	865
824	964
857	·1708063
891	162
925	261
959	360
992	458
1482026	557
060	656
094	755
128	854
161	952
195	·1709051
229	150
263	249
296	348
330	447
364	546
397	645
431	743
465	842

1482499	·1709941
532	·1710040
566	139
600	238
633	336
667	435
701	534
734	633
768	731
802	830
836	929
869	·1711027
903	126
937	225
970	323
1483004	422
038	521
071	620
105	718
139	817
173	916
206	·1712014
240	113
274	212
307	310
341	409
375	508
408	607
442	705
476	804
510	903
543	·1713001
577	100
611	199
644	297
678	396
712	495
745	594
779	692
813	791
847	890
880	988
914	·1714087
948	186
981	284
1484015	383
049	481
082	580
116	678
150	777
184	875
217	974
251	·1715072
285	171
318	269
352	368
386	466
419	565
453	664
487	762
521	861
554	959
588	·1716058
622	157
655	255

1484689	·1716354
723	452
756	551
790	649
824	748
858	846
891	945
925	·1717043
959	141
992	240
1485026	338
059	437
093	535
127	633
160	732
194	830
227	929
261	·1718027
295	126
328	224
362	323
396	421
429	520
463	618
497	717
531	815
564	914
598	·1719012
632	110
665	209
699	307
733	405
766	504
800	602
834	700
868	798
901	897
935	995
969	·1720093
1486002	192
036	290
069	388
103	487
137	585
170	683
204	781
237	880
271	978
305	·1721076
338	175
372	273
405	371
439	470
473	568
506	666
540	764
573	863
607	961
641	·1722059
674	158
708	256
742	354
775	453
809	551
843	649

1486877	·1722747
910	846
944	944
978	·1723042
1487011	140
045	238
078	336
112	435
146	533
179	631
213	729
246	827
280	925
314	·1724023
347	121
381	220
414	318
448	416
482	514
515	612
549	711
582	809
616	907
650	·1725005
683	103
717	201
750	299
784	398
818	496
851	594
885	692
918	790
952	888
986	986
1488019	·1726084
053	182
086	280
120	378
154	476
187	574
221	672
254	770
288	868
322	966
355	·1727064
389	162
422	260
456	359
490	457
523	555
557	653
590	751
624	849
658	947
691	·1728045
725	143
758	241
792	339
826	436
859	534
893	632
926	730
960	828
994	926
1489027	·1729024

1489061	·1729122
094	220
128	318
162	416
195	514
229	612
262	710
296	808
329	906
363	·1730004
396	101
430	199
463	297
497	395
530	493
564	590
597	688
631	786
665	884
698	982
732	·1731080
765	178
799	276
833	373
866	471
900	569
933	667
967	765

Excesses

10 = 29	1 = 3
20 = 58	2 = 6
30 = 88	3 = 9
	4 = 12
	5 = 15
	6 = 18
	7 = 20
	8 = 23
	9 = 26

17 = 49

Anti Log Excesses

10 = 3	1 = 0
20 = 7	2 = 1
30 = 10	3 = 1
40 = 14	4 = 1
50 = 17	5 = 2
60 = 21	6 = 2
70 = 24	7 = 2
80 = 27	8 = 3
90 = 31	9 = 3

49 = 17

1490000	·1731863
034	961
067	·1732058
101	156
134	254
168	352
201	450
235	547
268	645

1490302	·1732743
336	841
369	938
403	·1733036
436	134
470	232
504	329
537	427
571	525
604	622
638	720
671	818
705	915
738	·1734013
772	111
805	209
839	306
872	404
906	502
939	599
973	697
1491007	795
040	892
074	990
107	·1735088
141	186
175	283
208	381
242	479
275	576
309	674
342	772
376	869
409	967
443	·1736064
476	162
510	260
543	357
577	455
610	552
644	650
677	748
711	845
744	943
778	·1737040
811	138
845	236
878	333
912	431
945	528
979	626
1492012	724
046	821
079	919
113	·1738016
146	114
180	211
213	309
247	406
280	504
314	601
347	699
381	796
414	894
448	991

1492481	·1739089
515	186
548	284
582	381
615	479
649	576
682	673
716	771
749	868
783	966
816	·1740063
850	160
883	258
917	355
950	453
984	550
1493017	647
051	745
084	842
118	940
151	·1741037
185	134
218	232
252	329
285	427
319	524
352	621
386	719
419	816
453	914
486	·1742011
520	108
553	206
587	303
620	401
654	498
687	595
721	693
754	790
788	887
821	985
854	·1743082
888	179
921	276
955	374
988	471
1494021	568
055	666
088	763
122	860
155	958
189	·1744055
222	152
256	249
289	347
323	444
356	541
390	638
423	736
457	833
490	930
524	·1745027
557	124
591	222
624	319

1494658	·1745416
691	513
725	610
758	708
792	805
825	902
858	999
892	·1746096
925	194
959	291
992	388
1495025	485
059	582
092	679
126	776
159	874
193	971
226	·1747068
260	165
293	262
327	359
360	456
394	553
427	650
461	747
494	845
527	942
561	·1748039
594	136
628	233
661	330
694	427
728	524
761	621
795	718
828	815
861	912
895	·1749009
928	106
962	203
995	300
1496028	397
062	494
095	591
129	688
162	786
195	883
229	980
262	·1750077
296	174
329	271
362	368
396	465
429	562
463	659
496	756
529	852
563	949
596	·1751046
630	143
663	240
696	337
730	434
763	531
797	628

1496830	·1751725
863	821
897	918
930	·1752015
964	112
997	209
1497030	306
064	403
097	500
131	597
164	694
197	790
231	887
264	984
298	·1753081
331	178
364	275
398	372
431	468
465	565
498	662
531	759
565	856
598	952
632	·1754049
665	146
698	243
732	340
765	436
799	533
832	630
865	727
899	824
932	920
966	·1755017
999	114
1498032	211
066	308
099	404
133	501
166	598
199	695
233	792
266	888
300	985
333	·1756082
366	179
400	275
433	372
467	469
500	566
533	662
567	759
600	856
634	952
667	·1757049
700	146
734	242
767	339
800	435
833	532
867	629
900	725
933	822
967	918

Number	Log
1499000	·1758015
033	112
067	208
100	305
134	402
167	499
200	595
234	692
267	789
301	885
334	982
367	·1759079
401	175
434	272
467	368
500	465
534	561
567	658
600	754
634	851
667	947
700	·1760044
734	140
767	237
800	333
833	430
867	527
900	623
933	720
967	816

Excesses

10 = 29	1 = 3
20 = 58	2 = 6
30 = 87	3 = 9
	4 = 12
	5 = 15
	6 = 17
	7 = 20
	8 = 23
	9 = 26

17 = 49

Anti Log Excesses

10 = 3	1 = 0
20 = 7	2 = 1
30 = 10	3 = 1
40 = 14	4 = 1
50 = 17	5 = 2
60 = 21	6 = 2
70 = 24	7 = 2
80 = 28	8 = 3
90 = 31	9 = 3

49 = 17

Number	Log
1500000	·1760913
033	·1761010
067	106
100	203
134	299
167	396

Number	Log	Number	Log	Number	Log	Number	Log
1500200	·1761492	1502365	·1767754	1504527	·1773998	1506685	·1780225
234	589	399	850	560	·1774094	718	320
267	685	432	947	594	190	752	416
301	782	465	·1768043	627	286	785	512
334	878	498	139	660	382	818	608
367	974	532	235	693	478	851	703
401	·1762071	565	331	726	574	884	799
434	167	598	428	760	670	918	895
467	264	632	524	793	766	951	990
500	360	665	620	826	862	984	·1781086
534	456	698	716	859	957	1507017	182
567	553	732	812	892	·1775053	050	277
600	649	765	908	926	149	084	373
634	746	798	·1769004	959	245	117	468
667	842	831	101	992	341	150	564
700	938	865	197	1505025	437	183	660
734	·1763035	898	293	058	533	216	755
767	131	931	389	092	628	250	851
800	228	965	485	125	724	283	946
833	324	998	581	158	820	316	·1782042
867	420	1503031	677	191	916	349	138
900	517	064	773	224	·1776012	382	233
933	613	098	869	258	107	416	329
967	710	131	965	291	203	449	424
1501000	806	164	·1770062	324	299	482	520
033	902	197	158	357	395	515	615
067	999	230	254	391	491	548	711
100	·1764095	264	350	424	587	582	806
133	192	297	446	457	683	615	902
166	288	330	542	490	779	648	997
200	384	363	638	524	874	681	·1783093
233	481	397	734	557	970	714	188
266	577	430	830	590	·1777066	747	284
300	674	463	926	624	162	780	379
333	770	496	·1771023	657	258	813	475
366	866	530	119	690	354	847	570
400	963	563	215	723	449	880	666
433	·1765059	596	311	757	545	913	761
466	155	630	407	790	641	946	857
499	252	663	503	823	737	979	952
533	348	696	599	856	832	1508012	·1784048
566	444	729	695	889	928	045	143
599	540	763	791	923	·1778024	079	239
633	637	796	887	956	119	112	334
666	733	829	983	989	215	145	430
699	829	862	·1772079	1506022	311	178	525
733	925	895	175	055	406	211	621
766	·1766022	929	271	089	502	245	716
799	118	962	367	122	598	278	812
832	214	995	463	155	694	311	907
866	310	1504028	559	188	789	344	·1785002
899	406	062	655	221	885	377	098
932	503	095	751	255	981	410	193
966	599	128	847	288	·1779076	443	289
999	695	161	943	321	172	476	384
1502032	791	195	·1773039	354	268	510	479
066	888	228	135	387	363	543	575
099	984	261	231	420	459	576	670
132	·1767080	295	327	453	555	609	766
165	177	328	423	486	651	642	861
199	273	361	519	520	746	675	956
232	369	394	615	553	842	708	·1786052
265	465	428	711	586	938	742	147
299	562	461	807	619	·1780033	775	243
332	658	494	903	652	129	808	338

1508841	·1786433
874	529
908	624
941	720
974	815
1509007	910
040	·1787006
073	101
106	197
139	292
173	387
206	483
239	578
272	674
305	769
338	864
371	960
404	·1788055
437	150
470	246
504	341
537	436
570	531
603	627
636	722
669	817
702	912
735	·1789008
768	103
801	198
835	293
868	388
901	484
934	579
967	674

Excesses

10 = 29	1 = 3
20 = 58	2 = 6
30 = 87	3 = 9
	4 = 12
	5 = 14
	6 = 17
	7 = 20
	8 = 23
	9 = 26

17 = 48

Anti Log Excesses

10 = 3	1 = 0
20 = 7	2 = 1
30 = 10	3 = 1
40 = 14	4 = 1
50 = 17	5 = 2
60 = 21	6 = 2
70 = 24	7 = 2
80 = 28	8 = 3
90 = 31	9 = 3

48 = 17

1510000 ·1789769

1510033	·1789864	1512185	·1796046	1514332	·1802209	1516477	·1808356
066	960	218	141	365	304	510	450
099	·1790055	251	236	398	399	543	545
132	150	284	331	431	494	576	639
166	245	317	426	464	588	609	733
199	340	350	521	497	683	642	828
232	436	383	616	530	778	675	922
265	531	416	711	563	872	708	·1809017
298	626	449	806	596	967	740	111
331	721	482	900	629	·1803062	773	205
364	816	515	995	662	156	806	300
398	912	548	·1797090	695	251	839	394
431	·1791007	581	185	728	345	872	489
464	102	614	280	761	440	905	583
497	197	647	375	794	535	938	677
530	292	680	470	827	629	971	772
564	388	713	565	860	724	1517004	866
597	483	746	660	893	818	037	960
630	578	779	755	926	913	070	·1810055
663	673	813	849	959	·1804008	103	149
696	768	846	944	992	102	136	243
729	863	879	·1798039	1515025	197	169	337
762	958	912	134	058	291	202	432
795	·1792054	945	229	091	386	235	526
828	149	978	324	124	481	268	620
861	244	1513011	419	157	575	301	715
894	339	044	514	190	670	334	809
927	434	077	609	223	764	367	904
960	529	110	704	256	859	399	998
993	624	143	798	289	954	432	·1811092
1511026	719	176	893	322	·1805048	465	187
059	814	209	988	355	143	498	281
092	909	242	·1799083	388	237	531	376
125	·1793005	275	178	421	332	564	470
159	100	308	273	454	427	597	564
192	195	341	367	487	521	630	658
225	290	374	462	520	616	663	753
258	385	407	557	553	710	696	847
291	480	440	652	586	805	729	941
324	575	473	746	619	900	762	·1812035
357	670	506	841	652	994	795	129
390	765	539	936	685	·1806089	828	224
423	860	572	·1800030	718	183	861	318
456	956	605	125	751	278	894	412
490	·1794051	638	220	784	372	927	506
523	146	671	315	817	467	960	600
556	241	704	409	850	561	993	695
589	336	737	504	883	656	1518026	789
622	431	770	599	916	750	058	883
655	526	804	694	949	845	091	977
688	621	837	789	982	939	124	·1813071
721	716	870	883	1516015	·1807034	157	166
754	811	903	978	048	128	190	260
787	906	936	·1801073	081	223	223	354
821	·1795001	969	168	114	317	256	448
854	096	1514002	262	147	412	289	542
887	191	035	357	180	506	322	637
920	286	068	452	213	601	355	731
953	381	101	547	246	695	387	825
986	476	134	641	279	789	420	919
1512019	571	167	736	312	884	453	·1814013
052	666	200	831	345	978	486	108
085	761	233	925	378	·1808073	519	202
118	856	266	·1802020	411	167	552	296
152	951	299	115	444	261	585	390

1518618	·1814484
651	579
684	673
717	767
750	861
783	955
816	·1815050
849	144
882	238
915	332
948	426
981	520
1519014	614
046	709
079	803
112	897
145	991
178	·1816085
211	179
244	273
277	367
310	461
343	555
375	650
408	744
441	838
474	932
507	·1817026
540	120
573	214
606	308
639	402
672	496
704	590
737	684
770	778
803	872
836	966
869	·1818060
902	154
935	248
968	342

Excesses

10 = 29	1 = 3
20 = 57	2 = 6
30 = 86	3 = 9
	4 = 11
	5 = 14
	6 = 17
	7 = 20
	8 = 23
	9 = 26

17 = 48

Anti Log Excesses

10 = 3	1 = 0
20 = 7	2 = 1
30 = 10	3 = 1
40 = 14	4 = 1
50 = 17	5 = 2
60 = 21	6 = 2
70 = 24	7 = 2
80 = 28	8 = 3
90 = 31	9 = 3

48 = 17

1520001	·1818436
033	530
066	624
099	718
132	812
165	906
198	·1819000
231	094
264	188
297	282
330	376
362	470
395	563
428	657
461	751
494	845
527	939
560	·1820033
593	127
626	221
659	315
691	409
724	502
757	596
790	690
823	784
856	878
889	972
922	·1821066
954	160
987	254
1521020	348
053	441
086	535
118	629
151	723
184	817
217	911
250	·1822005
283	098
316	192
348	286
381	380
414	474
447	567
480	661
513	755
546	849
579	943
611	·1823036
644	130

1521677	·1823224
710	318
743	412
775	505
808	599
841	693
874	787
907	880
940	974
973	·1824068
1522005	162
038	255
071	349
104	443
137	536
170	630
203	724
236	817
268	911
301	·1825005
334	099
367	192
400	286
432	380
465	473
498	567
531	661
564	754
597	848
630	942
662	·1826036
695	129
728	223
761	317
794	410
827	504
860	598
893	691
925	785
958	878
991	972
1523024	·1827066
057	159
089	253
122	346
155	440
188	534
221	627
253	721
286	814
319	908
352	·1828002
385	095
417	189
450	282
483	376
516	470
549	563
582	657
615	750
647	844
680	937
713	·1829031
746	124
779	218

1523812	·1829311
845	405
878	498
910	592
943	685
976	779
1524009	872
042	966
074	·1830059
107	153
140	246
173	340
206	433
238	527
271	620
304	714
337	807
370	901
402	994
435	·1831088
468	181
501	274
534	368
566	461
599	555
632	648
665	741
698	835
730	928
763	·1832022
796	115
829	208
861	302
894	395
927	489
959	582
992	675
1525025	769
058	862
090	956
123	·1833049
156	142
189	236
221	329
254	422
287	516
320	609
353	702
385	795
418	889
451	982
484	·1834075
517	169
549	262
582	355
615	449
648	542
681	635
713	728
746	822
779	915
812	·1835008
845	101
877	195
910	288

1525943	·1835381
976	474
1526009	567
041	661
074	754
107	847
140	940
172	·1836034
205	127
238	220
270	314
303	407
336	500
369	593
401	687
434	780
467	873
500	966
532	·1837059
565	152
598	246
631	339
664	432
696	525
729	618
762	711
795	804
827	897
860	991
893	·1838084
925	177
958	270
991	363
1527024	457
056	550
089	643
122	736
155	829
187	922
220	·1839015
253	109
286	202
319	295
351	388
384	481
417	574
450	667
482	760
515	853
548	946
580	·1840039
613	132
646	225
679	318
711	411
744	504
777	597
809	690
842	783
875	876
907	969
940	·1841062
973	155
1528006	248
038	341

1528071	·1841434
104	527
136	620
169	713
202	806
234	899
267	992
300	·1842085
333	178
365	271
398	364
431	457
463	550
496	643
529	736
561	829
594	922
627	·1843015
660	108
692	201
725	294
758	387
790	480
823	573
856	666
888	759
921	851
954	944
987	·1844037
1529019	130
052	223
085	316
117	409
150	501
183	594
215	687
248	780
281	873
314	965
346	·1845058
379	151
412	244
444	337
477	429
510	522
542	615
575	708
608	801
641	893
673	986
706	·1846079
739	172
771	265
804	357
837	450
869	543
902	636
935	729
968	821

Excesses

10 = 28, 20 = 57, 30 = 85

1 = 3, 2 = 6, 3 = 9, 4 = 11, 5 = 14, 6 = 17, 7 = 20, 8 = 23, 9 = 26

16 = 47

Anti Log Excesses

10 = 4, 20 = 7, 30 = 11, 40 = 14, 50 = 18, 60 = 21, 70 = 25, 80 = 28, 90 = 32

1 = 0, 2 = 1, 3 = 1, 4 = 1, 5 = 2, 6 = 2, 7 = 2, 8 = 3, 9 = 3

47 = 16

1530000	·1846914
033	·1847007
066	100
098	192
131	285
164	378
196	471
229	563
262	656
295	749
327	841
360	934
393	·1848027
425	119
458	212
491	305
523	398
556	490
589	583
622	676
654	768
687	861
720	954
752	·1849046
785	139
817	232
850	325
883	417
915	510
948	603
980	695
1531013	788
046	881
078	973
111	·1850066
144	158
176	251

1531209	·1850344
242	436
275	529
307	621
340	714
373	807
405	899
438	992
470	·1851084
503	177
536	270
568	362
601	455
633	547
666	640
699	733
731	825
764	918
797	·1852010
829	103
862	195
895	288
928	380
960	473
993	565
1532026	658
058	750
091	842
123	935
156	·1853028
189	120
221	213
254	305
286	398
319	490
352	583
384	675
417	768
449	860
482	953
515	·1854045
547	138
580	230
612	323
645	415
678	507
710	600
743	692
775	785
808	877
841	969
873	·1855062
906	154
938	247
971	339
1533004	431
036	524
069	616
102	709
134	801
167	893
200	986
233	·1856078
265	171
298	263

1533331	·1856355
363	448
396	540
428	633
461	725
494	817
526	910
559	·1857002
591	095
624	187
657	279
689	372
722	464
754	556
787	649
820	741
852	833
885	925
917	·1858018
950	110
983	202
1534015	294
048	387
080	479
113	571
146	663
178	755
211	848
243	940
276	·1859032
308	124
341	217
373	309
406	401
438	494
471	586
503	678
536	770
568	863
601	955
634	·1860047
666	139
699	232
731	324
764	416
797	508
829	600
862	693
894	785
927	877
960	969
992	·1861061
1535025	153
057	245
090	338
123	430
155	522
188	614
220	706
253	798
286	890
318	982
351	·1862074
383	166
416	259

1535449	·1862351
481	443
514	535
546	627
579	719
611	811
644	903
676	995
709	·1863087
741	180
774	272
806	364
839	456
871	548
904	640
937	732
969	824
1536002	916
034	·1864008
067	100
100	192
132	284
165	376
197	468
230	560
262	652
295	744
327	836
360	928
392	·1865020
425	112
457	204
490	296
522	388
555	480
587	572
620	664
652	756
685	848
717	940
750	·1866032
782	124
815	216
847	308
880	400
913	492
945	584
978	676
1537010	768
043	860
076	951
108	·1867043
141	135
173	227
206	319
238	411
271	503
303	594
336	686
368	778
401	870
433	962
466	·1868053
498	145
531	237

Number	Log
1537563	·1868329
596	421
628	513
661	605
693	697
726	788
758	880
791	972
823	·1869064
856	156
888	248
921	340
953	431
986	523
1538018	615
051	707
083	799
116	890
148	982
181	·1870074
213	166
246	257
278	349
311	441
343	533
376	624
408	716
441	808
473	899
506	991
538	·1871083
571	175
603	266
636	358
668	450
701	542
733	634
766	725
798	817
831	909
863	·1872001
896	092
928	184
961	275
993	367
1539026	459
058	550
091	642
123	733
156	825
188	917
221	·1873008
253	100
286	192
318	284
351	375
383	467
416	559
448	650
481	742
513	834
546	925
578	·1874017
611	108
643	200

Number	Log
1539676	·1874292
708	383
741	475
773	566
806	658
838	750
871	841
903	933
936	·1875024
968	116

Excesses

10 = 28	1 = 3
20 = 57	2 = 6
30 = 85	3 = 8
	4 = 11
	5 = 14
	6 = 17
	7 = 20
	8 = 23
	9 = 25

16 = 46

Anti Log Excesses

10 = 4	1 = 0
20 = 7	2 = 1
30 = 11	3 = 1
40 = 14	4 = 1
50 = 18	5 = 2
60 = 21	6 = 2
70 = 25	7 = 2
80 = 28	8 = 3
90 = 32	9 = 3

46 = 16

Number	Log
1540000	·1875208
033	299
065	391
098	482
130	574
162	666
195	757
227	849
260	940
292	·1876032
325	123
357	215
390	306
422	398
455	489
487	581
520	672
552	764
585	855
617	947
650	·1877038
682	130
715	221
747	313
780	404
812	495

Number	Log
1540845	·1877587
877	678
910	770
942	861
974	952
1541007	·1878044
039	135
072	227
104	318
136	409
169	501
201	592
234	684
266	775
298	866
331	958
363	·1879049
396	141
428	232
460	323
493	415
525	506
558	598
590	689
623	780
655	872
688	963
720	·1880055
753	146
785	237
818	329
850	420
883	511
915	603
947	694
980	785
1542012	876
045	968
077	·1881059
109	150
142	242
174	333
207	424
239	516
271	607
304	698
336	789
369	881
401	972
433	·1882063
466	155
498	246
531	337
563	429
595	520
628	611
660	702
693	794
725	885
757	976
790	·1883067
822	159
855	250
887	341
919	432

Number	Log
1542952	·1883523
984	615
1543017	706
049	797
081	888
114	979
146	·1884071
179	162
211	253
243	344
276	435
308	527
341	618
373	709
405	800
438	891
470	982
503	·1885073
535	165
567	256
600	347
632	438
665	529
697	620
729	711
762	802
794	893
827	984
859	·1886076
891	167
924	258
956	349
989	440
1544021	531
053	622
086	713
118	804
151	895
183	987
215	·1887078
248	169
280	260
313	351
345	442
377	533
410	624
442	715
475	806
507	897
539	988
572	·1888079
604	170
637	261
669	352
701	443
734	534
766	625
798	716
830	807
863	898
895	989
927	·1889080
960	171
992	262
1545024	353

Number	Log
1545057	·1889444
089	535
122	626
154	717
186	808
219	899
251	990
284	·1890081
316	172
348	263
381	354
413	445
446	536
478	627
510	717
543	808
575	899
608	990
640	·1891081
672	172
705	263
737	353
769	444
801	535
834	626
866	717
898	807
931	898
963	989
995	·1892080
1546028	171
060	262
092	353
124	444
157	534
189	625
221	716
254	807
286	898
318	989
351	·1893080
383	170
416	261
448	352
480	443
513	534
545	624
578	715
610	806
642	897
675	988
707	·1894078
739	169
771	260
804	351
836	442
868	532
901	623
933	714
965	805
998	895
1547030	986
062	·1895077
094	168
127	258

1547159	·1895349
191	440
224	530
256	621
288	712
321	802
353	893
385	984
417	·1896075
450	165
482	256
514	347
547	437
579	528
611	619
644	709
676	800
708	890
740	981
773	·1897072
805	162
837	253
870	343
902	434
934	525
967	615
999	706
1548031	796
063	887
096	978
128	·1898068
160	159
193	249
225	340
257	431
290	521
322	612
354	702
386	793
419	884
451	974
483	·1899065
516	155
548	246
580	337
613	427
645	518
677	608
709	699
742	789
774	880
806	970
839	·1900061
871	151
903	242
936	332
968	423
1549000	513
032	604
065	694
097	785
129	875
162	966
194	·1901056
226	147

1549259	·1901237
291	328
323	418
356	509
388	599
420	690
452	780
485	871
517	961
549	·1902051
581	142
614	232
646	323
678	413
710	503
742	594
775	684
807	775
839	865
871	955
904	·1903046
936	136
968	227

Excesses

10 = 28	1 = 3
20 = 56	2 = 6
30 = 84	3 = 8
	4 = 11
	5 = 14
	6 = 17
	7 = 20
	8 = 22
	9 = 25

16 = 46

Anti Log Excesses

10 = 4	1 = 0
20 = 7	2 = 1
30 = 11	3 = 1
40 = 14	4 = 1
50 = 18	5 = 2
60 = 21	6 = 2
70 = 25	7 = 2
80 = 28	8 = 3
90 = 32	9 = 3

46 = 16

1550000	·1903317
033	407
065	498
097	588
130	679
162	769
194	859
226	950
259	·1904040
291	130
323	221
355	311
387	401

1550420	·1904491
452	582
484	672
516	762
549	853
581	943
613	·1905033
645	124
678	214
710	304
742	394
775	485
807	575
839	665
871	756
904	846
936	936
968	·1906027
1551000	117
032	207
065	297
097	388
129	478
161	568
193	659
226	749
258	839
290	930
322	·1907020
354	110
387	200
419	291
451	381
483	471
516	561
548	652
580	742
612	832
645	922
677	·1908012
709	103
742	193
774	283
806	373
838	463
871	553
903	643
935	734
967	824
999	914
1552032	·1909004
064	094
096	184
128	274
160	364
193	454
225	544
257	635
289	725
321	815
354	905
386	995
418	·1910085
450	175
482	265

1552515	·1910355
547	445
579	536
611	626
643	716
676	806
708	896
740	986
772	·1911076
804	166
837	256
869	346
901	437
933	527
965	617
998	707
1553030	797
062	887

Excesses

16 = 50

Anti Log Excesses

50 = 16

1553098	·1911987
134	·1912087
169	187
205	287
241	387
277	487
312	587
348	687
384	787
420	887
456	987
491	·1913087
527	187
563	286
599	386
634	486
670	586
706	686
742	786
778	886
813	986
849	·1914086
885	186
921	286
956	386
992	486
1554028	586
064	686
099	786
135	886
171	986
206	·1915085
242	185
278	285
313	385
349	485
385	585
421	685

1554456	·1915784
492	884
528	984
564	·1916084
599	183
635	283
671	383
707	483
742	583
778	683
814	783
849	882
885	982
921	·1917082
956	182
992	282
1555028	382
064	481
099	581
135	681
171	780
207	880
242	980
278	·1918079
314	179
350	279
385	379
421	478
457	578
492	678
528	778
564	877
599	977
635	·1919077
671	177
707	276
742	376
778	476
814	575
850	675
885	775
921	874
957	974
993	·1920074
1556028	173
064	273
100	373
135	472
171	572
207	672
242	771
278	871
314	971
349	·1921070
385	170
421	269
456	369
492	468
528	568
563	667
599	767
635	867
671	966
706	·1922066
742	165

Number	Log
1556778	·1922265
814	364
849	464
885	563
921	663
957	763
992	862
1557028	962
064	·1923061
099	161
135	260
171	360
206	459
242	559
278	658
313	758
349	857
385	957
420	·1924056
456	156
492	255
527	355
563	454
599	553
634	653
670	752
706	852
741	951
777	·1925051
813	150
848	250
884	349
920	449
955	548
991	648
1558027	747
062	847
098	946
134	·1926046
169	145
205	245
241	344
276	443
312	543
347	642
383	741
418	840
454	940
489	·1927039
525	138
561	237
596	336
632	436
668	535
703	634
739	733
775	833
810	932
846	·1928031
882	130
917	230
953	329
989	428
1559024	528
060	627

Number	Log
1559096	·1928726
131	826
167	925
203	·1929024
238	123
274	223
310	322
345	421
381	520
417	620
452	719
488	818
524	917
559	·1930016
595	116
630	215
666	314
701	413
737	513
772	612
808	711
844	810
879	909
915	·1931008
951	107
986	207

Excesses

10 = 28	1 = 3
20 = 56	2 = 6
30 = 84	3 = 8
	4 = 11
	5 = 14
	6 = 17
	7 = 20
	8 = 22
	9 = 25

18 = 50

Anti Log Excesses

10 = 4	1 = 0
20 = 7	2 = 1
30 = 11	3 = 1
40 = 14	4 = 1
50 = 18	5 = 2
60 = 21	6 = 2
70 = 25	7 = 3
80 = 29	8 = 3
90 = 32	9 = 3

50 = 18

Number	Log
1560022	·1931306
058	405
093	504
129	603
165	702
200	801
236	900
271	999
307	·1932099
342	198

Number	Log
1560378	·1932297
413	396
449	495
485	594
520	693
556	792
592	891
627	990
663	·1933089
699	188
734	287
770	386
806	485
841	584
877	683
912	782
948	882
983	981
1561019	·1934080
054	179
090	278
126	377
161	476
197	575
232	674
268	773
303	872
339	971
374	·1935070
410	169
446	268
481	367
517	466
552	565
588	663
623	762
659	861
694	960
730	·1936059
766	158
801	257
837	356
872	455
908	553
943	652
979	751
1562014	850
050	949
086	·1937048
121	147
157	246
192	345
228	443
263	542
299	641
334	740
370	839
406	938
441	·1938037
477	135
512	234
548	333
583	432
619	530
654	629

Number	Log
1562690	·1938728
726	827
761	926
797	·1939024
832	123
868	222
903	321
939	419
974	518
1563010	617
046	716
081	815
117	913
152	·1940012
188	111
223	210
259	308
294	407
330	506
366	605
401	703
437	802
472	901
508	999
543	·1941098
579	197
614	295
650	394
686	493
721	591
757	690
792	789
828	887
863	986
899	·1942085
934	183
970	282
1564005	381
041	479
076	578
112	677
147	775
183	874
218	973
254	·1943071
289	170
325	269
360	367
396	466
431	564
467	663
502	761
538	860
573	958
609	·1944057
645	156
680	254
716	353
751	451
787	550
822	648
858	747
893	845
929	944
964	·1945042

Number	Log
1565000	·1945141
035	239
071	338
106	436
142	535
177	633
213	732
248	830
283	929
319	·1946027
354	126
390	224
425	323
461	421
496	520
532	618
567	717
603	815
638	914
674	·1947012
709	110
745	209
780	307
816	405
851	504
887	602
922	700
958	799
993	897
1566029	996
064	·1948094
100	193
135	291
171	390
206	488
241	586
277	685
312	783
348	881
383	980
419	·1949078
454	176
490	275
525	373
560	471
596	569
631	668
667	766
702	864
738	962
773	·1950061
809	159
844	257
880	355
915	454
951	552
986	650
1567022	749
057	847
093	945
128	·1951044
164	142
199	240
235	338
270	437

1567306	·1951535
341	633
377	731
412	830
448	928
483	·1952026
518	124
554	222
589	320
625	418
660	517
696	615
731	713
767	811
802	909
837	·1953007
873	105
908	204
943	302
979	400
1568014	498
049	597
085	695
120	793
155	891
191	989
226	·1954087
262	185
297	284
333	382
368	480
404	578
439	676
474	774
510	872
545	970
581	·1955068
616	166
652	264
687	362
723	460
758	558
793	656
829	754
864	852
900	950
935	·1956048
971	146
1569006	244
042	342
077	440
112	538
148	636
183	734
218	832
254	930
289	·1957028
324	126
360	224
395	322
430	420
466	518
501	616
537	714
572	812

1569608	·1957909
643	·1958007
679	105
714	203
749	301
785	399
820	497
855	595
891	693
926	791
961	889
997	987

Excesses

10 = 28	1 = 3
20 = 56	2 = 6
30 = 83	3 = 8
	4 = 11
	5 = 14
	6 = 17
	7 = 19
	8 = 22
	9 = 25

18 = 49

Anti Log Excesses

10 = 4	1 = 0
20 = 7	2 = 1
30 = 11	3 = 1
40 = 14	4 = 1
50 = 18	5 = 2
60 = 22	6 = 2
70 = 25	7 = 3
80 = 29	8 = 3
90 = 32	9 = 3

49 = 18

1570032	·1959085
067	183
103	281
138	378
174	476
209	574
245	672
280	769
316	867
351	965
386	·1960063
422	161
457	259
492	357
528	454
563	552
598	650
634	748
669	846
704	944
740	·1961042
775	139
810	237
846	335

1570881	·1961433
916	530
952	628
987	726
1571022	824
058	921
093	·1962019
129	117
164	214
200	312
235	410
271	507
306	605
341	703
377	801
412	898
447	996
483	·1963094
518	192
553	289
589	387
624	485
659	583
695	680
730	778
765	876
801	973
836	·1964071
871	169
907	266
942	364
977	462
1572013	559
048	657
083	754
119	852
154	949
189	·1965047
225	144
260	242
295	340
331	437
366	535
401	633
437	730
472	828
507	926
543	·1966023
578	121
613	218
649	316
684	413
719	511
755	608
790	706
825	803
861	901
896	998
931	·1967096
967	193
1573002	291
037	388
073	486
108	583
143	681

1573179	·1967778
214	876
249	973
285	·1968071
320	168
355	266
391	363
426	461
461	558
497	656
532	753
567	850
602	948
638	·1969045
673	143
708	240
743	338
779	435
814	533
849	630
884	727
920	825
955	922
990	·1970019
1574026	117
061	214
096	311
132	409
167	506
202	603
238	701
273	798
308	896
344	993
379	·1971091
414	188
450	286
485	383
520	480
555	577
591	675
626	772
661	869
696	966
732	·1972064
767	161
802	258
837	355
873	453
908	550
943	647
979	745
1575014	842
049	939
085	·1973037
120	134
155	231
190	328
226	426
261	523
296	620
331	717
367	815
402	912
437	·1974009

1575472	·1974106
507	203
543	301
578	398
613	495
648	592
684	690
719	787
754	884
789	981
825	·1975078
860	175
895	272
931	370
966	467
1576001	564
037	661
072	758
107	855
142	952
178	·1976049
213	146
248	244
283	341
319	438
354	535
389	632
424	729
459	826
495	923
530	·1977020
565	118
600	215
636	312
671	409
706	506
741	603
776	700
812	797
847	894
882	991
917	·1978088
953	185
988	282
1577023	379
058	476
093	573
129	670
164	767
199	864
234	961
270	·1979058
305	155
340	252
375	349
410	446
446	543
481	640
516	737
551	834
587	931
622	·1980028
657	125
692	222
727	319

Number	Log
1577763	·1980416
798	513
833	609
868	706
904	803
939	900
974	997
1578009	·1981094
044	191
080	288
115	385
150	481
185	578
221	675
256	772
291	869
326	966
361	·1982063
396	159
431	256
467	353
502	450
537	546
572	643
607	740
642	837
677	934
713	·1983031
748	128
783	224
818	321
854	418
889	515
924	612
959	709
994	805
1579030	902
065	999
100	·1984095
135	192
171	289
206	385
241	482
276	579
311	676
346	772
381	869
417	966
452	·1985063
487	159
522	256
557	353
592	450
627	546
663	643
698	740
733	836
768	933
804	·1986030
839	126
874	223
909	320
944	416
979	513

Excesses

10 = 28	1 = 3
20 = 55	2 = 6
30 = 83	3 = 8
	4 = 11
	5 = 14
	6 = 17
	7 = 19
	8 = 22
	9 = 25

18 = 50

Anti Log Excesses

10 = 4	1 = 0
20 = 7	2 = 1
30 = 11	3 = 1
40 = 14	4 = 1
50 = 18	5 = 2
60 = 22	6 = 2
70 = 25	7 = 3
80 = 29	8 = 3
90 = 33	9 = 3

50 = 18

Number	Log
1580014	·1986610
050	706
085	803
120	900
155	996
190	·1987093
225	190
260	286
296	383
331	479
366	576
401	672
437	769
472	865
507	962
542	·1988059
577	155
612	252
647	348
683	445
718	541
753	638
788	734
823	831
858	928
893	·1989024
928	121
963	217
999	314
1581034	410
069	507
104	603
139	700
174	797
209	893
245	990
280	·1990086

Number	Log
1581315	·1990183
350	279
386	376
421	472
456	569
491	665
526	762
561	858
596	955
632	·1991051
667	148
702	244
737	341
772	437
807	533
842	630
877	726
912	822
948	919
983	·1992015
1582018	111
053	208
088	304
123	400
158	497
193	593
228	690
264	786
299	883
334	979
369	·1993076
404	172
439	268
474	365
509	461
544	557
580	654
615	750
650	846
685	943
720	·1994039
755	135
790	232
825	328
860	424
896	521
931	617
966	713
1583001	810
036	906
071	·1995002
106	098
141	195
176	291
211	387
246	483
281	580
316	676
351	772
386	868
421	964
456	·1996061
491	157
527	253
562	349

Number	Log
1583597	·1996446
632	542
667	638
702	734
737	830
772	926
807	·1997022
843	119
878	215
913	311
948	407
983	503
1584018	599
053	695
088	792
123	888
158	984
193	·1998080
228	177
263	273
298	369
333	465
368	561
403	657
438	753
474	850
509	946
544	·1999042
579	138
614	234
649	330
684	426
719	522
754	618
790	714
825	810
860	906
895	·2000002
930	098
965	194
1585000	290
035	386
070	482
105	579
140	675
175	771
210	867
245	963
280	·2001059
315	155
350	251
385	347
420	443
455	539
490	635
525	731
560	827
595	923
630	·2002019
665	115
700	211
736	306
771	402
806	498
841	594

Number	Log
1585876	·2002690
911	786
946	882
981	978
1586016	·2003074
051	169
086	265
121	361
156	457
191	553
226	649
261	745
296	841
331	937
366	·2004032
401	128
436	224
471	320
506	416
541	512
576	608
611	704
646	800
681	895
716	991
751	·2005087
786	183
821	279
856	375
891	471
926	566
961	662
996	758
1587031	854
066	949
101	·2006045
136	141
171	237
206	333
241	428
276	524
311	620
346	716
381	811
416	907
451	·2007003
486	099
521	194
556	290
591	386
626	481
661	577
696	673
731	768
766	864
801	960
836	·2008056
871	151
906	247
941	343
976	439
1588011	534
046	630
081	726
116	822

1588151	·2008917
186	·2009013
221	108
256	204
291	299
326	395
361	490
396	586
431	682
466	777
501	873
536	969
571	·2010064
606	160
641	256
676	351
711	447
746	543
781	638
816	734
851	829
885	925
920	·2011020
955	116
990	211
1589025	307
060	403
095	498
130	594
165	689
200	785
235	880
270	976
305	·2012071
340	167
375	263
410	358
445	454
480	549
515	645
550	740
585	836
620	931
655	·2013027
690	122
725	218
760	313
795	409
829	504
864	600
899	695
934	791
969	886

Excesses

10 = 27	1 = 3
20 = 54	2 = 5
30 = 82	3 = 8
	4 = 11
	5 = 14
	6 = 16
	7 = 19
	8 = 22
	9 = 25

18 = 48

Anti Log Excesses

10 = 4	1 = 0
20 = 7	2 = 1
30 = 11	3 = 1
40 = 15	4 = 1
50 = 18	5 = 2
60 = 22	6 = 2
70 = 26	7 = 3
80 = 29	8 = 3
90 = 33	9 = 3

48 = 18

1590004	·2013981
039	·2014077
074	172
109	267
143	363
178	458
213	553
248	649
283	744
318	839
353	935
388	·2015030
423	126
458	221
493	317
528	412
563	508
598	603
633	698
668	794
703	889
738	984
772	·2016080
807	175
842	270
877	366
912	461
947	556
982	652
1591017	747
052	842
086	938
121	·2017033
156	128
191	224
226	319
261	414

1591296	·2017509
331	605
366	700
401	795
436	890
471	986
506	·2018081
541	176
576	271
611	366
646	462
681	557
715	652
750	747
785	843
820	938
855	·2019033
890	128
925	223
960	319
995	414
1592029	509
064	604
099	700
134	795
169	890
204	985
239	·2020080
274	176
309	271
343	366
378	461
413	557
448	652
483	747
518	842
553	937
588	·2021032
623	127
657	223
692	318
727	413
762	508
797	603
832	698
867	793
902	888
937	983
971	·2022078
1593006	173
041	268
076	363
111	458
146	553
181	648
215	743
250	838
285	934
320	·2023029
354	124
389	219
424	314
459	409
494	504
529	599

1593564	·2023694
598	789
633	884
668	979
703	·2024074
738	169
773	264
808	359
843	454
878	549
912	644
947	739
982	834
1594017	929
052	·2025024
087	119
122	214
157	309
192	404
226	498
261	593
296	688
331	783
366	878
401	973
436	·2026068
470	163
505	258
540	352
575	447
609	542
644	637
679	732
714	827
749	922
784	·2027017
819	112
853	206
888	301
923	396
958	491
993	586
1595028	681
063	776
097	870
132	965
167	·2028060
202	155
236	249
271	344
306	439
341	534
376	629
411	723
446	818
480	913
515	·2029008
550	102
585	197
620	292
655	387
690	482
724	576
759	671
794	766

1595829	·2029861
863	955
898	·2030050
933	145
968	240
1596003	334
037	429
072	524
107	618
142	713
176	808
211	902
246	997
281	·2031092
316	186
350	281
385	376
420	470
455	565
489	660
524	754
559	849
594	944
629	·2032038
663	133
698	228
733	322
768	417
802	512
837	606
872	701
907	796
942	890
977	985
1597012	·2033079
046	174
081	268
116	363
151	457
186	552
221	647
256	741
290	836
325	930
360	·2034025
395	119
429	214
464	308
499	403
534	498
569	592
603	687
638	781
673	876
708	970
742	·2035065
777	159
812	254
847	348
881	443
916	537
951	632
985	726
1598020	821
055	915

1598089	·2036010
124	104
159	199
194	293
228	388
263	482
298	577
333	671
367	766
402	860
437	955
472	·2037049
507	144
541	238
576	332
611	427
646	521
680	615
715	710
750	804
785	898
820	993
854	·2038087
889	182
924	276
959	371
993	465
1599028	560
063	654
098	748
132	843
167	937
202	·2039031
236	126
271	220
306	314
340	409
375	503
410	597
445	691
479	786
514	880
549	974
584	·2040068
618	163
653	257
688	351
723	445
757	540
792	634
827	728
861	823
896	917
931	·2041011
965	106

Excesses

10 = 27	1 = 3
20 = 54	2 = 5
30 = 82	3 = 8
	4 = 11
	5 = 14
	6 = 16
	7 = 19
	8 = 22
	9 = 25

18 = 48

Anti Log Excesses

10 = 4	1 = 0
20 = 7	2 = 1
30 = 11	3 = 1
40 = 15	4 = 1
50 = 18	5 = 2
60 = 22	6 = 2
70 = 26	7 = 3
80 = 29	8 = 3
90 = 33	9 = 3

48 = 18

1600000	·2041200
035	294
070	388
104	483
139	577
174	671
209	765
243	860
278	954
313	·2042048
348	142
382	236
417	331
452	425
486	519
521	613
556	708
590	802
625	896
660	990
695	·2043084
729	178
764	272
799	367
834	461
868	555
903	649
938	743
973	837
1601007	931
042	·2044025
077	119
111	214
146	308
181	402
215	496
250	590

1601285	·2044684
319	778
354	872
389	966
423	·2045061
458	155
493	249
527	343
562	437
597	531
631	625
666	719
701	813
735	907
770	·2046001
805	095
839	189
874	283
909	377
944	471
978	565
1602013	659
048	753
083	847
117	941
152	·2047035
187	129
222	223
256	317
291	411
326	505
360	599
395	693
430	787
464	881
499	975
534	·2048069
568	163
603	257
638	351
672	444
707	538
742	632
776	726
811	820
846	914
880	·2049008
915	102
949	196
984	290
1603018	384
053	478
087	572
122	666
157	760
191	854
226	947
261	·2050041
295	135
330	229
365	322
399	416
434	510
469	604
503	698

1603538	·2050792
573	886
607	979
642	·2051073
677	167
711	261
746	355
781	449
815	543
850	636
885	730
919	824
954	918
989	·2052011
1604023	105
058	199
093	293
127	387
162	480
197	574
231	668
266	762
301	855
335	949
370	·2053043
405	137
439	230
474	324
508	418
543	511
577	605
612	699
646	792
681	886
716	980
750	·2054073
785	167
820	261
854	354
889	448
924	542
958	635
993	729
1605028	823
062	916
097	·2055010
131	104
166	197
200	291
235	385
269	478
304	572
339	666
373	759
408	853
443	947
477	·2056040
512	134
547	228
581	321
616	415
651	509
685	602
720	696
754	789

1605789	·2056883
823	976
858	·2057070
892	163
927	257
962	351
996	444
1606031	538
065	631
100	725
134	818
169	912
203	·2058005
238	099
273	192
307	286
342	379
377	473
411	566
446	660
481	753
515	847
550	940
585	·2059033
619	127
654	220
688	314
723	407
757	501
792	594
826	688
861	781
896	874
930	968
965	·2060061
999	155
1607034	248
068	342
103	435
137	529
172	622
207	715
241	809
276	902
310	996
345	·2061089
379	183
414	276
448	370
483	463
518	556
552	650
587	743
621	836
656	930
690	·2062023
725	116
759	210
794	303
829	396
863	490
898	583
932	676
967	770
1608001	863

1608036	·2062956
070	·2063050
105	143
140	236
174	329
209	423
243	516
278	609
312	702
347	796
381	889
416	982
451	·2064075
485	168
520	262
554	355
589	448
623	541
658	635
692	728
727	821
761	914
796	·2065007
830	101
865	194
899	287
934	380
968	474
1609003	567
037	660
072	753
106	846
141	940
175	·2066033
210	126
244	219
279	313
313	406
348	499
383	592
417	685
452	778
486	871
521	965
555	·2067058
590	151
624	244
659	337
693	430
728	523
762	616
797	709
831	803
866	896
900	989
935	·2068082
969	175

Excesses

10 = 27	1 = 3
20 = 54	2 = 5
30 = 81	3 = 8
	4 = 11
	5 = 14
	6 = 16
	7 = 19
	8 = 22
	9 = 24

17 = 47

Anti Log Excesses

10 = 4	1 = 0
20 = 7	2 = 1
30 = 11	3 = 1
40 = 15	4 = 1
50 = 18	5 = 2
60 = 22	6 = 2
70 = 26	7 = 3
80 = 30	8 = 3
90 = 33	9 = 3

47 = 17

1610004	·2068268
038	361
073	454
107	547
142	641
176	734
211	827
245	920
280	·2069013
314	106
349	199
383	292
418	385
452	478
487	571
521	664
556	757
590	850
625	943
659	·2070036
694	129
728	222
763	315
797	408
832	501
866	594
901	687
935	780
970	873
1611004	966
039	·2071059
073	152
108	245
142	338
177	431
211	524
246	617

1611280	·2071710
315	803
349	896
384	988
418	·2072081
453	174
487	267
522	360
556	453
591	546
625	639
660	732
694	824
729	917
763	·2073010
798	103
832	196
866	289
901	382
935	475
970	568
1612004	660
039	753
073	846
108	939
142	·2074032
176	125
211	218
245	310
280	403
314	496
349	589
383	681
418	774
452	867
486	960
521	·2075053
555	145
590	238
624	331
659	424
693	516
728	609
762	702
796	795
831	888
865	980
900	·2076073
934	166
969	259
1613003	351
038	444
072	537
106	630
141	722
175	815
210	908
244	·2077000
279	093
313	186
348	278
382	371
416	464
451	556
485	649

1613520	·2077742
554	834
589	927
623	·2078020
658	112
692	205
726	298
761	390
795	483
830	576
864	668
899	761
933	854
968	946
1614002	·2079039
036	132
071	224
105	317
139	409
174	502
208	594
242	687
277	779
311	872
345	965
380	·2080057
414	150
449	242
483	335
518	427
552	520
587	612
621	705
655	798
690	890
724	983
759	·2081075
793	168
828	260
862	353
897	445
931	538
965	630
1615000	723
034	815
068	908
103	·2082000
137	093
171	185
206	278
240	370
274	463
309	555
343	648
378	740
412	833
447	925
481	·2083018
516	110
550	203
584	295
619	388
653	480
687	572
722	665

1615756	·2083757
790	849
825	942
859	·2084034
893	126
928	219
962	311
997	404
1616031	496
066	589
100	681
135	774
169	866
203	958
238	·2085051
272	143
306	235
341	328
375	420
409	512
444	605
478	697
512	789
547	882
581	974
615	·2086066
650	159
684	251
718	343
753	436
787	528
821	620
856	713
890	805
925	897
959	990
994	·2087082
1617028	174
063	267
097	359
131	451
166	543
200	636
234	728
269	820
303	912
337	·2088005
372	097
406	189
440	281
475	373
509	466
543	558
578	650
612	742
646	835
681	927
715	·2089019
749	111
784	203
818	296
852	388
887	480
921	572
955	665

Number	Log
1617990	·2089757
1618024	849
058	941
093	·2090033
127	125
161	217
196	310
230	402
264	494
299	586
333	678
367	770
402	862
436	954
470	·2091046
505	139
539	231
573	323
608	415
642	507
676	599
711	691
745	783
779	875
814	967
848	·2092059
882	151
917	243
951	335
985	427
1619020	519
054	611
088	703
123	796
157	888
191	980
226	·2093072
260	164
294	256
328	348
363	440
397	532
431	624
465	716
500	808
534	900
568	992
602	·2094084
637	176
671	268
705	360
740	451
774	543
808	635
843	727
877	819
911	911
946	·2095003
980	095

Excesses

10 = 27	1 = 3
20 = 54	2 = 5
30 = 81	3 = 8
	4 = 11
	5 = 13
	6 = 16
	7 = 19
	8 = 22
	9 = 24

17 = 46

Anti Log Excesses

10 = 4	1 = 0
20 = 7	2 = 1
30 = 11	3 = 1
40 = 15	4 = 1
50 = 19	5 = 2
60 = 22	6 = 2
70 = 26	7 = 3
80 = 30	8 = 3
90 = 33	9 = 3

46 = 17

Number	Log
1620014	·2095187
049	279
083	371
117	463
152	555
186	647
220	739
254	831
289	923
323	·2096015
357	106
391	198
426	290
460	382
494	474
528	566
563	658
597	750
631	842
666	933
700	·2097025
734	117
769	209
803	301
837	393
871	485
906	576
940	668
974	760
1621008	852
043	943
077	·2098035
111	127
145	219
180	311
214	402
248	494

Number	Log
1621283	·2098586
317	678
351	769
386	861
420	953
454	·2099045
488	137
523	228
557	320
591	412
625	504
660	595
694	687
728	779
762	871
796	962
831	·2100054
865	146
899	237
933	329
968	421
1622002	512
036	604
070	696
104	788
139	879
173	971
207	·2101063
241	155
276	246
310	338
344	430
378	522
413	613
447	705
481	797
516	888
550	980
584	·2102072
619	163
653	255
687	347
721	438
756	530
790	621
824	713
858	804
893	896
927	987
961	·2103079
995	171
1623029	262
064	354
098	445
132	537
166	628
201	720
235	811
269	903
303	995
337	·2104086
372	178
406	269
440	361
474	452

Number	Log
1623509	·2104544
543	635
577	727
611	819
645	910
680	·2105002
714	093
748	185
782	276
817	368
851	459
885	551
919	642
953	734
988	825
1624022	917
056	·2106008
090	100
125	191
159	283
193	374
227	465
261	557
295	648
329	740
364	831
398	923
432	·2107014
466	106
500	197
534	288
568	380
603	471
637	563
671	654
705	746
740	837
774	929
808	·2108020
842	111
876	203
911	294
945	385
979	477
1625013	568
048	659
082	751
116	842
150	933
184	·2109025
219	116
253	207
287	299
321	390
356	481
390	573
424	664
458	755
492	847
526	938
560	·2110029
595	121
629	212
663	303
697	395

Number	Log
1625731	·2110486
765	577
799	668
834	760
868	851
902	942
936	·2111033
971	125
1626005	216
039	307
073	398
107	490
141	581
175	672
210	764
244	855
278	946
312	·2112038
346	129
380	220
414	311
449	402
483	493
517	585
551	676
586	767
620	858
654	949
688	·2113040
722	131
756	223
790	314
825	405
859	496
893	588
927	679
961	770
995	861
1627029	952
063	·2114043
097	134
132	226
166	317
200	408
234	499
268	590
302	681
336	772
370	863
404	954
439	·2115046
473	137
507	228
541	319
575	410
609	501
643	592
678	683
712	774
746	866
780	957
815	·2116048
849	139
883	230
917	321

1627951	·2116412
985	503
1628019	594
054	685
088	776
122	867
156	958
190	·2117049
224	140
258	231
292	322
326	413
361	504
395	595
429	686
463	777
497	868
531	959
565	·2118050
599	141
633	232
668	322
702	413
736	504
770	595
804	686
838	777
872	868
906	959
940	·2119050
975	141
1629009	232
043	323
077	414
111	505
145	596
179	687
213	778
247	869
282	959
316	·2120050
350	141
384	232
418	323
452	414
486	505
520	595
554	686
588	777
622	868
656	958
690	·2121049
724	140
758	231
792	322
826	413
860	504
895	594
929	685
963	776
997	867

Excesses

10 = 27	1 = 3
20 = 53	2 = 5
30 = 80	3 = 8
	4 = 11
	5 = 13
	6 = 16
	7 = 19
	8 = 21
	9 = 24

17 = 46

Anti Log Excesses

10 = 4	1 = 0
20 = 7	2 = 1
30 = 11	3 = 1
40 = 15	4 = 1
50 = 19	5 = 2
60 = 22	6 = 2
70 = 26	7 = 3
80 = 30	8 = 3
90 = 34	9 = 3

46 = 17

1630031	·2121958
065	·2122049
099	140
133	230
167	321
202	412
236	503
270	593
304	684
338	775
372	866
406	957
440	·2123047
474	138
509	229
543	320
577	410
611	501
645	592
679	683
713	773
747	864
781	955
815	·2124045
849	136
883	227
917	317
951	408
985	499
1631019	589
053	680
087	771
122	861
156	952
190	·2125043
224	133
258	224

1631292	·2125315
326	405
360	496
394	587
428	677
462	768
496	859
530	949
564	·2126040
598	131
632	221
666	312
700	403
735	493
769	584
803	675
837	765
871	856
905	947
939	·2127037
973	128
1632007	218
041	309
075	399
109	490
143	580
177	671
211	762
245	852
279	943
313	·2128033
347	124
381	214
415	305
449	395
483	486
517	577
551	667
585	758
619	848
654	939
688	·2129029
722	120
756	210
790	301
824	391
858	482
892	572
926	663
960	753
994	844
1633028	934
062	·2130025
096	115
130	205
164	296
198	386
232	477
266	567
300	658
334	748
368	839
402	929
436	·2131019
470	110

1633504	·2131200
538	291
572	381
606	472
640	562
674	653
708	743
742	833
776	924
810	·2132014
844	105
878	195
912	286
946	376
980	467
1634014	557
048	647
082	738
116	828
150	918
184	·2133009
218	099
252	189
286	280
320	370
354	460
388	550
422	641
456	731
490	821
524	911
558	·2134002
592	092
626	182
660	272
694	363
728	453
762	543
796	634
830	724
864	814
898	905
932	995
966	·2135085
1635000	175
034	266
068	356
101	446
135	536
169	627
203	717
237	807
271	897
305	987
339	·2136078
373	168
407	258
441	348
475	439
509	529
543	619
577	709
611	799
645	890
679	980

1635713	·2137070
747	160
781	251
815	341
849	431
883	521
917	611
951	701
985	791
1636019	882
053	972
087	·2138062
121	152
155	242
189	332
223	422
257	512
291	602
324	693
358	783
392	873
426	963
460	·2139053
494	143
528	233
562	323
596	413
630	504
664	594
698	684
732	774
766	864
800	954
834	·2140044
868	134
902	224
935	314
969	404
1637003	494
037	584
071	674
105	764
139	854
173	944
207	·2141034
240	124
274	214
308	304
342	394
376	484
410	574
444	664
478	754
512	844
546	934
580	·2142024
614	114
648	204
682	294
716	384
750	474
784	564
818	654
851	744
885	834

1637919	·2142924
953	·2143014
987	104
1638021	194
055	284
089	374
123	464
156	553
190	643
224	733
258	823
292	913
326	·2144003
360	093
394	183
428	273
461	362
495	452
529	542
563	632
597	722
631	812
665	902
699	991
733	·2145081
767	171
801	261
835	350
869	440
903	530
937	620
971	710
1639005	799
039	889
072	979
106	·2146069
140	158
174	248
208	338
242	428
276	518
310	608
344	698
377	787
411	877
445	967
479	·2147057
513	147
547	237
581	326
615	416
649	506
682	595
716	685
750	775
784	864
818	954
852	·2148044
886	134
919	223
953	313
987	403

Excesses

10 = 27	1 = 3
20 = 53	2 = 5
30 = 80	3 = 8
	4 = 11
	5 = 13
	6 = 16
	7 = 19
	8 = 21
	9 = 24

17 = 45

Anti Log Excesses

10 = 4	1 = 0
20 = 8	2 = 1
30 = 11	3 = 1
40 = 15	4 = 2
50 = 19	5 = 2
60 = 23	6 = 2
70 = 26	7 = 3
80 = 30	8 = 3
90 = 34	9 = 3

45 = 17

1640021	·2148493
054	582
088	672
122	762
156	852
190	941
224	·2149031
258	121
291	210
325	300
359	390
393	479
427	569
461	659
495	748
529	838
563	927
596	·2150017
630	106
664	196
698	285
732	375
766	465
800	554
834	644
868	734
901	823
935	913
969	·2151003
1641003	092
037	182
071	272
105	361
138	451
172	540
206	630
240	719

1641273	·2151809
307	898
341	988
375	·2152078
409	167
443	257
477	346
510	436
544	525
578	615
612	704
646	794
680	884
714	973
748	·2153063
782	152
815	242
849	331
883	421
917	510
951	600
985	689
1642019	779
052	868
086	958
120	·2154047
154	137
187	226
221	316
255	405
289	494
323	584
356	673
390	763
424	852
458	942
491	·2155031
525	121
559	210
593	299
627	389
661	478
695	568
728	657
762	747
796	836
830	926
864	·2156015
898	104
932	194
965	283
999	372
1643033	462
067	551
100	640
134	730
168	819
202	908
236	998
269	·2157087
303	176
337	266
371	355
404	444
438	534

1643472	·2157623
506	712
540	802
574	891
608	980
641	·2158070
675	159
709	248
743	338
777	427
811	516
845	605
878	695
912	784
946	873
980	962
1644013	·2159052
047	141
081	230
115	319
149	409
182	498
216	587
250	677
284	766
317	855
351	945
385	·2160034
419	123
453	212
486	301
520	390
554	480
588	569
621	658
655	747
689	836
723	925
757	·2161014
790	104
824	193
858	282
892	371
925	461
959	550
993	639
1645027	728
061	817
094	906
128	995
162	·2162085
196	174
229	263
263	352
297	441
331	530
365	619
398	708
432	797
466	887
500	976
533	·2163065
567	154
601	243
635	332

1645669	·2163421
702	510
736	599
770	689
804	778
837	867
871	956
905	·2164045
939	134
972	223
1646006	312
040	401
073	491
107	580
141	669
174	758
208	847
242	936
276	·2165025
309	114
343	203
377	292
411	381
444	470
478	559
512	648
546	737
580	826
613	915
647	·2166004
681	093
715	182
748	271
782	360
816	449
850	538
883	627
917	716
951	805
984	893
1647018	982
052	·2167071
085	160
119	249
153	338
187	427
220	516
254	605
288	694
322	783
355	872
389	961
423	·2168050

Excesses

17 = 45

Anti Log Excesses

45 = 17

1647461	·2168150
499	250
537	350
574	450
612	550
650	650
688	750
726	850
764	950
802	·2169050
840	150
878	250
916	349
954	449
992	549
1648030	649
068	749
106	849
144	949
181	·2170049
219	149
257	249
295	349
333	449
371	549
409	649
447	749
484	849
522	948
560	·2171048
598	148
636	248
674	348
712	448
750	547
788	647
826	747
864	847
902	946
940	·2172046
978	146
1649016	246
054	346
091	446
129	545
167	645
205	745
243	845
281	945
319	·2173045
357	144
394	244
432	344
470	444
508	543
546	643
584	743

1649622	·2173843
660	942
697	·2174042
735	142
773	242
811	341
849	441
887	541
925	640
963	740

Excesses

10 = 26   1 = 3
20 = 53   2 = 5
30 = 79   3 = 8
4 = 11
5 = 13
6 = 16
7 = 18
8 = 21
9 = 24

19 = 50

Anti Log Excesses

10 = 4   1 = 0
20 = 8   2 = 1
30 = 11   3 = 1
40 = 15   4 = 2
50 = 19   5 = 2
60 = 23   6 = 2
70 = 27   7 = 3
80 = 30   8 = 3
90 = 34   9 = 3

50 = 19

1650000	·2174839
038	939
076	·2175039
114	138
152	238
190	338
228	438
266	537
303	637
341	737
379	837
417	936
455	·2176036
493	136
531	235
569	335
606	434
644	534
682	634
720	733
758	833
796	933
834	·2177032
872	132
909	231
947	331

1650985	·2177430
1651023	530
061	629
099	729
137	828
175	928
212	·2178027
250	127
288	227
326	326
364	426
402	526
440	625
477	725
515	824
553	924
591	·2179023
628	123
666	222
704	322
742	421
780	521
817	620
855	720
893	819
931	919
969	·2180018
1652007	118
045	217
083	317
120	416
158	516
196	615
234	715
272	814
310	913
347	·2181013
385	112
423	212
461	311
498	410
536	510
574	609
612	708
650	808
688	907
725	·2182007
763	106
801	205
839	305
877	404
915	503
953	603
990	702
1653028	801
066	900
103	·2183000
141	099
179	198
217	297
255	397
293	496
330	596
368	695
406	794

1653444	·2183894
482	993
520	·2184092
557	192
595	291
633	390
671	489
708	589
746	688
784	787
822	886
859	986
897	·2185085
935	184
973	283
1654010	383
048	482
086	581
124	680
162	779
200	878
237	977
275	·2186077
313	176
351	275
389	374
427	473
464	572
502	671
540	770
578	870
615	969
653	·2187068
691	167
729	266
766	365
804	464
842	563
880	663
917	762
955	861
993	960
1655031	·2188059
068	158
106	257
144	356
182	456
219	555
257	654
295	753
333	852
370	951
408	·2189050
446	149
484	248
521	347
559	446
597	545
635	644
672	743
710	842
748	941
786	·2190040
823	139
861	238

1655899	·2190337
937	436
974	535
1656012	634
050	733
088	832
125	931
163	·2191030
201	129
239	228
276	327
314	426
352	524
390	623
427	722
465	821
503	920
541	·2192019
578	118
616	217
654	316
692	415
729	514
767	613
805	712
843	811
880	910
918	·2193009
955	107
993	206
1657031	305
068	404
106	503
144	602
181	701
219	799
257	898
295	997
332	·2194096
370	194
408	293
446	392
483	491
521	590
559	688
597	787
634	886
672	985
710	·2195084
748	183
785	282
823	380
860	479
898	578
936	677
973	775
1658011	874
049	973
086	·2196071
124	170
162	268
200	367
237	466
275	564
313	663

1658351	·2196762
388	861
426	959
463	·2197058
501	157
539	256
576	354
614	453
652	552
689	650
727	749
765	847
803	946
840	·2198045
878	143
916	242
954	341
991	439
1659029	538
066	636
104	735
142	834
179	932
217	·2199031
255	130
292	228
330	327
368	425
406	524
443	623
481	721
519	820
557	919
594	·2200017
632	116
669	214
707	313
745	411
782	510
820	608
858	707
895	805
933	904
970	·2201002

Excesses

10 = 26	1 = 3
20 = 52	2 = 5
30 = 79	3 = 8
	4 = 10
	5 = 13
	6 = 16
	7 = 18
	8 = 21
	9 = 24

19 = 50

Anti Log Excesses

10 = 4	1 = 0
20 = 8	2 = 1
30 = 11	3 = 1
40 = 15	4 = 2
50 = 19	5 = 2
60 = 23	6 = 2
70 = 27	7 = 3
80 = 30	8 = 3
90 = 34	9 = 3

50 = 19

1660008	·2201101
046	199
083	298
121	396
159	495
196	593
234	692
271	790
309	889
347	987
384	·2202086
422	184
460	282
497	381
535	479
572	577
610	676
648	774
685	873
723	971
761	·2203069
798	168
836	266
873	364
911	463
949	561
986	660
1661024	758
062	856
099	955
137	·2204053
174	151
212	250
250	348
287	447
325	545
363	643

1661400	·2204742
438	840
475	938
513	·2205037
551	135
588	234
626	332
664	430
701	529
739	627
776	725
814	823
852	922
889	·2206020
927	118
965	216
1662002	315
040	413
077	511
115	609
153	708
190	806
228	904
266	·2207002
303	101
341	199
378	297
416	395
454	494
491	592
529	690
566	788
604	887
641	985
679	·2208083
716	181
754	280
791	378
829	476
867	574
904	672
942	770
979	868
1663017	967
055	·2209065
092	163
130	261
168	359
205	457
243	555
280	653
318	752
356	850
393	948
431	·2210046
468	144
506	242
543	340
581	438
618	536
656	634
693	732
731	830
769	928
806	·2211026

1663844	·2211124
881	222
919	321
957	419
994	517
1664032	615
069	713
107	811
144	909
182	·2212007
219	105
257	203
294	301
332	399
370	497
407	595
445	693
482	790
520	888
558	986
595	·2213084
633	182
670	280
708	378
745	476
783	574
820	672
858	770
895	868
933	966
970	·2214064
1665008	162
045	260
083	357
120	455
158	553
195	651
233	749
270	847
308	945
345	·2215043
383	140
420	238
458	336
495	434
533	532
571	630
608	728
646	825
683	923
721	·2216021
759	119
796	216
834	314
871	412
909	510
946	608
984	705
1666021	803
059	901
096	999
134	·2217097
171	195
209	293
246	390

1666284	·2217488
321	586
359	684
396	781
434	879
471	977
509	·2218075
546	172
584	270
621	368
659	466
696	563
734	661
771	759
809	857
846	954
884	·2219052
921	150
959	248
996	345
1667034	443
071	541
109	638
146	736
184	833
221	931
259	·2220029
296	126
334	224
371	322
409	419
446	517
484	614
521	712
559	810
596	907
634	·2221005
671	103
709	200
746	298
783	395
821	493
858	591
896	688
933	786
970	884
1668008	981
045	·2222079
083	176
120	274
158	371
195	469
233	566
270	664
308	761
345	859
383	956
420	·2223054
458	151
495	249
533	346
570	444
608	541
645	639
682	736

1668720	·2223834
757	931
795	·2224029
832	126
869	224
907	321
944	419
982	516
1669019	614
057	711
094	809
132	906
169	·2225003
207	101
244	198
282	295
319	393
357	490
394	588
432	685
469	782
507	880
544	977
581	·2226074
619	172
656	269
694	367
731	464
768	561
806	659
843	756
880	853
918	951
955	·2227048
993	146

Excesses

10 = 26	1 = 3
20 = 52	2 = 5
30 = 78	3 = 8
	4 = 10
	5 = 13
	6 = 16
	7 = 18
	8 = 21
	9 = 23

19 = 49

Anti Log Excesses

10 = 4	1 = 0
20 = 8	2 = 1
30 = 12	3 = 1
40 = 15	4 = 2
50 = 19	5 = 2
60 = 23	6 = 2
70 = 27	7 = 3
80 = 31	8 = 3
90 = 34	9 = 3

49 = 19

1670030	·2227243

1670067	·2227340
105	438
142	535
180	632
217	729
255	827
292	924
330	·2228021
367	118
405	216
442	313
479	410
517	507
554	605
592	702
629	799
666	896
704	994
741	·2229091
778	188
816	285
853	383
891	480
928	577
965	674
1671003	772
040	869
078	966
115	·2230063
153	161
190	258
228	355
265	452
303	549
340	646
377	743
415	841
452	938
490	·2231035
527	132
564	229
602	326
639	423
676	520
714	618
751	715
789	812
826	909
863	·2232006
901	103
938	200
975	297
1672013	395
050	492
088	589
125	686
162	783
200	880
237	977
274	·2233074
312	171
349	268
387	365
424	462
461	559

1672499	·2233656
536	753
573	850
611	947
648	·2234044
686	141
723	238
760	335
798	432
835	529
872	626
910	723
947	820
985	917
1673022	·2235014
059	111
097	208
134	305
171	402
209	499
246	596
284	693
321	790
358	887
395	984
433	·2236081
470	178
507	275
544	372
582	469
619	566
656	663
694	760
731	857
768	953
806	·2237050
843	147
881	244
918	341
955	438
993	535
1674030	631
067	728
105	825
142	922
180	·2238018
217	115
254	212
291	309
329	406
366	502
403	599
440	696
478	793
515	890
552	987
590	·2239084
627	180
664	277
702	374
739	471
777	567
814	664
851	761
888	858

1674926	·2239954
963	·2240051
1675000	148
037	245
075	341
112	438
149	535
187	632
224	728
261	825
299	922
336	·2241019
374	115
411	212
448	309
485	405
523	502
560	598
597	695
634	792
672	888
709	985
746	·2242082
784	179
821	275
858	372
896	469
933	566
971	662
1676008	759
045	856
082	952
120	·2243049
157	145
194	242
231	338
269	435
306	531
343	628
380	724
418	821
455	917
492	·2244014
529	111
567	207
604	304
641	401
678	497
716	594
753	690
790	787
827	884
865	980
902	·2245077
939	174
977	270
1677014	367
051	463
089	560
126	656
164	753
201	849
238	946
275	·2246042
313	139

1677350	·2246235
387	332
424	428
462	525
499	621
536	717
573	814
611	910
648	·2247006
685	103
722	199
760	296
797	392
834	488
871	585
909	681
946	777
983	874
1678020	970
058	·2248067
095	163
132	259
169	356
207	452
244	548
281	645
318	741
356	838
393	934
430	·2249030
467	127
504	223
541	319
579	416
616	512
653	609
690	705
727	801
764	898
802	994
839	·2250090
876	186
913	283
951	379
988	475
1679025	571
062	668
100	764
137	860
174	957
211	·2251053
249	150
286	246
323	342
360	439
398	535
435	631
472	727
509	824
547	920
584	·2252016
621	112
658	208
695	304
732	400

1679770	·2252497
807	593
844	689
881	785
918	881
955	977
993	·2253073

Excesses

10 = 26	1 = 3
20 = 52	2 = 5
30 = 78	3 = 8
	4 = 10
	5 = 13
	6 = 16
	7 = 18
	8 = 21
	9 = 23

19 = 48

Anti Log Excesses

10 = 4	1 = 0
20 = 8	2 = 1
30 = 12	3 = 1
40 = 15	4 = 2
50 = 19	5 = 2
60 = 23	6 = 2
70 = 27	7 = 3
80 = 31	8 = 3
90 = 35	9 = 3

48 = 19

1680030	·2253169
067	266
104	362
142	458
179	554
216	650
253	746
291	842
328	938
365	·2254035
402	131
440	227
477	323
514	419
551	515
588	611
625	707
663	804
700	900
737	996
774	·2255092
811	188
848	284
886	380
923	476
960	573
997	669
1681035	765
072	861

1681109	·2255957
146	·2256053
183	149
220	245
258	341
295	437
332	533
369	629
406	725
443	821
480	917
517	·2257013
555	109
592	205
629	301
666	397
703	493
740	589
778	685
815	781
852	877
889	973
927	·2258069
964	165
1682001	261
038	357
075	453
112	548
150	644
187	740
224	836
261	932
298	·2259028
335	124
372	220
409	315
447	411
484	507
521	603
558	699
595	795
632	891
669	987
706	·2260082
744	178
781	274
818	370
855	466
892	562
929	658
966	754
1683003	849
041	945
078	·2261041
115	137
152	233
189	329
226	425
263	520
300	616
338	712
375	808
412	903
449	999
486	·2262095

1683523	·2262191
560	286
597	382
635	478
672	574
709	669
746	765
783	861
820	957
857	·2263052
894	148
932	244
969	340
1684006	435
043	531
080	627
117	723
154	818
191	914
229	·2264010
266	106
303	201
340	297
377	393
414	488
451	584
488	679
526	775
563	871
600	966
637	·2265062
674	158
711	253
748	349
785	444
823	540
860	636
897	731
934	827
971	923
1685008	·2266018
045	114
082	209
119	305
156	401
193	496
230	592
267	688
304	783
341	879
378	974
416	·2267070
453	165
490	261
527	356
564	452
601	547
638	643
675	738
713	834
750	930
787	·2268025
824	121
861	216
898	312

1685935	·2268407
972	502
1686009	598
046	693
083	789
120	884
157	980
194	·2269075
231	171
268	266
306	362
343	457
380	553
417	648
454	743
491	839
528	934
565	·2270029
602	125
639	220
676	316
713	411
750	507
787	602
824	698
861	793
899	889
936	984
973	·2271080
1687010	175
047	270
084	366
121	461
158	556
195	651
232	747
269	842
306	937
343	·2272032
380	128
417	223
454	318
491	414
528	509
565	605
602	700
639	795
676	891
713	986
750	·2273081
788	176
825	272
862	367
899	462
936	557
973	653
1688010	748
047	843
084	938
121	·2274034
158	129
195	224
232	319
269	415
306	510

1688343	·2274605
380	700
417	796
454	891
491	986
528	·2275081
565	177
602	272
639	367
676	462
713	558
750	653
787	748
824	843
861	938
898	·2276033
935	128
972	224
1689009	319
046	414
083	509
120	604
157	699
194	794
231	889
268	985
305	·2277080
342	175
379	270
416	365
453	460
490	555
527	650
564	746
601	841
638	936
675	·2278031
712	126
749	221
786	316
823	411
860	506
897	601
934	696
971	791

Excesses

10 = 26	1 = 3
20 = 52	2 = 5
30 = 77	3 = 8
	4 = 10
	5 = 13
	6 = 15
	7 = 18
	8 = 21
	9 = 23

19 = 48

Anti Log Excesses

10 = 4	1 = 0
20 = 8	2 = 1
30 = 12	3 = 1
40 = 16	4 = 2
50 = 19	5 = 2
60 = 23	6 = 2
70 = 27	7 = 3
80 = 31	8 = 3
90 = 35	9 = 3

48 = 19

1690008	·2278886
045	981
082	·2279076
119	171
156	266
193	361
230	456
267	551
304	646
341	741
378	836
415	931
452	·2280026
489	121
526	216
563	311
600	406
637	501
674	596
710	691
747	786
784	881
821	976
858	·2281071
895	166
932	261
969	356
1691006	450
043	545
080	640
117	735
154	830
191	925
228	·2282020
265	115
302	209
339	304

1691376	·2282399
413	494
450	589
487	684
524	779
561	874
597	968
634	·2283063
671	158
708	253
745	348
782	443
819	538
856	633
893	727
930	822
967	917
1692004	·2284012
041	107
078	202
115	297
152	391
188	486
225	581
262	676
299	770
336	865
373	960
410	·2285055
447	149
484	244
521	339
558	434
595	528
632	623
669	718
706	813
743	907
779	·2286002
816	097
853	192
890	286
927	381
964	476
1693001	571
038	665
074	760
111	855
148	950
185	·2287044
222	139
259	234
296	328
333	423
370	517
407	612
444	707
481	801
518	896
555	991
592	·2288085
629	180
665	274
702	369
739	464

1693776	·2288558
813	653
850	748
887	842
924	937
960	·2289031
997	126
1694034	221
071	315
108	410
145	505
182	599
219	694
255	788
292	883
329	977
366	·2290072
403	166
440	261
477	355
514	450
550	544
587	639
624	733
661	828
698	922
735	·2291017
772	111
809	206
845	300
882	395
919	489
956	584
993	678
1695030	773
067	867
104	962
140	·2292056
177	151
214	245
251	340
288	434
325	528
362	623
399	717
435	811
472	906
509	·2293000
546	095
583	189
620	284
657	378
694	473
730	567
767	662
804	756
841	851
878	945
915	·2294039
952	134
989	228
1696025	322
062	417
099	511
136	606

1696173	·2294700
210	794
246	889
283	983
320	·2295077
357	171
393	266
430	360
467	454
504	548
541	643
578	737
614	831
651	925
688	·2296020
725	114
762	208
799	302
836	397
873	491
909	585
946	680
983	774
1697020	869
057	963
094	·2297057
130	151
167	245
204	339
241	434
277	528
314	622
351	716
388	810
425	905
462	999
498	·2298093
535	187
572	282
609	376
646	470
683	564
719	658
756	752
793	846
830	941
866	·2299035
903	129
940	223
977	317
1698014	411
051	505
087	599
124	694
161	788
198	882
235	976
272	·2300070
308	164
345	258
382	352
419	447
455	541
492	635
529	729

1698566	·2300823
603	917
640	·2301011
676	105
713	200
750	294
787	388
824	482
861	576
897	670
934	764
971	858
1699008	952
044	·2302046
081	140
118	234
155	328
191	422
228	516
265	610
302	704
338	798
375	892
412	986
449	·2303080
485	174
522	268
559	362
596	456
632	550
669	644
706	738
743	832
779	926
816	·2304020
853	113
890	207
926	301
963	395

Excesses

10 = 26	1 = 3
20 = 51	2 = 5
30 = 77	3 = 8
	4 = 10
	5 = 13
	6 = 15
	7 = 18
	8 = 20
	9 = 23

18 = 47

Anti Log Excesses

10 = 4	1 = 0
20 = 8	2 = 1
30 = 12	3 = 1
40 = 16	4 = 2
50 = 20	5 = 2
60 = 23	6 = 2
70 = 27	7 = 3
80 = 31	8 = 3
90 = 35	9 = 4

47 = 18

1700000	·2304489
037	583
074	677
111	771
147	865
184	959
221	·2305053
258	147
295	241
332	335
368	429
405	523
442	616
479	710
515	804
552	898
589	992
626	·2306086
662	180
699	273
736	367
773	461
809	555
846	648
883	742
920	836
956	930
993	·2307024
1701030	117
067	211
103	305
140	399
177	493
214	587
250	681
287	774
323	868

1701360	·2307962
397	·2308056
433	149
470	243
507	337
543	431
580	524
617	618
654	712
690	806
727	899
764	993
801	·2309087
837	181
874	274
911	368
948	462
984	556
1702021	649
058	743
095	837
131	930
168	·2310024
205	117
242	211
278	305
315	398
352	492
389	586
425	679
462	773
499	866
536	960
572	·2311054
609	147
646	241
683	335
719	428
756	522
792	615
829	709
866	803
902	896
939	990
976	·2312084
1703012	177
049	271
086	364
123	458
159	552
196	645
233	739
270	833
306	926
343	·2313020
379	113
416	207
453	300
489	394
526	487
563	581
599	674
636	768
673	861
710	955

1703746	·2314048
783	142
820	235
857	329
893	422
930	516
966	609
1704003	703
040	796
076	890
113	983
150	·2315077
186	170
223	264
260	357
297	451
333	544
370	638
407	731
444	824
480	918
517	·2316011
553	104
590	198
627	291
663	385
700	478
737	571
773	665
810	758
846	851
883	945
920	·2317038
956	132
993	225
1705030	318
066	412
103	505
139	598
176	692
213	785
249	879
286	972
323	·2318065
359	159
396	252
433	345
470	439
506	532
543	626
580	719
617	812
653	906
690	999
726	·2319092
763	185
800	279
836	372
873	465
910	558
946	652
983	745
1706019	838
056	931
093	·2320025

1706129	·2320118
166	211
203	304
239	398
276	491
312	584
349	677
386	771
422	864
459	957
496	·2321050
532	143
569	236
605	329
642	423
679	516
715	609
752	702
789	795
825	889
862	982
898	·2322075
935	168
972	262
1707008	355
045	448
082	541
118	634
155	727
191	820
228	914
265	·2323007
301	100
338	193
374	286
411	379
447	472
484	565
520	659
557	752
593	845
630	938
667	·2324031
703	124
740	217
776	310
813	403
850	496
886	589
923	682
960	775
996	868
1708033	961
069	·2325054
106	148
143	241
179	334
216	427
253	520
289	613
326	706
362	799
399	892
436	985
472	·2326078

1708509	·2326171
545	264
582	357
618	450
655	542
691	635
728	728
764	821
801	914
838	·2327007
874	100
911	193
947	286
984	379
1709021	472
057	565
094	658
130	751
167	844
203	937
240	·2328029
276	122
313	215
349	308
386	401
423	494
459	587
496	680
532	772
569	865
606	958
642	·2329051
679	144
715	237
752	330
788	423
825	515
861	608
898	701
934	794
971	887

Excesses

10 = 25	1 = 3
20 = 51	2 = 5
30 = 76	3 = 8
	4 = 10
	5 = 13
	6 = 15
	7 = 18
	8 = 20
	9 = 23

18 = 47

Anti Log Excesses

10 = 4	1 = 0
20 = 8	2 = 1
30 = 12	3 = 1
40 = 16	4 = 2
50 = 20	5 = 2
60 = 24	6 = 2
70 = 27	7 = 3
80 = 31	8 = 3
90 = 35	9 = 4

47 = 18

1710008	·2329980
044	·2330073
081	166
117	258
154	351
191	444
227	537
264	630
300	723
337	816
373	908
410	·2331001
446	094
483	187
519	279
556	372
592	465
629	558
665	650
702	743
738	836
775	929
811	·2332021
848	114
884	207
921	300
957	392
994	485
1711030	578
067	671
103	763
140	856
177	949
213	·2333041
250	134
286	226
323	319

1711360	·2333412
396	504
433	597
469	690
506	782
542	875
579	967
615	·2334060
652	153
688	245
725	338
761	431
798	523
834	616
871	708
907	801
944	894
980	986
1712017	·2335079
053	172
090	264
126	357
163	449
199	542
236	635
272	727
309	820
345	913
382	·2336005
418	098
455	190
491	283
528	375
564	468
601	560
637	653
674	745
710	838
747	930
783	·2337023
820	116
856	208
893	301
929	394
966	486
1713002	579
039	671
075	764
112	856
148	949
185	·2338041
221	133
258	226
294	318
330	410
367	503
403	595
440	688
476	780
512	873
549	965
585	·2339058
622	150
658	243
695	335

1713731	·2339428
768	520
804	612
841	705
877	797
914	889
950	982
987	·2340074
1714023	167
060	259
096	351
133	444
169	536
206	628
242	721
279	813
315	906
352	998
388	·2341090
425	183
461	275
497	367
534	460
570	552
607	645
643	737
679	829
716	922
752	·2342014
789	106
825	198
862	291
898	383
935	475
971	567
1715008	660
044	752
080	844
117	936
153	·2343029
190	121
226	213
262	305
299	398
335	490
372	582
408	674
445	767
481	859
518	951
554	·2344043
591	136
627	228
663	320
700	412
736	505
773	597
809	689
845	781
882	874
918	966
955	·2345058
991	150
1716028	243
064	335

1716101	·2345427
137	519
174	611
210	703
246	795
283	888
319	980
356	·2346072
392	164
428	256
465	348
501	440
537	532
574	625
610	717
647	809
683	901
719	993
756	·2347085
792	177
829	269
865	361
902	453
938	545
975	637
1717011	729
048	821
084	913
120	·2348005
157	098
193	190
230	282
266	374
302	466
339	558
375	650
411	742
448	834
484	926
521	·2349018
557	110
593	202
630	294
666	386
702	478
739	570
775	662
812	754
848	846
884	938
921	·2350030
957	122
993	214
1718030	306
066	398
103	490
139	582
175	674
212	766
248	858
284	949
321	·2351041
357	133
394	225
430	317

1718466	·2351409
503	501
539	593
575	684
612	776
648	868
685	960
721	·2352052
757	144
794	236
830	328
866	419
903	511
939	603
976	695
1719012	787
048	879
085	971
121	·2353063
157	154
194	246
230	338
267	430
303	522
339	614
375	706
412	798
448	890
484	981
520	·2354073
557	165
593	257
629	349
666	441
702	532
738	624
775	716
811	808
848	899
884	991
920	·2355083
957	175
993	266

Excesses

10 = 25	1 = 3
20 = 51	2 = 5
30 = 76	3 = 8
	4 = 10
	5 = 13
	6 = 15
	7 = 18
	8 = 20
	9 = 23

18 = 46

Anti Log Excesses

10 = 4	1 = 0
20 = 8	2 = 1
30 = 12	3 = 1
40 = 16	4 = 2
50 = 20	5 = 2
60 = 24	6 = 2
70 = 28	7 = 3
80 = 32	8 = 3
90 = 36	9 = 4

46 = 18

1720029	·2355358
066	450
102	542
139	633
175	725
211	817
248	909
284	·2356000
320	092
357	184
393	276
430	367
466	459
502	551
538	642
575	734
611	825
647	917
683	·2357009
720	100
756	192
792	284
829	375
865	467
901	558
938	650
974	742
1721011	833
047	925
083	·2358017
119	108
156	200
192	291
228	383
264	475
301	566
337	658

1721373	·2358750
410	841
446	933
482	·2359024
519	116
555	208
592	299
628	391
664	483
700	574
737	666
773	757
809	849
845	940
882	·2360032
918	123
954	215
990	306
1722027	398
063	489
099	581
135	673
172	764
208	856
244	948
281	·2361039
317	131
353	222
390	314
426	405
463	497
499	588
535	679
571	771
608	862
644	953
680	·2362045
716	136
753	228
789	319
825	411
861	502
898	594
934	685
970	777
1723006	868
043	960
079	·2363051
115	143
151	234
188	326
224	417
260	509
296	600
333	692
369	783
405	874
441	966
478	·2364057
514	148
550	240
586	331
623	423
659	514
695	605

1723731	·2364697
768	788
804	879
840	970
876	·2365062
913	153
949	244
985	335
1724021	427
058	518
094	609
130	701
166	792
203	884
239	975
275	·2366066
311	158
348	249
384	340
420	431
456	523
493	614
529	705
565	796
601	888
638	979
674	·2367070
710	161
746	253
783	344
819	435
855	526
891	618
928	709
964	800
1725000	891
036	983
073	·2368074
109	165
145	256
181	347
218	438
254	529
290	621
326	712
363	803
399	894
435	985
471	·2369077
508	168
544	259
580	350
616	442
653	533
689	624
725	715
761	806
797	897
833	988
870	·2370080
906	171
942	262
978	353
1726014	444
050	535

1726087	·2370626
123	717
159	808
195	899
232	990
268	·2371081
304	172
340	263
377	354
413	445
449	537
485	628
522	719
558	810
594	901
630	992
666	·2372083
702	174
739	265
775	356
811	447
847	538
883	629
919	720
956	811
992	902
1727028	993
064	·2373084
101	175
137	266
173	357
209	448
245	539
281	630
318	721
354	812
390	903
426	994
462	·2374085
498	176
535	267
571	358
607	449
643	540
680	631
716	722
752	813
788	904
824	995
860	·2375085
897	176
933	267
969	358
1728005	449
041	540
077	631
113	722
149	812
186	903
222	994
258	·2376085
294	176
330	267
366	358
403	449

1728439	·2376539
475	630
511	721
548	812
584	903
620	994
656	·2377085
692	176
728	266
765	357
801	448
837	539
873	630
909	721
945	812
981	902
1729017	993
054	·2378084
090	175
126	265
162	356
198	447
234	538
270	628
306	719
343	810
379	901
415	991
451	·2379082
487	173
523	264
559	354
595	445
632	536
668	627
704	717
740	808
776	899
812	990
848	·2380080
884	171
921	262
957	353
993	443

Excesses

10 = 25	1 = 3
20 = 50	2 = 5
30 = 76	3 = 8
	4 = 10
	5 = 13
	6 = 15
	7 = 18
	8 = 20
	9 = 23
18 = 46	

Anti Log Excesses

10 = 4	1 = 0
20 = 8	2 = 1
30 = 12	3 = 1
40 = 16	4 = 2
50 = 20	5 = 2
60 = 24	6 = 2
70 = 28	7 = 3
80 = 32	8 = 3
90 = 36	9 = 4
46 = 18	

1730029	·2380534
065	625
101	715
137	806
173	896
210	987
246	·2381078
282	168
318	259
354	350
390	440
426	531
462	621
499	712
535	803
571	893
607	984
643	·2382075
679	165
715	256
751	346
788	437
824	528
860	618
896	709
932	800
968	890
1731004	981
040	·2383071
077	162
113	253
149	343
185	434
221	525
257	615
293	706
329	796

1731366	·2383887
402	977
438	·2384068
474	158
510	249
546	339
582	430
618	520
655	611
691	702
727	792
763	883
799	973
835	·2385064
871	154
907	244
943	335
979	425
1732015	516
051	606
087	697
123	787
159	878
195	968
232	·2386059
268	149
304	240
340	330
376	421
412	511
448	602
484	692
521	783
557	873
593	964
629	·2387054
665	144
701	235
737	325
773	415
809	506
845	596
881	687
917	777
953	867
989	958
1733025	·2388048
061	138
098	229
134	319
170	410
206	500
242	590
278	681
314	771
350	861
386	951
422	·2389042
458	132
494	222
530	312
566	403
602	493
638	583
674	674

1733710	·2389764
746	855
782	945
818	·2390035
854	126
890	216
926	306
963	396
999	487
1734035	577
071	667
107	757
143	848
179	938
215	·2391028
251	118
287	209
323	299
359	389
395	479
431	570
467	660
503	750
539	840
575	931
611	·2392021
647	111
683	201
719	292
755	382
791	472
828	562
864	653
900	743
936	833
972	923
1735008	·2393013
044	103
080	193
116	284
152	374
188	464
224	554
260	644
296	734
332	824
368	914
404	·2394005
440	095
476	185
512	275
548	365
584	455
620	545
656	635
692	726
728	816
764	906
800	996
836	·2395086
872	176
908	266
944	356
980	446
1736016	536

1736052	·2395626
088	716
124	806
160	896
196	986
232	·2396076
268	167
304	257
340	347
376	437
412	527
448	617
484	707
520	797
556	887
592	977
628	·2397067
664	157
700	247
736	337
772	427
808	517
844	607
880	697
916	787
952	877
988	967
1737024	·2398057
060	147
096	236
132	326
168	416
204	506
240	596
276	686
312	776
348	866
383	956
419	·2399046
455	136
491	226
527	316
563	406
599	496
635	586
671	675
707	765
743	855
779	945
815	·2400035
851	125
887	215
923	305
959	394
995	484
1738031	574
067	664
103	754
139	844
175	934
211	·2401023
247	113
283	203
319	293
355	382

1738391	·2401472
427	562
463	652
499	742
534	831
570	921
606	·2402011
642	101
678	191
714	281
750	371
786	460
822	550
858	640
894	730
930	819
966	909
1739002	999
038	·2403089
074	178
109	268
145	358
181	448
217	537
253	627
289	717
325	807
361	896
397	986
433	·2404076
469	166
505	255
541	345
577	435
613	524
649	614
684	703
720	793
756	883
792	972
828	·2405062
864	152
900	242
936	331
971	421

Excesses

10 = 25	1 = 3
20 = 50	2 = 5
30 = 75	3 = 8
	4 = 10
	5 = 13
	6 = 15
	7 = 18
	8 = 20
	9 = 23

18 = 45

Anti Log Excesses

10 = 4	1 = 0
20 = 8	2 = 1
30 = 12	3 = 1
40 = 16	4 = 2
50 = 20	5 = 2
60 = 24	6 = 2
70 = 28	7 = 3
80 = 32	8 = 3
90 = 36	9 = 4

45 = 18

1740007	·2405511
043	601
079	690
115	780
151	870
187	959
223	·2406049
259	138
295	228
331	318
367	407
403	497
439	587
475	676
511	766
546	855
582	945
618	·2407034
654	124
690	213
726	303
762	392
798	482
833	571
869	661
905	751
941	840
977	930
1741013	·2408020
049	109
085	199
120	288
156	378
192	467
228	557
264	646
300	736

1741336	·2408825
372	915
408	·2409004
444	094
480	183
516	273
552	362
588	452
624	541
660	631
695	720
731	810
767	899
803	989
839	·2410078
875	168
911	257
947	347
982	436
1742018	526
054	615
090	705
126	794
162	883
198	973
234	·2411062
269	151
305	241
341	330
377	420
413	509
449	598
485	688
521	777
556	866
592	956
628	·2412045
664	135
700	224
736	313
771	403
807	492
843	581
879	671
914	760
950	850
986	939
1743022	·2413028
058	118
094	207
129	296
165	386
201	475
237	565
273	654
309	743
345	833
381	922
416	·2414011
452	100
488	190
524	279
560	368
596	457
632	547

1743668	·2414636
703	725
739	814
775	904
811	993
847	·2415082
883	171
919	261
955	350
990	439
1744026	528
062	618
098	707
134	796
170	885
205	975
241	·2416064
277	153
313	242
348	332
384	421
420	510
456	599
492	688
528	777
563	866
599	956
635	·2417045
671	134
707	223
743	312
778	402
814	491
850	580
886	669
921	759
957	848
993	937
1745029	·2418026
065	115
101	204
136	293
172	383
208	472
244	561
280	650
316	739
351	828
387	917
423	·2419006
459	095
494	184
530	273
566	362
602	451
638	540
674	629
709	718
745	808
781	897
817	986
853	·2420075
889	164
924	253
960	342

1745996	·2420431
1746032	520
067	609
103	698
139	787
175	876
210	965
246	·2421054
282	143
318	232
353	321
389	410
425	499
461	588
497	677
533	766
568	855
604	944
640	·2422033
676	122
712	211
748	300
783	389
819	478
855	567
891	656
926	745
962	834
998	923
1747034	·2423012
069	101
105	190
141	278
177	367
212	456
248	545
284	634
320	723
355	812
391	901
427	989
463	·2424078
498	167
534	256
570	345
606	434
641	523
677	612
713	700
749	789
784	878
820	967
856	·2425056
892	145
927	234
963	323
999	411
1748035	500
070	589
106	678
142	767
178	856
213	945
249	·2426033
285	122

1748321	·2426211
356	300
392	388
428	477
464	566
499	655
535	744
571	832
607	921
642	·2427010
678	099
714	188
750	277
785	366
821	454
857	543
893	632
928	721
964	809
1749000	898
036	987
071	·2428075
107	164
143	252
179	341
214	430
250	518
286	607
322	696
357	785
393	873
429	962
465	·2429051
500	140
536	228
572	317
608	406
643	494
679	583
715	671
751	760
786	849
822	937
858	·2430026
894	115
929	203
965	292

Excesses

10 = 25	1 = 2
20 = 50	2 = 5
30 = 75	3 = 7
	4 = 10
	5 = 12
	6 = 15
	7 = 17
	8 = 20
	9 = 22

18 = 45

Anti Log Excesses

10 = 4	1 = 0
20 = 8	2 = 1
30 = 12	3 = 1
40 = 16	4 = 2
50 = 20	5 = 2
60 = 24	6 = 2
70 = 28	7 = 3
80 = 32	8 = 3
90 = 36	9 = 4

45 = 18

1750000	·2430380
036	469
072	558
107	646
143	735
179	824
214	912
250	·2431001
286	089
322	178
357	267
393	355
429	444
465	533
500	621
536	710
572	798
608	887
643	976
679	·2432064
715	153
751	242
786	330
822	419
857	507
893	596
929	684
964	773
1751000	861
036	950
071	·2433038
107	127
143	215
179	304
214	392
250	481
286	569

1751322	·2433658
357	746
393	835
428	923
464	·2434012
500	100
535	189
571	277
607	366
642	454
678	543
714	631
750	720
785	808
821	897
857	985
893	·2435073
928	162
964	250
999	338
1752035	427
071	515
106	604
142	692
178	780
213	869
249	957
284	·2436045
320	134
356	222
391	311
427	399
463	487
498	576
534	664
570	752
606	841
641	929
677	·2437018
713	106
749	194
784	283
820	371
855	459
891	548
927	636
962	725
998	813
1753034	901
069	990
105	·2438078
140	166
176	254
212	343
247	431
283	519
319	607
354	696
390	784
425	872
461	961
497	·2439049
532	138
568	226
604	314

1753639	·2439403
675	491
710	579
746	667
782	756
817	844
853	932
889	·2440020
924	109
960	197
995	285
1754031	373
067	462
102	550
138	638
174	726
209	814
245	902
280	990
316	·2441079
352	167
387	255
423	343
459	431
494	520
530	608
565	696
601	784
637	873
672	961
708	·2442049
744	137
779	225
815	313
850	401
886	490
922	578
957	666
993	754
1755029	842
064	930
100	·2443018
135	106
171	195
207	283
242	371
278	459
314	547
349	635
385	723
420	811
456	899
492	987
527	·2444075
563	163
599	251
634	339
670	427
705	515
741	604
777	692
812	780
848	868
884	956
919	·2445044

1755955	·2445132
990	220
1756026	308
062	396
097	484
133	572
168	660
204	748
239	836
275	924
310	·2446012
346	100
381	188
417	276
453	364
488	452
524	540
559	628
595	716
631	804
666	892
702	980
738	·2447068
773	156
809	244
844	331
880	419
916	507
951	595
987	683
1757022	771
058	859
093	947
129	·2448035
164	123
200	211
235	299
271	387
307	475
342	563
378	651
413	738
449	826
485	914
520	·2449002
556	090
591	178
627	266
662	354
698	441
733	529
769	617
804	705
840	793
875	881
911	969
946	·2450056
982	144
1758017	232
053	320
088	407
124	495
160	583
195	671
231	759

1758266	·2450846
302	934
338	·2451022
373	110
409	198
444	286
480	374
515	461
551	549
586	637
622	725
657	812
693	900
728	988
764	·2452076
799	163
835	251
870	339
906	427
941	514
977	602
1759013	690
048	777
084	865
119	952
155	·2453040
191	128
226	215
262	303
297	391
333	479
368	566
404	654
439	742
475	830
510	917
546	·2454005
581	093
617	180
652	268
688	355
723	443
759	531
794	618
830	706
865	794
901	881
936	969
972	·2455056

Excesses

10 = 25	1 = 2
20 = 49	2 = 5
30 = 74	3 = 7
	4 = 10
	5 = 12
	6 = 15
	7 = 17
	8 = 20
	9 = 22

18 = 44

Anti Log Excesses

10 = 4	1 = 0
20 = 8	2 = 1
30 = 12	3 = 1
40 = 16	4 = 2
50 = 20	5 = 2
60 = 24	6 = 2
70 = 28	7 = 3
80 = 32	8 = 3
90 = 36	9 = 4

44 = 18

1760007	·2455144
043	232
078	319
114	407
149	495
185	582
220	670
256	757
291	845
327	933
362	·2456020
398	108
433	196
469	283
504	371
540	458
575	546
611	634
646	721
682	809

Excesses

18 = 44

Anti Log Excesses

44 = 18

1760723	·2456909
763	·2457009
804	109
844	209
885	309
925	409
966	509
1761007	609
1761047	·2457709
088	809
128	909
169	·2458009
209	109
250	209
291	309
331	409
372	509
412	609
453	709
493	809
534	909
575	·2459009
615	109
656	209
696	309
737	409
777	509
818	609
859	709
899	809
940	909
980	·2460008
1762021	108
061	208
102	308
142	408
183	508
223	608
264	708
304	808
345	908
385	·2461008
426	108
466	208
507	308
547	407
588	507
628	607
669	707
710	807
750	906
791	·2462006
831	106
872	206
912	305
953	405
993	505
1763034	605
074	705
115	804
155	904
196	·2463004
236	104
277	204
317	303
358	403
398	503
439	603
479	702
520	802
560	902
601	·2464001
641	101
1763682	·2464201
722	301
763	400
803	500
844	600
884	699
925	799
965	899
1764006	999
046	·2465098
087	198
127	298
168	397
208	497
249	597
289	697
330	796
370	896
410	996
451	·2466095
491	195
532	294
572	394
613	493
653	593
694	693
734	792
775	892
815	991
856	·2467091
896	190
937	290
977	390
1765018	489
058	589
099	688
139	788
180	887
220	987
260	·2468087
301	186
341	286
382	385
422	485
463	584
503	684
543	784
584	883
624	983
665	·2469082
705	182
746	281
786	381
827	480
867	580
908	679
948	779
989	878
1766029	978
070	·2470077
110	176
151	276
191	375
232	475
272	574
1766313	·2470674
353	773
393	872
434	972
474	·2471071
515	171
555	270
596	370
636	469
676	568
717	668
757	767
798	867
838	966
879	·2472066
919	165
959	264
1767000	364
040	463
081	562
121	661
162	761
202	860
242	959
283	·2473059
323	158
364	257
404	356
445	456
485	555
525	654
566	754
606	853
646	952
686	·2474051
727	151
767	250
807	349
848	449
888	548
929	647
969	746
1768010	846
050	945
090	·2475044
131	143
171	242
212	342
252	441
293	540
333	639
373	738
414	838
454	937
495	·2476036
535	135
576	235
616	334
656	433
697	532
737	631
777	731
817	830
858	929
898	·2477028
1768938	·2477127
979	226
1769019	325
060	425
100	524
141	623
181	722
221	821
262	920
302	·2478019
343	118
383	217
424	316
464	415
504	514
545	613
585	712
625	812
665	911
706	·2479010
746	109
786	208
827	307
867	406
908	505
948	604
989	703

Excesses

10 = 25	1 = 2
20 = 49	2 = 5
30 = 74	3 = 7
40 = 98	4 = 10
	5 = 12
	6 = 15
	7 = 17
	8 = 20
	9 = 22

20 = 50

Anti Log Excesses

10 = 4	1 = 0
20 = 8	2 = 1
30 = 12	3 = 1
40 = 16	4 = 2
50 = 20	5 = 2
60 = 24	6 = 2
70 = 28	7 = 3
80 = 32	8 = 3
90 = 37	9 = 4

50 = 20

1770029	·2479802
069	901
110	·2480000
150	099
190	198
230	297
271	396
311	495
351	594

1770392	·2480693
432	792
473	891
513	990
554	·2481089
594	188
634	287
675	386
715	485
755	584
795	683
836	782
876	881
916	980
957	·2482079
997	178
1771037	276
077	375
118	474
158	573
198	672
239	771
279	870
320	968
360	·2483067
401	166
441	265
481	364
522	463
562	562
602	660
642	759
683	858
723	957
763	·2484056
804	155
844	254
884	352
924	451
965	550
1772005	649
045	748
086	846
126	945
166	·2485044
206	143
247	241
287	340
327	439
368	537
408	636
448	735
488	834
529	932
569	·2486031
609	130
650	228
690	327
730	426
770	525
811	623
851	722
891	821
932	919
972	·2487018

1773012	·2487117
052	216
093	314
133	413
173	512
214	610
254	709
294	808
334	907
375	·2488005
415	104
455	203
496	301
536	400
576	498
616	597
657	695
697	794
737	893
778	991
818	·2489090
858	188
898	287
939	385
979	484
1774019	583
060	681
100	780
140	878
180	977
221	·2490075
261	174
301	273
342	371
382	470
422	568
462	667
503	765
543	864
583	962
623	·2491061
663	159
704	258
744	356
784	455
824	553
864	651
905	750
945	848
985	947
1775025	·2492045
066	144
106	242
146	340
187	439
227	537
267	636
307	734
348	833
388	931
428	·2493029
468	128
508	226
549	325
589	423

1775629	·2493522
669	620
709	718
750	817
790	915
830	·2494014
870	112
911	211
951	309
991	407
1776031	506
071	604
112	702
152	800
192	899
232	997
272	·2495095
313	194
353	292
393	390
433	488
474	587
514	685
554	783
594	882
634	980
675	·2496078
715	176
755	275
795	373
835	471
876	570
916	668
956	766
996	864
1777037	963
077	·2497061
117	159
157	257
197	355
238	454
278	552
318	650
358	748
398	846
438	945
478	·2498043
519	141
559	239
599	338
639	436
679	534
720	632
760	730
800	829
840	927
881	·2499025
921	123
961	221
1778001	319
041	417
082	516
122	614
162	712
202	810

1778242	·2499908
282	·2500006
322	104
363	202
403	300
443	398
483	496
523	594
563	692
603	790
644	888
684	986
724	·2501084
764	182
804	280
844	378
884	476
925	575
965	673
1779005	771
045	869
085	967
125	·2502065
165	163
206	261
246	359
286	457
326	555
366	653
406	751
446	849
487	946
527	·2503044
567	142
607	240
647	338
687	436
727	534
768	632
808	730
848	828
888	926
928	·2504024
968	122

Excesses

10 = 24	1 = 2
20 = 49	2 = 5
30 = 73	3 = 7
40 = 98	4 = 10
	5 = 12
	6 = 15
	7 = 17
	8 = 20
	9 = 22

20 = 49

Anti Log Excesses

10 = 4	1 = 0
20 = 8	2 = 1
30 = 12	3 = 1
40 = 16	4 = 2
50 = 20	5 = 2
60 = 25	6 = 2
70 = 29	7 = 3
80 = 33	8 = 3
90 = 37	9 = 4

49 = 20

1780008	·2504220
049	317
089	415
129	513
169	611
209	709
249	807
289	905
330	·2505002
370	100
410	198
450	296
490	394
530	492
570	590
611	687
651	785
691	883
731	981
771	·2506079
811	177
851	275
891	372
931	470
971	568
1781011	666
051	764
091	861
131	959
172	·2507057
212	155
252	252
292	350
332	448
372	546
412	644
453	741

1781493	·2507839	1784097	·2514183	1786697	·2520509	1789294	·2526816	1790730	·2530302
533	937	137	280	737	606	334	913	770	398
573	·2508035	177	378	777	704	374	·2527010	810	495
613	133	217	475	817	801	414	107	850	592
653	230	257	573	857	898	454	204	890	688
693	328	297	670	897	995	494	301	930	785
733	426	337	768	937	·2521092	534	398	970	882
773	524	377	865	977	189	573	494	1791010	979
813	621	417	962	1787017	286	613	591	050	·2531075
853	719	457	·2515060	057	384	653	688	090	172
893	817	497	157	097	481	693	785	130	269
933	914	537	255	137	578	733	882	170	365
973	·2509012	577	352	177	675	773	979	210	462
1782014	110	617	450	217	772	813	·2528076	249	559
054	208	657	547	257	869	853	172	289	656
094	305	697	644	297	966	893	269	329	752
134	403	737	742	337	·2522063	933	366	369	849
174	501	777	839	377	160	973	463	409	946
214	598	817	937	417	257			449	·2532042
254	696	857	·2516034	457	354	Excesses		489	139
295	793	897	132	497	451			528	235
335	891	937	229	537	548	10 = 24	1 = 2	568	332
375	988	977	326	577	645	20 = 49	2 = 5	608	428
415	·2510086	1785017	424	616	743	30 = 73	3 = 7	648	525
455	184	057	521	656	840	40 = 97	4 = 10	688	622
495	281	098	618	696	937		5 = 12	728	718
535	379	138	715	736	·2523034		6 = 15	768	815
575	476	178	813	776	131		7 = 17	807	912
615	574	218	910	816	228		8 = 19	847	·2533009
655	671	258	·2517007	856	325		9 = 22	887	105
695	769	298	105	896	422			927	202
735	867	338	202	936	519	20 = 49		967	299
775	964	378	299	976	616			1792007	395
815	·2511062	418	396	1788016	713	Anti Log Excesses		047	492
856	160	458	494	056	810			086	588
896	258	498	591	096	907	10 = 4	1 = 0	126	685
936	355	538	688	136	·2524004	20 = 8	2 = 1	166	781
976	453	578	786	176	101	30 = 12	3 = 1	206	878
1783016	551	618	883	216	198	40 = 16	4 = 2	246	975
056	648	658	980	256	295	50 = 21	5 = 2	286	·2534071
096	746	698	·2518077	296	392	60 = 25	6 = 2	326	168
136	843	738	175	336	489	70 = 29	7 = 3	365	264
176	941	778	272	376	586	80 = 33	8 = 3	405	361
216	·2512038	818	369	416	683	90 = 37	9 = 4	445	457
256	136	858	467	455	780			485	554
296	233	898	564	495	877	49 = 20		525	650
336	331	938	661	535	974			565	747
376	428	978	758	575	·2525071	1790013	·2528560	605	843
416	526	1786018	856	615	168	053	656	644	940
456	623	058	953	655	265	093	753	684	·2535036
496	721	098	·2519050	695	362	132	850	724	133
536	818	138	148	735	459	172	947	764	229
576	916	178	245	775	556	212	·2529043	804	326
616	·2513013	218	342	815	653	252	140	844	422
656	111	258	439	855	750	292	237	884	519
697	208	298	537	895	847	332	334	923	615
737	306	338	634	935	944	372	431	963	712
777	403	378	731	975	·2526041	411	527	1793003	808
817	501	418	829	1789014	137	451	624	043	905
857	598	458	926	054	234	491	721	083	·2536001
897	696	497	·2520023	094	331	531	818	122	098
937	793	537	120	134	428	571	915	162	194
977	891	577	218	174	525	611	·2530011	202	291
1784017	988	617	315	214	622	651	108	242	387
057	·2514086	657	412	254	719	690	205	281	484

1793321	·2536580
361	676
401	773
441	869
480	966
520	·2537062
560	159
600	255
640	351
680	448
720	544
759	641
799	737
839	834
879	930
919	·2538026
959	123
999	219
1794038	316
078	412
118	509
158	605
198	701
237	798
277	894
317	990
357	·2539086
396	183
436	279
476	375
516	472
556	568
595	665
635	761
675	858
715	954
755	·2540050
795	147
835	243
874	339
914	435
954	532
994	628
1795034	724
073	821
113	917
153	·2541013
193	109
232	206
272	302
312	398
352	494
392	590
431	687
471	783
511	879
551	975
591	·2542071
630	168
670	264
710	360
750	456
789	553
829	649
869	745

1795908	·2542841
948	937
988	·2543034
1796028	130
067	226
107	322
147	418
187	514
227	610
266	707
306	803
346	899
386	995
426	·2544091
465	187
505	283
545	380
585	476
624	572
664	668
704	764
743	860
783	956
823	·2545052
863	148
902	244
942	340
982	436
1797022	532
062	628
101	725
141	821
181	917
221	·2546013
261	109
300	205
340	301
380	397
420	493
459	589
499	685
539	781
578	877
618	973
658	·2547069
698	165
737	261
777	357
817	453
856	549
896	645
936	741
976	837
1798015	933
055	·2548029
095	125
134	221
174	317
214	412
254	508
293	604
333	700
373	796
412	892
452	988

1798492	·2549083
532	179
571	275
611	371
651	467
690	563
730	659
770	755
810	851
849	947
889	·2550043
929	139
968	235
1799008	331
048	426
088	522
127	618
167	714
207	810
246	905
286	·2551001
326	097
366	193
405	288
445	384
485	480
524	576
564	672
604	767
644	863
683	959
723	·2552055
763	151
802	246
842	342
881	438
921	534
960	629

Excesses

10 = 24	1 = 2
20 = 48	2 = 5
30 = 73	3 = 7
	4 = 10
	5 = 12
	6 = 15
	7 = 17
	8 = 19
	9 = 22

20 = 48

Anti Log Excesses

10 = 4	1 = 0
20 = 8	2 = 1
30 = 12	3 = 1
40 = 17	4 = 2
50 = 21	5 = 2
60 = 25	6 = 2
70 = 29	7 = 3
80 = 33	8 = 3
90 = 37	9 = 4

48 = 20

1800000	·2552725
040	821
079	916
119	·2553012
159	108
199	204
238	299
278	395
318	491
357	586
397	682
437	778
477	874
516	969
556	·2554065
596	161
635	256
675	352
715	448
755	544
794	639
834	735
874	831
913	926
953	·2555022
992	117
1801032	213
071	308
111	404
151	500
190	595
230	691
270	786
310	882
349	977
389	·2556073
429	169

1801468	·2556264
508	360
547	455
587	551
626	646
666	742
706	838
745	933
785	·2557029
825	124
865	220
904	315
944	411
984	507
1802023	602
063	698
102	793
142	889
181	984
221	·2558080
261	176
300	271
340	367
380	462
420	558
459	653
499	749
539	844
578	940
618	·2559035
657	131
697	226
736	322
776	417
816	512
855	608
895	703
934	799
974	894
1803013	990
053	·2560085
093	180
132	276
172	371
212	467
252	562
291	658
331	753
371	848
410	944
450	·2561039
489	134
529	229
568	325
608	420
648	515
687	611
727	706
766	802
806	897
845	993
885	·2562088
925	183
964	279
1804004	374

1804043	·2562469	1806615	·2568656	1809183	·2574825	1810604	·2578234	1813167	·2584376
083	564	655	751	223	919	643	329	206	470
122	660	695	846	262	·2575014	683	423	246	565
162	755	734	941	302	109	722	518	285	659
202	850	774	·2569036	341	204	762	612	324	753
241	946	813	131	381	298	801	707	364	848
281	·2563041	853	226	420	393	841	802	403	942
320	136	892	321	460	488	880	896	443	·2585036
360	231	932	416	499	583	920	991	482	130
399	327	971	511	539	678	959	·2579085	522	225
439	422	1807011	606	578	772	999	180	561	319
479	517	050	701	618	867	1811038	274	600	413
518	613	090	796	657	962	078	369	640	508
558	708	129	891	697	·2576057	117	464	679	602
598	803	169	986	736	152	157	558	718	696
638	898	208	·2570081	776	246	196	653	757	790
677	994	248	176	815	341	236	747	797	885
717	·2564089	287	271	855	436	275	842	836	979
757	184	327	366	894	531	315	936	875	·2586073
796	279	366	461	934	625	354	·2580031	915	168
836	374	406	556	973	720	393	126	954	262
875	470	445	651			433	220	994	356
915	565	485	746	Excesses		472	315	1814033	450
954	660	525	841			512	409	073	545
994	755	564	936	10 = 24	1 = 2	551	504	112	639
1805034	850	604	·2571031	20 = 48	2 = 5	591	598	151	733
073	946	643	125	30 = 72	3 = 7	630	693	191	828
113	·2565041	683	220		4 = 10	669	787	230	922
152	136	722	315		5 = 12	709	882	269	·2587016
192	231	762	410		6 = 14	748	976	308	110
231	327	801	505		7 = 17	788	·2581071	348	205
271	422	841	600		8 = 19	827	165	387	299
311	517	880	695		9 = 22	867	260	426	393
350	612	920	790			906	354	466	488
390	707	959	885	20 = 48		945	449	505	582
429	803	999	980			985	543	545	676
469	898	1808038	·2572075	Anti Log Excesses		1812024	638	584	770
508	993	078	170			064	732	624	865
548	·2566088	117	265	10 = 4	1 = 0	103	827	663	959
587	183	157	360	20 = 8	2 = 1	143	921	702	·2588053
627	278	196	454	30 = 12	3 = 1	182	·2582016	742	147
666	373	236	549	40 = 17	4 = 2	221	110	781	241
706	469	275	644	50 = 21	5 = 2	261	205	821	336
745	564	315	739	60 = 25	6 = 2	300	299	860	430
785	659	354	834	70 = 29	7 = 3	339	394	900	524
824	754	394	929	80 = 33	8 = 3	378	488	939	618
864	849	433	·2573024	90 = 37	9 = 4	418	583	978	712
903	944	473	118			457	677	1815018	806
943	·2567039	512	213	48 = 20		496	771	057	900
982	135	552	308			536	866	096	995
1806022	230	591	403	1810012	·2576815	575	960	135	·2589089
061	325	631	498	052	909	615	·2583055	175	183
101	420	670	593	091	·2577004	654	149	214	277
141	515	710	688	131	098	694	244	253	371
180	610	749	782	170	193	733	338	293	465
220	705	789	877	210	287	772	432	332	559
259	801	828	972	249	382	812	527	371	654
299	896	868	·2574067	288	477	851	621	410	748
338	991	907	162	328	571	891	716	450	842
378	·2568086	947	256	367	666	930	810	489	936
418	181	986	351	407	761	970	905	528	·2590030
457	276	1809026	446	446	856	1813009	999	568	124
497	371	065	541	486	950	048	·2584093	607	218
536	466	105	635	525	·2578045	088	188	647	313
576	561	144	730	564	140	127	282	686	407

1815726	·2590501
765	595
804	689
844	783
883	877
922	971
961	·2591065
1816001	159
040	253
079	347
119	441
158	535
197	630
236	724
276	818
315	912
354	·2592006
394	100
433	194
473	288
512	382
552	476
591	570
630	664
670	758
709	852
748	946
787	·2593040
827	134
866	228
905	322
945	416
984	510
1817023	603
062	697
102	791
141	885
180	979
220	·2594073
259	167
298	261
337	355
377	449
416	543
455	637
495	731
534	825
573	918
612	·2595012
652	106
691	200
730	294
770	388
809	482
848	575
887	669
927	763
966	857
1818005	951
045	·2596045
084	139
123	232
162	326
202	420
241	514

1818280	·2596608
320	702
359	796
398	889
437	983
477	·2597077
516	171
555	265
595	359
634	453
673	546
712	640
752	734
791	828
830	922
870	·2598015
909	109
948	203
987	297
1819027	390
066	484
105	578
145	671
184	765
223	859
262	953
302	·2599046
341	140
380	234
420	327
459	421
498	515
537	609
577	702
616	796
655	890
695	983
734	·2600077
773	171
812	265
852	358
891	452
930	546
969	639

Excesses

10 = 24	1 = 2
20 = 48	2 = 5
30 = 72	3 = 7
	4 = 10
	5 = 12
	6 = 14
	7 = 17
	8 = 19
	9 = 22

20 = 47

Anti Log Excesses

10 = 4	1 = 0
20 = 8	2 = 1
30 = 13	3 = 1
40 = 17	4 = 2
50 = 21	5 = 2
60 = 25	6 = 3
70 = 29	7 = 3
80 = 33	8 = 3
90 = 38	9 = 4

47 = 20

1820008	·2600733
048	826
087	920
126	·2601013
165	107
204	201
244	294
283	388
322	482
361	576
401	669
440	763
479	857
519	950
558	·2602044
597	137
636	231
676	324
715	418
754	512
793	605
832	699
872	792
911	886
950	979
989	·2603073
1821028	167
068	260
107	354
146	447
185	541
225	634
264	728
303	821
342	915
381	·2604008
421	102

1821460	·2604195
499	289
538	382
577	475
617	569
656	662
695	756
734	849
774	943
813	·2605036
852	130
891	223
930	317
970	410
1822009	504
048	597
087	691
126	784
166	878
205	971
244	·2606065
283	158
323	252
362	345
401	438
440	532
479	625
519	718
558	811
597	905
636	998
675	·2607091
714	185
753	278
793	372
832	465
871	559
910	652
949	745
989	839
1823028	932
067	·2608025
106	118
146	212
185	305
224	398
263	492
302	585
342	678
381	771
420	865
459	958
498	·2609051
537	145
576	238
616	331
655	424
694	518
733	611
772	704
811	798
850	891
890	984
929	·2610077
968	171

1824007	·2610264
046	357
085	451
124	544
164	637
203	730
242	824
281	917
320	·2611010
359	103
398	196
438	290
477	383
516	476
555	569
594	662
633	755
672	848
712	942
751	·2612035
790	128
829	221
868	314
907	407
946	500
986	594
1825025	687
064	780
103	873
142	966
181	·2613059
220	152
260	246
299	339
338	432
377	525
416	618
455	711
494	804
534	898
573	991
612	·2614084
651	177
690	270
729	363
768	456
808	549
847	642
886	735
925	828
964	921
1826003	·2615014
042	107
082	200
121	293
160	386
199	479
238	572
277	665
316	758
356	851
395	944
434	·2616037
473	130
512	223

1826551	·2616316
590	409
629	502
668	595
707	688
746	781
785	874
824	967
863	·2617060
903	153
942	246
981	339
1827020	432
059	525
098	618
137	711
177	803
216	896
255	989
294	·2618082
333	175
372	268
411	361
450	454
489	547
528	640
567	733
606	826
645	919
684	·2619012
724	104
763	197
802	290
841	383
880	476
919	568
958	661
997	754
1828036	847
075	939
114	·2620032
153	125
192	218
231	311
271	403
310	496
349	589
388	682
427	775
466	867
505	960
544	·2621053
583	146
622	238
661	331
700	424
739	517
778	610
818	702
857	795
896	888
935	981
974	·2622074
1829013	166
052	259

1829091	·2622352
130	445
169	537
208	630
247	723
286	815
325	908
364	·2623001
403	094
442	186
481	279
520	372
559	464
598	557
638	649
677	742
716	834
755	927
794	·2624020
833	112
872	205
911	298
950	391
989	483

Excesses

10 = 24	1 = 2
20 = 48	2 = 5
30 = 71	3 = 7
	4 = 10
	5 = 12
	6 = 14
	7 = 17
	8 = 19
	9 = 21

20 = 47

Anti Log Excesses

10 = 4	1 = 0
20 = 8	2 = 1
30 = 13	3 = 1
40 = 17	4 = 2
50 = 21	5 = 2
60 = 25	6 = 3
70 = 29	7 = 3
80 = 34	8 = 3
90 = 38	9 = 4

47 = 20

1830028	·2624576
067	669
106	761
145	854
184	946
223	·2625039
262	131
301	224
340	317
379	409
418	502
457	594

1830496	·2625687
535	779
574	872
613	965
652	·2626057
691	150
730	242
769	335
808	427
847	520
886	613
925	705
964	798
1831003	890
042	983
081	·2627075
120	168
159	260
198	353
237	445
276	538
315	630
354	723
393	815
432	908
471	·2628000
510	093
549	185
588	278
627	370
666	463
705	555
744	648
783	740
822	833
861	925
900	·2629018
939	110
978	202
1832017	295
056	387
095	479
134	571
173	664
212	756
251	848
290	941
329	·2630033
368	126
407	218
446	311
485	403
524	495
563	588
602	680
641	773
680	865
719	958
758	·2631050
797	142
836	235
875	327
914	419
953	511
992	604

1833031	·2631696
070	788
109	881
148	973
187	·2632066
226	158
265	251
304	343
343	435
382	527
421	619
459	712
498	804
537	896
576	988
615	·2633080
654	173
693	265
732	357
771	449
810	542
849	634
888	726
927	819
966	911
1834005	·2634003
044	095
083	188
122	280
161	372
200	464
239	556
277	649
316	741
355	833
394	925
433	·2635017
472	109
511	201
550	294
589	386
628	478
667	570
706	662
745	754
784	846
822	939
861	·2636031
900	123
939	215
978	307
1835017	399
056	491
095	584
134	676
173	768
212	860
251	952
290	·2637044
329	136
367	228
406	320
445	412
484	504
523	596

1835562	·2637688
601	780
639	873
678	965
717	·2638057
756	149
795	241
834	333
873	425
912	517
951	609
990	701
1836029	793
068	885
107	977
146	·2639069
184	161
223	253
262	345
301	437
340	529
379	621
418	713
456	805
495	897
534	989
573	·2640081
612	173
651	265
690	357
729	448
768	540
807	632
846	724
885	816
924	908
963	·2641000
1837001	092
040	184
079	276
118	368
157	460
196	552
235	644
273	735
312	827
351	919
390	·2642011
429	103
468	195
507	287
545	378
584	470
623	562
662	654
701	746
740	838
779	930
817	·2643021
856	113
895	205
934	297
973	389
1838012	481
051	573

1838089	·2643664
128	756
167	848
206	940
245	·2644032
284	124
323	216
361	307
400	399
439	491
478	583
517	675
556	766
595	858
633	950
672	·2645042
711	133
750	225
789	317
828	408
867	500
905	592
944	684
983	775
1839022	867
061	959
100	·2646050
139	142
177	234
216	326
255	417
294	509
333	601
372	692
411	784
449	876
488	968
527	·2647059
566	151
605	243
643	334
682	426
721	517
760	609
798	700
837	792
876	884
915	975
954	·2648067
992	159

Excesses

10 = 24	1 = 2
20 = 48	2 = 5
30 = 71	3 = 7
	4 = 9
	5 = 12
	6 = 14
	7 = 17
	8 = 19
	9 = 21

19 = 46

Anti Log Excesses

10 = 4	1 = 0
20 = 8	2 = 1
30 = 13	3 = 1
40 = 17	4 = 2
50 = 21	5 = 2
60 = 25	6 = 3
70 = 30	7 = 3
80 = 34	8 = 3
90 = 38	9 = 4

46 = 19

1840031	·2648251
070	342
109	434
148	526
187	617
226	709
264	800
303	892
342	983
381	·2649075
420	167
459	258
498	350
536	441
575	533
614	624
653	716
692	808
730	899
769	991
808	·2650082
847	174
885	265
924	357
963	448
1841002	540
041	631
079	723
118	814
157	906
196	997
235	·2651089
273	180
312	272
351	363
390	455
428	546

1841467	·2651638
506	729
545	821
584	912
622	·2652004
661	095
700	187
739	278
778	369
816	461
855	552
894	644
933	735
971	827
1842010	918
049	·2653009
088	101
127	192
165	284
204	375
243	467
282	558
321	649
359	741
398	832
437	924
476	·2654015
514	107
553	198
592	289
630	381
669	472
708	563
747	654
785	746
824	837
863	928
902	·2655020
941	111
979	203
1843018	294
057	386
096	477
135	568
173	660
212	751
251	842
290	933
328	·2656025
367	116
406	207
444	299
483	390
522	482
561	573
599	665
638	756
677	847
715	939
754	·2657030
793	121
832	212
870	304
909	395
948	486

1843987	·2657577
1844026	668
064	760
103	851
142	942
181	·2658033
220	124
258	216
297	307
336	398
375	489
413	581
452	672
491	763
529	854
568	945
607	·2659037
646	128
684	219
723	310
762	401
800	492
839	583
878	675
917	766
955	857
994	948
1845033	·2660039
071	130
110	221
149	313
188	404
226	495
265	586
304	677
342	768
381	859
420	950
459	·2661041
497	132
536	223
575	314
613	405
652	496
691	588
730	679
768	770
807	861
846	952
884	·2662043
923	134
961	225
1846000	316
038	407
077	498
116	589
154	680
193	771
232	862
271	953
309	·2663044
348	135
387	226
425	317
464	408

1846503	·2663499
542	590
580	681
619	772
658	863
696	954
735	·2664045
774	136
813	227
851	318
890	409
929	500
967	591
1847006	682
044	772
083	863
121	954
160	·2665045
199	136
237	227
276	318
315	408
354	499
392	590
431	681
470	772
508	863
547	954
586	·2666045
625	136
663	227
702	318
741	409
779	500
818	591
856	681
895	772
933	863
972	954
1848011	·2667045
049	135
088	226
127	317
166	408
204	498
243	589
282	680
320	771
359	862
397	952
436	·2668043
474	134
513	225
552	316
590	406
629	497
668	588
707	679
745	769
784	860
823	951
861	·2669042
900	133
938	223
977	314

Number	Log
1849015	·2669405
054	496
093	587
131	677
170	768
209	859
248	950
286	·2670040
325	131
364	222
402	312
441	403
479	493
518	584
556	674
595	765
634	856
672	946
711	·2671037
749	128
788	219
826	309
865	400
904	491
942	581
981	672

Excesses

10 = 24	1 = 2
20 = 47	2 = 5
30 = 71	3 = 7
	4 = 9
	5 = 12
	6 = 14
	7 = 16
	8 = 19
	9 = 21
19 = 46	

Anti Log Excesses

10 = 4	1 = 0
20 = 8	2 = 1
30 = 13	3 = 1
40 = 17	4 = 2
50 = 21	5 = 2
60 = 25	6 = 3
70 = 30	7 = 3
80 = 34	8 = 3
90 = 38	9 = 4
46 = 19	

Number	Log
1850020	·2671763
059	854
097	944
136	·2672035
175	126
213	216
252	307
290	397
329	488
367	578

Number	Log	Number	Log	Number	Log	Number	Log
1850406	·2672669	1852913	·2678550	1855417	·2684415	1857918	·2690264
445	760	952	640	456	505	956	354
483	850	990	731	494	595	994	444
522	941	1853029	821	533	685	1858033	534
560	·2673031	067	912	571	775	071	624
599	122	106	·2679002	610	865	110	713
637	212	145	092	648	955	148	803
676	303	183	183	687	·2685046	187	893
715	394	222	273	725	136	225	983
753	484	260	363	764	226	264	·2691073
792	575	299	453	802	316	302	163
830	665	337	544	841	406	341	253
869	756	376	634	879	496	379	342
907	846	415	724	918	586	418	432
946	937	453	815	956	676	456	522
985	·2674028	492	905	995	766	495	612
1851023	118	530	995	1856033	856	533	702
062	209	569	·2680085	072	946	572	791
100	299	607	176	110	·2686036	610	881
139	390	646	266	149	126	649	971
177	480	684	356	187	216	687	·2692061
216	571	723	447	226	307	726	150
255	661	761	537	264	397	764	240
293	752	800	627	303	487	802	330
332	842	838	717	341	577	841	420
370	933	877	808	379	667	879	510
409	·2675023	915	898	418	757	918	599
447	114	954	988	456	847	956	689
486	204	992	·2681078	495	937	995	779
525	295	1854031	168	533	·2687027	1859033	869
563	385	069	259	572	117	071	959
602	476	108	349	610	207	110	·2693048
640	566	146	439	649	297	148	138
679	657	185	529	687	387	186	228
717	747	224	619	726	477	224	318
756	838	262	710	764	566	263	407
795	928	301	800	803	656	301	497
833	·2676019	339	890	841	746	339	587
872	109	378	980	880	836	378	676
910	200	416	·2682071	918	926	416	766
949	290	455	161	957	·2688016	455	856
987	381	493	251	995	106	493	946
1852026	471	532	341	1857034	196	532	·2694035
065	561	570	431	072	286	570	125
103	652	609	522	111	376	608	215
142	742	647	612	149	466	647	304
180	833	686	702	187	556	685	394
219	923	724	792	226	646	724	484
257	·2677014	763	882	264	736	762	574
296	104	801	973	303	826	801	663
335	194	840	·2683063	341	916	839	753
373	285	878	153	380	·2689006	877	843
412	375	917	243	418	096	916	932
450	466	955	334	456	186	954	·2695022
489	556	994	424	495	276	993	112
527	647	1855032	514	533	366		
566	737	071	604	572	455		
605	827	109	694	610	545		
643	918	148	784	649	635		
682	·2678008	186	874	687	725		
720	098	225	964	725	815		
759	188	263	·2684054	764	905		
797	279	302	144	802	995		
836	369	340	234	841	·2690084		
875	459	379	324	879	174		

Excesses

10 = 23	1 = 2
20 = 47	2 = 5
30 = 70	3 = 7
	4 = 9
	5 = 12
	6 = 14
	7 = 16
	8 = 19
	9 = 21

19 = 45

Anti Log Excesses

10 = 4	1 = 0
20 = 9	2 = 1
30 = 13	3 = 1
40 = 17	4 = 2
50 = 21	5 = 2
60 = 26	6 = 3
70 = 30	7 = 3
80 = 34	8 = 3
90 = 38	9 = 4

45 = 19

1860031	·2695202
070	291
108	381
146	471
185	560
223	650
262	739
300	829
339	918
377	·2696008
415	098
454	187
492	277
531	366
569	456
608	545
646	635
684	725
723	814
761	904
799	994
837	·2697084
876	173
914	263
952	353
991	442
1861029	532
068	621
106	711
145	800
183	890
221	979
260	·2698069
298	158
337	248
375	337
414	427

1861452	·2698516
490	606
529	695
567	785
605	874
643	964
682	·2699053
720	143
758	232
797	322
835	411
874	501
912	590
951	680
989	769
1862027	858
066	948
104	·2700037
142	127
180	216
219	306
257	395
295	485
334	574
372	664
411	753
449	843
488	932
526	·2701022
564	111
603	201
641	290
679	379
717	468
756	558
794	647
832	736
871	826
909	915
947	·2702005
985	094
1863024	184
062	273
100	362
139	452
177	541
216	631
254	720
293	810
331	899
369	988
408	·2703078
446	167
484	256
522	345
561	435
599	524
637	613
676	703
714	792
752	881
790	970
829	·2704060
867	149
905	238

1863944	·2704328
982	417
1864021	506
059	595
098	685
136	774
174	863
213	953
251	·2705042
289	131
327	220
366	310
404	399
442	488
481	578
519	667
557	756
595	845
634	935
672	·2706024
710	113
749	202
787	291
825	381
863	470
902	559
940	648
978	737
1865017	826
055	915
093	·2707005
131	094
170	183
208	272
246	361
285	450
323	539
361	629
399	718
438	807
476	896
514	985
553	·2708074
591	163
629	253
667	342
706	431
744	520
782	609
821	698
859	787
897	877
935	966
974	·2709055
1866012	144
050	233
089	322
127	411
165	500
203	589
242	678
280	767
318	856
357	945
395	·2710034

1866433	·2710124
471	213
510	302
548	391
586	480
625	569
663	658
701	747
739	836
778	925
816	·2711014
854	103
893	192
931	281
969	370
1867007	459
046	548
084	637
122	726
161	815
199	904
237	993
275	·2712082
314	171
352	260
390	349
428	438
466	527
505	615
543	704
581	793
619	882
657	971
696	·2713060
734	149
772	238
810	327
849	416
887	505
925	594
964	683
1868002	772
040	860
078	949
117	·2714038
155	127
193	216
231	305
269	394
308	482
346	571
384	660
422	749
460	838
499	927
537	·2715016
575	104
613	193
652	282
690	371
728	460
766	549
804	638
843	726
881	815

1868919	·2715904
957	993
995	·2716082
1869034	171
072	260
110	348
148	437
187	526
225	615
263	704
301	792
339	881
378	970
416	·2717059
454	147
492	236
530	325
569	413
607	502
645	591
683	680
722	768
760	857
798	946
836	·2718034
874	123
913	212
951	301
989	389

Excesses

10 = 23	1 = 2
20 = 47	2 = 5
30 = 70	3 = 7
	4 = 9
	5 = 12
	6 = 14
	7 = 16
	8 = 19
	9 = 21

19 = 45

Anti Log Excesses

10 = 4	1 = 0
20 = 9	2 = 1
30 = 13	3 = 1
40 = 17	4 = 2
50 = 21	5 = 2
60 = 26	6 = 3
70 = 30	7 = 3
80 = 34	8 = 3
90 = 39	9 = 4

45 = 19

1870027	·2718478
065	567
104	655
142	744
180	833
218	922
257	·2719010

1870295	·2719099	1872775	·2724856	1875253	·2730597	1877727	·2736324
333	188	813	944	291	685	765	412
371	276	852	·2725032	329	774	803	500
409	365	890	121	367	862	841	588
448	454	928	210	405	950	879	676
486	543	966	298	443	·2731038	917	763
524	631	1873004	386	481	126	955	851
562	720	042	475	520	215	993	939
600	809	080	563	558	303	1878031	·2737027
638	897	119	652	596	391	069	115
676	986	157	740	634	479	107	203
715	·2720074	195	829	672	567	145	291
753	163	233	917	710	655	184	379
791	251	271	·2726005	748	743	222	467
829	340	309	094	786	832	260	555
867	429	347	182	824	920	298	643
905	517	386	271	862	·2732008	336	731
943	606	424	359	900	096	374	819
982	695	462	448	938	184	412	907
1871020	784	500	536	976	272	450	994
058	872	538	624	1876014	360	488	·2738082
096	961	576	713	053	449	526	170
134	·2721050	614	801	091	537	564	258
173	138	653	889	129	625	602	346
211	227	691	977	167	713	640	434
249	315	729	·2727066	205	801	678	522
287	404	767	154	243	889	716	610
326	492	805	242	281	977	754	698
364	581	843	331	319	·2733066	792	786
402	670	881	419	357	154	830	874
440	758	920	508	395	242	868	962
478	847	958	596	433	330	906	·2739050
517	935	996	685	471	418	944	138
555	·2722024	1874034	773	509	506	982	225
593	112	072	861	547	594	1879020	313
631	201	110	950	586	683	058	401
669	290	148	·2728038	624	771	096	489
707	378	186	126	662	859	134	577
745	467	224	214	700	947	172	665
784	555	262	303	738	·2734035	210	753
822	644	300	391	776	123	248	840
860	732	338	479	814	211	286	928
898	821	376	568	852	299	324	·2740016
936	909	414	656	890	387	362	104
974	998	453	744	928	475	400	192
1872012	·2723086	491	832	966	563	438	279
051	175	529	921	1877004	651	476	367
089	263	567	·2729009	042	739	514	455
127	352	605	097	080	827	552	543
165	440	643	186	119	916	590	630
203	529	681	274	157	·2735004	628	718
241	617	720	362	195	092	666	806
279	706	758	450	233	180	704	894
318	794	796	539	271	268	742	982
356	883	834	627	309	356	780	·2741069
394	971	872	715	347	444	818	157
432	·2724060	910	803	385	532	856	245
470	148	948	891	423	620	894	333
508	237	986	980	461	708	932	421
546	325	1875024	·2730068	499	796	970	508
585	414	062	156	537	884		
623	502	100	244	575	972		
661	591	138	332	613	·2736060		
699	679	176	421	651	148		
737	767	214	509	689	236		

Excesses

10 = 23	1 = 2
20 = 46	2 = 5
30 = 69	3 = 7
	4 = 9
	5 = 12
	6 = 14
	7 = 16
	8 = 19
	9 = 21

19 = 44

Anti Log Excesses

10 = 4	1 = 0
20 = 9	2 = 1
30 = 13	3 = 1
40 = 17	4 = 2
50 = 22	5 = 2
60 = 26	6 = 3
70 = 30	7 = 3
80 = 35	8 = 3
90 = 39	9 = 4

44 = 19

1880008	·2741596
046	684
084	772
122	859
160	947
198	·2742035
236	122
274	210
312	298
350	386
388	473
426	561
464	649
502	736
540	824
578	912
616	·2743000
654	087
692	175
730	263
768	350
806	438
844	526
882	614
920	701
958	789
996	877
1881034	964
072	·2744052
110	140
148	228
186	315
224	403
262	491
300	578
338	666
375	753

1881413	·2744841
451	928
489	·2745016
527	104
565	191
603	279
641	367
679	455
717	542
755	630
793	718
831	805
869	893
907	980
945	·2746068
983	155
1882021	243
059	331
097	418
135	506
172	593
210	681
248	768
286	856
324	944
362	·2747031
400	119
438	206
476	294
514	381
552	469
590	556
628	644
666	731
704	819
742	906
780	994
818	·2748081
856	169
894	256
932	344
969	431
1883007	519
045	606
083	694
121	781
159	869
197	956
235	·2749044
273	131
311	219
349	306
387	393
425	481
463	568
500	656
538	743
576	831
614	918
652	·2750005
690	093
728	180
765	268
803	355
841	443

1883879	·2750530
917	617
955	705
993	792
1884031	880
069	967
107	·2751055
145	142
183	229
221	317
259	404
296	491
334	578
372	666
410	753
448	840
486	928
524	·2752015
562	103
600	190
638	278
676	365
714	452
752	540
790	627
827	714
865	801
903	889
941	976
979	·2753063
1885017	151
055	238
092	325
130	412
168	500
206	587
244	674
282	762
320	849
357	936
395	·2754023
433	111
471	198
509	285
547	373
585	460
622	547
660	634
698	722
736	809
774	896
812	984
850	·2755071
888	158
926	245
964	333
1886002	420
040	507
078	594
116	681
153	769
191	856
229	943
267	·2756030
305	117

1886343	·2756204
381	291
418	379
456	466
494	553
532	640
570	727
608	814
646	901
683	989
721	·2757076
759	163
797	250
835	337
873	424
911	511
948	599
986	686
1887024	773
062	860
100	947
138	·2758034
176	121
213	209
251	296
289	383
327	470
365	557
403	644
441	731
478	818
516	905
554	992
592	·2759079
630	166
667	253
705	340
743	428
781	515
818	602
856	689
894	776
932	863
970	950
1888007	·2760037
045	124
083	211
121	298
159	385
197	472
235	559
272	646
310	733
348	820
386	907
424	994
462	·2761081
500	168
537	255
575	342
613	429
651	516
689	603
727	690
765	777

1888802	·2761863
840	950
878	·2762037
916	124
954	211
991	298
1889029	385
067	472
105	559
142	646
180	733
218	820
256	907
294	994
331	·2763080
369	167
407	254
445	341
483	428
520	515
558	602
596	688
634	775
671	862
709	949
747	·2764036
785	123
823	210
860	296
898	383
936	470
974	557

Excesses

10 = 23	1 = 2
20 = 46	2 = 5
30 = 69	3 = 7
	4 = 9
	5 = 12
	6 = 14
	7 = 16
	8 = 18
	9 = 21

19 = 44

Anti Log Excesses

10 = 4	1 = 0
20 = 9	2 = 1
30 = 13	3 = 1
40 = 17	4 = 2
50 = 22	5 = 2
60 = 26	6 = 3
70 = 30	7 = 3
80 = 35	8 = 3
90 = 39	9 = 4

44 = 19

1890012	·2764644
050	731
088	818
125	904

1890163	·2764991
201	·2765078
239	165
277	252
314	339
352	426
390	512
428	599
465	686
503	773
541	860
579	946
617	·2766033
654	120
692	207
730	293
768	380
806	467
843	554
881	641
919	727
957	814
994	901
1891032	988
070	·2767075
107	161
145	248
183	335
221	422
258	508
296	595
334	682
372	768
410	855
447	942
485	·2768029
523	115
561	202
599	289
636	375
674	462
712	549
750	636
787	722
825	809
863	896
900	982
938	·2769069
976	155
1892014	242
051	328
089	415
127	502
165	588
203	675
240	762
278	849
316	935
354	·2770022
392	109
429	195
467	282
505	368
543	455
580	541

1892618	·2770628
656	715
693	801
731	888
769	974
807	·2771061
844	147
882	234
920	321
957	407
995	494
1893033	580
071	667
108	753
146	840
184	927
221	·2772013
259	100
297	186
335	273
372	359
410	446
448	533
485	619
523	706
561	792
599	879
636	965
674	·2773052
712	138
749	225
787	311
825	398
863	484
900	571
938	657
976	743
1894013	830
051	916
089	·2774003
127	089
164	176
202	262
240	349
277	435
315	522
353	608
391	695
428	781
466	868
504	954
541	·2775041
579	127
617	213
655	299
692	386
730	472
768	558
805	645
843	731
881	818
919	904
956	991
994	·2776077
1895032	163

1895069	·2776250
107	336
145	423
183	509
220	596
258	682
296	768
333	855
371	941
409	·2777027
447	113
484	200
522	286
560	372
597	459
635	545
672	631
710	717
747	804
785	890
823	976
860	·2778063
898	149
936	236
974	322
1896011	409
049	495
087	581
124	668
162	754
200	840
238	926
275	·2779013
313	099
351	185
388	271
426	357
463	444
501	530
538	616
576	702
614	788
651	875
689	961
727	·2780047
765	133
802	220
840	306
878	392
915	478
953	564
991	651
1897029	737
066	823
104	909
142	995
179	·2781082
217	168
254	254
292	340
329	427
367	513
405	599
442	685
480	771

1897518	·2781858
556	944
593	·2782030
631	116
669	202
706	288
744	374
781	461
819	547
856	633
894	719
932	805
969	891
1898007	977
045	·2783063
083	149
120	235
158	321
196	407
233	493
271	579
308	666
346	752
383	838
421	924
459	·2784010
496	096
534	182
571	268
609	354
646	440
684	526
722	612
759	698
797	784
835	871
873	957
910	·2785043
948	129
986	215
1899023	301
061	387
098	473
136	559
173	645
211	731
249	817
286	903
324	989
361	·2786075
399	161
436	247
474	333
512	419
549	505
587	591
624	677
662	763
699	849
737	935
775	·2787021
812	107
850	193
887	278
925	364

1899962	·2787450

Excesses

10 = 23	1 = 2
20 = 46	2 = 5
30 = 69	3 = 7
	4 = 9
	5 = 11
	6 = 14
	7 = 16
	8 = 18
	9 = 21

19 = 43

Anti Log Excesses

10 = 4	1 = 0
20 = 9	2 = 1
30 = 13	3 = 1
40 = 17	4 = 2
50 = 22	5 = 2
60 = 26	6 = 3
70 = 31	7 = 3
80 = 35	8 = 3
90 = 39	9 = 4

43 = 19

1900000	·2787536
038	622
075	708
113	794
151	879
189	965
226	·2788051
264	137
302	223
339	309
377	395
414	481
452	567
489	653
527	739
565	825
602	911
640	997
677	·2789082
715	168
752	254
790	340
828	426
865	512
903	598
940	683
978	769
1901015	855
053	941
091	·2790027
128	113
166	199
203	284
241	370
278	456

1901316	·2790542
354	628
391	713
429	799
466	885
504	971
541	·2791056
579	142
617	228
654	314
692	400
729	485
767	571
804	657
842	743

Excesses

19 = 43

Anti Log Excesses

43 = 19

1901886	·2791843
930	943
973	·2792043
1902017	143
061	243
105	343
149	443
192	543
236	643
280	743
323	843
367	943
411	·2793043
455	143
498	243
542	343
586	443
630	543
674	643
718	743
761	843
805	943
849	·2794043
893	143
937	243
981	343
1903024	443
068	543
112	643
156	743
200	843
243	943
287	·2795042
331	142
374	242
418	342
462	442
506	542
549	641
593	741
637	841

Number	Log
1903681	·2795941
725	·2796041
769	141
812	240
856	340
900	440
944	540
988	640
1904031	740
075	839
119	939
162	·2797039
206	139
250	239
294	339
337	438
381	538
425	638
469	738
513	838
556	938
600	·2798037
644	137
687	237
731	337
775	437
819	536
862	636
906	736
950	835
994	935
1905038	·2799035
081	134
125	234
169	334
212	433
256	533
300	633
344	732
387	832
431	932
475	·2800031
519	131
563	231
606	330
650	430
694	530
737	629
781	729
825	829
869	928
912	·2801028
956	128
1906000	227
044	327
088	427
131	526
175	626
219	726
262	825
306	925
350	·2802025
393	124
437	224
481	323

Number	Log
1906524	·2802423
568	522
612	622
655	721
699	821
743	921
786	·2803020
830	120
874	220
918	319
961	419
1907005	518
049	618
093	717
137	817
180	916
224	·2804016
268	115
311	215
355	314
399	413
442	513
486	612
530	711
573	811
617	910
661	·2805010
704	109
748	209
792	308
835	408
879	507
923	607
966	706
1908010	806
054	905
097	·2806005
141	104
185	203
228	303
272	402
316	501
359	601
403	700
447	799
490	899
534	998
578	·2807097
621	197
665	296
709	395
752	495
796	594
840	693
883	793
927	892
971	991
1909014	·2808091
058	190
102	289
145	389
189	488
233	587
276	686
320	786

Number	Log
1909364	·2808885
407	984
451	·2809083
495	182
538	282
582	381
626	480
669	580
713	679
756	778
800	877
843	977
887	·2810076
930	175
974	274

Excesses

10 = 23	1 = 2
20 = 46	2 = 5
30 = 68	3 = 7
40 = 91	4 = 9
	5 = 11
	6 = 14
	7 = 16
	8 = 18
	9 = 21

22 = 50

Anti Log Excesses

10 = 4	1 = 0
20 = 9	2 = 1
30 = 13	3 = 1
40 = 18	4 = 2
50 = 22	5 = 2
60 = 26	6 = 3
70 = 31	7 = 3
80 = 35	8 = 4
90 = 40	9 = 4

50 = 22

Number	Log
1910018	·2810373
061	472
105	572
149	671
192	770
236	869
280	968
323	·2811067
367	167
411	266
454	365
498	464
542	563
585	662
629	762
673	861
716	960
760	·2812059
803	158
847	257
890	357

Number	Log
1910934	·2812456
977	555
1911021	654
065	753
108	852
152	951
196	·2813050
239	149
283	248
326	347
370	446
413	546
457	645
500	744
544	843
588	942
631	·2814041
675	140
719	239
762	338
806	437
849	536
893	635
936	734
980	833
1912023	932
067	·2815031
111	130
154	229
198	328
242	427
285	526
329	625
372	724
416	823
459	921
503	·2816020
546	119
590	218
634	317
677	416
721	515
765	614
808	713
852	812
895	911
939	·2817010
982	108
1913026	207
069	306
113	405
156	504
200	603
243	702
287	801
330	900
374	999
418	·2818098
461	197
505	295
549	394
592	493
636	592
679	691
723	789

Number	Log
1913766	·2818888
810	987
853	·2819085
897	184
940	283
984	382
1914027	480
071	579
114	678
158	777
201	876
245	975
288	·2820073
332	172
375	271
419	370
463	469
506	567
550	666
594	765
637	863
681	962
724	·2821061
768	159
811	258
855	357
898	455
942	554
985	653
1915029	751
072	850
116	949
159	·2822047
203	146
246	245
290	343
333	442
377	541
420	639
464	738
507	837
551	935
594	·2823034
638	133
681	231
725	330
768	429
812	527
855	626
899	725
942	823
986	922
1916029	·2824021
073	119
116	218
160	316
203	415
247	513
290	612
334	710
377	809
421	908
464	·2825006
508	105
551	204

1916595	·2825302
638	401
682	499
725	598
769	696
812	795
856	893
899	992
942	·2826090
986	189
1917029	287
072	385
116	484
159	582
203	680
246	779
290	877
333	976
377	·2827074
420	173
464	271
507	370
551	468
594	566
638	665
681	763
725	861
768	960
812	·2828058
855	157
899	255
942	354
985	452
1918029	551
072	649
115	747
159	846
202	944
246	·2829042
289	141
333	239
376	337
420	436
463	534
507	632
550	731
594	829
637	927
681	·2830026
724	124
767	222
811	321
854	419
897	517
941	615
984	714
1919028	812
071	910
115	·2831008
158	106
202	205
245	303
288	401
332	500
375	598

1919418	·2831696
462	795
505	893
549	991
592	·2832090
636	188
679	286
723	384
766	483
809	581
853	679
896	777
939	875
983	973

Excesses

10 = 23	1 = 2
20 = 45	2 = 5
30 = 68	3 = 7
40 = 91	4 = 9
	5 = 11
	6 = 14
	7 = 16
	8 = 18
	9 = 20

22 = 49

Anti Log Excesses

10 = 4	1 = 0
20 = 9	2 = 1
30 = 13	3 = 1
40 = 18	4 = 2
50 = 22	5 = 2
60 = 26	6 = 3
70 = 31	7 = 3
80 = 35	8 = 4
90 = 40	9 = 4

49 = 22

1920026	·2833072
070	170
113	268
157	366
200	464
244	562
287	661
330	759
374	857
417	955
460	·2834053
504	151
547	249
590	347
634	445
677	543
720	641
764	739
807	838
851	936
894	·2835034
938	132

1920981	·2835230
1921025	328
068	426
111	524
155	622
198	720
241	818
285	916
328	·2836014
371	112
415	210
458	308
501	406
545	504
588	603
631	701
675	799
718	897
761	995
805	·2837093
848	190
892	288
935	386
979	484
1922022	582
066	680
109	778
152	876
196	974
239	·2838072
282	170
326	268
369	366
412	464
456	562
499	660
542	758
586	856
629	953
672	·2839051
716	149
759	247
802	345
846	443
889	541
932	639
976	737
1923019	835
062	933
106	·2840031
149	128
192	226
236	324
279	422
322	520
366	618
409	715
452	813
496	911
539	·2841009
582	107
626	204
669	302
712	400
756	497

1923799	·2841595
842	693
886	791
929	888
972	986
1924016	·2842084
059	182
102	280
146	378
189	475
232	573
276	671
319	769
362	867
405	964
449	·2843062
492	160
535	257
578	355
621	453
665	550
708	648
751	746
795	843
838	941
881	·2844039
925	136
968	234
1925011	332
055	429
098	527
141	625
185	722
228	820
271	918
315	·2845015
358	113
401	211
444	308
488	406
531	504
574	601
617	699
660	797
704	894
747	992
790	·2846089
834	187
877	284
920	382
964	479
1926007	577
050	674
094	772
137	869
180	967
223	·2847064
267	162
310	260
353	357
396	455
439	553
483	650
526	748
569	845

1926613	·2847943
656	·2848040
699	138
742	235
786	333
829	430
872	528
915	625
958	722
1927002	820
045	917
088	·2849014
132	112
175	209
218	307
261	404
305	502
348	599
391	697
434	794
477	891
520	989
564	·2850086
607	183
650	281
693	378
736	476
780	573
823	671
866	768
910	866
953	963
996	·2851060
1928039	158
083	255
126	352
169	450
212	547
255	644
298	742
342	839
385	936
428	·2852034
471	131
514	228
558	326
601	423
644	520
688	618
731	715
774	812
817	909
861	·2853007
904	104
947	201
990	298
1929033	395
076	493
120	590
163	687
206	785
249	882
292	979
335	·2854076
379	174

1929422	·2854271
465	368
508	465
551	562
594	659
638	757
681	854
724	951
767	·2855048
810	145
853	242
897	340
940	437
983	534

Excesses

10 = 23	1 = 2
20 = 45	2 = 5
30 = 68	3 = 7
40 = 90	4 = 9
	5 = 11
	6 = 14
	7 = 16
	8 = 18
	9 = 20

22 = 49

Anti Log Excesses

10 = 4	1 = 0
20 = 9	2 = 1
30 = 13	3 = 1
40 = 18	4 = 2
50 = 22	5 = 2
60 = 27	6 = 3
70 = 31	7 = 3
80 = 35	8 = 4
90 = 40	9 = 4

49 = 22

1930026	·2855631
069	728
112	825
156	923
199	·2856020
242	117
285	214
328	311
371	408
415	506
458	603
501	700
544	797
587	894
630	991
674	·2857088
717	185
760	282
803	379
846	476
889	573
933	671

1930976	·2857768
1931019	865
062	962
105	·2858059
148	156
192	253
235	350
278	447
321	544
364	641
407	738
451	835
494	932
537	·2859029
580	126
623	223
666	320
710	417
753	514
796	611
839	708
882	805
925	902
969	999
1932012	·2860096
055	193
098	290
141	387
184	484
228	580
271	677
314	774
357	871
400	968
443	·2861065
486	162
529	259
572	356
615	453
658	550
701	647
745	743
788	840
831	937
874	·2862034
917	131
960	228
1933004	324
047	421
090	518
133	615
176	712
219	809
262	905
305	·2863002
348	099
391	196
434	293
477	390
521	486
564	583
607	680
650	777
693	874
736	971

1933779	·2864067
822	164
865	261
908	358
951	455
994	551
1934038	648
081	745
124	841
167	938
210	·2865035
253	132
296	228
339	325
382	422
425	519
468	616
511	712
555	809
598	906
641	·2866002
684	099
727	196
770	292
813	389
856	486
899	582
942	679
985	776
1935028	872
072	969
115	·2867066
158	162
201	259
244	356
287	452
330	549
373	646
416	742
459	839
502	936
545	·2868032
588	129
631	225
674	322
717	418
760	515
803	611
847	708
890	805
933	901
976	998
1936019	·2869095
062	191
105	288
148	384
191	481
234	577
277	674
320	770
363	867
406	963
449	·2870060
492	156
535	253

1936578	·2870349
621	446
664	542
707	639
750	735
793	832
836	928
880	·2871025
923	121
966	218
1937009	314
052	411
095	507
138	604
181	700
224	797
267	893
310	989
353	·2872086
396	182
439	278
482	375
525	471
568	568
611	664
654	761
697	857
740	954
783	·2873050
826	146
869	243
912	339
955	435
998	532
1938041	628
084	724
127	821
170	917
213	·2874013
256	110
299	206
342	302
385	399
428	495
471	591
514	688
557	784
600	880
643	977
686	·2875073
729	169
772	266
815	362
858	458
901	554
943	651
986	747
1939029	843
072	939
115	·2876035
158	132
201	228
244	324
287	421
330	517

1939373	·2876613
416	709
459	806
502	902
545	998
588	·2877094
631	190
674	286
717	383
760	479
803	575
846	671
889	767
932	863
975	960

Excesses

10 = 22	1 = 2
20 = 45	2 = 4
30 = 67	3 = 7
40 = 90	4 = 9
	5 = 11
	6 = 13
	7 = 16
	8 = 18
	9 = 20

22 = 49

Anti Log Excesses

10 = 4	1 = 0
20 = 9	2 = 1
30 = 13	3 = 1
40 = 18	4 = 2
50 = 22	5 = 2
60 = 27	6 = 3
70 = 31	7 = 3
80 = 36	8 = 4
90 = 40	9 = 4

49 = 22

1940018	·2878056
061	152
104	248
147	344
190	440
232	537
275	633
318	729
361	825
404	921
447	·2879017
490	114
533	210
576	306
619	402
662	498
705	594
748	690
791	786
834	882
877	978

1940920	·2880074
963	170
1941005	266
048	362
091	458
134	554
177	650
220	746
263	843
306	939
349	·2881035
392	131
435	227
478	323
520	419
563	515
606	611
649	707
692	803
735	899
778	995
821	·2882091
864	187
907	283
950	379
993	475
1942035	570
078	666
121	762
164	858
207	954
250	·2883050
293	146
336	242
379	338
422	434
465	530
508	626
550	721
593	817
636	913
679	·2884009
722	105
765	201
807	297
850	393
893	489
936	585
979	681
1943022	777
065	872
108	968
151	·2885064
194	160
237	256
280	352
322	447
365	543
408	639
451	735
494	831
537	927
579	·2886022
622	118
665	214

1943708	·2886310
751	406
794	501
837	597
880	693
923	788
966	884
1944009	980
052	·2887076
094	171
137	267
180	363
223	459
266	555
309	650
351	746
394	842
437	937
480	·2888033
523	129
566	225
608	320
651	416
694	512
737	608
780	704
823	799
865	895
908	991
951	·2889086
994	182
1945037	278
080	373
122	469
165	565
208	660
251	756
294	852
337	947
379	·2890043
422	138
465	234
508	329
551	425
594	520
636	616
679	712
722	807
765	903
808	999
851	·2891094
893	190
936	285
979	381
1946022	476
065	572
108	667
150	763
193	859
236	954
279	·2892050
322	146
365	241
407	337
450	432

1946493	·2892528
536	623
579	719
622	814
664	910
707	·2893005
750	101
793	196
836	292
879	387
921	483
964	578
1947007	674
050	769
093	865
135	960
178	·2894056
221	151
263	247
306	342
349	437
392	533
434	628
477	723
520	819
563	914
606	·2895010
649	105
691	201
734	296
777	392
820	487
863	582
905	678
948	773
991	868
1948033	964
076	·2896059
119	154
162	250
204	345
247	440
290	536
333	631
376	726
419	822
461	917
504	·2897012
547	108
590	203
633	298
675	394
718	489
761	584
803	680
846	775
889	870
932	965
974	·2898061
1949017	156
060	251
103	346
146	441
188	537
231	632

1949274	·2898727
316	823
359	918
402	·2899013
445	108
487	204
530	299
573	394
616	489
659	584
701	679
744	775
787	870
829	965
872	·2900060
915	155
958	251

Excesses

10 = 22	1 = 2
20 = 45	2 = 4
30 = 67	3 = 7
40 = 89	4 = 9
	5 = 11
	6 = 13
	7 = 16
	8 = 18
	9 = 20

21 = 48

Anti Log Excesses

10 = 4	1 = 0
20 = 9	2 = 1
30 = 13	3 = 1
40 = 18	4 = 2
50 = 22	5 = 2
60 = 27	6 = 3
70 = 31	7 = 3
80 = 36	8 = 4
90 = 40	9 = 4

48 = 21

1950000	·2900346
043	441
086	537
129	632
172	727
214	822
257	917
300	·2901012
342	107
385	202
428	297
470	392
513	488
556	583
598	678
641	773
684	868
727	963
769	·2902059

1950812	·2902154
855	249
898	344
941	439
983	534
1951026	630
069	725
111	820
154	915
197	·2903010
239	105
282	200
325	295
367	390
410	485
453	580
495	675
538	770
581	865
623	960
666	·2904055
709	150
752	245
794	340
837	435
880	530
923	625
966	720
1952008	815
051	910
094	·2905005
136	100
179	195
222	290
264	385
307	479
350	574
392	669
435	764
478	859
520	954
563	·2906049
606	144
648	239
691	334
734	429
776	524
819	618
862	713
904	808
947	903
990	998
1953032	·2907093
075	187
118	282
160	377
203	472
246	567
288	662
331	757
374	852
416	947
459	·2908042
502	137
544	232

1953587	·2908326
630	421
672	516
715	611
758	706
800	801
843	895
886	990
928	·2909085
971	180
1954014	275
056	369
099	464
142	559
184	653
227	748
270	843
312	938
355	·2910032
398	127
440	222
483	317
525	412
568	506
610	601
653	696
695	790
738	885
781	980
823	·2911074
866	169
909	264
951	358
994	453
1955037	548
079	643
122	737
165	832
207	927
250	·2912022
293	117
335	211
378	306
421	400
463	495
506	589
548	684
591	778
633	873
676	968
718	·2913062
761	157
804	252
846	346
889	441
932	536
974	630
1956017	725
059	820
102	914
144	·2914009
187	103
229	198
272	292
315	387

1956357	·2914481
400	576
443	671
485	765
528	860
571	955
613	·2915049
656	144
699	238
741	333
784	427
826	522
869	616
911	711
954	805
996	900
1957039	994
082	·2916089
124	183
167	278
210	372
252	467
295	561
337	656
380	750
422	845
465	939
507	·2917034
550	128
592	222
635	317
677	411
720	505
762	600
805	694
848	789
890	883
933	978
976	·2918072
1958018	167
061	261
103	355
146	450
188	544
231	638
273	733
316	827
358	921
401	·2919016
443	110
486	204
528	299
571	393
614	487
656	582
699	676
742	770
784	865
827	959
869	·2920053
912	148
954	242
997	336
1959039	431
082	525

1959124	·2920619
167	714
209	808
252	902
294	997
337	·2921091
379	185
422	279
464	374
507	468
549	562
592	656
634	750
677	845
719	939
762	·2922033
804	128
847	222
890	316
932	410
975	505

Excesses

10 = 22	1 = 2
20 = 44	2 = 4
30 = 67	3 = 7
40 = 89	4 = 9
	5 = 11
	6 = 13
	7 = 16
	8 = 18
	9 = 20

21 = 47

Anti Log Excesses

10 = 5	1 = 0
20 = 9	2 = 1
30 = 14	3 = 1
40 = 18	4 = 2
50 = 23	5 = 2
60 = 27	6 = 3
70 = 32	7 = 3
80 = 36	8 = 4
90 = 41	9 = 4

47 = 21

1960018	·2922599
060	693
103	787
145	881
188	975
230	·2923070
273	164
315	258
358	352
400	446
443	540
485	635
528	729
570	823
613	917

1960655	·2924011
698	105
740	200
783	294
825	388
868	482
910	576
953	670
995	764
1961038	858
080	952
123	·2925046
165	140
208	234
250	329
293	423
335	517
378	611
420	705
463	799
505	893
547	987
590	·2926081
632	175
674	269
717	363
759	457
802	551
844	645
887	739
929	833
972	927
1962014	·2927022
057	116
099	210
142	304
184	398
227	492
269	586
312	680
354	774
397	868
439	962
482	·2928056
524	149
567	243
609	337
652	431
694	525
737	619
779	713
822	807
864	901
907	995
949	·2929089
992	183
1963034	277
076	370
119	464
161	558
203	652
246	746
288	840
331	934
373	·2930028

1963416	·2930122
458	216
501	310
543	403
586	497
628	591
671	685
713	779
756	873
798	966
840	·2931060
883	154
925	248
967	342
1964010	436
052	529
095	623
137	717
180	811
222	905
265	999
307	·2932092
349	186
392	280
434	374
476	468
519	562
561	655
604	749
646	843
689	937
731	·2933031
774	124
816	218
858	312
901	405
943	499
985	593
1965028	686
070	780
113	874
155	967
198	·2934061
240	155
283	249
325	342
367	436
410	530
452	624
494	718
537	811
579	905
621	999
664	·2935092
706	186
748	280
791	373
833	467
876	560
918	654
961	747
1966003	841
046	934
088	·2936028
130	122

1966173	·2936215	1968925	·2942292	1970449	·2945651	1973196	·2951701	1975939	·2957734
215	309	968	386	491	744	238	794	982	827
257	403	1969010	479	533	837	281	887	1976024	920
300	496	052	572	575	930	323	980	066	·2958012
342	590	095	666	618	·2946023	365	·2952073	108	105
384	684	137	759	660	117	407	166	150	198
427	777	179	852	702	210	450	258	192	290
469	871	222	946	745	303	492	351	235	383
511	965	264	·2943039	787	396	534	444	277	476
554	·2937058	306	132	829	489	576	537	319	568
596	152	349	226	872	582	618	630	361	661
639	245	391	319	914	676	660	723	403	754
681	339	433	412	956	769	703	816	445	846
724	432	476	506	999	862	745	909	488	939
766	526	518	599	1971041	955	787	·2953002	530	·2959032
809	619	560	692	083	·2947048	829	095	572	124
851	713	603	785	125	141	871	188	614	217
893	807	645	879	168	235	914	281	656	310
936	900	687	972	210	328	956	373	698	402
978	994	730	·2944065	252	421	998	466	741	495
1967020	·2938088	772	158	294	514	1974041	559	783	588
063	181	814	251	336	607	083	652	825	680
105	275	857	345	379	700	125	745	867	773
147	368	899	438	421	794	167	838	909	866
190	462	941	531	463	887	210	930	951	958
232	555	984	625	506	980	252	·2954023	994	·2960051
274	649			548	·2948073	294	116	1977036	143
317	742	Excesses		590	166	336	209	078	236
359	836			633	259	378	302	120	328
401	929	10 = 22	1 = 2	675	352	420	395	162	421
444	·2939023	20 = 44	2 = 4	717	445	463	487	204	513
486	116	30 = 66	3 = 7	760	538	505	580	246	606
528	210	40 = 88	4 = 9	802	631	547	673	288	699
571	303		5 = 11	844	724	589	766	330	791
613	397		6 = 13	886	817	631	859	372	884
655	490		7 = 15	929	911	673	952	414	977
698	584		8 = 18	971	·2949004	716	·2955044	456	·2961069
740	677		9 = 20	1972013	097	758	137	499	162
782	770			055	190	800	230	541	254
825	864	21 = 46		097	283	842	323	583	347
867	957			140	376	884	416	625	439
909	·2940050	Anti Log Excesses		182	469	926	509	667	532
952	144			224	562	969	601	709	624
994	237	10 = 5	1 = 0	267	655	1975011	694	752	717
1968036	331	20 = 9	2 = 1	309	748	053	787	794	809
079	424	30 = 14	3 = 1	351	841	095	880	836	902
121	518	40 = 18	4 = 2	393	934	137	973	878	994
163	611	50 = 23	5 = 2	436	·2950027	180	·2956065	920	·2962087
206	705	60 = 27	6 = 3	478	120	222	158	962	179
248	798	70 = 32	7 = 3	520	213	264	251	1978005	272
290	891	80 = 36	8 = 4	562	306	307	343	047	364
333	985	90 = 41	9 = 4	604	399	349	436	089	457
375	·2941078			647	492	391	529	131	549
417	171	46 = 21		689	585	433	622	173	642
460	265			731	678	476	714	215	734
502	358	1970026	·2944718	774	771	518	807	258	827
544	452	068	811	816	864	560	900	300	919
587	545	111	905	858	957	602	993	342	·2963012
629	639	153	998	900	·2951050	644	·2957086	384	104
671	732	195	·2945091	943	143	686	178	426	197
714	826	238	185	985	236	729	271	468	289
756	919	280	278	1973027	329	771	364	510	382
798	·2942012	322	371	069	422	813	456	552	474
841	106	364	464	111	515	855	549	594	567
883	199	407	558	154	608	897	642	636	659

1978678	·2963751
720	844
763	936
805	·2964028
847	121
889	213
931	306
973	398
1979016	491
058	583
100	676
142	768
184	860
226	953
268	·2965045
310	137
352	230
394	322
436	414
478	507
521	599
563	691
605	784
647	876
689	968
731	·2966061
773	153
815	245
857	338
899	430
941	522
983	615

Excesses

10 = 22	1 = 2
20 = 44	2 = 4
30 = 66	3 = 7
40 = 88	4 = 9
	5 = 11
	6 = 13
	7 = 15
	8 = 18
	9 = 20

21 = 46

Anti Log Excesses

10 = 5	1 = 0
20 = 9	2 = 1
30 = 14	3 = 1
40 = 18	4 = 2
50 = 23	5 = 2
60 = 27	6 = 3
70 = 32	7 = 3
80 = 36	8 = 4
90 = 41	9 = 4

46 = 21

1980026	·2966707
068	799
110	892
152	984
1980194	·2967076
236	169
278	261
320	353
362	446
404	538
446	630
488	723
531	815
573	907
615	·2968000
657	092
699	184
741	276
783	369
825	461
867	553
909	645
951	737
993	829
1981036	922
078	·2969014
120	106
162	198
204	290
246	383
288	475
330	567
372	660
414	752
456	844
498	936
540	·2970029
582	121
624	213
666	305
708	397
750	489
793	581
835	673
877	765
919	857
961	949
1982003	·2971041
045	134
087	226
129	318
171	410
213	502
255	594
297	687
339	779
381	871
423	963
465	·2972055
507	147
549	239
591	331
633	423
675	515
717	607
759	699
802	791
844	883
886	975
1982928	·2973067
970	159
1983012	251
054	344
096	436
138	528
180	620
222	712
264	804
306	896
348	988
390	·2974080
432	172
474	264
516	356
558	448
600	540
642	632
684	724
726	816
768	908
810	999
852	·2975091
894	183
936	275
978	367
1984020	459
062	551
104	643
146	735
188	827
230	919
272	·2976011
314	102
356	194
398	286
440	378
482	470
524	562
566	654
608	746
650	838
692	930
734	·2977022
776	114
818	205
860	297
902	389
944	481
986	573
1985028	665
070	756
112	848
154	940
196	·2978032
238	124
280	216
322	307
364	399
406	491
448	583
490	675
532	767
573	858
615	950
1985657	·2979042
699	134
741	226
783	317
825	409
867	501
909	592
951	684
993	776
1986035	868
077	959
119	·2980051
161	143
203	235
245	327
287	418
329	510
371	602
413	693
455	785
497	877
539	969
580	·2981060
622	152
664	244
706	336
748	428
790	519
832	611
874	703
916	794
958	886
1987000	978
042	·2982069
084	161
126	253
168	344
210	436
252	528
294	619
335	711
377	802
419	894
461	985
503	·2983077
545	168
587	260
629	352
671	443
713	535
755	627
797	718
838	810
880	902
922	993
964	·2984085
1988006	177
048	268
090	360
132	451
174	543
216	634
258	726
300	817
341	909
1988383	·2985000
425	092
467	183
509	275
551	366
593	458
635	549
677	641
719	732
761	824
803	915
844	·2986007
886	098
928	190
970	281
1989012	373
054	464
095	556
137	647
179	739
221	830
263	922
305	·2987013
347	105
389	196
431	288
473	379
515	470
557	562
598	653
640	744
682	836
724	927
766	·2988019
808	110
849	202
891	293
933	385
975	476

Excesses

10 = 22	1 = 2
20 = 44	2 = 4
30 = 66	3 = 7
40 = 88	4 = 9
	5 = 11
	6 = 13
	7 = 15
	8 = 18
	9 = 20

21 = 46

Anti Log Excesses

10 = 5	1 = 0
20 = 9	2 = 1
30 = 14	3 = 1
40 = 18	4 = 2
50 = 23	5 = 2
60 = 27	6 = 3
70 = 32	7 = 3
80 = 37	8 = 4
90 = 41	9 = 4

46 = 21

1990017	·2988567
059	659
101	750
143	841
185	933
227	·2989024
269	115
311	207
352	298
394	389
436	481
478	572
520	663
562	755
603	846
645	937
687	·2990029
729	120
771	211
813	303
854	394
896	485
938	577
980	668
1991022	759
064	851
105	942
147	·2991033
189	125
231	216
273	307
315	399
356	490
398	581
440	673
482	764
524	855

1991566	·2991946
607	·2992038
649	129
691	220
733	311
775	402
817	493
858	585
900	676
942	767
984	858
1992026	949
068	·2993041
109	132
151	223
193	315
235	406
277	497
319	588
360	680
402	771
444	862
486	953
528	·2994044
570	135
611	226
653	317
695	408
737	499
779	590
821	681
862	773
904	864
946	955
988	·2995046
1993030	137
072	228
113	320
155	411
197	502
239	593
281	684
323	775
364	866
406	957
448	·2996048
490	139
532	230
574	321
615	413
657	504
699	595
741	686
783	777
824	868
866	959
908	·2997050
949	141
991	232
1994033	323
075	414
116	505
158	596
200	687
242	778

1994284	·2997869
326	960
367	·2998051
409	142
451	233
493	324
535	415
577	506
618	597
660	688
702	779
744	870
786	961
827	·2999052
869	142
911	233
952	324
994	415
1995036	506
078	597
119	688
161	779
203	870
245	961
287	·3000052
328	143
370	233
412	324
453	415
495	506
537	597
579	688
620	779
662	870
704	961
746	·3001052
788	143
829	234
871	324
913	415
954	506
996	597
1996038	688
080	779
121	869
163	960
205	·3002051
247	142
289	233
330	323
372	414
414	505
455	595
497	686
539	777
581	868
622	958
664	·3003049
706	140
748	231
790	322
831	413
873	503
915	594
956	685

1996998	·3003776
1997040	867
082	957
123	·3004048
165	139
207	229
249	320
291	411
332	501
374	592
416	683
457	773
499	864
541	955
582	·3005046
624	136
666	227
707	318
749	409
791	500
832	590
874	681
916	772
957	862
999	953
1998041	·3006044
083	134
124	225
166	316
208	406
250	497
292	588
333	678
375	769
417	859
458	950
500	·3007040
542	131
583	221
625	312
667	403
708	493
750	584
792	675
833	765
875	856
917	947
958	·3008037
1999000	128
042	219
083	309
125	400
167	490
208	581
250	671
292	762
333	852
375	943
417	·3009033
458	124
500	214
542	305
583	395
625	486
667	576

1999708	·3009667
750	757
792	848
833	938
875	·3010029
917	119
958	210

Excesses

10 = 22	1 = 2
20 = 44	2 = 4
30 = 65	3 = 7
40 = 87	4 = 9
	5 = 11
	6 = 13
	7 = 15
	8 = 17
	9 = 20

21 = 45

Anti Log Excesses

10 = 5	1 = 0
20 = 9	2 = 1
30 = 14	3 = 1
40 = 18	4 = 2
50 = 23	5 = 2
60 = 28	6 = 3
70 = 32	7 = 3
80 = 37	8 = 4
90 = 41	9 = 4

45 = 21

2000000	·3010300
042	391
083	481
125	572
167	662
208	753
250	843
292	933
333	·3011024
375	114
417	204
458	295
500	385
542	476
583	566
625	657
667	747
708	838
750	928
792	·3012018
833	109
875	199
917	289
958	380
2001000	470
042	561
083	651
125	742
167	832

2001208	·3012923
250	·3013013
292	103
333	194
375	284
417	374
458	465
500	555
542	645
583	736
625	826
667	916
708	·3014007
750	097
792	187
833	277
875	368
917	458
958	548
2002000	638
041	728
083	819
124	909
166	999
207	·3015090
249	180
291	270
332	361
374	451
416	541
457	632
499	722
541	812
582	902
624	993
666	·3016083
707	173
749	263
790	353
832	444
873	534
915	624
956	715
998	805
2003040	895
081	985
123	·3017076
165	166
206	256
248	346
289	436
331	526
372	617
414	707
455	797
497	887
539	977
580	·3018067
622	158
664	248
705	338
747	428
789	518
830	608
872	698

2003914	·3018788
955	878
997	968
2004038	·3019058
080	148
121	239
163	329
204	419
246	509
287	599
329	689
370	780
412	870
453	960
495	·3020050
537	140
578	230
620	320
662	410
703	500
745	590
786	680
828	770
869	860
911	950
952	·3021040
994	130
2005036	220
077	310
119	400
161	490
202	580
244	670
285	760
327	850
368	940
410	·3022030
451	120
493	210
534	300
576	390
617	480
659	570
700	660
742	750
784	840
825	930
867	·3023020
909	110
950	200
992	290
2006033	380
075	470
116	559
158	649
199	739
241	829
282	919
324	·3024009
365	099
407	189
448	279
490	369
531	459
573	549

2006614	·3024638
656	728
697	818
739	908
780	998
822	·3025088
863	177
905	267
946	357
988	447
2007029	537
071	627
112	716
154	806
195	896
237	986
278	·3026076
320	166
361	256
403	345
444	435
486	525
528	615
569	705
611	794
653	884
694	974
736	·3027064
777	154
819	244
860	333
902	423
943	513
985	603
2008026	693
068	782
109	872
151	962
192	·3028051
234	141
275	231
317	321
358	410
400	500
441	590
483	680
524	770
566	859
607	949
648	·3029039
690	128
731	218
772	308
814	397
855	487
897	577
938	666
980	756
2009021	846
063	935
104	·3030025
146	115
187	204
229	294
270	384

2009312	·3030473
353	563
395	653
436	742
478	832
519	922
561	·3031011
602	101
644	190
685	280
727	369
768	459
810	548
851	638
893	728
934	817
976	907

Excesses

10 = 22	1 = 2
20 = 43	2 = 4
30 = 65	3 = 6
40 = 87	4 = 9
	5 = 11
	6 = 13
	7 = 15
	8 = 17
	9 = 19

21 = 45

Anti Log Excesses

10 = 5	1 = 0
20 = 9	2 = 1
30 = 14	3 = 1
40 = 18	4 = 2
50 = 23	5 = 2
60 = 28	6 = 3
70 = 32	7 = 3
80 = 37	8 = 4
90 = 42	9 = 4

45 = 21

2010017	·3031997
059	·3032086
100	176
141	265
183	355
224	444
265	534
307	623
348	713
390	803
431	892
473	982
514	·3033072
556	161
597	251
639	340
680	430
722	519
763	609

2010805	·3033698
846	788
887	877
929	967
970	·3034056
2011011	146
053	235
094	325
136	414
177	504
219	593
260	682
302	772
343	861
384	950
426	·3035040
467	129
508	219
550	308
591	398
633	487
674	577
716	666
757	756
799	845
840	935
882	·3036024
923	114
965	203
2012006	292
048	382
089	471
130	560
172	650
213	739
254	828
296	918
337	·3037007
379	096
420	186
462	275
503	364
545	454
586	543
627	632
669	722
710	811
751	900
793	990
834	·3038079
876	168
917	258
959	347
2013000	436
042	526
083	615
124	704
166	794
207	883
248	972
290	·3039062
331	151
372	240
414	330
455	419

2013496	·3039508
538	597
579	687
621	776
662	865
704	954
745	·3040043
787	133
828	222
869	311
911	401
952	490
993	579
2014035	668
076	758
117	847
159	936
200	·3041025
241	114
283	203
324	293
365	382
407	471
448	560
489	649
531	738
572	828
613	917
655	·3042006
696	095
737	184
779	273
820	363
862	452
903	541
945	630
986	719
2015028	808
069	898
110	987
152	·3043076
193	165
234	254
276	343
317	433
358	522
400	611
441	700
482	789
524	878
565	967
606	·3044056
648	145
689	234
730	323
772	412
813	502
854	591
896	680
937	769
978	858
2016020	947
061	·3045036
102	125
144	214

2016185	·3045303
226	392
268	481
309	570
350	659
392	748
433	837
474	926
516	·3046015
557	104
598	193
640	282
681	371
722	460
764	549
805	638
846	727
888	816
929	905
970	994
2017012	·3047083
053	172
094	261
136	350
177	439
218	528
260	617
301	705
342	794
384	883
425	972
466	·3048061
507	150
549	239
590	328
631	417
672	506
713	595
755	684
796	772
837	861
879	950
920	·3049039
961	128
2018003	217
044	305
085	394
127	483
168	572
209	661
251	750
292	838
333	927
375	·3050016
416	105
457	194
498	283
540	371
581	460
622	549
663	638
704	727
746	816
787	904
828	993

2018870	·3051082
911	171
952	260
994	349
2019035	437
076	526
118	615
159	704
200	793
241	881
283	970
324	·3052059
365	147
406	236
447	325
489	414
530	502
571	591
613	680
654	769
695	858
736	946
778	·3053035
819	124
860	212
901	301
942	390
984	478

Excesses

10 = 22	1 = 2
20 = 43	2 = 4
30 = 65	3 = 6
40 = 86	4 = 9
	5 = 11
	6 = 13
	7 = 15
	8 = 17
	9 = 19

21 = 45

Anti Log Excesses

10 = 5	1 = 0
20 = 9	2 = 1
30 = 14	3 = 1
40 = 19	4 = 2
50 = 23	5 = 2
60 = 28	6 = 3
70 = 32	7 = 3
80 = 37	8 = 4
90 = 42	9 = 4

45 = 21

2020025	·3053567
066	656
108	744
149	833
190	922
232	·3054010
273	099
314	188

2020356	·3054276
397	365
438	454
479	542
521	631
562	720
603	808
644	897
685	986
726	·3055074
768	163
809	252
850	340
891	429
932	518
974	606
2021015	695
056	783
098	872
139	960
180	·3056049
221	137
263	226
304	315
345	403
386	492
427	581
469	669
510	758
551	846
593	935
634	·3057023
675	112
716	200
758	289
799	377
840	466
881	554
922	643
963	731
2022005	820
046	909
087	997
128	·3058086
169	174
210	263
252	351
293	439
334	528
375	616
416	705
458	793
499	882
540	970
582	·3059059
623	147
664	236
705	324
747	413
788	501
829	590
870	678
911	767
952	855
994	944

2023035	·3060032
076	121
117	209
158	297
199	386
241	474
282	562
323	651
364	739
405	827
446	916
488	·3061004
529	092
570	181
611	269
652	358
693	446
735	535
776	623
817	712
858	800
899	888
940	977
982	·3062065
2024023	153
064	242
105	330
146	418
187	507
229	595
270	683
311	772
352	860
393	948
434	·3063036
476	125
517	213
558	301
599	389
640	477
681	566
723	654
764	742
805	831
846	919
887	·3064007
928	096
970	184
2025011	272
052	361
093	449
134	537
175	625
217	714
258	802
299	890
340	978
381	·3065066
422	154
464	243
505	331
546	419
587	507
628	595
669	684

2025711	·3065772
752	860
793	949
834	·3066037
875	125
916	213
957	302
998	390
2026039	478
080	566
121	654
162	742
204	831
245	919
286	·3067007
327	095
368	183
409	271
451	360
492	448
533	536
574	624
615	712
656	800
698	888
739	976
780	·3068064
821	152
862	240
903	328
944	417
985	505
2027026	593
067	681
108	769
149	857
191	945
232	·3069033
273	121
314	209
355	297
396	385
438	473
479	561
520	649
561	737
602	825
643	913
684	·3070002
725	090
766	178
807	266
848	354
889	442
931	530
972	618
2028013	706
054	794
095	882
136	970
177	·3071058
218	146
259	234
300	322
341	410

2028382	·3071498
424	585
465	673
506	761
547	849
588	937
629	·3072025
670	113
711	201
752	289
793	377
834	465
875	553
917	640
958	728
999	816
2029040	904
081	992
122	·3073080
163	168
204	256
245	344
286	432
327	520
368	608
409	695
450	783
491	871
532	959
573	·3074047
614	135
656	222
697	310
738	398
779	486
820	574
861	662
902	750
943	837
984	925

Excesses

10 = 21	1 = 2
20 = 42	2 = 4
30 = 64	3 = 6
40 = 86	4 = 9
	5 = 11
	6 = 13
	7 = 15
	8 = 17
	9 = 19
21 = 44	

Anti Log Excesses

10 = 5	1 = 0
20 = 9	2 = 1
30 = 14	3 = 1
40 = 19	4 = 2
50 = 23	5 = 2
60 = 28	6 = 3
70 = 33	7 = 3
80 = 37	8 = 4
	9 = 4
44 = 21	

2030025	·3075013
066	101
107	189
148	276
189	364
230	452
271	540
312	628
353	716
395	803
436	891
477	979
518	·3076067
559	155
600	242
641	330
682	418
723	505
764	593
805	681
846	769
887	856
928	944
969	·3077032
2031010	120
051	208
092	295
133	383
174	471
215	558
256	646
297	734
338	821
379	909
420	997
461	·3078084
502	172

2031543	·3078260
584	348
626	435
667	523
708	611
749	699
790	787
831	874
872	962
913	·3079050
954	137
995	225
2032036	313
077	400
118	488
159	575
200	663
241	750
282	838
323	925
364	·3080013
405	101
446	188
487	276
528	364
569	451
610	539
651	627
692	714
733	802
774	890
815	977
856	·3081065
897	152
938	240
979	327
2033020	415
061	502
102	590
143	677
184	765
225	852
266	940
307	·3082027
347	115
388	203
429	290
470	378
511	466
552	553
593	641
634	728
675	816
716	903
757	991
798	·3083078
839	166
880	253
921	341
962	428
2034003	515
044	603
085	690
126	777
167	865

2034208	·3083952
249	·3084040
290	127
331	215
372	302
413	390
454	477
495	565
536	652
576	740
617	827
658	915
699	·3085002
740	089
781	177
822	264
863	351
904	439
945	526
986	614
2035027	701
068	789
109	876
150	964
191	·3086051
232	138
273	226
314	313
355	400
396	488
437	575
478	662
519	750
559	837
600	924
641	·3087012
682	099
723	186
764	274
805	361
846	448
887	536
928	623
969	710
2036010	798
050	885
091	972
132	·3088060
173	147
214	234
255	321
296	409
337	496
378	583
419	670
460	757
501	845
541	932
582	·3089019
623	107
664	194
705	281
746	368
787	456
828	543

2036869	·3089630
910	717
951	804
992	891
2037032	979
073	·3090066
114	153
155	240
196	327
237	415
278	502
319	589
360	677
401	764
442	851
483	938
523	·3091026
564	113
605	200
646	287
687	374
728	461
769	549
810	636
851	723
892	810
933	897
974	984
2038014	·3092071
055	158
096	245
137	332
178	419
219	506
259	594
300	681
341	768
382	855
423	942
464	·3093029
504	117
545	204
586	291
627	378
668	465
709	552
750	639
791	726
832	813
873	900
914	987
955	·3094074
995	161
2039036	248
077	335
118	422
159	509
200	596
240	684
281	771
322	858
363	945
404	·3095032
445	119
485	206

2039526	·3095293
567	380
608	467
649	554
690	641
730	728
771	815
812	902
853	989
894	·3096076
935	163
976	250

Excesses

10 = 21	1 = 2
20 = 43	2 = 4
30 = 64	3 = 6
40 = 85	4 = 9
	5 = 11
	6 = 13
	7 = 15
	8 = 17
	9 = 19

21 = 44

Anti Log Excesses

10 = 5	1 = 0
20 = 9	2 = 1
30 = 14	3 = 1
40 = 19	4 = 2
50 = 23	5 = 2
60 = 28	6 = 3
70 = 33	7 = 3
80 = 37	8 = 4
	9 = 4

44 = 21

2040017	·3096336
058	423
099	510
140	597
181	684
221	771
262	858
303	945
344	·3097032
385	119
426	206
466	293
507	380
548	467
589	554
630	641
671	728
711	814
752	901
793	988
834	·3098075
875	162
916	249
956	335

2040997	·3098422
2041038	509
079	596
120	683
161	770
201	857
242	944
283	·3099031
324	118
365	205
406	292
446	378
487	465
528	552
569	639
610	726
650	812
691	899
732	986
772	·3100072
813	159
854	246
895	333
935	419
976	506
2042017	593
058	680
099	767
140	854
180	940
221	·3101027
262	114
303	201
344	288
385	374
425	461
466	548
507	634
548	721
589	808
630	895
670	981
711	·3102068
752	155
793	242
834	329
874	415
915	502
956	589
996	675
2043037	762
078	849
119	935
159	·3103022
200	109
241	195
282	282
323	369
364	455
404	542
445	629
486	715
527	802
568	889
609	975

2043649	·3104062
690	149
731	235
772	322
813	409
853	495
894	582
935	669
975	755
2044016	842
057	929
098	·3105015
138	102
179	189
220	275
261	362
302	449
342	535
383	622
424	709
464	795
505	882
546	969
587	·3106055
627	142
668	228
709	315
750	401
791	488
831	574
872	661
913	748
953	834
994	921
2045035	·3107008
076	094
116	181
157	267
198	354
239	440
280	527
320	613
361	700
402	786
442	873
483	959
524	·3108046
565	132
605	219
646	305
687	392
728	478
769	564
809	651
850	737
891	823
931	910
972	996
2046013	·3109083
054	169
094	256
135	342
176	429
217	515
258	601

2046298	·3109688
339	774
380	860
420	947
461	·3110033
502	120
542	206
583	293
624	379
664	466
705	552
746	638
786	725
827	811
868	897
908	984
949	·3111070
990	156
2047031	243
071	329
112	415
153	502
194	588
235	674
275	761
316	847
357	933
397	·3112020
438	106
479	192
519	279
560	365
601	451
641	538
682	624
723	710
763	797
804	883
845	969
885	·3113056
926	142
967	228
2048007	315
048	401
089	487
129	574
170	660
211	746
252	833
292	919
333	·3114005
374	092
415	178
456	264
496	350
537	437
578	523
618	609
659	695
700	781
740	867
781	954
822	·3115040
862	126
903	212

2048944	·3115298
984	385
2049025	471
066	557
106	644
147	730
188	816
228	902
269	989
310	·3116075
350	161
391	247
432	333
472	419
513	505
554	591
594	677
635	763
676	849
716	935
757	·3117022
798	108
838	194
879	280
919	366
960	452

Excesses

10 = 21	1 = 2
20 = 42	2 = 4
30 = 64	3 = 6
40 = 85	4 = 8
	5 = 11
	6 = 13
	7 = 15
	8 = 17
	9 = 19

20 = 43

Anti Log Excesses

10 = 5	1 = 0
20 = 9	2 = 1
30 = 14	3 = 1
40 = 19	4 = 2
50 = 24	5 = 2
60 = 28	6 = 3
70 = 33	7 = 3
80 = 38	8 = 4
	9 = 4

43 = 20

2050000	·3117539
041	625
081	711
122	797
163	883
203	969
244	·3118056
285	142
325	228
366	314

2050407	·3118400
447	486
488	572
529	658
569	744
610	830
651	916
691	·3119002
732	088
773	174
813	260
854	346
895	432
935	518
976	605
2051017	691
057	777
098	863
139	949
179	·3120035
220	121
261	207
301	293
342	379
382	465
423	551
463	637
504	723
544	809
585	895
626	981
666	·3121067
707	153
748	239
788	325
829	411
870	497
910	583
951	668
992	754
2052032	840
073	926
113	·3122012
154	098
194	184
235	270
275	356
316	442
357	528
397	614
438	699
479	785
519	871
560	957
600	·3123043
641	129
681	215
722	301
762	387
803	473
844	559
884	645
925	730
966	816
2053006	902

2053047	·3123988
088	·3124074
128	160
169	245
210	331
250	417
291	503
331	589
372	675
412	760
453	846
493	932
534	·3125018
575	104
615	190
656	275
697	361
737	447
778	533
818	619
859	705
899	790
940	876
980	962
2054021	·3126048
061	134
102	219
142	305
183	391
223	476
264	562
305	648
345	734
386	819
427	905
467	991
508	·3127077
548	163
589	248
629	334
670	420
710	505
751	591
791	677
832	762
872	848
913	934
953	·3128019
994	105
2055035	191
075	277
116	362
157	448
197	534
238	620
278	706
319	791
359	877
400	963
440	·3129048
481	134
521	220
562	305
602	391
643	476

2055683	·3129562
724	647
764	733
805	818
845	904
886	990
926	·3130075
967	161
2056008	247
048	332
089	418
130	504
170	589
211	675
251	761
292	846
332	932
373	·3131017
413	103
454	188
494	274
535	359
575	445
616	531
656	616
697	702
737	788
778	873
818	959
859	·3132044
899	130
940	215
980	301
2057021	386
061	472
102	557
142	643
183	728
223	814
264	899
304	985
345	·3133070
385	156
426	241
466	327
507	412
547	498
588	583
628	669
669	754
709	840
750	925
790	·3134011
831	096
871	182
912	267
952	353
993	438
2058033	524
074	609
114	695
155	780
195	865
236	951
276	·3135036

2058317	·3135121
357	207
398	292
438	378
479	463
519	549
560	634
600	720
641	805
681	890
722	976
762	·3136061
803	146
843	232
884	317
924	402
965	488
2059005	573
046	658
086	744
127	829
167	914
208	·3137000
248	085
288	170
329	256
369	341
409	426
450	512
490	597
531	682
571	768
612	853
652	938
693	·3138024
733	109
774	194
814	280
855	365
895	450
936	536
976	621

Excesses

10 = 21	1 = 2
20 = 42	2 = 4
30 = 63	3 = 6
40 = 84	4 = 8
	5 = 11
	6 = 13
	7 = 15
	8 = 17
	9 = 19

20 = 43

Anti Log Excesses

10 = 5	1 = 0
20 = 9	2 = 1
30 = 14	3 = 1
40 = 19	4 = 2
50 = 24	5 = 2
60 = 28	6 = 3
70 = 33	7 = 3
80 = 38	8 = 4
	9 = 4

43 = 20

2060017	·3138706
057	792
098	877
138	962
179	·3139047
219	133
259	218
300	303
340	388
380	473
421	559
461	644
502	729
542	815
583	900
623	985
664	·3140070
704	156
745	241
785	326
826	411
866	496
907	581
947	667
987	752
2061028	837
068	922
108	·3141007
149	093
189	178
230	263
270	349
311	434
351	519
392	604
432	690
472	775

2061513	·3141860
553	945
593	·3142030
634	115
674	200
715	285
755	370
796	455
836	540
877	625
917	711
957	796
998	881
2062038	966
078	·3143051
119	136
159	222
200	307
240	392
281	477
321	562
362	647
402	732
442	817
483	902
523	987
563	·3144072
604	157
644	243
685	328
725	413
766	498
806	583
847	668
887	753
927	838
968	923
2063008	·3145008
048	093
089	178
129	263
169	348
210	433
250	518
290	603
331	688
371	773
412	858
452	943
493	·3146028
533	113
574	198
614	283
654	368
695	453
735	538
775	623
816	708
856	793
896	878
937	963
977	·3147048
2064017	133
058	218
098	303

2064138	·3147388
179	473
219	558
259	643
300	728
340	812
381	897
421	982
462	·3148067
502	152
543	237
583	322
623	407
664	492
704	577
744	662
785	747
825	831
865	916
906	·3149001
946	086
986	171
2065027	256
067	340
107	425
148	510
188	595
228	680
269	765
309	849
349	934
390	·3150019
430	104
470	189
511	274
551	358
591	443
632	528
672	613
712	698
753	783
793	867
833	952
874	·3151037
914	122
954	207
995	292
2066035	376
075	461
116	546
156	631
196	716
237	800
277	885
317	970
358	·3152054
398	139
438	224
479	309
519	393
559	478
600	563
640	648
680	733
721	817

2066761	·3152902
801	987
842	·3153071
882	156
922	241
963	326
2067003	410
043	495
084	580
124	665
164	750
205	834
245	919
285	·3154004
326	088
366	173
406	258
447	342
487	427
527	512
568	596
608	681
648	766
689	850
729	935
769	·3155020
810	104
850	189
890	274
930	358
971	443
2068011	527
051	612
091	696
131	781
172	865
212	950
252	·3156035
293	119
333	204
373	289
414	373
454	458
494	542
535	627
575	711
615	796
656	880
696	965
736	·3157050
777	134
817	219
857	304
897	388
938	473
978	557
2069018	642
058	726
098	811
139	895
179	980
219	·3158064
260	149
300	233
340	318

2069381	·3158402
421	487
461	571
502	656
542	740
582	825
622	909
663	994
703	·3159078
743	163
783	247
823	332
864	416
904	501
944	585
985	670

Excesses

10 = 21	1 = 2
20 = 42	2 = 4
30 = 63	3 = 6
40 = 84	4 = 8
	5 = 11
	6 = 13
	7 = 15
	8 = 17
	9 = 19

20 = 42

Anti Log Excesses

10 = 5	1 = 0
20 = 10	2 = 1
30 = 14	3 = 1
40 = 19	4 = 2
50 = 24	5 = 2
60 = 29	6 = 3
70 = 33	7 = 3
80 = 38	8 = 4
	9 = 4

42 = 20

2070025	·3159754
065	839
105	923
146	·3160008
186	092
226	177
266	261
306	345
347	430
387	514
427	598
468	683
508	767
548	852
588	936
629	·3161021
669	105
709	190
749	274
789	358

2070830	·3161443
870	527
910	611
951	696
991	780
2071031	864
071	949
112	·3162033
152	117
192	202
232	286
272	371
312	455
353	540
393	624
433	709
473	793
513	877
554	962
594	·3163046
634	130
675	215
715	299
755	383
795	467
836	552
876	636
916	720
956	804
996	888
2072036	973
077	·3164057
117	141
157	226
197	310
237	394
278	479
318	563
358	647
399	732
439	816
479	900
519	984
560	·3165069
600	153
640	237
680	321
720	405
760	490
801	574
841	658
881	743
921	827
961	911
2073001	995
042	·3166080
082	164
122	248
162	332
202	416
242	500
283	585
323	669
363	753
403	837

2073443	·3166921
484	·3167005
524	090
564	174
605	258
645	342
685	426
725	510
766	595
806	679
846	763
886	847
926	931
966	·3168015
2074007	100
047	184
087	268
127	352
167	436
207	520
248	605
288	689
328	773
368	857
408	941
448	·3169025
489	109
529	193
569	277
609	361
649	445
689	529
730	614
770	698
810	782
850	866
890	950
930	·3170034
971	118
2075011	202
051	286
091	370
131	454
171	538
212	622
252	706
292	790
332	874
372	958
412	·3171042
453	127
493	211
533	295
573	379
613	463
653	547
694	630
734	714
774	798
814	882
854	966
894	·3172050
934	134
974	218
2076014	302

2076054	·3172386
094	470
134	554
175	638
215	722
255	806
295	890
335	974
375	·3173058
416	142
456	226
496	310
536	394
576	478
616	562
657	645
697	729
737	813
777	897
817	981
857	·3174065
898	149
938	233
978	317
2077018	401
058	485
098	569
138	652
178	736
218	820
258	904
298	988
338	·3175072
379	155
419	239
459	323
499	407
539	491
579	575
620	658
660	742
700	826
740	910
780	994
820	·3176078
860	161
900	245
940	329
980	413
2078020	497
060	581
101	664
141	748
181	832
221	916
261	·3177000
301	084
341	167
381	251
421	335
461	419
501	503
541	586
582	670
622	754

2078662	·3177837
702	921
742	·3178005
782	089
823	172
863	256
903	340
943	424
983	508
2079023	591
063	675
103	759
143	842
183	926
223	·3179010
263	093
303	177
343	261
383	344
423	428
463	512
503	595
544	679
584	763
624	846
664	930
704	·3180014
744	097
784	181
824	265
864	348
904	432
944	516
984	599

Excesses

10 = 21	1 = 2
20 = 42	2 = 4
30 = 63	3 = 6
40 = 84	4 = 8
	5 = 10
	6 = 13
	7 = 15
	8 = 17
	9 = 19

20 = 42

Anti Log Excesses

10 = 5	1 = 0
20 = 10	2 = 1
30 = 14	3 = 1
40 = 19	4 = 2
50 = 24	5 = 2
60 = 29	6 = 3
70 = 33	7 = 3
80 = 38	8 = 4
	9 = 4

42 = 20

2080025	·3180683
065	767

2080105	·3180850
145	934
185	·3181018
225	101
265	185
305	269
345	352
385	436
425	520
465	603
505	687
545	771
585	854
625	938
665	·3182022
705	105
746	189
786	272
826	356
866	439
906	523
946	606
986	690
2081026	774
066	857
106	941
146	·3183025
186	108
226	192
266	275
306	359
346	442
386	526
426	609
466	693
506	777
546	860
586	944
626	·3184028
666	111
707	195
747	278
787	362
827	445
867	529
907	612
947	696
987	779
2082027	863
067	946
107	·3185030
147	113
187	197
227	280
267	364
307	447
347	530
387	614
427	697
467	780
507	864
547	947
587	·3186031
627	114
667	198

2082707	·3186281
747	365
787	448
827	531
867	615
907	698
947	781
987	865
2083027	948
067	·3187032
107	115
147	199
187	282
227	366
267	449

Excesses

20 = 42

Anti Log Excesses

42 = 20

2083315	·3187549
363	649
411	749
459	849
507	949
555	·3188049
603	149
651	249
699	349
747	449
795	549
843	649
891	749
939	849
987	949
2084035	·3189049
083	149
131	249
179	349
227	449
275	549
323	649
371	749
419	849
467	949
515	·3190049
563	149
611	249
659	349
707	449
755	549
803	649
851	748
899	848
947	948
995	·3191048
2085043	148
090	248
138	348
186	448
234	548

2085282	·3191648
330	747
378	847
426	947
474	·3192047
522	147
570	246
618	346
666	446
714	546
762	646
810	745
858	845
906	945
954	·3193045
2086002	145
049	245
097	345
145	445
193	545
241	644
289	744
337	843
385	943
433	·3194043
481	143
529	242
577	342
625	442
673	542
721	642
768	741
816	841
864	941
912	·3195041
960	140
2087008	240
056	339
104	439
152	539
200	639
247	738
295	838
343	938
391	·3196038
439	137
487	237
535	336
583	436
631	536
679	635
726	735
774	834
822	934
870	·3197034
918	133
966	233
2088014	332
062	432
110	532
158	631
205	731
253	830
301	930
349	·3198030

2088397	·3198129
445	229
493	328
541	428
589	528
637	627
684	727
732	826
780	926
828	·3199026
876	125
923	225
971	324
2089019	424
067	523
115	623
163	722
211	822
259	921
307	·3200020
355	120
402	219
450	319
498	418
546	518
594	617
641	717
689	816
737	916
785	·3201015
833	115
881	214
929	314
977	413

Excesses

10 = 21	1 = 2
20 = 42	2 = 4
30 = 62	3 = 6
40 = 83	4 = 8
	5 = 10
	6 = 12
	7 = 15
	8 = 17
	9 = 19

24 = 50

Anti Log Excesses

10 = 5	1 = 0
20 = 10	2 = 1
30 = 14	3 = 1
40 = 19	4 = 2
50 = 24	5 = 2
60 = 29	6 = 3
70 = 34	7 = 3
80 = 38	8 = 4
90 = 43	9 = 4

50 = 24

2090025	·3201512
073	612

2090120	·3201711
168	811
216	910
264	·3202009
312	109
359	208
407	308
455	407
503	506
551	606
598	705
646	805
694	904
742	·3203003
790	103
837	202
885	302
933	401
981	500
2091029	600
076	699
124	799
172	898
220	997
268	·3204096
316	196
364	295
412	394
460	493
508	592
555	692
603	791
651	890
699	989
747	·3205089
794	188
842	288
890	387
938	486
986	585
2092033	685
081	784
129	883
177	982
225	·3206081
272	181
320	280
368	379
416	478
464	577
511	677
559	776
607	875
655	974
702	·3207073
750	173
797	272
845	371
893	470
941	569
988	668
2093036	767
084	866
132	965
180	·3208064

2093227	·3208164
275	263
323	362
371	461
419	560
466	659
514	758
562	857
610	956
658	·3209055
705	155
753	254
801	353
849	452
897	551
944	650
992	749
2094040	848
088	947
135	·3210046
183	145
230	244
278	343
326	442
374	541
421	640
469	739
517	838
565	937
613	·3211036
660	135
708	234
756	333
804	432
852	531
899	630
947	729
995	828
2095043	927
090	·3212026
138	125
185	224
233	323
281	422
329	521
376	619
424	718
472	817
520	916
567	·3213015
615	114
662	213
710	312
758	411
806	510
853	608
901	707
949	806
997	905
2096045	·3214004
092	102
140	201
188	300
236	399
283	498

2096331	·3214597
378	696
426	795
474	894
522	993
569	·3215091
617	190
665	289
713	388
760	486
808	585
855	683
903	782
951	881
998	980
2097046	·3216078
093	177
141	276
189	375
237	474
284	572
332	671
380	770
428	869
475	967
523	·3217066
570	164
618	263
666	362
714	461
761	559
809	658
857	757
905	856
952	954
2098000	·3218053
047	151
095	250
143	349
190	447
238	546
285	644
333	743
381	842
428	941
476	·3219039
523	138
571	237
619	336
667	434
714	533
762	631
810	730
858	828
905	927
953	·3220025
2099000	124
048	222
096	321
143	419
191	518
238	616
286	715
334	814
381	912

2099429	·3221011
476	109
524	208
572	306
619	405
667	503
714	602
762	700
810	799
857	897
905	996
952	·3222094

Excesses

10 = 21	1 = 2
20 = 41	2 = 4
30 = 62	3 = 6
40 = 83	4 = 8
	5 = 10
	6 = 12
	7 = 15
	8 = 17
	9 = 19

24 = 49

Anti Log Excesses

10 = 5	1 = 0
20 = 10	2 = 1
30 = 14	3 = 1
40 = 19	4 = 2
50 = 24	5 = 2
60 = 29	6 = 3
70 = 34	7 = 3
80 = 39	8 = 4
90 = 43	9 = 4

49 = 24

2100000	·3222193
048	292
096	390
143	489
191	587
239	686
287	784
334	883
382	981
429	·3223080
477	178
525	276
572	375
620	473
667	572
715	670
763	768
810	867
858	965
905	·3224064
953	162
2101001	260
048	359
096	457

2101143	·3224556
191	654
239	752
286	851
334	949
381	·3225048
429	146
477	244
524	342
572	441
619	539
667	637
714	735
762	834
809	932
857	·3226031
904	129
952	227
999	325
2102047	424
094	522
142	620
190	718
237	817
285	915
332	·3227014
380	112
428	210
475	308
523	407
570	505
618	603
666	701
713	799
761	898
808	996
856	·3228094
903	192
951	290
998	389
2103046	487
093	585
141	683
188	781
236	880
283	978
331	·3229076
379	174
426	272
474	370
521	468
569	566
617	664
664	762
712	861
759	959
807	·3230057
854	155
902	253
949	352
997	450
2104044	548
092	646
139	744
187	842

2104234	·3230940
282	·3231038
329	136
377	234
424	332
472	430
519	528
567	626
614	724
662	823
709	921
757	·3232019
805	117
852	215
900	313
947	411
995	509
2105042	607
090	705
137	803
185	901
232	999
280	·3233097
327	195
375	293
422	391
470	489
517	587
565	685
612	783
660	881
707	979
754	·3234077
802	175
849	272
897	370
944	468
992	566
2106039	664
087	762
134	860
182	958
229	·3235056
277	154
324	251
372	349
419	447
467	545
514	643
562	741
609	839
657	937
704	·3236035
752	133
799	230
847	328
894	426
941	524
989	622
2107036	719
084	817
131	915
179	·3237013
226	111
274	208

2107321	·3237306
369	404
416	502
464	600
511	697
559	795
606	893
653	991
701	·3238089
748	186
796	284
843	382
890	480
938	577
985	675
2108033	772
080	870
127	968
175	·3239066
222	163
270	261
317	359
365	457
412	554
460	652
507	749
555	847
602	945
650	·3240043
697	140
745	238
792	336
839	434
887	531
934	629
982	726
2109029	824
076	922
124	·3241019
171	117
219	214
266	312
313	410
361	507
408	605
456	702
503	800
550	898
598	995
645	·3242093
693	190
740	288
787	386
835	483
882	581
930	678
977	776

Excesses

10 = 21	1 = 2
20 = 41	2 = 4
30 = 62	3 = 6
40 = 83	4 = 8
	5 = 10
	6 = 12
	7 = 14
	8 = 17
	9 = 19

24 = 49

Anti Log Excesses

10 = 5	1 = 0
20 = 10	2 = 1
30 = 14	3 = 1
40 = 19	4 = 2
50 = 24	5 = 2
60 = 29	6 = 3
70 = 34	7 = 3
80 = 39	8 = 4
90 = 44	9 = 4

49 = 24

2110024	·3242874
072	971
119	·3243069
167	166
214	264
261	361
309	459
356	556
404	654
451	751
498	849
546	946
593	·3244044
641	141
688	239
735	336
782	434
830	531
877	629
924	726
971	823
2111019	921
066	·3245018
114	116
161	213
208	310
256	408
303	505
351	603
398	700
445	798
493	895
540	993
588	·3246090
635	188
682	285
730	382

2111777	·3246480
825	577
872	674
919	771
966	869
2112014	966
061	·3247064
108	161
155	258
203	356
250	453
298	551
345	648
392	745
440	842
487	940
535	·3248037
582	134
629	231
677	329
724	426
772	524
819	621
866	718
913	815
961	913
2113008	·3249010
055	107
102	204
150	302
197	399
245	497
292	594
339	691
386	788
434	886
481	983
528	·3250080
575	177
623	274
670	372
718	469
765	566
812	663
859	760
907	858
954	955
2114001	·3251052
048	149
096	246
143	344
191	441
238	538
285	635
332	732
380	829
427	926
474	·3252023
521	120
569	217
616	315
664	412
711	509
758	606
805	703

2114853	·3252801
900	898
947	995
994	·3253092
2115042	189
089	286
137	383
184	480
231	577
278	674
326	771
373	868
420	965
467	·3254062
514	159
562	256
609	353
656	450
703	547
751	644
798	742
846	839
893	936
940	·3255033
987	130
2116035	226
082	323
129	420
176	517
223	614
271	711
318	808
365	905
412	·3256002
460	099
507	196
555	293
602	390
649	487
696	584
744	681
791	778
838	875
885	972
932	·3257069
980	165
2117027	262
074	359
121	456
168	553
216	650
263	747
310	844
357	941
404	·3258038
452	134
499	231
546	328
593	425
640	522
688	618
735	715
782	812
829	909
876	·3259006

2117924	·3259102
971	199
2118018	296
065	393
112	490
160	586
207	683
254	780
301	877
349	974
396	·3260070
444	167
491	264
538	361
585	458
633	554
680	651
727	748
774	845
821	942
869	·3261038
916	135
963	232
2119010	329
057	426
104	522
151	619
198	716
245	813
292	909
340	·3262006
387	102
434	199
481	296
528	392
576	489
623	585
670	682
717	779
764	876
812	972
859	·3263069
906	166
953	263

Excesses

10 = 21	1 = 2
20 = 41	2 = 4
30 = 62	3 = 6
40 = 82	4 = 8
	5 = 10
	6 = 12
	7 = 14
	8 = 16
	9 = 18

24 = 49

Anti Log Excesses

10 = 5	1 = 0
20 = 10	2 = 1
30 = 15	3 = 1
40 = 19	4 = 2
50 = 24	5 = 2
60 = 29	6 = 3
70 = 34	7 = 3
80 = 39	8 = 4
90 = 44	9 = 4

49 = 24

2120000	·3263359
048	456
095	552
142	649
189	746
236	842
284	939
331	·3264035
378	132
425	229
472	325
520	422
567	518
614	615
661	711
708	808
755	904
802	·3265001
849	097
896	194
943	290
991	387
2121038	483
085	580
132	677
179	773
227	870
274	966
321	·3266063
368	159
415	256
463	352
510	449
557	545
604	642
651	738
698	835

2121745	·3266931
792	·3267028
839	124
886	221
934	317
981	414
2122028	510
075	606
122	703
169	799
216	896
263	992
310	·3268088
357	185
405	281
452	378
499	474
546	570
593	667
641	763
688	860
735	956
782	·3269052
829	149
876	245
923	342
970	438
2123017	534
064	631
112	727
159	824
206	920
253	·3270016
300	112
347	209
394	305
441	401
488	497
535	594
583	690
630	787
677	883
724	979
771	·3271075
818	172
865	268
912	364
959	460
2124006	557
053	653
100	750
147	846
194	942
241	·3272038
289	135
336	231
383	327
430	423
477	519
524	616
571	712
618	808
665	904
712	·3273000
759	097

2124806	·3273193
853	289
900	385
947	481
995	578
2125042	674
089	770
136	866
183	962
230	·3274059
277	155
324	251
371	347
418	443
465	539
512	635
559	731
606	827
653	923
700	·3275020
747	116
794	212
841	308
888	404
936	500
983	596
2126030	692
077	788
124	884
171	981
218	·3276077
265	173
312	269
359	365
406	461
453	557
500	653
547	749
594	845
641	941
688	·3277037
735	133
782	229
829	325
876	421
923	517
970	613
2127017	709
064	805
111	901
158	997
205	·3278093
252	189
299	285
346	381
393	477
440	573
487	669
534	765
581	861
628	957
675	·3279053
722	149
769	245
816	340

2127863	·3279436
910	532
957	628
2128004	724
051	820
098	916
145	·3280012
192	108
239	204
286	299
333	395
380	491
427	587
474	683
521	778
568	874
615	970
662	·3281066
709	162
756	258
803	354
850	450
897	546
944	642
991	737
2129038	833
085	929
132	·3282025
179	121
226	216
273	312
320	408
367	504
414	600
460	695
507	791
554	887
601	983
648	·3283078
695	174
742	269
789	365
836	461
883	557
930	652
977	748

Excesses

10 = 20	1 = 2
20 = 41	2 = 4
30 = 61	3 = 6
40 = 82	4 = 8
	5 = 10
	6 = 12
	7 = 14
	8 = 16
	9 = 18

24 = 48

Anti Log Excesses

10 = 5	1 = 0
20 = 10	2 = 1
30 = 15	3 = 1
40 = 20	4 = 2
50 = 24	5 = 2
60 = 29	6 = 3
70 = 34	7 = 3
80 = 39	8 = 4
90 = 44	9 = 4

48 = 24

2130024	·3283844
071	940
118	·3284036
165	131
212	227
259	323
306	419
353	514
399	610
446	705
493	801
540	897
587	992
634	·3285088
681	183
728	279
775	375
822	470
869	566
916	661
963	757
2131010	853
057	949
103	·3286044
150	140
197	236
244	332
291	427
338	523
385	618
432	714
479	810
526	905
573	·3287001
620	096
667	192
714	287

2131761	·3287383
807	478
854	574
901	669
948	765
995	860
2132042	956
089	·3288051
136	147
183	243
230	338
276	434
323	529
370	625
417	720
464	816
511	911
558	·3289007
605	102
652	198
699	293
745	389
792	484
839	580
886	675
933	771
979	866
2133026	962
073	·3290057
120	152
167	248
214	343
261	439
308	534
355	629
402	725
448	820
495	916
542	·3291011
589	106
636	202
682	297
729	393
776	488
823	583
870	679
917	774
964	870
2134011	965
058	·3292060
105	156
151	251
198	347
245	442
292	537
339	633
385	728
432	824
479	919
526	·3293014
573	109
620	205
667	300
714	395
761	490

2134808	·3293586
854	681
901	777
948	872
995	967
2135042	·3294062
088	158
135	253
182	348
229	443
276	538
322	634
369	729
416	824
463	919
510	·3295014
556	110
603	205
650	300
697	395
744	490
790	586
837	681
884	776
931	871
978	966
2136024	·3296062
071	157
118	252
165	347
212	442
259	538
306	633
353	728
400	823
447	918
493	·3297014
540	109
587	204
634	299
681	394
727	489
774	584
821	679
868	774
915	869
961	965
2137008	·3298060
055	155
102	250
149	345
195	440
242	535
289	630
336	725
382	820
429	916
475	·3299011
522	106
569	201
616	296
662	391
709	486
756	581
803	676

2137850	·3299771
896	866
943	961
990	·3300056
2138037	151
084	246
130	341
177	436
224	531
271	626
318	721
364	816
411	911
458	·3301006
505	101
552	196
598	291
645	386
692	481
739	576
785	671
832	766
878	861
925	956
972	·3302051
2139019	146
065	240
112	335
159	430
206	525
253	620
299	715
346	810
393	905
440	·3303000
487	095
533	189
580	284
627	379
674	474
720	569
767	663
813	758
860	853
907	948
954	·3304043

Excesses

10 = 20	1 = 2
20 = 41	2 = 4
30 = 61	3 = 6
40 = 81	4 = 8
	5 = 10
	6 = 12
	7 = 14
	8 = 16
	9 = 18

23 = 48

Anti Log Excesses

10 = 5	1 = 0
20 = 10	2 = 1
30 = 15	3 = 1
40 = 20	4 = 2
50 = 25	5 = 2
60 = 29	6 = 3
70 = 34	7 = 3
80 = 39	8 = 4
90 = 44	9 = 4

48 = 23

2140000	·3304138
047	233
094	328
141	423
188	518
234	612
281	707
328	802
375	897
421	992
468	·3305086
514	181
561	276
608	371
655	465
701	560
748	654
795	749
842	844
888	939
935	·3306033
981	128
2141028	223
075	318
122	413
168	507
215	602
262	697
309	792
355	886
402	981
448	·3307075
495	170
542	265
589	360
635	454
682	549

2141729	·3307644
776	739
822	833
869	928
915	·3308022
962	117
2142009	212
056	306
102	401
149	495
196	590
243	685
289	780
336	874
382	969
429	·3309064
476	159
522	253
569	348
615	442
662	537
709	632
756	726
802	821
849	915
896	·3310010
943	104
989	199
2143036	293
082	388
129	482
176	577
222	671
269	766
315	860
362	955
409	·3311050
456	144
502	239
549	333
596	428
643	522
689	617
736	711
782	806
829	900
876	995
922	·3312089
969	184
2144015	278
062	373
109	467
155	562
202	656
248	751
295	845
342	939
388	·3313034
435	128
481	223
528	317
575	411
622	506
668	600
715	695

2144762	·3313789
809	883
855	978
902	·3314072
948	167
995	261
2145042	355
088	450
135	544
181	639
228	733
275	827
321	922
368	·3315016
414	111
461	205
508	299
554	394
601	488
647	583
694	677
741	771
787	865
834	960
880	·3316054
927	148
974	242
2146020	337
067	431
113	526
160	620
207	714
253	808
300	903
346	997
393	·3317091
440	185
486	279
533	374
579	468
626	562
673	656
719	751
766	845
812	940
859	·3318034
906	128
952	222
999	317
2147045	411
092	505
138	599
185	693
231	788
278	882
324	976
371	·3319070
417	164
464	259
510	353
557	447
604	541
650	635
697	729
743	823

Number	Log
2147790	·3319917
837	·3320011
883	105
930	200
976	294
2148023	388
070	482
116	576
163	671
209	765
256	859
302	953
349	·3321047
395	141
442	235
488	329
535	423
581	517
628	612
674	706
721	800
768	894
814	988
861	·3322082
907	176
954	270
2149000	364
047	458
093	552
140	646
186	740
233	834
279	928
326	·3323022
372	116
419	210
466	304
512	398
559	492
605	586
652	680
698	774
745	868
791	962
838	·3324056
884	150
931	244
977	338

Excesses

10 = 20	1 = 2
20 = 40	2 = 4
30 = 61	3 = 6
40 = 81	4 = 8
	5 = 10
	6 = 12
	7 = 14
	8 = 16
	9 = 18

23 = 47

Anti Log Excesses

10 = 5	1 = 0
20 = 10	2 = 1
30 = 15	3 = 1
40 = 20	4 = 2
50 = 25	5 = 2
60 = 30	6 = 3
70 = 35	7 = 3
80 = 40	8 = 4
90 = 44	9 = 4

47 = 23

Number	Log
2150024	·3324432
070	526
117	620
163	714
210	808
256	901
303	995
349	·3325089
396	183
442	277
489	371
535	465
582	559
628	653
675	747
721	840
768	934
814	·3326028
861	122
907	216
954	309
2151000	403
047	497
093	591
140	685
186	779
233	873
279	967
326	·3327061
372	155
419	248
465	342
512	436
558	530
605	624
651	717
698	811

Number	Log
2151744	·3327905
790	999
837	·3328093
883	186
930	280
976	374
2152023	468
069	562
116	655
162	749
209	843
255	937
302	·3329031
348	124
395	218
441	312
487	406
534	499
580	593
627	686
673	780
720	874
766	968
813	·3330061
859	155
906	249
952	343
999	436
2153045	530
092	623
138	717
184	811
231	904
277	998
324	·3331091
370	185
416	279
463	373
509	466
556	560
602	654
648	748
695	841
741	935
788	·3332028
834	122
880	216
927	309
973	403
2154020	496
066	590
113	684
159	777
206	871
252	964
299	·3333058
345	152
392	245
438	339
485	432
531	526
577	619
624	713
670	806
717	900

Number	Log
2154763	·3333993
809	·3334087
856	180
902	274
949	367
995	461
2155041	555
088	648
134	742
181	835
227	929
273	·3335022
320	116
366	209
413	303
459	396
505	489
552	583
598	676
645	770
691	863
737	957
784	·3336050
830	144
877	237
923	331
969	424
2156015	518
062	611
108	705
154	798
200	891
247	985
293	·3337078
340	172
386	265
432	358
479	452
525	545
572	639
618	732
664	825
711	918
757	·3338012
804	105
850	198
896	291
943	385
989	478
2157036	572
082	665
128	758
175	852
221	945
268	·3339039
314	132
360	225
406	318
453	412
499	505
545	598
591	691
638	785
684	878
731	972

Number	Log
2157777	·3340065
823	158
870	251
916	345
963	438
2158009	531
055	624
102	718
148	811
195	905
241	998
287	·3341091
333	184
380	278
426	371
472	464
518	557
565	650
611	744
658	837
704	930
750	·3342023
796	116
843	210
889	303
935	396
981	489
2159028	582
074	675
121	768
167	861
213	954
260	·3343047
306	141
353	234
399	327
445	420
491	513
538	607
584	700
630	793
676	886
723	979
769	·3344072
816	165
862	258
908	351
954	444

Excesses

10 = 20	1 = 2
20 = 40	2 = 4
30 = 60	3 = 6
40 = 81	4 = 8
	5 = 10
	6 = 12
	7 = 14
	8 = 16
	9 = 18

23 = 47

Anti Log Excesses

10 = 5	1 = 0
20 = 10	2 = 1
30 = 15	3 = 1
40 = 20	4 = 2
50 = 25	5 = 2
60 = 30	6 = 3
70 = 35	7 = 3
80 = 40	8 = 4
90 = 45	9 = 4

47 = 23

2160001	·3344538
047	631
093	724
139	817
186	910
232	·3345003
279	096
325	189
371	282
417	375
464	468
510	561
556	654
602	747
648	840
695	933
741	·3346026
787	119
833	212
880	305
926	398
973	491
2161019	584
065	677
111	770
158	863
204	956
250	·3347049
296	142
342	235
389	328
435	421
481	514
527	607
574	700
620	793
667	886

2161713	·3347979
759	·3348072
805	165
852	258
898	351
944	444
990	537
2162036	630
083	722
129	815
175	908
221	·3349001
268	094
314	187
361	280
407	373
453	466
499	559
546	651
592	744
638	837
684	930
730	·3350023
777	115
823	208
869	301
915	394
961	487
2163008	579
054	672
100	765
146	858
192	951
239	·3351043
285	136
331	229
377	322
423	415
470	507
516	600
562	693
608	786
654	879
701	971
747	·3352064
793	157
839	250
886	343
932	435
979	528
2164025	621
071	714
117	807
164	899
210	992
256	·3353085
302	178
348	270
395	363
441	455
487	548
533	641
579	733
626	826
672	918

2164718	·3354011
764	104
810	197
857	289
903	382
949	475
995	568
2165041	660
087	753
133	845
179	938
225	·3355031
271	123
318	216
364	308
410	401
456	494
502	586
549	679
595	771
641	864
687	957
733	·3356049
780	142
826	234
872	327
918	420
964	512
2166011	605
057	697
103	790
149	883
195	975
242	·3357068
288	160
334	253
380	345
426	438
473	530
519	623
565	715
611	808
657	900
703	993
749	·3358085
795	178
841	271
887	363
934	456
980	548
2167026	641
072	733
118	826
165	918
211	·3359011
257	103
303	195
349	288
396	380
442	473
488	565
534	657
580	750
626	842
672	935

2167718	·3360027
764	119
810	212
857	304
903	397
949	489
995	581
2168041	674
087	766
133	859
179	951
225	·3361043
271	136
318	228
364	321
410	413
456	505
502	598
549	690
595	783
641	875
687	967
733	·3362060
779	152
825	245
871	337
917	429
963	521
2169010	614
056	706
102	798
148	890
194	983
240	·3363075
286	168
332	260
378	352
424	444
471	537
517	629
563	721
609	813
655	905
701	998
747	·3364090
793	182
839	274
885	367
932	459
978	552

Excesses

10 = 20	1 = 2
20 = 40	2 = 4
30 = 60	3 = 6
40 = 80	4 = 8
	5 = 10
	6 = 12
	7 = 14
	8 = 16
	9 = 18

23 = 46

Anti Log Excesses

10 = 5	1 = 0
20 = 10	2 = 1
30 = 15	3 = 1
40 = 20	4 = 2
50 = 25	5 = 2
60 = 30	6 = 3
70 = 35	7 = 3
80 = 40	8 = 4
90 = 45	9 = 4

46 = 23

2170024	·3364644
070	736
116	828
162	921
208	·3365013
254	105
300	197
346	289
392	382
438	474
484	566
530	658
576	750
623	842
669	934
715	·3366026
761	118
807	210
853	303
899	395
945	487
991	579
2171037	671
083	764
129	856
175	948
221	·3367040
267	132
314	225
360	317
406	409
452	501
498	593
544	685
590	777
636	869
682	961

2171728	·3368053
774	145
820	237
866	329
912	421
958	513
2172004	606
050	698
096	790
142	882
188	974
234	·3369066
280	158
326	250
372	342
418	434
465	526
511	618
557	710
603	802
649	894
695	986
741	·3370078
787	170
833	262
879	354
925	446
971	538
2173017	630
063	722
109	814
155	906
201	998
247	·3371090
293	182
339	274
385	365
431	457
477	549
523	641
569	733
615	825
661	917
707	·3372009
753	101
799	193
845	285
891	377
937	469
983	561
2174029	653
075	744
121	836
167	928
213	·3373020
259	112
305	203
351	295
397	387
443	479
489	571
535	663
581	755
627	847
673	939

2174719	·3374031
765	122
811	214
857	306
903	398
949	490
995	581
2175041	673
087	765
133	857
179	949
225	·3375040
271	132
317	224
363	316
409	408
454	499
500	591
546	683
592	775
638	866
684	958
730	·3376049
776	141
822	233
868	325
914	416
960	508
2176006	600
052	692
098	784
144	875
190	967
236	·3377059
282	151
328	242
374	334
420	425
466	517
512	609
558	700
603	792
649	883
695	975
741	·3378067
787	159
833	250
879	342
925	434
971	526
2177017	617
063	709
109	800
155	892
201	984
247	·3379075
292	167
338	258
384	350
430	442
476	533
522	625
568	716
614	808
660	900

2177706	·3379991
751	·3380083
797	174
843	266
889	358
935	449
981	541
2178027	632
073	724
119	815
165	907
211	998
257	·3381090
303	181
349	273
395	364
440	456
486	547
532	639
578	730
624	822
670	913
716	·3382005
762	096
808	188
854	279
899	371
945	462
991	554
2179037	645
083	737
129	828
175	920
221	·3383011
267	102
313	194
358	285
404	377
450	468
496	559
542	651
587	742
633	834
679	925
725	·3384016
771	108
817	199
863	291
909	382
955	473

Excesses

10 = 20	1 = 2
20 = 40	2 = 4
30 = 60	3 = 6
40 = 80	4 = 8
	5 = 10
	6 = 12
	7 = 14
	8 = 16
	9 = 18

23 = 46

Anti Log Excesses

10 = 5	1 = 1
20 = 10	2 = 1
30 = 15	3 = 2
40 = 20	4 = 2
50 = 25	5 = 3
60 = 30	6 = 3
70 = 35	7 = 4
80 = 40	8 = 4
90 = 45	9 = 5

46 = 23

2180001	·3384565
046	656
092	748
138	839
184	930
230	·3385022
275	113
321	205
367	296
413	387
459	479
505	570
551	662
597	753
643	844
689	935
734	·3386027
780	118
826	209
872	300
918	392
963	483
2181009	575
055	666
101	757
147	848
193	940
239	·3387031
285	122
331	213
377	305
422	396
468	488
514	579
560	670
606	761
651	853

2181697	·3387944
743	·3388035
789	126
835	217
880	309
926	400
972	491
2182018	582
064	673
109	765
155	856
201	947
247	·3389038
293	129
338	221
384	312
430	403
476	494
522	585
567	677
613	768
659	859
705	950
751	·3390041
796	133
842	224
888	315
934	406
980	497
2183026	588
072	679
118	770
164	861
210	952
255	·3391044
301	135
347	226
393	317
439	408
484	499
530	590
576	681
622	772
668	863
713	955
759	·3392046
805	137
851	228
896	319
942	410
987	501
2184033	592
079	683
125	774
170	865
216	956
262	·3393047
308	138
354	229
399	320
445	411
491	502
537	593
583	684
628	775

2184674	·3393866
720	957
766	·3394048
812	139
857	230
903	321
949	412
995	503
2185041	594
086	685
132	776
178	867
224	958
270	·3395049
315	140
361	231
407	322
453	413
498	504
544	595
589	686
635	777
681	868
727	959
772	·3396049
818	140
864	231
910	322
956	413
2186001	503
047	594
093	685
139	776
185	867
230	958
276	·3397049
322	140
368	231
413	322
459	412
504	503
550	594
596	685
642	776
687	866
733	957
779	·3398048
825	139
871	230
916	320
962	411
2187008	502
054	593
099	684
145	774
190	865
236	956
282	·3399047
328	138
373	228
419	319
465	410
511	501
556	592
602	682

2187647	·3399773
693	864
739	955
785	·3400045
830	136
876	226
922	317
968	408
2188013	499
059	589
104	680
150	771
196	862
242	952
287	·3401043
333	133
379	224
425	315
470	406
516	496
561	587
607	678
653	769
699	859
744	950
790	·3402040
836	131
882	222
927	312
973	403
2189018	493
064	584
110	675
156	765
201	856
247	946
293	·3403037
339	128
384	218
430	309
475	399
521	490
567	581
612	671
658	762
703	852
749	943
795	·3404034
841	124
886	215
932	305
978	396

Excesses

10 = 20	1 = 2
20 = 40	2 = 4
30 = 60	3 = 6
40 = 80	4 = 8
	5 = 10
	6 = 12
	7 = 14
	8 = 16
	9 = 18

23 = 46

Anti Log Excesses

10 = 5	1 = 1
20 = 10	2 = 1
30 = 15	3 = 2
40 = 20	4 = 2
50 = 25	5 = 3
60 = 30	6 = 3
70 = 35	7 = 4
80 = 40	8 = 4
90 = 45	9 = 5

46 = 23

2190024	·3404487
069	577
115	668
160	758
206	849
252	939
297	·3405030
343	120
388	211
434	301
480	392
526	482
571	573
617	663
663	754
709	844
754	935
800	·3406025
845	116
891	206
937	297
982	387
2191028	478
073	568
119	659
165	749
210	840
256	930
301	·3407021
347	111
393	201
438	292
484	382
529	473
575	563
621	653
666	744

2191712	·3407834
757	925
803	·3408015
849	105
895	196
940	286
986	377
2192032	467
078	557
123	648
169	738
214	829
260	919
306	·3409009
351	100
397	190
442	281
488	371
534	461
579	551
625	642
670	732
716	822
762	912
807	·3410003
853	093
898	184
944	274
990	364
2193035	455
081	545
126	636
172	726
218	816
263	906
309	997
354	·3411087
400	177
446	267
491	357
537	448
582	538
628	628
674	718
719	808
765	899
810	989
856	·3412079
901	169
947	259
992	350
2194038	440
083	530
129	620
174	711
220	801
265	892
311	982
357	·3413072
402	162
448	252
493	342
539	432
585	522
630	612

2194676	·3413703
721	793
767	883
813	973
858	·3414063
904	154
949	244
995	334
2195041	424
086	514
132	605
177	695
223	785
268	875
314	965
359	·3415055
405	145
450	235
496	325
541	415
587	506
632	596
678	686
724	776
769	866
815	956
860	·3416046
906	136
951	226
997	316
2196042	406
088	496
133	586
179	676
224	766
270	857
315	947
361	·3417037
407	127
452	217
498	307
543	397
589	487
634	577
680	667
725	757
771	847
816	937
862	·3418027
907	117
953	207
998	297
2197044	387
090	477
135	567
181	657
226	747
272	837
317	927
363	·3419017
408	106
454	196
499	286
545	376
590	466

2197636	·3419556
681	646
727	736
772	826
818	916
863	·3420006
909	096
954	186
2198000	276
045	366
091	455
136	545
182	635
227	725
273	815
318	904
364	994
409	·3421084
454	174
500	264
545	354
591	444
636	534
682	624
727	714
773	803
818	893
864	983
909	·3422073
955	163
2199000	252
046	342
091	432
137	522
182	612
228	701
273	791
319	881
364	971
410	·3423061
455	150
501	240
546	330
591	420
637	509
682	599
728	688
773	778
818	868
864	958
909	·3424047
955	137

Excesses

10 = 20	1 = 2
20 = 40	2 = 4
30 = 59	3 = 6
40 = 79	4 = 8
	5 = 10
	6 = 12
	7 = 14
	8 = 16
	9 = 18

23 = 45

Anti Log Excesses

10 = 5	1 = 1
20 = 10	2 = 1
30 = 15	3 = 2
40 = 20	4 = 2
50 = 25	5 = 3
60 = 30	6 = 3
70 = 35	7 = 4
80 = 40	8 = 4
90 = 45	9 = 5

45 = 23

2200000	·3424227
046	317
091	407
137	496
182	586
228	676
273	766
319	855
364	945
410	·3425034
455	124
500	214
546	304
591	393
637	483
682	573
727	663
773	752
818	842
864	931
909	·3426021
955	111
2201000	200
046	290
091	379
137	469
182	559
228	648
273	738
319	827
364	917
409	·3427007
455	096
500	186
546	275
591	365
636	455

2201682	·3427544
727	634
773	723
818	813
863	903
909	992
954	·3428082
2202000	171
045	261
090	351
136	440
181	530
227	619
272	709
317	799
363	888
408	978
454	·3429067
499	157
544	246
590	336
635	425
681	515
726	604
771	694
817	783
862	873
908	962
953	·3430052
998	141
2203044	231
089	320
135	410
180	499
225	588
271	678
316	767
362	857
407	946
452	·3431036
498	125
543	215
589	304
634	394
679	483
725	573
770	662
816	752
861	841
906	930
952	·3432020
997	109
2204043	199
088	288
133	377
178	467
224	556
269	646
314	735
359	824
405	914
450	·3433003
496	093
541	182
586	271

2204632	·3433360
677	450
723	539
768	628
813	717
859	807
904	896
950	986
995	·3434075
2205040	164
086	253
131	343
177	432
222	521
267	610
312	700
358	789
403	879
448	968
493	·3435057
539	146
584	236
630	325
675	414
720	503
766	593
811	682
857	772
902	861
947	950
992	·3436039
2206038	129
083	218
128	307
173	396
219	485
264	575
310	664
355	753
400	842
446	931
491	·3437021
537	110
582	199
627	288
672	377
718	467
763	556
808	645
853	734
899	823
944	913
990	·3438002
2207035	091
080	180
125	269
171	358
216	447
261	536
306	625
352	714
397	804
443	893
488	982
533	·3439071

2207578	·3439160
624	250
669	339
714	428
759	517
805	606
850	695
896	784
941	873
986	962
2208031	·3440051
077	141
122	230
167	319
212	408
258	497
303	586
349	675
394	764
439	853
484	942
530	·3441031
575	120
620	209
665	298
710	387
756	476
801	565
846	654
891	743
937	832
982	921
2209028	·3442010
073	099
118	188
163	277
209	366
254	455
299	544
344	633
389	722
435	811
480	900
525	989
570	·3443078
616	167
661	255
707	344
752	433
797	522
842	611
888	700
933	789
978	878

Excesses

10 = 20	1 = 2
20 = 39	2 = 4
30 = 59	3 = 6
40 = 79	4 = 8
	5 = 10
	6 = 12
	7 = 14
	8 = 16
	9 = 18

23 = 45

Anti Log Excesses

10 = 5	1 = 1
20 = 10	2 = 1
30 = 15	3 = 2
40 = 20	4 = 2
50 = 25	5 = 3
60 = 30	6 = 3
70 = 36	7 = 4
80 = 41	8 = 4
	9 = 5

45 = 23

2210023	·3443967
068	·3444056
114	145
159	234
204	323
249	412
294	501
340	589
385	678
430	767
475	856
521	945
566	·3445034
612	123
657	212
702	301
747	390
793	478
838	567
883	656
928	745
973	834
2211019	922
064	·3446011
109	100
154	189
199	278
245	366
290	455
335	544
380	633
425	722
471	810
516	899
561	988
606	·3447077
651	166
2211697	·3447254
742	343
787	432
832	521
877	610
923	698
968	787
2212013	876
058	965
103	·3448054
149	142
194	231
239	320
284	409
329	497
375	586
420	674
465	763
510	852
555	941
601	·3449029
646	118
691	207
736	296
781	384
827	473
872	561
917	650
962	739
2213007	828
053	916
098	·3450005
143	094
188	183
233	271
279	360
324	448
369	537
414	626
459	714
505	803
550	891
595	980
640	·3451069
685	157
731	246
776	334
821	423
866	512
911	600
957	689
2214002	777
047	866
092	955
137	·3452043
182	132
227	220
272	309
317	398
362	486
408	575
453	663
498	752
543	841
588	929
2214634	·3453018
679	106
724	195
769	283
814	372
860	460
905	549
950	637
995	726
2215040	814
085	903
130	991
175	·3454080
220	168
265	257
311	345
356	434
401	522
446	611
491	699
537	788
582	876
627	965
672	·3455053
717	142
762	230
807	319
852	407
897	496
942	584
988	673
2216033	761
078	850
123	938
168	·3456027
214	115
259	204
304	292
349	380
394	469
439	557
484	646
529	734
574	822
619	911
665	999
710	·3457088
755	176
800	264
845	353
890	441
935	530
980	618
2217025	706
070	795
116	883
161	972
206	·3458060
251	148
296	236
341	325
386	413
431	501
476	589
521	678
2217567	·3458766
612	855
657	943
702	·3459031
747	119
792	208
837	296
882	384
927	472
972	561
2218018	649
063	738
108	826
153	914
198	·3460002
243	091
288	179
333	267
378	355
423	443
469	532
514	620
559	708
604	796
649	884
694	973
739	·3461061
784	149
829	237
874	325
919	414
964	502
2219009	590
054	678
099	766
145	855
190	943
235	·3462031
280	119
325	207
370	296
415	384
460	472
505	560
550	648
595	737
640	825
685	913
730	·3463001
775	089
820	178
865	266
910	354
955	442

Excesses

10 = 20	1 = 2
20 = 39	2 = 4
30 = 59	3 = 6
40 = 78	4 = 8
	5 = 10
	6 = 12
	7 = 14
	8 = 16
	9 = 18

23 = 44

Anti Log Excesses

10 = 5	1 = 1
20 = 10	2 = 1
30 = 15	3 = 2
40 = 20	4 = 2
50 = 25	5 = 3
60 = 31	6 = 3
70 = 36	7 = 4
80 = 41	8 = 4
	9 = 5

44 = 23

2220000	·3463530
046	618
091	706
136	794
181	882
226	970
271	·3464059
316	147
361	235
406	323
451	411
496	499
541	587
586	675
631	763
676	851
721	939
766	·3465027
811	115
856	203
901	291
946	380
991	468
2221036	556
081	644
126	732
171	820
216	908
261	996
306	·3466084
351	172
396	260
441	348
486	436
531	524
576	612
622	700

2221667	·3466788
712	876
757	964
802	·3467052
847	140
892	228
937	316
982	404
2222027	492
072	579
117	667
162	755
207	843
252	931
297	·3468019
342	107
387	195
432	283
477	371
522	459
567	547
612	635
657	723
702	811
747	898
792	986
837	·3469074
882	162
927	250
971	338
2223016	426
061	514
106	602
151	690
196	777
241	865
286	953
331	·3470041
376	129
421	216
466	304
511	392
556	480
601	568
646	656
691	744
736	832
781	920
826	·3471008
871	095
916	183
961	271
2224006	359
051	446
096	534
141	621
186	709
231	797
276	885
321	972
366	·3472060
411	148
456	236
501	324
545	411

2224590	·3472499
635	587
680	675
725	763
770	850
815	938
860	·3473026
905	114
950	202
995	289
2225040	377
085	465
130	553
175	640
219	728
264	815
309	903
354	991
399	·3474079
444	166
489	254
534	342
579	430
624	517
669	605
714	692
759	780
804	868
849	955
893	·3475043
938	130
983	218
2226028	306
073	393
118	481
163	568
208	656
253	744
298	831
343	919
388	·3476006
433	094
478	182
523	269
567	357
612	444
657	532
702	620
747	708
792	795
837	883
882	971
927	·3477058
972	146
2227016	233
061	321
106	408
151	496
196	583
241	671
286	758
331	846
376	934
421	·3478021
465	109

2227510	·3478196
555	284
600	371
645	459
690	546
735	634
780	721
825	809
870	896
914	984
959	·3479071
2228004	159
049	246
094	334
138	421
183	509
228	596
273	684
318	771
363	859
408	946
453	·3480034
498	121
543	209
587	296
632	384
677	471
722	558
767	646
812	733
857	821
902	908
947	995
992	·3481083
2229036	170
081	258
126	345
171	432
216	520
260	607
305	695
350	782
395	869
440	957
484	·3482044
529	132
574	219
619	306
664	394
709	481
754	569
799	656
844	743
889	830
933	918
978	·3483005

Excesses

10 = 20	1 = 2
20 = 39	2 = 4
30 = 59	3 = 6
40 = 78	4 = 8
	5 = 10
	6 = 12
	7 = 14
	8 = 16
	9 = 18

22 = 44

Anti Log Excesses

10 = 5	1 = 1
20 = 10	2 = 1
30 = 15	3 = 2
40 = 20	4 = 2
50 = 26	5 = 3
60 = 31	6 = 3
70 = 36	7 = 4
80 = 41	8 = 4
	9 = 5

44 = 22

2230023	·3483092
068	179
113	267
157	354
202	442
247	529
292	616
337	703
381	791
426	878
471	965
516	·3484052
561	140
605	227
650	315
695	402
740	489
785	576
830	664
875	751
920	838
965	925
2231010	·3485012
054	100
099	187
144	274
189	361
234	449
278	536
323	624
368	711
413	798
458	885
502	973
547	·3486060
592	147
637	234

2231682	·3486321
726	409
771	496
816	583
861	670
906	757
950	845
995	932
2232040	·3487019
085	106
130	193
174	280
219	367
264	454
309	541
354	628
398	716
443	803
488	890
533	977
578	·3488064
622	152
667	239
712	326
757	413
802	500
846	587
891	674
936	761
981	848
2233026	935
070	·3489023
115	110
160	197
205	284
250	371
294	458
339	545
384	632
429	719
473	806
518	893
562	980
607	·3490067
652	154
697	241
741	329
786	416
831	503
876	590
921	677
965	764
2234010	851
055	938
100	·3491025
145	112
189	199
234	286
279	373
324	460
369	547
413	634
458	721
503	808
548	895

2234592	·3491982
637	·3492069
681	156
726	243
771	330
816	417
860	503
905	590
950	677
995	764
2235040	851
084	938
129	·3493025
174	112
219	199
264	286
308	373
353	460
398	547
443	634
487	721
532	807
576	894
621	981
666	·3494068
711	155
755	242
800	329
845	416
890	503
934	590
979	676
2236023	763
068	850
113	937
158	·3495024
202	110
247	197
292	284
337	371
382	458
426	544
471	631
516	718
561	805
605	892
650	979
694	·3496066
739	153
784	240
829	327
873	413
918	500
963	587
2237008	674
052	761
097	847
141	934
186	·3497021
231	108
276	194
320	281
365	367
410	454
455	541

2237499	·3497628
544	714
588	801
633	888
678	975
723	·3498062
767	148
812	235
857	322
902	409
946	495
991	582
2238035	668
080	755
125	842
169	929
214	·3499015
258	102
303	189
348	276
393	362
437	449
482	535
527	622
572	709
616	795
661	882
705	968
750	·3500055
795	142
839	228
884	315
928	401
973	488
2239018	575
063	662
107	748
152	835
197	922
242	·3501009
286	095
331	182
375	268
420	355
465	442
509	528
554	615
598	701
643	788
688	875
733	961
777	·3502048
822	134
867	221
912	307
956	394

Excesses

10 = 20	1 = 2
20 = 39	2 = 4
30 = 59	3 = 6
40 = 78	4 = 8
	5 = 10
	6 = 12
	7 = 14
	8 = 16
	9 = 18

22 = 43

Anti Log Excesses

10 = 5	1 = 1
20 = 10	2 = 1
30 = 15	3 = 2
40 = 21	4 = 2
50 = 26	5 = 3
60 = 31	6 = 3
70 = 36	7 = 4
80 = 41	8 = 4
	9 = 5

43 = 22

2240001	·3502480
045	567
090	653
135	740
179	826
224	913
268	999
313	·3503086
358	173
402	259
447	346
491	432
536	519
581	605
625	692
670	778
714	865
759	951
804	·3504038
848	124
893	211
937	297
982	384
2241027	470
072	557
116	643
161	730
206	816
251	902
295	989
340	·3505075
384	162
429	248
474	334
518	421
563	507
607	594

2241652	·3505680
697	766
741	853
786	939
830	·3506026
875	112
920	198
964	285
2242009	371
053	458
098	544
143	630
187	717
232	803
276	890
321	976
366	·3507062
410	149
455	235
499	322
544	408
589	494
633	581
678	667
722	754
767	840
812	926
856	·3508013
901	099
945	186
990	272
2243034	358
079	444
123	531
168	617
212	703
257	789
301	876
346	962
390	·3509049
435	135
480	221
524	307
569	394
613	480
658	566
703	652
747	738
792	825
836	911
881	997
926	·3510083
970	170
2244015	256
059	343
104	429
149	515
193	601
238	688
282	774
327	860
371	946
416	·3511032
460	119
505	205

2244549	·3511291
594	377
638	463
683	550
727	636
772	722
817	808
861	894
906	981
950	·3512067
995	153
2245040	239
084	325
129	411
173	497
218	583
262	669
307	755
351	842
396	928
440	·3513014
485	100
529	186
574	273
618	359
663	445
708	531
752	617
797	703
841	789
886	875
930	961
975	·3514047
2246019	134
064	220
108	306
153	392
197	478
242	564
286	650
331	736
375	822
420	908
464	994
509	·3515080
553	166
598	252
642	338
687	425
731	511
776	597
820	683
865	769
909	855
954	941
998	·3516027
2247043	113
087	199
132	285
176	371
221	457
265	543
310	629
354	715
399	801

2247443	·3516887
488	973
532	·3517059
577	145
621	231
666	317
710	403
755	489
799	574
844	660
888	746
933	832
977	918
2248022	·3518004
066	090
111	176
155	262
200	348
244	434
289	520
333	606
378	692
422	778
467	863
511	949
556	·3519035
600	121
645	207
689	293
734	379
778	465
822	551
867	637
911	722
956	808
2249000	894
045	980
089	·3520066
134	151
178	237
223	323
267	409
312	495
356	581
401	667
445	753
489	839
534	925
578	·3521010
623	096
667	182
711	268
756	354
800	439
845	525
889	611
934	697
978	783

Excesses

10 = 19	1 = 2
20 = 39	2 = 4
30 = 58	3 = 6
40 = 77	4 = 8
	5 = 10
	6 = 12
	7 = 14
	8 = 15
	9 = 17

22 = 43

Anti Log Excesses

10 = 5	1 = 1
20 = 10	2 = 1
30 = 16	3 = 2
40 = 21	4 = 2
50 = 26	5 = 3
60 = 31	6 = 3
70 = 36	7 = 4
80 = 41	8 = 4
	9 = 5

43 = 22

2250023	·3521868
067	954
112	·3522040
156	126
201	211
245	297
290	382
334	468
378	554
423	640
467	725
512	811
556	897
600	983
645	·3523069
689	154
734	240
778	326
822	412
867	497
911	583
956	668
2251000	754
044	840
089	926
133	·3524011
178	097
222	183
266	269
311	354
355	440
400	525
444	611
489	697
533	783
578	868
622	954

2251667	·3525040
711	126
756	211
800	297
845	382
889	468
933	554
978	639
2252022	725
067	810
111	896
155	982
200	·3526067
244	153
289	238
333	324
377	410
422	495
466	581
511	666
555	752
599	838
644	923
688	·3527009
733	094
777	180
821	266
866	351
910	437
955	522
999	608
2253043	694
087	779
132	865
176	950
220	·3528036
264	121
309	207
353	292
398	378
442	463
486	549
531	634
575	720
620	805
664	891
708	976
753	·3529062
797	147
842	233
886	318
930	404
975	489
2254019	575
064	660
108	746
152	831
197	917
241	·3530002
286	088
330	173
374	258
418	344
463	429
507	515

2254551	·3530600
595	686
640	771
684	857
729	942
773	·3531028
817	113
862	199
906	284
951	370
995	455
2255039	540
084	626
128	711
173	797
217	882
261	967
305	·3532052
350	138
394	223
438	308
482	393
527	479
571	564
616	650
660	735
704	820
749	906
793	991
838	·3533077
882	162
926	247
970	333
2256015	418
059	504
103	589
147	674
192	759
236	845
281	930
325	·3534015
369	100
414	186
458	271
503	357
547	442
591	527
635	612
680	698
724	783
768	868
812	953
857	·3535039
901	124
946	210
990	295
2257034	380
078	465
123	551
167	636
211	721
255	806
300	891
344	977
389	·3536062

2257433	·3536147
477	232
521	317
566	403
610	488
654	573
698	658
743	743
787	829
832	914
876	999
920	·3537084
964	169
2258009	255
053	340
097	425
141	510
185	595
230	681
274	766
318	851
362	936
407	·3538021
451	107
496	192
540	277
584	362
628	447
673	532
717	617
761	702
805	787
850	872
894	958
939	·3539043
983	128
2259027	213
071	298
116	384
160	469
204	554
248	639
292	724
337	809
381	894
425	979
469	·3540064
513	149
558	234
602	319
646	404
690	489
735	574
779	659
824	744
868	829
912	914
956	999

Excesses

10 = 19	1 = 2
20 = 39	2 = 4
30 = 58	3 = 6
40 = 77	4 = 8
	5 = 10
	6 = 12
	7 = 13
	8 = 15
	9 = 17

22 = 43

Anti Log Excesses

10 = 5	1 = 1
20 = 10	2 = 1
30 = 16	3 = 2
40 = 21	4 = 2
50 = 26	5 = 3
60 = 31	6 = 3
70 = 36	7 = 4
80 = 42	8 = 4
	9 = 5

43 = 22

2260001	·3541084
045	169
089	254
133	339
177	424
222	510
266	595
310	680
354	765
398	850
443	935
487	·3542020
531	105
575	190
620	275
664	360
709	445
753	530
797	615
841	700
886	784
930	869
974	954
2261018	·3543039
062	124
107	209
151	294
195	379
239	464
283	549
328	634
372	719
416	804
460	889
504	974
549	·3544058
593	143

2261637	·3544228
681	313
725	398
770	483
814	568
858	653
902	738
946	823
991	907
2262035	992
079	·3545077
123	162
167	247
212	332
256	417
300	502
344	587
388	672
433	756
477	841
521	926
565	·3546011
609	096
654	180
698	265
742	350
786	435
830	520
875	604
919	689
963	774
2263007	859
051	944
096	·3547028
140	113
184	198
228	283
272	368
317	452
361	537
405	622
449	707
493	792
538	876
582	961
626	·3548046
670	131
714	215
759	300
803	384
847	469
891	554
935	639
980	723
2264024	808
068	893
112	978
156	·3549063
200	147
244	232
288	317
332	402
376	486
421	571
465	655

2264509	·3549740
553	825
597	910
642	994
686	·3550079
730	164
774	249
818	333
863	418
907	502
951	587
995	672
2265039	756
084	841
128	925
172	·3551010
216	095
260	179
304	264
348	348
392	433
436	518
480	602
525	687
569	771
613	856
657	941
701	·3552025
746	110
790	194
834	279
878	364
922	448
966	533
2266010	617
054	702
098	787
142	871
187	956
231	·3553040
275	125
319	210
363	294
408	379
452	463
496	548
540	633
584	717
628	802
672	886
716	971
760	·3554055
804	140
849	224
893	309
937	393
981	478
2267025	562
069	647
113	731
157	816
201	900
245	985
290	·3555069
334	154

2267378	·3555238
422	322
466	407
510	491
554	576
598	660
642	745
686	829
731	914
775	998
819	·3556083
863	167
907	252
951	336
995	421
2268039	505
083	589
127	674
172	758
216	843
260	927
304	·3557011
348	096
392	180
436	265
480	349
524	433
568	518
613	602
657	687
701	771
745	855
789	940
833	·3558024
877	109
921	193
965	277
2269009	361
053	446
097	530
141	614
185	698
229	783
274	867
318	952
362	·3559036
406	120
450	205
494	289
538	374
582	458
626	542
670	626
714	711
758	795
802	879
846	963
890	·3560048
935	132
979	217

Excesses

10 = 19	1 = 2
20 = 38	2 = 4
30 = 58	3 = 6
40 = 77	4 = 8
	5 = 10
	6 = 12
	7 = 13
	8 = 15
	9 = 17

22 = 42

Anti Log Excesses

10 = 5	1 = 1
20 = 10	2 = 1
30 = 16	3 = 2
40 = 21	4 = 2
50 = 26	5 = 3
60 = 31	6 = 3
70 = 37	7 = 4
80 = 42	8 = 4
	9 = 5

42 = 22

2270023	·3560301
067	385
111	469
155	554
199	638
243	722
287	806
331	890
375	975
419	·3561059
463	143
507	227
551	312
595	396
639	481
683	565
727	649
771	733
815	818
859	902
903	986
947	·3562070
991	154
2271036	239
080	323
124	407
168	491
212	575
256	660
300	744
344	828
388	912
432	996
476	·3563081
520	165
564	249
608	333

2271652	·3563417
696	501
740	585
784	669
828	753
872	837
916	922
960	·3564006
2272004	090
048	174
092	258
136	343
180	427
224	511
268	595
312	679
356	763
400	847
444	931
488	·3565015
532	099
576	184
620	268
664	352
708	436
752	520
796	604
840	688
884	772
928	856
972	940
2273016	·3566024
060	108
104	192
148	276
192	360
236	445
280	529
324	613
368	697
412	781
456	865
500	949
544	·3567033
588	117
632	201
676	285
720	369
764	453
808	537
852	621
896	705
940	789
984	873
2274028	957
072	·3568041
116	124
160	208
204	292
248	376
292	460
336	544
380	628
424	712
468	796

2274512	·3568880
555	964
599	·3569048
643	132
687	216
731	300
775	384
819	468
863	552
907	636
951	720
995	803
2275039	887
083	971
127	·3570055
171	139
215	223
259	307
303	391
347	475
391	559
434	642
478	726
522	810
566	894
610	978
654	·3571061
698	145
742	229
786	313
830	397
874	481
918	565
962	649
2276006	733
050	817
094	900
138	984
182	·3572068
226	152
270	236
313	319
357	403
401	487
445	571
489	655
533	738
577	822
621	906
665	990
709	·3573074
752	157
796	241
840	325
884	409
928	493
972	576
2277016	660
060	744
104	828
148	912
192	995
236	·3574079
280	163
324	247

2277368	·3574330
411	414
455	497
499	581
543	665
587	749
631	832
675	916
719	·3575000
763	084
807	167
850	251
894	334
938	418
982	502
2278026	586
070	669
114	753
158	837
202	921
246	·3576004
289	088
333	171
377	255
421	339
465	423
509	506
553	590
597	674
641	758
685	841
728	925
772	·3577008
816	092
860	176
904	259
947	343
991	426
2279035	510
079	594
123	677
167	761
211	844
255	928
299	·3578012
343	095
386	179
430	262
474	346
518	430
562	513
605	597
649	680
693	764
737	848
781	931
825	·3579015
869	098
913	182
957	265

Excesses	
10 = 19	1 = 2
20 = 38	2 = 4
30 = 57	3 = 6
40 = 76	4 = 8
	5 = 10
	6 = 11
	7 = 13
	8 = 15
	9 = 17
22 = 42	

Anti Log Excesses	
10 = 5	1 = 1
20 = 10	2 = 1
30 = 16	3 = 2
40 = 21	4 = 2
50 = 26	5 = 3
60 = 31	6 = 3
70 = 37	7 = 4
80 = 42	8 = 4
	9 = 5
42 = 22	

2280001	·3579349
044	432
088	516
132	599
176	683
220	766
263	850
307	933
351	·3580017
395	100
439	184
483	267
527	351
571	434
615	518
659	601
702	685
746	768
790	852
834	935
878	·3581019
921	102
965	186
2281009	269
053	353
097	436
140	520
184	603
228	687
272	770
316	854
359	937
403	·3582021
447	104
491	187
535	271
579	354

2281623	·3582438
667	521
711	604
755	688
798	771
842	855
886	938
930	·3583021
974	105
2282017	188
061	272
105	355
149	438
193	522
236	605
280	689
324	772
368	855
412	939
455	·3584022
499	106
543	189
587	272
631	355
674	439
718	522
762	605
806	688
850	772
893	855
937	939
981	·3585022
2283025	105
069	189
112	272
156	356
200	439
244	522
288	605
331	689
375	772
419	855
463	938
507	·3586022
550	105
594	189
638	272
682	355
726	438
769	522
813	605
857	688
901	771
945	854
988	938
2284032	·3587021
076	104
120	187
164	270
207	354
251	437
295	520
339	603
383	686
426	770

2284470	·3587853
514	936
558	·3588019
601	102
645	186
688	269
732	352
776	435
820	518
863	602
907	685
951	768
995	851
2285039	934
082	·3589018
126	101
170	184
214	267
258	350
301	434
345	517
389	600
433	683
477	766
520	850
564	933
608	·3590016
652	099
695	182
739	265
782	348
826	431
870	514
914	597
957	681
2286001	764
045	847
089	930
133	·3591013
176	096
220	179
264	262
308	345
352	428
395	512
439	595
483	678
527	761
570	844
614	927
657	·3592010
701	093
745	176
789	259
832	342
876	425
920	508
964	591
2287007	674
051	758
094	841
138	924
182	·3593007
226	090
269	173

2287313	·3593256
357	339
401	422
445	505
488	588
532	671
576	754
620	837
663	920
707	·3594003
750	086
794	169
838	252
882	335
925	417
969	500
2288013	583
057	666
100	749
144	832
187	915
231	998
275	·3595081
319	164
362	247
406	330
450	413
494	496
537	579
581	661
624	744
668	827
712	910
756	993
799	·3596076
843	159
887	242
931	325
974	408
2289018	490
061	573
105	656
149	739
193	822
236	905
280	988
324	·3597071
368	154
411	237
455	319
498	402
542	485
586	568
629	651
673	733
716	816
760	899
804	982
848	·3598065
891	147
935	230
979	313

Excesses

10 = 19	1 = 2
20 = 38	2 = 4
30 = 57	3 = 6
40 = 76	4 = 8
	5 = 10
	6 = 11
	7 = 13
	8 = 15
	9 = 17

22 = 42

Anti Log Excesses

10 = 5	1 = 1
20 = 11	2 = 1
30 = 16	3 = 2
40 = 21	4 = 2
50 = 26	5 = 3
60 = 32	6 = 3
70 = 37	7 = 4
80 = 42	8 = 4
	9 = 5

42 = 22

2290023	·3598396
066	479
110	561
153	644
197	727
241	810
284	893
328	975
371	·3599058
415	141
459	224
503	307
546	389
590	472
634	555
678	638
721	721
765	803
808	886
852	969
896	·3600052
939	135
983	217
2291026	300
070	383
114	466
157	549
201	631
244	714
288	797
332	880
376	962
419	·3601045
463	127
507	210
551	293
594	376

2291638	·3601458
681	541
725	624
769	707
812	789
856	872
899	954
943	·3602037
987	120
2292030	202
074	285
117	367
161	450
205	533
248	616
292	698
335	781
379	864
423	947
466	·3603029
510	112
553	194
597	277
641	360
684	442
728	525
771	607
815	690
859	773
902	855
946	938
989	·3604020
2293033	103
077	186
120	268
164	351
207	433
251	516
295	599
338	681
382	764
425	846
469	929
513	·3605011
556	094
600	176
643	259
687	341
731	424
774	506
818	589
861	671
905	754
949	837
992	919
2294036	·3606002
079	084
123	167
167	249
210	332
254	414
297	497
341	579
385	662
428	744

2294472	·3606827
515	909
559	992
603	·3607074
646	157
690	239
733	322
777	404
821	487
864	569
908	652
951	734
995	817
2295039	899
082	982
126	·3608064
169	147
213	229
257	311
300	394
344	476
387	559
431	641
475	723
518	806
562	888
605	971
649	·3609053
692	135
736	218
779	300
823	383
866	465
910	547
953	630
997	712
2296040	795
084	877
128	959
171	·3610042
215	124
258	207
302	289
346	371
389	454
433	536
476	619
520	701
563	783
607	865
650	948
694	·3611030
737	112
781	194
824	277
868	359
911	442
955	524
999	606
2297042	688
086	771
129	853
173	935
216	·3612017
260	100

2297303	·3612182
347	265
390	347
434	429
477	511
521	594
564	676
608	758
652	840
695	923
739	·3613005
782	088
826	170
869	252
913	334
956	417
2298000	499
043	581
087	663
130	745
174	828
217	910
261	992
304	·3614074
348	156
391	239
435	321
478	403
522	485
565	567
609	650
652	732
696	814
739	896
783	978
826	·3615061
870	143
913	225
957	307
2299000	389
044	472
087	554
131	636
174	718
218	800
261	883
305	965
348	·3616047
392	129
435	211
479	293
522	375
566	457
609	539
653	621
696	704
740	786
783	868
826	950
870	·3617032
913	115
957	197

Excesses

10 = 19	1 = 2
20 = 38	2 = 4
30 = 57	3 = 6
40 = 76	4 = 8
	5 = 9
	6 = 11
	7 = 13
	8 = 15
	9 = 17

22 = 41

Anti Log Excesses

10 = 5	1 = 1
20 = 11	2 = 1
30 = 16	3 = 2
40 = 21	4 = 2
50 = 26	5 = 3
60 = 32	6 = 3
70 = 37	7 = 4
80 = 42	8 = 4
	9 = 5

41 = 22

2300000	·3617279
044	361
087	443
131	525
174	607
218	689
261	771
305	853
348	935
392	·3618017
435	099
479	181
522	263
566	346
609	428
653	510
696	592
740	674
783	756
827	838
870	920
913	·3619002
957	084
2301000	166
044	248
087	330
130	412
174	494
217	576
261	658
304	740
348	822
391	904
435	986
478	·3620068
522	150
565	232

2301609	·3620314
652	396
696	478
739	560
782	642
826	724
869	806
913	888
956	970
999	·3621052
2302043	134
086	215
130	297
173	379
217	461
260	543
304	625
347	707
391	789
434	871
478	953
521	·3622035
565	117
608	199
651	281
695	363
738	444
782	526
825	608
868	690
912	772
955	854
999	936
2303042	·3623018
085	100
129	182
172	263
216	345
259	427
302	509
346	591
389	672
433	754
476	836
519	918
563	·3624000
606	081
650	163
693	245
736	327
780	409
823	491
867	573
910	655
953	737
997	819
2304040	900
084	982
127	·3625064
170	146
214	228
257	309
301	391
344	473
387	555

2304431	·3625637
474	718
518	800
561	882
604	964
648	·3626045
691	127
735	208
778	290
821	372
865	454
908	535
952	617
995	699
2305038	781
082	863
125	944
169	·3627026
212	108
255	190
299	271
342	353
386	434
429	516
472	598
516	680
559	761
603	843
646	925
689	·3628007
733	088
776	170
820	251
863	333
906	415
949	497
993	578
2306036	660
079	742
122	824
166	905
209	987
253	·3629068
296	150
339	232
383	313
426	395
470	476
513	558
556	640
600	721
643	803
687	884
730	966
773	·3630048
817	129
860	211
904	292
947	374
990	456
2307033	537
077	619
120	700
163	782
206	864

2307250	·3630945
293	·3631027
337	108
380	190
423	272
467	353
510	435
554	516
597	598
640	680
683	761
727	843
770	924
813	·3632006
856	088
900	169
943	251
987	332
2308030	414
073	495
117	577
160	658
204	740
247	821
290	903
333	984
377	·3633066
420	147
463	229
506	310
550	392
593	473
637	555
680	636
723	718
766	799
810	881
853	962
896	·3634044
939	125
983	207
2309026	288
070	370
113	451
156	532
199	614
243	695
286	777
329	858
372	939
416	·3635021
459	102
503	184
546	265
589	346
632	428
676	509
719	591
762	672
805	753
849	835
892	916
936	998
979	·3636079

Excesses

10 = 19	1 = 2
20 = 38	2 = 4
30 = 57	3 = 6
40 = 75	4 = 8
	5 = 9
	6 = 11
	7 = 13
	8 = 15
	9 = 17

22 = 41

Anti Log Excesses

10 = 5	1 = 1
20 = 11	2 = 1
30 = 16	3 = 2
40 = 21	4 = 2
50 = 27	5 = 3
60 = 32	6 = 3
70 = 37	7 = 4
80 = 42	8 = 4
	9 = 5

41 = 22

2310022	·3636160
065	242
109	323
152	405
195	486
238	567
282	649
325	730
369	812
412	893
455	974
498	·3637056
542	137
585	219
628	300
671	381
715	462
758	544
802	625
845	706
888	787
931	869
975	950
2311018	·3638032
061	113
104	194
147	275
191	357
234	438
277	519
320	600
364	682
407	763
451	845
494	926
537	·3639007
580	088
2311624	·3639170
667	251
710	332
753	413
796	494
840	576
883	657
926	738
969	819
2312012	901
056	982
099	·3640064
142	145
185	226
229	307
272	389
316	470
359	551
402	632
445	713
489	795
532	876
575	957
618	·3641038
661	119
705	201
748	282
791	363
834	444
877	525
921	607
964	688
2313007	769
050	850
093	931
137	·3642013
180	094
223	175
266	256
309	337
353	419
396	500
439	581
482	662
526	743
569	824
613	905
656	986
699	·3643067
742	148
786	230
829	311
872	392
915	473
958	554
2314002	635
045	716
088	797
131	878
174	959
218	·3644041
261	122
304	203
347	284
390	365
2314434	·3644446
477	527
520	608
563	689
606	770
650	852
693	933
736	·3645014
779	095
822	176
866	257
909	338
952	419
995	500
2315038	581
082	662
125	743
168	824
211	905
254	986
298	·3646067
341	148
384	229
427	310
470	391
514	472
557	553
600	634
643	715
686	796
730	877
773	958
816	·3647039
859	120
902	201
946	282
989	363
2316032	444
075	525
118	606
161	687
204	768
247	849
290	930
333	·3648011
377	091
420	172
463	253
506	334
549	415
593	496
636	577
679	658
722	739
765	820
809	901
852	982
895	·3649063
938	144
981	225
2317025	305
068	386
111	467
154	548
197	629
2317241	·3649709
284	790
327	871
370	952
413	·3650033
456	114
499	195
542	276
585	357
628	438
672	518
715	599
758	680
801	761
844	842
888	922
931	·3651003
974	084
2318017	165
060	246
103	326
146	407
189	488
232	569
275	650
319	730
362	811
405	892
448	973
491	·3652054
535	134
578	215
621	296
664	377
707	458
750	538
793	619
836	700
879	781
922	862
966	942
2319009	·3653023
052	104
095	185
138	266
182	346
225	427
268	508
311	589
354	670
397	750
440	831
483	912
526	993
569	·3654073
613	154
656	234
699	315
742	396
785	477
828	557
871	638
914	719
957	800

Excesses

10 = 19	1 = 2
20 = 38	2 = 4
30 = 56	3 = 6
40 = 75	4 = 8
	5 = 9
	6 = 11
	7 = 13
	8 = 15
	9 = 17

22 = 41

Anti Log Excesses

10 = 5	1 = 1
20 = 11	2 = 1
30 = 16	3 = 2
40 = 21	4 = 2
50 = 27	5 = 3
60 = 32	6 = 3
70 = 37	7 = 4
80 = 43	8 = 4
	9 = 5

41 = 22

2320000	·3654880
044	961
087	·3655041
130	122
173	203
216	283
259	364
302	444
345	525
388	606
431	687
475	767
518	848
561	929
604	·3656010
647	090
690	171
733	251
776	332
819	413
862	493
906	574
949	654
992	735
2321035	816
078	896
121	977
164	·3657057
207	138
250	219
293	299
336	380
379	460
422	541
465	622
508	702
552	783

2321595	·3657863
638	944
681	·3658025
724	105
767	186
810	266
853	347
896	427
939	508
983	588
2322026	669
069	749
112	830
155	910
198	991
241	·3659071
284	152
327	233
370	313
413	394
456	474
499	555
542	635
585	716
628	796
671	877
714	957
757	·3660038
800	118
844	199
887	279
930	360
973	440
2323016	521
059	601
102	682
145	762
188	842
231	923
274	·3661003
317	084
360	164
403	245
446	325
489	406
532	486
575	567
618	647
661	728
705	808
748	889
791	969
834	·3662049
877	130
920	210
963	291
2324006	371
049	451
092	532
135	612
178	693
221	773
264	853
307	934
350	·3663014

2324393	·3663095
436	175
479	255
522	336
565	416
608	497
651	577
694	657
737	738
780	818
823	899
866	979
909	·3664059
952	139
995	220
2325038	300
081	380
124	460
167	541
210	621
253	702
296	782
339	862
382	943
425	·3665023
468	104
511	184
554	264
597	344
640	425
683	505
726	585
769	665
812	745
855	826
898	906
941	986
984	·3666066
2326027	147
070	227
113	308
156	388
199	468
242	548
285	629
328	709
371	789
414	869
457	949
500	·3667030
543	110
586	190
629	270
672	350
715	431
758	511
801	591
844	671
887	751
930	832
973	912
2327016	992
059	·3668072
102	152
145	233

2327188	·3668313
231	393
274	473
317	553
359	634
402	714
445	794
488	874
531	954
574	·3669035
617	115
660	195
703	275
746	355
789	436
832	516
875	596
918	676
961	756
2328004	837
047	917
090	997
133	·3670077
176	157
219	237
262	317
305	397
348	477
391	557
433	638
476	718
519	798
562	878
605	958
648	·3671038
691	118
734	198
777	278
820	358
863	439
906	519
949	599

Excesses

21 = 40

Anti Log Excesses

40 = 21

2329002	·3671699
056	799
109	899
163	999
217	·3672099
270	199
324	299
378	399
432	499
485	599
539	699
593	799
646	899
700	999

2329753	·3673099
807	199
861	299
914	399
968	499

Excesses

10 = 19	1 = 2
20 = 37	2 = 4
30 = 56	3 = 6
40 = 75	4 = 8
50 = 93	5 = 9
	6 = 11
	7 = 13
	8 = 15
	9 = 17

27 = 50

Anti Log Excesses

10 = 5	1 = 1
20 = 11	2 = 1
30 = 16	3 = 2
40 = 21	4 = 2
50 = 27	5 = 3
60 = 32	6 = 3
70 = 37	7 = 4
80 = 43	8 = 4
90 = 48	9 = 5

50 = 27

2330022	·3673599
076	699
129	799
183	899
237	999
290	·3674099
344	199
397	299
451	399
505	499
558	599
612	699
666	799
719	899
773	999
826	·3675099
880	199
934	299
987	399
2331041	498
095	598
148	698
202	798
255	898
309	998
363	·3676098
416	198
470	297
524	397
577	497
631	597

2331684	·3676697
738	797
791	897
845	997
898	·3677096
952	196
2332006	296
059	396
113	495
167	595
220	695
274	795
327	895
381	995
435	·3678095
488	195
542	294
596	394
649	494
703	594
756	693
810	793
863	893
917	993
970	·3679092
2333024	192
078	292
131	392
185	491
239	591
292	691
346	790
399	890
453	989
506	·3680089
560	189
613	288
667	388
720	488
774	588
827	687
881	787
935	887
988	986
2334042	·3681086
096	185
149	285
203	385
256	484
310	584
363	684
417	784
470	883
524	983
577	·3682083
631	182
684	282
738	381
791	481
845	580
898	680
952	779
2335005	879
059	978
112	·3683078

2335166	·3683177
220	277
273	377
327	476
381	576
434	676
488	775
541	875
595	974
648	·3684074
702	173
755	273
809	372
862	472
916	571
969	671
2336023	770
076	870
130	969
183	·3685069
237	168
290	267
344	367
397	466
451	565
504	665
558	764
611	864
665	963
718	·3686063
772	162
825	262
879	361
932	460
986	560
2337039	659
093	758
146	858
200	957
253	·3687057
307	156
360	255
414	355
467	454
521	553
574	653
627	752
681	852
734	951
787	·3688050
841	150
894	249
948	348
2338001	447
055	547
108	646
162	745
215	845
269	944
322	·3689044
376	143
429	242
483	342
536	441
590	540

2338643	·3689639
697	739
750	838
804	937
857	·3690036
910	136
964	235
2339017	334
070	433
124	533
177	632
231	731
284	830
338	930
391	·3691029
445	128
498	227
552	326
605	425
659	524
712	623
765	723
819	822
872	921
925	·3692020
979	119

Excesses

10 = 19	1 = 2
20 = 37	2 = 4
30 = 56	3 = 6
40 = 74	4 = 7
50 = 93	5 = 9
	6 = 11
	7 = 13
	8 = 15
	9 = 17

27 = 50

Anti Log Excesses

10 = 5	1 = 1
20 = 11	2 = 1
30 = 16	3 = 2
40 = 22	4 = 2
50 = 27	5 = 3
60 = 32	6 = 3
70 = 38	7 = 4
80 = 43	8 = 4
90 = 48	9 = 5

50 = 27

2340032	·3692218
086	317
139	416
193	516
246	615
300	714
353	813
406	912
460	·3693011
513	110

2340566	·3693209
620	309
673	408
727	507
780	606
833	705
887	804
940	903
993	·3694002
2341047	101
100	200
154	299
207	398
261	497
314	596
368	695
421	794
474	893
528	992
581	·3695091
634	190
688	289
741	388
795	487
848	586
901	685
955	784
2342008	883
061	982
115	·3696081
168	180
222	279
275	378
328	477
382	576
435	675
488	774
542	873
595	972
649	·3697071
702	170
755	269
809	367
862	466
915	565
968	664
2343022	763
075	862
128	961
182	·3698060
235	158
289	257
342	356
395	455
449	554
502	653
555	752
608	851
662	949
715	·3699048
768	147
822	246
875	344
929	443
982	542

2344035	·3699641
089	740
142	839
195	938
248	·3700037
302	135
355	234
408	333
462	432
515	530
569	629
622	728
675	827
729	925
782	·3701024
835	123
888	222
942	320
995	419
2345048	518
101	617
155	715
208	814
261	913
315	·3702011
368	110
422	208
475	307
528	406
582	504
635	603
688	702
741	801
795	899
848	998
901	·3703097
954	195
2346008	294
061	392
114	491
167	590
221	688
274	787
327	886
380	984
434	·3704083
487	181
540	280
593	379
647	477
700	576
753	675
806	773
860	872
913	970
966	·3705069
2347019	167
073	266
126	364
179	463
232	561
286	660
339	758
392	857
445	955

2347499	·3706054
552	152
605	251
658	349
712	448
765	546
818	645
871	743
925	842
978	940
2348031	·3707039
084	137
138	236
191	334
244	433
297	531
351	630
404	728
457	826
510	925
564	·3708023
617	121
670	220
723	318
777	417
830	515
883	614
936	712
990	811
2349043	909
096	·3709007
149	106
203	204
256	302
309	401
362	499
416	598
469	696
522	794
575	892
628	991
681	·3710089
734	187
787	286
841	384
894	482
947	580

Excesses

10 = 19	1 = 2
20 = 37	2 = 4
30 = 56	3 = 6
40 = 74	4 = 7
50 = 93	5 = 9
	6 = 11
	7 = 13
	8 = 15
	9 = 17

27 = 49

Anti Log Excesses

10 = 5	1 = 1
20 = 11	2 = 1
30 = 16	3 = 2
40 = 22	4 = 2
50 = 27	5 = 3
60 = 32	6 = 3
70 = 38	7 = 4
80 = 43	8 = 4
90 = 49	9 = 5

49 = 27

2350000	·3710679
054	777
107	875
160	974
213	·3711072
267	171
320	269
373	367
426	466
479	564
532	662
585	760
638	859
692	957
745	·3712055
798	153
851	252
905	350
958	448
2351011	546
064	644
117	742
170	840
223	938
276	·3713037
330	135
383	233
436	331
489	430
543	528
596	626
649	724
702	822
755	920
808	·3714018
861	116
914	215
968	313
2352021	411
074	509
127	608
180	706
233	804
286	902
339	·3715000
393	098
446	196
499	294
552	392
606	490
659	588

2352712	·3715686
765	785
818	883
871	981
924	·3716079
977	177
2353031	275
084	373
137	471
190	569
243	667
296	765
349	863
402	961
456	·3717059
509	157
562	255
615	353
668	451
721	549
774	647
827	745
880	843
933	941
986	·3718039
2354039	137
093	235
146	333
199	431
252	529
305	626
358	724
411	822
464	920
518	·3719018
571	116
624	214
677	312
730	410
783	508
836	606
889	704
942	801
995	899
2355048	997
101	·3720095
154	193
207	291
260	389
313	487
367	584
420	682
473	780
526	878
579	976
632	·3721074
685	172
738	270
791	367
844	465
897	563
950	661
2356004	758
057	856
110	954

2356163	·3722052
216	149
269	247
322	345
375	443
428	540
481	638
534	736
587	834
640	931
693	·3723029
746	127
799	225
852	322
905	420
958	518
2357011	616
064	713
117	811
170	909
223	·3724007
276	104
329	202
382	300
435	398
489	495
542	593
595	691
648	788
701	886
754	983
807	·3725081
860	179
913	276
966	374
2358019	472
072	570
125	667
178	765
231	863
284	960
337	·3726058
390	155
443	253
496	350
549	448
602	545
655	643
708	741
761	838
814	936
867	·3727034
920	131
973	229
2359026	326
079	424
132	521
185	619
238	716
291	814
344	911
397	·3728009
450	106
503	204
555	301

2359608	·3728399
661	496
714	594
767	691
820	789
873	886
926	984
979	·3729081

Excesses

10 = 18	1 = 2
20 = 37	2 = 4
30 = 55	3 = 6
40 = 74	4 = 7
50 = 92	5 = 9
	6 = 11
	7 = 13
	8 = 15
	9 = 17

27 = 49

Anti Log Excesses

10 = 5	1 = 1
20 = 11	2 = 1
30 = 16	3 = 2
40 = 22	4 = 2
50 = 27	5 = 3
60 = 33	6 = 3
70 = 38	7 = 4
80 = 43	8 = 4
90 = 49	9 = 5

49 = 27

2360032	·3729179
085	276
138	374
191	471
244	569
297	666
350	764
403	861
456	959
509	·3730056
562	153
615	251
668	348
721	445
774	543
827	640
880	738
933	835
986	933
2361038	·3731030
091	128
144	225
197	322
250	420
303	517
356	614
409	711
462	809

2361515	·3731906
568	·3732003
621	101
673	198
726	296
779	393
832	490
885	588
938	685
991	782
2362044	879
097	977
150	·3733074
203	171
256	268
309	366
362	463
415	560
468	657
520	755
573	852
626	949
679	·3734046
732	144
785	241
838	338
891	435
943	533
996	630
2363049	727
102	824
155	922
208	·3735019
261	116
314	213
367	311
420	408
473	505
526	602
578	700
631	797
684	894
737	991
790	·3736088
843	185
896	282
949	379
2364001	477
054	574
107	671
160	768
213	865
266	962
319	·3737059
372	156
424	254
477	351
530	448
583	545
636	642
689	739
742	836
795	933
847	·3738030
900	127

2364953	·3738224
2365006	321
058	419
111	516
164	613
217	710
270	807
323	904
376	·3739001
429	098
481	195
534	292
587	389
640	486
692	583
745	680
798	777
851	874
904	971
957	·3740068
2366010	165
063	262
115	359
168	456
221	553
274	650
326	747
379	843
432	940
485	·3741037
538	134
591	231
644	328
697	425
749	522
802	619
855	716
908	813
960	910
2367013	·3742006
066	103
119	200
171	297
224	394
277	491
330	588
383	685
436	781
489	878
542	975
594	·3743072
647	169
700	266
753	363
805	460
858	556
911	653
964	750
2368016	847
069	943
122	·3744040
175	137
227	234
280	330
333	427

2368386	·3744524
438	621
491	717
544	814
597	911
649	·3745008
702	104
755	201
808	298
861	395
914	491
967	588
2369020	685
072	782
125	878
178	975
231	·3746072
283	169
336	265
389	362
442	459
494	556
547	652
600	749
653	846
705	943
758	·3747039
811	136
864	233
916	329
969	426

Excesses

10 = 18	1 = 2
20 = 37	2 = 4
30 = 55	3 = 6
40 = 73	4 = 7
50 = 92	5 = 9
	6 = 11
	7 = 13
	8 = 15
	9 = 17

26 = 49

Anti Log Excesses

10 = 5	1 = 1
20 = 11	2 = 1
30 = 16	3 = 2
40 = 22	4 = 2
50 = 27	5 = 3
60 = 33	6 = 3
70 = 38	7 = 4
80 = 44	8 = 4
90 = 49	9 = 5

49 = 26

2370022	·3747522
075	619
127	716
180	812
233	909

2370285	·3748006
338	102
390	199
443	295
496	392
548	489
601	585
654	682
707	779
759	875
812	972
865	·3749068
918	165
970	261
2371023	358
076	454
129	551
181	647
234	744
287	840
340	937
392	·3750034
445	130
498	227
551	324
603	420
656	517
709	613
762	710
814	806
867	903
920	999
972	·3751095
2372025	192
077	288
130	384
183	481
235	577
288	674
341	770
394	867
446	963
499	·3752060
552	156
605	253
657	349
710	446
763	542
815	639
868	735
920	832
973	928
2373026	·3753024
078	121
131	217
184	313
237	410
289	506
342	603
395	699
447	795
500	892
552	988
605	·3754084
658	181

2373710	·3754277
763	374
816	470
869	566
921	663
974	759
2374027	855
079	951
132	·3755048
184	144
237	240
290	336
342	433
395	529
448	625
500	721
553	818
605	914
658	·3756010
711	106
763	203
816	299
869	395
921	491
974	588
2375026	684
079	780
132	876
184	973
237	·3757069
290	165
342	261
395	358
447	454
500	550
553	646
605	743
658	839
711	935
763	·3758031
816	128
868	224
921	320
974	416
2376026	512
079	608
132	704
184	800
237	897
289	993
342	·3759089
395	185
447	281
500	377
553	473
605	569
658	666
710	762
763	858
815	954
868	·3760050
920	146
973	242
2377026	338
078	434

2377131	·3760530
184	626
236	722
289	819
341	915
394	·3761011
446	107
499	203
551	299
604	395
657	491
709	587
762	683
815	779
867	875
920	971
972	·3762067
2378025	163
077	259
130	355
182	451
235	547
287	643
340	739
392	835
445	931
498	·3763027
550	123
603	219
656	315
708	411
761	507
813	602
866	698
918	794
971	890
2379023	986
076	·3764082
128	178
181	274
233	370
286	466
338	562
391	658
443	753
496	849
548	945
601	·3765041
653	137
706	233
758	329
811	425
863	520
916	616
969	712

Excesses

10 = 18	1 = 2
20 = 37	2 = 4
30 = 55	3 = 5
40 = 73	4 = 7
50 = 91	5 = 9
	6 = 11
	7 = 13
	8 = 15
	9 = 16

26 = 48

Anti Log Excesses

10 = 5	1 = 1
20 = 11	2 = 1
30 = 16	3 = 2
40 = 22	4 = 2
50 = 27	5 = 3
60 = 33	6 = 3
70 = 38	7 = 4
80 = 44	8 = 4
90 = 49	9 = 5

48 = 26

2380021	·3765808
074	903
127	999
179	·3766095
232	191
284	287
337	383
389	479
442	575
494	670
547	766
599	862
652	958
704	·3767053
757	149
809	245
862	341
914	436
967	532
2381019	628
072	724
124	819
177	915
229	·3768011
282	107
334	202
387	298
439	394
492	490
544	585
597	681
649	777
702	873
754	968
807	·3769064
859	160
911	256

2381964	·3769351
2382016	447
068	543
121	638
173	734
226	829
278	925
331	·3770021
383	116
436	212
488	308
541	404
593	499
646	595
698	691
751	786
803	882
856	977
908	·3771073
961	169
2383013	264
066	360
118	456
171	551
223	647
276	742
328	838
380	933
433	·3772029
485	124
537	220
590	315
642	411
695	506
747	602
800	697
852	793
905	888
957	984
2384010	·3773079
062	175
115	270
167	366
219	461
272	557
324	652
376	748
429	843
481	939
534	·3774034
586	130
639	225
691	321
744	416
796	512
848	607
901	703
953	798
2385005	894
058	989
110	·3775085
163	180
215	275
268	371
320	466

2385373	·3775561
425	657
477	752
530	848
582	943
634	·3776038
687	134
739	229
792	324
844	420
896	515
949	611
2386001	706
053	801
106	897
158	992
211	·3777087
263	183
315	278
368	374
420	469
472	564
525	660
577	755
630	850
682	945
734	·3778041
787	136
839	231
891	326
944	422
996	517
2387049	612
101	707
153	803
206	898
258	993
310	·3779088
363	184
415	279
468	374
520	469
572	565
625	660
677	755
729	850
782	946
834	·3780041
887	136
939	231
991	327
2388044	422
096	517
148	612
200	708
253	803
305	898
357	993
410	·3781088
462	183
515	278
567	373
619	469
672	564
724	659

2388776	·3781754
828	849
881	944
933	·3782039
985	134
2389038	230
090	325
143	420
195	515
247	610
300	705
352	800
404	895
456	990
509	·3783085
561	180
613	275
665	371
718	466
770	561
822	656
875	751
927	846
980	941

Excesses

10 = 18	1 = 2
20 = 36	2 = 4
30 = 55	3 = 5
40 = 73	4 = 7
50 = 91	5 = 9
	6 = 11
	7 = 13
	8 = 15
	9 = 16

26 = 48

Anti Log Excesses

10 = 5	1 = 1
20 = 11	2 = 1
30 = 16	3 = 2
40 = 22	4 = 2
50 = 27	5 = 3
60 = 33	6 = 3
70 = 38	7 = 4
80 = 44	8 = 4
90 = 49	9 = 5

48 = 26

2390032	·3784036
084	131
137	226
189	321
241	416
293	511
346	606
398	701
450	796
502	891
555	986
607	·3785081

2390659	·3785176
711	271
764	366
816	461
868	556
920	651
973	746
2391025	841
077	936
129	·3786031
182	126
234	221
286	316
339	411
391	505
444	600
496	695
548	790
601	885
653	980
705	·3787075
757	170
810	265
862	360
914	455
966	550
2392019	644
071	739
123	834
175	929
228	·3788024
280	119
332	214
384	309
437	403
489	498
541	593
593	688
646	782
698	877
750	972
802	·3789067
855	162
907	257
959	352
2393011	447
064	541
116	636
168	731
220	826
273	920
325	·3790015
377	110
429	205
481	299
533	394
585	489
637	584
690	678
742	773
794	868
846	963
899	·3791057
951	152
2394003	247

2394055	·3791342
108	436
160	531
212	626
264	721
317	815
369	910
421	·3792005
473	099
526	194
578	288
630	383
682	478
735	572
787	667
839	762
891	857
943	951
995	·3793046
2395047	141
099	235
152	330
204	424
256	519
308	614
361	708
413	803
465	898
517	992
570	·3794087
622	181
674	276
726	370
778	465
830	559
882	654
934	749
987	843
2396039	938
091	·3795033
143	127
196	222
248	316
300	411
352	505
404	600
456	694
508	789
560	883
613	978
665	·3796072
717	167
769	261
821	356
873	450
925	545
977	639
2397030	734
082	828
134	923
186	·3797017
239	112
291	206
343	301
395	395

2397447	·3797490
499	584
551	679
603	773
656	868
708	962
760	·3798056
812	151
864	245
916	339
968	434
2398020	528
073	623
125	717
177	811
229	906
281	·3799000
333	094
385	189
437	283
490	378
542	472
594	566
646	661
698	755
750	849
802	944
854	·3800038
907	133
959	227
2399011	321
063	416
115	510
167	604
219	698
271	793
323	887
375	981
427	·3801075
479	170
532	264
584	358
636	452
688	547
740	641
792	735
844	829
896	924
948	·3802018

Excesses

10 = 18	1 = 2
20 = 36	2 = 4
30 = 54	3 = 5
40 = 73	4 = 7
50 = 91	5 = 9
	6 = 11
	7 = 13
	8 = 15
	9 = 16

26 = 47

Anti Log Excesses

10 = 6	1 = 1
20 = 11	2 = 1
30 = 17	3 = 2
40 = 22	4 = 2
50 = 28	5 = 3
60 = 33	6 = 3
70 = 39	7 = 4
80 = 44	8 = 4
90 = 50	9 = 5

47 = 26

2400000	·3802112
052	206
104	301
157	395
209	489
261	583
313	678
365	772
417	866
469	960
521	·3803055
573	149
625	243
677	337
729	432
782	526
834	620
886	714
938	808
990	902
2401042	996
094	·3804090
146	185
198	279
250	373
302	467
354	562
406	656
458	750
510	844
562	938
615	·3805032
667	126
719	220
771	315
823	409
875	503
927	597
979	691
2402031	785
083	879
135	973
187	·3806067
239	161
291	255
343	349
395	443
447	537
499	631
551	725
603	820

2402655	·3806914
707	·3807008
759	102
811	196
863	290
915	384
967	478
2403019	572
071	666
123	760
175	854
227	948
280	·3808042
332	136
384	230
436	324
488	418
540	512
592	606
644	700
696	794
748	888
800	982
852	·3809076
904	170
956	264
2404008	358
060	452
112	545
164	639
216	733
268	827
320	921
372	·3810015
424	109
475	203
527	297
579	391
631	485
683	579
735	672
787	766
839	860
891	954
943	·3811048
995	142
2405047	236
099	330
151	423
203	517
255	611
307	705
359	798
411	892
463	986
515	·3812080
567	174
619	268
671	362
723	456
775	549
827	643
879	737
931	831
983	924

2406035	·3813018
087	112
138	206
190	299
242	393
294	487
346	581
398	674
450	768
502	862
554	956
606	·3814049
658	143
710	237
762	331
814	424
866	518
918	612
969	705
2407021	799
073	892
125	986
177	·3815080
229	173
281	267
333	361
385	455
437	548
489	642
541	736
592	830
644	923
696	·3816017
748	111
800	204
852	298
904	391
956	485
2408008	579
060	672
112	766
164	860
215	953
267	·3817047
319	140
371	234
423	327
475	421
527	514
579	608
631	702
683	795
735	889
787	983
838	·3818076
890	170
942	263
994	357
2409046	450
098	544
150	637
202	731
253	824
305	918
357	·3819011

2409409	·3819105
461	198
513	292
565	385
617	479
668	572
720	666
772	759
824	853
876	946
928	·3820040
980	133

Excesses

10 = 18	1 = 2
20 = 36	2 = 4
30 = 54	3 = 5
40 = 72	4 = 7
50 = 90	5 = 9
	6 = 11
	7 = 13
	8 = 14
	9 = 16

26 = 47

Anti Log Excesses

10 = 6	1 = 1
20 = 11	2 = 1
30 = 17	3 = 2
40 = 22	4 = 2
50 = 28	5 = 3
60 = 33	6 = 3
70 = 39	7 = 4
80 = 44	8 = 4
90 = 50	9 = 5

47 = 26

2410032	·3820227
083	320
135	414
187	507
239	601
291	694
343	788
395	881
447	975
498	·3821068
550	162
602	255
654	348
705	442
757	535
809	628
861	722
913	815
965	909
2411017	·3822002
069	095
120	189
172	282
224	375

2411276	·3822469
328	562
380	656
432	749
484	842
535	936
587	·3823029
639	122
691	216
742	309
794	403
846	496
898	589
950	683
2412002	776
054	869
106	962
157	·3824056
209	149
261	242
313	335
364	429
416	522
468	615
520	708
571	802
623	895
675	988
727	·3825081
779	175
831	268
883	361
935	454
986	548
2413038	641
090	734
142	827
193	921
245	·3826014
297	107
349	200
400	294
452	387
504	480
556	573
607	667
659	760
711	853
763	946
814	·3827039
866	132
918	225
970	318
2414022	412
074	505
126	598
178	691
229	784
281	877
333	970
385	·3828063
436	157
488	250
540	343
592	436

2414643	·3828529
695	622
747	715
799	808
850	902
902	995
954	·3829088
2415006	181
057	274
109	367
161	460
213	553
264	646
316	739
368	832
420	925
471	·3830018
523	111
575	204
627	297
678	390
730	483
782	576
834	669
885	762
937	855
989	948
2416041	·3831041
092	134
144	227
196	320
248	413
299	506
351	599
403	692
455	785
506	878
558	971
610	·3832064
662	157
713	250
765	343
817	436
868	529
920	622
971	715
2417023	808
075	901
126	994
178	·3833086
230	179
282	272
333	365
385	458
437	551
489	644
540	737
592	830
644	923
696	·3834016
747	109
799	201
851	294
903	387
954	480

2418006	·3834572
058	665
109	758
161	851
212	944
264	·3835037
316	130
367	223
419	315
471	408
523	501
574	594
626	686
678	779
730	872
781	965
833	·3836057
885	150
936	243
988	336
2419039	429
091	522
143	615
194	708
246	800
298	893
350	986
401	·3837079
453	171
505	264
556	357
608	449
659	542
711	634
763	727
814	820
866	912
918	·3838005
969	098

Excesses

10 = 18	1 = 2
20 = 36	2 = 4
30 = 54	3 = 5
40 = 72	4 = 7
50 = 90	5 = 9
	6 = 11
	7 = 13
	8 = 14
	9 = 16

26 = 47

Anti Log Excesses

10 = 6	1 = 1
20 = 11	2 = 1
30 = 17	3 = 2
40 = 22	4 = 2
50 = 28	5 = 3
60 = 33	6 = 3
70 = 39	7 = 4
80 = 44	8 = 4
90 = 50	9 = 5

47 = 26

2420021	·3838191
072	283
124	376
176	469
227	562
279	654
331	747
383	840
434	932
486	·3839025
538	117
589	210
641	303
692	395
744	488
796	581
847	673
899	766
951	858
2421002	951
054	·3840044
105	136
157	229
209	322
260	414
312	507
364	599
415	692
467	785
518	877
570	970
622	·3841063
673	155
725	248
777	340
829	433
880	525
932	618
984	710
2422035	803
087	895
138	988
190	·3842080
241	173
293	265
344	358
396	450
447	543
499	635
550	728
602	820

2422654	·3842913
705	·3843005
757	098
809	190
860	283
912	375
963	468
2423015	560
066	653
118	745
169	838
221	930
273	·3844023
324	115
376	208
428	300
479	393
531	485
582	578
634	670
685	762
737	855
788	947
840	·3845039
892	132
943	224
995	317
2424047	409
098	501
150	594
201	686
253	778
304	871
356	963
407	·3846056
459	148
510	240
562	333
613	425
665	517
717	610
768	702
820	795
872	887
923	979
975	·3847072
2425026	164
078	256
129	348
181	441
232	533
284	625
335	717
387	810
438	902
490	994
541	·3848086
593	179
644	271
696	363
747	455
799	548
850	640
902	732
953	824

2426005	·3848917
056	·3849009
108	101
160	193
211	286
263	378
315	470
366	562
418	655
469	747
521	839
572	931
624	·3850024
675	116
727	208
778	300
830	393
881	485
933	577
984	669
2427036	761
087	853
139	945
190	·3851037
242	130
293	222
345	314
396	406
448	498
499	590
551	682
602	774
654	867
705	959
757	·3852051
808	143
860	235
911	327
963	419
2428014	511
066	603
117	695
169	787
220	879
271	972
323	·3853064
374	156
425	248
477	340
528	432
580	524
631	616
683	708
734	800
786	892
837	984
889	·3854076
940	168
992	260
2429043	352
095	444
146	536
198	628
249	720
301	812

2429352	·3854904
404	996
455	·3855088
507	180
558	272
610	364
661	456
712	548
764	640
815	732
866	824
918	916
969	·3856008

Excesses

10 = 18	1 = 2
20 = 36	2 = 4
30 = 54	3 = 5
40 = 72	4 = 7
50 = 90	5 = 9
	6 = 11
	7 = 13
	8 = 14
	9 = 16

26 = 46

Anti Log Excesses

10 = 6	1 = 1
20 = 11	2 = 1
30 = 17	3 = 2
40 = 22	4 = 2
50 = 28	5 = 3
60 = 34	6 = 3
70 = 39	7 = 4
80 = 45	8 = 4
90 = 50	9 = 5

46 = 26

2430021	·3856100
072	192
124	284
175	375
227	467
278	559
330	651
381	743
433	835
484	927
535	·3857019
587	110
638	202
689	294
741	386
792	478
844	570
895	662
947	754
998	845
2431050	937
101	·3858029
152	121

2431204	·3858213
255	305
306	397
358	489
409	580
461	672
512	764
564	856
615	947
667	·3859039
718	131
769	223
821	315
872	407
923	499
975	591
2432026	682
078	774
129	866
180	958
232	·3860049
283	141
334	233
386	325
437	416
489	508
540	600
591	692
643	783
694	875
745	967
797	·3861059
848	150
900	242
951	334
2433003	426
054	517
106	609
157	701
208	792
260	884
311	975
362	·3862067
413	159
465	250
516	342
567	434
619	526
670	617
722	709
773	801
824	892
876	984
927	·3863075
978	167
2434030	259
081	350
133	442
184	534
235	625
287	717
338	808
389	900
441	992
492	·3864083

2434544	·3864175
595	267
646	358
698	450
749	541
800	633
851	724
903	816
954	907
2435005	999
057	·3865090
108	182
160	273
211	365
262	457
314	548
365	640
416	732
467	823
519	915
570	·3866006
621	098
673	189
724	281
776	372
827	464
878	555
930	647
981	738
2436032	830
083	921
135	·3867013
186	104
237	196
288	287
340	379
391	470
442	561
493	653
545	744
596	835
647	927
699	·3868018
750	110
802	201
853	293
904	384
956	476
2437007	567
058	658
109	750
161	841
212	932
263	·3869024
314	115
366	207
417	298
468	389
519	481
571	572
622	663
673	755
724	846
776	938
827	·3870029

2437878	·3870120
929	212
981	303
2438032	394
083	485
135	577
186	668
238	759
289	851
340	942
392	·3871034
443	125
494	216
545	308
597	399
648	490
699	581
750	673
802	764
853	855
904	946
955	·3872038
2439007	129
058	220
109	311
160	403
212	494
263	585
314	676
365	768
417	859
468	950
519	·3873041
570	133
622	224
673	315
724	406
775	498
827	589
878	680
929	771
980	862

Excesses

10 = 18	1 = 2
20 = 36	2 = 4
30 = 54	3 = 5
40 = 71	4 = 7
50 = 89	5 = 9
	6 = 11
	7 = 12
	8 = 14
	9 = 16

26 = 46

Anti Log Excesses

10 = 6	1 = 1
20 = 11	2 = 1
30 = 17	3 = 2
40 = 22	4 = 2
50 = 28	5 = 3
60 = 34	6 = 3
70 = 39	7 = 4
80 = 45	8 = 4
90 = 50	9 = 5

46 = 26

2440031	·3873953
082	·3874044
133	135
184	227
236	318
287	409
338	500
389	591
441	682
492	773
543	864
594	956
646	·3875047
697	138
748	229
799	321
851	412
902	503
953	594
2441004	685
056	776
107	867
158	958
209	·3876049
261	140
312	231
363	322
414	414
465	505
516	596
567	687
618	778
670	869
721	960
772	·3877051
823	142
875	233
926	324
977	415
2442028	506
080	597
131	688
182	779
233	870
284	961
335	·3878052
386	143
437	234
489	325
540	416
591	507

2442642	·3878598
694	689
745	780
796	871
847	962
899	·3879053
950	144
2443001	235
052	326
103	417
154	508
205	599
256	690
308	781
359	872
410	963
461	·3880054
512	144
563	235
614	326
665	417
717	508
768	599
819	690
870	781
922	872
973	963
2444024	·3881054
075	145
126	235
177	326
228	417
279	508
331	599
382	690
433	781
484	872
535	962
586	·3882053
637	144
688	235
740	325
791	416
842	507
893	598
944	688
995	779
2445046	870
097	961
149	·3883052
200	143
251	234
302	325
353	415
404	506
455	597
506	688
558	778
609	869
660	960
711	·3884051
762	141
813	232
864	323
915	414

2445967	·3884504
2446018	595
069	686
120	777
171	867
222	958
273	·3885049
324	140
375	230
426	321
477	412
528	502
580	593
631	683
682	774
733	865
784	955
835	·3886046
886	137
937	228
989	318
2447040	409
091	500
142	591
193	681
244	772
295	863
346	953
397	·3887044
448	134
499	225
550	316
601	406
652	497
703	588
754	678
806	769
857	859
908	950
959	·3888041
2448010	131
061	222
112	313
163	403
214	494
265	584
316	675
367	766
418	856
469	947
520	·3889037
571	128
623	218
674	308
725	399
776	489
827	580
878	670
929	761
980	851
2449031	942
082	·3890032
133	123
184	213
235	304

2449286	·3890394
337	485
388	575
439	666
490	756
541	847
592	937
643	·3891028
694	118
745	209
796	299
847	390
898	480
949	571

Excesses

10 = 18	1 = 2
20 = 36	2 = 4
30 = 53	3 = 5
40 = 71	4 = 7
50 = 89	5 = 9
	6 = 11
	7 = 12
	8 = 14
	9 = 16

26 = 45

Anti Log Excesses

10 = 6	1 = 1
20 = 11	2 = 1
30 = 17	3 = 2
40 = 23	4 = 2
50 = 28	5 = 3
60 = 34	6 = 3
70 = 39	7 = 4
80 = 45	8 = 5
90 = 51	9 = 5

45 = 26

2450000	·3891661
052	752
103	842
154	932
205	·3892023
256	113
307	203
358	294
409	384
460	475
511	565
562	656
613	746
664	837
715	927
766	·3893017
817	108
868	198
919	288
970	379
2451021	469
072	560

2451123	·3893650
174	740
225	831
276	921
327	·3894011
378	101
429	192
480	282
531	372
582	463
633	553
684	644
735	734
786	824
837	915
888	·3895005
939	095
990	185
2452040	276
091	366
142	456
193	546
244	637
295	727
346	817
397	907
448	998
499	·3896088
550	178
601	268
652	359
703	449
754	539
805	629
856	720
907	810
958	900
2453009	990
060	·3897081
111	171
162	261
213	351
264	442
315	532
366	622
417	712
467	803
518	893
569	983
620	·3898073
671	163
722	253
773	343
824	433
875	524
926	614
977	704
2454028	794
079	884
130	974
181	·3899064
232	154
282	245
333	335
384	425

2454435	·3899515
486	605
537	695
588	785
639	875
690	966
741	·3900056
792	146
843	236
893	326
944	416
995	506
2455046	596
097	686
148	776
199	866
250	956
301	·3901047
352	137
403	227
454	317
504	407
555	497
606	587
657	677
708	767
759	857
810	947
861	·3902037
912	127
963	217
2456014	307
065	397
115	487
166	577
217	667
268	757
319	847
370	937
421	·3903027
472	117
522	207
573	296
624	386
675	476
726	566
777	656
828	746
879	836
929	926
980	·3904016
2457031	106
082	196
133	286
184	376
235	466
286	556
336	646
387	735
438	825
489	915
540	·3905005
591	095
642	185
693	275

2457743	·3905365
794	454
845	544
896	634
947	724
998	814
2458049	904
100	994
150	·3906084
201	173
252	263
303	353
353	443
404	532
455	622
506	712
557	802
608	892
659	982
710	·3907072
760	162
811	251
862	341
913	431
963	521
2459014	610
065	700
116	790
167	880
218	969
269	·3908059
320	149
370	239
421	328
472	418
523	508
573	598
624	687
675	777
726	867
777	957
828	·3909046
879	136
930	226
980	316

Excesses

10 = 18	1 = 2
20 = 35	2 = 4
30 = 53	3 = 5
40 = 71	4 = 7
50 = 88	5 = 9
	6 = 11
	7 = 12
	8 = 14
	9 = 16

25 = 45

Anti Log Excesses

10 = 6	1 = 1
20 = 11	2 = 1
30 = 17	3 = 2
40 = 23	4 = 2
50 = 28	5 = 3
60 = 34	6 = 3
70 = 40	7 = 4
80 = 45	8 = 5
90 = 51	9 = 5

45 = 25

2460031	·3909405
082	495
133	585
183	674
234	764
285	853
336	943
386	·3910033
437	122
488	212
539	302
589	392
640	481
691	571
742	661
793	750
844	840
895	929
946	·3911019
996	109
2461047	198
098	288
149	378
199	467
250	557
301	646
352	736
402	826
453	915
504	·3912005
555	095
605	184
656	274
707	363
758	453
808	542
859	632
910	721
961	811
2462011	901
062	990
113	·3913080
164	170
214	259
265	349
316	438
367	528
418	617
469	707
520	796
571	886

2462621	·3913975
672	·3914065
723	154
774	244
824	333
875	423
926	512
977	602
2463027	691
078	781
129	870
180	960
230	·3915049
281	139
332	228
383	318
433	407
484	497
535	586
585	676
636	765
686	855
737	944
788	·3916033
838	123
889	212
940	301
991	391
2464041	480
092	570
143	659
194	748
244	838
295	927
346	·3917016
397	106
447	195
498	285
549	374
600	463
650	553
701	642
752	731
803	821
853	910
904	·3918000
955	089
2465006	178
056	268
107	357
158	446
209	535
259	625
310	714
361	803
411	893
462	982
512	·3919072
563	161
614	250
664	340
715	429
766	518
817	607
867	697

2465918	·3919786
969	875
2466020	964
070	·3920054
121	143
172	232
222	321
273	411
323	500
374	589
425	678
475	768
526	857
577	946
628	·3921035
678	125
729	214
780	303
830	392
881	482
931	571
982	660
2467033	749
083	838
134	927
185	·3922016
236	105
286	195
337	284
388	373
438	462
489	552
539	641
590	730
641	819
691	908
742	997
793	·3923086
844	175
894	265
945	354
996	443
2468046	532
097	621
147	710
198	799
249	888
299	978
350	·3924067
401	156
451	245
502	334
552	423
603	512
654	601
704	691
755	780
806	869
856	958
907	·3925047
957	136
2469008	225
059	314
109	403
160	492

2469211	·3925581
261	670
312	759
362	848
413	937
464	·3926026
514	115
565	204
616	293
666	382
717	471
767	560
818	649
869	738
919	827
970	916

Excesses

10 = 18	1 = 2
20 = 35	2 = 4
30 = 53	3 = 5
40 = 70	4 = 7
50 = 88	5 = 9
	6 = 11
	7 = 12
	8 = 14
	9 = 16

25 = 45

Anti Log Excesses

10 = 6	1 = 1
20 = 11	2 = 1
30 = 17	3 = 2
40 = 23	4 = 2
50 = 28	5 = 3
60 = 34	6 = 3
70 = 40	7 = 4
80 = 45	8 = 5
	9 = 5

45 = 25

2470021	·3927005
071	094
122	183
172	272
223	361
274	450
324	539
375	628
426	717
476	806
527	895
577	984
628	·3928073
678	162
729	251
779	339
830	428
881	517
931	606
982	695

2471033	·3928784
083	873
134	962
184	·3929051
235	140
285	229
336	318
386	406
437	495
488	584
538	673
589	762
640	851
690	940
741	·3930029
791	117
842	206
892	295
943	384
993	472
2472044	561
094	650
145	739
195	828
246	917
297	·3931006
347	095
398	183
449	272
499	361
550	450
600	538
651	627
701	716
752	805
802	893
853	982
903	·3932071
954	160
2473004	248
055	337
105	426
156	515
206	603
257	692
308	781
358	870
409	958
460	·3933047
510	136
561	225
611	313
662	402
712	491
763	580
813	668
864	757
914	846
965	935
2474015	·3934023
066	112
116	201
167	290
217	378
268	467

2474318	·3934556
369	644
419	733
470	821
520	910
571	999
621	·3935087
672	176
722	265
773	354
823	442
874	531
924	620
975	708
2475025	797
076	885
126	974
177	·3936062
227	151
278	239
328	328
379	417
429	505
480	594
530	683
581	771
631	860
682	948
732	·3937037
783	125
833	214
884	302
934	391
985	479
2476035	568
086	656
136	745
187	834
237	922
288	·3938011
338	100
389	188
439	277
490	365
540	454
591	542
641	631
692	719
742	808
793	896
843	985
894	·3939073
944	162
995	250
2477045	339
096	427
146	515
196	604
247	692
297	780
347	869
398	957
448	·3940046
499	134
549	223

2477600	·3940311
650	400
701	488
751	577
802	665
852	754
903	842
953	930
2478004	·3941019
054	107
105	195
155	284
205	372
256	461
306	549
356	637
407	726
457	814
508	902
558	991
609	·3942079
659	168
710	256
760	344
811	433
861	521
912	609
962	698
2479012	786
063	875
113	963
163	·3943051
214	140
264	228
315	316
365	404
416	493
466	581
517	669
567	757
617	846
668	934
718	·3944022
768	111
819	199
869	288
920	376
970	464

Excesses

10 = 18	1 = 2
20 = 35	2 = 4
30 = 53	3 = 5
40 = 70	4 = 7
50 = 88	5 = 9
	6 = 11
	7 = 12
	8 = 14
	9 = 16

25 = 44

Anti Log Excesses

10 = 6	1 = 1
20 = 11	2 = 1
30 = 17	3 = 2
40 = 23	4 = 2
50 = 28	5 = 3
60 = 34	6 = 3
70 = 40	7 = 4
80 = 46	8 = 5
	9 = 5

44 = 25

2480020	·3944553
071	641
121	729
171	817
222	906
272	994
323	·3945082
373	170
424	259
474	347
525	435
575	523
625	612
676	700
726	788
776	876
827	964
877	·3946052
928	140
978	228
2481028	317
079	405
129	493
179	581
230	670
280	758
331	846
381	934
431	·3947023
482	111
532	199
582	287
633	375
683	463
734	551
784	639
834	728
885	816
935	904
985	992
2482036	·3948080
086	168
137	256
187	344
237	433
288	521
338	609
388	697
439	785
489	873
540	961

2482590	·3949049
640	137
691	225
741	313
791	401
841	489
892	577
942	665
992	753
2483043	842
093	930
144	·3950018
194	106
244	194
295	282
345	370
395	458
445	546
496	634
546	722
596	810
647	898
697	986
748	·3951074
798	162
848	250
899	338
949	426
999	514
2484049	602
100	690
150	778
200	866
251	954
301	·3952042
352	130
402	218
452	306
503	393
553	481
603	569
653	657
704	745
754	833
804	921
854	·3953009
905	097
955	185
2485005	273
055	361
106	448
156	536
206	624
257	712
307	800
358	888
408	976
458	·3954064
509	151
559	239
609	327
659	415
710	503
760	591
810	679

2485860	·3954767
911	854
961	942
2486011	·3955030
061	118
112	206
162	294
212	382
262	470
313	557
363	645
413	733
463	821
514	908
564	996
614	·3956084
664	172
715	259
765	347
815	435
866	523
916	610
967	698
2487017	786
067	874
118	962
168	·3957050
218	138
268	225
319	313
369	400
419	488
469	576
520	664
570	752
620	840
670	927
721	·3958015
771	102
821	190
871	278
922	365
972	453
2488022	541
072	629
123	716
173	804
223	892
273	980
323	·3959067
373	155
423	243
473	330
524	418
574	505
624	593
674	681
725	768
775	856
825	944
875	·3960031
926	119
976	206
2489026	294
076	382

2489127	·3960469
177	557
227	645
277	732
328	820
378	907
428	995
478	·3961083
529	170
579	258
629	346
679	433
730	521
780	608
830	696
880	783
930	871
980	958

Excesses

10 = 17	1 = 2
20 = 35	2 = 3
30 = 52	3 = 5
40 = 70	4 = 7
50 = 87	5 = 9
	6 = 10
	7 = 12
	8 = 14
	9 = 16

25 = 44

Anti Log Excesses

10 = 6	1 = 1
20 = 11	2 = 1
30 = 17	3 = 2
40 = 23	4 = 2
50 = 29	5 = 3
60 = 34	6 = 3
70 = 40	7 = 4
80 = 46	8 = 5
	9 = 5

44 = 25

2490030	·3962046
080	133
131	221
181	308
231	396
281	484
332	571
382	659
432	747
482	834
533	922
583	·3963009
633	097
683	184
734	272
784	359
834	447
884	534

2490934	·3963622
984	709
2491034	797
084	884
135	972
185	·3964059
235	147
285	234
336	322
386	409
436	497
486	584
536	672
586	759
636	846
686	934
737	·3965021
787	108
837	196
887	283
938	371
988	458
2492038	546
088	633
138	721
188	808
238	895
288	983
339	·3966070
389	157
439	245
489	332
540	420
590	507
640	594
690	682
740	769
790	856
840	944
890	·3967031
941	119
991	206
2493041	293
091	381
141	468
191	555
241	642
291	730
342	817
392	904
442	992
492	·3968079
542	167
592	254
642	341
692	429
743	516
793	603
843	690
893	778
943	865
993	952
2494043	·3969039
093	127
144	214

2494194	·3969301
244	388
294	476
344	563
394	650
444	737
494	825
545	912
595	999
645	·3970086
695	174
745	261
795	348
845	435
895	523
945	610
995	697
2495045	784
095	872
146	959
196	·3971046
246	133
296	220
346	307
396	394
446	481
496	569
547	656
597	743
647	830
697	918
747	·3972005
797	092
847	179
897	266
947	353
997	440
2496047	527
097	615
148	702
198	789
248	876
298	963
348	·3973050
398	137
448	224
498	312
548	399
598	486
648	573
698	660
748	747
798	834
848	921
898	·3974008
949	095
999	182
2497049	269
099	357
149	444
199	531
249	618
299	705
349	792
399	879

2497449	·3974966
499	·3975053
549	140
599	227
649	314
699	401
750	488
800	575
850	662
900	749
950	836
2498000	923
050	·3976010
100	097
150	184
200	271
250	358
300	445
350	532
400	619
450	706
500	793
550	880
600	967
650	·3977054
700	141
750	228
800	315
850	402
900	489
950	575
2499000	662
050	749
100	836
150	923
200	·3978010
250	097
300	184
350	270
400	357
450	444
500	531
550	618
600	705
650	792
700	879
750	966
800	·3979053
850	140
900	227
950	313

Excesses

10 = 17	1 = 2
20 = 35	2 = 3
30 = 52	3 = 5
40 = 70	4 = 7
50 = 87	5 = 9
	6 = 10
	7 = 12
	8 = 14
	9 = 16

25 = 44

Anti Log Excesses

10 = 6	1 = 1
20 = 11	2 = 1
30 = 17	3 = 2
40 = 23	4 = 2
50 = 29	5 = 3
60 = 34	6 = 3
70 = 40	7 = 4
80 = 46	8 = 5
	9 = 5

44 = 25

2500000	·3979400
050	487
100	574
150	660
200	747
250	834
300	921
350	·3980008
400	095
450	182
500	269
550	355
600	442
650	529
700	616
750	702
800	789
850	876
900	963
950	·3981049
2501000	136
050	223
100	310
150	396
200	483
250	570
300	657
350	743
400	830
450	917
500	·3982004
550	091
600	178
650	265
700	352
750	438
800	525
850	612
900	699
950	785
2502000	872
050	959
100	·3983045
150	132
200	218
250	305
299	392
349	478
399	565
449	652
499	739

2502549	·3983825
599	912
649	999
699	·3984086
749	172
799	259
849	346
899	432
949	519
999	605
2503049	692
098	779
148	865
198	952
248	·3985039
298	126
348	212
398	299
448	386
498	472
548	559
598	645
648	732
698	818
748	905
798	991
848	·3986078
897	165
947	251
997	338
2504047	425
097	511
147	598
197	684
247	771
297	857
347	944
397	·3987030
447	117
496	203
546	290
596	376
646	463
696	550
746	636
796	723
846	810
896	896
946	983
996	·3988069
2505046	156
095	242
145	329
195	415
245	502
295	588
345	675
395	761
445	848
494	934
544	·3989021
594	107
644	193
694	280
744	366

2505794	·3989452
844	539
893	625
943	712
993	798
2506043	885
093	971
143	·3990058
193	144
243	231
292	317
342	404
392	490
442	576
492	663
542	749
592	835
642	922
691	·3991008
741	095
791	181
841	267
891	354
941	440
991	526
2507041	613
090	699
140	786
190	872
240	958
290	·3992045
340	131
390	217
440	304
489	390
539	477
589	563
639	649
688	736
738	822
788	908
838	994
888	·3993081
938	167
988	253
2508038	339
087	426
137	512
187	598
237	685
286	771
336	858
386	944
436	·3994030
486	117
536	203
586	289
636	375
685	462
735	548
785	634
835	720
884	807
934	893
984	979

2509034	·3995065
084	152
134	238
184	324
234	410
283	496
333	582
383	668
433	754
482	841
532	927
582	·3996013
632	099
681	186
731	272
781	358
831	444
881	531
931	617
981	703

Excesses

10 = 17	1 = 2
20 = 35	2 = 3
30 = 52	3 = 5
40 = 69	4 = 7
	5 = 9
	6 = 10
	7 = 12
	8 = 14
	9 = 16

25 = 43

Anti Log Excesses

10 = 6	1 = 1
20 = 12	2 = 1
30 = 17	3 = 2
40 = 23	4 = 2
50 = 29	5 = 3
60 = 35	6 = 3
70 = 40	7 = 4
80 = 46	8 = 5
	9 = 5

43 = 25

2510031	·3996789
080	875
130	961
180	·3997047
230	133
279	220
329	306
379	392
429	478
478	565
528	651
578	737
628	823
677	909
727	995
777	·3998081

2510827	·3998167
876	253
926	339
976	425
2511026	511
076	598
126	684
176	770
226	856
275	942
325	·3999028
375	114
425	200
474	286
524	372
574	458
624	544
673	631
723	717
773	803
823	889
872	975
922	·4000061
972	147
2512022	233
071	319
121	405
171	491
221	577
270	663
320	749
370	835
420	921
469	·4001007
519	093
569	179
619	265
668	351
718	437
768	523
818	609
867	695
917	781
967	867
2513017	953
066	·4002039
116	125
166	211
216	297
265	383
315	468
365	554
415	640
464	726
514	812
564	898
614	984
663	·4003070
713	156
763	242
813	328
862	414
912	499
962	585
2514011	671

2514061	·4003757
110	843
160	929
210	·4004015
259	101
309	187
359	273
409	359
458	445
508	530
558	616
608	702
657	788
707	874
757	960
807	·4005046
856	132
906	217
956	303
2515006	389
055	475
105	560
155	646
204	732
254	818
303	903
353	989
403	·4006075
452	161
502	247
552	333
602	419
651	505
701	590
751	676
801	762
850	848
900	933
950	·4007019
999	105
2516049	191
098	276
148	362
198	448
247	534
297	619
347	705
397	791
446	877
496	962
546	·4008048
595	134
645	219
694	305
744	390
794	476
843	562
893	647
943	733
993	819
2517042	905
092	990
142	·4009076
191	162
241	248

2517290	·4009333
340	419
390	505
439	590
489	676
539	761
589	847
638	933
688	·4010018
738	104
787	190
837	275
886	361
936	446
986	532
2518035	618
085	703
135	789
184	875
234	960
283	·4011046
333	131
383	217
432	303
482	388
532	474
581	560
631	645
680	731
730	816
780	902
829	987
879	·4012073
929	158
978	244
2519028	330
077	415
127	501
177	587
226	672
276	758
326	843
375	929
425	·4013014
474	100
524	185
574	271
623	356
673	442
723	527
772	613
822	698
871	784
921	869
971	955

Excesses

10 = 17	1 = 2
20 = 35	2 = 3
30 = 52	3 = 5
40 = 69	4 = 7
	5 = 9
	6 = 10
	7 = 12
	8 = 14
	9 = 16

25 = 43

Anti Log Excesses

10 = 6	1 = 1
20 = 12	2 = 1
30 = 17	3 = 2
40 = 23	4 = 2
50 = 29	5 = 3
60 = 35	6 = 3
70 = 41	7 = 4
80 = 46	8 = 5
	9 = 5

43 = 25

2520020	·4014040
070	126
120	211
169	297
219	382
268	468
318	553
367	638
417	724
466	809
516	894
566	980
615	·4015065
665	151
715	236
764	322
814	407
863	493
913	578
962	664
2521012	749
061	835
111	920
161	·4016005
210	091
260	176
310	261
359	347
409	432
458	518
508	603
557	688
607	774
656	859
706	944
756	·4017030
805	115

2521855	·4017201	2525075	·4022742	2528290	·4028269	2530069	·4031324	2533279	·4036830
905	286	124	827	340	354	119	409	328	914
954	371	174	912	389	439	168	493	378	999
2522004	457	223	997	439	524	218	578	427	·4037084
053	542	273	·4023082	488	609	267	663	476	168
103	627	322	167	537	694	316	748	526	253
152	712	372	253	587	778	366	833	575	337
202	798	421	338	636	863	415	918	624	422
251	883	471	423	685	948	464	·4032003	673	506
301	968	520	508	735	·4029033	514	088	723	591
350	·4018054	570	593	784	118	563	172	772	675
400	139	619	678	834	203	613	257	821	760
449	225	669	763	883	288	662	342	871	844
499	310	718	848	933	373	712	427	920	929
548	395	768	933	982	457	761	511	970	·4038013
598	481	817	·4024018	2529032	542	811	596	2534019	098
647	566	867	103	081	627	860	681	068	182
697	651	916	188	131	712	909	766	118	267
747	736	966	274	180	797	959	850	167	351
796	822	2526015	359	230	882	2531008	935	216	436
846	907	065	444	279	967	057	·4033020	265	521
896	992	114	529	328	·4030052	107	105	315	605
945	·4019077	164	614	378	136	156	189	364	690
995	163	213	699	427	221	206	274	413	775
2523044	248	263	784	476	306	255	359	463	859
094	333	312	869	526	391	304	444	512	944
143	418	361	954	575	476	354	528	562	·4039028
193	504	411	·4025039	625	561	403	613	611	113
242	589	460	124	674	646	452	698	660	197
292	674	509	209	724	731	502	782	710	282
341	759	559	294	773	815	551	867	759	366
391	845	608	379	823	900	601	951	808	451
440	930	658	464	872	985	650	·4034036	857	535
490	·4020015	707	549	921	·4031070	699	121	907	620
539	100	757	635	971	154	749	205	956	704
589	186	806	720			798	290	2535005	788
638	271	856	805	**Excesses**		847	375	054	873
688	356	905	890			897	460	104	957
737	441	955	975	10 = 17	1 = 2	946	544	153	·4040041
787	527	2527004	·4026060	20 = 34	2 = 3	996	629	202	126
836	612	054	145	30 = 52	3 = 5	2532045	714	252	210
886	697	103	230	40 = 69	4 = 7	094	799	301	295
935	782	153	315		5 = 9	144	883	351	379
985	868	202	400		6 = 10	193	968	400	464
2524034	953	252	485		7 = 12	242	·4035053	449	548
084	·4021038	301	570		8 = 14	292	137	499	633
134	123	351	655		9 = 15	341	222	548	717
183	209	400	740			391	306	597	802
233	294	450	825	25 = 43		440	391	646	886
283	379	499	910			489	476	696	971
332	464	549	995	**Anti Log Excesses**		539	560	745	·4041055
382	550	598	·4027080			588	645	794	140
431	635	648	165	10 = 6	1 = 1	637	730	843	224
481	720	697	250	20 = 12	2 = 1	687	814	893	309
530	805	746	335	30 = 17	3 = 2	736	899	942	393
580	890	796	419	40 = 23	4 = 2	786	983	991	477
629	975	845	504	50 = 29	5 = 3	835	·4036068	2536040	562
679	·4022060	894	589	60 = 35	6 = 3	884	153	090	646
728	145	944	674	70 = 41	7 = 4	934	237	139	730
778	231	993	759	80 = 47	8 = 5	983	322	188	815
827	316	2528043	844		9 = 5	2533032	407	237	899
877	401	092	929			081	491	287	984
926	486	142	·4028014	43 = 25		131	576	336	·4042068
976	572	191	099			180	660	385	153
2525025	657	241	184	2530020	·4031239	229	745	435	237

2536484	·4042322
534	406
583	490
632	575
682	659
731	743
780	827
829	912
879	996
928	·4043080
977	165
2537026	249
076	334
125	418
174	502
223	587
273	671
322	755
371	839
420	924
470	·4044008
519	092
568	176
617	261
667	345
716	429
765	514
814	598
864	683
913	767
962	851
2538011	936
061	·4045020
110	104
159	188
208	273
258	357
307	441
356	525
405	610
455	694
504	778
553	862
602	947
652	·4046031
701	115
750	199
799	284
849	368
898	452
947	536
996	621
2539046	705
095	789
144	873
193	957
243	·4047041
292	125
341	209
390	294
440	378
489	462
538	546
587	631
637	715

2539686	·4047799
735	883
784	967
833	·4048051
882	135
931	219
980	304

Excesses

10 = 17	1 = 2
20 = 34	2 = 3
30 = 51	3 = 5
40 = 69	4 = 7
	5 = 9
	6 = 10
	7 = 12
	8 = 14
	9 = 15

25 = 42

Anti Log Excesses

10 = 6	1 = 1
20 = 12	2 = 1
30 = 18	3 = 2
40 = 23	4 = 2
50 = 29	5 = 3
60 = 35	6 = 4
70 = 41	7 = 4
80 = 47	8 = 5
	9 = 5

42 = 25

2540030	·4048388
079	472
128	556
177	641
227	725
276	809
325	893
374	977
424	·4049061
473	145
522	229
571	314
621	398
670	482
719	566
768	650
817	734
866	818
915	902
964	986
2541014	·4050070
063	154
112	238
161	322
211	406
260	490
309	574
358	659
408	743

2541457	·4050827
506	911
555	995
605	·4051079
654	163
703	247
752	331
801	415
850	499
899	583
948	667
998	751
2542047	835
096	919
145	·4052003
195	087
244	171
293	255
342	339
391	423
440	507
489	591
538	675
588	759
637	843
686	927
735	·4053011
785	095
834	179
883	263
932	347
981	430
2543030	514
079	598
128	682
178	766
227	850
276	934
325	·4054018
374	102
423	186
472	270
521	354
571	437
620	521
669	605
718	689
768	773
817	857
866	941
915	·4055025
964	109
2544013	193
062	277
111	361
161	444
210	528
259	612
308	696
357	780
406	864
455	948
504	·4056032
554	115
603	199

2544652	·4056283
701	367
750	450
799	534
848	618
897	702
947	785
996	869
2545045	953
094	·4057037
143	121
192	205
241	289
290	373
339	456
388	540
437	624
486	708
536	791
585	875
634	959
683	·4058043
732	126
781	210
830	294
879	378
929	461
978	545
2546027	629
076	713
125	796
174	880
223	964
272	·4059048
321	131
370	215
419	299
468	383
518	466
567	550
616	634
665	718
714	801
763	885
812	969
861	·4060052
910	136
959	219
2547008	303
057	387
107	470
156	554
205	638
254	722
303	805
352	889
401	973
450	·4061056
499	140
548	223
597	307
646	391
696	474
745	558
794	642

2547843	·4061725
892	809
941	892
990	976
2548039	·4062060
088	143
137	227
186	311
235	394
284	478
333	561
382	645
431	729
480	812
529	896
578	980
627	·4063063
677	147
726	230
775	314
824	397
873	481
922	564
971	648
2549020	731
069	815
118	898
167	982
216	·4064066
265	149
314	233
363	317
412	400
461	484
510	567
559	651
608	734
657	818
706	901
755	985
804	·4065068
853	152
902	235
951	319

Excesses

10 = 17	1 = 2
20 = 34	2 = 3
30 = 51	3 = 5
40 = 68	4 = 7
	5 = 9
	6 = 10
	7 = 12
	8 = 14
	9 = 15

25 = 42

Anti Log Excesses

10 = 6	1 = 1
20 = 12	2 = 1
30 = 18	3 = 2
40 = 23	4 = 2
50 = 29	5 = 3
60 = 35	6 = 4
70 = 41	7 = 4
80 = 47	8 = 5
	9 = 5

42 = 25

2550000	·4065402
050	486
099	569
148	653
197	736
246	820
295	903
344	986
393	·4066070
442	153
491	236
540	320
589	403
638	487
687	570
736	654
785	737
834	821
883	904
932	988
981	·4067071
2551030	155
079	238
128	321
177	405
226	488
275	571
324	655
373	738
422	822
471	905
520	988
569	·4068072
618	155
667	238
716	322
765	405
814	489
863	572
912	655
960	739
2552009	822
058	905
107	989
156	·4069072
205	156
254	239
303	322
352	406
401	489
450	572

2552499	·4069655
548	739
597	822
646	905
695	988
744	·4070072
793	155
842	238
891	322
940	405
989	489
2553038	572
087	655
136	739
185	822
234	905
283	988
332	·4071072
381	155
430	238
479	321
527	405
576	488
625	571
674	654
723	738
772	821
821	904
870	987
919	·4072071
968	154
2554017	237
066	320
115	404
164	487
213	570
262	653
310	736
359	819
408	902
457	985
506	·4073069
555	152
604	235
653	318
702	402
751	485
800	568
849	651
898	735
947	818
996	901
2555045	984
093	·4074067
142	150
191	233
240	316
289	400
338	483
387	566
436	649
485	732
534	815
583	898
632	981

2555680	·4075065
729	148
778	231
827	314
876	397
925	480
974	563
2556023	646
072	729
121	812
170	895
219	978
267	·4076062
316	145
365	228
414	311
463	394
512	477
561	560
610	643
658	726
707	809
756	892
805	975
854	·4077058
903	141
952	224
2557001	307
049	390
098	473
147	556
196	639
245	723
294	806
343	889
392	972
440	·4078055
489	138
538	221
587	304
636	387
685	469
734	552
783	635
831	718
880	801
929	884
978	967
2558027	·4079050
076	133
125	216
174	299
222	382
271	465
320	548
369	631
418	714
467	797
516	880
565	963
613	·4080046
662	129
711	212
760	295
809	378

2558858	·4080460
907	543
956	626
2559004	709
053	792
102	875
151	958
199	·4081041
248	123
297	206
346	289
395	372
444	455
493	538
542	621
590	704
639	786
688	869
737	952
785	·4082035
834	118
883	201
932	284
981	367

Excesses

10 = 17	1 = 2
20 = 34	2 = 3
30 = 51	3 = 5
40 = 68	4 = 7
	5 = 8
	6 = 10
	7 = 12
	8 = 14
	9 = 15

24 = 42

Anti Log Excesses

10 = 6	1 = 1
20 = 12	2 = 1
30 = 18	3 = 2
40 = 24	4 = 2
50 = 29	5 = 3
60 = 35	6 = 4
70 = 41	7 = 4
80 = 47	8 = 5
	9 = 5

42 = 24

2560030	·4082449
079	532
128	615
176	698
225	780
274	863
323	946
371	·4083029
420	112
469	195
518	278
566	361

2560615	·4083443
664	526
713	609
762	692
811	774
860	857
909	940
957	·4084023
2561006	105
055	188
104	271
152	354
201	436
250	519
299	602
347	685
396	767
445	850
494	933
543	·4085016
592	098
641	181
690	264
738	347
787	429
836	512
885	595
933	678
982	760
2562031	843
080	926
128	·4086009
177	091
226	174
275	257
323	340
372	422
421	505
470	588
518	670
567	753
616	835
665	918
713	·4087001
762	083
811	166
860	249
909	331
958	414
2563007	496
056	579
104	662
153	744
202	827
251	910
299	992
348	·4088075
397	157
446	240
494	323
543	405
592	488
641	571
689	653
738	736

Number	Log
2563787	·4088818
836	901
884	983
933	·4089066
982	148
2564031	231
079	314
128	396
177	479
226	562
274	644
323	727
372	809
421	892
469	974
518	·4090057
567	139
616	222
664	304
713	387
762	469
811	552
859	634
908	717
957	799
2565006	882
054	964
103	·4091047
152	129
200	212
249	294
297	377
346	459
395	542
443	624
492	707
541	789
590	872
638	954
687	·4092037
736	119
785	202
833	284
882	367
931	449
980	532
2566028	614
077	697
126	779
175	862
223	944
272	·4093027
321	109
370	191
418	274
467	356
516	438
564	521
613	603
661	686
710	768
759	850
807	933
856	·4094015
905	097

Number	Log
2566954	·4094180
2567002	262
051	345
100	427
149	509
197	592
246	674
295	756
344	839
392	921
441	·4095004
490	086
538	168
587	251
635	333
684	415
733	498
781	580
830	663
879	745
928	827
976	910
2568025	992
074	·4096074
122	156
171	239
219	321
268	403
317	485
365	568
414	650
463	732
512	814
560	897
609	979
658	·4097061
706	144
755	226
803	309
852	391
901	473
949	556
998	638
2569047	720
096	802
144	885
193	967
242	·4098049
290	131
339	214
387	296
436	378
485	460
533	542
582	624
631	706
679	788
728	871
776	953
825	·4099035
874	117
922	200
971	282

Excesses

10 = 17	1 = 2
20 = 34	2 = 3
30 = 51	3 = 5
40 = 68	4 = 7
	5 = 8
	6 = 10
	7 = 12
	8 = 14
	9 = 15

24 = 41

Anti Log Excesses

10 = 6	1 = 1
20 = 12	2 = 1
30 = 18	3 = 2
40 = 24	4 = 2
50 = 30	5 = 3
60 = 35	6 = 4
70 = 41	7 = 4
80 = 47	8 = 5
	9 = 5

41 = 24

Number	Log
2570020	·4099364
068	446
117	529
165	611
214	693
263	775
311	857
360	939
409	·4100021
458	103
506	186
555	268
604	350
652	432
701	515
749	597
798	679
847	761
895	843
944	925
993	·4101007
2571041	089
090	172
138	254
187	336
235	418
284	500
332	582
381	664
430	746
478	829
527	911
576	993
624	·4102075
673	157
721	239
770	321

Number	Log
2571819	·4102403
867	485
916	567
965	649
2572013	731
062	814
110	896
159	978
208	·4103060
256	142
305	224
354	306
402	388
451	470
499	552
548	634
596	716
645	798
693	880
742	962
791	·4104044
839	126
888	208
937	290
985	372
2573034	454
082	536
131	618
179	700
228	782
276	864
325	946
373	·4105028
422	110
470	192
519	274
568	356
616	438
665	520
714	602
762	684
811	766
859	847
908	929
956	·4106011
2574005	093
053	175
102	257
151	339
199	421
248	503
297	585
345	667
394	749
442	830
491	912
539	994
588	·4107076
636	158
685	240
733	322
782	404
830	486
879	568
927	650

Number	Log
2574976	·4107732
2575024	813
073	895
121	977
170	·4108059
218	140
267	222
316	304
364	386
413	468
462	550
510	632
559	714
607	795
656	877
704	959
753	·4109041
801	123
850	205
898	287
947	369
995	450
2576044	532
092	614
141	696
189	777
238	859
286	941
335	·4110023
383	104
432	186
480	268
529	350
577	431
626	513
674	595
723	677
771	758
820	840
868	922
917	·4111004
965	085
2577014	167
062	249
111	331
159	412
208	494
256	576
305	658
353	739
402	821
450	903
499	985
547	·4112066
596	148
644	230
693	312
741	393
790	475
838	557
887	639
935	720
984	802
2578032	884
081	965

2578129	·4113047
178	128
226	210
275	292
323	373
372	455
420	537
469	619
517	700
566	782
614	864
663	945
711	·4114027
760	108
808	190
857	272
905	353
954	435
2579002	517
051	598
099	680
148	761
196	843
245	925
293	·4115006
342	088
390	170
438	251
487	333
535	414
583	496
632	577
680	659
729	740
777	822
826	903
874	985
923	·4116066
971	148

Excesses

10 = 17	1 = 2
20 = 34	2 = 3
30 = 51	3 = 5
40 = 67	4 = 7
	5 = 8
	6 = 10
	7 = 12
	8 = 13
	9 = 15

24 = 41

Anti Log Excesses

10 = 6	1 = 1
20 = 12	2 = 1
30 = 18	3 = 2
40 = 24	4 = 2
50 = 30	5 = 3
60 = 36	6 = 4
70 = 42	7 = 4
80 = 47	8 = 5
	9 = 5

41 = 24

2580020	·4116230
068	311
117	393
165	475
214	556
262	638
311	719
359	801
407	882
456	964
504	·4117045
552	127
601	208
649	290
698	371
746	453
795	534
843	616
892	697
940	779
989	860
2581037	942
086	·4118023
134	105
182	186
231	268
279	349
327	431
376	512
424	594
473	675
521	757
570	838
618	920
667	·4119001
715	082
763	164
812	245
860	326
908	408
957	489
2582005	571
054	652
102	734
151	815
199	897
248	978
296	·4120060
344	141
393	223
441	304

2582489	·4120385
538	467
586	548
635	629
683	711
731	792
780	874
828	955
876	·4121036
925	118
973	199
2583022	280
070	362
119	443
167	525
216	606
264	687
312	769
361	850
409	931
457	·4122012
506	094
554	175
603	256
651	338
699	419
748	501
796	582
844	663
893	745
941	826
990	907
2584038	988
086	·4123070
135	151
183	232
231	313
280	395
328	476
377	557
425	638
473	720
522	801
570	882
618	963
667	·4124045
715	126
764	207
812	288
860	370
909	451
957	532
2585005	613
053	695
102	776
150	857
198	938
247	·4125020
295	101
344	182
392	263
440	345
489	426
537	507
585	588

2585634	·4125670
682	751
731	832
779	913
827	995
876	·4126076
924	157
972	238
2586020	319
069	400
117	481
165	562
214	644
262	725
311	806
359	887
407	968
456	·4127049
504	130
552	211
600	293
649	374
697	455
745	536
794	618
842	699
891	780
939	861
987	942
2587036	·4128023
084	104
132	185
180	266
229	347
277	428
325	509
373	591
422	672
470	753
518	834
567	915
615	996
664	·4129077
712	158
760	239
809	320
857	401
905	482
953	564
2588002	645
050	726
098	807
146	888
195	969
243	·4130050
291	131
339	212
388	293
436	374
484	455
532	536
581	617
629	698
677	779
726	860

2588774	·4130941
823	·4131022
871	103
919	184
968	265
2589016	346
064	427
112	508
161	589
209	670
257	751
305	832
354	913
402	994
450	·4132075
498	156
547	237
595	318
643	399
691	480
740	561
788	642
836	723
884	804
933	884
981	965

Excesses

10 = 17	1 = 2
20 = 34	2 = 3
30 = 50	3 = 5
40 = 67	4 = 7
	5 = 8
	6 = 10
	7 = 12
	8 = 13
	9 = 15

24 = 41

Anti Log Excesses

10 = 6	1 = 1
20 = 12	2 = 1
30 = 18	3 = 2
40 = 24	4 = 2
50 = 30	5 = 3
60 = 36	6 = 4
70 = 42	7 = 4
80 = 48	8 = 5
	9 = 5

41 = 24

2590029	·4133046
077	127
126	208
174	289
222	370
270	451
319	532
367	613
415	694
463	775

2590512	·4133855
560	936
608	·4134017
656	098
705	179
753	260
801	341
849	422
898	502
946	583
994	664
2591042	745
091	826
139	907
187	988
235	·4135069
284	149
332	230
380	311
428	392
477	472
525	553
573	634
621	715
670	796
718	877
766	958
814	·4136039
863	119
911	200
959	281
2592007	362
056	442
104	523
152	604
200	685
249	765
297	846
345	927
393	·4137008
442	088
490	169
538	250
586	331
635	411
683	492
731	573
779	654
827	734
875	815
923	896
971	977
2593020	·4138057
068	138
116	219
164	300
213	380
261	461
309	542
357	623
406	703
454	784
502	865
550	946
599	·4139026

2593647	·4139107
695	188
743	269
791	349
839	430
887	511
935	592
984	672
2594032	753
080	834
128	914
177	995
225	·4140075
273	156
321	237
370	317
418	398
466	479
514	560
562	640
610	721
658	802
706	882
755	963
803	·4141043
851	124
899	205
948	285
996	366
2595044	447
092	527
140	608
188	688
236	769
284	849
333	930
381	·4142010
429	091
477	172
526	252
574	333
622	414
670	494
718	575
766	655
814	736
862	816
911	897
959	977
2596007	·4143058
055	138
103	219
151	299
199	380
247	461
296	541
344	622
392	703
440	783
489	864
537	944
585	·4144025
633	105
681	186
729	266

2596777	·4144347
825	427
874	508
922	588
970	669
2597018	749
066	830
114	910
162	991
210	·4145071
259	152
307	232
355	312
403	393
451	473
499	553
547	634
595	714
644	795
692	875
740	956
788	·4146036
836	117
884	197
932	278
980	358
2598029	439
077	519
125	599
173	680
221	760
269	840
317	921
365	·4147001
414	082
462	162
510	243
558	323
606	404
654	484
702	564
750	645
798	725
846	805
894	886
942	966
991	·4148047
2599039	127
087	207
135	288
183	368
231	448
279	528
327	609
376	689
424	769
472	850
520	930
568	·4149011
616	091
664	171
712	252
760	332
808	412
856	492

2599904	·4149573
952	653

Excesses

10 = 17	1 = 2
20 = 33	2 = 3
30 = 50	3 = 5
40 = 67	4 = 7
	5 = 8
	6 = 10
	7 = 12
	8 = 13
	9 = 15

24 = 40

Anti Log Excesses

10 = 6	1 = 1
20 = 12	2 = 1
30 = 18	3 = 2
40 = 24	4 = 2
50 = 30	5 = 3
60 = 36	6 = 4
70 = 42	7 = 4
80 = 48	8 = 5
	9 = 5

40 = 24

2600000	·4149733
048	814
096	894
145	975
193	·4150055
241	135
289	216
337	296
385	376
433	456
481	537
529	617
577	697
625	777
673	858
722	938
770	·4151018
818	098
866	179
914	259
962	339
2601010	419
058	500
106	580
154	660
202	740
250	821
298	901
346	981
394	·4152061
442	142
491	222
539	302
587	382

2601635	·4152463
683	543
731	623
779	703
827	783
875	863
923	943
971	·4153023
2602019	104
067	184
115	264
163	344
211	425
259	505
307	585
355	665
403	745
451	825
499	905
547	985
595	·4154066
643	146
691	226
739	306
787	386
836	466
884	546
932	626
980	707
2603028	787
076	867
124	947
172	·4155027
220	107
268	187
316	267
364	348
412	428
460	508
508	588
556	668
604	748
652	828
700	908
748	988
796	·4156068
844	148
892	228
940	308
988	388
2604036	468
084	548
132	629
180	709
228	789
276	869
324	949
372	·4157029
420	109
468	189
516	269
564	349
612	429
660	509
708	589

2604756	·4157669
804	749
852	829
900	909
948	989
996	·4158069
2605044	149
092	229
140	309
188	389
236	469
284	549
332	629
380	709
428	789
475	869
523	949
571	·4159029
619	109
667	189
715	269
763	349
811	429
859	509
907	588
955	668
2606003	748
051	828
099	908
147	988
195	·4160068
243	148
291	228
339	308
387	388
435	468
483	548
531	628
579	708
627	788
675	867
723	947
771	·4161027
818	107
866	187
914	267
962	347
2607010	427
058	506
106	586
154	666
202	746
250	826
298	906
346	986
394	·4162066
442	145
490	225
538	305
585	385
633	465
681	545
729	625
777	705
825	784

2607873	·4162864
921	944
969	·4163024
2608017	103
065	183
113	263
161	343
209	423
257	503
305	583
352	663
400	742
448	822
496	902
544	982
592	·4164061
640	141
688	221
736	301
784	380
832	460
880	540
927	620
975	699
2609023	779
071	859
119	939
167	·4165018
215	098
263	178
311	258
359	337
407	417
455	497
502	577
550	656
598	736
646	816
694	895
742	975
790	·4166054
838	134
885	214
933	293
981	373

Excesses

10 = 17	1 = 2
20 = 33	2 = 3
30 = 50	3 = 5
40 = 67	4 = 7
	5 = 8
	6 = 10
	7 = 12
	8 = 13
	9 = 15

24 = 40

Anti Log Excesses

10 = 6	1 = 1
20 = 12	2 = 1
30 = 18	3 = 2
40 = 24	4 = 2
50 = 30	5 = 3
60 = 36	6 = 4
70 = 42	7 = 4
80 = 48	8 = 5
	9 = 5

40 = 24

2610029	·4166453
077	533
125	612
173	692
221	772
268	852
316	931
364	·4167011
412	091
460	170
508	250
556	329
604	409
652	489
700	568
748	648
796	728
843	808
891	887
939	967
987	·4168047
2611035	126
083	206
131	285
179	365
226	445
274	524
322	604
370	684
417	763
465	843
513	922
561	·4169002
609	081
657	161
705	240
753	320
800	400
848	479
896	559
944	639
992	718
2612040	798
088	877
136	957
183	·4170036
231	116
279	195
327	275
375	354
423	434

2612471	·4170513
519	593
566	672
614	752
662	831
710	911
757	991
805	·4171070
853	150
901	230
949	309
997	389
2613045	468
093	548
140	627
188	707
236	786
284	865
331	945
379	·4172024
427	103
475	183
523	262
571	342
619	421
667	501
714	580
762	660
810	739
858	819
905	898
953	978
2614001	·4173057
049	137
097	216
145	296
193	375
241	455
288	534
336	614
384	693
432	772
479	852
527	931
575	·4174010
623	090
670	169
718	249
766	328
814	407
861	487
909	566
957	645
2615005	725
053	804
101	884
149	963
197	·4175042
244	122
292	201
340	280
388	360
435	439
483	519
531	598

2615579	·4175677
626	757
674	836
722	915
770	995
817	·4176074
865	154
913	233
961	312
2616008	392
056	471
104	550
152	629
200	709
248	788
296	867
344	946
391	·4177026
439	105
487	184
535	264
582	343
630	423
678	502
726	581
773	661
821	740
869	819
917	898
964	978
2617012	·4178057
060	136
108	215
155	295
203	374
251	453
299	532
346	612
394	691
442	770
490	849
537	929
585	·4179008
633	087
681	166
728	246
776	325
824	404
872	483
919	563
967	642
2618015	721
063	800
110	879
158	958
206	·4180037
254	116
301	196
349	275
397	354
445	433
492	513
540	592
588	671
636	750

Number	Log
2618683	·4180829
731	908
779	987
827	·4181066
874	146
922	225
970	304
2619018	383
065	463
113	542
161	621
209	700
256	779
304	858
352	937
399	·4182016
447	096
494	175
542	254
590	333
637	412
685	491
733	570
781	649
828	728
876	807
924	886
972	965

Excesses

10 = 17	1 = 2
20 = 33	2 = 3
30 = 50	3 = 5
40 = 66	4 = 7
	5 = 8
	6 = 10
	7 = 12
	8 = 13
	9 = 15

24 = 40

Anti Log Excesses

10 = 6	1 = 1
20 = 12	2 = 1
30 = 18	3 = 2
40 = 24	4 = 2
50 = 30	5 = 3
60 = 36	6 = 4
70 = 42	7 = 4
	8 = 5
	9 = 5

40 = 24

Number	Log
2620019	·4183045
067	124
115	203
163	282
210	361
258	440
306	519
354	598

Number	Log
2620401	·4183677
449	756
497	835
544	914
592	993
639	·4184072
687	151
735	230
782	310
830	389
878	468
926	547
973	626
2621021	705
069	784
117	863
164	942
212	·4185021
260	100
307	179
355	258
402	337
450	416
498	495
545	574
593	653
641	732
689	811
736	890
784	969
832	·4186048
880	127
927	206
975	285
2622023	364
070	443
118	522
165	600
213	679
261	758
308	837
356	916
404	995
452	·4187074
499	153
547	232
595	311
642	390
690	469
737	548
785	627
833	706
880	785
928	863
976	942
2623024	·4188021
071	100
119	179
167	258
214	337
262	416
309	494
357	573
405	652
452	731

Number	Log
2623500	·4188810
548	889
595	968
643	·4189047
690	125
738	204
786	283
833	362
881	441
929	520
976	599
2624024	678
071	756
119	835
167	914
214	993
262	·4190071
310	150
357	229
405	308
452	387
500	466
548	545
595	624
643	702
691	781
738	860
786	939
833	·4191017
881	096
929	175
976	254
2625024	332
072	411
119	490
167	569
214	647
262	726
310	805
357	884
405	962
453	·4192041
500	120
548	199
595	277
643	356
691	435
738	514
786	592
834	671
881	750
929	829
976	907
2626024	986
072	·4193065
119	144
167	222
215	301
262	380
310	459
357	537
405	616
452	695
500	774
547	852

Number	Log
2626595	·4193931
643	·4194010
690	088
738	167
786	245
833	324
881	403
928	481
976	560
2627023	639
071	718
118	796
166	875
214	954
261	·4195032
309	111
357	189
404	268
452	347
499	425
547	504
594	583
642	661
689	740
737	818
784	897
832	976
879	·4196054
927	133
975	212
2628022	290
070	369
118	447
165	526
213	604
260	683
308	761
355	840
403	919
450	997
498	·4197076
545	155
593	233
640	312
688	390
736	469
783	547
831	626
879	704
926	783
974	861
2629021	940
069	·4198018
116	097
164	176
211	254
259	333
306	412
354	490
401	569
449	647
496	726
544	804
591	883
639	961

Number	Log
2629686	·4199040
734	118
781	197
829	275
877	354
924	432
972	511

Excesses

10 = 17	1 = 2
20 = 33	2 = 3
30 = 50	3 = 5
40 = 66	4 = 7
	5 = 8
	6 = 10
	7 = 12
	8 = 13
	9 = 15

24 = 40

Anti Log Excesses

10 = 6	1 = 1
20 = 12	2 = 1
30 = 18	3 = 2
40 = 24	4 = 2
50 = 30	5 = 3
60 = 36	6 = 4
70 = 42	7 = 4
	8 = 5
	9 = 5

40 = 24

Number	Log
2630020	·4199589
067	668
115	746
162	825
210	903
257	982
305	·4200060
352	139
400	217
447	296
495	374
542	453
590	531
637	609
685	688
732	766
780	844
827	923
875	·4201001
922	080
970	158
2631017	237
065	315
112	394
160	472
207	550
255	629
302	707
350	785

2631397	·4201864
445	942
492	·4202021
540	099
587	178
635	256
682	335
730	413
777	491
825	570
872	648
920	726
967	805
2632015	883
062	962
110	·4203040
157	118
205	197
252	275
300	353
347	431
395	510
442	588
490	666
537	745
585	823
632	902
680	980
727	·4204058
775	137
822	215
870	293
917	371
965	450
2633012	528
060	606
107	684
154	763
202	841
249	919
296	997
344	·4205076
391	154
439	232
486	310
534	389
581	467
629	545
676	624
724	702
771	781
819	859
866	937
914	·4206016
961	094
2634009	172
056	250
104	328
151	406
199	484
246	562
293	641
341	719
388	797
435	875

2634483	·4206954
530	·4207032
578	110
625	188
673	267
720	345
768	423
815	501
863	580
910	658
958	736
2635005	814
053	893
100	971
148	·4208049
195	127
242	205
290	283
337	361
384	439
432	518
479	596
527	674
574	752
622	831
669	909
717	987
764	·4209065
811	143
859	221
906	299
953	377
2636001	456
048	534
096	612
143	690
191	768
238	846
286	924
333	·4210002
380	081
428	159
475	237
522	315
570	393
617	471
665	549
712	627
760	705
807	783
855	861
902	939
949	·4211018
997	096
2637044	174
091	252
139	330
186	408
234	486
281	564
328	642
376	720
423	798
470	876
518	954

2637565	·4212032
613	110
660	188
708	266
755	344
803	422
850	500
897	578
945	656
992	734
2638039	812
087	890
134	968
182	·4213046
229	124
276	202
324	280
371	358
418	436
466	514
513	592
561	670
608	748
655	826
703	903
750	981
797	·4214059
845	137
892	215
940	293
987	371
2639034	449
082	527
129	605
176	683
224	761
271	839
319	917
366	995
413	·4215073
461	150
508	228
555	306
602	384
650	462
697	540
744	618
792	696
839	774
887	852
934	930
981	·4216008

Excesses

10 = 16	1 = 2
20 = 33	2 = 3
30 = 49	3 = 5
40 = 66	4 = 7
	5 = 8
	6 = 10
	7 = 12
	8 = 13
	9 = 15

24 = 39

Anti Log Excesses

10 = 6	1 = 1
20 = 12	2 = 1
30 = 18	3 = 2
40 = 24	4 = 2
50 = 30	5 = 3
60 = 36	6 = 4
70 = 42	7 = 4
	8 = 5
	9 = 5

39 = 24

2640029	·4216085
076	163
123	241
171	319
218	397
266	475
313	553
360	631
408	708
455	786
502	864
549	942
597	·4217020
644	098
691	176
739	254
786	331
834	409
881	487
928	565
976	642
2641023	720
070	798
117	876
165	954
212	·4218032
259	110
307	188
354	265
402	343
449	421
496	499
544	576
591	654
638	732
685	810
733	888

2641780	·4218966
827	·4219044
874	122
922	199
969	277
2642016	355
064	433
111	510
159	588
206	666
253	744
301	821
348	899
395	977
442	·4220055
490	132
537	210
584	288
631	366
679	443
726	521
773	599
821	677
868	754
916	832
963	910
2643010	988
058	·4221065
105	143
152	221
199	299
247	376
294	454
341	532
388	609
436	687
483	764
530	842
577	920
625	997
672	·4222075
719	153
766	231
814	308
861	386
908	464
955	542
2644003	619
050	697
097	775
145	852
192	930
240	·4223007
287	085
334	163
382	240
429	318
476	396
523	473
571	551
618	628
665	706
712	783
760	861
807	938

Number	Log
2644854	·4224016
901	094
949	171
996	249
2645043	327
090	404
138	482
185	559
232	637
279	715
327	792
374	870
421	948
468	·4225025
516	103
563	180
610	258
657	335
705	413
752	490
799	568
846	645
894	723
941	800
988	878
2646035	955
083	·4226033
130	110
177	188
224	266
272	343
319	421
366	499
413	576
461	654
508	731
555	809
602	886
650	964
697	·4227041
744	119
791	196
838	274
885	351
932	428
979	506
2647027	583
074	660
121	738
168	815
216	893
263	970
310	·4228048
357	125
405	203
452	280
499	358
546	435
594	513
641	590
688	668
735	745
783	823
830	900
877	977

Number	Log
2647924	·4229055
972	132
2648019	209
066	287
113	364
160	442
207	519
254	597
301	674
349	752
396	829
443	906
490	984
538	·4230061
585	138
632	216
679	293
727	371
774	448
821	525
868	603
916	680
963	757
2649010	835
057	912
104	990
151	·4231067
198	144
245	222
293	299
340	376
387	453
434	531
482	608
529	685
576	763
623	840
670	918
717	995
764	·4232072
811	150
859	227
906	304
953	381

Excesses

10 = 16	1 = 2
20 = 33	2 = 3
30 = 49	3 = 5
40 = 66	4 = 7
	5 = 8
	6 = 10
	7 = 11
	8 = 13
	9 = 15

24 = 39

Anti Log Excesses

10 = 6	1 = 1
20 = 12	2 = 1
30 = 18	3 = 2
40 = 24	4 = 2
50 = 30	5 = 3
60 = 37	6 = 4
70 = 43	7 = 4
	8 = 5
	9 = 5

39 = 24

Number	Log
2650000	·4232459
048	536
095	613
142	690
189	768
237	845
284	922
331	·4233000
378	077
425	155
472	232
519	309
566	387
614	464
661	541
708	618
755	696
802	773
849	850
896	927
943	·4234005
991	082
2651038	159
085	236
132	314
180	391
227	468
274	545
321	622
368	699
415	776
462	853
509	931
557	·4235008
604	085
651	162
698	240
745	317
792	394
839	471
886	549
934	626
981	703
2652028	780
075	858
122	935
169	·4236012
216	089
263	166
311	243
358	320

Number	Log
2652405	·4236397
452	475
499	552
546	629
593	706
640	784
688	861
735	938
782	·4237015
829	092
876	169
923	246
970	323
2653017	401
065	478
112	555
159	632
206	709
253	786
300	863
347	940
394	·4238017
442	094
489	171
536	248
583	326
630	403
677	480
724	557
771	635
819	712
866	789
913	866
960	943
2654007	·4239020
054	097
101	174
148	251
195	328
242	405
289	482
336	559
384	636
431	713
478	790
525	867
572	944
619	·4240021
666	098
713	175
760	252
807	329
854	406
901	484
949	561
996	638
2655043	715
090	792
137	869
184	946
231	·4241023
278	100
325	177
372	254
419	331

Number	Log
2655466	·4241408
514	485
561	562
608	639
655	716
702	792
749	869
796	946
843	·4242023
890	100
937	177
984	254
2656031	331
078	408
125	485
172	562
219	639
267	716
314	793
361	870
408	947
455	·4243024
502	101
549	178
596	255
643	331
690	408
737	485
784	562
831	639
878	716
925	793
972	870
2657020	946
067	·4244023
114	100
161	177
208	254
255	331
302	408
349	485
396	561
443	638
490	715
537	792
584	869
631	946
678	·4245023
725	100
772	176
819	253
866	330
913	407
960	484
2658007	561
054	638
101	715
148	791
195	868
242	945
289	·4246022
336	098
383	175
430	252
477	329

2658525	·4246406
572	483
619	560
666	637
713	713
760	790
807	867
854	944
901	·4247020
948	097
995	174
2659042	251
089	327
136	404
183	481
230	558
277	634
324	711
371	788
418	865
465	941
512	·4248018
559	095
606	172
653	248
700	325
747	402
794	479
841	555
888	632
935	709
982	786

Excesses

10 = 16	1 = 2
20 = 33	2 = 3
30 = 49	3 = 5
40 = 65	4 = 7
	5 = 8
	6 = 10
	7 = 11
	8 = 13
	9 = 15

23 = 38

Anti Log Excesses

10 = 6	1 = 1
20 = 12	2 = 1
30 = 18	3 = 2
40 = 24	4 = 2
50 = 31	5 = 3
60 = 37	6 = 4
70 = 43	7 = 4
	8 = 5
	9 = 6

38 = 23

2660029	·4248862
076	939
123	·4249016
170	093

2660217	·4249169
264	246
311	323
358	400
405	476
452	553
499	630
546	707
593	783
640	860
687	937
733	·4250013
780	090
827	166
874	243
921	320
968	396
2661015	473
062	550
109	627
156	703
203	780
250	857
297	933
344	·4251010
391	086
438	163
485	240
532	316
579	393
626	470
673	546
720	623
767	699
814	776
861	853
908	929
955	·4252006
2662002	083
048	159
095	236
142	312
189	389
236	466
283	542
330	619
377	696
424	772
471	849
518	925
565	·4253002
612	078
659	155
706	231
753	308
800	384
847	461
894	537
941	614
987	691
2663034	767
081	844
128	921
175	997
222	·4254074

2663269	·4254150
316	227
363	303
410	380
457	456
504	533
551	609
598	686
645	762
692	839
738	915
785	992
832	·4255068
879	145
926	221
973	298
2664020	374
067	451
114	527
161	604
208	680
255	757
301	833
348	910
395	986
442	·4256063
489	139
536	216
583	292
630	368
677	445
724	521
771	597
818	674
864	750
911	827
958	903
2665005	980
052	·4257056
099	133
146	209
193	286
239	362
286	439
333	515
380	591
427	668
474	744
521	820
568	897
615	973
662	·4258050
709	126
756	203
802	279
849	356
896	432
943	508
990	585
2666037	661
084	737
131	814
177	890
224	967
271	·4259043

2666318	·4259119
365	196
412	272
459	348
506	424
552	501
599	577
646	653
693	730
740	806
787	883
834	959
881	·4260035
927	112
974	188
2667021	264
068	340
115	417
162	493
209	569
256	646
302	722
349	799
396	875
443	951
490	·4261028
537	104
584	180
631	256
677	333
724	409
771	485
818	561
865	638
912	714
959	790
2668006	866
052	943
099	·4262019
146	095
193	171
239	248
286	324
333	400
380	476
427	553
474	629
521	705
568	781
614	858
661	934
708	·4263010
755	086
801	163
848	239
895	315
942	391
989	467
2669036	543
083	619
130	695
176	772
223	848
270	924
317	·4264000

2669363	·4264077
410	153
457	229
504	305
551	382
598	458
645	534
692	610
738	686
785	762
832	838
879	914
925	991
972	·4265067

Excesses

10 = 16	1 = 2
20 = 33	2 = 3
30 = 49	3 = 5
40 = 65	4 = 7
	5 = 8
	6 = 10
	7 = 11
	8 = 13
	9 = 15

23 = 38

Anti Log Excesses

10 = 6	1 = 1
20 = 12	2 = 1
30 = 18	3 = 2
40 = 25	4 = 3
50 = 31	5 = 3
60 = 37	6 = 4
70 = 43	7 = 4
	8 = 5
	9 = 6

38 = 23

2670019	·4265143
066	219
112	296
159	372
206	448
253	524
300	600
347	676
394	752
441	828
487	905
534	981
581	·4266057
628	133
674	209
721	285
768	361
815	437
861	514
908	590
955	666
2671002	742

2671049	·4266818
096	894
143	970
190	·4267046
236	122
283	198
330	274
377	350
423	426
470	502
517	578
564	654
610	731
657	807
704	883
751	959
797	·4268035
844	111
891	187
938	263
984	339
2672031	415
078	491
125	567
171	643
218	719
265	795
312	871
359	947
406	·4269023
453	099
500	175
546	251
593	327
640	403
687	479
733	555
780	631
827	707
874	783
920	859
967	935
2673014	·4270011
061	087
107	163
154	239
201	315
248	391
294	467
341	543
388	619
435	695
481	771
528	847
575	923
622	999
668	·4271075
715	150
762	226
809	302
855	378
902	454
949	530
996	606
2674042	682

2674089	·4271758
136	834
183	910
229	986
276	·4272062
323	138
370	214
416	290
463	365
510	441
557	517
603	593
650	669
697	745
744	821
790	897
837	972
884	·4273048
931	124
977	200
2675024	276
071	352
117	428
164	504
210	579
257	655
304	731
350	807
397	882
444	958
491	·4274034
537	110
584	186
631	262
678	338
724	414
771	489
818	565
865	641
911	717
958	792
2676005	868
052	944
098	·4275020
145	095
192	171
239	247
285	323
332	399
379	475
425	551
472	627
518	702
565	778
612	854
658	930
705	·4276005
752	081
799	157
845	233
892	308
939	384
986	460
2677032	536
079	611

2677126	·4276687
173	763
219	839
266	914
313	990
359	·4277066
406	142
452	217
499	293
546	369
592	445
639	520
686	596
733	672
779	748
826	823
873	899
919	975
966	·4278050
2678012	126
059	201
106	277
152	353
199	428
246	504
293	580
339	656
386	731
433	807
480	883
526	959
573	·4279034
620	110
666	186
713	261
759	337
806	412
853	488
899	564
946	639
993	715
2679039	791
086	866
132	942
179	·4280017
226	093
272	169
319	244
366	320
413	396
459	471
506	547
553	622
599	698
646	774
692	849
739	925
786	·4281001
832	076
879	152
926	227
972	303

Excesses

10 = 16	1 = 2
20 = 32	2 = 3
30 = 49	3 = 5
40 = 65	4 = 7
	5 = 8
	6 = 10
	7 = 11
	8 = 13
	9 = 15

23 = 38

Anti Log Excesses

10 = 6	1 = 1
20 = 12	2 = 1
30 = 18	3 = 2
40 = 25	4 = 3
50 = 31	5 = 3
60 = 37	6 = 4
70 = 43	7 = 4
	8 = 5
	9 = 6

38 = 23

2680019	·4281378
065	454
112	529
159	605
205	681
252	756
299	832
346	908
392	983
439	·4282059
486	134
532	210
579	285
625	361
672	436
719	512
765	587
812	663
859	738
905	814
952	889
998	965
2681045	·4283040
092	116
138	191
185	267
232	342
278	418
325	494
371	569
418	645
464	721
511	796
557	872
604	947
651	·4284022
697	098

2681744	·4284173
791	248
837	324
884	399
930	475
977	550
2682024	626
070	701
117	777
164	852
210	928
257	·4285003
303	079
350	154
397	230
443	305
490	381
537	456
583	532
630	607
676	683
723	758
769	833
816	909
862	984
909	·4286059
956	135
2683002	210
049	286
096	361
142	436
189	512
235	587
282	662
328	738
375	813
421	889
468	964
515	·4287040
561	115
608	191
655	266
701	341
748	417
794	492
841	567
887	643
934	718
980	794
2684027	869
074	944
120	·4288020
167	095
214	170
260	245
307	321
353	396
400	471
446	547
493	622
539	698
586	773
632	848
679	924
725	999

Number	Log
2684772	·4289074
819	149
865	225
912	300
959	375
2685005	450
052	526
098	601
145	676
191	751
238	827
284	902
331	977
377	·4290053
424	128
470	204
517	279
563	354
610	430
656	505
703	580
750	655
796	731
843	806
890	881
936	956
983	·4291032
2686029	107
076	182
122	257
169	332
215	407
262	482
308	557
355	633
401	708
448	783
494	858
541	934
587	·4292009
634	084
680	159
727	235
773	310
820	385
866	460
913	536
959	611
2687006	686
052	761
099	837
145	912
192	987
238	·4293062
285	137
331	212
378	287
424	362
471	438
517	513
564	588
610	663
657	739
703	814
750	889

Number	Log
2687797	·4293964
843	·4294039
890	114
937	189
983	264
2688030	340
076	415
123	490
169	565
216	640
262	715
309	790
355	865
402	941
448	·4295016
495	091
541	166
587	241
634	316
680	391
726	466
773	541
819	616
866	691
912	766
959	842
2689005	917
052	992
098	·4296067
145	142
191	217
238	292
284	367
331	442
377	517
424	592
470	667
517	743
563	818
610	893

Excesses

23 = 38

Anti Log Excesses

38 = 23

Number	Log
2689672	·4296993
734	·4297093
796	193
858	293
920	393
982	493

Excesses

10 = 16	1 = 2
20 = 32	2 = 3
30 = 49	3 = 5
40 = 65	4 = 7
50 = 81	5 = 8
60 = 97	6 = 10
	7 = 11
	8 = 13
	9 = 15

31 = 50

Anti Log Excesses

10 = 6	1 = 1
20 = 12	2 = 1
30 = 19	3 = 2
40 = 25	4 = 3
50 = 31	5 = 3
60 = 37	6 = 4
70 = 43	7 = 4
80 = 49	8 = 5
90 = 56	9 = 6

50 = 31

Number	Log
2690044	·4297593
106	693
168	793
230	893
292	993
354	·4298093
416	193
477	293
539	393
601	493
663	593
725	693
787	793
849	893
911	993
973	·4299093
2691035	193
097	293
159	393
221	492
283	592
345	692
407	792
469	892
531	992
592	·4300092
654	192
716	292
778	392
840	492
902	592
964	691
2692026	791
088	891
149	991
211	·4301091
273	191

Number	Log
2692335	·4301290
397	390
459	490
521	590
583	690
645	790
707	889
769	989
831	·4302089
892	189
954	289
2693016	389
078	488
140	588
202	688
264	788
326	888
388	988
449	·4303087
511	187
573	287
635	386
697	486
759	586
820	686
882	786
944	886
2694006	985
068	·4304085
130	185
192	284
254	384
316	484
377	583
439	683
501	783
563	882
625	982
687	·4305082
748	181
810	281
872	381
934	480
996	580
2695058	680
119	779
181	879
243	979
305	·4306078
367	178
429	278
490	377
552	477
614	577
676	676
738	776
800	876
861	975
923	·4307075
985	174
2696047	274
109	373
171	473
232	572
294	672

Number	Log
2696356	·4307772
418	871
480	971
542	·4308070
603	170
665	269
727	369
789	468
851	568
913	667
974	767
2697036	866
098	966
159	·4309065
221	165
283	264
345	364
407	463
469	563
530	662
592	762
654	861
715	961
777	·4310060
839	159
901	259
963	358
2698025	458
086	557
148	657
210	756
271	856
333	955
395	·4311054
457	154
519	253
581	352
642	452
704	551
766	651
827	750
889	850
951	949
2699012	·4312049
074	148
136	247
198	347
260	446
322	545
383	645
445	744
507	843
568	943
630	·4313042
692	141
753	241
815	340
877	439
938	539

Excesses

10 = 16	1 = 2
20 = 32	2 = 3
30 = 48	3 = 5
40 = 64	4 = 6
50 = 81	5 = 8
60 = 97	6 = 10
	7 = 11
	8 = 13
	9 = 15

31 = 50

Anti Log Excesses

10 = 6	1 = 1
20 = 12	2 = 1
30 = 19	3 = 2
40 = 25	4 = 3
50 = 31	5 = 3
60 = 37	6 = 4
70 = 43	7 = 4
80 = 50	8 = 5
90 = 56	9 = 6

50 = 31

2700000	4313638
062	737
124	836
186	935
248	·4314034
309	134
371	233
433	332
494	432
556	531
618	630
679	730
741	829
803	928
864	·4315027
926	126
988	225
2701049	325
111	424
173	523
234	623
296	722
358	821
420	920
482	·4316019
544	118
605	218
667	317
729	416
790	515
852	614
914	713
975	813
2702037	912
099	·4317011
160	110
222	209

2702284	·4317308
345	408
407	507
469	606
530	705
592	804
654	903
715	·4318002
777	101
839	200
900	299
962	398
2703024	497
085	597
147	696
209	795
270	894
332	993
394	·4319092
455	191
517	290
579	389
640	488
702	587
764	686
825	785
887	884
949	983
2704010	·4320082
072	181
133	280
195	379
256	478
318	577
379	676
441	775
503	874
564	973
626	·4321072
688	171
749	270
811	369
873	468
934	566
996	665
2705058	764
119	863
181	962
243	·4322061
304	160
366	259
427	358
489	456
550	555
612	654
673	753
735	852
797	951
858	·4323050
920	149
982	248
2706043	346
105	445
166	544
228	643

2706289	·4323742
351	841
412	939
474	·4324038
536	137
597	236
659	335
721	434
782	532
844	631
905	730
967	828
2707028	927
090	·4325026
151	125
213	224
275	323
336	421
398	520
459	619
521	717
582	816
644	915
705	·4326013
767	112
829	211
890	309
952	408
2708013	507
075	606
136	705
198	804
259	902
321	·4327001
383	100
444	198
506	297
567	396
629	494
690	593
752	692
813	790
875	889
936	988
998	·4328086
2709059	185
121	283
182	382
244	480
305	579
367	677
428	776
490	875
551	973
613	·4329072
675	171
736	269
798	368
859	466
921	565
982	663

Excesses

10 = 16	1 = 2
20 = 32	2 = 3
30 = 48	3 = 5
40 = 64	4 = 6
50 = 80	5 = 8
60 = 96	6 = 10
	7 = 11
	8 = 13
	9 = 14

31 = 49

Anti Log Excesses

10 = 6	1 = 1
20 = 12	2 = 1
30 = 19	3 = 2
40 = 25	4 = 3
50 = 31	5 = 3
60 = 37	6 = 4
70 = 44	7 = 4
80 = 50	8 = 5
90 = 56	9 = 6

49 = 31

2710044	·4329762
105	860
167	959
228	·4330058
290	156
351	255
413	353
474	452
536	550
597	649
659	747
720	846
781	944
843	·4331043
904	141
966	240
2711027	338
089	437
150	535
212	634
273	732
335	830
396	929
458	·4332027
519	126
581	224
642	323
703	421
765	520
826	618
888	716
949	815
2712011	913
072	·4333011
134	110
195	208
257	307

2712318	·4333405
380	504
441	602
503	701
564	799
625	897
687	996
748	·4334094
809	192
871	291
932	389
994	487
2713055	586
117	684
178	782
240	881
301	979
362	·4335077
424	176
485	274
547	372
608	471
670	569
731	667
793	766
854	864
915	962
977	·4336061
2714038	159
099	257
161	355
222	453
284	551
345	650
407	748
468	846
530	945
591	·4337043
652	141
714	239
775	337
836	435
898	534
959	632
2715020	730
082	829
143	927
204	·4338025
266	123
327	221
389	319
450	418
512	516
573	614
635	712
696	810
757	908
819	·4339007
880	105
941	203
2716003	301
064	399
125	497
187	595
248	693

N	Log
2716309	·4339791
371	889
432	987
493	·4340085
555	184
616	282
677	380
739	478
800	576
861	674
923	772
984	870
2717045	968
107	·4341066
168	164
229	262
291	360
352	458
413	556
475	655
536	753
597	851
659	949
720	·4342047
781	145
843	243
904	341
965	439
2718027	536
088	634
149	732
211	830
272	928
333	·4343026
395	124
456	222
517	320
579	418
640	516
701	614
763	712
824	810
885	908
947	·4344005
2719008	103
069	201
131	299
192	397
253	495
314	593
375	691
436	789
498	886
559	984
620	·4345082
682	180
743	278
804	376
866	473
927	571
988	669

Excesses

10 = 16	1 = 2
20 = 32	2 = 3
30 = 48	3 = 5
40 = 64	4 = 6
50 = 80	5 = 8
60 = 96	6 = 10
	7 = 11
	8 = 13
	9 = 14

31 = 49

Anti Log Excesses

10 = 6	1 = 1
20 = 13	2 = 1
30 = 19	3 = 2
40 = 25	4 = 3
50 = 31	5 = 3
60 = 38	6 = 4
70 = 44	7 = 4
80 = 50	8 = 5
90 = 56	9 = 6

49 = 31

N	Log
2720050	·4345767
111	865
172	963
234	·4346061
295	159
356	257
417	354
478	452
539	550
601	647
662	745
723	843
785	941
846	·4347039
907	137
969	234
2721030	332
091	430
152	527
213	625
274	723
336	821
397	919
458	·4348017
520	114
581	212
642	310
704	407
765	505
826	603
887	700
948	798
2722009	896
071	993
132	·4349091
193	189
255	286

N	Log
2722316	·4349384
377	482
438	579
499	677
560	775
622	872
683	970
744	·4350068
806	165
867	263
928	361
989	458
2723050	556
111	654
173	751
234	849
295	946
356	·4351044
417	141
478	239
540	336
601	434
662	532
724	629
785	727
846	825
907	922
968	·4352020
2724029	117
091	215
152	312
213	410
274	507
335	605
396	702
458	800
519	897
580	995
641	·4353092
702	190
763	287
825	385
886	482
947	580
2725008	677
069	775
130	872
192	970
253	·4354067
314	164
375	262
436	359
497	457
559	554
620	652
681	749
742	847
803	944
864	·4355041
926	139
987	236
2726048	334
109	431
170	529
231	626

N	Log
2726292	·4355724
353	821
414	918
476	·4356016
537	113
598	210
659	308
720	405
781	502
843	600
904	697
965	794
2727026	892
087	989
148	·4357086
209	184
270	281
331	378
393	476
454	573
515	670
576	768
637	865
698	962
759	·4358060
820	157
881	254
942	352
2728003	449
064	546
126	643
187	740
248	837
309	935
370	·4359032
431	129
492	227
553	324
614	421
675	518
736	615
797	712
859	810
920	907
981	·4360004
2729042	102
103	199
164	296
225	393
286	490
347	587
408	684
469	781
530	878
591	976
652	·4361073
713	170
775	267
836	364
897	461
958	559

Excesses

10 = 16	1 = 2
20 = 32	2 = 3
30 = 48	3 = 5
40 = 64	4 = 6
50 = 80	5 = 8
60 = 96	6 = 10
	7 = 11
	8 = 13
	9 = 14

31 = 49

Anti Log Excesses

10 = 6	1 = 1
20 = 13	2 = 1
30 = 19	3 = 2
40 = 25	4 = 3
50 = 31	5 = 3
60 = 38	6 = 4
70 = 44	7 = 4
80 = 50	8 = 5
90 = 56	9 = 6

49 = 31

N	Log
2730019	·4361656
080	753
141	850
202	947
263	·4362044
324	141
385	238
446	335
507	432
568	529
629	626
690	724
751	821
812	918
873	·4363015
934	112
995	209
2731056	306
117	403
178	500
240	597
301	694
362	791
423	888
484	985
545	·4364082
606	179
667	276
728	373
789	470
850	567
911	664
972	761
2732033	858
094	955
155	·4365052
216	149

2732277	·4365246
338	343
399	440
460	537
521	634
582	731
643	828
704	925
765	·4366022
826	119
887	215
948	312
2733009	409
070	506
131	603
192	700
253	797
314	894
375	991
435	·4367087
496	184
557	281
618	378
679	475
740	572
801	669
862	766
923	863
984	959
2734045	·4368056
106	153
167	250
228	347
289	444
350	540
411	637
472	734
533	830
594	927
655	·4369024
716	121
777	218
838	315
898	411
959	508
2735020	605
081	701
142	798
203	895
264	992
325	·4370089
386	186
447	282
508	379
569	476
630	572
691	669
752	766
812	862
873	959
934	·4371056
995	152
2736056	249
117	346
178	442

2736239	·4371539
300	636
361	732
422	829
483	926
543	·4372022
604	119
665	216
726	312
787	409
848	506
909	602
970	699
2737031	796
091	892
152	989
213	·4373086
274	182
335	279
396	375
457	472
518	568
579	665
639	761
700	858
761	955
822	·4374051
883	148
944	244
2738005	341
066	437
127	534
187	630
248	727
309	824
370	920
431	·4375017
492	113
552	210
613	306
674	403
735	499
796	596
857	692
917	789
978	885
2739039	982
100	·4376078
161	175
222	271
282	368
343	464
404	560
465	657
526	753
587	850
648	946
709	·4377043
770	139
830	236
891	332
952	428

Excesses

10 = 16	1 = 2
20 = 32	2 = 3
30 = 48	3 = 5
40 = 64	4 = 6
50 = 79	5 = 8
60 = 95	6 = 10
	7 = 11
	8 = 13
	9 = 14

30 = 48

Anti Log Excesses

10 = 6	1 = 1
20 = 13	2 = 1
30 = 19	3 = 2
40 = 25	4 = 3
50 = 31	5 = 3
60 = 38	6 = 4
70 = 44	7 = 4
80 = 50	8 = 5
90 = 57	9 = 6

48 = 30

2740012	·4377525
073	621
134	718
195	814
256	911
317	·4378007
377	104
438	200
499	296
560	393
621	489
682	585
742	682
803	778
864	874
925	971
986	·4379067
2741047	163
107	260
168	356
229	452
290	549
351	645
412	741
472	838
533	934
594	·4380030
654	127
715	223
776	319
837	416
898	512
959	608
2742019	704
080	800
141	896
201	993

2742262	·4381089
323	185
384	282
445	378
506	474
566	571
627	667
688	763
748	859
809	955
870	·4382051
931	148
992	244
2743053	340
113	436
174	532
235	628
295	725
356	821
417	917
477	·4383014
538	110
599	206
660	302
721	398
782	494
842	590
903	686
964	782
2744024	879
085	975
146	·4384071
206	167
267	263
328	359
388	456
449	552
510	648
571	744
632	840
693	936
753	·4385032
814	128
875	224
935	320
996	416
2745057	512
117	608
178	704
239	800
299	897
360	993
421	·4386089
481	185
542	281
603	377
663	473
724	569
785	665
846	761
907	857
968	953
2746028	·4387049
089	145
150	241

2746210	·4387337
271	433
332	529
392	624
453	720
514	816
574	912
635	·4388008
696	104
756	200
817	296
878	392
938	488
999	584
2747060	680
120	776
181	872
242	968
302	·4389063
363	159
424	255
484	351
545	447
606	543
666	639
727	735
788	831
848	926
909	·4390022
970	118
2748030	214
091	310
152	406
212	501
273	597
334	693
394	789
455	885
516	981
576	·4391076
637	172
697	268
758	364
818	460
879	556
939	651
2749000	747
061	843
121	938
182	·4392034
243	130
303	226
364	322
425	418
485	513
546	609
607	705
667	800
728	896
789	992
849	·4393087
910	183
970	279

Excesses

10 = 16	1 = 2
20 = 32	2 = 3
30 = 47	3 = 5
40 = 63	4 = 6
50 = 79	5 = 8
60 = 95	6 = 10
	7 = 11
	8 = 13
	9 = 14

30 = 48

Anti Log Excesses

10 = 6	1 = 1
20 = 13	2 = 1
30 = 19	3 = 2
40 = 25	4 = 3
50 = 32	5 = 3
60 = 38	6 = 4
70 = 44	7 = 4
80 = 51	8 = 5
90 = 57	9 = 6

48 = 30

2750031	·4393374
091	470
152	566
212	662
273	758
334	854
394	949
455	·4394045
516	141
576	236
637	332
698	428
758	523
819	619
879	715
940	810
2751000	906
061	·4395002
121	097
182	193
243	288
303	384
364	479
425	575
485	670
546	766
606	862
667	957
727	·4396053
788	149
848	244
909	340
970	436
2752030	531
091	627
151	722
212	818

2752272	·4396913
333	·4397009
393	104
454	200
515	296
575	391
636	487
696	582
757	678
817	773
878	869
938	964
999	·4398060
2753059	155
120	251
180	346
241	442
301	537
362	633
423	728
483	824
544	919
604	·4399014
665	110
725	205
786	301
846	396
907	492
967	587
2754028	683
088	778
149	873
209	969
270	·4400064
330	159
391	255
451	350
512	446
572	541
633	637
693	732
754	828
814	923
875	·4401018
935	114
996	209
2755056	304
117	400
177	495
238	590
298	686
359	781
419	876
480	972
540	·4402067
601	162
661	258
722	353
782	448
843	544
903	639
964	734
2756024	830
085	925
145	·4403020

2756206	·4403116
266	211
326	306
387	401
447	496
508	591
568	687
629	782
689	877
750	973
810	·4404068
871	163
931	259
992	354
2757052	449
113	544
173	639
233	734
294	830
354	925
415	·4405020
475	116
536	211
596	306
657	401
717	496
777	591
838	687
898	782
959	877
2758019	972
080	·4406067
140	162
201	258
261	353
321	448
382	543
442	638
502	733
563	828
623	923
684	·4407018
744	114
805	209
865	304
926	399
986	494
2759046	589
107	684
167	779
227	874
288	969
348	·4408064
408	159
469	255
529	350
590	445
650	540
711	635
771	730
832	825
892	920
952	·4409015

Excesses

10 = 16	1 = 2
20 = 32	2 = 3
30 = 47	3 = 5
40 = 63	4 = 6
50 = 79	5 = 8
60 = 95	6 = 10
	7 = 11
	8 = 13
	9 = 14

30 = 48

Anti Log Excesses

10 = 6	1 = 1
20 = 13	2 = 1
30 = 19	3 = 2
40 = 25	4 = 3
50 = 32	5 = 3
60 = 38	6 = 4
70 = 44	7 = 4
80 = 51	8 = 5
90 = 57	9 = 6

48 = 30

2760013	·4409110
073	205
133	300
194	395
254	490
314	585
375	680
435	775
495	870
556	965
616	·4410060
676	155
737	250
797	345
858	440
918	535
979	630
2761039	725
100	819
160	914
220	·4411009
281	104
341	199
401	294
462	389
522	484
582	579
643	674
703	769
763	864
824	958
884	·4412053
944	148
2762005	243
065	338
125	433
186	528

2762246	·4412623
306	718
367	812
427	907
487	·4413002
548	097
608	192
668	287
729	381
789	476
849	571
910	666
970	761
2763030	856
091	950
151	·4414045
211	140
272	235
332	330
392	425
453	519
513	614
573	709
634	803
694	898
754	993
815	·4415088
875	183
935	278
995	372
2764055	467
115	562
176	656
236	751
296	846
357	940
417	·4416035
477	130
538	224
598	319
658	414
719	508
779	603
839	698
900	792
960	887
2765020	982
081	·4417076
141	171
201	266
261	360
321	455
381	550
442	644
502	739
562	834
623	928
683	·4418023
743	118
804	212
864	307
924	402
985	496
2766045	591
105	686

Number	Log
2766165	·4418780
225	875
285	970
346	·4419064
406	159
466	253
527	348
587	442
647	537
708	631
768	726
828	821
888	915
948	·4420010
2767008	104
069	199
129	293
189	388
250	482
310	577
370	671
430	766
490	860
550	955
611	·4421049
671	144
731	238
792	333
852	427
912	522
972	616
2768032	711
092	805
153	900
213	994
273	·4422088
334	183
394	277
454	372
514	466
574	561
634	655
695	750
755	844
815	938
875	·4423033
935	127
995	222
2769056	316
116	411
176	505
236	600
296	694
356	788
417	883
477	977
537	·4424071
598	166
658	260
718	354
778	449
838	543
898	637
959	732

Excesses

10 = 16	1 = 2
20 = 31	2 = 3
30 = 47	3 = 5
40 = 63	4 = 6
50 = 79	5 = 8
60 = 94	6 = 9
	7 = 11
	8 = 13
	9 = 14

30 = 47

Anti Log Excesses

10 = 6	1 = 1
20 = 13	2 = 1
30 = 19	3 = 2
40 = 25	4 = 3
50 = 32	5 = 3
60 = 38	6 = 4
70 = 45	7 = 5
80 = 51	8 = 5
90 = 57	9 = 6

47 = 30

Number	Log
2770019	·4424826
079	920
139	·4425015
199	109
259	203
320	298
380	392
440	486
500	581
560	675
620	769
680	864
740	958
800	·4426052
861	146
921	240
981	334
2771041	429
101	523
161	617
222	712
282	806
342	900
402	995
462	·4427089
522	183
583	277
643	371
703	465
763	560
823	654
883	748
943	843
2772003	937
063	·4428031
124	125
184	219

Number	Log
2772244	·4428313
304	408
364	502
424	596
484	690
544	784
604	878
665	973
725	·4429067
785	161
845	255
905	349
965	443
2773025	538
085	632
145	726
206	820
266	914
326	·4430008
386	102
446	196
506	290
566	385
626	479
686	573
747	667
807	761
867	855
927	949
987	·4431043
2774047	137
107	231
167	325
227	419
287	513
347	607
407	701
467	795
527	889
587	983
648	·4432078
708	172
768	266
828	360
888	454
948	548
2775008	642
068	736
128	830
188	924
248	·4433018
308	112
368	206
428	300
488	394
549	487
609	581
669	675
729	769
789	863
849	957
909	·4434051
969	145
2776029	239
089	333

Number	Log
2776149	·4434427
209	521
269	615
329	709
389	803
449	896
509	990
569	·4435084
629	178
689	272
749	366
809	460
869	554
929	648
989	741
2777049	835
109	929
169	·4436023
229	117
289	211
349	304
409	398
469	492
529	586
589	680
649	774
709	867
769	961
829	·4437055
889	149
949	243
2778009	337
069	430
129	524
189	618
249	711
309	805
369	899
429	993
489	·4438087
549	181
609	274
669	368
729	462
789	555
849	649
909	743
969	836
2779029	930
089	·4439024
149	117
209	211
269	305
329	398
389	492
449	586
509	679
569	773
629	867
689	961
749	·4440055
809	148
869	242
929	335
989	429

Excesses

10 = 16	1 = 2
20 = 31	2 = 3
30 = 47	3 = 5
40 = 63	4 = 6
50 = 78	5 = 8
60 = 94	6 = 9
	7 = 11
	8 = 13
	9 = 14

30 = 47

Anti Log Excesses

10 = 6	1 = 1
20 = 13	2 = 1
30 = 19	3 = 2
40 = 26	4 = 3
50 = 32	5 = 3
60 = 38	6 = 4
70 = 45	7 = 5
80 = 51	8 = 5
90 = 58	9 = 6

47 = 30

Number	Log
2780048	·4440522
108	616
168	710
228	803
288	897
348	991
408	·4441084
468	178
528	272
588	365
648	459
708	553
768	646
828	740
888	834
948	927
2781008	·4442021
068	115
127	208
187	302
247	395
307	489
367	582
427	676
487	769
547	863
607	957
667	·4443050
727	144
787	237
846	331
906	424
966	518
2782026	611
086	705
146	798
206	892

Number	Log
2782266	·4443985
326	·4444079
385	172
445	266
505	359
565	453
625	546
685	640
745	733
805	827
865	920
925	·4445014
985	107
2783045	201
104	294
164	388
224	481
284	575
344	668
404	761
463	855
523	948
583	·4446041
643	135
703	228
763	322
823	415
883	509
943	602
2784002	696
062	789
122	882
182	976
242	·4447069
302	162
361	256
421	349
481	442
541	536
601	629
661	722
721	816
781	909
841	·4448002
900	096
960	189
2785020	282
080	376
140	469
200	562
259	656
319	749
379	842
439	936
499	·4449029
559	122
618	216
678	309
738	402
798	495
858	588
918	681
977	775
2786037	868
097	961

Number	Log
2786157	·4450055
217	148
277	241
336	335
396	428
456	521
515	614
575	707
635	800
695	894
755	987
815	·4451080
874	173
934	266
994	359
2787054	453
114	546
174	639
233	733
293	826
353	919
412	·4452012
472	105
532	198
592	292
652	385
712	478
771	571
831	664
891	757
951	850
2788011	943
071	·4453036
130	130
190	223
250	316
309	409
369	502
429	595
489	688
549	781
609	874
668	968
728	·4454061
788	154
847	247
907	340
967	433
2789026	526
086	619
146	712
206	805
266	898
326	991
385	·4455084
445	177
505	270
564	363
624	456
684	549
743	642
803	735
863	828
923	921
983	·4456014

Excesses

10 = 16	1 = 2
20 = 31	2 = 3
30 = 47	3 = 5
40 = 62	4 = 6
50 = 78	5 = 8
	6 = 9
	7 = 11
	8 = 12
	9 = 14

30 = 47

Anti Log Excesses

10 = 6	1 = 1
20 = 13	2 = 1
30 = 19	3 = 2
40 = 26	4 = 3
50 = 32	5 = 3
60 = 38	6 = 4
70 = 45	7 = 5
80 = 51	8 = 5
90 = 58	9 = 6

47 = 30

Number	Log
2790043	·4456107
102	200
162	293
222	386
281	479
341	572
401	665
460	758
520	851
580	944
639	·4457037
699	130
759	223
818	315
878	408
938	501
998	594
2791058	687
118	780
177	873
237	966
297	·4458059
356	152
416	245
476	338
535	430
595	523
655	616
714	709
774	802
834	895
893	988
953	·4459081
2792013	174
072	266
132	359
192	452

Number	Log
2792251	·4459545
311	638
371	731
430	823
490	916
550	·4460009
609	102
669	195
729	288
789	380
849	473
909	566
968	658
2793028	751
088	844
147	937
207	·4461030
267	123
326	215
386	308
446	401
505	493
565	586
625	679
684	772
744	865
803	958
863	·4462050
922	143
982	236
2794041	328
101	421
161	514
220	606
280	699
340	792
399	884
459	977
519	·4463070
578	162
638	255
698	348
757	440
817	533
877	626
936	718
996	811
2795056	904
115	996
175	·4464089
235	182
294	274
354	367
414	460
473	552
533	645
593	738
652	830
712	923
771	·4465016
831	108
890	201
950	293
2796009	386
069	478

Number	Log
2796129	·4465571
188	663
248	756
308	849
367	941
427	·4466034
487	126
546	219
606	311
665	404
725	496
784	589
844	682
903	774
963	867
2797023	959
082	·4467052
142	144
202	237
261	329
321	422
380	514
440	607
499	699
559	792
618	884
678	977
738	·4468069
797	162
857	254
917	346
976	439
2798036	531
095	624
155	716
214	809
274	901
333	994
393	·4469086
453	178
512	271
572	363
631	455
691	548
750	640
810	733
869	825
929	918
989	·4470010
2799048	103
108	195
167	287
227	380
286	472
346	564
405	657
465	749
524	841
584	934
643	·4471026
703	118
762	211
822	303
881	395
941	488

Excesses

10 = 16	1 = 2
20 = 31	2 = 3
30 = 47	3 = 5
40 = 62	4 = 6
50 = 78	5 = 8
	6 = 9
	7 = 11
	8 = 12
	9 = 14

30 = 46

Anti Log Excesses

10 = 6	1 = 1
20 = 13	2 = 1
30 = 19	3 = 2
40 = 26	4 = 3
50 = 32	5 = 3
60 = 39	6 = 4
70 = 45	7 = 5
80 = 51	8 = 5
90 = 58	9 = 6

46 = 30

2800000	·4471580
060	672
119	765
179	857
239	949
298	·4472042
358	134
417	226
477	319
536	411
596	503
655	596
715	688
774	780
834	873
893	965
953	·4473057
2801012	150
072	242
131	334
191	427
250	519
310	611
369	703
429	795
488	887
548	980
607	·4474072
667	164
726	257
786	349
845	441
905	533
964	625
2802024	717
083	810
143	902

2802202	·4474994
262	·4475086
321	178
380	270
440	363
499	455
559	547
618	639
678	731
737	823
797	915
856	·4476007
916	099
975	192
2803035	284
094	376
154	468
213	560
272	652
332	744
391	836
451	928
510	·4477021
570	113
629	205
689	297
748	389
807	481
867	573
926	665
986	757
2804045	849
105	941
164	·4478033
224	126
283	218
342	310
402	402
461	494
521	586
580	678
640	770
699	862
759	954
818	·4479046
877	138
937	230
996	322
2805055	414
115	506
174	598
234	690
293	782
353	874
412	966
472	·4480058
531	150
590	242
650	334
709	426
768	518
828	609
887	701
946	793
2806006	885

2806065	·4480977
125	·4481069
184	161
244	253
303	345
363	437
422	529
481	621
541	712
600	804
659	896
719	988
778	·4482080
837	172
897	264
956	356
2807015	448
075	539
134	631
193	723
253	815
312	907
371	999
431	·4483090
490	182
550	274
609	366
669	458
728	550
788	641
847	733
906	825
966	917
2808025	·4484009
084	101
144	192
203	284
262	376
322	467
381	559
440	651
500	743
559	835
618	927
678	·4485018
737	110
796	202
856	293
915	385
974	477
2809034	568
093	660
152	752
212	844
271	936
330	·4486028
390	119
449	211
508	303
568	394
627	486
686	578
746	669
805	761
864	853

2809924	·4486944
983	·4487036

Excesses

10 = 15	1 = 2
20 = 31	2 = 3
30 = 46	3 = 5
40 = 62	4 = 6
50 = 77	5 = 8
	6 = 9
	7 = 11
	8 = 12
	9 = 14

30 = 46

Anti Log Excesses

10 = 6	1 = 1
20 = 13	2 = 1
30 = 19	3 = 2
40 = 26	4 = 3
50 = 32	5 = 3
60 = 39	6 = 4
70 = 45	7 = 5
80 = 52	8 = 5
90 = 58	9 = 6

46 = 30

2810042	·4487128
102	219
161	311
220	403
280	494
339	586
398	678
457	769
516	861
575	953
635	·4488044
694	136
753	227
813	319
872	410
931	502
991	593
2811050	685
109	777
169	868
228	960
287	·4489052
347	143
406	235
465	327
525	418
584	510
643	601
703	693
762	784
821	876
880	967
939	·4490059
998	150

2812058	·4490242
117	333
176	425
236	516
295	608
354	700
414	791
473	883
532	974
591	·4491066
650	157
709	249
769	340
828	432
887	523
947	615
2813006	706
065	798
125	889
184	981
243	·4492072
302	164
361	255
420	346
480	438
539	529
598	620
658	712
717	803
776	895
836	986
895	·4493078
954	169
2814013	261
072	352
131	443
191	535
250	626
309	717
369	809
428	900
487	991
546	·4494083
605	174
664	265
724	357
783	448
842	539
902	631
961	722
2815020	813
079	905
138	996
197	·4495087
257	179
316	270
375	361
434	453
493	544
552	635
612	727
671	818
730	909
790	·4496001
849	092

2815908	·4496183
967	275
2816026	366
085	457
145	549
204	640
263	731
322	823
381	914
440	·4497005
500	096
559	187
618	278
677	370
736	461
795	552
855	644
914	735
973	826
2817032	917
091	·4498008
150	099
210	191
269	282
328	373
387	465
446	556
505	647
565	738
624	829
683	920
742	·4499012
801	103
860	194
919	285
978	376
2818037	467
097	558
156	649
215	740
274	832
333	923
392	·4500014
452	105
511	196
570	287
629	379
688	470
747	561
807	652
866	743
925	834
984	925
2819043	·4501016
102	107
161	198
220	289
279	380
339	472
398	563
457	654
516	745
575	836
634	927
693	·4502018

2819752	·4502109
811	200
871	291
930	382
989	473

Excesses

10 = 15	1 = 2
20 = 31	2 = 3
30 = 46	3 = 5
40 = 62	4 = 6
50 = 77	5 = 8
	6 = 9
	7 = 11
	8 = 12
	9 = 14

30 = 46

Anti Log Excesses

10 = 6	1 = 1
20 = 13	2 = 1
30 = 19	3 = 2
40 = 26	4 = 3
50 = 32	5 = 3
60 = 39	6 = 4
70 = 45	7 = 5
80 = 52	8 = 5
90 = 58	9 = 6

46 = 30

2820048	·4502564
107	655
166	746
225	837
284	928
343	·4503019
402	110
461	201
520	292
580	383
639	474
698	565
757	656
816	747
875	838
934	929
993	·4504020
2821052	111
111	201
170	292
229	383
289	474
348	565
407	656
466	747
525	838
584	929
643	·4505020
702	111
761	202
820	293

2821879	·4505384
938	475
997	565
2822056	656
115	747
175	838
234	929
293	·4506020
352	110
411	201
470	292
529	383
588	474
647	565
706	656
765	747
824	838
883	928
942	·4507019
2823001	110
060	201
119	292
178	383
237	473
296	564
355	655
415	745
474	836
533	927
592	·4508018
651	109
710	200
769	290
828	381
887	472
946	562
2824005	653
064	744
123	835
182	926
241	·4509017
300	107
359	198
418	289
477	379
536	470
595	561
654	651
713	742
772	833
831	923
890	·4510014
949	105
2825008	195
067	286
126	377
185	467
244	558
303	649
362	739
421	830
480	921
539	·4511011
598	102
657	193

2825716	·4511283
775	374
834	465
893	555
952	646
2826011	737
070	828
129	919
188	·4512010
247	100
306	191
365	281
423	372
482	462
541	553
600	643
659	734
718	825
777	915
836	·4513006
895	097
954	187
2827013	278
072	368
131	459
190	549
249	640
308	730
367	821
426	912
485	·4514002
544	093
603	183
661	274
720	364
779	455
838	545
897	636
956	726
2828015	817
074	907
133	998
192	·4515088
251	179
310	269
369	360
428	450
487	541
545	631
604	722
663	812
722	903
781	993
840	·4516083
899	174
958	264
2829017	355
076	445
135	536
194	626
252	717
311	807
370	897
429	988
488	·4517078

2829547	·4517168
606	259
665	349
724	440
782	530
841	621
900	711
959	802

Excesses

10 = 15	1 = 2
20 = 31	2 = 3
30 = 46	3 = 5
40 = 62	4 = 6
50 = 77	5 = 8
	6 = 9
	7 = 11
	8 = 12
	9 = 14

30 = 46

Anti Log Excesses

10 = 7	1 = 1
20 = 13	2 = 1
30 = 20	3 = 2
40 = 26	4 = 3
50 = 33	5 = 3
60 = 39	6 = 4
70 = 46	7 = 5
80 = 52	8 = 5
90 = 59	9 = 6

46 = 30

2830018	·4517892
077	982
136	·4518073
195	163
254	253
313	344
372	434
431	524
489	615
548	705
607	795
666	886
725	976
784	·4519066
842	157
901	247
960	337
2831019	428
078	518
137	608
196	699
255	789
314	879
372	970
431	·4520060
490	150
549	240
608	330

2831667	·4520420
725	511
784	601
843	691
902	782
961	872
2832020	962
079	·4521053
138	143
197	233
255	324
314	414
373	504
432	594
491	684
550	774
608	865
667	955
726	·4522045
785	135
844	225
903	315
961	406
2833020	496
079	586
138	677
197	767
256	857
314	947
373	·4523037
432	127
491	218
550	308
609	398
667	488
726	578
785	668
843	758
902	848
961	938
2834020	·4524029
079	119
138	209
196	299
255	389
314	479
373	569
432	659
491	749
549	839
608	929
667	·4525019
725	110
784	200
843	290
902	380
961	470
2835020	560
078	650
137	740
196	830
255	920
314	·4526010
373	100
431	190

2835490	·4526280
549	370
607	460
666	550
725	640
784	730
843	820
902	910
960	·4527000
2836019	090
078	180
136	270
195	360
254	450
312	540
371	630
430	720
489	810
548	900
607	990
665	·4528080
724	170
783	260
841	350
900	440
959	530
2837017	619
076	709
135	799
194	889
253	979
312	·4529069
370	159
429	249
488	339
546	429
605	519
664	609
722	698
781	788
840	878
899	968
958	·4530058
2838017	148
075	238
134	328
193	418
251	507
310	597
369	687
427	777
486	867
545	957
603	·4531046
662	136
721	226
779	316
838	406
897	496
955	585
2839014	675
073	765
132	855
191	945
250	·4532035

2839308	·4532124
367	214
426	304
484	393
543	483
602	573
660	663
719	753
778	843
836	932
895	·4533022
954	112

Excesses

10 = 15	1 = 2
20 = 31	2 = 3
30 = 46	3 = 5
40 = 61	4 = 6
50 = 77	5 = 8
	6 = 9
	7 = 11
	8 = 12
	9 = 14

30 = 46

Anti Log Excesses

10 = 7	1 = 1
20 = 13	2 = 1
30 = 20	3 = 2
40 = 26	4 = 3
50 = 33	5 = 3
60 = 39	6 = 4
70 = 46	7 = 5
80 = 52	8 = 5
90 = 59	9 = 6

46 = 30

2840012	·4533201
071	291
130	381
188	470
247	560
306	650
364	740
423	830
482	920
540	·4534009
599	099
658	189
716	278
775	368
834	458
892	547
951	637
2841010	727
068	816
127	906
186	996
244	·4535085
303	175
362	265

2841420	·4535354
479	444
538	534
596	623
655	713
714	803
772	892
831	982
890	·4536071
948	161
2842007	250
066	340
124	429
183	519
242	609
300	698
359	788
418	878
476	967
535	·4537057
593	146
652	236
710	325
769	415
827	504
886	594
945	684
2843003	773
062	863
121	952
179	·4538042
238	131
297	221
355	310
414	400
473	490
531	579
590	669
649	758
707	848
766	937
824	·4539027
883	116
941	206
2844000	295
058	385
117	474
176	564
234	653
293	743
352	832
410	922
469	·4540011
528	100
586	190
645	279
703	369
762	458
820	548
879	637
937	727
996	816
2845055	905
113	995
172	·4541084

2845231	·4541174
289	263
348	353
406	442
465	532
523	621
582	710
640	800
699	889
758	978
816	·4542068
875	157
933	246
992	336
2846050	425
109	514
167	604
226	693
285	782
343	872
402	961
460	·4543050
519	140
577	229
636	318
694	408
753	497
812	586
870	676
929	765
987	854
2847046	944
104	·4544033
163	122
221	212
280	301
339	390
397	480
456	569
514	658
573	747
631	836
690	925
748	·4545015
807	104
865	193
924	283
982	372
2848041	461
099	551
158	640
216	729
275	818
333	907
392	996
450	·4546086
509	175
567	264
626	354
684	443
743	532
801	621
860	710
918	799
977	889

2849035	·4546978
094	·4547067
152	156
211	245
269	334
328	423
386	512
445	601
503	691
562	780
620	869
679	958
737	·4548047
796	136
854	226
913	315
971	404

Excesses

10 = 15	1 = 2
20 = 31	2 = 3
30 = 46	3 = 5
40 = 61	4 = 6
50 = 76	5 = 8
	6 = 9
	7 = 11
	8 = 12
	9 = 14

29 = 45

Anti Log Excesses

10 = 7	1 = 1
20 = 13	2 = 1
30 = 20	3 = 2
40 = 26	4 = 3
50 = 33	5 = 3
60 = 39	6 = 4
70 = 46	7 = 5
80 = 52	8 = 5
	9 = 6

45 = 29

2850030	·4548493
088	582
147	671
205	760
264	849
322	938
381	·4549028
439	117
497	206
556	295
614	384
673	473
731	562
790	651
848	740
907	829
965	918
2851024	·4550007
082	096

2851141	·4550185
199	274
258	364
316	453
374	542
433	631
491	720
550	809
608	898
667	987
725	·4551076
784	165
842	254
900	343
959	432
2852017	521
076	610
134	699
193	788
251	877
310	966
368	·4552055
426	144
485	233
543	322
602	411
660	500
719	589
777	678
836	766
894	855
952	944
2853011	·4553033
069	122
127	211
186	300
244	389
303	478
361	567
420	656
478	745
537	833
595	922
653	·4554011
712	100
770	189
828	278
887	367
945	456
2854003	545
062	633
120	722
179	811
237	900
296	989
354	·4555078
413	166
471	255
529	344
588	433
646	522
704	611
763	699
821	788
879	877

2854938	·4555966
996	·4556055
2855054	144
113	232
171	321
229	410
288	498
346	587
404	676
463	765
521	854
580	943
638	·4557031
697	120
755	209
814	297
872	386
930	475
989	563
2856047	652
105	741
164	830
222	919
280	·4558008
339	096
397	185
455	274
514	362
572	451
630	540
689	628
747	717
805	806
864	894
922	983
980	·4559072
2857039	160
097	249
155	338
214	426
272	515
330	604
389	692
447	781
505	870
564	958
622	·4560047
680	136
739	224
797	313
855	402
914	490
972	579
2858030	668
089	756
147	845
205	933
264	·4561022
322	110
380	199
439	287
497	376
555	465
614	553
672	642

2858730	·4561731
788	819
846	908
904	996
963	·4562085
2859021	173
079	262
138	350
196	439
254	527
313	616
371	704
429	793
488	881
546	970
604	·4563059
663	147
721	236
779	324
838	413
896	501
954	590

Excesses

10 = 15	1 = 2
20 = 30	2 = 3
30 = 46	3 = 5
40 = 61	4 = 6
50 = 76	5 = 8
	6 = 9
	7 = 11
	8 = 12
	9 = 14

29 = 44

Anti Log Excesses

10 = 7	1 = 1
20 = 13	2 = 1
30 = 20	3 = 2
40 = 26	4 = 3
50 = 33	5 = 3
60 = 39	6 = 4
70 = 46	7 = 5
80 = 53	8 = 5
	9 = 6

44 = 29

2860012	·4563678
070	767
128	855
187	944
245	·4564032
303	120
362	209
420	297
478	386
537	474
595	563
653	651
712	740
770	828

2860828	·4564916
886	·4565005
944	093
2861002	182
061	270
119	359
177	447
236	536
294	624
352	712
411	801
469	889
527	977
585	·4566066
643	154
701	242
760	331
818	419
876	508
935	596
993	685
2862051	773
110	862
168	950
226	·4567038
284	127
342	215
400	303
459	392
517	480
575	568
634	657
692	745
750	833
808	922
866	·4568010
924	098
983	187
2863041	275
099	363
157	452
215	540
273	628
332	716
390	804
448	892
507	981
565	·4569069
623	157
681	246
739	334
797	422
856	511
914	599
972	687
2864030	775
088	863
146	951
205	·4570040
263	128
321	216
380	305
438	393
496	481
554	570

2864612	·4570658
670	746
729	834
787	922
845	·4571010
903	099
961	187
2865019	275
078	363
136	451
194	539
252	628
310	716
368	804
427	892
485	980
543	·4572068
601	157
659	245
717	333
776	421
834	509
892	597
950	686
2866008	774
066	862
125	950
183	·4573038
241	126
299	214
357	302
415	390
474	478
532	566
590	654
648	743
706	831
764	919
822	·4574007
880	095
938	183
997	271
2867055	359
113	447
171	535
229	623
287	711
346	799
404	887
462	975
520	·4575063
578	151
636	239
694	328
752	416
810	504
869	592
927	680
985	768
2868043	855
101	943
159	·4576031
217	119
275	207
333	295

2868392	·4576383
450	471
508	559
566	647
624	735
682	823
740	911
798	999
856	·4577087
914	175
972	263
2869030	351
089	439
147	527
205	615
263	703
321	791
379	879
437	966
495	·4578054
553	142
611	230
669	318
727	406
786	494
844	582
902	670
960	757

Excesses

10 = 15	1 = 2
20 = 30	2 = 3
30 = 45	3 = 5
40 = 61	4 = 6
50 = 76	5 = 8
	6 = 9
	7 = 11
	8 = 12
	9 = 14

29 = 44

Anti Log Excesses

10 = 7	1 = 1
20 = 13	2 = 1
30 = 20	3 = 2
40 = 26	4 = 3
50 = 33	5 = 3
60 = 40	6 = 4
70 = 46	7 = 5
80 = 53	8 = 5
	9 = 6

44 = 29

2870018	·4578845
076	933
134	·4579021
192	109
250	197
308	284
366	372
424	460

2870483	·4579548
541	636
599	724
657	811
715	899
773	987
831	·4580075
889	163
947	251
2871005	339
063	427
121	515
179	602
237	690
295	778
353	865
411	953
469	·4581041
527	129
585	217
643	305
702	392
760	480
818	568
876	655
934	743
992	831
2872050	918
108	·4582006
166	094
224	182
282	270
340	358
398	445
456	533
514	621
572	708
630	796
688	884
746	971
804	·4583059
862	147
920	234
978	322
2873036	410
094	497
152	585
210	673
268	760
326	848
384	936
442	·4584023
500	111
558	199
616	286
674	374
732	462
790	549
848	637
906	725
964	812
2874022	900
080	988
138	·4585075
196	163

2874254	·4585251
312	338
370	426
428	514
486	601
544	689
602	777
660	864
718	952
776	·4586039
834	127
892	214
950	302
2875008	389
066	477
124	565
182	652
240	740
298	827
356	915
414	·4587002
472	090
529	177
587	265
645	352
703	440
761	527
819	615
877	702
935	790
993	878
2876051	965
109	·4588053
167	140
225	228
283	315
341	403
399	490
457	578
515	665
572	753
630	840
688	927
746	·4589015
804	102
862	190
920	277
978	365
2877036	452
094	540
152	627
210	715
268	802
326	890
384	977
441	·4590065
499	152
557	239
615	327
673	414
731	501
789	589
847	676
905	764
963	851

2878021	·4590939
079	·4591026
136	114
194	201
252	288
310	376
368	463
426	550
484	638
542	725
600	812
657	900
715	987
773	·4592074
831	162
889	249
947	336
2879005	424
063	511
121	598
179	686
237	773
295	860
352	948
410	·4593035
468	122
526	210
584	297
642	384
699	471
757	558
815	645
873	733
931	820
989	907

Excesses

10 = 15	1 = 2
20 = 30	2 = 3
30 = 45	3 = 5
40 = 60	4 = 6
50 = 76	5 = 8
	6 = 9
	7 = 11
	8 = 12
	9 = 14

29 = 44

Anti Log Excesses

10 = 7	1 = 1
20 = 13	2 = 1
30 = 20	3 = 2
40 = 26	4 = 3
50 = 33	5 = 3
60 = 40	6 = 4
70 = 46	7 = 5
80 = 53	8 = 5
	9 = 6

44 = 29

2880047	·4593995

2880105	·4594082
163	169
220	257
278	344
336	431
394	519
452	606
510	693
567	780
625	867
683	954
741	·4595042
799	129
857	216
915	304
973	391
2881031	478
088	565
146	652
204	739
262	827
320	914
378	·4596001
435	088
493	175
551	262
609	350
667	437
725	524
782	611
840	698
898	785
956	873
2882014	960
072	·4597047
129	134
187	221
245	308
303	396
361	483
419	570
476	657
534	744
592	831
650	918
708	·4598005
766	092
823	179
881	266
939	353
996	441
2883054	528
112	615
170	702
228	789
286	876
343	963
401	·4599050
459	137
517	224
575	311
633	398
690	485
748	572
806	659

2883863	·4599747
921	834
979	921
2884037	·4600008
095	095
153	182
210	269
268	356
326	443
384	530
442	617
500	704
557	791
615	878
673	965
730	·4601052
788	139
846	226
904	312
962	399
2885020	486
077	573
135	660
193	747
250	834
308	921
366	·4602008
423	095
481	182
539	269
597	356
655	443
713	530
770	617
828	704
886	791
943	877
2886001	964
059	·4603051
116	138
174	225
232	312
290	399
348	486
406	573
463	659
521	746
579	833
636	920
694	·4604007
752	094
809	180
867	267
925	354
983	441
2887041	528
099	615
156	702
214	789
272	876
329	962
387	·4605049
445	136
502	223
560	310

2887618	·4605397
675	483
733	570
791	657
848	743
906	830
964	917
2888021	·4606004
079	091
137	178
195	264
253	351
311	438
368	524
426	611
484	698
541	785
599	872
657	959
714	·4607045
772	132
830	219
887	305
945	392
2889003	479
060	565
118	652
176	739
233	825
291	912
349	999
406	·4608085
464	172
522	259
579	345
637	432
695	519
752	606
810	693
868	780
925	866
983	953

Excesses

10 = 15	1 = 2
20 = 30	2 = 3
30 = 45	3 = 5
40 = 60	4 = 6
50 = 75	5 = 8
	6 = 9
	7 = 11
	8 = 12
	9 = 14

29 = 43

Anti Log Excesses

10 = 7	1 = 1
20 = 13	2 = 1
30 = 20	3 = 2
40 = 27	4 = 3
50 = 33	5 = 3
60 = 40	6 = 4
70 = 46	7 = 5
80 = 53	8 = 5
	9 = 6

43 = 29

2890041	·4609040
098	126
156	213
214	299
271	386
329	472
387	559
444	645
502	732
560	819
617	905
675	992
733	·4610079
790	165
848	252
906	339
963	425
2891021	512
079	599
136	685
194	772
252	859
309	945
367	·4611032
425	118
482	205
540	291
598	378
655	464
713	551
771	638
828	724
886	811
944	897
2892001	984
059	·4612070
116	157
174	243
231	330
289	416
346	503
404	589
462	676
519	762
577	849
635	935
692	·4613022
750	108
808	195
865	281
923	368

2892981	·4613454
2893038	541
096	627
154	714
211	800
269	887
326	973
384	·4614060
441	146
499	232
556	319
614	405
672	492
729	578
787	665
845	751
902	838
960	924
2894017	·4615010
075	097
132	183
190	270
247	356
305	443
363	529
420	616
478	702
536	788
593	875
651	961
708	·4616047
766	134
823	220
881	306
938	393
996	479
2895054	565
111	652
169	738
227	824
284	911
342	997
399	·4617083
457	170
514	256
572	342
629	429
687	515
745	601
802	688
860	774
917	860
975	947
2896032	·4618033
090	119
147	206
205	292
263	378
320	465
378	551
435	637
493	724
550	810
608	896
665	983

2896723	·4619069
780	155
838	241
895	327
953	413
2897010	500
068	586
126	672
183	759
241	845
298	931
356	·4620018
413	104
471	190
528	276
586	362
643	448
701	535
758	621
816	707
873	794
931	880
988	966
2898046	·4621052
103	138
161	224
218	311
276	397
333	483
391	569
448	655
506	741
563	827
621	913
678	999
736	·4622086
793	172
851	258
908	344
966	430
2899023	516
081	603
138	689
196	775
253	861
311	947
368	·4623033
426	119
483	205
541	291
598	378
656	464
713	550
771	636
828	722
885	808
943	894

Excesses

10 = 15	1 = 2
20 = 30	2 = 3
30 = 45	3 = 5
40 = 60	4 = 6
50 = 75	5 = 8
	6 = 9
	7 = 11
	8 = 12
	9 = 14

29 = 43

Anti Log Excesses

10 = 7	1 = 1
20 = 13	2 = 1
30 = 20	3 = 2
40 = 27	4 = 3
50 = 33	5 = 3
60 = 40	6 = 4
70 = 47	7 = 5
80 = 53	8 = 5
	9 = 6

43 = 29

2900000	·4623980
058	·4624066
115	152
173	238
230	324
288	410
345	496
403	582
460	668
518	754
575	840
633	926
690	·4625012
747	098
805	185
862	271
920	357
977	443
2901035	529
092	615
150	701
207	787
264	873
322	959
379	·4626045
437	131
494	216
552	302
609	388
667	474
724	560
781	646
839	732
896	818
954	904
2902011	990
069	·4627076

2902126	·4627162
184	248
241	334
298	420
356	506
413	592
471	678
528	763
586	849
643	935
701	·4628021
758	107
815	193
873	279
930	365
987	451
2903045	536
102	622
160	708
217	794
275	880
332	966
390	·4629052
447	138
504	224
562	310
619	396
676	482
734	567
791	653
848	739
906	825
963	911
2904020	997
078	·4630082
135	168
193	254
250	340
308	426
365	512
423	597
480	683
537	769
595	854
652	940
709	·4631026
767	112
824	198
881	284
939	369
996	455
2905053	541
111	626
168	712
225	798
283	884
340	970
398	·4632056
455	141
513	227
570	313
628	398
685	484
742	570
800	655

2905857	·4632741
914	827
972	912
2906029	998
086	·4633084
144	169
201	255
258	341
316	426
373	512
430	598
488	683
545	769
602	855
660	941
717	·4634027
774	113
832	198
889	284
946	370
2907004	455
061	541
118	627
176	712
233	798
290	883
348	969
405	·4635054
462	140
520	225
577	311
634	397
692	482
749	568
806	654
864	739
921	825
978	911
2908036	996
093	·4636082
150	167
208	253
265	338
322	424
380	509
437	595
494	681
551	766
608	852
665	938
723	·4637023
780	109
837	194
895	280
952	365
2909009	451
067	536
124	622
181	707
239	793
296	878
353	964
411	·4638049
468	135
525	220

2909583	·4638306
640	391
697	477
754	562
811	648
868	733
926	819
983	904

Excesses

10 = 15	1 = 2
20 = 30	2 = 3
30 = 45	3 = 5
40 = 60	4 = 6
50 = 75	5 = 8
	6 = 9
	7 = 10
	8 = 12
	9 = 13

29 = 43

Anti Log Excesses

10 = 7	1 = 1
20 = 13	2 = 1
30 = 20	3 = 2
40 = 27	4 = 3
50 = 33	5 = 3
60 = 40	6 = 4
70 = 47	7 = 5
80 = 54	8 = 5
	9 = 6

43 = 29

2910040	·4638990
098	·4639075
155	161
212	246
270	332
327	417
384	503
442	588
499	674
556	759
614	845
671	930
728	·4640015
785	101
842	186
899	271
957	357
2911014	442
071	528
129	613
186	699
243	784
301	870
358	955
415	·4641040
472	126
529	211
586	296

Number	Log
2911644	·4641382
701	467
758	552
816	638
873	723
930	808
987	894
2912044	979
101	·4642064
159	150
216	235
273	320
331	406
388	491
445	576
503	662
560	747
617	832
674	918
731	·4643003
788	088
846	174
903	259
960	344
2913017	430
074	515
131	600
189	686
246	771
303	856
361	942
418	·4644027
475	112
532	198
589	283
646	368
704	454
761	539
818	624
876	709
933	794
990	879
2914047	965
104	·4645050
161	135
219	221
276	306
333	391
390	476
447	561
504	646
562	732
619	817
676	902
733	988
790	·4646073
847	158
905	243
962	328
2915019	413
076	499
133	584
190	669
248	754
305	839

Number	Log
2915362	·4646924
419	·4647010
476	095
533	180
591	265
648	350
705	435
762	521
819	606
876	691
934	776
991	861
2916048	946
105	·4648031
162	116
219	201
277	287
334	372
391	457
448	542
505	627
562	712
620	797
677	882
734	967
791	·4649053
848	138
905	223
963	308
2917020	393
077	478
134	563
191	648
248	733
305	818
362	903
419	988
477	·4650073
534	158
591	243
648	328
705	413
762	498
819	583
876	668
933	753
991	838
2918048	923
105	·4651008
162	093
219	178
276	263
333	348
390	433
447	518
505	603
562	688
619	773
676	858
733	943
790	·4652028
847	113
904	198
961	283
2919019	368

Number	Log
2919076	·4652453
133	538
190	623
247	708
304	793
361	877
418	962
475	·4653047
533	132
590	217
647	302
704	387
761	472
818	557
875	642
932	727
989	812

Excesses

10 = 15	1 = 2
20 = 30	2 = 3
30 = 45	3 = 5
40 = 60	4 = 6
50 = 74	5 = 7
	6 = 9
	7 = 10
	8 = 12
	9 = 13

29 = 43

Anti Log Excesses

10 = 7	1 = 1
20 = 13	2 = 1
30 = 20	3 = 2
40 = 27	4 = 3
50 = 34	5 = 3
60 = 40	6 = 4
70 = 47	7 = 5
80 = 54	8 = 5
	9 = 6

43 = 29

Number	Log
2920046	·4653896
103	981
160	·4654066
217	151
274	236
331	321
389	406
446	491
503	576
560	660
617	745
674	830
731	915
788	·4655000
845	085
902	169
959	254
2921016	339
073	424

Number	Log
2921130	·4655509
187	594
245	678
302	763
359	848
416	933
473	·4656018
530	103
587	187
644	272
701	357
758	441
815	526
872	611
929	696
986	781
2922043	866
100	950
157	·4657035
214	120
271	204
328	289
385	374
442	459
499	544
556	629
614	713
671	798
728	883
785	967
842	·4658052
899	137
956	221
2923013	306
070	391
127	475
184	560
241	645
298	730
355	815
412	900
469	984
526	·4659069
583	154
640	238
697	323
754	408
811	492
868	577
925	662
982	746
2924039	831
096	915
153	·4660000
210	084
267	169
324	253
381	338
438	423
495	507
552	592
609	677
666	761
723	846
780	931

Number	Log
2924837	·4661015
894	100
951	185
2925008	269
065	354
122	439
179	523
236	608
293	693
349	777
406	862
463	946
520	·4662031
577	115
634	200
691	284
748	369
805	454
862	538
919	623
976	707
2926033	792
090	876
147	961
204	·4663045
261	130
318	214
375	299
432	383
489	468
546	552
603	637
660	721
716	806
773	890
830	975
887	·4664059
944	144
2927001	228
058	313
115	397
172	482
229	566
286	651
343	735
400	820
457	904
514	988
571	·4665073
628	157
685	242
741	326
798	411
855	495
912	580
969	664
2928026	748
083	833
140	917
197	·4666002
254	086
311	171
368	255
424	340
481	424

2928538	·4666508
595	593
652	677
709	761
766	846
823	930
880	·4667015
937	099
994	184
2929051	268
107	353
164	437
221	521
278	606
335	690
392	774
449	859
506	943
563	·4668027
619	112
676	196
733	280
790	365
847	449
904	533
961	618

Excesses

10 = 15	1 = 2
20 = 30	2 = 3
30 = 45	3 = 5
40 = 59	4 = 6
50 = 74	5 = 7
	6 = 9
	7 = 10
	8 = 12
	9 = 13

29 = 43

Anti Log Excesses

10 = 7	1 = 1
20 = 13	2 = 1
30 = 20	3 = 2
40 = 27	4 = 3
50 = 34	5 = 3
60 = 40	6 = 4
70 = 47	7 = 5
80 = 54	8 = 5
	9 = 6

43 = 29

2930018	·4668702
075	786
131	871
188	955
245	·4669039
302	124
359	208
416	292
472	377
529	461

2930586	·4669545
643	629
700	713
757	797
814	882
871	966
928	·4670050
984	135
2931041	219
098	303
155	388
212	472
269	556
325	640
382	724
439	808
496	893
553	977
610	·4671061
667	146
724	230
781	314
837	398
894	482
951	566
2932008	651
065	735
122	819
178	903
235	987
292	·4672071
349	156
406	240
463	324
519	409
576	493
633	577
690	661
747	745
804	829
860	913
917	997
974	·4673081
2933031	166
088	250
145	334
201	418
258	502
315	586
372	670
429	754
486	838
542	923
599	·4674007
656	091
712	175
769	259
826	343
883	427
940	511
997	595
2934053	680
110	764
167	848
224	932

2934281	·4675016
338	100
394	184
451	268
508	352
564	436
621	520
678	604
735	688
792	772
849	856
905	940
962	·4676024
2935019	108
076	192
133	276
190	360
246	444
303	528
360	612
416	696
473	780
530	864
587	948
644	·4677032
701	116
757	200
814	284
871	368
927	452
984	536
2936041	620
097	704
154	788
211	872
268	956
325	·4678040
382	124
438	208
495	292
552	376
608	460
665	544
722	628
779	712
836	796
893	880
949	963
2937006	·4679047
063	131
119	215
176	299
233	383
289	467
346	551
403	635
459	719
516	803
573	887
630	970
687	·4680054
744	138
800	222
857	306
914	390

2937970	·4680473
2938027	557
084	641
140	725
197	809
254	893
310	976
367	·4681060
424	144
481	228
538	312
595	396
651	479
708	563
765	647
821	731
878	815
935	899
991	982
2939048	·4682066
105	150
161	233
218	317
275	401
331	485
388	569
445	653
501	736
558	820
615	904
671	987
728	·4683071
785	155
841	239
898	323
955	407

Excesses

10 = 15	1 = 2
20 = 30	2 = 3
30 = 44	3 = 4
40 = 59	4 = 6
50 = 74	5 = 7
	6 = 9
	7 = 10
	8 = 12
	9 = 13

28 = 42

Anti Log Excesses

10 = 7	1 = 1
20 = 14	2 = 1
30 = 20	3 = 2
40 = 27	4 = 3
50 = 34	5 = 3
60 = 41	6 = 4
70 = 47	7 = 5
80 = 54	8 = 5
	9 = 6

42 = 28

2940012	·4683490
069	574
126	658
182	741
239	825
296	909
352	992
409	·4684076
466	160
522	244
579	328
636	412
692	495
749	579
806	663
862	746
919	830
976	914
2941032	997
089	·4685081
146	165
202	248
259	332
316	416
372	499
429	583
486	667
542	750
599	834
656	918
712	·4686001
769	085
826	168
882	252
939	335
996	419
2942052	502
109	586
166	670
222	753
279	837
335	921
392	·4687004
448	088
505	172
561	255
618	339
675	423
731	506
788	590
845	673
901	757
958	840
2943015	924
071	·4688007
128	091
185	175
241	258
298	342
355	425
411	509
468	592
525	676
581	759
638	843

2943695	·4688926
751	·4689010
808	093
864	177
921	260
977	344
2944034	427
090	511
147	594
204	678
260	761
317	845
374	928
430	·4690012
487	095
544	179
600	262
657	346
713	429
770	513
826	596
883	680
939	763
996	847
2945053	930
109	·4691014
166	097
223	180
279	264
336	347
393	431
449	514
506	598
562	681
619	765
675	848
732	931
788	·4692015
845	098
902	181
958	265
2946015	348
071	432
128	515
184	599
241	682
297	766
354	849
411	932
467	·4693016
524	099
581	182
637	266
694	349
750	432
807	516
863	599
920	682
976	766
2947033	849
090	932
146	·4694016
203	099
259	182
316	266

2947372	·4694349
429	432
485	516
542	599
598	682
655	766
711	849
768	932
824	·4695016
881	099
938	182
994	266
2948051	349
107	432
164	516
220	599
277	682
333	765
390	848
446	931
503	·4696015
559	098
616	181
672	265
729	348
786	431
842	515
899	598
955	681
2949012	764
068	847
125	930
181	·4697014
238	097
294	180
351	264
407	347
464	430
520	513
577	596
633	679
690	763
746	846
803	929
859	·4698012
916	095
972	178

Excesses

10 = 15	1 = 2
20 = 29	2 = 3
30 = 44	3 = 4
40 = 59	4 = 6
50 = 74	5 = 7
	6 = 9
	7 = 10
	8 = 12
	9 = 13

28 = 42

Anti Log Excesses

10 = 7	1 = 1
20 = 14	2 = 1
30 = 20	3 = 2
40 = 27	4 = 3
50 = 34	5 = 3
60 = 41	6 = 4
70 = 47	7 = 5
80 = 54	8 = 5
	9 = 6

42 = 28

2950029	·4698262
085	345
142	428
198	511
255	594
311	677
368	761
424	844
481	927
537	·4699010
594	093
650	176
707	260
763	343
820	426
876	509
933	592
989	675
2951046	758
102	841
158	924
215	·4700008
271	091
328	174
384	257
441	340
497	423
554	506
610	589
667	672
723	755
780	838
836	921
893	·4701005
949	088
2952005	171
062	254
118	337
175	420
231	503
288	586
344	669
401	752
457	835
513	918
570	·4702001
626	084
683	167
739	250
796	333
852	416

2952909	·4702499
965	582
2953021	665
078	748
134	831
191	914
247	997
304	·4703080
360	163
417	246
473	329
529	412
586	495
642	578
699	661
755	744
812	827
868	910
925	993
981	·4704076
2954037	159
094	242
150	325
206	408
263	491
319	574
376	657
432	739
489	822
545	905
602	988
658	·4705071
714	154
771	237
827	320
883	403
940	485
996	568
2955052	651
109	734
165	817
222	900
278	983
335	·4706066
391	149
448	231
504	314
560	397
617	480
673	563
729	646
786	728
842	811
898	894
955	977
2956011	·4707060
067	143
124	226
180	309
237	392
293	474
350	557
406	640
463	723
519	806

2956575	·4707889
632	971
688	·4708054
744	137
801	219
857	302
913	385
970	468
2957026	551
082	634
139	716
195	799
251	882
308	964
364	·4709047
420	130
477	213
533	296
589	379
646	461
702	544
758	627
815	709
871	792
927	875
984	957
2958040	·4710040
096	123
153	206
209	289
265	372
322	454
378	537
434	620
491	702
547	785
603	868
660	950
716	·4711033
772	116
829	198
885	281
941	364
998	446
2959054	529
110	612
167	694
223	777
279	860
336	942
392	·4712025
448	108
505	190
561	273
617	355
674	438
730	520
786	603
843	685
899	768
955	851

Excesses

10 = 15	1 = 2
20 = 29	2 = 3
30 = 44	3 = 4
40 = 59	4 = 6
50 = 73	5 = 7
	6 = 9
	7 = 10
	8 = 12
	9 = 13

28 = 41

Anti Log Excesses

10 = 7	1 = 1
20 = 14	2 = 1
30 = 20	3 = 2
40 = 27	4 = 3
50 = 34	5 = 3
60 = 41	6 = 4
70 = 48	7 = 5
80 = 54	8 = 5
	9 = 6

41 = 28

2960012	·4712933
068	·4713016
124	099
181	181
237	264
293	347
350	429
406	512
462	594
519	677
575	759
631	842
688	924
744	·4714007
800	090
857	172
913	255
969	338
2961025	420
081	503
137	585
194	668
250	750
306	833
363	915
419	998
475	·4715081
532	163
588	246
644	328
701	411
757	493
813	576
870	658
926	741
982	823
2962038	906

2962094	·4715988
150	·4716071
207	153
263	236
319	318
376	401
432	483
488	566
545	648
601	731
657	813
713	896
769	978
825	·4717060
882	143
938	225
994	308
2963051	390
107	473
163	555
220	638
276	720
332	802
388	885
444	967
500	·4718049
557	132
613	214
669	297
726	379
782	462
838	544
895	627
951	709
2964007	791
063	874
119	956
175	·4719038
232	121
288	203
344	285
401	368
457	450
513	532
569	615
625	697
681	779
738	862
794	944
850	·4720026
907	109
963	191
2965019	273
075	356
131	438
187	520
244	603
300	685
356	767
412	850
468	932
524	·4721014
581	097
637	179
693	261

2965750	·4721344
806	426
862	508
918	591
974	673
2966030	755
087	838
143	920
199	·4722002
255	085
311	167
367	249
424	331
480	413
536	495
592	578
648	660
704	742
761	825
817	907
873	989
930	·4723071
986	153
2967042	235
098	318
154	400
210	482
267	565
323	647
379	729
435	811
491	893
547	975
604	·4724058
660	140
716	222
772	304
828	386
884	468
940	551
996	633
2968052	715
109	798
165	880
221	962
277	·4725044
333	126
389	208
446	290
502	372
558	454
614	537
670	619
726	701
783	783
839	865
895	947
951	·4726030
2969007	112
063	194
119	276
175	358
231	440
288	522
344	604

2969400	·4726686
456	769
512	851
568	933
625	·4727015
681	097
737	179
793	261
849	343
905	425
961	507

Excesses

10 = 15	1 = 2
20 = 29	2 = 3
30 = 44	3 = 4
40 = 59	4 = 6
50 = 73	5 = 7
	6 = 9
	7 = 10
	8 = 12
	9 = 13

28 = 41

Anti Log Excesses

10 = 7	1 = 1
20 = 14	2 = 1
30 = 20	3 = 2
40 = 27	4 = 3
50 = 34	5 = 3
60 = 41	6 = 4
70 = 48	7 = 5
80 = 55	8 = 6
	9 = 6

41 = 28

2970017	·4727589
073	671
130	753
186	835
242	917
298	999
354	·4728081
410	163
466	246
522	328
578	410
635	492
691	574
747	656
803	738
859	820
915	902
971	984
2971027	·4729066
083	148
140	230
196	312
252	394
308	476
364	558

2971420	·4729640
476	721
532	803
588	885
644	967
700	·4730049
756	131
813	213
869	295
925	377
981	459
2972037	541
093	623
149	705
205	787
261	869
317	951
373	·4731033
429	115
486	196
542	278
598	360
654	442
710	524
766	606
822	688
878	770
934	852
990	934
2973046	·4732016
102	098
158	179
214	261
270	343
326	425
382	507
438	589
495	671
551	753
607	835
663	916
719	998
775	·4733080
831	162
887	244
943	326
999	407
2974055	489
111	571
167	653
223	735
279	817
335	898
391	980
447	·4734062
503	143
559	225
615	307
671	389
727	471
783	553
839	634
895	716
951	798
2975008	880

2975064	·4734962
120	·4735044
176	125
232	207
288	289
344	370
400	452
456	534
512	615
568	697
624	779
680	861
736	943
792	·4736025
848	106
904	188
960	270
2976016	351
072	433
128	515
184	596
240	678
296	760
352	841
408	923
464	·4737005
520	087
576	169
632	251
688	332
744	414
800	496
856	577
912	659
968	741
2977024	822
080	904
136	986
192	·4738067
248	149
304	231
359	312
415	394
471	476
527	557
583	639
639	720
695	802
751	883
807	965
863	·4739046
919	128
975	210
2978031	291
087	373
143	455
199	536
255	618
311	700
367	781
423	863
479	945
535	·4740026
591	108
647	189

2978703	·4740271
759	352
815	434
870	515
926	597
982	679
2979038	760
094	842
150	923
206	·4741005
262	086
318	168
374	249
430	331
486	413
542	494
598	576
654	657
710	739
766	820
822	902
877	983
933	·4742065
989	146

Excesses

- 10 = 15, 1 = 2
- 20 = 29, 2 = 3
- 30 = 44, 3 = 4
- 40 = 58, 4 = 6
- 50 = 73, 5 = 7
- 6 = 9
- 7 = 10
- 8 = 12
- 9 = 13

28 = 41

Anti Log Excesses

- 10 = 7, 1 = 1
- 20 = 14, 2 = 1
- 30 = 21, 3 = 2
- 40 = 27, 4 = 3
- 50 = 34, 5 = 3
- 60 = 41, 6 = 4
- 70 = 48, 7 = 5
- 80 = 55, 8 = 6
- 9 = 6

41 = 28

2980045	·4742228
101	309
157	391
213	472
269	554
325	635
381	717
437	798
493	880
548	961
604	·4743043
660	124

2980716	·4743206
772	287
828	368
884	450
940	531
996	613
2981052	694
108	776
164	857
219	939
275	·4744020
331	101
387	183
443	264
499	346
555	427
611	509
667	590
722	672
778	753
834	834
890	916
946	997
2982002	·4745079
058	160
114	242
170	323
226	405
282	486
338	567
393	649
449	730
505	811
561	893
617	974
673	·4746055
728	137
784	218
840	299
896	381
952	462
2983008	543
064	625
120	706
176	787
231	869
287	950
343	·4747031
399	113
455	194
511	275
566	357
622	438
678	519
734	601
790	682
846	763
902	845
958	926
2984014	·4748007
069	089
125	170
181	251
237	333
293	414

2984349	·4748495
404	576
460	657
516	738
572	820
628	901
684	982
739	·4749064
795	145
851	226
907	308
963	389
2985019	470
074	551
130	632
186	713
242	795
298	876
354	957
409	·4750039
465	120
521	201
577	282
633	363
689	444
744	526
800	607
856	688
912	769
968	850
2986024	931
079	·4751013
135	094
191	175
247	256
303	337
359	418
414	500
470	581
526	662
581	743
637	824
693	905
749	987
805	·4752068
861	149
916	230
972	311
2987028	392
084	474
140	555
196	636
251	717
307	798
363	879
418	960
474	·4753041
530	122
586	203
642	284
698	365
753	447
809	528
865	609
920	690

2987976	·4753771
2988032	852
088	933
144	·4754014
200	095
255	176
311	257
367	338
422	419
478	500
534	581
590	662
646	743
702	824
757	906
813	987
869	·4755068
924	149
980	230
2989036	311
092	392
148	473
204	554
259	635
315	716
371	797
426	878
482	959
538	·4756040
593	121
649	202
705	283
761	364
817	445
873	526
928	607
984	688

Excesses

- 10 = 15, 1 = 2
- 20 = 29, 2 = 3
- 30 = 44, 3 = 4
- 40 = 58, 4 = 6
- 50 = 73, 5 = 7
- 6 = 9
- 7 = 10
- 8 = 12
- 9 = 13

28 = 41

Anti Log Excesses

- 10 = 7, 1 = 1
- 20 = 14, 2 = 1
- 30 = 21, 3 = 2
- 40 = 27, 4 = 3
- 50 = 34, 5 = 3
- 60 = 41, 6 = 4
- 70 = 48, 7 = 5
- 80 = 55, 8 = 6
- 9 = 6

41 = 28

Number	Log
2990040	·4756769
095	850
151	931
207	·4757012
262	092
318	173
374	254
429	335
485	416
541	497
597	578
653	659
709	740
764	821
820	902
876	983
931	·4758063
987	144
2991043	225
098	306
154	387
210	468
265	549
321	630
377	711
432	792
488	873
544	954
600	·4759034
656	115
712	196
767	277
823	358
879	439
934	519
990	600
2992046	681
101	762
157	843
213	924
268	·4760004
324	085
380	166
435	247
491	328
547	409
602	489
658	570
714	651
769	732
825	813
881	894
936	974
992	·4761055
2993048	136
103	217
159	298
215	379
270	459
326	540
382	621
437	701
493	782
549	863
604	944

Number	Log
2993660	·4762025
716	106
771	186
827	267
883	348
938	428
994	509
2994050	590
105	670
161	751
217	832
272	913
328	994
384	·4763075
439	155
495	236
551	317
606	397
662	478
718	559
773	639
829	720
885	801
940	881
996	962
2995052	·4764043
107	123
163	204
219	285
274	365
330	446
386	527
441	607
497	688
553	769
608	849
664	930
720	·4765011
775	091
831	172
887	253
942	333
998	414
2996054	495
109	575
165	656
221	737
276	817
332	898
387	979
443	·4766059
498	140
554	221
609	301
665	382
721	463
776	543
832	624
888	704
943	785
999	865
2997055	946
110	·4767026
166	107
222	188

Number	Log
2997277	·4767268
333	349
388	429
444	510
499	590
555	671
610	751
666	832
722	913
777	993
833	·4768074
889	154
944	235
2998000	315
056	396
111	476
167	557
222	638
278	718
333	799
389	879
444	960
500	·4769040
556	121
611	201
667	282
723	362
778	443
834	523
889	604
945	684
2999000	765
056	845
111	926
167	·4770006
223	087
278	167
334	248
389	328
445	409
500	489
556	569
611	650
667	730
723	810
778	891
834	971
889	·4771051
945	132

Excesses

10 = 15	1 = 2
20 = 29	2 = 3
30 = 44	3 = 4
40 = 58	4 = 6
50 = 73	5 = 7
	6 = 9
	7 = 10
	8 = 12
	9 = 13

28 = 40

Anti Log Excesses

10 = 7	1 = 1
20 = 14	2 = 1
30 = 21	3 = 2
40 = 28	4 = 3
50 = 34	5 = 3
60 = 41	6 = 4
70 = 48	7 = 5
80 = 55	8 = 6
	9 = 6

40 = 28

Number	Log
3000000	·4771212
056	293
111	373
167	454
223	534
278	615
334	695
389	775
445	856
500	936
556	·4772016
611	097
667	177
723	257
778	338
834	418
889	499
945	579
3001000	660
056	740
111	821
167	901
223	981
278	·4773062
334	142
389	222
445	303
500	383
556	463
611	544
667	624
722	704
778	785
833	865
889	945
944	·4774026
3002000	106
055	186
111	267
166	347
222	427
277	508
333	588
388	668
444	749
499	829
555	909
610	989
666	·4775069
721	149
777	230

Number	Log
3002832	·4775310
888	390
943	471
999	551
3003054	631
110	712
165	792
221	872
276	953
332	·4776033
387	113
443	193
498	273
554	353
609	434
665	514
720	594
776	675
831	755
887	835
942	915
998	995
3004053	·4777075
109	156
164	236
220	316
275	396
331	476
386	556
442	637
497	717
553	797
608	878
664	958
719	·4778038
775	118
830	198
885	278
941	359
996	439
3005052	519
107	599
163	679
218	759
274	840
329	920
384	·4779000
440	080
495	160
551	240
606	320
662	400
717	480
773	561
828	641
884	721
939	801
995	881
3006050	961
106	·4780041
161	121
216	201
272	282
327	362
382	442

Number	Log
3006438	·4780522
493	602
549	682
604	762
660	842
715	922
771	·4781002
826	082
881	162
937	242
992	322
3007048	402
103	482
159	562
214	642
270	722
325	802
380	882
436	963
491	·4782043
546	123
602	203
657	283
713	363
768	443
824	523
879	603
935	683
990	763
3008045	843
101	923
156	·4783003
211	083
267	163
322	243
377	323
433	402
488	482
544	562
599	642
655	722
710	802
766	882
821	962
876	·4784042
932	122
987	202
3009042	282
098	362
153	442
208	522
264	602
319	682
375	762
430	842
486	922
541	·4785002
597	082
652	162
707	242
763	321
818	401
873	481
929	561
984	641

Excesses

10 = 14	1 = 1
20 = 29	2 = 3
30 = 43	3 = 4
40 = 58	4 = 6
50 = 72	5 = 7
	6 = 9
	7 = 10
	8 = 12
	9 = 13

28 = 40

Anti Log Excesses

10 = 7	1 = 1
20 = 14	2 = 1
30 = 21	3 = 2
40 = 28	4 = 3
50 = 35	5 = 4
60 = 42	6 = 4
70 = 48	7 = 5
80 = 55	8 = 6
	9 = 6

40 = 28

Number	Log
3010039	·4785721
095	801
150	881
205	961
261	·4786041
316	121
371	201
427	280
482	360
537	440
593	520
648	600
703	680
759	759
814	839
869	919
925	999
980	·4787079
3011035	159
091	238
146	318
202	398
257	478
313	558
368	638
424	717
479	797
534	877
590	957
645	·4788037
700	117
756	196
811	276
866	356
922	435
977	515
3012032	595

Number	Log
3012088	·4788675
143	755
198	835
254	914
309	994
364	·4789074
420	153
475	233
530	313
586	392
641	472
696	552
752	632
807	712
862	792
917	871
972	951
3013027	·4790031
083	110
138	190
193	270
249	349
304	429
359	509
415	588
470	668
525	748
581	828
636	908
691	988
747	·4791067
802	147
857	227
913	306
968	386
3014023	466
079	545
134	625
189	705
245	784
300	864
355	944
411	·4792023
466	103
521	183
576	262
631	342
686	422
742	501
797	581
852	660
908	740
963	819
3015018	899
074	978
129	·4793058
184	138
240	217
295	297
350	377
406	456
461	536
516	616
571	695
626	775

Number	Log
3015681	·4793855
737	934
792	·4794014
847	093
903	173
958	252
3016013	332
069	411
124	491
179	571
235	650
290	730
345	809
400	889
455	968
510	·4795048
566	127
621	207
676	287
732	366
787	446
842	525
898	605
953	684
3017008	764
063	843
118	923
173	·4796002
229	082
284	161
339	241
395	320
450	400
505	479
560	559
615	638
670	718
726	797
781	877
836	956
892	·4797036
947	115
3018002	194
057	274
112	353
167	433
223	512
278	592
333	671
389	751
444	830
499	910
554	989
609	·4798069
664	148
720	228
775	307
830	386
886	466
941	545
996	624
3019051	704
106	783
161	863
217	942

Number	Log
3019272	·4799022
327	101
383	181
438	260
493	339
548	419
603	498
658	577
714	657
769	736
824	815
879	895
934	974
989	·4800053

Excesses

10 = 14	1 = 1
20 = 29	2 = 3
30 = 43	3 = 4
40 = 58	4 = 6
50 = 72	5 = 7
	6 = 9
	7 = 10
	8 = 12
	9 = 13

28 = 40

Anti Log Excesses

10 = 7	1 = 1
20 = 14	2 = 1
30 = 21	3 = 2
40 = 28	4 = 3
50 = 35	5 = 4
60 = 42	6 = 4
70 = 49	7 = 5
	8 = 6
	9 = 6

40 = 28

Number	Log
3020045	·4800133
100	212
155	291
210	371
265	450
320	529
376	609
431	688
486	767
541	847
596	926
651	·4801005
707	085
762	164
817	243
873	323
928	402
983	481
3021038	561
093	640
148	719
204	799

3021259	·4801878
314	957
369	·4802037
424	116
479	195
534	275
589	354
644	433
700	513
755	592
810	671
865	751
920	830
975	909
3022031	988
086	·4803067
141	146
196	226
251	305
306	384
362	464
417	543
472	622
527	702
582	781
637	860
693	939
748	·4804018
803	097
858	177
913	256
968	335
3023023	415
078	494
133	573
189	652
244	731
299	810
354	890
409	969
464	·4805048
520	127
575	206
630	285
685	365
740	444
795	523
850	602
905	681
960	760
3024016	840
071	919
126	998
181	·4806077
236	156
291	235
346	315
401	394
456	473
512	552
567	631
622	710
677	789
732	868
787	947

3024842	·4807027
897	106
952	185
3025007	264
062	343
117	422
173	501
228	580
283	659
338	738
393	817
448	896
503	976
558	·4808055
613	134
669	213
724	292
779	371
834	450
889	529
944	608
999	687
3026054	766
109	845
164	924
219	·4809003
274	082
329	161
384	240
439	319
495	398
550	477
605	556
660	635
715	714
770	793
825	872
880	951
935	·4810030
990	109
3027045	188
100	267
155	346
210	425
265	504
321	583
376	662
431	741
486	820
541	899
596	978
651	·4811057
706	136
761	215
816	294
871	373
926	452
981	531
3028036	610
091	689
146	768
201	847
256	926
311	·4812004
366	083

3028421	·4812162
476	241
531	320
586	399
642	478
697	557
752	636
807	715
862	794
917	873
972	951
3029027	·4813030
082	109
137	188
192	267
247	346
302	425
357	504
412	583
467	661
522	740
577	819
632	898
687	977
742	·4814056
797	134
852	213
907	292
962	371

Excesses

10 = 14	1 = 1
20 = 29	2 = 3
30 = 43	3 = 4
40 = 57	4 = 6
50 = 72	5 = 7
	6 = 9
	7 = 10
	8 = 11
	9 = 13

28 = 40

Anti Log Excesses

10 = 7	1 = 1
20 = 14	2 = 1
30 = 21	3 = 2
40 = 28	4 = 3
50 = 35	5 = 4
60 = 42	6 = 4
70 = 49	7 = 5
	8 = 6
	9 = 6

40 = 28

3030017	·4814450
072	529
127	608
182	687
237	766
292	844
347	923

3030402	·4815002
457	080
512	159
567	238
622	317
677	396
732	475
787	553
842	632
897	711
952	790
3031007	869
062	948
117	·4816026
172	105
227	184
282	262
337	341
392	420
447	499
502	578
557	657
612	735
667	814
722	893
777	971
832	·4817050
887	129
942	207
997	286
3032052	365
106	443
161	522
216	601
271	680
326	759
381	838
436	916
491	995
546	·4818074
601	152
656	231
711	310
766	388
821	467
876	546
931	624
986	703
3033041	782
096	860
151	939
206	·4819018
261	096
316	175
371	254
425	332
480	411
535	490
590	568
645	647
700	726
755	804
810	883
865	962
920	·4820040

3033975	·4820119
3034030	198
085	276
140	355
195	434
249	512
304	591
359	669
414	748
469	826
524	905
579	983
634	·4821062
689	141
744	219
799	298
854	377
908	455
963	534
3035018	613
073	691
128	770
183	848
238	927
293	·4822005
348	084
403	162
458	241
513	320
567	398
622	477
677	555
732	634
787	712
842	791
897	869
952	948
3036007	·4823027
061	105
116	184
171	262
226	341
281	419
336	498
391	576
446	655
501	733
555	812
610	890
665	969
720	·4824047
775	126
830	204
885	283
940	361
995	440
3037049	518
104	597
159	675
214	754
269	832
324	910
379	989
434	·4825067
489	146

Number	Log
3037543	·4825224
598	303
653	381
708	460
763	538
818	616
872	695
927	773
982	852
3038037	930
092	·4826009
147	087
202	166
257	244
312	322
366	401
421	479
476	557
531	636
586	714
641	793
695	871
750	950
805	·4827028
860	107
915	185
970	263
3039024	342
079	420
134	498
189	577
244	655
299	733
353	812
408	890
463	968
518	·4828047
573	125
628	203
682	282
737	360
792	438
847	517
902	595
957	673

Excesses

10 = 14	1 = 1
20 = 29	2 = 3
30 = 43	3 = 4
40 = 57	4 = 6
50 = 72	5 = 7
	6 = 9
	7 = 10
	8 = 11
	9 = 13

28 = 40

Anti Log Excesses

10 = 7	1 = 1
20 = 14	2 = 1
30 = 21	3 = 2
40 = 28	4 = 3
50 = 35	5 = 4
60 = 42	6 = 4
70 = 49	7 = 5
	8 = 6
	9 = 6

40 = 28

Number	Log
3040011	·4828752
066	830
121	908
176	987
231	·4829065
286	143
340	222
395	300
450	378
505	457
560	535
615	613
669	692
724	770
779	848
834	926
889	·4830004
944	082
998	161
3041053	239
108	317
162	396
217	474
272	552
327	631
382	709
437	787
491	866
546	944
601	·4831022
656	100
711	178
766	256
820	335
875	413
930	491
984	570
3042039	648
094	726
149	804
204	882
259	960
313	·4832039
368	117
423	195
477	273
532	351
587	429
642	508
697	586
752	664

Number	Log
3042806	·4832743
861	821
916	899
970	977
3043025	·4833055
080	133
135	212
190	290
245	368
299	446
354	524
409	602
463	681
518	759
573	837
627	915
682	993
737	·4834071
792	149
847	227
902	305
956	384
3044011	462
066	540
120	618
175	696
230	774
285	852
340	930
395	·4835008
449	086
504	164
559	242
613	321
668	399
723	477
777	555
832	633
887	711
941	789
996	867
3045051	945
106	·4836023
161	101
216	179
270	257
325	335
380	413
434	492
489	570
544	648
598	726
653	804
708	882
762	960
817	·4837038
872	116
926	194
981	272
3046036	350
091	428
146	506
201	584
255	662
310	740

Number	Log
3046365	·4837818
419	896
474	974
529	·4838052
583	130
638	208
693	286
747	364
802	442
857	520
911	598
966	676
3047021	754
075	832
130	910
185	988
239	·4839065
294	143
349	221
403	299
458	377
513	455
568	533
623	611
678	689
732	767
787	845
842	923
896	·4840001
951	079
3048006	157
060	234
115	312
170	390
224	468
279	546
334	624
388	702
443	780
498	858
552	935
607	·4841013
662	091
716	169
771	247
826	325
880	403
935	481
990	559
3049044	636
099	714
154	792
208	870
263	948
318	·4842026
372	103
427	181
482	259
536	337
591	415
646	493
700	570
755	648
810	726
864	804

Number	Log
3049919	·4842882
973	960

Excesses

10 = 14	1 = 1
20 = 29	2 = 3
30 = 43	3 = 4
40 = 57	4 = 6
50 = 71	5 = 7
	6 = 9
	7 = 10
	8 = 11
	9 = 13

27 = 39

Anti Log Excesses

10 = 7	1 = 1
20 = 14	2 = 1
30 = 21	3 = 2
40 = 28	4 = 3
50 = 35	5 = 4
60 = 42	6 = 4
70 = 49	7 = 5
	8 = 6
	9 = 6

39 = 27

Number	Log
3050028	·4843037
082	115
137	193
191	271
246	349
301	427
355	504
410	582
465	660
519	737
574	815
629	893
683	971
738	·4844049
793	127
847	204
902	282
957	360
3051011	437
066	515
121	593
175	670
230	748
285	826
339	904
394	982
448	·4845060
503	137
557	215
612	293
666	370
721	448
776	526
830	603

3051885	·4845681
940	759
994	836
3052049	914
104	992
158	·4846069
213	147
268	225
322	303
377	381
431	459
486	536
540	614
595	692
649	769
704	847
759	925
813	·4847002
868	080
923	158
977	235
3053032	313
087	391
141	468
196	546
250	623
305	701
359	778
414	856
468	933
523	·4848011
578	089
632	166
687	244
742	322
796	399
851	477
905	555
960	632
3054014	710
069	788
123	865
178	943
233	·4849020
287	098
342	175
396	253
451	330
505	408
560	486
614	563
669	641
724	719
778	796
833	874
887	951
942	·4850029
996	106
3055051	184
105	261
160	339
215	417
269	494
324	572
378	649

3055433	·4850727
487	804
542	882
596	959
651	·4851037
706	114
760	192
815	269
869	347
924	424
978	502
3056033	579
087	657
142	734
196	812
251	889
305	967
360	·4852044
414	122
469	199
524	277
578	354
633	432
687	509
742	587
796	664
851	742
905	819
960	897
3057014	974
069	·4853052
123	129
178	206
232	284
287	361
341	439
396	516
450	594
505	671
559	749
614	826
668	903
723	981
777	·4854058
832	135
886	213
941	290
995	367
3058050	445
104	522
159	600
213	677
268	755
322	832
377	910
431	987
486	·4855064
540	142
595	219
649	296
704	374
758	451
813	528
867	606
922	683

3058976	·4855760
3059031	838
085	915
140	992
194	·4856070
249	147
303	224
358	302
412	379
467	456
521	534
576	611
630	688
685	766
739	843
793	920
848	998
902	·4857075
957	152

Excesses

10 = 14	1 = 1
20 = 28	2 = 3
30 = 43	3 = 4
40 = 57	4 = 6
50 = 71	5 = 7
	6 = 9
	7 = 10
	8 = 11
	9 = 13

27 = 39

Anti Log Excesses

10 = 7	1 = 1
20 = 14	2 = 1
30 = 21	3 = 2
40 = 28	4 = 3
50 = 35	5 = 4
60 = 42	6 = 4
70 = 49	7 = 5
	8 = 6
	9 = 6

39 = 27

3060011	·4857230
066	307
120	384
175	462
229	539
284	616
338	694
393	771
447	848
502	926
556	·4858003
610	080
665	157
719	234
774	311
828	389
883	466

3060937	·4858543
992	621
3061046	698
100	775
155	853
209	930
264	·4859007
318	084
373	161
427	238
482	316
536	393
590	470
645	548
699	625
754	702
808	779
863	856
917	933
972	·4860011
3062026	088
080	165
135	242
189	319
244	396
298	474
353	551
407	628
462	706
516	783
570	860
625	937
679	·4861014
733	091
788	169
842	246
897	323
951	400
3063006	477
060	554
115	632
169	709
223	786
278	863
332	940
386	·4862017
441	094
495	171
549	248
604	326
658	403
713	480
767	557
822	634
876	711
931	788
985	865
3064039	942
094	·4863020
148	097
202	174
257	251
311	328
365	405
420	482

3064474	·4863559
529	636
583	714
638	791
692	868
747	945
801	·4864022
855	099
910	176
964	253
3065018	330
073	407
127	484
181	561
236	638
290	715
344	792
399	869
453	946
507	·4865023
562	101
616	178
670	255
725	332
779	409
833	486
888	563
942	640
997	717
3066051	793
106	870
160	947
215	·4866024
269	101
323	178
378	255
432	332
486	409
541	486
595	563
649	640
704	717
758	794
812	871
867	948
921	·4867025
975	102
3067030	179
084	256
138	333
193	410
247	487
301	564
356	640
410	717
464	794
519	871
573	948
627	·4868025
682	102
736	179
790	256
845	333
899	410
953	487

3068008	·4868563
062	640
116	717
171	794
225	871
279	948
334	·4869025
388	102
442	179
497	255
551	332
605	409
660	486
714	563
768	640
823	716
877	793
931	870
985	947
3069039	·4870024
093	101
148	177
202	254
256	331
311	408
365	485
419	562
474	639
528	716
582	793
637	869
691	946
745	·4871023
800	099
854	176
908	253
963	330

Excesses

10 = 14	1 = 1
20 = 28	2 = 3
30 = 42	3 = 4
40 = 57	4 = 6
50 = 71	5 = 7
	6 = 8
	7 = 10
	8 = 11
	9 = 13

27 = 39

Anti Log Excesses

10 = 7	1 = 1
20 = 14	2 = 1
30 = 21	3 = 2
40 = 28	4 = 3
50 = 35	5 = 4
60 = 42	6 = 4
70 = 49	7 = 5
	8 = 6
	9 = 6

39 = 27

3070017	·4871407
071	484
126	560
180	637
234	714
289	791
343	868
397	945
451	·4872021
505	098
559	175
614	251
668	328
722	405
777	482
831	559
885	636
940	712
994	789
3071048	866
103	942
157	·4873019
211	096
265	172
319	249
373	326
428	402
482	479
536	556
591	633
645	710
699	787
754	863
808	940
862	·4874017
917	093
971	170
3072025	247
079	323
133	400
187	477
242	553
296	630
350	707
405	783
459	860
513	937
568	·4875013
622	090
676	167
730	243
784	320
838	397
893	473
947	550
3073001	627
056	703
110	780
164	857
218	933
272	·4876010
326	087
381	163
435	240
489	317

3073544	·4876393
598	470
652	547
706	623
760	700
814	776
869	853
923	929
977	·4877006
3074032	082
086	159
140	236
194	312
248	389
302	466
357	542
411	619
465	696
520	772
574	849
628	925
682	·4878002
736	078
790	155
845	231
899	308
953	384
3075007	461
061	537
115	614
170	690
224	767
278	843
333	920
387	996
441	·4879073
495	149
549	226
603	303
658	379
712	456
766	532
820	609
874	685
928	762
983	838
3076037	915
091	991
145	·4880068
199	144
253	221
308	297
362	374
416	450
471	527
525	603
579	679
633	756
687	832
741	909
796	985
850	·4881062
904	138
958	215
3077012	291

3077066	·4881367
121	444
175	520
229	597
283	673
337	750
391	826
445	903
499	979
553	·4882055
608	132
662	208
716	285
770	361
824	438
878	514
933	591
987	667
3078041	743
095	820
149	896
203	972
258	·4883049
312	125
366	201
420	278
474	354
528	431
583	507
637	584
691	660
745	737
799	813
853	889
907	966
961	·4884042
3079015	118
070	195
124	271
178	347
232	424
286	500
340	576
395	653
449	729
503	805
557	882
611	958
665	·4885034
719	111
773	187
827	263
882	340
936	416
990	492

Excesses

10 = 14	1 = 1
20 = 28	2 = 3
30 = 42	3 = 4
40 = 56	4 = 6
50 = 71	5 = 7
	6 = 8
	7 = 10
	8 = 11
	9 = 13

27 = 38

Anti Log Excesses

10 = 7	1 = 1
20 = 14	2 = 1
30 = 21	3 = 2
40 = 28	4 = 3
50 = 35	5 = 4
60 = 42	6 = 4
70 = 50	7 = 5
	8 = 6
	9 = 6

38 = 27

3080044	·4885569
098	645
152	721
206	797
260	873
314	949
369	·4886026
423	102
477	178
531	255
585	331
639	407
693	484
747	560
801	636
855	713
909	789
963	865
3081018	941
072	·4887017
126	093
180	170
234	246
288	322
342	399
396	475
450	551
505	628
559	704
613	780
667	856
721	932
775	·4888008
829	085
883	161
937	237
991	313

Number	Log
3082045	·4888389
099	465
154	542
208	618
262	694
316	771
370	847
424	923
478	999
532	·4889075
586	151
640	228
694	304
748	380
802	456
856	532
910	608
965	685
3083019	761
073	837
127	913
181	989
235	·4890065
289	141
343	217
397	293
451	370
505	446
559	522
613	598
667	674
721	750
775	827
829	903
883	979
937	·4891055
991	131
3084045	207
100	283
154	359
208	435
262	512
316	588
370	664
424	740
478	816
532	892
586	968
640	·4892044
694	120
748	196
802	272
856	348
910	424
964	500
3085018	576
072	653
126	729
180	805
234	881
288	957
342	·4893033
396	109
450	185
504	261

Number	Log
3085558	·4893337
612	413
666	489
720	565
774	641
828	717
882	793
936	869
990	945
3086044	·4894021
098	097
152	173
206	249
260	325
314	401
368	477
422	553
476	629
530	705
584	781
638	857
692	933
746	·4895009
800	085
854	161
908	237
962	313
3087016	388
070	464
124	540
178	616
232	692
286	768
340	844
394	920
448	996
502	·4896072
556	148
610	224
664	300
718	376
772	452
826	527
880	603
934	679
988	755
3088042	831
096	907
150	983
204	·4897059
258	135
312	210
366	286
420	362
474	438
528	514
582	590
636	666
690	742
744	818
797	893
851	969
905	·4898045
959	121
3089013	197

Number	Log
3089067	·4898273
121	348
175	424
229	500
283	576
337	652
391	728
445	803
499	879
553	955
607	·4899031
661	107
715	183
769	258
823	334
877	410
930	486
984	562

Excesses

10 = 14	1 = 1
20 = 28	2 = 3
30 = 42	3 = 4
40 = 56	4 = 6
50 = 70	5 = 7
	6 = 8
	7 = 10
	8 = 11
	9 = 13

27 = 38

Anti Log Excesses

10 = 7	1 = 1
20 = 14	2 = 1
30 = 21	3 = 2
40 = 28	4 = 3
50 = 36	5 = 4
60 = 43	6 = 4
70 = 50	7 = 5
	8 = 6
	9 = 6

38 = 27

Number	Log
3090038	·4899638
092	713
146	789
200	865
254	941
308	·4900017
362	093
416	168
470	244
524	320
578	395
632	471
686	547
739	623
793	699
847	775
901	850
955	926

Number	Log
3091009	·4901002
063	077
117	153
171	229
225	304
279	380
333	456
386	532
440	608
494	684
548	759
602	835
656	911
710	986
764	·4902062
818	138
871	213
925	289
979	365
3092033	441
087	517
141	593
195	668
249	744
303	820
357	895
411	971
465	·4903047
518	122
572	198
626	274
680	349
734	425
788	501
842	576
896	652
950	728
3093003	803
057	879
111	955
165	·4904030
219	106
273	182
327	257
381	333
435	409
488	484
542	560
596	636
650	711
704	787
758	863
811	938
865	·4905014
919	089
973	165
3094027	240
081	316
135	391
189	467
243	543
296	618
350	694
404	770
458	845

Number	Log
3094512	·4905921
566	997
619	·4906072
673	148
727	223
781	299
835	374
889	450
942	525
996	601
3095050	677
104	752
158	828
212	903
266	979
320	·4907054
374	130
427	205
481	281
535	357
589	432
643	508
697	583
750	659
804	734
858	810
912	885
966	961
3096020	·4908036
073	112
127	187
181	263
235	338
289	414
343	489
396	565
450	640
504	716
558	791
612	867
666	942
719	·4909018
773	093
827	169
880	244
934	320
988	395
3097042	471
096	546
150	621
203	697
257	772
311	848
365	923
419	999
473	·4910074
526	150
580	225
634	300
688	376
742	451
796	527
849	602
903	678
957	753

3098010	·4910829
064	904
118	979
172	·4911055
226	130
280	205
333	281
387	356
441	431
495	507
549	582
603	658
656	733
710	809
764	884
817	960
871	·4912035
925	110
979	186
3099033	261
087	336
140	412
194	487
248	562
301	638
355	713
409	788
463	864
517	939
571	·4913014
624	090
678	165
732	240
785	316
839	391
893	466
946	542

Excesses

10 = 14	1 = 1
20 = 28	2 = 3
30 = 42	3 = 4
40 = 56	4 = 6
50 = 70	5 = 7
	6 = 8
	7 = 10
	8 = 11
	9 = 13

27 = 38

Anti Log Excesses

10 = 7	1 = 1
20 = 14	2 = 1
30 = 21	3 = 2
40 = 29	4 = 3
50 = 36	5 = 4
60 = 43	6 = 4
70 = 50	7 = 5
	8 = 6
	9 = 6

38 = 27

3100000	·4913617
054	692
108	768
162	843
216	918
269	994
323	·4914069
377	144
430	220
484	295
538	370
592	446
646	521
700	596
753	672
807	747
861	822
914	897
968	972
3101022	·4915047
075	123
129	198
183	273
237	349
291	424
345	499
398	575
452	650
506	725
559	801
613	876
667	951
720	·4916026
774	101
828	176
881	252
935	327
989	402
3102043	478
097	553
151	628
204	703
258	778
312	853
365	929
419	·4917004
473	079
526	154
580	229
634	304
687	380
741	455
795	530
848	605
902	680
956	755
3103010	831
064	906
118	981
171	·4918057
225	132
279	207
332	282
386	357
440	432

3103493	·4918508
547	583
601	658
654	733
708	808
762	883
815	958
869	·4919033
923	108
976	184
3104030	259
084	334
137	409
191	484
245	559
298	634
352	709
406	784
459	860
513	935
567	·4920010
620	085
674	160
728	235
781	310
835	385
889	460
942	536
996	611
3105050	686
104	761
158	836
212	911
265	986
319	·4921061
373	136
426	211
480	286
534	361
587	436
641	511
695	586
748	661
802	736
856	811
909	887
963	962
3106016	·4922037
070	112
123	187
177	262
230	337
284	412
338	487
391	562
445	637
499	712
552	787
606	862
660	937
713	·4923012
767	087
821	162
874	237
928	312

3106982	·4923387
3107035	461
089	536
143	611
196	686
250	761
304	836
357	911
411	986
465	·4924061
518	136
572	211
626	286
679	361
733	436
787	511
840	586
894	661
948	736
3108001	811
055	886
108	961
162	·4925035
215	110
269	185
322	260
376	335
430	410
483	485
537	560
591	635
644	710
698	785
752	860
805	934
859	·4926009
913	084
966	159
3109020	234
073	309
127	383
180	458
234	533
287	608
341	683
395	758
448	833
502	908
556	983
609	·4927057
663	132
717	207
770	282
824	357
877	432
931	506
984	581

Excesses

10 = 14	1 = 1
20 = 28	2 = 3
30 = 42	3 = 4
40 = 56	4 = 6
50 = 70	5 = 7
	6 = 8
	7 = 10
	8 = 11
	9 = 13

27 = 38

Anti Log Excesses

10 = 7	1 = 1
20 = 14	2 = 1
30 = 21	3 = 2
40 = 29	4 = 3
50 = 36	5 = 4
60 = 43	6 = 4
70 = 50	7 = 5
	8 = 6
	9 = 6

38 = 27

3110038	·4927656
091	731
145	806
199	881
252	955
306	·4928030
360	105
413	180
467	255
520	330
574	404
627	479
681	554
734	629
788	704
842	779
895	853
949	928
3111003	·4929003
056	077
110	152
163	227
217	302
270	377
324	452
377	526
431	601
485	676
538	750
592	825
645	900
699	975
752	·4930050
806	125
859	199
913	274
967	349

3112020	·4930423
074	498
127	573
181	647
234	722
288	797
341	871
395	946
449	·4931021
502	096
556	171
609	246
663	320
716	395
770	470
823	544
877	619
931	694
984	768
3113038	843
091	918
145	992
198	·4932067
252	142
305	216
359	291
412	366
466	440
519	515
573	590
626	664
680	739
734	814
787	888
841	963
894	·4933038
948	112
3114001	187
055	262
108	336
162	411
215	486
269	560
322	635
376	710
429	784
483	859
536	934
590	·4934008
643	083
697	157
750	232
804	306
857	381
911	455
964	530
3115018	605
071	679
125	754
178	829
232	903
285	978
339	·4935052
392	127
446	201

3115499	·4935276
553	350
606	425
660	500
713	574
767	649
820	723
874	798
927	872
981	947
3116034	·4936021
088	096
141	171
195	245
248	320
302	394
355	469
409	543
462	618
516	692
569	767
623	841
676	916
730	990
783	·4937065
837	139
890	214
943	288
997	363
3117050	437
104	512
157	586
211	661
264	735
318	810
371	884
425	959
478	·4938033
532	108
585	182
639	257
692	331
745	406
799	480
852	555
906	629
959	704
3118013	778
066	852
120	927
173	·4939001
226	076
280	150
333	225
387	299
440	374
494	448
547	522
601	597
654	671
707	745
761	820
814	894
868	969
921	·4940043

3118975	·4940118
3119028	192
082	267
135	341
188	415
242	490
295	564
349	638
402	713
456	787
509	861
563	936
616	·4941010
669	084
723	159
776	233
829	308
883	382
936	457
990	531

Excesses

10 = 14	1 = 1
20 = 28	2 = 3
30 = 42	3 = 4
40 = 56	4 = 6
50 = 70	5 = 7
	6 = 8
	7 = 10
	8 = 11
	9 = 13

27 = 37

Anti Log Excesses

10 = 7	1 = 1
20 = 14	2 = 1
30 = 22	3 = 2
40 = 29	4 = 3
50 = 36	5 = 4
60 = 43	6 = 4
70 = 50	7 = 5
	8 = 6
	9 = 6

37 = 27

3120043	·4941606
097	680
150	754
204	829
257	903
310	977
364	·4942052
417	126
470	200
524	275
577	349
631	423
684	498
738	572
791	646
845	721
898	795

3120951	·4942869
3121005	944
058	·4943018
111	092
165	167
218	241
271	315
325	390
378	464
432	538
485	612
539	686
592	760
646	835
699	909
752	983
806	·4944058
859	132
912	206
966	281
3122019	355
072	429
126	504
179	578
232	652
286	727
339	801
392	875
446	949
499	·4945023
553	097
606	172
660	246
713	320
767	395
820	469
873	543
927	617
980	691
3123033	765
087	840
140	914
193	988
247	·4946062
300	136
353	210
407	285
460	359
513	433
567	508
620	582
673	656
727	730
780	804
833	878
887	953
940	·4947027
993	101
3124047	175
100	249
153	323
207	398
260	472
313	546

3124367	·4947620
420	694
473	768
527	842
580	916
633	990
687	·4948065
740	139
793	213
847	287
900	361
953	435
3125007	510
060	584
113	658
167	732
220	806
273	880
327	954
380	·4949028
433	102
487	177
540	251
593	325
647	399
700	473
753	547
807	621
860	695
913	769
967	843
3126020	917
073	991
127	·4950066
180	140
233	214
287	288
340	362
393	436
447	510
500	584
553	658
607	732
660	806
713	880
767	954
820	·4951028
873	102
927	176
980	250
3127033	324
087	398
140	472
193	546
247	620
300	694
353	768
407	842
460	916
513	990
566	·4952064
619	138
672	212
726	286
779	360

3127832	·4952434
886	508
939	582
992	656
3128046	730
099	804
152	878
206	952
259	·4953026
312	100
366	174
419	248
472	322
525	396
578	470
632	544
685	618
738	692
791	766
845	840
898	914
951	988
3129005	·4954061
058	135
111	209
165	283
218	357
271	431
324	505
377	579
430	653
484	727
537	801
590	875
644	948
697	·4955022
750	096
804	170
857	244
910	318
963	392

Excesses

10 = 14	1 = 1
20 = 28	2 = 3
30 = 42	3 = 4
40 = 56	4 = 6
50 = 70	5 = 7
	6 = 8
	7 = 10
	8 = 11
	9 = 13

27 = 37

Anti Log Excesses

10 = 7	1 = 1
20 = 14	2 = 1
30 = 22	3 = 2
40 = 29	4 = 3
50 = 36	5 = 4
60 = 43	6 = 4
70 = 50	7 = 5
	8 = 6
	9 = 6

37 = 27

3130016	·4955466
069	540
123	613
176	687
229	761
283	835
336	909
389	983
443	·4956056
496	130
549	204
602	278
655	352
708	426
762	500
815	574
868	648
922	721
975	795
3131028	869
081	943
134	·4957017
187	091
241	164
294	238
347	312
401	385
454	459
507	533
561	607
614	681
667	755
720	828
773	902
826	976
880	·4958050
933	124
986	198
3132039	271
092	345
145	419
199	492
252	566
305	640
359	714
412	788
465	862
518	935
571	·4959009
624	083
678	156

3132731	·4959230
784	304
838	378
891	452
944	526
997	599
3133050	673
103	747
157	820
210	894
263	968
316	·4960041
369	115
422	189
476	262
529	336
582	410
635	483
688	557
741	631
795	704
848	778
901	852
955	926
3134008	·4961000
061	074
114	147
167	221
220	295
274	368
327	442
380	516
433	589
486	663
539	737
593	810
646	884
699	958
752	·4962031
805	105
858	179
912	252
965	326
3135018	400
071	473
124	547
177	621
231	694
284	768
337	842
390	915
443	989
496	·4963062
550	136
603	209
656	283
709	356
762	430
815	504
868	577
921	651
974	725
3136028	798
081	872
134	946

3136187	·4964019
240	093
293	166
347	240
400	313
453	387
506	460
559	534
612	608
666	681
719	755
772	829
825	902
878	976
931	·4965049
984	123
3137037	196
090	270
144	343
197	417
250	490
303	564
356	637
409	711
462	784
515	858
568	932
622	·4966005
675	079
728	152
781	226
834	299
887	373
940	446
993	520
3138046	593
100	667
153	740
206	814
259	887
312	961
365	·4967034
418	108
471	181
524	255
578	328
631	402
684	475
737	549
790	622
843	695
896	769
949	842
3139002	916
056	989
109	·4968063
162	136
215	210
268	283
321	357
374	430
427	504
480	577
533	651
586	724

3139639	·4968797
693	871
746	944
799	·4969017
852	091
905	164
958	238

Excesses

10 = 14	1 = 1
20 = 28	2 = 3
30 = 42	3 = 4
40 = 55	4 = 6
50 = 69	5 = 7
	6 = 8
	7 = 10
	8 = 11
	9 = 12

27 = 37

Anti Log Excesses

10 = 7	1 = 1
20 = 14	2 = 1
30 = 22	3 = 2
40 = 29	4 = 3
50 = 36	5 = 4
60 = 43	6 = 4
70 = 51	7 = 5
	8 = 6
	9 = 6

37 = 27

3140011	·4969311
064	385
117	458
170	532
223	605
276	678
330	752
383	825
436	898
489	972
542	·4970045
595	118
648	192
701	265
754	338
807	412
860	485
913	559
966	632
3141019	706
072	779
126	853
179	926
232	999
285	·4971073
338	146
391	219
444	293
497	366

3141550	·4971439	3144997	·4976201	3148439	·4980953	3150344	·4983580	3153781	·4988315
603	513	3145050	274	492	·4981026	397	653	834	388
656	586	103	348	545	099	450	726	887	461
709	659	156	421	598	172	503	798	940	534
762	733	209	494	651	245	556	871	993	606
815	806	261	567	704	318	609	944	3154046	679
868	879	314	640	757	391	662	·4984017	099	752
921	953	367	713	810	464	715	090	151	824
974	·4972026	420	787	863	537	768	163	204	897
3142027	099	473	860	916	610	820	236	257	970
081	173	526	933	969	683	873	309	310	·4989042
134	246	579	·4977006	3149022	756	926	382	363	115
187	319	632	079	074	829	979	455	416	188
240	393	685	152	127	902	3151032	528	468	260
293	466	738	226	180	975	085	601	521	333
346	539	791	299	233	·4982048	138	673	574	406
399	612	844	372	286	121	191	746	627	479
452	685	897	445	339	194	244	819	680	552
505	758	950	518	392	267	296	892	733	625
558	832	3146003	591	445	340	349	965	785	697
611	905	056	664	498	413	402	·4985038	838	770
664	978	109	737	551	486	455	111	891	843
717	·4973052	162	810	604	559	508	184	944	915
770	125	215	884	657	632	561	257	997	988
823	198	268	957	709	704	614	329	3155050	·4990061
876	272	321	·4978030	762	777	667	402	102	133
929	345	374	103	815	850	720	475	155	206
982	418	427	176	868	923	772	548	208	279
3143035	492	480	249	921	996	825	621	260	351
088	565	533	322	974	·4983069	878	694	313	424
141	638	586	395			931	766	366	497
194	712	639	468	Excesses		984	839	419	569
247	785	692	542			3152037	912	472	642
300	858	745	615	10 = 14	1 = 1	090	985	525	715
353	931	798	688	20 = 28	2 = 3	143	·4986058	577	788
406	·4974004	851	761	30 = 41	3 = 4	196	131	630	861
459	077	904	834	40 = 55	4 = 6	248	203	683	934
512	151	957	907	50 = 69	5 = 7	301	276	736	·4991006
565	224	3147010	980		6 = 8	354	349	789	079
618	297	063	·4979053		7 = 10	407	422	842	152
671	371	116	126		8 = 11	460	495	894	224
724	444	168	199		9 = 12	513	568	947	297
777	517	221	272			565	640	3156000	370
830	591	274	345	27 = 37		618	713	053	442
883	664	327	419			671	786	106	515
936	737	380	492	Anti Log Excesses		724	859	159	588
989	810	433	565			777	932	211	660
3144042	883	486	638	10 = 7	1 = 1	830	·4987005	264	733
095	956	539	711	20 = 14	2 = 1	883	077	317	806
149	·4975030	592	784	30 = 22	3 = 2	936	150	370	878
202	103	645	857	40 = 29	4 = 3	989	223	423	951
255	176	698	930	50 = 36	5 = 4	3153041	296	476	·4992023
308	250	751	·4980003	60 = 43	6 = 4	094	369	528	096
361	323	804	076	70 = 51	7 = 5	147	442	581	168
414	396	857	149		8 = 6	200	514	634	241
467	469	910	222		9 = 7	253	587	686	313
520	542	963	295			306	660	739	386
573	615	3148016	368	37 = 27		358	732	792	459
626	689	069	441			411	805	845	531
679	762	122	514	3150027	·4983142	464	878	898	604
732	835	175	587	080	215	517	951	951	677
785	908	228	660	133	288	570	4988024	3157003	749
838	981	280	733	186	361	623	097	056	822
891	·4976054	333	806	239	434	675	169	109	895
944	128	386	879	292	507	728	242	161	967

3157214	·4993040
267	113
320	185
373	258
426	331
478	403
531	476
584	548
637	621
690	693
743	766
795	838
848	911
901	984
953	·4994056
3158006	129
059	202
112	274
165	347
218	419
270	492
323	564
376	637
428	709
481	782
534	855
586	927
639	·4995000
692	072
745	145
798	217
851	290
903	362
956	435
3159009	507
061	580
114	652
167	725
220	797
273	870
326	943
378	·4996015
431	088
484	160
536	233
589	305
642	378
694	450
747	523
800	595
853	668
906	740
959	813

Excesses

10 = 14	1 = 1
20 = 28	2 = 3
30 = 41	3 = 4
40 = 55	4 = 6
50 = 69	5 = 7
	6 = 8
	7 = 10
	8 = 11
	9 = 12

26 = 36

Anti Log Excesses

10 = 7	1 = 1
20 = 15	2 = 2
30 = 22	3 = 2
40 = 29	4 = 3
50 = 36	5 = 4
60 = 44	6 = 4
70 = 51	7 = 5
	8 = 6
	9 = 7

36 = 26

3160011	·4996885
064	958
117	·4997030
169	103
222	175
275	247
327	320
380	392
433	465
485	537
538	610
591	682
644	755
697	827
750	900
802	972
855	·4998045
908	117
960	190
3161013	262
066	334
118	407
171	479
224	551
276	624
329	696
382	769
435	841
488	914
541	986
593	·4999059
646	131
699	204
751	276
804	349
857	421
909	494

3161962	·4999566
3162015	638
067	711
120	783
173	855
225	928
278	·5000000
331	072
384	144
436	217
489	289
542	361
594	434
647	506
700	578
752	651
805	723
858	795
911	868
964	940
3163017	·5001012
069	085
122	157
175	229
227	302
280	375
333	448
385	520
438	592
491	664
543	736
596	809
649	882
701	953
754	·5002025
807	097
859	170
912	242
965	314
3164017	387
070	459
123	531
175	604
228	676
281	748
333	821
386	893
439	965
491	·5003038
544	110
597	182
649	255
702	327
755	399
807	471
860	543
913	615
965	688
3165018	760
071	832
123	905
176	977
229	·5004049
281	122
334	194

3165387	·5004266
439	338
492	410
545	482
597	555
650	627
703	699
755	772
808	844
861	916
913	989
966	·5005061
3166019	133
071	205
124	277
177	349
229	422
281	494
333	566
386	638
439	710
492	782
544	855
597	927
650	999
702	·5006072
755	144
808	216
860	288
913	360
966	432
3167018	505
071	577
124	649
176	721
229	793
282	865
334	938
387	·5007010
440	082
492	154
544	226
596	298
649	370
702	442
755	514
807	587
860	659
913	731
965	803
3168018	875
071	947
123	·5008020
176	092
229	164
281	236
334	308
387	380
439	452
491	524
543	596
596	668
649	740
702	812
754	885

3168807	·5008957
860	·5009029
912	101
965	173
3169018	245
070	317
123	389
176	461
228	533
280	605
332	677
385	750
438	822
491	894
543	966
596	·5010038
649	110
701	182
754	254
807	326
859	398
912	470
964	542

Excesses

10 = 14	1 = 1
20 = 27	2 = 3
30 = 41	3 = 4
40 = 55	4 = 6
50 = 69	5 = 7
	6 = 8
	7 = 10
	8 = 11
	9 = 12

26 = 36

Anti Log Excesses

10 = 7	1 = 1
20 = 15	2 = 2
30 = 22	3 = 2
40 = 29	4 = 3
50 = 36	5 = 4
60 = 44	6 = 4
70 = 51	7 = 5
	8 = 6
	9 = 7

36 = 26

3170016	·5010614
069	686
122	758
174	830
227	902
280	974
332	·5011046
385	118
438	190
490	262
542	334
595	406
647	478

3170700	·5011550
753	622
805	694
857	766
909	838
962	910
3171015	982
068	·5012054
120	126
173	198
226	270
278	342
331	414
384	486
436	558
488	630
540	702
593	774
646	846
699	918
751	990
803	·5013062
855	134
908	206
961	278
3172014	350
066	422
119	494
172	566
224	637
276	709
328	781
381	853
434	925
487	997
539	·5014069
592	141
645	213
697	285
749	357
801	429
854	500
907	572
960	644
3173012	716
064	788
116	860
169	932
222	·5015004
275	076
327	147
379	219
431	291
484	363
537	435
590	507
642	579
694	651
746	723
799	794
852	866
905	938
957	·5016010
3174010	082
063	154

3174115	·5016225
167	297
219	369
272	441
325	513
378	585
430	656
482	728
534	800
587	872
640	944
693	·5017016
745	087
797	159
849	231
902	303
955	375
3175008	447
060	518
112	590
164	662
217	733
269	805
321	877
374	949
427	·5018021
480	093
532	164
584	236
636	308
689	379
742	451
795	523
847	595
899	667
951	739
3176004	810
057	882
110	954
162	·5019025
214	097
266	169
319	241
372	313
425	385
477	456
529	528
581	600
634	671
686	743
738	815
791	886
844	958
897	·5020030
949	101
3177001	173
053	245
106	317
158	389
210	461
263	532
316	604
369	676
421	747
473	819

3177525	·5020891
578	962
631	·5021034
684	106
736	177
788	249
840	321
893	392
945	464
997	536
3178050	607
103	679
156	751
208	822
260	894
312	966
365	·5022037
417	109
469	181
522	252
574	324
626	396
679	467
732	539
785	611
837	682
889	754
941	826
994	897
3179046	969
098	·5023041
151	112
204	183
257	254
309	326
361	398
413	470
466	541
518	613
570	685
623	756
675	828
727	900
780	971
833	·5024043
886	115
938	186
990	257

Excesses

10 = 14	1 = 1
20 = 27	2 = 3
30 = 41	3 = 4
40 = 55	4 = 6
50 = 68	5 = 7
	6 = 8
	7 = 10
	8 = 11
	9 = 12

26 = 36

Anti Log Excesses

10 = 7	1 = 1
20 = 15	2 = 2
30 = 22	3 = 2
40 = 29	4 = 3
50 = 37	5 = 4
60 = 44	6 = 4
70 = 51	7 = 5
	8 = 6
	9 = 7

36 = 26

3180042	·5024328
095	400
147	472
199	544
252	615
304	687
356	759
409	830
461	901
513	972
566	·5025044
619	116
672	188
724	259
776	331
828	403
881	474
933	545
985	616
3181038	688
090	760
142	832
195	903
247	974
299	·5026045
352	117
405	189
458	261
510	332
562	403
614	474
667	546
719	618
771	690
824	761
876	832
928	903
981	975
3182033	·5027047
085	119
138	190
190	261
242	332
295	404
347	476
399	548
452	619
504	690
556	761
609	833
662	904

3182715	·5027975
767	·5028047
819	119
871	191
924	262
976	333
3183028	404
081	476
133	547
185	618
238	690
290	762
342	834
395	905
447	976
499	·5029047
552	119
604	190
656	261
709	333
761	404
813	475
866	547
918	619
970	691
3184023	762
075	833
127	904
180	976
232	·5030047
284	118
337	190
389	261
441	332
494	404
546	475
598	546
651	618
703	689
755	760
808	832
860	903
912	974
965	·5031046
3185017	117
069	188
122	260
174	332
226	404
279	475
331	546
383	617
436	689
488	760
540	831
593	902
645	973
697	·5032044
750	116
802	187
854	258
907	330
959	401
3186011	472
064	544

Number	Log
3186116	·5032615
168	686
221	758
273	829
325	900
378	972
430	·5033043
482	114
535	186
587	257
639	328
692	400
744	471
796	542
849	614
901	685
953	756
3187006	828
058	899
110	970
162	·5034041
214	112
266	183
319	255
371	326
423	397
476	469
528	540
580	611
633	682
685	753
737	824
790	896
842	967
894	·5035038
947	110
999	181
3188051	252
103	324
155	395
207	466
260	537
312	608
364	679
417	751
469	822
521	893
574	964
626	·5036035
678	106
731	178
783	249
835	320
887	392
939	463
991	534
3189044	605
096	676
148	747
201	818
253	889
305	960
358	·5037032
410	103
462	174

Number	Log
3189515	·5037245
567	316
619	387
671	459
723	530
775	601
828	672
880	743
932	814
985	886

Excesses

10 = 14	1 = 1
20 = 27	2 = 3
30 = 41	3 = 4
40 = 55	4 = 6
50 = 68	5 = 7
	6 = 8
	7 = 10
	8 = 11
	9 = 12

26 = 36

Anti Log Excesses

10 = 7	1 = 1
20 = 15	2 = 2
30 = 22	3 = 2
40 = 29	4 = 3
50 = 37	5 = 4
60 = 44	6 = 4
70 = 51	7 = 5
	8 = 6
	9 = 7

36 = 26

Number	Log
3190037	·5037957
089	·5038028
142	099
194	170
246	241
298	313
350	384
402	455
455	526
507	597
559	668
612	739
664	810
716	881
768	953
820	·5039024
872	095
925	166
977	237
3191029	308
082	379
134	450
186	521
239	592
291	663
343	734

Number	Log
3191395	·5039805
447	876
499	947
552	·5040019
604	090
656	161
709	232
761	303
813	374
865	445
917	516
969	587
3192022	658
074	729
126	800
178	871
230	942
282	·5041013
335	084
387	155
439	226
492	297
544	368
596	439
648	510
700	581
752	652
805	723
857	794
909	865
962	936
3193014	·5042007
066	078
118	149
170	220
222	291
275	362
327	433
379	504
431	575
483	646
535	717
588	788
640	859
692	930
744	·5043001
796	072
848	143
901	214
953	285
3194005	356
057	427
109	498
161	569
214	640
266	711
318	782
371	852
423	923
475	994
527	·5044065
579	136
631	207
684	278
736	349

Number	Log
3194788	·5044420
840	491
892	562
944	633
997	703
3195049	774
101	845
153	916
205	987
257	·5045058
310	129
362	200
414	271
466	342
518	413
570	484
623	554
675	625
727	696
779	767
831	838
883	909
935	980
987	·5046051
3196039	122
092	192
144	263
196	334
248	405
300	476
352	547
405	617
457	688
509	759
561	830
613	901
665	972
718	·5047042
770	113
822	184
874	255
926	326
978	397
3197030	467
082	538
134	609
187	680
239	751
291	822
343	892
395	963
447	·5048034
500	104
552	175
604	246
656	317
708	388
760	459
812	529
864	600
916	671
969	742
3198021	813
073	884
125	954

Number	Log
3198177	·5049025
229	096
281	166
333	237
385	308
438	378
490	449
542	520
594	591
646	662
698	733
750	803
802	874
854	945
907	·5050015
959	086
3199011	157
063	227
115	298
167	369
219	439
271	510
323	581
375	652
427	723
479	794
532	864
584	935
636	·5051006
688	076
740	147
792	218
844	288
896	359
948	430

Excesses

10 = 14	1 = 1
20 = 27	2 = 3
30 = 41	3 = 4
40 = 54	4 = 5
50 = 68	5 = 7
	6 = 8
	7 = 10
	8 = 11
	9 = 12

26 = 35

Anti Log Excesses

10 = 7	1 = 1
20 = 15	2 = 2
30 = 22	3 = 2
40 = 29	4 = 3
50 = 37	5 = 4
60 = 44	6 = 4
70 = 51	7 = 5
	8 = 6
	9 = 7

35 = 26

Number	Log
3200000	·5051500

Number	Log
3200052	·5051571
104	642
157	712
209	783
261	854
313	924
365	995
417	·5052066
469	136
521	207
573	278
625	348
677	419
729	490
782	560
834	631
886	702
938	772
990	843
3201042	914
094	984
146	·5053055
198	126
250	196
302	267
354	338
406	408
458	479
510	550
563	620
615	690
667	760
719	831
771	902
823	973
875	·5054043
927	114
979	185
3202031	255
083	326
135	397
187	467
239	537
291	607
343	678
395	749
447	820
500	890
552	961
604	·5055032
656	102
708	172
760	242
812	313
864	384
916	455
968	525
3203020	596
072	667
124	737
176	807
228	877
280	948
332	·5056019
384	090

Number	Log
3203436	·5056160
488	231
540	302
592	372
644	442
696	512
748	583
800	654
852	725
904	795
956	865
3204008	935
060	·5057006
112	077
164	148
216	218
268	288
320	358
373	429
425	500
477	571
529	641
581	711
633	781
685	852
737	923
789	994
841	·5058064
893	134
945	204
997	275
3205049	345
101	415
153	486
205	557
257	628
309	698
361	768
413	838
465	909
517	980
569	·5059051
621	121
673	191
725	261
777	332
829	402
881	472
933	543
985	614
3206037	685
088	755
140	825
192	895
244	966
296	·5060036
348	106
400	177
452	247
504	317
556	388
608	458
660	528
712	599
764	670

Number	Log
3206816	·5060741
868	811
920	881
972	951
3207024	·5061022
076	092
128	162
180	233
232	303
284	373
336	444
388	514
440	584
492	655
544	725
596	795
648	866
700	936
752	·5062006
803	077
855	147
907	217
959	288
3208011	358
063	428
115	499
167	569
219	639
271	710
323	780
375	850
427	921
479	991
531	·5063061
583	132
635	202
687	272
739	343
791	413
843	483
894	554
946	624
998	694
3209050	765
102	835
154	905
206	976
258	·5064046
310	116
362	186
414	256
466	326
518	397
570	467
622	537
673	608
725	678
777	748
829	819
881	889
933	959
985	·5065029

Excesses

10 = 14	1 = 1
20 = 27	2 = 3
30 = 41	3 = 4
40 = 54	4 = 5
50 = 68	5 = 7
	6 = 8
	7 = 10
	8 = 11
	9 = 12

26 = 35

Anti Log Excesses

10 = 7	1 = 1
20 = 15	2 = 2
30 = 22	3 = 2
40 = 30	4 = 3
50 = 37	5 = 4
60 = 44	6 = 4
70 = 52	7 = 5
	8 = 6
	9 = 7

35 = 26

Number	Log
3210037	·5065099
089	169
141	240
193	310
245	380
296	451
348	521
400	591
452	662
504	732
556	802
608	872
660	942
712	·5066012
764	083
816	153
868	223
919	294
971	364
3211023	434
075	504
127	574
179	644
231	715
283	785
335	855
386	925
438	995
490	·5067065
542	136
594	206
646	276
698	346
750	416
802	486
854	557
906	627

Number	Log
3211958	·5067697
3212009	767
061	837
113	907
165	978
217	·5068048
269	118
321	188
373	258
425	328
476	399
528	469
580	539
632	609
684	679
736	749
787	819
839	889
891	959
943	·5069030
995	100
3213047	170
099	240
151	310
203	380
254	450
306	520
358	590
410	661
462	731
514	801
565	871
617	941
669	·5070011
721	081
773	151
825	221
877	291
929	361
981	431
3214032	502
084	572
136	642
188	712
240	782
292	852
343	922
395	992
447	·5071062
499	132
551	202
603	272
654	342
706	412
758	482
810	552
862	622
914	692
966	763
3215018	833
070	903
121	973
173	·5072043
225	113
277	183

3215329	·5072253
381	323
432	393
484	463
536	533
588	603
640	673
692	743
743	813
795	883
847	953
899	·5073023
951	093
3216003	163
054	233
106	303
158	373
209	442
261	512
313	582
365	652
417	722
469	792
520	862
572	932
624	·5074002
676	072
728	142
780	212
831	282
883	352
935	422
987	492
3217039	562
091	632
142	702
194	772
246	842
297	911
349	981
401	·5075051
453	121
505	191
557	261
608	331
660	401
712	471
764	541
816	611
868	681
919	751
971	821
3218023	891
074	960
126	·5076030
178	100
230	170
282	240
334	310
385	380
437	450
489	520
541	589
593	659
645	729

3218696	·5076799
748	869
800	939
851	·5077008
903	078
955	148
3219007	218
059	288
111	358
162	427
214	497
266	567
317	637
369	707
421	777
473	846
525	916
577	986
628	·5078056
680	126
732	196
783	266
835	336
887	406
938	475
990	545

Excesses

10 = 14	1 = 1
20 = 27	2 = 3
30 = 41	3 = 4
40 = 54	4 = 5
50 = 68	5 = 7
	6 = 8
	7 = 9
	8 = 11
	9 = 12

26 = 35

Anti Log Excesses

10 = 7	1 = 1
20 = 15	2 = 2
30 = 22	3 = 2
40 = 30	4 = 3
50 = 37	5 = 4
60 = 44	6 = 4
70 = 52	7 = 5
	8 = 6
	9 = 7

35 = 26

3220042	·5078615
094	684
146	754
198	824
249	894
301	964
353	·5079034
404	103
456	173
508	243

3220559	·5079313
611	383
663	453
715	522
767	592
819	662
870	731
922	801
974	871
3221025	941
077	·5080011
129	081
180	150
232	220
284	290
336	359
388	429
440	499
491	568
543	638
595	708
646	778
698	848
750	918
801	987
853	·5081057
905	127
956	196
3222008	266
060	336
112	405
164	475
216	545
267	614
319	684
371	754
422	823
474	893
526	963
577	·5082032
629	102
681	172
732	241
784	311
836	381
887	450
939	520
991	590
3223043	660
095	730
147	800
198	869
250	939
302	·5083009
353	078
405	148
457	218
508	287
560	356
612	425
663	495
715	565
767	635
818	704
870	774

3223922	·5083844
973	913
3224025	983
077	·5084053
128	122
180	192
232	262
283	331
335	401
387	471
438	540
490	610
542	680
593	749
645	819
697	889
749	958
801	·5085027
853	096
904	166
956	236
3225008	306
059	375
111	445
163	515
214	584
266	654
318	724
369	793
421	863
473	933
524	·5086002
576	071
628	140
679	210
731	280
783	350
834	419
886	489
938	559
989	628
3226041	697
093	766
144	836
196	906
248	976
299	·5087045
351	114
403	183
454	253
506	323
558	393
609	462
661	532
713	602
764	671
816	740
868	809
919	879
970	949
3227021	·5088019
073	088
125	157
177	226
228	296

3227280	·5088366
332	436
383	505
435	574
487	643
538	713
590	783
642	853
693	922
745	991
797	·5089060
848	130
900	200
952	270
3228003	339
055	408
107	477
158	547
210	616
262	685
313	755
365	825
417	895
468	964
519	·5090033
570	102
622	172
674	241
726	310
777	380
829	450
881	520
932	589
984	658
3229036	727
087	797
139	866
191	935
242	·5091005
294	074
346	143
397	213
449	283
501	353
552	422
603	491
654	560
706	630
758	699
810	768
861	838
913	907
965	976

Excesses

10 = 13	1 = 1
20 = 27	2 = 3
30 = 40	3 = 4
40 = 54	4 = 5
50 = 67	5 = 7
	6 = 8
	7 = 9
	8 = 11
	9 = 12

26 = 35

Anti Log Excesses

10 = 7	1 = 1
20 = 15	2 = 2
30 = 22	3 = 2
40 = 30	4 = 3
50 = 37	5 = 4
60 = 45	6 = 5
	7 = 5
	8 = 6
	9 = 7

35 = 26

3230016	·5092046
068	115
120	184
171	254
223	323
275	392
326	462
377	531
428	600
480	670
532	739
584	808
635	878
687	947
739	·5093016
790	086
842	155
894	224
945	294
996	363
3231047	432
099	502
151	571
203	640
254	710
306	779
358	848
409	918
461	987
513	·5094056
564	126
615	195
666	264
718	334
770	403
822	472
873	542

3231925	·5094611
977	680
3232028	750
079	819
130	888
182	958
234	·5095027
286	096
337	166
389	235
441	304
492	373
543	442
594	511
646	581
698	650
750	719
801	789
853	858
905	927
956	997
3233007	·5096066
058	135
110	205
162	274
214	343
265	412
317	481
369	550
420	620
471	689
522	758
574	828
626	897
678	966
729	·5097035
781	104
833	173
884	243
935	312
986	381
3234038	450
090	519
142	588
193	658
244	727
295	796
347	866
399	935
451	·5098004
502	074
554	143
606	212
657	281
708	350
759	419
811	489
863	558
915	627
966	696
3235017	765
068	834
120	904
172	973
224	·5099042

3235275	·5099111
326	180
377	249
429	319
481	388
533	457
584	526
635	595
686	664
738	733
790	802
842	871
893	941
944	·5100010
995	079
3236047	148
099	217
151	286
202	356
253	425
304	494
356	563
408	632
460	701
511	770
562	839
613	908
665	978
717	·5101047
769	116
820	185
871	254
922	323
974	392
3237026	461
078	530
129	599
180	668
231	737
283	807
335	876
387	945
438	·5102014
489	083
540	152
592	221
644	290
696	359
747	428
798	497
849	566
901	635
952	704
3238003	773
055	842
107	911
159	980
210	·5103049
261	118
312	187
364	257
416	326
468	395
519	464
570	533

3238621	·5103602
673	671
724	740
775	809
827	878
879	947
931	·5104016
982	085
3239033	154
084	223
136	292
188	361
240	430
291	498
342	567
393	636
445	705
496	774
547	843
599	912
651	981
703	·5105050
754	119
805	188
856	257
908	326
959	395

Excesses

10 = 13	1 = 1
20 = 27	2 = 3
30 = 40	3 = 4
40 = 54	4 = 5
50 = 67	5 = 7
	6 = 8
	7 = 9
	8 = 11
	9 = 12

26 = 35

Anti Log Excesses

10 = 7	1 = 1
20 = 15	2 = 2
30 = 22	3 = 2
40 = 30	4 = 3
50 = 37	5 = 4
60 = 45	6 = 5
	7 = 5
	8 = 6
	9 = 7

35 = 26

3240010	·5105464
062	533
114	602
166	671
217	740
268	809
319	878
371	947
422	·5106016

3240473	·5106085
525	154
576	223
627	292
679	360
731	429
783	498
834	567
885	636
936	705
988	774
3241039	843
090	912
142	980
194	·5107049
246	118
297	187
348	256
399	325
451	394
502	463
553	532
605	600
656	669
707	738
759	807
811	876
863	945
914	·5108014
965	083
3242016	152
068	220
119	289
170	358
222	427
273	496
324	565
376	633
427	702
478	771
530	840
582	909
634	978
685	·5109047
736	116
787	185
839	253
890	322
941	391
993	459
3243044	528
095	597
147	666
198	735
249	804
301	872
352	941
403	·5110010
455	079
507	148
559	217
610	285
661	354
712	423
764	492

3243815	·5110561
866	630
918	698
969	767
3244020	836
072	904
123	973
174	·5111042
226	111
277	180
328	249
380	317
431	386
482	455
534	523
585	592
636	661
688	729
739	798
790	867
842	936
894	·5112005
946	074
997	142
3245048	211
099	280
151	348
202	417
253	486
305	554
356	623
407	692
459	761
510	830
561	899
613	967
664	·5113036
715	105
767	173
818	242
869	311
921	379
972	448
3246023	517
075	585
126	654
177	723
229	791
280	860
331	929
383	997
434	·5114066
485	135
537	203
588	272
639	341
691	409
742	478
793	547
845	615
896	684
947	753
999	821
3247050	890
101	959

3247153	·5115027
204	096
255	165
307	233
358	302
409	371
461	439
512	508
563	577
615	645
666	714
717	783
769	851
820	920
871	989
923	·5116057
974	125
3248025	193
077	262
128	331
179	400
231	468
282	537
333	606
385	674
436	743
487	812
539	880
590	949
641	·5117018
693	086
744	154
795	222
846	291
897	360
948	429
3249000	497
051	566
102	635
154	703
205	772
256	841
308	909
359	977
410	·5118045
462	114
513	183
564	252
616	320
667	389
718	458
770	526
821	594
872	662
924	731
975	800

Excesses

10 = 13	1 = 1
20 = 27	2 = 3
30 = 40	3 = 4
40 = 54	4 = 5
50 = 67	5 = 7
	6 = 8
	7 = 9
	8 = 11
	9 = 12

26 = 35

Anti Log Excesses

10 = 7	1 = 1
20 = 15	2 = 2
30 = 22	3 = 2
40 = 30	4 = 3
50 = 37	5 = 4
60 = 45	6 = 5
	7 = 5
	8 = 6
	9 = 7

35 = 26

3250026	·5118869
077	937
128	·5119005
179	073
231	142
282	211
333	280
385	348
436	416
487	484
539	553
590	622
641	691
693	759
744	827
795	895
847	964
898	·5120033
949	102
3251000	170
051	238
102	306
154	375
205	444
256	513
308	581
359	649
410	717
462	786
513	855
564	924
615	992
666	·5121060
717	128
769	197
820	265
871	333

3251923	·5121402
974	471
3252025	540
077	608
128	676
179	744
231	813
282	881
333	949
384	·5122018
435	086
486	154
538	223
589	291
640	359
692	428
743	497
794	566
845	634
896	702
947	770
999	839
3253050	907
101	975
153	·5123044
204	112
255	180
307	249
358	317
409	385
460	454
511	523
562	592
614	660
665	728
716	796
768	865
819	933
870	·5124001
921	070
972	138
3254023	206
075	275
126	343
177	411
229	480
280	548
331	616
382	685
433	753
484	821
536	890
587	958
638	·5125026
689	095
740	163
791	231
843	300
894	368
945	436
997	505
3255048	573
099	641
150	710
201	778

3255252	·5125846
304	915
355	983
406	·5126051
457	120
508	188
559	256
611	325
662	393
713	461
765	529
816	597
867	665
918	734
969	802
3256020	870
072	939
123	·5127007
174	075
225	144
276	212
327	280
379	349
430	417
481	485
532	554
583	622
634	690
686	758
737	826
788	894
839	963
890	·5128031
941	099
993	168
3257044	236
095	304
146	373
197	441
248	509
300	577
351	645
402	713
453	782
504	850
555	918
607	987
658	·5129055
709	123
760	191
811	259
862	327
914	396
965	464
3258016	532
067	600
118	668
169	736
221	805
272	873
323	941
374	·5130009
425	077
476	145
528	214

Number	Log
3258579	·5130282
630	350
681	418
732	486
783	554
835	623
886	691
937	759
988	827
3259039	895
090	963
141	·5131032
192	100
243	168
295	236
346	304
397	372
448	440
499	508
550	576
602	645
653	713
704	781
755	849
806	917
857	985
908	·5132054
959	122

Excesses

10 = 13	1 = 1
20 = 27	2 = 3
30 = 40	3 = 4
40 = 53	4 = 5
50 = 67	5 = 7
	6 = 8
	7 = 9
	8 = 11
	9 = 12

26 = 34

Anti Log Excesses

10 = 7	1 = 1
20 = 15	2 = 2
30 = 22	3 = 2
40 = 30	4 = 3
50 = 37	5 = 4
60 = 45	6 = 5
	7 = 5
	8 = 6
	9 = 7

34 = 26

Number	Log
3260010	·5132190
062	258
113	326
164	394
215	462
266	530
317	598
369	667

Number	Log
3260420	·5132735
471	803
522	871
573	939
624	·5133007
675	075
726	143
777	211
829	279
880	347
931	415
982	484
3261033	552
084	620
135	688
186	756
237	824
289	892
340	960
391	·5134028
442	096
493	164
544	232
595	300
646	368
697	436
748	504
799	572
850	640
902	708
953	776
3262004	844
055	912
106	980
157	·5135048
208	116
259	184
310	252
362	321
413	389
464	457
515	525
566	593
617	661
668	728
719	796
770	864
821	932
872	·5136000
923	068
975	136
3263026	204
077	272
128	340
179	408
230	476
281	544
332	612
383	680
434	748
485	816
536	884
587	952
638	·5137020
689	088

Number	Log
3263741	·5137156
792	224
843	292
894	360
945	428
996	496
3264047	564
098	632
149	700
200	768
251	836
302	904
353	971
404	·5138039
455	107
507	175
558	243
609	311
660	379
711	447
762	515
813	583
864	651
915	719
966	786
3265017	854
068	922
119	990
170	·5139058
221	126
272	194
323	262
374	330
425	398
476	466
527	534
579	601
630	669
681	737
732	805
783	873
834	941
885	·5140008
936	076
987	144
3266038	212
089	280
140	348
191	415
242	483
293	551
344	619
395	687
446	755
497	823
548	891
599	959
650	·5141026
701	094
752	162
803	229
854	297
905	365
956	433
3267007	501

Number	Log
3267058	·5141569
109	636
160	704
211	772
262	840
313	908
364	976
415	·5142043
466	111
517	179
568	246
619	314
670	382
721	450
772	518
823	586
874	653
925	721
976	789
3268027	857
078	925
129	993
180	·5143060
231	128
282	196
333	263
384	331
435	399
486	466
537	534
588	602
639	670
690	738
741	806
792	873
843	941
894	·5144009
945	076
996	144
3269047	212
098	279
149	347
200	415
251	483
302	551
353	619
404	686
455	754
506	822
557	889
608	957
659	·5145025
710	092
761	160
812	228
863	295
914	363
965	431

Excesses

10 = 13	1 = 1
20 = 27	2 = 3
30 = 40	3 = 4
40 = 53	4 = 5
50 = 67	5 = 7
	6 = 8
	7 = 9
	8 = 11
	9 = 12

26 = 34

Anti Log Excesses

10 = 8	1 = 1
20 = 15	2 = 2
30 = 23	3 = 2
40 = 30	4 = 3
50 = 38	5 = 4
60 = 45	6 = 5
	7 = 5
	8 = 6
	9 = 7

34 = 26

Number	Log
3270016	·5145498
067	566
118	634
169	701
220	769
271	837
322	904
373	972
424	·5146040
474	107
525	175
576	243
627	310
678	378
729	446
780	513
831	581
882	649
933	716
984	784
3271035	852
086	919
137	987
188	·5147055
239	122
290	190
341	258
392	325
443	393
494	461
544	528
595	595
646	662
697	730
748	798
799	866
850	933

3271901	·5148001
952	069
3272003	136
054	204
105	272
156	339
207	407
258	475
309	542
360	609
411	676
461	744
512	812
563	880
614	947
665	·5149015
716	083
767	150
818	218
869	286
920	353
971	421
3273022	489
072	556
123	623
174	690
225	758
276	826
327	894
378	961
429	·5150028
480	095
531	163
582	231
633	299
683	366
734	434
785	502
836	569
887	636
938	703
989	771
3274040	839
091	907
142	974
193	·5151041
244	108
294	176
345	244
396	312
447	379
498	446
549	513
600	581
651	649
702	717
752	784
803	851
854	918
905	986
956	·5152054
3275007	122
058	189
109	256
160	323

3275210	·5152391
261	459
312	527
363	594
414	661
465	728
516	796
567	863
618	930
668	998
719	·5153066
770	134
821	201
872	268
923	335
974	403
3276025	470
076	537
126	605
177	673
228	741
279	808
330	875
381	942
431	·5154010
482	077
533	144
584	212
635	279
686	346
737	414
788	482
839	550
889	617
940	684
991	751
3277042	819
093	886
144	953
194	·5155021
245	088
296	155
347	223
398	290
449	357
500	425
551	493
602	561
652	628
703	695
754	762
805	830
856	897
907	964
957	·5156032
3278008	099
059	166
110	234
161	301
212	368
262	436
313	503
364	570
415	638
466	705

3278517	·5156772
567	840
618	907
669	974
720	·5157042
771	109
822	176
872	244
923	311
974	378
3279025	446
076	513
127	580
177	648
228	715
279	782
330	850
381	917
432	984
482	·5158052
533	119
584	186
635	254
686	321
737	388
787	456
838	523
889	590
940	658
991	725

Excesses

10 = 13	1 = 1
20 = 27	2 = 3
30 = 40	3 = 4
40 = 53	4 = 5
50 = 66	5 = 7
	6 = 8
	7 = 9
	8 = 11
	9 = 12

26 = 34

Anti Log Excesses

10 = 8	1 = 1
20 = 15	2 = 2
30 = 23	3 = 2
40 = 30	4 = 3
50 = 38	5 = 4
60 = 45	6 = 5
	7 = 5
	8 = 6
	9 = 7

34 = 26

3280042	·5158792
092	860
143	927
194	994
244	·5159061
295	128
346	195

3280397	·5159263
448	330
499	397
549	465
600	532
651	599
702	667
753	734
804	801
854	868
905	935
956	·5160002
3281006	070
057	137
108	204
159	272
210	339
261	406
311	474
362	541
413	608
464	675
515	742
566	809
616	877
667	944
718	·5161011
768	078
819	145
870	212
921	280
972	347
3282023	414
073	482
124	549
175	616
225	683
276	750
327	817
378	885
429	952
480	·5162019
530	087
581	154
632	221
682	288
733	355
784	422
835	490
886	557
937	624
987	691
3283038	758
089	825
139	892
190	959
241	·5163026
292	094
343	161
394	228
444	295
495	362
546	429
596	497
647	564

3283698	·5163631
748	698
799	765
850	832
901	899
952	966
3284003	·5164033
053	101
104	168
155	235
205	302
256	369
307	436
357	503
408	570
459	637
510	705
561	772
612	839
662	906
713	973
764	·5165040
814	107
865	174
916	241
966	308
3285017	375
068	442
118	510
169	577
220	644
271	711
322	778
373	845
423	912
474	979
525	·5166046
575	113
626	180
677	247
727	314
778	381
829	448
879	515
930	582
981	649
3286032	717
083	784
134	851
184	918
235	985
286	·5167052
336	119
387	186
438	253
488	320
539	387
590	454
640	521
691	588
742	655
792	722
843	789
894	856
944	923

3286995	·5167990
3287046	·5168057
096	124
147	191
198	258
249	325
300	392
351	459
401	526
452	593
503	660
553	727
604	794
655	861
705	928
756	995
807	·5169062
857	128
908	195
959	262
3288009	329
060	396
111	463
161	530
212	597
263	664
313	731
364	798
415	865
465	932
516	999
567	·5170066
617	132
668	199
719	266
769	333
820	400
871	467
921	534
972	601
3289023	668
073	735
124	802
175	869
225	935
276	·5171002
327	069
377	136
428	203
479	270
529	337
580	404
631	471
681	537
732	604
783	671
833	738
884	805
935	872
985	939

Excesses

10 = 13	1 = 1
20 = 26	2 = 3
30 = 40	3 = 4
40 = 53	4 = 5
50 = 66	5 = 7
	6 = 8
	7 = 9
	8 = 11
	9 = 12

26 = 34

Anti Log Excesses

10 = 8	1 = 1
20 = 15	2 = 2
30 = 23	3 = 2
40 = 30	4 = 3
50 = 38	5 = 4
60 = 45	6 = 5
	7 = 5
	8 = 6
	9 = 7

34 = 26

3290036	·5172006
087	073
137	139
188	206
239	273
289	340
340	407
391	474
441	540
492	607
543	674
593	741
644	808
695	875
745	941
796	·5173008
847	075
897	142
948	209
999	276
3291049	343
100	410
151	477
201	543
252	610
303	677
353	743
404	810
455	877
505	944
556	·5174011
607	078
657	144
708	211
759	278
809	345
859	412

3291909	·5174479
960	545
3292011	612
062	679
112	745
163	812
214	879
264	945
315	·5175012
366	079
416	146
467	213
518	280
568	346
619	413
670	480
720	546
771	613
822	680
872	747
922	814
972	881
3293023	948
074	·5176015
125	082
175	148
226	215
277	282
327	348
378	415
429	482
479	548
530	615
581	682
631	748
682	815
733	882
783	949
833	·5177016
883	083
934	149
985	216
3294036	283
086	349
137	416
188	483
238	549
289	616
340	683
390	749

Excesses

25 = 33

Anti Log Excesses

33 = 25

3294466	·5177849
541	949
617	·5178049
693	149
769	249
845	349

3294921	·5178449
997	549
3295073	649
148	749
224	849
300	949
376	·5179049
452	149
528	249
604	349
680	449
755	549
831	649
907	749
983	849
3296059	949
135	·5180049
210	148
286	248
362	348
438	448
514	548
590	648
665	748
741	848
817	948
893	·5181048
969	148
3297045	248
120	348
196	448
272	547
348	647
424	747
500	847
575	946
651	·5182046
727	146
803	246
879	346
955	446
3298030	545
106	645
182	745
258	845
333	945
409	·5183045
485	145
561	245
637	344
713	444
788	544
864	644
940	743
3299016	843
091	943
167	·5184043
243	142
319	242
394	342
470	442
546	541
622	641
697	740
773	840

3299849	·5184940
925	·5185040

Excesses

10 = 13	1 = 1
20 = 26	2 = 3
30 = 40	3 = 4
40 = 53	4 = 5
50 = 66	5 = 7
60 = 79	6 = 8
70 = 92	7 = 9
	8 = 11
	9 = 12

38 = 50

Anti Log Excesses

10 = 8	1 = 1
20 = 15	2 = 2
30 = 23	3 = 2
40 = 30	4 = 3
50 = 38	5 = 4
60 = 46	6 = 5
70 = 53	7 = 5
80 = 61	8 = 6
90 = 68	9 = 7

50 = 38

3300000	·5185139
076	239
152	338
228	438
304	538
380	638
455	738
531	838
607	937
683	·5186037
758	136
834	236
909	336
985	436
3301061	535
137	635
212	734
288	834
364	933
440	·5187033
515	132
591	232
667	332
743	432
818	531
894	631
970	730
3302046	830
121	929
197	·5188029
272	128
348	228
424	328
500	428

3302575	·5188527
651	627
727	726
803	826
878	925
954	·5189025
3303029	124
105	224
181	323
257	423
332	522
408	622
484	721
560	821
635	920
711	·5190020
786	118
862	218
938	317
3304014	417
089	517
165	617
240	716
316	816
392	914
468	·5191014
543	113
619	213
694	312
770	412
845	511
921	611
997	710
3305073	809
148	908
224	·5192008
299	107
375	207
451	306
527	405
602	504
678	604
753	703
829	803
904	902
980	·5193002
3306056	101
132	200
207	299
283	399
358	498
434	598
509	697
585	796
660	895
736	995
812	·5194094
888	193
963	292
3307039	392
114	491
190	590
265	689
341	789
416	888

3307492	·5194987
568	·5195086
644	186
719	285
795	385
870	484
946	583
3308021	682
097	781
172	880
248	980
323	·5196079
399	178
474	277
550	377
626	476
702	575
777	674
853	773
928	872
3309004	971
079	·5197070
155	170
230	269
306	368
381	467
457	566
532	665
608	765
683	864
759	963
834	·5198062
910	161
985	260

Excesses

10 = 13	1 = 1
20 = 26	2 = 3
30 = 39	3 = 4
40 = 53	4 = 5
50 = 66	5 = 7
60 = 79	6 = 8
70 = 92	7 = 9
	8 = 11
	9 = 12

38 = 50

Anti Log Excesses

10 = 8	1 = 1
20 = 15	2 = 2
30 = 23	3 = 2
40 = 30	4 = 3
50 = 38	5 = 4
60 = 46	6 = 5
70 = 53	7 = 5
80 = 61	8 = 6
90 = 68	9 = 7

50 = 38

3310061	·5198359
136	458

3310212	·5198558
287	657
363	756
439	855
515	954
590	·5199053
666	152
741	251
817	350
892	449
967	548
3311043	647
118	746
194	845
270	944
345	·5200043
421	142
496	241
572	340
647	439
723	538
798	637
874	736
949	835
3312025	934
100	·5201033
175	132
250	231
326	330
401	429
477	528
552	627
628	726
703	825
779	924
854	·5202022
930	121
3313005	220
081	319
156	418
232	517
307	616
383	715
458	814
534	913
609	·5203012
685	111
760	209
836	308
911	407
986	506
3314061	605
137	704
212	803
288	902
363	·5204000
439	099
514	198
590	297
665	396
741	495
816	593
892	692
967	791
3315042	890

3315117	·5204988
193	·5205087
268	186
344	285
419	384
495	483
570	581
646	680
721	779
796	878
871	976
947	·5206075
3316022	174
098	273
173	371
249	470
324	568
400	667
475	766
550	865
625	963
701	·5207062
776	160
852	259
927	358
3317003	457
078	556
153	655
228	753
304	852
379	950
455	·5208049
530	147
605	246
680	345
756	444
831	542
907	641
982	739
3318058	838
133	936
208	·5209035
283	134
359	233
434	331
510	430
585	528
660	627
735	725
811	824
886	922
962	·5210021
3319037	120
112	219
187	317
263	416
338	514
413	613
488	711
564	810
639	908
715	·5211007
790	105
865	204
940	302

Excesses

10 = 13	1 = 1
20 = 26	2 = 3
30 = 39	3 = 4
40 = 52	4 = 5
50 = 66	5 = 7
60 = 79	6 = 8
70 = 92	7 = 9
	8 = 10
	9 = 12

38 = 50

Anti Log Excesses

10 = 8	1 = 1
20 = 15	2 = 2
30 = 23	3 = 2
40 = 31	4 = 3
50 = 38	5 = 4
60 = 46	6 = 5
70 = 53	7 = 5
80 = 61	8 = 6
90 = 69	9 = 7

50 = 38

3320016	·5211401
091	499
166	598
241	696
317	795
392	893
468	992
543	·5212090
618	189
693	287
769	386
844	484
919	583
994	681
3321070	780
145	878
220	976
295	·5213074
371	173
446	271
522	370
597	468
672	567
747	665
823	764
898	862
973	960
3322048	·5214058
124	157
199	255
274	354
349	452
425	550
500	648
575	747
650	846
726	944

3322801	·5215042
876	141
951	239
3323026	338
101	436
177	534
252	632
327	731
402	829
478	927
553	·5216025
628	124
703	222
779	320
854	418
929	517
3324004	615
080	713
155	811
230	910
305	·5217008
380	106
455	204
531	303
606	401
681	499
756	597
832	696
907	794
982	892
3325057	990
132	:5218089
207	187
283	285
358	383
433	481
508	579
583	678
658	776
734	874
809	972
884	·5219070
959	168
3326034	266
109	364
184	462
260	561
335	659
410	757
485	856
560	954
635	·5220052
711	150
786	248
861	346
936	444
3327011	542
086	640
161	738
236	836
312	934
387	·5221032
462	130
537	228
612	326

3327687	·5221425
763	523
838	621
913	719
988	817
3328063	915
138	·5222013
213	111
288	209
363	307
438	405
514	503
589	601
664	699
739	797
814	895
889	993
964	·5223091
3329039	189
114	287
189	385
265	483
340	581
415	679
490	777
565	874
640	972
715	·5224070
790	168
865	266
940	364

Excesses

10 = 13	1 = 1
20 = 26	2 = 3
30 = 39	3 = 4
40 = 52	4 = 5
50 = 65	5 = 7
60 = 78	6 = 8
70 = 91	7 = 9
	8 = 10
	9 = 12

38 = 50

Anti Log Excesses

10 = 8	1 = 1
20 = 15	2 = 2
30 = 23	3 = 2
40 = 31	4 = 3
50 = 38	5 = 4
60 = 46	6 = 5
70 = 54	7 = 5
80 = 61	8 = 6
90 = 69	9 = 7

50 = 38

3330016	·5224462
091	560
166	658
241	756
316	854

3330391	·5224952
466	·5225049
541	147
616	245
691	343
766	441
841	539
916	637
991	735
3331066	832
141	930
216	·5226028
291	126
367	224
442	322
517	419
592	517
667	615
742	713
817	810
892	908
967	·5227006
3332042	104
117	202
192	300
267	397
342	495
417	593
492	691
567	788
642	886
717	984
792	·5228082
867	179
942	277
3333017	375
092	473
167	570
242	668
317	766
392	864
467	961
542	·5229059
617	157
692	255
767	352
842	450
917	547
992	645
3334067	743
142	841
217	938
292	·5230036
367	133
442	231
517	329
592	427
667	524
742	622
817	719
892	817
967	914
3335042	·5231012
117	110
192	208

3335267	·5231305
342	403
417	500
491	598
566	695
641	793
716	890
791	988
866	·5232086
941	184
3336016	281
091	379
166	476
241	574
316	671
391	769
466	866
541	964
616	·5233061
690	159
765	256
840	354
915	451
990	549
3337065	646
140	744
215	841
290	939
365	·5234036
440	134
515	231
589	329
664	426
739	524
814	621
889	719
964	816
3338039	914
114	·5235011
189	109
264	206
338	304
413	401
488	498
563	595
638	693
713	790
788	888
863	985
937	·5236083
3339012	180
087	277
162	374
237	472
312	569
387	667
462	764
536	862
611	959
686	·5237056
761	153
836	251
911	348
985	446

Excesses

10 = 13	1 = 1
20 = 26	2 = 3
30 = 39	3 = 4
40 = 52	4 = 5
50 = 65	5 = 7
60 = 78	6 = 8
70 = 91	7 = 9
	8 = 10
	9 = 12

38 = 50

Anti Log Excesses

10 = 8	1 = 1
20 = 15	2 = 2
30 = 23	3 = 2
40 = 31	4 = 3
50 = 38	5 = 4
60 = 46	6 = 5
70 = 54	7 = 5
80 = 61	8 = 6
90 = 69	9 = 7

50 = 38

3340060	·5237543
135	640
210	737
285	835
360	932
434	·5238029
509	126
584	224
659	321
734	419
809	516
883	613
958	710
3341033	808
108	905
183	·5239002
258	099
332	197
407	294
482	391
557	488
632	586
707	683
781	780
856	877
931	974
3342006	·5240071
080	169
155	266
230	363
305	460
380	558
455	655
529	752
604	849
679	946
754	·5241043

3342828	·5241141
903	238
978	335
3343053	432
128	529
203	626
277	724
352	821
427	918
502	·5242015
576	112
651	209
726	306
801	403
875	500
950	597
3344025	695
100	792
175	889
249	986
324	·5243083
399	180
474	277
548	374
623	471
698	568
773	665
847	762
922	859
997	956
3345072	·5244054
146	151
221	248
296	345
371	442
445	539
520	636
595	733
670	830
744	927
819	·5245024
894	121
969	218
3346043	315
118	412
192	508
267	605
342	702
417	799
491	896
566	993
641	·5246090
716	187
790	284
865	381
939	478
3347014	575
089	672
164	769
238	866
313	963
388	·5247059
463	156
537	253
612	350

3347686	·5247447
761	544
836	641
911	738
985	835
3348060	932
134	·5248028
209	125
284	222
359	319
433	416
508	513
582	609
657	706
732	803
807	900
881	996
956	·5249093
3349030	190
105	287
180	384
255	481
329	577
404	674
478	771
553	868
627	964
702	·5250061
777	158
852	255
926	351

Excesses

10 = 13	1 = 1
20 = 26	2 = 3
30 = 39	3 = 4
40 = 52	4 = 5
50 = 65	5 = 7
60 = 78	6 = 8
70 = 91	7 = 9
	8 = 10
	9 = 12

37 = 48

Anti Log Excesses

10 = 8	1 = 1
20 = 15	2 = 2
30 = 23	3 = 2
40 = 31	4 = 3
50 = 39	5 = 4
60 = 46	6 = 5
70 = 54	7 = 5
80 = 62	8 = 6
90 = 69	9 = 7

48 = 37

3350001	·5250448
075	545
150	642
224	738
299	835

3350374	·5250932
449	·5251029
523	125
598	222
672	319
747	416
821	512
896	609
971	706
3351046	803
120	899
195	996
269	·5252092
343	189
418	286
493	383
567	479
642	576
716	672
791	769
866	865
941	962
3352015	·5253059
090	156
164	252
239	349
313	445
388	542
462	638
537	735
611	832
686	929
761	·5254025
836	122
910	218
985	315
3353059	411
134	508
208	605
283	702
357	798
432	895
506	991
581	·5255088
655	184
730	281
804	377
879	474
953	570
3354028	667
102	763
177	860
252	956
327	·5256053
401	149
476	246
550	342
625	439
699	535
774	632
848	728
923	825
997	921
3355072	·5257018
146	114

3355221	·5257211
295	307
370	403
444	499
519	596
593	692
668	789
742	885
817	982
891	·5258078
966	175
3356040	271
115	367
189	463
264	560
338	656
412	753
487	849
561	946
636	·5259042
710	139
785	235
859	331
934	427
3357008	524
083	620
157	717
232	813
306	909
380	·5260005
454	102
529	198
603	294
678	390
752	487
827	583
901	680
976	776
3358050	872
125	968
199	·5261065
274	161
348	257
423	353
497	450
572	546
646	642
720	738
794	835
869	931
943	·5262027
3359018	123
092	220
167	316
241	412
316	508
390	605
465	701
539	797
614	893
688	989
762	·5263085
836	182
911	278
985	374

Excesses

10 = 13	1 = 1
20 = 26	2 = 3
30 = 39	3 = 4
40 = 52	4 = 5
50 = 65	5 = 7
60 = 78	6 = 8
70 = 91	7 = 9
	8 = 10
	9 = 12

37 = 48

Anti Log Excesses

10 = 8	1 = 1
20 = 15	2 = 2
30 = 23	3 = 2
40 = 31	4 = 3
50 = 39	5 = 4
60 = 46	6 = 5
70 = 54	7 = 5
80 = 62	8 = 6
90 = 70	9 = 7

48 = 37

3360060	·5263470
134	566
209	662
283	759
358	855
432	951
506	·5264047
580	143
655	239
729	335
804	431
878	528
953	624
3361027	720
102	816
176	912
250	·5265008
324	104
399	200
473	296
548	392
622	489
697	585
771	681
845	777
919	873
994	969
3362068	·5266065
143	161
217	257
291	353
365	449
440	545
514	641
589	737
663	833
737	929

3362811	·5267025
886	121
960	217
3363035	313
109	409
184	505
258	601
332	697
406	793
481	889
555	985
629	·5268081
703	177
778	273
852	369
927	465
3364001	561
075	657
149	753
224	849
298	945
373	·5269041
447	137
521	232
595	328
670	424
744	520
818	616
892	712
967	808
3365041	904
116	·5270000
190	096
264	191
338	287
413	383
487	479
561	575
635	671
710	766
784	862
858	958
932	·5271054
3366007	150
081	246
156	342
230	438
304	533
378	629
453	725
527	821
601	916
675	·5272012
750	108
824	204
898	300
972	396
3367047	491
121	587
195	683
269	779
344	874
418	970
492	·5273066
566	162

3367641	·5273257
715	353
789	449
863	545
938	640
3368012	736
086	831
160	927
234	·5274023
308	119
383	214
457	310
531	406
605	502
680	597
754	693
828	788
902	884
977	980
3369051	·5275076
125	171
199	267
273	362
347	458
422	553
496	649
570	745
644	841
719	936
793	·5276032
867	127
941	223

Excesses

10 = 13	1 = 1
20 = 26	2 = 3
30 = 39	3 = 4
40 = 52	4 = 5
50 = 65	5 = 7
60 = 77	6 = 8
70 = 90	7 = 9
	8 = 10
	9 = 12

37 = 48

Anti Log Excesses

10 = 8	1 = 1
20 = 15	2 = 2
30 = 23	3 = 2
40 = 31	4 = 3
50 = 39	5 = 4
60 = 46	6 = 5
70 = 54	7 = 5
80 = 62	8 = 6
90 = 70	9 = 7

48 = 37

3370015	·5276318
089	414
164	510
238	606

3370312	·5276701
386	797
460	892
534	988
609	·5277083
683	179
757	274
831	370
905	465
979	561
3371054	656
128	752
202	847
276	943
350	·5278038
424	134
499	230
573	325
647	420
721	516
795	611
869	707
944	802
3372018	898
092	993
166	·5279089
240	184
314	280
388	375
462	471
537	566
611	662
685	757
759	853
833	948
907	·5280044
981	139
3373055	235
130	330
204	426
278	521
352	616
426	711
500	807
574	902
648	998
722	·5281093
796	189
871	284
945	380
3374019	475
093	570
167	665
241	761
315	856
389	952
463	·5282047
537	142
612	237
686	333
760	428
834	524
908	619
982	714
3375056	809

3375130	·5282905
204	·5283000
278	096
352	191
426	286
500	381
574	477
649	572
723	667
797	762
871	858
945	953
3376019	·5284048
093	143
167	239
241	334
315	429
389	524
463	620
537	715
611	810
685	905
759	·5285001
833	096
907	191
981	286
3377055	382
129	477
203	572
278	667
352	762
426	857
500	953
574	·5286048
648	143
722	238
796	333
870	428
944	524
3378018	619
092	714
166	809
240	904
314	999
388	·5287094
462	189
536	285
610	380
684	475
758	570
832	665
906	760
980	855
3379054	950
128	·5288046
202	141
276	236
350	331
424	426
498	521
571	616
645	711
719	806
793	901
867	996

3379941	·5289091

Excesses

10 = 13	1 = 1
20 = 26	2 = 3
30 = 39	3 = 4
40 = 51	4 = 5
50 = 64	5 = 6
60 = 77	6 = 8
70 = 90	7 = 9
	8 = 10
	9 = 12

37 = 48

Anti Log Excesses

10 = 8	1 = 1
20 = 16	2 = 2
30 = 23	3 = 2
40 = 31	4 = 3
50 = 39	5 = 4
60 = 47	6 = 5
70 = 54	7 = 5
80 = 62	8 = 6
90 = 70	9 = 7

48 = 37

3380015	·5289186
089	281
163	376
237	471
311	566
385	661
459	756
533	851
607	946
681	·5290041
755	136
829	231
903	326
977	421
3381051	516
125	611
198	706
272	801
346	896
420	991
494	·5291086
568	181
642	276
716	371
790	466
864	561
938	656
3382012	751
086	845
160	940
233	·5292035
307	130
381	225
455	320
529	415

3382603	·5292510
677	605
751	700
825	794
899	889
972	984
3383046	·5293079
120	174
194	269
268	364
342	459
416	554
490	649
564	743
638	838
711	933
785	·5294028
859	123
933	218
3384007	312
081	407
155	502
229	597
302	691
376	786
450	881
524	976
598	·5295070
672	165
746	260
820	355
893	450
967	545
3385041	639
115	734
189	829
263	924
336	·5296018
410	113
484	208
558	303
632	397
706	492
779	586
853	681
927	776
3386001	871
075	965
149	·5297060
222	154
296	249
370	344
444	439
518	533
592	628
665	722
739	817
813	912
887	·5298007
961	101
3387035	196
108	290
182	385
256	479
330	574

3387403	·5298669
477	764
551	859
625	953
699	·5299048
773	142
846	237
920	331
994	426
3388068	520
141	615
215	710
289	805
363	899
436	994
510	·5300088
584	183
658	277
732	372
806	466
879	561
953	655
3389027	750
101	844
174	939
248	·5301033
322	128
396	222
469	317
543	411
617	506
691	600
764	695
838	789
912	884
986	978

Excesses

10 = 13	1 = 1
20 = 26	2 = 3
30 = 38	3 = 4
40 = 51	4 = 5
50 = 64	5 = 6
60 = 77	6 = 8
70 = 90	7 = 9
	8 = 10
	9 = 12

37 = 47

Anti Log Excesses

10 = 8	1 = 1
20 = 16	2 = 2
30 = 23	3 = 2
40 = 31	4 = 3
50 = 39	5 = 4
60 = 47	6 = 5
70 = 55	7 = 6
80 = 62	8 = 6
90 = 70	9 = 7

47 = 37

3390059	·5302073
133	167
207	262
281	356
354	451
428	545
502	640
576	734
649	829
723	923
797	·5303018
871	112
944	207
3391018	301
091	395
165	489
239	584
313	678
386	773
460	867
534	962
608	·5304056
681	151
755	245
829	339
903	433
976	528
3392050	622
123	717
197	811
271	905
345	999
418	·5305094
492	188
566	283
640	377
713	472
787	566
860	660
934	754
3393008	849
082	943
155	·5306037
229	131
302	226
376	320
450	414
524	508
597	603
671	697
744	791
818	885
892	980
966	·5307074
3394039	168
113	262
186	357
260	451
334	546
408	640
481	734
555	828
628	922
702	·5308016
776	111

3394850	·5308205
923	299
997	393
3395070	488
144	582
217	676
291	770
365	864
439	958
512	·5309053
586	147
659	241
733	335
806	429
880	523
954	617
3396028	711
101	806
175	900
248	994
322	·5310088
395	182
469	276
542	371
616	465
690	559
764	653
837	747
911	841
984	935
3397058	·5311029
131	123
205	217
278	311
352	405
426	500
500	594
573	688
647	782
720	876
794	970
867	·5312064
941	158
3398014	252
088	346
161	440
235	534
308	628
382	722
456	816
530	910
603	·5313004
677	098
750	192
824	286
897	380
971	474
3399044	568
118	662
191	756
265	850
338	944
412	·5314038
485	131
559	225

3399632	·5314319
706	413
779	507
853	601
926	695

Excesses

10 = 13	1 = 1
20 = 26	2 = 3
30 = 38	3 = 4
40 = 51	4 = 5
50 = 64	5 = 6
60 = 77	6 = 8
70 = 90	7 = 9
	8 = 10
	9 = 12

37 = 47

Anti Log Excesses

10 = 8	1 = 1
20 = 16	2 = 2
30 = 23	3 = 2
40 = 31	4 = 3
50 = 39	5 = 4
60 = 47	6 = 5
70 = 55	7 = 6
80 = 63	8 = 6
90 = 70	9 = 7

47 = 37

3400000	·5314789
074	883
148	977
221	·5315071
295	165
368	259
442	353
515	447
589	540
662	634
736	728
809	822
883	916
956	·5316010
3401030	104
103	198
177	291
250	385
324	479
397	573
471	667
544	761
618	854
691	948
765	·5317042
838	136
912	230
985	324
3402059	417
132	511
206	605

Number	Log
3402279	·5317699
353	793
426	887
500	980
573	·5318074
647	168
720	262
793	355
866	449
940	543
3403013	637
087	730
160	824
234	918
307	·5319012
381	105
454	199
528	293
601	387
675	480
748	574
822	668
895	762
969	855
3404042	949
116	·5320042
189	136
263	230
336	324
409	417
482	511
556	605
629	699
703	792
776	886
850	979
923	·5321073
997	166
3405070	260
144	354
217	448
290	541
363	635
437	728
510	822
584	915
657	·5322009
731	103
804	197
878	290
951	384
3406025	477
098	571
171	664
244	758
318	851
391	945
465	·5323039
538	133
612	226
685	320
758	413
831	507
905	600
978	694

Number	Log
3407052	·5323787
125	881
199	974
272	·5324068
346	161
419	255
492	348
565	442
639	535
712	629
786	722
859	816
932	909
3408005	·5325003
079	096
152	190
226	283
299	377
373	470
446	564
519	657
592	751
666	844
739	938
813	·5326031
886	124
959	217
3409032	311
106	404
179	498
253	591
326	685
399	778
472	872
546	965
619	·5327058
693	151
766	245
839	338
912	432
986	525

Excesses

10 = 13	1 = 1
20 = 26	2 = 3
30 = 38	3 = 4
40 = 51	4 = 5
50 = 64	5 = 6
60 = 77	6 = 8
70 = 89	7 = 9
	8 = 10
	9 = 11

37 = 47

Anti Log Excesses

10 = 8	1 = 1
20 = 16	2 = 2
30 = 24	3 = 2
40 = 31	4 = 3
50 = 39	5 = 4
60 = 47	6 = 5
70 = 55	7 = 6
80 = 63	8 = 6
90 = 71	9 = 7

47 = 37

Number	Log
3410059	·5327619
132	712
205	805
279	898
352	992
426	·5328085
499	179
572	272
645	365
719	458
792	552
865	645
938	738
3411012	831
085	925
159	·5329018
232	112
305	205
378	298
452	391
525	485
598	578
671	672
745	765
818	858
891	951
964	·5330045
3412038	138
111	231
185	324
258	418
331	511
404	604
478	697
551	791
624	884
697	977
771	·5331070
844	163
917	256
990	350
3413064	443
137	536
210	629
283	723
357	816
430	909
503	·5332002
576	096
650	189
723	282

Number	Log
3413796	·5332375
869	468
943	561
3414016	655
089	748
162	841
235	934
308	·5333027
382	120
455	214
528	307
601	400
675	493
748	586
821	679
894	772
968	865
3415041	958
114	·5334051
187	145
260	238
333	331
407	424
480	517
553	610
626	703
700	796
773	889
846	982
919	·5335075
992	168
3416065	261
139	354
212	447
285	540
358	633
431	726
504	819
578	912
651	·5336005
724	098
797	191
870	284
943	377
3417017	470
090	563
163	656
236	749
309	842
382	935
456	·5337028
529	121
602	214
675	307
748	400
821	493
895	586
968	679
3418041	772
114	865
187	958
260	·5338051
333	144
406	237
480	329

Number	Log
3418553	·5338422
626	515
699	608
772	701
845	794
918	887
991	980
3419065	·5339073
138	166
211	258
284	351
357	444
430	537
503	630
576	723
650	816
723	909
796	·5340001
869	094
942	187

Excesses

10 = 13	1 = 1
20 = 25	2 = 3
30 = 38	3 = 4
40 = 51	4 = 5
50 = 64	5 = 6
60 = 76	6 = 8
70 = 89	7 = 9
	8 = 10
	9 = 11

37 = 47

Anti Log Excesses

10 = 8	1 = 1
20 = 16	2 = 2
30 = 24	3 = 2
40 = 31	4 = 3
50 = 39	5 = 4
60 = 47	6 = 5
70 = 55	7 = 6
80 = 63	8 = 6
90 = 71	9 = 7

47 = 37

Number	Log
3420015	·5340280
088	373
161	466
234	558
307	651
381	744
454	837
527	929
600	·5341022
673	115
746	208
819	300
892	393
965	486
3421038	579
111	672

3421184	·5341765
258	857
331	950
404	·5342043
477	136
550	228
623	321
696	414
769	507
842	599
915	692
988	784
3422061	877
134	970
207	·5343063
280	155
353	248
427	341
500	434
573	526
646	619
719	711
792	804
865	897
938	990
3423011	·5344082
084	175
157	267
230	360
303	453
376	546
449	638
522	731
595	823
668	916
741	·5345009
814	102
887	194
960	287
3424033	379
106	472
179	564
252	657
325	749
398	842
471	934
544	·5346027
617	120
690	213
763	305
836	398
909	490
982	583
3425055	675
128	768
201	860
274	953
347	·5347045
420	138
493	230
566	323
639	415
712	508
785	600
858	693

3425931	·5347785
3426004	878
077	970
150	·5348063
223	155
296	248
369	340
442	433
515	525
588	618
661	710
734	803
807	895
879	988
952	·5349080
3427025	173
098	265
171	358
244	450
317	543
390	635
463	727
536	819
609	912
682	·5350004
755	097
828	189
901	282
974	374
3428047	467
120	559
192	651
265	743
338	836
411	928
484	·5351021
557	113
630	206
703	298
776	390
849	482
922	575
995	667
3429067	760
140	852
213	945
286	·5352037
359	129
432	221
505	314
578	406
651	498
724	590
796	683
869	775
942	868

Excesses

10 = 13	1 = 1
20 = 25	2 = 3
30 = 38	3 = 4
40 = 51	4 = 5
50 = 63	5 = 6
60 = 76	6 = 8
70 = 89	7 = 9
	8 = 10
	9 = 11

37 = 47

Anti Log Excesses

10 = 8	1 = 1
20 = 16	2 = 2
30 = 24	3 = 2
40 = 32	4 = 3
50 = 39	5 = 4
60 = 47	6 = 5
70 = 55	7 = 6
80 = 63	8 = 6
90 = 71	9 = 7

47 = 37

3430015	·5352960
088	·5353052
161	144
234	237
307	329
379	421
452	513
525	606
598	698
671	790
744	882
817	975
890	·5354067
962	159
3431035	251
108	344
181	436
254	528
327	620
399	713
472	805
545	897
618	989
691	·5355081
764	173
837	266
910	358
982	450
3432055	542
128	635
201	727
274	819
347	911
419	·5356003
492	095
565	187
638	279

3432711	·5356372
784	464
856	556
929	648
3433002	740
075	832
148	924
221	·5357016
293	109
366	201
439	293
512	385
584	477
657	569
730	661
803	753
876	845
949	937
3434021	·5358029
094	121
167	214
240	306
312	398
385	490
458	582
531	674
604	766
677	858
749	950
822	·5359042
895	134
968	226
3435040	318
113	410
186	502
259	594
331	686
404	778
477	870
550	962
622	·5360054
695	146
768	238
841	330
914	422
987	514
3436059	606
132	698
205	790
278	882
350	974
423	·5361066
496	158
569	249
641	341
714	433
787	525
860	617
932	709
3437005	801
077	893
150	985
223	·5362077
296	169
368	261

3437441	·5362352
514	444
587	536
659	628
732	720
805	812
878	904
950	996
3438023	·5363087
096	179
169	271
241	363
314	455
386	547
459	638
532	730
605	822
677	914
750	·5364006
823	098
896	189
968	281
3439041	373
113	465
186	557
259	649
332	740
404	832
477	924
550	·5365016
623	107
695	199
768	291
840	383
913	474
986	566

Excesses

10 = 13	1 = 1
20 = 25	2 = 3
30 = 38	3 = 4
40 = 51	4 = 5
50 = 63	5 = 6
60 = 76	6 = 8
70 = 88	7 = 9
	8 = 10
	9 = 11

36 = 46

Anti Log Excesses

10 = 8	1 = 1
20 = 16	2 = 2
30 = 24	3 = 2
40 = 32	4 = 3
50 = 40	5 = 4
60 = 47	6 = 5
70 = 55	7 = 6
80 = 63	8 = 6
90 = 71	9 = 7

46 = 36

Number	Log
3440059	·5365658
131	750
204	841
276	933
349	·5366025
422	117
495	208
567	300
640	392
712	484
785	575
858	667
931	759
3441003	851
076	942
148	·5367034
221	125
294	217
367	309
439	401
512	492
584	584
657	675
729	767
802	859
875	951
948	·5368042
3442020	134
093	225
165	317
238	408
310	500
383	592
456	684
529	775
601	867
674	958
746	·5369050
819	141
891	233
964	325
3443037	417
110	508
182	600
255	691
327	783
400	874
472	966
545	·5370057
617	149
690	240
763	332
836	423
908	515
981	606
3444053	698
126	789
198	881
271	972
343	·5371064
416	156
488	248
561	339
634	431
707	522

Number	Log
3444779	·5371614
852	705
924	797
997	888
3445069	980
142	·5372071
214	163
287	254
359	346
432	437
504	529
577	620
649	711
722	802
795	894
868	985
940	·5373077
3446013	168
085	260
158	351
230	443
303	534
375	626
448	717
520	809
593	900
665	991
738	·5374082
810	174
883	265
955	357
3447028	448
100	540
173	631
245	722
318	813
390	905
463	996
535	·5375088
608	179
680	271
753	362
825	453
898	544
970	636
3448043	727
115	819
188	910
260	·5376001
333	092
405	184
478	275
550	366
623	457
695	549
768	640
840	731
913	822
985	914
3449058	·5377005
130	097
203	188
275	279
348	370
420	462

Number	Log
3449493	·5377553
565	644
638	735
710	827
783	918
855	·5378009
928	100

Excesses

10 = 13	1 = 1
20 = 25	2 = 3
30 = 38	3 = 4
40 = 50	4 = 5
50 = 63	5 = 6
60 = 76	6 = 8
70 = 88	7 = 9
	8 = 10
	9 = 11

36 = 46

Anti Log Excesses

10 = 8	1 = 1
20 = 16	2 = 2
30 = 24	3 = 2
40 = 32	4 = 3
50 = 40	5 = 4
60 = 48	6 = 5
70 = 56	7 = 6
80 = 63	8 = 6
90 = 71	9 = 7

46 = 36

Number	Log
3450000	·5378191
073	282
145	374
218	465
290	556
363	647
435	739
508	830
580	921
653	·5379012
725	103
798	194
870	286
942	377
3451014	468
087	559
159	650
232	741
304	833
377	924
449	·5380015
522	106
594	197
667	288
739	379
812	470
884	562
956	653
3452028	744
101	835

Number	Log
3452173	·5380926
246	·5381017
318	108
391	199
463	291
536	382
608	473
681	564
753	655
825	746
897	837
970	928
3453042	·5382019
115	110
187	201
260	292
332	383
405	474
477	565
549	656
621	747
694	838
766	930
839	·5383021
911	112
984	203
3454056	294
128	385
200	476
273	567
345	658
418	748
490	839
563	930
635	·5384021
707	112
779	203
852	294
924	385
997	476
3455069	567
142	658
214	749
286	840
358	931
431	·5385022
503	113
576	204
648	295
720	386
792	477
865	567
937	658
3456010	749
082	840
154	931
226	·5386022
299	113
371	204
444	294
516	385
588	476
660	567
733	658
805	749

Number	Log
3456878	·5386840
950	931
3457022	·5387021
094	112
167	203
239	294
312	385
384	476
456	566
528	657
601	748
673	839
745	930
817	·5388021
890	111
962	202
3458035	293
107	384
179	474
251	565
324	656
396	747
468	838
540	929
613	·5389019
685	110
758	201
830	292
902	382
974	473
3459047	564
119	655
191	745
263	836
336	927
408	·5390018
480	108
552	199
625	289
697	380
769	471
841	562
914	652
986	743

Excesses

10 = 13	1 = 1
20 = 25	2 = 3
30 = 38	3 = 4
40 = 50	4 = 5
50 = 63	5 = 6
60 = 75	6 = 8
70 = 88	7 = 9
	8 = 10
	9 = 11

36 = 45

Anti Log Excesses

10 = 8	1 = 1
20 = 16	2 = 2
30 = 24	3 = 2
40 = 32	4 = 3
50 = 40	5 = 4
60 = 48	6 = 5
70 = 56	7 = 6
80 = 64	8 = 6
90 = 72	9 = 7

45 = 36

3460058	·5390834
130	925
203	·5391015
275	106
347	196
419	287
492	378
564	469
636	559
708	650
781	740
853	831
925	921
997	·5392012
3461070	103
142	194
214	284
286	375
359	465
431	556
503	646
575	737
648	828
720	919
792	·5393009
864	100
936	190
3462008	281
081	371
153	462
225	552
297	643
370	734
442	825
514	915
586	·5394006
658	096
730	187
803	277
875	368
947	458
3463019	549
092	639
164	730
236	820
308	911
380	·5395001
452	092
525	182
597	273
669	363

3463741	·5395454
813	544
885	635
958	725
3464030	816
102	906
174	997
246	·5396087
318	178
391	268
463	359
535	449
607	540
679	630
751	721
824	811
896	902
968	992
3465040	·5397083
112	173
184	263
257	353
329	444
401	534
473	625
545	715
617	806
689	896
761	987
834	·5398077
906	167
978	257
3466050	348
122	438
194	529
267	619
339	710
411	800
483	890
555	980
627	·5399071
699	161
771	252
843	342
915	432
988	522
3467060	613
132	703
204	794
276	884
348	974
420	·5400064
492	155
564	245
636	336
709	426
781	516
853	606
925	697
997	787
3468069	877
141	967
213	·5401058
285	148
357	238

3468430	·5401328
502	419
574	509
646	599
718	689
790	780
862	870
934	960
3469006	·5402050
078	141
150	231
222	321
294	411
366	501
438	591
510	682
583	772
655	862
727	952
799	·5403043
871	133
943	223

Excesses

10 = 13	1 = 1
20 = 25	2 = 3
30 = 38	3 = 4
40 = 50	4 = 5
50 = 63	5 = 6
60 = 75	6 = 8
70 = 88	7 = 9
	8 = 10
	9 = 11

36 = 45

Anti Log Excesses

10 = 8	1 = 1
20 = 16	2 = 2
30 = 24	3 = 2
40 = 32	4 = 3
50 = 40	5 = 4
60 = 48	6 = 5
70 = 56	7 = 6
80 = 64	8 = 6
90 = 72	9 = 7

45 = 36

3470015	·5403313
087	403
159	493
231	583
303	673
375	764
447	854
519	944
591	·5404034
663	124
735	214
807	305
879	395
951	485

3471023	·5404575
095	665
167	755
239	845
311	935
383	·5405025
456	115
528	206
600	296
672	386
744	476
816	566
888	656
960	746
3472032	836
104	926
176	·5406016
248	106
320	196
392	286
464	376
536	466
608	556
680	646
752	736
824	826
896	916
968	·5407006
3473040	096
112	187
183	277
255	367
327	457
399	547
471	637
543	727
615	817
687	907
759	997
831	·5408086
903	176
975	266
3474047	356
119	446
191	536
263	626
335	716
407	806
479	896
551	986
623	·5409076
695	166
767	256
839	346
911	436
982	526
3475054	616
126	706
198	795
270	885
342	975
414	·5410065
486	155
558	245
630	335

3475702	·5410425
774	515
846	605
918	694
989	784
3476061	874
133	964
205	·5411054
277	144
349	233
421	323
493	413
565	503
637	593
709	683
781	773
852	863
924	952
996	·5412042
3477068	132
140	222
212	311
284	401
356	491
428	581
500	670
571	760
643	850
715	940
787	·5413029
859	119
931	209
3478003	299
075	389
146	479
218	568
290	658
362	747
434	837
506	927
578	·5414017
650	106
721	196
793	286
865	376
937	466
3479009	556
081	645
153	735
225	824
296	914
368	·5415004
440	094
512	183
584	273
656	362
727	452
799	541
871	631
943	721

Excesses

10 = 12	1 = 1
20 = 25	2 = 3
30 = 37	3 = 4
40 = 50	4 = 5
50 = 62	5 = 6
60 = 75	6 = 8
70 = 87	7 = 9
	8 = 10
	9 = 11

36 = 45

Anti Log Excesses

10 = 8	1 = 1
20 = 16	2 = 2
30 = 24	3 = 2
40 = 32	4 = 3
50 = 40	5 = 4
60 = 48	6 = 5
70 = 56	7 = 6
80 = 64	8 = 6
90 = 72	9 = 7

45 = 36

3480015	·5415811
087	900
158	990
230	·5416079
302	169
374	259
446	349
518	438
589	528
661	617
733	707
805	796
877	886
949	976
3481020	·5417066
092	155
164	245
236	334
308	424
380	513
451	603
523	692
595	782
667	871
738	961
810	·5418051
882	141
954	230
3482026	320
098	409
169	499
241	588
313	678
385	767
456	857
528	946
600	·5419036
3482672	·5419125
743	215
815	304
887	394
959	483
3483031	573
103	662
174	752
246	841
318	931
390	·5420020
461	110
533	199
605	289
677	378
748	468
820	557
892	647
964	736
3484035	826
107	915
179	·5421004
251	093
322	183
394	272
466	362
538	451
609	541
681	630
753	720
825	809
896	899
968	988
3485040	·5422077
112	166
183	256
255	345
327	435
399	524
470	614
542	703
614	792
686	881
757	971
829	·5423060
900	150
972	239
3486044	328
116	417
187	507
259	596
331	686
403	775
474	864
546	953
617	·5424043
689	132
761	222
833	311
904	400
976	489
3487048	579
120	668
191	757
263	846
3487334	·5424936
406	·5425025
478	114
550	203
621	293
693	382
765	471
837	560
908	650
980	739
3488051	828
123	917
195	·5426007
267	096
338	185
410	274
481	364
553	453
625	542
697	631
768	721
840	810
911	899
983	988
3489054	·5427077
126	166
198	256
270	345
341	434
413	523
484	612
556	701
628	791
700	880
771	969
843	·5428058
914	147
986	236

Excesses

10 = 12	1 = 1
20 = 25	2 = 3
30 = 37	3 = 4
40 = 50	4 = 5
50 = 62	5 = 6
60 = 75	6 = 8
70 = 87	7 = 9
	8 = 10
	9 = 11

36 = 45

Anti Log Excesses

10 = 8	1 = 1
20 = 16	2 = 2
30 = 24	3 = 2
40 = 32	4 = 3
50 = 40	5 = 4
60 = 48	6 = 5
70 = 56	7 = 6
80 = 64	8 = 6
	9 = 7

45 = 36

3490057	·5428326
129	415
201	504
273	593
344	682
416	771
487	860
559	949
630	·5429039
702	128
774	217
846	306
917	395
989	484
3491060	573
132	662
203	751
275	840
347	929
419	·5430018
490	108
562	197
633	286
705	375
776	464
848	553
919	642
991	731
3492063	820
135	909
206	998
278	·5431087
349	176
421	265
492	354
564	443
635	532
707	621
778	710
850	799
921	888
993	977
3493065	·5432066
137	155
208	244
280	333
351	422
423	511
494	600
566	689
637	778
3493709	·5432867
780	956
852	·5433045
923	134
995	222
3494066	311
138	400
210	489
282	578
353	667
425	756
496	845
568	934
639	·5434023
711	112
782	201
854	289
925	378
997	467
3495068	556
140	645
211	734
283	823
354	912
426	·5435000
497	089
569	178
640	267
712	356
783	445
855	534
926	623
998	711
3496069	800
141	889
212	978
284	·5436066
355	155
427	244
498	333
570	422
641	511
713	599
784	688
856	777
927	866
999	955
3497070	·5437044
142	132
213	221
285	310
356	399
428	487
499	576
571	665
642	754
714	842
785	931
857	·5438020
928	109
3498000	197
071	286
143	375
214	464
286	552

3498357	·5438641
429	729
500	818
572	907
643	996
715	·5439084
786	173
857	262
928	351
3499000	439
071	528
143	616
214	705
286	794
357	883
429	971
500	·5440060
572	148
643	237
715	326
786	415
858	503
929	592

Excesses

10 = 12	1 = 1
20 = 25	2 = 3
30 = 37	3 = 4
40 = 50	4 = 5
50 = 62	5 = 6
60 = 75	6 = 8
70 = 87	7 = 9
	8 = 10
	9 = 11

36 = 45

Anti Log Excesses

10 = 8	1 = 1
20 = 16	2 = 2
30 = 24	3 = 2
40 = 32	4 = 3
50 = 40	5 = 4
60 = 48	6 = 5
70 = 56	7 = 6
80 = 64	8 = 6
	9 = 7

45 = 36

3500000	·5440680
071	769
143	857
214	946
286	·5441035
357	124
429	212
500	301
572	389
643	478
715	566
786	655
858	744

3500929	·5441833
3501000	921
071	·5442010
143	098
214	187
286	275
357	364
429	452
500	541
572	629
643	718
714	806
785	895
857	983
928	·5443072
3502000	161
071	250
143	338
214	427
285	515
356	604
428	692
499	781
571	869
642	958
714	·5444046
785	135
856	223
927	312
999	400
3503070	489
142	577
213	666
285	754
356	842
427	930
498	·5445019
570	107
641	196
713	284
784	373
856	461
927	550
998	638
3504069	727
141	815
212	904
284	992
355	·5446080
426	168
497	257
569	345
640	434
712	522
783	611
854	699
925	788
997	876
3505068	964
140	·5447052
211	141
282	229
353	318
425	406
496	495

3505568	·5447583
639	671
710	759
781	848
853	936
924	·5448025
995	113
3506066	201
138	289
209	378
281	466
352	554
423	642
494	731
566	819
637	908
708	996
779	·5449084
851	172
922	261
994	349
3507065	438
136	526
207	614
279	702
350	791
421	879
492	967
564	·5450055
635	144
706	232
777	320
849	408
920	497
991	585
3508062	673
134	761
205	849
276	937
347	·5451026
419	114
490	202
561	290
632	379
704	467
775	555
846	643
917	731
989	819
3509060	908
131	996
202	·5452084
274	172
345	260
416	348
487	437
559	525
630	613
701	701
772	789
844	877
915	965
986	·5453053

Excesses

10 = 12	1 = 1
20 = 25	2 = 3
30 = 37	3 = 4
40 = 50	4 = 5
50 = 62	5 = 6
60 = 74	6 = 7
70 = 87	7 = 9
	8 = 10
	9 = 11

36 = 44

Anti Log Excesses

10 = 8	1 = 1
20 = 16	2 = 2
30 = 24	3 = 2
40 = 32	4 = 3
50 = 40	5 = 4
60 = 48	6 = 5
70 = 56	7 = 6
80 = 65	8 = 7
	9 = 7

44 = 36

3510057	·5453142
129	230
200	318
271	406
342	494
414	582
485	670
556	758
627	847
698	935
769	·5454023
841	111
912	199
983	287
3511054	375
126	463
197	551
268	639
339	727
410	815
481	903
553	991
624	·5455079
695	167
766	255
838	343
909	432
980	520
3512051	608
122	696
193	784
265	872
336	960
407	·5456048
478	136
549	224
620	312

3512692	·5456400
763	488
834	576
905	664
976	752
3513047	840
119	928
190	·5457016
261	104
332	192
403	279
474	367
546	455
617	543
688	631
759	719
830	807
901	895
973	983
3514044	·5458071
115	159
186	247
257	335
328	423
399	511
470	599
542	686
613	774
684	862
755	950
826	·5459038
897	126
968	214
3515039	302
111	390
182	478
253	565
324	653
395	741
466	829
537	917
608	·5460005
680	092
751	180
822	268
893	356
964	444
3516035	532
106	619
177	707
249	795
320	883
391	971
462	·5461059
533	146
604	234
675	322
746	410
817	497
888	585
959	673
3517030	761
102	849
173	937
244	·5462024

3517315	·5462112
386	200
457	288
528	375
599	463
670	551
741	639
812	726
883	814
954	901
3518025	989
097	·5463077
168	165
239	252
310	340
381	428
452	516
523	603
594	691
665	779
736	867
807	954
878	·5464042
949	129
3519020	217
091	305
162	393
233	480
304	568
375	655
446	743
518	830
589	918
660	·5465006
731	094
802	181
873	269
944	356

Excesses

10 = 12	1 = 1
20 = 25	2 = 3
30 = 37	3 = 4
40 = 49	4 = 5
50 = 62	5 = 6
60 = 74	6 = 7
70 = 86	7 = 9
	8 = 10
	9 = 11

36 = 44

Anti Log Excesses

10 = 8	1 = 1
20 = 16	2 = 2
30 = 24	3 = 2
40 = 32	4 = 3
50 = 40	5 = 4
60 = 49	6 = 5
70 = 57	7 = 6
80 = 65	8 = 7
	9 = 7

44 = 36

3520015	·5465444
086	531
157	619
228	707
299	795
370	882
441	970
512	·5466057
583	145
654	232
725	320
796	408
867	496
938	583
3521009	671
080	758
151	846
222	933
293	·5467021
364	108
435	196
506	283
577	371
648	458
719	546
790	633
861	721
932	808
3522003	896
074	984
145	·5468072
216	159
287	247
358	334
429	422
500	509
571	597
642	684
713	772
783	859
854	947
925	·5469034
996	122
3523067	209
138	297
209	384
280	472
351	559
422	646
493	733
564	821

3523635	·5469908
706	996
777	·5470083
848	171
919	258
990	346
3524061	433
132	521
203	608
273	696
344	783
415	870
486	957
557	·5471045
628	132
699	220
770	307
841	395
912	482
983	570
3525054	657
125	744
196	831
266	919
337	·5472006
408	094
479	181
550	268
621	355
692	443
763	530
834	618
905	705
975	792
3526046	879
117	967
188	·5473054
259	142
330	229
401	316
472	403
543	491
614	578
684	666
755	753
826	840
897	927
968	·5474015
3527039	102
110	189
181	276
251	364
322	451
393	538
464	625
535	713
606	800
677	887
748	974
818	·5475062
889	149
960	236
3528031	323
102	411
173	498

3528244	·5475585
315	672
385	760
456	847
527	934
598	·5476021
669	108
740	195
810	283
881	370
952	457
3529023	544
094	632
165	719
235	806
306	893
377	980
448	·5477067
519	154
590	241
660	329
731	416
802	503
873	590
944	677

Excesses

10 = 12	1 = 1
20 = 25	2 = 3
30 = 37	3 = 4
40 = 49	4 = 5
50 = 62	5 = 6
60 = 74	6 = 7
70 = 86	7 = 9
	8 = 10
	9 = 11

36 = 44

Anti Log Excesses

10 = 8	1 = 1
20 = 16	2 = 2
30 = 24	3 = 2
40 = 32	4 = 3
50 = 41	5 = 4
60 = 49	6 = 5
70 = 57	7 = 6
80 = 65	8 = 7
	9 = 7

44 = 36

3530015	·5477764
085	852
156	939
227	·5478026
298	113
369	200
440	287
510	374
581	461
652	548
723	635

3530793	·5478723
864	810
935	897
3531006	984
077	·5479071
148	158
218	245
289	332
360	419
431	506
501	593
572	680
643	767
714	854
785	941
856	·5480028
926	116
997	203
3532068	290
139	377
209	464
280	551
351	638
422	725
492	812
563	899
634	986
705	·5481073
775	160
846	247
917	334
988	421
3533058	508
129	595
200	682
271	769
341	856
412	942
483	·5482029
554	116
624	203
695	290
766	377
837	464
907	551
978	638
3534049	725
120	812
190	899
261	986
332	·5483073
403	160
473	247
544	333
615	420
686	507
756	594
827	681
898	768
969	855
3535039	942
110	·5484029
181	116
252	202
322	289

3535393	·5484376
463	463
534	550
605	637
676	723
746	810
817	897
888	984
959	·5485071
3536029	158
100	245
170	332
241	418
312	505
383	592
453	679
524	766
595	853
666	939
736	·5486026
807	113
877	200
948	286
3537019	373
090	460
160	547
231	633
301	720
372	807
443	894
514	981
584	·5487068
655	154
725	241
796	328
867	415
938	501
3538008	588
079	675
149	762
220	848
291	935
362	·5488021
432	108
503	195
573	282
644	369
715	456
786	542
856	629
927	715
997	802
3539068	889
138	976
209	·5489062
280	149
351	235
421	322
492	409
562	496
633	582
704	669
775	755
845	842
916	929

3539986	·5490016

Excesses

10 = 12	1 = 1
20 = 25	2 = 3
30 = 37	3 = 4
40 = 49	4 = 5
50 = 61	5 = 6
60 = 74	6 = 7
70 = 86	7 = 9
	8 = 10
	9 = 11

35 = 43

Anti Log Excesses

10 = 8	1 = 1
20 = 16	2 = 2
30 = 24	3 = 2
40 = 33	4 = 3
50 = 41	5 = 4
60 = 49	6 = 5
70 = 57	7 = 6
80 = 65	8 = 7
	9 = 7

43 = 35

3540057	·5490102
127	189
198	275
268	362
339	448
410	535
481	622
551	709
622	795
692	882
763	968
833	·5491055
904	141
975	228
3541046	314
116	401
187	488
257	575
328	661
398	748
469	834
539	921
610	·5492007
680	094
751	180
822	267
893	353
963	440
3542034	526
104	613
175	699
245	786
316	872
386	959
457	·5493046

3542527	·5493133
598	219
669	306
740	392
810	479
881	565
951	652
3543022	738
092	825
163	911
233	997
304	·5494083
374	170
445	256
515	343
586	429
656	516
727	602
797	689
868	775
939	862
3544010	948
080	·5495035
151	121
221	208
292	294
362	381
433	467
503	554
574	640
644	726
715	812
785	899
856	985
926	·5496072
997	158
3545067	245
138	331
208	417
279	503
349	590
420	676
490	763
561	849
631	936
702	·5497022
772	108
843	194
913	281
984	367
3546054	454
125	540
195	626
266	712
336	799
407	885
477	972
548	·5498058
618	144
689	230
759	317
830	403
900	490
971	576
3547041	662

3547112	·5498748
182	835
253	921
323	·5499007
394	093
464	180
535	266
605	352
676	438
746	525
817	611
887	697
958	783
3548028	870
099	956
169	·5500042
239	128
309	215
380	301
450	387
521	473
591	560
662	646
732	732
803	818
873	905
944	991
3549014	·5501077
085	163
155	249
226	335
296	422
367	508
437	594
508	680
578	766
648	852
718	939
789	·5502025
859	111
930	197

Excesses

10 = 12	1 = 1
20 = 24	2 = 2
30 = 37	3 = 4
40 = 49	4 = 5
50 = 61	5 = 6
60 = 73	6 = 7
70 = 86	7 = 9
	8 = 10
	9 = 11

35 = 43

Anti Log Excesses

10 = 8	1 = 1
20 = 16	2 = 2
30 = 24	3 = 2
40 = 33	4 = 3
50 = 41	5 = 4
60 = 49	6 = 5
70 = 57	7 = 6
80 = 65	8 = 7
	9 = 7

43 = 35

3550000	·5502283
071	369
141	456
212	542
282	628
353	714
423	800
493	886
563	973
634	·5503059
704	145
775	231
845	317
916	403
986	489
3551057	575
127	661
197	747
267	834
338	920
408	·5504006
479	092
549	178
620	264
690	350
761	436
831	522
901	608
971	694
3552042	780
112	866
183	952
253	·5505038
324	124
394	211
464	297
534	383
605	469
675	555
746	641
816	727
887	813
957	899
3553027	985
097	·5506071
168	157
238	243
309	329
379	415
450	501
520	587

3553590	·5506673
660	759
731	845
801	931
872	·5507017
942	103
3554012	188
082	274
153	360
223	446
294	532
364	618
434	704
504	790
575	876
645	962
716	·5508048
786	134
856	220
926	306
997	392
3555067	478
138	564
208	650
278	735
348	821
419	907
489	993
560	·5509079
630	165
700	251
770	337
841	422
911	508
981	594
3556051	680
122	766
192	852
263	938
333	·5510024
403	109
473	195
544	281
614	367
684	453
754	539
825	624
895	710
965	796
3557035	882
106	968
176	·5511054
247	139
317	225
387	311
457	397
528	482
598	568
668	654
738	740
809	825
879	911
949	997
3558019	·5512083
090	169

3558160	·5512255
230	340
300	426
371	512
441	598
511	683
581	769
652	854
722	940
792	·5513026
862	112
933	197
3559003	283
073	369
143	455
214	540
284	626
354	712
424	798
495	883
565	969
635	·5514054
705	140
776	226
846	312
916	397
986	483

Excesses

10 = 12	1 = 1
20 = 24	2 = 2
30 = 37	3 = 4
40 = 49	4 = 5
50 = 61	5 = 6
60 = 73	6 = 7
70 = 85	7 = 9
	8 = 10
	9 = 11

35 = 43

Anti Log Excesses

10 = 8	1 = 1
20 = 16	2 = 2
30 = 25	3 = 3
40 = 33	4 = 3
50 = 41	5 = 4
60 = 49	6 = 5
70 = 57	7 = 6
80 = 65	8 = 7
	9 = 7

43 = 35

3560057	·5514569
127	655
197	740
267	826
338	911
408	997
478	·5515082
548	168
618	254

3560688	·5515340
759	425
829	511
899	596
969	682
3561040	767
110	853
180	939
250	·5516025
320	110
390	196
461	281
531	367
601	452
671	538
742	624
812	710
882	795
952	881
3562022	966
092	·5517052
163	137
233	223
303	308
373	394
443	479
513	565
584	650
654	736
724	821
794	907
865	993
935	·5518079
3563005	164
075	250
145	335
215	421
286	506
356	592
426	677
496	763
566	848
636	934
706	·5519019
776	105
847	190
917	276
987	361
3564057	447
127	532
197	618
268	703
338	788
408	873
478	959
548	·5520045
618	131
688	216
758	301
829	386
899	472
969	557
3565039	643
109	728
179	814

3565249	·5520899
319	985
390	·5521070
460	156
530	241
600	326
670	411
740	497
810	582
880	668
951	753
3566021	839
091	924
161	·5522010
231	095
301	180
371	265
441	351
511	436
581	521
652	606
722	692
792	777
862	863
932	948
3567002	·5523033
072	118
142	204
212	289
282	374
352	459
422	545
493	630
563	716
633	801
703	886
773	971
843	·5524057
913	142
983	228
3568053	313
123	398
193	483
263	569
333	654
403	739
474	824
544	910
614	995
684	·5525080
754	165
824	251
894	336
964	421
3569034	506
104	592
174	677
244	762
314	847
384	932
454	·5526017
524	103
594	188
664	273
734	358

3569804	·5526444
874	529
944	614

Excesses

10 = 12	1 = 1
20 = 24	2 = 2
30 = 37	3 = 4
40 = 49	4 = 5
50 = 61	5 = 6
60 = 73	6 = 7
70 = 85	7 = 9
	8 = 10
	9 = 11

35 = 43

Anti Log Excesses

10 = 8	1 = 1
20 = 16	2 = 2
30 = 25	3 = 3
40 = 33	4 = 3
50 = 41	5 = 4
60 = 49	6 = 5
70 = 57	7 = 6
80 = 66	8 = 7
	9 = 7

43 = 35

3570015	·5526699
085	784
155	869
225	955
295	·5527040
365	125
435	210
505	296
575	381
645	466
715	551
785	636
855	721
925	807
995	892
3571065	977
135	·5528062
205	147
275	232
345	317
415	402
485	488
555	573
625	658
695	743
765	828
835	913
905	998
975	·5529083
3572045	168
115	253
185	339
255	424

Number	Log
3572325	·5529509
395	594
465	679
535	764
605	849
675	934
745	·5530019
815	104
884	189
954	274
3573024	359
094	444
164	529
234	614
304	699
374	784
444	870
514	955
584	·5531040
654	125
724	210
794	295
864	380
934	465
3574004	550
074	635
144	720
214	805
284	890
354	975
423	·5532060
493	145
563	230
633	314
703	399
773	484
843	569
913	654
983	739
3575053	824
123	909
193	994
263	·5533079
333	164
402	249
472	334
542	419
612	503
682	588
752	673
822	758
892	843
962	928
3576032	·5534013
102	098
172	183
241	268
311	353
381	438
451	522
521	607
591	692
661	777
731	862
801	947

Number	Log
3576871	·5535032
940	117
3577010	201
080	286
150	371
220	456
290	541
360	626
430	710
500	795
570	880
639	965
709	·5536050
779	135
849	219
919	304
989	389
3578059	474
129	559
198	644
268	728
338	813
408	898
478	983
548	·5537067
617	152
687	237
757	322
827	407
897	492
967	576
3579037	661
107	746
176	831
246	915
316	·5538000
386	085
456	170
526	254
595	339
665	424
735	509
805	593
875	678
945	763

Excesses

10 = 12	1 = 1
20 = 24	2 = 2
30 = 36	3 = 4
40 = 49	4 = 5
50 = 61	5 = 6
60 = 73	6 = 7
	7 = 9
	8 = 10
	9 = 11

35 = 42

Anti Log Excesses

10 = 8	1 = 1
20 = 16	2 = 2
30 = 25	3 = 3
40 = 33	4 = 3
50 = 41	5 = 4
60 = 49	6 = 5
70 = 58	7 = 6
80 = 66	8 = 7
	9 = 7

42 = 35

Number	Log
3580014	·5538848
084	932
154	·5539017
224	101
294	186
364	271
433	356
503	440
573	525
643	610
713	695
783	779
852	864
922	948
992	·5540033
3581062	118
131	203
201	287
271	372
341	456
411	541
481	625
550	710
620	795
690	880
760	964
829	·5541049
899	133
969	218
3582039	303
109	388
179	472
248	557
318	641
388	726
458	810
527	895
597	979
667	·5542064
737	148
806	233
876	318
946	403
3583016	487
086	572
156	656
225	741
295	825
365	910
435	994
504	·5543079

Number	Log
3583574	·5543163
644	248
714	332
783	417
853	501
923	586
993	670
3584062	755
132	840
202	925
272	·5544009
341	094
411	178
481	263
551	347
620	432
690	516
760	601
830	685
899	770
969	854
3585039	938
109	·5545022
178	107
248	191
318	276
388	360
457	445
527	529
596	614
666	698
736	783
806	867
875	952
945	·5546036
3586015	121
085	205
154	290
224	374
294	458
364	542
433	627
503	711
572	796
642	880
712	965
782	·5547049
851	134
921	218
991	303
3587061	387
130	471
200	555
269	640
339	724
409	809
479	893
548	977
618	·5548061
687	146
757	230
827	315
897	399
966	484
3588036	568

Number	Log
3588106	·5548652
176	736
245	821
315	905
384	990
454	·5549074
524	158
594	242
663	327
733	411
802	495
872	579
942	664
3589012	748
081	833
151	917
220	·5550001
290	085
359	170
429	254
499	338
569	422
638	507
708	591
777	675
847	759
917	844
987	928

Excesses

10 = 12	1 = 1
20 = 24	2 = 2
30 = 36	3 = 4
40 = 48	4 = 5
50 = 61	5 = 6
60 = 73	6 = 7
	7 = 8
	8 = 10
	9 = 11

35 = 42

Anti Log Excesses

10 = 8	1 = 1
20 = 17	2 = 2
30 = 25	3 = 3
40 = 33	4 = 3
50 = 41	5 = 4
60 = 50	6 = 5
70 = 58	7 = 6
80 = 66	8 = 7
	9 = 7

42 = 35

Number	Log
3590056	·5551012
126	096
195	181
265	265
334	349
404	433
474	517
544	601

N	Log
3590613	·5551686
683	770
752	854
822	938
891	·5552023
961	107
3591031	191
101	275
170	360
240	444
309	528
379	612
448	696
518	780
587	865
657	949
727	·5553033
797	117
866	201
936	285
3592005	369
075	453
144	538
214	622
283	706
353	790
423	874
493	958
562	·5554043
632	127
701	211
771	295
840	379
910	463
979	547
3593049	631
118	715
188	799
258	884
328	968
397	·5555052
467	136
536	220
606	304
675	388
745	472
814	556
884	640
953	724
3594023	808
092	892
162	976
232	·5556061
302	145
371	229
441	313
510	397
580	481
649	565
719	649
788	733
858	817
927	901
997	985
3595066	·5557069

N	Log
3595136	·5557153
205	237
275	321
344	405
414	489
483	573
553	657
622	741
692	825
761	909
831	993
900	·5558077
970	161
3596040	245
110	328
179	412
249	496
318	580
388	664
457	748
527	832
596	916
666	·5559000
735	084
805	168
874	252
944	336
3597013	420
083	503
152	587
222	671
291	755
361	839
430	923
500	·5560007
569	091
639	175
708	259
778	342
847	426
917	510
986	594
3598056	678
125	762
194	846
263	930
333	·5561013
402	097
472	181
541	265
611	349
680	433
750	516
819	600
889	684
958	768
3599028	852
097	936
167	·5562020
236	104
306	187
375	271
445	355
514	439
584	522

N	Log
3599653	·5562606
723	690
792	774
862	857
931	941

Excesses

10 = 12 1 = 1
20 = 24 2 = 2
30 = 36 3 = 4
40 = 48 4 = 5
50 = 60 5 = 6
60 = 72 6 = 7
7 = 8
8 = 10
9 = 11

35 = 42

Anti Log Excesses

10 = 8 1 = 1
20 = 17 2 = 2
30 = 25 3 = 3
40 = 33 4 = 3
50 = 41 5 = 4
60 = 50 6 = 5
70 = 58 7 = 6
80 = 66 8 = 7
9 = 7

42 = 35

N	Log
3600000	·5563025
069	109
139	193
208	277
278	360
347	444
417	528
486	612
556	695
625	779
695	863
764	947
834	·5564030
903	114
973	198
3601042	282
111	365
180	449
250	533
319	617
389	700
458	784
528	867
597	951
667	·5565035
736	119
806	202
875	286
944	370
3602013	454
083	537

N	Log
3602152	·5565621
222	704
291	788
361	872
430	956
500	·5566039
569	123
638	206
707	290
777	374
846	458
916	541
985	625
3603055	708
124	792
194	875
263	959
332	·5567043
401	127
471	210
540	294
610	377
679	461
749	544
818	628
887	712
956	796
3604026	879
095	963
165	·5568046
234	130
303	213
372	297
442	380
511	464
581	547
650	631
720	715
789	799
858	882
927	966
997	·5569049
3605066	133
136	216
205	300
274	383
343	467
413	550
482	634
552	717
621	801
690	884
759	968
829	·5570051
898	135
968	218
3606037	302
106	385
175	469
245	552
314	636
384	719
453	803
522	886
591	970

N	Log
3606661	·5571053
730	137
800	220
869	304
938	387
3607007	470
077	553
146	637
215	720
284	804
354	887
423	971
493	·5572054
562	138
631	221
700	305
770	388
839	472
908	555
977	638
3608047	721
116	805
186	888
255	972
324	·5573055
393	139
463	222
532	306
601	389
670	472
740	555
809	639
878	722
947	806
3609017	889
086	972
156	·5574055
225	139
294	222
363	306
433	389
502	472
571	555
640	639
710	722
779	806
848	889
917	972
987	·5575055

Excesses

10 = 12 1 = 1
20 = 24 2 = 2
30 = 36 3 = 4
40 = 48 4 = 5
50 = 60 5 = 6
60 = 72 6 = 7
7 = 8
8 = 10
9 = 11

35 = 42

Anti Log Excesses

10 = 8	1 = 1
20 = 17	2 = 2
30 = 25	3 = 3
40 = 33	4 = 3
50 = 42	5 = 4
60 = 50	6 = 5
70 = 58	7 = 6
80 = 66	8 = 7
	9 = 7

42 = 35

3610056	·5575139
125	222
194	306
264	389
333	472
402	555
471	639
541	722
610	805
679	888
748	972
818	·5576055
887	139
956	222
3611025	305
095	388
164	472
233	555
302	638
371	721
440	805
510	888
579	971
648	·5577054
717	138
787	221
856	304
925	387
994	470
3612064	553
133	636
202	720
271	803
340	886
409	970
479	·5578053
548	136
617	219
686	302
756	385
825	469
894	552
963	635
3613032	718
101	801
171	884
240	968
309	·5579051
378	134
448	217
517	300

3613586	·5579383
655	467
724	550
793	633
863	716
932	799
3614001	882
070	965
139	·5580048
208	132
278	215
347	298
416	381
485	464
554	547
623	630
693	713
762	796
831	880
900	963
969	·5581046
3615038	129
108	212
177	295
246	378
315	461
384	544
453	627
523	710
592	793
661	876
730	960
799	·5582043
868	126
937	209
3616006	292
076	375
145	458
214	541
283	624
352	707
421	790
491	873
560	956
629	·5583039
698	122
767	205
836	288
905	371
974	454
3617043	537
112	620
182	703
251	786
320	869
389	952
458	·5584035
527	118
596	201
665	284
735	367
804	450
873	532
942	615
3618011	698

3618080	·5584781
149	864
218	947
287	·5585030
356	113
426	196
495	279
564	362
633	445
702	528
771	611
840	693
909	776
978	859
3619047	942
116	·5586025
185	108
255	191
324	274
393	357
462	440
531	522
600	605
669	688
738	771
807	854
876	937
945	·5587020

Excesses

10 = 12	1 = 1
20 = 24	2 = 2
30 = 36	3 = 4
40 = 48	4 = 5
50 = 60	5 = 6
60 = 72	6 = 7
	7 = 8
	8 = 10
	9 = 11

35 = 42

Anti Log Excesses

10 = 8	1 = 1
20 = 17	2 = 2
30 = 25	3 = 3
40 = 33	4 = 3
50 = 42	5 = 4
60 = 50	6 = 5
70 = 58	7 = 6
80 = 67	8 = 7
	9 = 7

42 = 35

3620014	·5587103
083	185
152	268
221	351
290	434
360	517
429	600
498	682

3620567	·5587765
636	848
705	931
774	·5588014
843	097
912	179
981	262
3621050	345
119	428
188	510
257	593
326	676
395	759
464	842
533	925
602	·5589007
671	090
740	173
809	256
878	338
947	421
3622017	504
086	587
155	669
224	752
293	835
362	918
431	·5590000
500	083
569	166
638	249
707	331
776	414
845	497
914	580
983	662
3623052	745
121	827
190	910
259	993
328	·5591076
397	158
466	241
535	324
604	407
673	489
742	572
811	654
880	737
949	820
3624018	903
087	985
156	·5592068
225	150
294	233
363	316
431	399
500	481
569	564
638	646
707	729
776	812
845	895
914	977
983	·5593060

3625052	·5593142
121	225
190	307
259	390
328	473
397	556
466	638
535	721
604	803
673	886
742	968
811	·5594051
880	133
949	216
3626018	298
086	381
155	464
224	547
293	629
362	712
431	794
500	877
569	959
638	·5595042
707	124
776	207
845	289
914	372
983	454
3627052	537
121	619
189	702
258	784
327	867
396	949
465	·5596032
534	114
603	197
672	279
741	362
810	444
879	527
948	609
3628016	692
085	774
154	857
223	939
292	·5597022
361	104
430	187
499	269
568	352
637	434
705	517
774	599
843	682
912	764
981	847
3629050	929
119	·5598012
188	094
257	177
326	259
394	341
463	423

3629532	·5598506
601	588
670	671
739	753
808	836
877	918
945	·5599001

Excesses

10 = 12	1 = 1
20 = 24	2 = 2
30 = 36	3 = 4
40 = 48	4 = 5
50 = 60	5 = 6
60 = 72	6 = 7
	7 = 8
	8 = 10
	9 = 11

35 = 42

Anti Log Excesses

10 = 8	1 = 1
20 = 17	2 = 2
30 = 25	3 = 3
40 = 33	4 = 3
50 = 42	5 = 4
60 = 50	6 = 5
70 = 58	7 = 6
80 = 67	8 = 7
	9 = 8

42 = 35

3630014	·5599083
083	165
152	247
221	330
290	412
358	495
427	577
496	660
565	742
634	825
703	907
772	989
841	·5600071
909	154
978	236
3631047	319
116	401
185	483
254	565
322	648
391	730
460	813
529	895
598	977
667	·5601059
735	142
804	224
873	306
942	388

3632011	·5601471
080	553
148	636
217	718
286	800
355	882
424	965
493	·5602047
561	129
630	211
699	294
768	376
837	458
906	540
974	623
3633043	705
112	787
181	869
250	952
319	·5603034
387	116
456	198
525	281
594	363
662	445
731	527
800	610
869	692
938	774
3634007	856
075	939
144	·5604021
213	103
282	185
350	267
419	349
488	432
557	514
625	596
694	678
763	761
832	843
901	925
970	·5605007
3635038	089
107	171
176	254
245	336
313	418
382	500
451	583
520	665
588	747
657	829
726	911
795	993
863	·5606075
932	157
3636001	240
070	322
138	404
207	486
276	568
345	650
413	732

3636482	·5606814
551	897
620	979
688	·5607061
757	143
826	225
895	307
963	389
3637032	471
101	553
170	635
238	717
307	799
376	882
445	964
513	·5608046
582	128
651	210
720	292
788	374
857	456
925	538
994	620
3638063	702
132	784
200	866
269	948
338	·5609030
407	112
475	194
544	276
613	358
682	440
750	522
819	604
887	686
956	768
3639025	850
094	932
162	·5610014
231	096
300	178
369	260
437	342
506	424
574	506
643	588
712	670
781	752
849	834
918	915
986	997

Excesses

10 = 12	1 = 1
20 = 24	2 = 2
30 = 36	3 = 4
40 = 48	4 = 5
50 = 60	5 = 6
60 = 72	6 = 7
	7 = 8
	8 = 10
	9 = 11

34 = 41

Anti Log Excesses

10 = 8	1 = 1
20 = 17	2 = 2
30 = 25	3 = 3
40 = 33	4 = 3
50 = 42	5 = 4
60 = 50	6 = 5
70 = 59	7 = 6
80 = 67	8 = 7
	9 = 8

41 = 34

3640055	·5611079
124	161
193	243
261	325
330	407
399	489
468	571
536	653
605	735
673	817
742	899
811	981
880	·5612062
948	144
3641017	226
085	308
154	390
222	472
291	554
360	636
429	718
497	800
566	881
634	963
703	·5613045
772	127
841	209
909	291
978	373
3642047	455
115	536
184	618
253	700
321	782
390	864
458	946
527	·5614027

3642595	·5614109
664	191
733	273
802	355
870	437
939	518
3643007	600
076	682
144	764
213	845
282	927
351	·5615009
419	091
488	173
556	255
625	336
693	418
762	500
831	582
900	663
968	745
3644037	827
105	909
174	990
242	·5616072
311	154
379	236
448	317
517	399
586	481
654	563
723	644
791	726
860	808
928	890
997	971
3645065	·5617053
134	135
202	217
271	298
340	380
409	462
477	544
546	625
614	707
683	788
751	870
820	952
888	·5618034
957	115
3646025	197
094	278
162	360
231	442
300	524
369	605
437	687
506	768
574	850
643	932
711	·5619014
780	095
848	177
917	258
985	340

3647054	·5619422
122	504
191	585
259	667
328	748
396	830
465	911
533	993
602	·5620074
671	156
740	238
808	320
877	401
945	483
3648014	564
082	646
151	727
219	809
288	890
356	972
425	·5621053
493	135
562	217
630	299
699	380
767	462
836	543
904	625
973	706
3649041	788
110	869
178	951
247	·5622032
315	114
384	195
452	277
521	358
589	440
658	521
726	603
795	684
863	766
932	847

Excesses

10 = 12	1 = 1
20 = 24	2 = 2
30 = 36	3 = 4
40 = 48	4 = 5
50 = 60	5 = 6
60 = 71	6 = 7
	7 = 8
	8 = 10
	9 = 11

34 = 41

Anti Log Excesses

10 = 8	1 = 1
20 = 17	2 = 2
30 = 25	3 = 3
40 = 34	4 = 3
50 = 42	5 = 4
60 = 50	6 = 5
70 = 59	7 = 6
80 = 67	8 = 7
	9 = 8

41 = 34

3650000	·5622929
069	·5623010
137	092
206	173
274	255
343	336
411	418
480	499
548	581
617	662
685	744
754	825
822	907
891	988
959	·5624070
3651028	151
096	233
165	314
233	396
302	477
370	558
439	639
507	721
576	802
644	884
712	965
780	·5625047
849	128
917	210
986	291
3652054	373
123	454
191	535
260	616
328	698
397	779
465	861
534	942
602	·5626024
671	105
739	187
808	268
876	349
945	430
3653013	512
081	593
149	675
218	756
286	837
355	918
423	·5627000

3653492	·5627081
560	163
629	244
697	325
766	406
834	488
903	569
971	651
3654039	732
107	813
176	894
244	976
313	·5628057
381	139
450	220
518	301
587	382
655	464
723	545
791	626
860	707
928	789
997	870
3655065	951
134	·5629032
202	114
271	195
339	276
407	357
475	439
544	520
612	602
681	683
749	764
818	845
886	926
955	·5630007
3656023	089
091	170
159	251
228	332
296	413
365	495
433	576
502	657
570	738
638	819
706	901
775	982
843	·5631063
912	144
980	226
3657048	307
116	388
185	469
253	550
322	631
390	713
459	794
527	875
595	956
663	·5632037
732	118
800	200
869	281

3657937	·5632362
3658005	443
073	524
142	605
210	686
279	767
347	849
415	930
483	·5633011
552	092
620	173
689	254
757	335
825	416
893	498
962	579
3659030	660
099	741
167	822
235	903
303	984
372	·5634065
440	146
509	227
577	308
645	389
713	471
782	552
850	633
919	714
987	795

Excesses

10 = 12	1 = 1
20 = 24	2 = 2
30 = 36	3 = 4
40 = 48	4 = 5
50 = 59	5 = 6
60 = 71	6 = 7
	7 = 8
	8 = 10
	9 = 11

34 = 41

Anti Log Excesses

10 = 8	1 = 1
20 = 17	2 = 2
30 = 25	3 = 3
40 = 34	4 = 3
50 = 42	5 = 4
60 = 50	6 = 5
70 = 59	7 = 6
80 = 67	8 = 7
	9 = 8

41 = 34

3660055	·5634876
123	957
192	·5635038
260	119
328	200

3660396	·5635281
465	362
533	443
602	524
670	605
738	686
806	767
875	848
943	929
3661011	·5636010
079	091
148	172
216	253
284	334
352	415
421	496
489	577
558	658
626	739
694	820
762	901
831	982
899	·5637063
967	144
3662035	225
104	306
172	387
240	468
308	549
377	630
445	711
513	792
581	873
650	953
718	·5638034
786	115
854	196
923	277
991	358
3663059	439
127	520
196	601
264	682
332	763
400	844
469	924
537	·5639005
605	086
673	167
742	248
810	329
878	410
946	491
3664015	572
083	653
151	733
219	814
288	895
356	976
424	·5640057
492	138
561	218
629	299
697	380
765	461

3664833	·5640542
901	623
970	703
3665038	784
106	865
174	946
243	·5641027
311	108
379	188
447	269
515	350
583	431
652	512
720	593
788	673
856	754
925	835
993	916
3666061	996
129	·5642077
197	158
265	239
334	319
402	400
470	481
538	562
607	642
675	723
743	804
811	885
879	965
947	·5643046
3667016	127
084	208
152	288
220	369
288	450
356	531
425	611
493	692
561	773
629	854
697	934
765	·5644015
834	096
902	177
970	257
3668038	338
106	418
174	499
243	580
311	661
379	741
447	822
515	903
583	984
651	·5645064
719	145
788	225
856	306
924	387
992	468
3669060	548
128	629
197	709
3669265	·5645790
333	871
401	952
469	·5646032
537	113
605	193
673	274
742	355
810	436
878	516
946	597

Excesses

10 = 12	1 = 1
20 = 24	2 = 2
30 = 36	3 = 4
40 = 47	4 = 5
50 = 59	5 = 6
60 = 71	6 = 7
	7 = 8
	8 = 9
	9 = 11

34 = 40

Anti Log Excesses

10 = 8	1 = 1
20 = 17	2 = 2
30 = 25	3 = 3
40 = 34	4 = 3
50 = 42	5 = 4
60 = 51	6 = 5
70 = 59	7 = 6
80 = 68	8 = 7
	9 = 8

40 = 34

3670014	·5646677
082	758
150	838
218	919
287	999
355	·5647080
423	160
491	241
559	322
627	403
695	483
763	564
831	644
899	725
968	805
3671036	886
104	966
172	·5648047
240	127
308	208
376	288
444	369
512	450
580	531
649	611
3671717	·5648692
785	772
853	853
921	933
989	·5649014
3672057	094
125	175
193	255
261	336
329	417
397	498
466	578
534	659
602	739
670	820
738	900
806	981
874	·5650061
942	142
3673010	222
078	303
146	383
214	463
282	543
350	624
419	704
487	785
555	865
623	946
691	·5651026
759	107
827	187
895	268
963	348
3674031	429
099	509
167	590
235	670
303	751
371	831
439	911
507	991
575	·5652072
643	152
711	233
779	313
847	394
915	474
983	555
3675052	635
120	715
188	795
256	876
324	956
392	·5653037
460	117
528	198
596	278
664	358
732	438
800	519
868	599
936	680
3676004	760
072	840
3676140	·5653920
208	·5654001
276	081
344	162
412	242
480	322
548	402
616	483
684	563
752	644
820	724
888	804
956	884
3677024	965
092	·5655045
160	126
228	206
296	286
364	366
432	447
500	527
568	607
636	687
704	768
772	848
840	928
908	·5656008
976	089
3678043	169
111	249
179	329
247	410
315	490
383	570
451	650
519	731
587	811
655	891
723	971
791	·5657052
859	132
927	212
995	292
3679063	373
131	453
199	533
267	613
335	693
403	773
471	854
539	934
607	·5658014
674	094
742	175
810	255
878	335
946	415

Excesses

10 = 12	1 = 1
20 = 24	2 = 2
30 = 35	3 = 4
40 = 47	4 = 5
50 = 59	5 = 6
60 = 71	6 = 7
	7 = 8
	8 = 9
	9 = 11

34 = 40

Anti Log Excesses

10 = 8	1 = 1
20 = 17	2 = 2
30 = 25	3 = 3
40 = 34	4 = 3
50 = 42	5 = 4
60 = 51	6 = 5
70 = 59	7 = 6
80 = 68	8 = 7
	9 = 8

40 = 34

3680014	·5658495
082	575
150	656
218	736
286	816
354	896
422	976
490	·5659056
558	136
625	216
693	297
761	377
829	457
897	537
965	617
3681033	697
101	777
169	857
237	938
305	·5660018
372	098
440	178
508	258
576	338
644	418
712	498
780	579
848	659
916	739
984	819
3682051	899
119	979
187	·5661059
255	139
323	219
391	299
459	379

3682527	·5661459
595	539
663	619
730	700
798	780
866	860
934	940
3683002	·5662020
070	100
138	180
206	260
273	340
341	420
409	500
477	580
545	660
613	740
681	820
749	900
816	980
884	·5663060
952	140
3684020	220
088	300
156	380
223	460
291	540
359	620
427	700
495	780
563	860
631	940
699	·5664020
766	100
834	179
902	259
970	339
3685038	419
106	499
173	579
241	659
309	739
377	819
445	899
513	979
580	·5665059
648	138
716	218
784	298
852	378
920	458
987	538
3686055	618
123	698
191	778
259	858
327	938
394	·5666018
462	097
530	177
598	257
665	337
733	417
801	497
869	577

3686937	·5666657
3687005	736
072	816
140	896
208	976
276	·5667056
344	136
412	215
479	295
547	375
615	455
683	535
750	615
818	695
886	775
954	854
3688021	934
089	·5668014
157	094
225	173
293	253
361	333
428	413
496	493
564	573
632	652
699	732
767	812
835	892
903	972
970	·5669052
3689038	131
106	211
174	291
241	371
309	450
377	530
445	610
512	690
580	769
648	849
716	929
783	·5670009
851	088
919	168
987	248

Excesses

10 = 12	1 = 1
20 = 24	2 = 2
30 = 35	3 = 4
40 = 47	4 = 5
50 = 59	5 = 6
60 = 71	6 = 7
	7 = 8
	8 = 9
	9 = 11

34 = 40

Anti Log Excesses

10 = 8	1 = 1
20 = 17	2 = 2
30 = 25	3 = 3
40 = 34	4 = 3
50 = 42	5 = 4
60 = 51	6 = 5
70 = 59	7 = 6
80 = 68	8 = 7
	9 = 8

40 = 34

3690054	·5670328
122	407
190	487
258	567
325	647
393	726
461	806
529	886
596	966
664	·5671045
732	125
800	204
867	284
935	364
3691003	444
071	523
138	603
206	683
274	763
342	842
409	922
477	·5672001
545	081
613	161
680	241
748	320
815	400
883	479
951	559
3692019	639
086	719
154	798
222	878
290	957
357	·5673037
425	116
493	196
561	276
628	356
696	435
763	515
831	594
899	674
967	753
3693034	833
102	913
170	993
238	·5674072
305	152
373	231
440	311

3693508	·5674390
576	470
644	549
711	629
779	709
846	789
914	868
982	948
3694050	·5675027
117	107
185	186
252	266
320	345
388	425
456	504
523	584
591	663
658	743
726	822
794	902
862	981
929	·5676061
997	141
3695064	221
132	300
200	380
268	459
335	539
403	618
470	698
538	777
606	857
674	936
741	·5677016
809	095
876	175
944	254
3696012	334
080	413
147	493
215	572
282	651
350	730
417	810
485	889
553	969
621	·5678048
688	128
756	207
823	287
891	366
958	446
3697026	525
094	605
162	684
229	763
297	842
364	922
432	·5679001
499	081
567	160
634	240
702	319
770	399
838	478

3697905	·5679558
973	637
3698040	716
108	795
175	875
243	954
311	·5680034
379	113
446	193
514	272
581	351
649	430
716	510
784	589
851	669
919	748
986	828
3699054	907
122	986
190	·5681065
257	145
325	224
392	304
460	383
527	462
595	541
662	621
730	700
797	780
865	859
932	938

Excesses

10 = 12	1 = 1
20 = 24	2 = 2
30 = 35	3 = 4
40 = 47	4 = 5
50 = 59	5 = 6
60 = 71	6 = 7
	7 = 8
	8 = 9
	9 = 11

34 = 40

Anti Log Excesses

10 = 9	1 = 1
20 = 17	2 = 2
30 = 26	3 = 3
40 = 34	4 = 3
50 = 43	5 = 4
60 = 51	6 = 5
70 = 60	7 = 6
	8 = 7
	9 = 8

40 = 34

3700000	·5682017
068	097
136	176
203	255
271	334

3700338	·5682414
406	493
473	572
541	651
608	731
676	810
743	890
811	969
878	·5683048
946	127
3701013	207
081	286
149	365
217	444
284	524
352	603
419	682
487	761
554	841
622	920
689	999
757	·5684078
824	158
892	237
959	316
3702027	395
094	475
162	554
229	633
297	712
364	792
432	871
499	950
567	·5685029
634	108
702	187
769	267
837	346
904	425
972	504
3703039	583
107	662
174	742
242	821
309	900
377	979
444	·5686058
512	137
579	217
647	296
714	375
782	454
849	533
917	612
984	692
3704052	771
119	850
187	929
254	·5687008
322	087
389	166
457	245
524	325
592	404
659	483

3704727	·5687562
794	641
862	720
929	799
997	878
3705064	957
132	·5688036
199	116
267	195
334	274
402	353
469	432
537	511
604	590
672	669
739	748
807	827
874	906
942	985
3706009	·5689064
077	143
144	222
212	301
279	381
346	460
413	539
481	618
548	697
616	776
683	855
751	934
818	·5690013
886	092
953	171
3707021	250
088	329
156	408
223	487
291	566
358	645
425	724
492	803
560	882
627	961
695	·5691040
762	119
830	198
897	277
965	356
3708032	435
100	514
167	593
235	672
302	751
369	829
436	908
504	987
571	·5692066
639	145
706	224
774	303
841	382
909	461
976	540
3709044	619

3709111	·5692698
178	777
245	856
313	934
380	·5693013
448	092
515	171
583	250
650	329
717	408
784	487
852	566
919	645
987	723

Excesses

10 = 12	1 = 1
20 = 23	2 = 2
30 = 35	3 = 4
40 = 47	4 = 5
50 = 59	5 = 6
60 = 70	6 = 7
	7 = 8
	8 = 9
	9 = 11

34 = 40

Anti Log Excesses

10 = 9	1 = 1
20 = 17	2 = 2
30 = 26	3 = 3
40 = 34	4 = 3
50 = 43	5 = 4
60 = 51	6 = 5
70 = 60	7 = 6
	8 = 7
	9 = 8

40 = 34

3710054	·5693802
122	881
189	960
257	·5694039
324	118
391	197
458	276
526	354
593	433
661	512
728	591
796	670
863	749
930	827
997	906
3711065	985
132	·5695064
200	143
267	222
334	300
401	379
469	458

3711536	·5695537
604	616
671	695
739	773
806	852
873	931
940	·5696010
3712008	089
075	168
143	246
210	325
277	404
344	483
412	561
479	640
547	719
614	798
681	876
748	955
816	·5697034
883	113
951	191
3713018	270
085	349
152	428
220	506
287	585
355	664
422	743
489	821
556	900
624	979
691	·5698058
759	136
826	215
893	294
960	373
3714028	451
095	530
163	609
230	688
297	766
364	845
432	923
499	·5699002
566	081
633	160
701	238
768	317
836	395
903	474
970	553
3715037	632
105	710
172	789
239	867
306	946
374	·5700025
441	104
509	182
576	261
643	339
710	418
778	497
845	576

3715912	·5700654
979	733
3716047	811
114	890
181	968
248	·5701047
316	126
383	205
450	283
517	362
585	440
652	519
719	597
786	676
854	754
921	833
989	912
3717056	991
123	·5702069
190	148
258	226
325	305
392	383
459	462
527	540
594	619
661	697
728	776
796	854
863	933
930	·5703012
997	091
3718065	169
132	248
199	326
266	405
333	483
400	562
468	640
535	719
602	797
669	875
737	954
804	·5704033
871	111
938	190
3719006	268
073	347
140	425
207	504
275	582
342	661
409	739
476	818
543	896
610	975
678	·5705053
745	132
812	210
879	289
947	367

Excesses

10 = 12	1 = 1
20 = 23	2 = 2
30 = 35	3 = 4
40 = 47	4 = 5
50 = 58	5 = 6
60 = 70	6 = 7
	7 = 8
	8 = 9
	9 = 11

34 = 39

Anti Log Excesses

10 = 9	1 = 1
20 = 17	2 = 2
30 = 26	3 = 3
40 = 34	4 = 3
50 = 43	5 = 4
60 = 51	6 = 5
70 = 60	7 = 6
	8 = 7
	9 = 8

39 = 34

3720014	·5705446
081	524
148	603
216	681
283	760
350	838
417	916
484	994
551	·5706073
619	151
686	230
753	308
820	387
887	465
954	544
3721022	622
089	701
156	779
223	857
291	935
358	·5707014
425	092
492	171
559	249
626	328
694	406
761	485
828	563
895	641
962	719
3722029	798
097	876
164	955
231	·5708033
298	112
365	190
432	268

3722500	·5708346
567	425
634	503
701	582
768	660
835	738
903	816
970	895
3723037	973
104	·5709052
171	130
238	208
305	286
372	365
440	443
507	522
574	600
641	678
708	756
775	835
843	913
910	991
977	·5710069
3724044	148
111	226
178	304
245	382
312	461
380	539
447	618
514	696
581	774
648	852
715	931
782	·5711009
849	087
917	165
984	244
3725051	322
118	400
185	478
252	557
319	635
386	713
453	791
520	869
588	947
655	·5712026
722	104
789	182
856	260
923	339
990	417
3726057	495
124	573
191	652
259	730
326	808
393	886
460	964
527	·5713042
594	121
661	199
728	277
795	355

3726862	·5713433
930	511
997	590
3727064	668
131	746
198	824
265	902
332	980
399	·5714059
466	137
533	215
600	293
667	371
734	449
801	527
869	605
936	684
3728003	762
070	840
137	918
204	996
271	·5715074
338	152
405	230
472	308
539	386
606	465
673	543
740	621
807	699
874	777
941	855
3729008	933
075	·5716011
142	089
210	167
277	245
344	323
411	401
478	479
545	558
612	636
679	714
746	792
813	870
880	948
947	·5717026

Excesses

10 = 12	1 = 1
20 = 23	2 = 2
30 = 35	3 = 4
40 = 47	4 = 5
50 = 58	5 = 6
60 = 70	6 = 7
	7 = 8
	8 = 9
	9 = 10

34 = 39

Anti Log Excesses

10 = 9	1 = 1
20 = 17	2 = 2
30 = 26	3 = 3
40 = 34	4 = 3
50 = 43	5 = 4
60 = 51	6 = 5
70 = 60	7 = 6
	8 = 7
	9 = 8

39 = 34

3730014	·5717104
081	182
148	260
215	338
282	416
349	494
416	572
483	650
550	728
617	806
684	884
751	962
818	·5718040
885	118
952	196
3731019	274
086	352
153	430
220	508
287	586
354	664
421	742
488	820
555	898
622	976
689	·5719054
756	132
823	210
890	288
957	366
3732024	444
091	522
158	600
225	678
292	756
359	834
426	912
493	990
560	·5720067
627	145
694	223
761	301
828	379
895	457
962	535
3733029	613
096	691
163	769
230	847
297	925
364	·5721002

3733431	·5721080
498	158
564	236
631	314
698	392
765	470
832	548
899	625
966	703
3734033	781
100	859
167	937
234	·5722015
301	093
368	171
435	248
502	326
569	404
636	482
703	560
770	638
837	716
903	794
970	871
3735037	949
104	·5723027
171	105
238	183
305	261
372	338
439	416
506	494
573	572
640	649
706	727
773	805
840	883
907	961
974	·5724039
3736041	116
108	194
175	272
242	350
309	427
376	505
443	583
509	661
576	738
643	816
710	894
777	972
844	·5725049
911	127
978	205
3737045	283
112	360
178	438
245	516
312	594
379	671
446	749
513	827
580	905
647	982
714	·5726060

3737781	·5726138
847	216
914	293
981	371
3738048	449
115	527
182	604
249	682
316	759
382	837
449	915
516	993
583	·5727070
650	148
717	225
784	303
851	381
917	459
984	536
3739051	614
118	691
185	769
252	847
318	925
385	·5728002
452	080
519	157
586	235
653	312
720	390
787	468
853	546
920	623
987	701

Excesses

10 = 12	1 = 1
20 = 23	2 = 2
30 = 35	3 = 4
40 = 46	4 = 5
50 = 58	5 = 6
60 = 70	6 = 7
	7 = 8
	8 = 9
	9 = 10

33 = 39

Anti Log Excesses

10 = 9	1 = 1
20 = 17	2 = 2
30 = 26	3 = 3
40 = 34	4 = 3
50 = 43	5 = 4
60 = 52	6 = 5
70 = 60	7 = 6
	8 = 7
	9 = 8

39 = 33

3740054	·5728778
121	856
3740188	·5728933
254	·5729011
321	089
388	167
455	244
522	322
589	399
655	477
722	554
789	632
856	709
923	787
990	865
3741056	943
123	·5730020
190	098
257	175
324	253
391	330
457	408
524	485
591	563
658	640
725	718
792	795
858	873
925	950
992	·5731028
3742059	105
125	183
192	261
259	339
326	416
393	494
460	571
526	649
593	726
660	804
727	881
793	959
860	·5732036
927	114
994	191
3743061	269
128	346
194	424
261	501
328	579
395	656
461	734
528	811
595	889
662	966
728	·5733044
795	121
862	198
929	275
995	353
3744062	430
129	508
196	585
263	663
330	740
396	818
463	895
3744530	·5733973
597	·5734050
663	128
730	205
797	283
864	360
930	437
997	514
3745064	592
131	669
197	747
264	824
331	902
398	979
464	·5735057
531	134
598	211
665	288
731	366
798	443
865	521
932	598
998	676
3746065	753
132	830
199	907
265	985
332	·5736062
399	140
466	217
532	295
599	372
665	449
732	526
799	604
866	681
932	759
999	836
3747066	913
133	990
199	·5737068
266	145
333	223
400	300
466	377
533	454
599	532
666	609
733	686
800	763
866	841
933	918
3748000	995
067	·5738072
133	150
200	227
266	305
333	382
400	459
467	536
533	614
600	691
667	768
734	845
800	923
3748867	·5739000
933	077
3749000	154
067	232
134	309
200	386
267	463
333	541
400	618
467	695
534	772
600	850
667	927
733	·5740004
800	081
867	158
934	235

Excesses

10 = 12	1 = 1
20 = 23	2 = 2
30 = 35	3 = 4
40 = 46	4 = 5
50 = 58	5 = 6
60 = 70	6 = 7
	7 = 8
	8 = 9
	9 = 10

33 = 39

Anti Log Excesses

10 = 9	1 = 1
20 = 17	2 = 2
30 = 26	3 = 3
40 = 34	4 = 3
50 = 43	5 = 4
60 = 52	6 = 5
70 = 60	7 = 6
	8 = 7
	9 = 8

39 = 33

3750000	·5740313
067	390
133	467
200	544
267	622
334	699
400	776
467	853
533	930
600	·5741007
667	085
734	162
800	239
867	316
933	393
3751000	470
067	548
134	625
200	702
3751267	·5741779
333	856
400	933
467	·5742011
534	088
600	165
667	242
733	320
800	397
866	474
933	551
3752000	628
067	705
133	782
200	859
266	937
333	·5743014
399	091
466	168
533	245
600	322
666	399
733	476
799	553
866	630
932	708
999	785
3753066	862
133	939
199	·5744016
266	093
332	170
399	247
465	324
532	401
598	478
665	555
732	632
799	709
865	786
932	863
998	941
3754065	·5745018
131	095
198	172
264	249
331	326
398	403
465	480
531	557
598	634
664	711
731	788
797	865
864	942
930	·5746019
997	096
3755063	173
130	250
196	327
263	404
330	481
397	558
463	635
530	712

Number	Log
3755596	·5746789
663	866
729	943
796	·5747020
862	097
929	173
995	250
3756062	327
128	404
195	481
261	558
328	635
395	712
462	789
528	866
595	943
661	·5748020
728	097
794	174
861	251
927	328
994	405
3757060	482
127	558
193	635
260	712
326	789
393	866
459	943
526	·5749020
592	097
659	174
725	251
792	327
858	404
925	481
992	558
3758059	635
125	712
192	789
258	866
325	942
391	·5750019
458	096
524	173
591	250
657	327
724	404
790	481
857	557
923	634
990	711
3759056	788
123	865
189	942
256	·5751018
322	095
389	172
455	249
522	326
588	403
655	479
721	556
788	633
854	710
3759921	·5751786
987	863

Excesses

10 = 12	1 = 1
20 = 23	2 = 2
30 = 35	3 = 4
40 = 46	4 = 5
50 = 58	5 = 6
60 = 69	6 = 7
	7 = 8
	8 = 9
	9 = 10

33 = 39

Anti Log Excesses

10 = 9	1 = 1
20 = 17	2 = 2
30 = 26	3 = 3
40 = 35	4 = 4
50 = 43	5 = 4
60 = 52	6 = 5
70 = 61	7 = 6
	8 = 7
	9 = 8

39 = 33

Number	Log
3760054	·5751940
120	·5752017
187	094
253	171
320	247
386	324
453	401
519	478
586	554
652	631
719	708
785	785
851	861
917	938
984	·5753015
3761050	092
117	168
183	245
250	322
316	399
383	475
449	552
516	629
582	706
649	782
715	859
782	936
848	·5754013
915	089
981	166
3762048	243
114	320
181	396
247	473
3762314	·5754550
380	627
447	703
513	780
579	856
645	933
712	·5755010
778	087
845	163
911	240
978	317
3763044	394
111	470
177	547
244	623
310	700
377	777
443	854
510	930
576	·5756007
642	083
708	160
775	236
841	313
908	390
974	467
3764041	543
107	620
174	696
240	773
307	849
373	926
439	·5757003
505	080
572	156
638	233
705	309
771	386
838	462
904	539
971	615
3765037	692
103	769
169	846
236	922
302	999
369	·5758075
435	152
502	228
568	305
635	381
701	458
767	534
833	611
900	687
966	764
3766033	840
099	917
166	994
232	·5759071
298	147
364	224
431	300
497	377
564	453
3766630	·5759530
697	606
763	683
829	759
895	836
962	912
3767028	989
095	·5760065
161	142
228	218
294	295
360	371
426	448
493	524
559	601
626	677
692	754
758	830
824	907
891	983
957	·5761060
3768024	136
090	213
157	289
223	366
289	442
355	519
422	595
488	672
555	748
621	825
687	901
753	978
820	·5762054
886	130
953	206
3769019	283
085	359
151	436
218	512
284	589
351	665
417	742
483	818
549	895
616	971
682	·5763047
749	123
815	200
881	276
947	353

Excesses

10 = 12	1 = 1
20 = 23	2 = 2
30 = 35	3 = 4
40 = 46	4 = 5
50 = 58	5 = 6
60 = 69	6 = 7
	7 = 8
	8 = 9
	9 = 10

33 = 39

Anti Log Excesses

10 = 9	1 = 1
20 = 17	2 = 2
30 = 26	3 = 3
40 = 35	4 = 4
50 = 43	5 = 4
60 = 52	6 = 5
70 = 61	7 = 6
	8 = 7
	9 = 8

39 = 33

Number	Log
3770014	·5763429
080	506
146	582
212	659
279	735
345	811
412	887
478	964
544	·5764040
610	117
677	193
743	270
809	346
875	422
942	498
3771008	575
075	651
141	728
207	804
273	880
340	956
406	·5765033
472	109
538	186
605	262
671	338
738	414
804	491
870	567
936	644
3772003	720
069	796
135	872
201	949
268	·5766025
334	101
400	177
466	254
533	330
599	407
665	483
731	559
798	635
864	712
930	788
996	864
3773063	940
129	·5767017
195	093
261	169
328	245

Number	Log
3773394	·5767321
460	397
526	474
593	550
659	627
725	703
791	779
858	855
924	932
990	·5768008
3774056	084
123	160
189	236
255	312
321	389
388	465
454	541
520	617
586	694
653	770
719	846
785	922
851	999
918	·5769075
984	151
3775050	227
116	303
183	379
249	456
315	532
381	608
447	684
513	760
580	836
646	913
712	989
778	·5770065
845	141
911	217
977	293
3776043	370
110	446
176	522
242	598
308	674
374	750
440	826
507	902
573	979
639	·5771055
705	131
772	207
838	283
904	359
970	435
3777036	511
102	588
169	664
235	740
301	816
367	892
433	968
499	·5772044
566	120
632	196

Number	Log
3777698	·5772272
764	348
831	424
897	501
963	577
3778029	653
095	729
161	805
228	881
294	957
360	·5773033
426	109
492	185
558	261
625	337
691	413
757	489
823	565
889	641
955	717
3779022	793
088	869
154	945
220	·5774021
286	097
352	174
418	250
484	326
551	402
617	478
683	554
749	630
815	706
881	782
948	857

Excesses

10 = 12	1 = 1
20 = 23	2 = 2
30 = 35	3 = 4
40 = 46	4 = 5
50 = 58	5 = 6
60 = 69	6 = 7
	7 = 8
	8 = 9
	9 = 10

33 = 39

Anti Log Excesses

10 = 9	1 = 1
20 = 17	2 = 2
30 = 26	3 = 3
40 = 35	4 = 4
50 = 43	5 = 4
60 = 52	6 = 5
70 = 61	7 = 6
	8 = 7
	9 = 8

39 = 33

Number	Log
3780014	·5774933
3780080	·5775009
146	085
212	161
278	237
344	313
410	389
477	465
543	541
609	617
675	693
741	769
807	845
873	921
939	997
3781006	·5776073
072	149
138	225
204	301
270	377
336	453
402	529
468	605
535	680
601	756
667	832
733	908
799	984
865	·5777060
931	136
997	212
3782063	288
129	364
196	439
262	515
328	591
394	667
460	743
526	819
592	895
658	971
724	·5778047
790	123
857	198
923	274
989	350
3783055	426
121	502
187	578
253	654
319	730
385	805
451	881
517	957
583	·5779033
650	109
716	185
782	260
848	336
914	412
980	488
3784046	564
112	640
178	715
244	791
310	867

Number	Log
3784376	·5779943
442	·5780019
508	095
574	170
640	246
707	322
773	398
839	473
905	549
971	625
3785037	701
103	777
169	853
235	928
301	·5781004
367	080
433	156
499	231
565	307
631	382
697	458
763	534
829	610
895	685
961	761
3786027	837
093	913
160	988
226	·5782064
292	140
358	216
424	291
490	367
556	443
622	519
688	594
754	670
820	746
886	822
952	897
3787018	973
084	·5783049
150	125
216	200
282	276
348	351
414	427
480	503
546	579
612	654
678	730
744	806
810	882
876	957
942	·5784033
3788008	108
074	184
140	260
206	336
272	411
338	487
404	562
470	638
536	713
602	789

Number	Log
3788668	·5784865
734	941
800	·5785016
866	092
932	167
998	243
3789064	319
130	395
196	470
262	546
328	621
394	697
460	772
526	848
592	923
657	999
723	·5786075
789	151
855	226
921	302
987	377

Excesses

10 = 11	1 = 1
20 = 23	2 = 2
30 = 34	3 = 3
40 = 46	4 = 5
50 = 57	5 = 6
60 = 69	6 = 7
	7 = 8
	8 = 9
	9 = 10

33 = 38

Anti Log Excesses

10 = 9	1 = 1
20 = 17	2 = 2
30 = 26	3 = 3
40 = 35	4 = 4
50 = 44	5 = 4
60 = 52	6 = 5
70 = 61	7 = 6
	8 = 7
	9 = 8

38 = 33

Number	Log
3790053	·5786453
119	528
185	604
251	679
317	755
383	830
449	906
515	982
581	·5787058
647	133
713	209
779	284
845	360
911	435
977	511

3791043	·5787586
108	662
174	737
240	813
306	888
372	964
438	·5788039
504	115
570	190
636	266
702	341
768	417
834	492
900	568
966	643
3792032	719
098	794
163	870
229	945
295	·5789021
361	096
427	172
493	247
559	323
625	398
691	474
757	549
823	625
889	700
954	776
3793020	851
086	927
152	·5790002
218	078
284	153
350	229
416	304
482	380
548	455
613	531
679	606
745	682
811	757
877	832
943	907
3794009	983
075	·5791058
141	134
207	209
272	285
338	360
404	436
470	511
536	587
602	662
668	738
734	813
799	888
865	963
931	·5792039
997	114
3795063	190
129	265
195	341
261	416

3795326	·5792491
392	566
458	642
524	717
590	793
656	868
722	944
788	·5793019
853	094
919	169
985	245
3796051	320
117	396
183	471
249	546
315	621
380	697
446	772
512	848
578	923
644	998
710	·5794073
775	149
841	224
907	300
973	375
3797039	450
105	525
170	601
236	676
302	751
368	826
434	902
500	977
565	·5795053
631	128
697	203
763	278
829	354
895	429
960	504
3798026	579
092	655
158	730
224	805
290	880
355	956
421	·5796031
487	106
553	181
618	257
684	332
750	407
816	482
882	558
948	633
3799013	708
079	783
145	859
211	934
277	·5797009
343	084
408	159
474	234
540	310

3799606	·5797385
671	460
737	535
803	610
869	685
934	761

Excesses

10 = 11	1 = 1
20 = 23	2 = 2
30 = 34	3 = 3
40 = 46	4 = 5
50 = 57	5 = 6
60 = 69	6 = 7
	7 = 8
	8 = 9
	9 = 10

33 = 38

Anti Log Excesses

10 = 9	1 = 1
20 = 17	2 = 2
30 = 26	3 = 3
40 = 35	4 = 4
50 = 44	5 = 4
60 = 52	6 = 5
70 = 61	7 = 6
	8 = 7
	9 = 8

38 = 33

3800000	·5797836
066	911
132	986
198	·5798062
264	137
329	212
395	287
461	362
527	437
592	513
658	588
724	663
790	738
855	813
921	888
987	964
3801053	·5799039
119	114
185	189
251	264
316	339
382	414
448	489
513	564
579	640
645	715
711	790
776	865
842	940
908	·5800015

3801974	·5800090
3802039	166
105	241
171	316
237	391
302	466
368	541
434	616
500	691
565	767
631	842
697	917
763	992
828	·5801067
894	142
960	217
3803026	292
091	367
157	442
223	517
289	592
354	667
420	742
485	818
551	893
617	968
683	·5802043
748	118
814	193
880	268
946	343
3804011	418
077	493
143	568
209	643
274	718
340	793
406	868
472	943
537	·5803018
603	093
668	168
734	243
800	318
866	393
931	468
997	543
3805063	618
129	693
194	768
260	843
325	918
391	993
457	·5804068
523	143
588	218
654	293
720	368
786	443
851	518
917	593
982	668
3806048	743
114	818
180	892

3806245	·5804967
311	·5805042
376	117
442	192
508	267
574	342
639	417
705	492
770	567
836	642
902	717
968	792
3807033	867
099	941
164	·5806016
230	091
296	166
362	241
427	316
493	391
558	466
624	541
690	616
756	690
821	765
887	840
952	915
3808018	990
084	·5807065
150	140
215	215
281	289
346	364
412	439
478	514
544	589
609	664
675	738
740	813
806	888
871	963
937	·5808038
3809003	113
069	187
134	262
200	337
265	412
331	487
396	562
462	636
528	711
594	786
659	861
725	936
790	·5809011
856	085
921	160
987	235

Excesses

10 = 11	1 = 1
20 = 23	2 = 2
30 = 34	3 = 3
40 = 46	4 = 5
50 = 57	5 = 6
60 = 68	6 = 7
	7 = 8
	8 = 9
	9 = 10
33 = 37	

Anti Log Excesses

10 = 9	1 = 1
20 = 18	2 = 2
30 = 26	3 = 3
40 = 35	4 = 4
50 = 44	5 = 4
60 = 53	6 = 5
70 = 61	7 = 6
	8 = 7
	9 = 8
37 = 33	

3810053	·5809310
119	384
184	459
250	534
315	609
381	684
446	759
512	833
577	908
643	983
709	·5810058
775	132
840	207
906	282
971	357
3811037	431
102	506
168	581
233	656
299	730
365	805
431	880
496	955
562	·5811029
627	104
693	178
758	253
824	328
889	403
955	478
3812021	553
087	627
152	702
218	776
283	851
349	926
414	·5812001

3812480	·5812075
545	150
611	225
676	300
742	374
807	449
873	523
939	598
3813005	673
070	748
136	822
201	897
267	972
332	·5813047
398	121
463	196
529	270
594	345
660	420
725	495
791	569
856	644
922	718
987	793
3814053	867
119	942
185	·5814017
250	092
316	166
381	241
447	315
512	390
578	464
643	539
709	614
774	689
840	763
905	838
971	912
3815036	987
102	·5815061
167	136
233	210
298	285
364	360
429	435
495	509
560	584
626	658
691	733
757	807
822	882
888	956
953	·5816031
3816019	105
084	180
150	254
215	329
281	403
346	478
412	552
477	627
543	702
608	777
674	851

3816739	·5816926
805	·5817000
870	075
936	149
3817001	224
067	298
132	373
198	447
263	522
329	596
394	671
460	745
525	820
591	894
656	969
722	·5818043
787	118
853	192
918	267
984	341
3818049	416
115	490
180	565
246	639
311	714
377	788
442	863
508	937
573	·5819012
639	086
704	160
770	234
835	309
901	383
966	458
3819032	532
097	607
163	681
228	756
294	830
359	905
425	979
490	·5820054
555	128
620	203
686	277
751	351
817	425
882	500
948	574

Excesses

10 = 11	1 = 1
20 = 23	2 = 2
30 = 34	3 = 3
40 = 46	4 = 5
50 = 57	5 = 6
60 = 68	6 = 7
	7 = 8
	8 = 9
	9 = 10
33 = 37	

Anti Log Excesses

10 = 9	1 = 1
20 = 18	2 = 2
30 = 26	3 = 3
40 = 35	4 = 4
50 = 44	5 = 4
60 = 53	6 = 5
70 = 61	7 = 6
	8 = 7
	9 = 8
37 = 33	

3820013	·5820649
079	723
144	798
210	872
275	947
341	·5821021
406	095
472	169
537	244
603	318
668	393
733	467
798	542
864	616
929	690
995	764
3821060	839
126	913
191	988
257	·5822062
322	137
388	211
453	285
519	359
584	434
649	508
714	583
780	657
845	731
911	805
976	880
3822042	954
107	·5823029
173	103
238	177
303	251
368	326
434	400
499	475
565	549
630	623
696	697
761	772
827	846
892	920
957	994
3823022	·5824069
088	143
153	217
219	291
284	366

3823350	·5824440
415	514
481	588
546	663
611	737
676	812
742	886
807	960
873	·5825034
938	109
3824004	183
069	257
134	331
199	406
265	480
330	554
396	628
461	702
527	776
592	851
657	925
722	999
788	·5826073
853	148
919	222
984	296
3825050	370
115	445
180	519
245	593
311	667
376	741
442	815
507	890
572	964
637	·5827038
703	112
768	187
834	261
899	335
964	409
3826029	483
095	557
160	632
226	706
291	780
356	854
421	928
487	·5828002
552	076
618	150
683	225
748	299
813	373
879	447
944	521
3827010	595
075	670
140	744
205	818
271	892
336	966
402	·5829040
467	114
532	188

3827597	·5829262
663	336
728	411
794	485
859	559
924	633
989	707
3828055	781
120	855
185	929
250	·5830003
316	077
381	152
447	226
512	300
577	374
642	448
708	522
773	596
838	670
903	744
969	818
3829034	892
100	966
165	·5831040
230	114
295	189
361	263
426	337
491	411
556	485
622	559
687	633
752	707
817	781
883	855
948	929

Excesses

10 = 11	1 = 1
20 = 23	2 = 2
30 = 34	3 = 3
40 = 45	4 = 5
50 = 57	5 = 6
60 = 68	6 = 7
	7 = 8
	8 = 9
	9 = 10

33 = 37

Anti Log Excesses

10 = 9	1 = 1
20 = 18	2 = 2
30 = 26	3 = 3
40 = 35	4 = 4
50 = 44	5 = 4
60 = 53	6 = 5
70 = 62	7 = 6
	8 = 7
	9 = 8

37 = 33

3830014	·5832003
079	077
144	151
209	225
275	299
340	373
405	447
470	521
536	595
601	669
666	743
731	817
797	891
862	965
927	·5833039
992	113
3831058	187
123	261
188	335
253	409
319	482
384	556
449	630
514	704
580	778
645	852
710	926
775	·5834000
841	074
906	148
971	222
3832036	296
102	370
167	444
232	518
297	592
363	666
428	740
493	813
558	887
624	961
689	·5835035
754	109
819	183
885	257
950	331
3833015	405
080	479
145	553
210	627
276	700
341	774
406	848
471	922
537	996
602	·5836070
667	144
732	218
798	291
863	365
928	439
993	513
3834058	587
123	661
189	734

3834254	·5836808
319	882
384	956
450	·5837030
515	104
580	178
645	252
710	325
775	399
841	473
906	547
971	621
3835036	695
102	768
167	842
232	916
297	990
362	·5838064
427	138
493	211
558	285
623	359
688	433
753	506
818	580
884	654
949	728
3836014	801
079	875
144	949
209	·5839023
275	097
340	171
405	244
470	318
535	392
600	466
666	539
731	613
796	687
861	761
926	834
991	908
3837057	982
122	·5840056
187	129
252	203
317	277
382	351
448	424
513	498
578	572
643	646
708	719
773	793
838	867
903	941
969	·5841014
3838034	088
099	161
164	235
229	309
294	383
360	456
425	530

3838490	·5841603
555	677
620	751
685	825
750	898
815	972
881	·5842046
946	120
3839011	193
076	267
141	340
206	414
271	488
336	562
401	635
466	709
532	782
597	856
662	930
727	·5843004
792	077
857	151
922	224
987	298

Excesses

10 = 11	1 = 1
20 = 23	2 = 2
30 = 34	3 = 3
40 = 45	4 = 5
50 = 57	5 = 6
60 = 68	6 = 7
	7 = 8
	8 = 9
	9 = 10

33 = 37

Anti Log Excesses

10 = 9	1 = 1
20 = 18	2 = 2
30 = 26	3 = 3
40 = 35	4 = 4
50 = 44	5 = 4
60 = 53	6 = 5
70 = 62	7 = 6
	8 = 7
	9 = 8

37 = 33

3840053	·5843371
118	445
183	519
248	593
313	666
378	740
443	813
508	887
573	960
638	·5844034
704	107
769	181

3840834	·5844255
899	329
964	402
3841029	476
094	549
159	623
224	696
289	770
354	843
419	917
485	991
550	·5845065
615	138
680	212
745	285
810	359
875	432
940	506
3842005	579
070	653
135	726
200	800
265	873
330	947
396	·5846020
461	094
526	167
591	241
656	314
721	388
786	461
851	535
916	609
981	683
3843046	756
111	830
176	903
241	977
306	·5847050
371	124
436	197
501	271
567	344
632	418
697	491
762	565
827	638
892	712
957	785
3844022	859
087	932
152	·5848005
217	078
282	152
347	225
412	299
477	372
542	446
607	519
672	593
737	666
802	740
867	813
932	887
997	960

3845062	·5849034
127	107
192	181
257	254
322	327
387	400
452	474
517	547
582	621
647	694
712	768
777	841
842	915
907	988
972	·5850062
3846037	135
102	209
167	282
232	355
297	428
362	502
427	575
492	649
557	722
622	796
687	869
752	942
817	·5851015
882	089
947	162
3847012	236
077	309
142	382
207	455
272	529
337	602
402	676
467	749
532	823
597	896
662	969
727	·5852042
792	116
857	189
922	263
987	336
3848052	409
117	482
182	556
247	629
312	703
377	776
442	849
507	922
572	996
637	·5853069
702	142
767	215
831	289
896	362
961	435
3849026	508
091	582
156	655
221	729

3849286	·5853802
351	875
416	948
481	·5854022
546	095
611	168
676	241
741	315
806	388
871	461
936	534

Excesses

10 = 11	1 = 1
20 = 23	2 = 2
30 = 34	3 = 3
40 = 45	4 = 5
50 = 56	5 = 6
60 = 68	6 = 7
	7 = 8
	8 = 9
	9 = 10

33 = 37

Anti Log Excesses

10 = 9	1 = 1
20 = 18	2 = 2
30 = 27	3 = 3
40 = 35	4 = 4
50 = 44	5 = 4
60 = 53	6 = 5
70 = 62	7 = 6
	8 = 7
	9 = 8

37 = 33

3850000	·5854608
065	681
130	754
195	827
260	901
325	974
390	·5855047
455	120
520	194
585	267
650	340
715	413
780	486
845	559
909	633
974	706
3851039	779
104	852
169	926
234	999
299	·5856072
364	145
429	218
494	291
559	365

3851624	·5856438
688	511
753	584
818	658
883	731
948	804
3852013	877
078	950
143	·5857023
208	097
273	170
337	243
402	316
467	389
532	462
597	536
662	609
727	682
792	755
856	828
921	901
986	974
3853051	·5858047
116	121
181	194
246	267
311	340
376	413
441	486
505	559
570	632
635	706
700	779
765	852
830	925
895	998
960	·5859071
3854024	144
089	217
154	290
219	363
284	437
349	510
413	583
478	656
543	729
608	802
673	875
738	948
803	·5860021
868	094
932	167
997	240
3855062	313
127	386
192	460
257	533
321	606
386	679
451	752
516	825
581	898
646	971
710	·5861044
775	117

3855840	·5861190
905	263
970	336
3856035	409
099	482
164	555
229	628
294	701
359	774
424	847
488	920
553	993
618	·5862066
683	139
748	212
813	285
877	358
942	431
3857007	504
072	577
137	650
202	723
266	796
331	869
396	942
461	·5863015
525	088
590	161
655	234
720	307
785	380
850	453
914	526
979	598
3858044	671
109	744
174	817
238	890
303	963
368	·5864036
433	109
498	182
562	255
627	328
692	401
757	474
821	547
886	620
951	693
3859016	765
080	838
145	911
210	984
275	·5865057
340	130
405	203
469	276
534	348
599	421
664	494
728	567
793	640
858	713
923	786
987	859

Excesses

10 = 11	1 = 1
20 = 23	2 = 2
30 = 34	3 = 3
40 = 45	4 = 5
50 = 56	5 = 6
60 = 68	6 = 7
	7 = 8
	8 = 9
	9 = 10

33 = 37

Anti Log Excesses

10 = 9	1 = 1
20 = 18	2 = 2
30 = 27	3 = 3
40 = 36	4 = 4
50 = 44	5 = 4
60 = 53	6 = 5
70 = 62	7 = 6
	8 = 7
	9 = 8

37 = 33

3860052	·5865931
117	·5866004
182	077
246	150
311	223
376	296
441	369
505	442
570	514
635	587
700	660
764	733
829	806
894	879
959	951
3861023	·5867024
088	097
153	170
218	243
282	316
347	388
412	461
477	534
541	607
606	680
671	753
736	825
800	898
865	971
930	·5868044
995	116
3862059	189
124	262
189	335
254	408
318	481
383	553

3862448	·5868626
513	699
577	772
642	844
707	917
772	990
836	·5869063
901	135
965	208
3863030	281
095	354
160	426
224	499
289	572
354	645
419	717
483	790
548	863
613	936
678	·5870008
742	081
807	154
871	227
936	299
3864001	372
066	444
130	517
195	590
260	663
325	735
389	808
454	881
518	954
583	·5871026
648	099
713	171
777	244
842	317
906	390
971	462
3865036	535
101	608
165	681
230	753
294	826
359	898
424	971
489	·5872044
553	117
618	189
683	262
748	334
812	407
877	480
941	553
3866006	625
071	698
136	770
200	843
265	915
329	988
394	·5873061
458	134
523	206
588	279

3866653	·5873351
717	424
782	496
846	569
911	642
976	715
3867041	787
105	860
170	932
234	·5874005
299	077
364	150
429	223
493	296
558	368
622	441
687	513
751	586
816	658
881	731
946	803
3868010	876
075	948
139	·5875021
204	093
268	166
333	239
398	312
463	384
527	457
592	529
656	602
721	674
785	747
850	819
915	892
980	964
3869044	·5876037
109	109
173	182
238	254
302	327
367	399
432	472
497	544
561	617
626	689
690	762
755	834
819	907
884	979
948	·5877052

Excesses

10 = 11	1 = 1
20 = 22	2 = 2
30 = 34	3 = 3
40 = 45	4 = 5
50 = 56	5 = 6
60 = 67	6 = 7
	7 = 8
	8 = 9
	9 = 10

32 = 36

Anti Log Excesses

10 = 9	1 = 1
20 = 18	2 = 2
30 = 27	3 = 3
40 = 36	4 = 4
50 = 44	5 = 4
60 = 53	6 = 5
70 = 62	7 = 6
	8 = 7
	9 = 8

36 = 32

3870013	·5877124
078	197
143	269
207	342
272	414
336	487
401	559
465	632
530	704
594	777
659	849
724	922
789	994
853	·5878067
918	139
982	212
3871047	284
111	357
176	429
240	502
305	574
369	646
434	718
498	791
563	863
628	936
693	·5879008
757	081
822	153
886	226
951	298
3872015	371
080	443
144	516
209	588
273	661
338	733

3872403	·5879805
468	877
532	950
597	·5880022
661	095
726	167
790	240
855	312
919	384
984	456
3873048	529
113	601
177	674
242	746
306	819
371	891
435	963
500	·5881035
564	108
629	180
693	253
758	325
822	398
887	470
952	542
3874017	614
081	687
146	759
210	832
275	904
339	976
404	·5882048
468	121
533	193
597	266
662	338
726	410
791	482
855	555
920	627
984	700
3875049	772
113	844
178	916
242	989
307	·5883061
371	133
436	205
500	278
565	350
629	422
694	494
758	567
823	639
887	712
952	784
3876016	856
081	928
145	·5884001
210	073
274	145
339	217
403	290
468	362
532	434

3876597	·5884506
661	579
726	651
790	723
855	795
919	868
984	940
3877048	·5885012
113	084
177	156
242	228
306	301
371	373
435	445
500	517
564	590
629	662
693	734
758	806
822	879
887	951
951	·5886023
3878015	095
079	167
144	239
208	312
273	384
337	456
402	528
466	601
531	673
595	745
660	817
724	889
789	961
853	·5887034
918	106
982	178
3879047	250
111	322
176	394
240	466
305	538
369	611
433	683
497	755
562	827
626	899
691	971
755	·5888043
820	115
884	188
949	260

Excesses

10 = 11	1 = 1
20 = 22	2 = 2
30 = 34	3 = 3
40 = 45	4 = 5
50 = 56	5 = 6
60 = 67	6 = 7
	7 = 8
	8 = 9
	9 = 10

32 = 36

Anti Log Excesses

10 = 9	1 = 1
20 = 18	2 = 2
30 = 27	3 = 3
40 = 36	4 = 4
50 = 45	5 = 5
60 = 54	6 = 5
70 = 62	7 = 6
	8 = 7
	9 = 8

36 = 32

3880013	·5888332
078	404
142	476
207	548
271	620
336	692
400	765
464	837
528	909
593	981
657	·5889053
722	125
786	197
851	269
915	341
980	413
3881044	486
109	558
173	630
237	702
301	774
366	846
430	918
495	990
559	·5890062
624	134
688	206
753	278
817	350
881	422
945	494
3882010	566
074	638
139	710
203	783
268	855
332	927

3882397	·5890999
461	·5891071
525	143
589	215
654	287
718	359
783	431
847	503
912	575
976	647
3883041	719
105	791
169	863
233	935
298	·5892007
362	079
427	151
491	223
556	295
620	367
684	439
748	511
813	583
877	655
942	727
3884006	799
071	871
135	942
199	·5893014
263	086
328	158
392	230
457	302
521	374
585	446
649	518
714	590
778	662
843	734
907	806
972	878
3885036	950
100	·5894022
164	094
229	166
293	237
358	309
422	381
486	453
550	525
615	597
679	669
744	741
808	813
872	885
936	957
3886001	·5895029
065	100
130	172
194	244
258	316
322	388
387	460
451	532
516	604

3886580	·5895675
644	747
708	819
773	891
837	963
902	·5896035
966	107
3887030	179
094	250
159	322
223	394
287	466
351	538
416	610
480	681
545	753
609	825
673	897
737	969
802	·5897041
866	112
931	184
995	256
3888059	328
123	400
188	472
252	543
316	615
380	687
445	759
509	831
574	903
638	974
702	·5898046
766	118
831	190
895	261
959	333
3889023	405
088	477
152	549
216	621
280	692
345	764
409	836
473	908
537	979
602	·5899051
666	123
731	195
795	266
859	338
923	410
988	482

Excesses

10 = 11	1 = 1
20 = 22	2 = 2
30 = 34	3 = 3
40 = 45	4 = 5
50 = 56	5 = 6
60 = 67	6 = 7
	7 = 8
	8 = 9
	9 = 10

32 = 36

Anti Log Excesses

10 = 9	1 = 1
20 = 18	2 = 2
30 = 27	3 = 3
40 = 36	4 = 4
50 = 45	5 = 5
60 = 54	6 = 5
70 = 63	7 = 6
	8 = 7
	9 = 8

36 = 32

3890052	·5899554
116	625
180	697
245	769
309	841
373	912
437	984
502	·5900056
566	128
630	199
694	271
759	343
823	415
887	486
951	558
3891016	630
080	702
144	773
208	845
273	916
337	988
401	·5901060
465	132
530	203
594	275
658	347
722	419
787	490
851	562
915	633
979	705
3892044	777
108	849
172	920
236	992
301	·5902063
365	135

3892429	·5902207
493	279
558	350
622	422
686	493
750	565
814	637
878	709
943	780
3893007	852
071	923
135	995
200	·5903067
264	139
328	210
392	282
457	353
521	425
585	496
649	568
713	639
777	711
842	783
906	855
970	926
3894034	998
099	·5904069
163	141
227	212
291	284
355	355
419	427
484	499
548	571
612	642
676	714
741	785
805	857
869	928
933	·5905000
997	071
3895061	143
126	214
190	286
254	357
318	429
382	501
446	573
511	644
575	716
639	787
703	859
767	930
831	·5906002
896	073
960	145
3896024	216
088	288
152	359
216	431
281	502
345	574
409	645
473	717
537	788

3896601	·5906860
666	931
730	·5907003
794	074
858	146
922	217
986	289
3897051	360
115	432
179	503
243	575
307	646
371	718
436	789
500	861
564	932
628	·5908004
692	075
756	147
820	218
884	290
949	361
3898013	433
077	504
141	576
205	647
269	718
333	789
397	860
462	932
526	·5909004
590	075
654	147
718	218
782	290
846	361
910	433
975	504
3899039	576
103	647
167	718
231	789
295	861
359	932
423	·5910004
488	075
552	147
616	218
680	290
744	361
808	432
872	503
936	575

Excesses

10 = 11	1 = 1
20 = 22	2 = 2
30 = 33	3 = 3
40 = 45	4 = 5
50 = 56	5 = 6
60 = 67	6 = 7
	7 = 8
	8 = 9
	9 = 10

32 = 36

Anti Log Excesses

10 = 9	1 = 1
20 = 18	2 = 2
30 = 27	3 = 3
40 = 36	4 = 4
50 = 45	5 = 5
60 = 54	6 = 5
70 = 63	7 = 6
	8 = 7
	9 = 8

36 = 32

3900000	·5910646
064	718
129	789
193	861
257	932
321	·5911004
385	075
449	146
513	217
577	289
641	360
705	432
770	503
834	574
898	645
962	717
3901026	788
090	860
154	931
218	·5912002
282	073
346	145
411	216
475	288
539	359
603	430
667	501
731	573
795	644
859	716
923	787
987	858
3902051	929
115	·5913001
179	072
243	143
308	214

3902372	·5913285
436	357
500	429
564	500
628	571
692	642
756	714
820	785
884	856
948	927
3903012	999
076	·5914070
140	141
204	212
268	284
332	355
396	426
461	497
525	569
589	640
653	711
717	782
781	854
845	925
909	996
973	·5915067
3904037	139
101	210
165	281
229	352
293	424
357	495
421	566
485	637
549	709
613	780
677	851
741	922
805	993
869	·5916064
933	136
997	207
3905061	278
125	349
189	421
253	492
317	563
381	634
445	705
509	776
573	848
637	919
701	990
765	·5917061
829	132
893	203
957	275
3906021	346
085	417
149	488
213	559
278	630
341	702
405	773
469	844

3906533	·5917915
597	987
661	·5918058
725	129
789	200
853	271
917	342
981	413
3907045	484
109	556
173	627
237	698
301	769
365	840
429	911
493	982
557	·5919053
621	124
685	195
749	267
813	338
877	409
941	480
3908005	551
069	622
133	693
197	764
261	835
325	906
389	978
453	·5920049
517	120
581	191
645	262
709	333
773	404
837	475
901	546
965	617
3909028	688
092	759
156	830
220	901
284	972
348	·5921043
412	114
476	185
540	257
604	328
668	399
732	470
796	541
860	612
924	683
988	754

Excesses

10 = 11	1 = 1
20 = 22	2 = 2
30 = 33	3 = 3
40 = 44	4 = 4
50 = 56	5 = 6
60 = 67	6 = 7
	7 = 8
	8 = 9
	9 = 10

32 = 36

Anti Log Excesses

10 = 9	1 = 1
20 = 18	2 = 2
30 = 27	3 = 3
40 = 36	4 = 4
50 = 45	5 = 5
60 = 54	6 = 5
70 = 63	7 = 6
	8 = 7
	9 = 8

36 = 32

3910052	·5921825
116	896
179	967
243	·5922038
307	109
371	180
435	251
499	322
563	393
627	464
691	535
755	606
819	677
883	748
947	819
3911011	890
074	961
138	·5923032
202	103
266	174
330	245
394	315
458	386
522	457
586	528
650	599
714	670
778	741
841	812
905	883
969	954
3912033	·5924025
097	096
161	167
225	238
289	309
353	380

3912417	·5924451
480	522
544	592
608	663
672	734
736	805
800	876
864	947
928	·5925018
992	089
3913056	160
119	231
183	302
247	373
311	443
375	514
439	585
503	656
567	727
630	798
694	869
758	940
822	·5926011
886	082
950	152
3914014	223
078	294
141	365
205	436
269	507
333	577
397	648
461	719
525	790
589	861
652	932
716	·5927003
780	074
844	144
908	215
972	286
3915035	357
099	428
163	499
227	569
291	640
355	711
419	782
483	853
546	924
610	994
674	·5928065
738	136
802	207
866	278
929	349
993	419
3916057	490
121	561
185	632
249	702
312	773
376	844
440	915
504	985

3916568	·5929056
632	127
695	198
759	269
823	340
887	410
951	481
3917015	552
078	623
142	693
206	764
270	835
334	906
398	976
461	·5930047
525	118
589	189
653	259
716	330
780	401
844	472
908	542
972	613
3918036	684
099	755
163	825
227	896
291	966
355	·5931037
419	108
482	179
546	249
610	320
674	391
737	462
801	532
865	603
929	673
993	744
3919057	815
120	886
184	956
248	·5932027
312	098
375	169
439	239
503	310
567	380
630	451
694	522
758	593
822	663
885	734
949	804

Excesses

10 = 11	1 = 1
20 = 22	2 = 2
30 = 33	3 = 3
40 = 44	4 = 4
50 = 55	5 = 6
60 = 67	6 = 7
	7 = 8
	8 = 9
	9 = 10

32 = 35

Anti Log Excesses

10 = 9	1 = 1
20 = 18	2 = 2
30 = 27	3 = 3
40 = 36	4 = 4
50 = 45	5 = 5
60 = 54	6 = 5
70 = 63	7 = 6
	8 = 7
	9 = 8

35 = 32

3920013	·5932875
077	946
141	·5933017
205	087
268	158
332	228
396	299
460	370
523	441
587	511
651	582
715	652
778	723
842	793
906	864
970	935
3921033	·5934006
097	076
161	147
225	217
288	288
352	358
416	429
480	499
543	570
607	641
671	712
735	782
798	853
862	923
926	994
990	·5935064
3922053	135
117	205
181	276
245	347
308	418

3922372	·5935488
436	559
500	629
563	700
627	770
691	841
755	911
818	982
882	·5936052
946	123
3923010	193
073	264
137	334
200	405
264	475
328	546
392	617
455	688
519	758
583	829
647	899
710	970
774	·5937040
838	111
902	181
965	252
3924029	322
092	393
156	463
220	534
284	604
347	675
411	745
475	816
539	886
602	957
666	·5938027
729	098
793	168
857	239
921	309
984	380
3925048	450
112	520
176	590
239	661
303	731
366	802
430	872
494	943
558	·5939013
621	084
685	154
749	225
813	295
876	366
940	436
3926003	507
067	577
131	648
195	718
258	789
322	859
385	929
449	999

3926513	·5940070
577	140
640	211
704	281
767	352
831	422
895	493
959	563
3927022	634
086	704
149	774
213	844
277	915
341	985
404	·5941056
468	126
531	197
595	267
658	337
722	407
786	478
850	548
913	619
977	689
3928040	760
104	830
168	900
232	970
295	·5942041
359	111
422	182
486	252
549	323
613	393
677	463
741	533
804	604
868	674
931	745
995	815
3929059	885
123	955
186	·5943026
250	096
313	167
377	237
440	307
504	377
568	448
632	518
695	588
759	658
822	729
886	799
949	870

Excesses

10 = 11	1 = 1
20 = 22	2 = 2
30 = 33	3 = 3
40 = 44	4 = 4
50 = 55	5 = 6
60 = 66	6 = 7
	7 = 8
	8 = 9
	9 = 10

32 = 35

Anti Log Excesses

10 = 9	1 = 1
20 = 18	2 = 2
30 = 27	3 = 3
40 = 36	4 = 4
50 = 45	5 = 5
60 = 54	6 = 5
70 = 63	7 = 6
	8 = 7
	9 = 8

35 = 32

3930013	·5943940
076	·5944010
140	080
204	151
268	221
331	291
395	361
458	432
522	502
585	573
649	643
713	713
777	783
840	854
904	924
967	994
3931031	·5945064
094	135
158	205
221	275
285	345
348	416
412	486
476	556
540	626
603	697
667	767
730	837
794	907
857	978
921	·5946048
984	118
3932048	188
112	258
176	328
239	399
303	469

3932366	·5946539
430	609
493	680
557	750
620	820
684	890
747	961
811	·5947031
874	101
938	171
3933001	241
065	311
129	382
193	452
256	522
320	592
383	662
447	732
510	803
574	873
637	943
701	·5948013
764	083
828	153
891	224
955	294
3934018	364
082	434
145	504
209	574
273	644
337	714
400	785
464	855
527	925
591	995
654	·5949065
718	135
781	205
845	275
908	346
972	416
3935035	486
099	556
162	626
226	696
289	766
353	836
416	907
480	977
543	·5950047
607	117
670	187
734	257
797	327
861	397
924	467
988	537
3936051	608
115	678
178	748
242	818
305	888
369	958
432	·5951028

3936496	·5951098
559	168
623	238
686	308
750	378
813	448
877	518
940	588
3937004	658
067	729
131	799
194	869
258	939
321	·5952009
385	079
448	149
512	219
575	289
639	359
702	429
766	499
829	569
893	639
956	709
3938020	779
083	849
147	919
210	989
274	·5953059
337	129
401	199
464	269
528	339
591	409
655	479
718	549
782	619
845	689
909	759
972	829
3939036	899
099	969
163	·5954039
226	109
290	178
353	248
417	318
480	388
544	458
607	528
671	598
734	668
797	738
860	808
924	878
987	948

Excesses

10 = 11	1 = 1
20 = 22	2 = 2
30 = 33	3 = 3
40 = 44	4 = 4
50 = 55	5 = 6
60 = 66	6 = 7
	7 = 8
	8 = 9
	9 = 10

32 = 35

Anti Log Excesses

10 = 9	1 = 1
20 = 18	2 = 2
30 = 27	3 = 3
40 = 36	4 = 4
50 = 45	5 = 5
60 = 54	6 = 5
70 = 63	7 = 6
	8 = 7
	9 = 8

35 = 32

3940051	·5955018
114	088
178	158
241	228
305	298
368	368
432	438
495	508
559	577
622	647
686	717
749	787
813	857
876	927
939	997
3941002	·5956067
066	137
129	207
193	276
256	346
320	416
383	486
447	556
510	626
574	696
637	766
701	836
764	906
827	976
890	·5957046
954	116
3942017	186
081	255
144	325
208	395
271	465
335	535

3942398	·5957605
462	674
525	744
588	814
651	884
715	954
778	·5958024
842	093
905	163
969	233
3943032	303
096	373
159	443
223	512
286	582
349	652
412	722
476	792
539	862
603	932
666	·5959002
730	071
793	141
856	211
919	281
983	350
3944046	420
110	490
173	560
237	629
300	699
364	769
427	839
490	909
553	979
617	·5960048
680	118
744	188
807	258
871	327
934	397
997	467
3945060	537
124	606
187	676
251	746
314	816
378	885
441	955
504	·5961025
567	095
631	164
694	234
758	304
821	374
884	443
947	513
3946011	583
074	653
138	722
201	792
265	861
328	931
391	·5962001
454	071

Number	Log
3946518	·5962140
581	210
645	280
708	350
771	419
834	489
898	559
961	629
3947025	698
088	768
151	837
214	907
278	977
341	·5963047
405	116
468	186
531	255
594	325
658	395
721	465
785	534
848	604
911	673
974	743
3948038	813
101	883
165	952
228	·5964022
291	091
354	161
418	231
481	301
545	370
608	440
671	509
734	579
798	649
861	719
924	788
987	858
3949051	927
114	997
178	·5965066
241	136
304	205
367	275
431	345
494	415
557	484
620	554
684	623
747	693
811	762
874	832
937	901

Excesses

10 = 11	1 = 1
20 = 22	2 = 2
30 = 33	3 = 3
40 = 44	4 = 4
50 = 55	5 = 6
60 = 66	6 = 7
	7 = 8
	8 = 9
	9 = 10

32 = 35

Anti Log Excesses

10 = 9	1 = 1
20 = 18	2 = 2
30 = 27	3 = 3
40 = 36	4 = 4
50 = 45	5 = 5
60 = 55	6 = 6
	7 = 6
	8 = 7
	9 = 8

35 = 32

Number	Log
3950000	·5965971
064	·5966041
127	111
190	180
253	250
317	319
380	389
444	458
507	528
570	597
633	667
697	736
760	806
823	876
886	946
950	·5967015
3951013	085
076	154
139	224
203	293
266	363
329	432
392	502
456	571
519	641
582	710
645	780
709	849
772	919
836	988
899	·5968058
962	127
3952025	197
089	266
152	336
215	405
278	475

Number	Log
3952342	·5968544
405	614
468	683
531	753
595	822
658	892
721	961
784	·5969031
848	100
911	170
974	239
3953037	309
101	378
164	448
227	517
290	587
354	656
417	726
480	795
543	865
606	934
669	·5970004
733	073
796	143
859	212
922	282
986	351
3954049	420
112	489
175	559
239	628
302	698
365	767
428	837
492	906
555	976
618	·5971045
681	115
744	184
807	254
871	323
934	392
997	461
3955060	531
124	600
187	670
250	739
313	809
377	878
440	948
503	·5972017
566	086
629	155
692	225
756	294
819	364
882	433
945	503
3956009	572
072	642
135	711
198	780
261	849
324	919
388	988

Number	Log
3956451	·5973058
514	127
577	196
640	265
703	335
767	404
830	474
893	543
956	612
3957020	681
083	751
146	820
209	890
272	959
335	·5974028
399	097
462	167
525	236
588	306
651	375
714	444
778	513
841	583
904	652
967	722
3958030	791
093	860
157	929
220	999
283	·5975068
346	137
409	206
472	276
536	345
599	415
662	484
725	553
788	622
851	692
914	761
977	830
3959041	899
104	969
167	·5976038
230	107
293	176
356	246
420	315
483	384
546	453
609	523
672	592
735	661
798	730
861	800
925	869
988	938

Excesses

10 = 11	1 = 1
20 = 22	2 = 2
30 = 33	3 = 3
40 = 44	4 = 4
50 = 55	5 = 6
60 = 66	6 = 7
	7 = 8
	8 = 9
	9 = 10

32 = 35

Anti Log Excesses

10 = 9	1 = 1
20 = 18	2 = 2
30 = 27	3 = 3
40 = 36	4 = 4
50 = 46	5 = 5
60 = 55	6 = 6
	7 = 6
	8 = 7
	9 = 8

35 = 32

Number	Log
3960051	·5977007
114	077
177	146
240	215
304	284
367	354
430	423
493	492
556	561
619	630
682	699
745	769
808	838
871	907
935	976
998	·5978046
3961061	115
124	184
187	253
250	322
313	391
376	461
440	530
503	599
566	668
629	738
692	807
755	876
818	945
881	·5979014
944	083
3962007	153
071	222
134	291
197	360
260	429
323	498

Number	Log
3962386	·5979568
449	637
512	706
575	775
638	844
702	913
765	982
828	·5980051
891	121
954	190
3963017	259
080	328
143	397
206	466
269	535
332	604
395	674
459	743
522	812
585	881
648	950
711	·5981019
774	088
837	157
900	227
963	296
3964026	365
089	434
152	503
215	572
278	641
342	710
405	779
468	848
531	917
594	986
657	·5982056
720	125
783	194
846	263
909	332
972	401
3965035	470
098	539
161	608
224	677
287	746
350	815
413	884
477	953
540	·5983022
603	091
666	160
729	229
792	298
855	367
918	436
981	505
3966044	575
107	644
170	713
233	782
296	851
359	920
422	989

Number	Log
3966485	·5984058
548	127
611	196
674	265
737	334
800	403
863	472
926	541
989	610
3967052	679
115	748
178	817
241	886
304	955
367	·5985024
430	093
493	162
556	231
619	300
682	369
745	437
808	506
871	575
934	644
997	713
3968060	782
123	851
186	920
249	989
312	·5986058
375	127
438	196
501	265
564	334
627	403
690	472
753	541
816	610
879	678
942	747
3969005	816
068	885
131	954
194	·5987023
257	092
320	161
383	230
446	299
509	368
572	437
635	505
698	574
761	643
824	712
887	781
950	850

Excesses

10 = 11	1 = 1
20 = 22	2 = 2
30 = 33	3 = 3
40 = 44	4 = 4
50 = 55	5 = 6
60 = 66	6 = 7
	7 = 8
	8 = 9
	9 = 10

32 = 35

Anti Log Excesses

10 = 9	1 = 1
20 = 18	2 = 2
30 = 27	3 = 3
40 = 37	4 = 4
50 = 46	5 = 5
60 = 55	6 = 6
	7 = 6
	8 = 7
	9 = 8

35 = 32

Number	Log
3970013	·5987919
076	988
139	·5988057
202	126
265	194
328	263
391	332
454	401
517	470
580	539
643	608
706	677
769	745
832	814
895	883
958	952
3971021	·5989021
083	090
146	158
209	227
272	296
335	365
398	434
461	503
524	571
587	640
650	709
713	778
776	847
839	916
902	984
965	·5990053
3972028	122
091	191
154	260
216	329
279	397

Number	Log
3972342	·5990466
405	535
468	604
531	673
594	742
657	810
720	879
783	948
846	·5991017
909	085
972	154
3973035	223
097	292
160	360
223	429
286	498
349	567
412	635
475	704
538	773
601	842
664	911
727	980
790	·5992048
852	117
915	186
978	255
3974041	323
104	392
167	461
230	530
293	598
356	667
419	735
481	804
544	873
607	942
670	·5993010
733	079
796	148
859	217
922	285
985	354
3975048	423
110	492
173	560
236	629
299	698
362	767
425	835
488	904
551	972
614	·5994041
677	110
739	179
802	247
865	316
928	385
991	454
3976054	522
117	591
180	659
242	728
305	797
368	866

Number	Log
3976431	·5994934
494	·5995003
557	071
620	140
683	209
745	278
808	346
871	415
934	483
997	552
3977060	620
122	689
185	758
248	827
311	895
374	964
437	·5996032
500	101
563	170
625	239
688	307
751	376
814	444
877	513
940	582
3978002	651
065	719
128	788
191	856
254	925
317	993
379	·5997062
442	130
505	199
568	268
631	337
694	405
756	474
819	542
882	611
945	679
3979008	748
071	816
133	885
196	953
259	·5998022
322	090
385	159
448	228
510	297
573	365
636	434
699	502
762	571
825	639
887	708
950	776

Excesses	
10 = 11	1 = 1
20 = 22	2 = 2
30 = 33	3 = 3
40 = 44	4 = 4
50 = 55	5 = 6
60 = 66	6 = 7
	7 = 8
	8 = 9
	9 = 10
32 = 35	

Anti Log Excesses	
10 = 9	1 = 1
20 = 18	2 = 2
30 = 27	3 = 3
40 = 37	4 = 4
50 = 46	5 = 5
60 = 55	6 = 6
	7 = 6
	8 = 7
	9 = 8
35 = 32	

3980013	·5998845
076	913
138	982
201	·5999050
264	119
327	187
390	256
453	324
515	393
578	461
641	530
704	598
767	667
830	735
892	804
955	872
3981018	941
081	·6000009
143	078
206	146
269	215
332	283
395	352
458	420
520	489
583	557
646	626
709	694
771	763
834	831
897	900
960	968
3982022	·6001037
085	105
148	174
211	242
273	311

3982336	·6001379
399	448
462	516
525	585
588	653
650	722
713	790
776	859
839	927
901	996
964	·6002064
3983027	132
090	200
152	269
215	337
278	406
341	474
403	543
466	611
529	680
592	748
654	817
717	885
780	954
843	·6003022
905	090
968	158
3984031	227
094	295
156	364
219	432
282	501
345	569
407	638
470	706
533	774
596	842
658	911
721	979
784	·6004048
847	116
909	185
972	253
3985035	321
098	389
160	458
223	526
286	595
349	663
411	731
474	799
537	868
600	936
662	·6005005
725	073
787	142
850	210
913	278
976	346
3986038	415
101	483
164	552
227	620
289	688
352	756

3986415	·6005825
478	893
540	962
603	·6006030
665	098
728	166
791	235
854	303
916	371
979	439
3987042	508
105	576
167	645
230	713
293	781
356	849
418	918
481	986
543	·6007054
606	122
669	191
732	259
794	327
857	395
919	464
982	532
3988045	600
108	668
170	737
233	805
296	873
359	941
421	·6008010
484	078
546	146
609	214
672	283
735	351
797	420
860	488
922	556
985	624
3989048	692
111	760
173	829
236	897
298	965
361	·6009033
424	102
487	170
549	238
612	306
674	375
737	443
800	511
863	579
925	647
988	715

Excesses	
10 = 11	1 = 1
20 = 22	2 = 2
30 = 33	3 = 3
40 = 44	4 = 4
50 = 54	5 = 5
60 = 65	6 = 7
	7 = 8
	8 = 9
	9 = 10
32 = 35	

Anti Log Excesses	
10 = 9	1 = 1
20 = 18	2 = 2
30 = 28	3 = 3
40 = 37	4 = 4
50 = 46	5 = 5
60 = 55	6 = 6
	7 = 6
	8 = 7
	9 = 8
35 = 32	

3990050	·6009784
113	852
176	920
239	988
301	·6010057
364	125
426	193
489	261
551	329
614	397
677	466
740	534
802	602
865	670
927	738
990	806
3991053	875
116	943
178	·6011011
241	079
303	147
366	215
428	284
491	352
554	420
617	488
679	556
742	624
804	692
867	760
929	829
992	897
3992055	965
118	·6012033
180	101
243	169
305	237

3992368	·6012305
430	374
493	442
556	510
619	578
681	646
744	714
806	782
869	850
931	918
994	986
3993057	·6013055
120	123
182	191
245	259
307	327
370	395
432	463
495	531
557	599
620	667
683	735
746	803
808	872
871	940
933	·6014008
996	076
3994058	144
121	212
183	280
246	348
309	416
372	484
434	552
497	620
559	688
622	756
684	824
747	892
809	960
872	·6015028
934	096
997	164
3995060	232
123	300
185	368
248	436
310	504
373	572
435	641
498	709
560	777
623	845
685	913
748	981
810	·6016049
873	117
935	185
998	253
3996061	321
124	389
186	457
249	525
311	593
374	661

3996436	·6016729
499	797
561	865
624	933
686	·6017001
749	069
811	137
874	205
937	273
999	340
3997061	408
124	476
187	544
250	612
312	680
375	748
437	816
500	884
562	952
625	·6018020
687	088
750	156
812	224
875	292
937	360
3998000	428
062	496
125	563
187	631
250	699
312	767
375	835
437	903
500	971
562	·6019039
625	107
687	175
750	242
812	310
875	378
937	446
3999000	514
062	582
125	650
187	718
250	786
312	854
375	921
437	989
500	·6020057
562	125
625	193
687	261
750	329
812	397
875	464
937	532

Excesses

10 = 11	1 = 1
20 = 22	2 = 2
30 = 33	3 = 3
40 = 43	4 = 4
50 = 54	5 = 5
60 = 65	6 = 7
	7 = 8
	8 = 9
	9 = 10

31 = 34

Anti Log Excesses

10 = 9	1 = 1
20 = 18	2 = 2
30 = 28	3 = 3
40 = 37	4 = 4
50 = 46	5 = 5
60 = 55	6 = 6
	7 = 6
	8 = 7
	9 = 8

34 = 31

4000000	·6020600
062	668
125	736
187	804
250	872
312	940
375	·6021007
437	075
500	143
562	211
625	279
687	347
750	414
812	482
875	550
937	618
4001000	686
062	754
125	821
187	889
250	957
312	·6022025
375	093
437	161
500	228
562	296
625	364
687	432
750	499
812	567
875	635
937	703
4002000	771
062	839
125	906
187	974
250	·6023042

4002312	·6023110
375	177
437	245
500	313
562	381
625	448
687	516
750	584
812	652
874	719
936	787
999	855
4003061	923
124	990
186	·6024058
249	126
311	194
374	261
436	329
499	397
561	465
624	532
686	600
749	668
811	736
874	803
936	871
999	939
4004061	·6025007
123	074
185	142
248	210
310	278
373	345
435	413
498	480
560	548
623	616
685	684
748	751
810	819
873	887
935	955
997	·6026022
4005059	090
122	157
184	225
247	293
309	361
372	428
434	496
497	563
559	631
622	699
684	767
746	834
808	902
871	969
933	·6027037
996	105
4006058	173
121	240
183	308
246	375
308	443

4006370	·6027511
432	579
495	646
557	714
620	781
682	849
745	916
807	984
870	·6028052
932	120
994	187
4007056	255
119	322
181	390
244	457
306	525
369	593
431	661
493	728
555	796
618	863
680	931
743	998
805	·6029066
868	133
930	201
993	269
4008055	337
117	404
179	472
242	539
304	607
367	674
429	742
491	809
553	877
616	944
678	·6030012
741	080
803	148
866	215
928	283
990	350
4009052	418
115	485
177	553
240	620
302	688
365	755
427	823
489	890
551	958
614	·6031025
676	093
739	160
801	228
863	295
925	363
988	430

Excesses

10 = 11	1 = 1
20 = 22	2 = 2
30 = 33	3 = 3
40 = 43	4 = 4
50 = 54	5 = 5
60 = 65	6 = 7
	7 = 8
	8 = 9
	9 = 10

31 = 34

Anti Log Excesses

10 = 9	1 = 1
20 = 18	2 = 2
30 = 28	3 = 3
40 = 37	4 = 4
50 = 46	5 = 5
60 = 55	6 = 6
	7 = 6
	8 = 7
	9 = 8

34 = 31

4010050	·6031498
113	565
175	633
237	700
299	768
362	835
424	903
487	970
549	·6032038
611	105
673	173
736	240
798	308
861	375
923	443
985	510
4011047	578
110	645
172	713
235	780
297	848
359	915
421	983
484	·6033050
546	118
609	185
671	253
733	320
795	388
858	455
920	523
983	590
4012045	658
107	725
169	793
232	860
294	928

4012356	·6033995
418	·6034062
481	129
543	197
606	264
668	332
730	399
792	467
855	534
917	602
979	669
4013041	737
104	804
166	872
229	939
291	·6035006
353	073
415	141
478	208
540	276
602	343
664	411
727	478
789	546
852	613
914	680
976	747
4014038	815
101	882
163	950
225	·6036017
287	085
350	152
412	219
474	286
536	354
599	421
661	489
723	556
785	624
848	691
910	758
972	825
4015034	893
097	960
159	·6037028
222	095
284	162
346	229
408	297
471	364
533	432
595	499
657	566
720	633
782	701
844	768
906	836
969	903
4016031	970
093	·6038037
155	105
218	172
280	240
342	307

4016404	·6038374
467	441
529	509
591	576
653	643
716	710
778	778
840	845
902	912
965	979
4017027	·6039047
089	114
151	182
213	249
275	316
338	383
400	451
462	518
524	585
587	652
649	720
711	787
773	854
836	921
898	989
960	·6040056
4018022	123
085	190
147	258
209	325
271	392
333	459
395	527
458	594
520	661
582	728
644	796
707	863
769	930
831	997
893	·6041065
956	132
4019018	199
080	266
142	333
204	400
266	468
329	535
391	602
453	669
515	737
578	804
640	871
702	938
764	·6042005
826	072
888	140
951	207

Excesses

10 = 11	1 = 1
20 = 22	2 = 2
30 = 32	3 = 3
40 = 43	4 = 4
50 = 54	5 = 5
60 = 65	6 = 7
	7 = 8
	8 = 9
	9 = 10

31 = 34

Anti Log Excesses

10 = 9	1 = 1
20 = 18	2 = 2
30 = 28	3 = 3
40 = 37	4 = 4
50 = 46	5 = 5
60 = 55	6 = 6
	7 = 6
	8 = 7
	9 = 8

34 = 31

4020013	·6042274
075	341
137	408
200	475
262	543
324	610
386	677
448	744
510	812
573	879
635	946
697	·6043013
759	080
821	147
883	214
946	281
4021008	349
070	416
132	483
194	550
256	617
319	684
381	752
443	819
505	886
567	953
629	·6044020
692	087
754	154
816	221
878	289
940	356
4022002	423
065	490
127	557
189	624
251	691

4022313	·6044758
375	826
438	893
500	960
562	·6045027
624	094
686	161
748	228
810	295
872	362
935	429
997	496
4023059	563
121	631
183	698
245	765
308	832
370	899
432	966
494	·6046033
556	100
618	167
680	234
742	301
805	368
867	435
929	502
991	569
4024053	636
115	704
177	771
239	838
302	905
364	972
426	·6047039
488	106
550	173
612	240
674	307
736	374
799	441
861	508
923	575
985	642
4025047	709
109	776
171	843
233	910
296	977
358	·6048044
420	111
482	178
544	245
606	312
668	379
730	446
792	513
854	580
917	647
979	714
4026041	781
103	848
165	915
227	982
289	·6049049

4026351	·6049116
413	183
475	250
537	317
599	384
662	451
724	518
786	585
848	652
910	719
972	786
4027034	852
096	919
158	986
220	·6050053
282	120
344	187
407	254
469	321
531	388
593	455
655	522
717	589
779	656
841	723
903	790
965	857
4028027	923
089	990
151	·6051057
213	124
276	191
338	258
400	325
462	392
524	459
586	526
648	592
710	659
772	726
834	793
896	860
958	927
4029020	994
082	·6052061
144	128
206	195
268	261
330	328
393	395
455	462
517	529
579	596
641	663
703	730
765	796
827	863
889	930
951	997

Excesses	
10 = 11	1 = 1
20 = 22	2 = 2
30 = 32	3 = 3
40 = 43	4 = 4
50 = 54	5 = 5
60 = 65	6 = 7
	7 = 8
	8 = 9
	9 = 10
31 = 33	

Anti Log Excesses	
10 = 9	1 = 1
20 = 19	2 = 2
30 = 28	3 = 3
40 = 37	4 = 4
50 = 46	5 = 5
60 = 56	6 = 6
	7 = 6
	8 = 7
	9 = 8
33 = 31	

4030013	·6053064
075	131
137	197
199	264
261	331
323	398
385	465
447	532
509	598
571	665
633	732
695	799
757	866
819	933
881	999
943	·6054066
4031005	133
067	200
129	267
191	334
253	400
315	467
377	534
439	601
501	667
563	734
625	801
687	868
750	935
812	·6055002
874	069
936	136
998	202
4032060	269
122	336
184	403
246	469

4032308	·6055536
370	603
432	670
494	736
556	803
618	870
680	937
742	·6056003
803	070
865	137
927	204
989	270
4033051	337
113	404
175	471
237	537
299	604
361	671
423	738
485	804
547	871
609	938
671	·6057005
733	071
795	138
857	205
919	272
981	338
4034043	405
105	472
167	539
229	605
291	672
353	738
415	805
477	872
539	939
601	·6058005
663	072
725	139
787	206
849	272
911	339
973	405
4035035	472
097	539
158	606
220	672
282	739
344	805
406	872
468	939
530	·6059006
592	072
654	139
716	205
778	272
840	339
902	406
964	472
4036026	539
088	605
150	672
212	739
273	806

4036335	·6059872
397	939
459	·6060005
521	072
583	139
645	206
707	272
769	339
831	405
893	472
955	538
4037017	605
079	671
141	738
203	805
264	872
326	938
388	·6061005
450	071
512	138
574	204
636	271
698	338
760	405
822	471
884	538
946	604
4038007	671
069	737
131	804
193	870
255	937
317	·6062003
379	070
441	137
503	204
565	270
626	337
688	403
750	470
812	536
874	603
936	669
998	736
4039060	802
122	869
184	935
245	·6063002
307	068
369	135
431	202
493	269
555	335
617	402
679	468
740	535
802	601
864	668
926	734
988	801

Excesses	
10 = 11	1 = 1
20 = 22	2 = 2
30 = 32	3 = 3
40 = 43	4 = 4
50 = 54	5 = 5
60 = 65	6 = 7
	7 = 8
	8 = 9
	9 = 10
31 = 33	

Anti Log Excesses	
10 = 9	1 = 1
20 = 19	2 = 2
30 = 28	3 = 3
40 = 37	4 = 4
50 = 46	5 = 5
60 = 56	6 = 6
	7 = 7
	8 = 7
	9 = 8
33 = 31	

4040050	·6063867
112	934
174	·6064000
236	067
298	133
359	200
421	266
483	333
545	399
607	466
669	532
731	599
793	665
854	732
916	798
978	865
4041040	931
102	998
164	·6065064
225	131
287	197
349	264
411	330
473	397
535	463
597	530
659	596
720	663
782	729
844	796
906	862
968	928
4042030	994
091	·6066061
153	127
215	194
277	260

4042339	·6066327
401	393
462	460
524	526
586	593
648	659
710	726
772	792
834	859
896	925
957	992
4043019	·6067058
081	124
143	190
205	257
267	323
328	390
390	456
452	523
514	589
575	656
637	722
699	788
761	854
823	921
885	987
946	·6068054
4044008	120
070	187
132	253
194	320
256	386
317	452
379	518
441	585
503	651
565	718
627	784
688	850
750	916
812	983
874	·6069049
935	116
997	182
4045059	249
121	315
183	381
245	447
306	514
368	580
430	647
492	713
553	779
615	845
677	912
739	978
801	·6070045
863	111
924	177
986	243
4046048	310
110	376
171	443
233	509
295	575

4046357	·6070641
418	708
480	774
542	840
604	906
666	973
728	·6071039
789	106
851	172
913	238
975	304
4047036	371
098	437
160	503
222	569
283	636
345	702
407	768
469	834
530	901
592	967
654	·6072034
716	100
777	166
839	232
901	299
963	365
4048024	431
086	497
148	564
210	630
271	696
333	762
395	829
457	895
518	961
580	·6073027
642	094
704	160
765	226
827	292
889	358
951	424
4049012	491
074	557
136	623
198	689
259	756
321	822
383	888
445	954
506	·6074021
568	087
630	153
692	219
753	286
815	352
877	418
939	484

Excesses

10 = 11	1 = 1
20 = 21	2 = 2
30 = 32	3 = 3
40 = 43	4 = 4
50 = 54	5 = 5
60 = 64	6 = 6
	7 = 7
	8 = 9
	9 = 10

31 = 33

Anti Log Excesses

10 = 9	1 = 1
20 = 19	2 = 2
30 = 28	3 = 3
40 = 37	4 = 4
50 = 47	5 = 5
60 = 56	6 = 6
	7 = 7
	8 = 7
	9 = 8

33 = 31

4050000	·6074550
062	616
124	683
186	749
247	815
309	881
371	948
433	·6075014
494	080
556	146
617	212
679	278
741	345
803	411
864	477
926	543
988	609
4051050	675
111	741
173	807
235	874
297	940
358	·6076006
420	072
481	138
543	204
605	271
667	337
728	403
790	469
852	535
914	601
975	668
4052037	734
098	800
160	866
222	932

4052284	·6076998
345	·6077064
407	130
469	196
531	262
592	329
654	395
715	461
777	527
839	593
901	659
962	725
4053024	791
085	857
147	923
209	990
271	·6078056
332	122
394	188
455	254
517	320
579	386
641	452
702	518
764	584
826	650
888	716
949	782
4054011	848
072	915
134	981
196	·6079047
258	113
319	179
381	245
442	311
504	377
565	443
627	509
689	575
751	641
812	707
874	773
935	839
997	905
4055059	971
121	·6080037
182	103
244	169
305	235
367	301
429	367
491	433
552	499
614	565
675	631
737	697
798	763
860	829
922	895
984	961
4056045	·6081027
107	093
168	159
230	225

4056292	·6081291
354	357
415	423
477	489
538	555
600	621
661	687
723	753
785	819
847	885
908	951
970	·6082017
4057031	083
093	149
154	215
216	281
278	347
340	413
401	479
463	545
524	611
586	677
647	743
709	809
770	875
832	940
894	·6083006
956	072
4058017	138
079	204
140	270
202	336
263	402
325	468
386	534
448	600
510	666
572	732
633	798
695	863
756	929
818	995
879	·6084061
941	127
4059002	193
064	259
126	325
188	391
249	457
311	523
372	589
434	654
495	720
557	786
618	852
680	918
741	984
803	·6085050
865	116
927	181
988	247

Excesses

10 = 11	1 = 1
20 = 21	2 = 2
30 = 32	3 = 3
40 = 43	4 = 4
50 = 54	5 = 5
60 = 64	6 = 6
	7 = 7
	8 = 9
	9 = 10

31 = 33

Anti Log Excesses

10 = 9	1 = 1
20 = 19	2 = 2
30 = 28	3 = 3
40 = 37	4 = 4
50 = 47	5 = 5
60 = 56	6 = 6
	7 = 7
	8 = 7
	9 = 8

33 = 31

4060050	·6085313
111	379
173	445
234	511
296	576
357	642
419	708
480	774
542	840
603	906
665	972
727	·6086038
789	103
850	169
912	235
973	301
4061035	367
096	433
158	498
219	564
281	630
342	696
404	762
465	828
527	893
588	959
650	·6087025
712	091
774	157
835	223
897	288
958	354
4062020	420
081	486
143	551
204	617
266	683

4062327	·6087749
389	815
450	881
512	946
573	·6088012
635	078
696	144
758	209
819	275
881	341
942	407
4063004	472
065	538
127	604
188	670
250	735
311	801
373	867
435	933
497	998
558	·6089064
620	130
681	196
743	261
804	327
866	393
927	459
989	524
4064050	590
112	656
173	722
235	787
296	853
358	919
419	985
481	·6090050
542	116
604	182
665	248
727	313
788	379
850	444
911	510
973	576
4065034	642
096	707
157	773
219	839
280	905
342	970
403	·6091036
465	101
526	167
588	233
649	299
711	364
772	430
834	496
895	562
957	627
4066018	693
080	758
141	824
203	890
264	956

4066326	·6092021
387	087
449	152
510	218
572	283
633	349
694	415
755	481
817	546
878	612
940	677
4067001	743
063	809
124	875
186	940
247	·6093006
309	071
370	137
432	202
493	268
555	334
616	400
678	465
739	531
801	596
862	662
924	727
985	793
4068047	859
108	925
170	990
231	·6094056
293	121
354	187
415	253
476	319
538	384
599	450
661	515
722	581
784	646
845	712
907	777
968	843
4069030	908
091	974
153	·6095040
214	106
276	171
337	237
398	302
459	368
521	433
582	499
644	564
705	630
767	695
828	761
890	826
951	892

Excesses

10 = 11	1 = 1
20 = 21	2 = 2
30 = 32	3 = 3
40 = 43	4 = 4
50 = 53	5 = 5
60 = 64	6 = 6
	7 = 7
	8 = 9
	9 = 10

31 = 33

Anti Log Excesses

10 = 9	1 = 1
20 = 19	2 = 2
30 = 28	3 = 3
40 = 37	4 = 4
50 = 47	5 = 5
60 = 56	6 = 6
	7 = 7
	8 = 7
	9 = 8

33 = 31

4070013	·6095957
074	·6096023
136	088
197	154
258	220
319	286
381	351
442	417
504	482
565	548
627	613
688	679
750	744
811	810
873	875
934	941
995	·6097006
4071056	072
118	137
179	203
241	268
302	334
364	399
425	465
487	530
548	596
610	661
671	727
732	792
793	858
855	923
916	989
978	·6098054
4072039	120
101	185
162	251
223	316

4072284	·6098382
346	447
407	513
469	578
530	643
592	708
653	774
715	839
776	905
837	970
898	·6099036
960	101
4073021	167
083	232
144	298
206	363
267	429
328	494
389	560
451	625
512	691
574	756
635	821
697	886
758	952
819	·6100017
880	083
942	148
4074003	214
065	279
126	345
188	410
249	476
310	541
371	606
433	671
494	737
556	802
617	868
678	933
739	999
801	·6101064
862	130
924	195
985	260
4075047	325
108	391
169	456
230	522
292	587
353	653
415	718
476	783
537	848
598	914
660	979
721	·6102045
783	110
844	175
905	240
966	306
4076028	371
089	437
151	502
212	567

4076273	·6102632
334	698
396	763
457	829
519	894
580	959
641	·6103024
702	090
764	155
825	221
887	286
948	351
4077009	416
070	482
132	547
193	613
254	678
315	744
377	809
438	874
500	939
561	·6104005
622	070
683	135
745	200
806	266
868	331
929	396
990	461
4078051	527
113	592
174	658
235	723
296	788
358	853
419	919
481	984
542	·6105049
603	114
664	180
726	245
787	310
848	375
909	441
971	506
4079032	571
094	636
155	702
216	767
277	832
339	897
400	963
461	·6106028
522	093
584	158
645	224
706	289
767	354
829	419
890	485
951	550

Excesses

10 = 11	1 = 1
20 = 21	2 = 2
30 = 32	3 = 3
40 = 43	4 = 4
50 = 53	5 = 5
60 = 64	6 = 6
	7 = 7
	8 = 9
	9 = 10

30 = 32

Anti Log Excesses

10 = 9	1 = 1
20 = 19	2 = 2
30 = 28	3 = 3
40 = 38	4 = 4
50 = 47	5 = 5
60 = 56	6 = 6
	7 = 7
	8 = 8
	9 = 8

32 = 30

4080012	·6106615
074	680
135	745
197	810
258	876
319	941
380	·6107006
442	071
503	137
564	202
625	267
687	332
748	398
809	463
870	528
932	593
993	658
4081054	723
115	789
177	854
238	919
299	984
360	·6108049
422	114
483	180
544	245
605	310
667	375
728	440
789	505
850	571
912	636
973	701
4082034	766
095	831
157	896
218	962

4082279	·6109027
340	092
402	157
463	222
524	287
585	353
647	418
708	483
769	548
830	613
892	678
953	743
4083014	808
075	874
137	939
198	·6110004
259	069
320	134
381	199
442	264
504	329
565	394
626	459
687	525
749	590
810	655
871	720
932	785
994	850
4084055	915
116	980
177	·6111046
238	111
299	176
361	241
422	306
483	371
544	436
606	501
667	566
728	631
789	696
851	761
912	826
973	891
4085034	957
095	·6112022
156	087
218	152
279	217
340	282
401	347
463	412
524	477
585	542
646	607
707	672
768	737
830	802
891	867
952	932
4086013	997
074	·6113062
135	127
197	192

4086258	·6113257
319	322
380	387
441	452
502	517
564	582
625	647
686	712
747	777
809	842
870	907
931	972
992	·6114037
4087053	102
114	167
176	232
237	297
298	362
359	427
420	492
481	557
543	622
604	687
665	752
726	817
787	882
848	947
910	·6115012
971	077
4088032	142
093	207
154	272
215	337
276	402
337	467
399	532
460	597
521	662
582	727
643	792
704	857
766	922
827	987
888	·6116052
949	117
4089010	181
071	246
132	311
193	376
255	441
316	506
377	571
438	636
499	701
560	766
622	831
683	896
744	961
805	·6117026
866	090
927	155
988	220

Excesses

10 = 11	1 = 1
20 = 21	2 = 2
30 = 32	3 = 3
40 = 43	4 = 4
50 = 53	5 = 5
60 = 64	6 = 6
	7 = 7
	8 = 9
	9 = 10

30 = 32

Anti Log Excesses

10 = 9	1 = 1
20 = 19	2 = 2
30 = 28	3 = 3
40 = 38	4 = 4
50 = 47	5 = 5
60 = 56	6 = 6
	7 = 7
	8 = 8
	9 = 8

32 = 30

4090049	·6117285
111	350
172	415
233	480
294	545
355	610
416	675
477	739
538	804
599	869
660	934
722	999
783	·6118064
844	129
905	194
966	258
4091027	323
088	388
149	453
211	518
272	583
333	648
394	713
455	777
516	842
577	907
638	972
699	·6119037
760	102
822	166
883	231
944	296
4092005	361
066	426
127	491
188	556
249	621

4092310	·6119685
371	750
433	815
494	880
555	944
616	·6120009
677	074
738	139
799	204
860	269
921	333
982	398
4093043	463
104	528
166	593
227	658
288	722
349	787
410	852
471	917
532	981
593	·6121046
654	111
715	176
776	241
837	306
898	370
959	435
4094021	500
082	565
143	629
204	694
265	759
326	824
387	888
448	953
509	·6122018
570	083
631	147
692	212
753	277
814	342
875	406
936	471
997	536
4095058	601
120	665
181	730
242	795
303	860
364	924
425	989
486	·6123054
547	119
608	183
669	248
730	313
791	378
852	442
913	507
974	572
4096035	637
096	701
157	766
218	830

Number	Log
4096279	·6123895
340	960
401	·6124025
462	089
523	154
584	219
645	284
706	348
767	413
829	477
890	542
951	607
4097012	672
073	736
134	801
195	865
256	930
317	995
378	·6125060
439	124
500	189
561	253
622	318
683	383
744	448
805	512
866	577
927	641
988	706
4098049	771
110	836
171	900
232	965
293	·6126029
354	094
415	159
476	224
537	288
598	353
659	417
720	482
781	546
842	611
903	676
964	741
4099025	805
086	870
147	934
208	999
269	·6127063
330	128
391	193
452	258
513	322
574	387
635	451
696	516
757	580
818	645
879	709
939	774

Excesses

10 = 11	1 = 1
20 = 21	2 = 2
30 = 32	3 = 3
40 = 42	4 = 4
50 = 53	5 = 5
60 = 64	6 = 6
	7 = 7
	8 = 8
	9 = 10

30 = 32

Anti Log Excesses

10 = 9	1 = 1
20 = 19	2 = 2
30 = 28	3 = 3
40 = 38	4 = 4
50 = 47	5 = 5
60 = 57	6 = 6
	7 = 7
	8 = 8
	9 = 8

32 = 30

Number	Log
4100000	·6127838
061	903
122	968
183	·6128033
244	097
305	162
366	226
427	291
488	355
549	420
610	484
671	549
732	613
793	678
854	743
915	808
976	872
4101037	937
098	·6129001
159	066
220	130
281	195
342	259
403	324
464	388
525	453
586	517
646	582
707	646
768	711
829	775
890	840
951	904
4102012	969
073	·6130033
134	098
195	162
256	227
4102317	·6130291
378	356
439	421
500	486
561	550
622	615
683	679
743	744
804	808
865	873
926	937
987	·6131002
4103048	066
109	131
170	195
231	259
292	323
353	388
414	452
475	517
536	581
596	646
657	710
718	775
779	839
840	904
901	968
962	·6132033
4104023	097
084	162
145	226
206	291
267	355
327	420
388	484
449	549
510	613
571	678
632	742
693	807
754	871
815	936
876	·6133000
936	065
997	129
4105058	193
119	257
180	322
241	386
302	451
363	515
424	580
485	644
545	709
606	773
667	838
728	902
789	966
850	·6134030
911	095
972	159
4106033	224
094	288
154	353
215	417

Number	Log
4106276	·6134481
337	545
398	610
459	674
520	739
581	803
641	868
702	932
763	997
824	·6135061
885	125
946	189
4107007	254
068	318
128	383
189	447
250	511
311	575
372	640
433	704
494	769
555	833
615	898
676	962
737	·6136026
798	090
859	155
920	219
980	284
4108041	348
102	412
163	476
224	541
285	605
345	670
407	734
467	798
528	862
589	927
650	991
711	·6137055
772	119
832	184
893	248
954	313
4109015	377
076	441
137	505
197	570
258	634
319	699
380	763
441	827
502	891
562	956
623	·6138020
684	084
745	148
806	213
867	277
927	341
988	405

Excesses

10 = 11	1 = 1
20 = 21	2 = 2
30 = 32	3 = 3
40 = 42	4 = 4
50 = 53	5 = 5
60 = 63	6 = 6
	7 = 7
	8 = 8
	9 = 10

30 = 32

Anti Log Excesses

10 = 9	1 = 1
20 = 19	2 = 2
30 = 28	3 = 3
40 = 38	4 = 4
50 = 47	5 = 5
60 = 57	6 = 6
	7 = 7
	8 = 8
	9 = 9

32 = 30

Number	Log
4110049	·6138470
110	534
171	598
232	662
292	727
353	791
414	856
475	920
536	984
597	·6139048
657	113
718	177
779	241
840	305
900	370
961	434
4111022	498
083	562
144	626
205	690
265	755
326	819
387	883
448	947
509	·6140012
570	076
630	140
691	204
752	269
813	333
873	397
934	461
995	526
4112056	590
117	654
178	718
238	782

4112299	·6140846
360	911
421	975
481	·6141039
542	103
603	168
664	232
724	296
785	360
846	424
907	488
968	553
4113029	617
089	681
150	745
211	809
272	873
332	938
393	·6142002
454	066
515	130
575	194
636	258
697	323
758	387
818	451
879	515
940	579
4114001	643
062	708
123	772
183	836
244	900
305	964
366	·6143028
426	093
487	157
548	221
609	285
669	349
730	413
791	477
852	541
912	606
973	670
4115034	734
095	798
155	862
216	926
277	990
338	·6144054
398	118
459	182
520	247
581	311
641	375
702	439
763	503
824	567
884	631
945	695
4116006	759
067	823
127	888
188	952

4116249	·6145016
310	080
370	144
431	208
491	272
552	336
613	400
674	464
734	528
795	592
856	656
917	720
977	785
4117038	849
099	913
160	977
220	·6146041
281	105
342	169
403	233
463	297
524	361
584	425
645	489
706	553
767	617
827	681
888	745
949	809
4118010	873
070	937
131	·6147001
192	065
253	129
313	193
374	257
434	321
495	385
556	449
617	513
677	577
738	641
799	705
860	769
920	833
981	897
4119041	961
102	·6148025
163	089
224	153
284	217
345	281
405	345
466	409
527	473
588	537
648	601
709	665
770	729
831	793
891	857
952	921

Excesses	
10 = 11	1 = 1
20 = 21	2 = 2
30 = 32	3 = 3
40 = 42	4 = 4
50 = 53	5 = 5
60 = 63	6 = 6
	7 = 7
	8 = 8
	9 = 9
30 = 32	

Anti Log Excesses	
10 = 9	1 = 1
20 = 19	2 = 2
30 = 28	3 = 3
40 = 38	4 = 4
50 = 47	5 = 5
60 = 57	6 = 6
	7 = 7
	8 = 8
	9 = 9
32 = 30	

4120012	·6148985
073	·6149049
134	113
195	177
255	241
316	305
376	369
437	433
498	497
559	561
619	625
680	688
740	752
801	816
862	880
923	944
983	·6150008
4121044	072
104	136
165	200
226	264
287	328
347	392
408	456
468	520
529	583
590	647
651	711
711	775
772	839
832	903
893	967
954	·6151031
4122015	095
075	159
136	223
196	287

4122257	·6151350
317	414
378	478
439	542
500	606
560	670
621	734
681	798
742	861
803	925
864	989
924	·6152053
985	117
4123045	181
106	245
166	309
227	373
288	437
349	500
409	564
470	628
530	692
591	756
651	820
712	883
773	947
834	·6153011
894	075
955	139
4124015	203
076	266
136	330
197	394
258	458
319	522
379	586
440	649
500	713
561	777
621	841
682	905
743	969
804	·6154032
864	096
925	160
985	224
4125046	288
106	352
167	415
227	479
288	543
349	607
410	670
470	734
531	798
591	862
652	926
712	990
773	·6155053
833	117
894	181
955	245
4126016	308
076	372
137	436

4126197	·6155500
258	563
318	627
379	691
439	755
500	818
560	882
621	946
682	·6156010
743	073
803	137
864	201
924	265
985	328
4127045	392
106	456
166	520
227	583
287	647
348	711
408	775
469	838
530	902
591	966
651	·6157030
712	093
772	157
833	221
893	285
954	348
4128014	412
075	475
135	539
196	603
256	667
317	730
377	794
438	858
499	922
560	985
620	·6158049
681	113
741	177
802	240
862	304
923	367
983	431
4129044	495
104	559
165	622
225	686
286	749
346	813
407	877
467	941
528	·6159004
588	068
649	131
709	195
770	259
831	323
892	386
952	450

Excesses

10 = 11	1 = 1
20 = 21	2 = 2
30 = 32	3 = 3
40 = 42	4 = 4
50 = 53	5 = 5
60 = 63	6 = 6
	7 = 7
	8 = 8
	9 = 9

30 = 32

Anti Log Excesses

10 = 9	1 = 1
20 = 19	2 = 2
30 = 28	3 = 3
40 = 38	4 = 4
50 = 47	5 = 5
60 = 57	6 = 6
	7 = 7
	8 = 8
	9 = 9

32 = 30

4130013	·6159513
073	577
134	641
194	705
255	768
315	832
376	895
436	959
497	·6160022
557	086
618	150
678	214
739	277
799	341
860	404
920	468
981	532
4131041	596
102	659
162	723
223	786
283	850
344	913
404	977
465	·6161040
525	104
586	168
646	232
707	295
767	359
828	422
888	486
949	549
4132009	613
070	677
130	741
191	804

4132251	·6161868
312	931
372	995
433	·6162058
493	122
554	185
614	249
675	312
735	376
796	439
856	503
917	567
977	631
4133038	694
098	758
159	821
219	885
280	948
340	·6163012
401	075
461	139
522	202
582	266
643	329
703	393
764	456
824	520
885	583
945	647
4134006	710
066	774
127	838
187	902
248	965
308	·6164029
369	092
429	156
490	219
550	283
611	346
671	410
731	473
791	537
852	600
912	664
973	727
4135033	791
094	854
154	918
215	981
275	·6165045
336	108
396	172
457	235
517	299
578	362
638	426
699	489
759	553
820	616
880	680
941	743
4136001	806
061	869
121	933

4136182	·6165996
242	·6166060
303	123
363	187
424	250
484	314
545	377
605	441
666	504
726	568
787	631
847	695
908	758
968	822
4137028	885
088	948
149	·6167011
209	075
270	138
330	202
391	265
451	329
512	392
572	456
633	519
693	583
754	646
814	710
874	773
934	836
995	899
4138055	963
116	·6168026
176	090
237	153
297	217
358	280
418	344
478	407
538	470
599	533
659	597
720	660
780	724
841	787
901	851
962	914
4139022	977
083	·6169040
143	104
203	167
263	231
324	294
384	358
445	421
505	484
566	547
626	611
686	674
746	738
807	801
867	864
928	927
988	991

Excesses

10 = 11	1 = 1
20 = 21	2 = 2
30 = 32	3 = 3
40 = 42	4 = 4
50 = 53	5 = 5
60 = 63	6 = 6
	7 = 7
	8 = 8
	9 = 9

30 = 32

Anti Log Excesses

10 = 10	1 = 1
20 = 19	2 = 2
30 = 29	3 = 3
40 = 38	4 = 4
50 = 48	5 = 5
60 = 57	6 = 6
	7 = 7
	8 = 8
	9 = 9

32 = 30

4140049	·6170054
109	118
170	181
230	245
290	308
350	371
411	434
471	498
532	561
592	625
653	688
713	751
773	814
833	878
894	941
954	·6171005
4141015	068
075	131
136	194
196	258
256	321
316	384
377	447
437	511
498	574
558	638
619	701
679	764
739	827
799	891
860	954
920	·6172017
981	080
4142041	144
101	207
161	270
222	333

4142282	·6172397
343	460
403	524
464	587
524	650
584	713
644	777
705	840
765	903
826	966
886	·6173030
946	093
4143006	156
067	219
127	283
188	346
248	409
308	472
368	536
429	599
489	662
550	725
610	789
670	852
730	915
791	978
851	·6174042
912	105
972	168
4144032	231
092	294
153	357
213	421
274	484
334	547
394	610
454	674
515	737
575	800
636	863
696	926
756	989
816	·6175053
877	116
937	179
997	242
4145057	306
118	369
178	432
239	495
299	558
359	621
419	684
480	747
540	811
601	874
661	937
721	·6176000
781	064
842	127
902	190
962	253
4146022	316
083	379
143	443

4146204	·6176506
264	569
324	632
384	695
445	758
505	822
565	885
625	948
686	·6177011
746	074
806	137
866	200
927	263
987	327
4147048	390
108	453
168	516
228	579
289	642
349	705
409	768
469	832
530	895
590	958
650	·6178021
710	084
771	147
831	210
892	273
952	337
4148012	400
072	463
133	526
193	589
253	652
313	715
374	778
434	841
494	904
554	967
615	·6179030
675	094
735	157
795	220
856	283
916	346
976	409
4149036	472
097	535
157	598
217	661
277	724
338	787
398	851
458	914
518	977
579	·6180040
639	103
699	166
759	229
820	292
880	355
940	418

Excesses

10 = 10	1 = 1
20 = 21	2 = 2
30 = 31	3 = 3
40 = 42	4 = 4
50 = 52	5 = 5
60 = 63	6 = 6
	7 = 7
	8 = 8
	9 = 9

30 = 31

Anti Log Excesses

10 = 10	1 = 1
20 = 19	2 = 2
30 = 29	3 = 3
40 = 38	4 = 4
50 = 48	5 = 5
60 = 57	6 = 6
	7 = 7
	8 = 8
	9 = 9

31 = 30

4150000	·6180481
061	544
121	607
181	670
241	733
302	796
362	859
422	922
482	985
543	·6181048
603	111
663	174
723	237
784	300
844	363
904	426
964	489
4151024	552
084	615
145	678
205	741
265	804
325	867
386	930
446	993
506	·6182056
566	119
627	182
687	245
747	308
807	371
868	434
928	497
988	560
4152048	623
108	686
168	749

4152229	·6182812
289	875
349	938
409	·6183001
470	064
530	127
590	190
650	253
710	316
770	379
831	442
891	505
951	568
4153011	631
072	694
132	757
192	820
252	883
312	945
372	·6184008
433	071
493	134
553	197
613	260
674	323
734	386
794	449
854	512
914	575
974	638
4154035	701
095	764
155	826
215	889
275	952
335	·6185015
396	078
456	141
516	204
576	267
636	330
696	393
757	456
817	519
877	581
937	644
997	707
4155057	770
118	833
178	896
238	959
298	·6186022
358	084
418	147
479	210
539	273
599	336
659	399
719	462
779	525
840	588
900	651
960	713
4156020	776
080	839

4156140	·6186902
201	965
261	·6187028
321	090
381	153
441	216
501	279
562	342
622	405
682	468
742	531
802	593
862	656
922	719
982	782
4157043	845
103	908
163	970
223	·6188033
283	096
343	159
404	221
464	284
524	347
584	410
644	473
704	536
764	598
824	661
885	724
945	787
4158005	850
065	913
125	975
185	·6189038
245	101
305	164
366	226
426	289
486	352
546	415
606	477
666	540
726	603
786	666
846	729
906	792
967	854
4159027	917
087	980
147	·6190043
207	105
267	168
327	231
387	294
448	356
508	419
568	482
628	545
688	607
748	670
808	733
868	796
928	858
988	921

Excesses

10 = 10	1 = 1
20 = 21	2 = 2
30 = 31	3 = 3
40 = 42	4 = 4
50 = 52	5 = 5
60 = 63	6 = 6
	7 = 7
	8 = 8
	9 = 9

30 = 31

Anti Log Excesses

10 = 10	1 = 1
20 = 19	2 = 2
30 = 29	3 = 3
40 = 38	4 = 4
50 = 48	5 = 5
60 = 57	6 = 6
	7 = 7
	8 = 8
	9 = 9

31 = 30

4160049	·6190984
109	·6191047
169	109
229	172
289	235
349	298
409	360
469	423
529	485
589	548
649	611
709	674
770	736
830	799
890	862
950	925
4161010	987
070	·6192050
130	113
190	176
250	238
310	301
370	363
430	426
491	489
551	552
611	614
671	677
731	740
791	803
851	865
911	928
971	990
4162031	·6193053
091	116
151	179
211	241

4162271	·6193304
332	366
392	429
452	492
512	555
572	617
632	680
692	742
752	805
812	868
872	931
932	993
992	·6194056
4163052	118
112	181
172	243
232	306
292	369
352	432
413	494
473	557
533	619
593	682
653	745
713	808
773	870
833	933
893	995
953	·6195058
4164013	120
073	183
133	246
193	309
253	371
313	434
373	496
433	559
493	621
553	684
613	746
673	809
733	872
793	935
853	997
913	·6196060
973	122
4165033	185
094	247
154	310
214	372
274	435
334	497
394	560
454	622
514	685
574	748
634	811
694	873
754	936
814	998
874	·6197061
934	123
994	186
4166054	248
114	311

4166174	·6197373
234	436
294	498
354	561
414	623
474	686
534	749
594	812
654	874
714	937
774	999
834	·6198062
894	124
954	187
4167014	249
074	312
134	374
194	437
254	499
314	562
374	624
434	687
494	749
554	812
614	874
674	937
734	999
794	·6199062
854	124
914	187
974	249
4168034	312
094	374
154	437
214	499
273	562
333	624
393	687
453	749
513	812
573	874
633	937
693	999
753	·6200062
813	124
873	187
933	249
993	312
4169053	374
113	437
173	499
233	562
293	624
353	686
413	748
473	811
533	873
593	936
653	998
713	·6201061
773	123
833	186
893	248
952	311

Excesses

10 = 10	1 = 1
20 = 21	2 = 2
30 = 31	3 = 3
40 = 42	4 = 4
50 = 52	5 = 5
60 = 63	6 = 6
	7 = 7
	8 = 8
	9 = 9

30 = 31

Anti Log Excesses

10 = 10	1 = 1
20 = 19	2 = 2
30 = 29	3 = 3
40 = 38	4 = 4
50 = 48	5 = 5
60 = 58	6 = 6
	7 = 7
	8 = 8
	9 = 9

31 = 30

4170012	·6201373
072	436
132	498
192	561
252	623
312	686
372	748
432	811
492	873
552	935
612	997
672	·6202060
732	122
792	185
852	247
912	310
972	372
4171031	435
091	497
151	559
211	621
271	684
331	746
391	809
451	871
511	934
571	996
631	·6203059
691	121
751	183
811	245
871	308
931	370
990	433
4172050	495
110	558
170	620

4172230	·6203682
290	744
350	807
410	869
470	932
530	994
590	·6204057
650	119
709	181
769	243
829	306
889	368
949	431
4173009	493
069	555
129	617
189	680
249	742
309	805
369	867
428	929
488	991
548	·6205054
608	116
668	179
728	241
788	304
848	366
907	428
967	490
4174027	553
087	615
147	677
207	739
267	802
327	864
387	926
447	988
506	·6206051
566	113
626	176
686	238
746	300
806	362
866	425
926	487
985	549
4175045	611
105	674
165	736
225	799
285	861
345	923
405	985
464	·6207048
524	110
584	172
644	234
704	297
764	359
824	421
884	483
943	546
4176003	608
063	670

4176123	·6207732
183	795
243	857
303	919
363	981
422	·6208044
482	106
542	168
602	230
662	293
722	355
781	417
841	479
901	542
961	604
4177021	666
081	728
141	791
201	853
260	915
320	977
380	·6209040
440	102
500	164
560	226
619	288
679	350
739	413
799	475
859	537
919	599
978	662
4178038	724
098	786
158	848
218	910
278	972
337	·6210035
397	097
457	159
517	221
577	284
637	346
696	408
756	470
816	532
876	594
936	657
996	719
4179055	781
115	843
175	905
235	967
295	·6211030
355	092
414	154
474	216
534	278
594	340
653	403
713	465
773	527
833	589
893	651
953	713

Excesses

10 = 10	1 = 1
20 = 21	2 = 2
30 = 31	3 = 3
40 = 42	4 = 4
50 = 52	5 = 5
	6 = 6
	7 = 7
	8 = 8
	9 = 9

30 = 31

Anti Log Excesses

10 = 10	1 = 1
20 = 19	2 = 2
30 = 29	3 = 3
40 = 38	4 = 4
50 = 48	5 = 5
60 = 58	6 = 6
	7 = 7
	8 = 8
	9 = 9

31 = 30

4180012	·6211776
072	838
132	900
192	962
251	·6212024
311	086
371	148
431	210
491	273
551	335
610	397
670	459
730	521
790	583
849	645
909	707
969	769
4181029	831
089	894
149	956
208	·6213018
268	080
328	142
388	204
447	266
507	328
567	391
627	453
687	515
747	577
806	639
866	701
926	763
986	825
4182045	887
105	949
165	·6214011

4182225	·6214073
284	136
344	198
404	260
464	322
523	384
583	446
643	508
703	570
763	632
823	694
882	756
942	818
4183002	880
062	942
121	·6215004
181	066
241	128
301	190
360	252
420	314
480	377
540	439
599	501
659	563
719	625
779	687
838	749
898	811
958	873
4184018	935
077	997
137	·6216059
197	121
257	183
316	245
376	307
436	369
496	431
555	493
615	555
675	617
735	679
794	741
854	803
914	865
974	927
4185033	989
093	·6217051
153	113
213	175
272	237
332	299
391	361
451	423
511	485
571	547
630	609
690	671
750	733
810	795
869	857
929	918
989	980
4186049	·6218042

4186108	·6218104
168	166
228	228
288	290
347	352
407	414
467	476
527	538
586	600
646	662
705	724
765	786
825	848
885	910
944	972
4187004	·6219034
064	096
124	157
183	219
243	281
302	343
362	405
422	467
482	529
541	591
601	653
661	715
721	777
780	839
840	900
899	962
959	·6220024
4188019	086
079	148
138	210
198	272
258	334
318	396
377	458
437	519
496	581
556	643
616	705
676	767
735	829
795	891
854	953
914	·6221015
974	077
4189034	138
093	200
153	262
213	324
273	386
332	448
392	510
451	572
511	633
571	695
631	757
690	819
750	881
809	943
869	·6222004
929	066

4189989	·6222128

Excesses

10 = 10	1 = 1
20 = 21	2 = 2
30 = 31	3 = 3
40 = 42	4 = 4
50 = 52	5 = 5
	6 = 6
	7 = 7
	8 = 8
	9 = 9

30 = 31

Anti Log Excesses

10 = 10	1 = 1
20 = 19	2 = 2
30 = 29	3 = 3
40 = 39	4 = 4
50 = 48	5 = 5
60 = 58	6 = 6
	7 = 7
	8 = 8
	9 = 9

31 = 30

4190048	·6222190
108	252
167	314
227	375
287	437
347	499
406	561
466	623
525	685
585	746
644	808
704	870
764	932
824	994
883	·6223056
943	117
4191002	179
062	241
122	303
182	365
241	427
301	488
360	550
420	612
480	674
540	735
599	797
659	859
718	921
778	983
837	·6224045
897	106
957	168
4192017	230
076	292

4192136	·6224353
195	415
255	477
315	539
375	600
434	662
494	724
553	786
613	848
672	910
732	971
792	·6225033
852	095
911	157
971	218
4193030	280
090	342
149	404
209	465
269	527
329	588
388	650
448	712
507	774
567	835
626	897
686	959
745	·6226021
805	082
865	144
925	206
984	268
4194044	329
103	391
163	453
222	515
282	576
342	638
402	700
461	762
521	823
580	885
640	946
699	·6227008
759	070
818	132
878	193
938	255
998	317
4195057	379
117	440
176	502
236	563
295	625
355	687
414	749
474	810
533	872
593	934
653	996
713	·6228057
772	119
832	180
891	242
951	304

4196010	·6228366
070	427
129	489
189	550
248	612
308	674
368	736
428	797
487	859
547	920
606	982
666	·6229044
725	106
785	167
844	229
904	290
963	352
4197023	413
082	475
142	537
202	599
262	660
321	722
381	783
440	845
500	907
559	969
619	·6230030
678	092
738	153
797	215
857	276
916	338
976	399
4198035	461
095	523
154	585
214	646
274	708
334	769
393	831
453	892
512	954
572	·6231016
631	078
691	139
750	201
810	262
869	324
929	385
988	447
4199048	508
107	570
167	631
226	693
286	755
345	817
405	878
464	940
524	·6232001
583	063
643	124
702	186
762	247
821	309

4199881	·6232370
940	432

Excesses

10 = 10	1 = 1
20 = 21	2 = 2
30 = 31	3 = 3
40 = 41	4 = 4
50 = 52	5 = 5
	6 = 6
	7 = 7
	8 = 8
	9 = 9

30 = 31

Anti Log Excesses

10 = 10	1 = 1
20 = 19	2 = 2
30 = 29	3 = 3
40 = 39	4 = 4
50 = 48	5 = 5
60 = 58	6 = 6
	7 = 7
	8 = 8
	9 = 9

31 = 30

4200000	·6232493
060	555
120	616
179	678
239	739
298	801
358	863
417	925
477	986
536	·6233048
596	109
655	171
715	232
774	294
834	355
893	417
953	478
4201012	540
072	601
131	663
191	724
250	786
310	847
369	909
429	970
488	·6234032
548	093
607	155
667	216
726	278
786	339
845	401
905	462
964	524

4202024	·6234585
083	647
143	708
202	770
262	831
321	893
381	954
440	·6235016
500	077
559	139
619	200
678	262
738	323
797	385
857	446
916	507
976	568
4203035	630
095	691
154	753
214	814
273	876
333	937
392	999
451	·6236060
510	122
570	183
629	245
689	306
748	368
808	429
867	491
927	552
986	614
4204046	675
105	736
165	797
224	859
284	920
343	982
403	·6237043
462	105
522	166
581	228
641	289
700	351
760	412
819	473
879	534
938	596
998	657
4205057	719
116	780
175	842
235	903
294	965
354	·6238026
413	087
473	148
532	210
592	271
651	333
711	394
770	456
830	517

4205889	·6238578
949	639
4206008	701
068	762
127	824
186	885
245	947
305	·6239008
364	069
424	130
483	192
543	253
602	315
662	376
721	437
781	498
840	560
900	621
959	683
4207018	744
077	806
137	867
196	928
256	989
315	·6240051
375	112
434	174
494	235
553	296
613	357
672	419
731	480
790	542
850	603
909	664
969	725
4208028	787
088	848
147	909
207	970
266	·6241032
326	093
385	155
444	216
503	277
563	338
622	400
682	461
741	522
801	583
860	645
920	706
979	768
4209038	829
097	890
157	951
216	·6242013
276	074
335	135
395	196
454	258
514	319
573	380
632	441
691	503

4209751	·6242564
810	625
870	686
929	748
989	809

Excesses

10 = 10	1 = 1
20 = 21	2 = 2
30 = 31	3 = 3
40 = 41	4 = 4
50 = 52	5 = 5
	6 = 6
	7 = 7
	8 = 8
	9 = 9

30 = 31

Anti Log Excesses

10 = 10	1 = 1
20 = 19	2 = 2
30 = 29	3 = 3
40 = 39	4 = 4
50 = 48	5 = 5
60 = 58	6 = 6
	7 = 7
	8 = 8
	9 = 9

31 = 30

4210048	·6242870
107	931
166	993
226	·6243054
285	115
345	176
404	238
464	299
523	360
582	421
641	483
701	544
760	605
820	666
879	728
939	789
998	850
4211057	911
116	973
176	·6244034
235	095
295	156
354	218
414	279
473	340
532	401
591	462
651	523
710	585
770	646
829	707

4211888	·6244768	4215744	·6248743	4219598	·6252710	4221730	·6254904	4225577	·6258860
947	830	804	804	657	771	789	965	637	921
4212007	891	863	865	716	832	848	·6255026	696	982
066	952	923	926	775	893	907	087	755	·6259042
126	·6245013	982	987	835	954	967	148	814	103
185	074	4216041	·6249048	894	·6253015	4222026	209	873	164
245	135	100	109	953	076	085	270	932	225
304	197	160	170			144	331	992	286
363	258	219	231	Excesses		204	391	4226051	347
422	319	278	292			263	452	110	407
482	380	337	354	10 = 10	1 = 1	322	513	169	468
541	442	397	415	20 = 21	2 = 2	381	574	228	529
601	503	456	476	30 = 31	3 = 3	441	635	287	590
660	564	516	537	40 = 41	4 = 4	500	696	346	650
719	625	575	598	50 = 52	5 = 5	559	757	405	711
778	686	634	659		6 = 6	618	818	465	772
838	747	693	720		7 = 7	677	878	524	833
897	809	753	781		8 = 8	736	939	583	894
957	870	812	842		9 = 9	796	·6256000	642	955
4213016	931	871	903			855	061	701	·6260015
075	992	930	964	30 = 31		914	122	760	076
134	·6246053	990	·6250025			973	183	820	137
194	114	4217049	086	Anti Log Excesses		4223033	244	879	198
253	176	109	147			092	305	938	258
313	237	168	208	10 = 10	1 = 1	151	366	997	319
372	298	227	269	20 = 19	2 = 2	210	427	4227056	380
431	359	286	330	30 = 29	3 = 3	269	487	115	441
490	420	346	391	40 = 39	4 = 4	328	548	174	501
550	481	405	453	50 = 49	5 = 5	388	609	233	562
609	543	464	514	60 = 58	6 = 6	447	670	293	623
669	604	523	575		7 = 7	506	731	352	684
728	665	583	636		8 = 8	565	792	411	744
787	726	642	697		9 = 9	625	852	470	805
846	787	701	758			684	913	529	866
906	848	760	819	31 = 30		743	974	588	927
965	909	820	880			802	·6257035	648	987
4214025	970	879	941	4220012	·6253137	861	096	707	·6261048
084	·6247032	938	·6251002	072	198	920	157	766	109
143	093	997	063	131	259	980	218	825	170
202	154	4218057	124	190	320	4224039	279	884	230
262	215	116	185	249	381	098	339	943	291
321	276	175	246	309	442	157	400	4228002	352
381	337	234	307	368	503	216	461	061	413
440	398	294	368	427	563	275	522	121	473
499	459	353	429	486	624	335	583	180	534
558	521	413	490	545	685	394	644	239	595
618	582	472	551	604	746	453	704	298	656
677	643	531	612	664	807	512	765	357	716
737	704	590	673	723	868	571	826	416	777
796	765	650	734	782	929	630	887	475	838
855	826	709	795	841	990	690	948	534	899
914	887	768	856	901	·6254051	749	·6258009	594	959
974	948	827	917	960	112	808	069	653	·6262020
4215033	·6248010	887	978	4221019	173	867	130	712	080
092	071	946	·6252039	078	234	927	191	771	141
151	132	4219005	100	138	295	986	252	830	202
211	193	064	161	197	356	4225045	313	889	263
270	254	124	222	256	417	104	374	948	323
330	315	183	283	315	478	163	434	4229007	384
389	376	242	344	375	538	222	495	067	445
448	437	301	405	434	599	282	556	126	506
507	498	361	466	493	660	341	617	185	566
567	559	420	527	552	721	400	678	244	627
626	621	479	588	612	782	459	739	303	687
685	682	538	649	671	843	518	799	362	748

4229421	·6262809
480	870
539	930
598	991
658	·6263052
717	113
776	173
835	234
894	294
953	355

Excesses

10 = 10	1 = 1
20 = 21	2 = 2
30 = 31	3 = 3
40 = 41	4 = 4
50 = 51	5 = 5
	6 = 6
	7 = 7
	8 = 8
	9 = 9

30 = 31

Anti Log Excesses

10 = 10	1 = 1
20 = 19	2 = 2
30 = 29	3 = 3
40 = 39	4 = 4
50 = 49	5 = 5
60 = 58	6 = 6
	7 = 7
	8 = 8
	9 = 9

31 = 30

4230012	·6263416
071	477
131	537
190	598
249	658
308	719
367	780
426	841
485	901
544	962
603	·6264022
662	083
721	144
780	205
840	265
899	326
958	386
4231017	447
076	508
135	569
194	629
253	690
312	750
371	811
431	871
490	932

4231549	·6264993
608	·6265054
667	114
726	175
785	235
844	296
903	357
962	418
4232021	478
080	539
139	599
198	660
258	720
317	781
376	842
435	903
494	963
553	·6266024
612	084
671	145
730	205
789	266
848	326
907	387
966	447
4233025	508
084	569
143	630
203	690
262	751
321	811
380	872
439	932
498	993
557	·6267053
616	114
675	175
734	236
793	296
852	357
911	417
970	478
4234029	538
088	599
147	659
206	720
265	780
324	841
384	901
443	962
502	·6268022
561	083
620	144
679	205
738	265
797	326
856	386
915	447
974	507
4235033	568
092	628
151	689
210	749
269	810
328	870

4235387	·6268931
446	991
505	·6269052
564	112
623	173
682	233
741	294
800	354
859	415
918	475
977	536
4236036	596
095	657
154	717
213	778
272	838
331	899
390	959
449	·6270020
508	080
567	141
627	201
686	262
745	322
804	383
863	443
922	504
981	564
4237040	625
099	685
158	746
217	806
276	867
335	927
394	988
453	·6271048
512	109
571	169
630	230
689	290
748	351
807	411
866	472
925	532
984	593
4238043	653
101	713
160	773
219	834
278	894
337	955
396	·6272015
455	076
514	136
573	197
632	257
691	318
750	378
809	439
868	499
927	560
986	620
4239045	680
104	740
163	801

4239222	·6272861
281	922
340	982
399	·6273043
458	103
517	164
576	224
635	285
694	345
753	405
812	465
871	526
930	586
989	647

Excesses

10 = 10	1 = 1
20 = 21	2 = 2
30 = 31	3 = 3
40 = 41	4 = 4
50 = 51	5 = 5
	6 = 6
	7 = 7
	8 = 8
	9 = 9

30 = 31

Anti Log Excesses

10 = 10	1 = 1
20 = 20	2 = 2
30 = 29	3 = 3
40 = 39	4 = 4
50 = 49	5 = 5
60 = 59	6 = 6
	7 = 7
	8 = 8
	9 = 9

31 = 30

4240048	·6273707
107	768
166	828
225	888
284	948
342	·6274009
401	069
460	130
519	190
578	251
637	311
696	371
755	431
814	492
873	552
932	613
991	673
4241050	734
109	794
168	854
227	914
286	975

4241345	·6275035
404	096
463	156
521	216
580	276
639	337
698	397
757	458
816	518
875	579
934	639
993	699
4242052	759
111	820
170	880
229	940
288	·6276000
346	061
405	121
464	182
523	242
582	302
641	362
700	423
759	483
818	544
877	604
936	664
995	724
4243054	785
113	845
171	906
230	966
289	·6277026
348	086
407	147
466	207
525	267
584	327
643	388
702	448
761	509
820	569
878	629
937	689
996	750
4244055	810
114	870
173	930
232	991
291	·6278051
350	111
409	171
467	232
526	292
585	352
644	412
703	473
762	533
821	593
880	653
939	714
998	774
4245056	834
115	894

4245174	·6278955
233	·6279015
292	075
351	135
410	196
469	256
527	316
586	376
645	437
704	497
763	557
822	617
881	678
940	738
998	798
4246057	858
116	919
175	979
234	·6280039
293	099
352	160
411	220
469	280
528	340
587	400
646	460
705	521
764	581
823	641
882	701
940	762
999	822
4247058	882
117	942
176	·6281002
235	062
294	123
353	183
411	243
470	303
529	363
588	423
647	484
706	544
764	604
823	664
882	724
941	784
4248000	845
059	905
118	965
177	·6282025
235	085
294	145
353	206
412	266
471	326
530	386
588	446
647	506
706	567
765	627
824	687
883	747
941	807

4249000	·6282867
059	928
118	988
177	·6283048
236	108
294	168
353	228
412	288
471	348
530	408
589	468
647	529
706	589
765	649
824	709
883	769
942	829

Excesses

10 = 10	1 = 1
20 = 20	2 = 2
30 = 31	3 = 3
40 = 41	4 = 4
50 = 51	5 = 5
	6 = 6
	7 = 7
	8 = 8
	9 = 9

30 = 31

Anti Log Excesses

10 = 10	1 = 1
20 = 20	2 = 2
30 = 29	3 = 3
40 = 39	4 = 4
50 = 49	5 = 5
60 = 59	6 = 6
	7 = 7
	8 = 8
	9 = 9

31 = 30

4250000	·6283889
059	949
118	·6284010
177	070
236	130
295	190
353	250
412	310
471	370
530	430
589	490
648	550
706	611
765	671
824	731
883	791
941	851
4251000	911
059	971

4251118	·6285031
177	091
236	151
294	211
353	271
412	331
471	391
529	452
588	512
647	572
706	632
765	692
824	752
882	812
941	872
4252000	932
059	992
117	·6286052
176	112
235	172
294	232
353	292
412	352
470	412
529	472
588	532
647	592
705	653
764	713
823	773
882	833
940	893
999	953
4253058	·6287013
117	073
176	133
235	193
293	253
352	313
411	373
470	433
528	493
587	553
646	613
705	673
763	733
822	793
881	853
940	913
998	973
4254057	·6288033
116	093
175	153
233	213
292	273
351	333
410	393
469	453
528	513
586	573
645	633
704	693
763	753
821	813
880	873

4254939	·6288933
998	993
4255056	·6289053
115	113
174	173
233	233
291	293
350	353
409	413
468	472
526	532
585	592
644	652
703	712
761	772
820	832
879	892
938	952
996	·6290012
4256055	072
113	132
172	192
231	252
290	312
348	372
407	431
466	491
525	551
583	611
642	671
701	731
760	791
818	851
877	911
936	971
995	·6291031
4257053	091
112	150
171	210
230	270
288	330
347	390
406	450
465	510
523	570
582	630
640	690
699	749
758	809
817	869
875	929
934	989
993	·6292049
4258052	109
110	169
169	229
228	289
287	348
345	408
404	468
462	528
521	588
580	648
639	708
697	768

4258756	·6292827
815	887
874	947
932	·6293007
991	067
4259049	127
108	186
167	246
226	306
284	366
343	426
402	486
461	545
519	605
578	665
636	725
695	785
754	845
813	905
871	965
930	·6294024
988	084

Excesses

10 = 10	1 = 1
20 = 20	2 = 2
30 = 31	3 = 3
40 = 41	4 = 4
50 = 51	5 = 5
	6 = 6
	7 = 7
	8 = 8
	9 = 9

30 = 31

Anti Log Excesses

10 = 10	1 = 1
20 = 20	2 = 2
30 = 29	3 = 3
40 = 39	4 = 4
50 = 49	5 = 5
60 = 59	6 = 6
	7 = 7
	8 = 8
	9 = 9

31 = 30

4260047	·6294144
106	204
165	263
223	323
282	383
341	443
400	503
458	563
517	622
575	682
634	742
693	802
752	862
810	922

4260869	·6294981
927	·6295041
986	101
4261045	161
104	220
162	280
221	340
279	400
338	460
397	520
456	579
514	639
573	699
631	759
690	818
749	878
808	938
866	998
925	·6296057
983	117
4262042	177
101	237
160	296
218	356
277	416
335	476
394	536
452	596
511	655
570	715
629	775
687	835
746	894
804	954
863	·6297014
922	074
981	133
4263039	193
098	253
156	313
215	372
274	432
333	491
391	551
450	611
508	671
567	731
625	791
684	850
743	910
802	969
860	·6298029
919	089
977	149
4264036	208
094	268
153	328
212	388
271	447
329	507
388	566
446	626
505	686
563	746
622	805

4264681	·6298865
740	925
798	985
857	·6299044
915	104
974	163
4265032	223
091	283
150	343
209	402
267	462
326	522
384	582
443	641
501	701
560	760
618	820
677	880
736	940
795	999
853	·6300059
912	118
970	178
4266029	238
087	298
146	357
205	417
264	476
322	536
381	595
439	655
498	715
556	775
615	834
673	894
732	953
790	·6301013
849	073
908	133
967	192
4267025	252
084	311
142	371
201	430
259	490
318	550
376	610
435	669
493	729
552	788
611	848
670	907
728	967
787	·6302027
845	087
904	146
962	206
4268021	265
079	325
138	384
196	444
255	503
314	563
373	623
431	683

4268490	·6302742
548	802
607	861
665	921
724	980
782	·6303040
841	099
899	159
958	218
4269016	278
075	337
133	397
192	457
251	517
310	576
368	636
427	695
485	755
544	814
602	874
661	933
719	993
778	·6304052
836	112
895	171
953	231

Excesses

10 = 10	1 = 1
20 = 20	2 = 2
30 = 31	3 = 3
40 = 41	4 = 4
50 = 51	5 = 5
	6 = 6
	7 = 7
	8 = 8
	9 = 9

29 = 30

Anti Log Excesses

10 = 10	1 = 1
20 = 20	2 = 2
30 = 29	3 = 3
40 = 39	4 = 4
50 = 49	5 = 5
	6 = 6
	7 = 7
	8 = 8
	9 = 9

30 = 29

4270012	·6304291
070	351
129	410
187	470
246	529
304	589
363	648
422	708
481	767
539	827

4270598	·6304886
656	946
715	·6305005
773	065
832	124
890	184
949	243
4271007	303
066	362
124	422
183	481
241	541
300	600
358	660
417	719
475	779
534	838
592	898
651	957
709	·6306017
768	076
826	136
885	195
943	255
4272002	314
060	374
119	433
177	493
236	552
294	612
353	671
412	731
471	790
529	850
588	909
646	969
705	·6307028
763	088
822	147
880	207
939	266
997	326
4273056	385
114	445
173	504
231	564
290	623
348	682
407	741
465	801
524	860
582	920
641	979
699	·6308039
758	098
816	158
875	217
933	277
992	336
4274050	396
109	455
167	515
226	574
284	634
343	693

4274401	·6308753
460	812
518	872
577	931
635	990
693	·6309049
751	109
810	168
868	228
927	287
985	347
4275044	406
102	466
161	525
219	584
278	643
336	703
395	762
453	822
512	881
570	941
629	·6310000
687	060
746	119
804	178
863	237
921	297
980	356
4276038	416
097	475
155	535
214	594
272	653
331	712
389	772
448	831
506	891
565	950
623	·6311010
681	069
739	128
798	187
856	247
915	306
973	366
4277032	425
090	484
149	543
207	603
266	662
324	722
383	781
441	841
500	900
558	959
617	·6312018
675	078
734	137
792	197
850	256
908	315
967	374
4278025	434
084	493
142	552

4278201	·6312611
259	671
318	730
376	790
435	849
493	908
552	967
610	·6313027
669	086
727	146
785	205
843	264
902	323
960	383
4279019	442
077	501
136	560
194	620
253	679
311	739
370	798
428	857
486	916
544	976
603	·6314035
661	094
720	153
778	213
837	272
895	331
954	390

Excesses

10 = 10	1 = 1
20 = 20	2 = 2
30 = 30	3 = 3
40 = 41	4 = 4
50 = 51	5 = 5
	6 = 6
	7 = 7
	8 = 8
	9 = 9

29 = 30

Anti Log Excesses

10 = 10	1 = 1
20 = 20	2 = 2
30 = 30	3 = 3
40 = 39	4 = 4
50 = 49	5 = 5
	6 = 6
	7 = 7
	8 = 8
	9 = 9

30 = 29

4280012	·6314450
071	509
129	568
187	627
245	687

4280304	·6314746
362	805
421	864
479	924
538	983
596	·6315042
655	101
713	161
771	220
829	279
888	338
946	398
4281005	457
063	516
122	575
180	635
239	694
297	753
355	812
413	872
472	931
530	990
589	·6316049
647	109
706	168
764	227
823	286
881	346
939	405
997	464
4282056	523
114	583
173	642
231	701
290	760
348	819
406	878
464	938
523	997
581	·6317056
640	115
698	174
757	233
815	293
873	352
931	411
990	470
4283048	530
107	589
165	648
224	707
282	766
340	825
398	885
457	944
515	·6318003
574	062
632	121
690	180
748	240
807	299
865	358
924	417
982	476
4284041	535

4284099	·6318595
157	654
215	713
274	772
332	831
391	890
449	950
507	·6319009
565	068
624	127
682	186
741	245
799	304
858	363
916	423
974	482
4285032	541
091	600
149	659
208	718
266	777
324	836
382	896
441	955
499	·6320014
558	073
616	132
674	191
732	250
791	309
849	369
908	428
966	487
4286024	546
082	605
141	664
199	723
258	782
316	841
374	900
432	960
491	·6321019
549	078
608	137
666	196
724	255
782	314
841	373
899	432
957	491
4287015	550
074	609
132	669
191	728
249	787
307	846
365	905
424	964
482	·6322023
541	082
599	141
657	200
715	259
774	318
832	377

4287890	·6322436
948	495
4288007	554
065	613
124	672
182	732
240	791
298	850
357	909
415	968
473	·6323027
531	086
590	145
648	204
707	263
765	322
823	381
881	440
940	499
998	558
4289056	617
114	676
173	735
231	794
289	853
347	912
406	971
464	·6324030
523	089
581	148
639	207
697	266
756	325
814	384
872	443
930	502
989	561

Excesses

10 = 10	1 = 1
20 = 20	2 = 2
30 = 30	3 = 3
40 = 41	4 = 4
50 = 51	5 = 5
	6 = 6
	7 = 7
	8 = 8
	9 = 9

29 = 30

Anti Log Excesses

10 = 10	1 = 1
20 = 20	2 = 2
30 = 30	3 = 3
40 = 39	4 = 4
50 = 49	5 = 5
	6 = 6
	7 = 7
	8 = 8
	9 = 9

30 = 29

4290047	·6324620
105	679
163	738
222	797
280	856
338	915
396	974
455	·6325033
513	092
572	151
630	210
688	269
746	328
805	387
863	446
921	505
979	564
4291038	623
096	682
154	741
212	800
271	859
329	918
387	977
445	·6326036
504	094
562	153
620	212
678	271
737	330
795	389
853	448
911	507
970	566
4292028	625
086	684
144	743
203	802
261	861
319	920
377	979
436	·6327037
494	096
552	155
610	214
669	273
727	332
785	391
843	450
902	509
960	568
4293018	627
076	686
135	744
193	803
251	862
309	921
367	980
425	·6328039
484	098
542	157
600	216
658	275
717	334
775	393

Number	Log
4293833	·6328451
891	510
950	569
4294008	628
066	687
124	746
183	805
241	864
299	922
357	981
415	·6329040
473	099
532	158
590	217
648	276
706	335
765	393
823	452
881	511
939	570
998	629
4295056	688
114	746
172	805
230	864
288	923
347	982
405	·6330041
463	100
521	159
580	217
638	276
696	335
754	394
812	453
870	512
929	570
987	629
4296045	688
103	747
162	806
220	865
278	923
336	982
394	·6331041
452	100
511	159
569	218
627	276
685	335
743	394
801	453
860	511
918	570
976	629
4297034	688
093	747
151	806
209	864
267	923
325	982
383	·6332041
442	099
500	158
558	217

Number	Log
4297616	·6332276
674	334
732	393
791	452
849	511
907	570
965	629
4298023	687
081	746
140	805
198	864
256	922
314	981
372	·6333040
430	099
489	157
547	216
605	275
663	334
721	392
779	451
838	510
896	569
954	627
4299012	686
070	745
128	804
186	862
244	921
303	980
361	·6334039
419	097
477	156
535	215
593	274
652	332
710	391
768	450
826	509
884	567
942	626

Excesses

10 = 10	1 = 1
20 = 20	2 = 2
30 = 30	3 = 3
40 = 40	4 = 4
50 = 51	5 = 5
	6 = 6
	7 = 7
	8 = 8
	9 = 9

29 = 29

Anti Log Excesses

10 = 10	1 = 1
20 = 20	2 = 2
30 = 30	3 = 3
40 = 40	4 = 4
50 = 49	5 = 5
	6 = 6
	7 = 7
	8 = 8
	9 = 9

29 = 29

Number	Log
4300000	·6334685
058	744
117	802
175	861
233	919
291	978
349	·6335037
407	096
466	154
524	213
582	272
640	331
698	389
756	448
814	507
872	566
931	624
989	683
4301047	741
105	800
163	859
221	918
279	976
337	·6336035
396	093
454	152
512	211
570	270
628	328
686	387
744	446
802	505
861	563
919	622
977	680
4302035	739
093	798
151	857
209	915
267	974
325	·6337032
383	091
442	149
500	208
558	267
616	326
674	384
732	443
790	501
848	560
906	619

Number	Log
4302964	·6337678
4303023	736
081	795
139	853
197	912
255	971
313	·6338030
371	088
429	147
487	205
545	264
604	322
662	381
720	440
778	499
836	557
894	616
952	674
4304010	733
068	791
126	850
185	908
243	967
301	·6339026
359	085
417	143
475	202
533	260
591	319
649	377
707	436
765	494
823	553
881	612
939	671
998	729
4305056	788
114	846
172	905
230	963
288	·6340022
346	080
404	139
462	197
520	256
578	314
636	373
694	432
752	491
811	549
869	608
927	666
985	725
4306043	783
101	842
159	900
217	959
275	·6341017
333	076
391	134
449	193
507	251
565	310
623	368
681	427

Number	Log
4306739	·6341486
797	545
856	603
914	662
972	720
4307030	779
088	837
146	896
204	954
262	·6342013
320	071
378	130
436	188
494	247
552	305
610	364
668	422
726	481
784	539
842	598
900	656
958	715
4308016	773
074	832
132	890
190	949
248	·6343007
306	066
364	124
422	183
481	241
539	300
597	358
655	417
713	475
771	534
829	592
887	651
945	709
4309003	768
061	826
119	885
177	943
235	·6344001
293	059
351	118
409	176
467	235
525	293
583	352
641	410
699	469
757	527
815	586
873	644
931	703
989	761

Excesses

10 = 10	1 = 1
20 = 20	2 = 2
30 = 30	3 = 3
40 = 40	4 = 4
50 = 50	5 = 5
	6 = 6
	7 = 7
	8 = 8
	9 = 9

29 = 29

Anti Log Excesses

10 = 10	1 = 1
20 = 20	2 = 2
30 = 30	3 = 3
40 = 40	4 = 4
50 = 50	5 = 5
	6 = 6
	7 = 7
	8 = 8
	9 = 9

29 = 29

4310047	·6344820
105	878
163	937
221	995
279	·6345054
337	112
395	171
453	229
511	287
569	345
627	404
685	462
743	521
801	579
859	638
917	696
975	755
4311033	813
091	872
149	930
207	988
265	·6346046
323	105
381	163
439	222
497	280
555	339
613	397
671	456
729	514
787	572
845	630
903	689
961	747
4312019	806
077	864
135	923

4312193	·6346981
250	·6347040
308	098
366	156
424	214
482	273
540	331
598	390
656	448
714	507
772	565
830	624
888	682
946	740
4313004	798
062	857
120	915
178	974
236	·6348032
294	091
352	149
410	207
468	265
526	324
584	382
642	441
700	499
758	557
816	615
873	674
931	732
989	791
4314047	849
105	907
163	965
221	·6349024
279	082
337	141
395	199
453	257
511	315
569	374
627	432
685	491
743	549
801	607
859	665
916	724
974	782
4315032	841
090	899
148	957
206	·6350015
264	074
322	132
380	190
438	248
496	307
554	365
612	424
670	482
727	540
785	598
843	657
901	715

4315959	·6350773
4316017	831
075	890
133	948
191	·6351007
249	065
307	123
365	181
423	240
481	298
538	356
596	414
654	473
712	531
770	589
828	647
886	706
944	764
4317002	822
060	880
118	939
176	997
233	·6352055
291	113
349	172
407	230
465	288
523	346
581	405
639	463
697	521
755	579
812	638
870	696
928	754
986	812
4318044	871
102	929
160	987
218	·6353045
276	103
334	161
391	220
449	278
507	336
565	394
623	453
681	511
739	569
797	627
854	686
912	744
970	802
4319028	860
086	918
144	976
202	·6354035
260	093
317	151
375	209
433	268
491	326
549	384
607	442
665	500

4319723	·6354558
780	617
838	675
896	733
954	791

Excesses

10 = 10	1 = 1
20 = 20	2 = 2
30 = 30	3 = 3
40 = 40	4 = 4
50 = 50	5 = 5
	6 = 6
	7 = 7
	8 = 8
	9 = 9

29 = 29

Anti Log Excesses

10 = 10	1 = 1
20 = 20	2 = 2
30 = 30	3 = 3
40 = 40	4 = 4
50 = 50	5 = 5
	6 = 6
	7 = 7
	8 = 8
	9 = 9

29 = 29

4320012	·6354849
070	907
128	966
186	·6355024
243	082
301	140
359	198
417	256
475	315
533	373
591	431
649	489
706	547
764	605
822	664
880	722
938	780
996	838
4321053	896
111	954
169	·6356013
227	071
285	129
343	187
401	245
459	303
516	362
574	420
632	478
690	536
748	594

4321806	·6356652
863	710
921	768
979	826
4322037	884
095	943
153	·6357001
210	059
268	117
326	175
384	233
442	291
500	349
557	408
615	466
673	524
731	582
789	640
847	698
904	756
962	814
4323020	873
078	931
136	989
194	·6358047
251	105
309	163
367	221
425	279
483	337
541	395
598	453
656	511
714	570
772	628
830	686
888	744
945	802
4324003	860
061	918
119	976
177	·6359034
235	092
292	150
350	208
408	266
466	324
523	382
581	440
639	498
697	556
755	615
813	673
870	731
928	789
986	847
4325044	905
101	963
159	·6360021
217	079
275	137
333	195
391	253
448	311
506	369

4325564	·6360427
622	485
679	543
737	601
795	659
853	717
911	775
969	833
4326026	891
084	949
142	·6361007
200	065
257	123
315	181
373	239
431	297
488	355
546	413
604	471
662	529
720	587
778	645
835	703
893	761
951	819
4327009	877
066	935
124	993
182	·6362051
240	109
297	167
355	225
413	283
471	341
528	399
586	457
644	515
702	573
760	631
818	689
875	747
933	805
991	863
4328049	921
106	979
164	·6363037
222	095
280	153
337	211
395	269
453	327
511	385
568	443
626	500
684	558
742	616
799	674
857	732
915	790
973	848
4329030	906
088	964
146	·6364022
204	080
261	138

4329319	·6364196
377	254
435	312
492	370
550	427
608	485
666	543
723	601
781	659
839	717
897	775
954	833

Excesses

10 = 10	1 = 1
20 = 20	2 = 2
30 = 30	3 = 3
40 = 40	4 = 4
50 = 50	5 = 5
	6 = 6
	7 = 7
	8 = 8
	9 = 9

29 = 29

Anti Log Excesses

10 = 10	1 = 1
20 = 20	2 = 2
30 = 30	3 = 3
40 = 40	4 = 4
50 = 50	5 = 5
	6 = 6
	7 = 7
	8 = 8
	9 = 9

29 = 29

4330012	·6364891
070	949
128	·6365007
185	065
243	122
300	180
358	238
416	296
474	354
531	412
589	470
647	528
705	586
762	644
820	701
878	759
936	817
993	875
4331051	933
109	991
167	·6366049
224	107
282	164
340	222

4331398	·6366280
455	338
513	396
570	454
628	512
686	570
744	627
801	685
859	743
917	801
975	859
4332032	917
090	974
147	·6367032
205	090
263	148
321	206
378	264
436	322
494	380
552	437
609	495
667	553
725	611
783	669
840	727
898	784
955	842
4333013	900
071	958
129	·6368016
186	074
244	131
301	189
359	247
417	305
475	362
532	420
590	478
648	536
706	594
763	652
821	709
878	767
936	825
994	883
4334052	941
109	999
167	·6369056
224	114
282	172
340	230
398	287
455	345
513	403
571	461
629	518
686	576
744	634
801	692
859	750
917	808
975	865
4335032	923
090	981

4335147	·6370039
205	096
263	154
321	212
378	270
436	327
493	385
551	443
609	501
667	558
724	616
782	674
839	732
897	789
954	847
4336012	905
070	963
128	·6371020
185	078
243	136
300	194
358	251
416	309
474	367
531	425
589	482
646	540
704	598
762	656
820	713
877	771
935	829
992	887
4337050	944
107	·6372002
165	059
223	117
281	175
338	233
396	290
453	348
511	406
569	464
627	521
684	579
742	637
799	695
857	752
914	810
972	867
4338030	925
088	983
145	·6373041
203	098
260	156
318	214
375	272
433	329
491	387
549	444
606	502
664	560
721	618
779	675
836	733

4338894	·6373790
952	848
4339010	906
067	964
125	·6374021
182	079
240	136
297	194
355	252
413	310
471	367
528	425
586	482
643	540
701	598
758	656
816	713
873	771
931	828
989	886

Excesses

10 = 10	1 = 1
20 = 20	2 = 2
30 = 30	3 = 3
40 = 40	4 = 4
50 = 50	5 = 5
	6 = 6
	7 = 7
	8 = 8
	9 = 9

29 = 29

Anti Log Excesses

10 = 10	1 = 1
20 = 20	2 = 2
30 = 30	3 = 3
40 = 40	4 = 4
50 = 50	5 = 5
	6 = 6
	7 = 7
	8 = 8
	9 = 9

29 = 29

4340047	·6374943
104	·6375001
162	059
219	117
277	174
334	232
392	289
449	347
507	404
565	462
623	520
680	578
738	635
795	693
853	750
910	808

4340968	·6375865
4341025	923
083	981
140	·6376039
198	096
256	154
314	211
371	269
429	326
486	384
544	442
601	500
659	557
716	615
774	672
831	730
889	787
947	845
4342005	902
062	960
120	·6377018
177	076
235	133
292	191
350	248
407	306
465	363
522	421
580	478
638	536
696	593
753	651
811	708
868	766
926	824
983	882
4343041	939
098	997
156	·6378054
213	112
271	169
328	227
386	284
443	342
501	399
559	457
617	514
674	572
732	629
789	687
847	744
904	802
962	859
4344019	917
077	975
134	·6379033
192	090
251	148
307	205
364	263
422	320
479	378
537	435
594	493
652	550

4344709	·6379608
767	665
825	723
883	780
940	838
998	895
4345055	953
113	·6380010
170	068
228	125
285	183
343	240
400	298
458	355
515	413
573	470
630	528
688	585
745	643
803	700
860	758
918	815
975	873
4346033	930
090	988
148	·6381045
205	103
263	160
320	218
378	275
435	333
493	390
550	448
608	505
665	563
723	620
780	677
838	734
895	792
953	849
4347010	907
068	964
125	·6382022
183	079
241	137
299	194
356	252
414	309
471	367
529	424
586	482
644	539
701	597
759	654
816	712
874	769
931	826
989	883
4348046	941
104	998
161	·6383056
219	113
276	171
334	228
391	286

4348448	·6383343
505	401
563	458
620	515
678	572
735	630
793	687
850	745
908	802
965	860
4349023	917
080	975
138	·6384032
195	089
253	146
310	204
368	261
425	319
483	376
540	434
598	491
655	549
713	606
770	663
828	720
885	778
943	835

Excesses

10 = 10	1 = 1
20 = 20	2 = 2
30 = 30	3 = 3
40 = 40	4 = 4
50 = 50	5 = 5
	6 = 6
	7 = 7
	8 = 8
	9 = 9

29 = 29

Anti Log Excesses

10 = 10	1 = 1
20 = 20	2 = 2
30 = 30	3 = 3
40 = 40	4 = 4
50 = 50	5 = 5
	6 = 6
	7 = 7
	8 = 8
	9 = 9

29 = 29

4350000	·6384893
058	950
115	·6385008
173	065
230	122
288	179
345	237
403	294
460	352

4350518	·6385409
575	467
633	524
690	581
748	638
805	696
862	753
919	811
977	868
4351034	925
092	982
149	·6386040
207	097
264	155
322	212
379	269
437	326
494	384
552	441
609	499
667	556
724	613
782	670
839	728
897	785
954	843
4352012	900
069	958
126	·6387015
183	072
241	129
298	187
356	244
413	302
471	359
528	416
586	473
643	531
701	588
758	645
816	702
873	760
931	817
988	875
4353045	932
102	989
160	·6388046
217	104
275	161
332	218
390	275
447	333
505	390
562	448
620	505
677	562
735	619
792	677
849	734
906	791
964	848
4354021	906
079	963
136	·6389020
194	077

4354251	·6389135
309	192
366	249
424	306
481	364
538	421
595	478
653	535
710	593
768	650
825	707
883	764
940	822
998	879
4355055	936
112	993
169	·6390051
227	108
284	165
342	222
399	280
457	337
514	394
572	451
629	509
686	566
743	623
801	680
858	738
916	795
973	852
4356031	909
088	966
146	·6391023
203	081
260	138
317	195
375	252
432	310
490	367
547	424
605	481
662	539
719	596
776	653
834	710
891	767
949	824
4357006	882
064	939
121	996
178	·6392053
235	110
293	167
350	225
408	282
465	339
523	396
580	454
637	511
694	568
752	625
809	682
867	739
924	797

4357982	·6392854
4358039	911
096	968
153	·6393025
211	082
268	140
326	197
383	254
441	311
498	368
555	425
612	482
670	539
727	597
785	654
842	711
899	768
956	825
4359014	882
071	940
129	997
186	·6394054
244	111
301	168
358	225
415	282
473	339
530	397
588	454
645	511
702	568
759	625
817	682
875	739
932	796
989	854

Excesses

10 = 10	1 = 1
20 = 20	2 = 2
30 = 30	3 = 3
40 = 40	4 = 4
50 = 50	5 = 5
	6 = 6
	7 = 7
	8 = 8
	9 = 9
29 = 29	

Anti Log Excesses

10 = 10	1 = 1
20 = 20	2 = 2
30 = 30	3 = 3
40 = 40	4 = 4
50 = 50	5 = 5
	6 = 6
	7 = 7
	8 = 8
	9 = 9
29 = 29	

4360046	·6394911
103	968
161	·6395025
218	082
276	139
333	196
390	253
447	310
505	367
562	425
620	482
677	539
734	596
791	653
849	710
906	767
964	824
4361021	882
078	939
135	996
193	·6396053
250	110
308	167
365	224
422	281
479	338
537	395
594	452
652	509
709	566
766	623
823	681
881	738
938	795
995	852
4362052	909
110	966
167	·6397023
225	080
282	137
339	194
396	251
454	308
511	365
569	422
626	479
683	536
740	593
798	650
855	707
912	764
969	822
4363027	879
084	936
142	993
199	·6398050
256	107
313	164
371	221
428	278
485	335
542	392
600	449
657	506
715	563

4363772	·6398620
829	677
886	734
944	791
4364001	848
058	905
115	962
173	·6399019
230	076
287	133
344	190
402	247
459	304
517	361
574	418
631	475
688	532
746	589
803	646
860	703
917	760
975	817
4365032	874
089	931
146	988
204	·6400045
261	102
318	159
375	216
433	273
490	330
547	387
604	444
662	501
719	558
776	614
833	671
891	728
948	785
4366006	842
063	899
120	956
177	·6401013
235	070
292	127
349	184
406	241
464	298
521	355
578	412
635	469
693	526
750	583
807	640
864	697
922	753
979	810
4367036	867
093	924
151	981
208	·6402038
265	095
322	152
380	209
437	266

4367494	·6402323
551	380
609	437
666	494
723	550
780	607
837	664
894	721
952	778
4368009	835
066	892
123	949
181	·6403006
238	063
295	119
352	176
410	233
467	290
524	347
581	404
639	461
696	518
753	575
810	632
868	688
925	745
982	802
4369039	859
097	916
154	973
211	·6404030
268	087
325	143
382	200
440	257
497	314
554	371
611	428
669	485
726	542
783	598
840	655
898	712
955	769

Excesses

10 = 10	1 = 1
20 = 20	2 = 2
30 = 30	3 = 3
40 = 40	4 = 4
50 = 50	5 = 5
	6 = 6
	7 = 7
	8 = 8
	9 = 9
29 = 29	

Anti Log Excesses

10 = 10	1 = 1
20 = 20	2 = 2
30 = 30	3 = 3
40 = 40	4 = 4
50 = 50	5 = 5
	6 = 6
	7 = 7
	8 = 8
	9 = 9
29 = 29	

4370012	·6404826
069	883
126	939
183	996
241	·6405053
298	110
355	167
412	224
470	281
527	338
584	394
641	451
698	508
755	565
813	622
870	679
927	735
984	792
4371042	849
099	906
156	963
213	·6406020
270	076
327	133
385	190
442	247
499	304
556	361
613	417
670	474
728	531
785	588
842	644
899	701
957	758
4372014	815
071	872
128	929
185	985
242	·6407042
300	099
357	156
414	212
471	269
528	326
585	383
643	440
700	497
757	553
814	610
871	667

4372928	·6407724
986	780
4373043	837
100	894
157	951
214	·6408007
271	064
329	121
386	178
443	234
500	291
557	348
614	405
672	461
729	518
786	575
843	632
900	688
957	745
4374015	802
072	859
129	915
186	972
243	·6409029
300	086
358	142
415	199
472	256
529	313
586	369
643	426
700	483
757	540
815	596
872	653
929	710
986	767
4375043	823
100	880
158	937
215	994
272	·6410050
329	107
386	164
443	221
500	277
557	334
615	390
672	447
729	504
786	561
843	617
900	674
958	731
4376015	788
072	844
129	901
186	957
243	·6411014
300	071
357	128
415	184
472	241
529	297
586	354

4376643	·6411411
700	468
757	524
814	581
872	637
929	694
986	751
4377043	808
100	864
157	921
214	978
271	·6412035
328	091
385	148
443	204
500	261
557	318
614	375
671	431
728	488
785	544
842	601
900	657
957	714
4378014	771
071	828
128	884
185	941
242	997
299	·6413054
356	111
413	168
471	224
528	281
585	337
642	394
699	450
756	507
813	564
870	621
927	677
984	734
4379041	790
098	847
156	903
213	960
270	·6414017
327	074
384	130
441	187
498	243
555	300
612	356
669	413
727	469
784	526
841	583
898	640
955	696

Excesses

10 = 10	1 = 1
20 = 20	2 = 2
30 = 30	3 = 3
40 = 40	4 = 4
50 = 50	5 = 5
	6 = 6
	7 = 7
	8 = 8
	9 = 9

29 = 29

Anti Log Excesses

10 = 10	1 = 1
20 = 20	2 = 2
30 = 30	3 = 3
40 = 40	4 = 4
50 = 50	5 = 5
	6 = 6
	7 = 7
	8 = 8
	9 = 9

29 = 29

4380012	·6414753
069	809
126	866
183	922
240	979
297	·6415035
354	092
411	149
468	206
526	262
583	319
640	375
697	432
754	488
811	545
868	601
925	658
982	714
4381039	771
096	827
153	884
210	941
267	998
325	·6416054
382	111
439	167
496	224
553	280
610	337
667	393
724	450
781	506
838	563
895	619
952	676
4382009	732
066	789

4382123	·6416845
180	902
237	959
294	·6417016
351	072
408	129
466	185
523	242
580	298
637	355
694	411
751	468
808	524
865	581
922	637
979	694
4383036	750
093	807
150	863
207	920
264	976
321	·6418033
378	089
435	146
492	202
549	259
606	315
663	372
720	428
777	485
834	541
891	598
949	654
4384006	711
063	767
120	824
177	880
234	937
291	993
348	·6419050
405	106
462	163
519	219
576	276
633	332
690	389
747	445
804	502
861	558
918	615
975	671
4385032	728
089	784
146	841
203	897
260	953
317	·6420009
374	066
431	122
488	179
545	235
602	292
659	348
716	405
773	461

4385830	·6420518
887	574
944	631
4386001	687
058	744
115	800
172	856
229	912
286	969
343	·6421025
400	082
457	138
514	195
571	251
628	308
685	364
742	421
799	477
856	534
913	590
970	646
4387027	702
084	759
141	815
198	872
255	928
312	985
369	·6422041
426	098
483	154
540	210
597	266
654	323
711	379
768	436
825	492
882	549
939	605
995	661
4388052	717
109	774
166	830
223	887
280	943
337	·6423000
394	056
451	113
508	169
565	225
622	281
679	338
736	394
793	451
850	507
907	564
964	620
4389021	676
078	732
135	789
192	845
249	902
306	958
363	·6424014
420	070
477	127

4389534	·6424183
590	240
647	296
704	352
761	408
818	465
875	521
932	578
989	634

Excesses

10 = 10	1 = 1
20 = 20	2 = 2
30 = 30	3 = 3
40 = 40	4 = 4
50 = 50	5 = 5
	6 = 6
	7 = 7
	8 = 8
	9 = 9

29 = 29

Anti Log Excesses

10 = 10	1 = 1
20 = 20	2 = 2
30 = 30	3 = 3
40 = 40	4 = 4
50 = 50	5 = 5
	6 = 6
	7 = 7
	8 = 8
	9 = 9

29 = 29

4390046	·6424691
103	747
160	803
217	859
274	916
331	972
388	·6425028
445	084
502	141
559	197
615	254
672	310
729	366
786	422
843	479
900	535
957	592
4391014	648
071	704
128	760
185	817
242	873
299	930
356	986
413	·6426042
470	098
526	155

4391583	·6426211
640	268
697	324
754	380
811	436
868	493
925	549
982	605
4392039	661
096	718
153	774
209	830
266	886
323	943
380	999
437	·6427055
494	111
551	168
608	224
665	281
722	337
779	393
836	449
892	506
949	562
4393006	618
063	674
120	731
177	787
234	843
291	899
348	956
405	·6428012
461	068
518	124
575	181
632	237
689	293
746	349
803	406
860	462
917	518
974	574
4394030	631
087	687
144	743
201	799
258	856
315	912
372	968
429	·6429024
486	080
543	136
599	193
656	249
713	305
770	361
827	418
884	474
941	530
998	586
4395054	642
111	698
168	755
225	811

4395282	·6429867
339	923
396	980
453	·6430036
509	092
566	148
623	204
680	260
737	317
794	373
851	429
908	485
964	541
4396021	597
078	654
135	710
192	766
249	822
306	879
363	935
419	991
476	·6431047
533	103
590	159
647	216
704	272
761	328
818	384
874	440
931	496
988	553
4397045	609
102	665
159	721
215	777
272	833
329	889
386	945
443	·6432002
500	058
556	114
613	170
670	226
727	282
784	339
841	395
898	451
955	507
4398011	563
068	619
125	675
182	731
239	788
296	844
352	900
409	956
466	·6433012
523	068
580	124
637	180
693	237
750	293
807	349
864	405
921	461

4398978	·6433517
4399034	573
091	629
148	686
205	742
262	798
319	854
375	910
432	966
489	·6434022
546	078
603	134
660	190
716	246
773	302
830	359
887	415
943	471

Excesses

10 = 10	1 = 1
20 = 20	2 = 2
30 = 30	3 = 3
40 = 40	4 = 4
50 = 49	5 = 5
	6 = 6
	7 = 7
	8 = 8
	9 = 9

28 = 28

Anti Log Excesses

10 = 10	1 = 1
20 = 20	2 = 2
30 = 30	3 = 3
40 = 40	4 = 4
50 = 51	5 = 5
	6 = 6
	7 = 7
	8 = 8
	9 = 9

28 = 28

4400000	·6434527
057	583
114	639
171	695
228	751
284	807
341	863
398	919
455	975
512	·6435032
569	088
625	144
682	200
739	256
796	312
852	368
909	424
966	480

4401023	6435536
080	592
137	648
193	704
250	760
307	816
364	872
421	928
478	984
534	·6436040
591	096
648	153
705	209
761	265
818	321
875	377
932	433
988	489
4402045	545
102	601
159	657
216	713
273	769
329	825
386	881
443	937
500	993
556	·6437049
613	105
670	161
727	217
783	273
840	329
897	385
954	441
4403011	497
068	553
124	609
181	665
238	721
295	777
351	833
408	889
465	945
522	·6438001
578	057
635	113
692	169
749	225
805	281
862	337
919	393
976	449
4404033	505
090	561
146	617
203	673
260	729
317	785
373	841
430	897
487	953
544	·6439009
600	065
657	121

Number	Log
4404714	·6439177
771	233
827	288
884	344
941	400
998	456
4405054	512
111	568
168	624
225	680
281	736
338	792
395	848
452	904
508	960
565	·6440016
622	072
679	128
735	184
792	240
849	295
906	351
962	407
4406019	463
076	519
133	575
189	631
246	687
303	743
360	799
416	855
473	911
529	966
586	·6441022
643	078
700	134
756	190
813	246
870	302
927	358
983	414
4407040	470
097	526
154	582
210	637
267	693
324	749
381	805
437	861
494	917
551	973
608	·6442029
664	084
721	140
777	196
834	252
891	308
948	364
4408004	420
061	476
118	532
175	588
231	643
288	699
345	755

Number	Log
4408402	·6442811
458	867
515	923
571	978
628	·6443034
685	090
742	146
798	202
855	258
912	314
969	370
4409025	425
082	481
138	537
195	593
252	649
309	705
365	760
422	816
479	872
536	928
592	984
649	·6444040
705	096
762	151
819	207
876	263
932	319
989	375

Excesses

10 = 10	1 = 1
20 = 20	2 = 2
30 = 30	3 = 3
40 = 39	4 = 4
50 = 49	5 = 5
	6 = 6
	7 = 7
	8 = 8
	9 = 9

28 = 28

Anti Log Excesses

10 = 10	1 = 1
20 = 20	2 = 2
30 = 30	3 = 3
40 = 41	4 = 4
50 = 51	5 = 5
	6 = 6
	7 = 7
	8 = 8
	9 = 9

28 = 28

Number	Log
4410046	·6444431
103	486
159	542
216	598
272	654
329	710
386	766

Number	Log
4410443	·6444821
499	877
556	933
612	989
669	·6445045
726	101
783	156
839	212
896	268
952	324
4411009	379
066	435
123	491
179	547
236	603
293	659
350	714
406	770
463	826
519	882
576	937
633	993
690	·6446049
746	105
803	161
859	217
916	272
973	328
4412030	384
086	440
143	495
199	551
256	607
312	663
369	718
426	774
483	830
539	886
596	941
652	997
709	·6447053
766	109
823	164
879	220
936	276
992	332
4413049	387
106	443
163	499
219	555
276	610
332	666
389	722
446	778
503	833
559	889
616	945
672	·6448001
729	056
785	112
842	168
899	224
956	279
4414012	335
069	391

Number	Log
4414125	·6448447
182	502
238	558
295	614
352	670
409	725
465	781
522	836
578	892
635	948
692	·6449004
749	059
805	115
862	171
918	227
975	282
4415031	338
088	393
145	449
202	505
258	561
315	616
371	672
428	728
484	784
541	839
597	895
654	950
711	·6450006
768	062
824	118
881	173
937	229
994	284
4416050	340
107	396
164	452
221	507
277	563
334	618
390	674
447	730
503	786
560	841
616	897
673	952
730	·6451008
787	064
843	120
900	175
956	231
4417013	286
069	342
126	398
182	454
239	509
296	565
353	620
409	676
466	731
522	787
579	843
635	899
692	954
748	·6452010

Number	Log
4417805	·6452065
862	121
919	176
975	232
4418032	288
088	344
145	399
201	455
258	510
314	566
371	621
427	677
484	733
541	789
598	844
654	900
711	955
767	·6453011
824	066
880	122
937	178
993	234
4419050	289
106	345
163	400
219	456
276	511
333	567
390	622
446	678
503	734
559	790
616	845
672	901
729	956
785	·6454012
842	067
898	123
955	178

Excesses

10 = 10	1 = 1
20 = 20	2 = 2
30 = 30	3 = 3
40 = 39	4 = 4
50 = 49	5 = 5
	6 = 6
	7 = 7
	8 = 8
	9 = 9

28 = 28

Anti Log Excesses

10 = 10	1 = 1
20 = 20	2 = 2
30 = 30	3 = 3
40 = 41	4 = 4
50 = 51	5 = 5
	6 = 6
	7 = 7
	8 = 8
	9 = 9
28 = 28	

4420011	·6454234
068	289
124	345
181	401
238	457
295	512
351	568
408	623
464	679
521	734
577	790
634	845
690	901
747	956
803	·6455012
860	067
916	123
973	178
4421029	234
086	289
142	345
199	401
256	457
313	512
369	568
426	623
482	679
539	734
595	790
652	845
708	901
765	956
821	·6456012
878	067
934	123
991	178
4422047	234
104	289
160	345
217	400
273	456
330	511
386	567
443	622
499	678
556	733
612	789
669	844
725	900
782	955
838	·6457011

4422895	·6457066
952	122
4423009	177
065	233
122	288
178	344
235	399
291	455
348	510
404	566
461	621
517	677
574	732
630	788
687	843
743	899
800	954
856	·6458010
913	065
969	121
4424026	176
082	232
139	287
195	343
252	398
308	454
365	509
421	565
478	620
534	676
591	731
647	787
704	842
760	897
817	952
873	·6459008
930	063
986	119
4425043	174
099	230
156	285
212	341
269	396
325	452
382	507
438	563
495	618
551	674
608	729
664	785
721	840
777	895
834	950
890	·6460006
947	061
4426003	117
060	172
116	228
173	283
229	339
286	394
342	450
399	505
455	560
512	615

4426568	·6460671
624	726
680	782
737	837
793	893
850	948
906	·6461004
963	059
4427019	114
076	169
132	225
189	280
245	336
302	391
358	447
415	502
471	558
528	613
584	668
641	723
697	779
754	834
810	890
867	945
923	·6462001
980	056
4428036	111
093	166
149	222
206	277
262	333
318	388
374	444
431	499
487	554
544	609
600	665
657	720
713	776
770	831
826	886
883	941
939	997
996	·6463052
4429052	108
109	163
165	219
222	274
278	329
335	384
391	440
447	495
503	551
560	606
616	661
673	716
729	772
786	827
842	883
899	938
955	993

Excesses

10 = 10	1 = 1
20 = 20	2 = 2
30 = 29	3 = 3
40 = 39	4 = 4
50 = 49	5 = 5
	6 = 6
	7 = 7
	8 = 8
	9 = 9
28 = 28	

Anti Log Excesses

10 = 10	1 = 1
20 = 20	2 = 2
30 = 31	3 = 3
40 = 41	4 = 4
50 = 51	5 = 5
	6 = 6
	7 = 7
	8 = 8
	9 = 9
28 = 28	

4430012	·6464048
068	104
125	159
181	214
238	269
294	325
350	380
406	436
463	491
519	546
576	601
632	657
689	712
745	768
802	823
858	878
915	933
971	989
4431027	·6465044
083	100
140	155
196	210
253	265
309	321
366	376
422	431
479	486
535	542
592	597
648	653
704	708
760	763
817	818
873	874
930	929
986	984
4432043	·6466039

4432099	·6466095
156	150
212	205
269	260
325	316
381	371
437	426
494	481
550	537
607	592
663	647
720	702
776	758
833	813
889	868
945	923
4433001	979
058	·6467034
114	089
171	144
227	200
284	255
340	310
396	365
452	421
509	476
565	531
622	586
678	642
735	697
791	752
848	807
904	863
960	918
4434016	973
073	·6468028
129	083
186	138
242	194
299	249
355	304
411	359
467	415
524	470
580	525
637	580
693	636
750	691
806	746
862	801
918	856
975	911
4435031	967
088	·6469022
144	077
201	132
257	188
313	243
369	298
426	353
482	408
539	463
595	519
652	574
708	629

Number	Log
4435764	·6469684
820	739
877	794
933	850
990	905
4436046	960
102	·6470015
158	070
215	125
271	181
328	236
384	291
441	346
497	401
553	456
609	512
666	567
722	622
779	677
835	732
891	787
947	843
4437004	898
060	953
117	·6471008
173	063
229	118
285	174
342	229
398	284
455	339
511	394
567	449
623	504
680	559
736	615
793	670
849	725
905	780
961	835
4438018	890
074	945
131	·6472000
187	056
243	111
299	166
356	221
412	276
469	331
525	386
581	441
637	497
694	552
750	607
807	662
863	717
919	772
975	827
4439032	882
088	937
145	992
201	·6473048
257	103
313	158
370	213

Number	Log
4439426	·6473268
482	323
538	378
595	433
651	488
708	543
764	598
820	653
876	709
933	764
989	819

Excesses

10 = 10	1 = 1
20 = 20	2 = 2
30 = 29	3 = 3
40 = 39	4 = 4
50 = 49	5 = 5
	6 = 6
	7 = 7
	8 = 8
	9 = 9

28 = 28

Anti Log Excesses

10 = 10	1 = 1
20 = 20	2 = 2
30 = 31	3 = 3
40 = 41	4 = 4
50 = 51	5 = 5
	6 = 6
	7 = 7
	8 = 8
	9 = 9

28 = 28

Number	Log
4440046	·6473874
102	929
158	984
214	·6474039
271	094
327	149
383	204
439	259
496	314
552	369
609	424
665	480
721	535
777	590
834	645
890	700
946	755
4441002	810
059	865
115	920
172	975
228	·6475030
284	085
340	140
397	195

Number	Log
4441453	·6475250
509	305
565	360
622	415
678	470
734	525
790	580
847	635
903	691
960	746
4442016	801
072	856
128	911
185	966
241	·6476021
297	076
353	131
410	186
466	241
522	296
578	351
635	406
691	461
747	516
803	571
860	626
916	681
972	736
4443028	791
085	846
141	901
198	956
254	·6477011
310	066
366	121
423	176
479	231
535	286
591	341
648	396
704	451
760	506
816	561
873	616
929	671
985	726
4444041	781
098	836
154	891
210	945
266	·6478000
323	055
379	110
435	165
491	220
548	275
604	330
660	385
716	440
773	495
829	550
885	605
941	660
998	715
4445054	770

Number	Log
4445110	·6478825
166	880
223	935
279	990
335	·6479045
391	100
448	154
504	209
560	264
616	319
673	374
729	429
785	484
841	539
897	594
953	649
4446010	704
066	759
122	814
178	869
235	923
291	978
347	·6480033
403	088
460	143
516	198
572	253
628	308
685	363
741	418
797	473
853	528
910	582
966	637
4447022	692
078	747
134	802
190	857
247	912
303	967
359	·6481022
415	077
472	131
528	186
584	241
640	296
697	351
753	406
809	461
865	516
921	570
977	625
4448034	680
090	735
146	790
202	845
259	900
315	955
371	·6482009
427	064
483	119
539	174
596	229
652	284
708	339

Number	Log
4448764	·6482394
821	448
877	503
933	558
989	613
4449045	668
101	723
158	777
214	832
270	887
326	942
382	997
438	·6483052
495	106
551	161
607	216
663	271
720	326
776	381
832	436
888	491
944	545

Excesses

10 = 10	1 = 1
20 = 20	2 = 2
30 = 29	3 = 3
40 = 39	4 = 4
50 = 49	5 = 5
	6 = 6
	7 = 7
	8 = 8
	9 = 9

28 = 28

Anti Log Excesses

10 = 10	1 = 1
20 = 20	2 = 2
30 = 31	3 = 3
40 = 41	4 = 4
50 = 51	5 = 5
	6 = 6
	7 = 7
	8 = 8
	9 = 9

28 = 28

Number	Log
4450000	·6483600
057	655
113	710
169	764
225	819
281	874
337	929
394	984
450	·6484039
506	093
562	148
618	203
674	258
731	313

4450787	·6484368
843	423
899	478
955	532
4451011	587
068	642
124	697
180	751
236	806
292	861
348	916
405	971
461	·6485026
517	080
573	135
629	190
685	245
742	299
798	354
854	409
910	464
966	518
4452022	573
079	628
135	683
191	738
247	793
303	847
359	902
416	957
472	·6486012
528	066
584	121
640	176
696	231
752	285
808	340
865	395
921	450
977	504
4453033	559
089	614
145	669
202	723
258	778
314	833
370	888
426	942
482	997
538	·6487052
594	107
651	161
707	216
763	271
819	326
875	380
931	435
987	490
4454043	545
100	599
156	654
212	709
268	764
324	818
380	873

4454436	·6487927
492	982
549	·6488037
605	092
661	146
717	201
773	256
829	311
885	365
941	420
998	474
4455054	529
110	584
166	639
222	693
278	748
334	803
390	858
447	912
503	967
559	·6489021
615	076
671	131
727	186
783	240
839	295
895	350
951	405
4456008	459
064	514
120	568
176	623
232	678
288	733
344	787
400	842
456	896
512	951
569	·6490006
625	061
681	115
737	170
793	224
849	279
905	334
961	389
4457017	443
073	498
130	552
186	607
242	662
298	717
354	771
410	826
466	880
522	935
578	989
634	·6491044
690	099
746	154
803	208
859	263
915	317
971	372
4458027	427

4458083	·6491482
139	536
195	591
251	645
307	700
363	754
419	809
476	863
532	918
588	973
644	·6492028
700	082
756	137
812	191
868	246
924	300
980	355
4459036	410
092	465
148	519
204	574
261	628
317	683
373	737
429	792
485	847
541	902
597	956
653	·6493011
709	065
765	120
821	174
877	229
933	283
989	338

Excesses

10 = 10	1 = 1
20 = 19	2 = 2
30 = 29	3 = 3
40 = 39	4 = 4
50 = 49	5 = 5
	6 = 6
	7 = 7
	8 = 8
	9 = 9

28 = 27

Anti Log Excesses

10 = 10	1 = 1
20 = 21	2 = 2
30 = 31	3 = 3
40 = 41	4 = 4
50 = 51	5 = 5
	6 = 6
	7 = 7
	8 = 8
	9 = 9

27 = 28

4460045	·6493392

4460101	·6493447
157	502
213	557
270	611
326	666
382	720
438	775
494	829
550	884
606	938
662	993
718	·6494047
774	102
830	157
886	212
942	266
998	321
4461054	375
110	430
166	484
222	539
278	593
334	648
390	702
446	757
502	811
558	866
615	920
671	975
727	·6495029
783	084
839	138
895	193
951	248
4462007	303
063	357
119	412
175	466
231	521
287	575
343	630
399	684
455	739
511	793
567	848
623	902
679	957
735	·6496011
791	066
847	120
903	175
959	229
4463015	284
071	338
127	393
183	447
239	502
295	556
351	611
407	665
463	720
519	774
575	829
631	883
687	938

4463743	·6496992
799	·6497047
855	101
911	156
967	210
4464023	265
079	319
135	374
191	428
247	483
303	537
359	592
415	646
471	700
527	754
583	809
639	863
695	918
751	972
807	·6498027
863	081
919	136
975	190
4465031	245
087	299
143	354
199	408
255	463
311	517
367	572
423	626
479	681
535	735
591	790
647	844
703	898
759	952
815	·6499007
871	061
927	116
983	170
4466039	225
095	279
151	334
207	388
263	443
319	497
375	552
431	606
487	661
543	715
599	769
655	823
711	878
767	932
823	987
879	·6500041
935	096
991	150
4467046	205
102	259
158	313
214	367
270	422
326	476

4467382	·6500531
438	585
494	640
550	694
606	749
662	803
718	857
774	911
830	966
886	·6501020
942	075
998	129
4468054	184
110	238
166	293
222	347
278	401
334	455
389	510
445	564
501	619
557	673
613	728
669	782
725	836
781	890
837	945
893	999
949	·6502054
4469005	108
061	163
117	217
173	271
229	325
284	380
340	434
396	489
452	543
508	597
564	651
620	706
676	760
732	815
788	869
844	924
900	978
956	·6503032

Excesses

10 = 10	1 = 1
20 = 19	2 = 2
30 = 29	3 = 3
40 = 39	4 = 4
50 = 49	5 = 5
	6 = 6
	7 = 7
	8 = 8
	9 = 9

28 = 27

Anti Log Excesses

10 = 10	1 = 1
20 = 21	2 = 2
30 = 31	3 = 3
40 = 41	4 = 4
50 = 51	5 = 5
	6 = 6
	7 = 7
	8 = 8
	9 = 9

27 = 28

4470012	·6503086
068	141
124	195
179	250
235	304
291	358
347	412
403	467
459	521
515	576
571	630
627	684
683	738
739	793
795	847
850	902
906	956
962	·6504010
4471018	064
074	119
130	173
186	227
242	281
298	336
354	390
410	445
466	499
521	553
577	607
633	662
689	716
745	770
801	824
857	879
913	933
969	988
4472025	·6505042
080	096
136	150
192	205
248	259
304	313
360	367
416	422
472	476
528	530
584	584
639	639
695	693
751	748
807	802

4472863	·6505856
919	910
975	965
4473031	·6506019
087	073
143	127
198	182
254	236
310	290
366	344
422	399
478	453
534	507
590	561
645	616
701	670
757	724
813	778
869	833
925	887
981	941
4474037	995
092	·6507050
148	104
204	158
260	212
316	267
372	321
428	375
484	429
539	483
595	537
651	592
707	646
763	700
819	754
875	809
931	863
986	917
4475042	971
098	·6508026
154	080
210	134
266	188
322	243
378	297
433	351
489	405
545	459
601	513
657	568
713	622
768	676
824	730
880	784
936	838
992	893
4476048	947
104	·6509001
160	055
215	110
271	164
327	218
383	272
439	326

4476495	·6509380
550	435
606	489
662	543
718	597
774	651
830	705
885	760
941	814
997	868
4477053	922
109	976
165	·6510030
220	085
276	139
332	193
388	247
444	301
500	355
555	410
611	464
667	518
723	572
779	626
835	680
890	734
946	788
4478002	843
058	897
114	951
170	·6511005
225	059
281	113
337	168
393	222
449	276
505	330
560	384
616	438
672	492
728	546
784	601
840	655
895	709
951	763
4479007	817
063	871
119	925
175	979
230	·6512034
286	088
342	142
398	196
453	250
509	304
565	358
621	412
677	466
733	520
788	575
844	629
900	683
956	737

Excesses

10 = 10	1 = 1
20 = 19	2 = 2
30 = 29	3 = 3
40 = 39	4 = 4
50 = 49	5 = 5
	6 = 6
	7 = 7
	8 = 8
	9 = 9

28 = 27

Anti Log Excesses

10 = 10	1 = 1
20 = 21	2 = 2
30 = 31	3 = 3
40 = 41	4 = 4
50 = 52	5 = 5
	6 = 6
	7 = 7
	8 = 8
	9 = 9

27 = 28

4480011	·6512791
067	845
123	899
179	953
235	·6513007
291	061
346	116
402	170
458	224
514	278
569	332
625	386
681	440
737	494
793	548
849	602
904	657
960	711
4481016	765
072	819
127	873
183	927
239	981
295	·6514035
351	089
407	143
462	197
518	251
574	305
630	359
685	414
741	468
797	522
853	576
908	630
964	684
4482020	738

4482076	·6514792
131	846
187	900
243	954
299	·6515008
355	062
411	116
466	170
522	224
578	278
634	332
689	386
745	440
801	494
857	548
912	602
968	656
4483024	710
080	764
135	818
191	872
247	927
303	981
358	·6516035
414	089
470	143
526	197
581	251
637	305
693	359
749	413
804	467
860	521
916	575
972	629
4484027	683
083	737
139	791
195	845
250	899
306	953
362	·6517007
418	061
473	115
529	169
585	223
641	277
696	331
752	385
808	439
864	492
919	546
975	600
4485031	654
087	708
142	762
198	816
254	870
310	924
365	978
421	·6518032
477	086
533	140
588	194
644	248

4485700	·6518302
756	356
811	410
867	464
923	518
979	572
4486034	626
090	680
146	734
202	787
257	841
313	895
368	949
424	·6519003
480	057
536	111
591	165
647	219
703	273
759	327
814	381
870	435
926	489
982	543
4487037	597
093	650
149	704
205	758
260	812
316	866
371	920
427	974
483	·6520028
539	082
594	136
650	190
706	244
762	297
817	351
873	405
928	459
984	513
4488040	567
096	621
151	675
207	729
263	783
319	836
374	890
430	944
485	998
541	·6521052
597	106
653	160
708	214
764	268
820	322
876	375
931	429
987	483
4489042	537
098	591
154	645
210	699
265	753

4489321	·6521806
377	860
433	914
488	968
544	·6522022
599	076
655	130
711	184
767	237
822	291
878	345
933	399
989	453

Excesses

10 = 10	1 = 1
20 = 19	2 = 2
30 = 29	3 = 3
40 = 39	4 = 4
50 = 48	5 = 5
	6 = 6
	7 = 7
	8 = 8
	9 = 9

28 = 27

Anti Log Excesses

10 = 10	1 = 1
20 = 21	2 = 2
30 = 31	3 = 3
40 = 41	4 = 4
50 = 52	5 = 5
	6 = 6
	7 = 7
	8 = 8
	9 = 9

27 = 28

4490045	·6522507
101	560
156	614
212	668
267	722
323	776
379	830
435	884
490	938
546	991
601	·6523045
657	099
713	153
769	207
824	261
880	314
935	368
991	422
4491047	476
103	530
158	584
214	637
269	691

4491325	·6523745
381	799
437	853
492	907
548	960
603	·6524014
659	068
715	122
771	176
826	230
882	283
937	337
993	391
4492049	445
105	499
160	553
216	606
271	660
327	714
383	768
439	821
494	875
550	929
605	983
661	·6525037
716	091
772	144
828	198
884	252
939	306
995	359
4493050	413
106	467
162	521
218	574
273	628
329	682
384	736
440	789
495	843
551	897
607	951
663	·6526005
718	059
774	112
829	166
885	220
940	274
996	327
4494052	381
108	435
163	489
219	542
274	596
330	650
386	704
442	757
497	811
553	865
608	919
664	972
719	·6527026
775	080
830	134
886	187

4494942	·6527241
998	295
4495053	349
109	402
164	456
220	509
275	563
331	617
387	671
443	724
498	778
554	832
609	886
665	939
720	993
776	·6528047
832	101
888	154
943	208
999	262
4496054	316
110	369
165	423
221	476
276	530
332	584
388	638
444	691
499	745
555	799
610	853
666	906
721	960
777	·6529013
832	067
888	121
943	175
999	228
4497055	282
111	336
166	390
222	443
277	497
333	550
388	604
444	658
499	712
555	765
611	819
667	872
722	926
778	980
833	·6530034
889	087
944	141
4498000	194
055	248
111	302
166	356
222	409
278	463
334	516
389	570
445	624
500	678

4498556	·6530731
611	785
667	838
722	892
778	945
833	999
889	·6531053
944	107
4499000	160
056	214
112	267
167	321
223	375
278	429
334	482
389	536
445	589
500	643
556	696
611	750
667	804
722	858
778	911
833	965
889	·6532018
944	072

Excesses

10 = 10	1 = 1
20 = 19	2 = 2
30 = 29	3 = 3
40 = 39	4 = 4
50 = 48	5 = 5
	6 = 6
	7 = 7
	8 = 8
	9 = 9

28 = 27

Anti Log Excesses

10 = 10	1 = 1
20 = 21	2 = 2
30 = 31	3 = 3
40 = 41	4 = 4
50 = 52	5 = 5
	6 = 6
	7 = 7
	8 = 8
	9 = 9

27 = 28

4500000	·6532125
056	179
112	233
167	287
223	340
278	394
334	447
389	501
445	554
500	608

4500556	·6532661
611	715
667	769
722	823
778	876
833	930
889	983
944	·6533037
4501000	090
055	144
111	197
167	251
223	305
278	359
334	412
389	466
445	519
500	573
556	626
611	680
667	733
722	787
778	840
833	894
889	948
944	·6534002
4502000	055
055	109
111	162
166	216
222	269
277	323
333	376
388	430
444	483
499	537
555	590
610	644
666	697
721	751
777	804
832	858
888	912
944	966
4503000	·6535019
055	073
111	126
166	180
222	233
277	287
333	340
388	394
444	447
499	501
555	554
610	608
666	661
721	715
777	768
832	822
888	875
943	929
999	982
4504054	·6536036
110	089

4504165	·6536143
221	196
276	250
332	303
387	357
443	410
498	464
554	517
609	571
665	624
720	678
776	731
831	785
887	839
942	892
998	945
4505053	999
109	·6537052
164	106
220	159
275	213
331	266
386	320
442	373
497	427
553	480
608	534
664	587
719	641
775	694
830	748
886	801
941	855
997	908
4506052	962
107	·6538015
162	069
218	122
273	176
329	229
384	283
440	336
495	390
551	443
606	497
662	550
717	604
773	657
828	711
884	764
939	818
995	871
4507050	924
106	977
161	·6539031
217	084
272	138
328	191
383	245
439	298
494	352
550	405
605	459
661	512
716	566

4507772	·6539619
827	673
882	726
937	779
993	832
4508048	886
104	939
159	993
215	·6540046
270	100
326	153
381	207
437	260
492	314
548	367
603	420
659	473
714	527
770	580
825	634
881	687
936	741
992	794
4509047	848
102	901
157	954
213	·6541007
268	061
324	114
379	168
435	221
490	275
546	328
601	382
657	435
712	488
768	541
823	595
879	648
934	702
989	755

Excesses

10 = 10	1 = 1
20 = 19	2 = 2
30 = 29	3 = 3
40 = 39	4 = 4
50 = 48	5 = 5
	6 = 6
	7 = 7
	8 = 8
	9 = 9

28 = 27

Anti Log Excesses

10 = 10	1 = 1
20 = 21	2 = 2
30 = 31	3 = 3
40 = 41	4 = 4
50 = 52	5 = 5
	6 = 6
	7 = 7
	8 = 8
	9 = 9

27 = 28

4510044	·6541809
100	862
155	915
211	968
266	·6542022
322	075
377	129
433	182
488	236
544	289
599	342
655	395
710	449
765	502
820	556
876	609
931	663
987	716
4511042	769
098	822
153	876
209	929
264	983
320	·6543036
375	089
430	142
485	196
541	249
596	303
652	356
707	410
763	463
818	516
874	569
929	623
985	676
4512040	730
095	783
150	836
206	889
261	943
317	996
372	·6544049
428	102
483	156
539	209
594	263
649	316
704	369
760	422
815	476

4512871	·6544529
926	583
982	636
4513037	689
093	742
148	796
203	849
258	902
314	955
369	·6545009
425	062
480	116
536	169
591	222
646	275
701	329
757	382
812	435
868	488
923	542
979	595
4514034	649
090	702
145	755
200	808
255	862
311	915
366	968
422	·6546021
477	075
533	128
588	181
643	234
698	288
754	341
809	394
865	447
920	501
976	554
4515031	607
086	660
141	714
197	767
252	820
308	873
363	927
419	980
474	·6547033
529	086
584	140
640	193
695	246
751	299
806	353
861	406
916	459
972	512
4516027	566
083	619
138	672
194	725
249	779
304	832
359	885
415	938

4516470	·6547992
526	·6548045
581	098
636	151
691	204
747	257
802	311
858	364
913	417
969	470
4517024	524
079	577
134	630
190	683
245	737
301	790
356	843
411	896
466	949
522	·6549002
577	056
633	109
688	162
743	215
798	269
854	322
909	375
965	428
4518020	481
075	534
130	588
186	641
241	694
297	747
352	800
407	853
462	906
518	960
573	·6550013
629	066
684	120
739	173
794	226
850	279
905	332
961	385
4519016	438
071	491
126	545
182	598
237	651
292	704
347	757
403	810
458	864
514	917
569	970
624	·6551023
679	076
735	129
790	183
846	236
901	289
956	342

Excesses

10 = 10	1 = 1
20 = 19	2 = 2
30 = 29	3 = 3
40 = 38	4 = 4
50 = 48	5 = 5
	6 = 6
	7 = 7
	8 = 8
	9 = 9

28 = 27

Anti Log Excesses

10 = 10	1 = 1
20 = 21	2 = 2
30 = 31	3 = 3
40 = 42	4 = 4
50 = 52	5 = 5
	6 = 6
	7 = 7
	8 = 8
	9 = 9

27 = 28

4520011	·6551395
067	448
122	502
177	555
232	608
288	661
343	714
399	767
454	820
509	873
564	926
620	980
675	·6552033
731	086
786	139
841	192
896	245
952	298
4521007	352
062	405
117	458
173	511
228	564
284	617
339	670
394	723
449	776
505	830
560	883
615	936
670	989
726	·6553042
781	095
836	148
891	201
947	254
4522002	308

4522058	·6553361
113	414
168	467
223	520
279	573
334	626
389	679
444	732
500	785
555	838
610	891
665	945
721	998
776	·6554051
831	104
886	157
942	210
997	263
4523053	316
108	369
163	422
218	475
274	528
329	582
384	635
439	688
495	741
550	794
605	847
661	900
716	953
771	·6555006
826	059
881	112
937	165
992	218
4524047	271
102	324
158	377
213	430
268	483
323	536
379	589
434	643
489	696
544	749
600	802
655	855
710	908
765	961
821	·6556014
876	067
931	120
986	173
4525042	226
097	279
152	332
207	385
263	438
318	491
373	544
428	597
484	650
539	703
594	756

4525649	·6556809
705	862
760	915
815	968
870	·6557021
926	074
981	127
4526036	180
091	233
147	286
202	339
257	392
312	445
368	498
423	551
478	604
533	657
589	710
644	763
699	816
754	869
810	922
865	975
920	·6558028
975	081
4527030	134
085	187
141	240
196	293
251	346
306	399
362	452
417	505
472	558
527	611
583	664
638	717
693	769
748	822
803	875
858	928
914	981
969	·6559034
4528024	087
079	140
135	193
190	246
245	299
300	352
356	405
411	458
466	511
521	564
576	617
631	670
687	723
742	776
797	828
852	881
908	934
963	987
4529018	·6560040
073	093
128	146
183	199

4529239	·6560252
294	305
349	358
404	411
460	464
515	517
570	569
625	622
680	675
735	728
791	781
846	834
901	887
956	940

Excesses

10 = 10	1 = 1
20 = 19	2 = 2
30 = 29	3 = 3
40 = 38	4 = 4
50 = 48	5 = 5
	6 = 6
	7 = 7
	8 = 8
	9 = 9

28 = 27

Anti Log Excesses

10 = 10	1 = 1
20 = 21	2 = 2
30 = 31	3 = 3
40 = 42	4 = 4
50 = 52	5 = 5
	6 = 6
	7 = 7
	8 = 8
	9 = 9

27 = 28

4530012	·6560993
067	·6561046
122	099
177	152
232	204
287	257
343	310
398	363
453	416
508	469
563	522
618	575
674	628
729	681
784	733
839	786
894	839
949	892
4531005	945
060	998
115	·6562051
170	104

4531225	·6562156
281	209
336	262
391	315
446	368
501	421
557	474
612	527
667	579
722	632
777	685
832	738
888	791
943	844
998	896
4532053	949
108	·6563002
163	055
219	108
274	161
329	214
384	267
439	319
494	372
549	425
604	478
660	531
715	584
770	636
825	689
880	742
935	795
991	848
4533046	901
101	953
156	·6564006
211	059
266	112
322	165
377	218
432	270
487	323
542	376
597	429
652	482
707	535
763	587
818	640
873	693
928	746
983	799
4534038	852
094	904
149	957
204	·6565010
259	063
314	115
369	168
424	221
479	274
535	327
590	380
645	432
700	485
755	538

4534810	·6565591
865	644
920	697
976	749
4535031	802
086	855
141	908
196	960
251	·6566013
306	066
361	119
417	171
472	224
527	277
582	330
637	383
692	436
747	488
802	541
858	594
913	647
968	699
4536023	752
078	805
133	858
188	910
243	963
299	·6567016
354	069
409	121
464	174
519	227
574	280
629	332
684	385
739	438
794	491
850	543
905	596
960	649
4537015	702
070	754
125	807
180	860
235	913
290	965
345	·6568018
401	071
456	124
511	176
566	229
621	282
676	335
731	387
786	440
841	492
896	545
952	598
4538007	651
062	703
117	756
172	809
227	862
282	914
337	967

4538392	·6569020
447	073
502	125
557	178
613	230
668	283
723	336
778	389
833	441
888	494
943	547
998	600
4539053	652
108	705
163	757
218	810
274	863
329	916
384	968
439	·6570021
494	074
549	127
604	179
659	232
714	284
769	337
824	390
879	443
934	495
989	548

Excesses

10 = 10	1 = 1
20 = 19	2 = 2
30 = 29	3 = 3
40 = 38	4 = 4
50 = 48	5 = 5
	6 = 6
	7 = 7
	8 = 8
	9 = 9

28 = 27

Anti Log Excesses

10 = 10	1 = 1
20 = 21	2 = 2
30 = 31	3 = 3
40 = 42	4 = 4
50 = 52	5 = 5
	6 = 6
	7 = 7
	8 = 8
	9 = 9

27 = 28

4540045	·6570601
100	654
155	706
210	759
265	811
320	864

4540375	·6570917
430	970
485	·6571022
540	075
595	127
650	180
705	233
760	286
815	338
870	391
926	443
981	496
4541036	548
091	601
146	654
201	707
256	759
311	812
366	864
421	917
476	970
531	·6572023
586	075
641	128
696	180
751	233
806	285
861	338
916	391
971	444
4542026	496
081	549
137	601
192	654
247	706
302	759
357	812
412	865
467	917
522	970
577	·6573022
632	075
687	127
742	180
797	233
852	286
907	338
962	391
4543017	443
072	496
127	548
182	601
237	654
292	707
347	759
402	812
457	864
512	917
567	969
622	·6574022
677	074
732	127
787	179
842	232
898	285

Number	Log
4543953	·6574338
4544008	390
063	443
118	495
173	548
228	600
283	653
338	705
393	758
448	810
503	863
558	916
613	969
668	·6575021
723	074
778	126
833	179
888	231
943	284
998	336
4545053	389
108	441
163	494
218	546
273	599
328	651
383	704
438	757
493	810
548	862
603	915
658	967
713	·6576020
768	072
823	125
878	177
933	230
988	282
4546043	335
098	387
153	440
208	492
263	545
318	597
373	650
428	702
483	755
538	807
593	860
648	912
703	965
758	·6577017
813	070
868	122
923	175
978	227
4547032	280
087	332
142	385
197	437
252	490
307	542
362	595
417	647
472	700

Number	Log
4547527	·6577752
582	805
637	857
692	910
747	962
802	·6578015
857	067
912	120
967	172
4548022	225
077	277
132	330
187	382
242	435
297	487
352	540
407	592
462	645
517	697
572	750
627	802
682	855
737	907
792	960
847	·6579012
901	065
956	117
4549011	170
066	222
121	275
176	327
231	380
286	432
341	485
396	537
451	590
506	642
561	695
616	747
671	800
726	852
781	905
836	956
891	·6580009
946	061

Excesses

10 = 10	1 = 1
20 = 19	2 = 2
30 = 29	3 = 3
40 = 38	4 = 4
50 = 48	5 = 5
	6 = 6
	7 = 7
	8 = 8
	9 = 9

28 = 27

Anti Log Excesses

10 = 10	1 = 1
20 = 21	2 = 2
30 = 31	3 = 3
40 = 42	4 = 4
50 = 52	5 = 5
	6 = 6
	7 = 7
	8 = 8
	9 = 9

27 = 28

Number	Log
4550000	·6580114
055	166
110	219
165	271
220	324
275	376
330	429
385	481
440	534
495	586
550	639
605	691
660	743
715	795
770	848
825	900
879	953
934	·6581005
989	058
4551044	110
099	163
154	215
209	268
264	320
319	372
374	424
429	477
484	529
539	582
594	634
648	687
703	739
758	792
813	844
868	897
923	949
978	·6582002
4552033	054
088	106
143	158
198	211
253	263
308	316
363	368
417	421
472	473
527	526
582	578
637	630
692	682
747	735

Number	Log
4552802	·6582787
857	840
912	892
966	945
4553021	997
076	·6583049
131	101
186	154
241	206
296	259
351	311
406	363
461	415
516	468
571	520
625	573
680	625
735	678
790	730
845	782
900	834
955	887
4554010	939
065	992
120	·6584044
174	096
229	148
284	201
339	253
394	306
449	358
504	410
559	462
613	515
668	567
723	620
778	672
833	725
888	777
943	829
998	881
4555053	934
108	986
162	·6585038
217	090
272	143
327	195
382	248
437	300
492	352
547	404
601	457
656	509
711	562
766	614
821	666
876	718
931	771
986	823
4556040	876
095	928
150	980
205	·6586032
260	085
315	137

Number	Log
4556370	·6586189
425	241
479	294
534	346
589	399
644	451
699	503
754	555
808	608
863	660
918	712
973	764
4557028	817
083	869
138	921
193	973
247	·6587026
302	078
357	130
412	182
467	235
522	287
576	340
631	392
686	444
741	496
796	549
851	601
906	653
961	705
4558015	758
070	810
125	862
180	914
235	967
290	·6588019
344	071
399	123
454	176
509	228
564	280
619	332
673	385
728	437
783	489
838	541
893	594
948	646
4559002	698
057	750
112	803
167	855
222	907
277	959
331	·6589012
386	064
441	116
496	168
551	221
606	273
660	325
715	377
770	429
825	481
880	534

4559935	·6589586
989	638

Excesses

10 = 10	1 = 1
20 = 19	2 = 2
30 = 29	3 = 3
40 = 38	4 = 4
50 = 48	5 = 5
	6 = 6
	7 = 7
	8 = 8
	9 = 9
28 = 27	

Anti Log Excesses

10 = 10	1 = 1
20 = 21	2 = 2
30 = 31	3 = 3
40 = 42	4 = 4
50 = 52	5 = 5
	6 = 6
	7 = 7
	8 = 8
	9 = 9
27 = 28	

4560044	·6589690
099	743
154	795
209	847
264	899
318	951
373	·6590003
428	056
483	108
538	160
593	212
647	265
702	317
757	369
812	421
866	473
921	525
976	578
4561031	630
086	682
141	734
195	787
250	839
305	891
360	943
415	995
470	·6591047
524	100
579	152
634	204
689	256
743	308
798	360
853	413

4561908	·6591465
963	517
4562018	569
072	621
127	673
182	726
237	778
291	830
346	882
401	934
456	986
511	·6592039
566	091
620	143
675	195
730	247
785	299
839	352
894	404
949	456
4563004	508
058	560
113	612
168	664
223	716
278	769
333	821
387	873
442	925
497	977
552	·6593029
606	082
661	134
716	186
771	238
825	290
880	342
935	394
990	446
4564045	499
100	551
154	603
209	655
264	707
319	759
373	811
428	863
483	915
538	967
592	·6594020
647	072
702	124
757	176
811	228
866	280
921	332
976	384
4565030	436
085	488
140	541
195	593
249	645
304	697
359	749
414	801

4565468	·6594853
523	905
578	957
633	·6595009
687	062
742	114
797	166
852	218
907	270
962	322
4566016	374
071	426
126	478
181	530
235	582
290	634
345	686
400	738
454	791
509	843
564	895
619	947
673	999
728	·6596051
782	103
837	155
892	207
947	259
4567001	311
056	363
111	415
166	467
220	519
275	571
330	623
385	675
439	728
494	780
549	832
604	884
658	936
713	988
768	·6597040
823	092
877	144
932	196
987	248
4568042	300
096	352
151	404
206	456
261	508
315	560
370	612
424	664
479	716
534	768
589	820
643	872
698	924
753	976
808	·6598028
862	080
917	132
972	184

4569027	·6598236
081	288
136	340
191	392
246	444
300	496
355	548
409	600
464	652
519	704
574	756
628	808
683	860
738	912
793	964
847	·6599016
902	068
956	120

Excesses

10 = 10	1 = 1
20 = 19	2 = 2
30 = 29	3 = 3
40 = 38	4 = 4
50 = 48	5 = 5
	6 = 6
	7 = 7
	8 = 8
	9 = 9
27 = 26	

Anti Log Excesses

10 = 11	1 = 1
20 = 21	2 = 2
30 = 32	3 = 3
40 = 42	4 = 4
50 = 53	5 = 5
	6 = 6
	7 = 7
	8 = 8
	9 = 9
26 = 27	

4570011	·6599172
066	224
121	276
175	328
230	380
285	432
340	484
394	536
449	588
503	640
558	692
613	744
668	796
722	848
777	900
832	952
887	·6600004
941	056

4570996	·6600108
4571050	160
105	212
160	264
215	316
269	368
324	419
379	471
434	523
488	575
543	627
597	679
652	731
707	783
762	835
816	887
871	939
925	991
980	·6601043
4572035	095
090	147
144	199
199	250
253	302
308	354
363	406
418	458
472	510
527	562
582	614
637	666
691	718
746	770
800	822
855	874
910	926
965	978
4573019	·6602030
074	082
128	134
183	185
238	237
293	289
347	341
402	393
456	445
511	497
566	549
621	601
675	653
730	705
784	757
839	808
893	860
948	912
4574003	964
058	·6603016
112	068
167	120
221	172
276	223
331	275
386	327
440	379
495	431

4574549	·6603483
604	535
659	587
714	639
768	691
823	742
877	794
932	846
986	898
4575041	950
096	·6604002
151	053
205	105
260	157
314	209
369	261
424	313
479	365
533	417
588	468
642	520
697	572
751	624
806	676
861	728
916	780
970	832
4576025	883
079	935
134	987
189	·6605039
244	091
298	143
353	194
407	246
462	298
516	350
571	402
626	454
681	505
735	557
790	609
844	661
899	713
953	765
4577008	816
062	868
117	920
172	972
227	·6606024
281	076
336	127
390	179
445	231
499	283
554	335
609	387
664	438
718	490
773	542
827	594
882	645
936	697
991	749
4578046	801

4578101	·6606853
155	905
210	956
264	·6607008
319	060
373	112
428	163
482	215
537	267
592	319
647	371
701	423
756	474
810	526
865	578
919	630
974	681
4579028	733
083	785
138	837
193	888
247	940
302	992
356	·6608044
411	096
465	148
520	199
574	251
629	303
683	355
738	406
793	458
848	510
902	562
957	613

Excesses

10 = 9	1 = 1
20 = 19	2 = 2
30 = 28	3 = 3
40 = 38	4 = 4
50 = 47	5 = 5
	6 = 6
	7 = 7
	8 = 8
	9 = 9

27 = 26

Anti Log Excesses

10 = 11	1 = 1
20 = 21	2 = 2
30 = 32	3 = 3
40 = 42	4 = 4
50 = 53	5 = 5
	6 = 6
	7 = 7
	8 = 8
	9 = 9

26 = 27

4580011	·6608665

4580066	·6608717
120	769
175	820
229	872
284	924
338	976
393	·6609027
448	079
503	131
557	183
612	234
666	286
721	338
775	390
830	441
884	493
939	545
993	597
4581048	648
103	700
158	752
212	804
267	855
321	907
376	958
430	·6610010
485	062
539	114
594	165
648	217
703	269
757	321
812	372
866	424
921	476
976	528
4582031	579
085	631
140	683
194	735
249	786
303	838
358	889
412	941
467	993
521	·6611045
576	096
630	148
685	200
739	252
794	303
849	355
904	407
958	459
4583013	510
067	562
122	613
176	665
231	717
285	769
340	820
394	872
449	923
503	975
558	·6612027

4583612	·6612079
667	130
721	182
776	233
830	285
885	337
939	389
994	440
4584048	492
103	543
158	595
213	647
267	699
322	750
376	802
431	853
485	905
540	957
594	·6613009
649	060
703	112
758	163
812	215
867	267
921	319
976	370
4585030	422
085	473
139	525
194	576
248	628
303	680
357	732
412	783
466	835
521	886
575	938
630	990
684	·6614042
739	093
793	145
848	196
902	248
957	299
4586011	351
066	403
120	455
175	506
229	558
284	609
338	661
393	712
447	764
502	816
556	868
611	919
665	971
720	·6615022
774	074
829	125
883	177
938	228
992	280
4587047	332
101	384

4587156	·6615435
210	487
265	538
319	590
374	641
428	693
483	744
537	796
592	848
646	900
701	951
755	·6616003
810	054
864	106
919	157
973	209
4588028	260
082	312
137	363
191	415
246	467
300	519
355	570
409	622
464	673
518	725
573	776
627	828
682	879
736	931
791	982
845	·6617034
900	085
954	137
4589009	188
063	240
118	292
172	344
227	395
281	447
336	498
390	550
445	601
499	653
554	704
608	756
663	807
717	859
772	910
826	962
881	·6618013
935	065
990	116

Excesses

10 = 9	1 = 1
20 = 19	2 = 2
30 = 28	3 = 3
40 = 38	4 = 4
50 = 47	5 = 5
	6 = 6
	7 = 7
	8 = 8
	9 = 9

27 = 26

Anti Log Excesses

10 = 11	1 = 1
20 = 21	2 = 2
30 = 32	3 = 3
40 = 42	4 = 4
50 = 53	5 = 5
	6 = 6
	7 = 7
	8 = 8
	9 = 10

26 = 27

4590044	·6618168
099	219
153	271
207	322
261	374
316	425
370	477
425	528
479	580
534	632
588	684
643	735
697	787
752	838
806	890
861	941
915	993
970	·6619044
4591024	096
079	147
133	199
188	250
242	302
297	353
351	405
406	456
460	508
514	559
568	611
623	662
677	714
732	765
786	817
841	868
895	920
950	971
4592004	·6620023

4592059	·6620074
113	126
168	177
222	229
277	280
331	332
386	383
440	435
494	486
548	538
603	589
657	640
712	691
766	743
821	794
875	846
930	897
984	949
4593039	·6621000
093	052
148	104
202	155
256	206
310	258
365	309
419	361
474	412
528	464
583	515
637	567
692	618
746	670
801	721
855	773
910	824
964	876
4594018	927
072	979
127	·6622030
181	082
236	133
290	184
345	235
399	287
454	338
508	390
562	441
616	493
671	544
725	596
780	647
834	699
889	750
943	802
998	853
4595052	904
107	955
161	·6623007
215	058
269	110
324	161
378	213
433	264
487	316
542	367

4595596	·6623419
651	470
705	521
759	572
813	624
868	675
922	727
977	778
4596031	830
086	881
140	933
195	984
249	·6624035
303	086
357	138
412	189
466	241
521	292
575	344
630	395
684	446
738	497
792	549
847	600
901	652
956	703
4597010	755
065	806
119	858
174	909
228	960
282	·6625011
336	063
391	114
445	166
500	217
554	268
609	319
663	371
717	422
771	474
826	525
880	577
935	628
989	679
4598044	730
098	782
152	833
206	885
261	936
315	987
370	·6626038
424	090
479	141
533	193
587	244
641	295
696	346
750	398
805	449
859	501
913	552
967	603
4599022	654
076	706

4599131	·6626757
185	809
240	860
294	911
348	962
402	·6627014
457	065
511	117
566	168
620	219
674	270
728	322
783	373
837	425
892	476
946	527

Excesses

10 = 9	1 = 1
20 = 19	2 = 2
30 = 28	3 = 3
40 = 38	4 = 4
50 = 47	5 = 5
	6 = 6
	7 = 7
	8 = 8
	9 = 9

27 = 26

Anti Log Excesses

10 = 11	1 = 1
20 = 21	2 = 2
30 = 32	3 = 3
40 = 42	4 = 4
50 = 53	5 = 5
	6 = 6
	7 = 7
	8 = 8
	9 = 10

26 = 27

4600000	·6627578
054	630
109	681
163	733
218	784
272	835
327	886
381	938
435	989
489	·6628040
544	091
598	143
653	194
707	245
761	296
815	348
870	399
924	451
979	502
4601033	553

4601087	·6628604
141	656
196	707
250	758
305	809
359	861
413	912
467	964
522	·6629015
576	066
631	117
685	169
739	220
793	271
848	322
902	374
957	425
4602011	476
065	527
119	579
174	630
228	681
283	732
337	784
391	835
445	886
500	937
554	989
608	·6630040
662	091
717	142
771	194
826	245
880	296
934	347
988	399
4603043	450
097	501
152	552
206	604
260	655
314	706
369	757
423	809
477	860
531	911
586	962
640	·6631014
695	065
749	116
803	167
857	219
912	270
966	321
4604020	372
074	423
129	474
183	526
238	577
292	628
346	679
400	731
455	782
509	833
563	884

4604617	·6631936
672	987
726	·6632038
781	089
835	140
889	191
943	243
998	294
4605052	345
106	396
160	448
215	499
269	550
324	601
378	652
432	703
486	755
541	806
595	857
649	908
703	959
758	·6633010
812	062
866	113
920	164
975	215
4606029	266
083	317
137	369
192	420
246	471
301	522
355	573
409	624
463	676
518	727
572	778
626	829
680	880
735	931
789	983
843	·6634034
897	085
952	136
4607006	187
060	238
114	290
169	341
223	392
277	443
331	494
386	545
440	597
494	648
548	699
603	750
657	801
711	852
765	903
820	954
874	·6635006
928	057
982	108
4608037	159
091	210

4608145	·6635261
199	312
254	363
308	415
362	466
416	517
471	568
525	619
579	670
633	721
688	772
742	824
796	875
850	926
905	977
959	·6636028
4609013	079
067	130
122	181
176	233
230	284
284	335
339	386
393	437
447	488
501	539
556	590
610	641
664	692
718	744
773	795
827	846
881	897
935	948
990	999

Excesses

10 = 9	1 = 1
20 = 19	2 = 2
30 = 28	3 = 3
40 = 38	4 = 4
50 = 47	5 = 5
	6 = 6
	7 = 7
	8 = 8
	9 = 8

27 = 26

Anti Log Excesses

10 = 11	1 = 1
20 = 21	2 = 2
30 = 32	3 = 3
40 = 42	4 = 4
50 = 53	5 = 5
	6 = 6
	7 = 7
	8 = 8
	9 = 10

26 = 27

4610044	·6637050
4610098	·6637101
152	152
207	203
261	254
315	305
369	357
423	408
477	459
532	510
586	561
640	612
694	663
749	714
803	765
857	816
911	867
966	918
4611020	969
074	·6638020
128	071
183	122
237	174
291	225
345	276
399	327
453	378
508	429
562	480
616	531
670	582
725	633
779	684
833	735
887	786
942	837
996	888
4612050	939
104	990
158	·6639041
212	092
267	143
321	195
375	246
429	297
484	348
538	399
592	450
646	501
700	552
754	603
809	654
863	705
917	756
971	807
4613026	858
080	909
134	960
188	·6640011
242	062
296	113
351	164
405	215
459	266
513	317
567	368

4613621	·6640419
676	470
730	521
784	572
838	623
893	674
947	725
4614001	776
055	827
109	878
163	929
218	980
272	·6641031
326	082
380	133
434	184
488	235
543	286
597	337
651	388
705	439
759	490
813	541
868	592
922	643
976	694
4615030	745
084	796
138	847
193	898
247	949
301	·6642000
355	051
409	102
463	153
518	204
572	255
626	306
680	357
734	408
788	459
843	510
897	561
951	612
4616005	663
059	713
113	764
168	815
222	866
276	917
330	968
384	·6643019
438	070
493	121
547	172
601	223
655	274
709	325
763	376
818	427
872	478
926	529
980	580
4617034	630
088	681

4617142	·6643732
196	783
251	834
305	885
359	936
413	987
467	·6644038
521	089
576	140
630	191
684	242
738	293
792	343
846	394
900	445
954	496
4618009	547
063	598
117	649
171	700
225	751
279	802
334	852
388	903
442	954
496	·6645005
550	056
604	107
658	158
712	209
767	260
821	311
875	361
929	412
983	463
4619037	514
091	565
145	616
200	667
254	718
308	768
362	819
416	870
470	921
524	972
578	·6646023
633	074
687	125
741	175
795	226
849	277
903	328
957	379

Excesses

10 = 9	1 = 1
20 = 19	2 = 2
30 = 28	3 = 3
40 = 38	4 = 4
50 = 47	5 = 5
	6 = 6
	7 = 7
	8 = 8
	9 = 8

27 = 26

Anti Log Excesses

10 = 11	1 = 1
20 = 21	2 = 2
30 = 32	3 = 3
40 = 42	4 = 4
50 = 53	5 = 5
	6 = 6
	7 = 7
	8 = 8
	9 = 10

26 = 27

4620011	·6646430
065	481
119	532
174	582
228	633
282	684
336	735
390	786
444	837
498	888
552	939
607	989
661	·6647040
715	091
769	142
823	193
877	244
931	294
985	345
4621039	396
093	447
148	498
202	549
256	599
310	650
364	701
418	752
472	803
526	854
580	904
634	955
689	·6648006
743	057
797	108
851	159
905	209
959	260

4622013	·6648311
067	362
121	413
175	464
229	514
283	565
338	616
392	667
446	718
500	769
554	819
608	870
662	921
716	972
770	·6649023
824	074
878	124
932	175
987	226
4623041	277
095	327
149	378
203	429
257	480
311	531
365	582
419	632
473	683
527	734
581	785
635	835
689	886
744	937
798	988
852	·6650038
906	089
960	140
4624014	191
068	241
122	292
176	343
230	394
284	445
338	496
392	546
446	597
500	648
554	699
609	749
663	800
717	851
771	902
825	952
879	·6651003
933	054
987	105
4625041	155
095	206
149	257
203	308
257	358
311	409
365	460
419	511
473	561

4625527	·6651612
582	663
636	714
690	764
744	815
798	866
852	917
906	967
960	·6652018
4626014	069
068	120
122	170
176	221
230	272
284	323
338	373
392	424
446	475
500	526
554	576
608	627
662	677
716	728
770	779
824	830
878	880
932	931
987	982
4627041	·6653033
095	083
149	134
203	185
257	236
311	286
365	337
419	388
473	439
527	489
581	540
635	590
689	641
743	692
797	743
851	793
905	844
959	895
4628013	946
067	996
121	·6654047
175	097
229	148
283	199
337	250
391	300
445	351
499	401
553	452
607	503
661	554
715	604
769	655
823	705
877	756
931	807
985	858

4629039	·6654908
093	959
147	·6655009
201	060
255	111
309	162
363	212
417	263
471	313
525	364
579	415
633	466
687	516
741	567
795	617
849	668
903	719
957	770

Excesses

10 = 9	1 = 1
20 = 19	2 = 2
30 = 28	3 = 3
40 = 38	4 = 4
50 = 47	5 = 5
	6 = 6
	7 = 7
	8 = 8
	9 = 8

27 = 25

Anti Log Excesses

10 = 11	1 = 1
20 = 21	2 = 2
30 = 32	3 = 3
40 = 43	4 = 4
50 = 53	5 = 5
	6 = 6
	7 = 7
	8 = 9
	9 = 10

25 = 27

4630011	·6655820
065	871
119	921
173	972
227	·6656023
281	074
335	124
389	175
443	225
497	276
551	326
605	377
659	428
713	479
767	529
821	580
875	630
929	681

4630983	·6656731
4631037	782
091	833
145	884
199	934
253	985
307	·6657035
361	086
415	136
469	187
523	238
577	289
631	339
685	390
739	440
793	491
847	541
901	592
955	642
4632009	693
063	744
117	795
171	845
225	896
279	946
332	997
386	·6658047
440	098
494	149
548	200
602	250
656	301
710	351
764	402
818	452
872	503
926	553
980	604
4633034	654
088	705
142	756
196	807
250	857
304	908
358	958
412	·6659009
466	059
520	110
574	160
628	211
682	261
735	312
789	362
843	413
897	464
951	515
4634005	565
059	616
113	666
167	717
221	767
275	818
329	868
383	919
437	969

4634491	·6660020
545	070
599	121
653	171
706	222
760	273
814	324
868	374
922	425
976	475
4635030	526
084	576
138	627
192	677
246	728
300	778
354	829
408	879
462	930
516	980
569	·6661031
623	081
677	132
731	182
785	233
839	283
893	334
947	384
4636001	435
055	485
109	536
163	586
217	637
271	687
324	738
378	788
432	839
486	889
540	940
594	990
648	·6662041
702	091
756	142
810	192
864	243
918	293
971	344
4637025	394
079	445
133	495
187	546
241	596
295	647
349	697
403	748
457	798
511	849
565	899
618	950
672	·6663000
726	051
780	101
834	152
888	202
942	253

4637996	·6663303
4638050	354
104	404
157	455
211	505
265	556
319	606
373	657
427	707
481	758
535	808
589	859
643	909
696	960
750	·6664010
804	060
858	110
912	161
966	211
4639020	262
074	312
127	363
181	413
235	464
289	514
343	565
397	615
451	666
505	716
559	767
613	817
666	867
720	917
774	968
828	·6665018
882	069
936	119
990	170

Excesses

10 = 9	1 = 1
20 = 19	2 = 2
30 = 28	3 = 3
40 = 37	4 = 4
50 = 47	5 = 5
	6 = 6
	7 = 7
	8 = 7
	9 = 8

27 = 25

Anti Log Excesses

10 = 11	1 = 1
20 = 21	2 = 2
30 = 32	3 = 3
40 = 43	4 = 4
50 = 53	5 = 5
	6 = 6
	7 = 7
	8 = 9
	9 = 10

25 = 27

4640044	·6665220
097	271
151	321
205	372
259	422
313	473
367	523
421	574
475	624
528	674
582	724
636	775
690	825
744	876
798	926
852	977
906	·6666027
959	078
4641013	128
067	178
121	228
175	279
229	329
283	380
337	430
390	481
444	531
498	582
552	632
606	682
660	732
713	783
767	833
821	884
875	934
929	985
983	·6667035
4642037	086
091	136
144	186
198	236
252	287
306	337
360	388
414	438
467	489
521	539
575	589
629	639
683	690
737	740

4642790	·6667791
844	841
898	892
952	942
4643006	992
060	·6668042
114	093
168	143
221	194
275	244
329	294
383	344
437	395
491	445
544	496
598	546
652	597
706	647
760	697
814	747
867	798
921	848
975	899
4644029	949
083	999
137	·6669049
190	100
244	150
298	201
352	251
406	301
460	351
513	402
567	452
621	503
675	553
729	603
783	653
836	704
890	754
944	805
998	855
4645051	905
105	955
159	·6670006
213	056
267	107
321	157
374	207
428	257
482	308
536	358
590	408
644	458
697	509
751	559
805	610
859	660
912	710
966	760
4646020	811
074	861
128	911
182	961
235	·6671012

4646289	·6671062
343	113
397	163
451	213
505	263
558	314
612	364
666	414
720	464
773	515
827	565
881	615
935	665
989	716
4647043	766
096	817
150	867
204	917
258	967
311	·6672018
365	068
419	118
473	168
527	219
581	269
634	319
688	369
742	420
796	470
849	520
903	570
957	621
4648011	671
064	721
118	771
172	822
226	872
280	922
334	972
387	·6673023
441	073
495	123
549	173
602	224
656	274
710	324
764	374
817	425
871	475
925	525
979	575
4649032	626
086	676
140	726
194	776
248	827
302	877
355	927
409	977
463	·6674027
517	077
570	128
624	178
678	228
732	278

4649785	·6674329
839	379
893	429
947	479

Excesses

10 = 9	1 = 1
20 = 19	2 = 2
30 = 28	3 = 3
40 = 37	4 = 4
50 = 47	5 = 5
	6 = 6
	7 = 7
	8 = 7
	9 = 8

27 = 25

Anti Log Excesses

10 = 11	1 = 1
20 = 21	2 = 2
30 = 32	3 = 3
40 = 43	4 = 4
50 = 53	5 = 5
	6 = 6
	7 = 7
	8 = 9
	9 = 10

25 = 27

4650000	·6674530
054	580
108	630
162	680
215	731
269	781
323	831
377	881
430	931
484	981
538	·6675032
592	082
645	132
699	182
753	233
807	283
860	333
914	383
968	433
4651022	483
075	534
129	584
183	634
237	684
290	734
344	784
398	835
452	885
505	935
559	985
613	·6676036
667	086

4651720	·6676136
774	186
828	236
882	286
935	337
989	387
4652043	437
097	487
150	537
204	587
258	638
312	688
365	738
419	788
473	838
527	888
580	939
634	989
688	·6677039
742	089
795	139
849	189
903	239
957	289
4653010	340
064	390
117	440
171	490
225	540
279	590
332	641
386	691
440	741
494	791
547	841
601	891
655	941
709	991
762	·6678042
816	092
870	142
924	192
977	242
4654031	292
084	342
138	392
192	443
246	493
299	543
353	593
407	643
461	693
514	743
568	793
622	844
676	894
729	944
783	994
836	·6679044
890	094
944	144
998	194
4655051	244
105	294
159	345

4655213	·6679395
266	445
320	495
373	545
427	595
481	645
535	695
588	746
642	796
696	846
750	896
803	946
857	996
910	·6680046
964	096
4656018	146
072	196
125	247
179	297
233	347
287	397
340	447
394	497
447	547
501	597
555	647
609	697
662	747
716	797
769	847
823	897
877	947
931	997
984	·6681048
4657038	098
092	148
146	198
199	248
253	298
306	348
360	398
414	448
468	498
521	548
575	598
628	648
682	698
736	748
790	798
843	848
897	898
950	948
4658004	998
058	·6682049
112	099
165	149
219	199
272	249
326	299
380	349
434	399
487	449
541	499
594	549
648	599

4658702	·6682649
756	699
809	749
863	799
916	849
970	899
4659024	949
078	999
131	·6683049
185	099
238	149
292	199
346	249
400	299
453	349
507	399
560	449
614	499
668	549
722	599
775	649
829	699
882	749
936	799
989	849

Excesses

10 = 9	1 = 1
20 = 19	2 = 2
30 = 28	3 = 3
40 = 37	4 = 4
50 = 47	5 = 5
	6 = 6
	7 = 7
	8 = 7
	9 = 8

27 = 25

Anti Log Excesses

10 = 11	1 = 1
20 = 21	2 = 2
30 = 32	3 = 3
40 = 43	4 = 4
50 = 54	5 = 5
	6 = 6
	7 = 8
	8 = 9
	9 = 10

25 = 27

4660043	·6683899
097	949
151	999
204	·6684049
258	099
311	149
365	199
419	249
473	299
526	349
580	399

4660633	·6684449
687	499
740	549
794	599
848	649
902	699
955	749
4661009	799
062	849
116	899
170	949
224	999
277	·6685049
331	099
384	149
438	199
491	249
545	299
599	349
653	399
706	449
760	498
813	548
867	598
920	648
974	698
4662028	748
082	798
135	848
189	898
242	948
296	998
349	·6686048
403	098
457	148
511	198
564	248
618	298
671	348
725	398
778	448
832	497
885	547
939	597
993	647
4663047	697
100	747
154	797
207	847
261	897
314	947
368	997
422	·6687047
476	097
529	147
583	196
636	246
690	296
743	346
797	396
850	446
904	496
958	546
4664012	596
065	646

4664119	·6687696
172	746
226	795
279	845
333	895
386	945
440	995
494	·6688045
548	095
601	145
655	195
708	245
762	294
815	344
869	394
922	444
976	494
4665030	544
084	594
137	644
191	694
244	744
298	793
351	843
405	893
458	943
512	993
565	·6689043
619	093
673	143
727	192
780	242
834	292
887	342
941	392
994	442
4666048	492
101	542
155	591
208	641
262	691
316	741
370	791
423	841
477	891
530	941
584	990
637	·6690040
691	090
744	140
798	190
851	240
905	289
958	339
4667012	389
066	439
120	489
173	539
227	588
280	638
334	688
387	738
441	788
494	838
548	888

4667601	·6690938
655	987
708	·6691037
762	087
815	137
869	186
923	236
977	286
4668030	336
084	386
137	436
191	485
244	535
298	585
351	635
405	685
458	735
512	784
565	834
619	884
672	934
726	984
779	·6692034
833	083
886	133
940	183
994	233
4669048	282
101	332
155	382
208	432
262	482
315	532
369	581
422	631
476	681
529	731
583	780
636	830
690	880
743	930
797	980
850	·6693030
904	079
957	129

Excesses

10 = 9	1 = 1
20 = 19	2 = 2
30 = 28	3 = 3
40 = 37	4 = 4
50 = 47	5 = 5
	6 = 6
	7 = 7
	8 = 7
	9 = 8

27 = 25

Anti Log Excesses

10 = 11	1 = 1
20 = 21	2 = 2
30 = 32	3 = 3
40 = 43	4 = 4
	5 = 5
	6 = 6
	7 = 8
	8 = 9
	9 = 10

25 = 27

4670011	·6693179
064	229
118	278
171	328
225	378
278	428
332	477
385	527
439	577
493	627
547	676
600	726
654	776
707	826
761	875
814	925
868	975
921	·6694025
975	075
4671028	125
082	174
135	224
189	274
242	324
296	373
349	423
403	473
456	523
510	572
563	622
617	672
670	722
724	771
777	821
831	871
884	921
938	970
991	·6695020
4672045	070
098	120
152	169
205	219
259	269
312	319
366	368
419	418
473	468
526	518
580	567
633	617
687	666

4672740	·6695716
794	766
847	816
901	865
954	915
4673008	965
061	·6696015
115	064
168	114
222	164
275	214
329	263
382	313
436	363
489	413
543	462
596	512
650	561
703	611
757	661
810	711
864	760
917	810
971	860
4674024	910
078	959
131	·6697009
185	058
238	108
292	158
345	208
399	257
452	307
506	357
559	407
613	456
666	506
720	555
773	605
827	655
880	705
934	754
987	804
4675041	853
094	903
148	953
201	·6698003
255	052
308	102
361	151
414	201
468	251
521	301
575	350
628	400
682	449
735	499
789	549
842	599
896	648
949	698
4676003	747
056	797
110	847
163	897

4676217	·6698946
270	996
324	·6699045
377	095
431	145
484	195
538	244
591	294
645	343
698	393
752	443
805	493
858	542
911	592
965	641
4677018	691
072	740
125	790
179	840
232	890
286	939
339	989
393	·6700038
446	088
500	138
553	188
607	237
660	287
714	336
767	386
821	435
874	485
927	534
980	584
4678034	634
087	684
141	733
194	783
248	832
301	882
355	931
408	981
462	·6701031
515	081
569	130
622	180
676	229
729	279
782	328
835	378
889	427
942	477
996	527
4679049	577
103	626
156	676
210	725
263	775
317	824
370	874
424	923
477	973
530	·6702022
583	072
637	122

4679690	·6702172
744	221
797	271
851	320
904	370
958	419

Excesses

10 = 9	1 = 1
20 = 19	2 = 2
30 = 28	3 = 3
40 = 37	4 = 4
50 = 46	5 = 5
	6 = 6
	7 = 7
	8 = 7
	9 = 8

27 = 25

Anti Log Excesses

10 = 11	1 = 1
20 = 22	2 = 2
30 = 32	3 = 3
40 = 43	4 = 4
	5 = 5
	6 = 6
	7 = 8
	8 = 9
	9 = 10

25 = 27

4680011	·6702469
065	518
118	568
171	617
224	667
278	716
331	766
385	816
438	866
492	915
545	965
599	·6703014
652	064
706	113
759	163
812	212
865	262
919	311
972	361
4681026	410
079	460
133	509
186	559
240	608
293	658
346	707
399	757
453	807
506	857
560	906

4681613	·6703956
667	·6704005
720	055
774	104
827	154
880	203
933	253
987	302
4682040	352
094	401
147	451
201	500
254	550
308	599
361	649
414	698
467	748
521	797
574	847
628	896
681	946
735	995
788	·6705045
842	094
895	144
948	193
4683001	243
055	292
108	342
162	391
215	441
269	490
322	540
375	589
428	639
482	688
535	738
589	787
642	837
696	886
749	936
802	985
855	·6706035
909	084
962	134
4684016	183
069	233
123	282
176	332
229	381
282	431
336	480
389	530
443	579
496	629
550	678
603	728
656	777
709	827
763	876
816	926
870	975
923	·6707025
976	074
4685029	124

4685083	·6707173
136	223
190	272
243	321
297	370
350	420
403	469
456	519
510	568
563	618
617	667
670	717
723	766
776	816
830	865
883	915
937	964
990	·6708014
4686044	063
097	113
150	162
203	212
257	261
310	310
364	359
417	409
470	458
523	508
577	557
630	607
684	656
737	706
790	755
843	805
897	854
950	904
4687004	953
057	·6709002
110	051
163	101
217	150
270	200
324	249
377	299
430	348
483	398
537	447
590	497
644	546
697	595
750	644
803	694
857	743
910	793
964	842
4688017	892
070	941
123	991
177	·6710040
230	090
284	139
337	188
390	237
443	287
497	336

4688550	·6710386
604	435
657	485
710	534
763	583
817	632
870	682
924	731
977	781
4689030	830
083	880
137	929
190	978
244	·6711027
297	077
350	126
403	176
457	225
510	275
563	324
616	373
670	422
723	472
777	521
830	571
883	620
936	670
990	719

Excesses

10 = 9	1 = 1
20 = 19	2 = 2
30 = 28	3 = 3
40 = 37	4 = 4
50 = 46	5 = 5
	6 = 6
	7 = 6
	8 = 7
	9 = 8

27 = 25

Anti Log Excesses

10 = 11	1 = 1
20 = 22	2 = 2
30 = 32	3 = 3
40 = 43	4 = 4
	5 = 5
	6 = 6
	7 = 8
	8 = 9
	9 = 10

25 = 27

4690043	·6711768
096	817
149	867
203	916
256	966
310	·6712015
363	064
416	113

4690469	·6712163
523	212
576	262
629	311
682	361
736	410
789	459
843	508
896	558
949	607
4691002	657
056	706
109	755
162	804
215	854
269	903
322	953
376	·6713002
429	051
482	100
535	150
589	199
642	249
695	298
748	347
802	396
855	446
908	495
961	544
4692015	593
068	643
122	692
175	742
228	791
281	840
335	889
388	939
441	988
494	·6714038
548	087
601	136
654	185
707	235
761	284
814	333
867	382
920	432
974	481
4693027	531
081	580
134	629
187	678
240	728
294	777
347	826
400	875
453	925
507	974
560	·6715023
613	072
666	122
720	171
773	221
826	270
879	319

Number	Log
4693933	·6715368
986	418
4694039	467
092	516
146	565
199	615
252	664
305	713
359	762
412	812
465	861
518	910
572	959
625	·6716009
678	058
731	107
785	156
838	206
891	255
944	304
998	353
4695051	403
104	452
157	501
211	550
264	600
317	649
370	698
424	747
477	797
530	846
583	895
637	944
690	994
743	·6717043
796	092
850	141
903	191
956	240
4696009	289
063	338
116	388
169	437
222	486
276	535
329	585
382	634
435	683
489	732
542	782
595	831
648	880
701	929
754	978
808	·6718027
861	077
914	126
967	175
4697021	224
074	274
127	323
180	372
234	421
287	470
340	519

Number	Log
4697393	·6718569
447	618
500	667
553	716
606	766
659	815
712	864
766	913
819	963
872	·6719012
925	061
979	110
4698032	159
085	208
138	258
192	307
245	356
298	405
351	455
404	504
457	553
511	602
564	651
617	700
670	750
724	799
777	848
830	897
883	946
937	995
990	·6720045
4699043	094
096	143
149	192
202	241
256	290
309	340
362	389
415	438
469	487
522	536
575	585
628	635
681	684
734	733
788	782
841	831
894	880
947	929

Excesses

10 = 9	1 = 1
20 = 18	2 = 2
30 = 28	3 = 3
40 = 37	4 = 4
50 = 46	5 = 5
	6 = 6
	7 = 6
	8 = 7
	9 = 8

27 = 25

Anti Log Excesses

10 = 11	1 = 1
20 = 22	2 = 2
30 = 32	3 = 3
40 = 43	4 = 4
	5 = 5
	6 = 6
	7 = 8
	8 = 9
	9 = 10

25 = 27

Number	Log
4700000	·6720978
053	·6721028
107	077
160	126
213	175
266	224
320	273
373	323
426	372
479	421
532	470
585	519
639	568
692	618
745	667
798	716
851	765
904	814
958	863
4701011	912
064	961
117	·6722011
171	060
224	109
277	158
330	207
383	256
436	305
490	354
543	404
596	453
649	502
702	551
755	600
809	649
862	698
915	747
968	796
4702021	845
074	895
128	944
181	993
234	·6723042
287	091
340	140
393	189
447	238
500	287
553	336
606	386
659	435

Number	Log
4702712	·6723484
766	533
819	582
872	631
925	680
978	729
4703031	778
085	827
138	876
191	925
244	975
297	·6724024
350	073
404	122
457	171
510	220
563	269
616	318
669	367
722	416
775	465
829	514
882	564
935	613
988	662
4704041	711
094	760
148	809
201	858
254	907
307	956
360	·6725005
413	054
466	103
519	152
573	201
626	250
679	299
732	348
785	397
838	447
892	496
945	545
998	594
4705051	643
104	692
157	741
210	790
263	839
317	888
370	937
423	986
476	·6726035
529	084
582	133
635	182
688	231
742	280
795	329
848	378
901	427
954	476
4706007	525
060	574
113	623

Number	Log
4706167	·6726672
220	721
273	770
326	819
379	868
432	917
485	966
538	·6727016
592	065
645	114
698	163
751	212
804	261
857	310
910	359
963	408
4707017	457
070	506
123	555
176	604
229	653
282	702
335	751
388	800
441	849
494	898
548	947
601	996
654	·6728045
707	094
760	142
813	191
866	240
919	289
972	338
4708025	387
079	436
132	486
185	535
238	584
291	633
344	682
397	730
450	779
503	828
556	877
610	926
663	975
716	·6729024
769	073
822	122
875	171
928	220
981	269
4709034	318
087	367
141	416
194	465
247	514
300	563
353	612
406	661
459	710
512	759
565	808

4709618	·6729857
671	906
724	955
778	·6730003
831	052
884	101
937	150
990	199

Excesses

10 = 9	1 = 1
20 = 18	2 = 2
30 = 28	3 = 3
40 = 37	4 = 4
50 = 46	5 = 5
	6 = 6
	7 = 6
	8 = 7
	9 = 8

26 = 24

Anti Log Excesses

10 = 11	1 = 1
20 = 22	2 = 2
30 = 33	3 = 3
40 = 43	4 = 4
	5 = 5
	6 = 7
	7 = 8
	8 = 9
	9 = 10

24 = 26

4710043	·6730248
096	297
149	346
202	395
255	444
308	493
361	542
414	591
467	640
521	689
574	738
627	786
680	835
733	884
786	933
839	982
892	·6731031
945	080
998	129
4711051	178
104	227
157	276
210	325
264	374
317	423
370	471
423	520
476	569

4711529	·6731618
582	667
635	716
688	765
741	814
794	863
847	912
900	960
953	·6732009
4712006	058
059	107
113	156
166	205
219	254
272	303
325	352
378	401
431	449
484	498
537	547
590	596
643	645
696	694
749	743
802	792
855	840
908	889
961	938
4713014	987
067	·6733036
120	085
174	134
227	183
280	231
333	280
386	329
439	378
492	427
545	476
598	525
651	574
704	622
757	671
810	720
863	769
916	818
969	867
4714022	916
075	965
128	·6734013
181	062
234	111
287	160
340	209
393	258
446	306
499	355
552	404
605	453
658	502
711	551
765	599
818	648
871	697
924	746

4714977	·6734795
4715030	844
083	893
136	942
189	990
242	·6735039
295	088
348	137
401	186
454	235
507	283
560	332
613	381
666	430
719	478
772	527
825	576
878	625
931	674
984	723
4716037	771
090	820
143	869
196	918
249	967
302	·6736016
355	064
408	113
461	162
514	211
567	259
620	308
673	357
726	406
779	455
832	504
885	552
938	601
991	650
4717044	699
097	747
150	796
203	845
256	894
309	943
362	992
415	·6737040
468	089
521	138
574	187
627	235
680	284
733	333
786	382
839	430
892	479
945	528
998	577
4718051	626
104	675
157	723
210	772
263	821
316	870
369	918

4718422	·6737967
475	·6738016
528	065
581	113
634	162
687	211
740	260
793	309
846	358
899	406
952	455
4719005	504
058	553
111	601
164	650
217	699
269	748
322	796
375	845
428	894
481	943
534	991
587	·6739040
640	089
693	138
746	186
799	235
852	284
905	333
958	381

Excesses

10 = 9	1 = 1
20 = 18	2 = 2
30 = 28	3 = 3
40 = 37	4 = 4
50 = 46	5 = 5
	6 = 6
	7 = 6
	8 = 7
	9 = 8

26 = 24

Anti Log Excesses

10 = 11	1 = 1
20 = 22	2 = 2
30 = 33	3 = 3
40 = 43	4 = 4
	5 = 5
	6 = 7
	7 = 8
	8 = 9
	9 = 10

24 = 26

4720011	·6739430
064	479
117	528
170	576
223	625
276	674

4720329	·6739723
382	771
435	820
488	868
541	917
594	966
647	·6740015
700	063
753	112
806	161
858	210
911	258
964	307
4721017	356
070	405
123	453
176	502
229	550
282	599
335	648
388	697
441	745
494	794
547	843
600	892
653	940
706	989
759	·6741037
812	086
865	135
917	184
970	232
4722023	281
076	330
129	379
182	427
235	476
288	524
341	573
394	622
447	671
500	719
553	768
606	817
659	866
712	914
764	963
817	·6742011
870	060
923	109
976	158
4723029	206
082	255
135	303
188	352
241	401
294	450
347	498
400	547
453	595
505	644
558	693
611	742
664	790
717	839

4723770	·6742887
823	936
876	985
929	·6743034
982	082
4724035	131
088	179
141	228
194	276
246	325
299	374
352	423
405	471
458	520
511	568
564	617
617	666
670	715
723	763
776	812
829	860
881	909
934	957
987	·6744006
4725040	055
093	104
146	152
199	201
252	249
305	298
358	346
410	395
463	444
516	493
569	541
622	590
675	638
728	687
781	735
834	784
887	833
939	882
992	930
4726045	979
098	·6745027
151	076
204	124
257	173
310	222
363	271
416	319
468	368
521	416
574	465
627	513
680	562
733	610
786	659
839	708
892	757
945	805
997	854
4727050	902
103	951
156	999

4727209	·6746048
262	096
315	145
368	193
420	242
473	291
526	340
579	388
632	437
685	485
738	534
791	582
843	631
896	679
949	728
4728002	776
055	825
108	873
161	922
214	971
266	·6747020
319	068
372	117
425	165
478	214
531	262
584	311
637	359
689	408
742	456
795	505
848	553
901	602
954	650
4729007	699
060	747
112	796
165	845
218	894
271	942
324	991
377	·6748039
430	088
483	136
535	185
588	233
641	282
694	330
747	379
800	427
852	476
905	524
958	573

Excesses

10 = 9	1 = 1
20 = 18	2 = 2
30 = 28	3 = 3
40 = 37	4 = 4
50 = 46	5 = 5
	6 = 6
	7 = 6
	8 = 7
	9 = 8

26 = 24

Anti Log Excesses

10 = 11	1 = 1
20 = 22	2 = 2
30 = 33	3 = 3
40 = 44	4 = 4
	5 = 5
	6 = 7
	7 = 8
	8 = 9
	9 = 10

24 = 26

4730011	·6748621
064	670
117	718
169	767
222	815
275	864
328	912
381	961
434	·6749009
487	058
540	106
592	155
645	203
698	252
751	300
804	349
857	398
909	447
962	495
4731015	544
068	592
121	641
174	689
226	738
279	786
332	835
385	883
438	932
491	980
543	·6750029
596	077
649	126
702	174
755	223
808	271
860	320
913	368

4731966	·6750417
4732019	465
072	513
125	561
177	610
230	658
283	707
336	755
389	804
442	852
494	901
547	949
600	998
653	·6751046
706	095
759	143
811	192
864	240
917	289
970	337
4733023	386
076	434
128	483
181	531
234	580
287	628
340	677
393	725
445	774
498	822
551	871
604	919
656	968
709	·6752016
762	064
815	112
868	161
921	209
973	258
4734026	306
079	355
132	403
185	452
238	500
290	549
343	597
396	646
449	694
501	743
554	791
607	840
660	888
713	936
766	984
818	·6753033
871	081
924	130
977	178
4735029	227
082	275
135	324
188	372
241	421
294	469
346	517

4735399	·6753565
452	614
505	662
557	711
610	759
663	808
716	856
768	905
821	953
874	·6754002
927	050
980	098
4736033	146
085	195
138	243
191	292
244	340
296	389
349	437
402	486
455	534
507	582
560	630
613	679
666	727
719	776
772	824
824	873
877	921
930	970
983	·6755018
4737035	066
088	114
141	163
194	211
246	260
299	308
352	357
405	405
457	453
510	501
563	550
616	598
668	647
721	695
774	744
827	792
880	840
933	888
985	937
4738038	985
091	·6756034
144	082
196	130
249	178
302	227
355	275
407	324
460	372
513	421
566	469
618	517
671	565
724	614
777	662

Number	Log
4738829	·6756711
882	759
935	807
988	855
4739040	904
093	952
146	·6757001
199	049
251	097
304	145
357	194
410	242
462	291
515	339
568	387
621	435
673	484
726	532
779	581
832	629
884	677
937	725
990	774

Excesses

10 = 9	1 = 1
20 = 18	2 = 2
30 = 28	3 = 3
40 = 37	4 = 4
50 = 46	5 = 5
	6 = 6
	7 = 6
	8 = 7
	9 = 8

26 = 24

Anti Log Excesses

10 = 11	1 = 1
20 = 22	2 = 2
30 = 33	3 = 3
40 = 44	4 = 4
	5 = 5
	6 = 7
	7 = 8
	8 = 9
	9 = 10

24 = 26

Number	Log
4740043	·6757822
095	871
148	919
201	967
254	·6758015
306	064
359	112
412	161
465	209
517	257
570	305
623	354
676	402
4740728	·6758451
781	499
833	547
886	595
939	644
992	692
4741044	740
097	788
150	837
203	885
255	934
308	982
361	·6759030
414	078
466	127
519	175
572	223
625	271
677	320
730	368
783	417
836	465
888	513
941	561
993	610
4742046	658
099	706
152	754
204	803
257	851
310	899
363	947
415	996
468	·6760044
521	092
574	140
626	189
679	237
731	285
784	333
837	382
890	430
942	479
995	527
4743048	575
101	623
153	672
206	720
259	768
312	816
364	865
417	913
469	961
522	·6761009
575	058
628	106
680	154
733	202
786	251
839	299
891	347
944	395
996	444
4744049	492
102	540
4744155	·6761588
207	637
260	685
313	733
366	781
418	830
471	878
523	926
576	974
629	·6762022
682	070
734	119
787	167
839	215
892	263
945	312
998	360
4745050	408
103	456
156	505
209	553
261	601
314	649
366	697
419	745
472	794
525	842
577	890
630	938
682	987
735	·6763035
788	083
841	131
893	180
946	228
998	276
4746051	324
104	372
157	420
209	469
262	517
315	565
368	613
420	662
473	710
525	758
578	806
631	854
684	902
736	951
789	999
841	·6764047
894	095
947	143
4747000	191
052	240
105	288
157	336
210	384
263	432
316	480
368	529
421	577
473	625
526	673
4747578	·6764722
631	770
684	818
737	866
789	914
842	962
894	·6765011
947	059
4748000	107
053	155
105	203
158	251
210	299
263	347
316	396
369	444
421	492
474	540
526	588
579	636
632	685
685	733
737	781
790	829
842	877
895	925
947	974
4749000	·6766022
053	070
106	118
158	166
211	214
263	262
316	310
369	359
422	407
474	455
527	503
579	551
632	599
684	647
737	695
790	744
843	792
895	840
948	888

Excesses

10 = 9	1 = 1
20 = 18	2 = 2
30 = 27	3 = 3
40 = 37	4 = 4
50 = 46	5 = 5
	6 = 5
	7 = 6
	8 = 7
	9 = 8

26 = 24

Anti Log Excesses

10 = 11	1 = 1
20 = 22	2 = 2
30 = 33	3 = 3
40 = 44	4 = 4
	5 = 5
	6 = 7
	7 = 8
	8 = 9
	9 = 10

24 = 26

Number	Log
4750000	·6766936
053	984
105	·6767032
158	080
211	129
264	177
316	225
369	273
421	321
474	369
526	417
579	465
632	513
685	561
737	610
790	658
842	706
895	754
947	802
4751000	850
053	898
106	946
158	995
211	·6768043
263	091
316	139
368	187
421	235
474	283
527	331
579	379
632	427
684	475
737	523
789	572
842	620
894	668
947	716
4752000	764
053	812
105	860
158	908
210	956
263	·6769004
315	052
368	100
421	149
474	197
526	245
579	293
631	341

N	Log
4752684	·6769389
736	437
789	485
841	533
894	581
947	629
4753000	677
052	725
105	773
157	821
210	869
262	918
315	966
367	·6770014
420	062
473	110
526	158
578	206
631	254
683	302
736	350
788	398
841	446
893	494
946	542
998	590
4754051	638
104	686
157	734
209	782
262	830
314	878
367	926
419	974
472	·6771022
524	071
577	119
629	167
682	215
735	263
788	311
840	359
893	407
945	455
998	503
4755050	551
103	599
155	647
208	695
260	743
313	791
365	839
418	887
471	935
524	983
576	·6772031
629	079
681	127
734	175
786	223
839	271
891	319
944	367
996	415
4756049	463

N	Log
4756101	·6772511
154	559
207	607
260	655
312	703
365	751
417	799
470	847
522	895
575	943
627	991
680	·6773039
732	087
785	135
837	183
890	231
942	279
995	327
4757047	375
100	423
153	471
206	519
258	567
311	615
363	663
416	711
468	758
521	806
573	854
626	902
678	950
731	998
783	·6774046
836	094
888	142
941	190
993	238
4758046	286
098	334
151	382
203	430
256	478
309	526
362	574
414	622
467	670
519	718
572	766
624	814
677	862
729	909
782	957
834	·6775005
887	053
939	101
992	149
4759044	197
097	245
149	293
202	341
254	389
307	437
359	485
412	533
464	581

N	Log
4759517	·6775629
569	676
622	724
674	772
727	820
779	868
832	916
884	964
937	·6776012
990	060

Excesses

10 = 9	1 = 1
20 = 18	2 = 2
30 = 27	3 = 3
40 = 37	4 = 4
50 = 46	5 = 5
	6 = 5
	7 = 6
	8 = 7
	9 = 8

26 = 24

Anti Log Excesses

10 = 11	1 = 1
20 = 22	2 = 2
30 = 33	3 = 3
40 = 44	4 = 4
	5 = 5
	6 = 7
	7 = 8
	8 = 9
	9 = 10

24 = 26

N	Log
4760043	·6776108
095	156
148	204
200	251
253	299
305	347
358	395
410	443
463	491
515	539
568	587
620	635
673	683
725	731
778	779
830	826
883	874
935	922
988	970
4761040	·6777018
093	066
145	114
198	162
250	210
303	258
355	305

N	Log
4761408	·6777353
460	401
513	449
565	497
618	545
670	593
723	641
775	689
828	737
880	785
933	832
985	880
4762038	928
090	976
143	·6778024
195	072
248	120
300	168
353	215
405	263
458	311
510	359
563	407
615	455
668	503
720	551
773	598
825	646
878	694
930	742
983	790
4763035	838
088	885
140	933
193	981
245	·6779029
298	077
350	125
403	172
455	220
508	268
560	316
613	364
665	412
718	460
770	508
822	555
874	603
927	651
979	699
4764032	747
084	795
137	842
189	890
242	938
294	986
347	·6780034
399	082
452	129
504	177
557	225
609	273
662	321
714	369
767	416

N	Log
4764819	·6780464
872	512
924	560
977	608
4765029	656
082	703
134	751
187	799
239	847
292	894
344	942
397	990
449	·6781038
502	086
554	134
606	181
658	229
711	277
763	325
816	373
868	421
921	468
973	516
4766026	564
078	612
131	659
183	707
236	755
288	803
341	850
393	898
446	946
498	994
551	·6782042
603	090
656	137
708	185
760	233
812	281
865	328
917	376
970	424
4767022	472
075	519
127	567
180	615
232	663
285	711
337	759
390	806
442	854
495	902
547	950
599	997
651	·6783045
704	093
756	141
809	188
861	236
914	284
966	332
4768019	379
071	427
124	475
176	523

4768229	·6783570
281	618
334	666
386	714
438	761
490	809
543	857
595	905
648	952
700	·6784000
753	048
805	096
858	143
910	191
963	239
4769015	287
067	334
119	382
172	430
224	478
277	525
329	573
382	621
434	669
487	716
539	764
592	812
644	860
697	907
749	955
801	·6785002
853	050
906	098
958	146

Excesses

10 = 9	1 = 1
20 = 18	2 = 2
30 = 27	3 = 3
40 = 36	4 = 4
50 = 46	5 = 5
	6 = 5
	7 = 6
	8 = 7
	9 = 8

26 = 24

Anti Log Excesses

10 = 11	1 = 1
20 = 22	2 = 2
30 = 33	3 = 3
40 = 44	4 = 4
	5 = 5
	6 = 7
	7 = 8
	8 = 9
	9 = 10

24 = 26

4770011	·6785193
063	241

4770116	·6785289
168	337
221	384
273	432
325	480
377	528
430	575
482	623
535	670
587	718
640	766
692	814
745	861
797	909
849	957
901	·6786005
954	052
4771006	100
059	147
111	195
164	243
216	291
269	338
321	386
373	434
425	482
478	529
530	577
583	624
635	672
688	720
740	768
793	815
845	863
897	911
949	959
4772002	·6787006
054	054
107	101
159	149
212	197
264	245
316	292
368	340
421	387
473	435
526	483
578	531
631	578
683	626
736	673
788	721
840	769
892	817
945	864
997	912
4773050	959
102	·6788007
155	054
207	102
259	150
311	198
364	245
416	293
469	340

4773521	·6788388
574	436
626	484
678	531
730	579
783	626
835	674
888	722
940	770
993	817
4774045	865
097	912
149	960
202	·6789007
254	055
307	103
359	151
411	198
463	246
516	293
568	341
621	388
673	436
726	484
778	532
830	579
882	627
935	674
987	722
4775040	769
092	817
144	865
196	913
249	960
301	·6790008
354	055
406	103
459	150
511	198
563	246
615	294
668	341
720	389
773	436
825	484
877	531
929	579
982	626
4776034	674
087	722
139	770
191	817
243	865
296	912
348	960
401	·6791007
453	055
505	102
557	150
610	197
662	245
715	293
767	341
820	388
872	436

4776924	·6791483
976	531
4777029	578
081	626
134	673
186	721
238	768
290	816
343	864
395	912
448	959
500	·6792007
552	054
604	102
657	149
709	197
761	244
813	292
866	339
918	387
971	434
4778023	482
075	529
127	577
180	625
232	673
285	720
337	768
389	815
441	863
494	910
546	958
599	·6793005
651	053
703	100
755	148
808	195
860	243
913	290
965	338
4779017	385
069	433
122	480
174	528
226	576
278	624
331	671
383	719
436	766
488	814
540	861
592	909
645	956
697	·6794004
749	051
801	099
854	146
906	194
959	241

Excesses

10 = 9	1 = 1
20 = 18	2 = 2
30 = 27	3 = 3
40 = 36	4 = 4
50 = 45	5 = 5
	6 = 5
	7 = 6
	8 = 7
	9 = 8

26 = 24

Anti Log Excesses

10 = 11	1 = 1
20 = 22	2 = 2
30 = 33	3 = 3
40 = 44	4 = 4
	5 = 5
	6 = 7
	7 = 8
	8 = 9
	9 = 10

24 = 26

4780011	·6794289
063	336
115	384
168	431
220	479
272	526
324	574
377	621
429	669
482	716
534	764
586	811
638	859
691	906
743	954
795	·6795001
847	049
900	096
952	144
4781005	191
057	239
109	286
161	334
214	381
266	429
318	476
370	524
423	571
475	619
527	666
579	714
632	761
684	809
737	856
789	904
841	951
893	999

4781946	·6796046
998	094
4782050	141
102	189
155	236
207	284
259	331
311	379
364	426
416	474
468	521
520	569
573	616
625	664
678	711
730	759
782	806
834	853
887	900
939	948
991	995
4783043	·6797043
096	090
148	138
200	185
252	233
305	280
357	328
409	375
461	423
514	470
566	518
618	565
670	613
723	660
775	708
827	755
879	803
932	850
984	897
4784036	944
088	992
141	·6798039
193	087
245	134
297	182
350	229
402	277
454	324
506	372
559	419
611	467
663	514
715	561
768	608
820	656
872	703
924	751
977	798
4785029	846
081	893
133	941
186	988
238	·6799036
290	083

4785342	·6799131
395	178
447	225
499	272
551	320
604	367
656	415
708	462
760	510
813	557
865	604
917	651
969	699
4786022	746
074	794
126	841
178	889
231	936
283	984
335	·6800031
387	078
440	125
492	173
544	220
596	268
649	315
701	363
753	410
805	458
858	505
910	552
962	599
4787014	647
066	694
118	742
171	789
223	837
275	884
327	931
380	978
432	·6801026
484	073
536	121
589	168
641	216
693	263
745	310
797	357
849	405
902	452
954	500
4788006	547
058	594
111	641
163	689
215	736
267	784
320	831
372	879
424	926
476	973
528	·6802020
580	068
633	115
685	163

4788737	·6802210
789	257
842	304
894	352
946	399
998	447
4789051	494
103	541
155	588
207	636
259	683
311	731
364	778
416	825
468	872
520	920
573	967
625	·6803015
677	062
729	109
781	156
833	204
886	251
938	299
990	346

Excesses

10 = 9	1 = 1
20 = 18	2 = 2
30 = 27	3 = 3
40 = 36	4 = 4
50 = 45	5 = 5
	6 = 5
	7 = 6
	8 = 7
	9 = 8

26 = 24

Anti Log Excesses

10 = 11	1 = 1
20 = 22	2 = 2
30 = 33	3 = 3
40 = 44	4 = 4
	5 = 6
	6 = 7
	7 = 8
	8 = 9
	9 = 10

24 = 26

4790042	·6803393
094	440
146	488
199	535
251	583
303	630
355	677
408	724
460	772
512	819
564	866

4790616	·6803913
668	961
721	·6804008
773	056
825	103
877	150
929	197
981	245
4791034	292
086	339
138	386
190	434
243	481
295	529
347	576
399	623
451	670
503	718
556	765
608	812
660	859
712	907
764	954
816	·6805002
869	049
921	096
973	143
4792025	191
077	238
129	285
182	332
234	380
286	427
338	474
390	521
442	569
495	616
547	663
599	710
651	758
703	805
755	852
808	899
860	947
912	994
964	·6806042
4793016	089
068	136
121	183
173	231
225	278
277	325
329	372
381	420
434	467
486	514
538	561
590	609
642	656
694	703
746	750
798	798
851	845
903	892
955	939

4794007	·6806987
059	·6807034
111	081
164	128
216	175
268	222
320	270
372	317
424	364
476	411
528	459
581	506
633	553
685	600
737	648
789	695
841	742
894	789
946	837
998	884
4795050	931
102	978
154	·6808026
206	073
258	120
311	167
363	214
415	261
467	309
519	356
571	403
624	450
676	498
728	545
780	592
832	639
884	686
936	733
988	781
4796041	828
093	875
145	922
197	970
249	·6809017
301	064
353	111
405	159
458	206
510	253
562	300
614	347
666	394
718	442
770	489
822	536
874	583
926	630
979	677
4797031	725
083	772
135	819
187	866
239	913
291	960
343	·6810008

4797396	·6810055
448	102
500	149
552	196
604	243
656	291
708	338
760	385
813	432
865	479
917	526
969	574
4798021	621
073	668
125	715
177	762
229	809
281	857
334	904
386	951
438	998
490	·6811045
542	092
594	140
646	187
698	234
750	281
802	328
855	375
907	423
959	470
4799011	517
063	564
115	611
167	658
219	705
271	752
323	800
375	847
427	894
480	941
532	988
584	·6812035
636	082
688	129
740	177
792	224
844	271
896	318
948	365

Excesses

10 = 9	1 = 1
20 = 18	2 = 2
30 = 27	3 = 3
40 = 36	4 = 4
50 = 45	5 = 5
	6 = 5
	7 = 6
	8 = 7
	9 = 8

26 = 24

Anti Log Excesses

10 = 11	1 = 1
20 = 22	2 = 2
30 = 33	3 = 3
40 = 44	4 = 4
	5 = 6
	6 = 7
	7 = 8
	8 = 9
	9 = 10

24 = 26

4800000	·6812412
052	459
105	506
157	554
209	601
261	648
313	695
365	742
417	789
469	836
521	883
573	931
625	978
677	·6813025
730	072
782	119
834	166
886	213
938	260
990	307
4801042	354
094	402
146	449
198	496
250	543
302	590
354	637
406	684
459	731
511	778
563	825
615	873
667	920
719	967
771	·6814014
823	061
875	108
927	155
979	202
4802031	249
083	296
135	343
188	390
240	438
292	485
344	532
396	579
448	626
500	673
552	720
604	767

4802656	·6814814
708	861
760	908
812	955
864	·6815003
916	050
968	097
4803020	144
072	191
124	238
176	285
229	332
281	379
333	426
385	473
437	520
489	567
541	614
593	661
645	708
697	755
749	802
801	850
853	897
905	944
957	991
4804009	·6816038
061	085
113	132
165	179
217	226
269	273
321	320
374	367
426	414
478	461
530	508
582	555
634	603
686	650
738	697
790	744
842	791
894	838
946	885
998	932
4805050	979
102	·6817026
154	073
206	120
258	167
310	214
362	261
414	308
466	355
518	402
570	449
622	496
674	543
726	590
778	637
830	684
882	731
934	778
986	825

4806038	·6817872
090	919
142	966
194	·6818013
246	060
298	107
350	154
402	201
454	248
506	295
559	342
611	389
663	436
715	483
767	530
819	577
871	624
923	671
975	718
4807027	765
079	812
131	859
183	906
235	953
287	·6819000
339	047
391	094
443	141
495	188
547	235
599	282
651	329
703	376
755	423
807	470
859	517
911	564
963	611
4808015	657
067	704
119	751
171	798
223	845
275	892
327	939
379	986
431	·6820033
483	080
535	127
587	174
639	221
691	268
743	315
795	362
847	409
898	456
950	503
4809002	550
054	597
106	644
158	690
210	737
262	784
314	831
366	878

4809418	·6820925
470	972
522	·6821019
574	066
626	113
678	160
730	207
782	254
834	301
886	348
938	395
990	441

Excesses

10 = 9	1 = 1
20 = 18	2 = 2
30 = 27	3 = 3
40 = 36	4 = 4
50 = 45	5 = 5
	6 = 5
	7 = 6
	8 = 7
	9 = 8

26 = 23

Anti Log Excesses

10 = 11	1 = 1
20 = 22	2 = 2
30 = 33	3 = 3
40 = 44	4 = 4
	5 = 6
	6 = 7
	7 = 8
	8 = 9
	9 = 10

23 = 26

4810042	·6821488
094	535
146	582
198	629
250	676
302	723
354	770
406	817
458	864
510	911
562	958
614	·6822004
666	051
718	098
770	145
822	192
874	239
926	286
977	333
4811029	380
081	427
133	474
185	521
237	567

4811289	·6822614
341	661
393	708
445	755
497	802
549	849
601	896
653	943
705	990
757	·6823036
809	083
861	130
913	177
965	224
4812017	271
069	318
121	365
173	412
224	459
276	505
328	552
380	599
432	646
484	693
536	740
588	787
640	834
692	880
744	927
796	974
848	·6824021
900	068
952	115
4813004	162
056	209
108	255
159	302
211	349
263	396
315	443
367	490
419	536
471	583
523	630
575	677
627	724
679	771
731	818
783	865
835	911
887	958
939	·6825005
990	052
4814042	099
094	146
146	192
198	239
250	286
302	333
354	380
406	427
458	474
510	521
562	567
614	614

4814666	·6825661
717	708
769	755
821	802
873	848
925	895
977	942
4815029	989
081	·6826036
133	083
185	129
237	176
289	223
340	270
392	316
444	363
496	410
548	457
600	504
652	551
704	597
756	644
808	691
860	738
912	785
963	832
4816015	878
067	925
119	972
171	·6827019
223	066
275	113
327	159
379	206
431	253
483	300
535	346
586	393
638	440
690	487
742	534
794	581
846	627
898	674
950	721
4817002	768
054	814
105	861
157	908
209	955
261	·6828001
313	048
365	095
417	142
469	189
521	236
573	282
624	329
676	376
728	423
780	469
832	516
884	563
936	610
988	656

4818039	·6828703
091	750
143	797
195	844
247	891
299	937
351	984
403	·6829031
454	078
506	124
558	171
610	218
662	265
714	311
766	358
818	405
870	452
922	498
973	545
4819025	592
077	639
129	685
181	732
233	779
285	826
337	872
388	919
440	966
492	·6830013
544	059
596	106
648	153
700	200
752	246
803	293
855	340
907	387
959	433

Excesses

10 = 9	1 = 1
20 = 18	2 = 2
30 = 27	3 = 3
40 = 36	4 = 4
50 = 45	5 = 5
	6 = 5
	7 = 6
	8 = 7
	9 = 8

26 = 23

Anti Log Excesses

10 = 11	1 = 1
20 = 22	2 = 2
30 = 33	3 = 3
40 = 44	4 = 4
	5 = 6
	6 = 7
	7 = 8
	8 = 9
	9 = 10

23 = 26

4820011	·6830480
063	527
114	574
166	620
218	667
270	713
322	760
374	807
426	854
478	900
529	947
581	994
633	·6831041
685	087
737	134
789	181
841	228
893	274
944	321
996	367
4821048	414
100	461
152	508
204	554
255	601
307	648
359	695
411	741
463	788
515	835
566	882
618	928
670	975
722	·6832021
774	068
826	115
878	162
930	208
981	255
4822033	302
085	349
137	395
189	442
241	488
292	535
344	582
396	629
448	675
500	722
552	768
603	815

4822655	·6832862
707	909
759	955
811	·6833002
863	049
914	096
966	142
4823018	189
070	235
122	282
174	329
225	376
277	422
329	469
381	515
433	562
485	609
536	656
588	702
640	749
692	795
744	842
796	889
847	936
899	982
951	·6834029
4824003	075
055	122
107	169
158	216
210	262
262	309
314	355
366	402
418	448
469	495
521	542
573	589
625	635
676	682
728	728
780	775
832	822
884	869
936	915
987	962
4825039	·6835008
091	055
143	101
195	148
247	195
298	242
350	288
402	335
454	381
505	428
557	474
609	521
661	568
713	615
765	661
816	708
868	754
920	801
972	847

4826024	·6835894
076	941
127	988
179	·6836034
231	081
283	127
334	174
386	220
438	267
490	314
542	361
594	407
645	454
697	500
749	547
801	593
852	640
904	687
956	734
4827008	780
059	827
111	873
163	920
215	966
267	·6837013
319	059
370	106
422	152
474	199
526	246
577	293
629	339
681	386
733	432
784	479
836	525
888	572
940	618
992	665
4828044	711
095	758
147	805
199	852
251	898
302	945
354	991
406	·6838038
458	084
509	131
561	177
613	224
665	270
716	317
768	363
820	410
872	456
924	503
976	550
4829027	597
079	643
131	690
183	736
234	783
286	829
338	876

4829390	·6838922
441	969
493	·6839015
545	062
597	108
648	155
700	201
752	248
804	294
855	341
907	388
959	435

Excesses

10 = 9	1 = 1
20 = 18	2 = 2
30 = 27	3 = 3
40 = 36	4 = 4
50 = 45	5 = 5
	6 = 5
	7 = 6
	8 = 7
	9 = 8

26 = 23

Anti Log Excesses

10 = 11	1 = 1
20 = 22	2 = 2
30 = 33	3 = 3
40 = 44	4 = 4
	5 = 6
	6 = 7
	7 = 8
	8 = 9
	9 = 10

23 = 26

4830011	·6839481
062	528
114	574
166	621
218	667
269	714
321	760
373	807
425	853
476	900
528	946
580	993
632	·6840039
683	086
735	132
787	179
839	225
890	272
942	318
994	365
4831046	411
097	458
149	504
201	551

4831253	·6840597
304	644
356	690
408	737
460	783
511	830
563	876
615	923
667	969
718	·6841016
770	062
822	109
874	155
925	202
977	248
4832029	295
081	341
132	388
184	434
236	481
288	527
339	574
391	620
443	667
495	713
546	760
598	806
650	853
702	899
753	946
805	992
857	·6842039
909	085
960	132
4833012	178
063	225
115	271
167	318
219	364
270	411
322	457
374	504
426	550
477	597
529	643
581	690
633	736
684	783
736	829
788	876
840	922
891	969
943	·6843015
994	062
4834046	108
098	155
150	201
201	248
253	294
305	341
357	387
408	433
460	479
512	526
564	572

4834615	·6843619
667	665
718	712
770	758
822	805
874	851
925	898
977	944
4835029	991
081	·6844037
132	084
184	130
235	177
287	223
339	269
391	315
442	362
494	408
546	455
598	501
649	548
701	594
752	641
804	687
856	734
908	780
959	827
4836011	873
063	919
115	965
166	·6845012
218	058
269	105
321	151
373	198
425	244
476	291
528	337
580	384
632	430
683	476
735	522
786	569
838	615
890	662
942	708
993	755
4837045	801
096	848
148	894
200	940
252	986
303	·6846033
355	079
407	126
459	172
510	219
562	265
613	312
665	358
717	404
769	450
820	497
872	543
923	590

4837975	·6846636
4838027	683
079	729
130	776
182	822
233	868
285	914
337	961
389	·6847007
440	054
492	100
543	147
595	193
647	239
699	285
750	332
802	378
853	425
905	471
957	518
4839009	564
060	610
112	656
163	703
215	749
267	796
319	842
370	888
422	934
473	981
525	·6848027
577	074
629	120
680	167
732	213
783	259
835	305
887	352
939	398
990	445

Excesses

10 = 9	1 = 1
20 = 18	2 = 2
30 = 27	3 = 3
40 = 36	4 = 4
50 = 45	5 = 4
	6 = 5
	7 = 6
	8 = 7
	9 = 8

26 = 23

Anti Log Excesses

10 = 11   1 = 1
20 = 22   2 = 2
30 = 33   3 = 3
40 = 45   4 = 4
5 = 6
6 = 7
7 = 8
8 = 9
9 = 10

23 = 26

4840042	·6848491
093	537
145	583
196	630
248	676
300	723
352	769
403	815
455	861
506	908
558	954
610	·6849001
662	047
713	093
765	139
816	186
868	232
919	279
971	325
4841023	371
075	417
126	464
178	510
229	557
281	603
333	649
385	695
436	742
488	788
539	835
591	881
642	927
694	973
746	·6850020
798	066
849	113
901	159
952	205
4842004	251
055	298
107	344
159	390
211	436
262	483
314	529
365	576
417	622
469	668
521	714
572	761
624	807

4842675	·6850853
727	899
778	946
830	992
881	·6851038
933	084
985	131
4843037	177
088	224
140	270
191	316
243	362
294	409
346	455
398	501
450	547
501	594
553	640
604	687
656	733
707	779
759	825
811	872
863	918
914	964
966	·6852010
4844017	057
069	103
120	149
172	195
223	242
275	288
327	334
379	380
430	427
482	473
533	519
585	565
636	612
688	658
740	704
792	750
843	797
895	843
946	889
998	935
4845049	982
101	·6853028
152	074
204	120
256	167
308	213
359	259
411	305
462	352
514	398
565	444
617	490
668	537
720	583
771	629
823	675
875	722
927	768
978	814

4846030	·6853860
081	907
133	953
184	999
236	·6854045
287	092
339	138
390	184
442	230
494	277
546	323
597	369
649	415
700	461
752	507
803	554
855	600
906	646
958	692
4847009	739
061	785
113	831
165	877
216	924
268	970
319	·6855016
371	062
422	108
474	154
525	201
577	247
628	293
680	339
731	386
783	432
835	478
887	524
938	571
990	617
4848041	663
093	709
144	755
196	801
247	848
299	894
350	940
402	986
453	·6856033
505	079
557	125
609	171
660	217
712	263
763	310
815	356
866	402
918	448
969	494
4849021	540
072	587
124	633
175	679
227	725
278	771
330	817

4849381	·6856864
433	910
485	956
537	·6857002
588	048
640	094
691	141
743	187
794	233
846	279
897	325
949	371

Excesses

10 = 9   1 = 1
20 = 18   2 = 2
30 = 27   3 = 3
40 = 36   4 = 4
50 = 45   5 = 4
6 = 5
7 = 6
8 = 7
9 = 8

26 = 23

Anti Log Excesses

10 = 11   1 = 1
20 = 22   2 = 2
30 = 33   3 = 3
40 = 45   4 = 4
5 = 6
6 = 7
7 = 8
8 = 9
9 = 10

23 = 26

4850000	·6857418
052	464
103	510
155	556
206	602
258	648
309	695
361	741
412	787
464	833
515	879
567	925
619	972
671	·6858018
722	064
774	110
825	156
877	202
928	248
980	294
4851031	341
083	387
134	433
186	479

4851237	·6858525
289	571
340	617
392	663
443	710
495	756
546	802
598	848
649	894
701	940
752	986
804	·6859032
855	079
907	125
958	171
4852010	217
061	263
113	309
164	355
216	401
268	448
320	494
371	540
423	586
474	632
526	678
577	724
629	770
680	816
732	862
783	909
835	955
886	·6860001
938	047
989	093
4853041	139
092	185
144	231
195	277
247	323
298	370
350	416
401	462
453	508
504	554
556	600
607	646
659	692
710	738
762	784
813	831
865	877
916	923
968	969
4854019	·6861015
071	061
122	107
174	153
225	199
277	245
328	291
380	337
431	384
483	430
534	476

4854586	·6861522
637	568
689	614
740	660
792	706
843	752
895	798
946	844
998	890
4855049	936
101	982
152	·6862028
204	074
255	121
307	167
358	213
410	259
461	305
513	351
564	397
616	443
667	489
719	535
770	581
822	627
873	673
925	719
976	765
4856028	811
079	857
131	903
182	949
234	995
285	·6863041
337	087
388	134
440	180
491	226
542	272
593	318
645	364
696	410
748	456
799	502
851	548
902	594
954	640
4857005	686
057	732
108	778
160	824
211	870
263	916
314	962
366	·6864008
417	054
469	100
520	146
572	192
623	238
675	284
726	330
778	376
829	422
881	468

4857932	·6864514
984	560
4858035	606
087	652
138	698
189	744
240	790
292	836
343	882
395	928
446	974
498	·6865020
549	066
601	112
652	158
704	204
755	250
807	296
858	342
910	388
961	434
4859013	480
064	526
116	572
167	618
219	664
270	710
321	756
372	802
424	848
475	894
527	940
578	986
630	·6866032
681	078
733	124
784	170
836	216
887	262
939	308
990	354

Excesses

10 = 9	1 = 1
20 = 18	2 = 2
30 = 27	3 = 3
40 = 36	4 = 4
50 = 45	5 = 4
	6 = 5
	7 = 6
	8 = 7
	9 = 8

26 = 23

Anti Log Excesses

10 = 11	1 = 1
20 = 22	2 = 2
30 = 34	3 = 3
40 = 45	4 = 4
	5 = 6
	6 = 7
	7 = 8
	8 = 9
	9 = 10

23 = 26

4860042	·6866400
093	446
145	492
196	538
247	583
298	629
350	675
401	721
453	767
504	813
556	859
607	905
659	951
710	997
762	·6867043
813	089
865	135
916	181
967	227
4861018	273
070	319
121	365
173	411
224	457
276	502
327	548
379	594
430	640
482	686
533	732
585	778
636	824
687	870
738	916
790	962
841	·6868008
893	054
944	100
996	146
4862047	192
099	237
150	283
202	329
253	375
304	421
355	467
407	513
458	559
510	605
561	651
613	697

4862664	·6868743
716	789
767	835
819	880
870	926
921	972
972	·6869018
4863024	064
075	110
127	156
178	202
230	248
281	294
333	339
384	385
436	431
487	477
538	523
589	569
641	615
692	661
744	707
795	753
847	798
898	844
950	890
4864001	936
052	982
103	·6870028
155	074
206	120
258	166
309	212
361	257
412	303
463	349
514	395
566	441
617	487
669	533
720	579
772	624
823	670
875	716
926	762
977	808
4865028	854
080	900
131	946
183	991
234	·6871037
286	083
337	129
388	175
439	221
491	267
542	313
594	358
645	404
697	450
748	496
800	542
851	588
902	633
953	679

4866005	·6871725
056	771
108	817
159	863
211	909
262	955
313	·6872000
364	046
416	092
467	138
519	184
570	230
622	275
673	321
724	367
775	413
827	459
878	505
930	550
981	596
4867032	642
083	688
135	734
186	780
238	825
289	871
341	917
392	963
443	·6873009
494	055
546	100
597	146
649	192
700	238
752	284
803	330
854	375
905	421
957	467
4868008	513
060	559
111	605
162	650
213	696
265	742
316	788
368	833
419	879
470	925
521	971
573	·6874017
624	063
676	108
727	154
779	200
830	246
881	292
932	338
984	383
4869035	429
087	475
138	521
189	566
240	612
292	658

4869343	·6874704
395	750
446	796
497	841
548	887
600	933
651	979
703	·6875024
754	070
805	116
856	162
908	207
959	253

Excesses

10 = 9	1 = 1
20 = 18	2 = 2
30 = 27	3 = 3
40 = 36	4 = 4
50 = 45	5 = 4
	6 = 5
	7 = 6
	8 = 7
	9 = 8

26 = 23

Anti Log Excesses

10 = 11	1 = 1
20 = 22	2 = 2
30 = 34	3 = 3
40 = 45	4 = 4
	5 = 6
	6 = 7
	7 = 8
	8 = 9
	9 = 10

23 = 26

4870011	·6875299
062	345
113	390
164	436
216	482
267	528
319	574
370	620
421	665
472	711
524	757
575	803
627	848
678	894
729	940
780	986
832	·6876031
883	077
935	123
986	169
4871037	214
088	260
140	306

4871191	·6876352
243	397
294	443
345	489
396	535
448	580
499	626
551	672
602	718
653	763
704	809
756	855
807	901
858	946
909	992
961	·6877038
4872012	084
064	129
115	175
166	221
217	267
269	312
320	358
372	404
423	450
474	495
525	541
577	587
628	633
679	678
730	724
782	769
833	815
885	861
936	907
987	952
4873038	998
090	·6878044
141	090
192	135
243	181
295	227
346	273
398	318
449	364
500	410
551	456
603	501
654	547
705	592
756	638
808	684
859	730
911	775
962	821
4874013	867
064	913
116	958
167	·6879004
218	049
269	095
321	141
372	187
424	232
475	278

4874526	·6879324
577	370
629	415
680	461
731	506
782	552
834	598
885	644
936	689
987	735
4875039	781
090	827
142	872
193	918
244	963
295	·6880009
347	055
398	101
449	146
500	192
552	237
603	283
654	329
705	375
757	420
808	466
859	511
910	557
962	603
4876013	649
064	694
115	740
167	785
218	831
270	877
321	923
372	968
423	·6881014
475	059
526	105
577	151
628	197
680	242
731	288
782	333
833	379
885	424
936	470
987	516
4877038	562
090	607
141	653
192	698
243	744
295	790
346	836
397	881
448	927
500	972
551	·6882018
602	063
653	109
705	155
756	201
807	246

4877858	·6882292
910	337
961	383
4878012	429
063	475
115	520
166	566
217	611
268	657
320	702
371	748
422	794
473	840
525	885
576	931
627	976
678	·6883022
730	067
781	113
832	159
883	205
935	250
986	296
4879037	341
088	387
140	432
191	478
242	523
293	569
345	615
396	661
447	706
498	752
550	797
601	843
652	888
703	934
755	979
806	·6884025
857	071
908	117
960	162

Excesses

10 = 9	1 = 1
20 = 18	2 = 2
30 = 27	3 = 3
40 = 36	4 = 4
50 = 45	5 = 4
	6 = 5
	7 = 6
	8 = 7
	9 = 8

26 = 23

Anti Log Excesses

10 = 11	1 = 1
20 = 22	2 = 2
30 = 34	3 = 3
40 = 45	4 = 4
	5 = 6
	6 = 7
	7 = 8
	8 = 9
	9 = 10

23 = 26

4880011	·6884208
062	253
113	299
164	344
215	390
267	435
318	481
369	526
420	572
472	618
523	664
574	709
625	755
677	800
728	846
779	891
830	937
882	982
933	·6885028
984	073
4881035	119
086	165
137	211
189	256
240	302
291	347
342	393
394	438
445	484
496	529
547	575
599	620
650	666
701	711
752	757
803	803
854	849
906	894
957	940
4882008	985
059	·6886031
111	076
162	122
213	167
264	213
316	258
367	304
418	349
469	395
520	440
571	486

4882623	·6886531
674	577
725	622
776	668
828	714
879	760
930	805
981	851
4883032	896
083	942
135	987
186	·6887033
237	078
288	124
340	169
391	215
442	260
493	306
544	351
595	397
647	442
698	488
749	533
800	579
851	624
902	670
954	715
4884005	761
056	806
107	852
159	897
210	943
261	988
312	·6888034
363	079
414	125
466	170
517	216
568	261
619	307
670	352
721	398
773	443
824	489
875	534
926	580
977	625
4885028	671
080	716
131	762
182	807
233	853
285	898
336	944
387	989
438	·6889035
489	080
540	126
592	171
643	217
694	262
745	308
796	353
847	399
899	444

4885950	·6889490
4886001	535
052	581
103	626
154	672
206	717
257	763
308	808
359	854
410	899
461	945
513	990
564	·6890036
615	081
666	127
717	172
768	218
820	263
871	308
922	353
973	399
4887024	444
075	490
126	535
177	581
229	626
280	672
331	717
382	763
433	808
484	854
536	899
587	945
638	990
689	·6891036
740	081
791	127
843	172
894	218
945	263
996	308
4888047	353
098	399
149	444
200	490
252	535
303	581
354	626
405	672
456	717
507	763
559	808
610	854
661	899
712	944
763	989
814	·6892035
865	080
916	126
968	171
4889019	217
070	262
121	308
172	353
223	399

4889274	·6892444
325	490
377	535
428	580
479	625
530	671
581	716
632	762
684	807
735	853
786	898
837	944
888	989
939	·6893035
990	080

Excesses

10 = 9	1 = 1
20 = 18	2 = 2
30 = 27	3 = 3
40 = 36	4 = 4
50 = 44	5 = 4
	6 = 5
	7 = 6
	8 = 7
	9 = 8

26 = 23

Anti Log Excesses

10 = 11	1 = 1
20 = 22	2 = 2
30 = 34	3 = 3
40 = 45	4 = 4
	5 = 6
	6 = 7
	7 = 8
	8 = 9
	9 = 10

23 = 26

4890041	·6893125
093	170
144	216
195	261
246	307
297	352
348	398
399	443
450	488
501	533
552	579
604	624
655	670
706	715
757	761
808	806
859	852
910	897
961	942
4891013	987
064	·6894033

4891115	·6894078
166	124
217	169
268	215
319	260
370	305
422	350
473	396
524	441
575	487
626	532
677	578
728	623
779	669
830	714
881	759
933	804
984	850
4892035	895
086	941
137	986
188	·6895032
239	077
290	122
341	167
392	213
444	258
495	304
546	349
597	394
648	439
699	485
750	530
801	576
852	621
903	667
955	712
4893006	757
057	802
108	848
159	893
210	939
261	984
312	·6896029
363	074
414	120
466	165
517	211
568	256
619	301
670	346
721	392
772	437
823	483
874	528
925	573
976	618
4894027	664
079	709
130	755
181	800
232	845
283	890
334	936
385	981

4894436	·6897027
487	072
538	117
589	162
640	208
692	253
743	299
794	344
845	389
896	434
947	480
998	525
4895049	570
100	615
151	661
202	706
253	752
304	797
355	842
407	887
458	933
509	978
560	·6898024
611	069
662	114
713	159
764	205
815	250
866	295
917	340
968	386
4896019	431
070	476
121	521
172	566
224	612
275	658
326	703
377	748
428	793
479	839
530	884
581	929
632	974
683	·6899020
734	065
785	110
836	155
887	201
938	246
989	291
4897040	336
091	382
143	427
194	473
245	518
296	563
347	608
398	654
449	699
500	744
551	789
602	835
653	880
704	925

4897755	·6899970
806	·6900016
857	061
908	106
959	151
4898010	197
061	242
112	287
163	332
214	378
265	423
316	468
368	513
419	559
470	604
521	649
572	694
623	740
674	785
725	830
776	875
827	921
878	966
929	·6901011
980	056
4899031	102
082	147
133	192
184	237
235	283
286	328
337	373
388	418
439	464
490	509
541	554
592	599
643	645
694	690
745	735
796	780
847	825
898	870
949	916

Excesses

10 = 9	1 = 1
20 = 18	2 = 2
30 = 27	3 = 3
40 = 35	4 = 4
50 = 44	5 = 4
	6 = 5
	7 = 6
	8 = 7
	9 = 8

26 = 23

Anti Log Excesses

10 = 11	1 = 1
20 = 23	2 = 2
30 = 34	3 = 3
40 = 45	4 = 5
	5 = 6
	6 = 7
	7 = 8
	8 = 9
	9 = 10

23 = 26

4900000	·6901961
051	·6902006
103	051
154	097
205	142
256	187
307	232
358	278
409	323
460	368
511	413
562	458
613	503
664	549
715	594
766	639
817	684
868	730
919	775
970	820
4901021	865
072	911
123	956
174	·6903001
225	046
276	091
327	136
378	182
429	227
480	272
531	317
582	362
633	407
684	453
735	498
786	543
837	588
888	634
939	679
990	724
4902041	769
092	814
143	859
194	905
245	950
296	995
347	·6904040
398	085
449	130
500	176
551	221

4902602	·6904266
653	311
704	356
755	401
806	447
857	492
908	537
959	582
4903010	627
061	672
112	718
163	763
214	808
265	853
316	898
367	943
418	989
469	·6905034
520	079
571	124
622	169
673	214
724	260
775	305
826	350
877	395
927	440
978	485
4904029	530
080	575
131	621
182	666
233	711
284	756
335	801
386	846
437	892
488	937
539	982
590	·6906027
641	072
692	117
743	162
794	207
845	253
896	298
947	343
998	388
4905049	433
100	478
151	523
202	568
253	614
304	659
355	704
406	749
457	794
508	839
559	884
610	929
660	975
711	·6907020
762	065
813	110
864	155

4905915	·6907200
966	245
4906017	290
068	336
119	381
170	426
221	471
272	516
323	561
374	606
425	651
476	696
527	741
578	787
629	832
680	877
731	922
781	967
832	·6908012
883	057
934	102
985	147
4907036	192
087	238
138	283
189	328
240	373
291	418
342	463
393	508
444	553
495	598
546	643
597	688
648	733
698	779
749	824
800	869
851	914
902	959
953	·6909004
4908004	049
055	094
106	139
157	184
208	229
259	274
310	319
361	364
412	409
463	454
513	500
564	545
615	590
666	635
717	680
768	725
819	770
870	815
921	860
972	905
4909023	950
074	995
125	·6910040
176	085

4909226	·6910130
277	175
328	220
379	265
430	311
481	356
532	401
583	446
634	491
685	536
736	581
787	626
837	671
888	716
939	761
990	806

Excesses

10 = 9	1 = 1
20 = 18	2 = 2
30 = 27	3 = 3
40 = 35	4 = 4
50 = 44	5 = 4
	6 = 5
	7 = 6
	8 = 7
	9 = 8

26 = 23

Anti Log Excesses

10 = 11	1 = 1
20 = 23	2 = 2
30 = 34	3 = 3
40 = 45	4 = 5
	5 = 6
	6 = 7
	7 = 8
	8 = 9
	9 = 10

23 = 26

4910041	·6910851
092	896
143	941
194	986
245	·6911031
296	076
347	121
398	166
448	211
499	256
550	301
601	346
652	391
703	436
754	481
805	526
856	571
907	616
958	661
4911009	706

4911059	·6911751
110	796
161	842
212	887
263	932
314	977
365	·6912022
416	067
467	112
518	157
568	202
619	247
670	292
721	337
772	382
823	427
874	472
925	517
976	562
4912027	607
077	652
128	697
179	742
230	787
281	832
332	877
383	922
434	967
484	·6913012
535	057
586	102
637	147
688	192
739	237
790	282
841	326
892	371
943	416
993	461
4913044	506
095	551
146	596
197	641
248	686
299	731
350	776
400	821
451	866
502	911
553	956
604	·6914001
655	046
706	091
757	136
807	181
858	226
909	271
960	316
4914011	361
062	406
113	451
164	496
214	541
265	586
316	631

4914367	·6914676
418	721
469	766
520	811
571	856
621	901
672	945
723	990
774	·6915035
825	080
876	125
927	170
978	215
4915028	260
079	305
130	350
181	395
232	440
283	485
334	530
385	575
435	620
486	665
537	710
588	754
639	799
690	844
740	889
791	934
842	979
893	·6916024
944	069
995	114
4916046	159
097	204
147	249
198	294
249	339
300	383
351	428
402	473
452	518
503	563
554	608
605	653
656	698
707	743
757	788
808	833
859	878
910	922
961	967
4917012	·6917012
062	057
113	102
164	147
215	192
266	237
317	282
368	327
419	371
469	416
520	461
571	506
622	551

4917673	·6917596
724	641
774	686
825	731
876	776
927	820
978	865
4918029	910
079	955
130	·6918000
181	045
232	090
283	135
334	179
384	224
435	269
486	314
537	359
588	404
639	449
689	494
740	539
791	584
842	628
892	673
943	718
994	763
4919045	808
096	853
147	897
197	942
248	987
299	·6919032
350	077
401	122
452	167
502	212
553	256
604	301
655	346
706	391
757	436
807	481
858	525
909	570
960	615

Excesses

10 = 9	1 = 1
20 = 18	2 = 2
30 = 27	3 = 3
40 = 35	4 = 4
50 = 44	5 = 4
	6 = 5
	7 = 6
	8 = 7
	9 = 8

26 = 23

Anti Log Excesses

10 = 11	1 = 1
20 = 23	2 = 2
30 = 34	3 = 3
40 = 45	4 = 5
	5 = 6
	6 = 7
	7 = 8
	8 = 9
	9 = 10

23 = 26

4920010	·6919660
061	705
112	750
163	795
214	840
265	884
315	929
366	974
417	·6920019
468	064
519	109
570	153
620	198
671	243
722	288
773	333
823	378
874	422
925	467
976	512
4921027	557
078	602
128	647
179	691
230	736
281	781
331	826
382	871
433	916
484	960
535	·6921005
586	050
636	095
687	140
738	185
789	229
839	274
890	319
941	364
992	409
4922043	454
094	498
144	543
195	588
246	633
297	677
347	722
398	767
449	812
500	857
550	902

4922601	·6921946
652	991
703	·6922036
754	081
805	126
855	171
906	215
957	260
4923008	305
058	350
109	394
160	439
211	484
261	529
312	574
363	619
414	663
465	708
516	753
566	798
617	842
668	887
719	932
769	977
820	·6923021
871	066
922	111
972	156
4924023	201
074	246
125	290
175	335
226	380
277	425
328	469
378	514
429	559
480	604
531	648
581	693
632	738
683	783
734	827
785	872
836	917
886	962
937	·6924006
988	051
4925039	096
089	141
140	186
191	231
242	275
292	320
343	365
394	410
445	454
495	499
546	544
597	589
648	633
698	678
749	723
800	768
851	812

4925901	·6924857
952	902
4926003	947
054	991
104	·6925036
155	081
206	126
257	170
307	215
358	260
409	305
460	349
510	394
561	439
612	484
663	528
713	573
764	617
815	662
866	707
916	752
967	796
4927018	841
069	886
119	931
170	975
221	·6926020
272	065
322	110
373	154
423	199
474	244
525	289
576	333
626	378
677	422
728	467
779	512
829	557
880	601
931	646
982	691
4928032	736
083	780
134	825
185	869
235	914
286	959
337	·6927004
388	048
438	093
489	138
540	183
591	227
641	272
692	316
742	361
793	406
844	451
895	495
945	540
996	585
4929047	630
098	674
148	719

4929199	·6927763
250	808
301	853
351	898
402	942
453	987
504	·6928031
554	076
605	121
655	166
706	210
757	255
808	300
858	345
909	389
960	434

Excesses

10 = 9	1 = 1
20 = 18	2 = 2
30 = 26	3 = 3
40 = 35	4 = 4
50 = 44	5 = 4
	6 = 5
	7 = 6
	8 = 7
	9 = 8

25 = 22

Anti Log Excesses

10 = 11	1 = 1
20 = 23	2 = 2
30 = 34	3 = 3
40 = 45	4 = 5
	5 = 6
	6 = 7
	7 = 8
	8 = 9
	9 = 10

22 = 25

4930011	·6928478
061	523
112	567
162	612
213	657
264	702
315	746
365	791
416	835
467	880
518	925
568	970
619	·6929014
670	059
721	103
771	148
822	193
872	238
923	282
974	327

4931025	·6929371
075	416
126	461
177	506
228	550
278	595
329	639
379	684
430	729
481	774
532	818
582	863
633	907
683	952
734	996
785	·6930041
836	086
886	131
937	175
988	220
4932039	264
089	309
140	353
190	398
241	443
292	488
343	532
393	577
444	621
494	666
545	711
596	756
647	800
697	845
748	889
799	934
850	978
900	·6931023
951	067
4933001	112
052	157
103	202
154	246
204	291
255	335
305	380
356	424
407	469
458	514
508	559
559	603
609	648
660	692
711	737
762	781
812	826
863	871
913	916
964	960
4934015	·6932005
066	049
116	094
167	138
217	183
268	227

4934319	·6932272
370	316
420	361
471	406
521	451
572	495
623	540
674	584
724	629
775	673
825	718
876	762
927	807
978	851
4935028	896
079	941
129	986
180	·6933030
231	075
282	119
332	164
383	208
433	253
484	297
535	342
586	386
636	431
687	475
737	520
788	565
838	610
889	654
940	699
991	743
4936041	788
092	832
142	877
193	921
244	966
295	·6934010
345	055
396	100
446	145
497	189
548	234
599	278
649	323
700	367
750	412
801	456
851	501
902	545
953	590
4937004	634
054	679
105	723
155	768
206	812
256	857
307	901
358	946
409	991
459	·6935036
510	080
560	125

4937611	·6935169
662	214
713	258
763	303
814	347
864	392
915	436
965	481
4938016	525
067	570
118	614
168	659
219	703
269	748
320	792
370	837
421	881
472	926
523	970
573	·6936015
624	059
674	104
725	148
775	193
826	237
877	282
928	326
978	371
4939029	415
079	460
130	504
180	549
231	593
281	638
332	682
383	727
434	771
484	816
535	860
585	905
636	949
686	994
737	·6937038
788	083
839	127
889	172
940	216
990	261

Excesses

10 = 9	1 = 1
20 = 18	2 = 2
30 = 26	3 = 3
40 = 35	4 = 4
50 = 44	5 = 4
	6 = 5
	7 = 6
	8 = 7
	9 = 8

25 = 22

Anti Log Excesses

10 = 11	1 = 1
20 = 23	2 = 2
30 = 34	3 = 3
40 = 45	4 = 5
	5 = 6
	6 = 7
	7 = 8
	8 = 9
	9 = 10

22 = 25

4940041	·6937305
091	350
142	394
192	439
243	483
294	528
345	572
395	617
446	661
496	706
547	750
597	795
648	839
698	884
749	928
800	973
851	·6938017
901	062
952	106
4941002	151
053	195
103	240
154	284
204	329
255	373
306	418
357	462
407	507
458	551
508	595
559	639
609	684
660	728
710	773
761	817
811	862
862	906
913	951
964	995
4942014	·6939040
065	084
115	129
166	173
216	218
267	262
317	307
368	351
419	396
470	440
520	485
571	529

4942621	·6939573
672	617
722	662
773	706
823	751
874	795
924	840
975	884
4943025	929
076	973
127	·6940018
178	062
228	107
279	151
329	196
380	240
430	284
481	328
531	373
582	417
632	462
683	506
733	551
784	595
835	640
886	684
936	729
987	773
4944037	817
088	861
138	906
189	950
239	995
290	·6941039
340	084
391	128
441	173
492	217
543	262
594	306
644	350
695	394
745	439
796	483
846	528
897	572
947	617
998	661
4945048	706
099	750
149	794
200	838
250	883
301	927
351	972
402	·6942016
453	061
504	105
554	150
605	194
655	238
706	282
756	327
807	371
857	416

4945908	·6942460
958	505
4946009	549
059	593
110	637
160	682
211	726
261	771
312	815
362	860
413	904
463	948
514	992
565	·6943037
616	081
666	126
717	170
767	215
818	259
868	303
919	347
969	392
4947020	436
070	481
121	525
171	570
222	614
272	658
323	702
373	747
424	791
474	836
525	880
575	924
626	968
676	·6944013
727	057
777	102
828	146
878	191
929	235
979	279
4948030	323
080	368
131	412
182	457
233	501
283	545
334	589
384	634
435	678
485	723
536	767
586	811
637	855
687	900
738	944
788	989
839	·6945033
889	077
940	121
990	166
4949041	210
091	255
142	299

4949192	·6945343
243	387
293	432
344	476
394	521
445	565
495	609
546	653
596	698
647	742
697	786
748	830
798	875
849	919
899	964
950	·6946008

Excesses

10 = 9	1 = 1
20 = 18	2 = 2
30 = 26	3 = 3
40 = 35	4 = 4
50 = 44	5 = 4
	6 = 5
	7 = 6
	8 = 7
	9 = 8

25 = 22

Anti Log Excesses

10 = 11	1 = 1
20 = 23	2 = 2
30 = 34	3 = 3
40 = 46	4 = 5
	5 = 6
	6 = 7
	7 = 8
	8 = 9
	9 = 10

22 = 25

4950000	·6946052
051	096
101	141
152	185
202	230
253	274
303	318
354	362
404	407
455	451
505	495
556	539
606	584
657	628
707	673
758	717
808	761
859	805
909	850
960	894

4951010	·6946938
061	982
111	·6947027
162	071
212	116
263	160
313	204
364	248
414	293
465	337
515	381
566	425
616	470
667	514
717	558
768	602
818	647
869	691
919	735
970	779
4952020	824
071	868
121	913
172	957
222	·6948001
273	045
323	090
374	134
424	178
475	222
525	267
576	311
626	355
677	399
727	444
777	488
827	532
878	576
928	621
979	665
4953029	709
080	753
130	798
181	842
231	886
282	930
332	975
383	·6949019
433	063
484	107
534	152
585	196
635	240
686	284
736	329
787	373
837	417
888	461
938	506
989	550
4954039	594
090	638
140	683
191	727
241	771

4954292	·6949815
342	860
392	904
442	948
493	992
543	·6950037
594	081
644	125
695	169
745	213
796	257
846	302
897	346
947	390
998	434
4955048	479
099	523
149	567
200	611
250	656
301	700
351	744
402	788
452	833
502	877
552	921
603	965
653	·6951009
704	053
754	098
805	142
855	186
906	230
956	275
4956007	319
057	363
108	407
158	451
209	495
259	540
310	584
360	628
410	672
460	717
511	761
561	805
612	849
662	893
713	937
763	982
814	·6952026
864	070
915	114
965	158
4957016	202
066	247
117	291
167	335
217	379
267	424
318	468
368	512
419	556
469	600
520	644

4957570	·6952689
621	733
671	777
722	821
772	865
823	909
873	954
923	998
973	·6953042
4958024	086
074	131
125	175
175	219
226	263
276	307
327	351
377	396
428	440
478	484
529	528
579	572
629	616
679	660
730	704
780	749
831	793
881	837
932	881
982	925
4959033	969
083	·6954014
133	058
183	102
234	146
284	190
335	234
385	279
436	323
486	367
537	411
587	455
638	499
688	543
738	587
788	632
839	676
889	720
940	764
990	808

Excesses

10 = 9	1 = 1
20 = 18	2 = 2
30 = 26	3 = 3
40 = 35	4 = 4
50 = 44	5 = 4
	6 = 5
	7 = 6
	8 = 7
	9 = 8

25 = 22

Anti Log Excesses

10 = 11	1 = 1
20 = 23	2 = 2
30 = 34	3 = 3
40 = 46	4 = 5
	5 = 6
	6 = 7
	7 = 8
	8 = 9
	9 = 10

22 = 25

4960041	·6954852
091	896
142	940
192	985
242	·6955029
292	073
343	117
393	161
444	205
494	249
545	293
595	338
646	382
696	426
746	470
796	514
847	558
897	602
948	646
998	691
4961049	735
099	779
150	823
200	867
250	911
300	955
351	999
401	·6956044
452	088
502	132
553	176
603	220
653	264
703	308
754	352
804	396
855	440
905	485
956	529
4962006	573
057	617
107	661
157	705
207	749
258	793
308	837
359	881
409	925
460	969
510	·6957014
560	058

4962610	·6957102
661	146
711	190
762	234
812	278
863	322
913	366
963	410
4963013	454
064	498
114	543
165	587
215	631
266	675
316	719
366	763
416	807
467	851
517	895
568	939
618	983
668	·6958027
718	071
769	115
819	159
870	203
920	248
971	292
4964021	336
071	380
121	424
172	468
222	512
273	556
323	600
374	644
424	688
474	732
524	776
575	820
625	864
676	908
726	953
776	997
826	·6959041
877	085
927	129
978	173
4965028	217
079	261
129	305
179	349
229	393
280	437
330	481
381	525
431	569
481	613
531	657
582	701
632	745
683	789
733	833
783	877
833	921

4965884	·6959965
934	·6960009
985	053
4966035	097
085	141
135	185
186	229
236	273
287	317
337	361
388	405
438	450
488	494
538	538
589	582
639	626
690	670
740	714
790	758
840	802
891	846
941	890
992	934
4967042	978
092	·6961022
142	066
193	110
243	154
294	198
344	242
394	286
444	330
495	374
545	418
595	462
645	506
696	550
746	594
797	638
847	682
897	726
947	770
998	814
4968048	858
099	902
149	946
199	990
249	·6962034
300	078
350	122
401	166
451	210
501	254
551	298
602	342
652	386
703	429
753	473
803	517
853	561
904	605
954	649
4969004	693
054	737
105	781

4969155	·6962825
206	869
256	913
306	957
356	·6963001
407	045
457	089
508	133
558	177
608	221
658	265
709	309
759	353
809	397
859	441
910	485
960	529

Excesses

10 = 9	1 = 1
20 = 17	2 = 2
30 = 26	3 = 3
40 = 35	4 = 3
50 = 44	5 = 4
	6 = 5
	7 = 6
	8 = 7
	9 = 8

25 = 22

Anti Log Excesses

10 = 11	1 = 1
20 = 23	2 = 2
30 = 34	3 = 3
40 = 46	4 = 5
	5 = 6
	6 = 7
	7 = 8
	8 = 9
	9 = 10

22 = 25

4970011	·6963573
061	617
111	661
161	705
212	748
262	792
312	836
362	880
413	924
463	968
514	·6964012
564	056
614	100
664	144
715	188
765	232
815	276
865	320
916	364

4970966	·6964408
4971016	452
066	496
117	539
167	583
218	627
268	671
318	715
368	759
419	803
469	847
519	891
569	935
620	979
670	·6965023
721	067
771	111
821	154
871	198
922	242
972	286
4972022	330
072	374
123	418
173	462
223	506
273	550
324	594
374	638
424	682
474	726
525	769
575	813
626	857
676	901
726	945
776	989
827	·6966033
877	077
927	121
977	165
4973028	208
078	252
128	296
178	340
229	384
279	428
329	472
379	516
430	560
480	604
530	647
580	691
631	735
681	779
731	823
781	867
832	911
882	955
933	998
983	·6967042
4974033	086
083	130
134	174
184	218

4974234	·6967262
284	306
335	349
385	393
435	437
485	481
536	525
586	569
636	613
686	657
737	700
787	744
837	788
887	832
938	876
988	920
4975038	964
088	·6968008
139	051
189	095
239	139
289	183
340	227
390	271
440	315
490	359
541	402
591	446
641	490
691	534
742	578
792	622
842	665
892	709
943	753
993	797
4976043	841
093	885
144	929
194	973
244	·6969016
294	060
345	104
395	148
445	192
495	236
545	279
595	323
646	367
696	411
746	455
796	499
847	542
897	586
947	630
997	674
4977048	718
098	762
148	805
198	849
249	893
299	937
349	981
399	·6970025
450	068

4977500	·6970112
550	156
600	200
651	244
701	288
751	331
801	375
851	419
901	463
952	506
4978002	550
052	594
102	638
153	682
203	726
253	769
303	813
354	857
404	901
454	945
504	989
555	·6971032
605	076
655	120
705	164
755	207
805	251
856	295
906	339
956	383
4979006	427
057	470
107	514
157	558
207	602
257	645
307	689
358	733
408	777
458	821
508	865
559	908
609	952
659	996
709	·6972040
760	083
810	127
860	171
910	215
960	258

Excesses

10 = 9	1 = 1
20 = 17	2 = 2
30 = 26	3 = 3
40 = 35	4 = 3
50 = 44	5 = 4
	6 = 5
	7 = 6
	8 = 7
	9 = 8

25 = 22

Anti Log Excesses

10 = 11	1 = 1
20 = 23	2 = 2
30 = 34	3 = 3
40 = 46	4 = 5
	5 = 6
	6 = 7
	7 = 8
	8 = 9
	9 = 10

22 = 25

4980010	·6972302
061	346
111	390
161	433
211	477
262	521
312	565
362	609
412	653
462	696
512	740
563	784
613	828
663	872
713	916
763	959
813	·6973003
864	047
914	091
964	134
4981014	178
065	222
115	266
165	309
215	353
265	397
315	441
366	484
416	528
466	572
516	616
567	659
617	703
667	747
717	791
767	834
817	878
868	922
918	966
968	·6974009
4982018	053
068	097
118	141
169	184
219	228
269	272
319	316
369	359
419	403
470	447
520	491

4982570	·6974534
620	578
670	622
720	666
771	709
821	753
871	796
921	840
972	884
4983022	928
072	971
122	·6975015
172	059
222	103
273	146
323	190
373	234
423	278
473	321
523	365
574	409
624	453
674	496
724	540
774	583
824	627
874	671
924	715
975	758
4984025	802
075	846
125	890
175	933
225	977
276	·6976020
326	064
376	108
426	152
476	195
526	239
577	283
627	327
677	370
727	414
777	457
827	501
878	545
928	589
978	632
4985028	676
078	720
128	764
179	807
229	851
279	894
329	938
379	982
429	·6977026
479	069
529	113
580	156
630	200
680	244
730	288
780	331

4985830	·6977375
881	419
931	463
981	506
4986031	550
081	593
131	637
181	680
231	724
282	768
332	812
382	855
432	899
482	942
532	986
582	·6978030
632	074
683	117
733	161
783	204
833	248
883	292
933	336
984	379
4987034	423
084	466
134	510
184	554
234	598
284	641
334	685
385	728
435	772
485	816
535	860
585	903
635	947
685	990
735	·6979034
786	077
836	121
886	165
936	209
986	252
4988036	296
086	339
136	383
187	427
237	471
287	514
337	558
387	601
437	645
487	688
537	732
587	776
637	820
688	863
738	907
788	950
838	994
888	·6980037
938	081
988	125
4989038	169

4989089	·6980212
139	256
189	299
239	343
289	386
339	430
389	474
439	518
489	561
539	605
590	648
640	692
690	735
740	779
790	822
840	866
890	910
940	954
990	997

Excesses

10 = 9	1 = 1
20 = 17	2 = 2
30 = 26	3 = 3
40 = 35	4 = 3
50 = 44	5 = 4
	6 = 5
	7 = 6
	8 = 7
	9 = 8

25 = 22

Anti Log Excesses

10 = 11	1 = 1
20 = 23	2 = 2
30 = 34	3 = 3
40 = 46	4 = 5
	5 = 6
	6 = 7
	7 = 8
	8 = 9
	9 = 10

22 = 25

4990040	·6981041
091	084
141	128
191	171
241	215
291	258
341	302
391	346
441	390
491	433
541	477
592	520
642	564
692	607
742	651
792	694
842	738

4990892	·6981782
942	826
992	869
4991042	913
093	956
143	·6982000
193	043
243	087
293	130
343	174
393	217
443	261
493	305
543	349
593	392
643	436
694	479
744	523
794	566
844	610
894	653
944	697
994	740
4992044	784
094	828
144	872
194	915
244	959
295	·6983002
345	046
395	089
445	133
495	176
545	220
595	263
645	307
695	350
745	394
795	437
845	481
895	524
945	568
996	612
4993046	656
096	699
146	743
196	786
246	830
296	873
346	917
396	960
446	·6984004
496	047
546	091
596	134
646	178
697	221
747	265
797	308
847	352
897	395
947	439
997	482
4994047	526
097	570

4994147	·6984614
197	657
247	701
297	744
347	788
397	831
447	875
497	918
547	962
598	·6985005
648	049
698	092
748	136
798	179
848	223
898	266
948	310
998	353
4995048	397
098	440
148	484
198	527
248	571
298	614
348	658
398	701
448	745
498	788
548	832
599	875
649	919
699	962
749	·6986006
799	049
849	093
899	136
949	180
999	223
4996049	267
099	310
149	354
199	397
249	441
299	484
349	528
399	571
449	615
499	658
549	702
599	745
649	789
699	832
749	876
799	919
849	963
900	·6987006
950	050
4997000	093
050	137
100	180
150	224
200	267
250	311
300	354
350	398

Number	Log
4997400	·6987441
450	485
500	528
550	572
600	615
650	659
700	702
750	745
800	788
850	832
900	875
950	919
4998000	962
050	·6988006
100	049
150	093
200	136
250	180
300	223
350	267
400	310
450	354
500	397
550	441
600	484
650	528
700	571
750	615
800	658
850	702
900	745
950	788
4999000	831
050	875
100	918
150	962
200	·6989005
250	049
300	092
350	136
400	179
450	223
500	266
550	310
600	353
650	397
700	440
750	483
800	526
850	570
900	613
950	657
5000000	700

Excesses

10 = 9	1 = 1
20 = 17	2 = 2
30 = 26	3 = 3
40 = 35	4 = 3
50 = 43	5 = 4
	6 = 5
	7 = 6
	8 = 7
	9 = 8

25 = 22

Anti Log Excesses

10 = 12	1 = 1
20 = 23	2 = 2
30 = 35	3 = 3
40 = 46	4 = 5
	5 = 6
	6 = 7
	7 = 8
	8 = 9
	9 = 10

22 = 25

Number	Log
5000100	·6989787
200	874
300	961
400	·6990048
500	134
600	221
700	308
800	395
900	482
5001000	569
100	655
200	742
300	829
400	916
500	·6991003
600	090
700	176
800	263
900	350
5002000	437
100	523
200	610
300	697
400	784
500	871
600	957
700	·6992044
800	131
900	218
5003000	304
100	391
199	478
299	565
399	651
499	738
599	825
699	912

Number	Log
5003799	·6992998
899	·6993085
999	172
5004099	259
199	345
299	432
399	519
498	605
598	692
698	779
798	865
898	952
998	·6994039
5005098	126
198	212
298	299
398	386
497	472
597	559
697	646
797	732
897	819
997	906
5006097	992
197	·6995079
297	165
396	252
496	339
596	425
696	512
796	599
896	685
996	772
5007095	858
195	945
295	·6996032
395	118
495	205
595	291
695	378
794	465
894	551
994	638
5008094	724
194	811
294	897
393	984
493	·6997071
593	157
693	244
793	330
893	417
992	503
5009092	590
192	677
292	763
392	850
491	936
591	·6998023
691	109
791	196
891	282
991	369

Excesses

10 = 9	1 = 1
20 = 17	2 = 2
30 = 26	3 = 3
40 = 35	4 = 3
50 = 43	5 = 4
60 = 52	6 = 5
70 = 61	7 = 6
80 = 69	8 = 7
90 = 78	9 = 8

50 = 43

Anti Log Excesses

10 = 12	1 = 1
20 = 23	2 = 2
30 = 35	3 = 3
40 = 46	4 = 5
50 = 58	5 = 6
60 = 69	6 = 7
70 = 81	7 = 8
80 = 92	8 = 9
	9 = 10

43 = 50

Number	Log
5010090	·6998455
190	542
290	628
390	715
489	801
589	888
689	974
789	·6999061
889	147
988	234
5011088	320
188	407
288	493
388	580
487	666
587	753
687	839
787	925
886	·7000012
986	098
5012086	185
186	271
285	358
385	444
485	531
585	617
684	703
784	790
884	876
984	963
5013083	·7001049
183	135
283	222
383	308
482	395
582	481
682	567

Number	Log
5013782	·7001654
881	740
981	827
5014081	913
180	999
280	·7002086
380	172
480	258
579	345
679	431
779	517
878	604
978	690
5015078	776
177	863
277	949
377	·7003035
477	122
576	208
676	294
776	381
875	467
975	553
5016075	640
174	726
274	812
374	899
473	985
573	·7004071
673	157
772	244
872	330
972	416
5017071	502
171	589
271	675
370	761
470	848
570	934
669	·7005020
769	106
869	193
968	279
5018068	365
167	451
267	537
367	624
466	710
566	796
666	882
765	969
865	·7006055
965	141
5019064	227
164	313
263	400
363	486
463	572
562	658
662	744
761	830
861	917
961	·7007003

Excesses

10 = 9	1 = 1
20 = 17	2 = 2
30 = 26	3 = 3
40 = 35	4 = 3
50 = 43	5 = 4
60 = 52	6 = 5
70 = 61	7 = 6
80 = 69	8 = 7
90 = 78	9 = 8

50 = 43

Anti Log Excesses

10 = 12	1 = 1
20 = 23	2 = 2
30 = 35	3 = 3
40 = 46	4 = 5
50 = 58	5 = 6
60 = 69	6 = 7
70 = 81	7 = 8
80 = 92	8 = 9
	9 = 10

43 = 50

5020060	·7007089
160	175
259	261
359	347
459	434
558	520
658	606
757	692
857	778
957	864
5021056	950
156	·7008037
255	123
355	209
454	295
554	381
654	467
753	553
853	639
952	726
5022052	812
151	898
251	984
351	·7009070
450	156
550	242
649	328
749	414
848	500
948	586
5023047	672
147	758
246	845
346	931
446	·7010017
545	103
645	189

5023744	·7010275
844	361
943	447
5024043	533
142	619
242	705
341	791
441	877
540	963
640	·7011049
739	135
839	221
938	307
5025038	393
137	479
237	565
336	651
436	737
535	823
635	909
734	995
834	·7012081
933	167
5026033	253
132	339
232	425
331	511
431	597
530	683
630	769
729	855
829	941
928	·7013027
5027027	112
127	198
226	284
326	370
425	456
525	542
624	628
724	714
823	800
923	886
5028022	972
121	·7014057
221	143
320	229
420	315
519	401
619	487
718	573
817	659
917	744
5029016	830
116	916
215	·7015002
315	088
414	174
513	260
613	345
712	431
812	517
911	603

Excesses

10 = 9	1 = 1
20 = 17	2 = 2
30 = 26	3 = 3
40 = 35	4 = 3
50 = 43	5 = 4
60 = 52	6 = 5
70 = 61	7 = 6
80 = 69	8 = 7
90 = 78	9 = 8

50 = 43

Anti Log Excesses

10 = 12	1 = 1
20 = 23	2 = 2
30 = 35	3 = 3
40 = 46	4 = 5
50 = 58	5 = 6
60 = 69	6 = 7
70 = 81	7 = 8
80 = 93	8 = 9
	9 = 10

43 = 50

5030010	·7015689
110	775
209	860
309	946
408	·7016032
507	118
607	204
706	289
806	375
905	461
5031004	547
104	633
203	718
303	804
402	890
501	976
601	·7017061
700	147
799	233
899	319
998	404
5032097	490
197	576
296	662
396	747
495	833
594	919
694	·7018005
793	090
892	176
992	262
5033091	348
190	433
290	519
389	605
488	690
588	776

5033687	·7018862
786	948
886	·7019033
985	119
5034084	205
184	290
283	376
382	462
482	547
581	633
680	719
780	804
879	890
978	976
5035077	·7020061
177	147
276	233
375	318
475	404
574	490
673	575
773	661
872	747
971	832
5036070	918
170	·7021003
269	089
368	175
468	260
567	346
666	431
765	517
865	603
964	688
5037063	774
162	859
262	945
361	·7022031
460	116
559	202
659	287
758	373
857	458
956	544
5038056	629
155	715
254	801
353	886
453	972
552	·7023057
651	143
750	228
850	314
949	399
5039048	485
147	570
246	656
346	741
445	827
544	912
643	998
743	·7024083
842	169
941	254

Excesses

10 = 9	1 = 1
20 = 17	2 = 2
30 = 26	3 = 3
40 = 35	4 = 3
50 = 43	5 = 4
60 = 52	6 = 5
70 = 60	7 = 6
80 = 69	8 = 7
90 = 78	9 = 8

50 = 43

Anti Log Excesses

10 = 12	1 = 1
20 = 23	2 = 2
30 = 35	3 = 3
40 = 46	4 = 5
50 = 58	5 = 6
60 = 70	6 = 7
70 = 81	7 = 8
80 = 93	8 = 9
	9 = 10

43 = 50

5040040	·7024340
139	425
239	511
338	596
437	682
536	767
635	853
735	938
834	·7025024
933	109
5041032	194
131	280
231	365
330	451
429	536
528	622
627	707
726	793
826	878
925	963
5042024	·7026049
123	134
222	220
321	305
421	390
520	476
619	561
718	647
817	732
916	817
5043015	903
115	988
214	·7027074
313	159
412	244
511	330
610	415

Number	Log
5043709	·7027500
809	586
908	671
5044007	756
106	842
205	927
304	·7028012
403	098
502	183
602	268
701	354
800	439
899	524
998	610
5045097	695
196	780
295	866
394	951
494	·7029036
593	122
692	207
791	292
890	377
989	463
5046088	548
187	633
286	719
385	804
484	889
584	974
683	·7030060
782	145
881	230
980	315
5047079	401
178	486
277	571
376	656
475	742
574	827
673	912
772	997
871	·7031082
970	168
5048069	253
168	338
268	423
367	509
466	594
565	679
664	764
763	849
862	935
961	·7032020
5049060	105
159	190
258	275
357	360
456	446
555	531
654	616
753	701
852	786
951	871

Excesses

10 = 9	1 = 1
20 = 17	2 = 2
30 = 26	3 = 3
40 = 34	4 = 3
50 = 43	5 = 4
60 = 52	6 = 5
70 = 60	7 = 6
80 = 69	8 = 7
90 = 77	9 = 8

50 = 43

Anti Log Excesses

10 = 12	1 = 1
20 = 23	2 = 2
30 = 35	3 = 3
40 = 46	4 = 5
50 = 58	5 = 6
60 = 70	6 = 7
70 = 81	7 = 8
80 = 93	8 = 9
	9 = 10

43 = 50

Number	Log
5050050	·7032957
149	·7033042
248	127
347	212
446	297
545	382
644	467
743	552
842	638
941	723
5051040	808
139	893
238	978
337	·7034063
436	148
535	233
634	318
733	403
832	489
931	574
5052030	659
129	744
228	829
327	914
426	999
525	·7035084
624	169
723	254
821	339
920	424
5053019	509
118	594
217	679
316	764
415	849
514	934
613	·7036019

Number	Log
5053712	·7036104
811	190
910	275
5054009	360
108	445
207	530
306	615
405	700
503	785
602	870
701	955
800	·7037040
899	125
998	209
5055097	294
196	379
295	464
394	549
493	634
591	719
690	804
789	889
888	974
987	·7038059
5056086	144
185	229
284	314
383	399
481	484
580	569
679	654
778	739
877	823
976	908
5057075	993
174	·7039078
272	163
371	248
470	333
569	418
668	503
767	588
866	673
965	757
5058063	842
162	927
261	·7040012
360	097
459	182
558	267
656	351
755	436
854	521
953	606
5059052	691
151	776
249	861
348	945
447	·7041030
546	115
645	200
744	285
842	370
941	454

Excesses

10 = 9	1 = 1
20 = 17	2 = 2
30 = 26	3 = 3
40 = 34	4 = 3
50 = 43	5 = 4
60 = 52	6 = 5
70 = 60	7 = 6
80 = 69	8 = 7
90 = 77	9 = 8

50 = 43

Anti Log Excesses

10 = 12	1 = 1
20 = 23	2 = 2
30 = 35	3 = 3
40 = 46	4 = 5
50 = 58	5 = 6
60 = 70	6 = 7
70 = 81	7 = 8
80 = 93	8 = 9
	9 = 10

43 = 50

Number	Log
5060040	·7041539
139	624
238	709
336	794
435	878
534	963
633	·7042048
732	133
831	218
929	302
5061028	387
127	472
226	557
325	641
423	726
522	811
621	896
720	980
818	·7043065
917	150
5062016	235
115	319
213	404
312	489
411	574
510	658
609	743
707	828
806	912
905	997
5063004	·7044082
102	167
201	251
300	336
399	421
497	505
596	590

Number	Log
5063695	·7044675
794	759
892	844
991	929
5064090	·7045014
188	098
287	183
386	268
485	352
583	437
682	522
781	606
880	691
978	776
5065077	860
176	945
274	·7046029
373	114
472	199
571	283
669	368
768	452
867	537
965	622
5066064	706
163	791
261	875
360	960
459	·7047045
558	129
656	214
755	298
854	383
952	468
5067051	552
150	637
248	721
347	806
446	890
544	975
643	·7048060
742	144
840	229
939	313
5068038	398
136	482
235	567
334	651
432	736
531	820
630	905
728	989
827	·7049074
925	159
5069024	243
123	328
221	412
320	497
419	581
517	666
616	750
714	835
813	919
912	·7050004

Excesses

10 = 9	1 = 1
20 = 17	2 = 2
30 = 26	3 = 3
40 = 34	4 = 3
50 = 43	5 = 4
60 = 51	6 = 5
70 = 60	7 = 6
80 = 69	8 = 7
90 = 77	9 = 8

49 = 42

Anti Log Excesses

10 = 12	1 = 1
20 = 23	2 = 2
30 = 35	3 = 3
40 = 47	4 = 5
50 = 58	5 = 6
60 = 70	6 = 7
70 = 82	7 = 8
80 = 93	8 = 9
	9 = 10

42 = 49

5070010	·7050088
109	173
208	257
306	341
405	426
503	510
602	595
701	679
799	764
898	848
996	933
5071095	·7051017
194	102
292	186
391	271
489	355
588	439
687	524
785	608
884	693
982	777
5072081	861
180	946
278	·7052030
377	115
475	199
574	283
672	368
771	452
870	537
968	621
5073067	705
165	790
264	874
362	958
461	·7053043
559	127
5073658	·7053212
757	296
855	380
954	465
5074052	549
151	633
249	718
348	802
446	886
545	971
643	·7054055
742	139
840	224
939	308
5075037	392
136	477
234	561
333	645
432	729
530	814
629	898
727	982
826	·7055067
924	151
5076023	235
121	319
220	404
318	488
417	572
515	657
614	741
712	825
811	909
909	994
5077007	·7056078
106	162
204	246
303	330
401	415
500	499
598	583
697	667
795	752
894	836
992	920
5078091	·7057004
189	088
288	173
386	257
485	341
583	425
681	509
780	594
878	678
977	762
5079075	846
174	930
272	·7058015
371	099
469	183
567	267
666	351
764	435
863	519
961	604

Excesses

10 = 9	1 = 1
20 = 17	2 = 2
30 = 26	3 = 3
40 = 34	4 = 3
50 = 43	5 = 4
60 = 51	6 = 5
70 = 60	7 = 6
80 = 68	8 = 7
90 = 77	9 = 8

49 = 42

Anti Log Excesses

10 = 12	1 = 1
20 = 23	2 = 2
30 = 35	3 = 4
40 = 47	4 = 5
50 = 58	5 = 6
60 = 70	6 = 7
70 = 82	7 = 8
80 = 93	8 = 9
	9 = 11

42 = 49

5080060	·7058688
158	772
256	856
355	940
453	·7059024
552	108
650	193
748	277
847	361
945	445
5081044	529
142	613
241	697
339	781
437	865
536	949
634	·7060034
732	118
831	202
929	286
5082028	370
126	454
224	538
323	622
421	706
520	790
618	874
716	958
815	·7061042
913	126
5083011	211
110	295
208	379
307	463
405	547
503	631
602	715
5083700	·7061799
798	883
897	967
995	·7062051
5084093	135
192	219
290	303
388	387
487	471
585	555
683	639
782	723
880	807
978	891
5085077	975
175	·7063059
273	143
372	227
470	311
568	395
667	479
765	563
863	646
962	730
5086060	814
158	898
256	982
355	·7064066
453	150
551	234
650	318
748	402
846	486
945	570
5087043	654
141	737
239	821
338	905
436	989
534	·7065073
633	157
731	241
829	325
927	409
5088026	493
124	576
222	660
320	744
419	828
517	912
615	996
714	·7066080
812	163
910	247
5089008	331
107	415
205	499
303	583
401	667
499	750
598	834
696	918
794	·7067002
892	086
991	169

Excesses

10 = 9	1 = 1
20 = 17	2 = 2
30 = 26	3 = 3
40 = 34	4 = 3
50 = 43	5 = 4
60 = 51	6 = 5
70 = 60	7 = 6
80 = 68	8 = 7
90 = 77	9 = 8

49 = 42

Anti Log Excesses

10 = 12	1 = 1
20 = 23	2 = 2
30 = 35	3 = 4
40 = 47	4 = 5
50 = 59	5 = 6
60 = 70	6 = 7
70 = 82	7 = 8
80 = 94	8 = 9
	9 = 11

42 = 49

5090089	·7067253
187	337
285	421
384	505
482	589
580	672
678	756
776	840
875	924
973	·7068007
5091071	091
169	175
268	259
366	343
464	426
562	510
660	594
759	678
857	761
955	845
5092053	929
151	·7069013
250	096
348	180
446	264
544	348
642	431
740	515
839	599
937	683
5093035	766
133	850
231	934
329	·7070017
428	101
526	185
624	269

5093722	·7070352
820	436
918	520
5094017	603
115	687
213	771
311	854
409	938
507	·7071022
605	105
704	189
802	273
900	356
998	440
5095096	524
194	607
292	691
391	775
489	858
587	942
685	·7072025
783	109
881	193
979	276
5096077	360
176	444
274	527
372	611
470	694
568	778
666	862
764	945
862	·7073029
960	112
5097058	196
157	280
255	363
353	447
451	530
549	614
647	697
745	781
843	864
941	948
5098039	·7074032
137	115
235	199
334	282
432	366
530	449
628	533
726	616
824	700
922	783
5099020	867
118	950
216	·7075034
314	117
412	201
510	284
608	368
706	451
804	535
902	618

Excesses

10 = 9	1 = 1
20 = 17	2 = 2
30 = 26	3 = 3
40 = 34	4 = 3
50 = 43	5 = 4
60 = 51	6 = 5
70 = 60	7 = 6
80 = 68	8 = 7
90 = 77	9 = 8

49 = 42

Anti Log Excesses

10 = 12	1 = 1
20 = 23	2 = 2
30 = 35	3 = 4
40 = 47	4 = 5
50 = 59	5 = 6
60 = 70	6 = 7
70 = 82	7 = 8
80 = 94	8 = 9
	9 = 11

42 = 49

5100000	·7075702
099	785
197	869
295	952
393	·7076036
491	119
589	203
687	286
785	370
883	453
981	537
5101079	620
177	704
275	787
373	870
471	954
569	·7077037
667	121
765	204
863	288
961	371
5102059	454
157	538
255	621
353	705
451	788
549	871
647	955
745	·7078038
843	122
941	205
5103039	288
137	372
235	455
333	539
431	622
529	705

5103627	·7078789
725	872
823	955
921	·7079039
5104019	122
116	205
214	289
312	372
410	455
508	539
606	622
704	705
802	789
900	872
998	955
5105096	·7080039
194	122
292	205
390	289
488	372
586	455
684	539
782	622
879	705
977	789
5106075	872
173	955
271	·7081038
369	122
467	205
565	288
663	371
761	455
859	538
957	621
5107054	704
152	788
250	871
348	954
446	·7082037
544	121
642	204
740	287
838	370
936	454
5108033	537
131	620
229	703
327	787
425	870
523	953
621	·7083036
719	119
816	203
914	286
5109012	369
110	452
208	535
306	619
404	702
501	785
599	868
697	951
795	·7084034
893	118
5109991	·7084201

Excesses

10 = 9	1 = 1
20 = 17	2 = 2
30 = 26	3 = 3
40 = 34	4 = 3
50 = 43	5 = 4
60 = 51	6 = 5
70 = 60	7 = 6
80 = 68	8 = 7
90 = 77	9 = 8

49 = 42

Anti Log Excesses

10 = 12	1 = 1
20 = 24	2 = 2
30 = 35	3 = 4
40 = 47	4 = 5
50 = 59	5 = 6
60 = 71	6 = 7
70 = 82	7 = 8
80 = 94	8 = 9
	9 = 11

42 = 49

5110089	·7084284
186	367
284	450
382	533
480	616
578	700
676	783
773	866
871	949
969	·7085032
5111067	115
165	198
263	282
360	365
458	448
556	531
654	614
752	697
849	780
947	863
5112045	946
143	·7086030
241	113
339	196
436	279
534	362
632	445
730	528
828	611
925	694
5113023	777
121	860
219	943
316	·7087026
414	109

5113512	·7087192
610	275
708	359
805	442
903	525
5114001	608
099	691
196	774
294	857
392	940
490	·7088023
587	106
685	189
783	272
881	355
979	438
5115076	521
174	604
272	687
369	770
467	853
565	936
663	·7089019
760	102
858	185
956	268
5116054	351
151	434
249	517
347	600
445	683
542	766
640	849
738	931
835	·7090014
933	097
5117031	180
129	263
226	346
324	429
422	512
519	595
617	678
715	761
813	844
910	926
5118008	·7091009
106	092
203	175
301	258
399	341
496	424
594	507
692	590
789	672
887	755
985	838
5119082	921
180	·7092004
278	087
375	170
473	252
571	335
668	418
766	501

5119864	·7092584
961	667

Excesses

10 = 8	1 = 1
20 = 17	2 = 2
30 = 25	3 = 3
40 = 34	4 = 3
50 = 42	5 = 4
60 = 51	6 = 5
70 = 59	7 = 6
80 = 68	8 = 7
90 = 76	9 = 8

49 = 42

Anti Log Excesses

10 = 12	1 = 1
20 = 24	2 = 2
30 = 35	3 = 4
40 = 47	4 = 5
50 = 59	5 = 6
60 = 71	6 = 7
70 = 82	7 = 8
80 = 94	8 = 9
	9 = 11

42 = 49

5120059	·7092750
157	832
254	915
352	998
450	·7093081
547	164
645	246
743	329
840	412
938	495
5121036	578
133	661
231	743
328	826
426	909
524	992
621	·7094074
719	157
817	240
914	323
5122012	405
109	488
207	571
305	654
402	737
500	819
597	902
695	985
793	·7095068
890	150
988	233
5123086	316
183	399
281	481

5123378	·7095564
476	647
573	729
671	812
769	895
866	978
964	·7096060
5124061	143
159	226
257	308
354	391
452	474
549	556
647	639
744	722
842	805
940	887
5125037	970
135	·7097053
232	135
330	218
427	301
525	383
622	466
720	549
818	631
915	714
5126013	796
110	879
208	962
305	·7098044
403	127
500	210
598	292
695	375
793	458
890	540
988	623
5127085	705
183	788
280	871
378	953
476	·7099036
573	118
671	201
768	284
866	366
963	449
5128061	531
158	614
256	696
353	779
451	862
548	944
646	·7100027
743	109
841	192
938	274
5129035	357
133	439
230	522
328	604
425	687
523	770
620	852

5129718	·7100935
815	·7101017
913	100

Excesses

10 = 8	1 = 1
20 = 17	2 = 2
30 = 25	3 = 3
40 = 34	4 = 3
50 = 42	5 = 4
60 = 51	6 = 5
70 = 59	7 = 6
80 = 68	8 = 7
90 = 76	9 = 8

49 = 41

Anti Log Excesses

10 = 12	1 = 1
20 = 24	2 = 2
30 = 35	3 = 4
40 = 47	4 = 5
50 = 59	5 = 6
60 = 71	6 = 7
70 = 83	7 = 8
80 = 94	8 = 9
	9 = 11

41 = 49

5130010	·7101182
108	265
205	347
303	430
400	512
498	595
595	677
692	760
790	842
887	925
985	·7102007
5131082	090
180	172
277	255
375	337
472	419
569	502
667	584
764	667
862	749
959	832
5132057	914
154	997
251	·7103079
349	161
446	244
544	326
641	409
739	491
836	574
933	656
5133031	738
128	821

5133226	·7103903
323	986
420	·7104068
518	150
615	233
713	315
810	398
907	480
5134005	562
102	645
200	727
297	810
394	892
492	974
589	·7105057
686	139
784	221
881	304
979	386
5135076	468
173	551
271	633
368	715
465	798
563	880
660	962
758	·7106045
855	127
952	209
5136050	292
147	374
244	456
342	539
439	621
536	703
634	786
731	868
828	950
926	·7107033
5137023	115
120	197
218	279
315	362
412	444
510	526
607	608
704	691
802	773
899	855
996	938
5138094	·7108020
191	102
288	184
385	267
483	349
580	431
677	513
775	596
872	678
969	760
5139067	842
164	924
261	·7109007
358	089
456	171

5139553	·7109253
650	335
748	418
845	500
942	582

Excesses

10 = 8	1 = 1
20 = 17	2 = 2
30 = 25	3 = 3
40 = 34	4 = 3
50 = 42	5 = 4
60 = 51	6 = 5
70 = 59	7 = 6
80 = 68	8 = 7
90 = 76	9 = 8

49 = 41.

Anti Log Excesses

10 = 12	1 = 1
20 = 24	2 = 2
30 = 35	3 = 4
40 = 47	4 = 5
50 = 59	5 = 6
60 = 71	6 = 7
70 = 83	7 = 8
80 = 95	8 = 9
	9 = 11

41 = 49

5140039	·7109664
137	746
234	829
331	911
428	993
526	·7110075
623	157
720	240
818	322
915	404
5141012	486
109	568
207	650
304	733
401	815
498	897
596	979
693	·7111061
790	143
887	225
985	307
5142082	390
179	472
276	554
373	636
471	718
568	800
665	882
762	964
860	·7112046
957	129

5143054	·7112211
151	293
249	375
346	457
443	539
540	621
637	703
735	785
832	867
929	949
5144026	·7113031
123	113
221	196
318	278
415	360
512	442
609	524
707	606
804	688
901	770
998	852
5145095	934
192	·7114016
290	098
387	180
484	262
581	344
678	426
775	508
873	590
970	672
5146067	754
164	836
261	918
358	·7115000
456	082
553	164
650	246
747	328
844	410
941	492
5147038	574
136	656
233	738
330	820
427	902
524	984
621	·7116066
718	148
816	230
913	311
5148010	393
107	475
204	557
301	639
398	721
495	803
593	885
690	967
787	·7117049
884	131
981	213
5149078	294
175	376
272	458

5149369	·7117540
466	622
564	704
661	786
758	868
855	950
952	·7118031

Excesses

10 = 8	1 = 1
20 = 17	2 = 2
30 = 25	3 = 3
40 = 34	4 = 3
50 = 42	5 = 4
60 = 51	6 = 5
70 = 59	7 = 6
80 = 68	8 = 7
90 = 76	9 = 8

48 = 41

Anti Log Excesses

10 = 12	1 = 1
20 = 24	2 = 2
30 = 36	3 = 4
40 = 47	4 = 5
50 = 59	5 = 6
60 = 71	6 = 7
70 = 83	7 = 8
80 = 95	8 = 9
	9 = 11

41 = 48

5150049	·7118113
146	195
243	277
340	359
437	441
534	523
632	604
729	686
826	768
923	850
5151020	932
117	·7119014
214	095
311	177
408	259
505	341
602	423
699	505
796	587
893	668
990	750
5152087	832
185	914
282	996
379	·7120077
476	159
573	241
670	323
767	405

5152864	·7120486
961	568
5153058	650
155	732
252	813
349	895
446	977
543	·7121059
640	140
737	222
834	304
931	386
5154028	467
125	549
222	631
319	713
416	794
513	876
610	958
707	·7122040
804	121
901	203
998	285
5155095	366
192	448
289	530
386	612
483	693
580	775
677	857
774	938
871	·7123020
968	102
5156065	183
162	265
259	347
356	428
453	510
550	592
647	673
744	755
841	837
938	918
5157035	·7124000
131	082
228	163
325	245
422	327
519	408
616	490
713	572
810	653
907	735
5158004	816
101	898
198	980
295	·7125061
392	143
489	224
586	306
683	388
779	469
876	551
973	632
5159070	714

5159167	·7125796
264	877
361	959
458	·7126040
555	122
652	204
749	285
845	367
942	448

Excesses

10 = 8	1 = 1
20 = 17	2 = 2
30 = 25	3 = 3
40 = 34	4 = 3
50 = 42	5 = 4
60 = 51	6 = 5
70 = 59	7 = 6
80 = 67	8 = 7
90 = 76	9 = 8

48 = 41

Anti Log Excesses

10 = 12	1 = 1
20 = 24	2 = 2
30 = 36	3 = 4
40 = 47	4 = 5
50 = 59	5 = 6
60 = 71	6 = 7
70 = 83	7 = 8
80 = 95	8 = 9
	9 = 11

41 = 48

5160039	·7126530
136	611
233	693
330	774
427	856
524	937
621	·7127019
718	101
814	182
911	264
5161008	345
105	427
202	508
299	590
396	671
493	753
589	834
686	916
783	997
880	·7128079
977	160
5162074	242
171	323
267	405
364	486
461	568
558	649

5162655	·7128731
752	812
849	894
945	975
5163042	·7129057
139	138
236	220
333	301
430	382
526	464
623	545
720	627
817	708
914	790
5164011	871
107	953
204	·7130034
301	115
398	197
495	278
591	360
688	441
785	522
882	604
979	685
5165076	767
172	848
269	929
366	·7131011
463	092
560	174
656	255
753	336
850	418
947	499
5166043	580
140	662
237	743
334	824
431	906
527	987
624	·7132068
721	150
818	231
914	313
5167011	394
108	475
205	557
302	638
398	719
495	800
592	882
689	963
785	·7133044
882	126
979	207
5168076	288
172	370
269	451
366	532
463	613
559	695
656	776
753	857
850	939

5168946	·7134020
5169043	101
140	182
236	264
333	345
430	426
527	507
623	589
720	670
817	751
913	832

Excesses

10 = 8	1 = 1
20 = 17	2 = 2
30 = 25	3 = 3
40 = 34	4 = 3
50 = 42	5 = 4
60 = 50	6 = 5
70 = 59	7 = 6
80 = 67	8 = 7
90 = 76	9 = 8

48 = 41

Anti Log Excesses

10 = 12	1 = 1
20 = 24	2 = 2
30 = 36	3 = 4
40 = 48	4 = 5
50 = 59	5 = 6
60 = 71	6 = 7
70 = 83	7 = 8
80 = 95	8 = 10
	9 = 11

41 = 48

5170010	·7134914
107	995
204	·7135076
300	157
397	239
494	320
590	401
687	482
784	564
881	645
977	726
5171074	807
171	888
267	970
364	·7136051
461	132
557	213
654	294
751	376
847	457
944	538
5172041	619
137	700
234	782
331	863

5172427	·7136944
524	·7137025
621	106
717	187
814	268
911	350
5173007	431
104	512
201	593
297	674
394	755
491	837
587	918
684	999
781	·7138080
877	161
974	242
5174070	323
167	404
264	486
360	567
457	648
554	729
650	810
747	891
843	972
940	·7139053
5175037	134
133	215
230	296
327	377
423	459
520	540
616	621
713	702
810	783
906	864
5176003	945
099	·7140026
196	107
293	188
389	269
486	350
582	431
679	512
776	593
872	674
969	755
5177065	836
162	917
258	998
355	·7141079
452	160
548	241
645	322
741	403
838	484
934	565
5178031	646
128	727
224	808
321	889
417	970
514	·7142051
610	132

5178707	·7142213
803	294
900	375
997	456
5179093	537
190	618
286	699
383	780
479	861
576	942
672	·7143023
769	104
865	184
962	265

Excesses

10 = 8	1 = 1
20 = 17	2 = 2
30 = 25	3 = 3
40 = 34	4 = 3
50 = 42	5 = 4
60 = 50	6 = 5
70 = 59	7 = 6
80 = 67	8 = 7
90 = 76	9 = 8

48 = 41

Anti Log Excesses

10 = 12	1 = 1
20 = 24	2 = 2
30 = 36	3 = 4
40 = 48	4 = 5
50 = 60	5 = 6
60 = 71	6 = 7
70 = 83	7 = 8
80 = 95	8 = 10
	9 = 11

41 = 48

5180058	·7143346
155	427
251	508
348	589
444	670
541	751
638	832
734	913
831	994
927	·7144075
5181024	155
120	236
217	317
313	398
410	479
506	560
603	641
699	722
796	802
892	883
989	964
5182085	·7145045

5182182	·7145126
278	207
374	288
471	368
567	449
664	530
760	611
857	692
953	773
5183050	854
146	934
243	·7146015
339	096
436	177
532	258
629	338
725	419
821	500
918	581
5184014	662
111	743
207	823
304	904
400	985
497	·7147066
593	147
689	227
786	308
882	389
979	470
5185075	550
172	631
268	712
365	793
461	873
557	954
654	·7148035
750	116
847	196
943	277
5186039	358
136	439
232	519
329	600
425	681
521	762
618	842
714	923
811	·7149004
907	085
5187003	165
100	246
196	327
293	407
389	488
485	569
582	649
678	730
775	811
871	891
967	972
5188064	·7150053
160	134
256	214
353	295

5188449	·7150375
546	456
642	537
738	617
835	698
931	779
5189027	859
124	940
220	·7151021
316	101
413	182
509	263
605	343
702	424
798	504
895	585
991	666

Excesses

10 = 8	1 = 1
20 = 17	2 = 2
30 = 25	3 = 3
40 = 33	4 = 3
50 = 42	5 = 4
60 = 50	6 = 5
70 = 59	7 = 6
80 = 67	8 = 7
90 = 75	9 = 8

48 = 40

Anti Log Excesses

10 = 12	1 = 1
20 = 24	2 = 2
30 = 36	3 = 4
40 = 48	4 = 5
50 = 60	5 = 6
60 = 72	6 = 7
70 = 84	7 = 8
80 = 95	8 = 10
	9 = 11

40 = 48

5190087	·7151746
184	827
280	908
376	988
473	·7152069
569	149
665	230
762	311
858	391
954	472
5191050	552
147	633
243	713
339	794
436	875
532	955
628	·7153036
725	116
821	197

5191917	·7153277
5192014	358
110	438
206	519
302	600
399	680
495	761
591	841
688	922
784	·7154002
880	083
977	163
5193073	244
169	324
265	405
362	485
458	566
554	646
650	727
747	807
843	888
939	968
5194036	·7155049
132	129
228	210
324	290
421	371
517	451
613	532
709	612
806	693
902	773
998	854
5195094	934
191	·7156015
287	095
383	175
479	256
576	336
672	417
768	497
864	578
960	658
5196057	739
153	819
249	899
345	980
442	·7157060
538	141
634	221
730	301
826	382
923	462
5197019	543
115	623
211	703
307	784
404	864
500	945
596	·7158025
692	105
788	186
885	266
981	347
5198077	427

5198173	·7158507
269	588
366	668
462	748
558	829
654	909
750	989
847	·7159070
943	150
5199039	230
135	311
231	391
327	471
424	552
520	632
616	712
712	793
808	873
904	953

Excesses

10 = 8	1 = 1
20 = 17	2 = 2
30 = 25	3 = 3
40 = 33	4 = 3
50 = 42	5 = 4
60 = 50	6 = 5
70 = 59	7 = 6
80 = 67	8 = 7
90 = 75	9 = 8

48 = 40

Anti Log Excesses

10 = 12	1 = 1
20 = 24	2 = 2
30 = 36	3 = 4
40 = 48	4 = 5
50 = 60	5 = 6
60 = 72	6 = 7
70 = 84	7 = 8
80 = 96	8 = 10
	9 = 11

40 = 48

5200000	·7160034
097	114
193	194
289	275
385	355
481	435
577	515
674	596
770	676
866	756
962	836
5201058	917
154	997
250	·7161077
346	158
443	238
539	318

5201635	·7161398
731	479
827	559
923	639
5202019	719
115	800
212	880
308	960
404	·7162040
500	120
596	201
692	281
788	361
884	441
980	522
5203077	602
173	682
269	762
365	842
461	923
557	·7163003
653	083
749	163
845	243
941	324
5204037	404
133	484
230	564
326	644
422	724
518	805
614	885
710	965
806	·7164045
902	125
998	205
5205094	286
190	366
286	446
382	526
478	606
574	686
670	766
767	847
863	927
959	·7165007
5206055	087
151	167
247	247
343	327
439	407
535	488
631	568
727	648
823	728
919	808
5207015	888
111	968
207	·7166048
303	128
399	208
495	288
591	369
687	449
783	529

5207879	·7166609
975	689
5208071	769
167	849
263	929
359	·7167009
455	089
551	169
647	249
743	329
839	409
935	489
5209031	569
127	649
223	729
319	809
415	889
511	970
607	·7168050
703	130
799	210
895	290
991	370

Excesses

10 = 8	1 = 1
20 = 17	2 = 2
30 = 25	3 = 3
40 = 33	4 = 3
50 = 42	5 = 4
60 = 50	6 = 5
70 = 58	7 = 6
80 = 67	8 = 7
90 = 75	9 = 8

48 = 40

Anti Log Excesses

10 = 12	1 = 1
20 = 24	2 = 2
30 = 36	3 = 4
40 = 48	4 = 5
50 = 60	5 = 6
60 = 72	6 = 7
70 = 84	7 = 8
	8 = 10
	9 = 11

40 = 48

5210087	·7168450
183	530
279	610
375	690
471	770
567	850
663	929
759	·7169009
855	089
951	169
5211046	249
142	329
238	409

5211334	·7169489
430	569
526	649
622	729
718	809
814	889
910	969
5212006	·7170049
102	129
198	209
294	289
390	369
486	449
581	528
677	608
773	688
869	768
965	848
5213061	928
157	·7171008
253	088
349	168
445	248
541	327
636	407
732	487
828	567
924	647
5214020	727
116	807
212	887
308	966
404	·7172046
500	126
595	206
691	286
787	366
883	446
979	525
5215075	605
171	685
267	765
362	845
458	925
554	·7173004
650	084
746	164
842	244
938	324
5216033	404
129	483
225	563
321	643
417	723
513	803
609	882
704	962
800	·7174042
896	122
992	201
5217088	281
184	361
279	441
375	521
471	600

5217567	·7174680
663	760
759	840
854	919
950	999
5218046	·7175079
142	159
238	238
334	318
429	398
525	478
621	557
717	637
813	717
908	797
5219004	876
100	956
196	·7176036
292	115
387	195
483	275
579	355
675	434
771	514
866	594
962	673

Excesses

10 = 8	1 = 1
20 = 17	2 = 2
30 = 25	3 = 2
40 = 33	4 = 3
50 = 42	5 = 4
60 = 50	6 = 5
70 = 58	7 = 6
80 = 67	8 = 7
90 = 75	9 = 7

48 = 40

Anti Log Excesses

10 = 12	1 = 1
20 = 24	2 = 2
30 = 36	3 = 4
40 = 48	4 = 5
50 = 60	5 = 6
60 = 72	6 = 7
70 = 84	7 = 8
	8 = 10
	9 = 11

40 = 48

5220058	·7176753
154	833
250	912
345	992
441	·7177072
537	152
633	231
728	311
824	391
920	470

5221016	·7177550
111	629
207	709
303	789
399	869
495	948
590	·7178028
686	107
782	187
878	267
973	346
5222069	426
165	506
261	585
356	665
452	744
548	824
644	904
739	983
835	·7179063
931	143
5223026	222
122	302
218	381
314	461
409	541
505	620
601	700
697	779
792	859
888	938
984	·7180018
5224079	098
175	177
271	257
366	336
462	416
558	495
654	575
749	654
845	734
941	814
5225036	893
132	973
228	·7181052
323	132
419	211
515	291
611	370
706	450
802	529
898	609
993	688
5226089	768
185	847
280	927
376	·7182006
472	086
567	165
663	245
759	324
854	404
950	483
5227046	563
141	642

5227237	·7182722
333	801
428	881
524	960
619	·7183040
715	119
811	198
906	278
5228002	357
098	437
193	516
289	596
385	675
480	755
576	834
671	913
767	993
863	·7184072
958	152
5229054	231
150	310
245	390
341	469
436	549
532	628
628	707
723	787
819	866
914	946

Excesses

10 = 8	1 = 1
20 = 17	2 = 2
30 = 25	3 = 2
40 = 33	4 = 3
50 = 42	5 = 4
60 = 50	6 = 5
70 = 58	7 = 6
80 = 66	8 = 7
90 = 75	9 = 7

48 = 40

Anti Log Excesses

10 = 12	1 = 1
20 = 24	2 = 2
30 = 36	3 = 4
40 = 48	4 = 5
50 = 60	5 = 6
60 = 72	6 = 7
70 = 84	7 = 8
	8 = 10
	9 = 11

40 = 48

5230010	·7185025
106	104
201	184
297	263
392	343
488	422
584	501

5230679	·7185581
775	660
870	739
966	819
5231062	898
157	977
253	·7186057
348	136
444	215
539	295
635	374
731	453
826	533
922	612
5232017	691
113	771
208	850
304	929
400	·7187009
495	088
591	167
686	247
782	326
877	405
973	484
5233068	564
164	643
260	722
355	802
451	881
546	960
642	·7188039
737	119
833	198
928	277
5234024	357
119	436
215	515
310	594
406	674
501	753
597	832
692	911
788	991
883	·7189070
979	149
5235075	228
170	307
266	387
361	466
457	545
552	624
648	704
743	783
839	862
934	941
5236030	·7190020
125	100
221	179
316	258
411	337
507	416
602	496
698	575
793	654

5236889	·7190733
984	812
5237080	892
175	971
271	·7191050
366	129
462	208
557	287
653	366
748	446
844	525
939	604
5238034	683
130	762
225	841
321	920
416	·7192000
512	079
607	158
703	237
798	316
894	395
989	474
5239084	553
180	633
275	712
371	791
466	870
562	949
657	·7193028
752	107
848	186
943	265

Excesses

10 = 8	1 = 1
20 = 17	2 = 2
30 = 25	3 = 2
40 = 33	4 = 3
50 = 41	5 = 4
60 = 50	6 = 5
70 = 58	7 = 6
80 = 66	8 = 7
90 = 75	9 = 7

48 = 40

Anti Log Excesses

10 = 12	1 = 1
20 = 24	2 = 2
30 = 36	3 = 4
40 = 48	4 = 5
50 = 60	5 = 6
60 = 72	6 = 7
70 = 84	7 = 8
	8 = 10
	9 = 11

40 = 48

5240039	·7193344
134	424
230	503

5240325	·7193582
420	661
516	740
611	819
707	898
802	977
897	·7194056
993	135
5241088	214
184	293
279	372
374	451
470	530
565	609
661	688
756	767
851	847
947	926
5242042	·7195005
137	084
233	163
328	242
424	321
519	400
614	479
710	558
805	637
900	716
996	795
5243091	874
187	953
282	·7196032
377	111
473	190
568	269
663	348
759	427
854	505
949	584
5244045	663
140	742
235	821
331	900
426	979
521	·7197058
617	137
712	216
807	295
903	374
998	453
5245093	532
189	611
284	690
379	769
475	848
570	926
665	·7198005
761	084
856	163
951	242
5246047	321
142	400
237	479
333	558
428	637

5246523	·7198715
618	794
714	873
809	952
904	·7199031
5247000	110
095	189
190	268
286	346
381	425
476	504
571	583
667	662
762	741
857	820
952	898
5248048	977
143	·7200056
238	135
334	214
429	293
524	371
619	450
715	529
810	608
905	687
5249000	766
096	844
191	923
286	·7201002
381	081
477	160
572	238
667	317
762	396
858	475
953	553

Excesses

10 = 8	1 = 1
20 = 17	2 = 2
30 = 25	3 = 2
40 = 33	4 = 3
50 = 41	5 = 4
60 = 50	6 = 5
70 = 58	7 = 6
80 = 66	8 = 7
90 = 75	9 = 7

48 = 40

Anti Log Excesses

10 = 12	1 = 1
20 = 24	2 = 2
30 = 36	3 = 4
40 = 48	4 = 5
50 = 60	5 = 6
60 = 72	6 = 7
70 = 85	7 = 8
	8 = 10
	9 = 11

40 = 48

5250048	·7201632
143	711
239	790
334	869
429	947
524	·7202026
620	105
715	184
810	262
905	341
5251000	420
096	499
191	577
286	656
381	735
476	814
572	892
667	971
762	·7203050
857	129
953	207
5252048	286
143	365
238	444
333	522
429	601
524	680
619	758
714	837
809	916
904	994
5253000	·7204073
095	152
190	230
285	309
380	388
476	467
571	545
666	624
761	703
856	781
951	860
5254047	939
142	·7205017
237	096
332	175
427	253
522	332
618	411
713	489
808	568
903	647
998	725
5255093	804
188	882
284	961
379	·7206040
474	118
569	197
664	275
759	354
854	433
950	511
5256045	590
140	668

5256235	·7206747
330	826
425	904
520	983
615	·7207062
710	140
806	219
901	298
996	376
5257091	455
186	533
281	612
376	690
471	769
566	847
662	926
757	·7208005
852	083
947	162
5258042	240
137	319
232	397
327	476
422	554
517	633
612	711
708	790
803	868
898	947
993	·7209025
5259088	104
183	182
278	261
373	339
468	418
563	497
658	575
753	654
848	732
943	811

Excesses

10 = 8	1 = 1
20 = 17	2 = 2
30 = 25	3 = 2
40 = 33	4 = 3
50 = 41	5 = 4
60 = 50	6 = 5
70 = 58	7 = 6
80 = 66	8 = 7
90 = 74	9 = 7

48 = 40

Anti Log Excesses

10 = 12	1 = 1
20 = 24	2 = 2
30 = 36	3 = 4
40 = 48	4 = 5
50 = 60	5 = 6
60 = 73	6 = 7
70 = 85	7 = 8
	8 = 10
	9 = 11

40 = 48

5260039	·7209889
134	967
229	·7210046
324	124
419	203
514	281
609	360
704	438
799	517
894	595
989	674
5261084	752
179	830
274	909
369	987
464	·7211066
559	144
654	223
749	301
844	379
939	458
5262034	536
129	615
224	693
319	772
414	850
509	929
604	·7212007
699	085
794	164
889	242
984	320
5263079	399
174	477
269	556
364	634
459	712
554	791
649	869
744	947
839	·7213026
934	104
5264029	182
124	261
219	339
314	418
409	496
504	574
599	653
694	731
789	809

5264884	·7213888
979	966
5265074	·7214044
169	123
264	201
359	279
454	358
549	436
644	514
739	592
834	671
929	749
5266024	827
119	906
214	984
309	·7215062
403	141
498	219
593	297
688	375
783	454
878	532
973	610
5267068	688
163	767
258	845
353	923
448	·7216001
543	080
638	158
732	236
827	314
922	393
5268017	471
112	549
207	627
302	706
397	784
492	863
587	941
682	·7217019
776	097
871	176
966	254
5269061	332
156	410
251	488
346	567
441	645
536	723
630	801
725	879
820	958
915	·7218036

Excesses

10 = 8	1 = 1
20 = 16	2 = 2
30 = 25	3 = 2
40 = 33	4 = 3
50 = 41	5 = 4
60 = 49	6 = 5
70 = 58	7 = 6
80 = 66	8 = 7
90 = 74	9 = 7

48 = 40

Anti Log Excesses

10 = 12	1 = 1
20 = 24	2 = 2
30 = 36	3 = 4
40 = 48	4 = 5
50 = 61	5 = 6
60 = 73	6 = 7
70 = 85	7 = 8
	8 = 10
	9 = 11

40 = 48

5270010	·7218114
105	192
200	270
295	349
389	427
484	505
579	583
674	661
769	739
864	818
959	896
5271054	974
148	·7219052
243	130
338	208
433	287
528	365
623	443
717	521
812	599
907	677
5272002	755
097	833
192	912
287	990
381	·7220068
476	146
571	224
666	302
761	380
856	458
950	536
5273045	615
140	693
235	771
330	849
424	927

5273519	·7221005
614	083
709	161
804	239
898	317
993	395
5274088	473
183	551
278	629
373	708
467	786
562	864
657	942
752	·7222020
846	098
941	176
5275036	254
131	332
226	410
320	488
415	566
510	644
605	722
700	800
794	878
889	956
984	·7223034
5276079	112
173	190
268	268
363	346
458	424
552	502
647	580
742	658
837	736
931	814
5277026	892
121	970
216	·7224048
310	126
405	204
500	282
595	360
689	438
784	516
879	594
974	672
5278068	750
163	828
258	906
353	984
447	·7225061
542	139
637	217
731	295
826	373
921	451
5279016	529
110	607
205	685
300	763
394	841
489	919
584	997

5279679	·7226074
773	152
868	230
963	308

Excesses

10 = 8	1 = 1
20 = 16	2 = 2
30 = 25	3 = 2
40 = 33	4 = 3
50 = 41	5 = 4
60 = 49	6 = 5
70 = 58	7 = 6
80 = 66	8 = 7
90 = 74	9 = 7

47 = 39

Anti Log Excesses

10 = 12	1 = 1
20 = 24	2 = 2
30 = 36	3 = 4
40 = 49	4 = 5
50 = 61	5 = 6
60 = 73	6 = 7
70 = 85	7 = 9
	8 = 10
	9 = 11

39 = 47

5280057	·7226386
152	464
247	542
341	620
436	698
531	775
625	853
720	931
815	·7227009
910	087
5281004	165
099	243
194	320
288	398
383	476
478	554
572	632
667	710
762	788
856	865
951	943
5282046	·7228021
140	099
235	177
330	255
424	332
519	410
613	488
708	566
803	644
897	721
992	799

5283087	·7228877
181	955
276	·7229033
371	110
465	188
560	266
655	344
749	422
844	499
938	577
5284033	655
128	733
222	810
317	888
412	966
506	·7230044
601	121
695	199
790	277
885	355
979	432
5285074	510
168	588
263	666
358	743
452	821
547	899
641	977
736	·7231054
831	132
925	210
5286020	287
114	365
209	443
304	520
398	598
493	676
587	753
682	831
776	909
871	987
966	·7232064
5287060	142
155	220
249	297
344	375
438	453
533	530
628	608
722	686
817	763
911	841
5288006	919
100	996
195	·7233074
289	152
384	229
479	307
573	385
668	462
762	540
857	617
951	695
5289046	773
140	850

5289235	·7233928
329	·7234006
424	083
518	161
613	238
707	316
802	394
897	471
991	549

Excesses

10 = 8 1 = 1
20 = 16 2 = 2
30 = 25 3 = 2
40 = 33 4 = 3
50 = 41 5 = 4
60 = 49 6 = 5
70 = 58 7 = 6
80 = 66 8 = 7
90 = 74 9 = 7

47 = 39

Anti Log Excesses

10 = 12 1 = 1
20 = 24 2 = 2
30 = 37 3 = 4
40 = 49 4 = 5
50 = 61 5 = 6
60 = 73 6 = 7
70 = 85 7 = 9
8 = 10
9 = 11

39 = 47

5290086	·7234626
180	704
275	782
369	859
464	937
558	·7235014
653	092
747	170
842	247
936	325
5291031	402
125	480
220	557
314	635
409	713
503	790
598	868
692	945
787	·7236023
881	100
976	178
5292070	255
165	333
259	411
353	488
448	566
542	643

5292637	·7236721
731	798
826	876
920	953
5293015	·7237031
109	108
204	186
298	263
393	341
487	418
582	496
676	573
770	651
865	728
959	806
5294054	883
148	961
243	·7238038
337	115
432	193
526	270
620	348
715	425
809	503
904	580
998	658
5295093	735
187	813
281	890
376	967
470	·7239045
565	122
659	200
754	277
848	355
942	432
5296037	509
131	587
226	664
320	742
414	819
509	896
603	974
698	·7240051
792	129
886	206
981	283
5297075	361
170	438
264	516
358	593
453	670
547	748
641	825
736	902
830	980
925	·7241057
5298019	135
113	212
208	289
302	367
396	444
491	521
585	599
680	676

5298774	·7241753
868	831
963	908
5299057	985
151	·7242063
246	140
340	217
434	295
529	372
623	449
717	527
812	604
906	681

Excesses

10 = 8 1 = 1
20 = 16 2 = 2
30 = 25 3 = 2
40 = 33 4 = 3
50 = 41 5 = 4
60 = 49 6 = 5
70 = 57 7 = 6
80 = 66 8 = 7
90 = 74 9 = 7

47 = 39

Anti Log Excesses

10 = 12 1 = 1
20 = 24 2 = 2
30 = 37 3 = 4
40 = 49 4 = 5
50 = 61 5 = 6
60 = 73 6 = 7
70 = 85 7 = 9
8 = 10
9 = 11

39 = 47

5300000	·7242759
095	836
189	913
284	990
378	·7243068
472	145
567	222
661	300
755	377
849	454
944	531
5301038	609
132	686
227	763
321	841
415	918
510	995
604	·7244073
698	150
793	227
887	304
981	382
5302076	459

5302170	·7244536
264	613
358	691
453	768
547	845
641	922
736	·7245000
830	077
924	154
5303019	231
113	308
207	386
301	463
396	540
490	617
584	695
678	772
773	849
867	926
961	·7246003
5304056	081
150	158
244	235
338	312
433	389
527	466
621	543
715	621
810	698
904	775
998	852
5305092	929
187	·7247006
281	084
375	161
469	238
564	315
658	392
752	469
846	546
941	624
5306035	701
129	778
223	855
317	932
412	·7248009
506	086
600	163
694	241
789	318
883	395
977	472
5307071	549
165	626
260	703
354	780
448	857
542	934
637	·7249011
731	089
825	166
919	243
5308013	320
108	397
202	474

5308296	·7249551
390	628
484	705
578	782
673	859
767	936
861	·7250013
955	090
5309049	167
144	244
238	321
332	398
426	475
520	552
614	629
709	706
803	783
897	860
991	938

Excesses

10 = 8 1 = 1
20 = 16 2 = 2
30 = 25 3 = 2
40 = 33 4 = 3
50 = 41 5 = 4
60 = 49 6 = 5
70 = 57 7 = 6
80 = 65 8 = 7
90 = 74 9 = 7

47 = 39

Anti Log Excesses

10 = 12 1 = 1
20 = 24 2 = 2
30 = 37 3 = 4
40 = 49 4 = 5
50 = 61 5 = 6
60 = 73 6 = 7
70 = 85 7 = 9
8 = 10
9 = 11

39 = 47

5310085	·7251015
179	092
274	169
368	246
462	323
556	400
650	477
744	554
838	631
933	708
5311027	785
121	862
215	939
309	·7252016
403	093
497	169
592	246

5311686	·7252323
780	400
874	477
968	554
5312062	631
156	708
250	785
345	862
439	939
533	·7253016
627	093
721	170
815	247
909	324
5313003	401
098	478
192	554
286	631
380	708
474	785
568	862
662	939
756	·7254016
850	093
944	170
5314039	247
133	323
227	400
321	477
415	554
509	631
603	708
697	785
791	862
885	939
979	·7255015
5315073	092
167	169
262	246
356	323
450	400
544	477
638	553
732	630
826	707
920	784
5316014	861
108	938
202	·7256015
296	091
390	168
484	245
578	322
672	399
766	475
860	552
955	629
5317049	706
143	783
237	860
331	936
425	·7257013
519	090
613	167
707	244

5317801	·7257320
895	397
989	474
5318083	551
177	627
271	704
365	781
459	858
553	935
647	·7258011
741	088
835	165
929	242
5319023	318
117	395
211	472
305	549
399	625
493	702
587	779
681	856
775	932
869	·7259009
963	086

Excesses

10 = 8	1 = 1
20 = 16	2 = 2
30 = 25	3 = 2
40 = 33	4 = 3
50 = 41	5 = 4
60 = 49	6 = 5
70 = 57	7 = 6
80 = 65	8 = 7
90 = 74	9 = 7

47 = 38

Anti Log Excesses

10 = 12	1 = 1
20 = 24	2 = 2
30 = 37	3 = 4
40 = 49	4 = 5
50 = 61	5 = 6
60 = 73	6 = 7
70 = 86	7 = 9
	8 = 10
	9 = 11

38 = 47

5320057	·7259162
151	239
245	316
339	393
433	469
527	546
621	623
715	699
809	776
903	853
997	930
5321091	·7260006

5321185	·7260083
279	160
373	236
466	313
560	390
654	466
748	543
842	620
936	696
5322030	773
124	850
218	926
312	·7261003
406	080
500	156
594	233
688	310
782	386
876	463
970	539
5323064	616
157	693
251	769
345	846
439	923
533	999
627	·7262076
721	152
815	229
909	306
5324003	382
097	459
191	535
284	612
378	689
472	765
566	842
660	918
754	995
848	·7263072
942	148
5325036	225
130	301
224	378
317	455
411	531
505	608
599	684
693	761
787	837
881	914
975	991
5326068	·7264067
162	144
256	220
350	297
444	373
538	450
632	526
726	603
819	679
913	756
5327007	833
101	909
195	986

5327289	·7265062
383	139
476	215
570	292
664	368
758	445
852	521
946	598
5328040	674
133	751
227	827
321	904
415	980
509	·7266057
603	133
696	209
790	286
884	362
978	439
5329072	515
166	592
259	668
353	745
447	821
541	898
635	974
728	·7267051
822	127
916	203

Excesses

10 = 8	1 = 1
20 = 16	2 = 2
30 = 24	3 = 2
40 = 33	4 = 3
50 = 41	5 = 4
60 = 49	6 = 5
70 = 57	7 = 6
80 = 65	8 = 7
90 = 73	9 = 7

47 = 38

Anti Log Excesses

10 = 12	1 = 1
20 = 25	2 = 2
30 = 37	3 = 4
40 = 49	4 = 5
50 = 61	5 = 6
60 = 74	6 = 7
70 = 86	7 = 9
	8 = 10
	9 = 11

38 = 47

5330010	·7267280
104	356
197	433
291	509
385	586
479	662
573	738

5330666	·7267815
760	891
854	968
948	·7268044
5331042	121
135	197
229	273
323	350
417	426
511	503
604	579
698	655
792	732
886	808
979	884
5332073	961
167	·7269037
261	114
355	190
448	266
542	343
636	419
730	495
823	572
917	648
5333011	724
105	801
198	877
292	954
386	·7270030
480	106
573	183
667	259
761	335
855	411
948	488
5334042	564
136	640
230	717
323	793
417	869
511	946
605	·7271022
698	098
792	175
886	251
979	327
5335073	403
167	480
261	556
354	632
448	709
542	785
635	861
729	937
823	·7272014
917	090
5336010	166
104	243
198	319
291	395
385	471
479	547
572	624
666	700

5336760	·7272776
854	853
947	929
5337041	·7273005
135	081
228	157
322	234
416	310
509	386
603	462
697	539
790	615
884	691
978	767
5338071	843
165	920
259	996
352	·7274072
446	148
540	224
633	301
727	377
821	453
914	529
5339008	605
102	682
195	758
289	834
382	910
476	986
570	·7275062
663	139
757	215
851	291
944	367

Excesses

10 = 8 1 = 1
20 = 16 2 = 2
30 = 24 3 = 2
40 = 33 4 = 3
50 = 41 5 = 4
60 = 49 6 = 5
70 = 57 7 = 6
80 = 65 8 = 7
90 = 73 9 = 7

47 = 38

Anti Log Excesses

10 = 12 1 = 1
20 = 25 2 = 2
30 = 37 3 = 4
40 = 49 4 = 5
50 = 61 5 = 6
60 = 74 6 = 7
70 = 86 7 = 9
8 = 10
9 = 11

38 = 47

5340038	·7275443
5340132	·7275519
225	595
319	672
412	748
506	824
600	900
693	976
787	·7276052
881	128
974	204
5341068	281
161	357
255	433
349	509
442	585
536	661
629	737
723	813
817	890
910	966
5342004	·7277042
097	118
191	194
285	270
378	346
472	422
565	498
659	574
753	650
846	726
940	803
5343033	879
127	955
221	·7278031
314	107
408	183
501	259
595	335
688	411
782	487
875	563
969	639
5344063	715
156	791
250	867
343	943
437	·7279019
530	095
624	171
718	247
811	323
905	399
998	475
5345092	551
185	627
279	703
372	779
466	855
559	931
653	·7280007
746	083
840	159
934	235
5346027	311
121	387

5346214	·7280463
308	539
401	615
495	691
588	767
682	843
775	919
869	995
962	·7281071
5347056	147
149	223
243	299
336	375
430	450
523	526
617	602
710	678
804	754
897	830
991	906
5348084	982
178	·7282058
271	134
365	210
458	286
552	362
645	437
739	513
832	589
926	665
5349019	741
113	817
206	893
300	969
393	·7283045
486	120
580	196
673	272
767	348
860	424
954	500

Excesses

10 = 8 1 = 1
20 = 16 2 = 2
30 = 24 3 = 2
40 = 32 4 = 3
50 = 41 5 = 4
60 = 49 6 = 5
70 = 57 7 = 6
80 = 65 8 = 6
90 = 73 9 = 7

47 = 38

Anti Log Excesses

10 = 12 1 = 1
20 = 25 2 = 2
30 = 37 3 = 4
40 = 49 4 = 5
50 = 62 5 = 6
60 = 74 6 = 7
70 = 86 7 = 9
8 = 10
9 = 11

38 = 47

5350047	·7283576
141	652
234	727
328	803
421	879
514	955
608	·7284031
701	107
795	182
888	258
982	334
5351075	410
169	486
262	562
355	638
449	713
542	789
636	865
729	941
823	·7285017
916	093
5352009	169
103	244
196	320
290	396
383	472
477	548
570	623
663	699
757	775
850	851
944	926
5353037	·7286002
130	078
224	154
317	230
411	305
504	381
597	457
691	533
784	608
878	684
971	760
5354064	836
158	911
251	987
345	·7287063
438	139
531	214
625	290
718	366

5354811	·7287442
905	517
998	593
5355092	669
185	744
278	820
372	896
465	972
558	·7288047
652	123
745	199
838	274
932	350
5356025	426
118	502
212	577
305	653
399	729
492	804
585	880
679	956
772	·7289031
865	107
959	183
5357052	258
145	334
239	410
332	485
425	561
519	637
612	712
705	788
799	864
892	939
985	·7290015
5358079	090
172	166
265	242
358	317
452	393
545	469
638	544
732	620
825	695
918	771
5359012	847
105	922
198	998
291	·7291074
385	149
478	225
571	300
665	376
758	452
851	527
945	603

Excesses

10 = 8	1 = 1
20 = 16	2 = 2
30 = 24	3 = 2
40 = 32	4 = 3
50 = 41	5 = 4
60 = 49	6 = 5
70 = 57	7 = 6
80 = 65	8 = 6
90 = 73	9 = 7

47 = 38

Anti Log Excesses

10 = 12	1 = 1
20 = 25	2 = 2
30 = 37	3 = 4
40 = 49	4 = 5
50 = 62	5 = 6
60 = 74	6 = 7
70 = 86	7 = 9
	8 = 10
	9 = 11

38 = 47

5360038	·7291678
131	754
224	830
318	905
411	981
504	·7292056
597	132
691	207
784	283
877	359
971	434
5361064	510
157	585
250	661
344	736
437	812
530	887
623	963
717	·7293038
810	114
903	189
996	265
5362090	340
183	416
276	492
369	567
463	643
556	718
649	794
742	869
836	945
929	·7294020
5363022	096
115	171
208	247
302	322
395	398

5363488	·7294473
581	549
675	624
768	700
861	775
954	850
5364047	926
141	·7295001
234	077
327	152
420	228
514	303
607	379
700	454
793	530
886	605
980	680
5365073	756
166	831
259	907
352	982
445	·7296058
539	133
632	208
725	284
818	359
911	435
5366005	510
098	586
191	661
284	736
377	812
470	887
564	963
657	·7297038
750	113
843	189
936	264
5367029	339
123	415
216	490
309	566
402	641
495	716
588	792
682	867
775	943
868	·7298018
961	093
5368054	169
147	244
240	319
334	395
427	470
520	545
613	621
706	696
799	771
892	847
986	922
5369079	997
172	·7299073
265	148
358	223
451	299

5369544	·7299374
637	449
730	525
824	600
917	675

Excesses

10 = 8	1 = 1
20 = 16	2 = 2
30 = 24	3 = 2
40 = 32	4 = 3
50 = 40	5 = 4
60 = 49	6 = 5
70 = 57	7 = 6
80 = 65	8 = 6
90 = 73	9 = 7

47 = 38

Anti Log Excesses

10 = 12	1 = 1
20 = 25	2 = 2
30 = 37	3 = 4
40 = 49	4 = 5
50 = 62	5 = 6
60 = 74	6 = 7
70 = 86	7 = 9
	8 = 10
	9 = 11

38 = 47

5370010	·7299750
103	826
196	901
289	976
382	·7300052
475	127
568	202
662	278
755	353
848	428
941	503
5371034	579
127	654
220	729
313	804
406	880
499	955
592	·7301030
686	105
779	181
872	256
965	331
5372058	406
151	482
244	557
337	632
430	707
523	783
616	858
709	933
802	·7302008

5372895	·7302083
988	159
5373082	234
175	309
268	384
361	460
454	535
547	610
640	685
733	760
826	836
919	911
5374012	986
105	·7303061
198	136
291	212
384	287
477	362
570	437
663	512
756	587
849	663
942	738
5375035	813
128	888
221	963
314	·7304038
407	114
500	189
593	264
687	339
780	414
873	489
966	564
5376059	640
152	715
245	790
338	865
431	940
524	·7305015
617	090
710	166
803	241
896	316
989	391
5377081	466
174	541
267	616
360	691
453	766
546	841
639	917
732	992
825	·7306067
918	142
5378011	217
104	292
197	367
290	442
383	517
476	592
569	667
662	742
755	817
848	893

5378941	·7306968
5379034	·7307043
127	118
220	193
313	268
406	343
499	418
592	493
685	568
777	643
870	718
963	793

Excesses

10 = 8	1 = 1
20 = 16	2 = 2
30 = 24	3 = 2
40 = 32	4 = 3
50 = 40	5 = 4
60 = 48	6 = 5
70 = 56	7 = 6
80 = 65	8 = 6
90 = 73	9 = 7

46 = 37

Anti Log Excesses

10 = 12	1 = 1
20 = 25	2 = 2
30 = 37	3 = 4
40 = 50	4 = 5
50 = 62	5 = 6
60 = 74	6 = 7
70 = 87	7 = 9
	8 = 10
	9 = 11

37 = 46

5380056	·7307868
149	943
242	·7308018
335	093
428	168
521	243
614	318
707	393
800	468
893	543
986	618
5381078	693
171	768
264	843
357	918
450	993
543	·7309068
636	143
729	218
822	293
915	368
5382008	443
100	518
193	593

5382286	·7309668
379	743
472	818
565	893
658	967
751	·7310042
844	117
937	192
5383029	267
122	342
215	417
308	492
401	567
494	642
587	717
680	792
772	867
865	942
958	·7311016
5384051	091
144	166
237	241
330	316
422	391
515	466
608	541
701	616
794	691
887	765
980	840
5385072	915
165	990
258	·7312065
351	140
444	215
537	290
630	364
722	439
815	514
908	589
5386001	664
094	739
187	814
279	888
372	963
465	·7313038
558	113
651	188
743	263
836	337
929	412
5387022	487
115	562
208	637
300	712
393	786
486	861
579	936
672	·7314011
764	086
857	160
950	235
5388043	310
136	385
228	460

5388321	·7314534
414	609
507	684
600	759
692	834
785	908
878	983
971	·7315058
5389063	133
156	207
249	282
342	357
435	432
527	506
620	581
713	656
806	731
898	805
991	880

Excesses

10 = 8	1 = 1
20 = 16	2 = 2
30 = 24	3 = 2
40 = 32	4 = 3
50 = 40	5 = 4
60 = 48	6 = 5
70 = 56	7 = 6
80 = 65	8 = 6
90 = 73	9 = 7

46 = 37

Anti Log Excesses

10 = 12	1 = 1
20 = 25	2 = 2
30 = 37	3 = 4
40 = 50	4 = 5
50 = 62	5 = 6
60 = 74	6 = 7
70 = 87	7 = 9
	8 = 10
	9 = 11

37 = 46

5390084	·7315955
177	·7316030
270	104
362	179
455	254
548	329
641	403
733	478
826	553
919	628
5391012	702
104	777
197	852
290	926
383	·7317001
475	076
568	150

5391661	·7317225
753	300
846	375
939	449
5392032	524
124	599
217	673
310	748
403	823
495	897
588	972
681	·7318047
773	121
866	196
959	271
5393052	345
144	420
237	495
330	569
422	644
515	719
608	793
701	868
793	942
886	·7319017
979	092
5394071	166
164	241
257	316
349	390
442	465
535	540
627	614
720	689
813	763
906	838
998	913
5395091	987
184	·7320062
276	136
369	211
462	286
554	360
647	435
740	509
832	584
925	659
5396018	733
110	808
203	882
296	957
388	·7321031
481	106
573	181
666	255
759	330
851	404
944	479
5397037	553
129	628
222	702
315	777
407	851
500	926
593	·7322001

5397685	·7322075
778	150
870	224
963	299
5398056	373
148	448
241	522
334	597
426	671
519	746
611	820
704	895
797	969
889	·7323044
982	118
5399074	193
167	267
260	342
352	416
445	491
538	565
630	640
723	714
815	789
908	863

Excesses

10 = 8	1 = 1
20 = 16	2 = 2
30 = 24	3 = 2
40 = 32	4 = 3
50 = 40	5 = 4
60 = 48	6 = 5
70 = 56	7 = 6
80 = 64	8 = 6
90 = 72	9 = 7

46 = 37

Anti Log Excesses

10 = 12	1 = 1
20 = 25	2 = 2
30 = 37	3 = 4
40 = 50	4 = 5
50 = 62	5 = 6
60 = 75	6 = 7
70 = 87	7 = 9
	8 = 10
	9 = 11

37 = 46

5400000	·7323938
093	·7324012
186	086
278	161
371	236
463	310
556	385
649	459
741	533
834	608
926	682

5401019	·7324757
111	831
204	906
297	980
389	·7325055
482	129
574	203
667	278
759	352
852	427
945	501
5402037	575
130	650
222	724
315	799
407	873
500	947
592	·7326022
685	096
778	171
870	245
963	319
5403055	394
148	468
240	543
333	617
425	691
518	766
610	840
703	914
795	989
888	·7327063
981	138
5404073	212
166	286
258	361
351	435
443	509
536	584
628	658
721	732
813	807
906	881
998	955
5405091	·7328030
183	104
276	178
368	253
461	327
553	401
646	475
738	550
831	624
923	698
5406016	773
108	847
201	921
293	996
386	·7329070
478	144
571	219
663	293
756	367
848	441
940	516

5407033	·7329590
125	664
218	738
310	813
403	887
495	961
588	·7330036
680	110
773	184
865	258
958	333
5408050	407
143	481
235	555
327	630
420	704
512	778
605	852
697	926
790	·7331001
882	075
975	149
5409067	223
159	298
252	372
344	446
437	520
529	594
622	669
714	743
806	817
899	891
991	965

Excesses

10 = 8	1 = 1
20 = 16	2 = 2
30 = 24	3 = 2
40 = 32	4 = 3
50 = 40	5 = 4
60 = 48	6 = 5
70 = 56	7 = 6
80 = 64	8 = 6
90 = 72	9 = 7

46 = 37

Anti Log Excesses

10 = 12	1 = 1
20 = 25	2 = 2
30 = 37	3 = 4
40 = 50	4 = 5
50 = 62	5 = 6
60 = 75	6 = 7
70 = 87	7 = 9
	8 = 10
	9 = 11

37 = 46

5410084	·7332040
176	114
269	188
5410361	·7332262
453	336
546	410
638	485
731	559
823	633
915	707
5411008	781
100	856
193	930
285	·7333004
377	078
470	152
562	226
655	300
747	375
839	449
932	523
5412024	597
117	671
209	745
301	819
394	894
486	968
578	·7334042
671	116
763	190
856	264
948	338
5413040	412
133	487
225	561
317	635
410	709
502	783
595	857
687	931
779	·7335005
872	079
964	153
5414056	227
149	302
241	376
333	450
426	524
518	598
610	672
703	746
795	820
887	894
980	968
5415072	·7336042
164	116
257	190
349	264
441	338
534	412
626	486
718	560
811	635
903	709
995	783
5416088	857
180	931
272	·7337005
5416365	·7337079
457	153
549	227
642	301
734	375
826	449
918	523
5417011	597
103	671
195	745
288	819
380	893
472	967
565	·7338041
657	115
749	189
841	263
934	337
5418026	411
118	485
211	559
303	633
395	706
487	780
580	854
672	928
764	·7339002
856	076
949	150
5419041	224
133	298
226	372
318	446
410	520
502	594
595	668
687	742
779	816
871	889
964	963

Excesses

10 = 8	1 = 1
20 = 16	2 = 2
30 = 24	3 = 2
40 = 32	4 = 3
50 = 40	5 = 4
60 = 48	6 = 5
70 = 56	7 = 6
80 = 64	8 = 6
90 = 72	9 = 7

46 = 37

Anti Log Excesses

10 = 12	1 = 1
20 = 25	2 = 2
30 = 37	3 = 4
40 = 50	4 = 5
50 = 62	5 = 6
60 = 75	6 = 7
70 = 87	7 = 9
	8 = 10
	9 = 11

37 = 46

5420056	·7340037
148	111
240	185
333	259
425	333
517	407
609	481
702	555
794	629
886	702
978	776
5421071	850
163	924
255	998
347	·7341072
439	146
532	220
624	294
716	367
808	441
901	515
993	589
5422085	663
177	737
269	811
362	884
454	958
546	·7342032
638	106
730	180
823	254
915	328
5423007	401
099	475
191	549
284	623
376	697
468	771
560	844
652	918
745	992
837	·7343066
929	140
5424021	213
113	287
206	361
298	435
390	509
482	582
574	656
666	730
5424759	·7343804
851	878
943	951
5425035	·7344025
127	099
219	173
312	247
404	320
496	394
588	468
680	542
772	615
864	689
957	763
5426049	837
141	910
233	984
325	·7345058
417	132
510	205
602	279
694	353
786	427
878	500
970	574
5427062	648
154	722
247	795
339	869
431	943
523	·7346016
615	090
707	164
799	238
891	311
984	385
5428076	459
168	532
260	606
352	680
444	753
536	827
628	901
720	974
813	·7347048
905	122
997	196
5429089	269
181	343
273	417
365	490
457	564
549	637
641	711
733	785
826	858
918	932

Excesses

10 = 8	1 = 1
20 = 16	2 = 2
30 = 24	3 = 2
40 = 32	4 = 3
50 = 40	5 = 4
60 = 48	6 = 5
70 = 56	7 = 6
80 = 64	8 = 6
90 = 72	9 = 7

46 = 37

Anti Log Excesses

10 = 12	1 = 1
20 = 25	2 = 2
30 = 37	3 = 4
40 = 50	4 = 5
50 = 62	5 = 6
60 = 75	6 = 8
70 = 87	7 = 9
	8 = 10
	9 = 11

37 = 46

5430010	·7348006
102	079
194	153
286	227
378	300
470	374
562	448
654	521
746	595
838	668
930	742
5431023	816
115	889
207	963
299	·7349037
391	110
483	184
575	257
667	331
759	405
851	478
943	552
5432035	625
127	699
219	773
311	846
403	920
495	993
587	·7350067
679	140
771	214
863	288
955	361
5433048	435
140	508
232	582
324	655

5433416	·7350729
508	803
600	876
692	950
784	·7351023
876	097
968	170
5434060	244
152	317
244	391
336	464
428	538
520	611
612	685
704	758
796	832
888	905
980	979
5435072	·7352052
164	126
256	199
348	273
440	346
532	420
624	493
716	567
808	640
900	714
992	787
5436084	861
176	934
268	·7353008
360	081
452	155
544	228
635	302
727	375
819	449
911	522
5437003	596
095	669
187	743
279	816
371	889
463	963
555	·7354036
647	110
739	183
831	257
923	330
5438015	404
107	477
199	550
291	624
383	697
475	771
566	844
658	917
750	991
842	·7355064
934	138
5439026	211
118	285
210	358
302	431

5439394	·7355505
486	578
578	652
670	725
762	798
853	872
945	945

Excesses

10 = 8	1 = 1
20 = 16	2 = 2
30 = 24	3 = 2
40 = 32	4 = 3
50 = 40	5 = 4
60 = 48	6 = 5
70 = 56	7 = 6
80 = 64	8 = 6
90 = 72	9 = 7

46 = 37

Anti Log Excesses

10 = 13	1 = 1
20 = 25	2 = 3
30 = 38	3 = 4
40 = 50	4 = 5
50 = 63	5 = 6
60 = 75	6 = 8
70 = 88	7 = 9
	8 = 10
	9 = 11

37 = 46

5440037	·7356018
129	092
221	165
313	238
405	312
497	385
589	459
681	532
773	605
864	679
956	752
5441048	825
140	899
232	972
324	·7357045
416	119
508	192
600	265
691	339
783	412
875	485
967	559
5442059	632
151	705
243	779
335	852
426	925
518	999
610	·7358072

5442702	·7358145
794	219
886	292
978	365
5443069	438
161	512
253	585
345	658
437	731
529	805
621	878
712	951
804	·7359025
896	098
988	171
5444080	244
172	318
264	391
355	464
447	537
539	611
631	684
723	757
815	830
906	904
998	977
5445090	·7360050
182	123
274	197
366	270
457	343
549	416
641	490
733	563
825	636
916	709
5446008	783
100	856
192	929
284	·7361002
375	075
467	149
559	222
651	295
743	368
834	441
926	514
5447018	588
110	661
202	734
293	807
385	880
477	954
569	·7362027
661	100
752	173
844	246
936	319
5448028	393
119	466
211	539
303	612
395	685
487	758
578	831

5448670	·7362905
762	978
854	·7363051
945	124
5449037	197
129	270
221	344
312	417
404	490
496	563
588	636
679	709
771	783
863	856
955	929

Excesses

10 = 8	1 = 1
20 = 16	2 = 2
30 = 24	3 = 2
40 = 32	4 = 3
50 = 40	5 = 4
60 = 48	6 = 5
70 = 56	7 = 6
80 = 64	8 = 6
90 = 72	9 = 7

46 = 37

Anti Log Excesses

10 = 13	1 = 1
20 = 25	2 = 3
30 = 38	3 = 4
40 = 50	4 = 5
50 = 63	5 = 6
60 = 75	6 = 8
70 = 88	7 = 9
	8 = 10
	9 = 11

37 = 46

5450046	·7364002
138	075
230	148
322	221
413	294
505	367
597	440
689	514
780	587
872	660
964	733
5451055	806
147	879
239	952
331	·7365025
422	098
514	171
606	244
697	317
789	390
881	464

5451973	·7365537
5452064	610
156	683
248	756
339	829
431	902
523	975
615	·7366048
706	121
798	194
890	267
981	340
5453073	413
165	486
256	559
348	632
440	705
531	778
623	851
715	924
807	997
898	·7367070
990	143
5454082	216
173	289
265	362
357	435
448	508
540	581
632	654
723	727
815	800
907	873
998	946
5455090	·7368019
182	092
273	165
365	238
456	311
548	384
640	457
731	530
823	603
915	676
5456006	748
098	821
190	894
281	967
373	·7369040
465	113
556	186
648	259
739	332
831	405
923	478
5457014	551
106	624
198	697
289	769
381	842
472	915
564	988
656	·7370061
747	134
839	207

5457931	·7370280
5458022	353
114	426
205	498
297	571
389	644
480	717
572	790
663	863
755	936
847	·7371009
938	081
5459030	154
121	227
213	300
304	373
396	446
488	519
579	591
671	664
762	737
854	810
946	883

Excesses

10 = 8	1 = 1
20 = 16	2 = 2
30 = 24	3 = 2
40 = 32	4 = 3
50 = 40	5 = 4
60 = 48	6 = 5
70 = 56	7 = 6
80 = 64	8 = 6
90 = 72	9 = 7

46 = 36

Anti Log Excesses

10 = 13	1 = 1
20 = 25	2 = 3
30 = 38	3 = 4
40 = 50	4 = 5
50 = 63	5 = 6
60 = 75	6 = 8
70 = 88	7 = 9
	8 = 10
	9 = 11

36 = 46

5460037	·7371956
129	·7372029
220	101
312	174
403	247
495	320
587	393
678	466
770	538
861	611
953	684
5461044	757
136	830

5461227	·7372902
319	975
411	·7373048
502	121
594	194
685	266
777	339
868	412
960	485
5462051	558
143	630
234	703
326	776
418	849
509	921
601	994
692	·7374067
784	140
875	213
967	285
5463058	358
150	431
241	504
333	576
424	649
516	722
607	795
699	867
790	940
882	·7375013
973	086
5464065	158
156	231
248	304
339	376
431	449
522	522
614	595
705	667
797	740
888	813
980	885
5465071	958
163	·7376031
254	104
346	176
437	249
529	322
620	394
712	467
803	540
895	612
986	685
5466078	758
169	830
261	903
352	976
444	·7377048
535	121
627	194
718	266
809	339
901	412
992	484
5467084	557

5467175	·7377630
267	702
358	775
450	848
541	920
633	993
724	·7378066
815	138
907	211
998	283
5468090	356
181	429
273	501
364	574
455	647
547	719
638	792
730	864
821	937
913	·7379010
5469004	082
095	155
187	227
278	300
370	373
461	445
553	518
644	590
735	663
827	736
918	808

Excesses

10 = 8	1 = 1
20 = 16	2 = 2
30 = 24	3 = 2
40 = 32	4 = 3
50 = 40	5 = 4
60 = 48	6 = 5
70 = 56	7 = 6
80 = 64	8 = 6
90 = 72	9 = 7

46 = 36

Anti Log Excesses

10 = 13	1 = 1
20 = 25	2 = 3
30 = 38	3 = 4
40 = 50	4 = 5
50 = 63	5 = 6
60 = 75	6 = 8
70 = 88	7 = 9
	8 = 10
	9 = 11

36 = 46

5470010	·7379881
101	953
192	·7380026
284	098
375	171

5470467	·7380244
558	316
649	389
741	461
832	534
924	606
5471015	679
106	751
198	824
289	896
381	969
472	·7381042
563	114
655	187
746	259
837	332
929	404
5472020	477
112	549
203	622
294	694
386	767
477	839
568	912
660	984
751	·7382057
843	129
934	202
5473025	274
117	347
208	419
299	492
391	564
482	637
573	709
665	782
756	854
847	927
939	999
·5474030	·7383072
121	144
213	217
304	289
395	361
487	434
578	506
669	579
761	651
852	724
943	796
5475035	869
126	941
217	·7384013
309	086
400	158
491	231
583	303
674	376
765	448
857	520
948	593
5476039	665
131	738
222	810
313	883

5476404	·7384955
496	·7385027
587	100
678	172
770	245
861	317
952	389
5477043	462
135	534
226	606
317	679
409	751
500	824
591	896
682	968
774	·7386041
865	113
956	185
5478048	258
139	330
230	402
321	475
413	547
504	619
595	692
686	764
778	837
869	909
960	981
5479052	·7387054
143	126
234	198
325	271
417	343
508	415
599	487
690	560
782	632
873	704
964	777

Excesses

10 = 8	1 = 1
20 = 16	2 = 2
30 = 24	3 = 2
40 = 32	4 = 3
50 = 40	5 = 4
60 = 48	6 = 5
70 = 56	7 = 6
80 = 63	8 = 6
90 = 71	9 = 7

46 = 36

Anti Log Excesses

10 = 13	1 = 1
20 = 25	2 = 3
30 = 38	3 = 4
40 = 50	4 = 5
50 = 63	5 = 6
60 = 76	6 = 8
70 = 88	7 = 9
	8 = 10
	9 = 11

36 = 46

5480055	·7387849
146	921
238	994
329	·7388066
420	138
511	211
603	283
694	355
785	427
876	500
968	572
5481059	644
150	717
241	789
332	861
424	933
515	·7389006
606	078
697	150
789	222
880	295
971	367
5482062	439
153	511
245	584
336	656
427	728
518	800
609	873
701	945
792	·7390017
883	089
974	162
5483065	234
157	306
248	378
339	451
430	523
521	595
612	667
704	739
795	812
886	884
977	956
5484068	·7391028
160	100
251	173
342	245
433	317
524	389
615	461

5484707	·7391534
798	606
889	678
980	750
5485071	822
162	894
253	967
345	·7392039
436	111
527	183
618	255
709	327
800	400
891	472
983	544
5486074	616
165	688
256	760
347	833
438	905
529	977
621	·7393049
712	121
803	193
894	265
985	338
5487076	410
167	482
258	554
350	626
441	698
532	770
623	842
714	914
805	987
896	·7394059
987	131
5488078	203
170	275
261	347
352	419
443	491
534	563
625	635
716	707
807	780
898	852
989	924
5489081	996
172	·7395068
263	140
354	212
445	284
536	356
627	428
718	500
809	572
900	644
991	716

Excesses

10 = 8	1 = 1
20 = 16	2 = 2
30 = 24	3 = 2
40 = 32	4 = 3
50 = 40	5 = 4
60 = 47	6 = 5
70 = 55	7 = 6
80 = 63	8 = 6
90 = 71	9 = 7

46 = 36

Anti Log Excesses

10 = 13	1 = 1
20 = 25	2 = 3
30 = 38	3 = 4
40 = 51	4 = 5
50 = 63	5 = 6
60 = 76	6 = 8
70 = 88	7 = 9
	8 = 10
	9 = 11

36 = 46

5490082	·7395788
174	860
265	932
356	·7396004
447	076
538	149
629	221
720	293
811	365
902	437
993	509
5491084	581
175	653
266	725
357	797
448	869
539	941
630	·7397013
722	085
813	157
904	229
995	301
5492086	373
177	445
268	517
359	589
450	661
541	733
632	805
723	877
814	949
905	·7398021
996	092
5493087	164
178	236
269	308
360	380

5493451	·7398452
542	524
633	596
724	668
815	740
906	812
997	884
5494088	956
179	·7399028
270	100
361	172
452	244
543	316
634	388
725	459
816	531
907	603
998	675
5495089	747
180	819
271	891
362	963
453	·7400035
544	107
635	179
726	250
817	322
908	394
999	466
5496090	538
181	610
272	682
363	754
454	826
545	897
636	969
727	·7401041
818	113
909	185
5497000	257
091	329
182	401
273	472
364	544
454	616
545	688
636	760
727	832
818	904
909	975
5498000	·7402047
091	119
182	191
273	263
364	335
455	406
546	478
637	550
728	622
819	694
909	766
5499000	837
091	909
182	981
273	·7403053

5499364	·7403125
455	196
546	268
637	340
728	412
819	484
910	555

Excesses

10 = 8	1 = 1
20 = 16	2 = 2
30 = 24	3 = 2
40 = 32	4 = 3
50 = 40	5 = 4
60 = 47	6 = 5
70 = 55	7 = 6
80 = 63	8 = 6
90 = 71	9 = 7

46 = 36

Anti Log Excesses

10 = 13	1 = 1
20 = 25	2 = 3
30 = 38	3 = 4
40 = 51	4 = 5
50 = 63	5 = 6
60 = 76	6 = 8
70 = 89	7 = 9
	8 = 10
	9 = 11

36 = 46

5500000	·7403627
091	699
182	771
273	842
364	914
455	986
546	·7404058
637	130
728	201
819	273
910	345
5501000	417
091	488
182	560
273	632
364	704
455	775
546	847
637	919
728	991
818	·7405062
909	134
5502000	206
091	277
182	349
273	421
364	493
454	564
545	636

5502636	·7405708
727	780
818	851
909	923
5503000	995
091	·7406066
181	138
272	210
363	281
454	353
545	425
636	497
727	568
817	640
908	712
999	783
5504090	855
181	927
272	998
362	·7407070
453	142
544	213
635	285
726	357
817	428
907	500
998	572
5505089	643
180	715
271	787
362	858
452	930
543	·7408001
634	073
725	145
816	216
906	288
997	360
5506088	431
179	503
270	575
360	646
451	718
542	789
633	861
724	933
814	·7409004
905	076
996	147
5507087	219
178	291
268	362
359	434
450	505
541	577
632	649
722	720
813	792
904	863
995	935
5508085	·7410006
176	078
267	150
358	221
449	293

5508539	·7410364
630	436
721	507
812	579
902	651
993	722
5509084	794
175	865
265	937
356	·7411008
447	080
538	151
628	223
719	294
810	366
901	437
991	509

Excesses

10 = 8	1 = 1
20 = 16	2 = 2
30 = 24	3 = 2
40 = 32	4 = 3
50 = 39	5 = 4
60 = 47	6 = 5
70 = 55	7 = 6
80 = 63	8 = 6
90 = 71	9 = 7

45 = 36

Anti Log Excesses

10 = 13	1 = 1
20 = 25	2 = 3
30 = 38	3 = 4
40 = 51	4 = 5
50 = 63	5 = 6
60 = 76	6 = 8
70 = 89	7 = 9
	8 = 10
	9 = 11

36 = 45

5510082	·7411581
173	652
264	724
354	795
445	867
536	938
627	·7412010
717	081
808	153
899	224
990	296
5511080	367
171	439
262	510
352	582
443	653
534	725
625	796
715	868

5511806	·7412939
897	·7413011
987	082
5512078	153
169	225
260	296
350	368
441	439
532	511
622	582
713	654
804	725
894	797
985	868
5513076	940
167	·7414011
257	082
348	154
439	225
529	297
620	368
711	440
801	511
892	582
983	654
5514073	725
164	797
255	868
345	939
436	·7415011
527	082
617	154
708	225
799	296
889	368
980	439
5515071	511
161	582
252	653
343	725
433	796
524	868
615	939
705	·7416010
796	082
887	153
977	224
5516068	296
159	367
249	438
340	510
431	581
521	653
612	724
702	795
793	867
884	938
974	·7417009
5517065	081
156	152
246	223
337	295
427	366
518	437
609	509

5517699	·7417580
790	651
881	723
971	794
5518062	865
152	937
243	·7418008
334	079
424	151
515	222
605	293
696	364
787	436
877	507
968	578
5519058	650
149	721
240	792
330	863
421	935
511	·7419006
602	077
693	149
783	220
874	291
964	362

Excesses

10 = 8	1 = 1
20 = 16	2 = 2
30 = 24	3 = 2
40 = 32	4 = 3
50 = 39	5 = 4
60 = 47	6 = 5
70 = 55	7 = 6
80 = 63	8 = 6
90 = 71	9 = 7

45 = 36

Anti Log Excesses

10 = 13	1 = 1
20 = 25	2 = 3
30 = 38	3 = 4
40 = 51	4 = 5
50 = 64	5 = 6
60 = 76	6 = 8
70 = 89	7 = 9
	8 = 10
	9 = 11

36 = 45

5520055	·7419434
145	505
236	576
327	647
417	719
508	790
598	861
689	932
779	·7420004
870	075

Number	Log
5520961	·7420146
5521051	217
142	289
232	360
323	431
413	502
504	574
594	645
685	716
776	787
866	859
957	930
5522047	·7421001
138	072
228	143
319	215
409	286
500	357
590	428
681	499
772	571
862	642
953	713
5523043	784
134	855
224	927
315	998
405	·7422069
496	140
586	211
677	282
767	354
858	425
948	496
5524039	567
129	638
220	709
310	781
401	852
491	923
582	994
672	·7423065
763	136
853	207
944	279
5525034	350
125	421
215	492
306	563
396	634
487	705
577	777
668	848
758	919
849	990
939	·7424061
5526030	132
120	203
211	274
301	345
392	416
482	488
573	559
663	630
754	701

Number	Log
5526844	·7424772
935	843
5527025	914
115	985
206	·7425056
296	127
387	198
477	269
568	341
658	412
749	483
839	554
930	625
5528020	696
110	767
201	838
291	909
382	980
472	·7426051
563	122
653	193
744	264
834	335
924	406
5529015	477
105	548
196	619
286	690
377	761
467	832
557	904
648	975
738	·7427046
829	117
919	188

Excesses

10 = 8	1 = 1
20 = 16	2 = 2
30 = 24	3 = 2
40 = 31	4 = 3
50 = 39	5 = 4
60 = 47	6 = 5
70 = 55	7 = 6
80 = 63	8 = 6
90 = 71	9 = 7

45 = 36

Anti Log Excesses

10 = 13	1 = 1
20 = 25	2 = 3
30 = 38	3 = 4
40 = 51	4 = 5
50 = 64	5 = 6
60 = 76	6 = 8
70 = 89	7 = 9
	8 = 10
	9 = 11

36 = 45

Number	Log
5530010	·7427259

Number	Log
5530100	·7427330
190	401
281	472
371	543
462	614
552	685
642	756
733	827
823	898
914	969
5531004	·7428040
094	110
185	181
275	252
366	323
456	394
546	465
637	536
727	607
818	678
908	749
998	820
5532089	891
179	962
269	·7429033
360	104
450	175
541	246
631	317
721	388
812	459
902	530
992	601
5533083	671
173	742
264	813
354	884
444	955
535	·7430026
625	097
715	168
806	239
896	310
986	381
5534077	452
167	522
257	593
348	664
438	735
528	806
619	877
709	948
800	·7431019
890	090
980	160
5535071	231
161	302
251	373
342	444
432	515
522	586
612	657
703	727
793	798
883	869

Number	Log
5535974	·7431940
5536064	·7432011
154	082
245	153
335	223
425	294
516	365
606	436
696	507
787	578
877	648
967	719
5537057	790
148	861
238	932
328	·7433003
419	073
509	144
599	215
690	286
780	357
870	427
960	498
5538051	569
141	640
231	711
322	781
412	852
502	923
592	994
683	·7434065
773	135
863	206
953	277
5539044	348
134	418
224	489
315	560
405	631
495	701
585	772
676	843
766	914
856	985
946	·7435055

Excesses

10 = 8	1 = 1
20 = 16	2 = 2
30 = 24	3 = 2
40 = 31	4 = 3
50 = 39	5 = 4
60 = 47	6 = 5
70 = 55	7 = 5
80 = 63	8 = 6
90 = 71	9 = 7

45 = 35

Anti Log Excesses

10 = 13	1 = 1
20 = 25	2 = 3
30 = 38	3 = 4
40 = 51	4 = 5
50 = 64	5 = 6
60 = 76	6 = 8
70 = 89	7 = 9
	8 = 10
	9 = 11

35 = 45

Number	Log
5540037	·7435126
127	197
217	267
307	338
398	409
488	480
578	550
668	621
759	692
849	763
939	833
5541029	904
120	975
210	·7436045
300	116
390	187
480	258
571	328
661	399
751	470
841	541
932	611
5542022	682
112	753
202	823
292	894
383	965
473	·7437035
563	106
653	177
744	247
834	318
924	389
5543014	459
104	530
195	601
285	671
375	742
465	813
555	883
646	954
736	·7438025
826	095
916	166
5544006	237
096	307
187	378
277	449
367	519
457	590
547	661

5544638	·7438731
728	802
818	872
908	943
998	·7439014
5545088	084
179	155
269	226
359	296
449	367
539	437
629	508
720	579
810	649
900	720
990	790
5546080	861
170	932
261	·7440002
351	073
441	143
531	214
621	285
711	355
801	426
892	496
982	567
5547072	638
162	708
252	779
342	849
432	920
522	990
613	·7441061
703	131
793	202
883	273
973	343
5548063	414
153	485
243	555
334	626
424	696
514	767
604	837
694	908
784	978
874	·7442049
964	119
5549054	190
145	260
235	331
325	401
415	472
505	542
595	613
685	683
775	754
865	824
955	895

Excesses

10 = 8	1 = 1
20 = 16	2 = 2
30 = 23	3 = 2
40 = 31	4 = 3
50 = 39	5 = 4
60 = 47	6 = 5
70 = 55	7 = 5
80 = 63	8 = 6
90 = 70	9 = 7

45 = 35

Anti Log Excesses

10 = 13	1 = 1
20 = 26	2 = 3
30 = 38	3 = 4
40 = 51	4 = 5
50 = 64	5 = 6
60 = 77	6 = 8
70 = 89	7 = 9
	8 = 10
	9 = 11

35 = 45

5550046	·7442965
136	·7443036
226	106
316	177
406	247
496	318
586	388
676	459
766	529
856	600
946	670
5551036	741
127	811
217	882
307	952
397	·7444023
487	093
577	163
667	234
757	304
847	375
937	445
5552027	516
117	586
207	657
297	727
387	797
477	868
567	938
658	·7445009
748	079
838	150
928	220
5553018	290
108	361
198	431
288	502

5553378	·7445572
468	642
558	713
648	783
738	854
828	924
918	994
5554008	·7446065
098	135
188	206
278	276
368	346
458	417
548	487
638	558
728	628
818	698
908	769
998	839
5555088	909
178	980
268	·7447050
358	121
448	191
538	261
628	332
718	402
808	472
898	543
988	613
5556078	683
168	754
258	824
348	894
438	965
528	·7448035
618	105
708	176
798	246
888	316
978	387
5557068	457
158	527
248	598
338	668
428	738
518	809
608	879
698	949
788	·7449019
878	090
968	160
5558058	230
148	301
238	371
328	441
418	511
507	582
597	652
687	722
777	793
867	863
957	933
5559047	·7450003
137	074

5559227	·7450144
317	214
407	284
497	355
587	425
677	495
767	565
857	636
947	706

Excesses

10 = 8	1 = 1
20 = 16	2 = 2
30 = 23	3 = 2
40 = 31	4 = 3
50 = 39	5 = 4
60 = 47	6 = 5
70 = 55	7 = 5
80 = 63	8 = 6
90 = 70	9 = 7

45 = 35

Anti Log Excesses

10 = 13	1 = 1
20 = 26	2 = 3
30 = 38	3 = 4
40 = 51	4 = 5
50 = 64	5 = 6
60 = 77	6 = 8
70 = 90	7 = 9
	8 = 10
	9 = 12

35 = 45

5560036	·7450776
126	846
216	917
306	987
396	·7451057
486	127
576	198
666	268
756	338
846	408
936	479
5561026	549
115	619
205	689
295	759
385	830
475	900
565	970
655	·7452040
745	110
835	181
925	251
5562015	321
104	391
194	461
284	532
374	602

5562464	·7452672
554	742
644	812
734	882
824	953
913	·7453023
5563003	093
093	163
183	233
273	303
363	374
453	444
543	514
632	584
722	654
812	724
902	795
992	865
5564082	935
172	·7454005
261	075
351	145
441	215
531	285
621	356
711	426
801	496
890	566
980	636
5565070	706
160	776
250	846
340	917
430	987
519	·7455057
609	127
699	197
789	267
879	337
969	407
5566058	477
148	547
238	618
328	688
418	758
507	828
597	898
687	968
777	·7456038
867	108
957	178
5567046	248
136	318
226	388
316	458
406	528
495	598
585	669
675	739
765	809
855	879
944	949
5568034	·7457019
124	089
214	159

5568304	·7457229
393	299
483	369
573	439
663	509
753	579
842	649
932	719
5569022	789
112	859
202	929
291	999
381	·7458069
471	139
561	209
650	279
740	349
830	419
920	489

Excesses

10 = 8	1 = 1
20 = 16	2 = 2
30 = 23	3 = 2
40 = 31	4 = 3
50 = 39	5 = 4
60 = 47	6 = 5
70 = 55	7 = 5
80 = 62	8 = 6
	9 = 7

45 = 35

Anti Log Excesses

10 = 13	1 = 1
20 = 26	2 = 3
30 = 38	3 = 4
40 = 51	4 = 5
50 = 64	5 = 6
60 = 77	6 = 8
70 = 90	7 = 9
	8 = 10
	9 = 12

35 = 45

5570009	·7458559
099	629
189	699
279	769
369	839
458	909
548	979
638	·7459049
728	119
817	189
907	259
997	329
5571087	399
176	469
266	539
356	609
446	679

5571535	·7459749
625	819
715	889
805	958
894	·7460028
984	098
5572074	168
163	238
253	308
343	378
433	448
522	518
612	588
702	658
792	728
881	798
971	868
5573061	938
150	·7461008
240	077
330	147
420	217
509	287
599	357
689	427
778	497
868	567
958	637
5574048	707
137	776
227	846
317	916
406	986
496	·7462056
586	126
675	196
765	266
855	335
944	405
5575034	475
124	545
214	615
303	685
393	755
483	824
572	894
662	964
752	·7463034
841	104
931	174
5576021	244
110	313
200	383
290	453
379	523
469	593
559	663
648	732
738	802
828	872
917	942
5577007	·7464012
097	081
186	151
276	221

5577365	·7464291
455	361
545	430
634	500
724	570
814	640
903	710
993	779
5578083	849
172	919
262	989
352	·7465059
441	128
531	198
620	268
710	338
800	407
889	477
979	547
5579069	617
158	686
248	756
337	826
427	896
517	966
606	·7466035
696	105
785	175
875	245
965	314

Excesses

10 = 8	1 = 1
20 = 16	2 = 2
30 = 23	3 = 2
40 = 31	4 = 3
50 = 39	5 = 4
60 = 47	6 = 5
70 = 55	7 = 5
80 = 62	8 = 6
	9 = 7

45 = 35

Anti Log Excesses

10 = 13	1 = 1
20 = 26	2 = 3
30 = 39	3 = 4
40 = 51	4 = 5
50 = 64	5 = 6
60 = 77	6 = 8
	7 = 9
	8 = 10
	9 = 12

35 = 45

5580054	·7466384
144	454
233	523
323	593
413	663
502	733

5580592	·7466802
681	872
771	942
861	·7467012
950	081
5581040	151
129	221
219	290
309	360
398	430
488	500
577	569
667	639
756	709
846	778
936	848
5582025	918
115	987
204	·7468057
294	127
384	196
473	266
563	336
652	405
742	475
831	545
921	615
5583010	684
100	754
190	823
279	893
369	963
458	·7469032
548	102
637	172
727	241
816	311
906	381
995	450
5584085	520
175	590
264	659
354	729
443	798
533	868
622	938
712	·7470007
801	077
891	147
980	216
5585070	286
159	355
249	425
338	495
428	564
517	634
607	704
697	773
786	843
876	912
965	982
5586055	·7471052
144	121
234	191
323	260

5586413	·7471330
502	399
592	469
681	539
771	608
860	678
950	747
5587039	817
129	886
218	956
308	·7472026
397	095
487	165
576	234
665	304
755	373
844	443
934	512
5588023	582
113	651
202	721
292	790
381	860
471	930
560	999
650	·7473069
739	138
829	208
918	277
5589008	347
097	416
186	486
276	555
365	625
455	694
544	764
634	833
723	903
813	972
902	·7474042
992	111

Excesses

10 = 8	1 = 1
20 = 16	2 = 2
30 = 23	3 = 2
40 = 31	4 = 3
50 = 39	5 = 4
60 = 47	6 = 5
70 = 54	7 = 5
80 = 62	8 = 6
	9 = 7

45 = 35

Anti Log Excesses

10 = 13	1 = 1
20 = 26	2 = 3
30 = 39	3 = 4
40 = 51	4 = 5
50 = 64	5 = 6
60 = 77	6 = 8
	7 = 9
	8 = 10
	9 = 12

35 = 45

5590081	·7474181
170	250
260	320
349	389
439	459
528	528
618	598
707	667
797	737
886	806
975	876
5591065	945
154	·7475015
244	084
333	153
423	223
512	292
601	362
691	431
780	501
870	570
959	639
5592048	709
138	778
227	848
317	917
406	987
495	·7476056
585	126
674	195
764	264
853	334
942	403
5593032	473
121	542
211	611
300	681
389	750
479	820
568	889
658	958
747	·7477028
836	097
926	167
5594015	236
105	305
194	375
283	444
373	514
462	583
551	652

5594641	·7477722
730	791
820	860
909	930
998	999
5595088	·7478069
177	138
266	207
356	277
445	346
534	415
624	485
713	554
802	624
892	693
981	762
5596071	832
160	901
249	970
339	·7479040
428	109
517	178
607	248
696	317
785	386
875	456
964	525
5597053	594
143	664
232	733
321	802
411	871
500	941
589	·7480010
679	079
768	149
857	218
947	287
5598036	357
125	426
215	495
304	565
393	634
482	703
572	772
661	842
750	911
840	980
929	·7481050
5599018	119
108	188
197	257
286	327
375	396
465	465
554	534
643	604
733	673
822	742
911	811

Excesses

10 = 8	1 = 1
20 = 16	2 = 2
30 = 23	3 = 2
40 = 31	4 = 3
50 = 39	5 = 4
60 = 47	6 = 5
70 = 54	7 = 5
80 = 62	8 = 6
	9 = 7

45 = 35

Anti Log Excesses

10 = 13	1 = 1
20 = 26	2 = 3
30 = 39	3 = 4
40 = 52	4 = 5
50 = 64	5 = 6
60 = 77	6 = 8
	7 = 9
	8 = 10
	9 = 12

35 = 45

5600000	·7481881
090	950
179	·7482019
268	088
358	158
447	227
536	296
625	365
715	434
804	504
893	573
983	642
5601072	711
161	781
250	850
340	919
429	988
518	·7483057
607	127
697	196
786	265
875	334
964	403
5602054	473
143	542
232	611
321	680
411	749
500	819
589	888
678	957
768	·7484026
857	095
946	164
5603035	234
125	303
214	372

5603303	·7484441
392	510
482	579
571	649
660	718
749	787
838	856
928	925
5604017	994
106	·7485063
195	133
285	202
374	271
463	340
552	409
641	478
731	547
820	616
909	686
998	755
5605087	824
177	893
266	962
355	·7486031
444	100
533	169
623	239
712	308
801	377
890	446
979	515
5606069	584
158	653
247	722
336	791
425	860
515	929
604	999
693	·7487068
782	137
871	206
960	275
5607050	344
139	413
228	482
317	551
406	620
495	689
585	758
674	827
763	896
852	965
941	·7488034
5608030	104
120	173
209	242
298	311
387	380
476	449
565	518
655	587
744	656
833	725
922	794
5609011	863

5609100	·7488932
189	·7489001
279	070
368	139
457	208
546	277
635	346
724	415
813	484
902	553
992	622

Excesses

10 = 8	1 = 1
20 = 15	2 = 2
30 = 23	3 = 2
40 = 31	4 = 3
50 = 39	5 = 4
60 = 46	6 = 5
70 = 54	7 = 5
80 = 62	8 = 6
	9 = 7

45 = 35

Anti Log Excesses

10 = 13	1 = 1
20 = 26	2 = 3
30 = 39	3 = 4
40 = 52	4 = 5
50 = 65	5 = 6
60 = 77	6 = 8
	7 = 9
	8 = 10
	9 = 12

35 = 45

5610081	·7489691
170	760
259	829
348	898
437	967
526	·7490036
615	105
705	174
794	243
883	312
972	381
5611061	450
150	519
239	588
328	657
417	726
507	795
596	864
685	933
774	·7491002
863	070
952	139
5612041	208
130	277
219	346

5612308	·7491415
397	484
487	553
576	622
665	691
754	760
843	829
932	898
5613021	967
110	·7492036
199	104
288	173
377	242
466	311
556	380
645	449
734	518
823	587
912	656
5614001	725
090	794
179	862
268	931
357	·7493000
446	069
535	138
624	207
713	276
802	345
891	414
980	482
5615070	551
159	620
248	689
337	758
426	827
515	896
604	964
693	·7494033
782	102
871	171
960	240
5616049	309
138	378
227	446
316	515
405	584
494	653
583	722
672	791
761	859
850	928
939	997
5617028	·7495066
117	135
206	204
295	272
384	341
473	410
562	479
651	548
740	616
829	685
918	754
5618007	823

5618096	·7495892
185	960
274	·7496029
363	098
452	167
541	236
630	304
719	373
808	442
897	511
986	580
5619075	648
164	717
253	786
342	855
431	923
520	992
609	·7497061
698	130
787	198
876	267
965	336

Excesses

10 = 8	1 = 1
20 = 15	2 = 2
30 = 23	3 = 2
40 = 31	4 = 3
50 = 39	5 = 4
60 = 46	6 = 5
70 = 54	7 = 5
80 = 62	8 = 6
	9 = 7

45 = 35

Anti Log Excesses

10 = 13	1 = 1
20 = 26	2 = 3
30 = 39	3 = 4
40 = 52	4 = 5
50 = 65	5 = 6
60 = 78	6 = 8
	7 = 9
	8 = 10
	9 = 12

35 = 45

5620054	·7497405
143	473
232	542
321	611
410	680
499	748
588	817
677	886
766	955
855	·7498023
943	092
5621032	161
121	229
210	298

5621299	·7498367
388	436
477	504
566	573
655	642
744	710
833	779
922	848
5622011	917
100	985
189	·7499054
278	123
367	191
455	260
544	329
633	397
722	466
811	535
900	604
989	672
5623078	741
167	810
256	878
345	947
434	·7500015
523	084
611	153
700	221
789	290
878	359
967	427
5624056	496
145	565
234	633
323	702
412	771
500	839
589	908
678	977
767	·7501045
856	114
945	182
5625034	251
123	320
212	388
300	457
389	526
478	594
567	663
656	731
745	800
834	869
923	937
5626012	·7502006
100	075
189	143
278	212
367	280
456	349
545	418
634	486
722	555
811	623
900	692
989	760

5627078	·7502829
167	897
256	966
344	·7503035
433	103
522	172
611	240
700	309
789	377
878	446
966	515
5628055	583
144	652
233	720
322	789
411	857
499	926
588	994
677	·7504063
766	132
855	200
944	269
5629032	337
121	406
210	474
299	543
388	611
476	680
565	748
654	817
743	885
832	954
921	·7505022

Excesses

10 = 8	1 = 1
20 = 15	2 = 2
30 = 23	3 = 2
40 = 31	4 = 3
50 = 39	5 = 4
60 = 46	6 = 5
70 = 54	7 = 5
80 = 62	8 = 6
	9 = 7

45 = 35

Anti Log Excesses

10 = 13	1 = 1
20 = 26	2 = 3
30 = 39	3 = 4
40 = 52	4 = 5
50 = 65	5 = 6
60 = 78	6 = 8
	7 = 9
	8 = 10
	9 = 12

35 = 45

5630009	·7505091
098	159
187	228

5630276	·7505296
365	365
453	433
542	502
631	570
720	639
809	707
897	776
986	844
5631075	913
164	981
253	·7506050
341	118
430	187
519	255
608	324
697	392
785	460
874	529
963	597
5632052	666
140	734
229	803
318	871
407	940
496	·7507008
584	077
673	145
762	213
851	282
939	350
5633028	419
117	487
206	556
294	624
383	692
472	761
561	829
649	898
738	966
827	·7508035
916	103
5634004	171
093	240
182	308
271	377
359	445
448	513
537	582
626	650
714	719
803	787
892	855
981	924
5635069	992
158	·7509061
247	129
335	197
424	266
513	334
602	403
690	471
779	539
868	608
956	676

5636045	·7509744
134	813
223	881
311	949
400	·7510018
489	086
577	154
666	223
755	291
844	359
932	428
5637021	496
110	564
198	633
287	701
376	769
464	838
553	906
642	974
731	·7511043
819	111
908	179
997	248
5638085	316
174	384
263	453
351	521
440	589
529	658
617	726
706	794
795	862
883	931
972	999
5639061	·7512067
149	136
238	204
327	272
415	340
504	409
593	477
681	545
770	613
859	682
947	750

Excesses

10 = 8	1 = 1
20 = 15	2 = 2
30 = 23	3 = 2
40 = 31	4 = 3
50 = 39	5 = 4
60 = 46	6 = 5
70 = 54	7 = 5
80 = 62	8 = 6
	9 = 7

45 = 35

Anti Log Excesses

10 = 13	1 = 1
20 = 26	2 = 3
30 = 39	3 = 4
40 = 52	4 = 5
50 = 65	5 = 6
60 = 78	6 = 8
	7 = 9
	8 = 10
	9 = 12

35 = 45

5640036	·7512818
125	887
213	955
302	·7513023
391	091
479	160
568	228
656	296
745	364
834	433
922	501
5641011	569
100	637
188	705
277	774
366	842
454	910
543	978
631	·7514047
720	115
809	183
897	251
986	319
5642075	388
163	456
252	524
340	592
429	661
518	729
606	797
695	865
784	933
872	·7515002
961	070
5643049	138
138	206
227	274
315	342
404	411
492	479
581	547
670	615
758	683
847	752
935	820
5644024	888
112	956
201	·7516024
290	092
378	161
467	229

5644555	·7516297
644	365
733	433
821	501
910	569
998	638
5645087	706
175	774
264	842
353	910
441	978
530	·7517046
618	115
707	183
795	251
884	319
973	387
5646061	455
150	523
238	591
327	659
415	728
504	796
592	864
681	932
769	·7518000
858	068
947	136
5647035	204
124	272
212	340
301	409
389	477
478	545
566	613
655	681
743	749
832	817
920	885
5648009	953
098	·7519021
186	089
275	157
363	225
452	294
540	362
629	430
717	498
806	566
894	634
983	702
5649071	770
160	838
248	906
337	974
425	·7520042
514	111
602	179
691	247
779	315
868	383
956	451

Excesses

10 = 8	1 = 1
20 = 15	2 = 2
30 = 23	3 = 2
40 = 31	4 = 3
50 = 38	5 = 4
60 = 46	6 = 5
70 = 54	7 = 5
80 = 62	8 = 6
	9 = 7

44 = 34

Anti Log Excesses

10 = 13	1 = 1
20 = 26	2 = 3
30 = 39	3 = 4
40 = 52	4 = 5
50 = 65	5 = 6
60 = 78	6 = 8
	7 = 9
	8 = 10
	9 = 12

34 = 44

5650045	·7520519
133	587
222	655
310	723
399	791
487	859
576	927
664	995
753	·7521063
841	131
930	199
5651018	267
107	335
195	403
284	471
372	539
460	607
549	675
637	743
726	811
814	879
903	947
991	·7522015
5652080	083
168	151
257	219
345	286
434	354
522	422
611	490
699	558
787	626
876	694
964	762
5653053	830
141	898
230	966

5653318	·7523034
407	102
495	170
583	238
672	306
760	374
849	442
937	510
5654026	577
114	645
202	713
291	781
379	849
468	917
556	985
645	·7524053
733	121
821	189
910	257
998	324
5655087	392
175	460
264	528
352	596
440	664
529	732
617	800
706	868
794	935
882	·7525003
971	071
5656059	139
148	207
236	275
324	343
413	411
501	478
590	546
678	614
766	682
855	750
943	818
5657032	886
120	953
208	·7526021
297	089
385	157
473	225
562	293
650	361
739	428
827	496
915	564
5658004	632
092	700
180	768
269	835
357	903
446	971
534	·7527039
622	107
711	174
799	242
887	310
976	378

5659064	·7527446
152	514
241	581
329	649
417	717
506	785
594	853
682	920
771	988
859	·7528056
947	124

Excesses

10 = 8	1 = 1
20 = 15	2 = 2
30 = 23	3 = 2
40 = 31	4 = 3
50 = 38	5 = 4
60 = 46	6 = 5
70 = 54	7 = 5
80 = 61	8 = 6
	9 = 7

44 = 34

Anti Log Excesses

10 = 13	1 = 1
20 = 26	2 = 3
30 = 39	3 = 4
40 = 52	4 = 5
50 = 65	5 = 7
60 = 78	6 = 8
	7 = 9
	8 = 10
	9 = 12

34 = 44

5660036	·7528191
124	259
213	327
301	395
389	463
478	530
566	598
654	666
743	734
831	801
919	869
5661007	937
096	·7529005
184	072
272	140
361	208
449	276
537	344
626	411
714	479
802	547
891	614
979	682
5662067	750
156	818

5662244	·7529885
332	953
420	·7530021
509	089
597	156
685	224
774	292
862	359
950	427
5663039	495
127	563
215	630
303	698
392	766
480	833
568	901
657	969
745	·7531037
833	104
921	172
5664010	240
098	307
186	375
275	443
363	510
451	578
539	646
628	713
716	781
804	849
892	916
981	984
5665069	·7532052
157	119
245	187
334	255
422	322
510	390
598	458
687	525
775	593
863	660
951	728
5666040	796
128	863
216	931
304	999
393	·7533066
481	134
569	202
657	269
746	337
834	404
922	472
5667010	540
099	607
187	675
275	742
363	810
451	878
540	945
628	·7534013
716	080
804	148
893	216

5667981	·7534283
5668069	351
157	418
245	486
334	554
422	621
510	689
598	756
686	824
775	892
863	959
951	·7535027
5669039	094
127	162
216	229
304	297
392	364
480	432
568	500
657	567
745	635
833	702
921	770

Excesses

10 = 8	1 = 1
20 = 15	2 = 2
30 = 23	3 = 2
40 = 31	4 = 3
50 = 38	5 = 4
60 = 46	6 = 5
70 = 54	7 = 5
80 = 61	8 = 6
	9 = 7

44 = 34

Anti Log Excesses

10 = 13	1 = 1
20 = 26	2 = 3
30 = 39	3 = 4
40 = 52	4 = 5
50 = 65	5 = 7
60 = 78	6 = 8
	7 = 9
	8 = 10
	9 = 12

34 = 44

5670009	·7535837
098	905
186	972
274	·7536040
362	108
450	175
538	243
627	310
715	378
803	445
891	513
979	580
5671067	648

5671156	·7536715
244	783
332	850
420	918
508	985
596	·7537053
685	120
773	188
861	255
949	323
5672037	390
125	458
213	525
302	593
390	660
478	728
566	795
654	863
742	930
830	998
919	·7538065
5673007	133
095	200
183	268
271	335
359	403
447	470
536	537
624	605
712	672
800	740
888	807
976	875
5674064	942
152	·7539010
241	077
329	145
417	212
505	279
593	347
681	414
769	482
857	549
945	617
5675034	684
122	751
210	819
298	886
386	954
474	·7540021
562	089
650	156
738	223
826	291
915	358
5676003	426
091	493
179	560
267	628
355	695
443	763
531	830
619	897
707	965
795	·7541032

5676883	·7541099
972	167
5677060	234
148	302
236	369
324	436
412	504
500	571
588	638
676	706
764	773
852	841
940	908
5678028	975
116	·7542043
204	110
292	177
381	245
469	312
557	379
645	447
733	514
821	581
909	649
997	716
5679085	783
173	851
261	918
349	985
437	·7543053
525	120
613	187
701	255
789	322
877	389
965	457

Excesses

10 = 8	1 = 1
20 = 15	2 = 2
30 = 23	3 = 2
40 = 31	4 = 3
50 = 38	5 = 4
60 = 46	6 = 5
70 = 54	7 = 5
80 = 61	8 = 6
	9 = 7

44 = 34

Anti Log Excesses

10 = 13	1 = 1
20 = 26	2 = 3
30 = 39	3 = 4
40 = 52	4 = 5
50 = 65	5 = 7
60 = 78	6 = 8
	7 = 9
	8 = 10
	9 = 12

34 = 44

5680053	·7543524
141	591
229	658
317	726
405	793
493	860
581	928
669	995
757	·7544062
846	129
934	197
5681022	264
110	331
198	399
286	466
374	533
462	600
550	668
638	735
726	802
814	869
902	937
990	·7545004
5682078	071
166	138
254	206
342	273
430	340
518	408
606	475
694	542
782	609
869	676
957	744
5683045	811
133	878
221	945
309	·7546013
397	080
485	147
573	214
661	282
749	349
837	416
925	483
5684013	550
101	618
189	685
277	752
365	819
453	886
541	954
629	·7547021
717	088
805	155
893	222
981	290
5685069	357
157	424
245	491
333	558
420	626
508	693
596	760
684	827

5685772	·7547894
860	961
948	·7548029
5686036	096
124	163
212	230
300	297
388	364
476	432
564	499
652	566
739	633
827	700
915	767
5687003	834
091	902
179	969
267	·7549036
355	103
443	170
531	237
619	304
707	371
794	439
882	506
970	573
5688058	640
146	707
234	774
322	841
410	908
498	976
586	·7550043
673	110
761	177
849	244
937	311
5689025	378
113	445
201	512
289	579
377	646
464	714
552	781
640	848
728	915
816	982
904	·7551049
992	116

Excesses

10 = 8	1 = 1
20 = 15	2 = 2
30 = 23	3 = 2
40 = 31	4 = 3
50 = 38	5 = 4
60 = 46	6 = 5
70 = 53	7 = 5
80 = 61	8 = 6
	9 = 7

44 = 34

Anti Log Excesses

10 = 13	1 = 1
20 = 26	2 = 3
30 = 39	3 = 4
40 = 52	4 = 5
50 = 65	5 = 7
60 = 79	6 = 8
	7 = 9
	8 = 10
	9 = 12

34 = 44

5690080	·7551183
167	250
255	317
343	384
431	451
519	518
607	585
695	652
783	719
870	787
958	854
5691046	921
134	988
222	·7552055
310	122
398	189
485	256
573	323
661	390
749	457
837	524
925	591
5692012	658
100	725
188	792
276	859
364	926
452	993
539	·7553060
627	127
715	194
803	261
891	328
979	395
5693066	462
154	529
242	596
330	663
418	730
506	797
593	864
681	931
769	998
857	·7554065
945	132
5694032	199
120	266
208	333
296	400
384	467
472	534

5694559	·7554601
647	668
735	735
823	802
911	869
998	935
5695086	·7555002
174	069
262	136
349	203
437	270
525	337
613	404
701	471
788	538
876	605
964	672
5696052	739
140	806
227	873
315	939
403	·7556006
491	073
578	140
666	207
754	274
842	341
929	408
5697017	475
105	542
193	609
281	676
368	742
456	809
544	876
632	943
719	·7557010
807	077
895	144
983	211
5698070	278
158	344
246	411
334	478
421	545
509	612
597	679
685	746
772	813
860	879
948	946
5699036	·7558013
123	080
211	147
299	214
386	280
474	347
562	414
650	481
737	548
825	615
913	682

Excesses

10 = 8	1 = 1
20 = 15	2 = 2
30 = 23	3 = 2
40 = 31	4 = 3
50 = 38	5 = 4
60 = 46	6 = 5
70 = 53	7 = 5
80 = 61	8 = 6
	9 = 7

44 = 34

Anti Log Excesses

10 = 13	1 = 1
20 = 26	2 = 3
30 = 39	3 = 4
40 = 52	4 = 5
50 = 66	5 = 7
60 = 79	6 = 8
	7 = 9
	8 = 10
	9 = 12

34 = 44

5700000	·7558749
088	816
176	882
264	949
351	·7559016
439	083
527	150
615	217
702	283
790	350
878	417
965	484
5701053	551
141	617
228	684
316	751
404	818
492	885
579	951
667	·7560018
755	085
842	152
930	219
5702018	285
105	352
193	419
281	486
368	552
456	619
544	686
631	753
719	820
807	886
895	953
982	·7561020
5703070	087
158	153

5703245	·7561220
333	287
421	354
508	420
596	487
684	554
771	621
859	687
947	754
5704034	821
122	888
209	954
297	·7562021
385	088
472	155
560	221
648	288
735	355
823	422
911	488
998	555
5705086	622
174	688
261	755
349	822
437	888
524	955
612	·7563022
699	089
787	155
875	222
962	289
5706050	355
138	422
225	489
313	555
400	622
488	689
576	756
663	822
751	889
839	956
926	·7564022
5707014	089
101	156
189	222
277	289
364	356
452	422
539	489
627	556
715	622
802	689
890	755
977	822
5708065	889
153	955
240	·7565022
328	089
415	155
503	222
591	289
678	355
766	422
853	488

5708941	·7565555
5709028	622
116	688
204	755
291	822
379	888
466	955
554	·7566022
641	088
729	155
817	221
904	288
992	355

Excesses

10 = 8	1 = 1
20 = 15	2 = 2
30 = 23	3 = 2
40 = 30	4 = 3
50 = 38	5 = 4
60 = 46	6 = 5
70 = 53	7 = 5
80 = 61	8 = 6
	9 = 7

44 = 33

Anti Log Excesses

10 = 13	1 = 1
20 = 26	2 = 3
30 = 39	3 = 4
40 = 53	4 = 5
50 = 66	5 = 7
60 = 79	6 = 8
	7 = 9
	8 = 11
	9 = 12

33 = 44

5710079	·7566421
167	488
254	554
342	621
430	688
517	754
605	821
692	887
780	954
867	·7567020
955	087
5711042	154
130	220
218	287
305	353
393	420
480	487
568	553
655	620
743	686
830	753
918	819
5712005	886

5712093	·7567952
180	·7568019
268	086
356	152
443	219
531	285
618	352
706	418
793	485
881	551
968	618
5713056	684
143	751
231	817
318	884
406	950
493	·7569017
581	083
668	150
756	216
843	283
931	350
5714018	416
106	483
193	549
281	616
368	682
456	749
543	815
631	881
718	948
806	·7570014
893	081
981	147
5715068	214
156	280
243	347
331	413
418	480
506	546
593	613
681	679
768	746
856	812
943	879
5716031	945
118	·7571012
206	078
293	145
380	211
468	277
555	344
643	410
730	477
818	543
905	610
993	676
5717080	742
168	809
255	875
343	942
430	·7572008
517	075
605	141
692	207

5717780	·7572274
867	340
955	407
5718042	473
130	540
217	606
304	672
392	739
479	805
567	872
654	938
742	·7573004
829	071
916	137
5719004	204
091	270
179	336
266	403
354	469
441	536
528	602
616	668
703	735
791	801
878	867
966	934

Excesses

10 = 8	1 = 1
20 = 15	2 = 2
30 = 23	3 = 2
40 = 30	4 = 3
50 = 38	5 = 4
60 = 46	6 = 5
70 = 53	7 = 5
80 = 61	8 = 6
	9 = 7

44 = 33

Anti Log Excesses

10 = 13	1 = 1
20 = 26	2 = 3
30 = 39	3 = 4
40 = 53	4 = 5
50 = 66	5 = 7
60 = 79	6 = 8
	7 = 9
	8 = 11
	9 = 12

33 = 44

5720053	·7574000
140	067
228	133
315	199
403	266
490	332
577	398
665	465
752	531
840	597

5720927	·7574664
5721014	730
102	796
189	863
277	929
364	995
451	·7575062
539	128
626	194
714	261
801	327
888	393
976	460
5722063	526
150	592
238	659
325	725
413	791
500	858
587	924
675	990
762	·7576057
849	123
937	189
5723024	255
112	322
199	388
286	454
374	521
461	587
548	653
636	720
723	786
810	852
898	918
985	985
5724072	·7577051
160	117
247	183
335	250
422	316
509	382
597	449
684	515
771	581
859	647
946	714
5725033	780
121	846
208	912
295	979
383	·7578045
470	111
557	177
645	244
732	310
819	376
907	442
994	509
5726081	575
169	641
256	707
343	773
430	840
518	906

5726605	·7578972
692	·7579038
780	105
867	171
954	237
5727042	303
129	369
216	436
304	502
391	568
478	634
565	700
653	767
740	833
827	899
915	965
5728002	·7580031
089	097
176	164
264	230
351	296
438	362
526	428
613	495
700	561
787	627
875	693
962	759
5729049	825
137	891
224	958
311	·7581024
398	090
486	156
573	222
660	288
747	354
835	421
922	487

Excesses

10 = 8	1 = 1
20 = 15	2 = 2
30 = 23	3 = 2
40 = 30	4 = 3
50 = 38	5 = 4
60 = 46	6 = 5
70 = 53	7 = 5
80 = 61	8 = 6
	9 = 7

44 = 33

Anti Log Excesses

10 = 13	1 = 1
20 = 26	2 = 3
30 = 40	3 = 4
40 = 53	4 = 5
50 = 66	5 = 7
60 = 79	6 = 8
	7 = 9
	8 = 11
	9 = 12

33 = 44

5730009	·7581553
096	619
184	685
271	751
358	817
446	884
533	950
620	·7582016
707	082
795	148
882	214
969	280
5731056	346
143	412
231	479
318	545
405	611
492	677
580	743
667	809
754	875
841	941
929	·7583007
5732016	073
103	140
190	206
278	272
365	338
452	404
539	470
626	536
714	602
801	668
888	734
975	800
5733063	866
150	932
237	999
324	·7584065
411	131
499	197
586	263
673	329
760	395
847	461
935	527
5734022	593
109	659
196	725
283	791
371	857

5734458	·7584923
545	989
632	·7585055
719	121
807	187
894	253
981	319
5735068	385
155	451
242	517
330	583
417	649
504	715
591	781
678	847
765	913
853	979
940	·7586045
5736027	111
114	177
201	243
288	309
376	375
463	441
550	507
637	573
724	639
811	705
899	771
986	837
5737073	903
160	969
247	·7587035
334	101
422	167
509	233
596	299
683	365
770	431
857	497
944	563
5738032	629
119	695
206	761
293	827
380	893
467	959
554	·7588025
641	091
729	157
816	222
903	288
990	354
5739077	420
164	486
251	552
338	618
426	684
513	750
600	816
687	882
774	948
861	·7589013
948	079

Excesses

10 = 8	1 = 1
20 = 15	2 = 2
30 = 23	3 = 2
40 = 30	4 = 3
50 = 38	5 = 4
60 = 45	6 = 5
70 = 53	7 = 5
80 = 61	8 = 6
	9 = 7

44 = 33

Anti Log Excesses

10 = 13	1 = 1
20 = 26	2 = 3
30 = 40	3 = 4
40 = 53	4 = 5
50 = 66	5 = 7
60 = 79	6 = 8
	7 = 9
	8 = 11
	9 = 12

33 = 44

5740035	·7589145
122	211
210	277
297	343
384	409
471	475
558	541
645	607
732	672
819	738
906	804
993	870
5741081	936
168	·7590002
255	068
342	134
429	199
516	265
603	331
690	397
777	463
864	529
951	595
5742038	661
126	726
213	792
300	858
387	924
474	990
561	·7591056
648	122
735	187
822	253
909	319
996	385
5743083	451
170	517

5743257	·7591582
344	648
432	714
519	780
606	846
693	912
780	977
867	·7592043
954	109
5744041	175
128	241
215	306
302	372
389	438
476	504
563	570
650	635
737	701
824	767
911	833
998	899
5745085	964
172	·7593030
259	096
346	162
433	228
520	293
608	359
695	425
782	491
869	556
956	622
5746043	688
130	754
217	820
304	885
391	951
478	·7594017
565	083
652	148
739	214
826	280
913	346
5747000	411
087	477
174	543
261	609
348	674
435	740
522	806
609	872
696	937
783	·7595003
870	069
957	134
5748044	200
131	266
218	332
305	397
392	463
479	529
566	594
653	660
739	726
826	792

Number	Log
5748913	·7595857
5749000	923
087	989
174	·7596054
261	120
348	186
435	251
522	317
609	383
696	449
783	514
870	580
957	646

Excesses

10 = 8	1 = 1
20 = 15	2 = 2
30 = 23	3 = 2
40 = 30	4 = 3
50 = 38	5 = 4
60 = 45	6 = 5
70 = 53	7 = 5
80 = 60	8 = 6
	9 = 7

44 = 33

Anti-Log Excesses

10 = 13	1 = 1
20 = 26	2 = 3
30 = 40	3 = 4
40 = 53	4 = 5
50 = 66	5 = 7
60 = 79	6 = 8
	7 = 9
	8 = 11
	9 = 12

33 = 44

Number	Log
5750044	·7596711
131	777
218	843
305	908
392	974
479	·7597040
566	105
653	171
740	237
827	302
913	368
5751000	434
087	499
174	565
261	631
348	696
435	762
522	828
609	893
696	959
783	·7598025
870	090
957	156

Number	Log
5752044	·7598222
131	287
217	353
304	418
391	484
478	550
565	615
652	681
739	747
826	812
913	878
5753000	943
087	·7599009
174	075
260	140
347	206
434	271
521	337
608	403
695	468
782	534
869	599
956	665
5754043	731
129	796
216	862
303	927
390	993
477	·7600058
564	124
651	190
738	255
825	321
911	386
998	452
5755085	517
172	583
259	649
346	714
433	780
520	845
606	911
693	976
780	·7601042
867	107
954	173
5756041	238
128	304
215	370
301	435
388	501
475	566
562	632
649	697
736	763
823	828
909	894
996	959
5757083	·7602025
170	090
257	156
344	221
430	287
517	352
604	418

Number	Log
5757691	·7602483
778	549
865	614
952	680
5758038	745
125	811
212	876
299	942
386	·7603007
473	073
559	138
646	204
733	269
820	335
907	400
993	466
5759080	531
167	597
254	662
341	728
428	793
514	858
601	924
688	989
775	·7604055
862	120
948	186

Excesses

10 = 8	1 = 1
20 = 15	2 = 2
30 = 23	3 = 2
40 = 30	4 = 3
50 = 38	5 = 4
60 = 45	6 = 5
70 = 53	7 = 5
80 = 60	8 = 6
	9 = 7

44 = 33

Anti-Log Excesses

10 = 13	1 = 1
20 = 27	2 = 3
30 = 40	3 = 4
40 = 53	4 = 5
50 = 66	5 = 7
60 = 80	6 = 8
	7 = 9
	8 = 11
	9 = 12

33 = 44

Number	Log
5760035	·7604251
122	317
209	382
296	448
382	513
469	578
556	644
643	709
730	775

Number	Log
5760816	·7604840
903	906
990	971
5761077	·7605036
164	102
250	167
337	233
424	298
511	364
598	429
684	494
771	560
858	625
945	691
5762031	756
118	821
205	887
292	952
378	·7606018
465	083
552	148
639	214
726	279
812	345
899	410
986	475
5763073	541
159	606
246	671
333	737
420	802
506	868
593	933
680	998
767	·7607064
853	129
940	194
5764027	260
114	325
200	390
287	456
374	521
461	587
547	652
634	717
721	783
808	848
894	913
981	979
5765068	·7608044
154	109
241	175
328	240
415	305
501	371
588	436
675	501
762	567
848	632
935	697
5766022	762
108	828
195	893
282	958
369	·7609024

Number	Log
5766455	·7609089
542	154
629	220
715	285
802	350
889	415
975	481
5767062	546
149	611
236	677
322	742
409	807
496	873
582	938
669	·7610003
756	068
842	134
929	199
5768016	264
102	329
189	395
276	460
362	525
449	590
536	656
623	721
709	786
796	851
883	917
969	982
5769056	·7611047
143	112
229	178
316	243
403	308
489	373
576	439
663	504
749	569
836	634
923	699

Excesses

10 = 8	1 = 1
20 = 15	2 = 2
30 = 23	3 = 2
40 = 30	4 = 3
50 = 38	5 = 4
60 = 45	6 = 5
70 = 53	7 = 5
80 = 60	8 = 6
	9 = 7

44 = 33

Anti-Log Excesses

10 = 13	1 = 1
20 = 27	2 = 3
30 = 40	3 = 4
40 = 53	4 = 5
50 = 66	5 = 7
60 = 80	6 = 8
	7 = 9
	8 = 11
	9 = 12

33 = 44

5770009	·7611765
096	830
182	895
269	960
356	·7612026
442	091
529	156
616	221
702	286
789	352
876	417
962	482
5771049	547
136	612
222	678
309	743
395	808
482	873
569	938
655	·7613004
742	069
829	134
915	199
5772002	264
089	330
175	395
262	460
348	525
435	590
522	655
608	721
695	786
781	851
868	916
955	981
5773041	·7614046
128	111
215	177
301	242
388	307
474	372
561	437
648	502
734	568
821	633
907	698
994	763
5774081	828
167	893
254	958
340	·7615023

5774427	·7615089
513	154
600	219
687	284
773	349
860	414
946	479
5775033	544
120	610
206	675
293	740
379	805
466	870
552	935
639	·7616000
726	065
812	130
899	195
985	261
5776072	326
158	391
245	456
332	521
418	586
505	651
591	716
678	781
764	846
851	911
937	976
5777024	·7617041
111	107
197	172
284	237
370	302
457	367
543	432
630	497
716	562
803	627
889	692
976	757
5778062	822
149	887
236	952
322	·7618017
409	082
495	147
582	212
668	277
755	342
841	407
928	472
5779014	537
101	602
187	668
274	733
360	798
447	863
533	928
620	993
706	·7619058
793	123
879	188
966	253

Excesses

10 = 8	1 = 1
20 = 15	2 = 2
30 = 23	3 = 2
40 = 30	4 = 3
50 = 38	5 = 4
60 = 45	6 = 5
70 = 53	7 = 5
80 = 60	8 = 6
	9 = 7

44 = 33

Anti-Log Excesses

10 = 13	1 = 1
20 = 27	2 = 3
30 = 40	3 = 4
40 = 53	4 = 5
50 = 66	5 = 7
60 = 80	6 = 8
	7 = 9
	8 = 11
	9 = 12

33 = 44

5780052	·7619318
139	383
225	448
312	513
398	578
485	642
571	707
658	772
744	837
831	902
917	967
5781004	·7620032
090	097
177	162
263	227
350	292
436	357
523	422
609	487
696	552
782	617
869	682
955	747
5782042	812
128	877
215	942
301	·7621007
388	072
474	137
560	202
647	266
733	331
820	396
906	461
993	526
5783079	591
166	656

5783252	·7621721
339	786
425	851
512	916
598	981
684	·7622046
771	111
857	175
944	240
5784030	305
117	370
203	435
290	500
376	565
462	630
549	695
635	760
722	824
808	889
895	954
981	·7623019
5785067	084
154	149
240	214
327	279
413	344
500	408
586	473
672	538
759	603
845	668
932	733
5786018	798
105	863
191	927
277	992
364	·7624057
450	122
537	187
623	252
709	317
796	381
882	446
969	511
5787055	576
141	641
228	705
314	770
401	835
487	900
573	965
660	·7625030
746	094
833	159
919	224
5788005	289
092	354
178	419
264	483
351	548
437	613
524	678
610	743
696	807
783	872

5788869	·7625937
956	·7626002
5789042	067
128	131
215	196
301	261
387	326
474	391
560	455
646	520
733	585
819	650
906	714
992	779

Excesses

10 = 8	1 = 1
20 = 15	2 = 2
30 = 23	3 = 2
40 = 30	4 = 3
50 = 38	5 = 4
60 = 45	6 = 5
70 = 53	7 = 5
80 = 60	8 = 6
	9 = 7

43 = 32

Anti-Log Excesses

10 = 13	1 = 1
20 = 27	2 = 3
30 = 40	3 = 4
40 = 53	4 = 5
50 = 67	5 = 7
60 = 80	6 = 8
	7 = 9
	8 = 11
	9 = 12

32 = 43

5790078	·7626844
165	909
251	974
337	·7627038
424	103
510	168
596	233
683	297
769	362
855	427
942	492
5791028	556
114	621
201	686
287	751
373	815
460	880
546	945
632	·7628009
719	074
805	139
891	204

5791978	·7628268
5792064	333
150	398
237	463
323	527
409	592
496	657
582	721
668	786
755	851
841	915
927	980
5793014	·7629045
100	110
186	174
272	239
359	304
445	368
531	433
618	498
704	562
790	627
877	692
963	756
5794049	821
135	886
222	951
308	·7630015
394	080
481	145
567	209
653	274
739	339
826	403
912	468
998	533
5795085	597
171	662
257	726
343	791
430	856
516	920
602	985
689	·7631050
775	114
861	179
947	244
5796034	308
120	373
206	438
292	502
379	567
465	631
551	696
637	761
724	825
810	890
896	955
982	·7632019
5797069	084
155	149
241	213
327	278
414	343
500	407

5797586	·7632472
672	536
759	601
845	666
931	730
5798017	795
104	859
190	924
276	988
362	·7633053
449	118
535	182
621	247
707	311
793	376
880	441
966	505
5799052	570
138	634
225	699
311	763
397	828
483	893
569	957
656	·7634022
742	086
828	151
914	215

Excesses

10 = 7	1 = 1
20 = 15	2 = 1
30 = 22	3 = 2
40 = 30	4 = 3
50 = 37	5 = 4
60 = 45	6 = 4
70 = 52	7 = 5
80 = 60	8 = 6
	9 = 7

43 = 32

Anti-Log Excesses

10 = 13	1 = 1
20 = 27	2 = 3
30 = 40	3 = 4
40 = 53	4 = 5
50 = 67	5 = 7
60 = 80	6 = 8
	7 = 9
	8 = 11
	9 = 12

32 = 43

5800000	·7634280
087	344
173	409
259	474
345	538
432	603
518	667
604	732

5800690	·7634796
776	861
863	925
949	990
5801035	·7635054
121	119
207	183
293	248
380	312
466	377
552	441
638	506
724	570
811	635
897	699
983	764
5802069	828
155	893
241	957
328	·7636022
414	086
500	151
586	215
672	280
758	344
845	409
931	473
5803017	538
103	602
189	667
275	731
362	795
448	860
534	924
620	988
706	·7637053
792	117
879	182
965	246
5804051	311
137	375
223	440
309	504
395	569
482	633
568	697
654	762
740	826
826	891
912	955
998	·7638020
5805084	084
171	149
257	213
343	277
429	342
515	406
601	471
687	535
773	600
860	664
946	728
5806032	793
118	857
204	922

5806290	·7638986
376	·7639050
462	115
549	179
635	244
721	308
807	373
893	437
979	501
5807065	566
151	630
237	694
323	759
410	823
496	888
582	952
668	·7640016
754	081
840	145
926	210
5808012	274
098	338
184	403
270	467
357	531
443	596
529	660
615	724
701	789
787	853
873	918
959	982
5809045	·7641046
131	111
217	175
303	239
389	304
476	368
562	432
648	497
734	561
820	625
906	690
992	754

Excesses

10 = 7	1 = 1
20 = 15	2 = 1
30 = 22	3 = 2
40 = 30	4 = 3
50 = 37	5 = 4
60 = 45	6 = 4
70 = 52	7 = 5
80 = 60	8 = 6
	9 = 7

43 = 32

Anti-Log Excesses

10 = 13	1 = 1
20 = 27	2 = 3
30 = 40	3 = 4
40 = 53	4 = 5
50 = 67	5 = 7
60 = 80	6 = 8
	7 = 9
	8 = 11
	9 = 12

32 = 43

5810078	·7641818
164	883
250	947
336	·7642011
422	076
508	140
594	204
680	268
766	333
852	397
938	461
5811025	526
111	590
197	654
283	719
369	783
455	847
541	911
627	976
713	·7643040
799	104
885	169
971	233
5812057	297
143	361
229	426
315	490
401	554
487	619
573	683
659	747
745	811
831	876
917	940
5813003	·7644004
089	069
175	133
261	197
347	261
433	326
519	390
605	454
691	518
777	583
863	647
949	711
5814035	775
121	839
207	904
293	968
379	·7645032

Number	Log
5814465	.7645096
551	161
637	225
723	289
809	353
895	418
981	482
5815067	546
153	610
239	674
325	739
411	803
497	867
583	931
669	995
755	·7646060
841	124
927	188
5816013	252
099	317
185	381
271	445
357	509
443	573
529	637
615	702
701	766
787	830
873	894
958	958
5817044	·7647022
130	087
216	151
302	215
388	279
474	343
560	407
646	472
732	536
818	600
904	664
990	728
5818076	792
162	856
248	921
334	985
420	·7648049
505	113
591	177
677	241
763	305
849	370
935	434
5819021	498
107	562
193	626
279	690
365	754
451	818
537	883
622	947
708	·7649011
794	075
880	139
966	203

Excesses

10 = 7	1 = 1
20 = 15	2 = 1
30 = 22	3 = 2
40 = 30	4 = 3
50 = 37	5 = 4
60 = 45	6 = 4
70 = 52	7 = 5
80 = 60	8 = 6
	9 = 7

43 = 32

Anti-Log Excesses

10 = 13	1 = 1
20 = 27	2 = 3
30 = 40	3 = 4
40 = 54	4 = 5
50 = 67	5 = 7
60 = 80	6 = 8
	7 = 9
	8 = 11
	9 = 12

32 = 43

Number	Log
5820052	·7649267
138	331
224	395
310	460
396	524
482	588
567	652
653	716
739	780
825	844
911	908
997	972
5821083	·7650036
169	100
255	165
341	229
426	293
512	357
598	421
684	485
770	549
856	613
942	677
5822028	741
114	805
199	869
285	933
371	997
457	·7651061
543	126
629	190
715	254
801	318
886	382
972	447
5823058	510
144	574

Number	Log
5823230	·7651638
316	702
402	766
487	830
573	894
659	958
745	·7652022
831	086
917	150
5824003	214
088	278
174	342
260	406
346	470
432	534
518	598
603	662
689	726
775	790
861	854
947	918
5825033	982
119	·7653046
204	110
290	174
376	238
462	302
548	366
634	430
719	494
805	558
891	622
977	686
5826063	750
148	814
234	878
320	942
406	·7654006
492	070
578	134
663	198
749	262
835	326
921	390
5827007	454
092	518
178	582
264	646
350	709
436	773
521	837
607	901
693	965
779	·7655029
865	093
950	157
5828036	221
122	285
208	349
294	413
379	477
465	541
551	605
637	668
722	732

Number	Log
5828808	·7655796
894	860
980	924
5829066	988
151	·7656052
237	116
323	180
409	244
494	309
580	373
666	436
752	500
838	564
923	628

Excesses

10 = 7	1 = 1
20 = 15	2 = 1
30 = 22	3 = 2
40 = 30	4 = 3
50 = 37	5 = 4
60 = 45	6 = 4
70 = 52	7 = 5
80 = 60	8 = 6
	9 = 7

43 = 32

Anti-Log Excesses

10 = 13	1 = 1
20 = 27	2 = 3
30 = 40	3 = 4
40 = 54	4 = 5
50 = 67	5 = 7
60 = 80	6 = 8
	7 = 9
	8 = 11
	9 = 12

32 = 43

Number	Log
5830009	·7656692
095	756
181	820
266	884
352	948
438	·7657011
524	075
609	139
695	203
781	267
867	331
952	395
5831038	459
124	522
210	586
295	650
381	714
467	778
553	842
638	906
724	969
810	·7658033

Number	Log
5831896	·7658097
981	161
5832067	225
153	289
238	352
324	416
410	480
496	544
581	608
667	672
753	735
839	799
924	863
5833010	927
096	991
181	·7659055
267	118
353	182
439	246
524	310
610	374
696	437
781	501
867	565
953	629
5834039	693
124	756
210	820
296	884
381	948
467	·7660012
553	075
638	139
724	203
810	267
896	330
981	394
5835067	458
153	522
238	586
324	649
410	713
495	777
581	841
667	904
752	968
838	·7661032
924	096
5836009	159
095	223
181	287
266	351
352	414
438	478
523	542
609	606
695	669
780	733
866	797
952	861
5837037	924
123	988
209	·7662052
294	116
380	179

5837466	·7662243
551	307
637	370
723	434
808	498
894	562
980	625
5838065	689
151	753
237	816
322	880
408	944
493	·7663008
579	071
665	135
750	199
836	262
922	326
5839007	390
093	453
179	517
264	581
350	645
435	708
521	772
607	836
692	899
778	963
864	·7664027
949	090

Excesses

10 = 7	1 = 1
20 = 15	2 = 1
30 = 22	3 = 2
40 = 30	4 = 3
50 = 37	5 = 4
60 = 45	6 = 4
70 = 52	7 = 5
80 = 60	8 = 6
	9 = 7

43 = 32

Anti-Log Excesses

10 = 13	1 = 1
20 = 27	2 = 3
30 = 40	3 = 4
40 = 54	4 = 5
50 = 67	5 = 7
60 = 81	6 = 8
	7 = 9
	8 = 11
	9 = 12

32 = 43

5840035	·7664154
120	218
206	281
292	345
377	409
463	472

5840548	·7664536
634	600
720	663
805	727
891	791
976	854
5841062	918
148	981
233	·7665045
319	109
404	172
490	236
576	300
661	363
747	427
832	491
918	554
5842004	618
089	681
175	745
260	809
346	872
432	936
517	·7666000
603	063
688	127
774	191
859	254
945	318
5843031	381
116	445
202	509
287	572
373	636
458	699
544	763
630	826
715	890
801	954
886	·7667017
972	081
5844057	144
143	208
228	272
314	335
400	399
485	462
571	526
656	589
742	653
827	717
913	780
998	844
5845084	907
169	971
255	·7668034
341	098
426	161
512	225
597	288
683	352
768	416
854	479
939	543
5846025	606

5846110	·7668670
196	733
281	797
367	860
452	924
538	987
623	·7669051
709	114
794	178
880	242
966	305
5847051	369
137	432
222	496
308	559
393	623
479	686
564	750
650	813
735	877
821	940
906	·7670004
992	067
5848077	131
163	194
248	258
334	321
419	385
505	448
590	511
676	575
761	638
847	702
932	765
5849018	829
103	892
188	956
274	·7671019
359	083
445	146
530	210
616	273
701	337
787	400
872	464
958	527

Excesses

10 = 7	1 = 1
20 = 15	2 = 1
30 = 22	3 = 2
40 = 30	4 = 3
50 = 37	5 = 4
60 = 45	6 = 4
70 = 52	7 = 5
80 = 59	8 = 6
	9 = 7

43 = 32

Anti-Log Excesses

10 = 13	1 = 1
20 = 27	2 = 3
30 = 40	3 = 4
40 = 54	4 = 5
50 = 67	5 = 7
60 = 81	6 = 8
	7 = 9
	8 = 11
	9 = 12

32 = 43

5850043	·7671590
129	654
214	717
300	781
385	844
471	908
556	971
641	·7672035
727	098
812	161
898	225
983	288
5851069	352
154	415
240	479
325	542
411	605
496	669
581	732
667	796
752	859
838	922
923	986
5852009	·7673049
094	113
180	176
265	240
350	303
436	366
521	430
607	493
692	557
778	620
863	683
948	747
5853034	810
119	873
205	937
290	·7674000
376	064
461	127
546	190
632	254
717	317
803	380
888	444
974	507
5854059	571
144	634
230	697
315	761

5854401	·7674824
486	887
571	951
657	·7675014
742	077
828	141
913	204
998	267
5855084	331
169	394
255	458
340	521
425	584
511	648
596	711
682	774
767	838
852	901
938	964
5856023	·7676028
108	091
194	154
279	218
365	281
450	344
535	407
621	471
706	534
791	597
877	661
962	724
5857048	787
133	851
218	914
304	977
389	·7677040
474	104
560	167
645	230
730	294
816	357
901	420
987	483
5858072	547
157	610
243	673
328	737
413	800
499	863
584	926
669	990
755	·7678053
840	116
925	179
5859011	243
096	306
181	369
267	432
352	496
437	559
523	622
608	685
693	749
779	812
864	875

5859949	·7678938

Excesses

10 = 7	1 = 1
20 = 15	2 = 1
30 = 22	3 = 2
40 = 30	4 = 3
50 = 37	5 = 4
60 = 45	6 = 4
70 = 52	7 = 5
80 = 59	8 = 6
	9 = 7

43 = 32

Anti-Log Excesses

10 = 13	1 = 1
20 = 27	2 = 3
30 = 40	3 = 4
40 = 54	4 = 5
50 = 67	5 = 7
60 = 81	6 = 8
	7 = 9
	8 = 11
	9 = 12

32 = 43

5860035	·7679002
120	065
205	128
291	191
376	255
461	318
547	381
632	444
717	507
802	571
888	634
973	697
5861058	760
144	823
229	887
314	950
400	·7680013
485	076
570	139
656	203
741	266
826	329
911	392
997	455
5862082	519
167	582
253	645
338	708
423	771
508	835
594	898
679	961
764	·7681024
850	087
935	150
5863020	·7681214
105	277
191	340
276	403
361	466
447	529
532	593
617	656
702	719
788	782
873	845
958	908
5864043	972
129	·7682035
214	098
299	161
385	224
470	287
555	350
640	414
726	477
811	540
896	603
981	666
5865067	729
152	792
237	855
322	919
408	982
493	·7683045
578	108
663	171
749	234
834	297
919	360
5866004	424
089	487
175	550
260	613
345	676
430	739
516	802
601	865
686	928
771	991
857	·7684054
942	118
5867027	181
112	244
197	307
283	370
368	433
453	496
538	559
624	622
709	685
794	748
879	811
964	874
5868050	937
135	·7685001
220	064
305	127
390	190
476	253
5868561	·7685316
646	379
731	442
816	505
902	568
987	631
5869072	694
157	757
242	820
328	883
413	946
498	·7686009
583	072
668	135
753	198
839	261
924	324

Excesses

10 = 7	1 = 1
20 = 15	2 = 1
30 = 22	3 = 2
40 = 30	4 = 3
50 = 37	5 = 4
60 = 44	6 = 4
70 = 52	7 = 5
80 = 59	8 = 6
	9 = 7

43 = 32

Anti-Log Excesses

10 = 14	1 = 1
20 = 27	2 = 3
30 = 41	3 = 4
40 = 54	4 = 5
50 = 68	5 = 7
60 = 81	6 = 8
	7 = 9
	8 = 11
	9 = 12

32 = 43

5870009	·7686387
094	450
179	513
265	577
350	640
435	703
520	766
605	829
690	892
776	955
861	·7687018
946	081
5871031	144
116	207
201	270
287	333
372	396
457	459
542	521
5871627	·7687584
712	647
797	710
883	773
968	836
5872053	899
138	962
223	·7688025
308	088
394	151
479	214
564	277
649	340
734	403
819	466
904	529
990	592
5873075	655
160	718
245	781
330	844
415	907
500	970
585	·7689033
671	096
756	158
841	221
926	284
5874011	347
096	410
181	473
266	536
352	599
437	662
522	725
607	788
692	851
777	914
862	976
947	·7690039
5875032	102
118	165
203	228
288	291
373	354
458	417
543	480
628	543
713	606
798	668
883	731
969	794
5876054	857
139	920
224	983
309	·7691046
394	109
479	172
564	234
649	297
734	360
819	423
904	486
990	549
5877075	612
5877160	·7691675
245	737
330	800
415	863
500	926
585	989
670	·7692052
755	115
840	177
925	240
5878010	303
095	366
180	429
266	492
351	554
436	617
521	680
606	743
691	806
776	869
861	931
946	994
5879031	·7693057
116	120
201	183
286	246
371	308
456	371
541	434
626	497
711	560
796	623
881	685
966	748

Excesses

10 = 7	1 = 1
20 = 15	2 = 1
30 = 22	3 = 2
40 = 30	4 = 3
50 = 37	5 = 4
60 = 44	6 = 4
70 = 52	7 = 5
80 = 59	8 = 6
	9 = 7

42 = 31

Anti-Log Excesses

10 = 14	1 = 1
20 = 27	2 = 3
30 = 41	3 = 4
40 = 54	4 = 5
50 = 68	5 = 7
60 = 81	6 = 8
	7 = 9
	8 = 11
	9 = 12

31 = 42

5880052	·7693811
137	874

5880222	·7693937
307	999
392	·7694062
477	125
562	188
647	251
732	313
817	376
902	439
987	502
5881072	565
157	627
242	690
327	753
412	816
497	878
582	941
667	·7695004
752	067
837	130
922	192
5882007	255
092	318
177	381
262	443
347	506
432	569
517	632
602	694
687	757
772	820
857	883
942	945
5883027	·7696008
112	071
197	134
282	196
367	259
452	322
537	385
622	447
707	510
792	573
877	635
962	698
5884047	761
132	824
217	886
302	949
387	·7697012
472	074
557	137
642	200
726	263
811	325
896	388
981	451
5885066	513
151	576
236	639
321	701
406	764
491	827
576	890
661	952

5885746	·7698015
831	078
916	140
5886001	203
086	266
171	328
256	391
341	454
426	516
511	579
595	642
680	704
765	767
850	830
935	892
5887020	955
105	·7699018
190	080
275	143
360	206
445	268
530	331
615	393
700	456
784	519
869	581
954	644
5888039	707
124	769
209	832
294	895
379	957
464	·7700020
549	082
634	145
719	208
803	270
888	333
973	395
5889058	458
143	521
228	583
313	646
398	709
483	771
568	834
652	896
737	959
822	·7701022
907	084
992	147

Excesses

10 = 7	1 = 1
20 = 15	2 = 1
30 = 22	3 = 2
40 = 30	4 = 3
50 = 37	5 = 4
60 = 44	6 = 4
70 = 52	7 = 5
80 = 59	8 = 6
	9 = 7

42 = 31

Anti-Log Excesses

10 = 14	1 = 1
20 = 27	2 = 3
30 = 41	3 = 4
40 = 54	4 = 5
50 = 68	5 = 7
60 = 81	6 = 8
	7 = 9
	8 = 11
	9 = 12

31 = 42

5890077	·7701209
162	272
247	334
332	397
416	460
501	522
586	585
671	647
756	710
841	773
926	835
5891011	898
095	960
180	·7702023
265	085
350	148
435	210
520	273
605	336
690	398
774	461
859	523
944	586
5892029	648
114	711
199	773
284	836
368	899
453	961
538	·7703024
623	086
708	149
793	211
878	274
962	336
5893047	399
132	461
217	524
302	586
387	649
471	711
556	774
641	837
726	899
811	962
896	·7704024
980	087
5894065	149
150	212
235	274
320	337

5894405	·7704399
489	462
574	524
659	587
744	649
829	711
914	774
998	836
5895083	899
168	961
253	·7705024
338	087
422	149
507	211
592	274
677	336
762	399
846	461
931	524
5896016	586
101	649
186	711
271	774
355	836
440	899
525	961
610	·7706024
694	086
779	148
864	211
949	273
5897034	336
118	398
203	461
288	523
373	586
458	648
542	710
627	773
712	835
797	898
881	960
966	·7707022
5898051	085
136	147
221	210
305	272
390	335
475	397
560	459
644	522
729	584
814	647
899	709
983	771
5899068	834
153	896
238	959
322	·7708021
407	083
492	146
577	208
662	271
746	333
831	395

5899916	·7708458

Excesses

10 = 7	1 = 1
20 = 15	2 = 1
30 = 22	3 = 2
40 = 29	4 = 3
50 = 37	5 = 4
60 = 44	6 = 4
70 = 52	7 = 5
80 = 59	8 = 6
	9 = 7

42 = 31

Anti-Log Excesses

10 = 14	1 = 1
20 = 27	2 = 3
30 = 41	3 = 4
40 = 54	4 = 5
50 = 68	5 = 7
60 = 81	6 = 8
	7 = 10
	8 = 11
	9 = 12

31 = 42

5900000	·7708520
085	583
170	645
255	707
339	770
424	832
509	894
594	957
678	·7709019
763	081
848	144
933	206
5901017	268
102	331
187	393
272	456
356	518
441	580
526	643
610	705
695	767
780	830
865	892
949	954
5902034	·7710017
119	079
203	141
288	204
373	266
458	328
542	391
627	453
712	515
796	578
881	640

Number	Log
5902966	·7710702
5903051	765
135	827
220	889
305	952
389	·7711014
474	076
559	138
643	201
728	263
813	325
898	388
982	450
5904067	512
152	575
236	637
321	699
406	761
490	824
575	886
660	948
744	·7712011
829	073
914	135
998	197
5905083	260
168	322
252	384
337	446
422	509
506	571
591	633
676	696
760	758
845	820
930	882
5906014	945
099	·7713007
184	069
268	131
353	194
438	256
522	318
607	380
692	443
776	505
861	567
946	629
5907030	691
115	754
199	816
284	878
369	940
453	·7714003
538	065
623	127
707	189
792	251
877	314
961	376
5908046	438
130	500
215	563
300	625
384	687

Number	Log
5908469	·7714749
554	811
638	874
723	936
807	998
892	·7715060
977	123
5909061	185
146	247
231	309
315	371
400	434
484	496
569	558
654	620
738	682
823	745
907	807
992	869

Excesses

10 = 7	1 = 1
20 = 15	2 = 1
30 = 22	3 = 2
40 = 29	4 = 3
50 = 37	5 = 4
60 = 44	6 = 4
70 = 51	7 = 5
80 = 59	8 = 6
	9 = 7

42 = 31

Anti-Log Excesses

10 = 14	1 = 1
20 = 27	2 = 3
30 = 41	3 = 4
40 = 54	4 = 5
50 = 68	5 = 7
60 = 82	6 = 8
	7 = 10
	8 = 11
	9 = 12

31 = 42

Number	Log
5910077	·7715931
161	993
246	·7716055
330	118
415	180
500	242
584	304
669	366
753	428
838	490
923	553
5911007	615
092	677
176	739
261	801
346	863
430	925

Number	Log
5911515	·7716988
599	·7717050
684	112
768	174
853	236
938	298
5912022	360
107	422
191	485
276	547
360	609
445	671
530	733
614	795
699	857
783	919
868	982
952	·7718044
5913037	106
122	168
206	230
291	292
375	354
460	416
544	478
629	540
713	603
798	665
882	727
967	789
5914052	851
136	913
221	975
305	·7719037
390	099
474	161
559	223
643	285
728	347
812	410
897	472
981	534
5915066	596
151	658
235	720
320	782
404	844
489	906
573	968
658	·7720030
742	092
827	154
911	216
996	278
5916080	340
165	402
249	464
334	527
418	589
503	651
587	713
672	775
756	837
841	899
925	961

Number	Log
5917010	·7721023
094	085
179	147
263	209
348	271
432	333
517	395
601	457
686	519
770	581
855	643
939	705
5918024	767
108	829
193	891
277	953
362	·7722015
446	077
531	139
615	201
700	263
784	325
869	387
953	449
5919038	511
122	573
207	635
291	697
375	759
460	821
544	883
629	945
713	·7723007
798	069
882	130
967	192

Excesses

10 = 7	1 = 1
20 = 15	2 = 1
30 = 22	3 = 2
40 = 29	4 = 3
50 = 37	5 = 4
60 = 44	6 = 4
70 = 51	7 = 5
80 = 59	8 = 6
	9 = 7

42 = 31

Anti-Log Excesses

10 = 14	1 = 1
20 = 27	2 = 3
30 = 41	3 = 4
40 = 54	4 = 5
50 = 68	5 = 7
60 = 82	6 = 8
	7 = 10
	8 = 11
	9 = 12

31 = 42

Number	Log
5920051	·7723254
136	316
220	378
305	440
389	502
473	564
558	626
642	688
727	750
811	812
896	874
980	936
5921065	998
149	·7724060
233	122
318	184
402	246
487	307
571	369
656	431
740	493
825	555
909	617
993	679
5922078	741
162	803
247	865
331	927
416	989
500	·7725050
584	112
669	174
753	236
838	298
922	360
5923006	422
091	484
175	546
260	608
344	669
429	731
513	793
597	855
682	917
766	979
851	·7726041
935	103
5924019	165
104	226
188	288
273	350
357	412
441	474
526	536
610	598
695	659
779	721
863	783
948	845
5925032	907
117	969
201	·7727031
285	092
370	154
454	216

Number	Log
5925538	·7727278
623	340
707	402
792	463
876	525
960	587
5926045	649
129	711
213	773
298	834
382	896
467	958
551	·7728020
635	082
720	144
804	205
888	267
973	329
5927057	391
141	453
226	514
310	576
395	638
479	700
563	762
648	823
732	885
816	947
901	·7729009
985	071
5928069	132
154	194
238	256
322	318
407	380
491	441
575	503
660	565
744	627
828	688
913	750
997	812
5929081	874
166	936
250	997
334	·7730059
419	121
503	183
587	244
672	306
756	368
840	430
925	491

Excesses

10 = 7	1 = 1
20 = 15	2 = 1
30 = 22	3 = 2
40 = 29	4 = 3
50 = 37	5 = 4
60 = 44	6 = 4
70 = 51	7 = 5
80 = 59	8 = 6
	9 = 7

42 = 31

Anti-Log Excesses

10 = 14	1 = 1
20 = 27	2 = 3
30 = 41	3 = 4
40 = 55	4 = 5
50 = 68	5 = 7
60 = 82	6 = 8
	7 = 10
	8 = 11
	9 = 12

31 = 42

Number	Log
5930009	·7730553
093	615
178	677
262	738
346	800
431	862
515	924
599	985
683	·7731047
768	109
852	171
936	232
5931021	294
105	356
189	417
274	479
358	541
442	602
526	664
611	726
695	788
779	849
864	911
948	973
5932032	·7732035
116	096
201	158
285	220
369	282
454	343
538	405
622	467
706	528
791	590
875	652
959	713
5933044	775

Number	Log
5933128	·7732837
212	898
296	960
381	·7733022
465	084
549	145
633	207
718	269
802	330
886	392
971	454
5934055	515
139	577
223	639
308	700
392	762
476	824
560	885
645	947
729	·7734009
813	070
897	132
982	194
5935066	255
150	317
234	378
319	440
403	502
487	563
571	625
655	687
740	748
824	810
908	872
992	933
5936077	995
161	·7735056
245	118
329	180
414	241
498	303
582	365
666	426
750	488
835	549
919	611
5937003	673
087	734
172	796
256	857
340	919
424	981
508	·7736042
593	104
677	165
761	227
845	289
929	350
5938014	412
098	473
182	535
266	597
350	658
435	720
519	781

Number	Log
5938603	·7736843
687	904
771	966
856	·7737027
940	089
5939024	151
108	212
192	274
277	335
361	397
445	458
529	520
613	581
697	643
782	705
866	766
950	828

Excesses

10 = 7	1 = 1
20 = 15	2 = 1
30 = 22	3 = 2
40 = 29	4 = 3
50 = 37	5 = 4
60 = 44	6 = 4
70 = 51	7 = 5
80 = 59	8 = 6
	9 = 7

42 = 31

Anti-Log Excesses

10 = 14	1 = 1
20 = 27	2 = 3
30 = 41	3 = 4
40 = 55	4 = 5
50 = 68	5 = 7
60 = 82	6 = 8
	7 = 10
	8 = 11
	9 = 12

31 = 42

Number	Log
5940034	·7737889
118	951
203	·7738012
287	074
371	135
455	197
539	258
623	320
708	381
792	443
876	504
960	566
5941044	628
128	689
212	751
297	812
381	874
465	935
549	997

Number	Log
5941633	·7739058
717	120
802	181
886	243
970	304
5942054	366
138	427
222	489
306	550
391	612
475	673
559	735
643	796
727	858
811	919
895	981
980	·7740042
5943064	104
148	165
232	227
316	288
400	350
484	411
568	472
653	534
737	595
821	657
905	718
989	780
5944073	841
157	903
241	964
326	·7741026
410	087
494	149
578	210
662	271
746	333
830	394
914	456
998	517
5945083	579
167	640
251	701
335	763
419	824
503	886
587	947
671	·7742009
755	070
839	131
923	193
5946008	254
092	316
176	377
260	439
344	500
428	561
512	623
596	684
680	746
764	807
848	868
932	930
5947017	991

5947101	·7743053
185	114
269	175
353	237
437	298
521	360
605	421
689	482
773	544
857	605
941	666
5948025	728
109	789
194	851
278	912
362	973
446	·7744035
530	096
614	157
698	219
782	280
866	342
950	403
5949034	464
118	526
202	587
286	648
370	710
454	771
538	832
622	894
706	955
790	·7745016
874	078
958	139

Excesses

10 = 7	1 = 1
20 = 15	2 = 1
30 = 22	3 = 2
40 = 29	4 = 3
50 = 37	5 = 4
60 = 44	6 = 4
70 = 51	7 = 5
80 = 58	8 = 6
	9 = 7

42 = 31

Anti-Log Excesses

10 = 14	1 = 1
20 = 27	2 = 3
30 = 41	3 = 4
40 = 55	4 = 5
50 = 68	5 = 7
60 = 82	6 = 8
	7 = 10
	8 = 11
	9 = 12

31 = 42

5950043	·7745200
5950127	·7745262
211	323
295	384
379	446
463	507
547	568
631	630
715	691
799	752
883	814
967	875
5951051	936
135	998
219	·7746059
303	120
387	181
471	243
555	304
639	365
723	427
807	488
891	549
975	611
5952059	672
143	733
227	795
311	856
395	917
479	978
563	·7747040
647	101
731	162
815	224
899	285
983	346
5953067	407
151	469
235	530
319	591
403	652
487	714
571	775
655	836
739	897
823	959
907	·7748020
991	081
5954075	142
159	204
243	265
327	326
411	387
495	449
579	510
663	571
747	632
830	694
914	755
998	816
5955082	877
166	939
250	·7749000
334	061
418	122
502	184
5955586	·7749245
670	306
754	367
838	428
922	490
5956006	551
090	612
174	673
258	734
342	796
426	857
510	918
593	979
677	·7750040
761	102
845	163
929	224
5957013	285
097	346
181	408
265	469
349	530
433	591
517	652
601	714
685	775
769	836
852	897
936	958
5958020	·7751019
104	081
188	142
272	203
356	264
440	325
524	386
608	448
692	509
776	570
859	631
943	692
5959027	753
111	814
195	876
279	937
363	998
447	·7752059
531	120
615	181
698	243
782	304
866	365
950	426

Excesses

10 = 7	1 = 1
20 = 15	2 = 1
30 = 22	3 = 2
40 = 29	4 = 3
50 = 36	5 = 4
60 = 44	6 = 4
70 = 51	7 = 5
80 = 58	8 = 6
	9 = 7

42 = 31

Anti-Log Excesses

10 = 14	1 = 1
20 = 27	2 = 3
30 = 41	3 = 4
40 = 55	4 = 5
50 = 69	5 = 7
60 = 82	6 = 8
	7 = 10
	8 = 11
	9 = 12

31 = 42

5960034	·7752487
118	548
202	609
286	671
370	732
454	793
537	854
621	915
705	976
789	·7753037
873	098
957	159
5961041	221
125	282
208	343
292	404
376	465
460	526
544	587
628	648
712	709
796	771
879	832
963	893
5962047	954
131	·7754015
215	076
299	137
383	198
466	259
550	320
634	381
718	443
802	504
886	565
970	626
5963053	687
5963137	·7754748
221	809
305	870
389	931
473	992
556	·7755053
640	114
724	175
808	236
892	297
976	358
5964060	420
143	481
227	542
311	603
395	664
479	725
563	786
646	847
730	908
814	969
898	·7756030
982	091
5965065	152
149	213
233	274
317	335
401	396
485	457
568	518
652	579
736	640
820	701
904	762
987	823
5966071	884
155	945
239	·7757006
323	067
406	128
490	189
574	250
658	311
742	372
825	433
909	494
993	555
5967077	616
161	677
244	738
328	799
412	860
496	921
580	982
663	·7758043
747	104
831	165
915	226
998	287
5968082	348
166	409
250	470
334	531
417	592
501	653

5968585	·7758713
669	774
752	835
836	896
920	957
5969004	·7759018
088	079
171	140
255	201
339	262
423	323
506	384
590	445
674	506
758	567
841	628
925	689

Excesses

10 = 7	1 = 1
20 = 15	2 = 1
30 = 22	3 = 2
40 = 29	4 = 3
50 = 36	5 = 4
60 = 44	6 = 4
70 = 51	7 = 5
80 = 58	8 = 6
	9 = 7

42 = 31

Anti-Log Excesses

10 = 14	1 = 1
20 = 27	2 = 3
30 = 41	3 = 4
40 = 55	4 = 5
50 = 69	5 = 7
60 = 82	6 = 8
	7 = 10
	8 = 11
	9 = 12

31 = 42

5970009	·7759750
093	810
176	871
260	932
344	993
428	·7760054
511	115
595	176
679	237
763	298
846	359
930	420
5971014	481
098	541
181	602
265	663
349	724
432	785
516	846

5971600	·7760907
684	968
767	·7761029
851	090
935	150
5972019	211
102	272
186	333
270	394
353	455
437	516
521	577
605	637
688	698
772	759
856	820
939	881
5973023	942
107	·7762003
191	063
274	124
358	185
442	246
525	307
609	368
693	429
777	489
860	550
944	611
5974028	672
111	733
195	794
279	855
362	915
446	976
530	·7763037
613	098
697	159
781	219
865	280
948	341
5975032	402
116	463
199	524
283	584
367	645
450	706
534	767
618	828
701	889
785	949
869	·7764010
952	071
5976036	132
120	193
203	253
287	314
371	375
454	436
538	497
622	557
705	618
789	679
873	740
956	801

5977040	·7764861
124	922
207	983
291	·7765044
375	104
458	165
542	226
625	287
709	347
793	408
876	469
960	530
5978044	591
127	651
211	712
295	773
378	834
462	894
545	955
629	·7766016
713	077
796	137
880	198
964	259
5979047	320
131	380
214	441
298	502
382	563
465	623
549	684
633	745
716	805
800	866
883	927
967	988

Excesses

10 = 7	1 = 1
20 = 15	2 = 1
30 = 22	3 = 2
40 = 29	4 = 3
50 = 36	5 = 4
60 = 44	6 = 4
70 = 51	7 = 5
80 = 58	8 = 6
	9 = 7

42 = 30

Anti-Log Excesses

10 = 14	1 = 1
20 = 28	2 = 3
30 = 41	3 = 4
40 = 55	4 = 6
50 = 69	5 = 7
60 = 83	6 = 8
	7 = 10
	8 = 11
	9 = 12

30 = 42

5980051	·7767048
134	109
218	170
301	230
385	291
469	352
552	413
636	473
720	534
803	595
887	655
970	716
5981054	777
138	837
221	898
305	959
388	·7768020
472	080
555	141
639	202
723	262
806	323
890	384
973	444
5982057	505
141	566
224	627
308	687
391	748
475	809
558	869
642	930
726	991
809	·7769051
893	112
976	173
5983060	233
143	294
227	355
311	415
394	476
478	536
561	597
645	658
728	718
812	779
896	840
979	900
5984063	961
146	·7770022
230	082
313	143
397	204
480	264
564	325
648	385
731	446
815	507
898	567
982	628
5985065	688
149	749
232	810
316	870
399	931

5985483	·7770992
566	·7771052
650	113
734	173
817	234
901	295
984	355
5986068	416
151	476
235	537
318	598
402	658
485	719
569	779
652	840
736	901
819	961
903	·7772022
986	082
5987070	143
153	204
237	264
320	325
404	385
487	446
571	506
654	567
738	628
821	688
905	749
988	809
5988072	870
155	930
239	991
322	·7773052
406	112
489	173
573	233
656	294
740	354
823	415
907	475
990	536
5989074	597
157	657
241	718
324	778
408	839
491	899
575	960
658	·7774020
742	081
825	141
909	202
992	262

Excesses

10 = 7	1 = 1
20 = 15	2 = 1
30 = 22	3 = 2
40 = 29	4 = 3
50 = 36	5 = 4
60 = 44	6 = 4
70 = 51	7 = 5
80 = 58	8 = 6
	9 = 7

42 = 30

Anti-Log Excesses

10 = 14	1 = 1
20 = 28	2 = 3
30 = 41	3 = 4
40 = 55	4 = 6
50 = 69	5 = 7
60 = 83	6 = 8
	7 = 10
	8 = 11
	9 = 12

30 = 42

5990076	·7774323
159	383
243	444
326	504
410	565
493	626
576	686
660	747
743	807
827	868
910	928
994	989
5991077	·7775049
161	110
244	170
328	231
411	291
494	351
578	412
661	472
745	533
828	593
912	654
995	714
5992079	775
162	835
245	896
329	956
412	·7776017
496	077
579	138
663	198
746	259
830	319
913	380
996	440
5993080	500

5993163	·7776561
247	621
330	682
414	742
497	803
580	863
664	924
747	984
831	·7777045
914	105
997	165
5994081	226
164	286
248	347
331	407
415	468
498	528
581	588
665	649
748	709
832	770
915	830
998	890
5995082	951
165	·7778011
249	072
332	132
415	193
499	253
582	313
666	374
749	434
832	495
916	555
999	615
5996083	676
166	736
249	797
333	857
416	917
499	978
583	·7779038
666	099
750	159
833	219
916	280
5997000	340
083	400
166	461
250	521
333	582
417	642
500	702
583	763
667	823
750	883
833	944
917	·7780004
5998000	065
084	125
167	185
250	246
334	306
417	366
500	427

5998584	·7780487
667	547
750	608
834	668
917	728
5999000	789
084	849
167	909
250	970
334	·7781030
417	090
500	151
584	211
667	271
750	332
834	392
917	452

Excesses

10 = 7	1 = 1
20 = 14	2 = 1
30 = 22	3 = 2
40 = 29	4 = 3
50 = 36	5 = 4
60 = 43	6 = 4
70 = 51	7 = 5
80 = 58	8 = 6
	9 = 7

42 = 30

Anti-Log Excesses

10 = 14	1 = 1
20 = 28	2 = 3
30 = 41	3 = 4
40 = 55	4 = 6
50 = 69	5 = 7
60 = 83	6 = 8
	7 = 10
	8 = 11
	9 = 12

30 = 42

6000000	·7781513
084	573
167	633
250	694
334	754
417	814
500	874
584	935
667	995
750	·7782055
834	116
917	176
6001000	236
084	297
167	357
250	417
334	477
417	538
500	598

6001584	·7782658
667	719
750	779
834	839
917	899
6002000	960
083	·7783020
167	080
250	141
333	201
417	261
500	321
583	382
667	442
750	502
833	562
916	623
6003000	683
083	743
166	804
250	864
333	924
416	984
499	·7784045
583	105
666	165
749	225
833	285
916	346
999	406
6004082	466
166	526
249	587
332	647
416	707
499	767
582	828
665	888
749	948
832	·7785008
915	068
998	129
6005082	189
165	249
248	309
331	369
415	430
498	490
581	550
664	610
748	671
831	731
914	791
998	851
6006081	911
164	972
247	·7786032
330	092
414	152
497	212
580	273
663	333
747	393
830	453
913	513

6006996	·7786573
6007080	634
163	694
246	754
329	814
413	874
496	935
579	995
662	·7787055
746	115
829	175
912	235
995	295
6008078	356
162	416
245	476
328	536
411	596
494	656
578	717
661	777
744	837
827	897
911	957
994	·7788017
6009077	077
160	137
243	198
327	258
410	318
493	378
576	438
659	498
743	558
826	618
909	679
992	739

Excesses

10 = 7	1 = 1
20 = 14	2 = 1
30 = 22	3 = 2
40 = 29	4 = 3
50 = 36	5 = 4
60 = 43	6 = 4
70 = 51	7 = 5
80 = 58	8 = 6
	9 = 7

42 = 30

Anti-Log Excesses

10 = 14	1 = 1
20 = 28	2 = 3
30 = 41	3 = 4
40 = 55	4 = 6
50 = 69	5 = 7
60 = 83	6 = 8
	7 = 10
	8 = 11
	9 = 12

30 = 42

Number	Log
6010075	·7788799
159	859
242	919
325	979
408	·7789039
491	099
575	160
658	220
741	280
824	340
907	400
990	460
6011074	520
157	580
240	640
323	700
406	760
490	821
573	881
656	941
739	·7790001
822	061
905	121
989	181
6012072	241
155	301
238	361
321	421
404	481
488	542
571	602
654	662
737	722
820	782
903	842
986	902
6013070	962
153	·7791022
236	082
319	142
402	202
485	262
568	322
652	382
735	442
818	502
901	562
984	622
6014067	683
150	743
234	803
317	863
400	923
483	983
566	·7792043
649	103
732	163
816	223
899	283
982	343
6015065	403
148	463
231	523
314	583
397	643

Number	Log
6015481	·7792703
564	763
647	823
730	883
813	943
896	·7793003
979	063
6016062	123
145	183
229	243
312	303
395	363
478	423
561	483
644	543
727	603
810	663
893	723
976	783
6017060	843
143	903
226	963
309	·7794023
392	083
475	143
558	203
641	263
724	323
807	383
890	443
974	503
6018057	563
140	623
223	683
306	743
389	803
472	862
555	922
638	982
721	·7795042
804	102
887	162
971	222
6019054	282
137	342
220	402
303	462
386	522
469	582
552	642
635	702
718	761
801	821
884	881
967	941

Excesses

10 = 7	1 = 1
20 = 14	2 = 1
30 = 22	3 = 2
40 = 29	4 = 3
50 = 36	5 = 4
60 = 43	6 = 4
70 = 51	7 = 5
80 = 58	8 = 6
	9 = 6

42 = 30

Anti-Log Excesses

10 = 14	1 = 1
20 = 28	2 = 3
30 = 42	3 = 4
40 = 55	4 = 6
50 = 69	5 = 7
60 = 83	6 = 8
	7 = 10
	8 = 11
	9 = 12

30 = 42

Number	Log
6020050	·7796001
133	061
216	121
299	181
383	241
466	301
549	361
632	421
715	480
798	540
881	600
964	660
6021047	720
130	780
213	840
296	900
379	960
462	·7797020
545	079
628	139
711	199
794	259
877	319
960	379
6022043	439
126	499
209	558
292	618
375	678
458	738
541	798
625	858
708	918
791	977
874	·7798037
957	097
6023040	157

Number	Log
6023123	·7798217
206	277
289	337
372	396
455	456
538	516
621	576
704	636
787	696
870	755
953	815
6024036	875
119	935
202	995
285	·7799055
368	114
451	174
534	234
617	294
700	354
783	414
866	473
949	533
6025032	593
115	653
198	713
281	772
364	832
447	892
530	952
613	·7800012
695	071
778	131
861	191
944	251
6026027	311
110	371
193	430
276	490
359	550
442	610
525	670
608	729
691	789
774	849
857	909
940	968
6027023	·7801028
106	088
189	148
272	208
355	267
438	327
521	387
604	447
687	506
770	566
853	626
935	686
6028018	745
101	805
184	865
267	925
350	984
433	·7802044

Number	Log
6028516	·7802104
599	164
682	223
765	283
848	343
931	403
6029014	462
097	522
180	582
262	642
345	701
428	761
511	821
594	881
677	940
760	·7803000
843	060
926	120

Excesses

10 = 7	1 = 1
20 = 14	2 = 1
30 = 22	3 = 2
40 = 29	4 = 3
50 = 36	5 = 4
60 = 43	6 = 4
70 = 50	7 = 5
80 = 58	8 = 6
	9 = 6

42 = 30

Anti-Log Excesses

10 = 14	1 = 1
20 = 28	2 = 3
30 = 42	3 = 4
40 = 55	4 = 6
50 = 69	5 = 7
	6 = 8
	7 = 10
	8 = 11
	9 = 12

30 = 42

Number	Log
6030009	·7803179
092	239
175	299
258	358
340	418
423	478
506	538
589	597
672	657
755	717
838	776
921	836
6031004	896
087	956
170	·7804015
252	075
335	135
418	194

6031501	·7804254
584	314
667	373
750	433
833	493
916	552
999	612
6032081	672
164	731
247	791
330	851
413	910
496	970
579	·7805030
662	089
744	149
827	209
910	268
993	328
6033076	388
159	447
242	507
325	567
407	626
490	686
573	746
656	805
739	865
822	925
905	984
988	·7806044
6034070	104
153	163
236	223
319	282
402	342
485	402
568	461
650	521
733	581
816	640
899	700
982	759
6035065	819
148	879
230	938
313	998
396	·7807058
479	117
562	177
645	236
727	296
810	356
893	415
976	475
6036059	534
142	594
224	654
307	713
390	773
473	832
556	892
639	951
721	·7808011
804	071

6036887	·7808130
970	190
6037053	249
136	309
218	369
301	428
384	488
467	547
550	607
632	666
715	726
798	786
881	845
964	905
6038047	964
129	·7809024
212	083
295	143
378	203
461	262
543	322
626	381
709	441
792	500
875	560
957	619
6039040	679
123	738
206	798
289	858
371	917
454	977
537	·7810036
620	096
702	155
785	215
868	274
951	334

Excesses

10 = 7	1 = 1
20 = 14	2 = 1
30 = 22	3 = 2
40 = 29	4 = 3
50 = 36	5 = 4
60 = 43	6 = 4
70 = 50	7 = 5
80 = 58	8 = 6
	9 = 6

41 = 30

Anti-Log Excesses

10 = 14	1 = 1
20 = 28	2 = 3
30 = 42	3 = 4
40 = 56	4 = 6
50 = 69	5 = 7
	6 = 8
	7 = 10
	8 = 11
	9 = 13

30 = 41

6040034	·7810393
116	453
199	512
282	572
365	631
448	691
530	750
613	810
696	869
779	929
861	988
944	·7811048
6041027	107
110	167
192	226
275	286
358	345
441	405
523	464
606	524
689	583
772	643
855	702
937	762
6042020	821
103	881
186	940
268	·7812000
351	059
434	119
517	178
599	238
682	297
765	356
848	416
930	475
6043013	535
096	594
178	654
261	713
344	773
427	832
509	892
592	951
675	·7813011
758	070
840	129
923	189
6044006	248
089	308
171	367

6044254	·7813427
337	486
419	546
502	605
585	664
668	724
750	783
833	843
916	902
998	962
6045081	·7814021
164	080
247	140
329	199
412	259
495	318
577	377
660	437
743	496
826	556
908	615
991	674
6046074	734
156	793
239	853
322	912
404	972
487	·7815031
570	090
652	150
735	209
818	268
901	328
983	387
6047066	447
149	506
231	565
314	625
397	684
479	744
562	803
645	862
727	922
810	981
893	·7816040
975	100
6048058	159
141	218
223	278
306	337
389	396
471	456
554	515
637	575
719	634
802	693
885	753
967	812
6049050	871
133	931
215	990
298	·7817049
381	109
463	168
546	227

6049629	·7817287
711	346
794	405
877	465
959	524

Excesses

10 = 7	1 = 1
20 = 14	2 = 1
30 = 22	3 = 2
40 = 29	4 = 3
50 = 36	5 = 4
60 = 43	6 = 4
70 = 50	7 = 5
80 = 57	8 = 6
	9 = 6

41 = 30

Anti-Log Excesses

10 = 14	1 = 1
20 = 28	2 = 3
30 = 42	3 = 4
40 = 56	4 = 6
50 = 70	5 = 7
	6 = 8
	7 = 10
	8 = 11
	9 = 13

30 = 41

6050042	·7817583
124	643
207	702
290	761
372	821
455	880
538	939
620	999
703	·7818058
786	117
868	176
951	236
6051033	295
116	354
199	414
281	473
364	532
447	592
529	651
612	710
694	769
777	829
860	888
942	947
6052025	·7819007
108	066
190	125
273	184
355	244
438	303
521	362

6052603	·7819421
686	481
768	540
851	599
934	659
6053016	718
099	777
181	836
264	896
347	955
429	·7820014
512	073
594	133
677	192
760	251
842	310
925	370
6054007	429
090	488
173	547
255	607
338	666
420	725
503	784
586	844
668	903
751	962
833	·7821021
916	081
998	140
6055081	199
164	258
246	317
329	377
411	436
494	495
576	554
659	614
742	673
824	732
907	791
989	850
6056072	910
154	969
237	·7822028
320	087
402	146
485	206
567	265
650	324
732	383
815	442
897	502
980	561
6057062	620
145	679
228	738
310	798
393	857
475	916
558	975
640	·7823034
723	094
805	153
888	212

6057970	·7823271
6058053	330
136	389
218	449
301	508
383	567
466	626
548	685
631	744
713	804
796	863
878	922
961	981
6059043	·7824040
126	099
208	158
291	218
373	277
456	336
538	395
621	454
703	513
786	572
868	632
951	691

Excesses

10 = 7	1 = 1
20 = 14	2 = 1
30 = 22	3 = 2
40 = 29	4 = 3
50 = 36	5 = 4
60 = 43	6 = 4
70 = 50	7 = 5
80 = 57	8 = 6
	9 = 6

41 = 30

Anti-Log Excesses

10 = 14	1 = 1
20 = 28	2 = 3
30 = 42	3 = 4
40 = 56	4 = 6
50 = 70	5 = 7
	6 = 8
	7 = 10
	8 = 11
	9 = 13

30 = 41

6060034	·7824750
116	809
199	868
281	927
364	986
446	·7825045
529	105
611	164
694	223
776	282
859	341

6060941	·7825400
6061024	459
106	518
189	577
271	636
353	696
436	755
518	814
601	873
683	932
766	991
848	·7826050
931	109
6062013	168
096	227
178	287
261	346
343	405
426	464
508	523
591	582
673	641
756	700
838	759
921	818
6063003	877
086	936
168	995
250	·7827054
333	114
415	173
498	232
580	291
663	350
745	409
828	468
910	527
993	586
6064075	645
157	704
240	763
322	822
405	881
487	940
570	999
652	·7828058
735	117
817	176
900	236
982	295
6065064	354
147	413
229	472
312	531
394	590
477	649
559	708
641	767
724	826
806	885
889	944
971	·7829003
6066054	062
136	121
218	180

6066301	·7829239
383	298
466	357
548	416
631	475
713	534
795	593
878	652
960	711
6067043	770
125	829
207	888
290	947
372	·7830006
455	065
537	124
619	183
702	242
784	301
867	360
949	419
6068031	478
114	537
196	596
279	655
361	714
443	773
526	832
608	890
691	949
773	·7831008
855	067
938	126
6069020	185
103	244
185	303
267	362
350	421
432	480
514	539
597	598
679	657
762	716
844	775
926	834

Excesses

10 = 7	1 = 1
20 = 14	2 = 1
30 = 21	3 = 2
40 = 29	4 = 3
50 = 36	5 = 4
60 = 43	6 = 4
70 = 50	7 = 5
80 = 57	8 = 6
	9 = 6

41 = 30

Anti-Log Excesses

10 = 14	1 = 1
20 = 28	2 = 3
30 = 42	3 = 4
40 = 56	4 = 6
50 = 70	5 = 7
	6 = 8
	7 = 10
	8 = 11
	9 = 13

30 = 41

6070009	·7831893
091	952
173	·7832011
256	070
338	129
421	188
503	246
585	305
668	364
750	423
832	482
915	541
997	600
6071079	659
162	718
244	777
327	836
409	895
491	953
574	·7833012
656	071
738	130
821	189
903	248
985	307
6072068	366
150	425
232	484
315	542
397	601
479	660
562	719
644	778
726	837
809	896
891	955
973	·7834014
6073056	072
138	131
220	190
303	249
385	308
467	367
550	426
632	485
714	543
797	602
879	661
961	720
6074044	779
126	838

6074208	·7834896
291	955
373	·7835014
455	073
538	132
620	191
702	250
784	308
867	367
949	426
6075031	485
114	544
196	603
278	661
361	720
443	779
525	838
607	897
690	956
772	·7836014
854	073
937	132
6076019	191
101	250
184	308
266	367
348	426
430	485
513	544
595	602
677	661
760	720
842	779
924	838
6077006	897
089	955
171	·7837014
253	073
335	132
418	191
500	249
582	308
665	367
747	426
829	484
911	543
994	602
6078076	661
158	720
240	778
323	837
405	896
487	955
569	·7838013
652	072
734	131
816	190
898	248
981	307
6079063	366
145	425
227	483
310	542
392	601
474	660

6079556	·7838719
639	777
721	836
803	895
885	953
968	·7839012

Excesses

10 = 7	1 = 1
20 = 14	2 = 1
30 = 21	3 = 2
40 = 29	4 = 3
50 = 36	5 = 4
60 = 43	6 = 4
70 = 50	7 = 5
80 = 57	8 = 6
	9 = 6

41 = 30

Anti-Log Excesses

10 = 14	1 = 1
20 = 28	2 = 3
30 = 42	3 = 4
40 = 56	4 = 6
50 = 70	5 = 7
	6 = 8
	7 = 10
	8 = 11
	9 = 13

30 = 41

6080050	·7839071
132	130
214	188
297	247
379	306
461	365
543	423
625	482
708	541
790	600
872	658
954	717
6081037	776
119	835
201	893
283	952
365	·7840011
448	069
530	128
612	187
694	245
777	304
859	363
941	422
6082023	480
105	539
188	598
270	656
352	715
434	774

6082516	·7840832
599	891
681	950
763	·7841009
845	067
927	126
6083010	185
092	243
174	302
256	361
338	419
421	478
503	537
585	595
667	654
749	713
832	771
914	830
996	889
6084078	947
160	·7842006
242	065
325	123
407	182
489	241
571	299
653	358
735	417
818	475
900	534
982	593
6085064	651
146	710
229	769
311	827
393	886
475	944
557	·7843003
639	062
721	120
804	179
886	238
968	296
6086050	355
132	413
214	472
297	531
379	589
461	648
543	707
625	765
707	824
789	882
872	941
954	·7844000
6087036	058
118	117
200	175
282	234
364	293
447	351
529	410
611	468
693	527
775	586

6087857	·7844644
939	703
6088022	761
104	820
186	879
268	937
350	996
432	·7845054
514	113
596	171
679	230
761	289
843	347
925	406
6089007	464
089	523
171	581
253	640
335	699
418	757
500	816
582	874
664	933
746	991
828	·7846050
910	108
992	167

Excesses

10 = 7	1 = 1
20 = 14	2 = 1
30 = 21	3 = 2
40 = 29	4 = 3
50 = 36	5 = 4
60 = 43	6 = 4
70 = 50	7 = 5
80 = 57	8 = 6
	9 = 6

41 = 30

Anti-Log Excesses

10 = 14	1 = 1
20 = 28	2 = 3
30 = 42	3 = 4
40 = 56	4 = 6
50 = 70	5 = 7
	6 = 8
	7 = 10
	8 = 11
	9 = 13

30 = 41

6090074	·7846225
156	284
239	343
321	401
403	460
485	518
567	577
649	635
731	694

6090813	·7846752
895	811
977	869
6091060	928
142	987
224	·7847045
306	104
388	162
470	221
552	279
634	338
716	396
798	455
880	513
962	572
6092044	630
127	689
209	747
291	806
373	864
455	923
537	981
619	·7848040
701	098
783	157
865	215
947	274
6093029	332
111	391
193	449
275	508
358	566
440	625
522	683
604	741
686	800
768	858
850	917
932	975
6094014	·7849034
096	092
178	151
260	209
342	268
424	326
506	385
588	443
670	502
752	560
834	619
916	677
998	736
6095080	794
163	852
245	911
327	969
409	·7850028
491	086
573	145
655	203
737	262
819	320
901	378
983	437
6096065	495

6096147	·7850554
229	612
311	671
393	729
475	787
557	846
639	904
721	963
803	·7851021
885	080
967	138
6097049	196
131	255
213	313
295	372
377	430
459	488
541	547
623	605
705	664
787	722
869	781
951	839
6098033	897
115	956
197	·7852014
279	073
361	131
443	189
525	248
607	306
689	364
771	423
853	481
935	540
6099017	598
099	656
181	715
263	773
345	831
427	890
509	948
591	·7853007
673	065
755	123
837	182
919	240

Excesses

10 = 7	1 = 1
20 = 14	2 = 1
30 = 21	3 = 2
40 = 28	4 = 3
50 = 36	5 = 4
60 = 43	6 = 4
70 = 50	7 = 5
80 = 57	8 = 6
	9 = 6

41 = 29

Anti-Log Excesses

10 = 14	1 = 1
20 = 28	2 = 3
30 = 42	3 = 4
40 = 56	4 = 6
50 = 70	5 = 7
	6 = 8
	7 = 10
	8 = 11
	9 = 13

29 = 41

6100000	·7853298
082	357
164	415
246	473
328	532
410	590
492	648
574	707
656	765
738	823
820	882
902	940
984	999
6101066	·7854057
148	115
230	174
312	232
394	290
476	349
558	407
640	465
722	524
804	582
885	640
967	698
6102049	757
131	815
213	873
295	932
377	990
459	·7855048
541	107
623	165
705	223
787	282
869	340
951	398
6103033	456
114	515
196	573
278	631
360	690
442	748
524	806
606	865
688	923
770	981
852	·7856039
934	098
6104016	156
097	214

6104179	·7856272
261	331
343	389
425	447
507	506
589	564
671	622
753	680
835	739
917	797
998	855
6105080	913
162	972
244	·7857030
326	088
408	146
490	205
572	263
654	321
736	379
817	438
899	496
981	554
6106063	612
145	671
227	729
309	787
391	845
472	904
554	962
636	·7858020
718	078
800	137
882	195
964	253
6107046	311
127	369
209	428
291	486
373	544
455	602
537	660
619	719
701	777
782	835
864	893
946	952
6108028	·7859010
110	068
192	126
274	184
355	243
437	301
519	359
601	417
683	475
765	534
847	592
928	650
6109010	708
092	766
174	824
256	883
338	941
419	999

6109501	·7860057
583	115
665	174
747	232
829	290
910	348
992	406

Excesses

10 = 7	1 = 1
20 = 14	2 = 1
30 = 21	3 = 2
40 = 28	4 = 3
50 = 36	5 = 4
60 = 43	6 = 4
70 = 50	7 = 5
80 = 57	8 = 6
	9 = 6

41 = 29

Anti-Log Excesses

10 = 14	1 = 1
20 = 28	2 = 3
30 = 42	3 = 4
40 = 56	4 = 6
50 = 70	5 = 7
	6 = 8
	7 = 10
	8 = 11
	9 = 13

29 = 41

6110074	·7860464
156	523
238	581
320	639
401	697
483	755
565	813
647	871
729	930
811	988
892	·7861046
974	104
6111056	162
138	220
220	279
302	337
383	395
465	453
547	511
629	569
711	627
792	686
874	744
956	802
6112038	860
120	918
201	976
283	·7862034
365	092

6112447	·7862151
529	209
610	267
692	325
774	383
856	441
938	499
6113019	557
101	615
183	674
265	732
347	790
428	848
510	906
592	964
674	·7863022
755	080
837	138
919	196
6114001	254
083	313
164	371
246	429
328	487
410	545
491	603
573	661
655	719
737	777
819	835
900	893
982	951
6115064	·7864010
146	068
227	126
309	184
391	242
473	300
554	358
636	416
718	474
800	532
881	590
963	648
6116045	706
127	764
208	822
290	880
372	938
454	996
535	·7865054
617	112
699	171
781	229
862	287
944	345
6117026	403
108	461
189	519
271	577
353	635
435	693
516	751
598	809
680	867

6117762	·7865925
843	983
925	·7866041
6118007	099
088	157
170	215
252	273
334	331
415	389
497	447
579	505
660	563
742	621
824	679
906	737
987	795
6119069	853
151	911
232	969
314	·7867027
396	085
478	143
559	201
641	259
723	317
804	375
886	433
968	491

Excesses

10 = 7	1 = 1
20 = 14	2 = 1
30 = 21	3 = 2
40 = 28	4 = 3
50 = 36	5 = 4
60 = 43	6 = 4
70 = 50	7 = 5
80 = 57	8 = 6
	9 = 6

41 = 29

Anti-Log Excesses

10 = 14	1 = 1
20 = 28	2 = 3
30 = 42	3 = 4
40 = 56	4 = 6
50 = 70	5 = 7
	6 = 8
	7 = 10
	8 = 11
	9 = 13

29 = 41

6120050	·7867549
131	607
213	665
295	723
376	781
458	839
540	897
621	955
703	·7868013

6120785	·7868071
866	129
948	186
6121030	244
112	302
193	360
275	418
357	476
438	534
520	592
602	650
683	708
765	766
847	824
928	882
6122010	940
092	998
173	·7869056
255	114
337	172
418	230
500	287
582	345
663	403
745	461
827	519
908	577
990	635
6123072	693
153	751
235	809
317	867
398	925
480	983
562	·7870040
643	098
725	156
807	214
888	272
970	330
6124051	388
133	446
215	504
296	562
378	620
460	677
541	735
623	793
705	851
786	909
868	967
949	·7871025
6125031	083
113	141
194	199
276	256
358	314
439	372
521	430
603	488
684	546
766	604
847	662
929	719

6126011	·7871777
092	835
174	893
255	951
337	·7872009
419	067
500	124
582	182
664	240
745	298
827	356
908	414
990	471
6127072	529
153	587
235	645
316	703
398	761
480	819
561	876
643	934
724	992
806	·7873050
888	108
969	166
6128051	223
132	281
214	339
296	397
377	455
459	512
540	570
622	628
703	686
785	744
867	801
948	859
6129030	917
111	975
193	·7874033
275	090
356	148
438	206
519	264
601	322
682	379
764	437
846	495
927	553

Excesses

10 = 7	1 = 1
20 = 14	2 = 1
30 = 21	3 = 2
40 = 28	4 = 3
50 = 35	5 = 4
60 = 43	6 = 4
70 = 50	7 = 5
80 = 57	8 = 6
	9 = 6

41 = 29

Anti Log Excesses

10 = 14	1 = 1
20 = 28	2 = 3
30 = 42	3 = 4
40 = 56	4 = 6
50 = 71	5 = 7
	6 = 8
	7 = 10
	8 = 11
	9 = 13

29 = 41

6130009	·7874611
090	668
172	726
253	784
335	842
416	899
498	957
580	·7875015
661	073
743	131
824	188
906	246
987	304
6131069	362
150	419
232	477
314	535
395	593
477	650
558	708
640	766
721	824
803	881
884	939
966	997
6132047	·7876055
129	113
211	170
292	228
374	286
455	344
537	401
618	459
700	517
781	574
863	632
944	690
6133026	748
107	805
189	863
270	921
352	979
433	·7877036
515	094
597	152
678	209
760	267
841	325
923	383
6134004	440
086	498

6134167	·7877556
249	613
330	671
412	729
493	787
575	844
656	902
738	960
819	·7878017
901	075
982	133
6135064	190
145	248
227	306
308	364
390	421
471	479
553	537
634	594
716	652
797	710
879	767
960	825
6136042	883
123	940
205	998
286	·7879056
368	113
449	171
530	229
612	286
693	344
775	402
856	459
938	517
6137019	575
101	632
182	690
264	748
345	805
427	863
508	920
590	978
671	·7880036
753	093
834	151
915	209
997	266
6138078	324
160	382
241	439
323	497
404	554
486	612
567	670
649	727
730	785
811	843
893	900
974	958
6139056	·7881015
137	073
219	131
300	188
382	246

6139463	·7881304
544	361
626	419
707	476
789	534
870	592
952	649

Excesses

10 = 7	1 = 1
20 = 14	2 = 1
30 = 21	3 = 2
40 = 28	4 = 3
50 = 35	5 = 4
60 = 42	6 = 4
70 = 50	7 = 5
80 = 57	8 = 6
	9 = 6

41 = 29

Anti-Log Excesses

10 = 14	1 = 1
20 = 28	2 = 3
30 = 42	3 = 4
40 = 56	4 = 6
50 = 71	5 = 7
	6 = 8
	7 = 10
	8 = 11
	9 = 13

29 = 41

6140033	·7881707
115	764
196	822
277	880
359	937
440	995
522	·7882052
603	110
685	168
766	225
847	283
929	340
6141010	398
092	455
173	513
254	571
336	628
417	686
499	743
580	801
662	858
743	916
824	974
906	·7883031
987	089
6142069	146
150	204
231	261
313	319
6142394	·7883376
476	434
557	492
638	549
720	607
801	664
883	722
964	779
6143045	837
127	894
208	952
290	·7884009
371	067
452	124
534	182
615	240
696	297
778	355
859	412
941	470
6144022	527
103	585
185	642
266	700
347	757
429	815
510	872
592	930
673	987
754	·7885045
836	102
917	160
998	217
6145080	275
161	332
243	390
324	447
405	505
487	562
568	620
649	677
731	735
812	792
893	850
975	907
6146056	965
137	·7886022
219	080
300	137
382	195
463	252
544	310
626	367
707	425
788	482
870	540
951	597
6147032	655
114	712
195	769
276	827
358	884
439	942
520	999
602	·7887057
6147683	·7887114
764	172
846	229
927	287
6148008	344
090	401
171	459
252	516
334	574
415	631
496	689
578	746
659	804
740	861
822	918
903	976
984	·7888033
6149065	091
147	148
228	206
309	263
391	320
472	378
553	435
635	493
716	550
797	608
879	665
960	722

Excesses

10 = 7	1 = 1
20 = 14	2 = 1
30 = 21	3 = 2
40 = 28	4 = 3
50 = 35	5 = 4
60 = 42	6 = 4
70 = 49	7 = 5
80 = 57	8 = 6
	9 = 6

41 = 29

Anti-Log Excesses

10 = 14	1 = 1
20 = 28	2 = 3
30 = 42	3 = 4
40 = 57	4 = 6
50 = 71	5 = 7
	6 = 8
	7 = 10
	8 = 11
	9 = 13

29 = 41

6150041	·7888780
122	837
204	895
285	952
366	·7889009
448	067
529	124
6150610	·7889182
692	239
773	296
854	354
935	411
6151017	469
098	526
179	583
261	641
342	698
423	756
504	813
586	870
667	928
748	985
829	·7890043
911	100
992	157
6152073	215
155	272
236	329
317	387
398	444
480	502
561	559
642	616
723	674
805	731
886	788
967	846
6153049	903
130	960
211	·7891018
292	075
374	132
455	190
536	247
617	304
699	362
780	419
861	476
942	534
6154024	591
105	648
186	706
267	763
349	820
430	878
511	935
592	992
674	·7892050
755	107
836	164
917	222
998	279
6155080	336
161	394
242	451
323	508
405	566
486	623
567	680
648	737
730	795
811	852
6155892	·7892909
973	967
6156054	·7893024
136	081
217	139
298	196
379	253
461	311
542	368
623	425
704	483
785	540
867	597
948	654
6157029	712
110	769
191	826
273	883
354	941
435	998
516	·7894055
597	113
679	170
760	227
841	284
922	342
6158003	399
085	456
166	513
247	571
328	628
409	685
491	742
572	800
653	857
734	914
815	971
897	·7895029
978	086
6159059	143
140	200
221	258
302	315
384	372
465	429
546	487
627	544
708	601
789	658
871	716
952	773

Excesses

10 = 7	1 = 1
20 = 14	2 = 1
30 = 21	3 = 2
40 = 28	4 = 3
50 = 35	5 = 4
60 = 42	6 = 4
70 = 49	7 = 5
80 = 56	8 = 6
	9 = 6

41 = 29

Anti-Log Excesses

10 = 14	1 = 1
20 = 28	2 = 3
30 = 43	3 = 4
40 = 57	4 = 6
50 = 71	5 = 7
	6 = 9
	7 = 10
	8 = 11
	9 = 13

29 = 41

6160033	·7895830
114	887
195	944
276	·7896002
358	059
439	116
520	173
601	230
682	288
763	345
845	402
926	459
6161007	517
088	574
169	631
250	688
332	745
413	803
494	860
575	917
656	974
737	·7897031
818	088
900	146
981	203
6162062	260
143	317
224	374
305	432
386	489
468	546
549	603
630	660
711	717
792	775
873	832
954	889
6163035	946
117	·7898003
198	061
279	118
360	175
441	232
522	289
603	346
684	404
766	461
847	518
928	575
6164009	632
090	689

6164171	·7898746
252	804
333	861
415	918
496	975
577	·7899032
658	089
739	146
820	204
901	261
982	318
6165063	375
144	432
226	489
307	546
388	603
469	661
550	718
631	775
712	832
793	889
874	946
955	·7900003
6166037	060
118	117
199	175
280	232
361	289
442	346
523	403
604	460
685	517
766	574
847	632
928	689
6167009	746
091	803
172	860
253	917
334	974
415	·7901031
496	088
577	145
658	202
739	260
820	317
901	374
982	431
6168063	488
144	545
226	602
307	659
388	716
469	773
550	830
631	887
712	944
793	·7902001
874	059
955	116
6169036	173
117	230
198	287
279	344
360	401

6169441	·7902458
522	515
603	572
684	629
765	686
847	743
928	800

Excesses

10 = 7	1 = 1
20 = 14	2 = 1
30 = 21	3 = 2
40 = 28	4 = 3
50 = 35	5 = 4
60 = 42	6 = 4
70 = 49	7 = 5
80 = 56	8 = 6
	9 = 6

40 = 28

Anti-Log Excesses

10 = 14	1 = 1
20 = 28	2 = 3
30 = 43	3 = 4
40 = 57	4 = 6
50 = 71	5 = 7
	6 = 9
	7 = 10
	8 = 11
	9 = 13

28 = 40

6170009	·7902857
090	914
171	971
252	·7903028
333	085
414	142
495	199
576	257
657	314
738	371
819	428
900	485
981	542
6171062	599
143	656
224	713
305	770
386	827
467	884
548	941
629	998
710	·7904055
791	112
872	169
953	226
6172034	283
115	340
196	397
277	454

6172358	·7904511
439	568
520	625
601	682
682	739
763	796
844	853
925	910
6173006	967
087	·7905024
168	081
249	138
330	195
411	252
492	309
573	366
654	423
735	479
816	536
897	593
978	650
6174059	707
140	764
221	821
302	878
383	935
464	992
545	·7906049
626	106
707	163
788	220
869	277
950	334
6175031	391
112	448
193	505
274	562
355	619
436	676
517	732
598	789
679	846
760	903
841	960
921	·7907017
6176002	074
083	131
164	188
245	245
326	302
407	359
488	416
569	472
650	529
731	586
812	643
893	700
974	757
6177055	814
136	871
217	928
298	985
379	·7908042
460	099
540	156

6177621	·7908212
702	269
783	326
864	383
945	440
6178026	497
107	554
188	611
269	668
350	724
431	781
512	838
593	895
673	952
754	·7909009
835	066
916	123
997	180
6179078	236
159	293
240	350
321	407
402	464
483	521
564	578
645	634
725	691
806	748
887	805
968	862

Excesses

10 = 7	1 = 1
20 = 14	2 = 1
30 = 21	3 = 2
40 = 28	4 = 3
50 = 35	5 = 4
60 = 42	6 = 4
70 = 49	7 = 5
80 = 56	8 = 6
	9 = 6

40 = 28

Anti-Log Excesses

10 = 14	1 = 1
20 = 28	2 = 3
30 = 43	3 = 4
40 = 57	4 = 6
50 = 71	5 = 7
	6 = 9
	7 = 10
	8 = 11
	9 = 13

28 = 40

6180049	·7909919
130	976
211	·7910033
292	089
373	146
454	203

6180534	·7910260
615	317
696	374
777	431
858	487
939	544
6181020	601
101	658
182	715
263	772
343	828
424	885
505	942
586	999
667	·7911056
748	113
829	169
910	226
990	283
6182071	340
152	397
233	453
314	510
395	567
476	624
557	681
637	738
718	794
799	851
880	908
961	965
6183042	·7912022
123	078
204	135
284	192
365	249
446	305
527	362
608	419
689	476
770	533
850	589
931	646
6184012	703
093	760
174	817
255	873
336	930
416	987
497	·7913044
578	101
659	157
740	214
821	271
901	328
982	384
6185063	441
144	498
225	555
306	611
387	668
467	725
548	782
629	838
710	895

6185791	·7913952
871	·7914009
952	065
6186033	122
114	179
195	236
276	292
356	349
437	406
518	462
599	519
680	576
761	633
841	689
922	746
6187003	803
084	860
165	916
245	973
326	·7915030
407	087
488	143
569	200
649	257
730	313
811	370
892	427
973	484
6188054	540
134	597
215	654
296	710
377	767
457	824
538	881
619	937
700	994
781	·7916051
861	107
942	164
6189023	221
104	277
185	334
265	391
346	447
427	504
508	561
589	617
669	674
750	731
831	788
912	844
992	901

Excesses

10 = 7	1 = 1
20 = 14	2 = 1
30 = 21	3 = 2
40 = 28	4 = 3
50 = 35	5 = 4
60 = 42	6 = 4
70 = 49	7 = 5
80 = 56	8 = 6
	9 = 6

40 = 28

Anti-Log Excesses

10 = 14	1 = 1
20 = 28	2 = 3
30 = 43	3 = 4
40 = 57	4 = 6
50 = 71	5 = 7
	6 = 9
	7 = 10
	8 = 11
	9 = 13

28 = 40

6190073	·7916958
154	·7917014
235	071
316	128
396	184
477	241
558	298
639	354
719	411
800	468
881	524
962	581
6191042	637
123	694
204	751
285	807
365	864
446	921
527	977
608	·7918034
688	091
769	147
850	204
931	261
6192011	317
092	374
173	430
254	487
334	544
415	600
496	657
577	714
657	770
738	827
819	883
900	940
980	997

6193061	·7919053
142	110
223	167
303	223
384	280
465	336
546	393
626	450
707	506
788	563
868	619
949	676
6194030	733
111	789
191	846
272	902
353	959
433	·7920016
514	072
595	129
676	185
756	242
837	299
918	355
998	412
6195079	468
160	525
241	581
321	638
402	695
483	751
563	808
644	864
725	921
806	977
886	·7921034
967	091
6196048	147
128	204
209	260
290	317
370	373
451	430
532	486
612	543
693	600
774	656
855	713
935	769
6197016	826
097	882
177	939
258	995
339	·7922052
419	108
500	165
581	222
661	278
742	335
823	391
903	448
984	504
6198065	561
145	617
226	674

6198307	·7922730
387	787
468	843
549	900
629	956
710	·7923013
791	069
871	126
952	182
6199033	239
113	295
194	352
275	408
355	465
436	521
517	578
597	634
678	691
759	747
839	804
920	860

Excesses

10 = 7	1 = 1
20 = 14	2 = 1
30 = 21	3 = 2
40 = 28	4 = 3
50 = 35	5 = 4
60 = 42	6 = 4
70 = 49	7 = 5
80 = 56	8 = 6
	9 = 6

40 = 28

Anti-Log Excesses

10 = 14	1 = 1
20 = 29	2 = 3
30 = 43	3 = 4
40 = 57	4 = 6
50 = 71	5 = 7
	6 = 9
	7 = 10
	8 = 11
	9 = 13

28 = 40

6200000	·7923917
081	973
162	·7924030
242	086
323	143
404	199
484	256
565	312
646	369
726	425
807	482
888	538
968	595
6201049	651
129	707

Number	Log
6201210	·7924764
291	821
371	877
452	933
533	990
613	·7925046
694	103
774	159
855	216
936	272
6202016	329
097	385
178	442
258	498
339	554
419	611
500	667
581	724
661	780
742	837
822	893
903	950
984	·7926006
6203064	062
145	119
225	175
306	232
387	288
467	345
548	401
628	457
709	514
790	570
870	627
951	683
6204031	739
112	796
193	852
273	909
354	965
434	·7927022
515	078
596	134
676	191
757	247
837	304
918	360
998	416
6205079	473
160	529
240	586
321	642
401	698
482	755
563	811
643	868
724	924
804	980
885	·7928037
965	093
6206046	149
127	206
207	262
288	319
368	375

Number	Log
6206449	·7928431
529	488
610	544
690	600
771	657
852	713
932	770
6207013	826
093	882
174	939
254	995
335	·7929051
415	108
496	164
577	220
657	277
738	333
818	390
899	446
979	502
6208060	558
140	615
221	671
301	728
382	784
462	840
543	897
624	953
704	·7930009
785	066
865	122
946	178
6209026	235
107	291
187	347
268	404
348	460
429	516
509	573
590	629
670	685
751	742
831	798
912	854
992	911

Excesses

10 = 7	1 = 1
20 = 14	2 = 1
30 = 21	3 = 2
40 = 28	4 = 3
50 = 35	5 = 4
60 = 42	6 = 4
70 = 49	7 = 5
80 = 56	8 = 6
	9 = 6

40 = 28

Anti-Log Excesses

10 = 14	1 = 1
20 = 29	2 = 3
30 = 43	3 = 4
40 = 57	4 = 6
50 = 71	5 = 7
	6 = 9
	7 = 10
	8 = 11
	9 = 13

28 = 40

Number	Log
6210073	·7930967
153	·7931023
234	079
315	136
395	192
476	248
556	305
637	361
717	417
798	474
878	530
959	586
6211039	642
120	699
200	755
281	811
361	868
442	924
522	980
603	·7932036
683	093
764	149
844	205
925	261
6212005	318
085	374
166	430
246	487
327	543
407	599
488	655
568	712
649	768
729	824
810	880
890	937
971	993
6213051	·7933049
132	105
212	162
293	218
373	274
454	330
534	387
615	443
695	499
776	555
856	612
936	668
6214017	724
097	780

Number	Log
6214178	·7933836
258	893
339	949
419	·7934005
500	061
580	118
661	174
741	230
821	286
902	343
982	399
6215063	455
143	511
224	567
304	624
385	680
465	736
546	792
626	848
706	905
787	961
867	·7935017
948	073
6216028	129
109	186
189	242
269	298
350	354
430	410
511	467
591	523
672	579
752	635
832	691
913	747
993	804
6217074	860
154	916
235	972
315	·7936028
395	085
476	141
556	197
637	253
717	309
798	365
878	422
958	478
6218039	534
119	590
200	646
280	702
360	758
441	815
521	871
602	927
682	983
762	·7937039
843	095
923	152
6219004	208
084	264
164	320
245	376
325	432

Number	Log
6219406	·7937488
486	545
566	601
647	657
727	713
808	769
888	825
968	881

Excesses

10 = 7	1 = 1
20 = 14	2 = 1
30 = 21	3 = 2
40 = 28	4 = 3
50 = 35	5 = 3
60 = 42	6 = 4
70 = 49	7 = 5
80 = 56	8 = 6
	9 = 6

40 = 28

Anti-Log Excesses

10 = 14	1 = 1
20 = 29	2 = 3
30 = 43	3 = 4
40 = 57	4 = 6
50 = 72	5 = 7
	6 = 9
	7 = 10
	8 = 11
	9 = 13

28 = 40

Number	Log
6220049	·7937938
129	994
209	·7938050
290	106
370	162
451	218
531	274
611	330
692	386
772	443
853	499
933	555
6221013	611
094	667
174	723
254	779
335	835
415	891
495	948
576	·7939004
656	060
737	116
817	172
897	228
978	284
6222058	340
138	396
219	452

6222299	·7939508
379	565
460	621
540	677
621	733
701	789
781	845
862	901
942	957
6223022	·7940013
103	069
183	125
263	181
344	237
424	294
504	350
585	406
665	462
745	518
826	574
906	630
986	686
6224067	742
147	798
227	854
308	910
388	966
468	·7941022
549	078
629	134
709	190
790	246
870	302
950	358
6225031	415
111	471
191	527
272	583
352	639
432	695
513	751
593	807
673	863
753	919
834	975
914	·7942031
994	087
6226075	143
155	199
235	255
316	311
396	367
476	423
557	479
637	535
717	591
797	647
878	703
958	759
6227038	815
119	871
199	927
279	983
359	·7943039
440	095

6227520	·7943151
600	207
681	263
761	319
841	375
922	431
6228002	487
082	543
162	599
243	655
323	711
403	767
483	823
564	879
644	935
724	991
805	·7944046
885	102
965	158
6229045	214
126	270
206	326
286	382
366	438
447	494
527	550
607	606
687	662
768	718
848	774
928	830

Excesses

10 = 7	1 = 1
20 = 14	2 = 1
30 = 21	3 = 2
40 = 28	4 = 3
50 = 35	5 = 3
60 = 42	6 = 4
70 = 49	7 = 5
80 = 56	8 = 6
	9 = 6

40 = 28

Anti Log Excesses

10 = 14	1 = 1
20 = 29	2 = 3
30 = 43	3 = 4
40 = 57	4 = 6
50 = 72	5 = 7
	6 = 9
	7 = 10
	8 = 11
	9 = 13

28 = 40

6230009	·7944886
089	942
169	998
249	·7945054
330	110

6230410	·7945166
490	222
570	278
651	333
731	389
811	445
891	501
972	557
6231052	613
132	669
212	725
293	781
373	837
453	893
533	949
613	·7946005
694	061
774	116
854	172
934	228
6232015	284
095	340
175	396
255	452
336	508
416	564
496	620
576	676
656	731
737	787
817	843
897	899
977	955
6233058	·7947011
138	067
218	123
298	179
378	234
459	290
539	346
619	402
699	458
779	514
860	570
940	626
6234020	682
100	737
180	793
261	849
341	905
421	961
501	·7948017
581	073
662	129
742	184
822	240
902	296
982	352
6235063	408
143	464
223	520
303	575
383	631
464	687
544	743

6235624	·7948799
704	855
784	911
865	966
945	·7949022
6236025	078
105	134
185	190
265	246
346	302
426	357
506	413
586	469
666	525
746	581
827	637
907	692
987	748
6237067	804
147	860
227	916
308	971
388	·7950027
468	083
548	139
628	195
708	251
789	306
869	362
949	418
6238029	474
109	530
189	585
269	641
350	697
430	753
510	809
590	864
670	920
750	976
831	·7951032
911	087
991	143
6239071	199
151	255
231	311
311	366
391	422
472	478
552	534
632	590
712	645
792	701
872	757
952	813

Excesses

10 = 7	1 = 1
20 = 14	2 = 1
30 = 21	3 = 2
40 = 28	4 = 3
50 = 35	5 = 3
60 = 42	6 = 4
70 = 49	7 = 5
80 = 56	8 = 6
	9 = 6

40 = 28

Anti Log Excesses

10 = 14	1 = 1
20 = 29	2 = 3
30 = 43	3 = 4
40 = 57	4 = 6
50 = 72	5 = 7
	6 = 9
	7 = 10
	8 = 11
	9 = 13

28 = 40

6240033	·7951868
113	924
193	980
273	·7952036
353	091
433	147
513	203
593	259
674	314
754	370
834	426
914	482
994	537
6241074	593
154	649
234	705
314	760
395	816
475	872
555	928
635	983
715	·7953039
795	095
875	151
955	206
6242035	262
116	318
196	373
276	429
356	485
436	541
516	596
596	652
676	708
756	764
836	819
916	875

6242997	·7953931
6243077	986
157	·7954042
237	098
317	154
397	209
477	265
557	321
637	376
717	432
797	488
878	543
958	599
6244038	655
118	711
198	766
278	822
358	878
438	933
518	989
598	·7955045
678	100
758	156
838	212
918	267
998	323
6245079	379
159	434
239	490
319	546
399	602
479	657
559	713
639	769
719	824
799	880
879	936
959	991
6246039	·7956047
119	102
199	158
279	214
359	269
439	325
520	381
600	436
680	492
760	548
840	603
920	659
6247000	715
080	770
160	826
240	882
320	937
400	993
480	·7957048
560	104
640	160
720	215
800	271
880	327
960	382
6248040	438
120	493

6248200	·7957549
280	605
360	660
440	716
520	772
600	827
680	883
760	938
840	994
920	·7958050
6249000	105
080	161
160	216
240	272
320	328
400	383
480	439
560	494
640	550
720	606
800	661
880	717
960	772

Excesses

10 = 7	1 = 1
20 = 14	2 = 1
30 = 21	3 = 2
40 = 28	4 = 3
50 = 35	5 = 3
60 = 42	6 = 4
70 = 49	7 = 5
80 = 56	8 = 6
	9 = 6

40 = 28

Anti Log Excesses

10 = 14	1 = 1
20 = 29	2 = 3
30 = 43	3 = 4
40 = 58	4 = 6
50 = 72	5 = 7
	6 = 9
	7 = 10
	8 = 12
	9 = 13

28 = 40

6250040	·7958828
120	884
200	939
280	995
360	·7959050
440	106
520	162
600	217
680	273
760	328
840	384
920	439
6251000	495

6251080	·7959551
160	606
240	662
320	717
400	773
480	828
560	884
640	940
720	995
800	·7960051
880	106
960	162
6252040	217
120	273
200	328
280	384
360	440
440	495
520	551
600	606
680	662
760	717
840	773
920	828
6253000	884
080	939
160	995
240	·7961050
320	106
400	162
480	217
559	273
639	328
719	384
799	439
879	495
959	550
6254039	606
119	661
199	717
279	772
359	828
439	883
519	939
599	994
679	·7962050
759	105
839	161
919	216
999	272
6255078	327
158	383
238	438
318	494
398	549
478	605
558	660
638	716
718	771
798	827
878	882
958	938
6256038	993
118	·7963049
197	104

6256277	·7963160
357	215
437	271
517	326
597	381
677	437
757	492
837	548
917	603
997	659
6257076	714
156	770
236	825
316	881
396	936
476	992
556	·7964047
636	103
716	158
796	213
876	269
955	324
6258035	380
115	435
195	491
275	546
355	602
435	657
515	712
595	768
674	823
754	879
834	934
914	990
994	·7965045
6259074	100
154	156
234	211
314	267
393	322
473	378
553	433
633	488
713	544
793	599
873	655
953	710

Excesses

10 = 7	1 = 1
20 = 14	2 = 1
30 = 21	3 = 2
40 = 28	4 = 3
50 = 35	5 = 3
60 = 42	6 = 4
70 = 49	7 = 5
	8 = 6
	9 = 6

40 = 28

Anti Log Excesses

10 = 14	1 = 1
20 = 29	2 = 3
30 = 43	3 = 4
40 = 58	4 = 6
50 = 72	5 = 7
	6 = 9
	7 = 10
	8 = 12
	9 = 13

28 = 40

6260032	·7965765
112	821
192	876
272	932
352	987
432	·7966043
512	098
592	153
671	209
751	264
831	320
911	375
991	430
6261071	486
151	541
230	597
310	652
390	707
470	763
550	818
630	873
710	929
789	984
869	·7967040
949	095
6262029	150
109	206
189	261
268	316
348	372
428	427
508	483
588	538
668	593
748	649
827	704
907	759
987	815
6263067	870
147	926
227	981
306	·7968036
386	092
466	147
546	202
626	258
705	313
785	368
865	424
945	479
6264025	534

6264105	·7968590
184	645
264	700
344	756
424	811
504	866
583	922
663	977
743	·7969032
823	088
903	143
983	198
6265062	254
142	309
222	364
302	420
382	475
461	530
541	586
621	641
701	696
781	752
860	807
940	862
6266020	917
100	973
180	·7970028
259	083
339	139
419	194
499	249
579	305
658	360
738	415
818	470
898	526
977	581
6267057	636
137	692
217	747
297	802
376	857
456	913
536	968
616	·7971023
695	079
775	134
855	189
935	244
6268015	300
094	355
174	410
254	465
334	521
413	576
493	631
573	686
653	742
732	797
812	852
892	907
972	963
6269051	·7972018
131	073
211	128

6269291	·7972184
370	239
450	294
530	349
610	405
689	460
769	515
849	570
929	626

Excesses

10 = 7	1 = 1
20 = 14	2 = 1
30 = 21	3 = 2
40 = 28	4 = 3
50 = 35	5 = 3
60 = 42	6 = 4
70 = 49	7 = 5
	8 = 6
	9 = 6

40 = 28

Anti Log Excesses

10 = 14	1 = 1
20 = 29	2 = 3
30 = 43	3 = 4
40 = 58	4 = 6
50 = 72	5 = 7
	6 = 9
	7 = 10
	8 = 12
	9 = 13

28 = 40

6270008	·7972681
088	736
168	791
248	847
327	902
407	957
487	·7973012
567	067
646	123
726	178
806	233
886	288
965	344
6271045	399
125	454
205	509
284	564
364	620
444	675
523	730
603	785
683	841
763	896
842	951
922	·7974006
6272002	061
081	117

6272161	·7974172
241	227
321	282
400	337
480	392
560	448
639	503
719	558
799	613
879	668
958	724
6273038	779
118	834
197	889
277	944
357	·7975000
437	055
516	110
596	165
676	220
755	275
835	331
915	386
994	441
6274074	496
154	551
234	606
313	661
393	717
473	772
552	827
632	882
712	937
791	992
871	·7976048
951	103
6275030	158
110	213
190	268
269	323
349	378
429	434
508	489
588	544
668	599
747	654
827	709
907	764
986	820
6276066	875
146	930
225	985
305	·7977040
385	095
464	150
544	205
624	261
703	316
783	371
863	426
942	481
6277022	536
102	591
181	646
261	701

6277341	·7977757
420	812
500	867
580	922
659	977
739	·7978032
819	087
898	142
978	197
6278058	252
137	308
217	363
296	418
376	473
456	528
535	583
615	638
695	693
774	748
854	803
934	858
6279013	914
093	969
172	·7979024
252	079
332	134
411	189
491	244
571	299
650	354
730	409
809	464
889	519
969	574

Excesses

10 = 7	1 = 1
20 = 14	2 = 1
30 = 21	3 = 2
40 = 28	4 = 3
50 = 35	5 = 3
60 = 42	6 = 4
70 = 48	7 = 5
	8 = 6
	9 = 6

40 = 28

Anti Log Excesses

10 = 14	1 = 1
20 = 29	2 = 3
30 = 43	3 = 4
40 = 58	4 = 6
50 = 72	5 = 7
	6 = 9
	7 = 10
	8 = 12
	9 = 13

28 = 40

6280048	·7979629
128	684

6280208	·7979740
287	795
367	850
446	905
526	960
606	·7980015
685	070
765	125
844	180
924	235
6281004	290
083	345
163	400
242	455
322	510
402	565
481	620
561	675
640	730
720	785
800	840
879	895
959	950
6282038	·7981005
118	060
198	116
277	171
357	226
436	281
516	336
596	391
675	446
755	501
834	556
914	611
993	666
6283073	721
153	776
232	831
312	886
391	941
471	996
550	·7982051
630	106
710	161
789	216
869	271
948	326
6284028	381
107	436
187	491
267	546
346	601
426	656
505	711
585	766
664	820
744	875
823	930
903	985
983	·7983040
6285062	095
142	150
221	205
301	260

Number	Log
6285380	·7983315
460	370
539	425
619	480
699	535
778	590
858	645
937	700
6286017	755
096	810
176	865
255	920
335	975
414	·7984030
494	085
573	140
653	194
733	249
812	304
892	359
971	414
6287051	469
130	524
210	579
289	634
369	689
448	744
528	799
607	854
687	909
766	964
846	·7985018
925	073
6288005	128
084	183
164	238
243	293
323	348
403	403
482	458
562	513
641	568
721	623
800	677
880	732
959	787
6289039	842
118	897
198	952
277	·7986007
357	062
436	117
516	172
595	226
675	281
754	336
834	391
913	446
993	501

Excesses

10 = 7	1 = 1
20 = 14	2 = 1
30 = 21	3 = 2
40 = 28	4 = 3
50 = 35	5 = 3
60 = 41	6 = 4
70 = 48	7 = 5
	8 = 6
	9 = 6

40 = 27

Anti Log Excesses

10 = 14	1 = 1
20 = 29	2 = 3
30 = 43	3 = 4
40 = 58	4 = 6
50 = 72	5 = 7
	6 = 9
	7 = 10
	8 = 12
	9 = 13

27 = 40

Number	Log
6290072	·7986556
152	611
231	666
311	720
390	775
469	830
549	885
628	940
708	995
787	·7987050
867	105
946	159
6291026	214
105	269
185	324
264	379
344	434
423	489
503	543
582	598
662	653
741	708
821	763
900	818
980	873
6292059	927
138	982
218	·7988037
297	092
377	147
456	202
536	257
615	312
695	366
774	421
854	476
933	531

Number	Log
6293012	·7988586
092	641
171	695
251	750
330	805
410	860
489	915
569	970
648	·7989024
728	079
807	134
886	189
966	244
6294045	298
125	353
204	408
284	463
363	518
442	572
522	627
601	682
681	737
760	792
840	846
919	901
999	956
6295078	·7990011
157	066
237	120
316	175
396	230
475	285
554	340
634	394
713	449
793	504
872	559
952	614
6296031	668
110	723
190	778
269	833
349	887
428	942
507	997
587	·7991052
666	107
746	161
825	216
905	271
984	326
6297063	380
143	435
222	490
302	545
381	599
460	654
540	709
619	764
698	818
778	873
857	928
937	983
6298016	·7992037
095	092

Number	Log
6298175	·7992147
254	202
334	256
413	311
492	366
572	421
651	475
731	530
810	585
889	640
969	694
6299048	749
127	804
207	858
286	913
366	968
445	·7993023
524	077
604	132
683	187
762	241
842	296
921	351

Excesses

10 = 7	1 = 1
20 = 14	2 = 1
30 = 21	3 = 2
40 = 28	4 = 3
50 = 34	5 = 3
60 = 41	6 = 4
70 = 48	7 = 5
	8 = 6
	9 = 6

40 = 27

Anti Log Excesses

10 = 14	1 = 1
20 = 29	2 = 3
30 = 43	3 = 4
40 = 58	4 = 6
50 = 72	5 = 7
	6 = 9
	7 = 10
	8 = 12
	9 = 13

27 = 40

Number	Log
6300000	·7993406
080	460
159	515
239	570
318	624
397	679
477	734
556	788
635	843
715	898
794	953
873	·7994007
953	062

Number	Log
6301032	·7994117
112	171
191	226
270	281
350	335
429	390
508	445
588	499
667	554
746	609
826	663
905	718
984	773
6302064	827
143	882
222	937
302	991
381	·7995046
460	101
540	155
619	210
698	265
778	319
857	374
936	429
6303016	483
095	538
174	593
254	648
333	702
412	757
492	811
571	866
650	921
730	975
809	·7996030
888	085
968	139
6304047	194
126	249
205	303
285	358
364	413
443	467
523	522
602	576
681	631
761	686
840	740
919	795
999	850
6305078	904
157	959
236	·7997013
316	068
395	123
474	177
554	232
633	286
712	341
791	396
871	450
950	505
6306029	560
109	614

6306188	·7997669
267	723
347	778
426	833
505	887
584	942
664	996
743	·7998051
822	106
901	160
981	215
6307060	269
139	324
219	379
298	433
377	488
456	542
536	597
615	652
694	706
773	761
853	815
932	870
6308011	924
091	979
170	·7999034
249	088
328	142
408	197
487	252
566	306
645	361
725	415
804	470
883	525
962	579
6309042	634
121	688
200	743
279	797
359	852
438	906
517	961
596	·8000016
676	070
755	125
834	179
913	234
993	288

Excesses

10 = 7	1 = 1
20 = 14	2 = 1
30 = 21	3 = 2
40 = 28	4 = 3
50 = 34	5 = 3
60 = 41	6 = 4
70 = 48	7 = 5
	8 = 6
	9 = 6

40 = 27

Anti Log Excesses

10 = 15	1 = 1
20 = 29	2 = 3
30 = 44	3 = 4
40 = 58	4 = 6
50 = 73	5 = 7
	6 = 9
	7 = 10
	8 = 12
	9 = 13

27 = 40

6310072	·8000343
151	397
230	452
310	506
389	561
468	615
547	670
626	724
706	779
785	834
864	888
943	943
6311023	997
102	·8001052
181	106
260	161
340	215
419	270
498	324
577	379
656	433
736	488
815	542
894	597
973	651
6312052	706
132	760
211	815
290	869
369	924
449	978
528	·8002033
607	087
686	142
765	196
845	251
924	305
6313003	360
082	414
161	468
241	523
320	577
399	632
478	686
557	741
637	795
716	850
795	904
874	959
953	·8003013
6314032	068

6314112	·8003122
191	177
270	231
349	286
428	340
508	395
587	449
666	503
745	558
824	612
904	667
983	721
6315062	776
141	830
220	885
299	939
379	994
458	·8004048
537	102
616	157
695	211
774	266
854	320
933	375
6316012	429
091	483
170	538
249	592
329	647
408	701
487	756
566	810
645	864
724	919
803	973
883	·8005028
962	082
6317041	136
120	191
199	245
278	300
358	354
437	409
516	463
595	517
674	572
753	626
832	681
912	735
991	789
6318070	844
149	898
228	953
307	·8006007
386	061
466	116
545	170
624	225
703	279
782	333
861	388
940	442
6319019	497
099	551
178	605

6319257	·8006660
336	714
415	768
494	823
573	877
652	932
732	986
811	·8007040
890	095
969	149

Excesses

10 = 7	1 = 1
20 = 14	2 = 1
30 = 21	3 = 2
40 = 28	4 = 3
50 = 34	5 = 3
60 = 41	6 = 4
70 = 48	7 = 5
	8 = 6
	9 = 6

40 = 27

Anti Log Excesses

10 = 15	1 = 1
20 = 29	2 = 3
30 = 44	3 = 4
40 = 58	4 = 6
50 = 73	5 = 7
	6 = 9
	7 = 10
	8 = 12
	9 = 13

27 = 40

6320048	·8007203
127	258
206	312
285	366
364	421
444	475
523	530
602	584
681	638
760	693
839	747
918	801
997	856
6321076	910
155	964
235	·8008019
314	073
393	128
472	182
551	236
630	291
709	345
788	399
867	454
946	508
6322025	562

6322105	·8008617
184	671
263	725
342	780
421	834
500	888
579	942
658	997
737	·8009051
816	105
895	160
974	214
6323054	268
133	323
212	377
291	431
370	486
449	540
528	594
607	649
686	703
765	757
844	811
923	866
6324002	920
081	974
161	·8010029
240	083
319	137
398	192
477	246
556	300
635	354
714	409
793	463
872	517
951	571
6325030	626
109	680
188	734
267	789
346	843
425	897
504	951
583	·8011006
663	060
742	114
821	168
900	223
979	277
6326058	331
137	385
216	440
295	494
374	548
453	602
532	657
611	711
690	765
769	819
848	874
927	928
6327006	982
085	·8012036
164	091

6327243	·8012145
322	199
401	253
480	308
559	362
638	416
717	470
796	525
875	579
954	633
6328033	687
112	741
191	796
270	850
349	904
428	958
507	·8013013
586	067
665	121
744	175
823	229
902	284
981	338
6329060	392
139	446
218	500
297	555
376	609
455	663
534	717
613	771
692	826
771	880
850	934
929	988

Excesses

10 = 7	1 = 1
20 = 14	2 = 1
30 = 21	3 = 2
40 = 27	4 = 3
50 = 34	5 = 3
60 = 41	6 = 4
70 = 48	7 = 5
	8 = 5
	9 = 6

40 = 27

Anti Log Excesses

10 = 15	1 = 1
20 = 29	2 = 3
30 = 44	3 = 4
40 = 58	4 = 6
50 = 73	5 = 7
	6 = 9
	7 = 10
	8 = 12
	9 = 13

27 = 40

6330008	·8014042
6330087	·8014097
166	151
245	205
324	259
403	313
482	368
561	422
640	476
719	530
798	584
877	638
956	693
6331035	747
114	801
193	855
272	909
351	963
430	·8015018
509	072
588	126
667	180
746	234
825	289
904	343
983	397
6332062	451
141	505
220	559
299	613
378	668
457	722
536	776
614	830
693	884
772	938
851	993
930	·8016047
6333009	101
088	155
167	209
246	263
325	317
404	371
483	426
562	480
641	534
720	588
799	642
878	696
957	750
6334036	805
114	859
193	913
272	967
351	·8017021
430	075
509	129
588	183
667	237
746	292
825	346
904	400
983	454
6335062	508
141	562
6335220	·8017616
298	670
377	725
456	779
535	833
614	887
693	941
772	995
851	·8018049
930	103
6336009	157
088	211
167	265
245	320
324	374
403	428
482	482
561	536
640	590
719	644
798	698
877	752
956	806
6337035	860
113	914
192	968
271	·8019023
350	077
429	131
508	185
587	239
666	293
745	347
823	401
902	455
981	509
6338060	563
139	617
218	671
297	725
376	779
455	833
533	887
612	942
691	996
770	·8020050
849	104
928	158
6339007	212
086	266
164	320
243	374
322	428
401	482
480	536
559	590
638	644
717	698
795	752
874	806
953	860

Excesses

10 = 7	1 = 1
20 = 14	2 = 1
30 = 21	3 = 2
40 = 27	4 = 3
50 = 34	5 = 3
60 = 41	6 = 4
70 = 48	7 = 5
	8 = 5
	9 = 6

40 = 27

Anti Log Excesses

10 = 15	1 = 1
20 = 29	2 = 3
30 = 44	3 = 4
40 = 58	4 = 6
50 = 73	5 = 7
	6 = 9
	7 = 10
	8 = 12
	9 = 13

27 = 40

6340032	·8020914
111	968
190	·8021022
269	076
347	130
426	184
505	238
584	292
663	346
742	400
821	454
899	508
978	562
6341057	616
136	670
215	724
294	778
373	832
451	886
530	940
609	994
688	·8022048
767	102
846	156
924	210
6342003	264
082	318
161	372
240	426
319	480
398	534
476	588
555	642
634	696
713	750
792	804
871	858
6342949	·8022912
6343028	966
107	·8023020
186	074
265	128
343	182
422	236
501	290
580	344
659	398
738	452
816	506
895	559
974	613
6344053	667
132	721
210	775
289	829
368	883
447	937
526	991
605	·8024045
683	099
762	153
841	207
920	261
999	315
6345077	369
156	423
235	476
314	530
393	584
471	638
550	692
629	746
708	800
787	854
865	908
944	962
6346023	·8025016
102	070
180	124
259	178
338	232
417	285
496	339
574	393
653	447
732	501
811	555
890	609
968	663
6347047	717
126	771
205	824
283	878
362	932
441	986
520	·8026040
598	094
677	148
756	202
835	256
914	309
992	363

6348071	·8026417
150	471
229	525
307	579
386	633
465	687
544	740
622	794
701	848
780	902
859	956
937	·8027010
6349016	064
095	118
174	171
252	225
331	279
410	333
489	387
567	441
646	495
725	549
804	603
882	657
961	711

Excesses

10 = 7	1 = 1
20 = 14	2 = 1
30 = 21	3 = 2
40 = 27	4 = 3
50 = 34	5 = 3
60 = 41	6 = 4
70 = 48	7 = 5
	8 = 5
	9 = 6

40 = 27

Anti Log Excesses

10 = 15	1 = 1
20 = 29	2 = 3
30 = 44	3 = 4
40 = 58	4 = 6
50 = 73	5 = 7
	6 = 9
	7 = 10
	8 = 12
	9 = 13

27 = 40

6350040	·8027764
119	818
197	872
276	926
355	980
434	·8028034
512	088
591	141
670	195
748	249
827	303

6350906	·8028357
985	411
6351063	464
142	518
221	572
300	626
378	680
457	734
536	788
614	841
693	895
772	949
851	·8029003
929	057
6352008	111
087	164
165	218
244	272
323	326
402	380
480	433
559	487
638	541
716	595
795	649
874	703
953	756
6353031	810
110	864
189	917
267	971
346	·8030025
425	079
503	133
582	186
661	240
740	294
818	348
897	401
976	455
6354054	509
133	563
212	617
290	670
369	724
448	778
526	832
605	885
684	939
762	993
841	·8031047
920	101
999	154
6355077	208
156	262
235	316
313	369
392	423
471	477
549	531
628	584
707	638
785	692
864	746
943	799

6356021	·8031853
100	907
179	961
257	·8032014
336	068
415	122
493	176
572	229
651	283
729	337
808	391
887	444
965	498
6357044	552
122	606
201	659
280	713
358	767
437	820
516	874
594	928
673	982
752	·8033035
830	089
909	143
988	196
6358066	250
145	304
224	358
302	411
381	465
459	519
538	572
617	626
695	680
774	734
853	787
931	841
6359010	895
088	948
167	·8034002
246	056
324	109
403	163
482	217
560	270
639	324
717	378
796	432
875	485
953	539

Excesses

10 = 7	1 = 1
20 = 14	2 = 1
30 = 20	3 = 2
40 = 27	4 = 3
50 = 34	5 = 3
60 = 41	6 = 4
70 = 48	7 = 5
	8 = 5
	9 = 6

39 = 27

Anti Log Excesses

10 = 15	1 = 1
20 = 29	2 = 3
30 = 44	3 = 4
40 = 59	4 = 6
50 = 73	5 = 7
	6 = 9
	7 = 10
	8 = 12
	9 = 13

27 = 39

6360032	·8034593
111	646
189	700
268	754
346	807
425	861
504	915
582	968
661	·8035022
739	076
818	129
897	183
975	237
6361054	290
132	344
211	398
290	451
368	505
447	559
525	612
604	666
683	720
761	773
840	827
918	881
997	934
6362076	988
154	·8036042
233	095
311	149
390	203
469	256
547	310
626	363
704	417
783	471
861	524
940	578
6363019	632
097	685
176	739
254	793
333	846
412	900
490	953
569	·8037007
647	061
726	114
804	168
883	222
962	275

6364040	·8037329
119	382
197	436
276	490
354	543
433	597
511	651
590	704
669	758
747	811
826	865
904	919
983	972
6365061	·8038026
140	079
218	133
297	186
376	240
454	294
533	347
611	401
690	454
768	508
847	562
925	615
6366004	669
082	722
161	776
240	829
318	883
397	937
475	990
554	·8039044
632	097
711	151
789	205
868	258
946	312
6367025	365
103	419
182	472
261	526
339	580
418	633
496	687
575	740
653	794
732	847
810	901
889	954
967	·8040008
6368046	061
124	115
203	169
281	222
360	276
438	329
517	383
595	436
674	490
752	543
831	597
909	650
988	704
6369066	757

6369145	·8040811
223	865
302	918
380	972
459	·8041025
537	079
616	132
694	186
773	239
851	293
930	346

Excesses

10 = 7	1 = 1
20 = 14	2 = 1
30 = 20	3 = 2
40 = 27	4 = 3
50 = 34	5 = 3
60 = 41	6 = 4
70 = 48	7 = 5
	8 = 5
	9 = 6

39 = 27

Anti Log Excesses

10 = 15	1 = 1
20 = 29	2 = 3
30 = 44	3 = 4
40 = 59	4 = 6
50 = 73	5 = 7
	6 = 9
	7 = 10
	8 = 12
	9 = 13

27 = 39

6370008	·8041400
087	453
165	507
244	560
322	614
401	667
479	721
558	774
636	828
715	881
793	935
872	988
950	·8042042
6371029	095
107	149
186	202
264	256
343	309
421	363
500	416
578	470
656	523
735	577
813	630
892	684

6371970	·8042737
6372049	791
127	844
206	898
284	951
363	·8043005
441	058
520	111
598	165
677	218
755	272
833	325
912	379
990	432
6373069	486
147	539
226	593
304	646
383	700
461	753
540	806
618	860
696	913
775	967
853	·8044020
932	074
6374010	127
089	181
167	234
246	288
324	341
402	394
481	448
559	501
638	555
716	608
795	662
873	715
951	769
6375030	822
108	875
187	929
265	982
344	·8045036
422	089
500	142
579	196
657	249
736	303
814	356
893	410
971	463
6376049	516
128	570
206	623
285	677
363	730
442	783
520	837
598	890
677	944
755	997
834	·8046050
912	104
990	157

6377069	·8046211
147	264
226	317
304	371
382	424
461	478
539	531
618	584
696	638
774	692
853	745
931	798
6378010	851
088	905
166	958
245	·8047012
323	065
402	118
480	172
558	225
637	278
715	332
793	385
872	439
950	492
6379029	545
107	599
185	652
264	705
342	759
421	812
499	865
577	919
656	972
734	·8048026
812	079
891	132
969	186

Excesses

10 = 7	1 = 1
20 = 14	2 = 1
30 = 20	3 = 2
40 = 27	4 = 3
50 = 34	5 = 3
60 = 41	6 = 4
70 = 48	7 = 5
	8 = 5
	9 = 6

39 = 27

Anti Log Excesses

10 = 15	1 = 1
20 = 29	2 = 3
30 = 44	3 = 4
40 = 59	4 = 6
50 = 73	5 = 7
	6 = 9
	7 = 10
	8 = 12
	9 = 13

27 = 39

6380048	·8048239
126	292
204	346
283	399
361	452
439	506
518	559
596	612
674	666
753	719
831	772
910	826
988	879
6381066	932
145	986
223	·8049039
301	092
380	146
458	199
536	252
615	306
693	359
771	412
850	466
928	519
6382006	572
085	625
163	679
241	732
320	785
398	839
477	892
555	945
633	999
712	·8050052
790	105
868	159
947	212
6383025	265
103	318
182	372
260	425
338	478
417	532
495	585
573	638
651	691
730	745
808	798
886	851
965	905

6384043	·8050958
121	·8051011
200	064
278	118
356	171
435	224
513	278
591	331
670	384
748	437
826	491
905	544
983	597
6385061	650
139	704
218	757
296	810
374	863
453	917
531	970
609	·8052023
688	076
766	130
844	183
923	236
6386001	289
079	343
157	396
236	449
314	502
392	556
471	609
549	662
627	715
705	769
784	822
862	875
940	928
6387019	981
097	·8053035
175	088
253	141
332	194
410	248
488	301
567	354
645	407
723	460
801	514
880	567
958	620
6388036	673
114	727
193	780
271	833
349	886
428	939
506	993
584	·8054046
662	099
741	152
819	205
897	259
975	312
6389054	365

6389132	·8054418
210	471
288	525
367	578
445	631
523	684
601	737
680	791
758	844
836	897
914	950
993	·8055003

Excesses

10 = 7	1 = 1
20 = 14	2 = 1
30 = 20	3 = 2
40 = 27	4 = 3
50 = 34	5 = 3
60 = 41	6 = 4
70 = 48	7 = 5
	8 = 5
	9 = 6

39 = 27

Anti Log Excesses

10 = 15	1 = 1
20 = 29	2 = 3
30 = 44	3 = 4
40 = 59	4 = 6
50 = 74	5 = 7
	6 = 9
	7 = 10
	8 = 12
	9 = 13

27 = 39

6390071	·8055056
149	110
227	163
306	216
384	269
462	322
540	376
619	429
697	482
775	535
853	588
932	641
6391010	694
088	748
166	801
245	854
323	907
401	960
479	·8056013
557	067
636	120
714	173
792	226
870	279

6391949	·8056332
6392027	385
105	439
183	492
261	545
340	598
418	651
496	704
574	758
653	811
731	864
809	917
887	970
965	·8057023
6393044	076
122	129
200	183
278	236
356	289
435	342
513	395
591	448
669	501
747	554
826	608
904	661
982	714
6394060	767
138	820
217	873
295	926
373	979
451	·8058032
529	086
608	139
686	192
764	245
842	298
920	351
999	404
6395077	457
155	510
233	563
311	616
389	670
468	723
546	776
624	829
702	882
780	935
859	988
937	·8059041
6396015	094
093	147
171	200
249	254
328	307
406	360
484	413
562	466
640	519
718	572
797	625
875	678
953	731

6397031	·8059784
109	837
187	890
266	943
344	996
422	·8060050
500	103
578	156
656	209
734	262
813	315
891	368
969	421
6398047	474
125	527
203	580
282	633
360	686
438	739
516	792
594	845
672	898
750	951
829	·8061004
907	057
985	110
6399063	163
141	216
219	269
297	322
375	375
454	428
532	482
610	535
688	588
766	641
844	694
922	747

Excesses

10 = 7	1 = 1
20 = 14	2 = 1
30 = 20	3 = 2
40 = 27	4 = 3
50 = 34	5 = 3
60 = 41	6 = 4
70 = 48	7 = 5
	8 = 5
	9 = 6

39 = 27

Anti Log Excesses

10 = 15	1 = 1
20 = 29	2 = 3
30 = 44	3 = 4
40 = 59	4 = 6
50 = 74	5 = 7
	6 = 9
	7 = 10
	8 = 12
	9 = 13

27 = 39

6400000	·8061800
079	853
157	906
235	959
313	·8062012
391	065
469	118
547	171
625	224
704	277
782	330
860	383
938	436
6401016	489
094	542
172	595
250	648
328	701
407	754
485	807
563	860
641	913
719	966
797	·8063019
875	072
953	125
6402031	178
110	231
188	283
266	336
344	389
422	442
500	495
578	548
656	601
734	654
812	707
890	760
969	813
6403047	866
125	919
203	972
281	·8064025
359	078
437	131
515	184
593	237
671	290
749	343
827	396
906	449

6403984	·8064502
6404062	555
140	608
218	661
296	713
374	766
452	819
530	872
608	925
686	978
764	·8065031
842	084
920	137
999	190
6405077	243
155	296
233	349
311	402
389	454
467	507
545	560
623	613
701	666
779	719
857	772
935	825
6406013	878
091	931
169	984
247	·8066037
325	090
404	142
482	195
560	248
638	301
716	354
794	407
872	460
950	513
6407028	566
106	619
184	672
262	725
340	778
418	831
496	883
574	936
652	989
730	·8067042
808	095
886	148
964	201
6408042	254
120	306
198	359
276	412
354	465
432	518
510	571
588	624
667	677
745	729
823	782
901	835
979	888

6409057	·8067941
135	994
213	·8068047
291	099
369	152
447	205
525	258
603	311
681	364
759	417
837	469
915	522
993	575

Excesses

10 = 7	1 = 1
20 = 14	2 = 1
30 = 20	3 = 2
40 = 27	4 = 3
50 = 34	5 = 3
60 = 41	6 = 4
70 = 47	7 = 5
	8 = 5
	9 = 6

39 = 26

Anti Log Excesses

10 = 15	1 = 1
20 = 29	2 = 3
30 = 44	3 = 4
40 = 59	4 = 6
50 = 74	5 = 7
	6 = 9
	7 = 10
	8 = 12
	9 = 13

26 = 39

6410071	·8068628
149	681
227	734
305	787
383	839
461	892
539	945
617	998
695	·8069051
773	104
851	156
929	209
6411007	262
085	315
163	368
241	421
319	473
397	526
475	579
553	632
631	685
709	737
787	790

6411865	·8069843
942	896
6412020	949
098	·8070002
176	054
254	107
332	160
410	213
488	266
566	318
644	371
722	424
800	477
878	530
956	582
6413034	635
112	688
190	741
268	794
346	846
424	899
502	952
580	·8071005
658	058
736	110
814	163
892	216
970	269
6414048	322
126	374
203	427
281	480
359	533
437	585
515	638
593	691
671	744
749	797
827	849
905	902
983	955
6415061	·8072008
139	060
217	113
295	166
373	219
451	271
529	324
606	377
684	430
762	482
840	535
918	588
996	641
6416074	693
152	746
230	799
308	852
386	904
464	957
542	·8073010
620	063
697	115
775	168
853	221

6416931	·8073274
6417009	326
087	379
165	432
243	485
321	537
399	590
477	643
555	695
632	748
710	801
788	854
866	906
944	959
6418022	·8074012
100	064
178	117
256	170
334	223
412	275
489	328
567	381
645	433
723	486
801	539
879	592
957	644
6419035	697
113	750
190	802
268	855
346	908
424	960
502	·8075013
580	066
658	119
736	171
814	224
891	277
969	329

Excesses

10 = 7	1 = 1
20 = 14	2 = 1
30 = 20	3 = 2
40 = 27	4 = 3
50 = 34	5 = 3
60 = 41	6 = 4
70 = 47	7 = 5
	8 = 5
	9 = 6

39 = 26

Anti Log Excesses

10 = 15	1 = 1
20 = 30	2 = 3
30 = 44	3 = 4
40 = 59	4 = 6
50 = 74	5 = 7
	6 = 9
	7 = 10
	8 = 12
	9 = 13

26 = 39

6420047	·8075382
125	435
203	487
281	540
359	593
437	645
514	698
592	751
670	803
748	856
826	909
904	961
982	·8076014
6421060	067
137	119
215	172
293	225
371	277
449	330
527	383
605	435
683	488
760	541
838	593
916	646
994	699
6422072	751
150	804
228	857
305	909
383	962
461	·8077015
539	067
617	120
695	173
772	225
850	278
928	330
6423006	383
084	436
162	488
240	541
317	594
395	646
473	699
551	751
629	804
707	857
784	909
862	962
940	·8078015

6424018	·8078067
096	120
174	172
251	225
329	278
407	330
485	383
563	435
641	488
718	541
796	593
874	646
952	699
6425030	751
108	804
185	856
263	909
341	962
419	·8079014
497	067
574	119
652	172
730	225
808	277
886	330
963	382
6426041	435
119	487
197	540
275	593
352	645
430	698
508	750
586	803
664	855
742	908
819	961
897	·8080013
975	066
6427053	118
130	171
208	223
286	276
364	329
442	381
519	434
597	486
675	539
753	591
831	644
908	696
986	749
6428064	802
142	854
220	907
297	959
375	·8081012
453	064
531	117
608	169
686	222
764	275
842	327
920	380
997	432

6429075	·8081485
153	537
231	590
308	642
386	695
464	747
542	800
619	852
697	905
775	958
853	·8082010
931	063

Excesses

10 = 7	1 = 1
20 = 14	2 = 1
30 = 20	3 = 2
40 = 27	4 = 3
50 = 34	5 = 3
60 = 41	6 = 4
70 = 47	7 = 5
	8 = 5
	9 = 6

39 = 26

Anti Log Excesses

10 = 15	1 = 1
20 = 30	2 = 3
30 = 44	3 = 4
40 = 59	4 = 6
50 = 74	5 = 7
	6 = 9
	7 = 10
	8 = 12
	9 = 13

26 = 39

6430008	·8082115
086	168
164	220
242	273
319	325
397	378
475	430
553	483
630	535
708	588
786	640
864	693
941	745
6431019	798
097	850
175	903
252	955
330	·8083008
408	060
486	113
563	165
641	218
719	270
797	323

6431874	·8083375
952	428
6432030	480
107	533
185	585
263	638
341	690
418	743
496	795
574	848
652	900
729	952
807	·8084005
885	057
962	110
6433040	162
118	215
196	267
273	320
351	372
429	425
507	477
584	530
662	582
740	635
817	687
895	740
973	792
6434051	844
128	897
206	949
284	·8085002
361	054
439	107
517	159
595	212
672	264
750	316
828	369
905	421
983	474
6435061	526
138	579
216	631
294	684
372	736
449	788
527	841
605	893
682	946
760	998
838	·8086051
915	103
993	156
6436071	208
148	260
226	313
304	365
381	418
459	470
537	522
615	575
692	627
770	680
848	732

6436925	·8086784
6437003	837
081	889
158	942
236	994
314	·8087047
391	099
469	151
547	204
624	256
702	308
780	361
857	413
935	466
6438013	518
090	570
168	623
246	675
323	728
401	780
479	832
556	885
634	937
712	990
789	·8088042
867	094
945	147
6439022	199
100	251
177	304
255	356
333	409
410	462
488	514
566	566
643	618
721	670
799	723
876	775
954	827

Excesses

10 = 7	1 = 1
20 = 13	2 = 1
30 = 20	3 = 2
40 = 27	4 = 3
50 = 34	5 = 3
60 = 40	6 = 4
70 = 47	7 = 5
	8 = 5
	9 = 6

39 = 26

Anti Log Excesses

10 = 15	1 = 1
20 = 30	2 = 3
30 = 44	3 = 4
40 = 59	4 = 6
50 = 74	5 = 7
	6 = 9
	7 = 10
	8 = 12
	9 = 13

26 = 39

6440032	·8088880
109	932
187	984
264	·8089037
342	089
420	141
497	194
575	246
653	298
730	351
808	403
886	456
963	508
6441041	560
118	613
196	665
274	717
351	769
429	822
507	874
584	926
662	979
739	·8090031
817	083
895	136
972	188
6442050	240
127	293
205	345
283	397
360	450
438	502
516	554
593	607
671	659
748	711
826	764
904	816
981	868
6443059	921
136	973
214	·8091025
292	078
369	130
447	182
524	235
602	287
680	339
757	391
835	444
912	496

6443990	·8091548
6444068	601
145	653
223	705
300	757
378	810
455	862
533	914
611	967
688	·8092019
766	071
843	123
921	176
999	228
6445076	280
154	332
231	385
309	437
386	489
464	542
542	594
619	646
697	698
774	751
852	803
929	855
6446007	907
085	960
162	·8093012
240	064
317	116
395	169
472	221
550	273
628	326
705	378
783	430
860	482
938	535
6447015	587
093	639
170	691
248	743
326	796
403	848
481	900
558	952
636	·8094005
713	057
791	109
868	161
946	214
6448023	266
101	318
179	370
256	422
334	475
411	527
489	579
566	631
644	684
721	736
799	788
876	840
954	892

Number	Log
6449031	·8094945
109	997
186	·8095049
264	101
342	153
419	206
497	258
574	310
652	362
729	414
807	467
884	519
962	571

Excesses

10 = 7	1 = 1
20 = 13	2 = 1
30 = 20	3 = 2
40 = 27	4 = 3
50 = 34	5 = 3
60 = 40	6 = 4
70 = 47	7 = 5
	8 = 5
	9 = 6

39 = 26

Anti Log Excesses

10 = 15	1 = 1
20 = 30	2 = 3
30 = 45	3 = 4
40 = 59	4 = 6
50 = 74	5 = 7
	6 = 9
	7 = 10
	8 = 12
	9 = 13

26 = 39

Number	Log
6450039	·8095623
117	675
194	728
272	780
349	832
427	884
504	936
582	989
659	·8096041
737	093
814	145
892	197
969	250
6451047	302
124	354
202	406
279	458
357	510
434	563
512	615
589	667
667	719
744	771
6451822	·8096823
899	876
977	928
6452054	980
132	·8097032
209	084
287	136
364	189
442	241
519	293
597	345
674	397
752	449
829	502
907	554
984	606
6453062	658
139	710
217	762
294	814
372	867
449	919
527	971
604	·8098023
682	075
759	127
837	179
914	231
992	284
6454069	336
146	388
224	440
301	492
379	544
456	596
534	649
611	701
689	753
766	805
844	857
921	909
999	961
6455076	·8099013
153	065
231	118
308	170
386	222
463	274
541	326
618	378
696	430
773	482
851	534
928	586
6456005	639
083	691
160	743
238	795
315	847
393	899
470	951
548	·8100003
625	055
702	107
780	159
6456857	·8100212
935	264
6457012	316
090	368
167	420
244	472
322	524
399	576
477	628
554	680
632	732
709	784
786	836
864	889
941	941
6458019	993
096	·8101045
174	097
251	149
328	201
406	253
483	305
561	357
638	409
716	461
793	513
870	565
948	617
6459025	669
103	721
180	773
257	826
335	878
412	930
490	982
567	·8102034
644	086
722	138
799	190
877	242
954	294

Excesses

10 = 7	1 = 1
20 = 13	2 = 1
30 = 20	3 = 2
40 = 27	4 = 3
50 = 34	5 = 3
60 = 40	6 = 4
70 = 47	7 = 5
	8 = 5
	9 = 6

39 = 26

Anti Log Excesses

10 = 15	1 = 1
20 = 30	2 = 3
30 = 45	3 = 4
40 = 59	4 = 6
50 = 74	5 = 7
	6 = 9
	7 = 10
	8 = 12
	9 = 13

26 = 39

Number	Log
6460031	·8102346
109	398
186	450
264	502
341	554
418	606
496	658
573	710
651	762
728	814
805	866
883	918
960	970
6461038	·8103022
115	074
192	126
270	178
347	230
424	282
502	334
579	386
657	438
734	490
811	542
889	594
966	646
6462044	698
121	750
198	802
276	854
353	906
430	958
508	·8104010
585	062
662	114
740	166
817	218
895	270
972	322
6463049	374
127	426
204	478
281	530
359	582
436	634
513	686
591	738
668	790
746	842
823	894
900	946
6463978	·8104998
6464055	·8105050
132	102
210	154
287	206
364	258
442	310
519	362
596	414
674	466
751	518
828	570
906	622
983	674
6465060	726
138	778
215	830
292	882
370	934
447	986
524	·8106038
602	090
679	142
756	193
834	245
911	297
988	349
6466066	401
143	453
220	505
298	557
375	609
452	661
530	713
607	765
684	817
762	869
839	921
916	973
994	·8107025
6467071	077
148	128
226	180
303	232
380	284
457	336
535	388
612	440
689	492
767	544
844	596
921	648
999	700
6468076	751
153	803
231	855
308	907
385	959
462	·8108011
540	062
617	114
694	166
772	218
849	270
926	322

6469004	·8108374
081	426
158	478
235	529
313	581
390	633
467	685
545	737
622	789
699	841
776	893
854	944
931	996

Excesses

10 = 7	1 = 1
20 = 13	2 = 1
30 = 20	3 = 2
40 = 27	4 = 3
50 = 34	5 = 3
60 = 40	6 = 4
70 = 47	7 = 5
	8 = 5
	9 = 6

39 = 26

Anti Log Excesses

10 = 15	1 = 1
20 = 30	2 = 3
30 = 45	3 = 4
40 = 60	4 = 6
50 = 74	5 = 7
	6 = 9
	7 = 10
	8 = 12
	9 = 13

26 = 39

6470008	·8109048
086	100
163	152
240	204
317	256
395	308
472	359
549	411
626	463
704	515
781	567
858	619
936	671
6471013	722
090	774
167	826
245	878
322	930
399	982
476	·8110033
554	085
631	137
708	189

6471785	·8110241
863	293
940	345
6472017	396
094	448
172	500
249	552
326	604
403	656
481	707
558	759
635	811
712	863
790	915
867	967
944	·8111018
6473021	070
099	122
176	174
253	226
330	278
408	329
485	381
562	433
639	485
717	537
794	588
871	640
948	692
6474026	744
103	796
180	847
257	899
334	951
412	·8112003
489	055
566	106
643	158
721	210
798	262
875	314
952	365
6475029	417
107	469
184	521
261	573
338	624
416	676
493	728
570	780
647	832
724	883
802	935
879	987
956	·8113039
6476033	090
110	142
188	194
265	246
342	298
419	349
496	401
574	453
651	505
728	556

6476805	·8113608
882	660
960	712
6477037	763
114	815
191	867
268	919
346	971
423	·8114022
500	074
577	126
654	178
732	229
809	281
886	333
963	384
6478040	436
118	488
195	540
272	591
349	643
426	695
503	747
581	798
658	850
735	902
812	954
889	·8115005
966	057
6479044	109
121	161
198	212
275	264
352	316
429	368
507	419
584	471
661	523
738	574
815	626
892	678
970	730

Excesses

10 = 7	1 = 1
20 = 13	2 = 1
30 = 20	3 = 2
40 = 27	4 = 3
50 = 34	5 = 3
60 = 40	6 = 4
70 = 47	7 = 5
	8 = 5
	9 = 6

39 = 26

Anti Log Excesses

10 = 15	1 = 1
20 = 30	2 = 3
30 = 45	3 = 4
40 = 60	4 = 6
50 = 75	5 = 7
	6 = 9
	7 = 10
	8 = 12
	9 = 13

26 = 39

6480047	·8115781
124	833
201	885
278	936
355	988
433	·8116040
510	091
587	143
664	195
741	247
818	298
895	350
973	402
6481050	453
127	505
204	557
281	608
358	660
436	712
513	763
590	815
667	867
744	919
821	970
898	·8117022
976	074
6482053	125
130	177
207	229
284	280
361	332
438	384
515	435
593	487
670	539
747	590
824	642
901	694
978	745
6483055	797
132	849
210	900
287	952
364	·8118004
441	055
518	107
595	159
672	210
749	262
827	314
904	365

6483981	·8118417
6484058	469
135	520
212	572
289	623
366	675
443	727
521	778
598	830
675	882
752	933
829	985
906	·8119037
983	088
6485060	140
137	191
214	243
292	295
369	346
446	398
523	450
600	501
677	553
754	604
831	656
908	708
985	759
6486062	811
140	863
217	914
294	966
371	·8120017
448	069
525	121
602	172
679	224
756	276
833	327
910	379
987	430
6487065	482
142	534
219	585
296	637
373	688
450	740
527	792
604	843
681	895
758	946
835	998
912	·8121049
989	101
6488066	153
144	204
221	256
298	307
375	359
452	411
529	462
606	514
683	565
760	617
837	668
914	720

6488991	·8121772
6489068	823
145	875
222	926
299	978
376	·8122029
453	081
531	132
608	184
685	236
762	287
839	339
916	390
993	442

Excesses

10 = 7	1 = 1
20 = 13	2 = 1
30 = 20	3 = 2
40 = 27	4 = 3
50 = 33	5 = 3
60 = 40	6 = 4
70 = 47	7 = 5
	8 = 5
	9 = 6

38 = 25

Anti Log Excesses

10 = 15	1 = 1
20 = 30	2 = 3
30 = 45	3 = 4
40 = 60	4 = 6
50 = 75	5 = 7
	6 = 9
	7 = 10
	8 = 12
	9 = 13

25 = 38

6490070	·8122493
147	545
224	597
301	648
378	700
455	751
532	803
609	854
686	906
763	957
840	·8123009
917	061
994	112
6491071	164
148	215
225	267
302	318
379	370
456	421
533	473
610	524
687	576

6491765	·8123627
842	679
919	730
996	782
6492073	833
150	885
227	937
304	988
381	·8124040
458	091
535	143
612	194
689	246
766	297
843	349
920	400
997	452
6493074	503
151	555
228	606
305	658
382	709
459	761
536	813
613	864
690	915
767	967
844	·8125018
921	070
998	121
6494075	173
152	224
229	276
306	327
383	379
460	430
537	481
614	533
691	584
768	636
845	687
922	739
999	790
6495076	842
153	893
230	945
306	996
383	·8126048
460	099
537	151
614	202
691	254
768	305
845	356
922	408
999	459
6496076	511
153	562
230	614
307	665
384	717
461	768
538	820
615	871
692	922

6496769	·8126974
846	·8127025
923	077
6497000	128
077	180
154	231
231	283
308	334
385	385
462	437
538	488
615	540
692	591
769	643
846	694
923	745
6498000	797
077	848
154	900
231	951
308	·8128003
385	054
462	105
539	157
616	208
693	260
770	311
847	363
923	414
6499000	465
077	517
154	568
231	620
308	671
385	722
462	774
539	825
616	877
693	928
770	979
847	·8129031
924	082

Excesses

10 = 7	1 = 1
20 = 13	2 = 1
30 = 20	3 = 2
40 = 27	4 = 3
50 = 33	5 = 3
60 = 40	6 = 4
70 = 47	7 = 5
	8 = 5
	9 = 6

38 = 25

Anti Log Excesses

10 = 15	1 = 1
20 = 30	2 = 3
30 = 45	3 = 4
40 = 60	4 = 6
50 = 75	5 = 7
	6 = 9
	7 = 10
	8 = 12
	9 = 13

25 = 38

6500000	·8129134
077	185
154	236
231	288
308	339
385	390
462	442
539	493
616	545
693	596
770	647
847	699
924	750
6501000	802
077	853
154	904
231	956
308	·8130007
385	058
462	110
539	161
616	213
693	264
769	315
846	367
923	418
6502000	469
077	521
154	572
231	624
308	675
385	726
462	778
538	829
615	880
692	932
769	983
846	·8131034
923	086
6503000	137
077	188
154	240
230	291
307	342
384	394
461	445
538	496
615	548
692	599
769	650
846	702

6503922	·8131753
999	804
6504076	856
153	907
230	958
307	·8132010
384	061
461	112
537	164
614	215
691	266
768	318
845	369
922	420
999	472
6505075	523
152	574
229	626
306	677
383	728
460	780
537	831
613	882
690	933
767	985
844	·8133036
921	087
998	139
6506075	190
151	241
228	293
305	344
382	395
459	446
536	498
613	549
689	600
766	652
843	703
920	754
997	805
6507074	857
150	908
227	959
304	·8134011
381	062
458	113
535	164
611	216
688	267
765	318
842	370
919	421
996	472
6508072	523
149	575
226	626
303	677
380	728
457	780
533	831
610	882
687	934
764	985
841	·8135036

6508917	·8135087
994	139
6509071	190
148	241
225	292
302	344
378	395
455	446
532	497
609	549
686	600
762	651
839	702
916	753
993	805

Excesses

10 = 7	1 = 1
20 = 13	2 = 1
30 = 20	3 = 2
40 = 27	4 = 3
50 = 33	5 = 3
60 = 40	6 = 4
70 = 47	7 = 5
	8 = 5
	9 = 6

38 = 25

Anti Log Excesses

10 = 15	1 = 1
20 = 30	2 = 3
30 = 45	3 = 4
40 = 60	4 = 6
50 = 75	5 = 7
	6 = 9
	7 = 10
	8 = 12
	9 = 13

25 = 38

6510070	·8135856
146	907
223	958
300	·8136010
377	061
454	112
530	163
607	215
684	266
761	317
838	368
914	419
991	471
6511068	522
145	573
222	624
298	676
375	727
452	778
529	829
606	880

6511682	·8136932
759	983
836	·8137034
913	085
989	137
6512066	188
143	239
220	290
297	341
373	392
450	444
527	495
604	546
680	597
757	648
834	700
911	751
988	802
6513064	853
141	904
218	956
295	·8138007
371	058
448	109
525	160
602	211
678	263
755	314
832	365
909	416
985	467
6514062	519
139	570
216	621
292	672
369	723
446	774
523	826
599	877
676	928
753	979
830	·8139030
906	081
983	133
6515060	184
137	235
213	286
290	337
367	388
444	439
520	491
597	542
674	593
751	644
827	695
904	746
981	798
6516058	849
134	900
211	951
288	·8140002
365	053
441	104
518	156
595	207

6516671	·8140258
748	309
825	360
902	411
978	462
6517055	513
132	565
208	616
285	667
362	718
439	769
515	820
592	871
669	922
746	974
822	·8141025
899	076
976	127
6518052	178
129	229
206	280
282	331
359	382
436	434
513	485
589	536
666	587
743	638
819	689
896	740
973	791
6519050	842
126	893
203	945
280	996
356	·8142047
433	098
510	149
586	200
663	251
740	302
816	353
893	404
970	455

Excesses

10 = 7	1 = 1
20 = 13	2 = 1
30 = 20	3 = 2
40 = 27	4 = 3
50 = 33	5 = 3
60 = 40	6 = 4
70 = 47	7 = 5
	8 = 5
	9 = 6

38 = 25

Anti Log Excesses

10 = 15	1 = 2
20 = 30	2 = 3
30 = 45	3 = 5
40 = 60	4 = 6
50 = 75	5 = 8
	6 = 9
	7 = 11
	8 = 12
	9 = 14

25 = 38

6520047	·8142506
123	557
200	609
277	660
353	711
430	762
507	813
583	864
660	915
737	966
813	·8143017
890	068
967	119
6521043	170
120	221
197	272
273	324
350	375
427	426
503	477
580	528
657	579
733	630
810	681
887	732
963	783
6522040	834
117	885
193	936
270	987
347	·8144038
423	089
500	140
577	191
653	242
730	293
807	345
883	396
960	447
6523037	498
113	549
190	600
267	651
343	702
420	753
496	804
573	855
650	906
726	957
803	·8145008
880	059

6523956	·8145110
6524033	161
110	212
186	263
263	314
340	365
416	416
493	467
569	518
646	569
723	620
799	671
876	722
953	773
6525029	824
106	875
182	926
259	977
336	·8146028
412	079
489	130
566	181
642	232
719	283
795	334
872	385
949	436
6526025	487
102	538
179	589
255	640
332	691
408	742
485	793
562	844
638	895
715	946
791	997
868	·8147048
945	099
6527021	150
098	201
174	252
251	303
328	354
404	405
481	456
557	507
634	558
711	609
787	660
864	711
940	762
6528017	813
094	864
170	914
247	965
323	·8148016
400	067
477	118
553	169
630	220
706	271
783	322
860	373

6528936	·8148424
6529013	475
089	526
166	577
242	628
319	679
396	730
472	780
549	831
625	882
702	933
778	984
855	·8149035
932	086

Excesses

10 = 7	1 = 1
20 = 13	2 = 1
30 = 20	3 = 2
40 = 27	4 = 3
50 = 33	5 = 3
60 = 40	6 = 4
70 = 47	7 = 5
	8 = 5
	9 = 6

38 = 25

Anti Log Excesses

10 = 15	1 = 2
20 = 30	2 = 3
30 = 45	3 = 5
40 = 60	4 = 6
50 = 75	5 = 8
	6 = 9
	7 = 11
	8 = 12
	9 = 14

25 = 38

6530008	·8149137
085	188
161	239
238	290
314	341
391	392
468	442
544	493
621	544
697	595
774	646
850	697
927	748
6531003	799
080	850
157	901
233	952
310	·8150002
386	053
463	104
539	155
616	206

6531692	·8150257
769	308
846	359
922	410
999	461
6532075	511
152	562
228	613
305	664
381	715
458	766
534	817
611	868
688	919
764	969
841	·8151020
917	071
994	122
6533070	173
147	224
223	275
300	326
376	376
453	427
529	478
606	529
682	580
759	631
836	682
912	733
989	783
6534065	834
142	885
218	936
295	987
371	·8152038
448	089
524	139
601	190
677	241
754	292
830	343
907	394
983	445
6535060	495
136	546
213	597
289	648
366	699
442	750
519	800
595	851
672	902
748	953
825	·8153004
901	055
978	105
6536054	156
131	207
207	258
284	309
360	360
437	410
513	461
590	512

6536666	·8153563
743	614
819	665
896	715
972	766
6537049	817
125	868
202	919
278	969
355	·8154020
431	071
508	122
584	173
661	224
737	274
814	325
890	376
967	427
6538043	477
119	528
196	579
272	630
349	681
425	731
502	782
578	833
655	884
731	935
808	985
884	·8155036
961	087
6539037	138
114	189
190	239
267	290
343	341
419	392
496	442
572	493
649	544
725	595
802	646
878	696
955	747

Excesses

10 = 7	1 = 1
20 = 13	2 = 1
30 = 20	3 = 2
40 = 27	4 = 3
50 = 33	5 = 3
60 = 40	6 = 4
70 = 46	7 = 5
	8 = 5
	9 = 6

38 = 25

Anti Log Excesses

10 = 15	1 = 2
20 = 30	2 = 3
30 = 45	3 = 5
40 = 60	4 = 6
50 = 75	5 = 8
	6 = 9
	7 = 11
	8 = 12
	9 = 14

25 = 38

6540031	·8155798
108	849
184	899
260	950
337	·8156001
413	052
490	102
566	153
643	204
719	255
796	305
872	356
948	407
6541025	458
101	508
178	559
254	610
331	661
407	711
484	762
560	813
636	864
713	914
789	965
866	·8157016
942	067
6542019	117
095	168
171	219
248	270
324	320
401	371
477	422
554	473
630	523
706	574
783	625
859	675
936	726
6543012	777
088	828
165	878
241	929
318	980
394	·8158030
471	081
547	132
623	183
700	233
776	284
853	335

6543929	·8158385
6544005	436
082	487
158	537
235	588
311	639
387	690
464	740
540	791
617	842
693	892
769	943
846	994
922	·8159044
999	095
6545075	146
151	197
228	247
304	298
381	349
457	399
533	450
610	501
686	551
762	602
839	653
915	703
992	754
6546068	805
144	855
221	906
297	957
374	·8160007
450	058
526	109
603	159
679	210
755	261
832	311
908	362
985	413
6547061	463
137	514
214	565
290	615
366	666
443	717
519	767
595	818
672	869
748	919
825	970
901	·8161021
977	071
6548054	122
130	173
206	223
283	274
359	324
435	375
512	426
588	476
664	527
741	578
817	628

6548894	·8161679
970	730
6549046	780
123	831
199	881
275	932
352	983
428	·8162033
504	084
581	135
657	185
733	236
810	286
886	337
962	388

Excesses

10 = 7	1 = 1
20 = 13	2 = 1
30 = 20	3 = 2
40 = 27	4 = 3
50 = 33	5 = 3
60 = 40	6 = 4
70 = 46	7 = 5
	8 = 5
	9 = 6

38 = 25

Anti Log Excesses

10 = 15	1 = 2
20 = 30	2 = 3
30 = 45	3 = 5
40 = 60	4 = 6
50 = 75	5 = 8
	6 = 9
	7 = 11
	8 = 12
	9 = 14

25 = 38

6550039	·8162438
115	489
191	540
268	590
344	641
420	691
497	742
573	793
649	843
726	894
802	944
878	995
955	·8163046
6551031	096
107	147
184	197
260	248
336	299
413	349
489	400
565	450

6551642	·8163501
718	551
794	602
870	653
947	703
6552023	754
099	804
176	855
252	906
328	956
405	·8164007
481	057
557	108
634	159
710	209
786	260
862	310
939	361
6553015	411
091	462
168	513
244	563
320	614
397	664
473	715
549	765
625	816
702	866
778	917
854	968
931	·8165018
6554007	069
083	119
159	170
236	220
312	271
388	321
465	372
541	422
617	473
693	524
770	574
846	625
922	675
999	726
6555075	776
151	827
227	877
304	928
380	978
456	·8166029
533	079
609	130
685	180
761	231
838	281
914	332
990	383
6556066	433
143	484
219	534
295	585
371	635
448	686
524	736

6556600	·8166787
676	837
753	888
829	938
905	989
982	·8167039
6557058	090
134	140
210	191
287	241
363	292
439	342
515	393
592	443
668	494
744	544
820	595
897	645
973	696
6558049	746
125	797
201	847
278	898
354	948
430	999
506	·8168049
583	100
659	150
735	201
811	251
888	302
964	352
6559040	403
116	453
193	503
269	554
345	604
421	655
497	705
574	756
650	806
726	857
802	907
879	958
955	·8169008

Excesses

10 = 7	1 = 1
20 = 13	2 = 1
30 = 20	3 = 2
40 = 26	4 = 3
50 = 33	5 = 3
60 = 40	6 = 4
70 = 46	7 = 5
	8 = 5
	9 = 6

38 = 25

Anti Log Excesses

10 = 15	1 = 2
20 = 30	2 = 3
30 = 45	3 = 5
40 = 60	4 = 6
50 = 75	5 = 8
	6 = 9
	7 = 11
	8 = 12
	9 = 14

25 = 38

6560031	·8169059
107	109
183	160
260	210
336	260
412	311
488	361
565	412
641	462
717	513
793	563
869	614
946	664
6561022	714
098	765
174	815
250	866
327	916
403	967
479	·8170017
555	068
631	118
708	168
784	219
860	269
936	320
6562012	370
089	421
165	471
241	521
317	572
393	622
470	673
546	723
622	774
698	824
774	874
850	925
927	975
6563003	·8171026
079	076
155	126
231	177
308	227
384	278
460	328
536	379
612	429
688	479
765	530
841	580

6563917	·8171630
993	681
6564069	731
146	782
222	832
298	882
374	933
450	983
526	·8172034
603	084
679	134
755	185
831	235
907	286
983	336
6565060	386
136	437
212	487
288	538
364	588
440	638
516	689
593	739
669	789
745	840
821	890
897	941
973	991
6566050	·8173041
126	092
202	142
278	192
354	243
430	293
506	343
583	394
659	444
735	495
811	545
887	595
963	646
6567039	696
116	746
192	797
268	847
344	897
420	948
496	998
572	·8174048
648	099
725	149
801	199
877	250
953	300
6568029	350
105	401
181	451
257	501
334	552
410	602
486	652
562	703
638	753
714	803
790	854

6568866	·8174904
943	954
6569019	·8175005
095	055
171	105
247	156
323	206
399	256
475	307
551	357
628	407
704	457
780	508
856	558
932	608

Excesses

10 = 7	1 = 1
20 = 13	2 = 1
30 = 20	3 = 2
40 = 26	4 = 3
50 = 33	5 = 3
60 = 40	6 = 4
70 = 46	7 = 5
	8 = 5
	9 = 6

38 = 25

Anti Log Excesses

10 = 15	1 = 2
20 = 30	2 = 3
30 = 45	3 = 5
40 = 60	4 = 6
50 = 76	5 = 8
	6 = 9
	7 = 11
	8 = 12
	9 = 14

25 = 38

6570008	·8175659
084	709
160	759
236	810
313	860
389	910
465	960
541	·8176011
617	061
693	111
769	162
845	212
921	262
997	313
6571073	363
150	413
226	463
302	514
378	564
454	614
530	665

6571606	·8176715
682	765
758	815
834	866
910	916
987	966
6572063	·8177017
139	067
215	117
291	167
367	218
443	268
519	318
595	368
671	419
747	469
823	519
899	569
975	620
6573052	670
128	720
204	770
280	821
356	871
432	921
508	971
584	·8178022
660	072
736	122
812	172
888	223
964	273
6574040	323
116	373
192	424
269	474
345	524
421	574
497	625
573	675
649	725
725	775
801	826
877	876
953	926
6575029	976
105	·8179027
181	077
257	127
333	177
409	227
485	278
561	328
637	378
713	428
789	479
865	529
941	579
6576018	629
094	679
170	730
246	780
322	830
398	880
474	930

6576550	·8179981
626	·8180031
702	081
778	131
854	181
930	232
6577006	282
082	332
158	382
234	432
310	483
386	533
462	583
538	633
614	683
690	733
766	784
842	834
918	884
994	934
6578070	984
146	·8181035
222	085
298	135
374	185
450	235
526	286
602	336
678	386
754	436
830	486
906	537
982	587
6579058	637
134	687
210	737
286	787
362	838
438	888
514	938
590	988
666	·8182038
742	088
818	138
894	189
970	239

Excesses

10 = 7	1 = 1
20 = 13	2 = 1
30 = 20	3 = 2
40 = 26	4 = 3
50 = 33	5 = 3
60 = 40	6 = 4
70 = 46	7 = 5
	8 = 5
	9 = 6

38 = 25

Anti Log Excesses

10 = 15	1 = 2
20 = 30	2 = 3
30 = 45	3 = 5
40 = 61	4 = 6
50 = 76	5 = 8
	6 = 9
	7 = 11
	8 = 12
	9 = 14

25 = 38

6580046	·8182289
122	339
198	389
274	439
350	489
426	540
502	590
578	640
654	690
730	740
806	790
882	841
958	891
6581034	941
110	991
186	·8183041
262	091
338	141
414	191
490	242
566	292
642	342
718	392
794	442
870	492
946	542
6582021	592
097	643
173	693
249	743
325	793
401	843
477	894
553	944
629	994
705	·8184044
781	094
857	144
933	194
6583009	244
085	294
161	345
237	395
313	445
389	495
465	545
541	595
617	645
692	695
768	745
844	795

6583920	·8184845
996	896
6584072	946
148	996
224	·8185046
300	096
376	146
452	196
528	246
604	296
680	346
756	396
832	447
907	497
983	547
6585059	597
135	647
211	697
287	747
363	797
439	847
515	897
591	947
667	997
743	·8186047
819	098
895	148
970	198
6586046	248
122	298
198	348
274	398
350	448
426	498
502	548
578	598
654	648
730	698
805	748
881	798
957	848
6587033	898
109	948
185	998
261	·8187048
337	098
413	149
489	199
565	249
640	299
716	349
792	399
868	449
944	499
6588020	549
096	599
172	649
248	699
324	749
399	799
475	849
551	899
627	949
703	999
779	·8188049

6588855	·8188099
931	149
6589006	199
082	249
158	299
234	349
310	399
386	449
462	499
538	549
614	599
689	649
765	699
841	749
917	799
993	849

Excesses

10 = 7	1 = 1
20 = 13	2 = 1
30 = 20	3 = 2
40 = 26	4 = 3
50 = 33	5 = 3
60 = 40	6 = 4
70 = 46	7 = 5
	8 = 5
	9 = 6

38 = 25

Anti Log Excesses

10 = 15	1 = 2
20 = 30	2 = 3
30 = 45	3 = 5
40 = 61	4 = 6
50 = 76	5 = 8
	6 = 9
	7 = 11
	8 = 12
	9 = 14

25 = 38

6590069	·8188899
145	949
221	999
296	·8189049
372	099
448	149
524	199
600	249
676	299
752	349
827	399
903	449
979	499
6591055	549
131	599
207	649
283	699
358	749
434	799
510	849

6591586	·8189899
662	949
738	999
814	·8190049
889	099
965	149
6592041	199
117	249
193	299
269	349
345	399
420	449
496	499
572	549
648	598
724	648
800	698
875	748
951	798
6593027	848
103	898
179	948
255	998
330	·8191048
406	098
482	148
558	198
634	248
710	298
785	348
861	398
937	448
6594013	498
089	548
165	598
240	647
316	697
392	747
468	797
544	847
620	897
695	947
771	997
847	·8192047
923	097
999	147
6595074	197
150	247
226	297
302	346
378	396
453	446
529	496
605	546
681	596
757	646
833	696
908	746
984	796
6596060	846
136	895
212	945
287	995
363	·8193045
439	095

6596515	·8193145
591	195
666	245
742	295
818	345
894	395
970	445
6597045	494
121	544
197	594
273	644
348	694
424	744
500	794
576	844
652	894
727	943
803	993
879	·8194043
955	093
6598031	143
106	193
182	243
258	293
334	342
409	392
485	442
561	492
637	542
712	592
788	642
864	692
940	741
6599016	791
091	841
167	891
243	941
319	991
394	·8195041
470	090
546	140
622	190
697	240
773	290
849	340
925	390

Excesses

10 = 7	1 = 1
20 = 13	2 = 1
30 = 20	3 = 2
40 = 26	4 = 3
50 = 33	5 = 3
60 = 40	6 = 4
70 = 46	7 = 5
	8 = 5
	9 = 6

38 = 25

Anti Log Excesses

10 = 15	1 = 2
20 = 30	2 = 3
30 = 46	3 = 5
40 = 61	4 = 6
	5 = 8
	6 = 9
	7 = 11
	8 = 12
	9 = 14

25 = 38

6600000	·8195439
076	489
152	539
228	589
304	639
379	689
455	739
531	788
607	838
682	888
758	938
834	988
910	·8196038
985	087
6601061	137
137	187
213	237
288	287
364	337
440	386
515	436
591	486
667	536
743	586
818	635
894	685
970	735
6602046	785
121	835
197	885
273	934
349	984
424	·8197034
500	084
576	134
651	183
727	233
803	283
879	333
954	383
6603030	432
106	482
182	532
257	582
333	632
409	681
484	731
560	781
636	831
712	881
787	930

6603863	·8197980
939	·8198030
6604014	080
090	129
166	179
242	229
317	279
393	329
469	379
544	428
620	478
696	528
772	578
847	627
923	677
999	727
6605074	777
150	827
226	876
301	926
377	976
453	·8199026
528	075
604	125
680	175
756	225
831	274
907	324
983	374
6606058	424
134	473
210	523
285	573
361	623
437	672
512	722
588	772
664	822
739	871
815	921
891	971
967	·8200021
6607042	070
118	120
194	170
269	220
345	269
421	319
496	369
572	418
648	468
723	518
799	568
875	617
950	667
6608026	717
102	767
177	817
253	866
329	916
404	966
480	·8201015
556	065
631	115
707	165

6608783	·8201214
858	264
934	314
6609010	363
085	413
161	463
236	512
312	562
388	612
463	662
539	711
615	761
690	811
766	860
842	910
917	960
993	·8202010

Excesses

10 = 7	1 = 1
20 = 13	2 = 1
30 = 20	3 = 2
40 = 26	4 = 3
50 = 33	5 = 3
60 = 39	6 = 4
70 = 46	7 = 5
	8 = 5
	9 = 6

38 = 25

Anti Log Excesses

10 = 15	1 = 2
20 = 30	2 = 3
30 = 46	3 = 5
40 = 61	4 = 6
	5 = 8
	6 = 9
	7 = 11
	8 = 12
	9 = 14

25 = 38

6610069	·8202059
144	109
220	159
296	208
371	258
447	308
522	357
598	407
674	457
749	507
825	556
901	606
976	656
6611052	705
127	755
203	805
279	854
354	904
430	954

6611506	·8203003
581	053
657	103
732	153
808	202
884	252
959	302
6612035	351
111	401
186	451
262	500
337	550
413	600
489	649
564	699
640	749
716	798
791	848
867	898
942	947
6613018	997
094	·8204046
169	096
245	146
320	195
396	245
472	295
547	344
623	394
698	444
774	493
850	543
925	593
6614001	642
076	692
152	741
228	791
303	841
379	890
454	940
530	990
606	·8205039
681	089
757	138
832	188
908	238
983	287
6615059	337
135	387
210	436
286	486
361	536
437	585
513	635
588	684
664	734
739	784
815	833
890	883
966	933
6616042	982
117	·8206032
193	081
268	131
344	181

6616419	·8206230
495	280
571	329
646	379
722	429
797	478
873	528
948	577
6617024	627
100	676
175	726
251	776
326	825
402	875
477	924
553	974
628	·8207024
704	073
780	123
855	172
931	222
6618006	272
082	321
157	371
233	420
308	470
384	519
460	569
535	619
611	668
686	718
762	767
837	817
913	866
988	916
6619064	966
139	·8208015
215	065
290	114
366	164
442	213
517	263
593	312
668	362
744	411
819	461
895	511
970	560

Excesses

10 = 7	1 = 1
20 = 13	2 = 1
30 = 20	3 = 2
40 = 26	4 = 3
50 = 33	5 = 3
60 = 39	6 = 4
70 = 46	7 = 5
	8 = 5
	9 = 6

38 = 25

Anti Log Excesses

10 = 15	1 = 2
20 = 30	2 = 3
30 = 46	3 = 5
40 = 61	4 = 6
	5 = 8
	6 = 9
	7 = 11
	8 = 12
	9 = 14

25 = 38

6620046	·8208610
121	659
197	709
272	758
348	808
423	857
499	907
574	956
650	·8209006
726	056
801	105
877	155
952	204
6621028	254
103	303
179	353
254	402
330	452
405	501
481	551
556	600
632	650
707	699
783	749
858	798
934	848
6622009	898
085	947
160	997
236	·8210046
311	096
387	145
462	195
538	244
613	294
689	343
764	393
840	442
915	492
991	541
6623066	591
142	640
217	690
293	739
368	789
444	838
519	888
595	937
670	987
746	·8211036
821	086

6623897	·8211135
972	185
6624048	234
123	284
199	333
274	383
350	432
425	482
500	531
576	581
651	630
727	680
802	729
878	778
953	828
6625029	877
104	927
180	976
255	·8212026
331	075
406	125
482	174
557	224
633	273
708	323
783	372
859	422
934	471
6626010	520
085	570
161	619
236	669
312	718
387	768
463	817
538	867
614	916
689	966
764	·8213015
840	065
915	114
991	163
6627066	213
142	262
217	312
293	361
368	411
443	460
519	510
594	559
670	608
745	658
821	707
896	757
972	806
6628047	856
122	905
198	954
273	·8214004
349	053
424	103
500	152
575	202
650	251
726	300

6628801	·8214350
877	399
952	449
6629028	498
103	547
178	597
254	646
329	696
405	745
480	795
556	844
631	893
706	943
782	992
857	·8215042
933	091

Excesses

10 = 7	1 = 1
20 = 13	2 = 1
30 = 20	3 = 2
40 = 26	4 = 3
50 = 33	5 = 3
60 = 39	6 = 4
70 = 46	7 = 5
	8 = 5
	9 = 6

38 = 25

Anti Log Excesses

10 = 15	1 = 2
20 = 31	2 = 3
30 = 46	3 = 5
40 = 61	4 = 6
	5 = 8
	6 = 9
	7 = 11
	8 = 12
	9 = 14

25 = 38

6630008	·8215140
083	190
159	239
234	289
310	338
385	387
461	437
536	486
611	535
687	585
762	634
838	684
913	733
988	782
6631064	832
139	881
215	931
290	980
365	·8216029
441	079
6631516	·8216128
592	177
667	227
742	276
818	326
893	375
969	424
6632044	474
119	523
195	572
270	622
345	671
421	720
496	770
572	819
647	869
722	918
798	967
873	·8217017
949	066
6633024	115
099	165
175	214
250	263
325	313
401	362
476	411
552	461
627	510
702	559
778	609
853	658
928	707
6634004	757
079	806
155	856
230	905
305	954
381	·8218004
456	053
531	102
607	152
682	201
757	250
833	300
908	349
984	398
6635059	447
134	497
210	546
285	595
360	645
436	694
511	743
586	793
662	842
737	891
812	941
888	990
963	·8219039
6636038	089
114	138
189	187
265	237
340	286
6636415	·8219335
491	384
566	434
641	483
717	532
792	582
867	631
943	680
6637018	730
093	779
169	828
244	877
319	927
395	976
470	·8220025
545	075
621	124
696	173
771	222
847	272
922	321
997	370
6638073	420
148	469
223	518
298	567
374	617
449	666
524	715
600	765
675	814
750	863
826	912
901	962
976	·8221011
6639052	060
127	109
202	159
278	208
353	257
428	306
503	356
579	405
654	454
729	503
805	553
880	602
955	651

Excesses

10 = 7	1 = 1
20 = 13	2 = 1
30 = 20	3 = 2
40 = 26	4 = 3
50 = 33	5 = 3
60 = 39	6 = 4
70 = 46	7 = 5
	8 = 5
	9 = 6

38 = 25

Anti Log Excesses

10 = 15	1 = 2
20 = 31	2 = 3
30 = 46	3 = 5
40 = 61	4 = 6
	5 = 8
	6 = 9
	7 = 11
	8 = 12
	9 = 14

25 = 38

6640031	·8221700
106	750
181	799
257	848
332	897
407	947
482	996
558	·8222045
633	094
708	144
784	193
859	242
934	291
6641009	341
085	390
160	439
235	488
311	538
386	587
461	636
536	686
612	735
687	784
762	833
838	882
913	932
988	981
6642063	·8223030
139	079
214	129
289	178
365	227
440	276
515	325
590	375
666	424
741	473
816	522
891	571
967	621
6643042	670
117	719
193	768
268	817
343	867
418	916
494	965
569	·8224014
644	063
719	113
795	162
6643870	·8224211
945	260
6644020	309
096	359
171	408
246	457
321	506
397	555
472	605
547	654
622	703
698	752
773	801
848	850
923	900
999	949
6645074	998
149	·8225047
224	096
300	146
375	195
450	244
525	293
601	342
676	391
751	441
826	490
901	539
977	588
6646052	637
127	686
202	736
278	785
353	834
428	883
503	932
579	981
654	·8226030
729	080
804	129
879	178
955	227
6647030	276
105	325
180	374
256	424
331	473
406	522
481	571
556	620
632	669
707	718
782	768
857	817
933	866
6648008	915
083	964
158	·8227013
233	062
309	112
384	161
459	210
534	259
609	308
685	357

6648760	·8227406
835	455
910	505
985	554
6649061	603
136	652
211	701
286	750
361	799
437	848
512	897
587	947
662	996
737	·8228045
813	094
888	143
963	192

Excesses

10 = 7	1 = 1
20 = 13	2 = 1
30 = 20	3 = 2
40 = 26	4 = 3
50 = 33	5 = 3
60 = 39	6 = 4
70 = 46	7 = 5
	8 = 5
	9 = 6

38 = 25

Anti Log Excesses

10 = 15	1 = 2
20 = 31	2 = 3
30 = 46	3 = 5
40 = 61	4 = 6
	5 = 8
	6 = 9
	7 = 11
	8 = 12
	9 = 14

25 = 38

6650038	·8228241
113	290
188	339
264	388
339	438
414	487
489	536
564	585
640	634
715	683
790	732
865	781
940	830
6651015	879
091	928
166	978
241	·8229027
316	076
391	125

6651467	·8229174
542	223
617	272
692	321
767	370
842	419
918	468
993	517
6652068	566
143	616
218	665
293	714
368	763
444	812
519	861
594	910
669	959
744	·8230008
819	057
895	106
970	155
6653045	204
120	253
195	302
270	351
346	401
421	450
496	499
571	548
646	597
721	646
796	695
872	744
947	793
6654022	842
097	891
172	940
247	989
322	·8231038
398	087
473	136
548	185
623	234
698	283
773	332
848	381
923	430
999	479
6655074	528
149	578
224	627
299	676
374	725
449	774
525	823
600	872
675	921
750	970
825	·8232019
900	068
975	117
6656050	166
125	215
201	264
276	313

6656351	·8232362
426	411
501	460
576	509
651	558
726	607
802	656
877	705
952	754
6657027	803
102	852
177	901
252	950
327	999
402	·8233048
477	097
553	146
628	195
703	244
778	293
853	342
928	391
6658003	440
078	489
153	538
228	587
304	636
379	685
454	734
529	783
604	832
679	881
754	930
829	979
904	·8234028
979	077
6659054	126
130	174
205	223
280	272
355	321
430	370
505	419
580	468
655	517
730	566
805	615
880	664
955	713

Excesses

10 = 7	1 = 1
20 = 13	2 = 1
30 = 20	3 = 2
40 = 26	4 = 3
50 = 33	5 = 3
60 = 39	6 = 4
70 = 46	7 = 5
	8 = 5
	9 = 6

38 = 25

Anti Log Excesses

10 = 15	1 = 2
20 = 31	2 = 3
30 = 46	3 = 5
40 = 61	4 = 6
	5 = 8
	6 = 9
	7 = 11
	8 = 12
	9 = 14

25 = 38

6660031	·8234762
106	811
181	860
256	909
331	958
406	·8235007
481	056
556	105
631	154
706	203
781	251
856	300
931	349
6661006	398
081	447
157	496
232	545
307	594
382	643
457	692
532	741
607	790
682	839
757	888
832	937
907	986
982	·8236034
6662057	083
132	132
207	181
282	230
357	279
432	328
508	377
583	426
658	475
733	524
808	573
883	621
958	670
6663033	719
108	768
183	817
258	866
333	915
408	964
483	·8237013
558	062
633	111
708	159
783	208

6663858	·8237257
933	306
6664008	355
083	404
158	453
233	502
308	551
383	600
458	648
533	697
609	746
684	795
759	844
834	893
909	942
984	991
6665059	·8238039
134	088
209	137
284	186
359	235
434	284
509	333
584	382
659	431
734	479
809	528
884	577
959	626
6666034	675
109	724
184	773
259	821
334	870
409	919
484	968
559	·8239017
634	066
709	115
784	163
859	212
934	261
6667009	310
084	359
159	408
234	457
309	505
384	554
459	603
534	652
609	701
684	750
759	799
834	847
909	896
984	945
6668059	994
134	·8240043
209	092
284	140
359	189
434	238
509	287
584	336
659	385

6668734	·8240433
808	482
883	531
958	580
6669033	629
108	678
183	726
258	775
333	824
408	873
483	922
558	970
633	·8241019
708	068
783	117
858	166
933	214

Excesses

10 = 7	1 = 1
20 = 13	2 = 1
30 = 20	3 = 2
40 = 26	4 = 3
50 = 33	5 = 3
60 = 39	6 = 4
70 = 46	7 = 5
	8 = 5
	9 = 6

37 = 24

Anti Log Excesses

10 = 15	1 = 2
20 = 31	2 = 3
30 = 46	3 = 5
40 = 61	4 = 6
	5 = 8
	6 = 9
	7 = 11
	8 = 12
	9 = 14

24 = 37

6670008	·8241263
083	312
158	361
233	410
308	458
383	507
458	556
533	605
608	654
683	703
758	751
833	800
907	849
982	898
6671057	946
132	995
207	·8242044
282	093
357	142

6671432	·8242191
507	239
582	288
657	337
732	386
807	435
882	483
957	532
6672032	581
107	630
182	678
256	727
331	776
406	825
481	874
556	922
631	971
706	·8243020
781	069
856	117
931	166
6673006	215
081	264
156	312
231	361
306	410
380	459
455	507
530	556
605	605
680	654
755	702
830	751
905	800
980	849
6674055	897
130	946
205	995
279	·8244044
354	092
429	141
504	190
579	239
654	287
729	336
804	385
879	434
954	482
6675029	531
103	580
178	629
253	677
328	726
403	775
478	824
553	872
628	921
703	970
778	·8245018
853	067
927	116
6676002	165
077	213
152	262
227	311

6676302	·8245359
377	408
452	457
527	506
601	554
676	603
751	652
826	700
901	749
976	798
6677051	847
126	895
201	944
275	993
350	·8246041
425	090
500	139
575	188
650	236
725	285
800	334
874	382
949	431
6678024	480
099	528
174	577
249	626
324	674
399	723
473	772
548	820
623	869
698	918
773	967
848	·8247015
923	064
997	113
6679072	161
147	210
222	259
297	307
372	356
447	405
521	453
596	502
671	551
746	599
821	648
896	697
971	745

Excesses

10 = 7	1 = 1
20 = 13	2 = 1
30 = 20	3 = 2
40 = 26	4 = 3
50 = 33	5 = 3
60 = 39	6 = 4
70 = 46	7 = 5
	8 = 5
	9 = 6

37 = 24

Anti Log Excesses

10 = 15	1 = 2
20 = 31	2 = 3
30 = 46	3 = 5
40 = 61	4 = 6
	5 = 8
	6 = 9
	7 = 11
	8 = 12
	9 = 14

24 = 37

6680045	·8247794
120	843
195	891
270	940
345	989
420	·8248037
494	086
569	134
644	183
719	232
794	280
869	329
944	378
6681018	426
093	475
168	524
243	572
318	621
393	670
467	718
542	767
617	816
692	864
767	913
842	961
916	·8249010
991	059
6682066	107
141	156
216	205
291	253
365	302
440	351
515	399
590	448
665	496
739	545
814	594
889	642
964	691
6683039	740
114	788
188	837
263	885
338	934
413	983
488	·8250031
562	080
637	128
712	177
787	226

6683862	·8250274
936	323
6684011	371
086	420
161	469
236	517
310	566
385	614
460	663
535	712
610	760
684	809
759	857
834	906
909	955
984	·8251003
6685058	052
133	100
208	149
283	198
358	246
432	295
507	343
582	392
657	440
732	489
806	538
881	586
956	635
6686031	683
105	732
180	780
255	829
330	878
405	926
479	975
554	·8252023
629	072
704	121
778	169
853	218
928	266
6687003	315
078	363
152	412
227	461
302	509
377	558
451	606
526	655
601	703
676	752
750	800
825	849
900	897
975	946
6688050	995
124	·8253043
199	092
274	140
349	189
423	237
498	286
573	334
648	383

6688722	·8253431
797	480
872	528
947	577
6689021	626
096	674
171	723
246	771
320	820
395	868
470	917
545	965
619	·8254014
694	062
769	111
844	159
918	208
993	256

Excesses

10 = 6	1 = 1
20 = 13	2 = 1
30 = 19	3 = 2
40 = 26	4 = 3
50 = 32	5 = 3
60 = 39	6 = 4
70 = 45	7 = 5
	8 = 5
	9 = 6

37 = 24

Anti Log Excesses

10 = 15	1 = 2
20 = 31	2 = 3
30 = 46	3 = 5
40 = 61	4 = 6
	5 = 8
	6 = 9
	7 = 11
	8 = 12
	9 = 14

24 = 37

6690068	·8254305
143	353
217	402
292	450
367	499
441	548
516	596
591	645
666	693
740	742
815	790
890	839
965	887
6691039	936
114	984
189	·8255033
263	081
338	130

6691413	·8255178
488	227
562	275
637	323
712	372
787	420
861	469
936	517
6692011	566
085	614
160	663
235	711
310	760
384	808
459	857
534	905
608	954
683	·8256002
758	051
832	099
907	148
982	196
6693057	245
131	293
206	342
281	390
355	439
430	487
505	535
580	584
654	632
729	681
804	729
878	778
953	826
6694028	875
102	923
177	972
252	·8257020
326	069
401	117
476	165
551	214
625	262
700	311
775	359
849	408
924	456
999	505
6695073	553
148	601
223	650
297	698
372	747
447	795
521	844
596	892
671	940
745	989
820	·8258037
895	086
969	134
6696044	183
119	231
193	279

6696268	·8258328
343	376
417	425
492	473
567	522
641	570
716	618
791	667
865	715
940	764
6697015	812
089	861
164	909
239	957
313	·8259006
388	054
463	103
537	151
612	199
687	248
761	296
836	345
911	393
985	442
6698060	490
135	538
209	587
284	635
359	684
433	732
508	780
582	829
657	877
732	925
806	974
881	·8260022
956	071
6699030	119
105	167
180	216
254	264
329	313
403	361
478	409
553	458
627	506
702	554
777	603
851	651
926	700

Excesses

10 = 6	1 = 1
20 = 13	2 = 1
30 = 19	3 = 2
40 = 26	4 = 3
50 = 32	5 = 3
60 = 39	6 = 4
70 = 45	7 = 5
	8 = 5
	9 = 6

37 = 24

Anti Log Excesses

10 = 15	1 = 2
20 = 31	2 = 3
30 = 46	3 = 5
40 = 62	4 = 6
	5 = 8
	6 = 9
	7 = 11
	8 = 12
	9 = 14

24 = 37

6700000	·8260748
075	796
150	845
224	893
299	941
374	990
448	·8261038
523	086
597	135
672	183
747	232
821	280
896	328
971	377
6701045	425
120	473
194	522
269	570
344	618
418	667
493	716
567	764
642	812
717	861
791	909
866	957
941	·8262006
6702015	054
090	102
164	151
239	199
314	247
388	296
463	344
537	392
612	441
687	489
761	537
836	586
910	634
985	682
6703060	731
134	779
209	827
283	876
358	924
432	972
507	·8263021
582	069
656	117
731	166

6703805	·8263214
880	262
955	311
6704029	359
104	407
178	456
253	504
327	552
402	601
477	649
551	697
626	745
700	794
775	842
849	890
924	939
999	987
6705073	·8264035
148	084
222	132
297	180
371	228
446	277
521	325
595	373
670	422
744	470
819	518
893	566
968	615
6706043	663
117	711
192	760
266	808
341	856
415	904
490	953
564	·8265001
639	049
714	097
788	146
863	194
937	242
6707012	291
086	339
161	387
235	435
310	484
384	532
459	580
534	628
608	677
683	725
757	773
832	821
906	870
981	918
6708055	966
130	·8266015
204	063
279	111
354	159
428	208
503	256
577	304

6708652	·8266352
726	401
801	449
875	497
950	545
6709024	594
099	642
173	690
248	738
322	786
397	835
471	883
546	931
620	979
695	·8267028
769	076
844	124
919	172
993	221

Excesses

10 = 6	1 = 1
20 = 13	2 = 1
30 = 19	3 = 2
40 = 26	4 = 3
50 = 32	5 = 3
60 = 39	6 = 4
70 = 45	7 = 5
	8 = 5
	9 = 6

37 = 24

Anti Log Excesses

10 = 15	1 = 2
20 = 31	2 = 3
30 = 46	3 = 5
40 = 62	4 = 6
	5 = 8
	6 = 9
	7 = 11
	8 = 12
	9 = 14

24 = 37

6710068	·8267269
142	317
217	365
291	413
366	462
440	510
515	558
589	606
664	655
738	703
813	751
887	799
962	847
6711036	896
111	944
185	992
260	·8268040

6711334	·8268089
409	137
483	185
558	233
632	281
707	330
781	378
856	426
930	474
6712005	522
079	571
154	619
228	667
303	715
377	763
452	812
526	860
601	908
675	956
750	·8269004
824	053
899	101
973	149
6713047	197
122	245
196	293
271	342
345	390
420	438
494	486
569	534
643	583
718	631
792	679
867	727
941	775
6714016	823
090	872
165	920
239	968
314	·8270016
388	064
463	112
537	161
611	209
686	257
760	305
835	353
909	401
984	449
6715058	498
133	546
207	594
282	642
356	690
430	738
505	787
579	835
654	883
728	931
803	979
877	·8271027
952	075
6716026	124
101	172

6716175	·8271220
249	268
324	316
398	364
473	412
547	461
622	509
696	557
771	605
845	653
919	701
994	749
6717068	798
143	846
217	894
292	942
366	990
440	·8272038
515	086
589	134
664	182
738	231
813	279
887	327
962	375
6718036	423
110	471
185	519
259	567
334	615
408	664
482	712
557	760
631	808
706	856
780	904
855	952
929	·8273000
6719003	048
078	096
152	145
227	193
301	241
375	289
450	337
524	385
599	433
673	481
748	529
822	577
896	625
971	674

Excesses

10 = 6	1 = 1
20 = 13	2 = 1
30 = 19	3 = 2
40 = 26	4 = 3
50 = 32	5 = 3
60 = 39	6 = 4
70 = 45	7 = 5
	8 = 5
	9 = 6

37 = 24

Anti Log Excesses

10 = 15	1 = 2
20 = 31	2 = 3
30 = 46	3 = 5
40 = 62	4 = 6
	5 = 8
	6 = 9
	7 = 11
	8 = 12
	9 = 14

24 = 37

6720045	·8273722
120	770
194	818
268	866
343	914
417	962
492	·8274010
566	058
640	106
715	154
789	203
864	251
938	299
6721012	347
087	395
161	443
236	491
310	539
384	587
459	635
533	683
607	731
682	779
756	827
831	875
905	923
979	972
6722054	·8275020
128	068
203	116
277	164
351	212
426	260
500	308
574	356
649	404
723	452

6722798	·8275500
872	548
946	596
6723021	644
095	692
169	740
244	788
318	836
393	884
467	932
541	980
616	·8276028
690	077
764	125
839	173
913	221
987	269
6724062	317
136	365
210	413
285	461
359	509
434	557
508	605
582	653
657	701
731	749
805	797
880	845
954	893
6725028	941
103	989
177	·8277037
251	085
326	133
400	181
474	229
549	277
623	325
697	373
772	421
846	469
921	517
995	565
6726069	613
144	661
218	709
292	757
367	805
441	853
515	901
590	949
664	997
738	·8278045
813	093
887	141
961	189
6727036	237
110	285
184	333
258	381
333	429
407	477
481	525
556	573

6727630	·8278621
704	669
779	717
853	765
927	813
6728002	861
076	909
150	957
225	·8279005
299	053
373	101
448	149
522	196
596	244
671	292
745	340
819	388
893	436
968	484
6729042	532
116	580
191	628
265	676
339	724
414	772
488	820
562	868
636	916
711	964
785	·8280012
859	060
934	108

Excesses

10 = 6	1 = 1
20 = 13	2 = 1
30 = 19	3 = 2
40 = 26	4 = 3
50 = 32	5 = 3
60 = 39	6 = 4
70 = 45	7 = 5
	8 = 5
	9 = 6

37 = 24

Anti Log Excesses

10 = 15	1 = 2
20 = 31	2 = 3
30 = 46	3 = 5
40 = 62	4 = 6
	5 = 8
	6 = 9
	7 = 11
	8 = 12
	9 = 14

24 = 37

6730008	·8280156
082	203
157	251
231	299
6730305	·8280347
379	395
454	443
528	491
602	539
677	587
751	635
825	683
899	731
974	779
6731048	827
122	875
197	922
271	970
345	·8281018
419	066
494	114
568	162
642	210
716	258
791	306
865	354
939	402
6732014	450
088	498
162	545
236	593
311	641
385	689
459	737
533	785
608	833
682	881
756	929
831	977
905	·8282024
979	072
6733053	120
128	168
202	216
276	264
350	312
425	360
499	408
573	456
647	503
722	551
796	599
870	647
944	695
6734019	743
093	791
167	839
241	887
316	934
390	982
464	·8283030
538	078
613	126
687	174
761	222
835	270
910	317
984	365
6735058	413
6735132	·8283461
207	509
281	557
355	605
429	652
503	700
578	748
652	796
726	844
800	892
875	940
949	988
6736023	·8284035
097	083
172	131
246	179
320	227
394	275
468	323
543	370
617	418
691	466
765	514
840	562
914	610
988	657
6737062	705
136	753
211	801
285	849
359	897
433	944
507	992
582	·8285040
656	088
730	136
804	184
879	231
953	279
6738027	327
101	375
175	423
250	471
324	518
398	566
472	614
546	662
621	710
695	757
769	805
843	853
917	901
992	949
6739066	997
140	·8286044
214	092
288	140
362	188
437	236
511	283
585	331
659	379
733	427
808	475
882	523
6739956	·8286570

Excesses

10 = 6	1 = 1
20 = 13	2 = 1
30 = 19	3 = 2
40 = 26	4 = 3
50 = 32	5 = 3
60 = 39	6 = 4
70 = 45	7 = 5
	8 = 5
	9 = 6

37 = 24

Anti Log Excesses

10 = 16	1 = 2
20 = 31	2 = 3
30 = 47	3 = 5
40 = 62	4 = 6
	5 = 8
	6 = 9
	7 = 11
	8 = 12
	9 = 14

24 = 37

6740030	·8286618
104	666
179	714
253	762
327	809
401	857
475	905
549	953
624	·8287000
698	048
772	096
846	144
920	192
994	239
6741069	287
143	335
217	383
291	431
365	478
440	526
514	574
588	622
662	669
736	717
810	765
885	813
959	861
6742033	908
107	956
181	·8288004
255	052
329	099
404	147
478	195
552	243
6742626	·8288290
700	338
774	386
849	434
923	482
997	529
6743071	577
145	625
219	673
293	720
368	768
442	816
516	864
590	911
664	959
738	·8289007
812	055
887	102
961	150
6744035	198
109	245
183	293
257	341
331	389
406	436
480	484
554	532
628	580
702	627
776	675
850	723
925	771
999	818
6745073	866
147	914
221	961
295	·8290009
369	057
443	105
518	152
592	200
666	248
740	296
814	343
888	391
962	439
6746036	486
110	534
185	582
259	630
333	677
407	725
481	773
555	820
629	868
703	916
778	963
852	·8291011
926	059
6747000	107
074	154
148	202
222	250
296	297
370	345

6747444	·8291393
519	441
593	488
667	536
741	584
815	631
889	679
963	727
6748037	774
111	822
185	870
260	917
334	965
408	·8292013
482	060
556	108
630	156
704	204
778	251
852	299
926	347
6749000	394
075	442
149	490
223	537
297	585
371	633
445	680
519	728
593	776
667	823
741	871
815	919
889	966
963	·8293014

Excesses

10 = 6	1 = 1
20 = 13	2 = 1
30 = 19	3 = 2
40 = 26	4 = 3
50 = 32	5 = 3
60 = 39	6 = 4
70 = 45	7 = 5
	8 = 5
	9 = 6

37 = 24

Anti Log Excesses

10 = 16	1 = 2
20 = 31	2 = 3
30 = 47	3 = 5
40 = 62	4 = 6
	5 = 8
	6 = 9
	7 = 11
	8 = 12
	9 = 14

24 = 37

6750038	·8293062
6750112	·8293109
186	157
260	204
334	252
408	300
482	347
556	395
630	443
704	490
778	538
852	586
926	633
6751000	681
074	729
149	776
223	824
297	872
371	919
445	967
519	·8294014
593	062
667	110
741	157
815	205
889	253
963	300
6752037	348
111	396
185	443
259	491
333	538
407	586
482	634
556	681
630	729
704	776
778	824
852	872
926	919
6753000	967
074	·8295015
148	062
222	110
296	157
370	205
444	253
518	300
592	348
666	395
740	443
814	491
888	538
962	586
6754036	633
110	681
184	729
258	776
332	824
406	871
480	919
555	967
629	·8296014
703	062
777	109
851	157
6754925	·8296205
999	252
6755073	300
147	347
221	395
295	443
369	490
443	538
517	585
591	633
665	681
739	728
813	776
887	823
961	871
6756035	918
109	966
183	·8297014
257	061
331	109
405	156
479	204
553	251
627	299
701	347
775	394
849	442
923	489
997	537
6757071	584
145	632
219	679
293	727
367	775
441	822
515	870
589	917
663	965
737	·8298012
811	060
885	107
959	155
6758033	203
107	250
181	298
255	345
329	393
403	440
477	488
551	535
625	583
699	630
773	678
847	725
921	773
995	820
6759068	868
142	916
216	963
290	·8299011
364	058
438	106
512	153
586	201
660	248
6759734	·8299296
808	343
882	391
956	438

Excesses

10 = 6	1 = 1
20 = 13	2 = 1
30 = 19	3 = 2
40 = 26	4 = 3
50 = 32	5 = 3
60 = 39	6 = 4
70 = 45	7 = 5
	8 = 5
	9 = 6

37 = 24

Anti Log Excesses

10 = 16	1 = 2
20 = 31	2 = 3
30 = 47	3 = 5
40 = 62	4 = 6
	5 = 8
	6 = 9
	7 = 11
	8 = 12
	9 = 14

24 = 37

6760030	·8299486
104	533
178	581
252	628
326	676
400	723
474	771
548	818
622	866
696	913
770	961
844	·8300008
918	056
992	103
6761066	151
139	198
213	246
287	293
361	341
435	388
509	436
583	484
657	531
731	578
805	626
879	673
953	721
6762027	768
101	816
175	864
249	911
323	959
6762397	·8301006
470	054
544	101
618	149
692	196
766	244
840	291
914	339
988	386
6763062	433
136	481
210	528
284	576
358	623
432	671
506	718
579	766
653	813
727	861
801	908
875	956
949	·8302003
6764023	051
097	098
171	145
245	193
319	240
393	288
466	335
540	383
614	430
688	478
762	525
836	573
910	620
984	667
6765058	715
132	762
206	810
280	857
353	905
427	952
501	·8303000
575	047
649	094
723	142
797	189
871	237
945	284
6766019	332
092	379
166	427
240	474
314	521
388	569
462	616
536	664
610	711
684	759
757	806
831	853
905	901
979	948
6767053	996
127	·8304043

Number	Log
6767201	·8304090
275	138
349	185
422	233
496	280
570	328
644	375
718	422
792	470
866	517
940	565
6768014	612
087	659
161	707
235	754
309	802
383	849
457	896
531	944
604	991
678	·8305039
752	086
826	133
900	181
974	228
6769048	276
122	323
195	370
269	418
343	465
417	513
491	560
565	607
639	655
712	702
786	749
860	797
934	844

Excesses

10 = 6	1 = 1
20 = 13	2 = 1
30 = 19	3 = 2
40 = 26	4 = 3
50 = 32	5 = 3
60 = 39	6 = 4
70 = 45	7 = 4
	8 = 5
	9 = 6

37 = 24

Anti Log Excesses

10 = 16	1 = 2
20 = 31	2 = 3
30 = 47	3 = 5
40 = 62	4 = 6
	5 = 8
	6 = 9
	7 = 11
	8 = 12
	9 = 14

24 = 37

Number	Log
6770008	·8305892
082	939
156	986
229	·8306034
303	081
377	128
451	176
525	223
599	271
673	318
746	365
820	413
894	460
968	507
6771042	555
116	602
189	649
263	697
337	744
411	792
485	839
559	886
633	934
706	981
780	·8307028
854	076
928	123
6772002	170
076	218
149	265
223	312
297	360
371	407
445	454
518	502
592	549
666	596
740	644
814	691
888	738
961	786
6773035	833
109	880
183	928
257	975
331	·8308022
404	070
478	117
552	165
626	212
700	259

Number	Log
6773773	·8308306
847	354
921	401
995	448
6774069	496
143	543
216	590
290	638
364	685
438	732
512	780
585	827
659	874
733	922
807	969
881	·8309016
954	064
6775028	111
102	158
176	206
250	253
323	300
397	347
471	395
545	442
619	489
692	537
766	584
840	631
914	679
988	726
6776061	773
135	820
209	868
283	915
356	962
430	·8310010
504	057
578	104
652	151
725	199
799	246
873	293
947	341
6777020	388
094	435
168	482
242	530
316	577
389	624
463	671
537	719
611	766
684	813
758	861
832	908
906	955
980	·8311002
6778053	050
127	097
201	144
275	191
348	239
422	286
496	333

Number	Log
6778570	·8311380
643	428
717	475
791	522
865	569
938	617
6779012	664
086	711
160	758
233	806
307	853
381	900
455	947
529	995
602	·8312042
676	089
750	136
824	184
897	231
971	278

Excesses

10 = 6	1 = 1
20 = 13	2 = 1
30 = 19	3 = 2
40 = 26	4 = 3
50 = 32	5 = 3
60 = 38	6 = 4
70 = 45	7 = 4
	8 = 5
	9 = 6

37 = 24

Anti Log Excesses

10 = 16	1 = 2
20 = 31	2 = 3
30 = 47	3 = 5
40 = 62	4 = 6
	5 = 8
	6 = 9
	7 = 11
	8 = 12
	9 = 14

24 = 37

Number	Log
6780045	·8312325
118	373
192	420
266	467
340	514
413	562
487	609
561	656
635	703
708	750
782	798
856	845
930	892
6781003	939
077	987
151	·8313034

Number	Log
6781225	·8313081
298	128
372	176
446	223
520	270
593	317
667	364
741	412
814	459
888	506
962	553
6782036	600
109	648
183	695
257	742
330	789
404	836
478	884
552	931
625	978
699	·8314025
773	072
847	120
920	167
994	214
6783068	261
141	308
215	356
289	403
362	450
436	497
510	544
584	592
657	639
731	686
805	733
878	780
952	827
6784026	875
100	922
173	969
247	·8315016
321	063
394	111
468	158
542	205
615	252
689	299
763	346
837	394
910	441
984	488
6785058	535
131	582
205	629
279	677
352	724
426	771
500	818
573	865
647	912
721	960
794	·8316007
868	054
942	101

6786016	·8316148
089	195
163	243
237	290
310	337
384	384
458	431
531	478
605	525
679	573
752	620
826	667
900	714
973	761
6787047	808
121	855
194	903
268	950
342	997
415	·8317044
489	091
563	138
636	185
710	233
784	280
857	327
931	374
6788005	421
078	468
152	515
226	562
299	610
373	657
447	704
520	751
594	798
668	845
741	892
815	939
888	986
962	·8318034
6789036	081
109	128
183	175
257	222
330	269
404	316
478	363
551	410
625	458
699	505
772	552
846	599
919	646
993	693

Excesses

10 = 6	1 = 1
20 = 13	2 = 1
30 = 19	3 = 2
40 = 26	4 = 3
50 = 32	5 = 3
60 = 38	6 = 4
70 = 45	7 = 4
	8 = 5
	9 = 6

37 = 24

Anti Log Excesses

10 = 16	1 = 2
20 = 31	2 = 3
30 = 47	3 = 5
40 = 62	4 = 6
	5 = 8
	6 = 9
	7 = 11
	8 = 12
	9 = 14

24 = 37

6790067	·8318740
140	787
214	834
288	881
361	929
435	976
509	·8319023
582	070
656	117
729	164
803	211
877	258
950	305
6791024	352
098	400
171	447
245	494
318	541
392	588
466	635
539	682
613	729
687	776
760	823
834	870
907	917
981	965
6792055	·8320012
128	059
202	106
276	153
349	200
423	247
496	294
570	341
644	388
717	435

6792791	·8320482
864	529
938	576
6793012	623
085	670
159	718
232	765
306	812
380	859
453	906
527	953
600	·8321000
674	047
748	094
821	141
895	188
968	235
6794042	282
116	329
189	376
263	423
336	470
410	517
484	564
557	611
631	658
704	705
778	753
851	800
925	847
999	894
6795072	941
146	988
219	·8322035
293	082
367	129
440	176
514	223
587	270
661	317
734	364
808	411
882	458
955	505
6796029	552
102	599
176	646
249	693
323	740
397	787
470	834
544	881
617	928
691	975
764	·8323022
838	069
912	116
985	163
6797059	210
132	257
206	304
279	351
353	398
426	445
500	492

6797574	·8323539
647	586
721	633
794	680
868	727
941	774
6798015	821
088	868
162	915
236	962
309	·8324009
383	056
456	103
530	150
603	197
677	244
750	291
824	338
897	385
971	432
6799045	479
118	526
192	573
265	620
339	667
412	714
486	760
559	807
633	854
706	901
780	948
853	995
927	·8325042

Excesses

10 = 6	1 = 1
20 = 13	2 = 1
30 = 19	3 = 2
40 = 26	4 = 3
50 = 32	5 = 3
60 = 38	6 = 4
70 = 45	7 = 4
	8 = 5
	9 = 6

37 = 24

Anti Log Excesses

10 = 16	1 = 2
20 = 31	2 = 3
30 = 47	3 = 5
40 = 63	4 = 6
	5 = 8
	6 = 9
	7 = 11
	8 = 13
	9 = 14

24 = 37

6800000	·8325089
074	136
148	183

6800221	·8325230
295	277
368	324
442	371
515	418
589	465
662	512
736	559
809	606
883	653
956	700
6801030	746
103	793
177	840
250	887
324	934
397	981
471	·8326028
544	075
618	122
691	169
765	216
838	263
912	310
986	357
6802059	404
132	451
206	497
280	544
353	591
427	638
500	685
574	732
647	779
721	826
794	873
868	920
941	967
6803015	·8327014
088	061
162	107
235	154
309	201
382	248
456	295
529	342
602	389
676	436
749	483
823	530
896	577
970	623
6804043	670
117	717
190	764
264	811
337	858
411	905
484	952
558	999
631	·8328046
705	093
778	139
852	186
925	233

6804999	·8328280
6805072	327
146	374
219	421
293	468
366	515
440	561
513	608
586	655
660	702
733	749
807	796
880	843
954	890
6806027	937
101	983
174	·8329030
248	077
321	124
395	171
468	218
541	265
615	312
688	358
762	405
835	452
909	499
982	546
6807056	593
129	640
203	686
276	733
349	780
423	827
496	874
570	921
643	968
717	·8330014
790	061
864	108
937	155
6808010	202
084	249
157	296
231	342
304	389
378	436
451	483
525	530
598	577
671	623
745	670
818	717
892	764
965	811
6809039	858
112	905
185	951
259	998
332	·8331045
406	092
479	139
553	185
626	232
699	279

6809773	·8331326
846	373
920	420
993	466

Excesses

10 = 6	1 = 1
20 = 13	2 = 1
30 = 19	3 = 2
40 = 26	4 = 3
50 = 32	5 = 3
60 = 38	6 = 4
70 = 45	7 = 4
	8 = 5
	9 = 6

37 = 24

Anti Log Excesses

10 = 16	1 = 2
20 = 31	2 = 3
30 = 47	3 = 5
40 = 63	4 = 6
	5 = 8
	6 = 9
	7 = 11
	8 = 13
	9 = 14

24 = 37

6810067	·8331513
140	560
213	607
287	654
360	700
434	747
507	794
581	841
654	888
727	935
801	981
874	·8332028
948	075
6811021	122
094	169
168	215
241	262
315	309
388	356
461	403
535	449
608	496
682	543
755	590
828	637
902	684
975	730
6812049	777
122	824
195	871
269	918
342	964

6812416	·8333011
489	058
562	105
636	151
709	198
783	245
856	292
929	339
6813003	385
076	432
150	479
223	526
296	573
370	619
443	666
516	713
590	760
663	806
737	853
810	900
883	947
957	993
6814030	·8334040
104	087
177	134
250	181
324	227
397	274
470	321
544	368
617	414
691	461
764	508
837	555
911	601
984	648
6815057	695
131	742
204	788
277	835
351	882
424	929
498	975
571	·8335022
644	069
718	116
791	162
864	209
938	256
6816011	303
084	349
158	396
231	443
304	489
378	536
451	583
525	630
598	676
671	723
745	770
818	817
891	863
965	910
6817038	957
111	·8336003

6817185	·8336050
258	097
331	144
405	190
478	237
551	284
625	331
698	377
771	424
845	471
918	517
991	564
6818065	611
138	658
211	704
285	751
358	798
431	844
505	891
578	938
651	984
725	·8337031
798	078
871	125
945	171
6819018	218
091	265
165	311
238	358
311	405
385	452
458	498
531	545
605	592
678	638
751	685
825	732
898	778
971	825

Excesses

10 = 6	1 = 1
20 = 13	2 = 1
30 = 19	3 = 2
40 = 25	4 = 3
50 = 32	5 = 3
60 = 38	6 = 4
70 = 45	7 = 4
	8 = 5
	9 = 6

37 = 23

Anti Log Excesses

10 = 16	1 = 2
20 = 31	2 = 3
30 = 47	3 = 5
40 = 63	4 = 6
	5 = 8
	6 = 9
	7 = 11
	8 = 13
	9 = 14

23 = 37

6820044	·8337872
118	918
191	965
264	·8338012
338	058
411	105
484	152
558	198
631	245
704	292
778	339
851	385
924	432
997	479
6821071	525
144	572
217	619
291	665
364	712
437	759
511	805
584	852
657	899
730	945
804	992
877	·8339039
950	085
6822024	132
097	179
170	225
244	272
317	319
390	365
463	412
537	459
610	505
683	552
757	598
830	645
903	692
976	738
6823050	785
123	832
196	879
270	925
343	972
416	·8340018
489	065
563	112
636	158
709	205

6823782	·8340252
856	298
929	345
6824002	391
076	438
149	485
222	531
295	578
369	625
442	671
515	718
588	764
662	811
735	858
808	904
881	951
955	998
6825028	·8341044
101	091
174	137
248	184
321	231
394	277
468	324
541	370
614	417
687	464
761	510
834	557
907	603
980	650
6826054	697
127	743
200	790
273	836
347	883
420	930
493	976
566	·8342023
639	069
713	116
786	163
859	209
932	256
6827006	303
079	349
152	396
225	442
299	489
372	535
445	582
518	629
592	675
665	722
738	768
811	815
885	861
958	908
6828031	955
104	·8343001
177	048
251	094
324	141
397	187
470	234

6828544	·8343281
617	327
690	374
763	420
836	467
910	513
983	560
6829056	607
129	653
203	700
276	746
349	793
422	839
495	886
569	932
642	979
715	·8344025
788	072
861	119
935	165

Excesses

10 = 6	1 = 1
20 = 13	2 = 1
30 = 19	3 = 2
40 = 25	4 = 3
50 = 32	5 = 3
60 = 38	6 = 4
70 = 45	7 = 4
	8 = 5
	9 = 6

37 = 23

Anti Log Excesses

10 = 16	1 = 2
20 = 31	2 = 3
30 = 47	3 = 5
40 = 63	4 = 6
	5 = 8
	6 = 9
	7 = 11
	8 = 13
	9 = 14

23 = 37

6830008	·8344212
081	258
154	305
227	351
301	398
374	444
447	491
520	537
593	584
667	631
740	677
813	724
886	770
959	817
6831033	863
106	910

6831179	·8344956
252	·8345003
325	049
399	096
472	143
545	189
618	236
691	282
765	329
838	375
911	422
984	468
6832057	515
130	561
204	608
277	654
350	701
423	747
496	794
570	840
643	887
716	933
789	980
862	·8346026
935	073
6833009	119
082	166
155	212
228	259
301	305
374	352
448	398
521	445
594	491
667	538
740	584
813	631
887	677
960	724
6834033	770
106	817
179	863
252	910
326	956
399	·8347003
472	049
545	096
618	142
691	189
765	235
838	282
911	328
984	375
6835057	421
130	468
203	514
277	561
350	607
423	654
496	700
569	747
642	793
716	840
789	886
862	933

6835935	·8347979
6836008	·8348025
081	072
154	118
228	165
301	211
374	258
447	304
520	351
593	397
666	444
739	490
812	537
886	583
959	629
6837032	676
105	722
178	769
251	815
325	862
398	908
471	955
544	·8349001
617	047
690	094
763	140
836	187
910	233
983	280
6838056	326
129	373
202	419
275	465
348	512
421	558
494	605
568	651
641	698
714	744
787	791
860	837
933	883
6839006	930
079	976
152	·8350023
226	069
299	115
372	162
445	208
518	255
591	301
664	348
737	394
810	440
884	487
957	533

Excesses

10 = 6	1 = 1
20 = 13	2 = 1
30 = 19	3 = 2
40 = 25	4 = 3
50 = 32	5 = 3
60 = 38	6 = 4
70 = 44	7 = 4
	8 = 5
	9 = 6

37 = 23

Anti Log Excesses

10 = 16	1 = 2
20 = 31	2 = 3
30 = 47	3 = 5
40 = 63	4 = 6
	5 = 8
	6 = 9
	7 = 11
	8 = 13
	9 = 14

23 = 37

6840030	·8350580
103	626
176	672
249	719
322	765
395	812
468	858
541	904
615	951
688	997
761	·8351044
834	090
907	136
980	183
6841053	229
126	276
199	322
272	368
345	415
418	461
492	508
565	554
638	600
711	647
784	693
857	740
930	786
6842003	832
076	879
149	925
222	971
295	·8352018
369	064
442	111
515	157
588	204
661	250

Number	Log
6842734	·8352296
807	343
880	389
953	435
6843026	482
099	528
172	575
245	621
318	667
391	714
465	760
538	806
611	853
684	899
757	945
830	992
903	·8353038
976	085
6844049	131
122	177
195	224
268	270
341	316
414	363
487	409
560	455
633	502
706	548
780	594
853	641
926	687
999	733
6845072	780
145	826
218	872
291	919
364	965
437	·8354011
510	058
583	104
656	151
729	197
802	243
875	290
948	336
6846021	382
094	429
167	475
240	521
313	567
386	614
459	660
532	706
605	753
679	799
752	845
825	892
898	938
971	984
6847044	·8355031
117	077
190	123
263	170
336	216
409	262

Number	Log
6847482	·8355309
555	355
628	401
701	448
774	494
847	540
920	586
993	633
6848066	679
139	725
212	772
285	818
358	864
431	911
504	957
577	·8356003
650	049
723	096
796	142
869	188
942	235
6849015	281
088	327
161	374
234	420
307	466
380	512
453	559
526	605
599	651
672	698
745	744
818	790
891	836
964	883

Excesses

10 = 6	1 = 1
20 = 13	2 = 1
30 = 19	3 = 2
40 = 25	4 = 3
50 = 32	5 = 3
60 = 38	6 = 4
70 = 44	7 = 4
	8 = 5
	9 = 6

37 = 23

Anti Log Excesses

10 = 16	1 = 2
20 = 32	2 = 3
30 = 47	3 = 5
40 = 63	4 = 6
	5 = 8
	6 = 9
	7 = 11
	8 = 13
	9 = 14

23 = 37

Number	Log
6850037	·8356929

Number	Log
6850110	·8356975
183	·8357021
256	068
329	114
402	160
475	207
548	253
621	299
694	345
767	392
840	438
913	484
986	530
6851059	577
132	623
205	669
278	715
351	762
424	808
497	854
570	901
643	947
716	993
789	·8358039
862	086
935	132
6852008	178
080	224
153	271
226	317
299	363
372	409
445	456
518	502
591	548
664	594
737	641
810	687
883	733
956	779
6853029	826
102	872
175	918
248	964
321	·8359010
394	057
467	103
540	149
613	195
686	242
759	288
832	334
904	380
977	427
6854050	473
123	519
196	565
269	611
342	658
415	704
488	750
561	796
634	843
707	889
780	935

Number	Log
6854853	·8359981
926	·8360027
999	074
6855072	120
145	166
217	212
290	258
363	305
436	351
509	397
582	443
655	489
728	536
801	582
874	628
947	674
6856020	720
093	767
166	813
239	859
311	905
384	951
457	998
530	·8361044
603	090
676	136
749	182
822	229
895	275
968	321
6857041	367
114	413
187	459
259	506
332	552
405	598
478	644
551	690
624	737
697	783
770	829
843	875
916	921
989	967
6858061	·8362014
134	060
207	106
280	152
353	198
426	244
499	291
572	337
645	383
718	429
790	475
863	521
936	568
6859009	614
082	660
155	706
228	752
301	798
374	844
447	891
519	937

Number	Log
6859592	·8362983
665	·8363029
738	075
811	121
884	167
957	214

Excesses

10 = 6	1 = 1
20 = 13	2 = 1
30 = 19	3 = 2
40 = 25	4 = 3
50 = 32	5 = 3
60 = 38	6 = 4
70 = 44	7 = 4
	8 = 5
	9 = 6

37 = 23

Anti Log Excesses

10 = 16	1 = 2
20 = 32	2 = 3
30 = 47	3 = 5
40 = 63	4 = 6
	5 = 8
	6 = 9
	7 = 11
	8 = 13
	9 = 14

23 = 37

Number	Log
6860030	·8363260
103	306
175	352
248	398
321	444
394	490
467	537
540	583
613	629
686	675
758	721
831	767
904	813
977	859
6861050	906
123	952
196	998
269	·8364044
341	090
414	136
487	182
560	228
633	275
706	321
779	367
852	413
924	459
997	505
6862070	551
143	597

6862216	·8364644
289	690
362	736
434	782
507	828
580	874
653	920
726	966
799	·8365012
872	058
944	105
6863017	151
090	197
163	243
236	289
309	335
382	381
454	427
527	473
600	519
673	566
746	612
819	658
892	704
964	750
6864037	796
110	842
183	888
256	934
329	980
401	·8366026
474	073
547	119
620	165
693	211
766	257
838	303
911	349
984	395
6865057	441
130	487
203	533
275	579
348	625
421	672
494	718
567	764
640	810
712	856
785	902
858	948
931	994
6866004	·8367040
077	086
149	132
222	178
295	224
368	270
441	316
513	363
586	409
659	455
732	501
805	547
878	593

6866950	·8367639
6867023	685
096	731
169	777
242	823
314	869
387	915
460	961
533	·8368007
606	053
678	099
751	145
824	191
897	237
970	283
6868042	329
115	376
188	422
261	468
334	514
406	560
479	606
552	652
625	698
698	744
770	790
843	836
916	882
989	928
6869062	974
134	·8369020
207	066
280	112
353	158
426	204
498	250
571	296
644	342
717	388
789	434
862	480
935	526

Excesses

10 = 6	1 = 1
20 = 13	2 = 1
30 = 19	3 = 2
40 = 25	4 = 3
50 = 32	5 = 3
60 = 38	6 = 4
70 = 44	7 = 4
	8 = 5
	9 = 6

36 = 23

Anti Log Excesses

10 = 16	1 = 2
20 = 32	2 = 3
30 = 47	3 = 5
40 = 63	4 = 6
	5 = 8
	6 = 9
	7 = 11
	8 = 13
	9 = 14

23 = 36

6870008	·8369572
081	618
153	664
226	710
299	756
372	802
444	848
517	894
590	940
663	986
736	·8370032
808	078
881	124
954	170
6871027	216
099	262
172	308
245	354
318	400
390	446
463	492
536	538
609	584
682	630
754	676
827	722
900	768
973	814
6872045	860
118	906
191	952
264	998
336	·8371044
409	090
482	136
555	182
627	228
700	274
773	320
846	366
918	412
991	457
6873064	503
137	549
209	596
282	642
355	687
428	733
500	779
573	825
646	871

6873719	·8371917
791	963
864	·8372009
937	055
6874010	101
082	147
155	193
228	239
300	285
373	331
446	377
519	423
591	469
664	515
737	561
810	606
882	652
955	698
6875028	744
100	790
173	836
246	882
319	928
391	974
464	·8373020
537	066
610	112
682	158
755	204
828	250
900	296
973	341
6876046	387
119	433
191	479
264	525
337	571
409	617
482	663
555	709
628	755
700	801
773	847
846	893
918	938
991	984
6877064	·8374030
137	076
209	122
282	168
355	214
427	260
500	306
573	352
645	398
718	443
791	489
864	535
936	581
6878009	627
082	673
154	719
227	765
300	811
372	857

6878445	·8374902
518	948
590	994
663	·8375040
736	086
809	132
881	178
954	224
6879027	270
099	316
172	361
245	407
317	453
390	499
463	545
535	591
608	637
681	683
753	728
826	774
899	820
971	866

Excesses

10 = 6	1 = 1
20 = 13	2 = 1
30 = 19	3 = 2
40 = 25	4 = 3
50 = 32	5 = 3
60 = 38	6 = 4
70 = 44	7 = 4
	8 = 5
	9 = 6

36 = 23

Anti Log Excesses

10 = 16	1 = 2
20 = 32	2 = 3
30 = 47	3 = 5
40 = 63	4 = 6
	5 = 8
	6 = 9
	7 = 11
	8 = 13
	9 = 14

23 = 36

6880044	·8375912
117	958
189	·8376004
262	050
335	095
407	141
480	187
553	233
625	279
698	325
771	371
843	417
916	462
989	508

6881061	·8376554
134	600
207	646
279	692
352	738
425	783
497	829
570	875
643	921
715	967
788	·8377013
861	059
933	104
6882006	150
079	196
151	242
224	288
297	334
369	379
442	425
515	471
587	517
660	563
733	609
805	655
878	700
950	746
6883023	792
096	838
168	884
241	930
314	975
386	·8378021
459	067
532	113
604	159
677	205
749	250
822	296
895	342
967	388
6884040	434
113	479
185	525
258	571
331	617
403	663
476	709
548	754
621	800
694	846
766	892
839	938
912	984
984	·8379029
6885057	075
129	121
202	167
275	213
347	258
420	304
492	350
565	396
638	442
710	487

6885783	·8379533
856	579
928	625
6886001	671
073	716
146	762
219	808
291	854
364	900
436	945
509	991
582	·8380037
654	083
727	129
799	174
872	220
945	266
6887017	312
090	357
162	403
235	449
308	495
380	541
453	586
525	632
598	678
671	724
743	769
816	815
888	861
961	907
6888034	952
106	998
179	·8381044
251	090
324	136
397	181
469	227
542	273
614	319
687	364
759	410
832	456
905	502
977	547
6889050	593
122	639
195	685
268	730
340	776
413	822
485	868
558	913
630	959
703	·8382005
776	051
848	096
921	142
993	188

Excesses

10 = 6	1 = 1
20 = 13	2 = 1
30 = 19	3 = 2
40 = 25	4 = 3
50 = 32	5 = 3
60 = 38	6 = 4
70 = 44	7 = 4
	8 = 5
	9 = 6

36 = 23

Anti Log Excesses

10 = 16	1 = 2
20 = 32	2 = 3
30 = 48	3 = 5
40 = 63	4 = 6
	5 = 8
	6 = 10
	7 = 11
	8 = 13
	9 = 14

23 = 36

6890066	·8382234
138	279
211	325
284	371
356	417
429	462
501	508
574	554
646	600
719	645
791	691
864	737
937	782
6891009	828
082	874
154	920
227	965
299	·8383011
372	057
444	103
517	148
590	194
662	240
735	285
807	331
880	377
952	423
6892025	468
097	514
170	560
243	605
315	651
388	697
460	743
533	788
605	834
678	880

6892750	·8383925
823	971
895	·8384017
968	062
6893040	108
113	154
186	200
258	245
331	291
403	337
476	382
548	428
621	474
693	519
766	565
838	611
911	656
983	702
6894056	748
128	794
201	839
273	885
346	931
419	976
491	·8385022
564	068
636	113
709	159
781	205
854	250
926	296
999	342
6895071	387
144	433
216	479
289	524
361	570
434	616
506	661
579	707
651	753
724	798
796	844
869	890
941	935
6896014	981
086	·8386027
159	072
231	118
304	164
376	209
449	255
521	301
594	346
666	392
739	438
811	483
884	529
956	575
6897029	620
101	666
174	712
246	757
319	803
391	849

6897464	·8386894
536	940
609	985
681	·8387031
754	077
826	122
899	168
971	214
6898044	259
116	305
189	350
261	396
334	442
406	487
479	533
551	579
624	624
696	670
769	716
841	761
913	807
986	852
6899058	898
131	944
203	989
276	·8388035
348	080
421	126
493	172
566	217
639	263
711	309
783	354
856	400
928	445

Excesses

10 = 6	1 = 1
20 = 13	2 = 1
30 = 19	3 = 2
40 = 25	4 = 3
50 = 31	5 = 3
60 = 38	6 = 4
70 = 44	7 = 4
	8 = 5
	9 = 6

36 = 23

Anti Log Excesses

10 = 16	1 = 2
20 = 32	2 = 3
30 = 48	3 = 5
40 = 64	4 = 6
	5 = 8
	6 = 10
	7 = 11
	8 = 13
	9 = 14

23 = 36

6900000	·8388491

Number	Log
6900073	·8388537
145	582
218	628
290	674
363	719
435	765
508	810
580	856
653	902
725	947
798	993
870	·8389038
942	084
6901015	130
087	175
160	221
232	266
305	312
377	358
450	403
522	449
595	494
667	540
739	585
812	631
884	677
957	722
6902029	768
102	813
174	859
247	905
319	950
391	996
464	·8390041
536	087
609	132
681	178
754	224
826	269
898	315
971	360
6903043	406
116	451
188	497
261	543
333	588
405	634
478	679
550	725
623	770
695	816
768	861
840	907
912	953
985	998
6904057	·8391044
130	089
202	135
275	180
347	226
419	272
492	317
564	363
637	408
709	454

Number	Log
6904781	·8391499
854	545
926	590
999	636
6905071	681
144	727
216	772
288	818
361	864
433	909
506	955
578	·8392000
650	046
723	091
795	137
868	182
940	228
6906012	273
085	319
157	364
230	410
302	455
374	501
447	546
519	592
592	637
664	683
736	729
809	774
881	820
954	865
6907026	911
098	956
171	·8393002
243	047
315	093
388	138
460	184
533	229
605	275
677	320
750	366
822	411
895	457
967	502
6908039	548
112	593
184	639
256	684
329	730
401	775
474	821
546	866
618	912
691	957
763	·8394003
835	048
908	094
980	139
6909053	185
125	230
197	276
270	321
342	367
414	412

Number	Log
6909487	·8394458
559	503
631	549
704	594
776	640
849	685
921	730
993	776

Excesses

10 = 6	1 = 1
20 = 13	2 = 1
30 = 19	3 = 2
40 = 25	4 = 3
50 = 31	5 = 3
60 = 38	6 = 4
70 = 44	7 = 4
	8 = 5
	9 = 6

36 = 23

Anti Log Excesses

10 = 16	1 = 2
20 = 32	2 = 3
30 = 48	3 = 5
40 = 64	4 = 6
	5 = 8
	6 = 10
	7 = 11
	8 = 13
	9 = 14

23 = 36

Number	Log
6910066	·8394821
138	867
210	912
283	958
355	·8395003
427	049
500	094
572	140
644	185
717	231
789	276
862	322
934	367
6911006	413
079	458
151	503
223	549
296	594
368	640
440	685
513	731
585	777
657	822
730	867
802	913
874	958
947	·8396004
6912019	049

Number	Log
6912091	·8396095
164	140
236	185
308	231
381	276
453	322
525	367
598	413
670	458
742	503
815	549
887	594
959	640
6913032	685
104	731
176	776
249	822
321	867
393	912
466	958
538	·8397003
610	049
683	094
755	140
827	185
900	230
972	276
6914044	321
116	367
189	412
261	458
333	503
406	548
478	594
550	639
623	685
695	730
767	775
840	821
912	866
984	912
6915057	957
129	·8398002
201	048
273	093
346	139
418	184
490	230
563	275
635	320
707	366
780	411
852	457
924	502
996	548
6916069	593
141	638
213	684
286	729
358	775
430	820
503	865
575	911
647	956
719	·8399001

Number	Log
6916792	·8399047
864	092
936	138
6917009	183
081	228
153	274
225	319
298	365
370	410
442	455
515	501
587	546
659	591
731	637
804	682
876	728
948	773
6918020	818
093	864
165	909
237	954
310	·8400000
382	045
454	090
526	136
599	181
671	227
743	272
815	317
888	363
960	408
6919032	453
104	499
177	544
249	589
321	635
394	680
466	726
538	771
610	816
683	862
755	907
827	952
899	998
972	·8401043

Excesses

10 = 6	1 = 1
20 = 13	2 = 1
30 = 19	3 = 2
40 = 25	4 = 3
50 = 31	5 = 3
60 = 38	6 = 4
70 = 44	7 = 4
	8 = 5
	9 = 6

36 = 23

Anti Log Excesses

10 = 16	1 = 2
20 = 32	2 = 3
30 = 48	3 = 5
40 = 64	4 = 6
	5 = 8
	6 = 10
	7 = 11
	8 = 13
	9 = 14

23 = 36

6920044	·8401088
116	134
188	179
261	224
333	270
405	315
477	360
550	406
622	451
694	496
766	542
839	587
911	632
983	678
6921055	723
128	768
200	814
272	859
344	904
417	950
489	995
561	·8402040
633	086
705	131
778	176
850	222
922	267
994	312
6922067	358
139	403
211	448
283	494
356	539
428	584
500	630
572	675
645	720
717	765
789	811
861	856
933	901
6923006	947
078	992
150	·8403037
222	083
295	128
367	173
439	219
511	264
583	309
656	354

6923728	·8403400
800	445
872	490
944	536
6924017	581
089	626
161	671
233	717
306	762
378	807
450	853
522	898
594	943
667	989
739	·8404034
811	079
883	124
955	170
6925028	215
100	260
172	305
244	351
316	396
389	441
461	487
533	532
605	577
677	622
750	668
822	713
894	758
966	803
6926038	849
111	894
183	939
255	985
327	·8405030
399	075
471	120
544	166
616	211
688	256
760	301
832	347
905	392
977	437
6927049	482
121	528
193	573
265	618
338	663
410	709
482	754
554	799
626	844
699	890
771	935
843	980
915	·8406025
987	071
6928059	116
132	161
204	206
276	252
348	297

6928420	·8406342
492	387
565	433
637	478
709	523
781	568
853	613
925	659
998	704
6929070	749
142	794
214	840
286	885
358	930
430	975
503	·8407021
575	066
647	111
719	156
791	201
863	247
936	292

Excesses

10 = 6	1 = 1
20 = 13	2 = 1
30 = 19	3 = 2
40 = 25	4 = 3
50 = 31	5 = 3
60 = 38	6 = 4
70 = 44	7 = 4
	8 = 5
	9 = 6

36 = 23

Anti Log Excesses

10 = 16	1 = 2
20 = 32	2 = 3
30 = 48	3 = 5
40 = 64	4 = 6
	5 = 8
	6 = 10
	7 = 11
	8 = 13
	9 = 14

23 = 36

6930008	·8407337
080	382
152	427
224	473
296	518
368	563
441	608
513	654
585	699
657	744
729	789
801	834
873	880
946	925

6931018	·8407970
090	·8408015
162	060
234	106
306	151
378	196
451	241
523	286
595	332
667	377
739	422
811	467
883	512
955	557
6932028	603
100	648
172	693
244	738
316	783
388	829
460	874
533	919
605	964
677	·8409009
749	054
821	100
893	145
965	190
6933037	235
109	280
182	326
254	371
326	416
398	461
470	506
542	551
614	597
686	642
758	687
831	732
903	777
975	822
6934047	867
119	913
191	958
263	·8410003
335	048
407	093
480	138
552	184
624	229
696	274
768	319
840	364
912	409
984	454
6935056	500
128	545
201	590
273	635
345	680
417	726
489	771
561	816
633	861

6935705	·8410906
777	951
849	996
921	·8411042
994	087
6936066	132
138	177
210	222
282	267
354	312
426	357
498	403
570	448
642	493
714	538
786	583
859	628
931	673
6937003	718
075	764
147	809
219	854
291	899
363	944
435	989
507	·8412034
579	079
651	125
723	170
796	215
868	260
940	305
6938012	350
084	395
156	440
228	485
300	530
372	576
444	621
516	666
588	711
660	756
732	801
804	846
876	891
949	936
6939021	982
093	·8413027
165	072
237	117
309	162
381	207
453	252
525	297
597	342
669	387
741	432
813	478
885	523
957	568

Excesses	
10 = 6	1 = 1
20 = 13	2 = 1
30 = 19	3 = 2
40 = 25	4 = 3
50 = 31	5 = 3
60 = 38	6 = 4
70 = 44	7 = 4
	8 = 5
	9 = 6
36 = 23	

Anti Log Excesses	
10 = 16	1 = 2
20 = 32	2 = 3
30 = 48	3 = 5
40 = 64	4 = 6
	5 = 8
	6 = 10
	7 = 11
	8 = 13
	9 = 14
23 = 36	

6940029	·8413613
101	658
173	703
245	748
317	793
390	838
462	883
534	928
606	973
678	·8414018
750	064
822	109
894	154
966	199
6941038	244
110	289
182	334
254	379
326	424
398	469
470	514
542	559
614	604
686	649
758	695
830	740
902	785
974	830
6942046	875
118	920
190	965
262	·8415010
334	055
406	100
478	145
550	190
622	235

6942695	·8415280
767	325
839	370
911	415
983	460
6943055	505
127	551
199	596
271	641
343	686
415	731
487	776
559	821
631	866
703	911
775	956
847	·8416001
919	046
991	091
6944063	136
135	181
207	226
279	271
351	316
423	361
495	406
567	451
639	496
711	541
783	586
855	631
927	676
999	721
6945071	766
143	811
215	856
287	901
359	946
431	991
503	·8417037
575	082
647	127
719	172
791	217
863	262
935	307
6946007	352
079	397
151	442
222	487
294	532
366	577
438	622
510	667
582	712
654	757
726	802
798	847
870	892
942	937
6947014	982
086	·8418027
158	072
230	117
302	162

6947374	·8418207
446	252
518	297
590	342
662	387
734	432
806	477
878	522
950	567
6948022	612
094	657
166	702
238	747
310	792
382	837
454	882
526	927
597	972
669	·8419017
741	062
813	107
885	151
957	196
6949029	241
101	286
173	331
245	376
317	421
389	466
461	511
533	556
605	601
677	646
749	691
821	736
893	781
965	826

Excesses	
10 = 6	1 = 1
20 = 13	2 = 1
30 = 19	3 = 2
40 = 25	4 = 3
50 = 31	5 = 3
60 = 38	6 = 4
70 = 44	7 = 4
	8 = 5
	9 = 6
36 = 23	

Anti Log Excesses	
10 = 16	1 = 2
20 = 32	2 = 3
30 = 48	3 = 5
40 = 64	4 = 6
	5 = 8
	6 = 10
	7 = 11
	8 = 13
	9 = 14
23 = 36	

6950036	·8419871
108	916
180	961
252	·8420006
324	051
396	096
468	141
540	186
612	230
684	275
756	320
828	365
900	410
972	455
6951044	500
116	545
187	590
259	635
331	680
403	725
475	770
547	815
619	860
691	905
763	950
835	994
907	·8421039
979	084
6952051	129
122	174
194	219
266	264
338	309
410	354
482	399
554	444
626	489
698	534
770	579
842	623
914	668
985	713
6953057	758
129	803
201	848
273	893
345	938
417	983
489	·8422028
561	073
633	117
705	162
776	207
848	252
920	297
992	342
6954064	387
136	432
208	477
280	522
352	567
424	611
495	656
567	701
639	746

6954711	·8422791
783	836
855	881
927	926
999	970
6955071	·8423015
142	060
214	105
286	150
358	195
430	240
502	285
574	330
646	374
718	419
789	464
861	509
933	554
6956005	599
077	644
149	689
221	734
293	778
364	823
436	868
508	913
580	958
652	·8424003
724	048
796	093
868	137
939	182
6957011	227
083	272
155	317
227	362
299	407
371	451
443	496
514	541
586	586
658	631
730	676
802	721
874	765
946	810
6958017	855
089	900
161	945
233	990
305	·8425035
377	079
449	124
520	169
592	214
664	259
736	304
808	348
880	393
952	438
6959023	483
095	528
167	573
239	617
311	662

6959383	·8425707
455	752
526	797
598	842
670	886
742	931
814	976
886	·8426021
957	066

Excesses

10 = 6	1 = 1
20 = 12	2 = 1
30 = 19	3 = 2
40 = 25	4 = 2
50 = 31	5 = 3
60 = 37	6 = 4
70 = 44	7 = 4
	8 = 5
	9 = 6

36 = 22

Anti Log Excesses

10 = 16	1 = 2
20 = 32	2 = 3
30 = 48	3 = 5
40 = 64	4 = 6
	5 = 8
	6 = 10
	7 = 11
	8 = 13
	9 = 14

22 = 36

6960029	·8426111
101	155
173	200
245	245
317	290
388	335
460	380
532	424
604	469
676	514
748	559
819	604
891	648
963	693
6961035	738
107	783
179	828
250	873
322	917
394	962
466	·8427007
538	052
610	097
681	141
753	186
825	231
897	276

6961969	·8427321
6962040	365
112	410
184	455
256	500
328	545
400	589
471	634
543	679
615	724
687	769
759	813
830	858
902	903
974	948
6963046	992
118	·8428037
189	082
261	127
333	172
405	216
477	261
548	306
620	351
692	396
764	440
836	485
907	530
979	575
6964051	619
123	664
195	709
266	754
338	798
410	843
482	888
554	933
625	978
697	·8429022
769	067
841	112
913	157
984	201
6965056	246
128	291
200	336
271	380
343	425
415	470
487	515
559	559
630	604
702	649
774	694
846	738
918	783
989	828
6966061	873
133	917
205	962
276	·8430007
348	052
420	096
492	141
563	186

6966635	·8430231
707	275
779	320
851	365
922	410
994	454
6967066	499
138	544
209	589
281	633
353	678
425	723
496	767
568	812
640	857
712	902
784	946
855	991
927	·8431036
999	081
6968071	125
142	170
214	215
286	259
358	304
429	349
501	394
573	438
645	483
716	528
788	572
860	617
932	662
6969003	707
075	751
147	796
219	841
290	885
362	930
434	975
506	·8432019
577	064
649	109
721	154
792	198
864	243
936	288

Excesses

10 = 6	1 = 1
20 = 12	2 = 1
30 = 19	3 = 2
40 = 25	4 = 2
50 = 31	5 = 3
60 = 37	6 = 4
70 = 44	7 = 4
	8 = 5
	9 = 6

36 = 22

Anti Log Excesses

10 = 16	1 = 2
20 = 32	2 = 3
30 = 48	3 = 5
40 = 64	4 = 6
	5 = 8
	6 = 10
	7 = 11
	8 = 13
	9 = 14

22 = 36

6970008	·8432332
079	377
151	422
223	466
295	511
366	556
438	601
510	645
582	690
653	735
725	779
797	824
868	869
940	913
6971012	958
084	·8433003
155	047
227	092
299	137
371	182
442	226
514	271
586	316
657	360
729	405
801	450
873	494
944	539
6972016	584
088	628
159	673
231	718
303	762
375	807
446	852
518	896
590	941
661	986
733	·8434030
805	075
877	120
948	164
6973020	209
092	254
163	298
235	343
307	387
378	432
450	477
522	521
594	566

6973665	·8434611
737	655
809	700
880	745
952	789
6974024	834
095	879
167	923
239	968
311	·8435013
382	057
454	102
526	147
597	191
669	236
741	280
812	325
884	370
956	414
6975027	459
099	504
171	548
242	593
314	638
386	682
457	727
529	771
601	816
673	861
744	905
816	950
888	995
959	·8436039
6976031	084
103	128
174	173
246	218
318	262
389	307
461	351
533	396
604	441
676	485
748	530
819	575
891	619
963	664
6977034	708
106	753
178	798
249	842
321	887
393	931
464	976
536	·8437021
608	065
679	110
751	154
823	199
894	244
966	288
6978037	333
109	377
181	422
252	467

6978324	·8437511
396	556
467	600
539	645
611	689
682	734
754	779
826	823
897	868
969	912
6979041	957
112	·8438001
184	046
255	091
327	135
399	180
470	224
542	269
614	314
685	358
757	403
829	447
900	492
972	536

Excesses

10 = 6	1 = 1
20 = 12	2 = 1
30 = 19	3 = 2
40 = 25	4 = 2
50 = 31	5 = 3
60 = 37	6 = 4
70 = 44	7 = 4
	8 = 5
	9 = 6
36 = 22	

Anti Log Excesses

10 = 16	1 = 2
20 = 32	2 = 3
30 = 48	3 = 5
40 = 64	4 = 6
	5 = 8
	6 = 10
	7 = 11
	8 = 13
	9 = 14
22 = 36	

6980043	·8438581
115	626
187	670
258	715
330	759
402	804
473	848
545	893
617	937
688	982
760	·8439027
831	071

6980903	·8439116
975	160
6981046	205
118	249
190	294
261	338
333	383
404	428
476	472
548	517
619	561
691	606
762	650
834	695
906	739
977	784
6982049	828
121	873
192	918
264	962
335	·8440007
407	051
479	096
550	140
622	185
693	229
765	274
837	319
908	363
980	408
6983051	452
123	497
195	541
266	586
338	630
409	675
481	719
553	764
624	808
696	853
767	897
839	942
911	986
982	·8441031
6984054	075
125	120
197	164
269	209
340	253
412	298
483	342
555	387
626	432
698	476
770	521
841	565
913	610
984	654
6985056	699
128	743
199	788
271	832
342	877
414	921
485	966

6985557	·8442010
629	055
700	099
772	144
843	188
915	233
986	277
6986058	321
130	366
201	410
273	455
344	499
416	544
487	588
559	633
631	677
702	722
774	766
845	811
917	855
988	900
6987060	944
132	989
203	·8443033
275	078
346	122
418	167
489	211
561	256
632	300
704	345
776	389
847	433
919	478
990	522
6988062	567
133	611
205	656
276	700
348	745
419	789
491	834
563	878
634	923
706	967
777	·8444011
849	056
920	100
992	145
6989063	189
135	234
206	278
278	323
350	367
421	412
493	456
564	500
636	545
707	589
779	634
850	678
922	723
993	767

Excesses

10 = 6	1 = 1
20 = 12	2 = 1
30 = 19	3 = 2
40 = 25	4 = 2
50 = 31	5 = 3
60 = 37	6 = 4
70 = 44	7 = 4
	8 = 5
	9 = 6
36 = 22	

Anti Log Excesses

10 = 16	1 = 2
20 = 32	2 = 3
30 = 48	3 = 5
40 = 64	4 = 6
	5 = 8
	6 = 10
	7 = 11
	8 = 13
	9 = 14
22 = 36	

6990065	·8444812
136	856
208	901
279	945
351	989
423	·8445034
494	078
566	123
637	167
709	212
780	256
852	300
923	345
995	389
6991066	434
138	478
209	523
281	567
352	611
424	656
495	700
567	745
638	789
710	834
781	878
853	922
924	967
996	·8446011
6992067	056
139	100
210	144
282	189
353	233
425	278
496	322
568	366
639	411

6992711	·8446455
782	500
854	544
925	589
997	633
6993068	677
140	722
211	766
283	811
354	855
426	899
497	944
569	988
640	·8447032
712	077
783	121
855	166
926	210
998	254
6994069	299
141	343
212	388
284	432
355	477
427	521
498	565
570	610
641	654
713	698
784	743
856	787
927	832
999	876
6995070	920
142	965
213	·8448009
285	053
356	098
428	142
499	187
571	231
642	275
713	320
785	364
856	408
928	453
999	497
6996071	542
142	586
214	630
285	675
357	719
428	763
500	808
571	852
643	896
714	941
785	985
857	·8449029
928	074
6997000	118
071	163
143	207
214	251
286	296

6997357	·8449340
429	384
500	429
572	473
643	517
714	562
786	606
857	650
929	695
6998000	739
072	783
143	828
215	872
286	916
357	961
429	·8450005
500	049
572	094
643	138
715	182
786	227
858	271
929	315
6999000	360
072	404
143	448
215	493
286	537
358	581
429	626
500	670
572	714
643	759
715	803
786	847
858	892
929	936

Excesses

10 = 6	1 = 1
20 = 12	2 = 1
30 = 19	3 = 2
40 = 25	4 = 2
50 = 31	5 = 3
60 = 37	6 = 4
70 = 43	7 = 4
	8 = 5
	9 = 6

36 = 22

Anti Log Excesses

10 = 16	1 = 2
20 = 32	2 = 3
30 = 48	3 = 5
40 = 64	4 = 6
	5 = 8
	6 = 10
	7 = 11
	8 = 13
	9 = 14

22 = 36

7000000	·8450980
072	·8451025
143	069
215	113
286	158
358	202
429	246
500	290
572	335
643	379
715	423
786	468
858	512
929	556
7001000	601
072	645
143	689
215	733
286	778
358	822
429	866
500	911
572	955
643	999
715	·8452044
786	088
857	132
929	176
7002000	221
072	265
143	309
214	354
286	398
357	442
429	486
500	531
571	575
643	619
714	664
786	708
857	752
928	796
7003000	841
071	885
143	929
214	973
285	·8453017
357	062
428	106
500	151
571	195
642	239
714	283
785	328
857	372
928	416
999	460
7004071	505
142	549
214	593
285	637
356	682
428	726
499	770
570	815

7004642	·8453859
713	903
785	947
856	992
927	·8454036
999	080
7005070	125
141	169
213	213
284	257
356	301
427	345
498	390
570	434
641	478
712	522
784	567
855	611
927	655
998	700
7006069	744
141	788
212	832
283	877
355	921
426	965
497	·8455009
569	053
640	098
712	142
783	186
854	230
926	275
997	319
7007068	363
140	407
211	452
282	496
354	540
425	584
496	628
568	673
639	717
711	761
782	805
853	850
925	894
996	938
7008067	982
139	·8456026
210	071
281	115
353	159
424	203
495	247
567	292
638	336
709	380
781	424
852	469
923	513
995	557
7009066	601
137	645
209	690

7009280	·8456734
351	778
423	822
494	866
565	911
637	955
708	999
779	·8457043
851	088
922	132
993	176

Excesses

10 = 6	1 = 1
20 = 12	2 = 1
30 = 19	3 = 2
40 = 25	4 = 2
50 = 31	5 = 3
60 = 37	6 = 4
70 = 43	7 = 4
	8 = 5
	9 = 6

36 = 22

Anti Log Excesses

10 = 16	1 = 2
20 = 32	2 = 3
30 = 48	3 = 5
40 = 65	4 = 6
	5 = 8
	6 = 10
	7 = 11
	8 = 13
	9 = 15

22 = 36

7010065	·8457220
136	264
207	308
279	353
350	397
421	441
493	485
564	529
635	574
707	618
778	662
849	706
921	750
992	794
7011063	839
135	883
206	927
277	971
348	·8458015
420	060
491	104
562	148
634	192
705	236
776	280

7011848	·8458325
919	369
990	413
7012062	457
133	501
204	545
275	590
347	634
418	678
489	722
561	766
632	810
703	855
775	899
846	943
917	987
988	·8459031
7013060	075
131	119
202	164
274	208
345	252
416	296
488	340
559	384
630	428
701	473
773	517
844	561
915	605
987	649
7014058	693
129	737
200	781
272	826
343	870
414	914
486	958
557	·8460002
628	046
699	090
771	135
842	179
913	223
984	267
7015056	311
127	355
198	399
270	443
341	488
412	532
483	576
555	620
626	664
697	708
768	752
840	796
911	841
982	885
7016054	929
125	973
196	·8461017
267	061
339	105
410	149

7016481	·8461193
552	238
624	282
695	326
766	370
837	414
909	458
980	502
7017051	546
122	590
194	635
265	679
336	723
407	767
479	811
550	855
621	899
692	943
764	987
835	·8462031
906	075
977	119
7018049	163
120	208
191	252
262	296
334	340
405	384
476	428
547	472
619	516
690	560
761	604
832	648
904	693
975	737
7019046	781
117	825
188	869
260	913
331	957
402	·8463001
473	045
545	089
616	133
687	177
758	221
830	265
901	310
972	354

Excesses

10 = 6	1 = 1
20 = 12	2 = 1
30 = 19	3 = 2
40 = 25	4 = 2
50 = 31	5 = 3
60 = 37	6 = 4
70 = 43	7 = 4
	8 = 5
	9 = 6

36 = 22

Anti Log Excesses

10 = 16	1 = 2
20 = 32	2 = 3
30 = 48	3 = 5
40 = 65	4 = 6
	5 = 8
	6 = 10
	7 = 11
	8 = 13
	9 = 15

22 = 36

7020043	·8463398
114	442
186	486
257	530
328	574
399	618
471	662
542	706
613	750
684	794
755	838
827	882
898	926
969	970
7021040	·8464014
112	059
183	103
254	147
325	191
396	235
468	279
539	323
610	367
681	411
752	455
824	499
895	543
966	587
7022037	631
108	675
180	719
251	763
322	807
393	851
464	895
536	939
607	983
678	·8465027
749	071
820	115
892	159
963	203
7023034	247
105	291
176	336
248	380
319	424
390	468
461	512
532	556
604	600

7023675	·8465644
746	688
817	732
888	776
959	820
7024031	864
102	908
173	952
244	996
315	·8466040
387	084
458	128
529	172
600	216
671	260
742	304
814	348
885	392
956	436
7025027	480
098	524
170	568
241	612
312	656
383	700
454	744
525	788
597	832
668	876
739	920
810	964
881	·8467008
952	052
7026024	096
095	140
166	184
237	228
308	272
379	316
451	360
522	404
593	448
664	492
735	536
806	580
877	624
949	668
7027020	712
091	755
162	799
233	843
304	887
376	931
447	975
518	·8468019
589	063
660	107
731	151
802	195
874	239
945	283
7028016	327
087	371
158	415
229	459

7028300	·8468503
372	547
443	591
514	635
585	679
656	723
727	767
798	811
870	855
941	899
7029012	942
083	986
154	·8469030
225	074
296	118
367	162
439	206
510	250
581	294
652	338
723	382
794	426
865	470
936	514

Excesses

10 = 6	1 = 1
20 = 12	2 = 1
30 = 19	3 = 2
40 = 25	4 = 2
50 = 31	5 = 3
60 = 37	6 = 4
70 = 43	7 = 4
	8 = 5
	9 = 6

36 = 22

Anti Log Excesses

10 = 16	1 = 2
20 = 32	2 = 3
30 = 49	3 = 5
40 = 65	4 = 6
	5 = 8
	6 = 10
	7 = 11
	8 = 13
	9 = 15

22 = 36

7030008	·8469558
079	602
150	646
221	690
292	733
363	777
434	821
505	865
577	909
648	953
719	997
790	·8470041

7030861	·8470085
932	129
7031003	173
074	217
145	261
217	305
288	348
359	392
430	436
501	480
572	524
643	568
714	612
785	656
857	700
928	744
999	788
7032070	832
141	875
212	919
283	963
354	·8471007
425	051
497	095
568	139
639	183
710	227
781	271
852	314
923	358
994	402
7033065	446
136	490
207	534
279	578
350	622
421	666
492	709
563	753
634	797
705	841
776	885
847	929
918	973
989	·8472017
7034060	061
132	105
203	148
274	192
345	236
416	280
487	324
558	368
629	412
700	456
771	499
842	543
913	587
985	631
7035056	675
127	719
198	763
269	807
340	850
411	894

7035482	·8472938
553	982
624	·8473026
695	070
766	114
837	157
908	201
979	245
7036051	289
122	333
193	377
264	421
335	465
406	508
477	552
548	596
619	640
690	684
761	728
832	772
903	815
974	859
7037045	903
116	947
187	991
258	·8474035
330	078
401	122
472	166
543	210
614	254
685	298
756	342
827	385
898	429
969	473
7038040	517
111	561
182	605
253	648
324	692
395	736
466	780
537	824
608	868
679	911
750	955
821	999
892	·8475043
963	087
7039035	131
106	174
177	218
248	262
319	306
390	350
461	393
532	437
603	481
674	525
745	569
816	613
887	656
958	700

Excesses

10 = 6	1 = 1
20 = 12	2 = 1
30 = 19	3 = 2
40 = 25	4 = 2
50 = 31	5 = 3
60 = 37	6 = 4
70 = 43	7 = 4
	8 = 5
	9 = 6

36 = 22

Anti Log Excesses

10 = 16	1 = 2
20 = 32	2 = 3
30 = 49	3 = 5
40 = 65	4 = 6
	5 = 8
	6 = 10
	7 = 11
	8 = 13
	9 = 15

22 = 36

7040029	·8475744
100	788
171	832
242	875
313	919
384	963
455	·8476007
526	050
597	094
668	138
739	182
810	226
881	269
952	313
7041023	357
094	401
165	445
236	488
307	532
378	576
449	620
520	664
591	708
662	752
733	795
804	839
875	883
946	927
7042017	971
088	·8477014
159	058
230	102
301	146
372	190
443	233
514	277
585	321

7042656	·8477365
727	408
798	452
869	496
940	540
7043011	584
082	627
153	671
224	715
295	759
366	802
437	846
508	890
579	934
650	977
721	·8478021
792	065
863	109
934	152
7044005	196
076	240
147	284
218	328
289	371
360	415
431	459
502	503
573	546
644	590
715	634
786	678
857	721
928	765
999	809
7045070	853
141	896
212	940
283	984
354	·8479028
425	071
496	115
566	159
637	203
708	246
779	290
850	334
921	378
992	421
7046063	465
134	509
205	553
276	596
347	640
418	684
489	727
560	771
631	815
702	859
773	902
844	946
915	990
986	·8480034
7047057	077
128	121
199	165

7047269	·8480208
340	252
411	296
482	340
553	383
624	427
695	471
766	514
837	558
908	602
979	646
7048050	689
121	733
192	777
263	820
334	864
405	908
476	952
546	995
617	·8481039
688	083
759	126
830	170
901	214
972	257
7049043	301
114	345
185	389
256	432
327	476
398	520
469	563
539	607
610	651
681	695
752	738
823	782
894	826
965	869

Excesses

10 = 6	1 = 1
20 = 12	2 = 1
30 = 18	3 = 2
40 = 25	4 = 2
50 = 31	5 = 3
60 = 37	6 = 4
70 = 43	7 = 4
	8 = 5
	9 = 6

35 = 22

Anti Log Excesses

10 = 16	1 = 2
20 = 32	2 = 3
30 = 49	3 = 5
40 = 65	4 = 6
	5 = 8
	6 = 10
	7 = 11
	8 = 13
	9 = 15

22 = 35

7050036	·8481913
107	957
178	·8482000
249	044
320	088
391	131
461	175
532	219
603	263
674	306
745	350
816	394
887	437
958	481
7051029	525
100	568
171	612
242	656
312	699
383	743
454	787
525	830
596	874
667	918
738	961
809	·8483005
880	049
951	092
7052021	136
092	180
163	223
234	267
305	311
376	354
447	398
518	442
589	485
660	529
730	573
801	616
872	660
943	704
7053014	747
085	791
156	835
227	878
298	922
368	965
439	·8484009
510	053
581	096

7053652	·8484140
723	184
794	227
865	271
936	315
7054006	358
077	402
148	445
219	489
290	533
361	576
432	620
503	664
573	707
644	751
715	795
786	838
857	882
928	925
999	969
7055070	·8485013
140	056
211	100
282	144
353	187
424	231
495	274
566	318
637	362
707	405
778	449
849	493
920	536
991	580
7056062	623
133	667
203	711
274	754
345	798
416	841
487	885
558	929
629	972
699	·8486016
770	059
841	103
912	147
983	190
7057054	234
125	277
195	321
266	365
337	408
408	452
479	496
550	539
620	583
691	626
762	670
833	713
904	757
975	801
7058046	844
116	888
187	931

7058258	·8486975
329	·8487019
400	062
471	106
541	149
612	193
683	237
754	280
825	324
896	367
966	411
7059037	454
108	498
179	542
250	585
321	629
391	672
462	716
533	759
604	803
675	847
746	890
816	934
887	977
958	·8488021

Excesses

10 = 6	1 = 1
20 = 12	2 = 1
30 = 18	3 = 2
40 = 25	4 = 2
50 = 31	5 = 3
60 = 37	6 = 4
70 = 43	7 = 4
	8 = 5
	9 = 6

35 = 22

Anti Log Excesses

10 = 16	1 = 2
20 = 32	2 = 3
30 = 49	3 = 5
40 = 65	4 = 6
	5 = 8
	6 = 10
	7 = 11
	8 = 13
	9 = 15

22 = 35

7060029	·8488064
100	108
170	151
241	195
312	239
383	282
454	326
525	369
595	413
666	456
737	500

7060808	·8488544
879	587
949	631
7061020	674
091	718
162	761
233	805
303	848
374	892
445	936
516	979
587	·8489023
658	066
728	110
799	153
870	197
941	240
7062012	284
082	327
153	371
224	414
295	458
366	502
436	545
507	589
578	632
649	676
720	719
790	763
861	806
932	850
7063003	893
073	937
144	980
215	·8490024
286	067
357	111
427	154
498	198
569	242
640	285
711	329
781	372
852	416
923	459
994	503
7064064	546
135	590
206	633
277	677
348	720
418	764
489	807
560	851
631	894
701	938
772	981
843	·8491025
914	068
985	112
7065055	155
126	199
197	242
268	286
338	329

7065409	·8491373
480	416
551	460
621	503
692	547
763	590
834	634
905	677
975	721
7066046	764
117	808
188	851
258	895
329	938
400	982
471	·8492025
541	069
612	112
683	156
754	199
824	243
895	286
966	330
7067037	373
107	417
178	460
249	503
320	547
390	590
461	634
532	677
603	721
673	764
744	808
815	851
886	895
956	938
7068027	982
098	·8493025
169	069
239	112
310	156
381	199
452	242
522	286
593	329
664	373
734	416
805	460
876	503
947	547
7069017	590
088	634
159	677
230	720
300	764
371	807
442	851
513	894
583	938
654	981
725	·8494025
795	068
866	111
937	155

Excesses

10 = 6	1 = 1
20 = 12	2 = 1
30 = 18	3 = 2
40 = 25	4 = 2
50 = 31	5 = 3
60 = 37	6 = 4
70 = 43	7 = 4
	8 = 5
	9 = 6

35 = 22

Anti Log Excesses

10 = 16	1 = 2
20 = 33	2 = 3
30 = 49	3 = 5
40 = 65	4 = 7
	5 = 8
	6 = 10
	7 = 11
	8 = 13
	9 = 15

22 = 35

7070008	·8494198
078	242
149	285
220	329
290	372
361	416
432	459
503	502
573	546
644	589
715	633
785	676
856	720
927	763
998	807
7071068	850
139	893
210	937
280	980
351	·8495024
422	067
493	110
563	154
634	197
705	241
775	284
846	328
917	371
987	414
7072058	458
129	501
200	545
270	588
341	632
412	675
482	718
553	762

7072624	·8495805
694	849
765	892
836	935
907	979
977	·8496022
7073048	066
119	109
189	153
260	196
331	239
401	283
472	326
543	370
613	413
684	456
755	500
825	543
896	587
967	630
7074038	673
108	717
179	760
250	804
320	847
391	890
462	934
532	977
603	·8497021
674	064
744	107
815	151
886	194
956	237
7075027	281
098	324
168	368
239	411
310	454
380	498
451	541
522	584
592	628
663	671
734	715
804	758
875	801
946	845
7076016	888
087	931
158	975
228	·8498018
299	062
370	105
440	148
511	192
582	235
652	278
723	322
794	365
864	408
935	452
7077006	495
076	539
147	582

7077217	·8498625
288	669
359	712
429	755
500	799
571	842
641	885
712	929
783	972
853	·8499015
924	059
995	102
7078065	145
136	189
206	232
277	275
348	319
418	362
489	405
560	449
630	492
701	535
772	579
842	622
913	665
983	709
7079054	752
125	795
195	839
266	882
337	925
407	969
478	·8500012
549	055
619	099
690	142
760	185
831	229
902	272
972	315

Excesses

10 = 6	1 = 1
20 = 12	2 = 1
30 = 18	3 = 2
40 = 25	4 = 2
50 = 31	5 = 3
60 = 37	6 = 4
70 = 43	7 = 4
	8 = 5
	9 = 6

35 = 22

Anti Log Excesses

10 = 16	1 = 2
20 = 33	2 = 3
30 = 49	3 = 5
40 = 65	4 = 7
	5 = 8
	6 = 10
	7 = 11
	8 = 13
	9 = 15

22 = 35

7080043	·8500359
113	402
184	445
255	488
325	532
396	575
467	618
537	662
608	705
678	748
749	792
820	835
890	878
961	922
7081031	965
102	·8501008
173	052
243	095
314	138
385	182
455	225
526	268
596	311
667	355
738	398
808	441
879	485
949	528
7082020	571
091	615
161	658
232	701
302	744
373	788
444	831
514	874
585	918
655	961
726	·8502004
797	047
867	091
938	134
7083008	177
079	221
150	264
220	307
291	350
361	394
432	437
502	480
573	524

7083644	·8502567
714	610
785	653
855	697
926	740
997	783
7084067	826
138	870
208	913
279	956
349	·8503000
420	043
491	086
561	129
632	173
702	216
773	259
843	302
914	346
985	389
7085055	432
126	475
196	519
267	562
337	605
408	648
479	692
549	735
620	778
690	821
761	865
831	908
902	951
973	994
7086043	·8504038
114	081
184	124
255	167
325	211
396	254
466	297
537	340
608	384
678	427
749	470
819	513
890	557
960	600
7087031	643
101	686
172	729
243	773
313	816
384	859
454	902
525	946
595	989
666	·8505032
736	075
807	119
877	162
948	205
7088019	248
089	291
160	335

7088230	·8505378
301	421
371	464
442	508
512	551
583	594
653	637
724	680
794	724
865	767
936	810
7089006	853
077	896
147	940
218	983
288	·8506026
359	069
429	112
500	156
570	199
641	242
711	285
782	328
852	372
923	415
993	458

Excesses

10 = 6	1 = 1
20 = 12	2 = 1
30 = 18	3 = 2
40 = 25	4 = 2
50 = 31	5 = 3
60 = 37	6 = 4
70 = 43	7 = 4
	8 = 5
	9 = 6

35 = 22

Anti Log Excesses

10 = 16	1 = 2
20 = 33	2 = 3
30 = 49	3 = 5
40 = 65	4 = 7
	5 = 8
	6 = 10
	7 = 11
	8 = 13
	9 = 15

22 = 35

7090064	·8506501
134	544
205	588
276	631
346	674
417	717
487	760
558	804
628	847
699	890

7090769	·8506933
840	976
910	·8507020
981	063
7091051	106
122	149
192	192
263	235
333	279
404	322
474	365
545	408
615	451
686	495
756	538
827	581
897	624
968	667
7092038	710
109	754
179	797
250	840
320	883
391	926
461	969
532	·8508013
602	056
673	099
743	142
814	185
884	228
955	272
7093025	315
096	358
166	401
237	444
307	487
378	531
448	574
519	617
589	660
660	703
730	746
801	789
871	833
942	876
7094012	919
083	962
153	·8509005
224	048
294	092
364	135
435	178
505	221
576	264
646	307
717	350
787	393
858	437
928	480
999	523
7095069	566
140	609
210	652
281	695

7095351	·8509739
422	782
492	825
562	868
633	911
703	954
774	997
844	·8510040
915	084
985	127
7096056	170
126	213
197	256
267	299
338	342
408	385
478	429
549	472
619	515
690	558
760	601
831	644
901	687
972	730
7097042	773
113	817
183	860
253	903
324	946
394	989
465	·8511032
535	075
606	118
676	161
747	205
817	248
888	291
958	334
7098028	377
099	420
169	463
240	506
310	549
381	592
451	636
521	679
592	722
662	765
733	808
803	851
874	894
944	937
7099015	980
085	·8512023
155	066
226	110
296	153
367	196
437	239
508	282
578	325
648	368
719	411
789	454
860	497

7099930	·8512540

Excesses

10 = 6	1 = 1
20 = 12	2 = 1
30 = 18	3 = 2
40 = 24	4 = 2
50 = 31	5 = 3
60 = 37	6 = 4
70 = 43	7 = 4
	8 = 5
	9 = 6

35 = 22

Anti Log Excesses

10 = 16	1 = 2
20 = 33	2 = 3
30 = 49	3 = 5
40 = 65	4 = 7
	5 = 8
	6 = 10
	7 = 11
	8 = 13
	9 = 15

22 = 35

7100000	·8512583
071	627
141	670
212	713
282	756
353	799
423	842
493	885
564	928
634	971
705	·8513014
775	057
846	100
916	143
986	186
7101057	229
127	273
198	316
268	359
338	402
409	445
479	488
550	531
620	574
690	617
761	660
831	703
902	746
972	789
7102042	832
113	875
183	918
254	961
324	·8514004
394	047

7102465	·8514091
535	134
606	177
676	220
746	263
817	306
887	349
958	392
7103028	435
098	478
169	521
239	564
310	607
380	650
450	693
521	736
591	779
662	822
732	865
802	908
873	951
943	994
7104013	·8515037
084	080
154	123
225	166
295	209
365	252
436	296
506	339
576	382
647	425
717	468
788	511
858	554
928	597
999	640
7105069	683
139	726
210	769
280	812
351	855
421	898
491	941
562	984
632	·8516027
702	070
773	113
843	156
914	199
984	242
7106054	285
125	328
195	371
265	414
336	457
406	500
476	543
547	586
617	629
687	672
758	715
828	758
899	801
969	844

7107039	·8516887
110	930
180	973
250	·8517016
321	059
391	102
461	145
532	188
602	231
672	274
743	317
813	360
883	403
954	446
7108024	489
094	532
165	575
235	618
306	661
376	704
446	746
517	789
587	832
657	875
728	918
798	961
868	·8518004
939	047
7109009	090
079	133
150	176
220	219
290	262
361	305
431	348
501	391
572	434
642	477
712	520
782	563
853	606
923	649
993	692

Excesses

10 = 6	1 = 1
20 = 12	2 = 1
30 = 18	3 = 2
40 = 24	4 = 2
50 = 31	5 = 3
60 = 37	6 = 4
70 = 43	7 = 4
	8 = 5
	9 = 6

35 = 22

Anti Log Excesses

10 = 16	1 = 2
20 = 33	2 = 3
30 = 49	3 = 5
40 = 65	4 = 7
	5 = 8
	6 = 10
	7 = 11
	8 = 13
	9 = 15

22 = 35

7110064	·8518735
134	778
204	821
275	864
345	906
415	949
486	992
556	·8519035
626	078
697	121
767	164
837	207
908	250
978	293
7111048	336
119	379
189	422
259	465
329	508
400	551
470	594
540	636
611	679
681	722
751	765
822	808
892	851
962	894
7112033	937
103	980
173	·8520023
243	066
314	109
384	152
454	195
525	237
595	280
665	323
736	366
806	409
876	452
946	495
7113017	538
087	581
157	624
228	667
298	710
368	753
438	795
509	838
579	881

7113649	·8520924
720	967
790	·8521010
860	053
930	096
7114001	139
071	182
141	225
212	267
282	310
352	353
422	396
493	439
563	482
633	525
704	568
774	611
844	654
914	696
985	739
7115055	782
125	825
196	868
266	911
336	954
406	997
477	·8522040
547	082
617	125
687	168
758	211
828	254
898	297
968	340
7116039	383
109	426
179	468
250	511
320	554
390	597
460	640
531	683
601	726
671	769
741	811
812	854
882	897
952	940
7117022	983
093	·8523026
163	069
233	112
303	154
374	197
444	240
514	283
584	326
655	369
725	412
795	454
865	497
936	540
7118006	583
076	626
146	669

7118217	·8523712
287	754
357	797
427	840
498	883
568	926
638	969
708	·8524012
778	054
849	097
919	140
989	183
7119059	226
130	269
200	311
270	354
340	397
411	440
481	483
551	526
621	568
692	611
762	654
832	697
902	740
972	783

Excesses

10 = 6	1 = 1
20 = 12	2 = 1
30 = 18	3 = 2
40 = 24	4 = 2
50 = 31	5 = 3
60 = 37	6 = 4
70 = 43	7 = 4
	8 = 5
	9 = 5

35 = 21

Anti Log Excesses

10 = 16	1 = 2
20 = 33	2 = 3
30 = 49	3 = 5
40 = 66	4 = 7
	5 = 8
	6 = 10
	7 = 11
	8 = 13
	9 = 15

21 = 35

7120043	·8524826
113	868
183	911
253	954
324	997
394	·8525040
464	082
534	125
604	168
675	211

7120745	·8525254
815	297
885	339
955	382
7121026	425
096	468
166	511
236	554
307	597
377	639
447	682
517	725
587	768
658	811
728	853
798	896
868	939
938	982
7122009	·8526025
079	067
149	110
219	153
289	196
360	239
430	282
500	324
570	367
640	410
711	453
781	496
851	538
921	581
991	624
7123062	667
132	710
202	752
272	795
342	838
413	881
483	924
553	966
623	·8527009
693	052
764	095
834	137
904	180
974	223
7124044	266
114	309
185	351
255	394
325	437
395	480
465	523
536	565
606	608
676	651
746	694
816	736
886	779
957	822
7125027	865
097	907
167	950
237	993

7125308	·8528036
378	079
448	121
518	164
588	207
658	250
729	292
799	335
869	378
939	421
7126009	463
079	506
150	549
220	592
290	634
360	677
430	720
500	763
571	805
641	848
711	891
781	934
851	977
921	·8529019
991	062
7127062	105
132	148
202	190
272	233
342	276
412	318
483	361
553	404
623	447
693	489
763	532
833	575
903	618
974	660
7128044	703
114	746
184	789
254	831
324	874
394	917
465	960
535	·8530002
605	045
675	088
745	131
815	173
885	216
956	259
7129026	301
096	344
166	387
236	430
306	472
376	515
446	558
517	600
587	643
657	686
727	729
797	771

7129867	·8530814
937	857

Excesses

10 = 6	1 = 1
20 = 12	2 = 1
30 = 18	3 = 2
40 = 24	4 = 2
50 = 30	5 = 3
60 = 37	6 = 4
70 = 43	7 = 4
	8 = 5
	9 = 5

35 = 21

Anti Log Excesses

10 = 16	1 = 2
20 = 33	2 = 3
30 = 49	3 = 5
40 = 66	4 = 7
	5 = 8
	6 = 10
	7 = 11
	8 = 13
	9 = 15

21 = 35

7130008	·8530899
078	942
148	985
218	·8531028
288	070
358	113
428	156
498	198
568	241
639	284
709	326
779	369
849	412
919	455
989	497
7131059	540
129	583
200	625
270	668
340	711
410	753
480	796
550	839
620	882
690	924
760	967
831	·8532010
901	052
971	095
7132041	138
111	180
181	223
251	266
321	308

7132391	·8532351
462	394
532	436
602	479
672	522
742	565
812	607
882	650
952	693
7133022	735
092	778
162	821
233	863
303	906
373	949
443	991
513	·8533034
583	077
653	119
723	162
793	205
863	247
933	290
7134004	333
074	375
144	418
214	461
284	504
354	546
424	589
494	631
564	674
634	717
704	759
774	802
845	845
915	887
985	930
7135055	973
125	·8534015
195	058
265	101
335	143
405	186
475	229
545	271
615	314
685	357
756	399
826	442
896	485
966	527
7136036	570
106	612
176	655
246	698
316	740
386	783
456	826
526	868
596	911
666	954
736	996
807	·8535039
877	081

7136947	·8535124
7137017	167
087	210
157	252
227	295
297	338
367	380
437	423
507	465
577	508
647	551
717	593
787	636
857	678
927	721
997	764
7138067	806
138	849
208	892
278	934
348	977
418	·8536019
488	062
558	105
628	147
698	190
768	232
838	275
908	318
978	360
7139048	403
118	445
188	488
258	531
328	573
398	616
468	658
538	701
608	744
678	786
748	829
818	871
888	914
958	957

Excesses

10 = 6	1 = 1
20 = 12	2 = 1
30 = 18	3 = 2
40 = 24	4 = 2
50 = 30	5 = 3
60 = 37	6 = 4
70 = 43	7 = 4
	8 = 5
	9 = 5

35 = 21

Anti Log Excesses

10 = 16	1 = 2
20 = 33	2 = 3
30 = 49	3 = 5
40 = 66	4 = 7
	5 = 8
	6 = 10
	7 = 12
	8 = 13
	9 = 15

21 = 35

7140029	·8536999
099	·8537042
169	084
239	127
309	170
379	212
449	255
519	297
589	340
659	383
729	425
799	468
869	510
939	553
7141009	595
079	638
149	681
219	723
289	766
359	808
429	851
499	894
569	936
639	979
709	·8538021
779	064
849	107
919	149
989	192
7142059	234
129	277
199	319
269	362
339	404
409	447
479	490
549	532
619	575
689	617
759	660
829	702
899	745
969	788
7143039	830
109	873
179	915
249	958
319	·8539000
389	043
459	085
529	128

7143599	·8539171
669	213
739	256
809	298
879	341
949	383
7144019	426
089	468
159	511
229	554
299	596
369	639
439	681
509	724
579	766
649	809
719	851
789	894
859	936
929	979
999	·8540022
7145069	064
139	106
209	149
279	191
349	234
419	277
489	319
559	362
629	404
699	447
768	489
838	532
908	574
978	617
7146048	659
118	702
188	744
258	787
328	829
398	872
468	914
538	957
608	999
678	·8541042
748	084
818	127
888	169
958	212
7147028	254
098	297
168	340
238	382
308	425
378	467
448	510
518	552
588	595
657	637
727	680
797	722
867	765
937	807
7148007	850
077	892

7148147	·8541935
217	977
287	·8542020
357	062
427	105
497	147
567	190
637	232
707	275
777	317
847	359
917	402
986	444
7149056	487
126	530
196	572
266	614
336	657
406	699
476	742
546	784
616	827
686	869
756	912
826	954
896	997
966	·8543039

Excesses

10 = 6	1 = 1
20 = 12	2 = 1
30 = 18	3 = 2
40 = 24	4 = 2
50 = 30	5 = 3
60 = 36	6 = 4
	7 = 4
	8 = 5
	9 = 5

35 = 21

Anti Log Excesses

10 = 16	1 = 2
20 = 33	2 = 3
30 = 49	3 = 5
40 = 66	4 = 7
	5 = 8
	6 = 10
	7 = 12
	8 = 13
	9 = 15

21 = 35

7150035	·8543082
105	124
175	167
245	209
315	252
385	294
455	337
525	379
595	422

7150665	·8543464
735	506
805	549
875	591
944	634
7151014	676
084	719
154	761
224	804
294	846
364	889
434	931
504	974
574	·8544016
644	059
714	101
783	143
853	186
923	228
993	271
7152063	313
133	356
203	398
273	441
343	483
413	526
483	568
552	610
622	653
692	695
762	738
832	780
902	823
972	865
7153042	908
112	950
182	992
252	·8545035
321	077
391	120
461	162
531	204
601	247
671	289
741	332
811	374
881	417
950	459
7154020	502
090	544
160	586
230	629
300	671
370	714
440	756
510	798
579	841
649	883
719	926
789	968
859	·8546011
929	053
999	095
7155069	138
139	180

7155208	·8546223
278	265
348	307
418	350
488	392
558	435
628	477
698	520
767	562
837	604
907	647
977	689
7156047	732
117	774
187	816
257	859
326	901
396	944
466	986
536	·8547028
606	071
676	113
746	156
815	198
885	240
955	283
7157025	325
095	368
165	410
235	452
304	495
374	537
444	580
514	622
584	664
654	707
724	749
793	791
863	834
933	876
7158003	919
073	961
143	·8548003
213	046
282	088
352	130
422	173
492	215
562	258
632	300
701	342
771	385
841	427
911	469
981	512
7159051	554
121	596
190	639
260	681
330	724
400	766
470	808
540	851
609	893
679	935

7159749	·8548978
819	·8549020
889	062
959	105

Excesses

10 = 6	1 = 1
20 = 12	2 = 1
30 = 18	3 = 2
40 = 24	4 = 2
50 = 30	5 = 3
60 = 36	6 = 4
	7 = 4
	8 = 5
	9 = 5

35 = 21

Anti Log Excesses

10 = 16	1 = 2
20 = 33	2 = 3
30 = 49	3 = 5
40 = 66	4 = 7
	5 = 8
	6 = 10
	7 = 12
	8 = 13
	9 = 15

21 = 35

7160028	·8549147
098	189
168	232
238	274
308	317
378	359
447	401
517	444
587	486
657	528
727	571
797	613
866	655
936	698
7161006	740
076	782
146	825
215	867
285	909
355	952
425	994
495	·8550037
565	079
634	121
704	164
774	206
844	248
914	291
983	333
7162053	375
123	418
193	460

7162263	·8550502
333	545
402	587
472	629
542	671
612	714
682	756
751	798
821	841
891	883
961	925
7163031	968
100	·8551010
170	052
240	095
310	137
380	179
449	222
519	264
589	306
659	349
729	391
798	433
868	476
938	518
7164008	560
078	602
147	645
217	687
287	729
357	772
427	814
496	856
566	899
636	941
706	983
775	·8552026
845	068
915	110
985	152
7165055	195
124	237
194	279
264	322
334	364
403	406
473	449
543	491
613	533
683	575
752	618
822	660
892	702
962	745
7166031	787
101	829
171	871
241	914
311	956
380	998
450	·8553041
520	083
590	125
659	167
729	210

7166799	·8553252
869	294
938	337
7167008	379
078	421
148	463
218	506
287	548
357	590
427	632
497	675
566	717
636	759
706	802
776	844
845	886
915	928
985	971
7168055	·8554013
124	055
194	097
264	140
334	182
403	224
473	266
543	309
613	351
682	393
752	435
822	478
892	520
961	562
7169031	604
101	647
171	689
240	731
310	773
380	816
450	858
519	900
589	942
659	985
729	·8555027
798	069
868	111
938	154

Excesses

10 = 6	1 = 1
20 = 12	2 = 1
30 = 18	3 = 2
40 = 24	4 = 2
50 = 30	5 = 3
60 = 36	6 = 4
	7 = 4
	8 = 5
	9 = 5

35 = 21

Anti Log Excesses

10 = 16	1 = 2
20 = 33	2 = 3
30 = 49	3 = 5
40 = 66	4 = 7
	5 = 8
	6 = 10
	7 = 12
	8 = 13
	9 = 15

21 = 35

7170007	·8555196
077	238
147	280
217	323
286	365
356	407
426	449
496	492
565	534
635	576
705	618
775	660
844	703
914	745
984	787
7171053	829
123	872
193	914
263	956
332	998
402	·8556040
472	083
541	125
611	167
681	209
751	252
820	294
890	336
960	378
7172030	420
099	463
169	505
239	547
308	589
378	632
448	674
517	716
587	758
657	800
727	843
796	885
866	927
936	969
7173005	·8557011
075	054
145	096
215	138
284	180
354	222
424	265
493	307

7173563	·8557349
633	391
702	433
772	476
842	518
912	560
981	602
7174051	644
121	687
190	729
260	771
330	813
399	855
469	897
539	940
609	982
678	·8558024
748	066
818	108
887	151
957	193
7175027	235
096	277
166	319
236	361
305	404
375	446
445	488
514	530
584	572
654	614
723	657
793	699
863	741
932	783
7176002	825
072	867
142	910
211	952
281	994
351	·8559036
420	078
490	120
560	163
629	205
699	247
769	289
838	331
908	373
978	416
7177047	458
117	500
187	542
256	584
326	626
396	669
465	711
535	753
605	795
674	837
744	879
814	921
883	964
953	·8560006
7178023	048

7178092	·8560090
162	132
231	174
301	216
371	259
440	301
510	343
580	385
649	427
719	469
789	511
858	554
928	596
998	638
7179067	680
137	722
207	764
276	806
346	848
416	891
485	933
555	975
624	·8561017
694	059
764	101
833	143
903	185
973	228

Excesses

10 = 6	1 = 1
20 = 12	2 = 1
30 = 18	3 = 2
40 = 24	4 = 2
50 = 30	5 = 3
60 = 36	6 = 4
	7 = 4
	8 = 5
	9 = 5

35 = 21

Anti Log Excesses

10 = 17	1 = 2
20 = 33	2 = 3
30 = 50	3 = 5
40 = 66	4 = 7
	5 = 8
	6 = 10
	7 = 12
	8 = 13
	9 = 15

21 = 35

7180042	·8561270
112	312
182	354
251	396
321	438
390	480
460	522
530	565

7180599	·8561607
669	649
739	691
808	733
878	775
948	817
7181017	859
087	901
156	944
226	986
296	·8562028
365	070
435	112
505	154
574	196
644	238
713	280
783	322
853	365
922	407
992	449
7182061	491
131	533
201	575
270	617
340	659
410	701
479	743
549	786
618	828
688	870
758	912
827	954
897	996
966	·8563038
7183036	080
106	122
175	164
245	206
314	249
384	291
454	333
523	375
593	417
663	459
732	501
802	543
871	585
941	627
7184011	669
080	711
150	753
219	796
289	838
359	880
428	922
498	964
567	·8564006
637	048
706	090
776	132
846	174
915	216
985	258
7185054	300

7185124	·8564342
194	385
263	427
333	469
402	511
472	553
542	595
611	637
681	679
750	721
820	763
889	805
959	847
7186029	889
098	931
168	973
237	·8565015
307	057
376	099
446	141
516	183
585	226
655	268
724	310
794	352
864	394
933	436
7187003	478
072	520
142	562
211	604
281	646
350	688
420	730
490	772
559	814
629	856
698	898
768	940
837	982
907	·8566024
977	066
7188046	108
116	150
185	192
255	234
324	276
394	318
463	360
533	402
603	444
672	486
742	528
811	570
881	613
950	655
7189020	697
089	739
159	781
229	823
298	865
368	907
437	949
507	991
576	·8567033

7189646	·8567075
715	117
785	159
854	201
924	243
994	285

Excesses

10 = 6	1 = 1
20 = 12	2 = 1
30 = 18	3 = 2
40 = 24	4 = 2
50 = 30	5 = 3
60 = 36	6 = 4
	7 = 4
	8 = 5
	9 = 5

35 = 21

Anti Log Excesses

10 = 17	1 = 2
20 = 33	2 = 3
30 = 50	3 = 5
40 = 66	4 = 7
	5 = 8
	6 = 10
	7 = 12
	8 = 13
	9 = 15

21 = 35

7190063	·8567327
133	369
202	411
272	453
341	495
411	537
480	579
550	621
619	663
689	705
758	747
828	789
898	831
967	873
7191037	915
106	957
176	999
245	·8568041
315	083
384	125
454	167
523	209
593	250
662	292
732	334
801	376
871	418
940	460
7192010	502
079	544

7192149	·8568586
219	628
288	670
358	712
427	754
497	796
566	838
636	880
705	922
775	964
844	·8569006
914	048
983	090
7193053	132
122	174
192	216
261	258
331	300
400	342
470	384
539	426
609	468
678	510
748	552
817	594
887	636
956	678
7194026	720
095	761
165	803
234	845
304	887
373	929
443	971
512	·8570013
582	055
651	097
721	139
790	181
860	223
929	265
999	307
7195068	349
138	391
207	433
277	475
346	517
416	558
485	600
555	642
624	684
694	726
763	768
833	810
902	852
972	894
7196041	936
111	978
180	·8571020
250	062
319	104
388	146
458	187
527	229
597	271

7196666	·8571313
736	355
805	397
875	439
944	481
7197014	523
083	565
153	607
222	649
292	691
361	732
431	774
500	816
570	858
639	900
708	942
778	984
847	·8572026
917	068
986	110
7198056	152
125	194
195	236
264	277
334	319
403	361
473	403
542	445
611	487
681	529
750	571
820	613
889	655
959	696
7199028	738
098	780
167	822
237	864
306	906
375	948
445	990
514	·8573032
584	074
653	115
723	157
792	199
862	241
931	283

Excesses

10 = 6	1 = 1
20 = 12	2 = 1
30 = 18	3 = 2
40 = 24	4 = 2
50 = 30	5 = 3
60 = 36	6 = 4
	7 = 4
	8 = 5
	9 = 5

35 = 21

Anti Log Excesses

10 = 17	1 = 2
20 = 33	2 = 3
30 = 50	3 = 5
40 = 66	4 = 7
	5 = 8
	6 = 10
	7 = 12
	8 = 13
	9 = 15

21 = 35

7200000	·8573325
070	367
139	409
209	451
278	492
348	534
417	576
487	618
556	660
625	702
695	744
764	786
834	828
903	869
973	911
7201042	953
112	995
181	·8574037
250	079
320	121
389	163
459	205
528	247
598	288
667	330
736	372
806	414
875	456
945	498
7202014	540
084	581
153	623
222	665
292	707
361	749
431	791
500	833
569	874
639	916
708	958
778	·8575000
847	042
917	084
986	126
7203055	167
125	209
194	251
264	293
333	335
402	377
472	418

7203541	·8575460
611	502
680	544
750	586
819	628
888	670
958	711
7204027	753
097	795
166	837
235	879
305	921
374	962
444	·8576004
513	046
582	088
652	130
721	172
791	213
860	255
929	297
999	339
7205068	381
138	423
207	464
276	506
346	548
415	590
485	632
554	674
623	715
693	757
762	799
831	841
901	883
970	925
7206040	966
109	·8577008
178	050
248	092
317	134
387	175
456	217
525	259
595	301
664	343
733	384
803	426
872	468
942	510
7207011	552
080	594
150	635
219	677
288	719
358	761
427	803
497	844
566	886
635	928
705	970
774	·8578012
843	053
913	095
982	137

7208052	·8578179
121	221
190	262
260	304
329	346
398	388
468	429
537	471
606	513
676	555
745	597
815	638
884	680
953	722
7209023	764
092	806
161	847
231	889
300	931
369	973
439	·8579014
508	056
577	098
647	140
716	182
786	223
855	265
924	307
994	349

Excesses

10 = 6	1 = 1
20 = 12	2 = 1
30 = 18	3 = 2
40 = 24	4 = 2
50 = 30	5 = 3
60 = 36	6 = 4
	7 = 4
	8 = 5
	9 = 5

35 = 21

Anti Log Excesses

10 = 17	1 = 2
20 = 33	2 = 3
30 = 50	3 = 5
40 = 66	4 = 7
	5 = 8
	6 = 10
	7 = 12
	8 = 13
	9 = 15

21 = 35

7210063	·8579390
132	432
202	474
271	516
340	557
410	599
479	641

7210548	·8579683
618	725
687	766
756	808
826	850
895	892
964	933
7211034	975
103	·8580017
172	059
242	100
311	142
380	184
450	226
519	267
588	309
658	351
727	393
796	434
866	476
935	518
7212004	560
074	601
143	643
212	685
282	727
351	768
420	810
490	852
559	894
628	935
698	977
767	·8581019
836	061
906	102
975	144
7213044	186
114	228
183	269
252	311
322	353
391	395
460	436
529	478
599	520
668	561
737	603
807	645
876	687
945	728
7214015	770
084	812
153	853
223	895
292	937
361	979
430	·8582020
500	062
569	104
638	145
708	187
777	229
846	271
916	312
985	354

7215054	·8582396
124	437
193	479
262	521
331	563
401	604
470	646
539	688
609	729
678	771
747	813
816	855
886	896
955	938
7216024	980
094	·8583021
163	063
232	105
301	146
371	188
440	230
509	272
579	313
648	355
717	397
786	438
856	480
925	522
994	563
7217064	605
133	647
202	688
271	730
341	772
410	813
479	855
549	897
618	939
687	980
756	·8584022
826	064
895	105
964	147
7218033	189
103	230
172	272
241	314
311	355
380	397
449	439
518	480
588	522
657	564
726	605
795	647
865	689
934	730
7219003	772
072	814
142	855
211	897
280	939
350	980
419	·8585022
488	064

7219557	·8585105
627	147
696	189
765	230
834	272
904	314
973	355

Excesses

10 = 6	1 = 1
20 = 12	2 = 1
30 = 18	3 = 2
40 = 24	4 = 2
50 = 30	5 = 3
60 = 36	6 = 4
	7 = 4
	8 = 5
	9 = 5

35 = 21

Anti Log Excesses

10 = 17	1 = 2
20 = 33	2 = 3
30 = 50	3 = 5
40 = 66	4 = 7
	5 = 8
	6 = 10
	7 = 12
	8 = 13
	9 = 15

21 = 35

7220042	·8585397
111	439
181	480
250	522
319	564
388	605
458	647
527	689
596	730
665	772
735	814
804	855
873	897
942	939
7221012	980
081	·8586022
150	063
219	105
288	147
358	188
427	230
496	272
565	314
635	355
704	397
773	438
842	480
912	521
981	563

7222050	·8586605
119	646
189	688
258	730
327	771
396	813
465	854
535	896
604	938
673	979
742	·8587021
812	063
881	104
950	146
7223019	187
088	229
158	271
227	312
296	354
365	396
435	437
504	479
573	520
642	562
711	604
781	645
850	687
919	728
988	770
7224058	812
127	853
196	895
265	936
334	978
404	·8588020
473	061
542	103
611	144
680	186
750	228
819	269
888	311
957	352
7225026	394
096	436
165	477
234	519
303	560
372	602
442	644
511	685
580	727
649	768
718	810
788	852
857	893
926	935
995	976
7226064	·8589018
134	060
203	101
272	143
341	184
410	226
480	268

7226549	·8589309
618	351
687	392
756	434
826	475
895	517
964	559
7227033	600
102	642
171	683
241	725
310	766
379	808
448	850
517	891
587	933
656	974
725	·8590016
794	057
863	099
932	141
7228002	182
071	224
140	265
209	307
278	348
347	390
417	431
486	473
555	515
624	556
693	598
762	639
832	681
901	722
970	764
7229039	805
108	847
177	889
247	930
316	972
385	·8591013
454	055
523	096
592	138
662	179
731	221
800	262
869	304
938	346

Excesses

10 = 6	1 = 1
20 = 12	2 = 1
30 = 18	3 = 2
40 = 24	4 = 2
50 = 30	5 = 3
60 = 36	6 = 4
	7 = 4
	8 = 5
	9 = 5

35 = 21

Anti Log Excesses

10 = 17	1 = 2
20 = 33	2 = 3
30 = 50	3 = 5
40 = 67	4 = 7
	5 = 8
	6 = 10
	7 = 12
	8 = 13
	9 = 15

21 = 35

7230007	·8591387
077	429
146	470
215	512
284	553
353	595
422	636
491	678
561	719
630	761
699	803
768	844
837	886
906	927
976	969
7231045	·8592010
114	052
183	093
252	135
321	176
390	218
460	259
529	301
598	342
667	384
736	425
805	467
874	508
944	550
7232013	592
082	633
151	675
220	716
289	758
358	799
427	841
497	882
566	924
635	965
704	·8593007
773	048
842	090
911	131
981	173
7233050	214
119	256
188	297
257	339
326	380
395	422
464	463

7233534	·8593505
603	546
672	588
741	629
810	671
879	712
948	754
7234017	795
086	837
156	878
225	920
294	961
363	·8594003
432	044
501	086
570	127
639	169
709	210
778	252
847	293
916	335
985	376
7235054	418
123	459
192	501
261	542
330	583
400	625
469	666
538	708
607	749
676	791
745	832
814	874
883	915
952	957
7236022	998
091	·8595040
160	081
229	123
298	164
367	206
436	247
505	288
574	330
643	371
712	413
782	454
851	496
920	537
989	579
7237058	620
127	662
196	703
265	744
334	786
403	827
472	869
542	910
611	952
680	993
749	·8596035
818	076
887	118
956	159

7238025	·8596201
094	242
163	284
232	325
301	366
370	408
440	449
509	491
578	532
647	574
716	615
785	656
854	698
923	739
992	781
7239061	822
130	864
199	905
268	947
337	988
407	·8597029
476	071
545	112
614	154
683	195
752	237
821	278
890	319
959	361

Excesses

10 = 6	1 = 1
20 = 12	2 = 1
30 = 18	3 = 2
40 = 24	4 = 2
50 = 30	5 = 3
60 = 36	6 = 4
	7 = 4
	8 = 5
	9 = 5

35 = 21

Anti Log Excesses

10 = 17	1 = 2
20 = 33	2 = 3
30 = 50	3 = 5
40 = 67	4 = 7
	5 = 8
	6 = 10
	7 = 12
	8 = 13
	9 = 15

21 = 35

7240028	·8597402
097	444
166	485
235	527
304	568
373	609
442	651

Number	Log
7240512	·8597692
581	734
650	775
719	816
788	858
857	899
926	941
995	982
7241064	·8598024
133	065
202	106
271	148
340	189
409	231
478	272
547	313
616	355
685	396
754	438
823	479
893	520
962	562
7242031	603
100	645
169	686
238	727
307	769
376	810
445	852
514	893
583	934
652	976
721	·8599017
790	059
859	100
928	141
997	183
7243066	224
135	266
204	307
273	348
342	390
411	431
480	473
549	514
618	555
687	597
756	638
825	679
894	721
964	762
7244033	804
102	845
171	886
240	928
309	969
378	·8600011
447	052
516	093
585	135
654	176
723	217
792	259
861	300
930	341

Number	Log
7244999	·8600383
7245068	424
137	466
206	507
275	548
344	590
413	631
482	672
551	714
620	755
689	797
758	838
827	879
896	921
965	962
7246034	·8601003
103	045
172	086
241	127
310	169
379	210
448	251
517	293
586	334
655	375
724	417
793	458
862	500
931	541
7247000	582
069	624
138	665
207	706
276	748
345	789
414	830
483	872
552	913
621	954
690	996
759	·8602037
828	078
897	120
966	161
7248035	202
104	244
173	285
242	326
311	368
380	409
449	450
518	492
587	533
656	574
725	616
794	657
862	698
931	740
7249000	781
069	822
138	864
207	905
276	946
345	988
414	·8603029

Number	Log
7249483	·8603070
552	112
621	153
690	194
759	236
828	277
897	318
966	359

Excesses

10 = 6	1 = 1
20 = 12	2 = 1
30 = 18	3 = 2
40 = 24	4 = 2
50 = 30	5 = 3
60 = 36	6 = 4
	7 = 4
	8 = 5
	9 = 5

35 = 21

Anti Log Excesses

10 = 17	1 = 2
20 = 33	2 = 3
30 = 50	3 = 5
40 = 67	4 = 7
	5 = 8
	6 = 10
	7 = 12
	8 = 13
	9 = 15

21 = 35

Number	Log
7250035	·8603401
104	442
173	483
242	525
311	566
380	607
449	649
518	690
587	731
656	773
725	814
794	855
863	896
931	938
7251000	979
069	·8604020
138	062
207	103
276	144
345	185
414	227
483	268
552	309
621	351
690	392
759	433
828	475
897	516

Number	Log
7251966	·8604557
7252035	598
104	640
173	681
242	722
310	764
379	805
448	846
517	887
586	929
655	970
724	·8605011
793	052
862	094
931	135
7253000	176
069	218
138	259
207	300
276	341
345	383
413	424
482	465
551	507
620	548
689	589
758	630
827	672
896	713
965	754
7254034	795
103	837
172	878
241	919
310	961
378	·8606002
447	043
516	084
585	126
654	167
723	208
792	249
861	291
930	332
999	373
7255068	414
137	456
206	497
274	538
343	579
412	621
481	662
550	703
619	744
688	786
757	827
826	868
895	909
964	951
7256032	992
101	·8607033
170	074
239	116
308	157
377	198

Number	Log
7256446	·8607239
515	281
584	322
653	363
722	404
790	445
859	487
928	528
997	569
7257066	610
135	652
204	693
273	734
342	775
411	817
479	858
548	899
617	940
686	981
755	·8608023
824	064
893	105
962	146
7258031	188
099	229
168	270
237	311
306	352
375	394
444	435
513	476
582	517
651	558
719	600
788	641
857	682
926	723
995	765
7259064	806
133	847
202	888
270	929
339	971
408	·8609012
477	053
546	094
615	135
684	177
753	218
821	259
890	300
959	341

Excesses

10 = 6	1 = 1
20 = 12	2 = 1
30 = 18	3 = 2
40 = 24	4 = 2
50 = 30	5 = 3
60 = 36	6 = 4
	7 = 4
	8 = 5
	9 = 5

35 = 21

Anti Log Excesses

10 = 17	1 = 2
20 = 33	2 = 3
30 = 50	3 = 5
40 = 67	4 = 7
	5 = 8
	6 = 10
	7 = 12
	8 = 13
	9 = 15

21 = 35

7260028	·8609383
097	424
166	465
235	506
304	547
372	589
441	630
510	671
579	712
648	753
717	795
786	836
854	877
923	918
992	959
7261061	·8610001
130	042
199	083
268	124
336	165
405	206
474	248
543	289
612	330
681	371
750	412
818	454
887	495
956	536
7262025	577
094	618
163	659
232	701
300	742
369	783
438	824
507	865
7262576	·8610906
645	948
713	989
782	·8611030
851	071
920	112
989	153
7263058	195
127	236
195	277
264	318
333	359
402	400
471	442
540	483
608	524
677	565
746	606
815	647
884	688
953	730
7264021	771
090	812
159	853
228	894
297	935
366	977
434	·8612018
503	059
572	100
641	141
710	182
779	223
847	265
916	306
985	347
7265054	388
123	429
191	470
260	511
329	553
398	594
467	635
536	676
604	717
673	758
742	799
811	841
880	882
948	923
7266017	964
086	·8613005
155	046
224	087
293	129
361	170
430	211
499	252
568	293
637	334
705	375
774	417
843	458
912	499
981	540
7267049	·8613581
118	622
187	663
256	704
325	745
393	787
462	828
531	869
600	910
669	951
737	992
806	·8614033
875	074
944	115
7268013	157
081	198
150	239
219	280
288	321
357	362
425	403
494	444
563	485
632	526
701	568
769	609
838	650
907	691
976	732
7269044	773
113	814
182	855
251	896
320	937
388	979
457	·8615020
526	061
595	102
663	143
732	184
801	225
870	266
939	307

Excesses

10 = 6	1 = 1
20 = 12	2 = 1
30 = 18	3 = 2
40 = 24	4 = 2
50 = 30	5 = 3
60 = 36	6 = 4
	7 = 4
	8 = 5
	9 = 5

35 = 21

Anti Log Excesses

10 = 17	1 = 2
20 = 33	2 = 3
30 = 50	3 = 5
40 = 67	4 = 7
	5 = 8
	6 = 10
	7 = 12
	8 = 13
	9 = 15

21 = 35

7270007	·8615348
076	389
145	430
214	472
282	513
351	554
420	595
489	636
558	677
626	718
695	759
764	800
833	841
901	882
970	924
7271039	965
108	·8616006
176	047
245	088
314	129
383	170
452	211
520	252
589	293
658	334
727	375
795	416
864	457
933	498
7272002	540
070	581
139	622
208	663
277	704
345	745
414	786
483	827
552	868
620	909
689	950
758	991
827	·8617032
895	073
964	114
7273033	155
102	196
170	237
239	279
308	320
377	361
445	402
7273514	·8617443
583	484
652	525
720	566
789	607
858	648
927	689
995	730
7274064	771
133	812
201	853
270	894
339	935
408	976
476	·8618017
545	058
614	099
683	140
751	181
820	222
889	263
958	304
7275026	345
095	386
164	428
232	469
301	510
370	551
439	592
507	633
576	674
645	715
714	756
782	797
851	838
920	879
988	920
7276057	961
126	·8619002
195	043
263	084
332	125
401	166
469	207
538	248
607	289
676	330
744	371
813	412
882	453
950	494
7277019	535
088	576
157	617
225	658
294	699
363	740
431	781
500	822
569	863
637	904
706	945
775	986
844	·8620027
912	068

7277981	·8620109
7278050	150
118	191
187	232
256	273
324	314
393	355
462	396
531	437
599	478
668	519
737	560
805	601
874	642
943	683
7279011	724
080	765
149	806
217	847
286	888
355	929
424	970
492	·8621011
561	052
630	093
698	134
767	175
836	216
904	257
973	298

Excesses

10 = 6   1 = 1
20 = 12   2 = 1
30 = 18   3 = 2
40 = 24   4 = 2
50 = 30   5 = 3
60 = 36   6 = 4
7 = 4
8 = 5
9 = 5

35 = 21

Anti Log Excesses

10 = 17   1 = 2
20 = 33   2 = 3
30 = 50   3 = 5
40 = 67   4 = 7
5 = 8
6 = 10
7 = 12
8 = 13
9 = 15

21 = 35

7280042	·8621338
110	379
179	420
248	461
316	502
385	543

7280454	·8621584
522	625
591	666
660	707
728	748
797	789
866	830
935	871
7281003	912
072	953
141	994
209	·8622035
278	076
347	117
415	158
484	199
553	240
621	281
690	322
759	363
827	404
896	444
965	485
7282033	526
102	567
171	608
239	649
308	690
376	731
445	772
514	813
582	854
651	895
720	936
788	977
857	·8623018
926	059
994	100
7283063	141
132	181
200	222
269	263
338	304
406	345
475	386
544	427
612	468
681	509
750	550
818	591
887	632
955	673
7284024	714
093	754
161	795
230	836
299	877
367	918
436	959
505	·8624000
573	041
642	082
711	123
779	164
848	205

7284916	·8624245
985	286
7285054	327
122	368
191	409
260	450
328	491
397	532
465	573
534	614
603	655
671	695
740	736
809	777
877	818
946	859
7286014	900
083	941
152	982
220	·8625023
289	064
358	105
426	145
495	186
563	227
632	268
701	309
769	350
838	391
907	432
975	473
7287044	514
112	554
181	595
250	636
318	677
387	718
455	759
524	800
593	841
661	882
730	922
799	963
867	·8626004
936	045
7288004	086
073	127
142	168
210	209
279	250
347	290
416	331
485	372
553	413
622	454
690	495
759	536
828	577
896	617
965	658
7289033	699
102	740
171	781
239	822
308	863

7289376	·8626903
445	944
514	985
582	·8627026
651	067
719	108
788	149
856	190
925	230
994	271

Excesses

10 = 6   1 = 1
20 = 12   2 = 1
30 = 18   3 = 2
40 = 24   4 = 2
50 = 30   5 = 3
60 = 36   6 = 4
7 = 4
8 = 5
9 = 5

35 = 21

Anti Log Excesses

10 = 17   1 = 2
20 = 34   2 = 3
30 = 50   3 = 5
40 = 67   4 = 7
5 = 8
6 = 10
7 = 12
8 = 13
9 = 15

21 = 35

7290062	·8627312
131	353
199	394
268	435
337	476
405	516
474	557
542	598
611	639
679	680
748	721
817	761
885	802
954	843
7291022	884
091	925
160	966
228	·8628007
297	047
365	088
434	129
502	170
571	211
640	252
708	292
777	333

7291845	·8628374
914	415
982	456
7292051	497
120	537
188	578
257	619
325	660
394	701
462	742
531	782
599	823
668	864
737	905
805	946
874	987
942	·8629027
7293011	068
079	109
148	150
217	191
285	232
354	272
422	313
491	354
559	395
628	436
696	477
765	517
834	558
902	599
971	640
7294039	681
108	721
176	762
245	803
313	844
382	885
450	925
519	966
588	·8630007
656	048
725	089
793	130
862	170
930	211
999	252
7295067	293
136	334
204	374
273	415
341	456
410	497
479	538
547	578
616	619
684	660
753	701
821	742
890	782
958	823
7296027	864
095	905
164	946
232	986

7296301	·8631027
369	068
438	109
507	149
575	190
644	231
712	272
781	313
849	353
918	394
986	435
7297055	476
123	517
192	557
260	598
329	639
397	680
466	720
534	761
603	802
671	843
740	884
808	924
877	965
945	·8632006
7298014	047
082	087
151	128
219	169
288	210
356	250
425	291
493	332
562	373
631	414
699	454
768	495
836	536
905	577
973	617
7299042	658
110	699
179	740
247	780
316	821
384	862
453	903
521	944
590	984
658	·8633025
727	066
795	107
864	147
932	188

Excesses

10 = 6	1 = 1
20 = 12	2 = 1
30 = 18	3 = 2
40 = 24	4 = 2
50 = 30	5 = 3
60 = 36	6 = 4
	7 = 4
	8 = 5
	9 = 5

34 = 20

Anti Log Excesses

10 = 17	1 = 2
20 = 34	2 = 3
30 = 50	3 = 5
40 = 67	4 = 7
	5 = 8
	6 = 10
	7 = 12
	8 = 13
	9 = 15

20 = 34

7300000	·8633229
069	270
137	310
206	351
274	392
343	433
411	473
480	514
548	555
617	595
685	636
754	677
822	718
891	758
959	799
7301028	840
096	881
165	921
233	962
302	·8634003
370	044
439	084
507	125
576	166
644	207
713	247
781	288
850	329
918	369
987	410
7302055	451
123	492
192	532
260	573
329	614
397	655
466	695

7302534	·8634736
603	777
671	817
740	858
808	899
877	940
945	980
7303014	·8635021
082	062
151	102
219	143
287	184
356	225
424	265
493	306
561	347
630	387
698	428
767	469
835	509
904	550
972	591
7304040	632
109	672
177	713
246	754
314	794
383	835
451	876
520	916
588	957
657	998
725	·8636038
793	079
862	120
930	161
999	201
7305067	242
136	283
204	323
273	364
341	405
409	445
478	486
546	527
615	567
683	608
752	649
820	689
889	730
957	771
7306025	812
094	852
162	893
231	934
299	974
368	·8637015
436	056
504	096
573	137
641	178
710	218
778	259
847	300
915	340

7306983	·8637381
7307052	422
120	462
189	503
257	544
326	584
394	625
462	666
531	706
599	747
668	788
736	828
805	869
873	910
941	950
7308010	991
078	·8638032
147	072
215	113
283	154
352	194
420	235
489	276
557	316
626	357
694	397
762	438
831	479
899	519
968	560
7309036	601
104	641
173	682
241	723
310	763
378	804
446	845
515	885
583	926
652	966
720	·8639007
788	048
857	088
925	129
994	170

Excesses

10 = 6	1 = 1
20 = 12	2 = 1
30 = 18	3 = 2
40 = 24	4 = 2
50 = 30	5 = 3
60 = 36	6 = 4
	7 = 4
	8 = 5
	9 = 5

34 = 20

Anti Log Excesses

10 = 17	1 = 2
20 = 34	2 = 3
30 = 50	3 = 5
40 = 67	4 = 7
	5 = 8
	6 = 10
	7 = 12
	8 = 13
	9 = 15

20 = 34

7310062	·8639210
130	251
199	292
267	332
336	373
404	413
472	454
541	495
609	535
678	576
746	617
814	657
883	698
951	738
7311020	779
088	820
156	860
225	901
293	942
362	982
430	·8640023
498	063
567	104
635	145
703	185
772	226
840	267
909	307
977	348
7312045	389
114	429
182	470
250	510
319	551
387	592
456	632
524	673
592	713
661	754
729	795
798	835
866	876
934	916
7313003	957
071	998
139	·8641038
208	079
276	119
344	160
413	201
481	241

7313550	·8641282
618	322
686	363
755	404
823	444
891	485
960	525
7314028	566
096	607
165	647
233	688
302	728
370	769
438	809
507	850
575	891
643	931
712	972
780	·8642012
848	053
917	094
985	134
7315053	175
122	215
190	256
259	296
327	337
395	378
464	418
532	459
600	499
669	540
737	581
805	621
874	662
942	702
7316010	743
079	783
147	824
215	865
284	905
352	946
420	986
489	·8643027
557	067
625	108
694	149
762	189
830	230
899	270
967	311
7317035	351
104	392
172	432
240	473
309	514
377	554
445	595
514	635
582	676
650	716
719	757
787	797
855	838
924	879

7317992	·8643919
7318060	960
129	·8644000
197	041
265	081
334	122
402	162
470	203
539	243
607	284
675	325
744	365
812	406
880	446
949	487
7319017	527
085	568
153	608
222	649
290	689
358	730
427	770
495	811
563	851
632	892
700	933
768	973
837	·8645014
905	054
973	095

Excesses

10 = 6	1 = 1
20 = 12	2 = 1
30 = 18	3 = 2
40 = 24	4 = 2
50 = 30	5 = 3
60 = 36	6 = 4
	7 = 4
	8 = 5
	9 = 5

34 = 20

Anti Log Excesses

10 = 17	1 = 2
20 = 34	2 = 3
30 = 51	3 = 5
40 = 67	4 = 7
	5 = 8
	6 = 10
	7 = 12
	8 = 13
	9 = 15

20 = 34

7320041	·8645135
110	176
178	216
246	257
315	297
383	338

7320451	·8645378
520	419
588	459
656	500
725	540
793	581
861	621
929	662
998	702
7321066	743
134	783
203	824
271	865
339	905
407	946
476	986
544	·8646027
612	067
681	108
749	148
817	189
886	229
954	270
7322022	310
090	351
159	391
227	432
295	472
364	513
432	553
500	594
568	634
637	675
705	715
773	756
841	796
910	837
978	877
7323046	918
115	958
183	999
251	·8647039
319	080
388	120
456	161
524	201
593	241
661	282
729	322
797	363
866	403
934	444
7324002	484
070	525
139	565
207	606
275	646
343	687
412	727
480	768
548	808
617	849
685	889
753	930
821	970

7324890	·8648010
958	051
7325026	091
094	132
163	172
231	213
299	253
367	294
436	334
504	375
572	415
640	456
709	496
777	537
845	577
913	617
982	658
7326050	698
118	739
186	779
255	820
323	860
391	901
459	941
528	982
596	·8649022
664	063
732	103
801	143
869	184
937	224
7327005	265
074	305
142	346
210	386
278	427
347	467
415	507
483	548
551	588
619	629
688	669
756	710
824	750
892	791
961	831
7328029	871
097	912
165	952
234	993
302	·8650033
370	074
438	114
506	154
575	195
643	235
711	276
779	316
848	357
916	397
984	437
7329052	478
121	518
189	559
257	599

7329325	·8650640
393	680
462	720
530	761
598	801
666	842
734	882
803	922
871	963
939	·8651003

Excesses

10 = 6	1 = 1
20 = 12	2 = 1
30 = 18	3 = 2
40 = 24	4 = 2
50 = 30	5 = 3
60 = 36	6 = 4
	7 = 4
	8 = 5
	9 = 5

34 = 20

Anti Log Excesses

10 = 17	1 = 2
20 = 34	2 = 3
30 = 51	3 = 5
40 = 67	4 = 7
	5 = 8
	6 = 10
	7 = 12
	8 = 13
	9 = 15

20 = 34

7330007	·8651044
076	084
144	125
212	165
280	205
348	246
417	286
485	327
553	367
621	407
689	448
758	488
826	529
894	569
962	609
7331030	650
099	690
167	731
235	771
303	812
371	852
440	892
508	933
576	973
644	·8652014
712	054

7331781	·8652094
849	135
917	175
985	216
7332053	256
122	296
190	337
258	377
326	418
394	458
463	498
531	539
599	579
667	620
735	660
804	700
872	741
940	781
7333008	822
076	862
144	902
213	943
281	983
349	·8653023
417	064
485	104
554	145
622	185
690	225
758	266
826	306
894	346
963	387
7334031	427
099	468
167	508
235	548
303	589
372	629
440	669
508	710
576	750
644	791
712	831
781	871
849	912
917	952
985	992
7335053	·8654033
121	073
190	113
258	154
326	194
394	235
462	275
530	315
599	356
667	396
735	436
803	477
871	517
939	557
7336008	598
076	638
144	678

7336212	·8654719
280	759
348	799
417	840
485	880
553	920
621	961
689	·8655001
757	041
825	082
894	122
962	162
7337030	203
098	243
166	283
234	324
302	364
371	404
439	445
507	485
575	525
643	566
711	606
779	646
848	687
916	727
984	767
7338052	808
120	848
188	888
256	929
325	969
393	·8656009
461	050
529	090
597	130
665	171
733	211
801	251
870	291
938	332
7339006	372
074	412
142	453
210	493
278	533
347	574
415	614
483	654
551	695
619	735
687	775
755	816
823	856
892	896
960	936

Excesses

10 = 6	1 = 1
20 = 12	2 = 1
30 = 18	3 = 2
40 = 24	4 = 2
50 = 30	5 = 3
60 = 36	6 = 4
	7 = 4
	8 = 5
	9 = 5

34 = 20

Anti Log Excesses

10 = 17	1 = 2
20 = 34	2 = 3
30 = 51	3 = 5
40 = 68	4 = 7
	5 = 8
	6 = 10
	7 = 12
	8 = 14
	9 = 15

20 = 34

7340028	·8656977
096	·8657017
164	057
232	098
300	138
368	178
436	219
505	259
573	299
641	340
709	380
777	420
845	460
913	501
981	541
7341049	581
118	622
186	662
254	702
322	742
390	783
458	823
526	863
594	904
662	944
731	984
799	·8658024
867	065
935	105
7342003	145
071	186
139	226
207	266
275	306
343	347
412	387
480	427

7342548	·8658468
616	508
684	548
752	588
820	629
888	669
956	709
7343024	749
092	790
161	830
229	870
297	910
365	951
433	991
501	·8659031
569	072
637	112
705	152
773	192
841	233
910	273
978	313
7344046	353
114	394
182	434
250	474
318	514
386	555
454	595
522	635
590	676
658	716
726	756
795	796
863	837
931	877
999	917
7345067	957
135	998
203	·8660038
271	078
339	118
407	159
475	199
543	239
611	279
680	320
748	360
816	400
884	440
952	480
7346020	521
088	561
156	601
224	641
292	682
360	722
428	762
496	802
564	843
632	883
700	923
768	963
837	·8661004
905	044

7346973	·8661084
7347041	124
109	164
177	205
245	245
313	285
381	325
449	366
517	406
585	446
653	486
721	526
789	567
857	607
925	647
993	687
7348061	727
130	768
198	808
266	848
334	888
402	929
470	969
538	·8662009
606	049
674	089
742	130
810	170
878	210
946	250
7349014	291
082	331
150	371
218	411
286	451
354	492
422	532
490	572
558	612
626	652
694	693
762	733
830	773
898	813
966	853

Excesses

10 = 6	1 = 1
20 = 12	2 = 1
30 = 18	3 = 2
40 = 24	4 = 2
50 = 30	5 = 3
60 = 35	6 = 4
	7 = 4
	8 = 5
	9 = 5

34 = 20

Anti Log Excesses

10 = 17	1 = 2
20 = 34	2 = 3
30 = 51	3 = 5
40 = 68	4 = 7
	5 = 8
	6 = 10
	7 = 12
	8 = 14
	9 = 15

20 = 34

7350035	·8662894
103	934
171	974
239	·8663014
307	054
375	095
443	135
511	175
579	215
647	255
715	295
783	336
851	376
919	416
987	456
7351055	496
123	537
191	577
259	617
327	657
395	697
463	737
531	778
599	818
667	858
735	898
803	938
871	979
939	·8664019
7352007	059
075	099
143	139
211	179
279	220
347	260
415	300
483	340
551	380
619	420
687	461
755	501
823	541
891	581
959	621
7353027	662
095	702
163	742
231	782
299	822
367	862
435	902

7353503	·8664943
571	983
639	·8665023
707	063
775	103
843	143
911	184
979	224
7354047	264
115	304
183	344
251	384
319	424
387	465
455	505
523	545
591	585
659	625
727	665
795	705
863	746
931	786
999	826
7355067	866
135	906
203	946
271	986
339	·8666027
407	067
475	107
543	147
611	187
679	227
747	267
815	307
883	348
950	388
7356018	428
086	468
154	508
222	548
290	588
358	628
426	669
494	709
562	749
630	789
698	829
766	869
834	909
902	949
970	990
7357038	·8667030
106	070
174	110
242	150
310	190
378	230
446	270
514	311
582	351
650	391
718	431
785	471
853	511

7357921	·8667551
989	591
7358057	631
125	672
193	712
261	752
329	792
397	832
465	872
533	912
601	952
669	992
737	·8668032
805	073
873	113
941	153
7359009	193
077	233
144	273
212	313
280	353
348	393
416	433
484	473
552	514
620	554
688	594
756	634
824	674
892	714
960	754

Excesses

10 = 6	1 = 1
20 = 12	2 = 1
30 = 18	3 = 2
40 = 24	4 = 2
50 = 30	5 = 3
60 = 35	6 = 4
	7 = 4
	8 = 5
	9 = 5

34 = 20

Anti Log Excesses

10 = 17	1 = 2
20 = 34	2 = 3
30 = 51	3 = 5
40 = 68	4 = 7
	5 = 8
	6 = 10
	7 = 12
	8 = 14
	9 = 15

20 = 34

7360028	·8668794
096	834
164	874
231	914
299	954

7360367	·8668995
435	·8669035
503	075
571	115
639	155
707	195
775	235
843	275
911	315
979	355
7361047	395
115	436
182	476
250	516
318	556
386	596
454	636
522	676
590	716
658	756
726	796
794	836
862	876
930	916
998	956
7362065	997
133	·8670037
201	077
269	117
337	157
405	197
473	237
541	277
609	317
677	357
745	397
812	437
880	477
948	517
7363016	557
084	597
152	637
220	678
288	718
356	758
424	798
492	838
559	878
627	918
695	958
763	998
831	·8671038
899	078
967	118
7364035	158
103	198
171	238
238	278
306	318
374	358
442	398
510	438
578	478
646	518
714	559

7364782	·8671599
849	639
917	679
985	719
7365053	759
121	799
189	839
257	879
325	919
393	959
460	999
528	·8672039
596	079
664	119
732	159
800	199
868	239
936	279
7366003	319
071	359
139	399
207	439
275	479
343	519
411	559
479	599
547	639
614	679
682	719
750	759
818	799
886	839
954	879
7367022	919
089	959
157	999
225	·8673039
293	079
361	119
429	159
497	199
565	239
632	279
700	319
768	359
836	399
904	439
972	479
7368040	519
107	559
175	599
243	639
311	679
379	719
447	759
515	799
582	839
650	879
718	919
786	959
854	999
922	·8674039
990	079
7369057	119
125	159

7369193	·8674199
261	239
329	279
397	319
465	359
532	399
600	439
668	479
736	519
804	559
872	599
939	639

Excesses

10 = 6	1 = 1
20 = 12	2 = 1
30 = 18	3 = 2
40 = 24	4 = 2
50 = 29	5 = 3
60 = 35	6 = 4
	7 = 4
	8 = 5
	9 = 5

34 = 20

Anti Log Excesses

10 = 17	1 = 2
20 = 34	2 = 3
30 = 51	3 = 5
40 = 68	4 = 7
	5 = 8
	6 = 10
	7 = 12
	8 = 14
	9 = 15

20 = 34

7370007	·8674679
075	719
143	759
211	799
279	839
346	879
414	919
482	959
550	999
618	·8675039
686	079
754	119
821	159
889	199
957	239
7371025	279
093	319
161	359
228	399
296	439
364	478
432	518
500	558
567	598

7371635	·8675638
703	678
771	718
839	758
907	798
974	838
7372042	878
110	918
178	958
246	998
314	·8676038
381	078
449	118
517	158
585	198
653	238
720	278
788	317
856	357
924	397
992	437
7373060	477
127	517
195	557
263	597
331	637
399	677
466	717
534	757
602	797
670	837
738	877
805	917
873	957
941	996
7374009	·8677036
077	076
145	116
212	156
280	196
348	236
416	276
484	316
551	356
619	396
687	436
755	476
823	516
890	555
958	595
7375026	635
094	675
162	715
229	755
297	795
365	835
433	875
500	915
568	955
636	995
704	·8678034
772	074
839	114
907	154
975	194

7376043	·8678234
111	274
178	314
246	354
314	394
382	434
450	473
517	513
585	553
653	593
721	633
788	673
856	713
924	753
992	793
7377060	833
127	872
195	912
263	952
331	992
398	·8679032
466	072
534	112
602	152
670	192
737	232
805	272
873	311
941	351
7378008	391
076	431
144	471
212	511
279	551
347	591
415	631
483	670
550	710
618	750
686	790
754	830
822	870
889	910
957	950
7379025	989
093	·8680029
160	069
228	109
296	149
364	189
431	229
499	269
567	308
635	348
702	388
770	428
838	468
906	508
973	548

Excesses

10 = 6	1 = 1
20 = 12	2 = 1
30 = 18	3 = 2
40 = 24	4 = 2
50 = 29	5 = 3
60 = 35	6 = 4
	7 = 4
	8 = 5
	9 = 5

34 = 20

Anti Log Excesses

10 = 17	1 = 2
20 = 34	2 = 3
30 = 51	3 = 5
	4 = 7
	5 = 8
	6 = 10
	7 = 12
	8 = 14
	9 = 15

20 = 34

7380041	·8680588
109	627
177	667
244	707
312	747
380	787
448	827
515	867
583	906
651	946
719	986
786	·8681026
854	066
922	106
990	146
7381057	185
125	225
193	265
261	305
328	345
396	385
464	425
532	464
599	504
667	544
735	584
802	624
870	664
938	704
7382006	744
073	783
141	823
209	863
277	903
344	943
412	983
480	·8682022

7382547	·8682062
615	102
683	142
751	182
818	222
886	261
954	301
7383022	341
089	381
157	421
225	461
292	500
360	540
428	580
496	620
563	660
631	700
699	739
766	779
834	819
902	859
970	899
7384037	939
105	978
173	·8683018
240	058
308	098
376	138
444	177
511	217
579	257
647	297
714	337
782	377
850	416
918	456
985	496
7385053	536
121	576
188	615
256	655
324	695
391	735
459	775
527	814
595	854
662	894
730	934
798	974
865	·8684013
933	053
7386001	093
068	133
136	173
204	213
272	252
339	292
407	332
475	372
542	412
610	451
678	491
745	531
813	571
881	611

7386948	·8684650
7387016	690
084	730
152	770
219	810
287	849
355	889
422	929
490	969
558	·8685009
625	048
693	088
761	128
828	168
896	207
964	247
7388031	287
099	327
167	367
234	406
302	446
370	486
437	526
505	565
573	605
640	645
708	685
776	725
843	764
911	804
979	844
7389046	884
114	923
182	963
249	·8686003
317	043
385	082
452	122
520	162
588	202
655	241
723	281
791	321
858	361
926	400
994	440

Excesses

10 = 6	1 = 1
20 = 12	2 = 1
30 = 18	3 = 2
40 = 23	4 = 2
50 = 29	5 = 3
60 = 35	6 = 4
	7 = 4
	8 = 5
	9 = 5

34 = 20

Anti Log Excesses

10 = 17	1 = 2
20 = 34	2 = 3
30 = 51	3 = 5
	4 = 7
	5 = 9
	6 = 10
	7 = 12
	8 = 14
	9 = 15

20 = 34

7390061	·8686480
129	520
197	560
264	599
332	639
400	679
467	719
535	758
603	798
670	838
738	878
806	917
873	957
941	997
7391009	·8687037
076	076
144	116
211	156
279	196
347	235
414	275
482	315
550	355
617	394
685	434
753	474
820	514
888	553
956	593
7392023	633
091	673
159	712
226	752
294	792
361	832
429	871
497	911
564	951
632	991
700	·8688030
767	070
835	110
903	150
970	189
7393038	229
105	269
173	308
241	348
308	388
376	428
444	467

7393511	·8688507
579	547
646	587
714	626
782	666
849	706
917	745
985	785
7394052	825
120	865
187	904
255	944
323	984
390	·8689023
458	063
526	103
593	143
661	182
728	222
796	262
864	301
931	341
999	381
7395066	421
134	460
202	500
269	540
337	579
404	619
472	659
540	699
607	738
675	778
743	818
810	857
878	897
945	937
7396013	976
081	·8690016
148	056
216	096
283	135
351	175
419	215
486	254
554	294
621	334
689	373
757	413
824	453
892	493
959	532
7397027	572
095	612
162	651
230	691
297	731
365	770
432	810
500	850
568	889
635	929
703	969
770	·8691008
838	048

7397906	·8691088
973	128
7398041	167
108	207
176	247
244	286
311	326
379	366
446	405
514	445
581	485
649	524
717	564
784	604
852	643
919	683
987	723
7399054	762
122	802
190	842
257	881
325	921
392	961
460	·8692000
528	040
595	080
663	119
730	159
798	199
865	238
933	278

Excesses

10 = 6	1 = 1
20 = 12	2 = 1
30 = 18	3 = 2
40 = 23	4 = 2
50 = 29	5 = 3
60 = 35	6 = 4
	7 = 4
	8 = 5
	9 = 5

34 = 20

Anti Log Excesses

10 = 17	1 = 2
20 = 34	2 = 3
30 = 51	3 = 5
	4 = 7
	5 = 9
	6 = 10
	7 = 12
	8 = 14
	9 = 15

20 = 34

7400000	·8692317
068	357
136	397
203	436
271	476

7400338	·8692516
406	555
473	595
541	635
609	674
676	714
744	754
811	793
879	833
946	873
7401014	912
082	952
149	992
217	·8693031
284	071
352	110
419	150
487	190
554	229
622	269
689	309
757	348
825	388
892	428
960	467
7402027	507
095	546
162	586
230	626
297	665
365	705
433	745
500	784
568	824
635	863
703	903
770	943
838	982
905	·8694022
973	062
7403040	101
108	141
175	180
243	220
311	260
378	299
446	339
513	378
581	418
648	458
716	497
783	537
851	577
918	616
986	656
7404053	695
121	735
189	775
256	814
324	854
391	893
459	933
526	973
594	·8695012
661	052

7404729	·8695091
796	131
864	171
931	210
999	250
7405066	289
134	329
201	369
269	408
336	448
404	487
471	527
539	566
606	606
674	646
742	685
809	725
877	764
944	804
7406012	844
079	883
147	923
214	962
282	·8696002
349	041
417	081
484	121
552	160
619	200
687	239
754	279
822	319
889	358
957	398
7407024	437
092	477
159	517
227	556
294	596
362	635
429	675
497	715
564	754
632	794
699	833
767	873
834	912
902	952
969	992
7408037	·8697031
104	071
172	110
239	150
307	189
374	229
442	268
509	308
577	348
644	387
712	427
779	466
847	506
914	545
982	585
7409049	625

7409117	·8697664
184	704
251	743
319	783
386	822
454	862
521	901
589	941
656	981
724	·8698020
791	060
859	099
926	139
994	178

Excesses

10 = 6	1 = 1
20 = 12	2 = 1
30 = 18	3 = 2
40 = 23	4 = 2
50 = 29	5 = 3
60 = 35	6 = 4
	7 = 4
	8 = 5
	9 = 5

34 = 20

Anti Log Excesses

10 = 17	1 = 2
20 = 34	2 = 3
30 = 51	3 = 5
	4 = 7
	5 = 9
	6 = 10
	7 = 12
	8 = 14
	9 = 15

20 = 34

7410061	·8698218
129	257
196	297
264	336
331	376
399	416
466	455
534	495
601	534
668	574
736	613
803	653
871	692
938	732
7411006	771
073	811
141	850
208	890
276	930
343	969
411	·8699009
478	048

7411546	·8699088
613	127
680	167
748	206
815	246
883	285
950	325
7412018	364
085	404
153	443
220	483
288	522
355	562
423	601
490	641
557	681
625	720
692	760
760	799
827	839
895	878
962	918
7413030	957
097	997
164	·8700036
232	076
299	115
367	155
434	194
502	234
569	273
637	313
704	352
771	392
839	431
906	471
974	510
7414041	550
109	589
176	629
244	668
311	708
378	747
446	787
513	826
581	866
648	905
716	945
783	984
850	·8701024
918	063
985	103
7415053	142
120	182
188	221
255	261
322	300
390	340
457	379
525	419
592	458
660	498
727	537
794	577
862	616

7415929	·8701656
997	695
7416064	735
132	774
199	814
266	853
334	893
401	932
469	972
536	·8702011
603	050
671	090
738	129
806	169
873	208
941	248
7417008	287
075	327
143	366
210	406
278	445
345	485
412	524
480	564
547	603
615	643
682	682
749	721
817	761
884	800
952	840
7418019	879
087	919
154	958
221	998
289	·8703037
356	077
424	116
491	155
558	195
626	234
693	274
761	313
828	353
895	392
963	432
7419030	471
097	511
165	550
232	589
300	629
367	668
434	708
502	747
569	787
637	826
704	866
771	905
839	945
906	984
974	·8704023

Excesses

10 = 6	1 = 1
20 = 12	2 = 1
30 = 18	3 = 2
40 = 23	4 = 2
50 = 29	5 = 3
60 = 35	6 = 4
	7 = 4
	8 = 5
	9 = 5

34 = 20

Anti Log Excesses

10 = 17	1 = 2
20 = 34	2 = 3
30 = 51	3 = 5
	4 = 7
	5 = 9
	6 = 10
	7 = 12
	8 = 14
	9 = 15

20 = 34

7420041	·8704063
108	102
176	142
243	181
310	221
378	260
445	299
513	339
580	378
647	418
715	457
782	497
850	536
917	576
984	615
7421052	654
119	694
186	733
254	773
321	812
389	852
456	891
523	930
591	970
658	·8705009
725	049
793	088
860	127
927	167
995	206
7422062	246
130	285
197	325
264	364
332	403
399	443
466	482

7422534	·8705522
601	561
668	600
736	640
803	679
871	719
938	758
7423005	798
073	837
140	876
207	916
275	955
342	995
409	·8706034
477	073
544	113
611	152
679	192
746	231
814	270
881	310
948	349
7424016	389
083	428
150	467
218	507
285	546
352	585
420	625
487	664
554	704
622	743
689	782
756	822
824	861
891	901
958	940
7425026	979
093	·8707019
160	058
228	098
295	137
362	176
430	216
497	255
564	294
632	334
699	373
766	413
834	452
901	491
968	531
7426036	570
103	609
170	649
238	688
305	728
372	767
440	806
507	846
574	885
642	924
709	964
776	·8708003
844	043

7426911	·8708082
978	121
7427046	161
113	200
180	239
248	279
315	318
382	357
450	397
517	436
584	476
652	515
719	554
786	594
853	633
921	672
988	712
7428055	751
123	791
190	830
257	869
325	909
392	948
459	987
527	·8709027
594	066
661	105
729	145
796	184
863	223
930	263
998	302
7429065	341
132	381
200	420
267	459
334	499
402	538
469	578
536	617
603	656
671	696
738	735
805	774
873	814
940	853

Excesses

10 = 6	1 = 1
20 = 12	2 = 1
30 = 18	3 = 2
40 = 23	4 = 2
50 = 29	5 = 3
60 = 35	6 = 4
	7 = 4
	8 = 5
	9 = 5

34 = 20

Anti Log Excesses

10 = 17	1 = 2
20 = 34	2 = 3
30 = 51	3 = 5
	4 = 7
	5 = 9
	6 = 10
	7 = 12
	8 = 14
	9 = 15

20 = 34

7430007	·8709892
075	932
142	971
209	·8710010
276	050
344	089
411	128
478	168
546	207
613	246
680	286
747	325
815	364
882	404
949	443
7431017	482
084	522
151	561
218	600
286	640
353	679
420	718
488	757
555	797
622	836
689	875
757	915
824	954
891	993
959	·8711033
7432026	072
093	111
160	151
228	190
295	229
362	269
429	308
497	347
564	387
631	426
699	465
766	504
833	544
900	583
968	622
7433035	662
102	701
169	740
237	780
304	819
371	858

7433438	·8711897
506	937
573	976
640	·8712015
708	055
775	094
842	133
909	173
977	212
7434044	251
111	290
178	330
246	369
313	408
380	448
447	487
515	526
582	565
649	605
716	644
784	683
851	723
918	762
985	801
7435053	840
120	880
187	919
254	958
322	998
389	·8713037
456	076
523	115
591	155
658	194
725	233
792	273
860	312
927	351
994	390
7436061	430
129	469
196	508
263	547
330	587
397	626
465	665
532	704
599	744
666	783
734	822
801	861
868	901
935	940
7437003	979
070	·8714018
137	058
204	097
272	136
339	176
406	215
473	254
540	293
608	333
675	372
742	411

7437809	·8714450
877	490
944	529
7438011	568
078	607
145	647
213	686
280	725
347	764
414	804
482	843
549	882
616	921
683	961
750	·8715000
818	039
885	078
952	118
7439019	157
086	196
154	235
221	274
288	314
355	353
423	392
490	431
557	471
624	510
691	549
759	588
826	628
893	667
960	706

Excesses

10 = 6	1 = 1
20 = 12	2 = 1
30 = 18	3 = 2
40 = 23	4 = 2
50 = 29	5 = 3
60 = 35	6 = 4
	7 = 4
	8 = 5
	9 = 5

34 = 20

Anti Log Excesses

10 = 17	1 = 2
20 = 34	2 = 3
30 = 51	3 = 5
	4 = 7
	5 = 9
	6 = 10
	7 = 12
	8 = 14
	9 = 15

20 = 34

7440027	·8715745
095	785
162	824

7440229	·8715863
296	902
363	941
431	981
498	·8716020
565	059
632	098
699	137
767	177
834	216
901	255
968	294
7441035	334
103	373
170	412
237	451
304	490
371	530
439	569
506	608
573	647
640	686
707	726
774	765
842	804
909	843
976	882
7442043	922
110	961
178	·8717000
245	039
312	079
379	118
446	157
514	196
581	235
648	275
715	314
782	353
849	392
917	431
984	471
7443051	510
118	549
185	588
252	627
320	667
387	706
454	745
521	784
588	823
656	862
723	902
790	941
857	980
924	·8718019
991	058
7444059	098
126	137
193	176
260	215
327	254
394	293
462	333
529	372

7444596	·8718411
663	450
730	489
797	529
865	568
932	607
999	646
7445066	685
133	724
200	764
267	803
335	842
402	881
469	920
536	959
603	999
670	·8719038
738	077
805	116
872	155
939	194
7446006	234
073	273
140	312
208	351
275	390
342	429
409	469
476	508
543	547
610	586
678	625
745	664
812	704
879	743
946	782
7447013	821
080	860
148	899
215	938
282	978
349	·8720017
416	056
483	095
550	134
618	173
685	213
752	252
819	291
886	330
953	369
7448020	408
088	447
155	487
222	526
289	565
356	604
423	643
490	682
557	722
625	761
692	800
759	839
826	878
893	917

7448960	·8720956
7449027	995
094	·8721035
162	074
229	113
296	152
363	191
430	230
497	269
564	309
631	348
698	387
766	426
833	465
900	504
967	543

Excesses

10 = 6	1 = 1
20 = 12	2 = 1
30 = 17	3 = 2
40 = 23	4 = 2
50 = 29	5 = 3
60 = 35	6 = 3
	7 = 4
	8 = 5
	9 = 5

34 = 20

Anti Log Excesses

10 = 17	1 = 2
20 = 34	2 = 3
30 = 51	3 = 5
	4 = 7
	5 = 9
	6 = 10
	7 = 12
	8 = 14
	9 = 15

20 = 34

7450034	·8721582
101	621
168	661
235	700
303	739
370	778
437	817
504	856
571	895
638	934
705	974
772	·8722013
839	052
906	091
974	130
7451041	170
108	208
175	247
242	286
309	326

7451376	·8722365
443	404
510	443
578	482
645	521
712	560
779	599
846	638
913	677
980	717
7452047	756
114	795
181	834
248	873
316	912
383	951
450	990
517	·8723029
584	068
651	108
718	147
785	186
852	225
919	264
986	303
7453054	342
121	381
188	420
255	459
322	499
389	538
456	577
523	616
590	655
657	694
724	733
791	772
859	811
926	850
993	889
7454060	928
127	968
194	·8724007
261	046
328	085
395	124
462	163
529	202
596	241
663	280
731	319
798	358
865	397
932	436
999	475
7455066	515
133	554
200	593
267	632
334	671
401	710
468	749
535	788
602	827
669	866

7455737	·8724905
804	944
871	983
938	·8725022
7456005	061
072	100
139	140
206	179
273	218
340	257
407	296
474	335
541	374
608	413
675	452
742	491
809	530
877	569
944	608
7457011	647
078	686
145	725
212	765
279	804
346	843
413	882
480	921
547	960
614	999
681	·8726038
748	077
815	116
882	155
949	194
7458016	233
083	272
150	311
217	350
284	389
352	428
419	467
486	506
553	545
620	584
687	623
754	662
821	701
888	741
955	780
7459022	819
089	858
156	897
223	936
290	975
357	·8727014
424	053
491	092
558	131
625	170
692	209
759	248
826	287
893	326
960	365

Excesses

10 = 6	1 = 1
20 = 12	2 = 1
30 = 17	3 = 2
40 = 23	4 = 2
50 = 29	5 = 3
60 = 35	6 = 3
	7 = 4
	8 = 5
	9 = 5

33 = 19

Anti Log Excesses

10 = 17	1 = 2
20 = 34	2 = 3
30 = 51	3 = 5
	4 = 7
	5 = 9
	6 = 10
	7 = 12
	8 = 14
	9 = 15

19 = 33

7460027	·8727404
094	443
161	482
228	521
295	560
362	599
429	638
496	677
563	716
631	755
698	794
765	833
832	872
899	911
966	950
7461033	989
100	·8728028
167	067
234	106
301	145
368	184
435	223
502	262
569	301
636	340
703	379
770	418
837	457
904	496
971	535
7462038	574
105	613
172	652
239	691
306	730
373	769
440	808

7462507	·8728847
574	886
641	925
708	964
775	·8729003
842	042
909	081
976	120
7463043	159
110	198
177	237
244	276
311	315
378	354
445	393
512	432
579	471
646	510
713	549
780	588
847	627
914	666
981	705
7464048	744
115	783
182	822
249	861
316	900
383	939
450	978
517	·8730017
584	056
651	095
718	134
784	172
851	211
918	250
985	289
7465052	328
119	367
186	406
253	445
320	485
387	523
454	562
521	601
588	640
655	679
722	718
789	757
856	796
923	835
990	874
7466057	913
124	952
191	991
258	·8731030
325	069
392	108
459	147
526	186
593	225
660	264
727	303
794	341

7466861	·8731380
928	419
995	458
7467062	497
128	536
195	575
262	614
329	653
396	692
463	731
530	770
597	809
664	848
731	887
798	926
865	965
932	·8732003
999	042
7468066	081
133	120
200	159
267	198
334	237
401	276
468	315
534	354
601	393
668	432
735	471
802	510
869	548
936	587
7469003	626
070	665
137	704
204	743
271	782
338	821
405	860
472	899
539	938
606	977
673	·8733016
739	055
806	093
873	132
940	171

Excesses

10 = 6	1 = 1
20 = 12	2 = 1
30 = 17	3 = 2
40 = 23	4 = 2
50 = 29	5 = 3
60 = 35	6 = 3
	7 = 4
	8 = 5
	9 = 5

33 = 19

Anti Log Excesses

10 = 17	1 = 2
20 = 34	2 = 3
30 = 52	3 = 5
	4 = 7
	5 = 9
	6 = 10
	7 = 12
	8 = 14
	9 = 15

19 = 33

7470007	·8733210
074	249
141	288
208	327
275	366
342	405
409	444
476	483
543	521
610	560
677	599
743	638
810	677
877	716
944	755
7471011	794
078	833
145	872
212	910
279	949
346	988
413	·8734027
480	066
547	105
613	144
680	183
747	222
814	261
881	299
948	338
7472015	377
082	416
149	455
216	494
283	533
350	572
416	611
483	649
550	688
617	727
684	766
751	805
818	844
885	883
952	922
7473019	961
086	999
152	·8735038
219	077
286	116
353	155

7473420	·8735194
487	233
554	272
621	310
688	349
755	388
821	427
888	466
955	505
7474022	544
089	583
156	622
223	660
290	699
357	738
424	777
490	816
557	855
624	894
691	932
758	971
825	·8736010
892	049
959	088
7475026	127
092	166
159	205
226	243
293	282
360	321
427	360
494	399
561	438
628	477
694	515
761	554
828	593
895	632
962	671
7476029	710
096	749
163	787
229	826
296	865
363	904
430	943
497	982
564	·8737020
631	059
698	098
765	137
831	176
898	215
965	254
7477032	292
099	331
166	370
233	409
299	448
366	487
433	525
500	564
567	603
634	642
701	681

7477768	·8737720
834	758
901	797
968	836
7478035	875
102	914
169	953
236	991
302	·8738030
369	069
436	108
503	147
570	186
637	224
704	263
770	302
837	341
904	380
971	418
7479038	457
105	496
172	535
238	574
305	613
372	651
439	690
506	729
573	768
640	807
706	845
773	884
840	923
907	962
974	·8739001

Excesses

10 = 6	1 = 1
20 = 12	2 = 1
30 = 17	3 = 2
40 = 23	4 = 2
50 = 29	5 = 3
60 = 35	6 = 3
	7 = 4
	8 = 5
	9 = 5

33 = 19

Anti Log Excesses

10 = 17	1 = 2
20 = 34	2 = 3
30 = 52	3 = 5
	4 = 7
	5 = 9
	6 = 10
	7 = 12
	8 = 14
	9 = 15

19 = 33

7480041	·8739039
107	078

7480174	·8739117
241	156
308	195
375	234
442	272
509	311
575	350
642	389
709	428
776	466
843	505
910	544
976	583
7481043	622
110	660
177	699
244	738
311	777
377	816
444	854
511	893
578	932
645	971
712	·8740009
778	048
845	087
912	126
979	165
7482046	203
113	242
179	281
246	320
313	359
380	397
447	436
513	475
580	514
647	553
714	591
781	630
848	669
914	708
981	746
7483048	785
115	824
182	863
248	902
315	940
382	979
449	·8741018
516	057
583	095
649	134
716	173
783	212
850	251
917	289
983	328
7484050	367
117	406
184	444
251	483
317	522
384	561
451	599

7484518	·8741638
585	677
651	716
718	754
785	793
852	832
919	871
985	909
7485052	948
119	987
186	·8742026
253	065
319	103
386	142
453	181
520	220
587	258
653	297
720	336
787	375
854	413
921	452
987	491
7486054	530
121	568
188	607
255	646
321	685
388	723
455	762
522	801
589	840
655	878
722	917
789	956
856	995
922	·8743033
989	072
7487056	111
123	149
190	188
256	227
323	266
390	304
457	343
523	382
590	421
657	459
724	498
791	537
857	575
924	614
991	653
7488058	692
124	730
191	769
258	808
325	847
392	885
458	924
525	963
592	·8744001
659	040
725	079
792	118

7488859	·8744156
926	195
992	234
7489059	272
126	311
193	350
259	389
326	427
393	466
460	505
527	543
593	582
660	621
727	660
794	698
860	737
927	776
994	814

Excesses

10 = 6	1 = 1
20 = 12	2 = 1
30 = 17	3 = 2
40 = 23	4 = 2
50 = 29	5 = 3
60 = 35	6 = 3
	7 = 4
	8 = 5
	9 = 5

33 = 19

Anti Log Excesses

10 = 17	1 = 2
20 = 34	2 = 3
30 = 52	3 = 5
	4 = 7
	5 = 9
	6 = 10
	7 = 12
	8 = 14
	9 = 16

19 = 33

7490061	·8744853
127	892
194	930
261	969
328	·8745008
394	047
461	085
528	124
595	163
661	201
728	240
795	279
862	318
928	356
995	395
7491062	434
129	472
195	511

7491262	·8745550
329	588
396	627
462	666
529	705
596	743
663	782
729	821
796	859
863	898
929	937
996	975
7492063	·8746014
130	053
196	091
263	130
330	169
397	207
463	246
530	285
597	324
664	362
730	401
797	440
864	478
931	517
997	556
7493064	594
131	633
197	672
264	710
331	749
398	788
464	826
531	865
598	904
664	942
731	981
798	·8747020
865	058
931	097
998	136
7494065	174
132	213
198	252
265	290
332	329
398	368
465	406
532	445
599	484
665	522
732	561
799	600
865	638
932	677
999	716
7495066	754
132	793
199	832
266	870
332	909
399	947
466	986
533	·8748025

Number	Log
7495599	·8748063
666	102
733	141
799	179
866	218
933	257
999	295
7496066	334
133	373
200	411
266	450
333	488
400	527
466	566
533	604
600	643
666	682
733	720
800	759
867	798
933	836
7497000	875
067	913
133	952
200	991
267	·8749029
333	068
400	107
467	145
533	184
600	223
667	261
733	300
800	338
867	377
934	416
7498000	454
067	493
134	532
200	570
267	609
334	647
400	686
467	725
534	763
600	802
667	840
734	879
800	918
867	956
934	995
7499000	·8750033
067	072
134	111
200	149
267	188
334	227
400	265
467	304
534	342
600	381
667	420
734	458
800	497
867	535
7499934	·8750574

Excesses

10 = 6	1 = 1
20 = 12	2 = 1
30 = 17	3 = 2
40 = 23	4 = 2
50 = 29	5 = 3
60 = 35	6 = 3
	7 = 4
	8 = 5
	9 = 5

33 = 19

Anti Log Excesses

10 = 17	1 = 2
20 = 35	2 = 3
30 = 52	3 = 5
	4 = 7
	5 = 9
	6 = 10
	7 = 12
	8 = 14
	9 = 16

19 = 33

Number	Log
7500000	·8750613
067	651
134	690
200	728
267	767
334	806
400	844
467	883
534	921
600	960
667	999
734	·8751037
800	076
867	114
934	153
7501000	192
067	230
134	269
200	307
267	346
334	384
400	423
467	462
534	500
600	539
667	577
734	616
800	655
867	693
934	732
7502000	770
067	809
134	847
200	886
267	925
7502333	·8751963
400	·8752002
467	040
533	079
600	117
667	156
733	195
800	233
867	272
933	310
7503000	349
067	387
133	426
200	465
266	503
333	542
400	581
466	619
533	658
600	696
666	735
733	773
800	812
866	850
933	889
999	928
7504066	966
133	·8753005
199	043
266	082
333	120
399	159
466	198
532	236
599	275
666	313
732	352
799	390
866	429
932	467
999	506
7505065	545
132	583
199	622
265	660
332	699
399	737
465	776
532	814
598	853
665	891
732	930
798	969
865	·8754007
931	046
998	084
7506065	123
131	161
198	200
265	238
331	277
398	315
464	354
531	392
598	431
7506664	·8754470
731	508
797	547
864	585
931	624
997	662
7507064	701
130	739
197	778
264	816
330	855
397	893
463	932
530	970
597	·8755009
663	047
730	086
796	124
863	163
930	202
996	240
7508063	279
129	317
196	356
263	394
329	433
396	471
462	510
529	548
596	587
662	625
729	664
795	702
862	741
929	779
995	818
7509062	856
128	895
195	933
261	972
328	·8756010
395	049
461	087
528	126
594	164
661	203
728	241
794	280
861	318
927	357
994	395

Excesses

10 = 6	1 = 1
20 = 12	2 = 1
30 = 17	3 = 2
40 = 23	4 = 2
50 = 29	5 = 3
60 = 35	6 = 3
	7 = 4
	8 = 5
	9 = 5

33 = 19

Anti Log Excesses

10 = 17	1 = 2
20 = 35	2 = 3
30 = 52	3 = 5
	4 = 7
	5 = 9
	6 = 10
	7 = 12
	8 = 14
	9 = 16

19 = 33

Number	Log
7510060	·8756434
127	472
194	511
260	549
327	588
393	626
460	665
526	703
593	742
660	780
726	819
793	857
859	896
926	934
992	973
7511059	·8757011
126	050
192	088
259	127
325	165
392	204
458	242
525	281
592	319
658	358
725	396
791	435
858	473
924	512
991	550
7512057	589
124	627
191	666
257	704
324	743
390	781
457	819
523	858
590	896
656	935
723	973
790	·8758012
856	050
923	089
989	127
7513056	166
122	204
189	243
255	281
322	320
389	358

Number	Log
7513455	·8758397
522	435
588	474
655	512
721	550
788	589
854	627
921	666
987	704
7514054	743
121	781
187	820
254	858
320	897
387	935
453	974
520	·8759012
586	050
653	089
719	127
786	166
852	204
919	243
986	281
7515052	320
119	358
185	397
252	435
318	473
385	512
451	550
518	589
584	627
651	666
717	704
784	743
850	781
917	819
983	858
7516050	896
117	935
183	973
250	·8760012
316	050
383	088
449	127
516	165
582	204
649	242
715	281
782	319
848	357
915	396
981	434
7517048	473
114	511
181	550
247	588
314	626
380	665
447	703
513	742
580	780
646	819
713	857
7517779	·8760895
846	934
912	972
979	·8761011
7518045	049
112	088
178	126
245	164
311	203
378	241
444	280
511	318
577	356
644	395
710	433
777	472
843	510
910	548
976	587
7519043	625
109	664
176	702
242	741
309	779
375	817
442	856
508	894
575	933
641	971
708	·8762009
774	048
841	086
907	125
974	163

Excesses

10 = 6	1 = 1
20 = 12	2 = 1
30 = 17	3 = 2
40 = 23	4 = 2
50 = 29	5 = 3
60 = 35	6 = 3
	7 = 4
	8 = 5
	9 = 5

33 = 19

Anti Log Excesses

10 = 17	1 = 2
20 = 35	2 = 3
30 = 52	3 = 5
	4 = 7
	5 = 9
	6 = 10
	7 = 12
	8 = 14
	9 = 16

19 = 33

Number	Log
7520040	·8762201
107	240
7520173	·8762278
240	317
306	355
373	393
439	432
506	470
572	508
639	547
705	585
772	624
838	662
905	700
971	739
7521038	777
104	816
171	854
237	892
304	931
370	969
437	·8763008
503	046
569	084
636	123
702	161
769	199
835	238
902	276
968	315
7522035	353
101	391
168	430
234	468
301	506
367	545
434	583
500	622
567	660
633	698
699	737
766	775
832	813
899	852
965	890
7523032	928
098	967
165	·8764005
231	044
298	082
364	120
431	159
497	197
563	235
630	274
696	312
763	350
829	389
896	427
962	466
7524029	504
095	542
162	581
228	619
294	657
361	696
427	734
7524494	·8764772
560	811
627	849
693	888
760	926
826	964
893	·8765003
959	041
7525025	079
092	118
158	156
225	194
291	233
358	271
424	309
491	348
557	386
623	424
690	463
756	501
823	539
889	578
956	616
7526022	654
088	693
155	731
221	769
288	808
354	846
421	884
487	923
553	961
620	999
686	·8766038
753	076
819	114
886	153
952	191
7527019	229
085	268
151	306
218	344
284	383
351	421
417	459
483	498
550	536
616	574
683	613
749	651
816	689
882	728
948	766
7528015	804
081	843
148	881
214	919
281	958
347	996
413	·8767034
480	072
546	111
613	149
679	187
745	226
7528812	·8767264
878	302
945	341
7529011	379
077	417
144	456
210	494
277	532
343	570
410	609
476	647
542	685
609	724
675	762
742	800
808	839
874	877
941	915

Excesses

10 = 6	1 = 1
20 = 12	2 = 1
30 = 17	3 = 2
40 = 23	4 = 2
50 = 29	5 = 3
60 = 35	6 = 3
	7 = 4
	8 = 5
	9 = 5

33 = 19

Anti Log Excesses

10 = 17	1 = 2
20 = 35	2 = 3
30 = 52	3 = 5
	4 = 7
	5 = 9
	6 = 10
	7 = 12
	8 = 14
	9 = 16

19 = 33

Number	Log
7530007	·8767953
074	992
140	·8768030
206	068
273	107
339	145
406	183
472	222
538	260
605	298
671	336
738	375
804	413
870	451
937	490
7531003	528
069	566
136	605

7531202	·8768643
269	681
335	719
401	758
468	796
534	834
601	873
667	911
733	949
800	987
866	·8769026
933	064
999	102
7532065	140
132	179
198	217
264	255
331	294
397	332
464	370
530	408
596	447
663	485
729	523
795	562
862	600
928	638
995	676
7533061	715
127	753
194	791
260	829
326	868
393	906
459	944
526	982
592	·8770021
658	059
725	097
791	136
857	174
924	212
990	250
7534057	289
123	327
189	365
256	403
322	442
388	480
455	518
521	556
587	595
654	633
720	671
787	709
853	748
919	786
986	824
7535052	862
118	901
185	939
251	977
317	·8771015
384	054
450	092

7535516	·8771130
583	168
649	207
715	245
782	283
848	321
915	360
981	398
7536047	436
114	474
180	512
246	551
313	589
379	627
445	665
512	704
578	742
644	780
711	818
777	857
843	895
910	933
976	971
7537042	·8772009
109	048
175	086
241	124
308	162
374	200
440	239
507	277
573	315
639	353
706	392
772	430
838	468
905	506
971	544
7538037	583
104	621
170	659
236	697
303	735
369	774
435	812
502	850
568	888
634	927
701	965
767	·8773003
833	041
900	079
966	118
7539032	156
099	194
165	232
231	270
298	309
364	347
430	385
497	423
563	461
629	500
695	538
762	576

7539828	·8773614
894	652
961	691

Excesses

10 = 6	1 = 1
20 = 12	2 = 1
30 = 17	3 = 2
40 = 23	4 = 2
50 = 29	5 = 3
60 = 35	6 = 3
	7 = 4
	8 = 5
	9 = 5

33 = 19

Anti Log Excesses

10 = 17	1 = 2
20 = 35	2 = 3
30 = 52	3 = 5
	4 = 7
	5 = 9
	6 = 10
	7 = 12
	8 = 14
	9 = 16

19 = 33

7540027	·8773729
093	767
160	805
226	843
292	882
359	920
425	958
491	996
558	·8774034
624	073
690	111
756	149
823	187
889	225
955	263
7541022	302
088	340
154	378
221	416
287	454
353	492
419	531
486	569
552	607
618	645
685	683
751	722
817	760
884	798
950	836
7542016	874
082	912
149	951

7542215	·8774989
281	·8775027
348	065
414	103
480	141
546	180
613	218
679	256
745	294
812	332
878	370
944	409
7543011	447
077	485
143	523
209	561
276	599
342	638
408	676
475	714
541	752
607	790
673	828
740	867
806	905
872	943
938	981
7544005	·8776019
071	057
137	095
204	134
270	172
336	210
402	248
469	286
535	324
601	362
667	401
734	439
800	477
866	515
933	553
999	591
7545065	629
131	668
198	706
264	744
330	782
396	820
463	858
529	897
595	935
662	973
728	·8777011
794	049
860	087
927	125
993	164
7546059	202
125	240
192	278
258	316
324	354
390	392
457	430

7546523	·8777469
589	507
655	545
722	583
788	621
854	659
920	697
987	735
7547053	774
119	812
185	850
252	888
318	926
384	964
450	·8778002
517	040
583	079
649	117
715	155
782	193
848	231
914	269
980	307
7548047	345
113	384
179	422
245	460
312	498
378	536
444	574
510	612
577	650
643	688
709	727
775	765
841	803
908	841
974	879
7549040	917
106	955
173	993
239	·8779031
305	069
371	108
438	146
504	184
570	222
636	260
702	298
769	336
835	374
901	412
967	450

Excesses

10 = 6	1 = 1
20 = 12	2 = 1
30 = 17	3 = 2
40 = 23	4 = 2
50 = 29	5 = 3
60 = 35	6 = 3
	7 = 4
	8 = 5
	9 = 5
33 = 19	

Anti Log Excesses

10 = 17	1 = 2
20 = 35	2 = 3
30 = 52	3 = 5
	4 = 7
	5 = 9
	6 = 10
	7 = 12
	8 = 14
	9 = 16
19 = 33	

7550034	·8779489
100	527
166	565
232	603
299	641
365	679
431	717
497	755
563	793
630	831
696	869
762	908
828	946
894	984
961	·8780022
7551027	060
093	098
159	136
226	174
292	212
358	250
424	288
490	326
557	365
623	403
689	441
755	479
821	517
888	555
954	593
7552020	631
086	669
153	707
219	745
285	783
351	821
417	859

7552484	·8780898
550	936
616	974
682	·8781012
748	050
815	088
881	126
947	164
7553013	202
079	240
146	278
212	316
278	354
344	392
410	430
477	468
543	507
609	545
675	583
741	621
807	659
874	697
940	735
7554006	773
072	811
138	849
205	887
271	925
337	963
403	·8782001
469	039
536	077
602	115
668	153
734	191
800	230
866	268
933	306
999	344
7555065	382
131	420
197	458
264	496
330	534
396	572
462	610
528	648
594	686
661	724
727	762
793	800
859	838
925	876
992	914
7556058	952
124	990
190	·8783028
256	066
322	104
389	142
455	180
521	218
587	256
653	294
719	332

7556786	·8783371
852	409
918	447
984	485
7557050	523
116	561
183	599
249	637
315	675
381	713
447	751
513	789
579	827
646	865
712	903
778	941
844	979
910	·8784017
976	055
7558043	093
109	131
175	169
241	207
307	245
373	283
439	321
506	359
572	397
638	435
704	473
770	511
836	549
903	587
969	625
7559035	663
101	701
167	739
233	777
299	815
366	853
432	891
498	929
564	967
630	·8785005
696	043
762	081
829	119
895	157
961	195

Excesses

10 = 6	1 = 1
20 = 11	2 = 1
30 = 17	3 = 2
40 = 23	4 = 2
50 = 29	5 = 3
60 = 34	6 = 3
	7 = 4
	8 = 5
	9 = 5
33 = 19	

Anti Log Excesses

10 = 17	1 = 2
20 = 35	2 = 3
30 = 52	3 = 5
	4 = 7
	5 = 9
	6 = 10
	7 = 12
	8 = 14
	9 = 16
19 = 33	

7560027	·8785233
093	271
159	309
225	347
291	385
358	423
424	461
490	499
556	537
622	575
688	613
754	651
821	689
887	727
953	765
7561019	803
085	841
151	879
217	917
283	955
350	993
416	·8786031
482	069
548	107
614	145
680	183
746	221
812	259
878	297
945	335
7562011	373
077	410
143	448
209	486
275	524
341	562
408	600
474	638
540	676
606	714
672	752
738	790
804	828
870	866
936	904
7563003	942
069	980
135	·8787018
201	056
267	094
333	132

7563399	·8787170
465	208
531	246
598	284
664	322
730	360
796	398
862	436
928	474
994	512
7564060	549
126	587
192	625
259	663
325	701
391	739
457	777
523	815
589	853
655	891
721	929
787	967
853	·8788005
920	043
986	081
7565052	119
118	157
184	195
250	233
316	270
382	308
448	346
514	384
580	422
647	460
713	498
779	536
845	574
911	612
977	650
7566043	688
109	726
175	764
241	802
307	840
373	877
440	915
506	953
572	991
638	·8789029
704	067
770	105
836	143
902	181
968	219
7567034	257
100	295
166	333
232	371
299	408
365	446
431	484
497	522
563	560
629	598

7567695	·8789636
761	674
827	712
893	750
959	788
7568025	826
091	863
157	901
224	939
290	977
356	·8790015
422	053
488	091
554	129
620	167
686	205
752	243
818	280
884	318
950	356
7569016	394
082	432
148	470
214	508
281	546
347	584
413	622
479	660
545	697
611	735
677	773
743	811
809	849
875	887
941	925

Excesses

10 = 6	1 = 1
20 = 11	2 = 1
30 = 17	3 = 2
40 = 23	4 = 2
50 = 29	5 = 3
60 = 34	6 = 3
	7 = 4
	8 = 5
	9 = 5

33 = 19

Anti Log Excesses

10 = 17	1 = 2
20 = 35	2 = 3
30 = 52	3 = 5
	4 = 7
	5 = 9
	6 = 10
	7 = 12
	8 = 14
	9 = 16

19 = 33

7570007	·8790963
7570073	·8791000
139	038
205	076
271	114
337	152
403	190
469	228
535	266
602	304
668	341
734	379
800	417
866	455
932	493
998	531
7571064	569
130	607
196	645
262	682
328	720
394	758
460	796
526	834
592	872
658	910
724	948
790	985
856	·8792023
922	061
988	099
7572054	137
120	175
186	213
252	251
319	288
385	326
451	364
517	402
583	440
649	478
715	516
781	553
847	591
913	629
979	667
7573045	705
111	743
177	781
243	818
309	856
375	894
441	932
507	970
573	·8793008
639	046
705	083
771	121
837	159
903	197
969	235
7574035	273
101	311
167	348
233	386
299	424

7574365	·8793462
431	500
497	538
563	576
629	613
695	651
761	689
827	727
893	765
959	803
7575025	840
091	878
157	916
223	954
289	992
355	·8794030
421	068
487	105
553	143
619	181
685	219
751	257
817	295
883	332
949	370
7576015	408
081	446
147	484
213	522
279	559
345	597
411	635
477	673
543	711
609	749
675	786
741	824
807	862
873	900
939	938
7577005	976
071	·8795013
137	051
203	089
269	127
335	165
401	202
467	240
533	278
599	316
665	354
731	392
797	429
863	467
929	505
995	543
7578061	581
127	618
193	656
259	694
325	732
391	770
457	807
523	845
589	883

7578655	·8795921
721	959
787	996
853	·8796034
919	072
985	110
7579051	148
117	185
183	223
248	261
314	299
380	337
446	374
512	412
578	450
644	488
710	526
776	563
842	601
908	639
974	677

Excesses

10 = 6	1 = 1
20 = 11	2 = 1
30 = 17	3 = 2
40 = 23	4 = 2
50 = 29	5 = 3
60 = 34	6 = 3
	7 = 4
	8 = 5
	9 = 5

33 = 19

Anti Log Excesses

10 = 17	1 = 2
20 = 35	2 = 3
30 = 52	3 = 5
	4 = 7
	5 = 9
	6 = 10
	7 = 12
	8 = 14
	9 = 16

19 = 33

7580040	·8796715
106	752
172	790
238	828
304	866
370	904
436	941
502	979
568	·8797017
634	055
700	093
766	130
832	168
898	206
963	244

7581029	·8797282
095	319
161	357
227	395
293	433
359	470
425	508
491	546
557	584
623	622
689	659
755	697
821	735
887	773
953	810
7582019	848
085	886
151	924
217	962
282	999
348	·8798037
414	075
480	113
546	150
612	188
678	226
744	264
810	301
876	339
942	377
7583008	415
074	453
140	490
206	528
272	566
337	604
403	641
469	679
535	717
601	755
667	792
733	830
799	868
865	906
931	943
997	981
7584063	·8799019
129	057
195	094
261	132
326	170
392	208
458	245
524	283
590	321
656	359
722	396
788	434
854	472
920	510
986	547
7585052	585
118	623
183	661
249	698

Number	Log
7585315	·8799736
381	774
447	812
513	849
579	887
645	925
711	962
777	·8800000
843	038
908	076
974	113
7586040	151
106	189
172	227
238	264
304	302
370	340
436	378
502	415
568	453
633	491
699	528
765	566
831	604
897	642
963	679
7587029	717
095	755
161	792
227	830
293	868
358	906
424	943
490	981
556	·8801019
622	056
688	094
754	132
820	170
886	207
951	245
7588017	283
083	321
149	358
215	396
281	434
347	471
413	509
479	547
544	584
610	622
676	660
742	698
808	735
874	773
940	811
7589006	848
072	886
137	924
203	962
269	999
335	·8802037
401	075
467	112
533	150

Number	Log
7589599	·8802188
665	225
730	263
796	301
862	338
928	376
994	414

Excesses

10 = 6	1 = 1
20 = 11	2 = 1
30 = 17	3 = 2
40 = 23	4 = 2
50 = 29	5 = 3
60 = 34	6 = 3
	7 = 4
	8 = 5
	9 = 5

33 = 19

Anti Log Excesses

10 = 17	1 = 2
20 = 35	2 = 3
30 = 52	3 = 5
	4 = 7
	5 = 9
	6 = 10
	7 = 12
	8 = 14
	9 = 16

19 = 33

Number	Log
7590060	·8802452
126	489
192	527
257	565
323	602
389	640
455	678
521	715
587	753
653	791
719	828
784	866
850	904
916	941
982	979
7591048	·8803017
114	055
180	092
245	130
311	168
377	205
443	243
509	281
575	318
641	356
707	394
772	431
838	469
904	507

Number	Log
7591970	·8803544
7592036	582
102	620
168	657
233	695
299	733
365	770
431	808
497	846
563	883
629	921
694	959
760	996
826	·8804034
892	072
958	109
7593024	147
089	185
155	222
221	260
287	298
353	335
419	373
485	411
550	448
616	486
682	524
748	561
814	599
880	637
945	674
7594011	712
077	750
143	787
209	825
275	863
340	900
406	938
472	975
538	·8805013
604	051
670	088
736	126
801	164
867	201
933	239
999	277
7595065	314
131	352
196	390
262	427
328	465
394	503
460	540
525	578
591	615
657	653
723	691
789	728
855	766
920	804
986	841
7596052	879
118	917
184	954

Number	Log
7596250	·8805992
315	·8806029
381	067
447	105
513	142
579	180
644	218
710	255
776	293
842	330
908	368
974	406
7597039	443
105	481
171	519
237	556
303	594
368	631
434	669
500	707
566	744
632	782
698	820
763	857
829	895
895	932
961	970
7598027	·8807008
092	045
158	083
224	120
290	158
356	196
421	233
487	271
553	309
619	346
685	384
750	421
816	459
882	497
948	534
7599014	572
079	609
145	647
211	685
277	722
343	760
408	797
474	835
540	873
606	910
672	948
737	985
803	·8808023
869	061
935	098

Excesses

10 = 6	1 = 1
20 = 11	2 = 1
30 = 17	3 = 2
40 = 23	4 = 2
50 = 29	5 = 3
60 = 34	6 = 3
	7 = 4
	8 = 5
	9 = 5

33 = 19

Anti Log Excesses

10 = 17	1 = 2
20 = 35	2 = 3
30 = 52	3 = 5
	4 = 7
	5 = 9
	6 = 10
	7 = 12
	8 = 14
	9 = 16

19 = 33

Number	Log
7600000	·8808136
066	173
132	211
198	249
264	286
329	324
395	361
461	399
527	437
593	474
658	512
724	549
790	587
856	625
921	662
987	700
7601053	737
119	775
185	813
250	850
316	888
382	925
448	963
514	·8809001
579	038
645	076
711	113
777	151
842	188
908	226
974	264
7602040	301
105	339
171	376
237	414
303	451
369	489

7602434	·8809527
500	564
566	602
632	639
697	677
763	714
829	752
895	790
960	827
7603026	865
092	902
158	940
224	977
289	·8810015
355	053
421	090
487	128
552	165
618	203
684	240
750	278
815	315
881	353
947	391
7604013	428
078	466
144	503
210	541
276	578
341	616
407	653
473	691
539	729
604	766
670	804
736	841
802	879
867	916
933	954
999	991
7605065	·8811029
130	066
196	104
262	142
328	179
393	217
459	254
525	292
591	329
656	367
722	404
788	442
854	479
919	517
985	554
7606051	592
116	630
182	667
248	705
314	742
379	780
445	817
511	855
577	892
642	930

7606708	·8811967
774	·8812005
840	042
905	080
971	117
7607037	155
102	193
168	230
234	268
300	305
365	343
431	380
497	418
563	455
628	493
694	530
760	568
825	605
891	643
957	680
7608023	718
088	755
154	793
220	830
285	868
351	905
417	943
483	980
548	·8813018
614	055
680	093
745	130
811	168
877	205
943	243
7609008	281
074	318
140	356
205	393
271	431
337	468
403	505
468	543
534	580
600	618
665	655
731	693
797	730
863	768
928	805
994	843

Excesses

10 = 6	1 = 1
20 = 11	2 = 1
30 = 17	3 = 2
40 = 23	4 = 2
50 = 29	5 = 3
60 = 34	6 = 3
	7 = 4
	8 = 5
	9 = 5

33 = 19

Anti Log Excesses

10 = 18	1 = 2
20 = 35	2 = 4
30 = 53	3 = 5
	4 = 7
	5 = 9
	6 = 11
	7 = 12
	8 = 14
	9 = 16

19 = 33

7610060	·8813880
125	918
191	955
257	993
322	·8814030
388	068
454	105
520	143
585	180
651	218
717	255
782	293
848	330
914	368
979	405
7611045	443
111	480
176	518
242	555
308	593
374	630
439	668
505	705
571	743
636	780
702	817
768	855
833	892
899	930
965	967
7612030	·8815005
096	042
162	080
228	117
293	155
359	192
425	230
490	267
556	305
622	342
687	380
753	417
819	455
884	492
950	529
7613016	567
081	604
147	642
213	679
278	717
344	754

7613410	·8815792
475	829
541	867
607	904
672	942
738	979
804	·8816017
869	054
935	091
7614001	129
066	166
132	204
198	241
263	279
329	316
395	354
460	391
526	429
592	466
657	503
723	541
789	578
854	616
920	653
986	691
7615051	728
117	766
183	803
248	840
314	878
380	915
445	953
511	990
577	·8817028
642	065
708	103
774	140
839	177
905	215
971	252
7616036	290
102	327
168	365
233	402
299	439
364	477
430	514
496	552
561	589
627	627
693	664
758	702
824	739
890	776
955	814
7617021	851
087	889
152	926
218	964
283	·8818001
349	038
415	076
480	113
546	151
612	188

7617677	·8818226
743	263
809	300
874	338
940	375
7618005	413
071	450
137	488
202	525
268	563
334	600
399	637
465	675
531	712
596	750
662	787
727	824
793	862
859	899
924	937
990	974
7619056	·8819011
121	049
187	086
252	124
318	161
384	198
449	236
515	273
581	311
646	348
712	385
777	423
843	460
909	498
974	535

Excesses

10 = 6	1 = 1
20 = 11	2 = 1
30 = 17	3 = 2
40 = 23	4 = 2
50 = 29	5 = 3
60 = 34	6 = 3
	7 = 4
	8 = 5
	9 = 5

33 = 19

Anti Log Excesses

10 = 18	1 = 2
20 = 35	2 = 4
30 = 53	3 = 5
	4 = 7
	5 = 9
	6 = 11
	7 = 12
	8 = 14
	9 = 16

19 = 33

7620040	·8819572
105	610
171	647
237	685
302	722
368	759
434	797
499	834
565	872
630	909
696	946
762	984
827	·8820021
893	059
958	096
7621024	133
090	171
155	208
221	245
286	283
352	320
418	358
483	395
549	432
615	470
680	507
746	544
811	582
877	619
943	657
7622008	694
074	731
139	769
205	806
271	844
336	881
402	918
467	956
533	993
598	·8821031
664	068
730	105
795	143
861	180
926	217
992	255
7623058	292
123	329
189	367
254	404
320	442
386	479
451	516
517	554
582	591
648	628
714	666
779	703
845	740
910	778
976	815
7624041	853
107	890
173	927
238	965

7624304	·8822002
369	039
435	077
500	114
566	151
632	189
697	226
763	263
828	301
894	338
960	375
7625025	413
091	450
156	488
222	525
287	562
353	600
419	637
484	674
550	712
615	749
681	786
746	824
812	861
877	898
943	936
7626009	973
074	·8823010
140	048
205	085
271	122
336	160
402	197
468	234
533	272
599	309
664	346
730	384
795	421
861	458
926	496
992	533
7627058	570
123	608
189	645
254	682
320	720
385	757
451	794
516	832
582	869
648	906
713	944
779	981
844	·8824018
910	056
975	093
7628041	130
106	167
172	205
238	242
303	279
369	317
434	354
500	391

7628565	·8824429
631	466
696	503
762	541
827	578
893	615
958	653
7629024	690
090	727
155	765
221	802
286	839
352	876
417	914
483	951
548	988
614	·8825026
679	063
745	100
810	138
876	175
942	212

Excesses

10 = 6	1 = 1
20 = 11	2 = 1
30 = 17	3 = 2
40 = 23	4 = 2
50 = 28	5 = 3
60 = 34	6 = 3
	7 = 4
	8 = 5
	9 = 5

33 = 19

Anti Log Excesses

10 = 18	1 = 2
20 = 35	2 = 4
30 = 53	3 = 5
	4 = 7
	5 = 9
	6 = 11
	7 = 12
	8 = 14
	9 = 16

19 = 33

7630007	·8825249
073	287
138	324
204	361
269	399
335	436
400	473
466	511
531	548
597	585
662	622
728	660
793	697
859	734

7630924	·8825771
990	809
7631055	846
121	883
187	921
252	958
318	995
383	·8826033
449	070
514	107
580	144
645	182
711	219
776	256
842	293
907	331
973	368
7632038	405
104	443
169	480
235	517
300	554
366	592
431	629
497	666
562	704
628	741
693	778
759	815
824	853
890	890
955	927
7633021	964
086	·8827002
152	039
217	076
283	113
348	151
414	188
479	225
545	263
610	300
676	337
741	374
807	412
872	449
938	486
7634003	523
069	561
134	598
200	635
265	672
331	710
396	747
462	784
527	821
593	859
658	896
724	933
789	970
855	·8828008
920	045
986	082
7635051	119
117	157

7635182	·8828194
248	231
313	268
379	306
444	343
510	380
575	417
641	455
706	492
772	529
837	566
903	604
968	641
7636034	678
099	715
164	753
230	790
295	827
361	864
426	901
492	939
557	976
623	·8829013
688	050
754	088
819	125
885	162
950	199
7637016	237
081	274
147	311
212	348
278	386
343	423
408	460
474	497
539	534
605	572
670	609
736	646
801	683
867	721
932	758
998	795
7638063	832
129	869
194	907
259	944
325	981
390	·8830018
456	056
521	093
587	130
652	167
718	204
783	242
849	279
914	316
979	353
7639045	390
110	428
176	465
241	502
307	539
372	576

7639438	·8830614
503	651
569	688
634	725
699	762
765	800
830	837
896	874
961	911

Excesses

10 = 6	1 = 1
20 = 11	2 = 1
30 = 17	3 = 2
40 = 23	4 = 2
50 = 28	5 = 3
60 = 34	6 = 3
	7 = 4
	8 = 5
	9 = 5

33 = 19

Anti Log Excesses

10 = 18	1 = 2
20 = 35	2 = 4
30 = 53	3 = 5
	4 = 7
	5 = 9
	6 = 11
	7 = 12
	8 = 14
	9 = 16

19 = 33

7640027	·8830948
092	986
158	·8831023
223	060
288	097
354	134
419	172
485	209
550	246
616	283
681	320
747	358
812	395
877	432
943	469
7641008	506
074	544
139	581
205	618
270	655
335	692
401	730
466	767
532	804
597	841
663	878
728	915

7641793	·8831953
859	990
924	·8832027
990	064
7642055	101
121	139
186	176
251	213
317	250
382	287
448	324
513	362
579	399
644	436
709	473
775	510
840	547
906	585
971	622
7643037	659
102	696
167	733
233	770
298	808
364	845
429	882
494	919
560	956
625	993
691	·8833031
756	068
822	105
887	142
952	179
7644018	217
083	254
149	291
214	328
279	365
345	402
410	440
476	477
541	514
606	551
672	588
737	625
803	662
868	700
933	737
999	774
7645064	811
130	848
195	885
260	923
326	960
391	997
457	·8834034
522	071
587	108
653	145
718	183
784	220
849	257
914	294
980	331

7646045	·8834368
111	405
176	443
241	480
307	517
372	554
438	591
503	628
568	665
634	703
699	740
765	777
830	814
895	851
961	888
7647026	925
091	962
157	·8835000
222	037
288	074
353	111
418	148
484	185
549	222
615	260
680	297
745	334
811	371
876	408
941	445
7648007	482
072	519
138	557
203	594
268	631
334	668
399	705
464	742
530	779
595	816
661	854
726	891
791	928
857	965
922	·8836002
987	039
7649053	076
118	113
183	150
249	188
314	225
380	262
445	299
510	336
576	373
641	410
706	447
772	484
837	522
902	559
968	596

Excesses

10 = 6	1 = 1
20 = 11	2 = 1
30 = 17	3 = 2
40 = 23	4 = 2
50 = 28	5 = 3
60 = 34	6 = 3
	7 = 4
	8 = 5
	9 = 5

33 = 19

Anti Log Excesses

10 = 18	1 = 2
20 = 35	2 = 4
30 = 53	3 = 5
	4 = 7
	5 = 9
	6 = 11
	7 = 12
	8 = 14
	9 = 16

19 = 33

7650033	·8836633
099	670
164	707
229	744
295	781
360	818
425	856
491	893
556	930
621	967
687	·8837004
752	041
817	078
883	115
948	152
7651014	189
079	226
144	264
210	301
275	338
340	375
406	412
471	449
536	486
602	523
667	560
732	597
798	634
863	672
928	709
994	746
7652059	783
124	820
190	857
255	894
320	931
386	968

7652451	·8838005
516	042
582	079
647	117
712	154
778	191
843	228
908	265
974	302
7653039	339
104	376
170	413
235	450
300	487
366	524
431	561
496	599
562	636
627	673
692	710
758	747
823	784
888	821
954	858
7654019	895
084	932
150	969
215	·8839006
280	043
346	080
411	117
476	154
542	192
607	229
672	266
738	303
803	340
868	377
934	414
999	451
7655064	488
129	525
195	562
260	599
325	636
391	673
456	710
521	747
587	784
652	821
717	858
783	896
848	933
913	970
979	·8840007
7656044	044
109	081
174	118
240	155
305	192
370	229
436	266
501	303
566	340
632	377

7656697	·8840414
762	451
828	488
893	525
958	562
7657023	599
089	636
154	673
219	711
285	748
350	785
415	822
481	859
546	896
611	933
676	970
742	·8841007
807	044
872	081
938	118
7658003	155
068	192
133	229
199	266
264	303
329	340
395	377
460	414
525	451
590	488
656	525
721	562
786	599
852	636
917	673
982	710
7659047	747
113	784
178	821
243	858
309	895
374	932
439	969
504	·8842006
570	043
635	080
700	117
766	154
831	191
896	228
961	265

Excesses

10 = 6	1 = 1
20 = 11	2 = 1
30 = 17	3 = 2
40 = 23	4 = 2
50 = 28	5 = 3
60 = 34	6 = 3
	7 = 4
	8 = 5
	9 = 5

33 = 19

Anti Log Excesses

10 = 18	1 = 2
20 = 35	2 = 4
30 = 53	3 = 5
	4 = 7
	5 = 9
	6 = 11
	7 = 12
	8 = 14
	9 = 16

19 = 33

7660027	·8842302
092	339
157	376
222	413
288	450
353	487
418	524
484	561
549	598
614	635
679	672
745	709
810	746
875	783
940	820
7661006	857
071	894
136	931
201	968
267	·8843005
332	042
397	079
463	116
528	153
593	190
658	227
724	264
789	301
854	338
919	375
985	412
7662050	449
115	486
180	523
246	560
311	597
376	634
441	671
507	708
572	745
637	782
702	819
768	856
833	893
898	930
963	967
7663029	·8844004
094	041
159	078
224	115
290	152

7663355	·8844189
420	226
485	263
551	300
616	337
681	374
746	411
812	448
877	485
942	522
7664007	559
073	596
138	633
203	670
268	707
333	744
399	781
464	817
529	854
594	891
660	928
725	965
790	·8845002
855	039
921	076
986	113
7665051	150
116	187
182	224
247	261
312	298
377	335
442	372
508	409
573	446
638	483
703	520
769	557
834	594
899	631
964	668
7666029	704
095	741
160	778
225	815
290	852
356	889
421	926
486	963
551	·8846000
616	037
682	074
747	111
812	148
877	185
943	222
7667008	259
073	296
138	333
203	369
269	406
334	443
399	480
464	517
529	554

7667595	·8846591
660	628
725	665
790	702
855	739
921	776
986	813
7668051	850
116	887
182	924
247	960
312	997
377	·8847034
442	071
508	108
573	145
638	182
703	219
768	256
834	293
899	330
964	367
7669029	404
094	440
160	477
225	514
290	551
355	588
420	625
485	662
551	699
616	736
681	773
746	810
811	847
877	884
942	921

Excesses

10 = 6	1 = 1
20 = 11	2 = 1
30 = 17	3 = 2
40 = 23	4 = 2
50 = 28	5 = 3
60 = 34	6 = 3
	7 = 4
	8 = 5
	9 = 5

33 = 18

Anti Log Excesses

10 = 18	1 = 2
20 = 35	2 = 4
30 = 53	3 = 5
	4 = 7
	5 = 9
	6 = 11
	7 = 12
	8 = 14
	9 = 16

18 = 33

7670007	·8847957
072	994
137	·8848031
203	068
268	105
333	142
398	179
463	216
529	253
594	290
659	327
724	363
789	400
854	437
920	474
985	511
7671050	548
115	585
180	622
246	659
311	696
376	732
441	769
506	806
571	843
637	880
702	917
767	954
832	991
897	·8849028
962	065
7672028	101
093	138
158	175
223	212
288	249
353	286
419	323
484	360
549	396
614	433
679	470
744	507
810	544
875	581
940	618
7673005	655
070	692
135	728
201	765
266	802
331	839
396	876
461	913
526	950
592	987
657	·8850023
722	060
787	097
852	134
917	171
983	208
7674048	245
113	282
178	319

7674243	·8850355
308	392
373	429
439	466
504	503
569	540
634	577
699	613
764	650
829	687
895	724
960	761
7675025	798
090	835
155	872
220	908
286	945
351	982
416	·8851019
481	056
546	093
611	130
676	166
742	203
807	240
872	277
937	314
7676002	351
067	388
132	424
197	461
263	498
328	535
393	572
458	609
523	646
588	682
653	719
719	756
784	793
849	830
914	867
979	903
7677044	940
109	977
174	·8852014
240	051
305	088
370	125
435	161
500	198
565	235
630	272
695	309
761	346
826	382
891	419
956	456
7678021	493
086	530
151	566
216	603
282	640
347	677
412	714

7678477	·8852751
542	787
607	824
672	861
737	898
802	935
868	972
933	·8853008
998	045
7679063	082
128	119
193	156
258	193
323	229
388	266
454	303
519	340
584	377
649	413
714	450
779	487
844	524
909	561
974	598

Excesses

10 = 6	1 = 1
20 = 11	2 = 1
30 = 17	3 = 2
40 = 23	4 = 2
50 = 28	5 = 3
60 = 34	6 = 3
	7 = 4
	8 = 5
	9 = 5

33 = 18

Anti Log Excesses

10 = 18	1 = 2
20 = 35	2 = 4
30 = 53	3 = 5
	4 = 7
	5 = 9
	6 = 11
	7 = 12
	8 = 14
	9 = 16

18 = 33

7680040	·8853634
105	671
170	708
235	745
300	782
365	818
430	855
495	892
560	929
625	966
691	·8854003
756	039

7680821	8854076
886	113
951	150
7681016	187
081	223
146	260
211	297
276	334
342	371
407	407
472	444
537	481
602	518
667	555
732	591
797	628
862	665
927	702
992	739
7682058	775
123	812
188	849
253	886
318	923
383	959
448	996
513	·8855033
578	070
643	106
708	143
773	180
839	217
904	254
969	290
7683034	327
099	364
164	401
229	438
294	474
359	511
424	548
489	585
554	621
619	658
685	695
750	732
815	769
880	805
945	842
7684010	879
075	916
140	952
205	989
270	·8856026
335	063
400	100
465	136
530	173
595	210
661	247
726	283
791	320
856	357
921	394
986	430

7685051	·8856467
116	504
181	541
246	578
311	614
376	651
441	688
506	725
571	761
636	798
702	835
767	872
832	908
897	945
962	982
7686027	·8857019
092	055
157	092
222	129
287	166
352	202
417	239
482	276
547	313
612	349
677	386
742	423
807	460
872	497
937	533
7687003	570
068	607
133	644
198	680
263	717
328	754
393	791
458	827
523	864
588	901
653	937
718	974
783	·8858011
848	048
913	084
978	121
7688043	158
108	195
173	231
238	268
303	305
368	342
433	378
498	415
563	452
628	489
693	525
759	562
824	599
889	635
954	672
7689019	709
084	746
149	782
214	819

7689279	·8858856
344	893
409	929
474	966
539	·8859003
604	039
669	076
734	113
799	150
864	186
929	223
994	260

Excesses

10 = 6	1 = 1
20 = 11	2 = 1
30 = 17	3 = 2
40 = 23	4 = 2
50 = 28	5 = 3
60 = 34	6 = 3
	7 = 4
	8 = 5
	9 = 5

33 = 18

Anti Log Excesses

10 = 18	1 = 2
20 = 35	2 = 4
30 = 53	3 = 5
	4 = 7
	5 = 9
	6 = 11
	7 = 12
	8 = 14
	9 = 16

18 = 33

7690059	·8859297
124	333
189	370
254	407
319	443
384	480
449	517
514	554
579	590
644	627
709	664
774	700
839	737
904	774
969	811
7691034	847
099	884
164	921
229	957
294	994
359	·8860031
424	067
489	104
554	141

7691619	·8860178
684	214
749	251
814	288
879	324
944	361
7692009	398
074	435
139	471
204	508
269	545
334	581
399	618
464	655
529	691
594	728
659	765
724	801
789	838
854	875
919	912
984	948
7693049	985
114	·8861022
179	058
244	095
309	132
374	168
439	205
504	242
569	278
634	315
699	352
764	389
829	425
894	462
959	499
7694024	535
089	572
154	609
219	645
284	682
349	719
414	755
479	792
544	829
609	865
674	902
739	939
804	975
869	·8862012
934	049
999	085
7695064	122
129	159
194	195
259	232
324	269
389	305
454	342
519	379
584	415
649	452
714	489
779	525

7695844	·8862562
909	599
973	635
7696038	672
103	709
168	745
233	782
298	819
363	855
428	892
493	929
558	965
623	·8863002
688	039
753	075
818	112
883	149
948	185
7697013	222
078	259
143	295
208	332
273	369
338	405
403	442
468	478
533	515
598	552
662	588
727	625
792	662
857	698
922	735
987	772
7698052	808
117	845
182	882
247	918
312	955
377	992
442	·8864028
507	065
572	101
637	138
702	175
767	211
832	248
897	285
961	321
7699026	358
091	395
156	431
221	468
286	504
351	541
416	578
481	614
546	651
611	688
676	724
741	761
806	797
871	834
936	871

Excesses

10 = 6	1 = 1
20 = 11	2 = 1
30 = 17	3 = 2
40 = 23	4 = 2
50 = 28	5 = 3
60 = 34	6 = 3
	7 = 4
	8 = 5
	9 = 5

32 = 18

Anti Log Excesses

10 = 18	1 = 2
20 = 35	2 = 4
30 = 53	3 = 5
	4 = 7
	5 = 9
	6 = 11
	7 = 12
	8 = 14
	9 = 16

18 = 32

7700000	·8864907
065	944
130	980
195	·8865017
260	054
325	090
390	127
455	164
520	200
585	237
650	273
715	310
780	347
845	383
910	420
974	457
7701039	493
104	530
169	566
234	603
299	640
364	676
429	713
494	750
559	786
624	823
689	859
754	896
818	933
883	969
948	·8866006
7702013	042
078	079
143	116
208	152
273	189
338	225

7702403	·8866262
468	299
533	335
597	372
662	408
727	445
792	482
857	518
922	555
987	591
7703052	628
117	665
182	701
247	738
311	774
376	811
441	848
506	884
571	921
636	957
701	994
766	·8867031
831	067
896	104
961	140
7704025	177
090	213
155	250
220	287
285	323
350	360
415	396
480	433
545	470
610	506
674	543
739	579
804	616
869	652
934	689
999	726
7705064	762
129	799
194	835
258	872
323	908
388	945
453	982
518	·8868018
583	055
648	091
713	128
778	164
842	201
907	238
972	274
7706037	311
102	347
167	384
232	420
297	457
362	494
426	530
491	567
556	603

7706621	·8868640
686	676
751	713
816	750
881	786
945	823
7707010	859
075	896
140	932
205	969
270	·8869005
335	042
400	079
464	115
529	152
594	188
659	225
724	261
789	298
854	334
919	371
983	408
7708048	444
113	481
178	517
243	554
308	590
373	627
437	663
502	700
567	736
632	773
697	810
762	846
827	883
891	919
956	956
7709021	992
086	·8870029
151	065
216	102
281	138
345	175
410	211
475	248
540	285
605	321
670	358
735	394
799	431
864	467
929	504
994	540

Excesses

10 = 6	1 = 1
20 = 11	2 = 1
30 = 17	3 = 2
40 = 23	4 = 2
50 = 28	5 = 3
60 = 34	6 = 3
	7 = 4
	8 = 5
	9 = 5

32 = 18

Anti Log Excesses

10 = 18	1 = 2
20 = 35	2 = 4
30 = 53	3 = 5
	4 = 7
	5 = 9
	6 = 11
	7 = 12
	8 = 14
	9 = 16

18 = 32

7710059	·8870577
124	613
189	650
253	686
318	723
383	759
448	796
513	832
578	869
642	906
707	942
772	979
837	·8871015
902	052
967	088
7711032	125
096	161
161	198
226	234
291	271
356	307
421	344
485	380
550	417
615	453
680	490
745	526
810	563
874	599
939	636
7712004	672
069	709
134	745
199	782
263	818
328	855
393	891

7712458	·8871928
523	964
588	·8872001
652	038
717	074
782	110
847	147
912	183
977	220
7713041	256
106	293
171	329
236	366
301	402
366	439
430	475
495	512
560	548
625	585
690	621
754	658
819	694
884	731
949	767
7714014	804
079	840
143	877
208	913
273	950
338	986
403	·8873023
467	059
532	096
597	132
662	169
727	205
791	242
856	278
921	315
986	351
7715051	388
116	424
180	461
245	497
310	534
375	570
440	607
504	643
569	680
634	716
699	752
764	789
828	825
893	862
958	898
7716023	935
088	971
152	·8874008
217	044
282	081
347	117
412	154
476	190
541	227
606	263

7716671	·8874300
736	336
800	373
865	409
930	445
995	482
7717060	518
124	555
189	591
254	628
319	664
383	701
448	737
513	773
578	810
643	846
707	883
772	919
837	956
902	992
967	·8875029
7718031	065
096	102
161	138
226	174
290	211
355	247
420	284
485	320
550	357
614	393
679	430
744	466
809	503
873	539
938	575
7719003	612
068	648
133	685
197	721
262	758
327	794
392	831
456	867
521	903
586	940
651	976
716	·8876013
780	049
845	086
910	122
975	159

Excesses

10 = 6	1 = 1
20 = 11	2 = 1
30 = 17	3 = 2
40 = 23	4 = 2
50 = 28	5 = 3
60 = 34	6 = 3
	7 = 4
	8 = 5
	9 = 5

32 = 18

Anti Log Excesses

10 = 18	1 = 2
20 = 36	2 = 4
30 = 53	3 = 5
	4 = 7
	5 = 9
	6 = 11
	7 = 12
	8 = 14
	9 = 16

18 = 32

7720039	·8876195
104	231
169	268
234	304
298	341
363	377
428	414
493	450
557	486
622	523
687	559
752	596
817	632
881	669
946	705
7721011	741
076	778
140	814
205	851
270	887
335	924
399	960
464	996
529	·8877033
594	069
658	106
723	142
788	178
853	215
917	251
982	288
7722047	324
112	360
176	397
241	433
306	470
371	506

7722435	·8877543
500	579
565	615
630	652
694	688
759	725
824	761
889	797
953	834
7723018	870
083	907
148	943
212	979
277	·8878016
342	052
406	089
471	125
536	161
601	198
665	234
730	271
795	307
860	343
924	380
989	416
7724054	453
119	489
183	525
248	562
313	598
377	635
442	671
507	707
572	744
636	780
701	817
766	853
831	889
895	926
960	962
7725025	998
089	·8879035
154	071
219	108
284	144
348	180
413	217
478	253
543	290
607	326
672	362
737	399
801	435
866	472
931	508
996	544
7726060	581
125	617
190	653
254	690
319	726
384	763
449	799
513	835
578	872

7726643	·8879908
707	944
772	981
837	·8880017
902	054
966	090
7727031	126
096	163
160	199
225	235
290	272
355	308
419	344
484	381
549	417
613	454
678	490
743	526
807	563
872	599
937	635
7728002	672
066	708
131	744
196	781
260	817
325	854
390	890
454	926
519	963
584	999
649	·8881035
713	072
778	108
843	144
907	181
972	217
7729037	253
101	290
166	326
231	362
295	399
360	435
425	471
489	508
554	544
619	581
684	617
748	653
813	690
878	726
942	762

Excesses

10 = 6	1 = 1
20 = 11	2 = 1
30 = 17	3 = 2
40 = 22	4 = 2
50 = 28	5 = 3
60 = 34	6 = 3
	7 = 4
	8 = 4
	9 = 5

32 = 18

Anti Log Excesses

10 = 18	1 = 2
20 = 36	2 = 4
30 = 53	3 = 5
	4 = 7
	5 = 9
	6 = 11
	7 = 12
	8 = 14
	9 = 16

18 = 32

7730007	·8881799
072	835
136	871
201	907
266	944
330	980
395	·8882017
460	053
524	089
589	126
654	162
718	198
783	235
848	271
912	307
977	344
7731042	380
107	416
171	453
236	489
301	525
365	562
430	598
495	634
559	670
624	707
689	743
753	779
818	816
883	852
947	888
7732012	925
077	961
141	997
206	·8883034
271	070
335	106

7732400	·8883143
465	179
529	215
594	252
659	288
723	324
788	361
852	397
917	433
982	470
7733046	506
111	542
176	578
240	615
305	651
370	687
434	724
499	760
564	796
628	833
693	869
758	905
822	942
887	978
952	·8884014
7734016	050
081	087
146	123
210	159
275	196
340	232
404	268
469	305
533	341
598	377
663	413
727	450
792	486
857	522
921	559
986	595
7735051	631
115	668
180	704
245	740
309	776
374	813
438	849
503	885
568	922
632	958
697	994
762	·8885030
826	067
891	103
956	139
7736020	176
085	212
149	248
214	284
279	321
343	357
408	393
473	430
537	466

7736602	·8885502
666	538
731	575
796	611
860	647
925	683
990	720
7737054	756
119	792
183	829
248	865
313	901
377	937
442	974
507	·8886010
571	046
636	082
700	119
765	155
830	191
894	228
959	264
7738024	300
088	336
153	373
217	409
282	445
347	481
411	518
476	554
540	590
605	627
670	663
734	699
799	736
863	772
928	808
993	844
7739057	881
122	917
187	953
251	989
316	·8887026
380	062
445	098
510	134
574	171
639	207
703	243
768	279
833	316
897	352
962	388

Excesses

10 = 6	1 = 1
20 = 11	2 = 1
30 = 17	3 = 2
40 = 22	4 = 2
50 = 28	5 = 3
60 = 34	6 = 3
	7 = 4
	8 = 4
	9 = 5

32 = 18

Anti Log Excesses

10 = 18	1 = 2
20 = 36	2 = 4
30 = 53	3 = 5
	4 = 7
	5 = 9
	6 = 11
	7 = 12
	8 = 14
	9 = 16

18 = 32

7740026	·8887424
091	461
156	497
220	533
285	569
349	606
414	642
479	678
543	714
608	751
672	787
737	823
801	859
866	896
931	932
995	968
7741060	·8888004
124	040
189	077
254	113
318	149
383	185
447	222
512	258
577	294
641	330
706	367
770	403
835	439
899	475
964	512
7742029	548
093	584
158	620
222	656
287	693
352	729

7742416	·8888765
481	801
545	838
610	874
674	910
739	947
804	983
868	·8889019
933	055
997	091
7743062	127
126	163
191	200
256	236
320	272
385	308
449	345
514	381
578	417
643	453
708	489
772	526
837	562
901	598
966	634
7744030	670
095	707
160	743
224	779
289	815
353	851
418	888
482	924
547	960
612	996
676	·8890032
741	069
805	105
870	141
934	177
999	213
7745063	250
128	286
193	322
257	358
322	394
386	431
451	467
515	503
580	539
644	575
709	612
774	648
838	684
903	720
967	756
7746032	793
096	829
161	865
225	901
290	937
354	973
419	·8891010
484	046
548	082

7746613	·8891118
677	154
742	191
806	227
871	263
935	299
7747000	335
064	371
129	408
194	444
258	480
323	516
387	552
452	589
516	625
581	661
645	697
710	733
774	769
839	806
903	842
968	878
7748033	914
097	950
162	986
226	·8892023
291	059
355	095
420	131
484	167
549	203
613	240
678	276
742	312
807	348
871	384
936	420
7749000	457
065	493
129	529
194	565
259	601
323	637
388	674
452	710
517	746
581	782
646	818
710	854
775	891
839	927
904	963
968	999

Excesses

10 = 6	1 = 1
20 = 11	2 = 1
30 = 17	3 = 2
40 = 22	4 = 2
50 = 28	5 = 3
60 = 34	6 = 3
	7 = 4
	8 = 4
	9 = 5

32 = 18

Anti Log Excesses

10 = 18	1 = 2
20 = 36	2 = 4
30 = 53	3 = 5
	4 = 7
	5 = 9
	6 = 11
	7 = 12
	8 = 14
	9 = 16

18 = 32

7750033	·8893035
097	071
162	107
226	144
291	180
355	216
420	252
484	288
549	324
613	360
678	397
742	433
807	469
871	505
936	541
7751000	577
065	613
129	650
194	686
258	722
323	758
387	794
452	830
516	866
581	903
645	939
710	975
774	·8894011
839	047
903	083
968	120
7752032	156
097	192
161	228
226	264
290	300
355	336

7752419	·8894372
484	409
548	445
613	481
677	517
742	553
806	589
871	625
935	661
7753000	698
064	734
129	770
193	806
258	842
322	878
387	914
451	950
516	987
580	·8895023
645	059
709	095
774	131
838	167
903	203
967	239
7754032	275
096	312
161	348
225	384
290	420
354	456
419	492
483	528
548	564
612	601
677	637
741	673
805	709
870	745
934	781
999	817
7755063	853
128	889
192	925
257	962
321	998
386	·8896034
450	070
515	106
579	142
644	178
708	214
773	250
837	286
901	323
966	359
7756030	395
095	431
159	467
224	503
288	539
353	575
417	611
482	648
546	684

7756611	·8896720
675	756
740	792
804	828
868	864
933	900
997	936
7757062	972
126	·8897008
191	045
255	081
320	117
384	153
449	189
513	225
577	261
642	297
706	333
771	369
835	405
900	441
964	478
7758029	514
093	550
157	586
222	622
286	658
351	694
415	730
480	766
544	802
609	838
673	874
738	910
802	946
866	983
931	·8898019
995	055
7759060	091
124	127
189	163
253	199
317	235
382	271
446	307
511	343
575	379
640	415
704	451
769	487
833	524
897	560
962	596

Excesses

10 = 6	1 = 1
20 = 11	2 = 1
30 = 17	3 = 2
40 = 22	4 = 2
50 = 28	5 = 3
60 = 34	6 = 3
	7 = 4
	8 = 4
	9 = 5

32 = 18

Anti Log Excesses

10 = 18	1 = 2
20 = 36	2 = 4
30 = 54	3 = 5
	4 = 7
	5 = 9
	6 = 11
	7 = 12
	8 = 14
	9 = 16

18 = 32

7760026	·8898632
091	668
155	704
220	740
284	776
348	812
413	848
477	884
542	920
606	956
671	992
735	·8899028
799	064
864	100
928	137
993	173
7761057	209
122	245
186	281
250	317
315	353
379	389
444	425
508	461
573	497
637	533
701	569
766	605
830	641
895	677
959	713
7762023	749
088	785
152	821
217	857
281	893
346	930

7762410	8899966
474	·8900002
539	038
603	074
668	110
732	146
796	182
861	218
925	254
990	290
7763054	326
118	362
183	398
247	434
312	470
376	506
440	542
505	578
569	614
634	650
698	686
762	722
827	758
891	794
956	830
7764020	866
084	902
149	938
213	974
278	·8901010
342	046
406	082
471	118
535	155
600	191
664	227
728	263
793	299
857	335
922	371
986	407
7765050	443
115	479
179	515
244	551
308	587
372	623
437	659
501	695
566	731
630	767
694	803
759	839
823	875
887	911
952	947
7766016	983
081	·8902019
145	055
209	091
274	127
338	163
402	199
467	235
531	271

7766596	·8902307
660	343
724	379
789	415
853	451
918	487
982	523
7767046	559
111	595
175	631
239	667
304	703
368	739
433	775
497	811
561	847
626	883
690	919
754	955
819	991
883	·8903027
947	063
7768012	099
076	135
141	171
205	207
269	242
334	278
398	314
462	350
527	386
591	422
655	458
720	494
784	530
849	566
913	602
977	638
7769042	674
106	710
170	746
235	782
299	818
363	854
428	890
492	926
556	962
621	998
685	·8904034
750	070
814	106
878	142
943	178

Excesses

10 = 6	1 = 1
20 = 11	2 = 1
30 = 17	3 = 2
40 = 22	4 = 2
50 = 28	5 = 3
60 = 34	6 = 3
	7 = 4
	8 = 4
	9 = 5

32 = 18

Anti Log Excesses

10 = 18	1 = 2
20 = 36	2 = 4
30 = 54	3 = 5
	4 = 7
	5 = 9
	6 = 11
	7 = 13
	8 = 14
	9 = 16

18 = 32

7770007	·8904214
071	250
136	286
200	322
264	358
329	394
393	430
457	466
522	502
586	537
650	573
715	609
779	645
843	681
908	717
972	753
7771036	789
101	825
165	861
229	897
294	933
358	969
423	·8905005
487	041
551	077
616	113
680	149
744	185
809	221
873	257
937	293
7772002	329
066	364
130	400
195	436
259	472
323	508

7772388	·8905544
452	580
516	616
581	652
645	688
709	724
773	760
838	796
902	832
966	868
7773031	904
095	940
159	976
224	·8906011
288	047
352	083
417	119
481	155
545	191
610	227
674	263
738	299
803	335
867	371
931	407
996	443
7774060	479
124	515
189	550
253	586
317	622
382	658
446	694
510	730
574	766
639	802
703	838
767	874
832	910
896	946
960	982
7775025	·8907018
089	053
153	089
218	125
282	161
346	197
410	233
475	269
539	305
603	341
668	377
732	413
796	449
861	484
925	520
989	556
7776053	592
118	628
182	664
246	700
311	736
375	772
439	808
504	844

7776568	·8907879
632	915
696	951
761	987
825	·8908023
889	059
954	095
7777018	131
082	167
147	203
211	239
275	274
339	310
404	346
468	382
532	418
597	454
661	490
725	526
789	562
854	598
918	633
982	669
7778047	705
111	741
175	777
239	813
304	849
368	885
432	920
496	956
561	992
625	·8909028
689	064
754	100
818	136
882	172
946	208
7779011	243
075	279
139	315
204	351
268	387
332	423
396	459
461	495
525	530
589	566
653	602
718	638
782	674
846	710
911	746
975	782

Excesses

10 = 6	1 = 1
20 = 11	2 = 1
30 = 17	3 = 2
40 = 22	4 = 2
50 = 28	5 = 3
60 = 33	6 = 3
	7 = 4
	8 = 4
	9 = 5

32 = 18

Anti Log Excesses

10 = 18	1 = 2
20 = 36	2 = 4
30 = 54	3 = 5
	4 = 7
	5 = 9
	6 = 11
	7 = 13
	8 = 14
	9 = 16

18 = 32

7780039	·8909817
103	853
168	889
232	925
296	961
360	997
425	·8910033
489	069
553	104
617	140
682	176
746	212
810	248
874	284
939	320
7781003	355
067	391
132	427
196	463
260	499
324	535
389	571
453	606
517	642
581	678
646	714
710	750
774	786
838	822
903	857
967	893
7782031	929
095	965
160	·8911001
224	037
288	073
352	108

7782417	·8911144
481	180
545	216
609	252
674	288
738	324
802	360
866	395
931	431
995	467
7783059	503
123	539
188	575
252	610
316	646
380	682
444	718
509	754
573	790
637	826
701	861
766	897
830	933
894	969
958	·8912005
7784023	041
087	076
151	112
215	148
280	184
344	220
408	256
472	291
536	327
601	363
665	399
729	435
793	471
858	506
922	542
986	578
7785050	614
115	650
179	686
243	721
307	757
371	793
436	829
500	865
564	901
628	936
693	972
757	·8913008
821	044
885	080
949	116
7786014	151
078	187
142	223
206	259
270	295
335	330
399	366
463	402
527	438

7786592	·8913474
656	510
720	545
784	581
848	617
913	653
977	688
7787041	724
105	760
169	796
234	832
298	868
362	903
426	939
490	975
555	·8914011
619	047
683	082
747	118
811	154
876	190
940	226
7788004	261
068	297
132	333
197	369
261	405
325	440
389	476
453	512
518	548
582	584
646	619
710	655
774	691
839	727
903	763
967	798
7789031	834
095	870
160	906
224	942
288	977
352	·8915013
416	049
481	085
545	121
609	156
673	192
737	228
802	264
866	299
930	335
994	371

Excesses

10 = 6	1 = 1
20 = 11	2 = 1
30 = 17	3 = 2
40 = 22	4 = 2
50 = 28	5 = 3
60 = 33	6 = 3
	7 = 4
	8 = 4
	9 = 5

32 = 18

Anti Log Excesses

10 = 18	1 = 2
20 = 36	2 = 4
30 = 54	3 = 5
	4 = 7
	5 = 9
	6 = 11
	7 = 13
	8 = 14
	9 = 16

18 = 32

7790058	·8915407
122	443
187	478
251	514
315	550
379	586
443	622
508	657
572	693
636	729
700	765
764	800
828	836
893	872
957	908
7791021	944
085	979
149	·8916015
213	051
278	086
342	122
406	158
470	194
534	230
599	265
663	301
727	337
791	373
855	408
919	444
984	480
7792048	516
112	551
176	587
240	623
304	659
369	694

7792433	·8916730
497	766
561	802
625	838
689	873
754	909
818	945
882	981
946	·8917016
7793010	052
074	088
139	124
203	159
267	195
331	231
395	267
459	302
523	338
588	374
652	410
716	445
780	481
844	517
908	553
973	588
7794037	624
101	660
165	696
229	731
293	767
357	803
422	838
486	874
550	910
614	946
678	981
742	·8918017
806	053
871	089
935	124
999	160
7795063	196
127	232
191	267
255	303
320	339
384	375
448	410
512	446
576	482
640	518
704	553
769	589
833	625
897	660
961	696
7796025	732
089	768
153	803
218	839
282	875
346	911
410	946
474	982
538	·8919018

7796602	·8919053
666	089
731	125
795	161
859	196
923	232
987	268
7797051	303
115	339
179	375
244	411
308	446
372	482
436	518
500	553
564	589
628	625
692	661
757	696
821	732
885	768
949	803
7798013	839
077	875
141	911
205	946
270	982
334	·8920018
398	053
462	089
526	125
590	160
654	196
718	232
782	268
847	303
911	339
975	375
7799039	410
103	446
167	482
231	517
295	553
359	589
424	625
488	660
552	696
616	732
680	767
744	803
808	839
872	874
936	910

Excesses

10 = 6	1 = 1
20 = 11	2 = 1
30 = 17	3 = 2
40 = 22	4 = 2
50 = 28	5 = 3
60 = 33	6 = 3
	7 = 4
	8 = 4
	9 = 5

32 = 18

Anti Log Excesses

10 = 18	1 = 2
20 = 36	2 = 4
30 = 54	3 = 5
	4 = 7
	5 = 9
	6 = 11
	7 = 13
	8 = 14
	9 = 16

18 = 32

7800000	·8920946
065	982
129	·8921017
193	053
257	089
321	124
385	160
449	196
513	231
577	267
641	303
706	338
770	374
834	410
898	445
962	481
7801026	517
090	553
154	588
218	624
282	660
347	695
411	731
475	767
539	802
603	838
667	874
731	909
795	945
859	981
923	·8922016
987	052
7802052	088
116	123
180	159
244	195
308	230

7802372	·8922266
436	302
500	337
564	373
628	409
692	444
756	480
821	516
885	551
949	587
7803013	623
077	658
141	694
205	730
269	765
333	801
397	837
461	872
525	908
589	944
653	979
718	·8923015
782	051
846	086
910	122
974	157
7804038	193
102	229
166	264
230	300
294	336
358	372
422	407
486	443
550	479
615	514
679	550
743	585
807	621
871	657
935	692
999	728
7805063	764
127	799
191	835
255	871
319	906
383	942
447	978
511	·8924013
575	049
639	085
704	120
768	156
832	191
896	227
960	263
7806024	298
088	334
152	370
216	405
280	441
344	477
408	512
472	548

7806536	·8924583
600	619
664	655
728	690
792	726
856	762
921	797
985	833
7807049	869
113	904
177	940
241	975
305	·8925011
369	047
433	082
497	118
561	154
625	189
689	225
753	260
817	296
881	332
945	367
7808009	403
073	439
137	474
201	510
265	545
329	581
393	617
457	652
521	688
586	723
650	759
714	795
778	830
842	866
906	902
970	937
7809034	973
098	·8926008
162	044
226	080
290	115
354	151
418	186
482	222
546	258
610	293
674	329
738	364
802	400
866	436
930	471
994	507

Excesses

10 = 6	1 = 1
20 = 11	2 = 1
30 = 17	3 = 2
40 = 22	4 = 2
50 = 28	5 = 3
60 = 33	6 = 3
	7 = 4
	8 = 4
	9 = 5

32 = 18

Anti Log Excesses

10 = 18	1 = 2
20 = 36	2 = 4
30 = 54	3 = 5
	4 = 7
	5 = 9
	6 = 11
	7 = 13
	8 = 14
	9 = 16

18 = 32

7810058	·8926542
122	578
186	614
250	649
314	685
378	720
442	756
506	792
570	827
634	863
698	898
762	934
826	970
890	·8927005
954	041
7811018	076
082	112
146	148
210	183
274	219
338	254
402	290
466	325
530	361
594	397
658	432
722	468
786	503
850	539
914	575
978	610
7812042	646
106	681
170	717
234	752
298	788
362	824

7812426	·8927859
490	895
554	930
618	966
682	·8928001
746	037
810	073
874	108
938	144
7813002	179
066	215
130	251
194	286
258	322
322	357
386	393
450	428
514	464
578	500
642	535
706	571
770	606
834	642
898	677
962	713
7814026	749
090	784
154	820
218	855
282	891
346	926
410	962
474	997
538	·8929033
602	069
666	104
730	140
794	175
858	211
922	246
986	282
7815050	317
114	353
178	389
242	424
306	460
370	495
434	531
498	566
562	602
626	637
690	673
754	708
818	744
882	780
946	815
7816010	851
074	886
138	922
202	957
266	993
330	·8930028
394	064
457	099
521	135

7816585	·8930170
649	206
713	242
777	277
841	313
905	348
969	384
7817033	419
097	455
161	490
225	526
289	561
353	597
417	632
481	668
545	704
609	739
673	775
737	810
801	846
865	881
929	917
993	952
7818057	988
120	·8931023
184	059
248	094
312	130
376	165
440	201
504	236
568	272
632	308
696	343
760	379
824	414
888	450
952	485
7819016	521
080	556
144	592
208	627
272	663
336	698
399	734
463	769
527	805
591	840
655	876
719	911
783	947
847	982
911	·8932018
975	053

Excesses

10 = 6	1 = 1
20 = 11	2 = 1
30 = 17	3 = 2
40 = 22	4 = 2
50 = 28	5 = 3
60 = 33	6 = 3
	7 = 4
	8 = 4
	9 = 5

32 = 18

Anti Log Excesses

10 = 18	1 = 2
20 = 36	2 = 4
30 = 54	3 = 5
	4 = 7
	5 = 9
	6 = 11
	7 = 13
	8 = 14
	9 = 16

18 = 32

7820039	·8932089
103	124
167	160
231	195
295	231
359	266
422	302
486	337
550	373
614	408
678	444
742	479
806	515
870	550
934	586
998	621
7821062	657
126	692
190	728
254	763
318	799
381	834
445	870
509	905
573	941
637	976
701	·8933012
765	047
829	083
893	118
957	154
7822021	189
085	225
149	260
212	296
276	331
340	367

7822404	8933402
468	438
532	473
596	509
660	544
724	580
788	615
852	651
916	686
979	722
7823043	757
107	793
171	828
235	864
299	899
363	935
427	970
491	·8934006
555	041
619	077
682	112
746	147
810	183
874	218
938	254
7824002	289
066	325
130	360
194	396
258	431
322	467
385	502
449	538
513	573
577	609
641	644
705	680
769	715
833	751
897	786
961	821
7825024	857
088	892
152	928
216	963
280	999
344	·8935034
408	070
472	105
536	141
600	176
663	212
727	247
791	282
855	318
919	353
983	389
7826047	424
111	460
175	495
238	531
302	566
366	601
430	637
494	672

Number	Log
7826558	·8935708
622	743
686	779
750	814
813	850
877	885
941	920
7827005	956
069	991
133	·8936027
197	062
261	098
324	133
388	169
452	204
516	239
580	275
644	310
708	346
772	381
835	417
899	452
963	487
7828027	523
091	558
155	594
219	629
283	665
346	700
410	735
474	771
538	806
602	842
666	877
730	913
794	948
857	983
921	·8937019
985	054
7829049	090
113	125
177	161
241	196
304	231
368	267
432	302
496	338
560	373
624	409
688	444
751	479
815	515
879	550
943	586

Excesses

10 = 6	1 = 1
20 = 11	2 = 1
30 = 17	3 = 2
40 = 22	4 = 2
50 = 28	5 = 3
60 = 33	6 = 3
	7 = 4
	8 = 4
	9 = 5

32 = 18

Anti Log Excesses

10 = 18	1 = 2
20 = 36	2 = 4
30 = 54	3 = 5
	4 = 7
	5 = 9
	6 = 11
	7 = 13
	8 = 14
	9 = 16

18 = 32

Number	Log
7830007	·8937621
071	656
135	692
198	727
262	763
326	798
390	834
454	869
518	904
582	940
645	975
709	·8938011
773	046
837	082
901	117
965	152
7831029	188
092	223
156	259
220	294
284	329
348	365
412	400
475	436
539	471
603	506
667	542
731	577
795	613
859	648
922	683
986	719
7832050	754
114	790
178	825
242	860
305	896

Number	Log
7832369	·8938931
433	967
497	·8939002
561	037
625	073
688	108
752	144
816	179
880	214
944	250
7833008	285
071	321
135	356
199	391
263	427
327	462
391	497
454	533
518	568
582	604
646	639
710	674
774	710
837	745
901	780
965	816
7834029	851
093	887
156	922
220	957
284	993
348	·8940028
412	064
476	099
539	134
603	170
667	205
731	240
795	276
859	311
922	347
986	382
7835050	417
114	453
178	488
241	523
305	559
369	594
433	630
497	665
560	700
624	736
688	771
752	806
816	842
880	877
943	912
7836007	948
071	983
135	·8941019
199	054
262	089
326	125
390	160
454	195

Number	Log
7836518	·8941231
581	266
645	301
709	337
773	372
837	407
900	443
964	478
7837028	514
092	549
156	584
219	620
283	655
347	690
411	726
475	761
538	796
602	832
666	867
730	902
794	938
857	973
921	·8942008
985	044
7838049	079
113	114
176	150
240	185
304	220
368	256
431	291
495	326
559	362
623	397
687	433
750	468
814	503
878	539
942	574
7839006	609
069	645
133	680
197	715
261	751
324	786
388	821
452	857
516	892
580	927
643	963
707	998
771	·8943033
835	069
898	104
962	139

Excesses

10 = 6	1 = 1
20 = 11	2 = 1
30 = 17	3 = 2
40 = 22	4 = 2
50 = 28	5 = 3
60 = 33	6 = 3
	7 = 4
	8 = 4
	9 = 5

32 = 18

Anti Log Excesses

10 = 18	1 = 2
20 = 36	2 = 4
30 = 54	3 = 5
	4 = 7
	5 = 9
	6 = 11
	7 = 13
	8 = 14
	9 = 16

18 = 32

Number	Log
7840026	·8943175
090	210
154	245
217	281
281	316
345	351
409	387
472	422
536	457
600	493
664	528
728	563
791	599
855	634
919	669
983	705
7841046	740
110	775
174	811
238	846
301	881
365	916
429	952
493	987
556	·8944022
620	058
684	093
748	128
812	164
875	199
939	234
7842003	270
067	305
130	340
194	376
258	411
322	446

7842385	·8944481
449	517
513	552
577	587
640	623
704	658
768	693
832	729
895	764
959	799
7843023	835
087	870
150	905
214	940
278	976
342	·8945011
405	046
469	082
533	117
597	152
660	188
724	223
788	258
852	293
915	329
979	364
7844043	399
107	435
170	470
234	505
298	541
362	576
425	611
489	646
553	682
616	717
680	752
744	788
808	823
871	858
935	893
999	929
7845063	964
126	999
190	·8946035
254	070
318	105
381	140
445	176
509	211
572	246
636	282
700	317
764	352
827	387
891	423
955	458
7846019	493
082	528
146	564
210	599
274	634
337	670
401	705
465	740

7846528	·8946775
592	811
656	846
720	881
783	916
847	952
911	987
974	·8947022
7847038	058
102	093
166	128
229	163
293	199
357	234
420	269
484	304
548	340
612	375
675	410
739	445
803	481
866	516
930	551
994	586
7848058	622
121	657
185	692
249	727
312	763
376	798
440	833
504	868
567	904
631	939
695	974
758	·8948009
822	045
886	080
949	115
7849013	150
077	186
141	221
204	256
268	291
332	327
395	362
459	397
523	432
586	468
650	503
714	538
778	573
841	609
905	644
969	679

Excesses

10 = 6	1 = 1
20 = 11	2 = 1
30 = 17	3 = 2
40 = 22	4 = 2
50 = 28	5 = 3
60 = 33	6 = 3
	7 = 4
	8 = 4
	9 = 5

32 = 18

Anti Log Excesses

10 = 18	1 = 2
20 = 36	2 = 4
30 = 54	3 = 5
	4 = 7
	5 = 9
	6 = 11
	7 = 13
	8 = 14
	9 = 16

18 = 32

7850032	·8948714
096	750
160	785
223	820
287	855
351	890
415	926
478	961
542	996
606	·8949031
669	067
733	102
797	137
860	172
924	208
988	243
7851051	278
115	313
179	348
242	384
306	419
370	454
433	489
497	525
561	560
625	595
688	630
752	665
816	701
879	736
943	771
7852007	806
070	842
134	877
198	912
261	947
325	982

7852389	·8950018
452	053
516	088
580	123
643	159
707	194
771	229
834	264
898	299
962	335
7853025	370
089	405
153	440
216	475
280	511
344	546
407	581
471	616
535	652
598	687
662	722
726	757
789	792
853	827
917	863
980	898
7854044	933
108	968
171	·8951003
235	039
299	074
362	109
426	144
490	179
553	215
617	250
681	285
744	320
808	355
872	391
935	426
999	461
7855063	496
126	531
190	567
254	602
317	637
381	672
444	707
508	743
572	778
635	813
699	848
763	883
826	918
890	954
954	989
7856017	·8952024
081	059
145	094
208	130
272	165
336	200
399	235
463	270

7856526	·8952305
590	341
654	376
717	411
781	446
845	481
908	517
972	552
7857036	587
099	622
163	657
226	692
290	728
354	763
417	798
481	833
545	868
608	903
672	939
736	974
799	·8953009
863	044
926	079
990	114
7858054	150
117	185
181	220
245	255
308	290
372	325
435	361
499	396
563	431
626	466
690	501
754	536
817	572
881	607
944	642
7859008	677
072	712
135	747
199	783
263	818
326	853
390	888
453	923
517	958
581	993
644	·8954029
708	064
771	099
835	134
899	169
962	204

Excesses

10 = 6	1 = 1
20 = 11	2 = 1
30 = 17	3 = 2
40 = 22	4 = 2
50 = 28	5 = 3
60 = 33	6 = 3
	7 = 4
	8 = 4
	9 = 5

32 = 18

Anti Log Excesses

10 = 18	1 = 2
20 = 36	2 = 4
30 = 54	3 = 5
	4 = 7
	5 = 9
	6 = 11
	7 = 13
	8 = 14
	9 = 16

18 = 32

7860026	·8954239
090	275
153	310
217	345
280	380
344	415
408	450
471	485
535	521
598	556
662	591
726	626
789	661
853	696
916	731
980	767
7861044	802
107	837
171	872
234	907
298	942
362	977
425	·8955013
489	048
553	083
616	118
680	153
743	188
807	223
871	259
934	294
998	329
7862061	364
125	399
188	434
252	469
316	504
7862379	·8955540
443	575
506	610
570	645
634	680
697	715
761	750
824	785
888	821
952	856
7863015	891
079	926
142	961
206	996
270	·8956031
333	066
397	102
460	137
524	172
587	207
651	242
715	277
778	312
842	347
905	382
969	418
7864033	453
096	488
160	523
223	558
287	593
350	628
414	663
478	698
541	734
605	769
668	804
732	839
795	874
859	909
923	944
986	979
7865050	·8957014
113	050
177	085
240	120
304	155
368	190
431	225
495	260
558	295
622	330
685	365
749	400
813	436
876	471
940	506
7866003	541
067	576
130	611
194	646
258	681
321	716
385	752
448	787
7866512	·8957822
575	857
639	892
702	927
766	962
830	997
893	·8958032
957	067
7867020	102
084	137
147	173
211	208
274	243
338	278
402	313
465	348
529	383
592	418
656	453
719	488
783	523
846	558
910	594
974	629
7868037	664
101	699
164	734
228	769
291	804
355	839
418	874
482	909
545	944
609	979
673	·8959014
736	049
800	085
863	120
927	155
990	190
7869054	225
117	260
181	295
244	330
308	365
372	400
435	435
499	470
562	505
626	540
689	575
753	610
816	646
880	681
943	716

Excesses

10 = 6	1 = 1
20 = 11	2 = 1
30 = 17	3 = 2
40 = 22	4 = 2
50 = 28	5 = 3
60 = 33	6 = 3
	7 = 4
	8 = 4
	9 = 5

32 = 18

Anti Log Excesses

10 = 18	1 = 2
20 = 36	2 = 4
30 = 54	3 = 5
	4 = 7
	5 = 9
	6 = 11
	7 = 13
	8 = 14
	9 = 16

18 = 32

7870007	·8959751
070	786
134	821
197	856
261	891
325	926
388	961
452	996
515	·8960031
579	066
642	101
706	136
769	171
833	206
896	241
960	277
7871023	312
087	347
150	382
214	417
277	452
341	487
404	522
468	557
531	592
595	627
659	662
722	697
786	732
849	767
913	802
976	837
7872040	872
103	907
167	942
230	977
294	·8961012
7872357	·8961047
421	082
484	118
548	153
611	188
675	223
738	258
802	293
865	328
929	363
992	398
7873056	433
119	468
183	503
246	538
310	573
373	608
437	643
500	678
564	713
627	748
691	783
754	818
818	853
881	888
945	923
7874008	958
072	993
135	·8962028
199	063
262	098
326	133
389	168
453	203
516	238
580	273
643	308
707	343
770	378
834	414
897	449
961	484
7875024	519
088	554
151	589
215	624
278	659
342	694
405	729
469	764
532	799
596	834
659	869
723	904
786	939
850	974
913	·8963009
977	044
7876040	079
104	114
167	149
231	184
294	219
358	254
421	289

7876484	8963324
548	359
611	394
675	429
738	464
802	499
865	534
929	569
992	604
7877056	639
119	674
183	709
246	744
310	779
373	814
437	849
500	884
564	919
627	954
691	989
754	·8964024
817	059
881	094
944	129
7878008	164
071	199
135	234
198	269
262	304
325	339
389	374
452	409
516	444
579	479
643	514
706	549
769	584
833	619
896	654
960	689
7879023	724
087	759
150	794
214	829
277	864
341	899
404	934
467	969
531	·8965004
594	038
658	073
721	108
785	143
848	178
912	213
975	248

Excesses

10 = 6	1 = 1
20 = 11	2 = 1
30 = 17	3 = 2
40 = 22	4 = 2
50 = 28	5 = 3
60 = 33	6 = 3
	7 = 4
	8 = 4
	9 = 5

32 = 18

Anti Log Excesses

10 = 18	1 = 2
20 = 36	2 = 4
30 = 54	3 = 5
	4 = 7
	5 = 9
	6 = 11
	7 = 13
	8 = 14
	9 = 16

18 = 32

7880039	·8965283
102	318
165	353
229	388
292	423
356	458
419	493
483	528
546	563
610	598
673	633
737	668
800	703
863	738
927	773
990	808
7881054	843
117	878
181	913
244	948
307	982
371	·8966017
434	052
498	087
561	122
625	157
688	192
752	227
815	262
878	297
942	332
7882005	367
069	402
132	437
196	472
259	507
322	542

7882386	·8966577
449	612
513	647
576	681
640	716
703	751
767	786
830	821
893	856
957	891
7883020	926
084	961
147	996
211	·8967031
274	066
337	101
401	136
464	171
528	206
591	241
654	276
718	311
781	346
845	380
908	415
972	450
7884035	485
098	520
162	555
225	590
289	625
352	660
416	695
479	730
542	765
606	800
669	835
733	869
796	904
859	939
923	974
986	·8968009
7885050	044
113	079
176	114
240	149
303	184
367	219
430	254
494	289
557	323
620	358
684	393
747	428
811	463
874	498
937	533
7886001	568
064	603
128	638
191	673
254	707
318	742
381	777
445	812

7886508	·8968847
571	882
635	917
698	952
762	987
825	·8969022
888	057
952	091
7887015	126
079	161
142	196
205	231
269	266
332	301
396	336
459	371
522	406
586	440
649	475
712	510
776	545
839	580
903	615
966	650
7888029	685
093	720
156	755
220	790
283	825
346	859
410	894
473	929
536	964
600	999
663	·8970034
727	069
790	104
853	139
917	174
980	208
7889044	243
107	278
170	313
234	348
297	383
360	418
424	453
487	488
551	522
614	557
677	592
741	627
804	662
867	697
931	732
994	767

Excesses

10 = 6	1 = 1
20 = 11	2 = 1
30 = 17	3 = 2
40 = 22	4 = 2
50 = 28	5 = 3
60 = 33	6 = 3
	7 = 4
	8 = 4
	9 = 5

32 = 17

Anti Log Excesses

10 = 18	1 = 2
20 = 36	2 = 4
30 = 54	3 = 5
	4 = 7
	5 = 9
	6 = 11
	7 = 13
	8 = 15
	9 = 16

17 = 32

7890058	·8970801
121	836
184	871
248	906
311	941
374	976
438	·8971011
501	046
564	081
628	115
691	150
755	185
818	220
881	255
945	290
7891008	325
071	360
135	394
198	429
261	464
325	499
388	534
452	569
515	604
578	638
642	673
705	708
768	743
832	778
895	813
958	848
7892022	883
085	917
148	952
212	987
275	·8972022
339	057

Number	Log
7892402	·8972092
465	127
529	161
592	196
655	231
719	266
782	301
845	336
909	371
972	405
7893035	440
099	475
162	510
225	545
289	580
352	615
415	649
479	684
542	719
606	754
669	789
732	824
796	858
859	893
922	928
986	963
7894049	998
112	·8973033
176	068
239	102
302	137
366	172
429	207
492	242
556	277
619	311
682	346
746	381
809	416
872	451
936	486
999	520
7895062	555
126	590
189	625
252	660
316	695
379	729
442	764
506	799
569	834
632	869
696	904
759	938
822	973
886	·8974008
949	043
7896012	078
075	113
139	147
202	182
265	217
329	252
392	287
455	322

Number	Log
7896519	·8974356
582	391
645	426
709	461
772	496
835	531
899	565
962	600
7897025	635
089	670
152	705
215	740
279	774
342	809
405	844
468	879
532	914
595	948
658	983
722	·8975018
785	053
848	088
912	122
975	157
7898038	192
102	227
165	262
228	297
291	331
355	366
418	401
481	436
545	471
608	505
671	540
735	575
798	610
861	645
924	679
988	714
7899051	749
114	784
178	819
241	853
304	888
368	923
431	958
494	993
557	·8976027
621	062
684	097
747	132
811	167
874	201
937	236

Excesses

10 = 6	1 = 1
20 = 11	2 = 1
30 = 17	3 = 2
40 = 22	4 = 2
50 = 28	5 = 3
60 = 33	6 = 3
	7 = 4
	8 = 4
	9 = 5

32 = 17

Anti Log Excesses

10 = 18	1 = 2
20 = 36	2 = 4
30 = 55	3 = 5
	4 = 7
	5 = 9
	6 = 11
	7 = 13
	8 = 15
	9 = 16

17 = 32

Number	Log
7900000	·8976271
064	306
127	340
190	375
254	410
317	445
380	480
444	514
507	549
570	584
633	619
697	654
760	688
823	723
887	758
950	793
7901013	827
076	862
140	897
203	932
266	967
330	·8977001
393	036
456	071
519	106
583	141
646	175
709	210
772	245
836	280
899	315
962	349
7902026	384
089	419
152	454
215	488
279	523

Number	Log
7902342	·8977558
405	593
468	627
532	662
595	697
658	732
722	767
785	801
848	836
911	871
975	906
7903038	940
101	975
164	·8978010
228	045
291	079
354	114
417	149
481	184
544	218
607	253
671	288
734	323
797	357
860	392
924	427
987	462
7904050	497
113	531
177	566
240	601
303	636
366	670
430	705
493	740
556	775
619	809
683	844
746	879
809	914
872	948
936	983
999	·8979018
7905062	053
125	087
189	122
252	157
315	191
378	226
442	261
505	296
568	330
631	365
695	400
758	435
821	469
884	504
948	539
7906011	574
074	608
137	643
201	678
264	713
327	747
390	782

Number	Log
7906454	·8979817
517	852
580	886
643	921
707	956
770	991
833	·8980025
896	060
959	095
7907023	129
086	164
149	199
212	234
276	268
339	303
402	338
465	373
529	407
592	442
655	477
718	511
781	546
845	581
908	616
971	650
7908034	685
098	720
161	754
224	789
287	824
351	859
414	893
477	928
540	963
603	997
667	·8981032
730	067
793	102
856	136
920	171
983	206
7909046	240
109	275
172	310
236	345
299	379
362	414
425	449
488	483
552	518
615	553
678	588
741	622
805	657
868	692
931	726
994	761

Excesses

10 = 5	1 = 1
20 = 11	2 = 1
30 = 16	3 = 2
40 = 22	4 = 2
50 = 27	5 = 3
60 = 33	6 = 3
	7 = 4
	8 = 4
	9 = 5

32 = 17

Anti Log Excesses

10 = 18	1 = 2
20 = 36	2 = 4
30 = 55	3 = 5
	4 = 7
	5 = 9
	6 = 11
	7 = 13
	8 = 15
	9 = 16

17 = 32

7910057	·8981796
121	830
184	865
247	900
310	935
373	970
437	·8982004
500	039
563	074
626	108
689	143
753	178
816	212
879	247
942	282
7911005	317
069	351
132	386
195	421
258	455
322	490
385	525
448	559
511	594
574	629
637	663
701	698
764	733
827	768
890	802
953	837
7912017	872
080	906
143	941
206	976
269	·8983010
333	045

7912396	·8983080
459	114
522	149
585	184
649	218
712	253
775	288
838	323
901	357
965	392
7913028	427
091	461
154	496
217	531
280	565
344	600
407	635
470	669
533	704
596	739
660	773
723	808
786	843
849	877
912	912
975	947
7914039	981
102	·8984016
165	051
228	085
291	120
355	155
418	189
481	224
544	259
607	293
670	328
734	363
797	397
860	432
923	467
986	501
7915049	536
113	571
176	605
239	640
302	675
365	709
428	744
492	779
555	813
618	848
681	883
744	917
807	952
871	987
934	·8985021
997	056
7916060	091
123	125
186	160
250	195
313	229
376	264
439	299

7916502	·8985333
565	368
629	403
692	437
755	472
818	506
881	541
944	576
7917008	610
071	645
134	680
197	714
260	749
323	784
386	818
450	853
513	888
576	922
639	957
702	991
765	·8986026
828	061
892	095
955	130
7918018	165
081	199
144	234
207	269
271	303
334	338
397	372
460	407
523	442
586	476
649	511
713	546
776	580
839	615
902	649
965	684
7919028	719
091	753
154	788
218	823
281	857
344	892
407	926
470	961
533	996
596	·8987030
660	065
723	100
786	134
849	169
912	204
975	238

Excesses

10 = 5	1 = 1
20 = 11	2 = 1
30 = 16	3 = 2
40 = 22	4 = 2
50 = 27	5 = 3
60 = 33	6 = 3
	7 = 4
	8 = 4
	9 = 5

32 = 17

Anti Log Excesses

10 = 18	1 = 2
20 = 36	2 = 4
30 = 55	3 = 5
	4 = 7
	5 = 9
	6 = 11
	7 = 13
	8 = 15
	9 = 16

17 = 32

7920038	·8987273
102	307
165	342
228	376
291	411
354	446
417	480
480	515
543	550
607	584
670	619
733	653
796	688
859	723
922	757
985	792
7921048	826
112	861
175	896
238	930
301	965
364	999
427	·8988034
490	069
553	103
616	138
680	172
743	207
806	242
869	276
932	311
995	345
7922058	380
121	415
185	449
248	484
311	518

7922374	·8988553
437	588
500	622
563	657
626	691
689	726
753	761
816	795
879	830
942	864
7923005	899
068	934
131	968
194	·8989003
257	037
321	072
384	107
447	141
510	176
573	210
636	245
699	280
762	314
825	349
888	383
952	418
7924015	452
078	487
141	522
204	556
267	591
330	625
393	660
456	694
519	729
583	764
646	798
709	833
772	867
835	902
898	936
961	971
7925024	·8990006
087	040
150	075
213	109
277	144
340	178
403	213
466	248
529	282
592	317
655	351
718	386
781	420
844	455
907	490
970	524
7926034	559
097	593
160	628
223	662
286	697
349	732
412	766

Number	Log
7926475	·8990801
538	835
601	870
664	904
727	939
791	973
854	·8991008
917	043
980	077
7927043	112
106	146
169	181
232	215
295	250
358	284
421	319
484	353
547	388
610	423
674	457
737	492
800	526
863	561
926	595
989	630
7928052	664
115	699
178	734
241	768
304	803
367	837
430	872
493	906
556	941
620	976
683	·8992010
746	045
809	079
872	114
935	148
998	183
7929061	217
124	252
187	286
250	321
313	355
376	390
439	425
502	459
565	494
628	528
692	563
755	597
818	632
881	666
944	701

Excesses

10 = 5	1 = 1
20 = 11	2 = 1
30 = 16	3 = 2
40 = 22	4 = 2
50 = 27	5 = 3
60 = 33	6 = 3
	7 = 4
	8 = 4
	9 = 5

32 = 17

Anti Log Excesses

10 = 18	1 = 2
20 = 36	2 = 4
30 = 55	3 = 5
	4 = 7
	5 = 9
	6 = 11
	7 = 13
	8 = 15
	9 = 16

17 = 32

Number	Log
7930007	·8992735
070	770
133	804
196	839
259	873
322	908
385	943
448	977
511	·8993012
574	046
637	081
700	115
763	150
826	184
889	219
953	253
7931016	288
079	322
142	357
205	391
268	426
331	460
394	495
457	529
520	564
583	598
646	633
709	668
772	702
835	737
898	771
961	806
7932024	840
087	875
150	909
213	944
276	978

Number	Log
7932339	·8994013
402	047
465	082
528	116
592	151
655	185
718	220
781	254
844	289
907	323
970	358
7933033	392
096	427
159	461
222	496
285	530
348	565
411	599
474	634
537	668
600	703
663	737
726	772
789	806
852	841
915	875
978	910
7934041	944
104	979
167	·8995013
230	048
293	082
356	117
419	151
482	186
545	220
608	255
671	289
734	324
797	358
860	393
923	427
986	462
7935049	496
112	531
175	565
238	600
301	634
364	668
427	703
490	737
553	772
616	806
679	841
742	875
805	910
868	944
931	979
994	·8996013
7936057	048
120	082
183	117
246	151
309	186
372	220

Number	Log
7936435	·8996255
498	289
561	324
624	358
687	392
750	427
813	461
876	496
939	530
7937002	565
065	599
128	634
191	668
254	703
317	737
380	772
443	806
506	841
569	875
632	910
695	944
758	978
821	·8997013
884	047
947	082
7938010	116
073	151
136	185
199	220
262	254
325	289
388	323
451	358
514	392
577	426
640	461
703	495
766	530
829	564
892	599
955	633
7939018	668
081	702
144	737
207	771
270	805
333	840
396	874
459	909
522	943
585	978
648	·8998012
711	047
774	081
837	115
900	150
963	184

Excesses

10 = 5	1 = 1
20 = 11	2 = 1
30 = 16	3 = 2
40 = 22	4 = 2
50 = 27	5 = 3
60 = 33	6 = 3
	7 = 4
	8 = 4
	9 = 5

32 = 17

Anti Log Excesses

10 = 18	1 = 2
20 = 37	2 = 4
30 = 55	3 = 5
	4 = 7
	5 = 9
	6 = 11
	7 = 13
	8 = 15
	9 = 16

17 = 32

Number	Log
7940026	·8998219
089	253
152	288
215	322
278	357
341	391
404	425
466	460
529	494
592	529
655	563
718	598
781	632
844	667
907	701
970	735
7941033	770
096	804
159	839
222	873
285	908
348	942
411	977
474	·8999011
537	045
600	080
663	114
726	148
789	183
852	217
915	252
978	286
7942041	321
103	355
166	389
229	424
292	458

7942355	·8999493
418	527
481	562
544	596
607	630
670	665
733	699
796	734
859	768
922	803
985	837
7943048	871
111	906
174	940
237	975
300	·9000009
363	043
425	078
488	112
551	147
614	181
677	215
740	250
803	284
866	319
929	353
992	388
7944055	422
118	456
181	491
244	525
307	560
370	594
433	628
495	663
558	697
621	732
684	766
747	800
810	835
873	869
936	904
999	938
7945062	972
125	·9001007
188	041
251	076
314	110
377	144
439	179
502	213
565	248
628	282
691	316
754	351
817	385
880	420
943	454
7946006	488
069	523
132	557
195	592
257	626
320	660
383	695

7946446	·9001729
509	764
572	798
635	832
698	867
761	901
824	935
887	970
950	·9002004
7947013	039
075	073
138	107
201	142
264	176
327	211
390	245
453	279
516	314
579	348
642	382
705	417
767	451
830	486
893	520
956	554
7948019	589
082	623
145	657
208	692
271	726
334	761
397	795
459	829
522	864
585	898
648	932
711	967
774	·9003001
837	036
900	070
963	104
7949026	139
088	173
151	207
214	242
277	276
340	311
403	345
466	379
529	414
592	448
655	482
717	517
780	551
843	585
906	620
969	654

Excesses

10 = 5	1 = 1
20 = 11	2 = 1
30 = 16	3 = 2
40 = 22	4 = 2
50 = 27	5 = 3
60 = 33	6 = 3
	7 = 4
	8 = 4
	9 = 5

31 = 17

Anti Log Excesses

10 = 18	1 = 2
20 = 37	2 = 4
30 = 55	3 = 5
	4 = 7
	5 = 9
	6 = 11
	7 = 13
	8 = 15
	9 = 16

17 = 31

7950032	·9003689
095	723
158	757
221	792
284	826
346	860
409	895
472	929
535	963
598	998
661	·9004032
724	067
787	101
850	135
912	170
975	204
7951038	238
101	273
164	307
227	341
290	376
353	410
415	444
478	479
541	513
604	547
667	582
730	616
793	650
856	685
919	719
981	753
7952044	788
107	822
170	857
233	891
296	925

7952359	·9004960
422	994
484	·9005028
547	063
610	097
673	131
736	166
799	200
862	234
924	269
987	303
7953050	337
113	372
176	406
239	440
302	475
365	509
427	543
490	578
553	612
616	646
679	681
742	715
805	749
867	784
930	818
993	852
7954056	886
119	921
182	955
245	989
308	·9006024
370	058
433	092
496	127
559	161
622	195
685	230
748	264
810	298
873	333
936	367
999	401
7955062	435
125	470
187	504
250	538
313	573
376	607
439	641
502	676
565	710
627	744
690	779
753	813
816	847
879	881
942	916
7956005	950
067	984
130	·9007019
193	053
256	087
319	122
382	156

7956444	·9007190
507	225
570	259
633	293
696	327
759	362
821	396
884	430
947	465
7957010	499
073	533
136	568
198	602
261	636
324	670
387	705
450	739
513	773
576	808
638	842
701	876
764	910
827	945
890	979
952	·9008013
7958015	048
078	082
141	116
204	150
267	185
329	219
392	253
455	288
518	322
581	356
644	390
706	425
769	459
832	493
895	528
958	562
7959021	596
083	630
146	665
209	699
272	733
335	768
397	802
460	836
523	870
586	905
649	939
712	973
774	·9009007
837	042
900	076
963	110

Excesses

10 = 5	1 = 1
20 = 11	2 = 1
30 = 16	3 = 2
40 = 22	4 = 2
50 = 27	5 = 3
60 = 33	6 = 3
	7 = 4
	8 = 4
	9 = 5

31 = 17

Anti Log Excesses

10 = 18	1 = 2
20 = 37	2 = 4
30 = 55	3 = 5
	4 = 7
	5 = 9
	6 = 11
	7 = 13
	8 = 15
	9 = 16

17 = 31

7960026	·9009144
088	179
151	213
214	247
277	282
340	316
402	350
465	384
528	419
591	453
654	487
717	521
779	556
842	590
905	624
968	658
7961031	693
093	727
156	761
219	795
282	830
345	864
407	898
470	933
533	967
596	·9010001
659	035
721	070
784	104
847	138
910	172
973	207
7962035	241
098	275
161	309
224	344
287	378

7962349	·9010412
412	446
475	481
538	515
601	549
663	583
726	618
789	652
852	686
915	720
977	755
7963040	789
103	823
166	857
228	892
291	926
354	960
417	994
480	·9011029
542	063
605	097
668	131
731	165
794	200
856	234
919	268
982	303
7964045	337
107	371
170	405
233	439
296	474
359	508
421	542
484	576
547	611
610	645
672	679
735	713
798	748
861	782
924	816
986	850
7965049	884
112	919
175	953
237	987
300	·9012021
363	056
426	090
489	124
551	158
614	193
677	227
740	261
802	295
865	329
928	364
991	398
7966053	432
116	466
179	501
242	535
305	569
367	603

7966430	·9012637
493	672
556	706
618	740
681	774
744	808
807	843
869	877
932	911
995	945
7967058	980
120	·9013014
183	048
246	082
309	116
371	151
434	185
497	219
560	253
622	287
685	322
748	356
811	390
873	424
936	458
999	493
7968062	527
124	561
187	595
250	630
313	664
375	698
438	732
501	766
564	800
626	835
689	869
752	903
815	937
877	971
940	·9014006
7969003	040
066	074
128	108
191	142
254	176
317	211
379	245
442	279
505	313
568	347
630	382
693	416
756	450
819	484
881	518
944	553

Excesses

10 = 5	1 = 1
20 = 11	2 = 1
30 = 16	3 = 2
40 = 22	4 = 2
50 = 27	5 = 3
60 = 33	6 = 3
	7 = 4
	8 = 4
	9 = 5

31 = 17

Anti Log Excesses

10 = 18	1 = 2
20 = 37	2 = 4
30 = 55	3 = 6
	4 = 7
	5 = 9
	6 = 11
	7 = 13
	8 = 15
	9 = 17

17 = 31

7970007	·9014587
070	621
132	655
195	689
258	723
320	758
383	792
446	826
509	860
571	894
634	929
697	963
760	997
822	·9015031
885	065
948	099
7971010	134
073	168
136	202
199	236
261	270
324	305
387	339
450	373
512	407
575	441
638	475
700	510
763	544
826	578
889	612
951	646
7972014	680
077	715
139	749
202	783
265	817

7972328	·9015851
390	885
453	920
516	954
579	988
641	·9016022
704	056
767	090
829	125
892	159
955	193
7973017	227
080	261
143	295
206	329
268	364
331	398
394	432
456	466
519	500
582	534
645	569
707	603
770	637
833	671
895	705
958	739
7974021	773
084	808
146	842
209	876
272	910
334	944
397	978
460	·9017012
522	047
585	081
648	115
711	149
773	183
836	217
899	251
961	286
7975024	320
087	354
149	388
212	422
275	456
337	490
400	525
463	559
526	593
588	627
651	661
714	695
776	729
839	764
902	798
964	832
7976027	866
090	900
152	934
215	968
278	·9018002
341	037

7976403	·9018071
466	105
529	139
591	173
654	207
717	241
779	276
842	310
905	344
967	378
7977030	412
093	446
155	480
218	514
281	549
343	583
406	617
469	651
531	685
594	719
657	753
719	787
782	821
845	856
907	890
970	924
7978033	958
096	992
158	·9019026
221	060
284	094
346	129
409	163
472	197
534	231
597	265
660	299
722	333
785	367
848	401
910	436
973	470
7979036	504
098	538
161	572
224	606
286	640
349	674
412	708
474	743
537	777
599	811
662	845
725	879
787	913
850	947
913	981
975	·9020015

Excesses

10 = 5 1 = 1
20 = 11 2 = 1
30 = 16 3 = 2
40 = 22 4 = 2
50 = 27 5 = 3
60 = 33 6 = 3
7 = 4
8 = 4
9 = 5

31 = 17

Anti Log Excesses

10 = 18 1 = 2
20 = 37 2 = 4
30 = 55 3 = 6
4 = 7
5 = 9
6 = 11
7 = 13
8 = 15
9 = 17

17 = 31

7980038	·9020049
101	084
163	118
226	152
289	186
351	220
414	254
477	288
539	322
602	356
665	390
727	425
790	459
853	493
915	527
978	561
7981041	595
103	629
166	663
228	697
291	731
354	766
416	800
479	834
542	868
604	902
667	936
730	970
792	·9021004
855	038
918	072
980	106
7982043	140
105	174
168	208
231	242
293	277

7982356	·9021311
419	345
481	379
544	413
607	447
669	481
732	515
794	549
857	583
920	617
982	651
7983045	685
108	720
170	754
233	788
296	822
358	856
421	890
483	924
546	958
609	992
671	·9022026
734	060
797	094
859	128
922	162
984	197
7984047	231
110	265
172	299
235	333
298	367
360	401
423	435
485	469
548	503
611	537
673	571
736	605
799	639
861	673
924	707
986	742
7985049	776
112	810
174	844
237	878
299	912
362	946
425	980
487	·9023014
550	048
613	082
675	116
738	150
800	184
863	218
926	252
988	286
7986051	320
113	355
176	389
239	423
301	457
364	491

7986426	·9023525
489	559
552	593
614	627
677	661
740	695
802	729
865	763
927	797
990	831
7987053	865
115	899
178	933
240	967
303	·9024001
366	035
428	069
491	103
553	138
616	172
679	206
741	240
804	274
866	308
929	342
991	376
7988054	410
117	444
179	478
242	512
304	546
367	580
430	614
492	648
555	682
617	716
680	750
743	784
805	818
868	852
930	886
993	920
7989056	954
118	988
181	·9025022
243	056
306	090
368	124
431	158
494	192
556	226
619	260
681	294
744	328
807	362
869	396
932	430
994	464

Excesses

10 = 5 1 = 1
20 = 11 2 = 1
30 = 16 3 = 2
40 = 22 4 = 2
50 = 27 5 = 3
60 = 33 6 = 3
7 = 4
8 = 4
9 = 5

31 = 17

Anti Log Excesses

10 = 18 1 = 2
20 = 37 2 = 4
30 = 55 3 = 6
4 = 7
5 = 9
6 = 11
7 = 13
8 = 15
9 = 17

17 = 31

7990057	·9025498
119	532
182	567
245	601
307	635
370	669
432	703
495	737
557	771
620	805
683	839
745	873
808	907
870	941
933	975
995	·9026009
7991058	043
121	077
183	111
246	145
308	179
371	213
433	247
496	281
559	315
621	349
684	383
746	417
809	451
871	485
934	519
996	553
7992059	587
122	621
184	655
247	689
309	723

7992372	·9026757
434	791
497	825
560	859
622	893
685	927
747	961
810	995
872	·9027029
935	063
997	097
7993060	131
123	165
185	199
248	233
310	267
373	301
435	335
498	369
560	403
623	437
686	470
748	504
811	538
873	572
936	606
998	640
7994061	674
123	708
186	742
248	776
311	810
374	844
436	878
499	912
561	946
624	980
686	·9028014
749	048
811	082
874	116
936	150
999	184
7995061	218
124	252
187	286
249	320
312	354
374	388
437	422
499	456
562	490
624	524
687	558
749	591
812	625
874	659
937	693
999	727
7996062	761
125	795
187	829
250	863
312	897
375	931

7996437	·9028965
500	999
562	·9029033
625	067
687	101
750	135
812	169
875	203
937	237
7997000	271
062	305
125	339
188	373
250	406
313	440
375	474
438	508
500	542
563	576
625	610
688	644
750	678
813	712
875	746
938	780
7998000	814
063	848
125	882
188	916
250	950
313	984
375	·9030018
438	052
500	085
563	119
625	153
688	187
750	221
813	255
875	289
938	323
7999000	357
063	391
125	425
188	459
250	493
313	527
375	561
438	595
500	628
563	662
625	696
688	730
750	764
813	798
875	832
938	866

Excesses

10 = 5	1 = 1
20 = 11	2 = 1
30 = 16	3 = 2
40 = 22	4 = 2
50 = 27	5 = 3
60 = 33	6 = 3
	7 = 4
	8 = 4
	9 = 5

31 = 17

Anti Log Excesses

10 = 18	1 = 2
20 = 37	2 = 4
30 = 55	3 = 6
	4 = 7
	5 = 9
	6 = 11
	7 = 13
	8 = 15
	9 = 17

17 = 31

8000000	·9030900
063	934
125	968
188	·9031002
250	036
313	070
375	103
438	137
500	171
563	205
625	239
688	273
750	307
813	341
875	375
938	409
8001000	443
063	477
125	511
188	545
250	578
313	612
375	646
438	680
500	714
563	748
625	782
688	816
750	850
813	884
875	918
938	952
8002000	985
063	·9032019
125	053
188	087
250	121

8002313	·9032155
375	189
438	223
500	257
563	291
625	325
688	359
750	392
813	426
875	460
937	494
8003000	528
062	562
125	596
187	630
250	664
312	698
375	731
437	765
500	799
562	833
625	867
687	901
750	935
812	969
875	·9033003
937	037
8004000	071
062	104
124	138
187	172
249	206
312	240
374	274
437	308
499	341
562	375
624	409
687	443
749	477
812	511
874	545
936	579
999	613
8005061	646
124	680
186	714
249	748
311	782
374	816
436	850
499	884
561	918
624	951
686	985
748	·9034019
811	053
873	087
936	121
998	155
8006061	189
123	222
186	256
248	290
311	324

8006373	·9034358
435	392
498	426
560	460
623	493
685	527
748	561
810	595
873	629
935	663
997	697
8007060	731
122	764
185	798
247	832
310	866
372	900
435	934
497	968
559	·9035002
622	035
684	069
747	103
809	137
872	171
934	205
997	239
8008059	272
121	306
184	340
246	374
309	408
371	442
434	476
496	510
558	543
621	577
683	611
746	645
808	679
871	713
933	746
995	780
8009058	814
120	848
183	882
245	916
308	949
370	983
432	·9036017
495	051
557	085
620	119
682	153
745	186
807	220
869	254
932	288
994	322

Excesses

10 = 5	1 = 1
20 = 11	2 = 1
30 = 16	3 = 2
40 = 22	4 = 2
50 = 27	5 = 3
60 = 33	6 = 3
	7 = 4
	8 = 4
	9 = 5

31 = 17

Anti Log Excesses

10 = 18	1 = 2
20 = 37	2 = 4
30 = 55	3 = 6
	4 = 7
	5 = 9
	6 = 11
	7 = 13
	8 = 15
	9 = 17

17 = 31

8010057	·9036356
119	389
182	423
244	457
306	491
369	525
431	559
494	593
556	626
618	660
681	694
743	728
806	762
868	796
931	829
993	863
8011055	897
118	931
180	965
243	999
305	·9037032
367	066
430	100
492	134
555	168
617	202
679	235
742	269
804	303
867	337
929	371
992	405
8012054	438
116	472
179	506
241	540
304	574

8012366	·9037608
428	641
491	675
553	709
616	743
678	777
740	810
803	844
865	878
928	912
990	946
8013052	980
115	·9038013
177	047
240	081
302	115
364	149
427	182
489	216
552	250
614	284
676	318
739	351
801	385
863	419
926	453
988	487
8014051	521
113	554
175	588
238	622
300	656
363	690
425	723
487	757
550	791
612	825
675	859
737	892
799	926
862	960
924	994
986	·9039028
8015049	061
111	095
174	129
236	163
298	197
361	231
423	264
486	298
548	332
610	366
673	400
735	433
797	467
860	501
922	535
985	569
8016047	602
109	636
172	670
234	704
296	737
359	771

8016421	·9039805
484	839
546	873
608	906
671	940
733	974
795	·9040008
858	041
920	075
982	109
8017045	143
107	177
170	211
232	244
294	278
357	312
419	346
481	379
544	413
606	447
668	481
731	515
793	548
856	582
918	616
980	650
8018043	683
105	717
167	751
230	785
292	818
354	852
417	886
479	920
542	954
604	987
666	·9041021
729	055
791	089
853	122
916	156
978	190
8019040	224
103	258
165	291
227	325
290	359
352	393
414	426
477	460
539	494
601	528
664	561
726	595
789	629
851	663
913	697
976	730

Excesses

10 = 5	1 = 1
20 = 11	2 = 1
30 = 16	3 = 2
40 = 22	4 = 2
50 = 27	5 = 3
60 = 32	6 = 3
	7 = 4
	8 = 4
	9 = 5

31 = 17

Anti Log Excesses

10 = 18	1 = 2
20 = 37	2 = 4
30 = 55	3 = 6
	4 = 7
	5 = 9
	6 = 11
	7 = 13
	8 = 15
	9 = 17

17 = 31

8020038	·9041764
100	798
163	832
225	865
287	899
350	933
412	967
474	·9042000
537	034
599	068
661	102
724	135
786	169
848	203
911	237
973	271
8021035	304
098	338
160	372
222	406
285	439
347	473
409	507
472	541
534	574
596	608
659	642
721	676
783	709
846	743
908	777
970	811
8022033	844
095	878
157	912
220	946
282	979

8022344	·9043013
407	047
469	080
531	114
594	148
656	182
718	215
781	249
843	283
905	317
968	350
8023030	384
092	418
154	451
217	485
279	519
341	553
404	586
466	620
528	654
591	688
653	721
715	755
778	789
840	822
902	856
965	890
8024027	924
089	957
152	991
214	·9044025
276	059
338	092
401	126
463	160
525	193
588	227
650	261
712	295
775	328
837	362
899	396
962	429
8025024	463
086	497
148	531
211	564
273	598
335	632
398	665
460	699
522	733
585	767
647	800
709	834
771	868
834	901
896	935
958	969
8026021	·9045003
083	036
145	070
208	104
270	137
332	171

8026394	·9045205
457	239
519	272
581	306
644	340
706	373
768	407
831	441
893	475
955	508
8027017	542
080	576
142	609
204	643
267	677
329	710
391	744
453	778
516	812
578	845
640	879
703	913
765	946
827	980
889	·9046014
952	047
8028014	081
076	115
139	149
201	182
263	216
325	250
388	283
450	317
512	351
574	384
637	418
699	452
761	485
824	519
886	553
948	586
8029010	620
073	654
135	688
197	721
259	755
322	789
384	822
446	856
509	890
571	923
633	957
695	991
758	·9047024
820	058
882	092
944	125

Excesses

10 = 5	1 = 1
20 = 11	2 = 1
30 = 16	3 = 2
40 = 22	4 = 2
50 = 27	5 = 3
60 = 32	6 = 3
	7 = 4
	8 = 4
	9 = 5

31 = 17

Anti Log Excesses

10 = 18	1 = 2
20 = 37	2 = 4
30 = 55	3 = 6
	4 = 7
	5 = 9
	6 = 11
	7 = 13
	8 = 15
	9 = 17

17 = 31

8030007	·9047159
069	193
131	226
194	260
256	294
318	327
380	361
443	395
505	428
567	462
629	496
692	530
754	563
816	597
878	631
941	664
8031003	698
065	732
127	765
190	799
252	833
314	866
376	900
439	933
501	967
563	·9048001
625	034
688	068
750	102
812	135
875	169
937	203
999	236
8032061	270
124	304
186	337
248	371

8032310	·9048405
373	438
435	472
497	506
559	539
621	573
684	607
746	640
808	674
870	708
933	741
995	775
8033057	809
119	842
182	876
244	910
306	943
368	977
431	·9049010
493	044
555	078
617	111
680	145
742	179
804	212
866	246
929	280
991	313
8034053	347
115	381
177	414
240	448
302	482
364	515
426	549
489	582
551	616
613	650
675	683
738	717
800	751
862	784
924	818
986	852
8035049	885
111	919
173	953
235	986
298	·9050020
360	053
422	087
484	121
547	154
609	188
671	222
733	255
795	289
858	322
920	356
982	389
8036044	423
107	457
169	490
231	524
293	558

8036355	·9050591
418	625
480	658
542	692
604	726
666	759
729	793
791	827
853	860
915	894
978	927
8037040	961
102	995
164	·9051028
226	062
289	095
351	129
413	163
475	196
537	230
600	264
662	297
724	331
786	364
848	398
911	432
973	465
8038035	499
097	532
159	566
222	600
284	633
346	667
408	701
471	734
533	768
595	801
657	835
719	869
782	902
844	936
906	969
968	·9052003
8039030	037
092	070
155	104
217	137
279	171
341	205
403	238
466	272
528	305
590	339
652	373
714	406
777	440
839	473
901	507
963	541

Excesses

10 = 5	1 = 1
20 = 11	2 = 1
30 = 16	3 = 2
40 = 22	4 = 2
50 = 27	5 = 3
60 = 32	6 = 3
	7 = 4
	8 = 4
	9 = 5

31 = 17

Anti Log Excesses

10 = 18	1 = 2
20 = 37	2 = 4
30 = 55	3 = 6
	4 = 7
	5 = 9
	6 = 11
	7 = 13
	8 = 15
	9 = 17

17 = 31

8040025	·9052574
088	608
150	641
212	675
274	709
336	742
399	776
461	809
523	843
585	876
647	910
709	944
772	977
834	·9053011
896	044
958	078
8041020	111
083	145
145	179
207	212
269	246
331	279
393	313
456	346
518	380
580	414
642	447
704	481
766	514
829	548
891	582
953	615
8042015	649
077	682
140	716
202	749
264	783

8042326	·9053817
388	850
450	884
513	917
575	951
637	984
699	·9054018
761	052
823	085
886	119
948	152
8043010	186
072	219
134	253
196	287
259	320
321	354
383	387
445	421
507	454
569	488
632	521
694	555
756	589
818	622
880	656
942	689
8044004	723
067	756
129	790
191	824
253	857
315	891
377	924
440	958
502	991
564	·9055025
626	058
688	092
750	126
812	159
875	193
937	226
999	260
8045061	293
123	327
185	360
248	394
310	427
372	461
434	495
496	528
558	562
620	595
683	629
745	662
807	696
869	729
931	763
993	796
8046055	830
118	864
180	897
242	931
304	964

8046366	·9055998
428	·9056031
490	065
553	098
615	132
677	165
739	199
801	232
863	266
925	299
988	333
8047050	367
112	400
174	434
236	467
298	501
360	534
422	568
485	601
547	635
609	668
671	702
733	735
795	769
857	802
919	836
982	869
8048044	903
106	936
168	970
230	·9057004
292	037
354	071
416	104
479	138
541	171
603	205
665	238
727	272
789	305
851	339
913	372
976	406
8049038	439
100	473
162	506
224	540
286	573
348	607
410	640
473	674
535	707
597	741
659	775
721	808
783	842
845	875
907	909
969	942

Excesses

10 = 5	1 = 1
20 = 11	2 = 1
30 = 16	3 = 2
40 = 22	4 = 2
50 = 27	5 = 3
60 = 32	6 = 3
	7 = 4
	8 = 4
	9 = 5

31 = 17

Anti Log Excesses

10 = 19	1 = 2
20 = 37	2 = 4
30 = 56	3 = 6
	4 = 7
	5 = 9
	6 = 11
	7 = 13
	8 = 15
	9 = 17

17 = 31

8050032	·9057976
094	·9058009
156	043
218	076
280	110
342	143
404	177
466	210
528	244
591	277
653	311
715	344
777	378
839	411
901	445
963	478
8051025	512
087	545
149	579
212	612
274	646
336	679
398	713
460	746
522	780
584	813
646	847
708	880
770	914
833	947
895	981
957	·9059014
8052019	048
081	081
143	115
205	148
267	182

8052329	·9059215
391	249
454	282
516	316
578	349
640	382
702	416
764	449
826	483
888	516
950	550
8053012	583
074	617
137	650
199	684
261	717
323	751
385	784
447	818
509	851
571	885
633	918
695	952
757	985
819	·9060019
882	052
944	086
8054006	119
068	153
130	186
192	219
254	253
316	287
378	320
440	353
502	387
564	420
627	454
689	487
751	521
813	554
875	588
937	621
999	655
8055061	688
123	721
185	755
247	788
309	822
371	855
433	889
496	922
558	956
620	989
682	·9061023
744	056
806	090
868	123
930	156
992	190
8056054	223
116	257
178	290
240	324
302	357

8056364	·9061391
427	424
489	458
551	491
613	524
675	558
737	591
799	625
861	658
923	692
985	725
8057047	759
109	792
171	826
233	859
295	892
357	926
419	960
481	993
544	·9062026
606	060
668	093
730	127
792	160
854	193
916	227
978	260
8058040	294
102	327
164	361
226	394
288	428
350	461
412	494
474	528
536	561
598	595
660	628
722	662
785	695
847	729
909	762
971	795
8059033	829
095	862
157	896
219	929
281	963
343	996
405	·9063029
467	063
529	096
591	130
653	163
715	197
777	230
839	263
901	297
963	330

Excesses

10 = 5	1 = 1
20 = 11	2 = 1
30 = 16	3 = 2
40 = 22	4 = 2
50 = 27	5 = 3
60 = 32	6 = 3
	7 = 4
	8 = 4
	9 = 5

31 = 17

Anti Log Excesses

10 = 19	1 = 2
20 = 37	2 = 4
30 = 56	3 = 6
	4 = 7
	5 = 9
	6 = 11
	7 = 13
	8 = 15
	9 = 17

17 = 31

8060025	·9063364
087	397
149	431
211	464
273	497
335	531
398	564
460	598
522	631
584	665
646	698
708	731
770	765
832	798
894	832
956	865
8061018	898
080	932
142	965
204	999
266	·9064032
328	066
390	099
452	132
514	166
576	199
638	233
700	266
762	299
824	333
886	366
948	400
8062010	433
072	467
134	500
196	533
258	567

8062320	·9064600
382	634
444	667
506	700
568	734
630	767
692	801
754	834
816	867
878	901
940	934
8063002	968
064	·9065001
126	034
188	068
250	101
312	135
374	168
436	201
499	235
561	268
623	302
685	335
747	368
809	402
871	435
933	468
995	502
8064057	535
119	569
181	602
243	635
305	669
367	702
429	736
491	769
553	802
615	836
677	869
739	903
801	936
863	969
925	·9066003
987	036
8065049	069
111	103
173	136
235	170
297	203
359	236
421	270
483	303
544	336
606	370
668	403
730	437
792	470
854	503
916	537
978	570
8066040	604
102	637
164	670
226	704
288	737

8066350	·9066770
412	804
474	837
536	870
598	904
660	937
722	971
784	·9067004
846	037
908	071
970	104
8067032	137
094	171
156	204
218	238
280	271
342	304
404	338
466	371
528	404
590	438
652	471
714	504
776	538
838	571
900	605
962	638
8068024	671
086	705
148	738
210	771
272	805
334	838
396	871
458	905
520	938
582	972
644	·9068005
705	038
767	072
829	105
891	138
953	172
8069015	205
077	238
139	272
201	305
263	338
325	372
387	405
449	438
511	472
573	505
635	538
697	572
759	605
821	639
883	672
945	705

Excesses

10 = 5	1 = 1
20 = 11	2 = 1
30 = 16	3 = 2
40 = 22	4 = 2
50 = 27	5 = 3
60 = 32	6 = 3
	7 = 4
	8 = 4
	9 = 5

31 = 17

Anti Log Excesses

10 = 19	1 = 2
20 = 37	2 = 4
30 = 56	3 = 6
	4 = 7
	5 = 9
	6 = 11
	7 = 13
	8 = 15
	9 = 17

17 = 31

8070007	·9068739
069	772
131	805
193	839
255	872
316	905
378	939
440	972
502	·9069005
564	039
626	072
688	105
750	139
812	172
874	205
936	239
998	272
8071060	305
122	339
184	372
246	405
308	439
370	472
432	505
494	539
555	572
617	605
679	639
741	672
803	705
865	739
927	772
989	805
8072051	839
113	872
175	905
237	939

8072299	·9069972
361	·9070005
423	039
485	072
547	105
609	139
670	172
732	205
794	238
856	272
918	305
980	338
8073042	372
104	405
166	438
228	472
290	505
352	538
414	572
476	605
538	638
599	672
661	705
723	738
785	772
847	805
909	838
971	872
8074033	905
095	938
157	971
219	·9071005
281	038
343	071
405	105
466	138
528	171
590	205
652	238
714	271
776	305
838	338
900	371
962	404
8075024	438
086	471
148	504
209	538
271	571
333	604
395	638
457	671
519	704
581	737
643	771
705	804
767	837
829	871
891	904
952	937
8076014	971
076	·9072004
138	037
200	070
262	104

Number	Log
8076324	·9072137
386	170
448	204
510	237
572	270
633	303
695	337
757	370
819	403
881	437
943	470
8077005	503
067	536
129	570
191	603
253	636
314	670
376	703
438	736
500	769
562	803
624	836
686	869
748	902
810	936
872	969
933	·9073002
995	036
8078057	069
119	102
181	135
243	169
305	202
367	235
429	269
490	302
552	335
614	368
676	402
738	435
800	468
862	501
924	535
986	568
8079047	601
109	634
171	668
233	701
295	734
357	768
419	801
481	834
543	867
604	901
666	934
728	967
790	·9074000
852	034
914	067
976	100

Excesses

10 = 5	1 = 1
20 = 11	2 = 1
30 = 16	3 = 2
40 = 22	4 = 2
50 = 27	5 = 3
60 = 32	6 = 3
	7 = 4
	8 = 4
	9 = 5

31 = 17

Anti Log Excesses

10 = 19	1 = 2
20 = 37	2 = 4
30 = 56	3 = 6
	4 = 7
	5 = 9
	6 = 11
	7 = 13
	8 = 15
	9 = 17

17 = 31

Number	Log
8080038	·9074133
100	167
161	200
223	233
285	267
347	300
409	333
471	366
533	400
595	433
656	466
718	499
780	533
842	566
904	599
966	632
8081028	666
090	699
151	732
213	765
275	799
337	832
399	865
461	898
523	932
585	965
646	998
708	·9075031
770	065
832	098
894	131
956	164
8082018	197
079	231
141	264
203	297
265	330

Number	Log
8082327	·9075364
389	397
451	430
512	463
574	497
636	530
698	563
760	596
822	630
884	663
946	696
8083007	729
069	763
131	796
193	829
255	862
317	895
379	929
440	962
502	995
564	·9076028
626	062
688	095
750	128
811	161
873	195
935	228
997	261
8084059	294
121	327
183	361
244	394
306	427
368	460
430	494
492	527
554	560
616	593
677	626
739	660
801	693
863	726
925	759
987	793
8085048	826
110	859
172	892
234	925
296	959
358	992
419	·9077025
481	058
543	091
605	125
667	158
729	191
791	224
852	258
914	291
976	324
8086038	357
100	390
162	424
223	457
285	490

Number	Log
8086347	·9077523
409	556
471	590
533	623
594	656
656	689
718	723
780	756
842	789
903	822
965	855
8087027	889
089	922
151	955
213	988
274	·9078021
336	055
398	088
460	121
522	154
584	187
645	221
707	254
769	287
831	320
893	353
955	387
8088016	420
078	453
140	486
202	519
264	552
325	586
387	619
449	652
511	685
573	718
635	752
696	785
758	818
820	851
882	884
944	917
8089005	951
067	984
129	·9079017
191	050
253	084
314	117
376	150
438	183
500	216
562	249
623	283
685	316
747	349
809	382
871	415
933	449
994	482

Excesses

10 = 5	1 = 1
20 = 11	2 = 1
30 = 16	3 = 2
40 = 21	4 = 2
50 = 27	5 = 3
60 = 32	6 = 3
	7 = 4
	8 = 4
	9 = 5

31 = 17

Anti Log Excesses

10 = 19	1 = 2
20 = 37	2 = 4
30 = 56	3 = 6
	4 = 7
	5 = 9
	6 = 11
	7 = 13
	8 = 15
	9 = 17

17 = 31

Number	Log
8090056	·9079515
118	548
180	581
242	614
303	648
365	681
427	714
489	747
551	780
612	813
674	847
736	880
798	913
860	946
921	979
983	·9080012
8091045	046
107	079
169	112
230	145
292	178
354	212
416	245
477	278
539	311
601	344
663	377
725	411
786	444
848	477
910	510
972	543
8092034	576
095	610
157	643
219	676
281	709

Number	Log
8092343	·9080742
404	775
466	808
528	842
590	875
651	908
713	941
775	974
837	·9081007
899	041
960	074
8093022	107
084	140
146	173
208	206
269	240
331	273
393	306
455	339
516	372
578	405
640	438
702	472
764	505
825	538
887	571
949	604
8094011	637
072	670
134	704
196	737
258	770
319	803
381	836
443	869
505	903
567	936
628	969
690	·9082002
752	035
814	068
875	101
937	135
999	168
8095061	201
122	234
184	267
246	300
308	333
370	367
431	400
493	433
555	466
617	499
678	532
740	565
802	598
864	632
925	665
987	698
8096049	731
111	764
172	797
234	830
296	863

Number	Log
8096358	·9082897
419	930
481	963
543	996
605	·9083029
666	062
728	095
790	128
852	162
913	195
975	228
8097037	261
099	294
160	327
222	360
284	393
346	427
407	460
469	493
531	526
593	559
654	592
716	625
778	658
840	691
901	725
963	758
8098025	791
087	824
148	857
210	890
272	923
334	956
395	989
457	·9084023
519	056
581	089
642	122
704	155
766	188
828	221
889	254
951	287
8099013	321
075	354
136	387
198	420
260	453
321	486
383	519
445	552
507	585
568	619
630	652
692	685
754	718
815	751
877	784
939	817

Excesses

10 = 5	1 = 1
20 = 11	2 = 1
30 = 16	3 = 2
40 = 21	4 = 2
50 = 27	5 = 3
60 = 32	6 = 3
	7 = 4
	8 = 4
	9 = 5

31 = 17

Anti Log Excesses

10 = 19	1 = 2
20 = 37	2 = 4
30 = 56	3 = 6
	4 = 7
	5 = 9
	6 = 11
	7 = 13
	8 = 15
	9 = 17

17 = 31

Number	Log
8100000	·9084850
062	883
124	916
186	950
247	983
309	·9085016
371	049
433	082
494	115
556	148
618	181
679	214
741	247
803	280
865	314
926	347
988	380
8101050	413
112	446
173	479
235	512
297	545
358	578
420	611
482	644
544	678
605	711
667	744
729	777
790	810
852	843
914	876
976	909
8102037	942
099	975
161	·9086008
222	041

Number	Log
8102284	·9086075
346	108
408	141
469	174
531	207
593	240
654	273
716	306
778	339
840	372
901	405
963	438
8103025	471
086	505
148	538
210	571
271	604
333	637
395	670
457	703
518	736
580	769
642	802
703	835
765	868
827	901
888	934
950	967
8104012	·908700
074	033
135	066
197	100
259	133
320	166
382	199
444	232
505	265
567	298
629	331
691	364
752	397
814	430
876	463
937	496
999	529
8105061	562
122	595
184	628
246	661
307	695
369	728
431	761
492	794
554	827
616	860
678	893
739	926
801	959
863	992
924	·9088025
986	058
8106048	091
109	124
171	157
233	190

Number	Log
8106294	·9088223
356	256
418	289
479	322
541	355
603	389
664	422
726	455
788	488
849	521
911	554
973	587
8107034	620
096	653
158	686
220	719
281	752
343	785
405	818
466	851
528	884
590	917
651	950
713	983
775	·9089016
836	049
898	082
960	115
8108021	148
083	181
145	214
206	247
268	280
330	314
391	347
453	380
515	413
576	446
638	479
700	512
761	545
823	578
885	611
946	644
8109008	677
069	710
131	743
193	776
254	809
316	842
378	875
439	908
501	941
563	974
624	·9090007
686	040
748	073
809	106
871	139
933	172
994	205

Excesses

10 = 5	1 = 1
20 = 11	2 = 1
30 = 16	3 = 2
40 = 21	4 = 2
50 = 27	5 = 3
60 = 32	6 = 3
	7 = 4
	8 = 4
	9 = 5

31 = 17

Anti Log Excesses

10 = 19	1 = 2
20 = 37	2 = 4
30 = 56	3 = 6
	4 = 7
	5 = 9
	6 = 11
	7 = 13
	8 = 15
	9 = 17

17 = 31

8110056	·9090238
118	271
179	304
241	337
303	370
364	403
426	436
488	469
549	502
611	535
672	568
734	601
796	634
857	667
919	700
981	733
8111042	766
104	799
166	832
227	865
289	898
351	931
412	964
474	997
535	·9091030
597	063
659	096
720	129
782	162
844	195
905	228
967	261
8112029	294
090	327
152	360
214	393
275	426

8112337	·9091459
398	492
460	525
522	558
583	591
645	624
707	657
768	690
830	723
891	756
953	789
8113015	822
076	855
138	888
200	921
261	954
323	987
385	·9092020
446	053
508	086
569	119
631	152
693	185
754	218
816	251
878	284
939	317
8114001	350
062	383
124	416
186	449
247	482
309	515
370	548
432	581
494	614
555	647
617	680
679	713
740	746
802	779
863	812
925	845
987	878
8115048	911
110	944
171	977
233	·9093010
295	043
356	076
418	109
480	141
541	174
603	207
664	240
726	273
788	306
849	339
911	372
972	405
8116034	438
096	471
157	504
219	537
280	570

8116342	·9093603
404	636
465	669
527	702
588	735
650	768
712	801
773	834
835	867
896	900
958	933
8117020	966
081	998
143	·9094031
204	064
266	097
328	130
389	163
451	196
512	229
574	262
636	295
697	328
759	361
820	394
882	427
944	460
8118005	493
067	526
128	559
190	592
252	625
313	658
375	691
436	724
498	756
559	789
621	822
683	855
744	888
806	921
867	954
929	987
991	·9095020
8119052	053
114	086
175	119
237	152
298	185
360	218
422	251
483	284
545	317
606	349
668	382
730	415
791	448
853	481
914	514
976	547

Excesses

10 = 5	1 = 1
20 = 11	2 = 1
30 = 16	3 = 2
40 = 21	4 = 2
50 = 27	5 = 3
60 = 32	6 = 3
	7 = 4
	8 = 4
	9 = 5

31 = 16

Anti Log Excesses

10 = 19	1 = 2
20 = 37	2 = 4
30 = 56	3 = 6
	4 = 7
	5 = 9
	6 = 11
	7 = 13
	8 = 15
	9 = 17

16 = 31

8120037	·9095580
099	613
161	646
222	679
284	712
345	745
407	778
468	811
530	843
592	876
653	909
715	942
776	975
838	·9096008
899	041
961	074
8121023	107
084	140
146	173
207	206
269	239
330	272
392	304
454	337
515	370
577	403
638	436
700	469
761	502
823	535
885	568
946	601
8122008	634
069	667
131	699
192	732
254	765

8122315	·9096798
377	831
439	864
500	897
562	930
623	963
685	996
746	·9097029
808	061
869	094
931	127
993	160
8123054	193
116	226
177	259
239	292
300	325
362	358
423	391
485	423
547	456
608	489
670	522
731	555
793	588
854	621
916	654
977	687
8124039	720
100	752
162	785
224	818
285	851
347	884
408	917
470	950
531	983
593	·9098016
654	049
716	081
777	114
839	147
900	180
962	213
8125024	246
085	279
147	312
208	345
270	378
331	410
393	443
454	476
516	509
577	542
639	575
700	608
762	641
824	674
885	706
947	739
8126008	772
070	805
131	838
193	871
254	904

Number	Log
8126316	·9098937
377	970
439	·9099002
500	035
562	068
623	101
685	134
746	167
808	200
870	232
931	265
993	298
8127054	331
116	364
177	397
239	430
300	463
362	495
423	528
485	561
546	594
608	627
669	660
731	693
792	726
854	758
915	791
977	824
8128038	857
100	890
161	923
223	956
284	989
346	·9100021
407	054
469	087
531	120
592	153
654	186
715	219
777	252
838	284
900	317
961	350
8129023	383
084	416
146	449
207	482
269	514
330	547
392	580
453	613
515	646
576	679
638	712
699	745
761	777
822	810
884	843
945	876

Excesses

10 = 5	1 = 1
20 = 11	2 = 1
30 = 16	3 = 2
40 = 21	4 = 2
50 = 27	5 = 3
60 = 32	6 = 3
	7 = 4
	8 = 4
	9 = 5

31 = 16

Anti Log Excesses

10 = 19	1 = 2
20 = 37	2 = 4
30 = 56	3 = 6
	4 = 7
	5 = 9
	6 = 11
	7 = 13
	8 = 15
	9 = 17

16 = 31

Number	Log
8130007	·9100909
068	942
130	975
191	·9101007
253	040
314	073
376	106
437	139
499	172
560	204
622	237
683	270
745	303
806	336
868	369
929	402
991	434
8131052	467
114	500
175	533
237	566
298	599
360	631
421	664
483	697
544	730
606	763
667	796
728	828
790	861
851	894
913	927
974	960
8132036	993
097	·9102025
159	058
220	091

Number	Log
8132282	·9102124
343	157
405	190
466	222
528	255
589	288
651	321
712	354
774	387
835	419
897	452
958	485
8133020	518
081	551
143	584
204	616
266	649
327	682
388	715
450	748
511	781
573	813
634	846
696	879
757	912
819	945
880	978
942	·9103010
8134003	043
065	076
126	109
188	142
249	174
311	207
372	240
433	273
495	306
556	339
618	371
679	404
741	437
802	470
864	503
925	535
987	568
8135048	601
110	634
171	667
233	700
294	732
355	765
417	798
478	831
540	864
601	896
663	929
724	962
786	995
847	·9104027
909	060
970	093
8136031	126
093	159
154	192
216	224

Number	Log
8136277	·9104257
339	290
400	323
462	356
523	388
585	421
646	454
707	487
769	519
830	552
892	585
953	618
8137015	651
076	683
138	716
199	749
260	782
322	815
383	847
445	880
506	913
568	946
629	979
691	·9105011
752	044
813	077
875	110
936	143
998	175
8138059	208
121	241
182	274
244	306
305	339
366	372
428	405
489	438
551	470
612	503
674	536
735	569
796	602
858	634
919	667
981	700
8139042	733
104	765
165	798
227	831
288	864
349	897
411	929
472	962
534	995
595	·9106028
657	060
718	093
779	126
841	159
902	191
964	224

Excesses

10 = 5	1 = 1
20 = 11	2 = 1
30 = 16	3 = 2
40 = 21	4 = 2
50 = 27	5 = 3
60 = 32	6 = 3
	7 = 4
	8 = 4
	9 = 5

31 = 16

Anti Log Excesses

10 = 19	1 = 2
20 = 37	2 = 4
30 = 56	3 = 6
	4 = 7
	5 = 9
	6 = 11
	7 = 13
	8 = 15
	9 = 17

16 = 31

Number	Log
8140025	·9106257
087	290
148	323
209	355
271	388
332	421
394	454
455	486
516	519
578	552
639	585
701	618
762	650
824	683
885	716
946	749
8141008	781
069	814
131	847
192	880
253	912
315	945
376	978
438	·9107011
499	043
561	076
622	109
683	142
745	174
806	207
868	240
929	273
990	306
8142052	338
113	371
175	404
236	437

8142297	·9107469
359	502
420	535
482	568
543	600
605	633
666	666
727	699
789	731
850	764
912	797
973	830
8143034	862
096	895
157	928
219	961
280	993
341	·9108026
403	059
464	091
526	124
587	157
648	190
710	222
771	255
833	288
894	321
955	353
8144017	386
078	419
139	452
201	484
262	517
324	550
385	583
446	615
508	648
569	681
631	714
692	746
753	779
815	812
876	844
938	877
999	910
8145060	943
122	975
183	·9109008
245	041
306	074
367	106
429	139
490	172
551	205
613	237
674	270
736	303
797	335
858	368
920	401
981	434
8146042	466
104	499
165	532
227	565

8146288	·9109597
349	630
411	663
472	695
534	728
595	761
656	794
718	826
779	859
840	892
902	924
963	957
8147024	990
086	·9110023
147	055
209	088
270	121
331	153
393	186
454	219
515	252
577	284
638	317
700	350
761	382
822	415
884	448
945	481
8148006	513
068	546
129	579
190	611
252	644
313	677
375	710
436	742
497	775
559	808
620	840
681	873
743	906
804	938
865	971
927	·9111004
988	037
8149050	069
111	102
172	135
234	167
295	200
356	233
418	265
479	298
540	331
602	364
663	396
724	429
786	462
847	494
908	527
970	560

Excesses

10 = 5	1 = 1
20 = 11	2 = 1
30 = 16	3 = 2
40 = 21	4 = 2
50 = 27	5 = 3
60 = 32	6 = 3
	7 = 4
	8 = 4
	9 = 5

31 = 16

Anti Log Excesses

10 = 19	1 = 2
20 = 38	2 = 4
30 = 56	3 = 6
	4 = 8
	5 = 9
	6 = 11
	7 = 13
	8 = 15
	9 = 17

16 = 31

8150031	·9111592
093	625
154	658
215	690
277	723
338	756
399	788
461	821
522	854
583	887
645	919
706	952
767	985
829	·9112017
890	050
951	083
8151013	115
074	148
135	181
197	213
258	246
319	279
381	311
442	344
503	377
565	409
626	442
687	475
749	507
810	540
871	573
933	606
994	638
8152055	671
117	704
178	736
239	769

8152301	·9112802
362	834
423	867
485	900
546	932
607	965
669	998
730	·9113030
791	063
853	096
914	128
975	161
8153037	194
098	226
159	259
221	292
282	324
343	357
405	389
466	422
527	455
589	488
650	520
711	553
773	586
834	618
895	651
957	683
8154018	716
079	749
141	782
202	814
263	847
325	880
386	912
447	945
508	978
570	·9114010
631	043
692	076
754	108
815	141
876	173
938	206
999	239
8155060	271
122	304
183	337
244	369
306	402
367	435
428	467
489	500
551	533
612	565
673	598
735	631
796	663
857	696
919	728
980	761
8156041	794
103	826
164	859
225	892

8156286	·9114924
348	957
409	990
470	·9115022
532	055
593	088
654	120
716	153
777	185
838	218
899	251
961	283
8157022	316
083	349
145	381
206	414
267	447
328	479
390	512
451	544
512	577
574	610
635	642
696	675
758	708
819	740
880	773
941	805
8158003	838
064	871
125	903
187	936
248	969
309	·9116001
370	034
432	066
493	099
554	132
616	164
677	197
738	230
799	262
861	295
922	327
983	360
8159045	393
106	425
167	458
228	490
290	523
351	556
412	588
474	621
535	654
596	686
657	719
719	751
780	784
841	817
902	849
964	882

Excesses

10 = 5	1 = 1
20 = 11	2 = 1
30 = 16	3 = 2
40 = 21	4 = 2
50 = 27	5 = 3
60 = 32	6 = 3
	7 = 4
	8 = 4
	9 = 5

31 = 16

Anti Log Excesses

10 = 19	1 = 2
20 = 38	2 = 4
30 = 56	3 = 6
	4 = 8
	5 = 9
	6 = 11
	7 = 13
	8 = 15
	9 = 17

16 = 31

8160025	·9116914
086	947
148	980
209	·9117012
270	045
331	078
393	110
454	143
515	175
576	208
638	241
699	273
760	306
822	338
883	371
944	404
8161005	436
067	469
128	501
189	534
250	567
312	599
373	632
434	664
495	697
557	730
618	762
679	795
741	827
802	860
863	893
924	925
986	958
8162047	990
108	·9118023
169	056
231	088

8162292	·9118121
353	153
414	186
476	219
537	251
598	284
659	316
721	349
782	382
843	414
904	447
966	479
8163027	512
088	545
149	577
211	610
272	642
333	675
394	707
456	740
517	773
578	805
639	838
701	870
762	903
823	935
884	968
946	·9119001
8164007	033
068	066
129	098
191	131
252	164
313	196
374	229
436	261
497	294
558	326
619	359
681	392
742	424
803	457
864	489
925	522
987	554
8165048	587
109	620
170	652
232	685
293	717
354	750
415	782
477	815
538	848
599	880
660	913
722	945
783	978
844	·9120010
905	043
966	076
8166028	108
089	141
150	173
211	206

8166273	·9120238
334	271
395	304
456	336
518	369
579	401
640	434
701	466
762	499
824	531
885	564
946	597
8167007	629
069	662
130	694
191	727
252	759
313	792
375	825
436	857
497	890
558	922
619	955
681	987
742	·9121020
803	052
864	085
926	118
987	150
8168048	183
109	215
170	248
232	280
293	313
354	345
415	378
476	410
538	443
599	476
660	508
721	541
783	573
844	606
905	638
966	671
8169027	703
089	736
150	768
211	801
272	833
333	866
395	899
456	931
517	964
578	996
639	·9122029
701	061
762	094
823	126
884	159
945	191

Excesses

10 = 5	1 = 1
20 = 11	2 = 1
30 = 16	3 = 2
40 = 21	4 = 2
50 = 27	5 = 3
60 = 32	6 = 3
	7 = 4
	8 = 4
	9 = 5

31 = 16

Anti Log Excesses

10 = 19	1 = 2
20 = 38	2 = 4
30 = 56	3 = 6
	4 = 8
	5 = 9
	6 = 11
	7 = 13
	8 = 15
	9 = 17

16 = 31

8170007	·9122224
068	256
129	289
190	321
251	354
313	387
374	419
435	452
496	484
557	517
619	549
680	582
741	614
802	647
863	679
925	712
986	744
8171047	777
108	809
169	842
231	874
292	907
353	939
414	972
475	·9123005
536	037
598	070
659	102
720	135
781	167
842	200
904	232
965	265
8172026	297
087	330
148	362
210	395

8172271	·9123427
332	460
393	492
454	525
515	557
577	590
638	622
699	655
760	687
821	720
882	752
944	785
8173005	817
066	850
127	882
188	915
250	947
311	980
372	·9124012
433	045
494	077
555	110
617	142
678	175
739	207
800	240
861	272
922	305
984	337
8174045	370
106	402
167	435
228	467
289	500
351	532
412	565
473	597
534	630
595	662
656	695
718	727
779	760
840	792
901	825
962	857
8175023	890
085	922
146	955
207	987
268	·9125020
329	052
390	085
452	117
513	150
574	182
635	215
696	247
757	280
819	312
880	345
941	377
8176002	410
063	442
124	475
185	507

8176247	·9125540
308	572
369	605
430	637
491	670
552	702
613	735
675	767
736	800
797	832
858	864
919	897
980	929
8177042	962
103	994
164	·9126027
225	059
286	092
347	124
408	157
470	189
531	222
592	254
653	286
714	319
775	351
836	384
898	416
959	449
8178020	481
081	514
142	546
203	579
264	611
326	644
387	676
448	709
509	741
570	773
631	806
692	838
753	871
815	903
876	936
937	968
998	·9127001
8179059	033
120	066
181	098
243	130
304	163
365	195
426	228
487	260
548	293
609	325
670	358
732	390
793	423
854	455
915	487
976	520

Excesses

10 = 5	1 = 1
20 = 11	2 = 1
30 = 16	3 = 2
40 = 21	4 = 2
50 = 27	5 = 3
60 = 32	6 = 3
	7 = 4
	8 = 4
	9 = 5

31 = 16

Anti Log Excesses

10 = 19	1 = 2
20 = 38	2 = 4
30 = 56	3 = 6
	4 = 8
	5 = 9
	6 = 11
	7 = 13
	8 = 15
	9 = 17

16 = 31

8180037	·9127552
098	585
159	617
221	650
282	682
343	715
404	747
465	780
526	812
587	844
648	877
710	909
771	942
832	974
893	·9128007
954	039
8181015	072
076	104
137	136
198	169
260	201
321	234
382	266
443	299
504	331
566	363
626	396
687	428
748	461
810	493
871	526
932	558
993	591
8182054	623
115	656
176	688
237	720

8182298	·9128753
360	785
421	818
482	850
543	883
604	915
665	947
726	980
787	·9129012
848	045
910	077
971	110
8183032	142
093	174
154	207
215	239
276	272
337	304
398	337
459	369
521	401
582	434
643	466
704	499
765	531
826	564
887	596
948	628
8184009	661
070	693
131	726
193	758
254	790
315	823
376	855
437	888
498	920
559	953
620	985
681	·9130017
742	050
803	082
865	115
926	147
987	179
8185048	212
109	244
170	277
231	309
292	342
353	374
414	406
475	439
537	471
598	504
659	536
720	568
781	601
842	633
903	666
964	698
8186025	730
086	763
147	795
208	828

8186269	·9130860
331	893
392	925
453	957
514	990
575	·9131022
636	055
697	087
758	119
819	152
880	184
941	217
8187002	249
063	281
125	314
186	346
247	379
308	411
369	443
430	476
491	508
552	540
613	573
674	605
735	638
796	670
857	702
918	735
979	767
8188041	800
102	832
163	864
224	897
285	929
346	962
407	994
468	·9132026
529	059
590	091
651	123
712	156
773	188
834	221
895	253
956	285
8189018	318
079	350
140	383
201	415
262	447
323	480
384	512
445	544
506	577
567	609
628	642
689	674
750	706
811	739
872	771
933	803
994	836

Excesses

10 = 5	1 = 1
20 = 11	2 = 1
30 = 16	3 = 2
40 = 21	4 = 2
50 = 27	5 = 3
60 = 32	6 = 3
	7 = 4
	8 = 4
	9 = 5

31 = 16

Anti Log Excesses

10 = 19	1 = 2
20 = 38	2 = 4
30 = 57	3 = 6
	4 = 8
	5 = 9
	6 = 11
	7 = 13
	8 = 15
	9 = 17

16 = 31

8190055	·9132868
116	900
178	933
239	965
300	998
361	·9133030
422	062
483	095
544	127
605	160
666	192
727	224
788	257
849	289
910	321
971	354
8191032	386
093	418
154	451
215	483
276	515
337	548
398	580
459	613
521	645
582	677
643	710
704	742
765	774
826	807
887	839
948	871
8192009	904
070	936
131	968
192	·9134001
253	033

8192314	·9134066
375	098
436	130
497	163
558	195
619	227
680	260
741	292
802	324
863	357
924	389
985	421
8193046	454
107	486
168	518
229	551
290	583
351	615
412	648
474	680
535	712
596	745
657	777
718	810
779	842
840	874
901	907
962	939
8194023	971
084	·9135004
145	036
206	068
267	101
328	133
389	165
450	198
511	230
572	262
633	295
694	327
755	359
816	392
877	424
938	456
999	489
8195060	521
121	553
182	586
243	618
304	650
365	683
426	715
487	747
548	780
609	812
670	844
731	877
792	909
853	941
914	974
975	·9136006
8196036	038
097	071
158	103
219	135

8196280	·9136167
341	200
402	232
463	265
524	297
585	329
646	361
707	394
768	426
829	458
890	491
951	523
8197012	555
073	588
134	620
195	652
256	685
317	717
378	749
439	782
500	814
561	846
622	879
683	911
744	943
805	975
866	·9137008
927	040
988	072
8198049	105
110	137
171	169
232	202
293	234
354	266
415	299
476	331
537	363
598	395
659	428
720	460
781	492
842	525
903	557
964	589
8199025	622
086	654
147	686
208	718
269	751
330	783
391	815
452	848
513	880
574	912
635	945
696	977
757	·9138009
818	041
879	074
940	106

Excesses

10 = 5	1 = 1
20 = 11	2 = 1
30 = 16	3 = 2
40 = 21	4 = 2
50 = 26	5 = 3
60 = 32	6 = 3
	7 = 4
	8 = 4
	9 = 5

31 = 16

Anti Log Excesses

10 = 19	1 = 2
20 = 38	2 = 4
30 = 57	3 = 6
	4 = 8
	5 = 9
	6 = 11
	7 = 13
	8 = 15
	9 = 17

16 = 31

8200000	·9138138
061	171
122	203
183	235
244	268
305	300
366	332
427	364
488	397
549	429
610	461
671	494
732	526
793	558
854	590
915	623
976	655
8201037	687
098	720
159	752
220	784
281	816
342	849
403	881
464	913
525	946
586	978
647	·9139010
708	042
769	075
830	107
891	139
951	172
8202012	204
073	236
134	268
195	301

8202256	·9139333
317	365
378	398
439	430
500	462
561	494
622	527
683	559
744	591
805	623
866	656
927	688
988	720
8203049	753
110	785
171	817
232	849
293	882
353	914
414	946
475	978
536	·9140011
597	043
658	075
719	107
780	140
841	172
902	204
963	236
8204024	269
085	301
146	333
207	366
268	398
329	430
390	462
451	495
511	527
572	559
633	591
694	624
755	656
816	688
877	721
938	753
999	785
8205060	817
121	850
182	882
243	914
304	946
365	979
426	·9141011
486	043
547	075
608	108
669	140
730	172
791	204
852	237
913	269
974	301
8206035	333
096	366
157	398

8206218	·9141430
279	462
340	495
400	527
461	559
522	591
583	624
644	656
705	688
766	720
827	753
888	785
949	817
8207010	849
071	882
132	914
192	946
253	978
314	·9142010
375	043
436	075
497	107
558	139
619	172
680	204
741	236
802	268
863	301
924	333
984	365
8208045	397
106	430
167	462
228	494
289	526
350	558
411	591
472	623
533	655
594	687
654	720
715	752
776	784
837	816
898	848
959	881
8209020	913
081	945
142	977
203	·9143010
264	042
324	074
385	106
446	138
507	171
568	203
629	235
690	267
751	299
812	332
873	364
934	396
994	428

Excesses

10 = 5	1 = 1
20 = 11	2 = 1
30 = 16	3 = 2
40 = 21	4 = 2
50 = 26	5 = 3
60 = 32	6 = 3
	7 = 4
	8 = 4
	9 = 5

30 = 16

Anti Log Excesses

10 = 19	1 = 2
20 = 38	2 = 4
30 = 57	3 = 6
	4 = 8
	5 = 9
	6 = 11
	7 = 13
	8 = 15
	9 = 17

16 = 30

8210055	·9143461
116	493
177	525
238	557
299	589
360	622
421	654
482	686
543	718
603	750
664	783
725	815
786	847
847	879
908	911
969	944
8211030	976
091	·9144008
151	040
212	073
273	105
334	137
395	169
456	201
517	234
578	266
639	298
699	330
760	362
821	395
882	427
943	459
8212004	491
065	523
126	555
187	588
247	620

8212308	·9144652
369	684
430	716
491	749
552	781
613	813
674	845
735	877
795	910
856	942
917	974
978	·9145006
8213039	038
100	071
161	103
222	135
282	167
343	199
404	232
465	264
526	296
587	328
648	360
709	392
769	425
830	457
891	489
952	521
8214013	553
074	586
135	618
196	650
256	682
317	714
378	747
439	779
500	811
561	843
622	875
682	907
743	940
804	972
865	·9146004
926	036
987	068
8215048	101
109	133
169	165
230	197
291	229
352	261
413	294
474	326
535	358
595	390
656	422
717	454
778	487
839	519
900	551
961	583
8216021	615
082	647
143	680
204	712

8216265	·9146744
326	776
387	808
447	840
508	873
569	905
630	937
691	969
752	·9147001
813	033
873	066
934	098
995	130
8217056	162
117	194
178	226
238	258
299	291
360	323
421	355
482	387
543	419
604	451
664	484
725	516
786	548
847	580
908	612
969	644
8218029	676
090	709
151	741
212	773
273	805
334	837
395	869
455	902
516	934
577	966
638	998
699	·9148030
760	062
820	095
881	127
942	159
8219003	191
064	223
125	255
185	287
246	320
307	352
368	384
429	416
490	448
550	480
611	512
672	545
733	577
794	609
855	641
915	673
976	705

Excesses

10 = 5	1 = 1
20 = 11	2 = 1
30 = 16	3 = 2
40 = 21	4 = 2
50 = 26	5 = 3
60 = 32	6 = 3
	7 = 4
	8 = 4
	9 = 5

30 = 16

Anti Log Excesses

10 = 19	1 = 2
20 = 38	2 = 4
30 = 57	3 = 6
	4 = 8
	5 = 9
	6 = 11
	7 = 13
	8 = 15
	9 = 17

16 = 30

8220037	·9148737
098	769
159	802
219	834
280	866
341	898
402	930
463	962
524	994
584	·9149027
645	059
706	091
767	123
828	155
889	187
949	219
8221010	251
071	284
132	316
193	348
253	380
314	412
375	444
436	476
497	508
558	541
618	573
679	605
740	637
801	669
862	701
922	733
983	765
8222044	798
105	830
166	862
226	894

8222287	·9149926
348	958
409	990
470	·9150022
531	055
591	087
652	119
713	151
774	183
835	215
895	247
956	279
8223017	311
078	344
139	376
199	408
260	440
321	472
382	504
443	536
503	568
564	601
625	633
686	665
747	697
807	729
868	761
929	793
990	825
8224051	857
111	890
172	922
233	954
294	986
355	·9151018
415	050
476	082
537	114
598	146
659	178
719	211
780	243
841	275
902	307
963	339
8225023	371
084	403
145	435
206	467
266	499
327	532
388	564
449	596
510	628
570	660
631	692
692	724
753	756
814	788
874	820
935	852
996	885
8226057	917
117	949
178	981

8226239	·9152013
300	045
361	077
421	109
482	141
543	173
604	205
664	238
725	270
786	302
847	334
908	366
968	398
8227029	430
090	462
151	494
211	526
272	558
333	590
394	623
455	655
515	687
576	719
637	751
698	783
758	815
819	847
880	879
941	911
8228001	943
062	975
123	·9153007
184	039
245	072
305	104
366	136
427	168
488	200
548	232
609	264
670	296
731	328
791	360
852	392
913	424
974	456
8229034	488
095	521
156	553
217	585
278	617
338	649
399	681
460	713
521	745
581	777
642	809
703	841
764	873
824	905
885	937
946	969

Excesses

10 = 5	1 = 1
20 = 11	2 = 1
30 = 16	3 = 2
40 = 21	4 = 2
50 = 26	5 = 3
60 = 32	6 = 3
	7 = 4
	8 = 4
	9 = 5

30 = 16

Anti Log Excesses

10 = 19	1 = 2
20 = 38	2 = 4
30 = 57	3 = 6
	4 = 8
	5 = 9
	6 = 11
	7 = 13
	8 = 15
	9 = 17

16 = 30

8230007	·9154002
067	034
128	066
189	098
250	130
310	162
371	194
432	226
493	258
553	290
614	322
675	354
736	386
796	418
857	450
918	482
979	514
8231039	546
100	579
161	611
222	643
282	675
343	707
404	739
465	771
525	803
586	835
647	867
707	899
768	931
829	963
890	995
950	·9155027
8232011	059
072	091
133	123
193	155

8232254	·9155187
315	219
376	252
436	283
497	315
558	347
619	379
679	412
740	444
801	476
861	508
922	540
983	572
8233044	604
104	636
165	668
226	700
287	732
347	764
408	796
469	828
530	860
590	892
651	924
712	956
772	988
833	·9156020
894	052
955	084
8234015	116
076	148
137	180
197	212
258	244
319	276
380	308
440	340
501	372
562	404
623	436
683	468
744	500
805	533
865	565
926	597
987	629
8235048	661
108	693
169	725
230	757
290	789
351	821
412	853
473	885
533	917
594	949
655	981
715	·9157013
776	045
837	077
898	109
958	141
8236019	173
080	205
140	237

8236201	·9157269
262	301
322	333
383	365
444	397
505	429
565	461
626	493
687	525
747	557
808	589
869	621
930	653
990	685
8237051	717
112	749
172	781
233	813
294	845
354	877
415	909
476	941
537	973
597	·9158005
658	037
719	069
779	101
840	133
901	165
961	197
8238022	229
083	261
143	293
204	325
265	357
326	389
386	421
447	453
508	485
568	517
629	549
690	581
750	613
811	645
872	677
932	709
993	741
8239054	773
115	805
175	837
236	869
297	901
357	933
418	965
479	997
539	·9159029
600	061
661	093
721	125
782	157
843	189
903	221
964	253

Excesses

10 = 5	1 = 1
20 = 11	2 = 1
30 = 16	3 = 2
40 = 21	4 = 2
50 = 26	5 = 3
60 = 32	6 = 3
	7 = 4
	8 = 4
	9 = 5

30 = 16

Anti Log Excesses

10 = 19	1 = 2
20 = 38	2 = 4
30 = 57	3 = 6
	4 = 8
	5 = 9
	6 = 11
	7 = 13
	8 = 15
	9 = 17

16 = 30

8240025	·9159285
085	317
146	349
207	381
267	413
328	445
389	477
450	509
510	540
571	572
632	604
692	636
753	668
814	700
874	732
935	764
996	796
8241056	828
117	860
178	892
238	924
299	956
360	988
420	·9160020
481	052
542	084
602	116
663	148
724	180
784	212
845	244
906	276
966	308
8242027	340
088	372
148	404
209	436

8242270	·9160468
330	500
391	532
452	564
512	596
573	628
634	660
694	692
755	724
816	756
876	788
937	820
998	852
8243058	884
119	915
179	947
240	979
301	·9161011
361	043
422	075
483	107
543	139
604	171
665	203
725	235
786	267
847	299
907	331
968	363
8244029	395
089	427
150	459
211	491
271	523
332	555
393	586
453	618
514	650
574	682
635	714
696	746
756	778
817	810
878	842
938	874
999	906
8245060	938
120	970
181	·9162002
242	034
302	066
363	098
423	130
484	161
545	193
605	225
666	257
727	289
787	321
848	353
909	385
969	417
8246030	449
090	481
151	513

8246212	·9162545
272	577
333	609
394	640
454	672
515	704
576	736
636	768
697	800
757	832
818	864
879	896
939	928
8247000	960
061	992
121	·9163024
182	056
242	088
303	119
364	151
424	183
485	215
546	247
606	279
667	311
727	343
788	375
849	407
909	439
970	471
8248031	503
091	534
152	566
212	598
273	630
334	662
394	694
455	726
516	758
576	790
637	822
697	854
758	886
819	917
879	949
940	981
8249000	·9164013
061	045
122	077
182	109
243	141
304	173
364	205
425	237
485	268
546	300
607	332
667	364
728	396
788	428
849	460
910	492
970	524

Excesses

10 = 5	1 = 1
20 = 11	2 = 1
30 = 16	3 = 2
40 = 21	4 = 2
50 = 26	5 = 3
60 = 32	6 = 3
	7 = 4
	8 = 4
	9 = 5

30 = 16

Anti Log Excesses

10 = 19	1 = 2
20 = 33	2 = 4
30 = 57	3 = 6
	4 = 8
	5 = 9
	6 = 11
	7 = 13
	8 = 15
	9 = 17

16 = 30

8250031	·9164556
091	587
152	619
213	651
273	683
334	715
394	747
455	779
516	811
576	843
637	875
697	907
758	938
819	970
879	·9165002
940	034
8251000	066
061	098
122	130
182	162
243	193
303	225
364	257
425	289
485	321
546	353
606	385
667	417
728	449
788	481
849	512
909	544
970	576
8252031	608
091	640
152	672
212	704

8252273	·9165736
334	768
394	799
455	831
515	863
576	895
636	927
697	959
758	991
818	·9166023
879	054
939	086
8253000	118
061	150
121	182
182	214
242	246
303	278
363	310
424	341
485	373
545	405
606	437
666	469
727	501
788	533
848	565
909	596
969	628
8254030	660
090	692
151	724
212	756
272	788
333	820
393	851
454	883
514	915
575	947
636	979
696	·9167011
757	043
817	074
878	106
938	138
999	170
8255060	202
120	234
181	266
241	298
302	329
362	361
423	393
484	425
544	457
605	489
665	520
726	552
786	584
847	616
907	648
968	680
8256029	712
089	744
150	775

8256210	·9167807
271	839
331	871
392	903
453	935
513	967
574	998
634	·9168030
695	062
755	094
816	126
876	158
937	189
998	221
8257058	253
119	285
179	317
240	349
300	381
361	412
421	444
482	476
543	508
603	540
664	572
724	604
785	635
845	667
906	699
966	731
8258027	763
087	795
148	826
209	858
269	890
330	922
390	954
451	986
511	·9169017
572	049
632	081
693	113
753	145
814	177
875	209
935	240
996	272
8259056	304
117	336
177	368
238	400
298	431
359	463
419	495
480	527
540	559
601	591
662	622
722	654
783	686
843	718
904	750
964	782

Excesses

10 = 5	1 = 1
20 = 11	2 = 1
30 = 16	3 = 2
40 = 21	4 = 2
50 = 26	5 = 3
60 = 32	6 = 3
	7 = 4
	8 = 4
	9 = 5

30 = 16

Anti Log Excesses

10 = 19	1 = 2
20 = 38	2 = 4
30 = 57	3 = 6
	4 = 8
	5 = 10
	6 = 11
	7 = 13
	8 = 15
	9 = 17

16 = 30

8260025	·9169813
085	845
146	877
206	909
267	941
327	972
388	·9170004
448	036
509	068
569	100
630	132
691	163
751	195
812	227
872	259
933	291
993	322
8261054	354
114	386
175	418
235	450
296	481
356	513
417	545
477	577
538	609
598	641
659	672
719	704
780	736
840	768
901	800
962	831
8262022	863
083	895
143	927
204	959
8262264	·9170990
325	·9171022
385	054
446	086
506	118
567	150
627	181
688	213
748	245
809	277
869	309
930	340
990	372
8263051	404
111	436
172	468
232	499
293	531
353	563
414	595
474	627
535	658
595	690
656	722
716	754
777	786
837	817
898	849
958	881
8264019	913
079	944
140	976
200	·9172008
261	040
321	072
382	103
442	135
503	167
563	199
624	231
684	262
745	294
805	326
866	358
926	390
987	421
8265047	453
108	485
168	517
229	549
289	580
350	612
410	644
471	676
531	708
592	739
652	771
713	803
773	835
834	866
894	898
955	930
8266015	962
076	994
136	·9173025
8266197	·9173057
257	089
318	121
378	152
439	184
499	216
560	248
620	280
681	311
741	343
802	375
862	407
923	438
983	470
8267044	502
104	534
164	566
225	597
285	629
346	661
406	693
467	724
527	756
588	788
648	820
709	851
769	883
830	915
890	947
951	978
8268011	·9174010
072	042
132	074
193	106
253	137
314	169
374	201
434	233
495	264
555	296
616	328
676	360
737	391
797	423
858	455
918	487
979	518
8269039	550
100	582
160	614
221	645
281	677
341	709
402	741
462	772
523	804
583	836
644	868
704	899
765	931
825	963
886	995
946	·9175026

Excesses

10 = 5	1 = 1
20 = 11	2 = 1
30 = 16	3 = 2
40 = 21	4 = 2
50 = 26	5 = 3
60 = 32	6 = 3
	7 = 4
	8 = 4
	9 = 5

30 = 16

Anti Log Excesses

10 = 19	1 = 2
20 = 38	2 = 4
30 = 57	3 = 6
	4 = 8
	5 = 10
	6 = 11
	7 = 13
	8 = 15
	9 = 17

16 = 30

8270007	·9175058
067	090
127	122
188	153
248	185
309	217
369	249
430	280
490	312
551	344
611	376
672	407
732	439
792	471
853	503
913	534
974	566
8271034	598
095	630
155	661
216	693
276	725
337	757
397	788
457	820
518	852
578	884
639	915
699	947
760	979
820	·9176011
881	042
941	074
8272001	106
062	137
122	169
183	201
8272243	·9176233
304	264
364	296
425	328
485	360
545	391
606	423
666	455
727	486
787	518
848	550
908	582
969	613
8273029	645
089	677
150	709
210	740
271	772
331	804
392	835
452	867
512	899
573	931
633	962
694	994
754	·9177026
815	058
875	089
935	121
996	153
8274056	184
117	216
177	248
238	280
298	311
358	343
419	375
479	406
540	438
600	470
661	502
721	533
781	565
842	597
902	628
963	660
8275023	692
084	724
144	755
204	787
265	819
325	850
386	882
446	914
507	946
567	977
627	·9178009
688	041
748	072
809	104
869	136
929	167
990	199
8276050	231
111	263

8276171	·9178294
232	326
292	358
352	389
413	421
473	453
534	485
594	516
654	548
715	580
775	611
836	643
896	675
956	706
8277017	738
077	770
138	801
198	833
259	865
319	897
379	928
440	960
500	992
561	·9179023
621	055
681	087
742	118
802	150
863	182
923	213
983	245
8278044	277
104	309
165	340
225	372
285	404
346	435
406	467
467	499
527	530
587	562
648	594
708	625
769	657
829	689
889	721
950	752
8279010	784
070	816
131	847
191	879
252	911
312	942
372	974
433	·9180006
493	037
554	069
614	101
674	132
735	164
795	196
856	227
916	259
976	291

Excesses

10 = 5	1 = 1
20 = 10	2 = 1
30 = 16	3 = 2
40 = 21	4 = 2
50 = 26	5 = 3
60 = 31	6 = 3
	7 = 4
	8 = 4
	9 = 5

30 = 16

Anti Log Excesses

10 = 19	1 = 2
20 = 38	2 = 4
30 = 57	3 = 6
	4 = 8
	5 = 10
	6 = 11
	7 = 13
	8 = 15
	9 = 17

16 = 30

8280037	·9180322
097	354
158	386
218	417
278	449
339	481
399	512
459	544
520	576
580	608
641	639
701	671
761	703
822	734
882	766
942	798
8281003	829
063	861
124	893
184	924
244	956
305	987
365	·9181019
425	051
486	082
546	114
607	146
667	177
727	209
788	241
848	272
908	304
969	336
8282029	367
090	399
150	431
210	462

8282271	·9181494
331	526
391	557
452	589
512	621
573	652
633	684
693	716
754	747
814	779
874	811
935	842
995	874
8283055	906
116	937
176	969
237	·9182000
297	032
357	064
418	095
478	127
538	159
599	190
659	222
719	254
780	285
840	317
901	349
961	380
8284021	412
082	443
142	475
202	507
263	538
323	570
383	602
444	633
504	665
564	697
625	728
685	760
746	792
806	823
866	855
927	886
987	918
8285047	950
108	981
168	·9183013
228	045
289	076
349	108
409	140
470	171
530	203
590	234
651	266
711	298
771	329
832	361
892	393
952	424
8286013	456
073	487
133	519

8286194	·9183551
254	582
315	614
375	646
435	677
496	709
556	740
616	772
677	804
737	835
797	867
858	899
918	930
978	962
8287039	993
099	·9184025
159	057
220	088
280	120
340	151
401	183
461	215
521	247
582	278
642	310
702	341
763	373
823	404
883	436
944	468
8288004	499
064	531
125	563
185	594
245	626
306	657
366	689
426	720
486	752
547	784
607	815
667	847
728	878
788	910
848	942
909	973
969	·9185005
8289029	037
090	068
150	100
210	131
271	163
331	195
391	226
452	258
512	289
572	321
633	353
693	384
753	416
814	447
874	479
934	511
994	542

Excesses

10 = 5	1 = 1
20 = 10	2 = 1
30 = 16	3 = 2
40 = 21	4 = 2
50 = 26	5 = 3
60 = 31	6 = 3
	7 = 4
	8 = 4
	9 = 5

30 = 16

Anti Log Excesses

10 = 19	1 = 2
20 = 38	2 = 4
30 = 57	3 = 6
	4 = 8
	5 = 10
	6 = 11
	7 = 13
	8 = 15
	9 = 17

16 = 30

8290055	·9185574
115	605
175	637
236	669
296	700
356	732
417	763
477	795
537	827
598	858
658	890
718	921
779	953
839	984
899	·9186016
959	048
8291020	079
080	111
140	142
201	174
261	206
321	237
382	269
442	300
502	332
562	364
623	395
683	427
743	458
804	490
864	521
924	553
985	585
8292045	616
105	648
165	679
226	711

8292286	·9186743
346	774
407	806
467	837
527	869
588	900
648	932
708	964
768	995
829	·9187027
889	058
949	090
8293010	121
070	153
130	185
190	216
251	248
311	279
371	311
432	342
492	374
552	406
612	437
673	469
733	500
793	532
854	563
914	595
974	627
8294035	658
095	690
155	721
215	753
276	784
336	816
396	848
456	879
517	911
577	942
637	974
698	·9188005
758	037
818	068
878	100
939	132
999	163
8295059	195
120	226
180	258
240	289
300	321
361	353
421	384
481	416
541	447
602	479
662	510
722	542
783	573
843	605
903	637
963	668
8296024	700
084	731
144	763

8296204	·9188794
265	826
325	857
385	889
446	920
506	952
566	984
626	·9189015
687	047
747	078
807	110
867	141
928	173
988	204
8297048	236
108	267
169	299
229	331
289	362
349	394
410	425
470	457
530	488
591	520
651	551
711	583
771	614
832	646
892	677
952	709
8298012	740
073	772
133	804
193	835
253	867
314	898
374	930
434	961
494	993
555	·9190024
615	056
675	087
735	119
796	150
856	182
916	213
976	245
8299037	276
097	308
157	340
217	371
278	403
338	434
398	466
458	497
519	529
579	560
639	592
699	623
760	655
820	686
880	718
940	749

Excesses

10 = 5	1 = 1
20 = 10	2 = 1
30 = 16	3 = 2
40 = 21	4 = 2
50 = 26	5 = 3
60 = 31	6 = 3
	7 = 4
	8 = 4
	9 = 5

30 = 16

Anti Log Excesses

10 = 19	1 = 2
20 = 38	2 = 4
30 = 57	3 = 6
	4 = 8
	5 = 10
	6 = 11
	7 = 13
	8 = 15
	9 = 17

16 = 30

8300000	·9190781
061	812
121	844
181	875
241	907
302	938
362	970
422	·9191001
482	033
543	064
603	096
663	127
723	159
784	190
844	222
904	254
964	285
8301025	317
085	348
145	380
205	411
265	443
326	474
386	506
446	537
506	569
567	600
627	632
687	663
747	695
808	726
868	758
928	789
988	821
8302048	852
109	884
169	915

8302229	·9191947
289	978
350	·9192010
410	041
470	073
530	104
590	136
651	167
711	199
771	230
831	262
892	293
952	325
8303012	356
072	388
132	419
193	451
253	482
313	514
373	545
434	577
494	608
554	640
614	671
674	703
735	734
795	766
855	797
915	829
975	860
8304036	892
096	923
156	955
216	986
277	·9193018
337	049
397	081
457	112
517	144
578	175
638	207
698	238
758	270
818	301
879	332
939	364
999	395
8305059	427
119	458
180	490
240	521
300	553
360	584
420	616
481	647
541	679
601	710
661	742
721	773
782	805
842	836
902	868
962	899
8306022	931
083	962

8306143	·9193994
203	·9194025
263	056
323	088
384	119
444	151
504	182
564	214
624	245
685	277
745	308
805	340
865	371
925	403
986	434
8307046	466
106	497
166	529
226	560
286	591
347	623
407	654
467	686
527	717
587	749
648	780
708	812
768	843
828	875
888	906
949	938
8308009	969
069	·9195000
129	032
189	063
249	095
310	126
370	158
430	189
490	221
550	252
610	284
671	315
731	346
791	378
851	409
911	441
972	472
8309032	504
092	535
152	567
212	598
272	630
333	661
393	692
453	724
513	755
573	787
633	818
694	850
754	881
814	913
874	944
934	975
994	·9196007

Excesses

10 = 5	1 = 1
20 = 10	2 = 1
30 = 16	3 = 2
40 = 21	4 = 2
50 = 26	5 = 3
60 = 31	6 = 3
	7 = 4
	8 = 4
	9 = 5

30 = 16

Anti Log Excesses

10 = 19	1 = 2
20 = 38	2 = 4
30 = 57	3 = 6
	4 = 8
	5 = 10
	6 = 11
	7 = 13
	8 = 15
	9 = 17

16 = 30

8310055	·9196038
115	070
175	101
235	133
295	164
355	196
416	227
476	258
536	290
596	321
656	353
716	384
777	416
837	447
897	479
957	510
8311017	541
077	573
138	604
198	636
258	667
318	699
378	730
438	761
499	793
559	824
619	856
679	887
739	919
799	950
859	982
920	·9197013
980	044
8312040	076
100	107
160	139
220	170

8312281	·9197202
341	233
401	264
461	296
521	327
581	359
641	390
702	421
762	453
822	484
882	516
942	547
8313002	579
063	610
123	641
183	673
243	704
303	736
363	767
423	798
484	830
544	861
604	893
664	924
724	956
784	987
844	·9198018
905	050
965	081
8314025	113
085	144
145	175
205	207
265	238
325	270
386	301
446	333
506	364
566	395
626	427
686	458
746	490
807	521
867	552
927	584
987	615
8315047	647
107	678
167	709
227	741
288	772
348	804
408	835
468	866
528	898
588	929
648	961
709	992
769	·9199023
829	055
889	086
949	118
8316009	149
069	180
129	212

8316190	·9199244
250	275
310	306
370	338
430	369
490	401
550	432
610	463
671	495
731	526
791	558
851	589
911	620
971	652
8317031	683
091	715
151	746
212	777
272	809
332	840
392	872
452	903
512	934
572	966
632	997
692	·9200028
753	060
813	091
873	123
933	154
993	185
8318053	217
113	248
173	280
233	311
294	342
354	374
414	405
474	436
534	468
594	499
654	531
714	562
774	593
835	625
895	656
955	688
8319015	719
075	750
135	782
195	813
255	844
315	876
375	907
436	939
496	970
556	·9201001
616	033
676	064
736	095
796	127
856	158
916	189
976	221

Excesses

10 = 5	1 = 1
20 = 10	2 = 1
30 = 16	3 = 2
40 = 21	4 = 2
50 = 26	5 = 3
60 = 31	6 = 3
	7 = 4
	8 = 4
	9 = 5

30 = 16

Anti Log Excesses

10 = 19	1 = 2
20 = 38	2 = 4
30 = 57	3 = 6
	4 = 8
	5 = 10
	6 = 11
	7 = 13
	8 = 15
	9 = 17

16 = 30

8320037	·9201252
097	284
157	315
217	346
277	378
337	409
397	440
457	472
517	503
577	535
637	566
698	597
758	629
818	660
878	691
938	723
998	754
8321058	785
118	817
178	848
238	880
298	911
359	942
419	974
479	·9202005
539	036
599	068
659	099
719	130
779	162
839	193
899	224
959	256
8322019	287
080	319
140	350
200	381

8322260	·9202413
320	444
380	475
440	507
500	538
560	569
620	601
680	632
740	663
801	695
861	726
921	757
981	789
8323041	820
101	851
161	883
221	914
281	946
341	977
401	·9203008
461	040
521	071
581	102
642	134
702	165
762	196
822	228
882	259
942	290
8324002	322
062	353
122	384
182	416
242	447
302	478
362	510
422	541
482	572
543	604
603	635
663	666
723	698
783	729
843	760
903	792
963	823
8325023	854
083	886
143	917
203	948
263	980
323	·9204011
383	042
443	074
503	105
564	136
624	168
684	199
744	230
804	262
864	293
924	324
984	355
8326044	387
104	418

Number	Log
8326164	·9204449
224	481
284	512
344	543
404	575
464	606
524	637
584	668
644	700
705	731
765	763
825	794
885	825
945	857
8327005	888
065	919
125	951
185	982
245	·9205013
305	044
365	076
425	107
485	138
545	170
605	201
665	232
725	264
785	295
845	326
905	358
965	389
8328025	420
086	452
146	483
206	514
266	545
326	577
386	608
446	639
506	671
566	702
626	733
686	765
746	796
806	827
866	859
926	890
986	921
8329046	952
106	984
166	·9206015
226	046
286	078
346	109
406	140
466	172
526	203
586	234
646	265
706	297
766	328
826	359
886	391
946	422

Excesses

10 = 5	1 = 1
20 = 10	2 = 1
30 = 16	3 = 2
40 = 21	4 = 2
50 = 26	5 = 3
60 = 31	6 = 3
	7 = 4
	8 = 4
	9 = 5

30 = 16

Anti Log Excesses

10 = 19	1 = 2
20 = 38	2 = 4
30 = 58	3 = 6
	4 = 8
	5 = 10
	6 = 12
	7 = 13
	8 = 15
	9 = 17

16 = 30

Number	Log
8330007	·9206453
067	484
127	516
187	547
247	578
307	610
367	641
427	672
487	704
547	735
607	766
667	797
727	829
787	860
847	891
907	923
967	954
8331027	985
087	·9207016
147	048
207	079
267	110
327	142
387	173
447	204
507	235
567	267
627	298
687	329
747	361
807	392
867	423
927	454
987	486
8332047	517
107	548
167	579

Number	Log
8332227	·9207611
287	642
347	673
407	705
467	736
527	767
587	798
647	830
707	861
767	892
827	923
887	955
947	986
8333007	·9208017
067	049
127	080
187	111
247	142
307	174
367	205
427	236
487	267
547	299
607	330
667	361
727	393
787	424
847	455
907	486
967	518
8334027	549
087	580
147	611
207	643
267	674
327	705
387	736
447	768
507	799
567	830
627	861
687	893
747	924
807	955
867	987
927	·9209018
987	049
8335047	080
107	112
167	143
227	174
287	205
347	237
407	268
467	299
527	330
587	362
647	393
707	424
767	455
827	487
887	518
947	549
8336007	580
067	612

Number	Log
8336127	·9209643
187	674
247	705
307	737
367	768
427	799
487	830
547	862
607	893
667	924
726	955
786	987
846	·9210018
906	049
966	080
8337026	112
086	143
146	174
206	205
266	237
326	268
386	299
446	330
506	361
566	393
626	424
686	455
746	486
806	518
866	549
926	580
986	611
8338046	643
106	674
166	705
226	736
286	768
346	799
406	830
466	861
526	892
586	924
645	955
705	986
765	·9211017
825	049
885	080
945	111
8339005	142
065	174
125	205
185	236
245	267
305	298
365	330
425	361
485	392
545	423
605	455
665	486
725	517
785	548
845	579
905	611
965	642

Excesses

10 = 5	1 = 1
20 = 10	2 = 1
30 = 16	3 = 2
40 = 21	4 = 2
50 = 26	5 = 3
	6 = 3
	7 = 4
	8 = 4
	9 = 5

30 = 16

Anti Log Excesses

10 = 19	1 = 2
20 = 38	2 = 4
30 = 58	3 = 6
	4 = 8
	5 = 10
	6 = 12
	7 = 13
	8 = 15
	9 = 17

16 = 30

Number	Log
8340024	·9211673
084	704
144	735
204	767
264	798
324	829
384	860
444	892
504	923
564	954
624	985
684	·9212016
744	048
804	079
864	110
924	141
984	172
8341044	204
104	235
163	266
223	297
283	329
343	360
403	391
463	422
523	453
583	485
643	516
703	547
763	578
823	609
883	641
943	672
8342003	703
063	734
123	765
182	797

8342242	·9212828
302	859
362	890
422	921
482	953
542	984
602	·9213015
662	046
722	077
782	109
842	140
902	171
962	202
8343022	233
081	265
141	296
201	327
261	358
321	389
381	421
441	452
501	483
561	514
621	545
681	577
741	608
801	639
861	670
920	701
980	733
8344040	764
100	795
160	826
220	857
280	888
340	919
400	950
460	981
520	·9214013
580	044
639	075
699	106
759	137
819	168
879	200
939	231
999	262
8345059	293
119	324
179	356
239	387
299	418
358	449
418	480
478	511
538	543
598	574
658	605
718	636
778	667
838	699
898	730
958	761
8346018	792
077	823

8346137	·9214854
197	886
257	917
317	948
377	979
437	·9215010
497	041
557	073
617	104
676	135
736	166
796	197
856	228
916	260
976	291
8347036	322
096	353
156	384
216	415
276	447
335	478
395	509
455	540
515	571
575	602
635	634
695	665
755	696
815	727
874	758
934	789
994	821
8348054	852
114	883
174	914
234	945
294	976
354	·9216008
414	039
473	070
533	101
593	132
653	163
713	194
773	226
833	257
893	288
953	319
8349012	350
072	381
132	412
192	444
252	475
312	506
372	537
432	568
492	599
551	631
611	662
671	693
731	724
791	755
851	786
911	817
971	849

Excesses

10 = 5	1 = 1
20 = 10	2 = 1
30 = 16	3 = 2
40 = 21	4 = 2
50 = 26	5 = 3
	6 = 3
	7 = 4
	8 = 4
	9 = 5

30 = 16

Anti Log Excesses

10 = 19	1 = 2
20 = 38	2 = 4
30 = 58	3 = 6
	4 = 8
	5 = 10
	6 = 12
	7 = 13
	8 = 15
	9 = 17

16 = 30

8350030	·9216880
090	911
150	942
210	973
270	·9217004
330	035
390	067
450	098
509	129
569	160
629	191
689	222
749	253
809	285
869	316
929	347
988	378
8351048	409
108	440
168	471
228	503
288	534
348	565
408	596
467	627
527	658
587	689
647	720
707	752
767	783
827	814
887	845
946	876
8352006	907
066	938
126	969
186	·9218001

8352246	·9218032
306	063
365	094
425	125
485	156
545	187
605	219
665	250
725	281
784	312
844	343
904	374
964	405
8353024	436
084	468
144	499
203	530
263	561
323	592
383	623
443	654
503	685
563	716
622	748
682	779
742	810
802	842
862	873
922	904
982	935
8354041	966
101	997
161	·9219028
221	059
281	090
341	122
401	153
460	184
520	215
580	246
640	277
700	308
760	339
819	370
879	402
939	433
999	464
8355059	495
119	526
179	557
238	588
298	619
358	650
418	682
478	713
538	744
597	775
657	806
717	837
777	868
837	899
897	930
956	961
8356016	993
076	·9220024

8356136	·9220055
196	086
256	117
315	148
375	179
435	210
495	241
555	272
615	304
674	335
734	366
794	397
854	428
914	459
974	490
8357033	521
093	552
153	583
213	614
273	646
333	677
392	708
452	739
512	770
572	801
632	832
692	863
751	894
811	925
871	956
931	988
991	·9221019
8358051	050
110	081
170	112
230	143
290	174
350	205
409	236
469	267
529	298
589	329
649	360
709	392
768	423
828	454
888	485
948	516
8359008	547
067	578
127	609
187	640
247	671
307	702
367	733
426	764
486	796
546	827
606	858
666	889
725	920
785	951
845	982
905	·9222013
965	044

Excesses

10 = 5	1 = 1
20 = 10	2 = 1
30 = 16	3 = 2
40 = 21	4 = 2
50 = 26	5 = 3
	6 = 3
	7 = 4
	8 = 4
	9 = 5

30 = 16

Anti Log Excesses

10 = 19	1 = 2
20 = 38	2 = 4
30 = 58	3 = 6
	4 = 8
	5 = 10
	6 = 12
	7 = 13
	8 = 15
	9 = 17

16 = 30

8360024	·9222075
084	106
144	137
204	168
264	199
323	231
383	262
443	293
503	324
563	355
622	386
682	417
742	448
802	479
862	510
922	541
981	572
8361041	603
101	634
161	665
221	697
280	728
340	759
400	790
460	821
520	852
579	883
639	914
699	945
759	976
818	·9223007
878	038
938	069
998	100
8362058	131
117	162
177	193

8362237	·9223224
297	256
357	287
416	318
476	349
536	380
596	411
656	442
715	473
775	504
835	535
895	566
955	597
8363014	628
074	659
134	690
194	721
253	752
313	783
373	814
433	845
493	876
552	907
612	938
672	970
732	·9224001
792	032
851	063
911	094
971	125
8364031	156
090	187
150	218
210	249
270	280
330	311
389	342
449	373
509	404
569	435
628	466
688	497
748	528
808	559
868	590
927	621
987	652
8365047	683
107	714
166	745
226	777
286	808
346	839
405	870
465	901
525	932
585	963
645	994
704	·9225025
764	056
824	087
884	118
943	149
8366003	180
063	211

8366123	·9225242
182	273
242	304
302	335
362	366
421	397
481	428
541	459
601	490
661	521
720	552
780	583
840	614
900	645
959	676
8367019	707
079	738
139	769
198	800
258	831
318	862
378	893
437	924
497	955
557	986
617	·9226017
676	048
736	079
796	110
856	141
915	172
975	203
8368035	234
095	265
154	296
214	327
274	359
334	390
393	421
453	452
513	483
573	514
632	545
692	576
752	607
812	638
871	669
931	700
991	731
8369051	762
110	793
170	824
230	855
290	886
349	917
409	948
469	979
529	·9227010
588	041
648	072
708	103
768	134
827	165
887	196
947	227

Excesses

10 = 5	1 = 1
20 = 10	2 = 1
30 = 16	3 = 2
40 = 21	4 = 2
50 = 26	5 = 3
	6 = 3
	7 = 4
	8 = 4
	9 = 5

30 = 16

Anti Log Excesses

10 = 19	1 = 2
20 = 39	2 = 4
30 = 58	3 = 6
	4 = 8
	5 = 10
	6 = 12
	7 = 13
	8 = 15
	9 = 17

16 = 30

8370006	·9227258
066	289
126	320
186	351
245	382
305	413
365	444
425	475
484	506
544	537
604	568
664	598
723	629
783	660
843	691
902	722
962	753
8371022	784
082	815
141	846
201	877
261	908
321	939
380	970
440	·9228001
500	032
559	063
619	094
679	125
739	156
798	187
858	218
918	249
978	280
8372037	311
097	342
157	373

8372216	·9228404
276	435
336	466
396	497
455	528
515	559
575	590
634	621
694	652
754	683
814	714
873	745
933	776
993	807
8373053	838
112	869
172	900
232	931
291	962
351	993
411	·9229024
471	055
530	086
590	117
650	148
709	178
769	209
829	240
888	271
948	302
8374008	333
068	364
127	395
187	426
247	457
306	488
366	519
426	550
486	581
545	612
605	643
665	674
724	705
784	736
844	767
903	798
963	829
8375023	860
083	891
142	922
202	953
262	984
321	·9230014
381	045
441	076
500	107
560	138
620	169
680	200
739	231
799	262
859	293
918	324
978	355
8376038	386

8376097	·9230417
157	448
217	479
277	510
336	541
396	572
456	603
515	634
575	664
635	695
694	726
754	757
814	788
873	819
933	850
993	881
8377052	912
112	943
172	974
232	·9231005
291	036
351	067
411	098
470	129
530	160
590	190
649	221
709	252
769	283
828	314
888	345
948	376
8378007	407
067	438
127	469
186	500
246	531
306	562
365	593
425	624
485	655
545	685
604	716
664	747
724	778
783	809
843	840
903	871
962	902
8379022	933
082	964
141	995
201	·9232026
261	057
320	088
380	118
440	149
499	180
559	211
619	242
678	273
738	304
798	335
857	366
917	397

8379977	·9232428

Excesses

10 = 5	1 = 1
20 = 10	2 = 1
30 = 16	3 = 2
40 = 21	4 = 2
50 = 26	5 = 3
	6 = 3
	7 = 4
	8 = 4
	9 = 5

30 = 15

Anti Log Excesses

10 = 19	1 = 2
20 = 39	2 = 4
30 = 58	3 = 6
	4 = 8
	5 = 10
	6 = 12
	7 = 14
	8 = 15
	9 = 17

15 = 30

8380036	·9232459
096	490
156	520
215	551
275	582
335	613
394	644
454	675
514	706
573	737
633	768
693	799
752	830
812	861
872	891
931	922
991	953
8381051	984
110	·9233015
170	046
230	077
289	108
349	139
408	170
468	201
528	232
587	262
647	293
707	324
766	355
826	386
886	417
945	448
8382005	479
065	510

8382124	·9233541
184	571
244	602
303	633
363	664
423	695
482	726
542	757
602	788
661	819
721	850
780	880
840	911
900	942
959	973
8383019	·9234004
079	035
138	066
198	097
258	128
317	159
377	189
437	220
496	251
556	282
615	313
675	344
735	375
794	406
854	437
914	468
973	498
8384033	529
093	560
152	591
212	622
271	653
331	684
391	715
450	746
510	776
570	807
629	838
689	869
749	900
808	931
868	962
927	993
987	·9235024
8385047	054
106	085
166	116
226	147
285	178
345	209
404	240
464	271
524	301
583	332
643	363
703	394
762	425
822	456
881	487
941	518

8386001	·9235548
060	579
120	610
180	641
239	672
299	703
358	734
418	765
478	796
537	826
597	857
657	888
716	919
776	950
835	981
895	·9236012
955	043
8387014	073
074	104
134	135
193	166
253	197
312	228
372	259
432	289
491	320
551	351
610	382
670	413
730	444
789	475
849	506
908	536
968	567
8388028	598
087	629
147	660
207	691
266	722
326	752
385	783
445	814
505	845
564	876
624	907
683	938
743	968
803	999
862	·9237030
922	061
981	092
8389041	123
101	153
160	184
220	215
279	246
339	277
399	308
458	339
518	369
577	400
637	431
697	462
756	493
816	524

8389875	·9237555
935	585
995	616

Excesses

10 = 5	1 = 1
20 = 10	2 = 1
30 = 16	3 = 2
40 = 21	4 = 2
50 = 26	5 = 3
	6 = 3
	7 = 4
	8 = 4
	9 = 5

30 = 15

Anti Log Excesses

10 = 19	1 = 2
20 = 39	2 = 4
30 = 58	3 = 6
	4 = 8
	5 = 10
	6 = 12
	7 = 14
	8 = 15
	9 = 17

15 = 30

8390054	·9237647
114	678
173	709
233	740
293	770
352	801
412	832
471	863
531	894
590	925
650	956
710	986
769	·9238017
829	048
888	079
948	110
8391008	141
067	171
127	202
186	233
246	264
306	295
365	326
425	357
484	388
544	419
603	449
663	480
723	511
782	542
842	573
901	604
961	634

8392021	·9238665
080	696
140	727
199	758
259	789
318	819
378	850
438	881
497	912
557	943
616	973
676	·9239004
735	035
795	066
855	097
914	128
974	158
8393033	189
093	220
152	251
212	282
272	313
331	343
391	374
450	405
510	436
569	467
629	498
689	528
748	559
808	590
867	621
927	652
986	682
8394046	713
106	744
165	775
225	806
284	836
344	867
403	898
463	929
523	960
582	991
642	·9240021
701	052
761	083
820	114
880	145
939	175
999	206
8395059	237
118	268
178	299
237	329
297	360
356	391
416	422
475	453
535	484
595	514
654	545
714	576
773	607
833	638

8395892	·9240668
952	699
8396011	730
071	761
131	792
190	822
250	853
309	884
369	915
428	946
488	976
547	·9241007
607	038
667	069
726	100
786	130
845	161
905	192
964	223
8397024	254
083	284
143	315
202	346
262	377
322	408
381	438
441	469
500	500
560	531
619	562
679	592
738	623
798	654
857	685
917	715
976	746
8398036	777
096	808
155	839
215	869
274	900
334	931
393	962
453	993
512	·9242023
572	054
631	085
691	116
750	146
810	177
869	208
929	239
989	270
8399048	300
108	331
167	362
227	393
286	424
346	454
405	485
465	516
524	547
584	577
643	608
703	639

8399762	·9242670
822	700
881	731
941	762

Excesses

10 = 5	1 = 1
20 = 10	2 = 1
30 = 16	3 = 2
40 = 21	4 = 2
50 = 26	5 = 3
	6 = 3
	7 = 4
	8 = 4
	9 = 5

30 = 15

Anti Log Excesses

10 = 19	1 = 2
20 = 39	2 = 4
30 = 58	3 = 6
	4 = 8
	5 = 10
	6 = 12
	7 = 14
	8 = 15
	9 = 17

15 = 30

8400000	·9242793
060	824
120	854
179	885
239	916
298	947
358	977
417	·9243008
477	039
536	070
596	101
655	131
715	162
774	193
834	224
893	254
953	285
8401012	316
072	347
131	377
191	408
250	439
310	470
369	501
429	531
488	562
548	593
607	624
667	654
727	685
786	716
846	747

8401905	·9243777
965	808
8402024	839
084	870
143	900
203	931
262	962
322	993
381	·9244023
441	054
500	085
560	116
619	146
679	177
738	208
798	239
857	269
917	300
976	331
8403036	362
095	392
155	423
214	454
274	485
333	515
393	546
452	577
512	608
571	638
631	669
690	700
750	731
809	761
869	792
928	823
988	854
8404047	884
107	915
166	946
226	977
285	·9245007
345	038
404	069
464	100
523	130
583	161
642	192
702	223
761	253
821	284
880	315
940	346
999	376
8405059	407
118	438
177	468
237	499
296	530
356	561
415	591
475	622
534	653
594	684
653	714
713	745

8405772	·9245776
832	807
891	837
951	868
8406010	899
070	930
129	960
189	991
248	·9246022
308	053
367	083
427	114
486	145
546	175
605	206
665	237
724	268
783	298
843	329
902	360
962	391
8407021	421
081	452
140	483
200	513
259	544
319	575
378	606
438	636
497	667
557	698
616	729
676	759
735	790
795	821
854	851
913	882
973	913
8408032	944
092	974
151	·9247005
211	036
270	066
330	097
389	128
449	159
508	189
568	220
627	251
686	281
746	312
805	343
865	374
924	404
984	435
8409043	466
103	496
162	527
222	558
281	589
341	619
400	650
459	681
519	711
578	742

8409638	·9247773
697	804
757	834
816	865
876	896
935	926
995	957

Excesses

10 = 5	1 = 1
20 = 10	2 = 1
30 = 16	3 = 2
40 = 21	4 = 2
50 = 26	5 = 3
	6 = 3
	7 = 4
	8 = 4
	9 = 5

30 = 15

Anti Log Excesses

10 = 19	1 = 2
20 = 39	2 = 4
30 = 58	3 = 6
	4 = 8
	5 = 10
	6 = 12
	7 = 14
	8 = 15
	9 = 17

15 = 30

8410054	·9247988
113	·9248018
173	049
232	080
292	110
351	141
411	172
470	203
530	233
589	264
649	295
708	325
767	356
827	387
886	417
946	448
8411005	479
065	510
124	540
184	571
243	602
302	632
362	663
421	694
481	724
540	755
600	786
659	816
719	847

8411778	·9248878
837	909
897	939
956	970
8412016	·9249001
075	031
135	062
194	093
253	123
313	154
372	185
432	215
491	246
551	277
610	307
670	338
729	369
788	399
848	430
907	461
967	492
8413026	522
086	553
145	584
204	614
264	645
323	676
383	706
442	737
502	768
561	798
620	829
680	860
739	890
799	921
858	952
918	982
977	·9250013
8414036	044
096	074
155	105
215	136
274	166
334	197
393	228
452	258
512	289
571	320
631	350
690	381
749	412
809	442
868	473
928	504
987	534
8415047	565
106	596
165	626
225	657
284	688
344	718
403	749
462	780
522	810
581	841

8415641	·9250872
700	902
760	933
819	964
878	994
938	·9251025
997	056
8416057	086
116	117
175	148
235	178
294	209
354	240
413	270
472	301
532	331
591	362
651	393
710	423
769	454
829	485
888	515
948	546
8417007	577
066	607
126	638
185	669
245	699
304	730
364	761
423	791
482	822
542	853
601	883
661	914
720	945
779	975
839	·9252006
898	036
957	067
8418017	098
076	128
136	159
195	190
254	220
314	251
373	282
433	312
492	343
551	374
611	404
670	435
730	465
789	496
848	527
908	557
967	588
8419027	619
086	649
145	680
205	711
264	741
324	772
383	802
442	833

8419502	·9252864
561	894
620	925
680	955
739	986
799	·9253017
858	047
917	078
977	109

Excesses

10 = 5	1 = 1
20 = 10	2 = 1
30 = 15	3 = 2
40 = 21	4 = 2
50 = 26	5 = 3
	6 = 3
	7 = 4
	8 = 4
	9 = 5

30 = 15

Anti Log Excesses

10 = 19	1 = 2
20 = 39	2 = 4
30 = 58	3 = 6
	4 = 8
	5 = 10
	6 = 12
	7 = 14
	8 = 15
	9 = 17

15 = 30

8420036	·9253139
096	170
155	200
214	231
274	262
333	292
392	323
452	354
511	384
571	415
630	445
689	476
749	507
808	537
867	568
927	599
986	629
8421046	660
105	690
164	721
224	752
283	782
342	813
402	844
461	874
521	905
580	935

8421639	·9253966
699	997
758	·9254027
817	058
877	089
936	119
996	150
8422055	180
114	211
174	242
233	272
292	303
352	333
411	364
470	395
530	425
589	456
649	486
708	517
767	548
827	578
886	609
945	640
8423005	670
064	701
123	731
183	762
242	793
302	823
361	854
420	884
480	915
539	946
598	976
658	·9255007
717	037
776	068
836	099
895	129
954	160
8424014	190
073	221
133	252
192	282
251	313
311	343
370	374
429	405
489	435
548	466
607	496
667	527
726	558
785	588
845	619
904	649
963	680
8425023	710
082	741
141	772
201	802
260	833
319	863
379	894
438	925

8425498	·9255955
557	986
616	·9256016
676	047
735	078
794	108
854	139
913	169
972	200
8426032	230
091	261
150	292
210	322
269	353
328	383
388	414
447	445
506	475
566	506
625	536
684	567
744	597
803	628
862	659
922	689
981	720
8427040	750
100	781
159	812
218	842
278	873
337	903
396	934
456	964
515	995
574	·9257026
634	056
693	087
752	117
812	148
871	178
930	209
990	240
8428049	270
108	301
168	331
227	362
286	392
346	423
405	454
464	484
524	515
583	545
642	576
701	606
761	637
820	668
879	698
939	729
998	759
8429057	790
117	820
176	851
235	882
295	912

8429354	·9257943
413	973
473	·9258004
532	034
591	065
651	096
710	126
769	157
828	187
888	218
947	248

Excesses

10 = 5	1 = 1
20 = 10	2 = 1
30 = 15	3 = 2
40 = 21	4 = 2
50 = 26	5 = 3
	6 = 3
	7 = 4
	8 = 4
	9 = 5

30 = 15

Anti Log Excesses

10 = 19	1 = 2
20 = 39	2 = 4
30 = 58	3 = 6
	4 = 8
	5 = 10
	6 = 12
	7 = 14
	8 = 16
	9 = 17

15 = 30

8430006	·9258279
066	309
125	340
184	371
244	401
303	432
362	462
422	493
481	523
540	554
600	584
659	615
718	645
777	676
837	707
896	737
955	768
8431015	798
074	829
133	859
193	890
252	920
311	951
370	982
430	·9259012

8431489	·9259043
548	073
608	104
667	134
726	165
786	195
845	226
904	256
963	287
8432023	317
082	348
141	379
201	409
260	440
319	470
379	501
438	531
497	562
556	592
616	623
675	653
734	684
794	714
853	745
912	776
972	806
8433031	837
090	867
149	898
209	928
268	959
327	989
387	·9260020
446	050
505	081
564	111
624	142
683	172
742	203
802	234
861	264
920	295
979	325
8434039	356
098	386
157	417
217	447
276	478
335	508
394	539
454	569
513	600
572	630
631	661
691	691
750	722
809	752
869	783
928	814
987	844
8435046	875
106	905
165	936
224	966
284	997

8435343	·9261027
402	058
461	088
521	119
580	149
639	180
698	210
758	241
817	271
876	302
936	332
995	363
8436054	393
113	424
173	454
232	485
291	515
350	546
410	576
469	607
528	637
587	668
647	698
706	729
765	759
825	790
884	820
943	851
8437002	881
062	912
121	942
180	973
239	·9262003
299	034
358	064
417	095
476	125
536	156
595	186
654	217
713	247
773	278
832	308
891	339
950	369
8438010	400
069	430
128	461
188	491
247	522
306	552
365	583
425	613
484	644
543	674
602	705
662	735
721	766
780	796
839	827
899	857
958	888
8439017	918
076	949
136	979

8439195	·9263010
254	040
313	071
373	101
432	132
491	162
550	193
609	223
669	254
728	284
787	315
846	345
906	376
965	406

Excesses

10 = 5	1 = 1
20 = 10	2 = 1
30 = 15	3 = 2
40 = 21	4 = 2
50 = 26	5 = 3
	6 = 3
	7 = 4
	8 = 4
	9 = 5

30 = 15

Anti Log Excesses

10 = 19	1 = 2
20 = 39	2 = 4
30 = 58	3 = 6
	4 = 8
	5 = 10
	6 = 12
	7 = 14
	8 = 16
	9 = 17

15 = 30

8440024	·9263437
083	467
143	498
202	528
261	558
320	589
380	619
439	650
498	680
557	711
617	741
676	772
735	802
794	833
854	863
913	894
972	924
8441031	955
090	985
150	·9264016
209	046
268	077

8441327	·9264107
387	138
446	168
505	198
564	229
624	259
683	290
742	320
801	351
860	381
920	412
979	442
8442038	473
097	503
157	534
216	564
275	595
334	625
394	655
453	686
512	716
571	747
630	777
690	808
749	838
808	869
867	899
927	930
986	960
8443045	991
104	·9265021
163	051
223	082
282	112
341	143
400	173
460	204
519	234
578	265
637	295
696	326
756	356
815	387
874	417
933	448
992	478
8444052	509
111	539
170	569
229	600
289	630
348	661
407	691
466	722
525	752
585	783
644	813
703	844
762	874
821	904
881	935
940	965
999	996
8445058	·9266026
117	057

8445177	·9266087
236	118
295	148
354	178
413	209
473	239
532	270
591	300
650	331
709	361
769	392
828	422
887	452
946	483
8446005	513
065	544
124	574
183	605
242	635
301	666
361	696
420	726
479	757
538	787
597	818
657	848
716	879
775	909
834	940
893	970
953	·9267000
8447012	031
071	061
130	092
189	122
249	153
308	183
367	213
426	244
485	274
545	305
604	335
663	366
722	396
781	426
840	457
900	487
959	518
8448018	548
077	578
136	609
196	639
255	670
314	700
373	731
432	761
491	791
551	822
610	852
669	883
728	913
787	944
847	974
906	·9268004
965	035

8449024	9268065
083	096
142	126
202	156
261	187
320	217
379	248
438	278
498	309
557	339
616	369
675	400
734	430
793	461
853	491
912	521
971	552

Excesses

10 = 5	1 = 1
20 = 10	2 = 1
30 = 15	3 = 2
40 = 21	4 = 2
50 = 26	5 = 3
	6 = 3
	7 = 4
	8 = 4
	9 = 5

30 = 15

Anti Log Excesses

10 = 19	1 = 2
20 = 39	2 = 4
30 = 58	3 = 6
	4 = 8
	5 = 10
	6 = 12
	7 = 14
	8 = 16
	9 = 17

15 = 30

8450030	·9268582
089	613
148	643
208	674
267	704
326	734
385	765
444	795
503	826
563	856
622	886
681	917
740	947
799	978
858	·9269008
918	038
977	069
8451036	099
095	130

8451154	·9269160
213	190
273	221
332	251
391	282
450	312
509	343
568	373
628	403
687	434
746	464
805	495
864	525
923	555
983	586
8452042	616
101	647
160	677
219	707
278	738
337	768
397	798
456	829
515	859
574	890
633	920
692	950
752	981
811	·9270011
870	041
929	072
988	102
8453047	133
106	163
166	193
225	224
284	254
343	285
402	315
461	345
520	376
580	406
639	437
698	467
757	497
816	528
875	558
934	588
994	619
8454053	649
112	680
171	710
230	740
289	771
348	801
408	832
467	862
526	892
585	923
644	953
703	983
762	·9271014
822	044
881	075
940	105

8454999	9271135
8455058	166
117	196
176	226
236	257
295	287
354	318
413	348
472	378
531	409
590	439
649	469
709	500
768	530
827	561
886	591
945	621
8456004	652
063	682
123	712
182	743
241	773
300	804
359	834
418	864
477	895
536	925
596	955
655	986
714	·9272016
773	046
832	077
891	107
950	138
8457009	168
069	198
128	229
187	259
246	289
305	320
364	350
423	380
482	411
542	441
601	471
660	502
719	532
778	562
837	593
896	623
955	653
8458014	684
074	714
133	745
192	775
251	805
310	836
369	866
428	896
487	927
546	957
606	987
665	·9273018
724	048
783	078

Number	Log
8458842	·9273109
901	139
960	169
8459019	200
078	230
138	260
197	291
256	321
315	352
374	382
433	412
492	443
551	473
610	503
670	534
729	564
788	594
847	625
906	655
965	685

Excesses

10 = 5	1 = 1
20 = 10	2 = 1
30 = 15	3 = 2
40 = 21	4 = 2
50 = 26	5 = 3
	6 = 3
	7 = 4
	8 = 4
	9 = 5

30 = 15

Anti Log Excesses

10 = 19	1 = 2
20 = 39	2 = 4
30 = 58	3 = 6
	4 = 8
	5 = 10
	6 = 12
	7 = 14
	8 = 16
	9 = 18

15 = 30

Number	Log
8460024	·9273716
083	746
142	776
201	807
261	837
320	867
379	898
438	928
497	958
556	989
615	·9274019
674	049
733	080
792	110
852	140
911	171

Number	Log
8460970	·9274201
8461029	231
088	262
147	292
206	322
265	353
324	383
383	413
442	444
502	474
561	504
620	535
679	565
738	595
797	626
856	656
915	686
974	717
8462033	747
092	777
152	808
211	838
270	868
329	899
388	929
447	959
506	990
565	·9275020
624	050
683	080
742	111
801	141
861	171
920	202
979	232
8463038	262
097	293
156	323
215	353
274	384
333	414
392	444
451	475
510	505
569	535
629	565
688	596
747	626
806	656
865	687
924	717
983	747
8464042	778
101	808
160	838
219	869
278	899
337	929
397	960
456	990
515	·9276020
574	050
633	081
692	111
751	141

Number	Log
8464810	·9276172
869	202
928	232
987	263
8465046	293
105	323
164	353
223	384
283	414
342	444
401	475
460	505
519	535
578	566
637	596
696	626
755	657
814	687
873	717
932	747
991	778
8466050	808
109	838
168	869
228	899
287	929
346	959
405	990
464	·9277020
523	050
582	081
641	111
700	141
759	172
818	202
877	232
936	262
995	293
8467054	323
113	354
172	384
231	414
291	444
350	475
409	505
468	535
527	566
586	596
645	626
704	656
763	687
822	717
881	747
940	778
999	808
8468058	838
117	868
176	899
235	929
294	959
353	990
412	·9278020
471	050
530	080
590	111

Number	Log
8468649	·9278141
708	171
767	201
826	232
885	262
944	292
8469003	323
062	353
121	383
180	413
239	444
298	474
357	504
416	534
475	565
534	595
593	625
652	656
711	686
770	716
829	746
888	777
947	807

Excesses

10 = 5	1 = 1
20 = 10	2 = 1
30 = 15	3 = 2
40 = 21	4 = 2
50 = 26	5 = 3
	6 = 3
	7 = 4
	8 = 4
	9 = 5

30 = 15

Anti Log Excesses

10 = 19	1 = 2
20 = 39	2 = 4
30 = 58	3 = 6
	4 = 8
	5 = 10
	6 = 12
	7 = 14
	8 = 16
	9 = 18

15 = 30

Number	Log
8470006	·9278837
065	867
124	898
183	928
243	958
302	988
361	·9279019
420	049
479	079
538	110
597	140
656	170
715	200

Number	Log
8470774	·9279231
833	261
892	291
951	321
8471010	352
069	382
128	412
187	442
246	473
305	503
364	533
423	563
482	594
541	624
600	654
659	684
718	715
777	745
836	775
895	806
954	836
8472013	866
072	896
131	927
190	957
249	987
308	·9280017
367	048
426	078
485	108
544	138
603	169
662	199
721	229
780	259
839	290
898	320
957	350
8473016	380
076	411
135	441
194	471
253	501
312	532
371	562
430	592
489	622
548	653
607	683
666	713
725	743
784	773
843	804
902	834
961	864
8474020	894
079	925
138	955
197	985
256	·9281015
315	046
374	076
433	106
492	136
551	167

8474610	·9281197
669	227
728	257
787	287
846	318
905	348
964	378
8475023	408
082	439
141	469
200	499
259	529
318	560
377	590
436	620
495	650
554	680
613	711
672	741
731	771
790	801
849	832
908	862
967	892
8476026	922
085	952
143	983
202	·9282013
261	043
320	073
379	104
438	134
497	164
556	194
615	224
674	255
733	285
792	315
851	345
910	376
969	406
8477028	436
087	466
146	497
205	527
264	557
323	587
382	617
441	648
500	678
559	708
618	738
677	768
736	799
795	829
854	859
913	889
972	920
8478031	950
090	980
149	·9283010
208	040
267	071
326	101
385	131

8478444	·9283161
503	191
562	222
621	252
680	282
739	312
798	342
857	373
916	403
974	433
8479033	463
092	494
151	524
210	554
269	584
328	614
387	645
446	675
505	705
564	735
623	765
682	796
741	826
800	856
859	886
918	916
977	947

Excesses

10 = 5	1 = 1
20 = 10	2 = 1
30 = 15	3 = 2
40 = 20	4 = 2
50 = 26	5 = 3
	6 = 3
	7 = 4
	8 = 4
	9 = 5

30 = 15

Anti Log Excesses

10 = 20	1 = 2
20 = 39	2 = 4
30 = 59	3 = 6
	4 = 8
	5 = 10
	6 = 12
	7 = 14
	8 = 16
	9 = 18

15 = 30

8480036	·9283977
095	·9284007
154	037
213	067
272	098
331	128
390	158
449	188
508	218

8480567	·9284248
625	279
684	309
743	339
802	369
861	399
920	430
979	460
8481038	490
097	520
156	550
215	581
274	611
333	641
392	671
451	701
510	731
569	762
628	792
687	822
746	852
805	882
864	913
922	943
981	973
8482040	·9285003
099	033
158	063
217	094
276	124
335	154
394	184
453	214
512	245
571	275
630	305
689	335
748	365
807	395
866	426
925	456
983	486
8483042	516
101	546
160	576
219	607
278	637
337	667
396	697
455	727
514	758
573	788
632	818
691	848
750	878
809	908
868	939
926	969
985	999
8484044	·9286029
103	059
162	089
221	120
280	150
339	180

8484398	·9286210
457	240
516	270
575	301
634	331
693	361
752	391
810	421
869	451
928	482
987	512
8485046	542
105	572
164	602
223	632
282	662
341	693
400	723
459	753
518	783
576	813
635	843
694	874
753	904
812	934
871	964
930	994
989	·9287024
8486048	055
107	085
166	115
225	145
284	175
342	205
401	235
460	266
519	296
578	326
637	356
696	386
755	416
814	447
873	477
932	507
991	537
8487049	567
108	597
167	627
226	658
285	688
344	718
403	748
462	778
521	808
580	838
639	869
697	899
756	929
815	959
874	989
933	·9288019
992	049
8488051	080
110	110
169	140

8488228	·9288170
287	200
345	230
404	260
463	290
522	321
581	351
640	381
699	411
758	441
817	471
876	501
934	532
993	562
8489052	592
111	622
170	652
229	682
288	712
347	742
406	773
465	803
523	833
582	863
641	893
700	923
759	953
818	983
877	·9289014
936	044
995	074

Excesses

10 = 5	1 = 1
20 = 10	2 = 1
30 = 15	3 = 2
40 = 20	4 = 2
50 = 26	5 = 3
	6 = 3
	7 = 4
	8 = 4
	9 = 5

30 = 15

Anti Log Excesses

10 = 20	1 = 2
20 = 39	2 = 4
30 = 59	3 = 6
	4 = 8
	5 = 10
	6 = 12
	7 = 14
	8 = 16
	9 = 18

15 = 30

8490054	·9289104
112	134
171	164
230	194
289	224

8490348	·9289255
407	285
466	315
525	345
584	375
642	405
701	435
760	465
819	496
878	526
937	556
996	586
8491055	616
114	646
172	676
231	706
290	737
349	767
408	797
467	827
526	857
585	887
643	917
702	947
761	977
820	·9290008
879	038
938	068
997	098
8492056	128
114	158
173	188
232	218
291	248
350	279
409	309
468	339
527	369
586	399
644	429
703	459
762	489
821	519
880	549
939	580
998	610
8493056	640
115	670
174	700
233	730
292	760
351	790
410	820
469	850
527	881
586	911
645	941
704	971
763	·9291001
822	031
881	061
940	091
998	121
8494057	151
116	182

8494175	·9291212
234	242
293	272
352	302
410	332
469	362
528	392
587	422
646	452
705	482
764	513
822	543
881	573
940	603
999	633
8495058	663
117	693
176	723
234	753
293	783
352	813
411	844
470	874
529	904
588	934
646	964
705	994
764	·9292024
823	054
882	084
941	114
8496000	144
058	175
117	205
176	235
235	265
294	295
353	325
412	355
470	385
529	415
588	445
647	475
706	505
765	536
823	566
882	596
941	626
8497000	656
059	686
118	716
177	746
235	776
294	806
353	836
412	866
471	896
530	927
588	957
647	987
706	·9293017
765	047
824	077
883	107
941	137

8498000	·9293167
059	197
118	227
177	257
236	287
294	317
353	348
412	378
471	408
530	438
589	468
647	498
706	528
765	558
824	588
883	618
942	648
8499000	678
059	708
118	738
177	768
236	799
295	829
353	859
412	889
471	919
530	949
589	979
648	·9294009
706	039
765	069
824	099
883	129
942	159

Excesses

10 = 5	1 = 1
20 = 10	2 = 1
30 = 15	3 = 2
40 = 20	4 = 2
50 = 26	5 = 3
	6 = 3
	7 = 4
	8 = 4
	9 = 5

30 = 15

Anti Log Excesses

10 = 20	1 = 2
20 = 39	2 = 4
30 = 59	3 = 6
	4 = 8
	5 = 10
	6 = 12
	7 = 14
	8 = 16
	9 = 18

15 = 30

8500000	·9294189
059	219

8500118	·9294249
177	279
236	310
295	340
353	370
412	400
471	430
530	460
589	490
648	520
706	550
765	580
824	610
883	640
942	670
8501000	700
059	730
118	760
177	790
236	820
295	850
353	880
412	910
471	940
530	970
589	·9295001
647	031
706	061
765	091
824	121
883	151
941	181
8502000	211
059	241
118	271
177	301
236	331
294	361
353	391
412	421
471	451
530	481
588	511
647	541
706	571
765	601
824	631
882	661
941	691
8503000	721
059	751
118	782
176	812
235	842
294	872
353	902
412	932
470	962
529	992
588	·9296022
647	052
706	082
764	112
823	142
882	172

8503941	·9296202
8504000	232
058	262
117	292
176	322
235	352
294	382
352	412
411	442
470	472
529	502
587	532
646	562
705	592
764	622
823	652
881	682
940	712
999	742
8505058	772
117	802
175	832
234	862
293	893
352	923
411	953
469	983
528	·9297013
587	043
646	073
704	103
763	133
822	163
881	193
940	223
998	253
8506057	283
116	313
175	343
234	373
292	403
351	433
410	463
469	493
527	523
586	553
645	583
704	613
763	643
821	673
880	703
939	733
998	763
8507056	793
115	823
174	853
233	883
291	913
350	943
409	973
468	·9298003
527	033
585	063
644	093
703	123

Number	Log
8507762	·9298153
820	183
879	213
938	243
997	273
8508056	303
114	333
173	363
232	393
291	423
349	453
408	483
467	513
526	543
584	573
643	603
702	633
761	663
819	693
878	723
937	753
996	783
8509055	813
113	843
172	873
231	903
290	933
348	963
407	993
466	·9299023
525	053
583	083
642	113
701	143
760	173
818	203
877	233
936	263
995	293

Excesses

10 = 5	1 = 1
20 = 10	2 = 1
30 = 15	3 = 2
40 = 20	4 = 2
50 = 26	5 = 3
	6 = 3
	7 = 4
	8 = 4
	9 = 5

30 = 15

Anti Log Excesses

10 = 20	1 = 2
20 = 39	2 = 4
30 = 59	3 = 6
	4 = 8
	5 = 10
	6 = 12
	7 = 14
	8 = 16
	9 = 18

15 = 30

Number	Log
8510053	·9299323
112	353
171	383
230	413
288	443
347	473
406	503
465	532
523	562
582	592
641	622
700	652
758	682
817	712
876	742
935	772
993	802
8511052	832
111	862
170	892
228	922
287	952
346	982
405	·9300012
463	042
522	072
581	102
640	132
698	162
757	192
816	222
875	252
933	282
992	312
8512051	342
110	372
168	402
227	432
286	462
344	492
403	522
462	552
521	582
579	611
638	641
697	671
756	701
814	731
873	761
932	791
991	821

Number	Log
8513049	·9300851
108	881
167	911
226	941
284	971
343	·9301001
402	031
460	061
519	091
578	121
637	151
695	181
754	211
813	241
872	271
930	301
989	330
8514048	360
106	390
165	420
224	450
283	480
341	510
400	540
459	570
518	600
576	630
635	660
694	690
752	720
811	750
870	780
929	810
987	840
8515046	870
105	900
163	930
222	960
281	989
340	·9302019
398	049
457	079
516	109
574	139
633	169
692	199
751	229
809	259
868	289
927	319
985	349
8516044	379
103	409
162	439
220	469
279	499
338	528
396	558
455	588
514	618
573	648
631	678
690	708
749	738
807	768

Number	Log
8516866	·9302798
925	828
984	858
8517042	888
101	918
160	948
218	978
277	·9303008
336	037
394	067
453	097
512	127
571	157
629	187
688	217
747	247
805	277
864	307
923	337
981	367
8518040	397
099	427
158	456
216	486
275	516
334	546
392	576
451	606
510	636
568	666
627	696
686	726
745	756
803	786
862	816
921	846
979	875
8519038	905
097	935
155	965
214	995
273	·9304025
331	055
390	085
449	115
508	145
566	175
625	204
684	234
742	264
801	294
860	324
918	354
977	384

Excesses

10 = 5	1 = 1
20 = 10	2 = 1
30 = 15	3 = 2
40 = 20	4 = 2
50 = 26	5 = 3
	6 = 3
	7 = 4
	8 = 4
	9 = 5

29 = 15

Anti Log Excesses

10 = 20	1 = 2
20 = 39	2 = 4
	3 = 6
	4 = 8
	5 = 10
	6 = 12
	7 = 14
	8 = 16
	9 = 18

15 = 29

Number	Log
8520036	·9304414
094	444
153	474
212	504
270	533
329	563
388	593
446	623
505	653
564	683
623	713
681	743
740	773
799	803
857	833
916	863
975	892
8521033	922
092	952
151	982
209	·9305012
268	042
327	072
385	102
444	132
503	162
561	191
620	221
679	251
737	281
796	311
855	341
913	371
972	401
8522031	431
089	461
148	490

Number	Log
8522207	·9305520
265	550
324	580
383	610
441	640
500	670
559	700
617	730
676	760
735	789
793	819
852	849
911	879
969	909
8523028	939
087	969
145	999
204	·9306029
263	058
321	088
380	118
439	148
497	178
556	208
615	238
673	268
732	298
791	327
849	357
908	387
967	417
8524025	447
084	477
143	507
201	537
260	567
319	596
377	626
436	656
495	686
553	716
612	746
671	776
729	806
788	836
847	865
905	895
964	925
8525022	955
081	985
140	·9307015
198	045
257	075
316	104
374	134
433	164
492	194
550	224
609	254
668	284
726	314
785	343
844	373
902	403
961	433

Number	Log
8526020	·9307463
078	493
137	523
195	553
254	582
313	612
371	642
430	672
489	702
547	732
606	762
665	791
723	821
782	851
840	881
899	911
958	941
8527016	971
075	·9308000
134	030
192	060
251	090
310	120
368	150
427	180
485	210
544	239
603	269
661	299
720	329
779	359
837	389
896	419
955	448
8528013	478
072	508
130	538
189	568
248	598
306	628
365	657
424	687
482	717
541	747
599	777
658	807
717	837
775	866
834	896
893	926
951	956
8529010	986
068	·9309016
127	045
186	075
244	105
303	135
362	165
420	195
479	225
537	254
596	284
655	314
713	344
772	374

Number	Log
8529831	·9309404
889	434
948	463

Excesses

10 = 5	1 = 1
20 = 10	2 = 1
30 = 15	3 = 2
40 = 20	4 = 2
50 = 25	5 = 3
	6 = 3
	7 = 4
	8 = 4
	9 = 5

29 = 15

Anti Log Excesses

10 = 20	1 = 2
20 = 39	2 = 4
	3 = 6
	4 = 8
	5 = 10
	6 = 12
	7 = 14
	8 = 16
	9 = 18

15 = 29

Number	Log
8530006	·9309493
065	523
124	553
182	583
241	613
299	642
358	672
417	702
475	732
534	762
593	792
651	822
710	851
768	881
827	911
886	941
944	971
8531003	·9310000
061	030
120	060
179	090
237	120
296	150
354	179
413	209
472	239
530	269
589	299
647	329
706	358
765	388
823	418
882	448

Number	Log
8531940	·9310478
999	508
8532058	537
116	567
175	597
234	627
292	657
351	687
409	716
468	746
527	776
585	806
644	836
702	865
761	895
819	925
878	955
937	985
995	·9311015
8533054	044
112	074
171	104
230	134
288	164
347	194
406	223
464	253
523	283
581	313
640	343
698	372
757	402
816	432
874	462
933	492
991	521
8534050	551
109	581
167	611
226	641
284	671
343	700
401	730
460	760
519	790
577	820
636	850
694	879
753	909
812	939
870	969
929	999
987	·9312028
8535046	058
104	088
163	118
222	148
280	177
339	207
397	237
456	267
515	297
573	326
632	356
690	386

Number	Log
8535749	·9312416
807	446
866	476
925	505
983	535
8536042	565
100	595
159	625
217	654
276	684
335	714
393	744
452	774
510	803
569	833
627	863
686	893
745	923
803	952
862	982
920	·9313012
979	042
8537037	072
096	101
155	131
213	161
272	191
330	221
389	250
447	280
506	310
565	340
623	369
682	399
740	429
799	459
857	489
916	518
974	548
8538033	578
092	608
150	638
209	667
267	697
326	727
384	757
443	787
502	816
560	846
619	876
677	906
736	935
794	965
853	995
911	·9314025
970	054
8539029	084
087	114
146	144
204	174
263	203
321	233
380	263
438	293
497	323

8539556	·9314352
614	382
673	412
731	442
790	471
848	501
907	531
965	561

Excesses

10 = 5	1 = 1
20 = 10	2 = 1
30 = 15	3 = 2
40 = 20	4 = 2
50 = 25	5 = 3
	6 = 3
	7 = 4
	8 = 4
	9 = 5

29 = 15

Anti Log Excesses

10 = 20	1 = 2
20 = 39	2 = 4
	3 = 6
	4 = 8
	5 = 10
	6 = 12
	7 = 14
	8 = 16
	9 = 18

15 = 29

8540024	·9314590
082	620
141	650
200	680
258	710
317	739
375	769
434	799
492	829
551	858
609	888
668	918
726	948
785	977
844	·9315007
902	037
961	067
8541019	097
078	126
136	156
195	186
253	216
312	245
370	275
429	305
487	335
546	364
605	394

8541663	·9315424
722	454
780	483
839	513
897	543
956	573
8542014	603
073	632
131	662
190	692
248	722
307	751
366	781
424	811
483	841
541	870
600	900
658	930
717	960
775	989
834	·9316019
892	049
951	079
8543009	108
068	138
126	168
185	198
243	227
302	257
360	287
419	317
478	346
536	376
595	406
653	436
712	466
770	495
829	525
887	555
946	585
8544004	614
063	644
121	674
180	704
238	733
297	763
355	793
414	823
472	852
531	882
589	912
648	942
706	971
765	·9317001
823	031
882	061
941	090
999	120
8545058	150
116	179
175	209
233	239
292	269
350	298
409	328

8545467	·9317358
526	388
584	417
643	447
701	477
760	507
818	536
877	566
935	596
994	625
8546052	655
111	685
169	715
228	744
286	774
345	804
403	834
462	863
520	893
579	923
637	952
696	982
754	·9318012
813	042
871	071
930	101
988	131
8547047	161
105	190
164	220
222	250
281	279
339	309
398	339
456	369
515	398
573	428
632	458
690	487
749	517
807	547
866	577
924	606
983	636
8548041	666
100	696
158	725
217	755
275	785
334	814
392	844
451	874
509	904
568	933
626	963
685	993
743	·9319023
802	052
860	082
919	112
977	141
8549036	171
094	201
153	230
211	260

8549269	·9319290
328	320
386	349
445	379
503	409
562	438
620	468
679	498
737	528
796	557
854	587
913	617
971	646

Excesses

10 = 5	1 = 1
20 = 10	2 = 1
30 = 15	3 = 2
40 = 20	4 = 2
50 = 25	5 = 3
	6 = 3
	7 = 4
	8 = 4
	9 = 5

29 = 15

Anti Log Excesses

10 = 20	1 = 2
20 = 39	2 = 4
	3 = 6
	4 = 8
	5 = 10
	6 = 12
	7 = 14
	8 = 16
	9 = 18

15 = 29

8550030	·9319676
088	706
147	736
205	765
264	795
322	825
381	854
439	884
498	914
556	943
615	973
673	·9320003
731	033
790	062
848	092
907	122
965	151
8551024	181
082	211
141	240
199	270
258	300
316	330

8551375	·9320359
433	389
492	419
550	448
609	478
667	508
725	537
784	567
842	597
901	627
959	656
8552018	686
076	716
135	745
193	775
252	805
310	834
369	864
427	894
486	923
544	953
602	983
661	·9321012
719	042
778	072
836	102
895	131
953	161
8553012	191
070	220
129	250
187	280
245	309
304	339
362	369
421	398
479	428
538	458
596	487
655	517
713	547
772	576
830	606
889	636
947	665
8554005	695
064	725
122	754
181	784
239	814
298	843
356	873
415	903
473	932
531	962
590	992
648	·9322021
707	051
765	081
824	110
882	140
941	170
999	199
8555057	229
116	259

8555174	·9322288
233	318
291	348
350	377
408	407
467	437
525	466
583	496
642	526
700	555
759	585
817	615
876	644
934	674
993	704
8556051	733
109	763
168	793
226	822
285	852
343	882
402	911
460	941
518	971
577	·9323000
635	030
694	060
752	089
811	119
869	149
928	178
986	208
8557044	238
103	267
161	297
220	327
278	356
337	386
395	416
453	445
512	475
570	505
629	534
687	564
746	593
804	623
862	653
921	682
979	712
8558038	742
096	771
155	801
213	831
271	860
330	890
388	920
447	949
505	979
563	·9324008
622	038
680	068
739	097
797	127
856	157
914	186

8558972	·9324216
8559031	246
089	275
148	305
206	335
264	364
323	394
381	423
440	453
498	483
557	512
615	542
673	572
732	601
790	631
849	661
907	690
965	720

Excesses

10 = 5 1 = 1
20 = 10 2 = 1
30 = 15 3 = 2
40 = 20 4 = 2
50 = 25 5 = 3
6 = 3
7 = 4
8 = 4
9 = 5

29 = 15

Anti Log Excesses

10 = 20 1 = 2
20 = 39 2 = 4
3 = 6
4 = 8
5 = 10
6 = 12
7 = 14
8 = 16
9 = 18

15 = 29

8560024	·9324749
082	779
141	809
199	838
258	868
316	898
374	927
433	957
491	987
550	·9325016
608	046
666	075
725	105
783	135
842	164
900	194
958	224
8561017	253

8561075	·9325283
134	312
192	342
250	372
309	401
367	431
426	461
484	490
542	520
601	549
659	579
718	609
776	638
834	668
893	698
951	727
8562010	757
068	786
126	816
185	846
243	875
302	905
360	934
418	964
477	994
535	·9326023
594	053
652	083
710	112
769	142
827	171
886	201
944	231
8563002	260
061	290
119	320
177	349
236	379
294	408
353	438
411	468
469	497
528	527
586	557
645	586
703	616
761	645
820	675
878	705
937	734
995	764
8564053	793
112	823
170	853
228	882
287	912
345	941
404	971
462	·9327001
520	030
579	060
637	089
695	119
754	149
812	178

8564871	·9327208
929	238
987	267
8565046	297
104	326
162	356
221	386
279	415
338	445
396	474
454	504
513	534
571	563
629	593
688	622
746	652
805	681
863	711
921	741
980	770
8566038	800
096	829
155	859
213	889
272	918
330	948
388	977
447	·9328007
505	037
563	066
622	096
680	125
739	155
797	185
855	214
914	244
972	273
8567030	303
089	333
147	362
205	392
264	421
322	451
380	480
439	510
497	540
556	569
614	599
672	628
731	658
789	687
847	717
906	747
964	776
8568022	806
081	835
139	865
198	894
256	924
314	954
373	983
431	·9329013
489	042
548	072
606	101

8568664	·9329131
723	161
781	190
839	220
898	249
956	279
8569014	308
073	338
131	368
189	397
248	427
306	456
365	486
423	515
481	545
540	575
598	604
656	634
715	663
773	693
831	722
890	752
948	782

Excesses

10 = 5 1 = 1
20 = 10 2 = 1
30 = 15 3 = 2
40 = 20 4 = 2
50 = 25 5 = 3
6 = 3
7 = 4
8 = 4
9 = 5

29 = 15

Anti Log Excesses

10 = 20 1 = 2
20 = 39 2 = 4
3 = 6
4 = 8
5 = 10
6 = 12
7 = 14
8 = 16
9 = 18

15 = 29

8570006	·9329811
065	841
123	870
181	900
240	929
298	959
356	989
415	·9330018
473	048
531	077
590	107
648	136
706	166

8570765	·9330195
823	225
881	255
940	284
998	314
8571056	343
115	373
173	402
231	432
290	462
348	491
406	521
465	550
523	580
581	609
640	639
698	668
756	698
815	728
873	757
931	787
990	816
8572048	846
106	875
165	905
223	934
281	964
340	993
398	·9331023
456	053
515	082
573	112
631	141
690	171
748	200
806	230
865	259
923	289
981	318
8573040	348
098	378
156	407
215	437
273	466
331	496
390	525
448	555
506	584
565	614
623	643
681	673
739	702
798	732
856	762
914	791
973	821
8574031	850
089	880
148	909
206	939
264	968
323	998
381	·9332027
439	057
498	086

8574556	·9332116
614	145
673	175
731	205
789	234
847	264
906	293
964	323
8575022	352
081	382
139	411
197	441
256	470
314	500
372	529
431	559
489	588
547	618
605	647
664	677
722	707
780	736
839	766
897	795
955	825
8576014	854
072	884
130	913
188	943
247	972
305	·9333002
363	031
422	061
480	090
538	120
597	149
655	179
713	208
771	238
830	267
888	297
946	326
8577005	356
063	385
121	415
180	445
238	474
296	504
354	533
413	563
471	592
529	622
588	651
646	681
704	710
762	740
821	769
879	799
937	828
996	858
8578054	887
112	917
170	946
229	976
287	·9334005

8578345	·9334035
404	064
462	094
520	123
578	153
637	182
695	212
753	242
812	271
870	301
928	330
986	360
8579045	389
103	419
161	448
220	478
278	507
336	537
394	566
453	596
511	625
569	655
628	684
686	714
744	743
802	773
861	802
919	832
977	861

Excesses

10 = 5	1 = 1
20 = 10	2 = 1
30 = 15	3 = 2
40 = 20	4 = 2
50 = 25	5 = 3
	6 = 3
	7 = 4
	8 = 4
	9 = 5

29 = 15

Anti Log Excesses

10 = 20	1 = 2
20 = 39	2 = 4
	3 = 6
	4 = 8
	5 = 10
	6 = 12
	7 = 14
	8 = 16
	9 = 18

15 = 29

8580035	·9334891
094	920
152	950
210	979
269	·9335009
327	038
385	068

8580443	·9335097
502	127
560	156
618	186
676	215
735	244
793	274
851	303
910	333
968	362
8581026	392
084	421
143	451
201	480
259	510
317	539
376	569
434	598
492	628
550	657
609	687
667	716
725	746
784	775
842	805
900	834
958	864
8582017	893
075	923
133	952
191	982
250	·9336011
308	041
366	070
424	100
483	129
541	159
599	188
657	218
716	247
774	277
832	306
890	336
949	365
8583007	394
065	424
123	453
182	483
240	512
298	542
356	571
415	601
473	630
531	660
589	689
648	719
706	748
764	778
822	807
881	837
939	866
997	896
8584055	925
114	954
172	984

8584230	·9337013
288	043
347	072
405	102
463	131
521	161
580	190
638	220
696	249
754	279
813	308
871	337
929	367
987	396
8585046	426
104	455
162	485
220	514
279	544
337	573
395	603
453	632
512	662
570	691
628	720
686	750
744	779
803	809
861	838
919	868
977	897
8586036	927
094	956
152	986
210	·9338015
269	044
327	074
385	103
443	133
502	162
560	192
618	221
676	251
734	280
793	310
851	339
909	368
967	398
8587026	427
084	457
142	486
200	516
259	545
317	575
375	604
433	634
491	663
550	692
608	722
666	751
724	781
783	810
841	840
899	869
957	898

8588015	·9338928
074	957
132	987
190	·9339016
248	046
307	075
365	105
423	134
481	163
539	193
598	222
656	252
714	281
772	311
830	340
889	369
947	399
8589005	428
063	458
122	487
180	517
238	546
296	576
354	605
413	634
471	664
529	693
587	723
645	752
704	782
762	811
820	840
878	870
936	899
995	929

Excesses

10 = 5	1 = 1
20 = 10	2 = 1
30 = 15	3 = 2
40 = 20	4 = 2
50 = 25	5 = 3
	6 = 3
	7 = 4
	8 = 4
	9 = 5

29 = 15

Anti Log Excesses

10 = 20	1 = 2
20 = 40	2 = 4
	3 = 6
	4 = 8
	5 = 10
	6 = 12
	7 = 14
	8 = 16
	9 = 18

15 = 29

8590053	·9339958
8590111	·9339987
169	·9340017
228	046
286	076
344	105
402	135
460	164
519	193
577	223
635	252
693	282
751	311
810	341
868	370
926	399
984	429
8591042	458
101	488
159	517
217	546
275	576
333	605
392	635
450	664
508	694
566	723
624	752
683	782
741	811
799	841
857	870
915	900
973	929
8592032	958
090	988
148	·9341017
206	047
264	076
323	106
381	135
439	164
497	194
555	223
614	253
672	282
730	311
788	341
846	370
905	400
963	429
8593021	458
079	488
137	517
195	547
254	576
312	605
370	635
428	664
486	694
545	723
603	753
661	782
719	811
777	841
835	870

8593894	·9341900
952	929
8594010	958
068	988
126	·9342017
185	047
243	076
301	105
359	135
417	164
475	194
534	223
592	252
650	282
708	311
766	341
825	370
883	399
941	429
999	458
8595057	487
115	517
174	546
232	576
290	605
348	634
406	664
464	693
523	723
581	752
639	781
697	811
755	840
813	870
872	899
930	928
988	958
8596046	987
104	·9343017
162	046
221	075
279	105
337	134
395	163
453	193
511	222
570	252
628	281
686	310
744	340
802	369
860	398
919	428
977	457
8597035	487
093	516
151	545
209	575
268	604
326	633
384	663
442	692
500	722
558	751
616	780

8597675	·9343810
733	839
791	868
849	898
907	927
965	957
8598024	986
082	·9344015
140	045
198	074
256	103
314	133
372	162
431	192
489	221
547	250
605	280
663	309
721	338
779	368
838	397
896	427
954	456
8599012	485
070	515
128	544
187	573
245	603
303	632
361	661
419	691
477	720
535	750
594	779
652	808
710	838
768	867
826	896
884	926
942	955

Excesses

10 = 5	1 = 1
20 = 10	2 = 1
30 = 15	3 = 2
40 = 20	4 = 2
50 = 25	5 = 3
	6 = 3
	7 = 4
	8 = 4
	9 = 5

29 = 15

Anti Log Excesses

10 = 20	1 = 2
20 = 40	2 = 4
	3 = 6
	4 = 8
	5 = 10
	6 = 12
	7 = 14
	8 = 16
	9 = 18

15 = 29

8600000	·9344984
059	·9345014
117	043
175	072
233	102
291	131
349	161
407	190
466	219
524	249
582	278
640	307
698	337
756	366
814	395
873	425
931	454
989	483
8601047	513
105	542
163	571
221	601
279	630
338	660
396	689
454	718
512	748
570	777
628	806
686	836
745	865
803	895
861	924
919	953
977	983
8602035	·9346012
093	041
151	071
210	100
268	129
326	159
384	188
442	217
500	247
558	276
616	305
675	335
733	364
791	393
849	423
907	452

8602965	·9346481
8603023	511
081	540
139	569
198	599
256	628
314	657
372	687
430	716
488	745
546	775
604	804
663	833
721	863
779	892
837	922
895	951
953	980
8604011	·9347010
069	039
127	068
186	098
244	127
302	156
360	186
418	215
476	244
534	273
592	303
650	332
709	361
767	391
825	420
883	449
941	479
999	508
8605057	537
115	567
173	596
231	625
290	655
348	684
406	713
464	743
522	772
580	801
638	831
696	860
754	889
812	919
871	948
929	977
987	·9348007
8606045	036
103	065
161	095
219	124
277	153
335	182
393	212
452	241
510	270
568	300
626	329
684	358

8606742	·9348388
800	417
858	446
916	476
974	505
8607033	534
091	563
149	593
207	622
265	651
323	681
381	710
439	739
497	769
555	798
613	827
671	857
730	886
788	915
846	945
904	974
962	·9349003
8608020	032
078	062
136	091
194	120
252	150
310	179
369	208
427	238
485	267
543	296
601	325
659	355
717	384
775	413
833	443
891	472
949	501
8609007	531
065	560
124	589
182	618
240	648
298	677
356	706
414	736
472	765
530	794
588	823
646	853
704	882
762	911
820	941
879	970
937	999
995	·9350029

Excesses

10 = 5	1 = 1
20 = 10	2 = 1
30 = 15	3 = 2
40 = 20	4 = 2
50 = 25	5 = 3
	6 = 3
	7 = 4
	8 = 4
	9 = 5

29 = 15

Anti Log Excesses

10 = 20	1 = 2
20 = 40	2 = 4
	3 = 6
	4 = 8
	5 = 10
	6 = 12
	7 = 14
	8 = 16
	9 = 18

15 = 29

8610053	·9350058
111	087
169	116
227	146
285	175
343	204
401	234
459	263
517	292
575	321
633	351
692	380
750	409
808	439
866	468
924	497
982	526
8611040	556
098	585
156	614
214	644
272	673
330	702
388	731
446	761
504	790
562	819
621	849
679	878
737	907
795	936
853	966
911	995
969	·9351024
8612027	054
085	083
143	112

8612201	·9351141
259	171
317	200
375	229
433	258
491	288
549	317
608	346
666	376
724	405
782	434
840	463
898	493
956	522
8613014	551
072	581
130	610
188	639
246	668
304	698
362	727
420	756
478	785
536	815
594	844
652	873
711	903
769	932
827	961
885	990
943	·9352020
8614001	049
059	078
117	107
175	137
233	166
291	195
349	224
407	254
465	283
523	312
581	341
639	371
697	400
755	429
813	459
871	488
929	517
987	546
8615045	576
104	605
162	634
220	663
278	693
336	722
394	751
452	780
510	810
568	839
626	868
684	897
742	927
800	956
858	985
916	·9353014

8615974	·9353043
8616032	073
090	102
148	131
206	161
264	190
322	219
380	248
438	278
496	307
554	336
612	365
670	394
728	424
786	453
844	482
902	511
961	541
8617019	570
077	599
135	628
193	658
251	687
309	716
367	745
425	775
483	804
541	833
599	862
657	892
715	921
773	950
831	979
889	·9354009
947	038
8618005	067
063	096
121	126
179	155
237	184
295	213
353	242
411	272
469	301
527	330
585	359
643	389
701	418
759	447
817	476
875	506
933	535
991	564
8619049	593
107	622
165	652
223	681
281	710
339	739
397	769
455	798
513	827
571	856
629	885
687	915

8619745	·9354944
803	973
861	·9355002
919	032
977	061

Excesses

10 = 5	1 = 1
20 = 10	2 = 1
30 = 15	3 = 2
40 = 20	4 = 2
50 = 25	5 = 3
	6 = 3
	7 = 4
	8 = 4
	9 = 5

29 = 15

Anti Log Excesses

10 = 20	1 = 2
20 = 40	2 = 4
	3 = 6
	4 = 8
	5 = 10
	6 = 12
	7 = 14
	8 = 16
	9 = 18

15 = 29

8620035	·9355090
093	119
151	149
209	178
267	207
325	236
383	265
441	295
499	324
557	353
615	382
673	412
731	441
789	470
847	500
905	529
963	558
8621021	587
079	616
137	646
195	675
253	704
311	733
369	762
427	792
485	821
543	850
601	879
659	908
717	938
775	967

8621833	·9355996
891	·9356025
949	055
8622007	084
065	113
123	142
181	171
239	201
297	230
355	259
413	288
471	317
529	347
587	376
645	405
703	434
761	463
819	493
877	522
935	551
993	580
8623051	609
109	639
167	668
225	697
283	726
341	755
399	785
457	814
515	843
573	872
631	901
689	931
747	960
805	989
863	·9357018
921	047
979	077
8624037	106
095	135
153	164
211	193
269	223
327	252
385	281
443	310
500	339
558	369
616	398
674	427
732	456
790	485
848	514
906	544
964	573
8625022	602
080	631
138	660
196	690
254	719
312	748
370	777
428	806
486	836
544	865

8625602	·9357894
660	923
718	952
776	982
834	·9358011
892	040
950	069
8626008	098
066	128
124	157
182	186
240	215
298	244
355	273
413	303
471	332
529	361
587	390
645	419
703	449
761	478
819	507
877	536
935	565
993	594
8627051	624
109	653
167	682
225	711
283	740
341	769
399	799
457	828
515	857
573	886
631	915
688	945
746	974
804	·9359003
862	032
920	061
978	090
8628036	120
094	149
152	178
210	207
268	236
326	265
384	295
442	324
500	353
558	382
616	411
674	440
732	470
790	499
847	528
905	557
963	586
8629021	615
079	645
137	674
195	703
253	732
311	761

8629369	·9359790
427	820
485	849
543	878
601	907
659	936
717	965
775	994
832	·9360024
890	053
948	082

Excesses

10 = 5	1 = 1
20 = 10	2 = 1
30 = 15	3 = 2
40 = 20	4 = 2
50 = 25	5 = 3
	6 = 3
	7 = 4
	8 = 4
	9 = 5

29 = 15

Anti Log Excesses

10 = 20	1 = 2
20 = 40	2 = 4
	3 = 6
	4 = 8
	5 = 10
	6 = 12
	7 = 14
	8 = 16
	9 = 18

15 = 29

8630006	·9360111
064	140
122	169
180	199
238	228
296	257
354	286
412	315
470	344
528	373
586	403
644	432
702	461
759	490
817	519
875	548
933	577
991	607
8631049	636
107	665
165	694
223	723
281	752
339	781
397	811

8631455	·9360840
513	869
570	898
628	927
686	956
744	985
802	·9361015
860	044
918	073
976	102
8632034	131
092	160
150	189
208	219
266	248
323	277
381	306
439	335
497	364
555	393
613	423
671	452
729	481
787	510
845	539
903	568
961	597
8633019	627
076	656
134	685
192	714
250	743
308	772
366	801
424	830
482	860
540	889
598	918
656	947
713	976
771	·9362005
829	034
887	063
945	093
8634003	122
061	151
119	180
177	209
235	238
293	267
351	297
408	326
466	355
524	384
582	413
640	442
698	471
756	500
814	529
872	559
930	588
987	617
8635045	646
103	675
161	704

8635219	·9362733
277	762
335	792
393	821
451	850
509	879
566	908
624	937
682	966
740	996
798	·9363025
856	054
914	083
972	112
8636030	141
088	170
145	199
203	228
261	258
319	287
377	316
435	345
493	374
551	403
609	432
667	461
724	491
782	520
840	549
898	578
956	607
8637014	636
072	665
130	694
188	723
245	752
303	782
361	811
419	840
477	869
535	898
593	927
651	956
709	985
766	·9364014
824	044
882	073
940	102
998	131
8638056	160
114	189
172	218
229	247
287	276
345	305
403	335
461	364
519	393
577	422
635	451
693	480
750	509
808	538
866	567
924	596

8638982	·9364626
8639040	655
098	684
156	713
213	742
271	771
329	800
387	829
445	858
503	887
561	917
619	946
676	975
734	·9365004
792	033
850	062
908	091
966	120

Excesses

10 = 5	1 = 1
20 = 10	2 = 1
30 = 15	3 = 2
40 = 20	4 = 2
50 = 25	5 = 3
	6 = 3
	7 = 4
	8 = 4
	9 = 5

29 = 15

Anti Log Excesses

10 = 20	1 = 2
20 = 40	2 = 4
	3 = 6
	4 = 8
	5 = 10
	6 = 12
	7 = 14
	8 = 16
	9 = 18

15 = 29

8640024	·9365149
082	178
139	207
197	236
255	266
313	295
371	324
429	353
487	382
544	411
602	440
660	469
718	498
776	527
834	556
892	585
950	615
8641007	644

8641065	·9365673
123	702
181	731
239	760
297	789
355	818
412	847
470	876
528	905
586	934
644	963
702	993
760	·9366022
817	051
875	080
933	109
991	138
8642049	167
107	196
165	225
222	254
280	283
338	312
396	341
454	371
512	400
570	429
627	458
685	487
743	516
801	545
859	574
917	603
975	632
8643032	661
090	690
148	719
206	748
264	777
322	807
379	836
437	865
495	894
553	923
611	952
669	981
727	·9367010
784	039
842	068
900	097
958	126
8644016	155
074	184
131	213
189	242
247	272
305	301
363	330
421	359
479	388
536	417
594	446
652	475
710	504
768	533

8644826	·9367562
883	591
941	620
999	649
8645057	678
115	707
173	737
230	766
288	795
346	824
404	853
462	882
520	911
577	940
635	969
693	998
751	·9368027
809	056
867	085
924	114
982	143
8646040	172
098	201
156	230
214	259
271	289
329	318
387	347
445	376
503	405
561	434
618	463
676	492
734	521
792	550
850	579
907	608
965	637
8647023	666
081	695
139	724
197	753
254	782
312	811
370	840
428	869
486	898
544	927
601	957
659	986
717	·9369015
775	044
833	073
890	102
948	131
8648006	160
064	189
122	218
179	247
237	276
295	305
353	334
411	363
469	392
526	421

8648584	·9369450
642	479
700	508
758	537
815	566
873	595
931	624
989	653
8649047	682
104	711
162	740
220	769
278	798
336	828
394	857
451	886
509	915
567	944
625	973
683	·9370002
740	031
798	060
856	089
914	118
972	147

Excesses

10 = 5	1 = 1
20 = 10	2 = 1
30 = 15	3 = 2
40 = 20	4 = 2
50 = 25	5 = 3
	6 = 3
	7 = 4
	8 = 4
	9 = 5

29 = 15

Anti Log Excesses

10 = 20	1 = 2
20 = 40	2 = 4
	3 = 6
	4 = 8
	5 = 10
	6 = 12
	7 = 14
	8 = 16
	9 = 18

15 = 29

8650029	·9370176
087	205
145	234
203	263
261	292
318	321
376	350
434	379
492	408
550	437
607	466

8650665	·9370495
723	524
781	553
839	582
896	611
954	640
8651012	669
070	698
128	727
185	756
243	785
301	814
359	843
417	872
474	901
532	930
590	959
648	988
706	·9371017
763	046
821	075
879	104
937	133
994	162
8652052	191
110	220
168	249
226	278
283	307
341	336
399	365
457	394
515	423
572	452
630	481
688	510
746	539
804	568
861	597
919	626
977	655
8653035	684
092	713
150	742
208	771
266	800
324	829
381	858
439	887
497	916
555	945
612	974
670	·9372003
728	032
786	061
844	090
901	119
959	148
8654017	177
075	206
132	235
190	264
248	293
306	322
364	351

8654421	·9372380
479	409
537	438
595	467
652	496
710	525
768	554
826	583
884	612
941	641
999	670
8655057	699
115	728
172	757
230	786
288	815
346	844
403	873
461	902
519	931
577	960
635	989
692	·9373018
750	047
808	076
866	105
923	134
981	163
8656039	192
097	221
154	250
212	279
270	308
328	337
385	366
443	394
501	423
559	452
616	481
674	510
732	539
790	568
848	597
905	626
963	655
8657021	684
079	713
136	742
194	771
252	800
310	829
367	858
425	887
483	916
541	945
598	974
656	·9374003
714	032
772	061
829	090
887	119
945	148
8658003	177
060	206
118	235

8658176	·9374264
234	293
291	322
349	350
407	379
465	408
522	437
580	466
638	495
696	524
753	553
811	582
869	611
927	640
984	669
8659042	698
100	727
158	756
215	785
273	814
331	843
388	872
446	901
504	930
562	959
619	988
677	·9375017
735	046
793	075
850	104
908	133
966	162

Excesses

10 = 5	1 = 1
20 = 10	2 = 1
30 = 15	3 = 2
40 = 20	4 = 2
50 = 25	5 = 3
	6 = 3
	7 = 4
	8 = 4
	9 = 5

29 = 14

Anti Log Excesses

10 = 20	1 = 2
20 = 40	2 = 4
	3 = 6
	4 = 8
	5 = 10
	6 = 12
	7 = 14
	8 = 16
	9 = 18

14 = 29

8660024	·9375191
081	219
139	248
197	277

8660255	·9375306
312	335
370	364
428	393
485	422
543	451
601	480
659	509
716	538
774	567
832	596
890	625
947	654
8661005	683
063	712
121	741
178	770
236	798
294	827
351	856
409	885
467	914
525	943
582	972
640	·9376001
698	030
756	059
813	088
871	117
929	146
986	175
8662044	204
102	233
160	262
217	291
275	319
333	348
390	377
448	406
506	435
564	464
621	493
679	522
737	551
795	580
852	609
910	638
968	667
8663025	696
083	725
141	753
199	782
256	811
314	840
372	869
429	898
487	927
545	956
603	985
660	·9377014
718	043
776	072
833	101
891	130
949	158

8664007	·9377187
064	216
122	245
180	274
237	303
295	332
353	361
410	390
468	419
526	448
584	477
641	506
699	535
757	563
814	592
872	621
930	650
988	679
8665045	708
103	737
161	766
218	795
276	824
334	853
391	882
449	911
507	939
565	968
622	997
680	·9378026
738	055
795	084
853	113
911	142
968	171
8666026	200
084	229
142	257
199	286
257	315
315	344
372	373
430	402
488	431
545	460
603	489
661	518
718	547
776	575
834	604
892	633
949	662
8667007	691
065	720
122	749
180	778
238	807
295	836
353	864
411	893
468	922
526	951
584	980
641	·9379009
699	038

8667757	·9379067
815	096
872	125
930	153
988	182
8668045	211
103	240
161	269
218	298
276	327
334	356
391	385
449	414
507	442
564	471
622	500
680	529
737	558
795	587
853	616
910	645
968	674
8669026	702
083	731
141	760
199	789
257	818
314	847
372	876
430	905
487	934
545	963
603	992
660	·9380020
718	049
776	078
833	107
891	136
949	165

Excesses

10 = 5	1 = 1
20 = 10	2 = 1
30 = 15	3 = 2
40 = 20	4 = 2
50 = 25	5 = 3
	6 = 3
	7 = 4
	8 = 4
	9 = 5

29 = 14

Anti Log Excesses

10 = 20	1 = 2
20 = 40	2 = 4
	3 = 6
	4 = 8
	5 = 10
	6 = 12
	7 = 14
	8 = 16
	9 = 18

14 = 29

8670006	·9380194
064	223
122	252
179	280
237	309
295	338
352	367
410	396
468	425
525	454
583	483
641	512
698	540
756	569
814	598
871	627
929	656
987	685
8671044	714
102	743
160	771
217	800
275	829
333	858
390	887
448	916
506	945
563	974
621	·9381002
679	031
736	060
794	089
852	118
909	147
967	176
8672024	205
082	233
140	262
197	291
255	320
313	349
370	378
428	407
486	436
543	464
601	493
659	522
716	551
774	580
832	609
889	638

8672947	·9381667
8673005	695
062	724
120	753
178	782
235	811
293	840
350	869
408	897
466	926
523	955
581	984
639	·9382013
696	042
754	071
812	099
869	128
927	157
985	186
8674042	215
100	244
158	273
215	302
273	330
330	359
388	388
446	417
503	446
561	475
619	504
676	532
734	561
792	590
849	619
907	648
964	677
8675022	706
080	734
137	763
195	792
253	821
310	850
368	879
426	908
483	936
541	965
598	994
656	·9383023
714	052
771	081
829	110
887	138
944	167
8676002	196
060	225
117	254
175	283
232	311
290	340
348	369
405	398
463	427
521	456
578	485
636	513

8676693	·9383542
751	571
809	600
866	629
924	658
982	686
8677039	715
097	744
154	773
212	802
270	831
327	860
385	888
443	917
500	946
558	975
615	·9384004
673	033
731	061
788	090
846	119
903	148
961	177
8678019	206
076	234
134	263
192	292
249	321
307	350
364	379
422	407
480	436
537	465
595	494
652	523
710	552
768	580
825	609
883	638
941	667
998	696
8679056	725
113	753
171	782
229	811
286	840
344	869
401	897
459	926
517	955
574	984
632	·9385013
689	042
747	070
805	099
862	128
920	157
977	186

Excesses

10 = 5	1 = 1
20 = 10	2 = 1
30 = 15	3 = 2
40 = 20	4 = 2
50 = 25	5 = 3
	6 = 3
	7 = 4
	8 = 4
	9 = 5

29 = 14

Anti Log Excesses

10 = 20	1 = 2
20 = 40	2 = 4
	3 = 6
	4 = 8
	5 = 10
	6 = 12
	7 = 14
	8 = 16
	9 = 18

14 = 29

8680035	·9385215
093	243
150	272
208	301
265	330
323	359
381	387
438	416
496	445
553	474
611	503
669	532
726	560
784	589
841	618
899	647
957	676
8681014	704
072	733
129	762
187	791
245	820
302	848
360	877
417	906
475	935
533	964
590	993
648	·9386021
705	050
763	079
821	108
878	137
936	165
993	194
8682051	223
109	252

8682166	·9386281
224	309
281	338
339	367
396	396
454	425
512	453
569	482
627	511
684	540
742	569
800	597
857	626
915	655
972	684
8683030	713
088	741
145	770
203	799
260	828
318	857
375	885
433	914
491	943
548	972
606	·9387001
663	029
721	058
778	087
836	116
894	145
951	173
8684009	202
066	231
124	260
182	289
239	317
297	346
354	375
412	404
469	433
527	461
585	490
642	519
700	548
757	577
815	605
872	634
930	663
988	692
8685045	720
103	749
160	778
218	807
275	836
333	864
391	893
448	922
506	951
563	979
621	·9388008
678	037
736	066
793	095
851	123

8685909	·9388152
966	181
8686024	210
081	238
139	267
196	296
254	325
312	354
369	382
427	411
484	440
542	469
599	497
657	526
714	555
772	584
830	613
887	641
945	670
8687002	699
060	728
117	756
175	785
233	814
290	843
348	872
405	900
463	929
520	958
578	987
635	·9389015
693	044
750	073
808	102
866	130
923	159
981	188
8688038	217
096	245
153	274
211	303
268	332
326	361
384	389
441	418
499	447
556	476
614	504
671	533
729	562
786	591
844	619
901	648
959	677
8689017	706
074	735
132	763
189	792
247	821
304	850
362	878
419	907
477	936
534	965
592	993

8689650	·9390022
707	051
765	080
822	108
880	137
937	166
995	195

Excesses

10 = 5	1 = 1
20 = 10	2 = 1
30 = 15	3 = 2
40 = 20	4 = 2
50 = 25	5 = 3
	6 = 3
	7 = 4
	8 = 4
	9 = 5

29 = 14

Anti Log Excesses

10 = 20	1 = 2
20 = 40	2 = 4
	3 = 6
	4 = 8
	5 = 10
	6 = 12
	7 = 14
	8 = 16
	9 = 18

14 = 29

8690052	·9390223
110	252
167	281
225	310
282	338
340	367
397	396
455	425
513	453
570	482
628	511
685	540
743	568
800	597
858	626
915	655
973	683
8691030	712
088	741
145	770
203	798
260	827
318	856
376	885
433	913
491	942
548	971
606	·9391000
663	028

8691721	·9391057
778	086
836	115
893	143
951	172
8692008	201
066	230
123	258
181	287
238	316
296	345
353	373
411	402
469	431
526	460
584	488
641	517
699	546
756	574
814	603
871	632
929	661
986	689
8693044	718
101	747
159	776
216	804
274	833
331	862
389	891
446	919
504	948
561	977
619	·9392005
676	034
734	063
791	092
849	120
906	149
964	178
8694021	207
079	235
136	264
194	293
251	322
309	350
366	379
424	408
482	436
539	465
597	494
654	523
712	551
769	580
827	609
884	638
942	666
999	695
8695057	724
114	752
172	781
229	810
287	839
344	867
402	896

8695459	·9392925
517	954
574	982
632	·9393011
689	040
747	068
804	097
862	126
919	155
977	183
8696034	212
092	241
149	269
207	298
264	327
322	356
379	384
437	413
494	442
552	470
609	499
667	528
724	557
782	585
839	614
896	643
954	671
8697011	700
069	729
126	758
184	786
241	815
299	844
356	872
414	901
471	930
529	958
586	987
644	·9394016
701	045
759	073
816	102
874	131
931	159
989	188
8698046	217
104	246
161	274
219	303
276	332
334	360
391	389
449	418
506	446
564	475
621	504
679	533
736	561
794	590
851	619
908	647
966	676
8699023	705
081	734
138	762

8699196	·9394791
253	820
311	848
368	877
426	906
483	934
541	963
598	992
656	·9395020
713	049
771	078
828	107
886	135
943	164

Excesses

10 = 5	1 = 0
20 = 10	2 = 1
30 = 15	3 = 1
40 = 20	4 = 2
50 = 25	5 = 2
	6 = 3
	7 = 3
	8 = 4
	9 = 4

29 = 14

Anti Log Excesses

10 = 20	1 = 2
20 = 40	2 = 4
	3 = 6
	4 = 8
	5 = 10
	6 = 12
	7 = 14
	8 = 16
	9 = 18

14 = 29

8700000	·9395193
058	221
115	250
173	279
230	307
288	336
345	365
403	393
460	422
518	451
575	479
633	508
690	537
748	566
805	594
863	623
920	652
977	680
8701035	709
092	738
150	766
207	795

8701265	·9395824
322	852
380	881
437	910
495	938
552	967
610	996
667	·9396024
724	053
782	082
839	110
897	139
954	168
8702012	196
069	225
127	254
184	282
242	311
299	340
357	368
414	397
471	426
529	454
586	483
644	512
701	541
759	569
816	598
874	627
931	655
988	684
8703046	713
103	741
161	770
218	799
276	827
333	856
391	885
448	913
506	942
563	971
620	999
678	·9397028
735	057
793	085
850	114
908	143
965	171
8704023	200
080	229
137	257
195	286
252	315
310	343
367	372
425	401
482	429
540	458
597	487
654	515
712	544
769	573
827	601
884	630
942	659

8704999	·9397687
8705057	716
114	744
171	773
229	802
286	830
344	859
401	888
459	916
516	945
573	974
631	·9398002
688	031
746	060
803	088
861	117
918	146
975	174
8706033	203
090	232
148	260
205	289
263	318
320	346
377	375
435	403
492	432
550	461
607	489
665	518
722	547
779	575
837	604
894	633
952	661
8707009	690
067	719
124	747
181	776
239	804
296	833
354	862
411	890
469	919
526	948
583	976
641	·9399005
698	034
756	062
813	091
871	120
928	148
985	177
8708043	205
100	234
158	263
215	291
272	320
330	349
387	377
445	406
502	434
560	463
617	492
674	520

8708732	·9399549
789	578
847	606
904	635
961	663
8709019	692
076	721
134	749
191	778
248	807
306	835
363	864
421	892
478	921
536	950
593	978
650	·9400007
708	036
765	064
823	093
880	121
937	150
995	179

Excesses

10 = 5	1 = 0
20 = 10	2 = 1
30 = 15	3 = 1
40 = 20	4 = 2
50 = 25	5 = 2
	6 = 3
	7 = 3
	8 = 4
	9 = 4

29 = 14

Anti Log Excesses

10 = 20	1 = 2
20 = 40	2 = 4
	3 = 6
	4 = 8
	5 = 10
	6 = 12
	7 = 14
	8 = 16
	9 = 18

14 = 29

8710052	·9400207
110	236
167	265
224	293
282	322
339	350
397	379
454	408
511	436
569	465
626	493
684	522
741	551

8710798	·9400579
856	608
913	637
971	665
8711028	694
085	722
143	751
200	780
258	808
315	837
372	865
430	894
487	923
545	951
602	980
659	·9401009
717	037
774	066
832	094
889	123
946	152
8712004	180
061	209
118	237
176	266
233	295
291	323
348	352
405	380
463	409
520	438
578	466
635	495
692	523
750	552
807	581
865	609
922	638
979	666
8713037	695
094	724
151	752
209	781
266	809
324	838
381	867
438	895
496	924
553	953
611	981
668	·9402010
725	038
783	067
840	096
897	124
955	153
8714012	181
070	210
127	239
184	267
242	296
299	324
356	353
414	382
471	410

8714529	·9402439
586	467
643	496
701	525
758	553
815	582
873	610
930	639
988	667
8715045	696
102	725
160	753
217	782
274	810
332	839
389	868
447	896
504	925
561	953
619	982
676	·9403011
733	039
791	068
848	096
906	125
963	153
8716020	182
078	211
135	239
192	268
250	296
307	325
364	354
422	382
479	411
537	439
594	468
651	496
709	525
766	554
823	582
881	611
938	639
995	668
8717053	696
110	725
167	754
225	782
282	811
340	839
397	868
454	897
512	925
569	954
626	982
684	·9404011
741	039
798	068
856	097
913	125
970	154
8718028	182
085	211
143	239
200	268

8718257	·9404297
315	325
372	354
429	382
487	411
544	439
601	468
659	497
716	525
773	554
831	582
888	611
945	639
8719003	668
060	697
117	725
175	754
232	782
289	811
347	839
404	868
461	896
519	925
576	954
634	982
691	·9405011
748	039
806	068
863	096
920	125
978	154

Excesses

10 = 5	1 = 0
20 = 10	2 = 1
30 = 15	3 = 1
40 = 20	4 = 2
50 = 25	5 = 2
	6 = 3
	7 = 3
	8 = 4
	9 = 4

29 = 14

Anti Log Excesses

10 = 20	1 = 2
20 = 40	2 = 4
	3 = 6
	4 = 8
	5 = 10
	6 = 12
	7 = 14
	8 = 16
	9 = 18

14 = 29

8720035	·9405182
092	211
150	239
207	268
264	296

8720322	·9405325
379	353
436	382
494	411
551	439
608	468
666	496
723	525
780	553
838	582
895	610
952	639
8721010	668
067	696
124	725
182	753
239	782
296	810
354	839
411	867
468	896
526	925
583	953
640	982
698	·9406010
755	039
812	067
870	096
927	124
984	153
8722042	181
099	210
156	239
214	267
271	296
328	324
385	353
443	381
500	410
557	438
615	467
672	495
729	524
787	553
844	581
901	610
959	638
8723016	667
073	695
131	724
188	752
245	781
303	809
360	838
417	866
475	895
532	923
589	952
647	980
704	·9407009
761	037
818	066
876	095
933	123
990	152

8724048	·9407180
105	209
162	237
220	266
277	294
334	323
392	351
449	380
506	408
564	437
621	465
678	494
735	522
793	551
850	579
907	608
965	637
8725022	665
079	694
137	722
194	751
251	779
309	808
366	836
423	865
480	893
538	922
595	950
652	979
710	·9408007
767	036
824	064
882	093
939	121
996	150
8726053	178
111	207
168	235
225	264
283	292
340	321
397	350
455	378
512	407
569	435
626	464
684	492
741	521
798	549
856	578
913	606
970	635
8727027	663
085	692
142	720
199	749
257	777
314	806
371	834
429	863
486	891
543	920
600	948
658	977
715	·9409005

8727772	·9409034
830	062
887	091
944	119
8728001	148
059	176
116	205
173	233
231	262
288	290
345	319
402	347
460	376
517	404
574	433
632	462
689	490
746	518
803	547
861	575
918	604
975	632
8729033	661
090	689
147	718
204	746
262	775
319	803
376	832
433	860
491	889
548	917
605	946
663	974
720	·9410003
777	031
834	060
892	088
949	117

Excesses

10 = 5	1 = 0
20 = 10	2 = 1
30 = 15	3 = 1
40 = 20	4 = 2
50 = 25	5 = 2
	6 = 3
	7 = 3
	8 = 4
	9 = 4

29 = 14

Anti Log Excesses

10 = 20	1 = 2
20 = 40	2 = 4
	3 = 6
	4 = 8
	5 = 10
	6 = 12
	7 = 14
	8 = 16
	9 = 18
14 = 29	

8730006	·9410145
064	174
121	202
178	231
235	259
293	288
350	316
407	345
464	373
522	402
579	430
636	459
693	487
751	516
808	544
865	573
923	601
980	630
8731037	658
094	687
152	715
209	744
266	772
323	801
381	829
438	857
495	886
552	914
610	943
667	971
724	·9411000
782	028
839	057
896	085
953	114
8732011	142
068	171
125	199
182	228
240	256
297	285
354	313
411	342
469	370
526	399
583	427
640	456
698	484
755	513
812	541
869	569

8732927	·9411598
984	626
8733041	655
098	683
156	712
213	740
270	769
327	797
385	826
442	854
499	883
556	911
614	940
671	968
728	996
785	·9412025
843	053
900	082
957	110
8734014	139
072	167
129	196
186	224
243	253
301	281
358	310
415	338
472	367
530	395
587	423
644	452
701	480
759	509
816	537
873	566
930	594
988	623
8735045	651
102	680
159	708
217	737
274	765
331	793
388	822
446	850
503	879
560	907
617	936
674	964
732	993
789	·9413021
846	050
903	078
961	107
8736018	135
075	163
132	192
190	220
247	249
304	277
361	306
419	334
476	363
533	391
590	420

8736647	·9413448
705	476
762	505
819	533
876	562
934	590
991	619
8737048	647
105	676
163	704
220	732
277	761
334	789
391	818
449	846
506	875
563	903
620	932
678	960
735	989
792	·9414017
849	045
906	074
964	102
8738021	131
078	159
135	187
193	216
250	245
307	273
364	301
421	330
479	358
536	387
593	415
650	443
707	472
765	500
822	529
879	557
936	586
994	614
8739051	642
108	671
165	699
222	728
280	756
337	785
394	813
451	841
508	870
566	898
623	927
680	955
737	984
795	·9415012
852	040
909	069
966	097

Excesses

10 = 5	1 = 0
20 = 10	2 = 1
30 = 15	3 = 1
40 = 20	4 = 2
50 = 25	5 = 2
	6 = 3
	7 = 3
	8 = 4
	9 = 4
29 = 14	

Anti Log Excesses

10 = 20	1 = 2
20 = 40	2 = 4
	3 = 6
	4 = 8
	5 = 10
	6 = 12
	7 = 14
	8 = 16
	9 = 18
14 = 29	

8740023	·9415126
081	154
138	183
195	211
252	239
309	268
367	296
424	325
481	353
538	382
595	410
653	438
710	467
767	495
824	524
881	552
939	581
996	609
8741053	637
110	666
167	694
225	723
282	751
339	779
396	808
453	836
511	865
568	893
625	922
682	950
739	978
797	·9416007
854	035
911	064
968	092
8742025	120
083	149

8742140	·9416177
197	206
254	234
311	263
369	291
426	319
483	348
540	376
597	405
655	433
712	461
769	490
826	518
883	547
941	575
998	603
8743055	632
112	660
169	689
226	717
284	745
341	774
398	802
455	831
512	859
570	887
627	916
684	944
741	973
798	·9417001
855	029
913	058
970	086
8744027	115
084	143
141	171
199	200
256	228
313	257
370	285
427	313
484	342
542	370
599	399
656	427
713	455
770	484
828	512
885	541
942	569
999	597
8745056	626
113	654
171	683
228	711
285	739
342	768
399	796
456	824
514	853
571	881
628	910
685	938
742	966
799	995

8745857	·9418023
914	052
971	080
8746028	108
085	137
143	165
200	194
257	222
314	250
371	279
428	307
486	335
543	364
600	392
657	421
714	449
771	477
828	506
886	534
943	562
8747000	591
057	619
114	648
171	676
229	704
286	733
343	761
400	789
457	818
514	846
572	875
629	903
686	931
743	960
800	988
857	·9419017
915	045
972	073
8748029	102
086	130
143	159
200	187
257	215
315	244
372	272
429	300
486	329
543	357
600	385
658	414
715	442
772	471
829	499
886	527
943	556
8749000	584
058	612
115	641
172	669
229	698
286	726
343	754
400	783
458	811
515	839

8749572	·9419868
629	896
686	924
743	953
800	981
858	·9420010
915	038
972	066

Excesses

10 = 5	1 = 0
20 = 10	2 = 1
30 = 15	3 = 1
40 = 20	4 = 2
50 = 25	5 = 2
	6 = 3
	7 = 3
	8 = 4
	9 = 4

29 = 14

Anti Log Excesses

10 = 20	1 = 2
20 = 40	2 = 4
	3 = 6
	4 = 8
	5 = 10
	6 = 12
	7 = 14
	8 = 16
	9 = 18

14 = 29

8750029	·9420095
086	123
143	151
200	180
258	208
315	236
372	265
429	293
486	322
543	350
600	378
658	407
715	435
772	463
829	492
886	520
943	548
8751000	577
058	605
115	633
172	662
229	690
286	718
343	747
400	775
458	804
515	832
572	860

8751629	·9420889
686	917
743	945
800	974
857	·9421002
915	030
972	059
8752029	087
086	115
143	144
200	172
257	200
314	229
372	257
429	285
486	314
543	342
600	371
657	399
714	427
771	456
829	484
886	512
943	541
8753000	569
057	597
114	626
171	654
228	682
286	711
343	739
400	767
457	796
514	824
571	852
628	881
685	909
743	937
800	966
857	994
914	·9422022
971	051
8754028	079
085	107
142	136
199	164
257	192
314	221
371	249
428	277
485	306
542	334
599	362
656	391
714	419
771	447
828	476
885	504
942	532
999	561
8755056	589
113	617
170	646
228	674
285	702

8755342	·9422731
399	759
456	787
513	816
570	844
627	872
684	901
741	929
799	957
856	986
913	·9423014
970	042
8756027	071
084	099
141	127
198	156
255	184
313	212
370	241
427	269
484	297
541	326
598	354
655	382
712	411
769	439
826	467
884	495
941	524
998	552
8757055	580
112	609
169	637
226	665
283	694
340	722
397	750
454	779
512	807
569	835
626	864
683	892
740	920
797	948
854	977
911	·9424005
968	033
8758025	062
082	090
140	118
197	147
254	175
311	203
368	232
425	260
482	288
539	316
596	345
653	373
710	401
768	430
825	458
882	486
939	515
996	543

8759053	·9424571
110	599
167	628
224	656
281	684
338	713
395	741
453	769
510	798
567	826
624	854
681	882
738	911
795	939
852	967
909	996
966	·9425024

Excesses

10 = 5	1 = 0
20 = 10	2 = 1
30 = 15	3 = 1
40 = 20	4 = 2
50 = 25	5 = 2
	6 = 3
	7 = 3
	8 = 4
	9 = 4

29 = 14

Anti-Log Excesses

10 = 20	1 = 2
20 = 40	2 = 4
	3 = 6
	4 = 8
	5 = 10
	6 = 12
	7 = 14
	8 = 16
	9 = 18

14 = 29

8760023	·9425052
080	081
137	109
195	137
252	165
309	194
366	222
423	250
480	279
537	307
594	335
651	363
708	392
765	420
822	448
879	477
937	505
994	533
8761051	562

8761108	·9425590	8764817	·9427428	8768524	·9429264	8770576	·9430281	8774281	·9432115
165	618	874	456	581	293	633	309	338	143
222	646	931	485	638	321	690	337	395	172
279	675	988	513	695	349	747	366	452	200
336	703	8765045	541	752	377	804	394	509	228
393	731	102	569	809	406	861	422	566	256
450	760	159	598	866	434	918	450	623	284
507	788	216	626	923	462	975	479	680	313
564	816	273	654	980	490	8771032	507	737	341
621	844	330	682	8769037	519	089	535	794	369
678	873	387	711	094	547	146	563	851	397
735	901	444	739	151	575	203	592	908	425
793	929	501	767	208	603	260	620	965	454
850	958	558	795	265	632	317	648	8775022	482
907	986	615	824	322	660	374	676	079	510
964	·9426014	672	852	379	688	431	704	136	538
8762021	042	729	880	436	716	488	733	193	566
078	071	786	908	493	745	545	761	250	595
135	099	843	937	550	773	602	789	307	623
192	127	900	965	607	801	659	817	364	651
249	155	957	993	664	829	716	846	421	679
306	184	8766014	·9428021	721	858	773	874	478	707
363	212	071	050	778	886	830	902	535	736
420	240	128	078	835	914	887	930	592	764
477	269	186	106	892	942	944	958	649	792
534	297	243	135	949	970	8772001	987	706	820
591	325	300	163			058	·9431015	763	848
649	354	357	191			115	043	820	877
706	382	414	219			172	071	877	905
763	410	471	248			229	100	933	933
820	438	528	276			286	128	990	961
877	467	585	304			343	156	8776047	989
934	495	642	332			400	184	104	·9433018
991	523	699	361			457	212	161	046
8763048	552	756	389			514	241	218	074
105	580	813	417			571	269	275	102
162	608	870	445			628	297	332	130
219	636	927	474			685	325	389	159
276	665	984	502			742	353	446	187
333	693	8767041	530			799	382	503	215
390	721	098	558			856	410	560	243
447	749	155	587			913	438	617	271
504	778	212	615			970	466	674	300
561	806	269	643			8773027	494	731	328
618	834	326	671			084	523	788	356
676	863	383	700			141	551	845	384
733	891	440	728			198	579	902	412
790	919	497	756			255	607	959	441
847	947	554	784			312	636	8777016	469
904	976	611	813			369	664	073	497
961	·9427004	668	841			426	692	130	525
8764018	032	725	869			483	720	187	553
075	060	782	897			540	748	244	581
132	089	839	926			597	777	301	610
189	117	896	954			654	805	358	638
246	145	954	982	8770006	·9429999	711	833	415	666
303	174	8768011	·9429010	063	·9430027	768	861	472	694
360	202	068	039	120	055	825	889	529	722
417	230	125	067	177	083	882	918	586	751
474	258	182	095	234	112	939	946	643	779
531	287	239	123	291	140	996	974	700	807
588	315	296	152	348	168	8774053	·9432002	756	835
645	343	353	180	405	196	110	031	813	863
702	371	410	208	462	225	167	059	870	892
759	400	467	236	519	253	224	087	927	920

Excesses

10 = 5	1 = 0
20 = 10	2 = 1
30 = 15	3 = 1
40 = 20	4 = 2
50 = 25	5 = 2
	6 = 3
	7 = 3
	8 = 4
	9 = 4

29 = 14

Anti-Log Excesses

10 = 20	1 = 2
20 = 40	2 = 4
	3 = 6
	4 = 8
	5 = 10
	6 = 12
	7 = 14
	8 = 16
	9 = 18

14 = 29

8777984	·9433948
8778041	976
098	·9434004
155	032
212	061
269	089
326	117
383	145
440	173
497	202
554	230
611	258
668	286
725	314
782	342
839	371
896	399
953	427
8779010	455
067	483
123	511
180	540
237	568
294	596
351	624
408	652
465	681
522	709
579	737
636	765
693	793
750	821
807	850
864	878
921	906
978	934

Excesses

10 = 5	1 = 0
20 = 10	2 = 1
30 = 15	3 = 1
40 = 20	4 = 2
50 = 25	5 = 2
	6 = 3
	7 = 3
	8 = 4
	9 = 4

28 = 14

Anti-Log Excesses

10 = 20	1 = 2
20 = 40	2 = 4
	3 = 6
	4 = 8
	5 = 10
	6 = 12
	7 = 14
	8 = 16
	9 = 18

14 = 28

8780035	·9434962
092	990
149	·9435019
206	047
262	075
319	103
376	131
433	159
490	188
547	216
604	244
661	272
718	300
775	328
832	357
889	385
946	413
8781003	441
060	469
117	497
174	525
230	554
287	582
344	610
401	638
458	666
515	694
572	723
629	751
686	779
743	807
800	835
857	863
914	892
971	920
8782028	948
085	976
141	·9436004
198	032
255	060
312	089
369	117
426	145
483	173
540	201
597	229
654	258
711	286
768	314
825	342
882	370
939	398
995	426
8783052	455
109	483
166	511
223	539
280	567
337	595
394	624
451	652
508	680
565	708
622	736
679	764

8783735	·9436792
792	821
849	849
906	877
963	905
8784020	933
077	961
134	989
191	·9437018
248	046
305	074
362	102
419	130
475	158
532	186
589	215
646	243
703	271
760	299
817	327
874	355
931	383
988	412
8785045	440
102	468
158	496
215	524
272	552
329	580
386	608
443	637
500	665
557	693
614	721
671	749
728	777
784	805
841	834
898	862
955	890
8786012	918
069	946
126	974
183	·9438002
240	030
297	059
354	087
410	115
467	143
524	171
581	199
638	227
695	255
752	284
809	312
866	340
923	368
980	396
8787036	424
093	452
150	480
207	508
264	537
321	565
378	593

8787435	·9438621
492	649
549	677
605	705
662	733
719	762
776	790
833	818
890	846
947	874
8788004	902
061	930
117	958
174	986
231	·9439015
288	043
345	071
402	099
459	127
516	155
573	183
630	211
686	239
743	268
800	296
857	324
914	352
971	380
8789028	408
085	436
142	464
198	492
255	521
312	549
369	577
426	605
483	633
540	661
597	689
654	717
710	745
767	773
824	802
881	830
938	858
995	886

Excesses

10 = 5	1 = 0
20 = 10	2 = 1
30 = 15	3 = 1
40 = 20	4 = 2
50 = 25	5 = 2
	6 = 3
	7 = 3
	8 = 4
	9 = 4

28 = 14

Anti-Log Excesses

10 = 20	1 = 2
20 = 40	2 = 4
	3 = 6
	4 = 8
	5 = 10
	6 = 12
	7 = 14
	8 = 16
	9 = 18

14 = 28

8790052	·9439914
109	942
165	970
222	998
279	·9440026
336	055
393	083
450	111
507	139
564	167
621	195
677	223
734	251
791	279
848	307
905	335
962	364
8791019	392
076	420
132	448
189	476
246	504
303	532
360	560
417	588
474	616
531	645
587	673
644	701
701	729
758	757
815	785
872	813
929	841
985	869
8792042	897
099	926
156	954
213	982
270	·9441010
327	038
384	066
440	094
497	122
554	150
611	178
668	206
725	235
782	263
838	291
895	319

Number	Log
8792952	·9441347
8793009	375
066	403
123	431
180	459
237	487
293	515
350	544
407	572
464	600
521	628
578	656
635	684
691	712
748	740
805	768
862	796
919	824
976	852
8794033	880
089	909
146	937
203	965
260	993
317	·9442021
374	049
431	077
487	105
544	133
601	161
658	189
715	217
772	245
829	274
885	302
942	330
999	358
8795056	386
113	414
170	442
226	470
283	498
340	526
397	554
454	582
511	610
568	638
624	667
681	695
738	723
795	751
852	779
909	807
965	835
8796022	863
079	891
136	919
193	947
250	975
307	·9443003
363	031
420	059
477	087
534	116
591	144

Number	Log
8796648	·9443172
704	200
761	228
818	256
875	284
932	312
989	340
8797045	368
102	396
159	424
216	452
273	480
330	508
386	537
443	565
500	593
557	621
614	649
671	677
727	705
784	733
841	761
898	789
955	817
8798012	845
068	873
125	901
182	929
239	957
296	985
353	·9444013
409	042
466	070
523	098
580	126
637	154
694	182
750	210
807	238
864	266
921	294
978	322
8799035	350
091	378
148	406
205	434
262	462
319	490
375	518
432	546
489	574
546	603
603	631
660	659
716	687
773	715
830	743
887	771
944	799

Excesses

10 = 5	1 = 0
20 = 10	2 = 1
30 = 15	3 = 1
40 = 20	4 = 2
50 = 25	5 = 2
	6 = 3
	7 = 3
	8 = 4
	9 = 4
28 = 14	

Anti-Log Excesses

10 = 20	1 = 2
20 = 41	2 = 4
	3 = 6
	4 = 8
	5 = 10
	6 = 12
	7 = 14
	8 = 16
	9 = 18
14 = 28	

Number	Log
8800000	·9444827
057	855
114	883
171	911
228	939
285	967
341	995
398	·9445023
455	051
512	079
569	107
625	135
682	163
739	191
796	219
853	247
910	275
966	304
8801023	332
080	360
137	388
194	416
250	444
307	472
364	500
421	528
478	556
534	584
591	612
648	640
705	668
762	696
818	724
875	752
932	780
989	808
8802046	836

Number	Log
8802103	·9445864
159	892
216	920
273	948
330	976
387	·9446004
443	032
500	060
557	088
614	116
671	144
727	172
784	200
841	228
898	256
955	284
8803011	312
068	340
125	368
182	396
239	424
295	452
352	481
409	509
466	537
523	565
579	593
636	621
693	649
750	677
806	705
863	733
920	761
977	789
8804034	817
090	845
147	873
204	901
261	929
318	957
374	985
431	·9447013
488	041
545	069
602	097
658	125
715	153
772	181
829	209
886	237
942	265
999	293
8805056	321
113	349
169	377
226	405
283	433
340	461
397	489
453	517
510	545
567	573
624	601
680	629
737	657

Number	Log
8805794	·9447685
851	713
908	741
964	769
8806021	797
078	825
135	853
192	881
248	909
305	937
362	965
419	993
475	·9448021
532	049
589	077
646	105
702	133
759	161
816	189
873	217
930	245
986	273
8807043	301
100	329
157	357
213	385
270	413
327	441
384	469
441	497
497	525
554	553
611	581
668	609
724	637
781	665
838	693
895	721
951	749
8808008	777
065	805
122	833
179	861
235	889
292	917
349	945
406	973
462	·9449001
519	029
576	057
633	085
689	113
746	141
803	169
860	197
916	225
973	253
8809030	281
087	309
143	337
200	365
257	392
314	420
371	448
427	476

8809484	·9449504
541	532
598	560
654	588
711	616
768	644
825	672
881	700
938	728
995	756

Excesses

10 = 5	1 = 0
20 = 10	2 = 1
30 = 15	3 = 1
40 = 20	4 = 2
50 = 25	5 = 2
	6 = 3
	7 = 3
	8 = 4
	9 = 4

28 = 14

Anti-Log Excesses

10 = 20	1 = 2
20 = 41	2 = 4
	3 = 6
	4 = 8
	5 = 10
	6 = 12
	7 = 14
	8 = 16
	9 = 18

14 = 28

8810052	·9449784
108	812
165	840
222	868
279	896
335	924
392	952
449	980
506	·9450008
562	036
619	064
676	092
733	120
789	148
846	176
903	204
960	232
8811016	260
073	288
130	316
187	344
243	372
300	400
357	428
414	455
470	483

8811527	·9450511
584	539
641	567
697	595
754	623
811	651
867	679
924	707
981	735
8812038	763
094	791
151	819
208	847
265	875
321	903
378	931
435	959
492	987
548	·9451015
605	043
662	071
719	099
775	127
832	155
889	183
946	211
8813002	239
059	266
116	294
172	322
229	350
286	378
343	406
399	434
456	462
513	490
570	518
626	546
683	574
740	602
797	630
853	658
910	686
967	714
8814023	742
080	770
137	798
194	826
250	854
307	881
364	909
421	937
477	965
534	993
591	·9452021
647	049
704	077
761	105
818	133
874	161
931	189
988	217
8815044	245
101	273
158	301

8815215	·9452329
271	357
328	385
385	412
441	440
498	468
555	496
612	524
668	552
725	580
782	608
839	636
895	664
952	692
8816009	720
065	748
122	776
179	804
236	832
292	860
349	887
406	915
462	943
519	971
576	999
633	·9453027
689	055
746	083
803	111
859	139
916	167
973	195
8817029	223
086	251
143	279
200	307
256	335
313	362
370	390
426	418
483	446
540	474
597	502
653	530
710	558
767	586
823	614
880	642
937	670
993	698
8818050	726
107	753
164	781
220	809
277	837
334	865
390	893
447	921
504	949
560	977
617	·9454005
674	033
731	061
787	089
844	117

8818901	·9454144
957	172
8819014	200
071	228
127	256
184	284
241	312
298	340
354	368
411	396
468	424
524	452
581	479
638	507
694	535
751	563
808	591
864	619
921	647
978	675

Excesses

10 = 5	1 = 0
20 = 10	2 = 1
30 = 15	3 = 1
40 = 20	4 = 2
50 = 25	5 = 2
	6 = 3
	7 = 3
	8 = 4
	9 = 4

28 = 14

Anti-Log Excesses

10 = 20	1 = 2
20 = 41	2 = 4
	3 = 6
	4 = 8
	5 = 10
	6 = 12
	7 = 14
	8 = 16
	9 = 18

14 = 28

8820035	·9454703
091	731
148	759
205	787
261	814
318	842
375	870
431	898
488	926
545	954
601	982
658	·9455010
715	038
771	066
828	094
885	121

8820941	·9455149
998	177
8821055	205
112	233
168	261
225	289
282	317
338	345
395	373
452	401
508	428
565	456
622	484
678	512
735	540
792	568
848	596
905	624
962	652
8822018	679
075	707
132	735
188	763
245	791
302	819
358	847
415	875
472	903
528	930
585	958
642	986
698	·9456014
755	042
812	070
869	098
925	126
982	154
8823039	182
095	209
152	237
209	265
265	293
322	321
379	349
435	377
492	405
549	433
605	460
662	488
719	516
775	544
832	572
889	600
945	628
8824002	656
059	684
115	711
172	739
229	767
285	795
342	823
398	851
455	879
512	907
568	935

8824625	·9456962
682	990
738	·9457018
795	046
852	074
908	102
965	130
8825022	158
078	185
135	213
192	241
248	269
305	297
362	325
418	353
475	381
532	408
588	436
645	464
702	492
758	520
815	548
872	576
928	604
985	631
8826042	659
098	687
155	715
211	743
268	771
325	799
381	827
438	854
495	882
551	910
608	938
665	966
721	994
778	·9458022
835	050
891	077
948	105
8827005	133
061	161
118	189
174	217
231	245
288	273
344	300
401	328
458	356
514	384
571	412
628	440
684	468
741	496
798	523
854	551
911	579
967	607
8828024	635
081	663
137	691
194	718
251	746

8828307	·9458774
364	802
421	830
477	858
534	886
590	913
647	941
704	969
760	997
817	·9459025
874	053
930	081
987	108
8829043	136
100	164
157	192
213	220
270	248
327	276
383	303
440	331
497	359
553	387
610	415
666	443
723	471
780	498
836	526
893	554
950	582

Excesses

10 = 5	1 = 0
20 = 10	2 = 1
30 = 15	3 = 1
40 = 20	4 = 2
50 = 25	5 = 2
	6 = 3
	7 = 3
	8 = 4
	9 = 4

28 = 14

Anti-Log Excesses

10 = 20	1 = 2
20 = 41	2 = 4
	3 = 6
	4 = 8
	5 = 10
	6 = 12
	7 = 14
	8 = 16
	9 = 18

14 = 28

8830006	·9459610
063	638
119	665
176	693
233	721
289	749

8830346	·9459777
403	805
459	833
516	860
572	888
629	916
686	944
742	972
799	·9460000
855	027
912	055
969	083
8831025	111
082	139
139	167
195	195
252	222
308	250
365	278
422	306
478	334
535	362
592	389
648	417
705	445
761	473
818	501
875	529
931	556
988	584
8832044	612
101	640
158	668
214	696
271	724
327	751
384	779
441	807
497	835
554	863
611	891
667	918
724	946
780	974
837	·9461002
894	030
950	058
8833007	085
063	113
120	141
177	169
233	197
290	225
346	252
403	280
460	308
516	336
573	364
629	392
686	419
743	447
799	475
856	503
912	531
969	558

8834026	·9461586
082	614
139	642
195	670
252	698
309	725
365	753
422	781
478	809
535	837
592	865
648	892
705	920
761	948
818	976
875	·9462004
931	031
988	059
8835044	087
101	115
158	143
214	171
271	198
327	226
384	254
441	282
497	310
554	337
610	365
667	393
723	421
780	449
837	477
893	504
950	532
8836006	560
063	588
120	616
176	643
233	671
289	699
346	727
402	755
459	782
516	810
572	838
629	866
685	894
742	922
799	949
855	977
912	·9463005
968	033
8837025	061
081	088
138	116
195	144
251	172
308	200
364	227
421	255
478	283
534	311
591	339
647	366

8837704	·9463394
760	422
817	450
874	478
930	505
987	533
8838043	561
100	589
156	617
213	644
270	672
326	700
383	728
439	756
496	783
552	811
609	839
666	867
722	895
779	922
835	950
892	978
948	·9464006
8839005	034
062	061
118	089
175	117
231	145
288	173
344	200
401	228
457	256
514	284
571	312
627	339
684	367
740	395
797	423
853	451
910	478
967	506

Excesses

10 = 5	1 = 0
20 = 10	2 = 1
30 = 15	3 = 1
40 = 20	4 = 2
50 = 25	5 = 2
	6 = 3
	7 = 3
	8 = 4
	9 = 4

28 = 14

Anti-Log Excesses

10 = 20	1 = 2
20 = 41	2 = 4
	3 = 6
	4 = 8
	5 = 10
	6 = 12
	7 = 14
	8 = 16
	9 = 18

14 = 28

8840023	·9464534
080	562
136	589
193	617
249	645
306	673
362	701
419	728
476	756
532	784
589	812
645	840
702	867
758	895
815	923
871	951
928	978
985	·9465006
8841041	034
098	062
154	090
211	117
267	145
324	173
380	201
437	228
494	256
550	284
607	312
663	339
720	367
776	395
833	423
889	451
946	478
8842003	506
059	534
116	562
172	589
229	617
285	645
342	673
398	700
455	728
511	756
568	784
625	812
681	839
738	867
794	895
851	923

8842907	·9465950
964	978
8843020	·9466006
077	034
133	061
190	089
247	117
303	145
360	173
416	200
473	228
529	256
586	284
642	311
699	339
755	367
812	395
868	422
925	450
982	478
8844038	506
095	533
151	561
208	589
264	617
321	644
377	672
434	700
490	728
547	756
603	783
660	811
716	839
773	867
829	894
886	922
943	950
999	978
8845056	·9467005
112	033
169	061
225	089
282	116
338	144
395	172
451	200
508	227
564	255
621	283
677	311
734	338
790	366
847	394
904	422
960	449
8846017	477
073	505
130	533
186	560
243	588
299	616
356	644
412	671
469	699
525	727

8846582	·9467755
638	782
695	810
751	838
808	866
864	893
921	921
977	949
8847034	977
090	·9468004
147	032
203	060
260	088
316	115
373	143
430	171
486	198
543	226
599	254
656	282
712	309
769	337
825	365
882	393
938	420
995	448
8848051	476
108	504
164	531
221	559
277	587
334	615
390	642
447	670
503	698
560	725
616	753
673	781
729	809
786	836
842	864
899	892
955	920
8849012	947
068	975
125	·9469003
181	031
238	058
294	086
351	114
407	141
464	169
520	197
577	225
633	252
690	280
746	308
803	336
859	363
916	391
972	419

Excesses

10 = 5	1 = 0
20 = 10	2 = 1
30 = 15	3 = 1
40 = 20	4 = 2
50 = 25	5 = 2
	6 = 3
	7 = 3
	8 = 4
	9 = 4

28 = 14

Anti Log Excesses

10 = 20	1 = 2
20 = 41	2 = 4
	3 = 6
	4 = 8
	5 = 10
	6 = 12
	7 = 14
	8 = 16
	9 = 18

14 = 28

8850029	·9469446
085	474
142	502
198	530
255	557
311	585
368	613
424	640
481	668
537	696
594	724
650	751
707	779
763	807
820	835
876	862
933	890
989	918
8851046	945
102	973
159	·9470001
215	029
272	056
328	084
385	112
441	139
498	167
554	195
611	223
667	250
723	278
780	306
836	333
893	361
949	389
8852006	417
062	444

8852119	·9470472
175	500
232	527
288	555
345	583
401	611
458	638
514	666
571	694
627	721
684	749
740	777
797	805
853	832
910	860
966	888
8853023	915
079	943
136	971
192	999
248	·9471026
305	054
361	082
418	109
474	137
531	165
587	192
644	220
700	248
757	276
813	303
870	331
926	359
983	386
8854039	414
096	442
152	469
209	497
265	525
321	553
378	580
434	608
491	636
547	663
604	691
660	719
717	746
773	774
830	802
886	830
943	857
999	885
8855056	913
112	940
168	968
225	996
281	·9472023
338	051
394	079
451	106
507	134
564	162
620	190
677	217
733	245

Number	Log
8855790	·9472273
846	300
902	328
959	356
8856015	383
072	411
128	439
185	466
241	494
298	522
354	549
411	577
467	605
524	632
580	660
636	688
693	716
749	743
806	771
862	799
919	827
975	854
8857032	882
088	910
145	937
201	965
257	993
314	·9473020
370	048
427	076
483	103
540	131
596	159
653	186
709	214
765	242
822	269
878	297
935	325
991	352
8858048	380
104	408
161	435
217	463
273	491
330	519
386	546
443	574
499	602
556	629
612	657
669	685
725	712
781	740
838	768
894	795
951	823
8859007	851
064	878
120	906
177	934
233	961
289	989
346	·9474017
402	044

Number	Log
8859459	·9474072
515	100
572	127
628	155
684	183
741	210
797	238
854	266
910	293
967	321

Excesses

10 = 5	1 = 0
20 = 10	2 = 1
30 = 15	3 = 1
40 = 20	4 = 2
50 = 25	5 = 2
	6 = 3
	7 = 3
	8 = 4
	9 = 4

28 = 14

Anti-Log Excesses

10 = 20	1 = 2
20 = 41	2 = 4
	3 = 6
	4 = 8
	5 = 10
	6 = 12
	7 = 14
	8 = 16
	9 = 18

14 = 28

Number	Log
8860023	·9474349
080	376
136	404
192	432
249	459
305	487
362	515
418	542
475	570
531	598
587	625
644	653
700	680
757	708
813	736
870	763
926	791
982	819
8861039	846
095	874
152	902
208	929
265	957
321	985
377	·9475012
434	040

Number	Log
8861490	·9475068
547	095
603	123
660	150
716	178
772	206
829	233
885	261
942	289
998	316
8862054	344
111	372
167	399
224	427
280	455
337	482
393	510
449	538
506	565
562	593
619	620
675	648
731	676
788	703
844	731
901	759
957	786
8863014	814
070	842
126	869
183	897
239	925
296	952
352	980
408	·9476007
465	035
521	063
578	090
634	118
690	146
747	173
803	201
860	229
916	256
973	284
8864029	311
085	339
142	367
198	394
255	422
311	450
367	477
424	505
480	533
537	560
593	588
649	615
706	643
762	671
819	698
875	726
931	754
988	781
8865044	809
101	836

Number	Log
8865157	·9476864
213	892
270	919
326	947
383	975
439	·9477002
495	030
552	057
608	085
665	113
721	140
777	168
834	196
890	223
947	251
8866003	278
059	306
116	334
172	361
229	389
285	417
341	444
398	472
454	499
511	527
567	555
623	582
680	610
736	638
792	665
849	693
905	720
962	748
8867018	776
074	803
131	831
187	859
244	886
300	914
356	941
413	969
469	997
526	·9478024
582	052
638	079
695	107
751	135
807	162
864	190
920	218
977	245
8868033	273
089	300
146	328
202	356
259	383
315	411
371	438
428	466
484	494
540	521
597	549
653	576
710	604
766	632

Number	Log
8868822	·9478659
879	687
935	715
991	742
8869048	770
104	797
161	825
217	853
273	880
330	908
386	935
442	963
499	991
555	·9479018
612	046
668	073
724	101
781	129
837	156
893	184
950	211

Excesses

10 = 5	1 = 0
20 = 10	2 = 1
30 = 15	3 = 1
40 = 20	4 = 2
50 = 24	5 = 2
	6 = 3
	7 = 3
	8 = 4
	9 = 4

28 = 14

Anti-Log Excesses

10 = 20	1 = 2
20 = 41	2 = 4
	3 = 6
	4 = 8
	5 = 10
	6 = 12
	7 = 14
	8 = 16
	9 = 18

14 = 28

Number	Log
8870006	·9479239
063	267
119	294
175	322
232	349
288	377
344	405
401	432
457	460
513	487
570	515
626	542
683	570
739	598
795	625

Number	Log
8870852	·9479653
908	680
964	708
8871021	736
077	763
133	791
190	818
246	846
303	874
359	901
415	929
472	956
528	984
584	·9480012
641	039
697	067
753	094
810	122
866	150
923	177
979	205
8872035	232
092	260
148	288
204	315
261	343
317	370
373	398
430	426
486	453
542	481
599	508
655	536
711	563
768	591
824	619
881	646
937	674
993	701
8873050	729
106	756
162	784
219	812
275	839
331	867
388	894
444	922
500	950
557	977
613	·9481005
669	032
726	060
782	087
838	115
895	143
951	170
8874007	198
064	225
120	253
177	280
233	308
289	336
346	363
402	391
458	418

Number	Log
8874515	·9481446
571	473
627	501
684	529
740	556
796	584
853	611
909	639
965	666
8875022	694
078	722
134	749
191	777
247	804
303	832
360	859
416	887
472	915
529	942
585	970
641	997
698	·9482025
754	052
810	080
867	107
923	135
979	163
8876036	190
092	218
148	245
205	273
261	300
317	328
374	356
430	383
486	411
543	438
599	466
655	493
712	521
768	549
824	576
881	604
937	631
993	659
8877050	686
106	714
162	741
219	769
275	797
331	824
388	852
444	879
500	907
556	934
613	962
669	989
725	·9483017
782	045
838	072
894	100
951	127
8878007	155
063	182
120	210

Number	Log
8878176	·9483237
232	265
289	292
345	320
401	348
458	375
514	403
570	430
627	458
683	485
739	513
795	540
852	568
908	596
964	623
8879021	651
077	678
133	706
190	733
246	761
302	788
359	816
415	843
471	871
528	898
584	926
640	954
696	981
753	·9484009
809	036
865	064
922	091
978	119

Excesses

10 = 5	1 = 0
20 = 10	2 = 1
30 = 15	3 = 1
40 = 20	4 = 2
50 = 24	5 = 2
	6 = 3
	7 = 3
	8 = 4
	9 = 4

28 = 14

Anti-Log Excesses

10 = 20	1 = 2
20 = 41	2 = 4
	3 = 6
	4 = 8
	5 = 10
	6 = 12
	7 = 14
	8 = 16
	9 = 18

14 = 28

Number	Log
8880034	·9484146
091	174
147	201

Number	Log
8880203	·9484229
260	256
316	284
372	312
428	339
485	367
541	394
597	422
654	449
710	477
766	504
823	532
879	559
935	587
991	614
8881048	642
104	669
160	697
217	725
273	752
329	780
386	807
442	835
498	862
554	890
611	917
667	945
723	972
780	·9485000
836	027
892	055
948	082
8882005	110
061	137
117	165
174	192
230	220
286	247
343	275
399	303
455	330
511	358
568	385
624	413
680	440
737	468
793	495
849	523
905	550
962	578
8883018	605
074	633
131	660
187	688
243	715
299	743
356	770
412	798
468	825
525	853
581	880
637	908
693	936
750	963
806	991

Number	Log
8883862	·9486018
919	046
975	073
8884031	101
087	128
144	156
200	183
256	211
313	238
369	266
425	293
481	321
538	348
594	376
650	403
706	431
763	458
819	486
875	513
932	541
988	568
8885044	596
100	623
157	651
213	678
269	706
325	733
382	761
438	788
494	816
551	843
607	871
663	898
719	926
776	953
832	981
888	·9487008
944	036
8886001	063
057	091
113	118
170	146
226	173
282	201
338	228
395	256
451	283
507	311
563	338
620	366
676	393
732	421
788	448
845	476
901	503
957	531
8887013	558
070	586
126	613
182	641
239	668
295	696
351	723
407	751
464	778

8887520	·9487806
576	833
632	861
689	888
745	916
801	943
857	971
914	998
970	·9488026
8888026	053
082	081
139	108
195	136
251	163
307	191
364	218
420	246
476	273
532	300
589	328
645	355
701	383
757	410
814	438
870	465
926	493
982	520
8889039	548
095	575
151	603
207	630
264	658
320	685
376	713
432	740
489	768
545	795
601	823
657	850
714	878
770	905
826	933
882	960
939	988
995	·9489015

Excesses

10 = 5	1 = 0
20 = 10	2 = 1
30 = 15	3 = 1
40 = 20	4 = 2
50 = 24	5 = 2
	6 = 3
	7 = 3
	8 = 4
	9 = 4

28 = 14

Anti-Log Excesses

10 = 20	1 = 2
20 = 41	2 = 4
	3 = 6
	4 = 8
	5 = 10
	6 = 12
	7 = 14
	8 = 16
	9 = 18

14 = 28

8890051	·9489043
107	070
164	097
220	125
276	152
332	180
389	208
445	235
501	262
557	290
614	317
670	345
726	372
782	400
838	427
895	455
951	482
8891007	510
063	537
120	565
176	592
232	620
288	647
345	674
401	702
457	729
513	757
570	784
626	812
682	839
738	866
794	894
851	921
907	949
963	976
8892019	·9490004
076	031
132	059
188	086
244	114
301	141
357	169
413	196
469	224
525	251
582	278
638	306
694	333
750	361
807	388
863	416

8892919	·9490443
975	471
8893031	498
088	526
144	553
200	581
256	608
313	635
369	663
425	690
481	718
537	745
594	773
650	800
706	828
762	855
819	883
875	910
931	937
987	965
8894043	992
100	·9491020
156	047
212	075
268	102
325	130
381	157
437	185
493	212
549	239
606	267
662	294
718	322
774	349
830	377
887	404
943	432
999	459
8895055	487
112	514
168	541
224	569
280	596
336	624
393	651
449	679
505	706
561	734
617	761
674	788
730	816
786	843
842	871
898	898
955	926
8896011	953
067	981
123	·9492008
179	035
236	063
292	090
348	118
404	145
460	173
517	200

8896573	·9492227
629	255
685	282
741	310
798	337
854	364
910	392
966	419
8897022	447
079	474
135	502
191	529
247	556
303	584
360	611
416	639
472	666
528	694
584	721
641	748
697	776
753	803
809	831
865	858
922	886
978	913
8898034	940
090	968
146	995
203	·9493023
259	050
315	078
371	105
427	132
484	160
540	187
596	215
652	242
708	270
764	297
821	324
877	352
933	379
989	407
8899045	434
102	462
158	489
214	516
270	544
326	571
383	599
439	626
495	653
551	681
607	708
663	736
720	763
776	791
832	818
888	845
944	873

Excesses

10 = 5	1 = 0
20 = 10	2 = 1
30 = 15	3 = 1
40 = 20	4 = 2
50 = 24	5 = 2
	6 = 3
	7 = 3
	8 = 4
	9 = 4

28 = 14

Anti Log Excesses

10 = 20	1 = 2
20 = 41	2 = 4
	3 = 6
	4 = 8
	5 = 10
	6 = 12
	7 = 14
	8 = 16
	9 = 18

14 = 28

8900000	·9493900
057	928
113	955
169	982
225	·9494010
281	037
338	065
394	092
450	119
506	147
562	174
618	202
675	229
731	257
787	284
843	311
899	339
956	366
8901012	394
068	421
124	448
180	476
236	503
293	531
349	558
405	585
461	613
517	640
573	668
630	695
686	722
742	750
798	777
854	805
910	832
967	859
8902023	887

Number	Log
8902079	·9494914
135	942
191	969
247	996
304	·9495024
360	051
416	079
472	106
528	133
584	161
641	188
697	216
753	243
809	270
865	298
921	325
978	353
8903034	380
090	407
146	435
202	462
258	490
314	517
371	544
427	572
483	599
539	627
595	654
651	681
708	709
764	736
820	764
876	791
932	818
988	846
8904045	873
101	901
157	928
213	955
269	983
325	·9496010
381	037
438	065
494	092
550	120
606	147
662	174
718	202
775	229
831	257
887	284
943	311
999	339
8905055	366
111	393
168	421
224	448
280	476
336	503
392	530
448	558
504	585
561	613
617	640
673	667

Number	Log
8905729	·9496695
785	722
841	749
897	777
954	804
8906010	832
066	859
122	886
178	914
234	941
290	968
347	996
403	·9497023
459	051
515	078
571	105
627	133
683	160
740	187
796	215
852	242
908	269
964	297
8907020	324
076	352
132	379
189	406
245	434
301	461
357	488
413	516
469	543
525	570
582	598
638	625
694	653
750	680
806	707
862	735
918	762
974	789
8908031	817
087	844
143	872
199	899
255	926
311	954
367	981
423	·9498008
480	036
536	063
592	090
648	118
704	145
760	173
816	200
872	227
929	255
985	282
8909041	309
097	337
153	364
209	391
265	419
321	446

Number	Log
8909378	·9498473
434	501
490	528
546	556
602	583
658	610
714	638
770	665
827	692
883	720
939	747
995	774

Excesses

10 = 5	1 = 0
20 = 10	2 = 1
30 = 15	3 = 1
40 = 20	4 = 2
50 = 24	5 = 2
	6 = 3
	7 = 3
	8 = 4
	9 = 4

28 = 14

Anti Log Excesses

10 = 21	1 = 2
20 = 41	2 = 4
	3 = 6
	4 = 8
	5 = 10
	6 = 12
	7 = 14
	8 = 16
	9 = 18

14 = 28

Number	Log
8910051	·9498802
107	829
163	856
219	884
275	911
332	939
388	966
444	993
500	·9499021
556	048
612	075
668	103
724	130
780	157
837	185
893	212
949	239
8911005	267
061	294
117	321
173	349
229	376
285	403
342	431

Number	Log
8911398	·9499458
454	485
510	513
566	540
622	567
678	595
734	622
790	649
847	677
903	704
959	731
8912015	759
071	786
127	813
183	841
239	868
295	896
351	923
408	950
464	978
520	·9500005
576	032
632	060
688	087
744	114
800	142
856	169
912	196
969	224
8913025	251
081	278
137	306
193	333
249	360
305	388
361	415
417	442
473	470
530	497
586	524
642	552
698	579
754	606
810	634
866	661
922	688
978	716
8914034	743
090	770
147	798
203	825
259	852
315	880
371	907
427	934
483	962
539	989
595	·9501016
651	044
707	071
764	098
820	125
876	153
932	180
988	207

Number	Log
8915044	·9501235
100	262
156	289
212	317
268	344
324	371
380	399
437	426
493	453
549	481
605	508
661	535
717	563
773	590
829	617
885	645
941	672
997	699
8916053	727
110	754
166	781
222	809
278	836
334	863
390	890
446	918
502	945
558	972
614	·9502000
670	027
726	054
782	081
839	109
895	136
951	163
8917007	191
063	218
119	245
175	273
231	300
287	327
343	355
399	382
455	409
511	436
567	464
624	491
680	518
736	546
792	573
848	600
904	628
960	655
8918016	682
072	709
128	737
184	764
240	791
296	819
352	846
408	873
464	901
521	928
577	955
633	982

8918689	·9503010
745	037
801	064
857	092
913	119
969	146
8919025	174
081	201
137	228
193	255
249	283
305	310
361	337
418	365
474	392
530	419
586	447
642	474
698	501
754	528
810	556
866	583
922	610
978	638

Excesses

10 = 5	1 = 0
20 = 10	2 = 1
30 = 15	3 = 1
40 = 19	4 = 2
50 = 24	5 = 2
	6 = 3
	7 = 3
	8 = 4
	9 = 4

28 = 14

Anti Log Excesses

10 = 21	1 = 2
20 = 41	2 = 4
	3 = 6
	4 = 8
	5 = 10
	6 = 12
	7 = 14
	8 = 16
	9 = 18

14 = 28

8920034	·9503665
090	692
146	719
202	747
258	774
314	801
370	829
426	856
483	883
539	910
595	938
651	965

8920707	·9503992
763	·9504020
819	047
875	074
931	101
987	129
8921043	156
099	183
155	211
211	238
267	265
323	292
379	320
435	347
491	374
547	402
603	429
660	456
716	484
772	511
828	538
884	565
940	593
996	620
8922052	647
108	674
164	702
220	729
276	756
332	784
388	811
444	838
500	865
556	893
612	920
668	947
724	975
780	·9505002
836	029
892	056
948	084
8923004	111
061	138
117	165
173	193
229	220
285	247
341	274
397	302
453	329
509	356
565	384
621	411
677	438
733	465
789	493
845	520
901	547
957	574
8924013	602
069	629
125	656
181	683
237	711
293	738

8924349	·9505765
405	793
461	820
517	847
573	874
629	902
685	929
741	956
797	983
853	·9506011
909	038
965	065
8925022	092
078	120
134	147
190	174
246	201
302	229
358	256
414	283
470	310
526	338
582	365
638	392
694	419
750	447
806	474
862	501
918	529
974	556
8926030	583
086	610
142	638
198	665
254	692
310	719
366	747
422	774
478	801
534	828
590	856
646	883
702	910
758	937
814	965
870	992
926	·9507019
982	046
8927038	074
094	101
150	128
206	155
262	183
318	210
374	237
430	264
486	292
542	319
598	346
654	373
710	400
766	428
822	455
878	482
934	509

8927990	·9507537
8928046	564
102	591
158	618
214	646
270	673
326	700
382	727
438	755
494	782
550	809
606	836
662	864
718	891
774	918
830	945
886	973
942	·9508000
998	027
8929054	054
110	081
166	109
222	136
278	163
334	190
390	218
446	245
502	272
558	299
614	327
670	354
726	381
782	408
838	436
894	463
950	490

Excesses

10 = 5	1 = 0
20 = 10	2 = 1
30 = 15	3 = 1
40 = 19	4 = 2
50 = 24	5 = 2
	6 = 3
	7 = 3
	8 = 4
	9 = 4

28 = 14

Anti Log Excesses

10 = 21	1 = 2
20 = 41	2 = 4
	3 = 6
	4 = 8
	5 = 10
	6 = 12
	7 = 14
	8 = 16
	9 = 18

14 = 28

8930006	·9508517
062	544
118	572
174	599
230	626
286	653
342	681
398	708
454	735
510	762
566	789
622	817
678	844
734	871
790	898
846	926
902	953
958	980
8931014	·9509007
070	034
126	062
182	089
238	116
294	143
350	171
406	198
462	225
518	252
574	280
630	307
686	334
742	361
798	388
854	416
910	443
966	470
8932022	497
078	524
134	552
189	579
245	606
301	633
357	661
413	688
469	715
525	742
581	769
637	797
693	824
749	851
805	878
861	905
917	933
973	960
8933029	987
085	·9510014
141	042
197	069
253	096
309	123
365	150
421	178
477	205
533	232
589	259

8933645	·9510286
701	314
757	341
813	368
869	395
925	422
981	450
8934037	477
093	504
148	531
204	558
260	586
316	613
372	640
428	667
484	695
540	722
596	749
652	776
708	803
764	830
820	858
876	885
932	912
988	939
8935044	966
100	994
156	·9511021
212	048
268	075
324	102
380	130
436	157
492	184
547	211
603	238
659	266
715	293
771	320
827	347
883	374
939	402
995	429
8936051	456
107	483
163	510
219	538
275	565
331	592
387	619
443	646
499	674
555	701
611	728
667	756
722	783
778	810
834	837
890	864
946	891
8937002	919
058	946
114	973
170	·9512000
226	027

8937282	·9512055
338	082
394	109
450	136
506	163
562	191
618	218
674	245
730	272
785	299
841	326
897	354
953	381
8938009	408
065	435
121	462
177	490
233	517
289	544
345	571
401	598
457	625
513	653
569	680
625	707
680	734
736	761
792	788
848	816
904	843
960	870
8939016	897
072	924
128	952
184	979
240	·9513006
296	033
352	060
408	087
464	115
520	142
575	169
631	196
687	223
743	250
799	278
855	305
911	332
967	359

Excesses

10 = 5	1 = 0
20 = 10	2 = 1
30 = 15	3 = 1
40 = 19	4 = 2
50 = 24	5 = 2
	6 = 3
	7 = 3
	8 = 4
	9 = 4

28 = 14

Anti Log Excesses

10 = 21	1 = 2
20 = 41	2 = 4
	3 = 6
	4 = 8
	5 = 10
	6 = 12
	7 = 14
	8 = 16
	9 = 19

14 = 28

8940023	·9513386
079	413
135	441
191	468
247	495
303	522
358	549
414	576
470	604
526	631
582	658
638	685
694	712
750	739
806	767
862	794
918	821
974	848
8941030	875
085	902
141	930
197	957
253	984
309	·9514011
365	038
421	065
477	093
533	120
589	147
645	174
701	201
756	228
812	256
868	283
924	310
980	337
8942036	364
092	391
148	418
204	446
260	473
316	500
372	527
427	554
483	581
539	609
595	636
651	663
707	690
763	717
819	744

8942875	·9514771
931	799
987	826
8943042	853
098	880
154	907
210	934
266	962
322	989
378	·9515016
434	043
490	070
546	097
602	124
657	151
713	179
769	206
825	233
881	260
937	287
993	314
8944049	342
105	369
161	396
217	423
272	450
328	477
384	504
440	532
496	559
552	586
608	613
664	640
720	667
776	694
831	722
887	749
943	776
999	803
8945055	830
111	857
167	884
223	912
279	939
334	966
390	993
446	·9516020
502	047
558	074
614	102
670	129
726	156
782	183
838	210
893	237
949	264
8946005	292
061	319
117	346
173	373
229	400
285	427
341	454
396	481
452	509

8946508	·9516536
564	563
620	590
676	617
732	644
788	671
844	698
899	725
955	753
8947011	780
067	807
123	834
179	861
235	888
291	915
346	942
402	970
458	997
514	·9517024
570	051
626	078
682	105
738	132
794	159
849	187
905	214
961	241
8948017	268
073	295
129	322
185	349
241	376
296	404
352	431
408	458
464	485
520	512
576	539
632	566
688	593
743	621
799	648
855	675
911	702
967	729
8949023	756
079	783
135	810
190	837
246	865
302	892
358	919
414	946
470	973
526	·9518000
581	027
637	054
693	081
749	109
805	136
861	163
917	190
973	217

Excesses

10 = 5	1 = 0
20 = 10	2 = 1
30 = 15	3 = 1
40 = 19	4 = 2
50 = 24	5 = 2
	6 = 3
	7 = 3
	8 = 4
	9 = 4

28 = 14

Anti Log Excesses

10 = 21	1 = 2
20 = 41	2 = 4
	3 = 6
	4 = 8
	5 = 10
	6 = 12
	7 = 14
	8 = 16
	9 = 19

14 = 28

8950028	·9518244
084	271
140	298
196	325
252	353
308	380
364	407
419	434
475	461
531	488
587	515
643	542
699	569
755	597
811	624
866	651
922	678
978	705
8951034	732
090	759
146	786
202	813
257	840
313	868
369	895
425	922
481	949
537	976
593	·9519003
648	030
704	057
760	084
816	111
872	138
928	166
984	193
8952039	220

8952095	·9519247
151	274
207	301
263	328
319	355
374	382
430	409
486	436
542	464
598	491
654	518
710	545
765	572
821	599
877	626
933	653
989	680
8953045	707
101	735
156	762
212	789
268	816
324	843
380	870
436	897
491	924
547	951
603	978
659	·9520005
715	033
771	060
826	087
882	114
938	141
994	168
8954050	195
106	222
162	249
217	276
273	303
329	330
385	358
441	385
497	412
552	439
608	466
664	493
720	520
776	547
832	574
887	601
943	628
999	655
8955055	683
111	710
167	737
222	764
278	791
334	818
390	845
446	872
502	899
557	926
613	953
669	980

8955725	·9521007
781	035
837	062
892	089
948	116
8956004	143
060	170
116	197
172	224
227	251
283	278
339	305
395	332
451	359
507	386
562	414
618	441
674	468
730	495
786	522
841	549
897	576
953	603
8957009	630
065	657
121	684
176	711
232	738
288	765
344	792
400	819
455	847
511	874
567	901
623	928
679	955
735	982
790	·9522009
846	036
902	063
958	090
8958014	117
069	144
125	171
181	198
237	225
293	252
349	280
404	307
460	334
516	361
572	388
628	415
683	442
739	469
795	496
851	523
907	550
962	577
8959018	604
074	631
130	658
186	685
242	712
297	740

8959353	·9522767
409	794
465	821
521	848
576	875
632	902
688	929
744	956
800	983
855	·9523010
911	037
967	064

Excesses

10 = 5	1 = 0
20 = 10	2 = 1
30 = 15	3 = 1
40 = 19	4 = 2
50 = 24	5 = 2
	6 = 3
	7 = 3
	8 = 4
	9 = 4

28 = 14

Anti Log Excesses

10 = 21	1 = 2
20 = 41	2 = 4
	3 = 6
	4 = 8
	5 = 10
	6 = 12
	7 = 14
	8 = 16
	9 = 19

14 = 28

8960023	·9523091
079	118
134	145
190	172
246	199
302	226
358	253
413	281
469	308
525	335
581	362
637	389
692	416
748	443
804	470
860	497
916	524
971	551
8961027	578
083	605
139	632
195	659
250	686
306	713

8961362	·9523740
418	767
474	794
529	821
585	848
641	876
697	903
753	929
808	956
864	983
920	·9524010
976	038
8962032	065
087	092
143	119
199	146
255	173
310	200
366	227
422	254
478	281
534	308
589	335
645	362
701	389
757	416
813	443
868	470
924	497
980	524
8963036	551
091	578
147	605
203	632
259	659
315	686
370	713
426	740
482	767
538	794
594	821
649	848
705	875
761	902
817	929
872	956
928	984
984	·9525011
8964040	038
096	065
151	092
207	119
263	146
319	173
374	200
430	227
486	254
542	281
598	308
653	335
709	362
765	389
821	416
876	443
932	470

8964988	·9525497
8965044	524
099	551
155	578
211	605
267	632
323	659
378	686
434	713
490	740
546	767
601	794
657	821
713	848
769	875
825	902
880	929
936	956
992	983
8966048	·9526010
103	037
159	064
215	091
271	118
326	145
382	172
438	199
494	226
549	253
605	280
661	307
717	334
772	361
828	388
884	415
940	442
996	469
8967051	496
107	523
163	550
219	577
274	604
330	631
386	658
442	685
497	712
553	739
609	766
665	793
720	820
776	847
832	874
888	901
943	928
999	955
8968055	982
111	·9527009
166	036
222	063
278	090
334	117
389	144
445	171
501	198
557	225

8968612	·9527252
668	279
724	306
780	333
835	360
891	387
947	414
8969003	441
058	468
114	495
170	522
226	549
281	576
337	603
393	630
449	657
504	684
560	711
616	738
672	765
727	792
783	819
839	846
895	873
950	900

Excesses

10 = 5	1 = 0
20 = 10	2 = 1
30 = 15	3 = 1
40 = 19	4 = 2
50 = 24	5 = 2
	6 = 3
	7 = 3
	8 = 4
	9 = 4

28 = 14

Anti Log Excesses

10 = 21	1 = 2
20 = 41	2 = 4
	3 = 6
	4 = 8
	5 = 10
	6 = 12
	7 = 14
	8 = 17
	9 = 19

14 = 28

8970006	·9527927
062	954
118	981
173	·9528008
229	035
285	062
341	089
396	116
452	143
508	170
563	197

8970619	·9528224
675	251
731	278
786	305
842	332
898	359
954	386
8971009	413
065	440
121	467
177	494
232	521
288	548
344	575
399	602
455	629
511	656
567	683
622	709
678	736
734	763
790	790
845	818
901	845
957	872
8972013	898
068	925
124	952
180	979
235	·9529006
291	033
347	060
403	087
458	114
514	141
570	168
626	195
681	222
737	249
793	276
848	303
904	330
960	357
8973016	384
071	411
127	438
183	465
238	492
294	519
350	546
406	573
461	600
517	627
573	654
629	681
684	708
740	735
796	761
851	788
907	815
963	842
8974019	869
074	896
130	923
186	950

8974241	·9529977
297	·9530004
353	031
409	058
464	085
520	112
576	139
631	166
687	193
743	220
799	247
854	274
910	301
966	328
8975021	355
077	382
133	408
189	435
244	462
300	489
356	516
411	543
467	570
523	597
578	624
634	651
690	678
746	705
801	732
857	759
913	786
968	813
8976024	840
080	867
136	894
191	921
247	948
303	974
358	·9531001
414	028
470	055
525	082
581	109
637	136
693	163
748	190
804	217
860	244
915	271
971	298
8977027	325
082	352
138	379
194	406
250	433
305	460
361	487
417	514
472	541
528	568
584	594
639	621
695	648
751	675
806	702

8977862	·9531729
918	756
974	783
8978029	810
085	837
141	864
196	891
252	918
308	945
363	972
419	999
475	·9532025
530	052
586	079
642	106
698	133
753	160
809	187
865	214
920	241
976	268
8979032	295
087	322
143	349
199	376
254	403
310	429
366	456
421	483
477	510
533	537
588	564
644	591
700	618
756	645
811	672
867	699
923	726
978	753

Excesses

10 = 5	1 = 0
20 = 10	2 = 1
30 = 15	3 = 1
40 = 19	4 = 2
50 = 24	5 = 2
	6 = 3
	7 = 3
	8 = 4
	9 = 4

28 = 13

Anti Log Excesses

10 = 21	1 = 2
20 = 41	2 = 4
	3 = 6
	4 = 8
	5 = 10
	6 = 12
	7 = 14
	8 = 17
	9 = 19

13 = 28

8980034	·9532780
090	806
145	833
201	860
257	887
312	914
368	941
424	968
479	995
535	·9533022
591	049
646	076
702	103
758	130
813	156
869	183
925	210
980	237
8981036	264
092	291
147	318
203	345
259	372
314	399
370	426
426	453
481	479
537	506
593	533
648	560
704	587
760	614
815	641
871	668
927	695
982	722
8982038	749
094	776
149	803
205	830
261	856
316	883
372	910
428	937
483	964
539	991
595	·9534018
650	045
706	072
762	099
817	126

8982873	·9534152
929	179
984	206
8983040	233
096	260
151	287
207	314
263	341
318	368
374	395
430	422
485	448
541	475
597	502
652	529
708	556
764	583
819	610
875	637
931	664
986	691
8984042	717
098	744
153	771
209	798
265	825
320	852
376	879
431	906
487	933
543	960
598	986
654	·9535013
710	040
765	067
821	094
877	121
932	148
988	175
8985044	202
099	229
155	255
211	282
266	309
322	336
378	363
433	390
489	417
544	444
600	471
656	498
711	524
767	551
823	578
878	605
934	632
990	659
8986045	686
101	713
157	740
212	766
268	793
323	820
379	847
435	874

8986490	·9535901
546	928
602	955
657	982
713	·9536008
769	035
824	062
880	089
935	116
991	143
8987047	170
102	197
158	224
214	250
269	277
325	304
381	331
436	358
492	385
547	412
603	439
659	465
714	492
770	519
826	546
881	573
937	600
992	627
8988048	654
104	680
159	707
215	734
271	761
326	788
382	815
438	842
493	869
549	896
604	922
660	949
716	976
771	·9537003
827	030
883	057
938	084
994	110
8989049	137
105	164
161	191
216	218
272	245
328	272
383	299
439	325
494	352
550	379
606	406
661	433
717	460
772	487
828	514
884	540
939	567
995	594

Excesses

10 = 5	1 = 0
20 = 10	2 = 1
30 = 14	3 = 1
40 = 19	4 = 2
50 = 24	5 = 2
	6 = 3
	7 = 3
	8 = 4
	9 = 4

28 = 13

Anti Log Excesses

10 = 21	1 = 2
20 = 41	2 = 4
	3 = 6
	4 = 8
	5 = 10
	6 = 12
	7 = 14
	8 = 17
	9 = 19

13 = 28

8990051	·9537621
106	648
162	675
217	702
273	728
329	755
384	782
440	809
495	836
551	863
607	890
662	917
718	943
774	970
829	997
885	·9538024
940	051
996	078
8991052	105
107	131
163	158
218	185
274	212
330	239
385	266
441	293
496	319
552	346
608	373
663	400
719	427
775	454
830	481
886	507
941	534
997	561
8992053	588

8992108	·9538615
164	642
219	669
275	696
331	722
386	749
442	776
497	803
553	830
609	857
664	883
720	910
775	937
831	964
887	991
942	·9539018
998	045
8993053	071
109	098
165	125
220	152
276	179
331	206
387	233
443	259
498	286
554	313
609	340
665	367
721	394
776	420
832	447
887	474
943	501
998	528
8994054	555
110	581
165	608
221	635
276	662
332	689
388	716
443	743
499	769
554	796
610	823
666	850
721	877
777	904
832	930
888	957
944	984
999	·9540011
8995055	038
110	065
166	091
221	118
277	145
333	172
388	199
444	226
499	252
555	279
611	306
666	333

Number	Log
8995722	·9540360
777	387
833	413
888	440
944	467
8996000	494
055	521
111	548
166	574
222	601
278	628
333	655
389	682
444	709
500	735
555	762
611	789
667	816
722	843
778	870
833	896
889	923
944	950
8997000	977
056	·9541004
111	031
167	057
222	084
278	111
333	138
389	165
445	192
500	218
556	245
611	272
667	299
722	326
778	353
834	379
889	406
945	433
8998000	460
056	487
111	513
167	540
223	567
278	594
334	621
389	648
445	674
500	701
556	728
612	755
667	782
723	808
778	835
834	862
889	889
945	916
8999000	942
056	969
112	996
167	·9542023
223	050
278	077

Number	Log
8999334	·9542103
389	130
445	157
500	184
556	211
612	237
667	264
723	291
778	318
834	345
889	371
945	398

Excesses

10 = 5	1 = 0
20 = 10	2 = 1
30 = 14	3 = 1
40 = 19	4 = 2
50 = 24	5 = 2
	6 = 3
	7 = 3
	8 = 4
	9 = 4

28 = 13

Anti Log Excesses

10 = 21	1 = 2
20 = 41	2 = 4
	3 = 6
	4 = 8
	5 = 10
	6 = 12
	7 = 14
	8 = 17
	9 = 19

13 = 28

Number	Log
9000000	·9542425
056	452
112	479
167	505
223	532
278	559
334	586
389	613
445	640
500	666
556	693
612	720
667	747
723	774
778	800
834	827
889	854
945	881
9001000	908
056	934
112	961
167	988
223	·9543015
278	042

Number	Log
9001334	·9543068
389	095
445	122
500	149
556	176
611	202
667	229
723	256
778	283
834	310
889	336
945	363
9002000	390
056	417
111	444
167	470
222	497
278	524
334	551
389	578
445	604
500	631
556	658
611	685
667	712
722	738
778	765
833	792
889	819
944	845
9003000	872
056	899
111	926
167	953
222	979
278	·9544006
333	033
389	060
444	087
500	113
555	140
611	167
666	194
722	220
777	247
833	274
889	301
944	328
9004000	354
055	381
111	408
166	435
222	462
277	488
333	515
388	542
444	569
499	595
555	622
610	649
666	676
721	703
777	729
833	756
888	783

Number	Log
9004944	·9544810
999	836
9005055	863
110	890
166	917
221	944
277	970
332	997
388	·9545024
443	051
499	077
554	104
610	131
665	158
721	185
776	211
832	238
887	265
943	292
999	318
9006054	345
110	372
165	399
221	426
276	452
332	479
387	506
443	533
498	559
554	586
609	613
665	640
720	666
776	693
831	720
887	747
942	773
998	800
9007053	827
109	854
164	881
220	907
275	934
331	961
386	988
442	·9546014
497	041
553	068
608	095
664	121
719	148
775	175
830	202
886	228
941	255
997	282
9008052	309
108	336
163	362
219	389
274	416
330	443
385	469
441	496
496	523

Number	Log
9008552	·9546550
607	576
663	603
718	630
774	657
830	683
885	710
941	737
996	764
9009052	790
107	817
163	844
218	871
273	897
329	924
384	951
440	978
495	·9547004
551	031
606	058
662	085
717	111
773	138
828	165
884	192
939	218
995	245

Excesses

10 = 5	1 = 0
20 = 10	2 = 1
30 = 14	3 = 1
40 = 19	4 = 2
50 = 24	5 = 2
	6 = 3
	7 = 3
	8 = 4
	9 = 4

28 = 13

Anti Log Excesses

10 = 21	1 = 2
20 = 41	2 = 4
	3 = 6
	4 = 8
	5 = 10
	6 = 12
	7 = 15
	8 = 17
	9 = 19

13 = 28

Number	Log
9010050	·9547272
106	299
161	325
217	352
272	379
328	406
383	432
439	459
494	486

9010550	·9547512
605	539
661	566
716	593
772	620
827	646
883	673
938	700
994	727
9011049	753
105	780
160	807
216	834
271	860
327	887
382	914
438	940
493	967
549	994
604	·9548021
660	047
715	074
771	101
826	128
882	154
937	181
993	208
9012048	235
103	261
159	288
214	315
270	342
325	368
381	395
436	422
492	449
547	475
603	502
658	529
714	555
769	582
825	609
880	636
936	662
991	689
9013047	716
102	743
158	769
213	796
268	823
324	849
379	876
435	903
490	930
546	956
601	983
657	·9549010
712	037
768	063
823	090
879	117
934	143
990	170
9014045	197
101	224

9014156	·9549250
212	277
267	304
322	331
378	357
433	384
489	411
544	437
600	464
655	491
711	518
766	544
822	571
877	598
933	624
988	651
9015043	678
099	705
154	731
210	758
265	785
321	812
376	838
432	865
487	892
543	918
598	945
654	972
709	998
764	·9550025
820	052
875	079
931	105
986	132
9016042	159
097	185
153	212
208	239
264	266
319	292
374	319
430	346
485	372
541	399
596	426
652	453
707	479
763	506
818	533
874	559
929	586
984	613
9017040	640
095	666
151	693
206	720
262	746
317	773
373	800
428	827
484	853
539	880
594	907
650	934
705	960

9017761	·9550987
816	·9551014
872	040
927	067
983	094
9018038	121
093	147
149	174
204	201
260	227
315	254
371	281
426	307
482	334
537	361
592	388
648	414
703	441
759	468
814	494
870	521
925	548
980	574
9019036	601
091	628
147	654
202	681
258	708
313	735
369	761
424	788
479	815
535	841
590	868
646	895
701	921
757	948
812	975
867	·9552001
923	028
978	055

Excesses

10 = 5	1 = 0
20 = 10	2 = 1
30 = 14	3 = 1
40 = 19	4 = 2
50 = 24	5 = 2
	6 = 3
	7 = 3
	8 = 4
	9 = 4

28 = 13

Anti Log Excesses

10 = 21	1 = 2
20 = 42	2 = 4
	3 = 6
	4 = 8
	5 = 10
	6 = 12
	7 = 15
	8 = 17
	9 = 19

13 = 28

9020034	·9552082
089	108
145	135
200	162
255	188
311	215
366	242
422	268
477	295
533	322
588	348
643	375
699	402
754	428
810	455
865	482
921	509
976	535
9021031	562
087	589
142	615
198	642
253	669
309	695
364	722
419	749
475	775
530	802
586	829
641	855
697	882
752	909
807	935
863	962
918	989
974	·9553015
9022029	042
085	069
140	095
195	122
251	149
306	175
362	202
417	229
472	256
528	282
583	309
639	336
694	362
750	389
805	416

9022860	·9553442
916	469
971	496
9023027	522
082	549
137	576
193	602
248	629
304	656
359	682
414	709
470	736
525	762
581	789
636	816
692	842
747	869
802	896
858	922
913	949
969	976
9024024	·9554002
079	029
135	056
190	082
246	109
301	136
356	162
412	189
467	216
523	242
578	269
633	296
689	322
744	349
800	376
855	402
910	429
966	456
9025021	482
077	509
132	536
187	562
243	589
298	616
354	642
409	669
464	696
520	722
575	749
631	775
686	802
741	829
797	855
852	882
908	909
963	935
9026018	962
074	989
129	·9555015
185	042
240	069
295	095
351	122
406	149

9026462	·9555175
517	202
572	229
628	255
683	282
739	309
794	335
849	362
905	388
960	415
9027016	442
071	468
126	495
182	522
237	548
292	575
348	602
403	628
459	655
514	682
569	708
625	735
680	761
736	788
791	815
846	841
902	868
957	895
9028012	921
068	948
123	975
179	·9556001
234	028
289	055
345	081
400	108
456	134
511	161
566	188
622	214
677	241
732	268
788	294
843	321
899	348
954	374
9029009	401
065	427
120	454
175	481
231	507
286	534
342	561
397	587
452	614
508	641
563	667
618	694
674	720
729	747
785	774
840	800
895	827
951	854

Excesses

10 = 5	1 = 0
20 = 10	2 = 1
30 = 14	3 = 1
40 = 19	4 = 2
50 = 24	5 = 2
	6 = 3
	7 = 3
	8 = 4
	9 = 4

28 = 13

Anti Log Excesses

10 = 21	1 = 2
20 = 42	2 = 4
	3 = 6
	4 = 8
	5 = 10
	6 = 12
	7 = 15
	8 = 17
	9 = 19

13 = 28

9030006	·9556880
061	907
117	933
172	960
228	987
283	·9557013
338	040
394	067
449	093
504	120
560	146
615	173
670	200
726	226
781	253
837	280
892	306
947	333
9031003	359
058	386
113	413
169	439
224	466
279	493
335	519
390	546
446	572
501	599
556	626
612	652
667	679
722	706
778	732
833	759
888	785
944	812
999	839

9032055	·9557865
110	892
165	919
221	945
276	972
331	998
387	·9558025
442	052
497	078
553	105
608	131
663	158
719	185
774	211
830	238
885	265
940	291
996	318
9033051	344
106	371
162	398
217	424
272	451
328	477
383	504
438	531
494	557
549	584
604	610
660	637
715	664
770	690
826	717
881	744
937	770
992	797
9034047	823
103	850
158	877
213	903
269	930
324	956
379	983
435	·9559010
490	036
545	063
601	089
656	116
711	143
767	169
822	196
877	222
933	249
988	276
9035043	302
099	329
154	355
209	382
265	409
320	435
375	462
431	488
486	515
541	542
597	568

9035652	·9559595
707	621
763	648
818	675
873	701
929	728
984	754
9036039	781
095	808
150	834
205	861
261	887
316	914
371	941
427	967
482	994
537	·9560020
593	047
648	073
703	100
759	127
814	153
869	180
925	206
980	233
9037035	260
091	286
146	313
201	339
257	366
312	393
367	419
423	446
478	472
533	499
589	525
644	552
699	579
755	605
810	632
865	658
921	685
976	712
9038031	738
087	765
142	791
197	818
253	845
308	871
363	898
419	924
474	951
529	977
584	·9561004
640	031
695	057
750	084
806	110
861	137
916	164
972	190
9039027	217
082	243
138	270
193	296

9039248	·9561323
304	350
359	376
414	403
470	429
525	456
580	482
635	509
691	536
746	562
801	589
857	615
912	642
967	668

Excesses

10 = 5	1 = 0
20 = 10	2 = 1
30 = 14	3 = 1
40 = 19	4 = 2
50 = 24	5 = 2
	6 = 3
	7 = 3
	8 = 4
	9 = 4

28 = 13

Anti Log Excesses

10 = 21	1 = 2
20 = 42	2 = 4
	3 = 6
	4 = 8
	5 = 10
	6 = 12
	7 = 15
	8 = 17
	9 = 19

13 = 28

9040023	·9561695
078	722
133	748
189	775
244	801
299	828
354	854
410	881
465	908
520	934
576	961
631	987
686	·9562014
742	040
797	067
852	094
908	120
963	147
9041018	173
073	200
129	226
184	253

9041239	·9562279	9044833	·9564005	9048426	·9565730	9050415	·9566685	9054005	·9568407
295	306	889	032	481	757	470	711	060	434
350	333	944	059	536	783	525	738	116	460
405	359	999	085	592	810	581	764	171	487
461	386	9045054	112	647	836	636	791	226	513
516	412	110	138	702	863	691	817	281	540
571	439	165	165	757	889	746	844	336	566
626	465	220	191	813	916	802	870	392	592
682	492	276	218	868	942	857	897	447	619
737	519	331	244	923	969	912	923	502	645
792	545	386	271	978	995	967	950	557	672
848	572	441	297	9049034	·9566022	9051023	976	613	698
903	598	497	324	089	048	078	·9567003	668	725
958	625	552	351	144	075	133	029	723	751
9042014	651	607	377	199	101	188	056	778	778
069	678	662	404	255	128	244	082	833	804
124	704	718	430	310	154	299	109	889	831
179	731	773	457	365	181	354	135	944	857
235	758	828	483	420	207	409	162	999	884
290	784	884	510	476	234	464	188	9055054	910
345	811	939	536	531	260	520	215	110	937
401	837	994	563	586	287	575	241	165	963
456	864	9046049	589	641	313	630	268	220	990
511	890	105	616	697	340	685	294	275	·9569016
567	917	160	642	752	366	741	321	330	043
622	943	215	669	807	393	796	347	386	069
677	970	270	695	862	419	851	374	441	096
732	997	326	722	918	446	906	400	496	122
788	·9563023	381	749	973	472	962	427	551	149
843	050	436	775			9052017	453	606	175
898	076	492	802			072	480	662	202
954	103	547	828			127	506	717	228
9043009	129	602	855			183	533	772	255
064	156	657	881			238	559	827	281
119	183	713	908			293	586	883	308
175	209	768	934			348	612	938	334
230	236	823	961			403	639	993	360
285	262	878	987			459	665	9056048	387
341	289	934	·9565014			514	692	103	413
396	315	989	040			569	718	159	440
451	342	9047044	067			624	745	214	466
506	368	099	093			680	771	269	493
562	395	155	120			735	798	324	519
617	421	210	146			790	824	379	546
672	448	265	173			845	851	435	572
728	475	321	200			901	877	490	599
783	501	376	226			956	904	545	625
838	528	431	253			9053011	930	600	652
893	554	486	279			066	957	656	678
949	581	542	306			122	983	711	705
9044004	607	597	332			177	·9568010	766	731
059	634	652	359			232	036	821	758
115	660	707	385			287	063	876	784
170	687	763	412			342	089	932	810
225	713	818	438			398	116	987	837
280	740	873	465			453	142	9057042	863
336	767	928	491			508	169	097	890
391	793	984	518			563	195	152	916
446	820	9048039	544	9050028	·9566499	619	222	208	943
502	846	094	571	083	526	674	248	263	969
557	873	149	597	139	552	729	275	318	996
612	899	205	624	194	579	784	301	373	·9570022
667	926	260	650	249	605	839	328	428	049
723	952	315	677	304	632	895	354	484	075
778	979	371	704	360	658	950	381	539	102

Excesses

10 = 5	1 = 0
20 = 10	2 = 1
30 = 14	3 = 1
40 = 19	4 = 2
50 = 24	5 = 2
	6 = 3
	7 = 3
	8 = 4
	9 = 4

28 = 13

Anti Log Excesses

10 = 21	1 = 2
20 = 42	2 = 4
	3 = 6
	4 = 8
	5 = 10
	6 = 12
	7 = 15
	8 = 17
	9 = 19

13 = 28

9057594	·9570128
649	155
704	181
760	208
815	234
870	260
925	287
980	314
9058036	340
091	367
146	393
201	420
256	446
312	472
367	499
422	525
477	552
532	578
588	605
643	631
698	658
753	684
808	711
864	737
919	764
974	790
9059029	816
084	843
140	869
195	896
250	922
305	949
360	975
415	·9571002
471	028
526	055
581	081
636	107
691	134
747	160
802	187
857	213
912	240
967	266

Excesses

10 = 5	1 = 0
20 = 10	2 = 1
30 = 14	3 = 1
40 = 19	4 = 2
50 = 24	5 = 2
	6 = 3
	7 = 3
	8 = 4
	9 = 4

28 = 13

Anti Log Excesses

10 = 21	1 = 2
20 = 42	2 = 4
	3 = 6
	4 = 8
	5 = 10
	6 = 13
	7 = 15
	8 = 17
	9 = 19

13 = 28

9060023	·9571293
078	319
133	346
188	372
243	398
299	425
354	451
409	478
464	504
519	531
574	557
630	584
685	610
740	636
795	663
850	689
906	716
961	742
9061016	769
071	795
126	822
181	848
237	875
292	901
347	927
402	954
457	980
513	·9572007
568	033
623	060
678	086
733	113
788	139
844	165
899	192
954	218
9062009	245
064	271
119	298
175	324
230	350
285	377
340	403
395	430
450	456
506	483
561	509
616	536
671	562
726	588
782	615

9062837	·9572641
892	668
947	694
9063002	721
057	747
113	774
168	800
223	826
278	853
333	879
388	906
444	932
499	959
554	985
609	·9573011
664	038
719	064
775	091
830	117
885	144
940	170
995	196
9064050	223
106	249
161	276
216	302
271	329
326	355
381	381
437	408
492	434
547	461
602	487
657	514
712	540
767	566
823	593
878	619
933	646
988	672
9065043	699
098	725
154	751
209	778
264	804
319	831
374	857
429	884
484	910
540	936
595	963
650	989
705	·9574016
760	042
815	068
871	095
926	121
981	148
9066036	174
091	201
146	227
201	253
257	280
312	306
367	333

9066422	·9574359
477	385
532	412
588	438
643	465
698	491
753	518
808	544
863	570
918	597
974	623
9067029	650
084	676
139	702
194	729
249	755
304	782
360	808
415	834
470	861
525	887
580	914
635	940
690	967
746	993
801	·9575019
856	046
911	072
966	099
9068021	125
076	151
132	178
187	204
242	231
297	257
352	283
407	310
462	336
517	363
573	389
628	416
683	442
738	468
793	495
848	521
903	548
959	574
9069014	600
069	627
124	653
179	680
234	706
289	732
344	759
400	785
455	812
510	838
565	864
620	891
675	917
730	943
786	970
841	996
896	·9576023
951	049

Excesses

10 = 5	1 = 0
20 = 10	2 = 1
30 = 14	3 = 1
40 = 19	4 = 2
50 = 24	5 = 2
	6 = 3
	7 = 3
	8 = 4
	9 = 4

28 = 13

Anti Log Excesses

10 = 21	1 = 2
20 = 42	2 = 4
	3 = 6
	4 = 8
	5 = 10
	6 = 13
	7 = 15
	8 = 17
	9 = 19

13 = 28

9070006	·9576075
061	102
116	128
171	155
227	181
282	207
337	234
392	260
447	287
502	313
557	339
612	366
668	392
723	419
778	445
833	471
888	498
943	524
998	550
9071053	577
108	603
164	630
219	656
274	682
329	709
384	735
439	762
494	788
549	814
605	841
660	867
715	894
770	920
825	946
880	973
935	999
990	·9577025

9072045	·9577052
101	078
156	105
211	131
266	157
321	184
376	210
431	236
486	263
541	289
597	316
652	342
707	368
762	395
817	421
872	447
927	474
982	500
9073037	527
093	553
148	579
203	606
258	632
313	659
368	685
423	711
478	738
533	764
589	791
644	817
699	843
754	870
809	896
864	922
919	949
974	975
9074029	·9578002
084	028
140	054
195	081
250	107
305	133
360	160
415	186
470	212
525	239
580	265
635	292
691	318
746	344
801	371
856	397
911	423
966	450
9075021	476
076	502
131	529
186	555
242	582
297	608
352	634
407	661
462	687
517	713
572	740

9075627	·9578766
682	792
737	819
792	845
848	872
903	898
958	924
9076013	951
068	977
123	·9579003
178	030
233	056
288	082
343	109
398	135
454	162
509	188
564	214
619	241
674	267
729	293
784	320
839	346
894	372
949	399
9077004	425
059	451
115	478
170	504
225	531
280	557
335	583
390	610
445	636
500	662
555	689
610	715
665	741
720	768
776	794
831	820
886	847
941	873
996	899
9078051	926
106	952
161	978
216	·9580005
271	031
326	058
381	084
436	110
492	137
547	163
602	189
657	216
712	242
767	268
822	295
877	321
932	347
987	374
9079042	400
097	427
152	453

9079208	·9580479
263	506
318	532
373	558
428	585
483	611
538	637
593	664
648	690
703	716
758	743
813	769
868	795
923	822
978	848

Excesses

10 = 5	1 = 0
20 = 10	2 = 1
30 = 14	3 = 1
40 = 19	4 = 2
50 = 24	5 = 2
	6 = 3
	7 = 3
	8 = 4
	9 = 4

28 = 13

Anti Log Excesses

10 = 21	1 = 2
20 = 42	2 = 4
	3 = 6
	4 = 8
	5 = 10
	6 = 13
	7 = 15
	8 = 17
	9 = 19

13 = 28

9080034	·9580874
089	901
144	927
199	953
254	980
309	·9581006
364	032
419	059
474	085
529	111
584	138
639	164
694	190
749	217
804	243
859	269
915	296
970	322
9081025	348
080	375
135	401

9081190	·9581427
245	454
300	480
355	506
410	533
465	559
520	585
575	611
630	638
685	664
740	690
795	717
851	743
906	769
961	796
9082016	822
071	848
126	875
181	901
236	927
291	954
346	980
401	·9582006
456	033
511	059
566	085
621	112
676	138
731	164
786	191
841	217
897	243
952	270
9083007	296
062	322
117	348
172	375
227	401
282	427
337	454
392	480
447	506
502	533
557	559
612	586
667	612
722	638
777	665
832	691
887	717
942	743
997	770
9084052	796
108	822
163	849
218	875
273	901
328	928
383	954
438	980
493	·9583007
548	033
603	059
658	085
713	112

9084768	·9583138
823	164
878	191
933	217
988	243
9085043	270
098	296
153	322
208	349
263	375
318	401
373	427
428	454
483	480
538	506
593	533
649	559
704	585
759	612
814	638
869	664
924	690
979	717
9086034	743
089	769
144	796
199	822
254	848
309	875
364	901
419	927
474	953
529	980
584	·9584006
639	032
694	059
749	085
804	111
859	138
914	164
969	190
9087024	216
079	243
134	269
189	295
244	322
299	348
354	374
409	400
464	427
519	453
574	479
629	506
684	532
739	558
795	585
850	611
905	637
960	663
9088015	690
070	716
125	742
180	769
235	795
290	821

9088345	·9584847
400	874
455	900
510	926
565	953
620	979
675	·9585005
730	032
785	058
840	084
895	110
950	137
9089005	163
060	189
115	216
170	242
225	268
280	294
335	321
390	347
445	373
500	400
555	426
610	452
665	478
720	505
775	531
830	557
885	584
940	610
995	636

Excesses

10 = 5	1 = 0
20 = 10	2 = 1
30 = 14	3 = 1
40 = 19	4 = 2
50 = 24	5 = 2
	6 = 3
	7 = 3
	8 = 4
	9 = 4

28 = 13

Anti Log Excesses

10 = 21	1 = 2
20 = 42	2 = 4
	3 = 6
	4 = 8
	5 = 10
	6 = 13
	7 = 15
	8 = 17
	9 = 19

13 = 28

9090050	·9585662
105	689
160	715
215	741
270	768

9090325	·9585794
380	820
435	846
490	873
545	899
600	925
655	952
710	978
765	·9586004
820	030
875	057
930	083
985	109
9091040	135
095	162
150	188
205	214
260	241
315	267
370	293
425	319
480	346
535	372
590	398
645	424
700	451
755	477
810	503
865	530
920	556
975	582
9092030	608
085	635
140	661
195	687
250	713
305	740
360	766
415	792
470	819
525	845
580	871
635	897
690	924
745	950
800	976
855	·9587002
910	029
965	055
9093020	081
075	107
130	134
185	160
240	186
295	213
350	239
405	265
460	291
515	318
570	344
625	370
680	396
735	422
790	449
845	475

9093900	·9587501
955	527
9094010	554
064	580
119	606
174	632
229	659
284	685
339	711
394	737
449	764
504	790
559	816
614	843
669	869
724	895
779	921
834	948
889	974
944	·9588000
999	026
9095054	053
109	079
164	105
219	131
274	158
329	184
384	210
439	236
494	263
549	289
604	315
659	341
714	368
769	394
824	420
879	446
934	473
989	499
9096044	525
099	551
154	578
209	604
263	630
318	656
373	683
428	709
483	735
538	761
593	787
648	814
703	840
758	866
813	892
868	919
923	945
978	971
9097033	997
088	·9589024
143	050
198	076
253	102
308	129
363	155
418	181

9097473	·9589207
528	234
583	260
638	286
693	312
747	339
802	365
857	391
912	417
967	443
9098022	470
077	496
132	522
187	548
242	575
297	601
352	627
407	653
462	680
517	706
572	732
627	758
682	785
737	811
792	837
847	863
902	890
956	916
9099011	942
066	968
121	995
176	·9590021
231	047
286	073
341	099
396	126
451	152
506	178
561	204
616	231
671	257
726	283
781	309
836	335
891	362
946	388

Excesses

10 = 5	1 = 0
20 = 10	2 = 1
30 = 14	3 = 1
40 = 19	4 = 2
50 = 24	5 = 2
	6 = 3
	7 = 3
	8 = 4
	9 = 4

27 = 13

Anti Log Excesses

10 = 21	1 = 2
20 = 42	2 = 4
	3 = 6
	4 = 8
	5 = 10
	6 = 13
	7 = 15
	8 = 17
	9 = 19

13 = 27

9100000	·9590414
055	440
110	467
165	493
220	519
275	545
330	571
385	598
440	624
495	650
550	676
605	703
660	729
715	755
770	781
825	807
880	834
935	860
989	886
9101044	912
099	939
154	965
209	991
264	·9591017
319	043
374	070
429	096
484	122
539	148
594	174
649	201
704	227
759	253
814	279
868	306
923	332
978	358
9102033	384
088	410
143	437
198	463
253	489
308	515
363	541
418	568
473	594
528	620
583	646
637	672
692	699
747	725

9102802	·9591751
857	777
912	804
967	830
9103022	856
077	882
132	908
187	935
242	961
297	987
352	·9592013
406	039
461	066
516	092
571	118
626	144
681	170
736	197
791	223
846	249
901	275
956	301
9104011	327
066	354
120	380
175	406
230	432
285	458
340	485
395	511
450	537
505	563
560	589
615	616
670	642
725	668
779	694
834	720
889	747
944	773
999	799
9105054	825
109	851
164	878
219	904
274	930
329	956
384	982
438	·9593009
493	035
548	061
603	087
658	113
713	140
768	166
823	192
878	218
933	244
988	270
9106042	297
097	323
152	349
207	375
262	401
317	428

9106372	·9593454
427	480
482	506
537	532
592	558
646	585
701	611
756	637
811	663
866	689
921	716
976	742
9107031	768
086	794
141	820
195	847
250	873
305	899
360	925
415	951
470	977
525	·9594004
580	030
635	056
690	082
744	108
799	134
854	161
909	187
964	213
9108019	239
074	265
129	292
184	318
239	344
293	370
348	396
403	422
458	449
513	475
568	501
623	527
678	553
733	579
787	606
842	632
897	658
952	684
9109007	710
062	736
117	763
172	789
227	815
281	841
336	867
391	893
446	920
501	946
556	972
611	998
666	·9595024
721	050
775	077
830	103
885	129

9109940	·9595155
995	181

Excesses

10 = 5	1 = 0
20 = 10	2 = 1
30 = 14	3 = 1
40 = 19	4 = 2
50 = 24	5 = 2
	6 = 3
	7 = 3
	8 = 4
	9 = 4

27 = 13

Anti Log Excesses

10 = 21	1 = 2
20 = 42	2 = 4
	3 = 6
	4 = 8
	5 = 10
	6 = 13
	7 = 15
	8 = 17
	9 = 19

13 = 27

9110050	·9595207
105	234
160	260
215	286
269	312
324	338
379	364
434	391
489	417
544	443
599	469
654	495
708	521
763	548
818	574
873	600
928	626
983	652
9111038	678
093	704
148	731
202	757
257	783
312	809
367	835
422	861
477	888
532	914
587	940
641	966
696	992
751	·9596018
806	044
861	071

9111916	·9596097
971	123
9112026	149
080	175
135	201
190	228
245	254
300	280
355	306
410	332
464	358
519	384
574	411
629	437
684	463
739	489
794	515
849	541
903	568
958	594
9113013	620
068	646
123	672
178	698
233	724
288	751
342	777
397	803
452	829
507	855
562	881
617	907
672	934
726	960
781	986
836	·9597012
891	038
946	064
9114001	090
056	117
110	143
165	169
220	195
275	221
330	247
385	273
440	300
494	326
549	352
604	378
659	404
714	430
769	456
824	483
878	509
933	535
988	561
9115043	587
098	613
153	639
208	665
262	692
317	718
372	744
427	770

9115482	·9597796
537	822
592	848
646	875
701	901
756	927
811	953
866	979
921	·9598005
975	031
9116030	057
085	084
140	110
195	136
250	162
305	188
359	214
414	240
469	266
524	293
579	319
634	345
688	371
743	397
798	423
853	449
908	475
963	502
9117018	528
072	554
127	580
182	606
237	632
292	658
347	684
401	711
456	737
511	763
566	789
621	815
676	841
730	867
785	893
840	920
895	946
950	972
9118005	998
060	·9599024
114	050
169	076
224	102
279	128
334	155
389	181
443	207
498	233
553	259
608	285
663	311
718	337
772	363
827	390
882	416
937	442
992	468

9119047	·9599494
101	520
156	546
211	572
266	599
321	625
375	651
430	677
485	703
540	729
595	755
650	781
704	807
759	834
814	860
869	886
924	912
979	938

Excesses

10 = 5	1 = 0
20 = 10	2 = 1
30 = 14	3 = 1
40 = 19	4 = 2
50 = 24	5 = 2
	6 = 3
	7 = 3
	8 = 4
	9 = 4

27 = 13

Anti Log Excesses

10 = 21	1 = 2
20 = 42	2 = 4
	3 = 6
	4 = 8
	5 = 10
	6 = 13
	7 = 15
	8 = 17
	9 = 19

13 = 27

9120033	·9599964
088	990
143	·9600016
198	042
253	069
308	095
362	121
417	147
472	173
527	199
582	225
636	251
691	277
746	303
801	330
856	356
911	382
965	408

9121020	·9600434
075	460
130	486
185	512
239	538
294	564
349	591
404	617
459	643
514	669
568	695
623	721
678	747
733	773
788	799
842	825
897	852
952	878
9122007	904
062	930
116	956
171	982
226	·9601008
281	034
336	060
391	086
445	112
500	139
555	165
610	191
665	217
719	243
774	269
829	295
884	321
939	347
993	373
9123048	399
103	426
158	452
213	478
267	504
322	530
377	556
432	582
487	608
541	634
596	660
651	686
706	713
761	739
815	765
870	791
925	817
980	843
9124035	869
089	895
144	921
199	947
254	973
309	999
363	·9602025
418	052
473	078
528	104

9124583	·9602130
637	156
692	182
747	208
802	234
857	260
911	286
966	313
9125021	339
076	365
131	391
185	417
240	443
295	469
350	495
405	521
459	547
514	573
569	599
624	625
679	651
733	678
788	704
843	730
898	756
953	782
9126007	808
062	834
117	860
172	886
226	912
281	938
336	964
391	990
446	·9603016
500	043
555	069
610	095
665	121
720	147
774	173
829	199
884	225
939	251
993	277
9127048	303
103	329
158	355
213	381
267	407
322	434
377	460
432	486
486	512
541	538
596	564
651	590
706	616
760	642
815	668
870	694
925	720
979	746
9128034	772
089	798

9128144	·9603824
199	850
253	877
308	903
363	929
418	955
472	981
527	·9604007
582	033
637	059
691	085
746	111
801	137
856	163
911	189
965	215
9129020	241
075	267
130	293
184	319
239	346
294	372
349	398
404	424
458	450
513	476
568	502
623	528
677	554
732	580
787	606
842	632
896	658
951	684

Excesses

10 = 5	1 = 0
20 = 10	2 = 1
30 = 14	3 = 1
40 = 19	4 = 2
50 = 24	5 = 2
	6 = 3
	7 = 3
	8 = 4
	9 = 4

27 = 13

Anti Log Excesses

10 = 21	1 = 2
20 = 42	2 = 4
	3 = 6
	4 = 8
	5 = 11
	6 = 13
	7 = 15
	8 = 17
	9 = 19

13 = 27

9130006	·9604710
061	736

9130116	·9604762
170	789
225	815
280	841
335	867
389	893
444	919
499	945
554	971
608	997
663	·9605023
718	049
773	075
827	101
882	127
937	153
992	179
9131046	205
101	231
156	257
211	283
265	310
320	336
375	362
430	388
485	414
539	440
594	466
649	492
704	518
758	544
813	570
868	596
923	622
977	648
9132032	674
087	700
142	726
196	752
251	778
306	804
361	830
415	856
470	882
525	908
580	934
634	961
689	987
744	·9606013
799	039
853	065
908	091
963	117
9133018	143
072	169
127	195
182	221
237	247
291	273
346	299
401	325
455	351
510	377
565	403
620	429

9133674	·9606455	9137232	·9608146	Anti Log Excesses		9142812	·9610798	9146366	·9612486
729	481	287	172	10 = 21	1 = 2	867	824	421	512
784	507	341	198	20 = 42	2 = 4	921	850	475	537
839	533	396	224		3 = 6	976	876	530	563
893	559	451	250		4 = 8	9143031	902	585	589
948	585	506	276		5 = 11	085	928	639	615
9134003	611	560	302		6 = 13	140	954	694	641
058	637	615	328		7 = 15	195	980	749	667
112	663	670	354		8 = 17	249	·9611006	803	693
167	689	725	380		9 = 19	304	032	858	719
222	715	779	406			359	058	913	745
277	742	834	432			413	083	967	771
331	768	889	458	13 = 27		468	109	9147022	797
386	794	943	485			523	135	077	823
441	820	998	511	9140022	·9609472	577	161	131	849
496	846	9138053	537	077	498	632	187	186	875
550	872	108	563	132	524	687	213	241	901
605	898	162	589	186	550	742	239	295	927
660	924	217	615	241	576	796	265	350	953
715	950	272	641	296	602	851	291	405	979
769	976	326	667	351	628	906	317	459	·9613005
824	·9607002	381	693	405	654	960	343	514	031
879	028	436	719	460	680	9144015	369	568	057
933	054	491	745	515	706	070	395	623	082
988	080	545	771	569	732	124	421	678	108
9135043	106	600	797	624	758	179	447	732	134
098	132	655	823	679	784	234	473	787	160
152	158	709	849	734	810	288	499	842	186
207	184	764	875	788	836	343	525	896	212
262	210	819	901	843	862	398	551	951	238
317	236	874	927	898	888	452	577	9148006	264
371	262	928	953	952	914	507	603	060	290
426	288	983	979	9141007	940	562	629	115	316
481	314	9139038	·9609005	062	966	616	655	170	342
536	340	092	031	116	992	671	681	224	368
590	366	147	057	171	·9610018	726	707	279	394
645	392	202	083	226	044	780	733	334	420
700	418	256	109	280	070	835	759	388	446
754	444	311	135	335	096	890	785	443	472
809	470	366	161	390	122	944	811	498	498
864	496	421	187	445	148	999	837	552	524
919	522	475	213	499	174	9145054	863	607	550
973	548	530	239	554	200	108	888	662	575
9136028	574	585	265	609	226	163	914	716	601
083	600	639	291	663	252	218	940	771	627
138	626	694	317	718	278	273	966	826	653
192	652	749	343	773	304	327	992	880	679
247	678	804	369	827	330	382	·9612018	935	705
302	704	858	395	882	356	437	044	990	731
356	730	913	420	937	382	491	070	9149044	757
411	756	968	446	992	408	546	096	099	783
466	782	Excesses		9142046	434	601	122	153	809
521	808			101	460	655	148	208	835
575	834	10 = 5	1 = 0	156	486	710	174	263	861
630	860	20 = 10	2 = 1	210	512	765	200	317	887
685	886	30 = 14	3 = 1	265	538	819	226	372	913
740	912	40 = 19	4 = 2	320	564	874	252	427	939
794	938	50 = 24	5 = 2	374	590	929	278	481	965
849	964		6 = 3	429	616	983	304	536	991
904	990		7 = 3	484	642	9146038	330	591	·9614016
958	·9608016		8 = 4	538	668	093	356	645	042
9137013	042		9 = 4	593	694	147	382	700	068
068	068			648	720	202	408	755	094
123	094			703	746	257	434	809	120
177	120	27 = 13		757	772	311	460	864	146

9149919	·9614172
973	198

Excesses

10 = 5	1 = 0
20 = 9	2 = 1
30 = 14	3 = 1
40 = 19	4 = 2
50 = 24	5 = 2
	6 = 3
	7 = 3
	8 = 4
	9 = 4

27 = 13

Anti Log Excesses

10 = 21	1 = 2
20 = 42	2 = 4
	3 = 6
	4 = 8
	5 = 11
	6 = 13
	7 = 15
	8 = 17
	9 = 19

13 = 27

9150028	·9614224
082	250
137	276
192	302
246	328
301	354
356	380
410	405
465	431
520	457
574	483
629	509
684	535
738	561
793	587
847	613
902	639
957	665
9151011	691
066	717
121	743
175	769
230	794
285	820
339	846
394	872
448	898
503	924
558	950
612	976
667	·9615002
722	028
776	054
831	080

9151886	·9615106
940	131
995	157
9152049	183
104	209
159	235
213	261
268	287
323	313
377	339
432	365
487	391
541	417
596	443
650	468
705	494
760	520
814	546
869	572
924	598
978	624
9153033	650
087	676
142	702
197	728
251	754
306	779
361	805
415	831
470	857
524	883
579	909
634	935
688	961
743	987
798	·9616013
852	039
907	065
961	090
9154016	116
071	142
125	168
180	194
234	220
289	246
344	272
398	298
453	324
508	350
562	375
617	401
671	427
726	453
781	479
835	505
890	531
945	557
999	583
9155054	609
108	635
163	660
218	686
272	712
327	738
381	764

9155436	·9616790
491	816
545	842
600	868
654	894
709	920
764	946
818	971
873	997
928	·9617023
982	049
9156037	075
091	101
146	127
201	153
255	179
310	205
364	231
419	256
474	282
528	308
583	334
637	360
692	386
747	412
801	438
856	464
910	490
965	515
9157020	541
074	567
129	593
183	619
238	645
293	671
347	697
402	723
456	749
511	774
566	800
620	826
675	852
729	878
784	904
839	930
893	956
948	982
9158002	·9618007
057	033
112	059
166	085
221	111
275	137
330	163
385	189
439	215
494	241
548	266
603	292
658	318
712	344
767	370
821	396
876	422
931	448

9158985	·9618474
9159040	499
094	525
149	551
204	577
258	603
313	629
367	655
422	681
476	706
531	732
586	758
640	784
695	810
749	836
804	862
859	888
913	914
968	939

Excesses

10 = 5	1 = 0
20 = 9	2 = 1
30 = 14	3 = 1
40 = 19	4 = 2
50 = 24	5 = 2
	6 = 3
	7 = 3
	8 = 4
	9 = 4

27 = 13

Anti Log Excesses

10 = 21	1 = 2
20 = 42	2 = 4
	3 = 6
	4 = 8
	5 = 11
	6 = 13
	7 = 15
	8 = 17
	9 = 19

13 = 27

9160022	·9618965
077	991
132	·9619017
186	043
241	069
295	095
350	121
404	146
459	172
514	198
568	224
623	250
677	276
732	302
786	327
841	353
896	379

9160950	·9619405
9161005	431
059	457
114	483
169	509
223	534
278	560
332	586
387	612
441	638
496	664
551	690
605	716
660	741
714	767
769	793
823	819
878	845
933	871
987	897
9162042	923
096	948
151	974
205	·9620000
260	026
315	052
369	078
424	104
478	129
533	155
587	181
642	207
697	233
751	259
806	285
860	311
915	336
969	362
9163024	388
079	414
133	440
188	466
242	492
297	517
351	543
406	569
461	595
515	621
570	647
624	673
679	698
733	724
788	750
842	776
897	802
952	828
9164006	854
061	880
115	905
170	931
224	957
279	983
334	·9621009
388	035
443	061

9164497	·9621086
552	112
606	138
661	164
715	190
770	216
825	242
879	267
934	293
988	319
9165043	345
097	371
152	397
206	422
261	448
316	474
370	500
425	526
479	552
534	578
588	603
643	629
697	655
752	681
807	707
861	732
916	758
970	784
9166025	810
079	836
134	862
188	888
243	913
297	939
352	965
407	991
461	·9622017
516	043
570	068
625	094
679	120
734	146
788	172
843	198
897	224
952	249
9167007	275
061	301
116	327
170	353
225	379
279	404
334	430
388	456
443	482
497	508
552	534
607	559
661	585
716	611
770	637
825	663
879	689
934	714
988	740

9168043	·9622766
097	792
152	818
206	844
261	869
316	895
370	921
425	947
479	973
534	999
588	·9623024
643	050
697	076
752	102
806	128
861	154
915	179
970	205
9169024	231
079	257
134	283
188	309
243	334
297	360
352	386
406	412
461	438
515	464
570	489
624	515
679	541
733	567
788	593
842	619
897	644
951	670

Excesses

10 = 5	1 = 0
20 = 9	2 = 1
30 = 14	3 = 1
40 = 19	4 = 2
50 = 24	5 = 2
	6 = 3
	7 = 3
	8 = 4
	9 = 4

27 = 13

Anti Log Excesses

10 = 21	1 = 2
20 = 42	2 = 4
	3 = 6
	4 = 8
	5 = 11
	6 = 13
	7 = 15
	8 = 17
	9 = 19

13 = 27

9170006	·9623696
060	722
115	748
170	773
224	799
279	825
333	851
388	877
442	903
497	928
551	954
606	980
660	·9624006
715	032
769	057
824	083
878	109
933	135
987	161
9171042	186
096	212
151	238
205	264
260	290
314	316
369	341
424	367
478	393
533	419
587	445
642	470
696	496
751	522
805	548
860	574
914	600
969	625
9172023	651
078	677
132	703
187	729
241	754
296	780
350	806
405	832
459	858
514	883
568	909
623	935
677	961
732	987
786	·9625012
841	038
895	064
950	090
9173004	116
059	142
113	167
168	193
222	219
277	245
331	271
386	296
440	322
495	348

9173549	·9625374
604	400
658	425
713	451
767	477
822	503
876	529
931	554
985	580
9174040	606
094	632
149	658
203	683
258	709
312	735
367	761
421	787
476	812
530	838
585	864
639	890
694	916
748	941
803	967
857	993
912	·9626019
966	045
9175021	070
075	096
130	122
184	148
239	174
293	199
348	225
402	251
457	277
511	302
566	328
620	354
675	380
729	406
784	431
838	457
893	483
947	509
9176002	535
056	560
111	586
165	612
220	638
274	664
329	689
383	715
438	741
492	767
547	792
601	818
656	844
710	870
765	896
819	921
874	947
928	973
983	999
9177037	·9627025

9177092	·9627050
146	076
201	102
255	128
309	153
364	179
418	205
473	231
527	257
582	282
636	308
691	334
745	360
800	386
854	411
909	437
963	463
9178018	489
072	514
127	540
181	566
236	592
290	617
345	643
399	669
454	695
508	721
562	746
617	772
671	798
726	824
780	849
835	875
889	901
944	927
998	953
9179053	978
107	·9628004
162	030
216	056
271	081
325	107
380	133
434	159
489	184
543	210
597	236
652	262
706	288
761	313
815	339
870	365
924	391
979	416

Excesses

10 = 5	1 = 0
20 = 9	2 = 1
30 = 14	3 = 1
40 = 19	4 = 2
50 = 24	5 = 2
	6 = 3
	7 = 3
	8 = 4
	9 = 4

27 = 13

Anti Log Excesses

10 = 21	1 = 2
20 = 42	2 = 4
	3 = 6
	4 = 8
	5 = 11
	6 = 13
	7 = 15
	8 = 17
	9 = 19

13 = 27

9180033	·9628442
088	468
142	494
197	519
251	545
306	571
360	597
414	622
469	648
523	674
578	700
632	726
687	751
741	777
796	803
850	829
905	854
959	880
9181014	906
068	932
122	958
177	984
231	·9629009
286	035
340	061
395	087
449	112
504	138
558	164
613	190
667	215
721	241
776	267
830	293
885	318
939	344
994	370
9182048	·9629396
103	421
157	447
212	473
266	499
320	524
375	550
429	576
484	602
538	627
593	653
647	679
702	705
756	730
811	756
865	782
919	808
974	833
9183028	859
083	885
137	911
192	936
246	962
301	988
355	·9630014
409	039
464	065
518	091
573	117
627	142
682	168
736	194
791	220
845	245
899	271
954	297
9184008	323
063	348
117	374
172	400
226	426
281	451
335	477
389	503
444	529
498	554
553	580
607	606
662	631
716	657
771	683
825	709
879	734
934	760
988	786
9185043	812
097	837
152	863
206	889
260	915
315	940
369	966
424	992
478	·9631018
533	043
9185587	·9631069
641	095
696	120
750	146
805	172
859	198
914	223
968	249
9186022	275
077	301
131	326
186	352
240	378
295	404
349	429
403	455
458	481
512	507
567	532
621	558
676	584
730	610
784	635
839	661
893	687
948	712
9187002	738
057	764
111	790
165	815
220	841
274	867
329	892
383	918
438	944
492	970
546	995
601	·9632021
655	047
710	073
764	098
819	124
873	150
927	175
982	201
9188036	227
091	253
145	278
199	304
254	330
308	355
363	381
417	407
472	433
526	458
580	484
635	510
689	535
744	561
798	587
852	613
907	638
961	664
9189016	690
070	715
9189125	·9632741
179	767
233	793
288	818
342	844
397	870
451	895
505	921
560	947
614	973
669	998
723	·9633024
777	050
832	075
886	101
941	127
995	153

Excesses

10 = 5	1 = 0
20 = 9	2 = 1
30 = 14	3 = 1
40 = 19	4 = 2
50 = 24	5 = 2
	6 = 3
	7 = 3
	8 = 4
	9 = 4

27 = 13

Anti Log Excesses

10 = 21	1 = 2
20 = 42	2 = 4
	3 = 6
	4 = 8
	5 = 11
	6 = 13
	7 = 15
	8 = 17
	9 = 19

13 = 27

9190049	·9633178
104	204
158	230
213	255
267	281
321	307
376	332
430	358
485	384
539	410
594	435
648	461
702	487
757	512
811	538
866	564
920	590
974	615
9191029	641
9191083	·9633667
138	692
192	718
246	744
301	769
355	795
410	821
464	847
518	872
573	898
627	924
682	950
736	975
790	·9634001
845	027
899	052
954	078
9192008	104
062	129
117	155
171	181
225	207
280	232
334	258
389	284
443	309
497	335
552	361
606	386
661	412
715	438
769	464
824	489
878	515
933	541
987	566
9193041	592
096	618
150	643
205	669
259	695
313	721
368	746
422	772
476	798
531	823
585	849
640	875
694	900
748	926
803	952
857	977
912	·9635003
966	029
9194020	054
075	080
129	106
183	132
238	157
292	183
347	209
401	234
455	260
510	286
564	311

9194618	·9635337
673	363
727	388
782	414
836	440
890	465
945	491
999	517
9195054	543
108	568
162	594
217	620
271	645
325	671
380	697
434	722
489	748
543	774
597	799
652	825
706	851
760	876
815	902
869	928
924	953
978	979
9196032	·9636005
087	030
141	056
195	082
250	107
304	133
358	159
413	185
467	210
522	236
576	261
630	287
685	313
739	338
793	364
848	390
902	416
957	441
9197011	467
065	493
120	518
174	544
228	570
283	595
337	621
391	647
446	672
500	698
555	724
609	749
663	775
718	801
772	826
826	852
881	878
935	903
989	929
9198044	955
098	980

9198152	·9637006
207	032
261	057
316	083
370	109
424	134
479	160
533	186
587	211
642	237
696	263
750	288
805	314
859	340
913	365
968	391
9199022	416
077	442
131	468
185	493
240	519
294	545
348	570
403	596
457	622
511	647
566	673
620	699
674	724
729	750
783	776
837	801
892	827
946	853

Excesses

10 = 5	1 = 0
20 = 9	2 = 1
30 = 14	3 = 1
40 = 19	4 = 2
50 = 24	5 = 2
	6 = 3
	7 = 3
	8 = 4
	9 = 4

27 = 13

Anti Log Excesses

10 = 21	1 = 2
20 = 42	2 = 4
	3 = 6
	4 = 8
	5 = 11
	6 = 13
	7 = 15
	8 = 17
	9 = 19

13 = 27

9200000	·9637878
055	904

9200109	·9637930
164	955
218	981
272	·9638007
327	032
381	058
435	084
490	109
544	135
598	161
653	186
707	212
761	237
816	263
870	289
924	314
979	340
9201033	366
087	391
142	417
196	443
250	468
305	494
359	520
413	545
468	571
522	597
576	622
631	648
685	674
739	699
794	725
848	751
902	776
957	802
9202011	827
065	853
120	879
174	904
228	930
283	956
337	981
391	·9639007
446	033
500	058
554	084
609	110
663	135
717	161
772	186
826	212
880	238
935	263
989	289
9203043	315
098	340
152	366
206	392
261	417
315	443
369	468
424	494
478	520
532	545
587	571

9203641	·9639597
695	622
750	648
804	674
858	699
913	725
967	750
9204021	776
076	802
130	827
184	853
239	879
293	904
347	930
402	955
456	981
510	·9640007
565	032
619	058
673	084
728	109
782	135
836	160
891	186
945	212
999	237
9205053	263
108	289
162	314
216	340
271	365
325	391
379	417
434	442
488	468
542	494
597	519
651	545
705	570
760	596
814	622
868	647
923	673
977	699
9206031	724
085	750
140	775
194	801
248	827
303	852
357	878
411	904
466	929
520	955
574	980
629	·9641006
683	032
737	057
791	083
846	108
900	134
954	160
9207009	185
063	211
117	237

9207172	·9641262
226	288
280	313
335	339
389	365
443	390
497	416
552	442
606	467
660	493
715	518
769	544
823	570
878	595
932	621
986	646
9208040	672
095	698
149	723
203	749
258	774
312	800
366	826
421	851
475	877
529	902
583	928
638	954
692	979
746	·9642005
801	031
855	056
909	082
964	107
9209018	133
072	159
126	184
181	210
235	235
289	261
344	287
398	312
452	338
506	363
561	389
615	415
669	440
724	466
778	491
832	517
886	543
941	568
995	594

Excesses

10 = 5	1 = 0
20 = 9	2 = 1
30 = 14	3 = 1
40 = 19	4 = 2
50 = 24	5 = 2
	6 = 3
	7 = 3
	8 = 4
	9 = 4

27 = 13

Anti Log Excesses

10 = 21	1 = 2
20 = 42	2 = 4
	3 = 6
	4 = 8
	5 = 11
	6 = 13
	7 = 15
	8 = 17
	9 = 19

13 = 27

9210049	·9642619
104	645
158	670
212	696
267	722
321	747
375	773
429	798
484	824
538	850
592	875
647	901
701	926
755	952
809	978
864	·9643003
918	029
972	054
9211027	080
081	106
135	131
189	157
244	182
298	208
352	234
406	259
461	285
515	310
569	336
624	361
678	387
732	413
786	438
841	464
895	489
949	515
9212004	541

9212058	·9643566
112	592
166	618
221	643
275	669
329	694
383	720
438	745
492	771
546	797
601	822
655	848
709	873
763	899
818	925
872	950
926	976
980	·9644001
9213035	027
089	053
143	078
198	104
252	129
306	155
360	180
415	206
469	232
523	257
577	283
632	308
686	334
740	359
795	385
849	411
903	436
957	462
9214012	487
066	513
120	539
174	564
229	590
283	615
337	641
391	666
446	692
500	718
554	743
608	769
663	794
717	820
771	845
826	871
880	897
934	922
988	948
9215043	973
097	999
151	·9645024
205	050
260	076
314	101
368	127
422	152
477	178
531	203

9215585	·9645229
639	255
694	280
748	306
802	331
856	357
911	382
965	408
9216019	434
073	459
128	485
182	510
236	536
290	561
345	587
399	613
453	638
507	664
562	689
616	715
670	740
724	766
779	791
833	817
887	843
941	868
996	894
9217050	919
104	945
158	970
213	996
267	·9646021
321	047
375	073
430	098
484	124
538	149
592	175
647	200
701	226
755	251
809	277
864	303
918	328
972	354
9218026	379
081	405
135	430
189	456
243	481
298	507
352	533
406	558
460	584
514	609
569	635
623	660
677	686
731	711
786	737
840	763
894	788
948	814
9219003	839
057	865

9219111	·9646890
165	916
220	941
274	967
328	992
382	·9647018
436	044
491	069
545	095
599	120
653	146
708	171
762	197
816	222
870	248
925	273
979	299

Excesses

10 = 5	1 = 0
20 = 9	2 = 1
30 = 14	3 = 1
40 = 19	4 = 2
50 = 24	5 = 2
	6 = 3
	7 = 3
	8 = 4
	9 = 4

27 = 13

Anti Log Excesses

10 = 21	1 = 2
20 = 42	2 = 4
	3 = 6
	4 = 8
	5 = 11
	6 = 13
	7 = 15
	8 = 17
	9 = 19

13 = 27

9220033	·9647325
087	350
141	376
196	401
250	427
304	452
358	478
413	503
467	529
521	554
575	580
630	606
684	631
738	657
792	682
846	708
901	733
955	759
9221009	784

9221063	·9647810
118	835
172	861
226	886
280	912
334	938
389	963
443	989
497	·9648014
551	040
606	065
660	091
714	116
768	142
822	167
877	193
931	218
985	244
9222039	269
094	295
148	321
202	346
256	372
310	397
365	423
419	448
473	474
527	499
581	525
636	550
690	576
744	602
798	627
853	653
907	678
961	704
9223015	729
069	755
124	780
178	806
232	831
286	857
340	882
395	908
449	933
503	959
557	984
612	·9649010
666	035
720	061
774	087
828	112
883	138
937	163
991	189
9224045	214
099	240
154	265
208	291
262	316
316	342
370	367
425	393
479	418
533	444

9224587	·9649469
641	495
696	520
750	546
804	571
858	597
912	622
967	648
9225021	673
075	699
129	724
183	750
238	775
292	801
346	827
400	852
454	878
509	903
563	929
617	954
671	980
725	·9650005
780	031
834	056
888	082
942	107
996	133
9226051	158
105	184
159	209
213	235
267	260
322	286
376	311
430	337
484	362
538	388
593	413
647	439
701	464
755	490
809	515
863	541
918	566
972	592
9227026	617
080	643
134	668
189	694
243	719
297	745
351	770
405	796
460	821
514	847
568	872
622	898
676	923
730	949
785	974
839	·9651000
893	025
947	051
9228001	076
056	102

9228110	·9651127
164	153
218	178
272	204
326	229
381	255
435	280
489	306
543	331
597	357
652	382
706	408
760	433
814	459
868	484
922	510
977	535
9229031	561
085	586
139	612
193	637
247	663
302	688
356	714
410	739
464	765
518	790
573	816
627	841
681	867
735	892
789	918
843	943
898	969
952	994

Excesses

10 = 5	1 = 0
20 = 9	2 = 1
30 = 14	3 = 1
40 = 19	4 = 2
50 = 24	5 = 2
	6 = 3
	7 = 3
	8 = 4
	9 = 4

27 = 13

Anti Log Excesses

10 = 21	1 = 2
20 = 42	2 = 4
	3 = 6
	4 = 8
	5 = 11
	6 = 13
	7 = 15
	8 = 17
	9 = 19

13 = 27

9230006	·9652020

9230060	·9652045
114	071
168	096
223	122
277	147
331	172
385	198
439	223
493	249
548	274
602	300
656	325
710	351
764	376
818	402
873	427
927	453
981	478
9231035	504
089	529
143	555
198	580
252	606
306	631
360	657
414	682
468	708
523	733
577	759
631	784
685	809
739	835
793	860
848	886
902	911
956	937
9232010	962
064	988
118	·9653013
173	039
227	064
281	090
335	115
389	141
443	166
497	192
552	217
606	243
660	268
714	293
768	319
822	345
877	370
931	396
985	421
9233039	447
093	472
147	497
201	523
256	548
310	574
364	599
418	625
472	650
526	676

9233581	·9653701
635	727
689	752
743	778
797	803
851	829
905	854
960	880
9234014	905
068	930
122	956
176	981
230	·9654007
284	032
339	058
393	083
447	109
501	134
555	160
609	185
663	211
718	236
772	262
826	287
880	312
934	338
988	363
9235042	389
097	414
151	440
205	465
259	491
313	516
367	542
421	567
476	593
530	618
584	643
638	669
692	694
746	720
800	745
855	771
909	796
963	822
9236017	847
071	873
125	898
179	923
233	949
288	974
342	·9655000
396	025
450	051
504	076
558	102
612	127
667	153
721	178
775	203
829	229
883	254
937	280
991	305
9237045	331

9237100	·9655356
154	382
208	407
262	433
316	458
370	483
424	509
479	534
533	560
587	585
641	611
695	636
749	662
803	687
857	712
912	738
966	763
9238020	789
074	814
128	840
182	865
236	890
290	916
345	941
399	967
453	992
507	·9656018
561	043
615	069
669	094
723	119
777	145
832	170
886	196
940	221
994	247
9239048	272
102	298
156	323
210	348
265	374
319	399
373	425
427	450
481	476
535	501
589	526
643	552
697	577
752	603
806	628
860	654
914	679
968	704

Excesses

10 = 5	1 = 0
20 = 9	2 = 1
30 = 14	3 = 1
40 = 19	4 = 2
50 = 24	5 = 2
	6 = 3
	7 = 3
	8 = 4
	9 = 4

27 = 13

Anti Log Excesses

10 = 21	1 = 2
20 = 43	2 = 4
	3 = 6
	4 = 9
	5 = 11
	6 = 13
	7 = 15
	8 = 17
	9 = 19

13 = 27

9240022	·9656730
076	755
130	781
184	806
239	832
293	857
347	883
401	908
455	933
509	959
563	984
617	·9657010
671	035
726	061
780	086
834	111
888	137
942	162
996	188
9241050	213
104	239
158	264
213	289
267	315
321	340
375	366
429	391
483	416
537	442
591	467
645	493
699	518
754	544
808	569
862	594
916	620
970	645

9242024	·9657671
078	696
132	722
186	747
240	772
295	798
349	823
403	849
457	874
511	899
565	925
619	950
673	976
727	·9658001
781	027
836	052
890	077
944	103
998	128
9243052	154
106	179
160	205
214	230
268	255
322	281
377	306
431	332
485	357
539	383
593	408
647	433
701	459
755	484
809	510
863	535
917	560
972	586
9244026	611
080	637
134	662
188	687
242	713
296	738
350	764
404	789
458	815
512	840
566	865
621	891
675	916
729	942
783	967
837	992
891	·9659018
945	043
999	069
9245053	094
107	119
161	145
215	170
270	196
324	221
378	246
432	272
486	297

9245540	·9659323
594	348
648	373
702	399
756	424
810	450
864	475
919	500
973	526
9246027	551
081	577
135	602
189	627
243	653
297	678
351	704
405	729
459	754
513	780
567	805
622	831
676	856
730	881
784	907
838	932
892	958
946	983
9247000	·9660008
054	034
108	059
162	084
216	110
270	135
324	161
379	186
433	211
487	237
541	262
595	288
649	313
703	338
757	364
811	389
865	415
919	440
973	465
9248027	491
081	516
135	541
190	567
244	592
298	618
352	643
406	669
460	694
514	719
568	745
622	770
676	796
730	821
784	846
838	872
892	897
946	922
9249000	948

9249055	·9660973
109	999
163	·9661024
217	049
271	075
325	100
379	126
433	151
487	176
541	202
595	227
649	252
703	278
757	303
811	329
865	354
919	379
973	405

Excesses

10 = 5	1 = 0
20 = 9	2 = 1
30 = 14	3 = 1
40 = 19	4 = 2
50 = 23	5 = 2
	6 = 3
	7 = 3
	8 = 4
	9 = 4

27 = 13

Anti Log Excesses

10 = 21	1 = 2
20 = 43	2 = 4
	3 = 6
	4 = 9
	5 = 11
	6 = 13
	7 = 15
	8 = 17
	9 = 19

13 = 27

9250028	·9661430
082	456
136	481
190	506
244	532
298	557
352	582
406	608
460	633
514	659
568	684
622	709
676	735
730	760
784	785
838	811
892	836
946	862

9251000	·9661887
054	912
109	938
163	963
217	988
271	·9662014
325	039
379	065
433	090
487	115
541	141
595	166
649	191
703	217
757	242
811	268
865	293
919	318
973	344
9252027	369
081	394
135	420
189	445
243	470
298	496
352	521
406	547
460	572
514	597
568	623
622	648
676	673
730	699
784	724
838	749
892	775
946	800
9253000	826
054	851
108	876
162	902
216	927
270	952
324	978
378	·9663003
432	028
486	054
540	079
594	104
648	130
702	155
756	180
811	206
865	231
919	257
973	282
9254027	307
081	333
135	358
189	383
243	409
297	434
351	459
405	485
459	510

9254513	·9663535
567	561
621	586
675	611
729	637
783	662
837	688
891	713
945	738
999	764
9255053	789
107	814
161	840
215	865
269	890
323	916
377	941
431	966
485	992
539	·9664017
593	042
647	068
701	093
755	118
809	144
864	169
918	195
972	220
9256026	245
080	271
134	296
188	321
242	347
296	372
350	397
404	423
458	448
512	473
566	499
620	524
674	549
728	575
782	600
836	625
890	651
944	676
998	701
9257052	727
106	752
160	777
214	803
268	828
322	853
376	879
430	904
484	929
538	955
592	980
646	·9665005
700	031
754	056
808	081
862	107
916	132
970	157

9258024	·9665183
078	208
132	233
186	259
240	284
294	309
348	335
402	360
456	385
510	411
564	436
618	461
672	487
726	512
780	537
834	563
888	588
942	613
996	639
9259050	664
104	689
158	715
212	740
266	765
320	791
374	816
428	841
482	867
536	892
590	917
644	943
698	968
752	993
806	·9666019
860	044
914	069
968	095

Excesses

10 = 5   1 = 0
20 = 9   2 = 1
30 = 14   3 = 1
40 = 19   4 = 2
50 = 23   5 = 2
6 = 3
7 = 3
8 = 4
9 = 4

27 = 13

Anti Log Excesses

10 = 21   1 = 2
20 = 43   2 = 4
3 = 6
4 = 9
5 = 11
6 = 13
7 = 15
8 = 17
9 = 19

13 = 27

9260022	·9666120
076	145
130	171
184	196
238	221
292	247
346	272
400	297
454	323
508	348
562	373
616	399
670	424
724	449
778	475
832	500
886	525
940	551
994	576
9261048	601
102	627
156	652
210	677
264	703
318	728
372	753
426	778
480	804
534	829
588	854
642	880
696	905
750	930
804	956
858	981
912	·9667006
966	032
9262020	057
074	082
128	108
182	133
236	158
290	184
344	209
398	234
452	259
506	285
560	310
614	335
668	361
721	386

9262775	·9667411
829	437
883	462
937	487
991	513
9263045	538
099	563
153	589
207	614
261	639
315	664
369	690
423	715
477	740
531	766
585	791
639	816
693	842
747	867
801	892
855	918
909	943
963	968
9264017	993
071	·9668019
125	044
179	069
233	095
287	120
341	145
395	171
449	196
503	221
557	246
611	272
665	297
719	322
772	348
826	373
880	398
934	424
988	449
9265042	474
096	499
150	525
204	550
258	575
312	601
366	626
420	651
474	677
528	702
582	727
636	752
690	778
744	803
798	828
852	854
906	879
960	904
9266014	929
068	955
122	980
176	·9669005
230	031

9266283	·9669056
337	081
391	107
445	132
499	157
553	182
607	208
661	233
715	258
769	284
823	309
877	334
931	359
985	385
9267039	410
093	435
147	461
201	486
255	511
309	536
363	562
417	587
470	612
524	638
578	663
632	688
686	713
740	739
794	764
848	789
902	815
956	840
9268010	865
064	890
118	916
172	941
226	966
280	992
334	·9670017
388	042
442	067
496	093
549	118
603	143
657	168
711	194
765	219
819	244
873	270
927	295
981	320
9269035	345
089	371
143	396
197	421
251	446
305	472
359	497
413	522
467	547
520	573
574	598
628	623
682	649
736	674

9269790	·9670699
844	724
898	750
952	775

Excesses

10 = 5	1 = 0
20 = 9	2 = 1
30 = 14	3 = 1
40 = 19	4 = 2
50 = 23	5 = 2
	6 = 3
	7 = 3
	8 = 4
	9 = 4

27 = 13

Anti Log Excesses

10 = 21	1 = 2
20 = 43	2 = 4
	3 = 6
	4 = 9
	5 = 11
	6 = 13
	7 = 15
	8 = 17
	9 = 19

13 = 27

9270006	·9670800
060	825
114	851
168	876
222	901
276	926
330	952
383	977
437	·9671002
491	028
545	053
599	078
653	103
707	129
761	154
815	179
869	204
923	230
977	255
9271031	280
085	305
139	331
192	356
246	381
300	407
354	432
408	457
462	482
516	508
570	533
624	558
678	583

9271732	·9671609
786	634
840	659
894	684
947	710
9272001	735
055	760
109	785
163	811
217	836
271	861
325	886
379	912
433	937
487	962
541	987
595	·9672013
648	038
702	063
756	088
810	114
864	139
918	164
972	190
9273026	215
080	240
134	265
188	291
242	316
295	341
349	366
403	392
457	417
511	442
565	467
619	493
673	518
727	543
781	568
835	594
889	619
942	644
996	669
9274050	695
104	720
158	745
212	770
266	795
320	821
374	846
428	871
482	896
536	922
589	947
643	972
697	997
751	·9673023
805	048
859	073
913	098
967	124
9275021	149
075	174
129	199
182	225

9275236	·9673250
290	275
344	300
398	325
452	351
506	376
560	401
614	426
668	452
721	477
775	502
829	527
883	553
937	578
991	603
9276045	628
099	654
153	679
207	704
261	729
314	754
368	780
422	805
476	830
530	855
584	881
638	906
692	931
746	956
800	982
853	·9674007
907	032
961	057
9277015	083
069	108
123	133
177	158
231	183
285	209
338	234
392	259
446	284
500	310
554	335
608	360
662	385
716	410
770	436
824	461
877	486
931	511
985	537
9278039	562
093	587
147	612
201	637
255	663
309	688
362	713
416	738
470	764
524	789
578	814
632	839
686	864

9278740	·9674890
794	915
847	940
901	965
955	991
9279009	·9675016
063	041
117	066
171	091
225	117
278	142
332	167
386	192
440	218
494	243
548	268
602	293
656	319
710	344
763	369
817	394
871	419
925	445
979	470

Excesses

10 = 5	1 = 0
20 = 9	2 = 1
30 = 14	3 = 1
40 = 19	4 = 2
50 = 23	5 = 2
	6 = 3
	7 = 3
	8 = 4
	9 = 4

27 = 13

Anti Log Excesses

10 = 21	1 = 2
20 = 43	2 = 4
	3 = 6
	4 = 9
	5 = 11
	6 = 13
	7 = 15
	8 = 17
	9 = 19

13 = 27

9280033	·9675495
087	520
141	545
194	571
248	596
302	621
356	646
410	671
464	697
518	722
572	747
625	772

9280679	·9675798
733	823
787	848
841	873
895	898
949	924
9281003	949
056	974
110	999
164	·9676024
218	050
272	075
326	100
380	125
434	150
487	176
541	201
595	226
649	251
703	276
757	302
811	327
865	352
918	377
972	402
9282026	428
080	453
134	478
188	503
242	528
295	554
349	579
403	604
457	629
511	654
565	680
619	705
673	730
726	755
780	780
834	806
888	831
942	856
996	881
9283050	906
103	932
157	957
211	982
265	·9677007
319	032
373	058
427	083
480	108
534	133
588	158
642	184
696	209
750	234
804	259
857	284
911	309
965	335
9284019	360
073	385
127	410

9284181	·9677435
234	461
288	486
342	511
396	536
450	561
504	587
558	612
611	637
665	662
719	687
773	712
827	738
881	763
935	788
988	814
9285042	839
096	864
150	889
204	914
258	940
311	965
365	990
419	·9678015
473	040
527	065
581	091
635	116
688	141
742	166
796	191
850	217
904	242
958	267
9286011	292
065	317
119	343
173	368
227	393
281	418
335	443
388	468
442	494
496	519
550	544
604	569
658	594
711	619
765	645
819	670
873	695
927	720
981	745
9287034	771
088	796
142	821
196	846
250	871
304	896
357	922
411	947
465	972
519	997
573	·9679022
627	047

9287681	·9679073
734	098
788	123
842	148
896	173
950	199
9288004	224
057	249
111	274
165	299
219	324
273	350
327	375
380	400
434	425
488	450
542	475
596	501
649	526
703	551
757	576
811	601
865	626
919	652
972	677
9289026	702
080	727
134	752
188	777
242	803
295	828
349	853
403	878
457	903
511	928
565	954
618	979
672	·9680004
726	029
780	054
834	079
887	104
941	130
995	155

Excesses

10 = 5	1 = 0
20 = 9	2 = 1
30 = 14	3 = 1
40 = 19	4 = 2
50 = 23	5 = 2
	6 = 3
	7 = 3
	8 = 4
	9 = 4

27 = 13

Anti Log Excesses

10 = 21	1 = 2
20 = 43	2 = 4
	3 = 6
	4 = 9
	5 = 11
	6 = 13
	7 = 15
	8 = 17
	9 = 19

13 = 27

9290049	·9680180
103	205
157	230
210	255
264	281
318	306
372	331
426	356
479	381
533	406
587	432
641	457
695	482
749	507
802	532
856	557
910	582
964	608
9291018	633
071	658
125	683
179	708
233	733
287	759
341	784
394	809
448	834
502	859
556	884
610	909
663	935
717	960
771	985
825	·9681010
879	035
932	060
986	086
9292040	111
094	136
148	161
202	186
255	211
309	236
363	262
417	287
471	312
524	337
578	362
632	387
686	412
740	438

9292793	·9681463
847	488
901	513
955	538
9293009	563
062	588
116	614
170	639
224	664
278	689
331	714
385	739
439	764
493	790
547	815
600	840
654	865
708	890
762	915
816	940
869	966
923	991
977	·9682016
9294031	041
085	066
138	091
192	116
246	141
300	167
354	192
407	217
461	242
515	267
569	292
623	317
676	343
730	368
784	393
838	418
892	443
945	468
999	493
9295053	519
107	544
161	569
214	594
268	619
322	644
376	669
429	694
483	720
537	745
591	770
645	795
698	820
752	845
806	870
860	896
914	921
967	946
9296021	971
075	996
129	·9683021
183	046
236	071

9296290	·9683097
344	122
398	147
451	172
505	197
559	222
613	247
667	272
720	298
774	323
828	348
882	373
935	398
989	423
9297043	448
097	473
151	499
204	524
258	549
312	574
366	599
419	624
473	649
527	674
581	700
635	725
688	750
742	775
796	800
850	825
903	850
957	875
9298011	901
065	926
119	951
172	976
226	·9684001
280	026
334	051
387	076
441	101
495	127
549	152
603	177
656	202
710	227
764	252
818	277
871	302
925	328
979	353
9299033	378
086	403
140	428
194	453
248	478
302	503
355	528
409	554
463	579
517	604
570	629
624	654
678	679
732	704

9299785	·9684729
839	754
893	780
947	805

Excesses

10 = 5	1 = 0
20 = 9	2 = 1
30 = 14	3 = 1
40 = 19	4 = 2
50 = 23	5 = 2
	6 = 3
	7 = 3
	8 = 4
	9 = 4

27 = 13

Anti Log Excesses

10 = 21	1 = 2
20 = 43	2 = 4
	3 = 6
	4 = 9
	5 = 11
	6 = 13
	7 = 15
	8 = 17
	9 = 19

13 = 27

9300000	·9684830
054	855
108	880
162	905
216	930
269	955
323	980
377	·9685005
431	031
484	056
538	081
592	106
646	131
699	156
753	181
807	206
861	231
914	256
968	281
9301022	307
076	332
129	357
183	382
237	407
291	432
344	457
398	482
452	507
506	532
560	558
613	583
667	608

9301721	·9685633
775	658
828	683
882	708
936	733
990	758
9302043	783
097	808
151	834
205	859
258	884
312	909
366	934
420	959
473	984
527	·9686009
581	034
635	059
688	085
742	110
796	135
850	160
903	185
957	210
9303011	235
065	260
118	285
172	310
226	335
279	361
333	386
387	411
441	436
494	461
548	486
602	511
656	536
709	561
763	586
817	611
871	636
924	662
978	687
9304032	712
086	737
139	762
193	787
247	812
301	837
354	862
408	887
462	912
516	937
569	963
623	988
677	·9687013
730	038
784	063
838	088
892	113
945	138
999	163
9305053	188
107	213
160	238

9305214	·9687264
268	289
322	314
375	339
429	364
483	389
536	414
590	439
644	464
698	489
751	514
805	539
859	564
913	590
966	615
9306020	640
074	665
128	690
181	715
235	740
289	765
342	790
396	815
450	840
504	865
557	890
611	915
665	940
719	966
772	991
826	·9688016
880	041
933	066
987	091
9307041	116
095	141
148	166
202	191
256	216
309	241
363	266
417	291
471	316
524	342
578	367
632	392
685	417
739	442
793	467
847	492
900	517
954	542
9308008	567
062	592
115	617
169	642
223	667
276	692
330	717
384	743
438	768
491	793
545	818
599	843
652	868

9308706	·9688893
760	918
814	943
867	968
921	993
975	·9689018
9309028	043
082	068
136	093
190	118
243	143
297	169
351	194
404	219
458	244
512	269
565	294
619	319
673	344
727	369
780	394
834	419
888	444
941	469
995	494

Excesses

10 = 5	1 = 0
20 = 9	2 = 1
30 = 14	3 = 1
40 = 19	4 = 2
50 = 23	5 = 2
	6 = 3
	7 = 3
	8 = 4
	9 = 4

27 = 13

Anti Log Excesses

10 = 21	1 = 2
20 = 43	2 = 4
	3 = 6
	4 = 9
	5 = 11
	6 = 13
	7 = 15
	8 = 17
	9 = 19

13 = 27

9310049	·9689519
103	544
156	569
210	594
264	619
317	644
371	670
425	695
478	720
532	745
586	770

9310640	·9689795
693	820
747	845
801	870
854	895
908	920
962	945
9311015	970
069	995
123	·9690020
177	045
230	070
284	095
338	121
391	146
445	171
499	196
552	221
606	246
660	271
714	296
767	321
821	346
875	371
928	396
982	421
9312036	446
089	471
143	496
197	521
250	546
304	571
358	596
412	621
465	646
519	672
573	697
626	722
680	747
734	772
787	797
841	822
895	847
948	872
9313002	897
056	922
110	947
163	972
217	997
271	·9691022
324	047
378	072
432	097
485	122
539	147
593	172
646	197
700	222
754	247
807	272
861	297
915	322
969	347
9314022	373
076	398

9314130	·9691423
183	448
237	473
291	498
344	523
398	548
452	573
505	598
559	623
613	648
666	673
720	698
774	723
827	748
881	773
935	798
988	823
9315042	848
096	873
149	898
203	923
257	948
310	973
364	998
418	·9692023
472	048
525	073
579	098
633	123
686	148
740	173
794	198
847	223
901	248
955	273
9316008	298
062	323
116	348
169	373
223	398
277	423
330	448
384	473
438	498
491	523
545	548
599	574
652	599
706	624
760	649
813	674
867	699
921	724
974	749
9317028	774
082	799
135	824
189	849
243	874
296	899
350	924
404	949
457	974
511	999
565	·9693024

9317618	·9693049
672	074
726	099
779	124
833	149
887	174
940	199
994	224
9318048	249
101	274
155	299
208	324
262	349
316	374
369	399
423	424
477	449
530	474
584	499
638	524
691	549
745	574
799	599
852	624
906	649
960	674
9319013	699
067	724
121	749
174	774
228	799
282	824
335	849
389	874
443	899
496	924
550	949
603	974
657	999
711	·9694024
764	049
818	074
872	099
925	124
979	149

Excesses

10 = 5	1 = 0
20 = 9	2 = 1
30 = 14	3 = 1
40 = 19	4 = 2
50 = 23	5 = 2
	6 = 3
	7 = 3
	8 = 4
	9 = 4

27 = 13

Anti Log Excesses

10 = 21	1 = 2
20 = 43	2 = 4
	3 = 6
	4 = 9
	5 = 11
	6 = 13
	7 = 15
	8 = 17
	9 = 19

13 = 27

9320033	·9694174
086	199
140	224
194	249
247	274
301	299
355	324
408	349
462	374
516	399
569	424
623	449
676	474
730	499
784	524
837	549
891	574
945	599
998	624
9321052	649
106	674
159	699
213	724
267	749
320	774
374	799
427	824
481	849
535	874
588	899
642	924
696	949
749	974
803	999
857	·9695024
910	049
964	074
9322017	099
071	124
125	149
178	174
232	199
286	224
339	249
393	274
447	299
500	324
554	349
607	374
661	399
715	424

9322768	·9695449
822	474
876	499
929	524
983	549
9323036	574
090	599
144	624
197	649
251	674
305	698
358	723
412	748
466	773
519	798
573	823
626	848
680	873
734	898
787	923
841	948
895	973
948	998
9324002	·9696023
055	048
109	073
163	098
216	123
270	148
324	173
377	198
431	223
484	248
538	273
592	298
645	323
699	348
753	373
806	398
860	423
913	448
967	473
9325021	498
074	523
128	548
181	573
235	598
289	623
342	648
396	673
450	698
503	723
557	748
610	773
664	798
718	822
771	847
825	872
878	897
932	922
986	947
9326039	972
093	997
147	·9697022
200	047

9326254	·9697072
307	097
361	122
415	147
468	172
522	197
575	222
629	247
683	272
736	297
790	322
843	347
897	372
951	397
9327004	421
058	446
112	471
165	496
219	521
272	546
326	571
380	596
433	621
487	646
540	671
594	696
648	721
701	746
755	771
808	796
862	821
916	846
969	871
9328023	896
076	921
130	946
184	971
237	996
291	·9698020
344	045
398	070
452	095
505	120
559	145
612	170
666	195
720	220
773	245
827	270
880	295
934	320
988	345
9329041	370
095	395
148	420
202	445
256	470
309	495
363	520
416	544
470	569
524	594
577	619
631	644
684	669

9329738	·9698694
791	719
845	744
899	769
952	794

Excesses

10 = 5	1 = 0
20 = 9	2 = 1
30 = 14	3 = 1
40 = 19	4 = 2
50 = 23	5 = 2
	6 = 3
	7 = 3
	8 = 4
	9 = 4

27 = 12

Anti Log Excesses

10 = 21	1 = 2
20 = 43	2 = 4
	3 = 6
	4 = 9
	5 = 11
	6 = 13
	7 = 15
	8 = 17
	9 = 19

12 = 27

9330006	·9698819
059	844
113	869
167	894
220	919
274	944
327	969
381	994
435	·9699018
488	043
542	068
595	093
649	118
703	143
756	168
810	193
863	218
917	243
970	268
9331024	293
078	318
131	343
185	368
238	393
292	418
346	442
399	467
453	492
506	517
560	542
613	567

9331667	·9699592
721	617
774	642
828	667
881	692
935	717
988	742
9332042	767
096	791
149	816
203	841
256	866
310	891
364	916
417	941
471	966
524	991
578	·9700016
631	041
685	066
739	091
792	116
846	141
899	165
953	190
9333006	215
060	240
114	265
167	290
221	315
274	340
328	365
381	390
435	415
489	440
542	465
596	490
649	514
703	539
756	564
810	589
864	614
917	639
971	664
9334024	689
078	714
131	739
185	764
239	789
292	814
346	838
399	863
453	888
506	913
560	938
614	963
667	988
721	·9701013
774	038
828	063
881	088
935	113
988	137
9335042	162
096	187

9335149	·9701212
203	237
256	262
310	287
363	312
417	337
470	362
524	387
578	412
631	436
685	461
738	486
792	511
845	536
899	561
953	586
9336006	611
060	636
113	661
167	686
220	710
274	735
327	760
381	785
435	810
488	835
542	860
595	885
649	910
702	935
756	960
809	984
863	·9702009
916	034
970	059
9337024	084
077	109
131	134
184	159
238	184
291	209
345	234
398	259
452	283
506	308
559	333
613	358
666	383
720	408
773	433
827	458
880	483
934	508
987	532
9338041	557
095	582
148	607
202	632
255	657
309	682
362	707
416	732
469	757
523	782
576	806

9338630	·9702831
683	856
737	881
791	906
844	931
898	956
951	981
9339005	·9703006
058	031
112	055
165	080
219	105
272	130
326	155
379	180
433	205
487	230
540	255
594	279
647	304
701	329
754	354
808	379
861	404
915	429
968	454

Excesses

10 = 5	1 = 0
20 = 9	2 = 1
30 = 14	3 = 1
40 = 19	4 = 2
50 = 23	5 = 2
	6 = 3
	7 = 3
	8 = 4
	9 = 4

27 = 12

Anti Log Excesses

10 = 21	1 = 2
20 = 43	2 = 4
	3 = 6
	4 = 9
	5 = 11
	6 = 13
	7 = 15
	8 = 17
	9 = 19

12 = 27

9340022	·9703479
075	504
129	528
183	553
236	578
290	603
343	628
397	653
450	678
504	703

9340557	·9703728
611	752
664	777
718	802
771	827
825	852
878	877
932	902
985	927
9341039	952
093	976
146	·9704001
200	026
253	051
307	076
360	101
414	126
467	151
521	176
574	200
628	225
681	250
735	275
788	300
842	325
895	350
949	375
9342002	400
056	424
109	449
163	474
217	499
270	524
324	549
377	574
431	599
484	624
538	648
591	673
645	698
698	723
752	748
805	773
859	797
912	822
966	847
9343019	872
073	897
126	922
180	947
233	972
287	996
340	·9705021
394	046
447	071
501	096
554	121
608	146
661	171
715	195
768	220
822	245
875	270
929	295
983	320

9344036	·9705345
090	369
143	394
197	419
250	444
304	469
357	494
411	519
464	544
518	568
571	593
625	618
678	643
732	668
785	693
839	718
892	743
946	767
999	792
9345053	817
106	842
160	867
213	892
267	917
320	941
374	966
427	991
481	·9706016
534	041
588	066
641	091
695	115
748	140
802	165
855	190
909	215
962	240
9346016	265
069	289
123	314
176	339
230	364
283	389
337	414
390	439
444	463
497	488
551	513
604	538
658	563
711	588
765	613
818	637
872	662
925	687
979	712
9347032	737
086	762
139	787
193	811
246	836
300	861
353	886
407	911
460	936

9347514	·9706961
567	985
621	·9707010
674	035
727	060
781	085
834	110
888	135
941	160
995	185
9348048	209
102	234
155	259
209	284
262	309
316	334
369	358
423	383
476	408
530	433
583	458
637	483
690	508
744	532
797	557
851	582
904	607
958	632
9349011	657
065	681
118	706
172	731
225	756
279	781
332	806
386	830
439	855
492	880
546	905
599	930
653	955
706	980
760	·9708004
813	029
867	054
920	079
974	104

Excesses

10 = 5	1 = 0
20 = 9	2 = 1
30 = 14	3 = 1
40 = 19	4 = 2
50 = 23	5 = 2
	6 = 3
	7 = 3
	8 = 4
	9 = 4

27 = 12

Anti Log Excesses

10 = 22	1 = 2
20 = 43	2 = 4
	3 = 6
	4 = 9
	5 = 11
	6 = 13
	7 = 15
	8 = 17
	9 = 19

12 = 27

9350027	·9708129
081	153
134	178
188	203
241	228
295	253
348	278
402	302
455	327
509	352
562	377
615	402
669	427
722	451
776	476
829	501
883	526
936	551
990	576
9351043	600
097	625
150	650
204	675
257	700
311	725
364	749
418	774
471	799
524	824
578	849
631	873
685	898
738	923
792	948
845	973
899	998
952	·9709022
9352006	047
059	072
113	097
166	122
219	147
273	171
326	196
380	221
433	246
487	271
540	296
594	320
647	345
701	370

9352754	·9709395
808	420
861	444
914	469
968	494
9353021	519
075	544
128	569
182	594
235	618
289	643
342	668
396	693
449	718
503	743
556	767
609	792
663	817
716	842
770	867
823	891
877	916
930	941
984	966
9354037	991
091	·9710016
144	040
197	065
251	090
304	115
358	140
411	164
465	189
518	214
572	239
625	264
678	289
732	313
785	338
839	363
892	388
946	413
999	437
9355053	462
106	487
160	512
213	537
266	561
320	586
373	611
427	636
480	661
534	686
587	710
641	735
694	760
747	785
801	810
854	834
908	859
961	884
9356015	909
068	934
121	958
175	983

9356228	·9711008
282	033
335	058
389	082
442	107
496	132
549	157
602	182
656	206
709	231
763	256
816	281
870	306
923	330
977	355
9357030	380
083	405
137	430
190	454
244	479
297	504
351	529
404	554
457	578
511	603
564	628
618	653
671	678
725	702
778	727
831	752
885	777
938	802
992	826
9358045	851
099	876
152	901
205	926
259	950
312	975
366	·9712000
419	025
473	049
526	074
579	099
633	124
686	149
740	173
793	198
847	223
900	248
953	273
9359007	297
060	322
114	347
167	372
221	397
274	421
327	446
381	471
434	496
488	520
541	545
595	570
648	595

9359701	·9712620
755	644
808	669
862	694
915	719
968	744

Excesses

10 = 5	1 = 0
20 = 9	2 = 1
30 = 14	3 = 1
40 = 19	4 = 2
50 = 23	5 = 2
	6 = 3
	7 = 3
	8 = 4
	9 = 4

27 = 12

Anti Log Excesses

10 = 22	1 = 2
20 = 43	2 = 4
	3 = 6
	4 = 9
	5 = 11
	6 = 13
	7 = 15
	8 = 17
	9 = 19

12 = 27

9360022	·9712768
075	793
129	818
182	843
236	867
289	892
342	917
396	942
449	967
503	991
556	·9713016
609	041
663	066
716	091
770	115
823	140
877	165
930	190
983	214
9361037	239
090	264
144	289
197	314
250	338
304	363
357	388
411	413
464	437
517	462
571	487

9361624	·9713512
678	537
731	561
785	586
838	611
891	636
945	660
998	685
9362052	710
105	735
158	759
212	784
265	809
319	834
372	859
425	883
479	908
532	933
586	958
639	982
692	·9714007
746	032
799	057
853	082
906	106
959	131
9363013	156
066	181
120	205
173	230
226	255
280	280
333	304
387	329
440	354
493	379
547	404
600	428
654	453
707	478
760	503
814	527
867	552
921	577
974	602
9364027	627
081	651
134	676
188	701
241	726
294	750
348	775
401	800
455	825
508	849
561	874
615	899
668	924
722	948
775	973
828	998
882	·9715023
935	048
988	072
9365042	097

9365095	·9715122
149	147
202	171
255	196
309	221
362	246
416	270
469	295
522	320
576	345
629	369
683	394
736	419
789	444
843	468
896	493
949	518
9366003	543
056	567
110	592
163	617
216	642
270	666
323	691
377	716
430	741
483	765
537	790
590	815
643	840
697	864
750	889
804	914
857	939
910	963
964	988
9367017	·9716013
070	038
124	062
177	087
231	112
284	137
337	161
391	186
444	211
497	236
551	260
604	285
658	310
711	335
764	359
818	384
871	409
924	434
978	458
9368031	483
085	508
138	533
191	557
245	582
298	607
351	632
405	656
458	681
512	706

9368565	·9716731
618	755
672	780
725	805
778	829
832	854
885	879
939	904
992	928
9369045	953
099	978
152	·9717003
205	027
259	052
312	077
365	102
419	126
472	151
526	176
579	201
632	225
686	250
739	275
792	299
846	324
899	349
952	374

Excesses

10 = 5	1 = 0
20 = 9	2 = 1
30 = 14	3 = 1
40 = 19	4 = 2
50 = 23	5 = 2
	6 = 3
	7 = 3
	8 = 4
	9 = 4

27 = 12

Anti Log Excesses

10 = 22	1 = 2
20 = 43	2 = 4
	3 = 6
	4 = 9
	5 = 11
	6 = 13
	7 = 15
	8 = 17
	9 = 19

12 = 27

9370006	·9717398
059	423
113	448
166	473
219	497
273	522
326	547
379	572
433	596

9370486	·9717621
539	646
593	670
646	695
700	720
753	745
806	769
860	794
913	819
966	844
9371020	868
073	893
126	918
180	942
233	967
286	992
340	·9718017
393	041
446	066
500	091
553	115
607	140
660	165
713	190
767	214
820	239
873	264
927	289
980	313
9372033	338
087	363
140	387
193	412
247	437
300	462
353	486
407	511
460	536
514	560
567	585
620	610
674	635
727	659
780	684
834	709
887	733
940	758
994	783
9373047	808
100	832
154	857
207	882
260	907
314	931
367	956
420	981
474	·9719005
527	030
580	055
634	079
687	104
740	129
794	154
847	178
900	203

9373954	·9719228
9374007	253
060	277
114	302
167	327
220	351
274	376
327	401
380	426
434	450
487	475
540	500
594	524
647	549
700	574
754	598
807	623
860	648
914	673
967	697
9375020	722
074	747
127	771
180	796
234	821
287	845
340	870
394	895
447	920
500	944
554	969
607	994
660	·9720018
714	043
767	068
820	092
874	117
927	142
980	167
9376034	191
087	216
140	241
194	265
247	290
300	315
354	339
407	364
460	389
514	414
567	438
620	463
674	488
727	512
780	537
834	562
887	586
940	611
994	636
9377047	660
100	685
154	710
207	735
260	759
314	784
367	809

9377420	·9720833
474	858
527	883
580	907
633	932
687	957
740	981
793	·9721006
847	031
900	056
953	080
9378007	105
060	130
113	154
167	179
220	204
273	228
327	253
380	278
433	302
487	327
540	352
593	376
646	401
700	426
753	451
806	475
860	500
913	525
966	549
9379020	574
073	599
126	623
180	648
233	673
286	698
339	722
393	747
446	772
499	796
553	821
606	846
659	870
713	895
766	920
819	944
873	969
926	994
979	·9722018

Excesses

10 = 5	1 = 0
20 = 9	2 = 1
30 = 14	3 = 1
40 = 19	4 = 2
50 = 23	5 = 2
	6 = 3
	7 = 3
	8 = 4
	9 = 4

27 = 12

Anti Log Excesses

10 = 22	1 = 2
20 = 43	2 = 4
	3 = 6
	4 = 9
	5 = 11
	6 = 13
	7 = 15
	8 = 17
	9 = 19

12 = 27

9380032	·9722043
086	068
139	092
192	117
246	142
299	167
352	191
406	216
459	241
512	265
566	290
619	315
672	339
725	364
779	389
832	413
885	438
939	463
992	487
9381045	512
099	537
152	561
205	586
258	611
312	635
365	660
418	685
472	709
525	734
578	759
631	783
685	808
738	833
791	857
845	882
898	907
951	931
9382005	956
058	981
111	·9723005
164	030
218	055
271	079
324	104
378	129
431	153
484	178
537	203
591	227
644	252
697	277

9382751	·9723301
804	326
857	351
910	375
964	400
9383017	425
070	449
124	474
177	499
230	523
284	548
337	573
390	597
443	622
497	647
550	671
603	696
657	721
710	745
763	770
816	795
870	819
923	844
976	869
9384029	893
083	918
136	943
189	967
243	992
296	·9724017
349	041
402	066
456	091
509	115
562	140
616	165
669	189
722	214
775	239
829	263
882	288
935	312
989	337
9385042	362
095	386
148	411
202	436
255	460
308	485
361	510
415	534
468	559
521	584
575	608
628	633
681	658
734	682
788	707
841	732
894	756
947	781
9386001	806
054	830
107	855
161	880

9386214	·9724904
267	929
320	953
374	978
427	·9725003
480	027
533	052
587	077
640	101
693	126
746	151
800	175
853	200
906	225
960	249
9387013	274
066	299
119	323
173	348
226	372
279	397
332	422
386	446
439	471
492	496
545	520
599	545
652	570
705	594
758	619
812	644
865	668
918	693
972	717
9388025	742
078	767
131	791
185	816
238	841
291	865
344	890
398	915
451	939
504	964
557	989
611	·9726013
664	038
717	062
770	087
824	112
877	136
930	161
983	186
9389037	210
090	235
143	260
196	284
250	309
303	333
356	358
409	383
463	407
516	432
569	457
622	481

9389676	·9726506
729	530
782	555
835	580
889	604
942	629
995	654

Excesses

10 = 5	1 = 0
20 = 9	2 = 1
30 = 14	3 = 1
40 = 19	4 = 2
50 = 23	5 = 2
	6 = 3
	7 = 3
	8 = 4
	9 = 4

27 = 12

Anti Log Excesses

10 = 22	1 = 2
20 = 43	2 = 4
	3 = 6
	4 = 9
	5 = 11
	6 = 13
	7 = 15
	8 = 17
	9 = 19

12 = 27

9390048	·9726678
102	703
155	727
208	752
261	777
315	801
368	826
421	851
474	875
528	900
581	924
634	949
687	974
741	998
794	·9727023
847	048
900	072
954	097
9391007	121
060	146
113	171
167	195
220	220
273	244
326	269
380	294
433	318
486	343
539	368

9391592	·9727392
646	417
699	441
752	466
805	491
859	515
912	540
965	565
9392018	589
072	614
125	638
178	663
231	688
285	712
338	737
391	761
444	786
498	811
551	835
604	860
657	885
710	909
764	934
817	958
870	983
923	·9728008
977	032
9393030	057
083	081
136	106
190	131
243	155
296	180
349	205
402	229
456	254
509	278
562	303
615	328
669	352
722	377
775	401
828	426
881	451
935	475
988	500
9394041	524
094	549
148	574
201	598
254	623
307	647
361	672
414	697
467	721
520	746
573	770
627	795
680	820
733	844
786	869
840	893
893	918
946	943
999	967

9395052	·9728992
106	·9729016
159	041
212	066
265	090
318	115
372	139
425	164
478	189
531	213
585	238
638	263
691	287
744	312
797	336
851	361
904	385
957	410
9396010	435
064	459
117	484
170	508
223	533
276	558
330	582
383	607
436	631
489	656
542	681
596	705
649	730
702	754
755	779
808	803
862	828
915	853
968	877
9397021	902
075	926
128	951
181	976
234	·9730000
287	025
341	049
394	074
447	099
500	123
553	148
607	172
660	197
713	221
766	246
819	271
873	295
926	320
979	344
9398032	369
085	394
139	418
192	443
245	467
298	492
351	516
405	541
458	566

9398511	·9730590
564	615
617	639
671	664
724	689
777	713
830	738
883	762
937	787
990	811
9399043	836
096	861
149	885
203	910
256	934
309	959
362	983
415	·9731008
469	033
522	057
575	082
628	106
681	131
735	155
788	180
841	205
894	229
947	254

Excesses

10 = 5	1 = 0
20 = 9	2 = 1
30 = 14	3 = 1
40 = 18	4 = 2
50 = 23	5 = 2
	6 = 3
	7 = 3
	8 = 4
	9 = 4

27 = 12

Anti Log Excesses

10 = 22	1 = 2
20 = 43	2 = 4
	3 = 6
	4 = 9
	5 = 11
	6 = 13
	7 = 15
	8 = 17
	9 = 19

12 = 27

9400000	·9731278
054	303
107	328
160	352
213	377
266	401
320	426
373	451

9400426	·9731475
479	500
532	524
586	549
639	573
692	598
745	623
798	647
852	672
905	696
958	721
9401011	745
064	770
117	795
171	819
224	844
277	868
330	893
383	917
437	942
490	966
543	991
596	·9732016
649	040
702	065
756	089
809	114
862	138
915	163
968	188
9402022	212
075	237
128	261
181	286
234	310
287	335
341	360
394	384
447	409
500	433
553	458
607	482
660	507
713	531
766	556
819	581
872	605
926	630
979	654
9403032	679
085	703
138	728
191	753
245	777
298	802
351	826
404	851
457	875
510	900
564	924
617	949
670	974
723	998
776	·9733023
830	047

Number	Log
9403883	·9733072
936	096
989	121
9404042	145
095	170
149	195
202	219
255	244
308	268
361	293
414	317
468	342
521	366
574	391
627	416
680	440
733	465
787	489
840	514
893	538
946	563
999	587
9405052	612
105	636
159	661
212	686
265	710
318	735
371	759
424	784
478	808
531	833
584	857
637	882
690	906
743	931
797	956
850	980
903	·9734005
956	029
9406009	054
062	078
116	103
169	127
222	152
275	176
328	201
381	225
434	250
488	275
541	299
594	324
647	348
700	373
753	397
807	422
860	446
913	471
966	495
9407019	520
072	544
125	569
179	594
232	618
285	643

Number	Log
9407338	·9734667
391	692
444	716
498	741
551	765
604	790
657	814
710	839
763	863
816	888
870	912
923	937
976	962
9408029	986
082	·9735011
135	035
188	060
242	084
295	109
348	133
401	158
454	182
507	207
560	231
614	256
667	280
720	305
773	329
826	354
879	379
932	403
986	428
9409039	452
092	477
145	501
198	526
251	550
304	575
358	599
411	624
464	648
517	673
570	697
623	722
676	746
730	771
783	795
836	820
889	845
942	869
995	894

Excesses

10 = 5	1 = 0
20 = 9	2 = 1
30 = 14	3 = 1
40 = 18	4 = 2
50 = 23	5 = 2
	6 = 3
	7 = 3
	8 = 4
	9 = 4

27 = 12

Anti Log Excesses

10 = 22	1 = 2
20 = 43	2 = 4
	3 = 6
	4 = 9
	5 = 11
	6 = 13
	7 = 15
	8 = 17
	9 = 19

12 = 27

Number	Log
9410048	·9735918
101	943
155	967
208	992
261	·9736016
314	041
367	065
420	090
473	114
527	139
580	163
633	188
686	212
739	237
792	261
845	286
898	310
952	335
9411005	360
058	384
111	409
164	433
217	458
270	482
323	507
377	531
430	556
483	580
536	605
589	629
642	654
695	678
748	703
802	727
855	752
908	776
961	801
9412014	825
067	850
120	874
173	899
227	923
280	948
333	973
386	997
439	·9737022
492	046
545	071
598	095
652	120
705	144

Number	Log
9412758	·9737169
811	193
864	218
917	242
970	267
9413023	291
077	316
130	340
183	365
236	389
289	414
342	438
395	463
448	487
501	512
555	536
608	561
661	585
714	610
767	634
820	659
873	683
926	708
979	732
9414033	757
086	781
139	806
192	830
245	855
298	879
351	904
404	928
457	953
511	977
564	·9738002
617	026
670	051
723	075
776	100
829	124
882	149
935	173
989	198
9415042	222
095	247
148	271
201	296
254	320
307	345
360	369
413	394
467	418
520	443
573	467
626	492
679	516
732	541
785	565
838	590
891	614
944	639
998	663
9416051	688
104	712
157	737

Number	Log
9416210	·9738761
263	786
316	810
369	835
422	859
475	883
529	908
582	932
635	957
688	981
741	·9739006
794	030
847	055
900	079
953	104
9417006	128
059	153
113	177
166	202
219	226
272	251
325	275
378	300
431	324
484	349
537	373
590	398
644	422
697	447
750	471
803	496
856	520
909	545
962	569
9418015	594
068	618
121	642
174	667
228	691
281	716
334	740
387	765
440	789
493	814
546	838
599	863
652	887
705	912
758	936
811	961
865	985
918	·9740010
971	034
9419024	059
077	083
130	108
183	132
236	156
289	181
342	205
395	230
448	254
502	279
555	303
608	328

9419661	·9740352
714	377
767	401
820	426
873	450
926	475
979	499

Excesses

10 = 5	1 = 0
20 = 9	2 = 1
30 = 14	3 = 1
40 = 18	4 = 2
50 = 23	5 = 2
	6 = 3
	7 = 3
	8 = 4
	9 = 4

27 = 12

Anti Log Excesses

10 = 22	1 = 2
20 = 43	2 = 4
	3 = 7
	4 = 9
	5 = 11
	6 = 13
	7 = 15
	8 = 17
	9 = 20

12 = 27

9420032	·9740524
085	548
139	572
192	597
245	621
298	646
351	670
404	695
457	719
510	744
563	768
616	793
669	817
722	842
775	866
828	891
882	915
935	939
988	964
9421041	988
094	·9741013
147	037
200	062
253	086
306	111
359	135
412	160
465	184
518	209

9421571	·9741233
625	258
678	282
731	306
784	331
837	355
890	380
943	404
996	429
9422049	453
102	478
155	502
208	527
261	551
314	576
368	600
421	624
474	649
527	673
580	698
633	722
686	747
739	771
792	796
845	820
898	845
951	869
9423004	894
057	918
110	942
163	967
217	991
270	·9742016
323	040
376	065
429	089
482	114
535	138
588	163
641	187
694	211
747	236
800	260
853	285
906	309
959	334
9424012	358
065	383
118	407
172	431
225	456
278	480
331	505
384	529
437	554
490	578
543	603
596	627
649	652
702	676
755	700
808	725
861	749
914	774
967	798

9425020	·9742823
073	847
126	872
180	896
233	920
286	945
339	969
392	994
445	·9743018
498	043
551	067
604	092
657	116
710	140
763	165
816	189
869	214
922	238
975	263
9426028	287
081	312
134	336
187	360
240	385
294	409
347	434
400	458
453	483
506	507
559	531
612	556
665	580
718	605
771	629
824	654
877	678
930	702
983	727
9427036	751
089	776
142	800
195	825
248	849
301	874
354	898
407	922
460	947
513	971
566	996
619	·9744020
673	045
726	069
779	093
832	118
885	142
938	167
991	191
9428044	216
097	240
150	264
203	289
256	313
309	338
362	362
415	387

9428468	·9744411
521	435
574	460
627	484
680	509
733	533
786	558
839	582
892	606
945	631
998	655
9429051	680
104	704
157	729
210	753
263	777
316	802
370	826
423	851
476	875
529	900
582	924
635	948
688	973
741	997
794	·9745022
847	046
900	071
953	095

Excesses

10 = 5	1 = 0
20 = 9	2 = 1
30 = 14	3 = 1
40 = 18	4 = 2
50 = 23	5 = 2
	6 = 3
	7 = 3
	8 = 4
	9 = 4

27 = 12

Anti Log Excesses

10 = 22	1 = 2
20 = 43	2 = 4
	3 = 7
	4 = 9
	5 = 11
	6 = 13
	7 = 15
	8 = 17
	9 = 20

12 = 27

9430006	·9745119
059	144
112	168
165	193
218	217
271	242
324	266

9430377	·9745290
430	315
483	339
536	364
589	388
642	412
695	437
748	461
801	486
854	510
907	535
960	559
9431013	583
066	608
119	632
172	657
225	681
278	705
331	730
384	754
437	779
490	803
543	828
596	852
649	876
702	901
755	925
808	950
861	974
914	998
967	·9746023
9432020	047
073	072
126	096
179	120
232	145
285	169
338	194
392	218
445	243
498	267
551	291
604	316
657	340
710	365
763	389
816	413
869	438
922	462
975	487
9433028	511
081	535
134	560
187	584
240	609
293	633
346	657
399	682
452	706
505	731
558	755
611	779
664	804
717	828
770	853

9433823	·9746877
876	901
929	926
982	950
9434035	975
088	999
141	·9747023
194	048
247	072
300	097
353	121
406	145
459	170
512	194
565	219
618	243
671	267
724	292
777	316
830	341
883	365
936	389
989	414
9435042	438
095	463
148	487
201	511
254	536
307	560
360	584
413	609
466	633
519	658
571	682
624	706
677	731
730	755
783	780
836	804
889	828
942	853
995	877
9436048	902
101	926
154	950
207	975
260	999
313	·9748023
366	048
419	072
472	097
525	121
578	145
631	170
684	194
737	219
790	243
843	267
896	292
949	316
9437002	340
055	365
108	389
161	414
214	438

9437267	·9748462
320	487
373	511
426	535
479	560
532	584
585	609
638	633
691	657
744	682
797	706
850	731
903	755
956	779
9438009	804
062	828
115	852
168	877
221	901
274	926
327	950
380	974
433	999
486	·9749023
539	047
591	072
644	096
697	120
750	145
803	169
856	194
909	218
962	242
9439015	267
068	291
121	315
174	340
227	364
280	389
333	413
386	437
439	462
492	486
545	510
598	535
651	559
704	584
757	608
810	632
863	657
916	681
969	705

Excesses

10 = 5	1 = 0
20 = 9	2 = 1
30 = 14	3 = 1
40 = 18	4 = 2
50 = 23	5 = 2
	6 = 3
	7 = 3
	8 = 4
	9 = 4

26 = 12

Anti Log Excesses

10 = 22	1 = 2
20 = 43	2 = 4
	3 = 7
	4 = 9
	5 = 11
	6 = 13
	7 = 15
	8 = 17
	9 = 20

12 = 26

9440022	·9749730
075	754
128	778
181	803
234	827
287	852
339	876
392	900
445	925
498	949
551	973
604	998
657	·9750022
710	046
763	071
816	095
869	120
922	144
975	168
9441028	193
081	217
134	241
187	266
240	290
293	314
346	339
399	363
452	387
505	412
558	436
611	461
663	485
716	509
769	534
822	558
875	582
928	607
981	631
9442034	655
087	680
140	704
193	728
246	753
299	777
352	801
405	826
458	850
511	875
564	899
617	923
670	948

9442723	·9750972
776	996
828	·9751021
881	045
934	070
987	094
9443040	118
093	143
146	167
199	191
252	216
305	240
358	264
411	289
464	313
517	337
570	362
623	386
676	410
729	435
782	459
834	483
887	508
940	532
993	557
9444046	581
099	605
152	630
205	654
258	678
311	703
364	727
417	751
470	776
523	800
576	824
629	849
682	873
734	897
787	922
840	946
893	970
946	995
999	·9752019
9445052	043
105	068
158	092
211	116
264	141
317	165
370	189
423	214
476	238
529	262
581	287
634	311
687	335
740	360
793	384
846	409
899	433
952	457
9446005	482
058	506
111	530

9446164	·9752555
217	579
270	603
323	628
375	652
428	676
481	701
534	725
587	749
640	774
693	798
746	822
799	847
852	871
905	895
958	920
9447011	944
064	968
116	993
169	·9753017
222	041
275	066
328	090
381	114
434	139
487	163
540	187
593	211
646	236
699	260
752	284
804	309
857	333
910	357
963	382
9448016	406
069	430
122	455
175	479
228	503
281	528
334	552
387	576
440	601
492	625
545	649
598	674
651	698
704	722
757	747
810	771
863	795
916	820
969	844
9449022	868
075	893
127	917
180	941
233	965
286	990
339	·9754014
392	038
445	063
498	087
551	111

9449604	·9754136
657	160
709	184
762	209
815	233
868	257
921	282
974	306

Excesses

10 = 5	1 = 0
20 = 9	2 = 1
30 = 14	3 = 1
40 = 18	4 = 2
50 = 23	5 = 2
	6 = 3
	7 = 3
	8 = 4
	9 = 4

26 = 12

Anti Log Excesses

10 = 22	1 = 2
20 = 43	2 = 4
	3 = 7
	4 = 9
	5 = 11
	6 = 13
	7 = 15
	8 = 17
	9 = 20

12 = 26

9450027	·9754330
080	355
133	379
186	403
239	427
292	452
344	476
397	500
450	525
503	549
556	573
609	598
662	622
715	646
768	671
821	695
873	719
926	743
979	768
9451032	792
085	816
138	841
191	865
244	889
297	914
350	938
403	962
455	987

9451508	·9755011
561	035
614	059
667	084
720	108
773	132
826	157
879	181
932	205
984	230
9452037	254
090	278
143	303
196	327
249	351
302	375
355	400
408	424
460	448
513	473
566	497
619	521
672	546
725	570
778	594
831	618
884	643
937	667
989	691
9453042	716
095	740
148	764
201	788
254	813
307	837
360	862
413	886
465	910
518	935
571	959
624	983
677	·9756007
730	032
783	056
836	080
889	105
941	129
994	153
9454047	178
100	202
153	226
206	250
259	275
312	299
365	323
417	348
470	372
523	396
576	420
629	445
682	469
735	493
788	518
841	542
893	566

9454946	·9756590
999	615
9455052	639
105	663
158	688
211	712
264	736
316	761
369	785
422	809
475	833
528	858
581	882
634	906
687	931
739	955
792	979
845	·9757003
898	028
951	052
9456004	076
057	101
110	125
163	149
215	173
268	198
321	222
374	246
427	271
480	295
533	319
586	343
638	368
691	392
744	416
797	440
850	465
903	489
956	513
9457008	538
061	562
114	586
167	610
220	635
273	659
326	683
379	708
431	732
484	756
537	780
590	805
643	829
696	853
749	877
802	902
854	926
907	950
960	975
9458013	999
066	·9758023
119	047
172	072
224	096
277	120
330	144

9458383	·9758169
436	193
489	217
542	242
594	266
647	290
700	314
753	339
806	363
859	387
912	411
965	436
9459017	460
070	484
123	509
176	533
229	557
282	581
335	606
387	630
440	654
493	678
546	703
599	727
652	751
705	775
757	800
810	824
863	848
916	873
969	897

Excesses

10 = 5	1 = 0
20 = 9	2 = 1
30 = 14	3 = 1
40 = 18	4 = 2
50 = 23	5 = 2
	6 = 3
	7 = 3
	8 = 4
	9 = 4

26 = 12

Anti Log Excesses

10 = 22	1 = 2
20 = 44	2 = 4
	3 = 7
	4 = 9
	5 = 11
	6 = 13
	7 = 15
	8 = 17
	9 = 20

12 = 26

9460022	·9758921
074	945
127	970
180	994

9460233	·9759018
286	042
339	067
392	091
444	115
497	139
550	164
603	188
656	212
709	236
762	261
814	285
867	309
920	333
973	358
9461026	382
079	406
132	430
184	455
237	479
290	503
343	527
396	552
449	576
501	600
554	625
607	649
660	673
713	697
766	722
819	746
871	770
924	794
977	819
9462030	843
083	867
136	891
188	916
241	940
294	964
347	988
400	·9760013
453	037
505	061
558	085
611	110
664	134
717	158
770	182
822	207
875	231
928	255
981	279
9463034	303
087	328
140	352
192	376
245	400
298	425
351	449
404	473
457	497
509	522
562	546
615	570

9463668	·9760594
721	619
774	643
826	667
879	691
932	716
985	740
9464038	764
091	789
143	813
196	837
249	861
302	886
355	910
408	934
460	958
513	983
566	·9761007
619	031
672	055
724	079
777	104
830	128
883	152
936	176
989	201
9465041	225
094	249
147	273
200	298
253	322
306	346
358	370
411	395
464	419
517	443
570	467
623	492
675	516
728	540
781	564
834	588
887	613
939	637
992	661
9466045	685
098	710
151	734
204	758
256	782
309	807
362	831
415	855
468	879
520	903
573	928
626	952
679	976
732	·9762000
785	025
837	049
890	073
943	097
996	122
9467049	146

9467101	·9762170
154	194
207	218
260	243
313	267
365	291
418	315
471	340
524	364
577	388
630	412
682	436
735	461
788	485
841	509
894	533
946	558
999	582
9468052	606
105	630
158	654
210	679
263	703
316	727
369	751
422	776
475	800
527	824
580	848
633	872
686	897
739	921
791	945
844	969
897	994
950	·9763018
9469003	042
055	066
108	090
161	115
214	139
267	163
319	187
372	212
425	236
478	260
531	284
583	308
636	333
689	357
742	381
795	405
847	430
900	454
953	478

Excesses

10 = 5	1 = 0
20 = 9	2 = 1
30 = 14	3 = 1
40 = 18	4 = 2
50 = 23	5 = 2
	6 = 3
	7 = 3
	8 = 4
	9 = 4

26 = 12

Anti Log Excesses

10 = 22	1 = 2
20 = 44	2 = 4
	3 = 7
	4 = 9
	5 = 11
	6 = 13
	7 = 15
	8 = 17
	9 = 20

12 = 26

9470006	·9763502
059	526
111	551
164	575
217	599
270	623
323	647
375	672
428	696
481	720
534	744
587	769
639	793
692	817
745	841
798	865
851	890
903	914
956	938
9471009	962
062	986
114	·9764011
167	035
220	059
273	083
326	107
378	132
431	156
484	180
537	204
590	229
642	253
695	277
748	301
801	325
854	350
906	374

9471959	·9764398
9472012	422
065	446
117	471
170	495
223	519
276	543
329	567
381	592
434	616
487	640
540	664
593	688
645	713
698	737
751	761
804	785
856	809
909	834
962	858
9473015	882
068	906
120	930
173	955
226	979
279	·9765003
331	027
384	051
437	076
490	100
543	124
595	148
648	172
701	197
754	221
806	245
859	269
912	293
965	318
9474018	342
070	366
123	390
176	414
229	438
281	463
334	487
387	511
440	535
493	559
545	584
598	608
651	632
704	656
756	680
809	705
862	729
915	753
968	777
9475020	801
073	826
126	850
179	874
231	898
284	922
337	946

9475390	·9765971
442	995
495	·9766019
548	043
601	067
654	092
706	116
759	140
812	164
865	188
917	212
970	237
9476023	261
076	285
128	309
181	333
234	358
287	382
339	406
392	430
445	454
498	478
551	503
603	527
656	551
709	575
762	599
814	624
867	648
920	672
973	696
9477025	720
078	744
131	769
184	793
236	817
289	841
342	865
395	890
447	914
500	938
553	962
606	986
658	·9767010
711	035
764	059
817	083
869	107
922	131
975	155
9478028	180
080	204
133	228
186	252
239	276
291	300
344	325
397	349
450	373
502	397
555	421
608	445
661	470
713	494
766	518

9478819	·9767542
872	566
924	590
977	615
9479030	639
083	663
135	687
188	711
241	735
294	760
346	784
399	808
452	832
505	856
557	880
610	905
663	929
716	953
768	977
821	·9768001
874	025
927	050
979	074

Excesses

10 = 5	1 = 0
20 = 9	2 = 1
30 = 14	3 = 1
40 = 18	4 = 2
50 = 23	5 = 2
	6 = 3
	7 = 3
	8 = 4
	9 = 4

26 = 12

Anti Log Excesses

10 = 22	1 = 2
20 = 44	2 = 4
	3 = 7
	4 = 9
	5 = 11
	6 = 13
	7 = 15
	8 = 17
	9 = 20

12 = 26

9480032	·9768098
085	122
138	146
190	170
243	195
296	219
349	243
401	267
454	291
507	315
560	340
612	364
665	388

9480718	·9768412
771	436
823	460
876	485
929	509
981	533
9481034	557
087	581
140	605
192	629
245	654
298	678
351	702
403	726
456	750
509	774
562	799
614	823
667	847
720	871
772	895
825	919
878	944
931	968
983	992
9482036	·9769016
089	040
142	064
194	088
247	113
300	137
353	161
405	185
458	209
511	233
563	257
616	282
669	306
722	330
774	354
827	378
880	402
933	427
985	451
9483038	475
091	499
143	523
196	547
249	571
302	596
354	620
407	644
460	668
513	692
565	716
618	740
671	765
723	789
776	813
829	837
882	861
934	885
987	909
9484040	934
092	958

9484145	·9769982
198	·9770006
251	030
303	054
356	078
409	103
461	127
514	151
567	175
620	199
672	223
725	248
778	272
830	296
883	320
936	344
989	368
9485041	392
094	416
147	440
199	465
252	489
305	513
358	537
410	561
463	585
516	609
568	633
621	658
674	682
727	706
779	730
832	754
885	778
937	802
990	826
9486043	851
096	875
148	899
201	923
254	947
306	971
359	995
412	·9771019
465	044
517	068
570	092
623	116
675	140
728	164
781	188
833	212
886	237
939	261
992	285
9487044	309
097	333
150	357
202	381
255	405
308	430
361	454
413	478
466	502
519	526

9487571	·9771550
624	574
677	598
729	623
782	647
835	671
888	695
940	719
993	743
9488046	767
098	791
151	816
204	840
256	864
309	888
362	912
414	936
467	960
520	984
573	·9772008
625	033
678	057
731	081
783	105
836	129
889	153
941	177
994	201
9489047	226
100	250
152	274
205	298
258	322
310	346
363	370
416	394
468	418
521	443
574	467
626	491
679	515
732	539
784	563
837	587
890	611
943	635
995	660

Excesses

10 = 5	1 = 0
20 = 9	2 = 1
30 = 14	3 = 1
40 = 18	4 = 2
50 = 23	5 = 2
	6 = 3
	7 = 3
	8 = 4
	9 = 4

26 = 12

Anti Log Excesses

10 = 22	1 = 2
20 = 44	2 = 4
	3 = 7
	4 = 9
	5 = 11
	6 = 13
	7 = 15
	8 = 17
	9 = 20

12 = 26

9490048	·9772684
101	708
153	732
206	756
259	780
311	804
364	828
417	852
469	876
522	901
575	925
627	949
680	973
733	997
786	·9773021
838	045
891	070
944	094
996	118
9491049	142
102	166
154	190
207	214
260	238
312	262
365	286
418	311
470	335
523	359
576	383
628	407
681	431
734	455
786	479
839	503
892	528
944	552
997	576
9492050	600
102	624
155	648
208	672
261	696
313	720
366	744
419	769
471	793
524	817
577	841
629	865
682	889

9492735	·9773913
787	937
840	961
893	985
945	·9774010
998	034
9493051	058
103	082
156	106
209	130
261	154
314	178
367	202
419	226
472	250
525	275
577	299
630	323
683	347
735	371
788	395
841	419
893	443
946	467
999	491
9494051	515
104	540
157	564
209	588
262	612
315	636
367	660
420	684
473	708
525	732
578	756
631	780
683	805
736	829
789	853
841	877
894	901
947	925
999	949
9495052	973
105	997
157	·9775021
210	045
262	069
315	094
368	118
420	142
473	166
526	190
578	214
631	238
684	262
736	286
789	310
842	334
894	358
947	383
9496000	407
052	431
105	455

9496158	·9775479
210	503
263	527
316	551
368	575
421	599
474	623
526	647
579	672
631	696
684	720
737	744
789	768
842	792
895	816
947	840
9497000	864
053	888
105	912
158	936
211	960
263	984
316	·9776009
369	033
421	057
474	081
526	105
579	129
632	153
684	177
737	201
790	225
842	249
895	273
948	297
9498000	321
053	346
106	370
158	394
211	418
263	442
316	466
369	490
421	514
474	538
527	562
579	586
632	610
685	634
737	658
790	682
843	707
895	731
948	755
9499000	779
053	803
106	827
158	851
211	875
264	899
316	923
369	947
422	971
474	995
527	·9777019

9499579	·9777043
632	067
685	091
737	116
790	140
843	164
895	188
948	212

Excesses

10 = 5	1 = 0
20 = 9	2 = 1
30 = 14	3 = 1
40 = 18	4 = 2
50 = 23	5 = 2
	6 = 3
	7 = 3
	8 = 4
	9 = 4

26 = 12

Anti Log Excesses

10 = 22	1 = 2
20 = 44	2 = 4
	3 = 7
	4 = 9
	5 = 11
	6 = 13
	7 = 15
	8 = 17
	9 = 20

12 = 26

9500000	·9777236
053	260
106	284
158	308
211	332
264	356
316	380
369	404
422	428
474	452
527	476
579	500
632	525
685	549
737	573
790	597
843	621
895	645
948	669
9501000	693
053	717
106	741
158	765
211	789
264	813
316	837
369	862
421	886

9501474	·9777910
527	934
579	958
632	982
685	·9778006
737	030
790	054
842	078
895	102
948	126
9502000	150
053	174
106	198
158	222
211	246
263	270
316	294
369	318
421	343
474	367
526	391
579	415
632	439
684	463
737	487
790	511
842	535
895	559
947	583
9503000	607
053	631
105	655
158	679
210	703
263	727
316	751
368	775
421	799
474	823
526	847
579	872
631	896
684	920
737	944
789	968
842	992
894	·9779016
947	040
9504000	064
052	088
105	112
157	136
210	160
263	184
315	208
368	232
421	256
473	280
526	304
578	328
631	352
684	376
736	400
789	424
841	448

9504894	·9779473
947	497
999	521
9505052	545
104	569
157	593
210	617
262	641
315	665
367	689
420	713
473	737
525	761
578	785
630	809
683	833
736	857
788	881
841	905
893	929
946	953
999	977
9506051	·9780001
104	025
156	049
209	073
262	097
314	121
367	145
419	169
472	193
525	218
577	242
630	266
682	290
735	314
788	338
840	362
893	386
945	410
998	434
9507051	458
103	482
156	506
208	530
261	554
313	578
366	602
419	626
471	650
524	674
576	698
629	722
682	746
734	770
787	794
839	818
892	842
945	866
997	890
9508050	914
102	938
155	962
207	986
260	·9781010

9508313	·9781034
365	058
418	082
470	106
523	130
576	154
628	178
681	202
733	226
786	250
838	274
891	298
944	322
996	346
9509049	370
101	394
154	418
207	442
259	466
312	490
364	514
417	538
469	562
522	586
575	610
627	634
680	658
732	682
785	706
838	730
890	754
943	778
995	802

Excesses

10 = 5	1 = 0
20 = 9	2 = 1
30 = 14	3 = 1
40 = 18	4 = 2
50 = 23	5 = 2
	6 = 3
	7 = 3
	8 = 4
	9 = 4

26 = 12

Anti Log Excesses

10 = 22	1 = 2
20 = 44	2 = 4
	3 = 7
	4 = 9
	5 = 11
	6 = 13
	7 = 15
	8 = 18
	9 = 20

12 = 26

9510048	·9781826
100	850
153	874

9510206	·9781898
258	922
311	946
363	970
416	994
468	·9782019
521	043
574	067
626	091
679	115
731	139
784	163
836	187
889	211
942	235
994	259
9511047	283
099	307
152	331
204	355
257	379
310	403
362	427
415	451
467	475
520	499
572	523
625	547
678	571
730	595
783	619
835	643
888	667
940	691
993	715
9512045	739
098	763
151	787
203	811
256	835
308	859
361	883
413	907
466	931
519	955
571	979
624	·9783003
676	027
729	051
781	075
834	099
886	123
939	147
992	171
9513044	195
097	219
149	243
202	267
254	291
307	315
360	339
412	363
465	387
517	411
570	435

9513622	·9783459
675	482
727	506
780	530
833	554
885	578
938	602
990	626
9514043	650
095	674
148	698
200	722
253	746
306	770
358	794
411	818
463	842
516	866
568	890
621	914
673	938
726	962
778	986
831	·9784010
884	034
936	058
989	082
9515041	106
094	130
146	154
199	178
251	202
304	226
357	250
409	274
462	298
514	322
567	346
619	370
672	394
724	418
777	442
829	466
882	490
935	514
987	538
9516040	562
092	586
145	610
197	634
250	658
302	682
355	706
407	730
460	754
512	778
565	802
618	826
670	850
723	873
775	897
828	921
880	945
933	969
985	993

9517038	·9785017
090	041
143	065
195	089
248	113
301	137
353	162
406	186
458	210
511	234
563	258
616	282
668	306
721	330
773	354
826	378
878	401
931	425
983	449
9518036	473
089	497
141	521
194	545
246	569
299	593
351	617
404	641
456	665
509	689
561	713
614	737
666	761
719	785
771	809
824	833
876	857
929	881
982	905
9519034	929
087	953
139	977
192	·9786001
244	025
297	049
349	073
402	097
454	121
507	144
559	168
612	192
664	216
717	240
769	264
822	288
874	312
927	336
979	360

Excesses

10 = 5	1 = 0
20 = 9	2 = 1
30 = 14	3 = 1
40 = 18	4 = 2
50 = 23	5 = 2
	6 = 3
	7 = 3
	8 = 4
	9 = 4

26 = 12

Anti Log Excesses

10 = 22	1 = 2
20 = 44	2 = 4
	3 = 7
	4 = 9
	5 = 11
	6 = 13
	7 = 15
	8 = 18
	9 = 20

12 = 26

9520032	·9786384
085	408
137	432
190	456
242	480
295	504
347	528
400	552
452	576
505	600
557	624
610	648
662	672
715	696
767	719
820	743
872	767
925	791
977	815
9521030	839
082	863
135	887
187	911
240	935
292	959
345	983
397	·9787007
450	031
502	055
555	079
608	103
660	127
713	151
765	175
818	199
870	223
923	246

9521975	·9787270
9522028	294
080	318
133	342
185	366
238	390
290	414
343	438
395	462
448	486
500	510
553	534
605	558
658	582
710	606
763	630
815	654
868	677
920	701
973	725
9523025	749
078	773
130	797
183	821
235	845
288	869
340	893
393	917
445	941
498	965
550	989
603	·9788013
655	037
708	060
760	084
813	108
865	132
918	156
970	180
9524023	204
075	228
128	252
180	276
233	300
285	324
338	348
390	372
443	396
495	420
548	443
600	467
653	491
705	515
758	539
810	563
863	587
915	611
968	635
9525020	659
073	683
125	707
178	731
230	755
283	779

9525335	·9788802
388	826
440	850
493	874
545	898
598	922
650	946
703	970
755	994
808	·9789018
860	042
913	066
965	090
9526018	114
070	137
122	161
175	185
227	209
280	233
332	257
385	281
437	305
490	329
542	353
595	377
647	401
700	425
752	448
805	472
857	496
910	520
962	544
9527015	568
067	592
120	616
172	640
225	664
277	688
330	712
382	736
435	759
487	783
540	807
592	831
644	855
697	879
749	903
802	927
854	951
907	975
959	999
9528012	·9790023
064	047
117	071
169	094
222	118
274	142
327	166
379	190
432	214
484	238
537	262
589	286
642	310
694	334

9528746	·9790358
799	381
851	405
904	429
956	453
9529009	477
061	501
114	525
166	549
219	573
271	597
324	621
376	645
429	668
481	692
534	716
586	740
638	764
691	788
743	812
796	836
848	860
901	884
953	908

Excesses

10 = 5	1 = 0
20 = 9	2 = 1
30 = 14	3 = 1
40 = 18	4 = 2
50 = 23	5 = 2
	6 = 3
	7 = 3
	8 = 4
	9 = 4

26 = 12

Anti Log Excesses

10 = 22	1 = 2
20 = 44	2 = 4
	3 = 7
	4 = 9
	5 = 11
	6 = 13
	7 = 15
	8 = 18
	9 = 20

12 = 26

9530006	·9790931
058	955
111	979
163	·9791003
216	027
268	051
321	075
373	099
425	123
478	147
530	171
583	194

9530635	·9791218
688	242
740	266
793	290
845	314
898	338
950	362
9531003	386
055	410
107	433
160	457
212	481
265	505
317	529
370	553
422	577
475	601
527	625
580	649
632	673
685	696
737	720
789	744
842	768
894	792
947	816
999	840
9532052	864
104	888
157	911
209	935
262	959
314	983
366	·9792007
419	031
471	055
524	079
576	103
629	127
681	150
734	174
786	198
838	222
891	246
943	270
996	294
9533048	318
101	342
153	366
206	389
258	413
311	437
363	461
415	485
468	509
520	533
573	557
625	581
678	604
730	628
783	652
835	676
887	700
940	724
992	748

9534045	·9792772
097	796
150	819
202	843
255	867
307	891
359	915
412	939
464	963
517	987
569	·9793011
622	034
674	058
727	082
779	106
831	130
884	154
936	178
989	202
9535041	226
094	249
146	273
198	297
251	321
303	345
356	369
408	393
461	417
513	441
566	464
618	488
670	512
723	536
775	560
828	584
880	608
933	632
985	655
9536037	679
090	703
142	727
195	751
247	775
300	799
352	823
404	847
457	870
509	894
562	918
614	942
667	966
719	990
771	·9794014
824	038
876	061
929	085
981	109
9537034	133
086	157
138	181
191	205
243	229
296	252
348	276
401	300

9537453	·9794324
505	348
558	372
610	396
663	420
715	443
768	467
820	491
872	515
925	539
977	563
9538030	587
082	611
134	634
187	658
239	682
292	706
344	730
397	754
449	778
501	801
554	825
606	849
659	873
711	897
764	921
816	945
868	968
921	992
973	·9795016
9539026	040
078	064
130	088
183	112
235	135
288	159
340	183
393	207
445	231
497	255
550	279
602	302
655	326
707	350
759	374
812	398
864	422
917	446
969	470

Excesses

10 = 5	1 = 0
20 = 9	2 = 1
30 = 14	3 = 1
40 = 18	4 = 2
50 = 23	5 = 2
	6 = 3
	7 = 3
	8 = 4
	9 = 4

26 = 12

Anti Log Excesses

10 = 22	1 = 2
20 = 44	2 = 4
	3 = 7
	4 = 9
	5 = 11
	6 = 13
	7 = 15
	8 = 18
	9 = 20

12 = 26

9540021	·9795493
074	517
126	541
179	565
231	589
284	613
336	636
388	660
441	684
493	708
546	732
598	756
650	780
703	803
755	827
808	851
860	875
912	899
965	923
9541017	947
070	970
122	994
174	·9796018
227	042
279	066
332	090
384	114
436	137
489	161
541	185
594	209
646	233
698	257
751	281
803	304
856	328
908	352
960	376
9542013	400
065	424
118	447
170	471
222	495
275	519
327	543
380	567
432	591
484	614
537	638
589	662
642	686

9542694	·9796710
746	734
799	757
851	781
904	805
956	829
9543008	853
061	877
113	900
166	924
218	948
270	972
323	996
375	·9797020
428	044
480	067
532	091
585	115
637	139
690	163
742	187
794	210
847	234
899	258
951	282
9544004	306
056	330
109	353
161	377
213	401
266	425
318	449
371	473
423	497
475	520
528	544
580	568
632	592
685	616
737	640
790	663
842	687
894	711
947	735
999	759
9545052	783
104	806
156	830
209	854
261	878
313	902
366	926
418	949
471	973
523	997
575	·9798021
628	045
680	069
733	092
785	116
837	140
890	164
942	188
994	212
9546047	235

9546099	·9798259
152	283
204	307
256	331
309	355
361	378
413	402
466	426
518	450
571	474
623	498
675	521
728	545
780	569
832	593
885	617
937	641
990	664
9547042	688
094	712
147	736
199	760
251	783
304	807
356	831
409	855
461	879
513	903
566	926
618	950
670	974
723	998
775	·9799022
827	045
880	069
932	093
985	117
9548037	141
089	165
142	188
194	212
246	236
299	260
351	284
404	308
456	331
508	355
561	379
613	403
665	427
718	450
770	474
822	498
875	522
927	546
980	569
9549032	593
084	617
137	641
189	665
241	689
294	712
346	736
398	760
451	784

9549503	·9799807
555	831
608	855
660	879
713	903
765	927
817	950
870	974
922	998
974	·9800022

Excesses

10 = 5	1 = 0
20 = 9	2 = 1
30 = 14	3 = 1
40 = 18	4 = 2
50 = 23	5 = 2
	6 = 3
	7 = 3
	8 = 4
	9 = 4

26 = 12

Anti Log Excesses

10 = 22	1 = 2
20 = 44	2 = 4
	3 = 7
	4 = 9
	5 = 11
	6 = 13
	7 = 15
	8 = 18
	9 = 20

12 = 26

9550027	·9800046
079	069
131	093
184	117
236	141
288	165
341	188
393	212
446	236
498	260
550	284
603	308
655	331
707	355
760	379
812	403
864	427
917	450
969	474
9551021	498
074	522
126	546
178	569
231	593
283	617
335	641
388	665

9551440	·9800688
493	712
545	736
597	760
650	784
702	807
754	831
807	855
859	879
911	903
964	926
9552016	950
068	974
121	998
173	·9801022
225	045
278	069
330	093
382	117
435	141
487	164
539	188
592	212
644	236
696	260
749	283
801	307
853	331
906	355
958	379
9553010	402
063	426
115	450
168	474
220	498
272	521
325	545
377	569
429	593
482	617
534	640
586	664
639	688
691	712
743	736
796	759
848	783
900	807
953	831
9554005	854
057	878
110	902
162	926
214	950
267	973
319	997
371	·9802021
424	045
476	069
528	092
581	116
633	140
685	164
738	188
790	211

9554842	·9802235
895	259
947	283
999	306
9555052	330
104	354
156	378
209	401
261	425
313	449
365	473
418	497
470	520
522	544
575	568
627	592
679	616
732	639
784	663
836	687
889	711
941	734
993	758
9556046	782
098	806
150	830
203	853
255	877
307	901
360	925
412	949
464	972
517	996
569	·9803020
621	044
674	067
726	091
778	115
831	139
883	163
935	186
987	210
9557040	234
092	258
144	281
197	305
249	329
301	353
354	377
406	400
458	424
511	448
563	472
615	495
668	519
720	543
772	567
825	590
877	614
929	638
981	662
9558034	686
086	709
138	733
191	757
243	781

9558295	·9803804
348	828
400	852
452	876
505	900
557	923
609	947
661	971
714	995
766	·9804018
818	042
871	066
923	090
975	113
9559028	137
080	161
132	185
185	209
237	232
289	256
341	280
394	304
446	327
498	351
551	375
603	399
655	422
708	446
760	470
812	494
865	517
917	541
969	565

Excesses

10 = 5	1 = 0
20 = 9	2 = 1
30 = 14	3 = 1
40 = 18	4 = 2
50 = 23	5 = 2
	6 = 3
	7 = 3
	8 = 4
	9 = 4

26 = 12

Anti Log Excesses

10 = 22	1 = 2
20 = 44	2 = 4
	3 = 7
	4 = 9
	5 = 11
	6 = 13
	7 = 15
	8 = 18
	9 = 20

12 = 26

9560021	·9804589
074	613
126	636
178	660

9560231	·9804684
283	707
335	731
388	755
440	779
492	802
544	826
597	850
649	874
701	897
754	921
806	945
858	969
910	992
963	·9805016
9561015	040
067	064
120	087
172	111
224	135
277	159
329	182
381	206
433	230
486	254
538	277
590	301
643	325
695	349
747	372
799	396
852	420
904	444
956	467
9562009	491
061	515
113	539
166	562
218	586
270	610
322	634
375	657
427	681
479	705
532	729
584	752
636	776
688	800
741	824
793	847
845	871
898	895
950	919
9563002	942
054	966
107	990
159	·9806014
211	037
264	061
316	085
368	109
420	132
473	156
525	180
577	204

9563630	·9806227
682	251
734	275
786	299
839	322
891	346
943	370
995	394
9564048	417
100	441
152	465
205	488
257	512
309	536
361	560
414	583
466	607
518	631
571	655
623	678
675	702
727	726
780	750
832	773
884	797
936	821
989	845
9565041	868
093	892
146	916
198	939
250	963
302	987
355	·9807011
407	034
459	058
511	082
564	106
616	129
668	153
721	177
773	201
825	224
877	248
930	272
982	295
9566034	319
086	343
139	367
191	390
243	414
295	438
348	461
400	485
452	509
505	533
557	556
609	580
661	604
714	628
766	651
818	675
870	699
923	722
975	746

9567027	·9807770
079	794
132	817
184	841
236	865
289	888
341	912
393	936
445	960
498	983
550	·9808007
602	031
654	055
707	078
759	102
811	126
863	149
916	173
968	197
9568020	221
072	244
125	268
177	292
229	315
281	339
334	363
386	387
438	410
490	434
543	458
595	481
647	505
699	529
752	553
804	576
856	600
908	624
961	647
9569013	671
065	695
117	719
170	742
222	766
274	790
326	813
379	837
431	861
483	885
535	908
588	932
640	956
692	979
744	·9809003
797	027
849	050
901	074
953	098

Excesses

10 = 5	1 = 0
20 = 9	2 = 1
30 = 14	3 = 1
40 = 18	4 = 2
50 = 23	5 = 2
	6 = 3
	7 = 3
	8 = 4
	9 = 4

26 = 12

Anti Log Excesses

10 = 22	1 = 2
20 = 44	2 = 4
	3 = 7
	4 = 9
	5 = 11
	6 = 13
	7 = 15
	8 = 18
	9 = 20

12 = 26

9570006	·9809122
058	145
110	169
162	193
215	216
267	240
319	264
371	288
424	311
476	335
528	359
580	382
633	406
685	430
737	453
789	477
842	501
894	525
946	548
998	572
9571051	596
103	620
155	643
207	667
260	691
312	714
364	738
416	762
469	786
521	809
573	833
625	857
677	880
730	904
782	928
834	951
886	975

9571939	·9809999
991	·9810023
9572043	046
095	070
148	094
200	117
252	141
304	165
357	188
409	212
461	236
513	260
565	283
618	307
670	331
722	354
774	378
827	402
879	425
931	449
983	473
9573036	496
088	520
140	544
192	568
244	591
297	615
349	639
401	662
453	686
506	710
558	733
610	757
662	781
715	804
767	828
819	852
871	876
923	899
976	923
9574028	947
080	970
132	994
185	·9811018
237	041
289	065
341	089
393	112
446	136
498	160
550	183
602	207
655	231
707	255
759	278
811	302
863	326
916	349
968	373
9575020	397
072	420
125	444
177	468
229	491
281	515
333	539

9575386	·9811562
438	586
490	610
542	633
594	657
647	681
699	704
751	728
803	752
856	776
908	799
960	823
9576012	847
064	870
117	894
169	918
221	941
273	965
325	989
378	·9812012
430	036
482	060
534	083
587	107
639	131
691	155
743	178
795	202
848	226
900	249
952	273
9577004	297
056	320
109	344
161	368
213	391
265	415
317	439
370	462
422	486
474	510
526	533
578	557
631	581
683	604
735	628
787	652
840	675
892	699
944	723
996	746
9578048	770
101	794
153	817
205	841
257	865
309	888
362	912
414	936
466	959
518	983
570	·9813007
623	030
675	054
727	078

9578779	·9813101
831	125
884	149
936	172
988	196
9579040	220
092	243
145	267
197	291
249	314
301	338
353	362
405	385
458	409
510	433
562	456
614	480
666	504
719	527
771	551
823	575
875	598
927	622
980	646

Excesses

10 = 5	1 = 0
20 = 9	2 = 1
30 = 14	3 = 1
40 = 18	4 = 2
50 = 23	5 = 2
	6 = 3
	7 = 3
	8 = 4
	9 = 4

26 = 12

Anti Log Excesses

10 = 22	1 = 2
20 = 44	2 = 4
	3 = 7
	4 = 9
	5 = 11
	6 = 13
	7 = 15
	8 = 18
	9 = 20

12 = 26

9580032	·9813669
084	693
136	717
188	740
241	764
293	788
345	811
397	835
449	859
502	882
554	906
606	930
658	953
710	977

9580762	·9814001
815	024
867	048
919	071
971	095
9581023	119
076	142
128	166
180	190
232	213
284	237
337	261
389	284
441	308
493	332
545	355
597	379
650	403
702	426
754	450
806	474
858	497
911	521
963	545
9582015	568
067	592
119	616
171	639
224	663
276	687
328	710
380	734
432	757
485	781
537	805
589	828
641	852
693	876
745	899
798	923
850	947
902	970
954	994
9583006	·9815018
058	041
111	065
163	089
215	112
267	136
319	159
372	183
424	207
476	230
528	254
580	278
632	301
685	325
737	349
789	372
841	396
893	420
945	443
998	467
9584050	490
102	514

9584154	·9815538
206	561
258	585
311	609
363	632
415	656
467	680
519	703
571	727
624	751
676	774
728	798
780	822
832	845
884	869
937	892
989	916
9585041	940
093	963
145	987
197	·9816011
250	034
302	058
354	081
406	105
458	129
510	152
563	176
615	200
667	223
719	247
771	271
823	294
876	318
928	341
980	365
9586032	389
084	412
136	436
188	460
241	483
293	507
345	531
397	554
449	578
501	601
554	625
606	649
658	672
710	696
762	720
814	743
867	767
919	790
971	814
9587023	838
075	861
127	885
179	908
232	932
284	956
336	979
388	·9817003
440	026
492	050

9587545	·9817074
597	097
649	121
701	145
753	168
805	192
857	215
910	239
962	263
9588014	286
066	310
118	334
170	357
222	381
275	404
327	428
379	452
431	475
483	499
535	522
587	546
640	570
692	593
744	617
796	641
848	664
900	688
952	711
9589005	735
057	759
109	782
161	806
213	829
265	853
317	877
370	900
422	924
474	948
526	971
578	995
630	·9818018
682	042
735	066
787	089
839	113
891	136
943	160
995	184

Excesses

10 = 5	1 = 0
20 = 9	2 = 1
30 = 14	3 = 1
40 = 18	4 = 2
50 = 23	5 = 2
	6 = 3
	7 = 3
	8 = 4
	9 = 4

26 = 12

Anti Log Excesses

10 = 22	1 = 2
20 = 44	2 = 4
	3 = 7
	4 = 9
	5 = 11
	6 = 13
	7 = 15
	8 = 18
	9 = 20

12 = 26

9590047	·9818207
100	231
152	254
204	278
256	302
308	325
360	349
412	373
465	396
517	420
569	443
621	467
673	491
725	514
777	538
829	561
882	585
934	609
986	632
9591038	656
090	679
142	703
194	727
247	750
299	774
351	797
403	821
455	845
507	868
559	892
611	915
664	939
716	963
768	986
820	·9819010
872	033
924	057
976	081
9592028	104
081	128
133	151
185	175
237	199
289	222
341	246
393	269
445	293
498	317
550	340
602	364
654	387

9592706	·9819411
758	435
810	458
862	482
915	505
967	529
9593019	553
071	576
123	600
175	623
227	647
279	671
332	694
384	718
436	741
488	765
540	789
592	812
644	836
696	859
748	883
801	907
853	930
905	954
957	977
9594009	·9820001
061	025
113	048
165	072
218	095
270	119
322	142
374	166
426	190
478	213
530	237
582	260
634	284
687	308
739	331
791	355
843	378
895	402
947	426
999	449
9595051	473
103	496
156	520
208	543
260	567
312	591
364	614
416	638
468	661
520	685
572	709
625	732
677	756
729	779
781	803
833	826
885	850
937	874
989	897
9596041	921
093	944

9596146	·9820968
198	992
250	·9821015
302	039
354	062
406	086
458	109
510	133
562	157
614	180
667	204
719	227
771	251
823	274
875	298
927	322
979	345
9597031	369
083	392
135	416
188	440
240	463
292	487
344	510
396	534
448	557
500	581
552	605
604	628
656	652
709	675
761	699
813	722
865	746
917	770
969	793
9598021	817
073	840
125	864
177	887
230	911
282	935
334	958
386	982
438	·9822005
490	029
542	052
594	076
646	100
698	123
750	147
803	170
855	194
907	217
959	241
9599011	265
063	288
115	312
167	335
219	359
271	382
323	406
375	430
428	453
480	477

9599532	·9822500
584	524
636	547
688	571
740	594
792	618
844	642
896	665
948	689

Excesses

10 = 5	1 = 0
20 = 9	2 = 1
30 = 14	3 = 1
40 = 18	4 = 2
50 = 23	5 = 2
	6 = 3
	7 = 3
	8 = 4
	9 = 4

26 = 12

Anti Log Excesses

10 = 22	1 = 2
20 = 44	2 = 4
	3 = 7
	4 = 9
	5 = 11
	6 = 13
	7 = 15
	8 = 18
	9 = 20

12 = 26

9600000	·9822712
053	736
105	759
157	783
209	807
261	830
313	854
365	877
417	901
469	924
521	948
573	971
625	995
678	·9823019
730	042
782	066
834	089
886	113
938	136
990	160
9601042	184
094	207
146	231
198	254
250	278
302	301
355	325

9601407	·9823348
459	372
511	396
563	419
615	443
667	466
719	490
771	513
823	537
875	560
927	584
979	608
9602032	631
084	655
136	678
188	702
240	725
292	749
344	772
396	796
448	819
500	843
552	867
604	890
656	914
708	937
761	961
813	984
865	·9824008
917	031
969	055
9603021	079
073	102
125	126
177	149
229	173
281	196
333	220
385	244
437	267
489	291
542	314
594	338
646	361
698	385
750	408
802	432
854	455
906	479
958	502
9604010	526
062	550
114	573
166	597
218	620
270	644
322	667
375	691
427	714
479	738
531	761
583	785
635	808
687	832
739	856

9604791	·9824879
843	903
895	926
947	950
999	973
9605051	997
103	·9825020
155	044
207	067
259	091
312	114
364	138
416	162
468	185
520	209
572	232
624	256
676	279
728	303
780	326
832	350
884	373
936	397
988	420
9606040	444
092	467
144	491
196	514
248	538
301	562
353	585
405	609
457	632
509	656
561	679
613	703
665	726
717	750
769	773
821	797
873	820
925	844
977	867
9607029	891
081	914
133	938
185	961
237	985
289	·9826009
341	032
393	056
446	079
498	103
550	126
602	150
654	173
706	197
758	220
810	244
862	267
914	291
966	314
9608018	338
070	361
122	385

9608174	·9826408
226	432
278	455
330	479
382	502
434	526
486	550
538	573
590	597
642	620
694	644
747	667
799	691
851	714
903	738
955	761
9609007	785
059	808
111	832
163	855
215	879
267	903
319	926
371	950
423	973
475	997
527	·9827020
579	044
631	067
683	091
735	114
787	138
839	161
891	185
943	208
995	232

Excesses

10 = 5	1 = 0
20 = 9	2 = 1
30 = 14	3 = 1
40 = 18	4 = 2
50 = 23	5 = 2
	6 = 3
	7 = 3
	8 = 4
	9 = 4

26 = 12

Anti Log Excesses

10 = 22	1 = 2
20 = 44	2 = 4
	3 = 7
	4 = 9
	5 = 11
	6 = 13
	7 = 15
	8 = 18
	9 = 20

12 = 26

9610047	·9827255
099	279
151	302
203	326
255	349
307	373
359	396
412	420
464	443
516	467
568	490
620	514
672	537
724	561
776	584
828	608
880	631
932	655
984	678
9611036	702
088	725
140	749
192	772
244	796
296	819
348	843
400	866
452	890
504	913
556	937
608	960
660	984
712	·9828007
764	031
816	054
868	078
920	101
972	125
9612024	148
076	172
128	195
180	219
232	242
284	266
336	289
388	313
440	336
492	360
544	383
596	407
648	430
700	454
752	477
804	501
856	524
908	548
961	571
9613013	595
065	618
117	642
169	665
221	689
273	712
325	736
377	759

9613429	·9828783
481	806
533	830
585	853
637	877
689	900
741	924
793	947
845	971
897	994
949	·9829018
9614001	041
053	065
105	088
157	112
209	135
261	159
313	182
365	206
417	229
469	253
521	276
573	300
625	323
677	347
729	370
781	394
833	417
885	441
937	464
989	488
9615041	511
093	535
145	558
197	582
249	605
301	629
353	652
405	676
457	699
509	723
561	746
613	770
665	793
717	817
769	840
821	864
873	887
925	910
977	934
9616029	957
081	981
133	·9830004
185	028
237	051
289	075
341	098
393	122
445	145
497	169
549	192
601	216
653	239
705	263
757	286

9616809	·9830310
861	333
913	357
965	380
9617017	404
069	427
121	451
173	474
225	498
277	521
329	544
381	568
433	591
485	615
537	638
589	662
641	685
693	709
745	732
797	756
848	779
900	803
952	826
9618004	850
056	873
108	897
160	920
212	944
264	967
316	991
368	·9831014
420	037
472	061
524	084
576	108
628	131
680	155
732	178
784	202
836	225
888	249
940	272
992	296
9619044	319
096	343
148	366
200	390
252	413
304	436
356	460
408	483
460	507
512	530
564	554
616	577
668	601
720	624
772	648
824	671
876	695
928	718
980	741

Excesses

10 = 5	1 = 0
20 = 9	2 = 1
30 = 14	3 = 1
40 = 18	4 = 2
50 = 23	5 = 2
	6 = 3
	7 = 3
	8 = 4
	9 = 4

26 = 12

Anti Log Excesses

10 = 22	1 = 2
20 = 44	2 = 4
	3 = 7
	4 = 9
	5 = 11
	6 = 13
	7 = 16
	8 = 18
	9 = 20

12 = 26

9620032	·9831765
084	788
136	812
188	835
240	859
292	882
344	906
396	929
447	953
499	976
551	·9832000
603	023
655	046
707	070
759	093
811	117
863	140
915	164
967	187
9621019	211
071	234
123	258
175	281
227	305
279	328
331	351
383	375
435	398
487	422
539	445
591	469
643	492
695	516
747	539
799	563
851	586
903	610

9621955	·9832633
9622007	656
058	680
110	703
162	727
214	750
266	774
318	797
370	821
422	844
474	868
526	891
578	914
630	938
682	961
734	985
786	·9833008
838	032
890	055
942	079
994	102
9623046	126
098	149
150	172
202	196
254	219
306	243
358	266
409	290
461	313
513	337
565	360
617	383
669	407
721	430
773	454
825	477
877	501
929	524
981	548
9624033	571
085	594
137	618
189	641
241	665
293	688
345	712
397	735
449	759
500	782
552	805
604	829
656	852
708	876
760	899
812	923
864	946
916	970
968	993
9625020	·9834016
072	040
124	063
176	087
228	110
280	133

9625332	·9834157
384	180
436	204
488	227
539	251
591	274
643	298
695	321
747	344
799	368
851	391
903	415
955	438
9626007	462
059	485
111	508
163	532
215	555
267	579
319	602
371	626
422	649
474	672
526	696
578	719
630	743
682	766
734	790
786	813
838	836
890	860
942	883
994	907
9627046	930
098	954
150	977
202	·9835000
253	024
305	047
357	071
409	094
461	118
513	141
565	164
617	188
669	211
721	235
773	258
825	282
877	305
929	328
981	352
9628032	375
084	399
136	422
188	446
240	469
292	492
344	516
396	539
448	563
500	586
552	610
604	633
656	656

9628708	·9835680
760	703
811	727
863	750
915	773
967	797
9629019	820
071	844
123	867
175	891
227	914
279	937
331	961
383	984
435	·9836008
486	031
538	054
590	078
642	101
694	125
746	148
798	172
850	195
902	218
954	242

Excesses

10 = 5	1 = 0
20 = 9	2 = 1
30 = 14	3 = 1
40 = 18	4 = 2
50 = 23	5 = 2
	6 = 3
	7 = 3
	8 = 4
	9 = 4

26 = 12

Anti Log Excesses

10 = 22	1 = 2
20 = 44	2 = 4
	3 = 7
	4 = 9
	5 = 11
	6 = 13
	7 = 16
	8 = 18
	9 = 20

12 = 26

9630006	·9836265
058	289
110	312
161	335
213	359
265	382
317	406
369	429
421	453
473	476
525	499

9630577	·9836523
629	546
681	570
733	593
784	617
836	640
888	663
940	687
992	710
9631044	734
096	757
148	780
200	804
252	827
304	851
356	874
407	897
459	921
511	944
563	968
615	991
667	·9837014
719	038
771	061
823	085
875	108
927	132
978	155
9632030	178
082	202
134	225
186	249
238	272
290	295
342	319
394	342
446	366
498	389
549	412
601	436
653	459
705	483
757	506
809	529
861	553
913	576
965	600
9633017	623
069	646
120	670
172	693
224	717
276	740
328	763
380	787
432	810
484	834
536	857
588	880
639	904
691	927
743	951
795	974
847	997
899	·9838021

9633951	·9838044
9634003	068
055	091
107	114
158	138
210	161
262	185
314	208
366	231
418	255
470	278
522	301
574	325
626	348
677	372
729	395
781	418
833	442
885	465
937	489
989	512
9635041	535
093	559
145	582
196	606
248	629
300	652
352	676
404	699
456	723
508	746
560	769
612	793
663	816
715	839
767	863
819	886
871	910
923	933
975	956
9636027	980
079	·9839003
130	027
182	050
234	073
286	097
338	120
390	143
442	167
494	190
546	214
597	237
649	260
701	284
753	307
805	331
857	354
909	377
961	401
9637012	424
064	447
116	471
168	494
220	518
272	541

9637324	·9839564
376	588
428	611
479	634
531	658
583	681
635	705
687	728
739	751
791	775
843	798
894	822
946	845
998	868
9638050	892
102	915
154	938
206	962
258	985
309	·9840009
361	032
413	055
465	079
517	102
569	125
621	149
673	172
724	196
776	219
828	242
880	266
932	289
984	312
9639036	336
088	359
139	383
191	406
243	429
295	453
347	476
399	499
451	523
503	546
554	569
606	593
658	616
710	640
762	663
814	686
866	710
918	733
969	756

Excesses

10 = 5	1 = 0
20 = 9	2 = 1
30 = 14	3 = 1
40 = 18	4 = 2
50 = 23	5 = 2
	6 = 3
	7 = 3
	8 = 4
	9 = 4
26 = 12	

Anti Log Excesses

10 = 22	1 = 2
20 = 44	2 = 4
	3 = 7
	4 = 9
	5 = 11
	6 = 13
	7 = 16
	8 = 18
	9 = 20
12 = 26	

9640021	·9840780
073	803
125	827
177	850
229	873
281	897
332	920
384	943
436	967
488	990
540	·9841013
592	037
644	060
695	084
747	107
799	130
851	154
903	177
955	200
9641007	224
059	247
110	270
162	294
214	317
266	340
318	364
370	387
422	411
473	434
525	457
577	481
629	504
681	527
733	551
785	574
836	597
888	621

9641940	·9841644
992	668
9642044	691
096	714
148	738
199	761
251	784
303	808
355	831
407	854
459	878
511	901
562	924
614	948
666	971
718	995
770	·9842018
822	041
874	065
925	088
977	111
9643029	135
081	158
133	181
185	205
236	228
288	251
340	275
392	298
444	321
496	345
548	368
599	392
651	415
703	438
755	462
807	485
859	508
910	532
962	555
9644014	578
066	602
118	625
170	648
222	672
273	695
325	718
377	742
429	765
481	788
533	812
584	835
636	858
688	882
740	905
792	929
844	952
896	975
947	999
999	·9843022
9645051	045
103	069
155	092
207	115
258	139

9645310	·9843162
362	185
414	209
466	232
518	255
569	279
621	302
673	325
725	349
777	372
829	395
880	419
932	442
984	465
9646036	489
088	512
140	535
191	559
243	582
295	605
347	629
399	652
451	675
502	699
554	722
606	745
658	769
710	792
762	815
813	839
865	862
917	885
969	909
9647021	932
073	955
124	979
176	·9844002
228	025
280	049
332	072
384	095
435	119
487	142
539	165
591	189
643	212
695	235
746	259
798	282
850	305
902	329
954	352
9648005	375
057	399
109	422
161	445
213	468
265	492
316	515
368	539
420	562
472	585
524	608
576	632
627	655

9648679	·9844678
731	702
783	725
835	748
886	772
938	795
990	818
9649042	842
094	865
146	888
197	912
249	935
301	958
353	982
405	·9845005
456	028
508	052
560	075
612	098
664	122
716	145
767	168
819	192
871	215
923	238
975	261

Excesses

10 = 5	1 = 0
20 = 9	2 = 1
30 = 14	3 = 1
40 = 18	4 = 2
50 = 23	5 = 2
	6 = 3
	7 = 3
	8 = 4
	9 = 4
26 = 12	

Anti Log Excesses

10 = 22	1 = 2
20 = 44	2 = 4
	3 = 7
	4 = 9
	5 = 11
	6 = 13
	7 = 16
	8 = 18
	9 = 20
12 = 26	

9650026	·9845285
078	308
130	331
182	355
234	378
285	401
337	425
389	448
441	471
493	495

9650545	·9845518
596	541
648	565
700	588
752	611
804	634
855	658
907	681
959	704
9651011	728
063	751
114	774
166	798
218	821
270	844
322	868
373	891
425	914
477	938
529	961
581	984
632	·9846007
684	031
736	054
788	077
840	101
892	124
943	147
995	171
9652047	194
099	217
151	241
202	264
254	287
306	311
358	334
410	357
461	381
513	404
565	427
617	450
669	474
720	497
772	520
824	544
876	567
928	590
979	614
9653031	637
083	660
135	683
187	707
238	730
290	753
342	777
394	800
445	823
497	847
549	870
601	893
653	917
704	940
756	963
808	986
860	·9847010

9653912	·9847033
963	056
9654015	080
067	103
119	126
171	149
222	173
274	196
326	219
378	243
430	266
481	289
533	313
585	336
637	359
688	382
740	406
792	429
844	452
896	476
947	499
999	522
9655051	546
103	569
155	592
206	615
258	639
310	662
362	685
413	709
465	732
517	755
569	778
621	802
672	825
724	848
776	872
828	895
880	918
931	941
983	965
9656035	988
087	·9848011
138	035
190	058
242	081
294	104
346	128
397	151
449	174
501	198
553	221
604	244
656	267
708	291
760	314
812	337
863	361
915	384
967	407
9657019	430
070	454
122	477
174	500
226	524

9657278	·9848547
329	570
381	593
433	617
485	640
536	663
588	687
640	710
692	733
744	756
795	780
847	803
899	826
951	850
9658002	873
054	896
106	919
158	943
209	966
261	989
313	·9849013
365	036
417	059
468	082
520	106
572	129
624	152
675	175
727	199
779	222
831	245
882	269
934	292
986	315
9659038	338
089	362
141	385
193	408
245	431
297	455
348	478
400	501
452	525
504	548
555	571
607	594
659	618
711	641
762	664
814	688
866	711
918	734
969	757

Excesses

10 = 4	1 = 0
20 = 9	2 = 1
30 = 13	3 = 1
40 = 18	4 = 2
50 = 22	5 = 2
	6 = 3
	7 = 3
	8 = 4
	9 = 4

26 = 12

Anti Log Excesses

10 = 22	1 = 2
20 = 44	2 = 4
	3 = 7
	4 = 9
	5 = 11
	6 = 13
	7 = 16
	8 = 18
	9 = 20

12 = 26

9660021	·9849781
073	804
125	827
176	850
228	874
280	897
332	920
384	943
435	967
487	990
539	·9850013
591	037
642	060
694	083
746	106
798	130
849	153
901	176
953	199
9661005	223
056	246
108	269
160	292
212	316
263	339
315	362
367	386
419	409
470	432
522	455
574	479
626	502
677	525
729	548
781	572
833	595
884	618

9661936	·9850641
988	665
9662040	688
091	711
143	734
195	758
247	781
298	804
350	828
402	851
454	874
505	897
557	921
609	944
661	967
712	990
764	·9851014
816	037
868	060
919	083
971	107
9663023	130
075	153
126	176
178	200
230	223
282	246
333	269
385	293
437	316
488	339
540	362
592	386
644	409
695	432
747	455
799	479
851	502
902	525
954	548
9664006	572
058	595
109	618
161	641
213	665
265	688
316	711
368	734
420	758
472	781
523	804
575	827
627	851
678	874
730	897
782	920
834	944
885	967
937	990
989	·9852013
9665041	037
092	060
144	083
196	106
248	130

9665299	·9852153
351	176
403	199
454	223
506	246
558	269
610	292
661	316
713	339
765	362
817	385
868	409
920	432
972	455
9666023	478
075	502
127	525
179	548
230	571
282	595
334	618
386	641
437	664
489	687
541	711
592	734
644	757
696	780
748	804
799	827
851	850
903	873
955	897
9667006	920
058	943
110	966
161	990
213	·9853013
265	036
317	059
368	083
420	106
472	129
523	152
575	175
627	199
679	222
730	245
782	268
834	292
885	315
937	338
989	361
9668041	385
092	408
144	431
196	454
247	478
299	501
351	524
403	547
454	570
506	594
558	617
610	640

9668661	·9853663
713	686
765	710
816	733
868	756
920	779
971	803
9669023	826
075	849
127	872
178	895
230	919
282	942
333	965
385	988
437	·9854012
489	035
540	058
592	081
644	104
695	128
747	151
799	174
851	197
902	221
954	244

Excesses

10 = 4	1 = 0
20 = 9	2 = 1
30 = 13	3 = 1
40 = 18	4 = 2
50 = 22	5 = 2
	6 = 3
	7 = 3
	8 = 4
	9 = 4

26 = 12

Anti Log Excesses

10 = 22	1 = 2
20 = 45	2 = 5
	3 = 7
	4 = 9
	5 = 11
	6 = 13
	7 = 16
	8 = 18
	9 = 20

12 = 26

9670006	·9854267
057	290
109	313
161	337
212	360
264	383
316	406
368	430
419	453
471	476

9670523	·9854499
574	522
626	546
678	569
730	592
781	615
833	638
885	662
936	685
988	708
9671040	731
091	755
143	778
195	801
247	824
298	847
350	871
402	894
453	917
505	940
557	963
608	987
660	·9855010
712	033
764	056
815	080
867	103
919	126
970	149
9672022	172
074	196
125	219
177	242
229	265
280	288
332	312
384	335
436	358
487	381
539	404
591	428
642	451
694	474
746	497
797	521
849	544
901	567
952	590
9673004	613
056	637
108	660
159	683
211	706
263	729
314	753
366	776
418	799
469	822
521	845
573	869
624	892
676	915
728	938
779	962
831	985

9673883	·9856008
935	031
986	054
9674038	078
090	101
141	124
193	147
245	170
296	194
348	217
400	240
451	263
503	286
555	310
606	333
658	356
710	379
761	402
813	426
865	449
917	472
968	495
9675020	518
072	542
123	565
175	588
227	611
278	634
330	658
382	681
433	704
485	727
537	750
588	774
640	797
692	820
743	843
795	866
847	890
898	913
950	936
9676002	959
053	982
105	·9857006
157	029
208	052
260	075
312	098
363	121
415	145
467	168
518	191
570	214
622	237
673	261
725	284
777	307
828	330
880	353
932	377
984	400
9677035	423
087	446
139	469
190	492

9677242	·9857516
294	539
345	562
397	585
449	608
500	632
552	655
604	678
655	701
707	724
759	748
810	771
862	794
913	817
965	840
9678017	863
068	887
120	910
172	933
223	956
275	979
327	·9858003
378	026
430	049
482	072
533	095
585	118
637	142
688	165
740	188
792	211
843	234
895	257
947	281
998	304
9679050	327
102	350
153	373
205	397
257	420
308	443
360	466
412	489
463	512
515	536
567	559
618	582
670	605
722	628
773	652
825	675
877	698
928	721
980	744

Excesses

10 = 4	1 = 0
20 = 9	2 = 1
30 = 13	3 = 1
40 = 18	4 = 2
50 = 22	5 = 2
	6 = 3
	7 = 3
	8 = 4
	9 = 4

26 = 12

Anti Log Excesses

10 = 22	1 = 2
20 = 45	2 = 5
	3 = 7
	4 = 9
	5 = 11
	6 = 13
	7 = 16
	8 = 18
	9 = 20

12 = 26

9680031	·9858767
083	791
135	814
186	837
238	860
290	883
341	906
393	930
445	953
496	976
548	999
600	·9859022
651	045
703	069
755	092
806	115
858	138
910	161
961	184
9681013	208
064	231
116	254
168	277
219	300
271	323
323	347
374	370
426	393
478	416
529	439
581	463
633	486
684	509
736	532
788	555
839	578
891	601

Number	Log
9681942	·9859625
994	648
9682046	671
097	694
149	717
201	740
252	764
304	787
356	810
407	833
459	856
511	879
562	903
614	926
665	949
717	972
769	995
820	·9860018
872	042
924	065
975	088
9683027	111
079	134
130	157
182	181
233	204
285	227
337	250
388	273
440	296
492	319
543	343
595	366
647	389
698	412
750	435
801	458
853	482
905	505
956	528
9684008	551
060	574
111	597
163	621
214	644
266	667
318	690
369	713
421	736
473	759
524	783
576	806
627	829
679	852
731	875
782	898
834	921
886	945
937	968
989	991
9685041	·9861014
092	037
144	060
195	084
247	107

Number	Log
9685299	·9861130
350	153
402	176
454	199
505	222
557	246
608	269
660	292
712	315
763	338
815	361
866	384
918	407
970	431
9686021	454
073	477
125	500
176	523
228	547
279	570
331	593
383	616
434	639
486	662
538	685
589	709
641	732
692	755
744	778
796	801
847	824
899	847
950	871
9687002	894
054	917
105	940
157	963
209	986
260	·9862009
312	033
363	056
415	079
467	102
518	125
570	148
621	171
673	194
725	218
776	241
828	264
880	287
931	310
983	333
9688034	356
086	380
138	403
189	426
241	449
292	472
344	495
396	518
447	542
499	565
550	588
602	611

Number	Log
9688654	·9862634
705	657
757	680
808	703
860	727
912	750
963	773
9689015	796
067	819
118	842
170	865
221	889
273	912
325	935
376	958
428	981
479	·9863004
531	027
583	050
634	074
686	097
737	120
789	143
841	166
892	189
944	212
995	235

Excesses

10 = 4	1 = 0
20 = 9	2 = 1
30 = 13	3 = 1
40 = 18	4 = 2
50 = 22	5 = 2
	6 = 3
	7 = 3
	8 = 4
	9 = 4

26 = 12

Anti Log Excesses

10 = 22	1 = 2
20 = 45	2 = 5
	3 = 7
	4 = 9
	5 = 11
	6 = 13
	7 = 16
	8 = 18
	9 = 20

12 = 26

Number	Log
9690047	·9863259
099	282
150	305
202	328
253	351
305	374
357	398
408	421
460	444

Number	Log
9690511	·9863467
563	490
615	513
666	536
718	559
769	583
821	606
872	629
924	652
976	675
9691027	698
079	721
130	744
182	768
234	791
285	814
337	837
388	860
440	883
492	906
543	929
595	953
646	976
698	999
750	·9864022
801	045
853	068
904	091
956	114
9692008	137
059	161
111	184
162	207
214	230
265	253
317	276
369	299
420	322
472	346
523	369
575	392
627	415
678	438
730	461
781	484
833	507
884	530
936	554
988	577
9693039	600
091	623
142	646
194	669
246	692
297	715
349	738
400	762
452	785
503	808
555	831
607	854
658	877
710	900
761	923
813	946

Number	Log
9693865	·9864970
916	993
968	·9865016
9694019	039
071	062
122	085
174	108
226	131
277	154
329	178
380	201
432	224
483	247
535	270
587	293
638	316
690	339
741	362
793	386
844	409
896	432
948	455
999	478
9695051	501
102	524
154	547
206	570
257	593
309	617
360	640
412	663
463	686
515	709
566	732
618	755
670	778
721	801
773	824
824	847
876	870
927	894
979	917
9696031	940
082	963
134	986
185	·9866009
237	032
288	055
340	078
392	101
443	124
495	148
546	171
598	194
649	217
701	240
753	263
804	286
856	309
907	332
959	355
9697010	378
062	402
113	425
165	448

9697217	·9866471
268	494
320	517
371	540
423	563
474	586
526	609
578	632
629	656
681	679
732	702
784	725
835	748
887	771
938	794
990	817
9698042	840
093	863
145	886
196	910
248	933
299	956
351	979
402	·9867002
454	025
506	048
557	071
609	094
660	117
712	140
763	163
815	187
866	210
918	233
970	256
9699021	279
073	302
124	325
176	348
227	371
279	394
330	417
382	440
433	463
485	487
537	510
588	533
640	556
691	579
743	602
794	625
846	648
897	671
949	694

Excesses

10 = 4	1 = 0
20 = 9	2 = 1
30 = 13	3 = 1
40 = 18	4 = 2
50 = 22	5 = 2
	6 = 3
	7 = 3
	8 = 4
	9 = 4

26 = 12

Anti Log Excesses

10 = 22	1 = 2
20 = 45	2 = 5
	3 = 7
	4 = 9
	5 = 11
	6 = 13
	7 = 16
	8 = 18
	9 = 20

12 = 26

9700000	·9867717
052	740
104	763
155	786
207	810
258	833
310	856
361	879
413	902
464	925
516	948
567	971
619	994
671	·9868017
722	040
774	063
825	087
877	110
928	133
980	156
9701031	179
083	202
134	225
186	248
238	271
289	294
341	317
392	340
444	364
495	387
547	410
598	433
650	456
701	479
753	502
804	525
856	548

9701908	·9868571
959	594
9702011	617
062	640
114	663
165	686
217	710
268	733
320	756
371	779
423	802
474	825
526	848
577	871
629	894
681	917
732	940
784	963
835	986
887	·9869009
938	032
990	056
9703041	079
093	102
144	125
196	148
247	171
299	194
350	217
402	240
453	263
505	286
557	309
608	332
660	355
711	378
763	401
814	425
866	448
917	471
969	494
9704020	517
072	540
123	563
175	586
226	609
278	632
329	655
381	678
432	701
484	724
536	747
587	770
639	793
690	816
742	840
793	863
845	886
896	909
948	932
999	955
9705051	978
102	·9870001
154	024
205	047

9705257	·9870070
308	093
360	116
411	139
463	162
514	185
566	208
617	231
669	254
720	278
772	301
823	324
875	347
927	370
978	393
9706030	416
081	439
133	462
184	485
236	508
287	531
339	554
390	577
442	600
493	623
545	646
596	669
648	692
699	716
751	739
802	762
854	785
905	808
957	831
9707008	854
060	877
111	900
163	923
214	946
266	969
317	992
369	·9871015
420	038
472	061
523	084
575	107
626	130
678	153
729	176
781	199
832	223
884	246
935	269
987	292
9708038	315
090	338
141	361
193	384
244	407
296	430
347	453
399	476
450	499
502	522
553	545

9708605	·9871568
656	591
708	614
759	637
811	660
862	683
914	706
965	729
9709017	752
068	775
120	798
171	822
223	845
274	868
326	891
377	914
429	937
480	960
532	983
583	·9872006
635	029
686	052
738	075
789	098
841	121
892	144
944	167
995	190

Excesses

10 = 4	1 = 0
20 = 9	2 = 1
30 = 13	3 = 1
40 = 18	4 = 2
50 = 22	5 = 2
	6 = 3
	7 = 3
	8 = 4
	9 = 4

26 = 12

Anti Log Excesses

10 = 22	1 = 2
20 = 45	2 = 5
	3 = 7
	4 = 9
	5 = 11
	6 = 13
	7 = 16
	8 = 18
	9 = 20

12 = 26

9710047	·9872213
098	236
150	259
201	282
253	305
304	328
356	351
407	374

9710459	·9872397
510	420
562	443
613	466
665	489
716	512
768	535
819	559
871	582
922	605
974	628
9711025	651
077	674
128	697
180	720
231	743
283	766
334	789
386	812
437	835
489	858
540	881
592	904
643	927
694	950
746	973
797	996
849	·9873019
900	042
952	065
9712003	088
055	111
106	134
158	157
209	180
261	203
312	226
364	249
415	272
467	295
518	318
570	341
621	364
673	387
724	410
776	433
827	456
879	479
930	502
982	526
9713033	549
084	572
136	595
187	618
239	641
290	664
342	687
393	710
445	733
496	756
548	779
599	802
651	825
702	848
754	871

9713805	·9873894
857	917
908	940
960	963
9714011	986
062	·9874009
114	032
165	055
217	078
268	101
320	124
371	147
423	170
474	193
526	216
577	239
629	262
680	285
732	308
783	331
835	354
886	377
937	400
989	423
9715040	446
092	469
143	492
195	515
246	538
298	561
349	584
401	607
452	630
504	653
555	676
607	699
658	722
709	745
761	768
812	791
864	814
915	837
967	860
9716018	883
070	906
121	929
173	952
224	975
276	998
327	·9875021
378	044
430	067
481	090
533	113
584	136
636	159
687	182
739	205
790	228
842	251
893	274
944	297
996	320
9717047	343
099	366

9717150	·9875389
202	412
253	435
305	458
356	481
408	504
459	527
510	550
562	573
613	596
665	619
716	642
768	665
819	688
871	711
922	734
974	757
9718025	780
076	803
128	826
179	849
231	872
282	895
334	918
385	941
437	964
488	987
539	·9876010
591	033
642	056
694	079
745	102
797	125
848	148
900	171
951	194
9719003	217
054	240
105	263
157	286
208	309
260	332
311	355
363	378
414	401
466	424
517	447
568	470
620	493
671	516
723	539
774	562
826	585
877	608
928	631
980	654

Excesses

10 = 4	1 = 0
20 = 9	2 = 1
30 = 13	3 = 1
40 = 18	4 = 2
50 = 22	5 = 2
	6 = 3
	7 = 3
	8 = 4
	9 = 4
26 = 12	

Anti Log Excesses

10 = 22	1 = 2
20 = 45	2 = 5
	3 = 7
	4 = 9
	5 = 11
	6 = 13
	7 = 16
	8 = 18
	9 = 20
12 = 26	

9720031	·9876677
083	700
134	723
186	746
237	769
289	792
340	814
391	837
443	860
494	883
546	906
597	929
649	952
700	975
751	998
803	·9877021
854	044
906	067
957	090
9721009	113
060	136
112	159
163	182
214	205
266	228
317	251
369	274
420	297
472	320
523	343
574	366
626	389
677	412
729	435
780	458
832	481
883	504

9721934	·9877527
986	550
9722037	573
089	596
140	619
192	642
243	665
294	688
346	711
397	734
449	757
500	780
552	803
603	826
654	849
706	872
757	894
809	917
860	940
912	963
963	986
9723014	·9878009
066	032
117	055
169	078
220	101
272	124
323	147
374	170
426	193
477	216
529	239
580	262
632	285
683	308
734	331
786	354
837	377
889	399
940	422
991	445
9724043	468
094	491
146	514
197	537
249	560
300	583
351	606
403	629
454	652
506	675
557	698
608	721
660	744
711	767
763	790
814	813
866	836
917	859
968	882
9725020	905
071	928
123	951
174	973
225	996

9725277	·9879019
328	042
380	065
431	088
482	111
534	134
585	157
637	180
688	203
740	226
791	249
842	272
894	295
945	318
997	341
9726048	364
099	387
151	410
202	433
254	456
305	478
356	501
408	524
459	547
511	570
562	593
613	616
665	639
716	662
768	685
819	708
871	731
922	754
973	777
9727025	800
076	823
128	846
179	869
230	892
282	915
333	937
385	960
436	983
487	·9880006
539	029
590	052
642	075
693	098
744	121
796	144
847	167
899	190
950	213
9728001	236
053	259
104	282
156	305
207	328
258	351
310	374
361	397
412	420
464	442
515	465
567	488

9728618	·9880511
669	534
721	557
772	580
824	603
875	626
926	649
978	672
9729029	695
081	718
132	741
183	764
235	787
286	810
338	832
389	855
440	878
492	901
543	924
595	947
646	970
697	993
749	·9881016
800	039
851	062
903	085
954	108

Excesses

10 = 4	1 = 0
20 = 9	2 = 1
30 = 13	3 = 1
40 = 18	4 = 2
50 = 22	5 = 2
	6 = 3
	7 = 3
	8 = 4
	9 = 4

26 = 11

Anti Log Excesses

10 = 22	1 = 2
20 = 45	2 = 5
	3 = 7
	4 = 9
	5 = 11
	6 = 13
	7 = 16
	8 = 18
	9 = 20

11 = 26

9730006	·9881131
057	154
108	176
160	199
211	222
263	245
314	268
365	291
417	314

9730468	·9881337
519	360
571	383
622	406
674	429
725	452
776	475
828	498
879	520
931	543
982	566
9731033	589
085	612
136	635
187	658
239	681
290	704
342	727
393	750
444	773
496	796
547	819
599	841
650	864
701	887
753	910
804	933
855	956
907	979
958	·9882002
9732010	025
061	048
112	071
164	094
215	117
266	140
318	162
369	185
421	208
472	231
523	254
575	277
626	300
677	323
729	346
780	369
832	392
883	415
934	438
986	460
9733037	483
088	506
140	529
191	552
243	575
294	598
345	621
397	644
448	667
499	690
551	713
602	735
653	758
705	781
756	804

9733808	·9882827
859	850
910	873
962	896
9734013	919
064	942
116	965
167	988
218	·9883011
270	034
321	057
373	080
424	102
475	125
527	148
578	171
629	194
681	217
732	240
783	263
835	286
886	309
938	332
989	354
9735040	377
092	400
143	423
194	446
246	469
297	492
348	515
400	538
451	561
503	584
554	607
605	629
657	652
708	675
759	698
811	721
862	744
913	767
965	790
9736016	813
067	836
119	859
170	881
222	904
273	927
324	950
376	973
427	996
478	·9884019
530	042
581	065
632	088
684	110
735	133
786	156
838	179
889	202
940	225
992	248
9737043	271
095	294

9737146	·9884317
197	339
249	362
300	385
351	408
403	431
454	454
505	477
557	500
608	523
659	546
711	569
762	591
813	614
865	637
916	660
967	683
9738019	706
070	729
121	752
173	775
224	797
276	820
327	843
378	866
430	889
481	912
532	935
584	958
635	981
686	·9885004
738	026
789	049
840	072
892	095
943	118
994	141
9739046	164
097	187
148	210
200	232
251	255
302	278
354	301
405	324
456	347
508	370
559	393
610	416
662	439
713	461
764	484
816	507
867	530
918	553
970	576

Excesses

10 = 4	1 = 0
20 = 9	2 = 1
30 = 13	3 = 1
40 = 18	4 = 2
50 = 22	5 = 2
	6 = 3
	7 = 3
	8 = 4
	9 = 4

26 = 11

Anti Log Excesses

10 = 22	1 = 2
20 = 45	2 = 5
	3 = 7
	4 = 9
	5 = 11
	6 = 13
	7 = 16
	8 = 18
	9 = 20

11 = 26

9740021	·9885599
072	622
124	645
175	667
226	690
278	713
329	736
380	759
432	782
483	805
534	828
586	851
637	873
688	896
740	919
791	942
842	965
894	988
945	·9886011
996	034
9741048	057
099	079
150	102
202	125
253	148
304	171
356	194
407	217
458	240
510	263
561	285
612	308
664	331
715	354
766	377
818	400
869	423

9741920	·9886446
972	468
9742023	491
074	514
126	537
177	560
228	583
279	606
331	629
382	651
433	674
485	697
536	720
587	743
639	766
690	789
741	812
793	834
844	857
895	880
947	903
998	926
9743049	949
101	972
152	995
203	·9887017
255	040
306	063
357	086
409	109
460	132
511	155
562	178
614	200
665	223
716	246
768	269
819	292
870	315
922	338
973	361
9744024	383
076	406
127	429
178	452
230	475
281	498
332	521
383	544
435	566
486	589
537	612
589	635
640	658
691	681
743	704
794	726
845	749
897	772
948	795
999	818
9745051	841
102	864
153	886
204	909

9745256	·9887932
307	955
358	978
410	·9888001
461	024
512	047
564	069
615	092
666	115
718	138
769	161
820	184
871	207
923	229
974	252
9746025	275
077	298
128	321
179	344
231	367
282	389
333	412
384	435
436	458
487	481
538	504
590	527
641	549
692	572
744	595
795	618
846	641
897	664
949	687
9747000	710
051	732
103	755
154	778
205	801
257	824
308	847
359	869
410	892
462	915
513	938
564	961
616	984
667	·9889007
718	029
769	052
821	075
872	098
923	121
975	144
9748026	167
077	189
129	212
180	235
231	258
282	281
334	304
385	327
436	349
488	372
539	395

9748590	·9889418
641	441
693	464
744	487
795	509
847	532
898	555
949	578
9749000	601
052	624
103	646
154	669
206	692
257	715
308	738
359	761
411	784
462	806
513	829
565	852
616	875
667	898
718	921
770	943
821	966
872	989
924	·9890012
975	035

Excesses

10 = 4	1 = 0
20 = 9	2 = 1
30 = 13	3 = 1
40 = 18	4 = 2
50 = 22	5 = 2
	6 = 3
	7 = 3
	8 = 4
	9 = 4

26 = 11

Anti Log Excesses

10 = 22	1 = 2
20 = 45	2 = 5
	3 = 7
	4 = 9
	5 = 11
	6 = 13
	7 = 16
	8 = 18
	9 = 20

11 = 26

9750026	·9890058
077	081
129	103
180	126
231	149
283	172
334	195
385	217

9750436	·9890240
488	263
539	286
590	309
642	332
693	355
744	377
795	400
847	423
898	446
949	469
9751000	492
052	514
103	537
154	560
206	583
257	606
308	629
359	651
411	674
462	697
513	720
564	743
616	766
667	788
718	811
770	834
821	857
872	880
923	903
975	925
9752026	948
077	971
128	994
180	·9891017
231	040
282	062
334	085
385	108
436	131
487	154
539	177
590	199
641	222
692	245
744	268
795	291
846	314
898	336
949	359
9753000	382
051	405
103	428
154	451
205	473
256	496
308	519
359	542
410	565
461	588
513	610
564	633
615	656
666	679
718	702

9753769	·9891725
820	747
872	770
923	793
974	816
9754025	839
077	861
128	884
179	907
230	930
282	953
333	976
384	998
435	·9892021
487	044
538	067
589	090
640	113
692	135
743	158
794	181
845	204
897	227
948	249
999	272
9755050	295
102	318
153	341
204	364
255	386
307	409
358	432
409	455
461	478
512	500
563	523
614	546
666	569
717	592
768	614
819	637
871	660
922	683
973	706
9756024	729
076	751
127	774
178	797
229	820
281	843
332	865
383	888
434	911
486	934
537	957
588	980
639	·9893002
691	025
742	048
793	071
844	094
895	116
947	139
998	162
9757049	185

9757100	·9893208
152	230
203	253
254	276
305	299
357	322
408	344
459	367
510	390
562	413
613	436
664	459
715	481
767	504
818	527
869	550
920	573
972	595
9758023	618
074	641
125	664
177	687
228	709
279	732
330	755
382	778
433	801
484	823
535	846
586	869
638	892
689	915
740	937
791	960
843	983
894	·9894006
945	029
996	051
9759048	074
099	097
150	120
201	143
253	165
304	188
355	211
406	234
457	257
509	279
560	302
611	325
662	348
714	371
765	393
816	416
867	439
919	462
970	485

Excesses

10 = 4
20 = 9
30 = 13
40 = 18
50 = 22

1 = 0
2 = 1
3 = 1
4 = 2
5 = 2
6 = 3
7 = 3
8 = 4
9 = 4

26 = 11

Anti Log Excesses

10 = 22
20 = 45

1 = 2
2 = 5
3 = 7
4 = 9
5 = 11
6 = 13
7 = 16
8 = 18
9 = 20

11 = 26

9760021	·9894507
072	530
123	553
175	576
226	599
277	621
328	644
380	667
431	690
482	713
533	735
584	758
636	781
687	804
738	827
789	849
841	872
892	895
943	918
994	941
9761046	963
097	986
148	·9895009
199	032
250	054
302	077
353	100
404	123
455	145
507	168
558	191
609	214
660	237
711	259
763	282
814	305
865	328

9761916	9895351
968	373
9762019	396
070	419
121	442
172	464
224	487
275	510
326	533
377	556
428	578
480	601
531	624
582	647
633	670
685	692
736	715
787	738
838	761
889	783
941	806
992	829
9763043	852
094	875
145	897
197	920
248	943
299	966
350	988
402	·9896011
453	034
504	057
555	080
606	102
658	125
709	148
760	171
811	193
862	216
914	239
965	262
9764016	285
067	307
118	330
170	353
221	376
272	398
323	421
375	444
426	467
477	489
528	512
579	535
631	558
682	581
733	603
784	626
835	649
887	672
938	694
989	717
9765040	740
091	763
143	785
194	808

9765245	·9896831
296	854
347	877
399	899
450	922
501	945
552	968
603	990
655	·9897013
706	036
757	059
808	081
859	104
911	127
962	150
9766013	173
064	195
115	218
167	241
218	264
269	286
320	309
371	332
423	355
474	377
525	400
576	423
627	446
679	469
730	491
781	514
832	537
883	560
935	582
986	605
9767037	628
088	651
139	673
190	696
242	719
293	742
344	765
395	787
446	810
498	833
549	856
600	878
651	901
702	924
754	947
805	969
856	992
907	·9898015
958	038
9768010	060
061	083
112	106
163	129
214	151
265	174
317	197
368	220
419	242
470	265
521	288

9768573	·9898311
624	333
675	356
726	379
777	402
828	424
880	447
931	470
982	493
9769033	515
084	538
136	561
187	584
238	606
289	629
340	652
391	675
443	697
494	720
545	743
596	766
647	788
699	811
750	834
801	857
852	879
903	902
954	925

Excesses

10 = 4	1 = 0
20 = 9	2 = 1
30 = 13	3 = 1
40 = 18	4 = 2
50 = 22	5 = 2
	6 = 3
	7 = 3
	8 = 4
	9 = 4

26 = 11

Anti Log Excesses

10 = 22	1 = 2
20 = 45	2 = 5
	3 = 7
	4 = 9
	5 = 11
	6 = 13
	7 = 16
	8 = 18
	9 = 20

11 = 26

9770006	·9898948
057	970
108	993
159	·9899016
210	039
261	061
313	084
364	107

9770415	·9899130
466	152
517	175
569	198
620	221
671	243
722	266
773	289
824	312
876	334
927	357
978	380
9771029	403
080	425
131	448
183	471
234	494
285	516
336	539
387	562
438	585
490	607
541	630
592	653
643	675
694	698
745	721
797	744
848	766
899	789
950	812
9772001	835
052	857
104	880
155	903
206	926
257	949
308	971
359	994
411	·9900017
462	040
513	062
564	085
615	108
666	130
718	153
769	176
820	199
871	221
922	244
973	267
9773025	290
076	312
127	335
178	358
229	381
280	403
332	426
383	449
434	471
485	494
536	517
587	540
639	562
690	585

9773741	·9900608
792	631
843	653
894	676
945	699
997	721
9774048	744
099	767
150	790
201	812
252	835
304	858
355	881
406	903
457	926
508	949
559	971
610	994
662	·9901017
713	040
764	062
815	085
866	108
917	131
969	153
9775020	176
071	199
122	221
173	244
224	267
275	290
327	312
378	335
429	358
480	380
531	403
582	426
633	449
685	471
736	494
787	517
838	540
889	562
940	585
991	608
9776043	630
094	653
145	676
196	699
247	721
298	744
350	767
401	789
452	812
503	835
554	858
605	880
656	903
708	926
759	948
810	971
861	994
912	·9902017
963	039
9777014	062

9777065	·9902085
117	108
168	130
219	153
270	176
321	198
372	221
423	244
475	267
526	289
577	312
628	335
679	358
730	381
781	403
833	426
884	449
935	471
986	494
9778037	517
088	539
139	562
191	585
242	608
293	630
344	653
395	676
446	698
497	721
548	744
600	767
651	789
702	812
753	835
804	857
855	880
906	903
957	926
9779009	948
060	971
111	994
162	·9903016
213	039
264	062
315	084
367	107
418	130
469	153
520	175
571	198
622	221
673	243
724	266
776	289
827	312
878	334
929	357
980	380

Excesses

10 = 4	1 = 0
20 = 9	2 = 1
30 = 13	3 = 1
40 = 18	4 = 2
50 = 22	5 = 2
	6 = 3
	7 = 3
	8 = 4
	9 = 4

26 = 11

Anti Log Excesses

10 = 23	1 = 2
20 = 45	2 = 5
	3 = 7
	4 = 9
	5 = 11
	6 = 14
	7 = 16
	8 = 18
	9 = 20

11 = 26

9780031	·9903402
082	425
133	448
185	470
236	493
287	516
338	539
389	561
440	584
491	607
542	629
594	652
645	675
696	697
747	720
798	743
849	766
900	788
951	811
9781002	834
054	856
105	879
156	902
207	924
258	947
309	970
360	992
411	·9904015
463	038
514	061
565	083
616	106
667	129
718	151
769	174
820	197
871	219

9781923	·9904242
974	265
9782025	288
076	310
127	333
178	356
229	378
280	401
332	424
383	446
434	469
485	492
536	514
587	537
638	560
689	582
740	605
792	628
843	651
894	673
945	696
996	719
9783047	741
098	764
149	787
200	809
251	832
303	855
354	878
405	900
456	923
507	946
558	968
609	991
660	·9905014
711	036
763	059
814	082
865	104
916	127
967	150
9784018	172
069	195
120	218
171	240
222	263
274	286
325	309
376	331
427	354
478	377
529	399
580	422
631	445
682	467
734	490
785	513
836	535
887	558
938	581
989	603
9785040	626
091	649
142	671
193	694

9785244	·9905717
296	739
347	762
398	785
449	807
500	830
551	853
602	875
653	898
704	921
755	944
807	966
858	989
909	·9906012
960	034
9786011	057
062	080
113	102
164	125
215	148
266	170
317	193
369	216
420	238
471	261
522	284
573	306
624	329
675	352
726	374
777	397
828	420
879	442
931	465
982	488
9787033	510
084	533
135	556
186	578
237	601
288	624
339	646
390	669
441	692
493	714
544	737
595	760
646	782
697	805
748	828
799	850
850	873
901	896
952	918
9788003	941
054	964
106	986
157	·9907009
208	032
259	054
310	077
361	100
412	122
463	145
514	168

9788565	·9907190
616	213
667	236
718	258
770	281
821	304
872	326
923	349
974	372
9789025	394
076	417
127	440
178	462
229	485
280	508
331	530
383	553
434	576
485	598
536	621
587	643
638	666
689	689
740	711
791	734
842	757
893	779
944	802
995	825

Excesses

10 = 4	1 = 0
20 = 9	2 = 1
30 = 13	3 = 1
40 = 18	4 = 2
50 = 22	5 = 2
	6 = 3
	7 = 3
	8 = 4
	9 = 4

26 = 11

Anti Log Excesses

10 = 23	1 = 2
20 = 45	2 = 5
	3 = 7
	4 = 9
	5 = 11
	6 = 14
	7 = 16
	8 = 18
	9 = 20

11 = 26

9790046	·9907847
098	870
149	893
200	915
251	938
302	961
353	983

9790404	·9908006
455	029
506	051
557	074
608	097
659	119
710	142
761	165
813	187
864	210
915	233
966	255
9791017	278
068	300
119	323
170	346
221	368
272	391
323	414
374	436
425	459
476	482
527	504
579	527
630	550
681	572
732	595
783	618
834	640
885	663
936	686
987	708
9792038	731
089	753
140	776
191	799
242	821
293	844
344	867
396	889
447	912
498	935
549	957
600	980
651	·9909003
702	025
753	048
804	070
855	093
906	116
957	138
9793008	161
059	184
110	206
161	229
212	252
263	274
315	297
366	320
417	342
468	365
519	387
570	410
621	433
672	455

9793723	·9909478
774	501
825	523
876	546
927	569
978	591
9794029	614
080	636
131	659
182	682
233	704
285	727
336	750
387	772
438	795
489	818
540	840
591	863
642	885
693	908
744	931
795	953
846	976
897	999
948	·9910021
999	044
9795050	067
101	089
152	112
203	134
254	157
305	180
357	202
408	225
459	248
510	270
561	293
612	315
663	338
714	361
765	383
816	406
867	429
918	451
969	474
9796020	497
071	519
122	542
173	564
224	587
275	610
326	632
377	655
428	678
479	700
530	723
582	745
633	768
684	791
735	813
786	836
837	859
888	881
939	904
990	926

9797041	·9910949
092	972
143	994
194	·9911017
245	040
296	062
347	085
398	107
449	130
500	153
551	175
602	198
653	221
704	243
755	266
806	288
857	311
908	334
959	356
9798011	379
062	402
113	424
164	447
215	469
266	492
317	515
368	537
419	560
470	582
521	605
572	628
623	650
674	673
725	696
776	718
827	741
878	763
929	786
980	809
9799031	831
082	854
133	876
184	899
235	922
286	944
337	967
388	990
439	·9912012
490	035
541	057
592	080
643	103
694	125
745	148
796	170
847	193
898	216
949	238

Excesses

10 = 4	1 = 0
20 = 9	2 = 1
30 = 13	3 = 1
40 = 18	4 = 2
50 = 22	5 = 2
	6 = 3
	7 = 3
	8 = 4
	9 = 4

26 = 11

Anti Log Excesses

10 = 23	1 = 2
20 = 45	2 = 5
	3 = 7
	4 = 9
	5 = 11
	6 = 14
	7 = 16
	8 = 18
	9 = 20

11 = 26

9800000	·9912261
052	283
103	306
154	329
205	351
256	374
307	397
358	419
409	442
460	464
511	487
562	510
613	532
664	555
715	577
766	600
817	623
868	645
919	668
970	690
9801021	713
072	736
123	758
174	781
225	803
276	826
327	849
378	871
429	894
480	917
531	939
582	962
633	984
684	·9913007
735	030
786	052
837	075

9801888	·9913097
939	120
990	143
9802041	165
092	188
143	210
194	233
245	256
296	278
347	301
398	323
449	346
500	369
551	391
602	414
653	436
704	459
755	482
806	504
857	527
908	549
959	572
9803010	595
061	617
112	640
163	662
214	685
265	708
316	730
367	753
418	775
469	798
520	820
571	843
622	866
673	888
724	911
775	933
826	956
877	979
928	·9914001
979	024
9804030	046
081	069
132	092
183	114
234	137
285	159
336	182
387	205
438	227
489	250
540	272
591	295
642	318
693	340
744	363
795	385
846	408
897	430
948	453
999	476
9805050	498
101	521
152	543

9805203	·9914566
254	588
305	611
356	634
407	656
458	679
509	701
560	724
611	747
662	769
713	792
764	814
815	837
866	859
917	882
968	905
9806019	927
070	950
121	972
172	995
223	·9915018
274	040
325	063
376	085
427	108
478	130
529	153
580	176
631	198
682	221
733	243
784	266
835	289
886	311
937	334
988	356
9807039	379
090	401
141	424
192	447
243	469
294	492
345	514
396	537
447	559
498	582
549	605
600	627
651	650
702	672
753	695
804	718
854	740
905	763
956	785
9808007	808
058	830
109	853
160	876
211	898
262	921
313	943
364	966
415	988
466	·9916011

9808517	·9916034
568	056
619	079
670	101
721	124
772	146
823	169
874	192
925	214
976	237
9809027	259
078	282
129	304
180	327
231	350
282	372
333	395
384	417
435	440
486	462
537	485
588	508
639	530
690	553
741	575
792	598
842	620
893	643
944	665
995	688

Excesses

10 = 4	1 = 0
20 = 9	2 = 1
30 = 13	3 = 1
40 = 18	4 = 2
50 = 22	5 = 2
	6 = 3
	7 = 3
	8 = 4
	9 = 4

25 = 11

Anti Log Excesses

10 = 23	1 = 2
20 = 45	2 = 5
	3 = 7
	4 = 9
	5 = 11
	6 = 14
	7 = 16
	8 = 18
	9 = 20

11 = 25

9810046	·9916711
097	733
148	756
199	778
250	801
301	823

9810352	·9916846
403	869
454	891
505	914
556	936
607	959
658	981
709	·9917004
760	026
811	049
862	071
913	094
964	117
9811015	139
066	162
117	184
168	207
219	229
270	252
320	274
371	297
422	320
473	342
524	365
575	387
626	410
677	432
728	455
779	477
830	500
881	523
932	545
983	568
9812034	590
085	613
136	635
187	658
238	680
289	703
340	726
391	748
442	771
493	793
543	816
594	838
645	861
696	883
747	906
798	929
849	951
900	974
951	996
9813002	·9918019
053	041
104	064
155	086
206	109
257	132
308	154
359	177
410	199
461	222
512	244
563	267
613	289

9813664	·9918312
715	334
766	357
817	380
868	402
919	425
970	447
9814021	470
072	492
123	515
174	537
225	560
276	583
327	605
378	628
429	650
480	673
531	695
581	718
632	740
683	763
734	785
785	808
836	830
887	853
938	876
989	898
9815040	921
091	943
142	966
193	988
244	·9919011
295	033
346	056
397	078
448	101
498	124
549	146
600	169
651	191
702	214
753	236
804	259
855	281
906	304
957	326
9816008	349
059	371
110	394
161	417
212	439
263	461
313	484
364	506
415	529
466	552
517	574
568	597
619	619
670	642
721	664
772	687
823	709
874	732
925	754

9816976	·9919777
9817027	799
077	822
128	845
179	867
230	890
281	912
332	935
383	957
434	980
485	·9920002
536	025
587	047
638	070
689	092
740	115
790	137
841	160
892	182
943	205
994	228
9818045	250
096	273
147	295
198	318
249	340
300	363
351	385
402	408
453	430
503	453
554	475
605	498
656	520
707	543
758	565
809	588
860	611
911	633
962	656
9819013	678
064	701
115	723
165	746
216	768
267	791
318	813
369	836
420	858
471	881
522	903
573	926
624	948
675	971
726	993
776	·9921016
827	038
878	061
929	083
980	106

Excesses

10 = 4	1 = 0
20 = 9	2 = 1
30 = 13	3 = 1
40 = 18	4 = 2
50 = 22	5 = 2
	6 = 3
	7 = 3
	8 = 4
	9 = 4

25 = 11

Anti Log Excesses

10 = 23	1 = 2
20 = 45	2 = 5
	3 = 7
	4 = 9
	5 = 11
	6 = 14
	7 = 16
	8 = 18
	9 = 20

11 = 25

9820031	·9921129
082	151
133	174
184	196
235	219
286	241
337	264
387	286
438	309
489	331
540	354
591	376
642	399
693	421
744	444
795	466
846	489
897	511
948	534
998	556
9821049	579
100	601
151	624
202	646
253	669
304	691
355	714
406	736
457	759
508	782
558	804
609	827
660	849
711	872
762	894
813	917
864	939

9821915	·9921962
966	984
9822017	·9922007
067	029
118	052
169	074
220	097
271	119
322	142
373	164
424	187
475	209
526	232
577	254
627	277
678	299
729	322
780	344
831	367
882	389
933	412
984	434
9823035	457
086	479
136	502
187	524
238	547
289	569
340	592
391	614
442	637
493	659
544	682
595	704
645	727
696	749
747	772
798	794
849	817
900	839
951	862
9824002	884
053	907
104	929
154	952
205	974
256	997
307	·9923019
358	042
409	064
460	087
511	109
562	132
612	154
663	177
714	199
765	222
816	244
867	267
918	289
969	312
9825020	334
070	356
121	379
172	401

9825223	·9923424
274	446
325	469
376	491
427	514
478	536
528	559
579	581
630	604
681	626
732	649
783	671
834	694
885	716
936	739
986	761
9826037	784
088	806
139	829
190	851
241	874
292	896
343	919
394	941
444	964
495	986
546	·9924009
597	031
648	054
699	076
750	099
801	121
851	144
902	166
953	189
9827004	211
055	234
106	256
157	278
208	301
259	324
309	346
360	369
411	391
462	414
513	436
564	459
615	481
666	504
716	526
767	549
818	571
869	593
920	616
971	638
9828022	661
073	683
123	706
174	728
225	751
276	773
327	796
378	818
429	841
480	863

9828530	·9924886
581	908
632	931
683	953
734	976
785	998
836	·9925021
886	043
937	066
988	088
9829039	110
090	133
141	155
192	178
243	200
293	223
344	245
395	268
446	290
497	313
548	335
599	358
650	380
700	403
751	425
802	448
853	470
904	492
955	515

Excesses

10 = 4	1 = 0
20 = 9	2 = 1
30 = 13	3 = 1
40 = 18	4 = 2
50 = 22	5 = 2
	6 = 3
	7 = 3
	8 = 4
	9 = 4

25 = 11

Anti Log Excesses

10 = 23	1 = 2
20 = 45	2 = 5
	3 = 7
	4 = 9
	5 = 11
	6 = 14
	7 = 16
	8 = 18
	9 = 20

11 = 25

9830006	·9925537
056	560
107	582
158	605
209	627
260	650
311	672

9830362	·9925695
412	717
463	740
514	762
565	785
616	807
667	829
718	852
769	874
819	897
870	919
921	942
972	964
9831023	987
074	·9926009
125	032
175	054
226	077
277	099
328	122
379	144
430	166
481	189
531	211
582	234
633	256
684	279
735	301
786	324
837	346
887	369
938	391
989	414
9832040	436
091	458
142	481
193	503
243	526
294	548
345	571
396	593
447	616
498	638
548	661
599	683
650	706
701	728
752	750
803	773
854	796
904	818
955	840
9833006	863
057	885
108	908
159	930
210	953
260	975
311	998
362	·9927020
413	043
464	065
515	087
565	110
616	132

9833667	·9927155
718	177
769	200
820	222
871	245
921	267
972	290
9834023	312
074	334
125	357
176	379
226	402
277	424
328	447
379	469
430	492
481	514
532	537
582	559
633	581
684	604
735	626
786	649
837	671
887	694
938	716
989	739
9835040	761
091	783
142	806
192	828
243	851
294	873
345	896
396	918
447	941
497	963
548	985
599	·9928008
650	030
701	053
752	075
802	098
853	120
904	143
955	165
9836006	187
057	210
107	232
158	255
209	277
260	300
311	322
362	345
412	367
463	389
514	412
565	434
616	457
667	479
717	502
768	524
819	546
870	569
921	591

9836972	·9928614
9837022	636
073	659
124	681
175	704
226	726
277	748
327	771
378	793
429	816
480	838
531	861
582	883
632	905
683	928
734	950
785	973
836	995
886	·9929018
937	040
988	063
9838039	085
090	107
141	130
191	152
242	175
293	197
344	220
395	242
445	265
496	287
547	309
598	332
649	354
700	377
750	399
801	422
852	444
903	467
954	489
9839005	511
055	534
106	556
157	579
208	601
259	624
309	646
360	668
411	691
462	713
513	736
563	758
614	781
665	803
716	825
767	848
818	870
868	893
919	915
970	937

Excesses

10 = 4	1 = 0
20 = 9	2 = 1
30 = 13	3 = 1
40 = 18	4 = 2
50 = 22	5 = 2
	6 = 3
	7 = 3
	8 = 4
	9 = 4
25 = 11	

Anti Log Excesses

10 = 23	1 = 2
20 = 45	2 = 5
	3 = 7
	4 = 9
	5 = 11
	6 = 14
	7 = 16
	8 = 18
	9 = 20
11 = 25	

9840021	·9929960
072	982
122	·9930005
173	027
224	050
275	072
326	094
377	117
427	139
478	162
529	184
580	207
631	229
681	251
732	274
783	296
834	319
885	341
935	364
986	386
9841037	408
088	431
139	453
189	476
240	498
291	520
342	543
393	565
443	588
494	610
545	633
596	655
647	677
698	700
748	722
799	745
850	767

9841901	·9930789
952	812
9842002	834
053	857
104	879
155	902
206	924
256	946
307	969
358	991
409	·9931014
460	036
510	058
561	081
612	103
663	126
714	148
764	171
815	193
866	215
917	238
968	260
9843018	283
069	305
120	327
171	350
222	372
272	395
323	417
374	439
425	462
475	484
526	507
577	529
628	551
679	574
729	596
780	619
831	641
882	664
933	686
983	709
9844034	731
085	753
136	776
187	798
237	821
288	843
339	865
390	888
441	910
491	933
542	955
593	977
644	·9932000
695	022
745	045
796	067
847	089
898	112
948	134
999	157
9845050	179
101	201
152	224

9845202	·9932246
253	269
304	291
355	313
406	336
456	358
507	381
558	403
609	425
659	448
710	470
761	493
812	515
863	537
913	560
964	582
9846015	605
066	627
116	649
167	672
218	694
269	717
320	739
370	761
421	784
472	806
523	829
574	851
624	873
675	896
726	918
777	941
827	963
878	985
929	·9933008
980	030
9847031	053
081	075
132	097
183	120
234	142
284	165
335	187
386	209
437	232
487	254
538	276
589	299
640	321
691	344
741	366
792	388
843	411
894	433
944	456
995	478
9848046	500
097	523
148	545
198	568
249	590
300	612
351	635
401	657
452	679

9848503	·9933702
554	724
604	747
655	769
706	791
757	814
808	836
858	859
909	881
960	903
9849011	926
061	948
112	970
163	993
214	·9934015
264	038
315	060
366	083
417	105
467	127
518	150
569	172
620	194
671	217
721	239
772	262
823	284
874	306
924	329
975	351

Excesses

10 = 4	1 = 0
20 = 9	2 = 1
30 = 13	3 = 1
40 = 18	4 = 2
50 = 22	5 = 2
	6 = 3
	7 = 3
	8 = 4
	9 = 4
25 = 11	

Anti Log Excesses

10 = 23	1 = 2
20 = 45	2 = 5
	3 = 7
	4 = 9
	5 = 11
	6 = 14
	7 = 16
	8 = 18
	9 = 20
11 = 25	

9850026	·9934374
077	396
127	418
178	441
229	463
280	485

9850330	·9934508
381	530
432	553
483	575
533	597
584	620
635	642
686	664
737	687
787	709
838	732
889	754
940	776
990	799
9851041	821
092	843
143	866
193	888
244	911
295	933
346	955
396	978
447	·9935000
498	022
549	045
599	067
650	090
701	112
752	134
802	157
853	179
904	201
955	224
9852005	246
056	269
107	291
158	313
208	336
259	358
310	380
361	403
411	425
462	447
513	470
564	492
614	515
665	537
716	559
767	582
817	604
868	626
919	649
970	671
9853020	693
071	716
122	738
173	761
223	783
274	805
325	828
376	850
426	872
477	895
528	917
579	939

9853629	·9935962
680	984
731	·9936007
782	029
832	051
883	074
934	096
984	118
9854035	141
086	163
137	185
187	208
238	230
289	253
340	275
390	297
441	320
492	342
543	364
593	387
644	409
695	431
746	454
796	476
847	499
898	521
948	543
999	566
9855050	588
101	610
151	633
202	655
253	677
304	700
354	722
405	745
456	767
507	789
557	812
608	834
659	856
710	879
760	901
811	923
862	946
912	968
963	990
9856014	·9937013
065	035
115	057
166	080
217	102
268	125
318	147
369	169
420	192
470	214
521	236
572	259
623	281
673	303
724	326
775	348
826	370
876	393

9856927	·9937415
978	437
9857028	460
079	482
130	504
181	527
231	549
282	572
333	594
384	616
434	639
485	661
536	683
586	706
637	728
688	750
739	773
789	795
840	817
891	840
941	862
992	884
9858043	907
094	929
144	951
195	974
246	996
296	·9938018
347	041
398	063
449	085
499	108
550	130
601	152
652	175
702	197
753	219
804	242
854	264
905	286
956	309
9859007	331
057	354
108	376
159	398
209	421
260	443
311	465
362	488
412	510
463	532
514	555
564	577
615	599
666	622
717	644
767	666
818	689
869	711
919	733
970	756

Excesses

10 = 4, 20 = 9, 30 = 13, 40 = 18, 50 = 22

1 = 0, 2 = 1, 3 = 1, 4 = 2, 5 = 2, 6 = 3, 7 = 3, 8 = 4, 9 = 4

25 = 11

Anti Log Excesses

10 = 23, 20 = 45

1 = 2, 2 = 5, 3 = 7, 4 = 9, 5 = 11, 6 = 14, 7 = 16, 8 = 18, 9 = 20

11 = 25

9860021	·9938778
071	800
122	823
173	845
224	867
274	890
325	912
376	934
426	957
477	979
528	·9939001
579	024
629	046
680	068
731	091
781	113
832	135
883	158
934	180
984	202
9861035	225
086	247
136	269
187	292
238	314
288	336
339	359
390	381
441	403
491	426
542	448
593	470
643	493
694	515
745	537
795	560
846	582

9861897	·9939604
948	627
998	649
9862049	671
100	694
150	716
201	738
252	761
302	783
353	805
404	828
455	850
505	872
556	894
607	917
657	939
708	961
759	984
809	·9940006
860	028
911	051
962	073
9863012	095
063	118
114	140
164	162
215	185
266	207
316	229
367	252
418	274
468	296
519	319
570	341
621	363
671	386
722	408
773	430
823	452
874	475
925	497
975	519
9864026	542
077	564
127	586
178	609
229	631
279	653
330	676
381	698
432	720
482	743
533	765
584	787
634	810
685	832
736	854
786	876
837	899
888	921
938	943
989	966
9865040	988
090	·9941010
141	033

9865192	·9941055
243	077
293	100
344	122
395	144
445	167
496	189
547	211
597	233
648	256
699	278
749	300
800	323
851	345
901	367
952	390
9866003	412
053	435
104	457
155	479
205	502
256	524
307	546
357	568
408	591
459	613
510	635
560	658
611	680
662	702
712	725
763	747
814	769
864	792
915	814
966	836
9867016	858
067	881
118	903
168	925
219	948
270	970
320	992
371	·9942015
422	037
472	059
523	081
574	104
624	126
675	148
726	171
776	193
827	215
878	238
928	260
979	282
9868030	304
080	327
131	349
182	371
232	394
283	416
334	438
384	461
435	483

9868486	·9942505
536	527
587	550
638	572
688	594
739	617
790	639
840	661
891	683
942	706
992	728
9869043	750
094	773
144	795
195	817
246	840
296	862
347	884
398	906
448	929
499	951
550	973
600	996
651	·9943018
702	040
752	062
803	085
854	107
904	129
955	152

Excesses

10 = 4	1 = 0
20 = 9	2 = 1
30 = 13	3 = 1
40 = 18	4 = 2
50 = 22	5 = 2
	6 = 3
	7 = 3
	8 = 4
	9 = 4

25 = 11

Anti Log Excesses

10 = 23	1 = 2
20 = 45	2 = 5
	3 = 7
	4 = 9
	5 = 11
	6 = 14
	7 = 16
	8 = 18
	9 = 20

11 = 25

9870006	·9943174
056	196
107	218
158	241
208	263
259	285
9870310	·9943308
360	330
411	352
461	374
512	397
563	419
613	441
664	464
715	486
765	508
816	530
867	553
917	575
968	597
9871019	620
069	642
120	664
171	686
221	709
272	731
323	753
373	776
424	798
475	820
525	842
576	865
627	887
677	909
728	932
778	954
829	976
880	999
930	·9944021
981	043
9872032	065
082	088
133	110
184	132
234	155
285	177
336	199
386	221
437	244
488	266
538	288
589	310
639	333
690	355
741	377
791	400
842	422
893	444
943	466
994	489
9873045	511
095	533
146	556
197	578
247	600
298	622
348	645
399	667
450	689
500	711
551	734
9873602	·9944756
652	778
703	801
754	823
804	845
855	867
906	890
956	912
9874007	934
057	956
108	979
159	·9945001
209	023
260	046
311	068
361	090
412	112
463	135
513	157
564	179
614	201
665	224
716	246
766	268
817	290
868	313
918	335
969	357
9875019	380
070	402
121	424
171	446
222	469
273	491
323	513
374	535
425	558
475	580
526	602
576	624
627	647
678	669
728	691
779	713
830	736
880	758
931	780
981	803
9876032	825
083	847
133	869
184	892
235	914
285	936
336	958
386	981
437	·9946003
488	025
538	047
589	070
640	092
690	114
741	136
791	159
842	181
9876893	·9946203
943	225
994	248
9877045	270
095	292
146	315
196	337
247	359
298	381
348	404
399	426
450	448
500	470
551	493
601	515
652	537
703	559
753	582
804	604
855	626
905	648
956	671
9878006	693
057	715
108	737
158	760
209	782
259	804
310	826
361	849
411	871
462	893
513	915
563	938
614	960
664	982
715	·9947004
766	027
816	049
867	071
917	093
968	116
9879019	138
069	160
120	182
171	205
221	227
272	249
322	271
373	294
424	316
474	338
525	360
575	383
626	405
677	427
727	449
778	472
828	494
879	516
930	538
980	561

Excesses

10 = 4	1 = 0
20 = 9	2 = 1
30 = 13	3 = 1
40 = 18	4 = 2
50 = 22	5 = 2
	6 = 3
	7 = 3
	8 = 4
	9 = 4

25 = 11

Anti Log Excesses

10 = 23	1 = 2
20 = 45	2 = 5
	3 = 7
	4 = 9
	5 = 11
	6 = 14
	7 = 16
	8 = 18
	9 = 20

11 = 25

9880031	·9947583
081	605
132	627
183	650
233	672
284	694
335	716
385	739
436	761
486	783
537	805
588	828
638	850
689	872
739	894
790	916
841	939
891	961
942	983
992	·9948005
9881043	028
094	050
144	072
195	094
245	117
296	139
347	161
397	183
448	206
498	228
549	250
600	272
650	295
701	317
751	339
802	361
853	383

9881903	·9948406
954	428
9882004	450
055	472
106	495
156	517
207	539
257	561
308	584
359	606
409	628
460	650
510	673
561	695
611	717
662	739
713	762
763	784
814	806
864	828
915	851
966	873
9883016	895
067	917
117	939
168	962
219	984
269	·9949006
320	028
370	051
421	073
472	095
522	117
573	140
623	162
674	184
724	206
775	228
826	251
876	273
927	295
977	317
9884028	340
079	362
129	384
180	406
230	428
281	451
332	473
382	495
433	517
483	540
534	562
584	584
635	606
686	628
736	651
787	673
837	695
888	717
939	740
989	762
9885040	784
090	806
141	828

9885191	·9949851
242	873
293	895
343	917
394	940
444	962
495	984
546	·9950006
596	028
647	051
697	073
748	095
798	117
849	140
900	162
950	184
9886001	206
051	228
102	251
152	273
203	295
254	317
304	339
355	362
405	384
456	406
506	428
557	451
608	473
658	495
709	517
759	539
810	562
860	584
911	606
962	628
9887012	650
063	673
113	695
164	717
214	739
265	762
316	784
366	806
417	828
467	850
518	873
568	895
619	917
670	939
720	961
771	984
821	·9951006
872	028
922	050
972	072
9888023	095
074	117
125	139
175	161
226	184
276	206
327	228
378	250
428	273

9888479	·9951295
529	317
580	339
630	361
681	384
731	406
782	428
833	450
883	472
934	495
984	517
9889035	539
085	561
136	583
187	606
237	628
288	650
338	672
389	694
439	717
490	739
540	761
591	783
642	805
692	828
743	850
793	872
844	894
894	916
945	939
995	961

Excesses

10 = 4	1 = 0
20 = 9	2 = 1
30 = 13	3 = 1
40 = 18	4 = 2
50 = 22	5 = 2
	6 = 3
	7 = 3
	8 = 4
	9 = 4

25 = 11

Anti Log Excesses

10 = 23	1 = 2
20 = 46	2 = 5
	3 = 7
	4 = 9
	5 = 11
	6 = 14
	7 = 16
	8 = 18
	9 = 20

11 = 25

9890046	·9951983
097	·9952005
147	027
198	050
248	072

9890299	·9952094
349	116
400	138
450	161
501	183
552	205
602	227
653	249
703	272
754	294
804	316
855	338
905	360
956	383
9891007	405
057	427
108	449
158	471
209	494
259	516
310	538
360	560
411	582
461	604
512	627
563	649
613	671
664	693
714	715
765	738
815	760
866	782
916	804
967	826
9892017	849
068	871
119	893
169	915
220	937
270	960
321	982
371	·9953004
422	026
472	048
523	070
573	093
624	115
675	137
725	159
776	181
826	204
877	226
927	248
978	270
9893028	292
079	315
129	337
180	359
231	381
281	403
332	425
382	448
433	470
483	492
534	514

9893584	·9953536
635	559
685	581
736	603
786	625
837	647
888	670
938	692
989	714
9894039	736
090	758
140	780
191	803
241	825
292	847
342	869
393	891
443	914
494	936
544	958
595	980
646	·9954002
696	024
747	047
797	069
848	091
898	113
949	135
999	158
9895050	180
100	202
151	224
201	246
252	268
302	291
353	313
403	335
454	357
505	379
555	401
606	424
656	446
707	468
757	490
808	512
858	535
909	557
959	579
9896010	601
060	623
111	645
161	668
212	690
262	712
313	734
363	756
414	778
465	801
515	823
566	845
616	867
667	889
717	911
768	934
818	956

9896869	·9954978
919	·9955000
970	022
9897020	044
071	067
121	089
172	111
222	133
273	155
323	177
374	200
424	222
475	244
525	266
576	288
626	310
677	333
728	355
778	377
829	399
879	421
930	443
980	466
9898031	488
081	510
132	532
182	554
233	576
283	599
334	621
384	643
435	665
485	687
536	709
586	732
637	754
687	776
738	798
788	820
839	842
889	864
940	887
990	909
9899041	931
091	953
142	975
192	997
243	·9956020
293	042
344	064
394	086
445	108
495	130
546	153
596	175
647	197
697	219
748	241
798	263
849	286
899	308
950	330

Excesses

10 = 4	1 = 0
20 = 9	2 = 1
30 = 13	3 = 1
40 = 18	4 = 2
50 = 22	5 = 2
	6 = 3
	7 = 3
	8 = 4
	9 = 4

25 = 11

Anti Log Excesses

10 = 23	1 = 2
20 = 46	2 = 5
	3 = 7
	4 = 9
	5 = 11
	6 = 14
	7 = 16
	8 = 18
	9 = 21

11 = 25

9900000	·9956352
051	374
102	396
152	418
203	441
253	463
304	485
354	507
405	529
455	551
506	574
556	596
607	618
657	640
708	662
758	684
809	706
859	729
910	751
960	773
9901011	795
061	817
112	839
162	861
213	884
263	906
314	928
364	950
415	972
465	994
516	·9957017
566	039
617	061
667	083
718	105
768	127
819	149

9901869	·9957172
920	194
970	216
9902021	238
071	260
121	282
172	304
222	327
273	349
323	371
374	393
424	415
475	437
525	459
576	482
626	504
677	526
727	548
778	570
828	592
879	614
929	637
980	659
9903030	681
081	703
131	725
182	747
232	769
283	792
333	814
384	836
434	858
485	880
535	902
586	924
636	946
687	969
737	991
788	·9958013
838	035
889	057
939	079
990	101
9904040	124
091	146
141	168
192	190
242	212
292	234
343	256
393	279
444	301
494	323
545	345
595	367
646	389
696	411
747	433
797	456
848	478
898	500
949	522
999	544
9905050	566
100	589

9905151	·9958611
201	633
252	655
302	677
353	699
403	721
454	743
504	766
554	788
605	810
655	832
706	854
756	876
807	898
857	920
908	943
958	965
9906009	987
059	·9959009
110	031
160	053
211	075
261	098
312	120
362	142
413	164
463	186
514	208
564	230
614	252
665	275
715	297
766	319
816	341
867	363
917	385
968	407
9907018	429
069	452
119	474
170	496
220	518
271	540
321	562
371	584
422	606
472	628
523	651
573	673
624	695
674	717
725	739
775	761
826	783
876	805
927	828
977	850
9908028	872
078	894
128	916
179	938
229	960
280	982
330	·9960004
381	027

9908431	·9960049
482	071
532	093
583	115
633	137
684	159
734	181
784	204
835	226
885	248
936	270
986	292
9909037	314
087	336
138	358
188	380
239	403
289	425
340	447
390	469
440	491
491	513
541	535
592	557
642	579
693	602
743	624
794	646
844	668
895	690
945	712
995	734

Excesses

10 = 4	1 = 0
20 = 9	2 = 1
30 = 13	3 = 1
40 = 18	4 = 2
50 = 22	5 = 2
	6 = 3
	7 = 3
	8 = 4
	9 = 4

25 = 11

Anti Log Excesses

10 = 23	1 = 2
20 = 46	2 = 5
	3 = 7
	4 = 9
	5 = 11
	6 = 14
	7 = 16
	8 = 18
	9 = 21

11 = 25

9910046	·9960756
096	778
147	801
197	823

9910248	·9960845
298	867
349	889
399	911
450	933
500	956
550	978
601	·9961000
651	022
702	044
752	066
803	088
853	110
904	132
954	155
9911004	177
055	199
105	221
156	243
206	265
257	287
307	309
358	331
408	354
459	376
509	398
559	420
610	442
660	464
711	486
761	508
812	530
862	552
913	575
963	597
9912013	619
064	641
114	663
165	685
215	707
266	729
316	751
367	773
417	796
467	818
518	840
568	862
619	884
669	906
720	928
770	950
820	972
871	994
921	·9962016
972	039
9913022	061
073	083
123	105
174	127
224	149
274	171
325	193
375	215
426	237
476	260

9913527	·9962282
577	304
627	326
678	348
728	370
779	392
829	414
880	436
930	458
981	480
9914031	503
081	525
132	547
182	569
233	591
283	613
334	635
384	657
434	679
485	701
535	723
586	746
636	768
687	790
737	812
787	834
838	856
888	878
939	900
989	922
9915040	944
090	966
140	989
191	·9963011
241	033
292	055
342	077
393	099
443	121
493	143
544	165
594	187
645	209
695	231
746	254
796	276
846	298
897	320
947	342
998	364
9916048	386
099	408
149	430
199	452
250	475
300	497
351	519
401	541
451	563
502	585
552	607
603	629
653	651
704	673
754	695

9916804	·9963717
855	739
905	762
956	784
9917006	806
057	828
107	850
157	872
208	894
258	916
309	938
359	960
409	982
460	·9964004
510	026
561	049
611	071
662	093
712	115
762	137
813	159
863	181
914	203
964	225
9918014	247
065	269
115	291
166	313
216	336
266	358
317	380
367	402
418	424
468	446
519	468
569	490
619	512
670	534
720	556
771	578
821	600
871	622
922	645
972	667
9919023	689
073	711
123	733
174	755
224	777
275	799
325	821
375	843
426	865
476	887
527	909
577	931
628	953
678	976
728	998
779	·9965020
829	042
880	064
930	086
980	108

Excesses

10 = 4	1 = 0
20 = 9	2 = 1
30 = 13	3 = 1
40 = 18	4 = 2
50 = 22	5 = 2
	6 = 3
	7 = 3
	8 = 4
	9 = 4

25 = 11

Anti Log Excesses

10 = 23	1 = 2
20 = 46	2 = 5
	3 = 7
	4 = 9
	5 = 11
	6 = 14
	7 = 16
	8 = 18
	9 = 21

11 = 25

9920031	·9965130
081	152
132	174
182	196
232	218
283	240
333	262
384	284
434	306
484	329
535	351
585	373
636	395
686	417
736	439
787	461
837	483
888	505
938	527
988	549
9921039	571
089	593
140	615
190	637
240	659
291	682
341	704
392	726
442	748
492	770
543	792
593	814
644	836
694	858
744	880
795	902
845	924

9921895	·9965946
946	968
996	990
9922047	·9966012
097	035
147	057
198	079
248	101
299	123
349	145
399	167
450	189
500	211
551	233
601	255
651	277
702	299
752	321
803	343
853	365
903	387
954	409
9923004	432
054	454
105	476
155	498
206	520
256	542
306	564
357	586
407	608
458	630
508	652
558	674
609	696
659	718
709	740
760	762
810	784
861	806
911	828
961	850
9924012	872
062	895
113	917
163	939
213	961
264	983
314	·9967005
364	027
415	049
465	071
516	093
566	115
616	137
667	159
717	181
767	203
818	225
868	247
919	269
969	291
9925019	313
070	335
120	357

9925171	·9967379
221	402
271	424
322	446
372	468
422	490
473	512
523	534
574	556
624	578
674	600
725	622
775	644
825	666
876	688
926	710
977	732
9926027	754
077	776
128	798
178	820
228	842
279	864
329	886
379	908
430	930
480	952
531	975
581	997
631	·9968019
682	041
732	063
782	085
833	107
883	129
934	151
984	173
9927034	195
085	217
135	239
185	261
236	283
286	305
336	327
387	349
437	371
488	393
538	415
588	437
639	459
689	481
739	504
790	526
840	548
891	570
941	592
991	614
9928042	636
092	658
142	680
193	702
243	724
293	746
344	768
394	790

9928444	·9968812
495	834
545	856
596	878
646	900
696	922
747	944
797	966
847	988
898	·9969010
948	032
998	054
9929049	076
099	098
150	120
200	142
250	164
301	186
351	208
401	230
452	252
502	274
552	296
603	318
653	340
703	363
754	385
804	407
854	429
905	451
955	473

Excesses

10 = 4	1 = 0
20 = 9	2 = 1
30 = 13	3 = 1
40 = 18	4 = 2
50 = 22	5 = 2
	6 = 3
	7 = 3
	8 = 4
	9 = 4

25 = 11

Anti Log Excesses

10 = 23	1 = 2
20 = 46	2 = 5
	3 = 7
	4 = 9
	5 = 11
	6 = 14
	7 = 16
	8 = 18
	9 = 21

11 = 25

9930006	·9969495
056	517
106	539
157	561
207	583

9930257	·9969605
308	627
358	649
408	671
459	693
509	715
559	737
610	759
660	781
710	803
761	825
811	847
861	869
912	891
962	913
9931013	935
063	957
113	979
164	·9970001
214	023
264	045
315	067
365	089
415	111
466	133
516	155
566	177
617	199
667	221
717	243
768	265
818	287
868	309
919	331
969	353
9932019	375
070	397
120	419
170	441
221	463
271	485
321	507
372	529
422	551
472	573
523	595
573	617
624	639
674	661
724	683
775	706
825	728
875	750
926	772
976	794
9933026	816
077	838
127	860
177	882
228	904
278	926
328	948
379	970
429	992
479	·9971014

9933530	·9971036
580	058
630	080
681	102
731	124
781	146
832	168
882	190
932	212
983	234
9934033	256
083	278
134	300
184	322
234	344
285	366
335	388
385	410
436	432
486	454
536	476
587	498
637	520
687	542
738	564
788	586
838	608
889	630
939	652
989	674
9935040	696
090	718
140	740
190	762
241	784
291	806
341	828
392	850
442	872
492	894
543	916
593	938
643	960
694	982
744	·9972004
794	026
845	048
895	070
945	092
996	114
9936046	136
096	158
147	180
197	202
247	224
298	246
348	268
398	290
449	312
499	334
549	356
600	378
650	400
700	422
750	444

9936801	·9972466
851	488
901	509
952	531
9937002	553
052	575
103	597
153	619
203	641
254	663
304	685
354	707
405	729
455	751
505	773
556	795
606	817
656	839
706	861
757	883
807	905
857	927
908	949
958	971
9938008	993
059	·9973015
109	037
159	059
210	081
260	103
310	125
361	147
411	169
461	191
511	213
562	235
612	257
662	279
713	301
763	323
813	345
864	367
914	389
964	411
9939015	433
065	455
115	477
165	499
216	521
266	543
316	565
367	587
417	609
467	631
518	653
568	675
618	697
669	719
719	741
769	763
819	785
870	807
920	829
970	851

Excesses

10 = 4	1 = 0
20 = 9	2 = 1
30 = 13	3 = 1
40 = 17	4 = 2
50 = 22	5 = 2
	6 = 3
	7 = 3
	8 = 3
	9 = 4

25 = 11

Anti Log Excesses

10 = 23	1 = 2
20 = 46	2 = 5
	3 = 7
	4 = 9
	5 = 11
	6 = 14
	7 = 16
	8 = 18
	9 = 21

11 = 25

9940021	·9973873
071	895
121	917
172	939
222	961
272	983
322	·9974004
373	026
423	048
473	070
524	092
574	114
624	136
675	158
725	180
775	202
825	224
876	246
926	268
976	290
9941027	312
077	334
127	356
177	378
228	400
278	422
328	444
379	466
429	488
479	510
530	532
580	554
630	576
680	598
731	620
781	642
831	664

9941882	·9974686
932	708
982	730
9942032	751
083	773
133	795
183	817
234	839
284	861
334	883
385	905
435	927
485	949
535	971
586	993
636	·9975015
686	037
737	059
787	081
837	103
887	125
938	147
988	169
9943038	191
089	213
139	235
189	257
239	279
290	301
340	323
390	344
441	366
491	388
541	410
591	432
642	454
692	476
742	498
793	520
843	542
893	564
943	586
994	608
9944044	630
094	652
145	674
195	696
245	718
295	740
346	762
396	784
446	806
496	828
547	850
597	872
647	894
698	916
748	938
798	960
848	981
899	·9976003
949	025
999	047
9945050	069
100	091

9945150	·9976113
200	135
251	157
301	179
351	201
401	223
452	245
502	267
552	289
603	311
653	333
703	355
753	377
804	399
854	421
904	443
954	464
9946005	486
055	508
105	530
156	552
206	574
256	596
306	618
357	640
407	662
457	684
507	706
558	728
608	750
658	772
708	794
759	816
809	838
859	860
910	882
960	903
9947010	925
060	947
111	969
161	991
211	·9977013
261	035
312	057
362	079
412	101
462	122
513	145
563	167
613	189
664	211
714	233
764	255
814	277
865	299
915	321
965	342
9948015	364
066	386
116	408
166	430
216	452
267	474
317	496
367	518

9948417	·9977540
468	562
518	584
568	606
618	628
669	650
719	672
769	694
820	715
870	737
920	759
970	781
9949021	803
071	825
121	847
171	869
222	891
272	913
322	935
372	957
423	979
473	·9978001
523	023
573	044
624	066
674	088
724	110
774	132
825	154
875	176
925	198
975	220

Excesses

10 = 4	1 = 0
20 = 9	2 = 1
30 = 13	3 = 1
40 = 17	4 = 2
50 = 22	5 = 2
	6 = 3
	7 = 3
	8 = 3
	9 = 4

25 = 11

Anti Log Excesses

10 = 23	1 = 2
20 = 46	2 = 5
	3 = 7
	4 = 9
	5 = 11
	6 = 14
	7 = 16
	8 = 18
	9 = 21

11 = 25

9950026	·9978242
076	264
126	286
176	308

9950227	·9978330
277	351
327	373
377	395
428	417
478	439
528	461
578	483
629	505
679	527
729	549
779	571
830	593
880	615
930	637
980	659
9951031	680
081	702
131	724
181	746
232	768
282	790
332	812
382	834
433	856
483	878
533	900
583	922
634	944
684	966
734	987
784	·9979009
835	031
885	053
935	075
985	097
9952035	119
086	141
136	163
186	185
236	207
287	229
337	251
387	272
437	294
488	316
538	338
588	360
638	382
689	404
739	426
789	448
839	470
890	492
940	514
990	536
9953040	557
090	579
141	601
191	623
241	645
291	667
342	689
392	711
442	733

9953492	·9979755
543	777
593	799
643	821
693	842
744	864
794	886
844	908
894	930
944	952
995	974
9954045	996
095	·9980018
145	040
196	062
246	083
296	105
346	127
397	149
447	171
497	193
547	215
597	237
648	259
698	281
748	303
798	325
849	347
899	368
949	390
999	412
9955049	434
100	456
150	478
200	500
250	522
301	544
351	566
401	587
451	609
501	631
552	653
602	675
652	697
702	719
753	741
803	763
853	785
903	806
953	828
9956004	850
054	872
104	894
154	916
205	938
255	960
305	982
355	·9981004
405	026
456	047
506	069
556	091
606	113
657	135
707	157

9956757	·9981179
807	201
857	223
908	245
958	267
9957008	288
058	310
109	332
159	354
209	376
259	398
309	420
360	442
410	464
460	486
510	507
560	529
611	551
661	573
711	595
761	617
812	639
862	661
912	683
962	705
9958012	726
063	748
113	770
163	792
213	814
263	836
314	858
364	880
414	902
464	924
514	945
565	967
615	989
665	·9982011
715	033
765	055
816	077
866	099
916	121
966	142
9959017	164
067	186
117	208
167	230
217	252
268	274
318	296
368	318
418	340
468	361
519	383
569	405
619	427
669	449
719	471
770	493
820	515
870	537
920	559
970	580

Excesses

10 = 4	1 = 0
20 = 9	2 = 1
30 = 13	3 = 1
40 = 17	4 = 2
50 = 22	5 = 2
	6 = 3
	7 = 3
	8 = 3
	9 = 4

25 = 11

Anti Log Excesses

10 = 23	1 = 2
20 = 46	2 = 5
	3 = 7
	4 = 9
	5 = 11
	6 = 14
	7 = 16
	8 = 18
	9 = 21

11 = 25

9960021	·9982602
071	624
121	646
171	668
221	690
272	712
322	734
372	756
422	777
472	799
523	821
573	843
623	865
673	887
723	909
774	930
824	952
874	974
924	996
974	·9983018
9961025	040
075	062
125	084
175	106
225	127
276	149
326	171
376	193
426	215
476	237
526	259
577	281
627	303
677	325
727	346
777	368
828	390

9961878	·9983412
928	434
978	456
9962028	478
079	499
129	521
179	543
229	565
279	587
330	609
380	631
430	653
480	674
530	696
580	718
631	740
681	762
731	784
781	806
831	828
882	850
932	871
982	893
9963032	915
082	937
133	959
183	981
233	·9984003
283	025
333	046
383	068
434	090
484	112
534	134
584	156
634	178
685	200
735	221
785	243
835	265
885	287
935	309
986	331
9964036	353
086	375
136	396
186	418
237	440
287	462
337	484
387	506
437	528
487	549
538	571
588	593
638	615
688	637
738	659
789	681
839	702
889	724
939	746
989	768
9965039	790
090	812

9965140	·9984834
190	856
240	877
290	899
340	921
391	943
441	965
491	987
541	·9985009
591	031
641	052
692	074
742	096
792	118
842	140
892	162
943	184
993	205
9966043	227
093	249
143	271
193	293
244	315
294	337
344	358
394	380
444	402
494	424
545	446
595	468
645	490
695	512
745	533
795	555
846	577
896	599
946	621
996	643
9967046	664
096	686
147	708
197	730
247	752
297	774
347	796
397	817
448	839
498	861
548	883
598	905
648	927
698	949
748	970
799	992
849	·9986014
899	036
949	058
999	080
9968049	102
100	123
150	145
200	167
250	189
300	211
350	233

9968401	·9986255
451	276
501	298
551	320
601	342
651	364
702	386
752	407
802	429
852	451
902	473
952	495
9969002	517
053	539
103	560
153	582
203	604
253	626
303	648
354	670
404	692
454	713
504	735
554	757
604	779
654	801
705	823
755	844
805	866
855	888
905	910
955	932

Excesses

10 = 4	1 = 0
20 = 9	2 = 1
30 = 13	3 = 1
40 = 17	4 = 2
50 = 22	5 = 2
	6 = 3
	7 = 3
	8 = 3
	9 = 4

25 = 11

Anti Log Excesses

10 = 23	1 = 2
20 = 46	2 = 5
	3 = 7
	4 = 9
	5 = 11
	6 = 14
	7 = 16
	8 = 18
	9 = 21

11 = 25

9970006	·9986954
056	976
106	997
156	·9987019

9970206	·9987041
256	063
306	085
357	107
407	128
457	150
507	172
557	194
607	216
657	238
708	260
758	281
808	303
858	325
908	347
958	369
9971008	391
059	412
109	434
159	456
209	478
259	500
309	522
359	543
410	565
460	587
510	609
560	631
610	653
660	675
710	696
761	718
811	740
861	762
911	783
961	805
9972011	827
061	849
112	871
162	893
212	915
262	937
312	958
362	980
412	·9988002
463	024
513	046
563	067
613	089
663	111
713	133
763	155
814	177
864	199
914	220
964	242
9973014	264
064	286
114	308
164	330
215	352
265	373
315	395
365	417
415	439

9973465	·9988461
515	482
566	504
616	526
666	548
716	570
766	592
816	613
866	635
916	657
967	679
9974017	701
067	722
117	744
167	766
217	788
267	810
318	832
368	853
418	875
468	897
518	919
568	941
618	963
668	984
719	·9989006
769	028
819	050
869	072
919	093
969	115
9975019	137
069	159
120	181
170	203
220	224
270	246
320	268
370	290
420	312
470	333
521	355
571	377
621	399
671	421
721	443
771	464
821	486
871	508
922	530
972	552
9976022	574
072	595
122	617
172	639
222	661
272	683
322	705
373	726
423	748
473	770
523	792
573	814
623	835
673	857

9976723	·9989879
774	901
824	923
874	944
924	966
974	988
9977024	·9990010
074	032
124	053
174	075
225	097
275	119
325	141
375	163
425	184
475	206
525	228
575	250
625	272
676	294
726	315
776	337
826	359
876	381
926	403
976	424
9978026	446
076	468
127	490
177	512
227	533
277	555
327	577
377	599
427	621
477	642
527	664
578	686
628	708
678	730
728	752
778	773
828	795
878	817
928	839
978	861
9979029	882
079	904
129	926
179	948
229	970
279	991
329	·9991013
379	035
429	057
479	079
530	100
580	122
630	144
680	166
730	188
780	209
830	231
880	253
930	275

9979980	·9991297

Excesses

10 = 4	1 = 0
20 = 9	2 = 1
30 = 13	3 = 1
40 = 17	4 = 2
50 = 22	5 = 2
	6 = 3
	7 = 3
	8 = 3
	9 = 4

25 = 11

Anti Log Excesses

10 = 23	1 = 2
20 = 46	2 = 5
	3 = 7
	4 = 9
	5 = 11
	6 = 14
	7 = 16
	8 = 18
	9 = 21

11 = 25

9980031	·9991318
081	340
131	362
181	384
231	406
281	427
331	449
381	471
431	493
481	515
532	536
582	558
632	580
682	602
732	624
782	645
832	667
882	689
932	711
982	733
9981033	754
083	776
133	798
183	820
233	842
283	863
333	885
383	907
433	929
483	951
533	972
584	994
634	·9992016
684	038
734	060

9981784	·9992081	9985039	·9993497	9988294	·9994913	9990096	·9995696	9993348	·9997110
834	103	089	519	344	934	146	718	398	131
884	125	139	541	394	956	196	740	448	153
934	147	190	563	444	978	246	761	498	175
984	168	240	585	494	·9995000	296	783	548	197
9982034	190	290	606	544	021	346	805	598	218
084	212	340	628	594	043	396	827	648	240
135	234	390	650	644	065	446	848	699	262
185	256	440	672	694	087	496	870	749	284
235	277	490	693	744	109	546	892	799	305
285	299	540	715	794	130	596	914	849	327
335	321	590	737	844	152	646	935	899	349
385	343	640	759	894	174	696	957	949	371
435	365	690	781	944	196	746	979	999	392
485	386	740	802	994	217	796	·9996001	9994049	414
535	408	790	824	9989044	239	846	022	099	436
585	430	840	846	095	261	896	044	149	458
635	452	891	868	145	283	946	066	199	479
686	474	941	889	195	304	996	088	249	501
736	495	991	911	245	326	9991046	109	299	523
786	517	9986041	933	295	348	097	131	349	545
836	539	091	955	345	370	147	153	399	566
886	561	141	977	395	391	197	175	449	588
936	583	191	998	445	413	247	196	499	610
986	604	241	·9994020	495	435	297	218	549	632
9983036	626	291	042	545	457	347	240	599	653
086	648	341	064	595	478	397	262	649	675
136	670	391	085	645	500	447	283	699	697
186	691	441	107	695	522	497	305	749	719
236	713	491	129	745	544	547	327	799	740
287	735	541	151	795	565	597	349	849	762
337	757	592	173	845	587	647	370	899	784
387	779	642	194	895	609	697	392	949	806
437	800	692	216	945	631	747	414	999	827
487	822	742	238	995	652	797	436	9995049	849
537	844	792	260			847	457	099	871
587	866	842	281			897	479	149	893
637	888	892	303			947	501	199	914
687	909	942	325			997	523	249	936
737	931	992	347			9992047	544	299	958
787	953	9987042	368			097	566	349	980
837	975	092	390			147	588	399	·9998001
888	996	142	412			197	610	449	023
938	·9993018	192	434			247	631	499	045
988	040	242	456			298	653	550	066
9984038	062	292	477			348	675	600	088
088	084	342	499			398	697	650	110
138	105	393	521			448	718	700	132
188	127	443	543			498	740	750	153
238	149	493	564			548	762	800	175
288	171	543	586			598	784	850	197
338	193	593	608			648	805	900	219
388	214	643	630			698	827	950	240
438	236	693	651			748	849	9996000	262
488	258	743	673			798	871	050	284
539	280	793	695			848	892	100	306
589	301	843	717			898	914	150	327
639	323	893	739			948	936	200	349
689	345	943	760			998	958	250	371
739	367	993	782			9993048	979	300	392
789	389	9988043	804			098	·9997001	350	414
839	410	093	826			148	023	400	436
889	432	143	847			198	045	450	457
939	454	194	869			248	066	500	479
989	476	244	891	9990046	·9995674	298	088	550	501

Excesses

10 = 4
20 = 9
30 = 13
40 = 17
50 = 22

1 = 0
2 = 1
3 = 1
4 = 2
5 = 2
6 = 3
7 = 3
8 = 3
9 = 4

25 = 11

Anti Log Excesses

10 = 23
20 = 46

1 = 2
2 = 5
3 = 7
4 = 9
5 = 11
6 = 14
7 = 16
8 = 18
9 = 21

11 = 25

9996600	·9998523
650	544
700	566
750	588
800	610
850	631
900	653
950	675
9997000	696
050	718
100	740
150	762
200	783
250	805
300	827
350	849
400	870
450	892
500	914
550	935
600	957
650	979
700	·9999001
750	023
800	044
850	066
900	088
950	110
9998000	131
050	153
100	175
150	196
200	218
250	240
300	262
350	283
400	305
450	327
500	349
550	370
600	392
650	414
700	435
750	457
800	479
850	501
900	522
950	544
9999000	566
050	587
100	609
150	631
200	653
250	674
300	696
350	718
400	740
450	761
500	783
550	805
600	826
650	848
700	870
750	892
800	913
9999850	·9999935
900	957
950	978

Excesses

10 = 4	1 = 0
20 = 9	2 = 1
30 = 13	3 = 1
40 = 17	4 = 2
50 = 22	5 = 2
	6 = 3
	7 = 3
	8 = 3
	9 = 4

25 = 11

Anti Log Excesses

10 = 23	1 = 2
20 = 46	2 = 5
	3 = 7
	4 = 9
	5 = 12
	6 = 14
	7 = 16
	8 = 18
	9 = 21

11 = 25